ELSEVIER'S
DICTIONARY OF
BIRD NAMES

ELSEVIER'S DICTIONARY OF BIRD NAMES

in

Latin, English, French, German and Italian

compiled by

MURRAY WROBEL
London, United Kingdom

2002
ELSEVIER
Amsterdam – Boston – London – New York – Oxford – Paris
San Diego – San Francisco – Singapore – Sydney – Tokyo

ELSEVIER SCIENCE B.V.
Sara Burgerhartstraat 25
P.O. Box 211, 1000 AE Amsterdam, The Netherlands

First edition 2002

Library of Congress Cataloging in Publication Data
A catalog record from the Library of Congress has been applied for.

British Library Cataloguing in Publication Data
A catalogue record from the British Library has been applied for.

ISBN: 0-444-50836-8

♾ The paper used in this publication meets the requirements of ANSI/NISO Z39.48-1992 (Permanence of Paper).
Printed in The Netherlands.

Preface

This dictionary has been compiled with the aim of giving an overview of the English, French, German and Italian names of birds. The Basic Table contains, in alphabetical order, the scientific names of orders, families species and some sub-species with, again in alphabetical order, their identified names in English, French, German and Italian, which are given in the singular for species and sub-species and in the plural for other terms. The relevant order and family are shown for each term and the taxonomy is that used by the majority of the sources consulted.

The criteria to determine which names are hyphenated or shown as two or more words are based on the frequency they are used by, and the relative importance I have attached to, the sources consulted.

My special thanks go to Marco Isiaia and Giovanni Soldato for their invaluable help with Italian names and sources, to Katayoun Asadollahzadeh-Sarab who prepared this work for publication and to my wife Myra for her support and encouragement.

Murray Wrobel

Contents

Abbreviations

The following abbreviations denote a general, but not necessarily exclusive, use in the various areas.

(Ants)	-	One or more of the French-speaking islands of the West Indies
(ANZ)	-	Australia and New Zealand
(CSA)	-	Central and Southern Africa
(ISC)	-	Indian Subcontinent
(NA)	-	North America
(Qué)	-	Quebec
(WI)	-	One or more of the English-speaking islands of the West Indies

Basic Table

A

1   *Abeillia abeillei*
Apodiformes - Trochilidae
e   Emerald-chinned Hummingbird;
Abeillé's Hummingbird
f   Colibri d'Abeillé
d   Abeillé-Kolibri
i   Colibrì gola di smeraldo

2   *Abroscopus albogularis*
Passeriformes - Sylviidae
e   Rufous-faced Warbler; Rufous-faced
Flycatcher-Warbler; White-throated
Flycatcher-Warbler; Fulvous-faced
Flycatcher-Warbler; Fulvous-faced
Warbler; White-throated Warbler
f   Pouillot à gorge blanche
d   Rostwangenlaubsänger
i   Luì pigliamosche golabianca

3   *Abroscopus schisticeps*
Passeriformes - Sylviidae
e   Black-faced Warbler; Blackfaced
Flycatcher-Warbler
f   Pouillot à face noire; Gobemouche à
face noire; Gobe-mouches à face
noire
d   Schieferkopflaubsänger
i   Luì pigliamosche faccianera

4   *Abroscopus superciliaris*
Passeriformes - Sylviidae
e   Yellow-bellied Warbler; Yellow-
bellied Flycatcher-Warbler; White-
throated Warbler; White-throated
Flycatcher-Warbler; Stock-Warbler;
Stock-Flycatcher-Warbler
f   Pouillot à sourcils blancs
d   Bambuslaubsänger
i   Luì pigliamosche ventregiallo

5   *Aburria aburri*
Galliformes - Cracidae
e   Wattled Guan; Wattled Piping-Guan

f   Pénélope aburri
d   Aburri
i   Penelope aburri

6   *Acanthagenys rufogularis*
Passeriformes - Meliphagidae
e   Spiny-cheeked Honeyeater; Spiny-
cheeked Wattlebird
f   Méliphage à bavette
d   Braunkehlhonigfresser
i   Mangiamiclc guancesetolose

7   *Acanthidops bairdii*
Passeriformes - Fringillidae
e   Peg-billed Finch; Peg billed Sparrow
f   Bec-en-cheville gris
d   Spitzschnabclämmerling
i   Passero becco a piolo

8   *Acanthisitta chloris*
Passeriformes - Xenicidae
e   Rifleman
f   Xénique grimpeur
d   Grenadier, Zwergschlüpfer;
Scheinkleiber
i   Acantisitta verde

9   *Acanthiza apicalis*
Passeriformes - Pardalotidae
e   Inland Thornbill (ANZ); Brown-
tailed Thornbill; Broad-tailed
Thornbill; White-tailed Thornbill
f   Acanthize troglodyte
d   Stelzschwanzdornschnabel

10   *Acanthiza chrysorrhoa*
Passeriformes - Pardalotidae
e   Yellow-rumped Thornbill; Yellow-
tailed Thornbill
f   Acanthize à croupion jaune
d   Gelbbürzeldornschnabel
i   Becco a spina dal groppone giallo;
Becco a spina gropponegiallo

11   *Acanthiza ewingii*
Passeriformes - Pardalotidae
e   Tasmanian Thornbill
f   Acanthize de Tasmanie
d   Tasman-Dornschnabel
i   Becco a spina della Tasmania

12   *Acanthiza inornata*
Passeriformes - Pardalotidae
e   Western Thornbill
f   Acanthize sobre
d   Walddornschnabel
i   Becco a spina occidentale

13   *Acanthiza iredalei*
Passeriformes - Pardalotidae
e   Slender-billed Thornbill; Slender
Thornbill; Samphire Thornbill; Dark
Thornbill
f   Acanthize d'Iredale
d   Grunddornschnabel
i   Becco a spina snello

14   *Acanthiza iredalei hedleyi*
Passeriformes - Pardalotidae
e   Eastern Slender-billed Thornbird
(ANZ)

15   *Acanthiza iredalei rosinae*
Passeriformes - Pardalotidae
e   St. Vincent's Gulf Slender-billed
Thornbird (ANZ)

16   *Acanthiza katherina*
Passeriformes - Pardalotidae
e   Mountain Thornbill
f   Acanthize des montagnes
d   Bergdornschnabel
i   Becco a spina montana

17   *Acanthiza lineata*
Passeriformes - Pardalotidae
e   Striated Thornbill
f   Acanthize ride
d   Stricheldornschnabel
i   Becco a spina striato

18   *Acanthiza murina*
Passeriformes - Pardalotidae
e   Papuan Thornbill; De Vis's
Thornbill; Bar-tailed Thornbill; De
Vis's Tree-Warbler; New Guinea
Mountain-Thornibill; New Guinea
Thornbill
f   Acanthize de Nouvelle-Guinée
d   Papua-Dornschnabel
i   Becco a spina

19   *Acanthiza nana*
Passeriformes - Pardalotidae
e   Yellow Thornbill; Little Thornbill
f   Acanthize nain
d   Gelbbauchdornschnabel
i   Becco a spina nano

20   *Acanthiza pusilla*
Passeriformes - Pardalotidae
e   Brown Thornbill
f   Acanthize mignon
d   Roststirndornschnabel
i   Becco a spina bruno

21   *Acanthiza pusilla archibaldi*
Passeriformes - Pardalotidae
e   King Island Brown Thornbird (ANZ)

22   *Acanthiza reguloides*
Passeriformes - Pardalotidae
e   Buff-rumped Thornbill; Buff-tailed
Thornbill
f   Acanthize à croupion beige
d   Goldhähnchendornschnabel
i   Becco a spina dal groppone nocciola;
Becco a spina gropponenocciola

23   *Acanthiza robustirostris*
Passeriformes - Pardalotidae
e   Slaty-backed Thornbill; Slate-backed
Thornbill; Robust Thornbill; Large-
billed Thornbill
f   Acanthize ardoisé
d   Graurückendornschnabel
i   Becco a spina robusto

24   *Acanthiza uropygialis*
Passeriformes - Pardalotidae
e   Chestnut-rumped Thornbill;
Chestnut-tailed Thornbill
f   Acanthize à croupion roux
d   Braunbürzeldornschnabel
i   Becco a spina codacastana

25   *Acanthorhynchus superciliosus*
Passeriformes - Meliphagidae
e   Western Spinebill
f   Méliphage festonné
d   Buntkopfhonigfresser; Westlicher
Dornschnabelhonigesser

i Mangiamiele becco a spina
occidentale

26 *Acanthorhynchus tenuirostris*
Passeriformes - Meliphagidae
e Eastern Spinebill
f Méliphage à bec grêle
d Östlicher Dornschnabelhonigesser;
Rotnackenhonigfresser
i Mangiamiele becco a spina orientale

27 *Acanthornis magnus*
Passeriformes - Pardalotidae
e Scrub Tit
f Séricorne de Tasmanie
d Stammhuscher
i Sericeo pettobianco

28 *Acanthornis magnus greenianus*
Passeriformes - Pardalotidae
e King Island Scrubtit (ANZ)

29 *Accipiter albogularis*
Falconiformes - Accipitridae
e Pied Goshawk; Pied Hawk; Pied
Sparrowhawk
f Autour pie
d Elsterhabicht
i Astore bianco e nero; Sparviere
variopinto; Sparviero variopinto

30 *Accipiter atricapillus*
Falconiformes - Accipitridae
e American Goshawk
f Autour américain
d Habicht

31 *Accipiter badius*
Falconiformes - Accipitridae
e Shikra; Little Banded-Goshawk
(CSA); Indian Sparrowhawk; Ceylon
Shikra
f Épervier shikra
d Shikra; Shikrasperber; Schikra;
Schikrasperber
i Scikra; Shikra; Astore baio

32 *Accipiter bicolor*
Falconiformes - Accipitridae
e Bicoloured Hawk; Two-coloured
Hawk; Bicoloured Sparrowhawk

f Épervier bicolore
d Zweifarbensperber
i Sparviero bicolore; Sparviere bicolor

33 *Accipiter brachyurus*
Falconiformes - Accipitridae
e New Britain Sparrowhawk
f Épervier de Nouvelle-Bretagne
d Dreifarbensperber
i Sparviero della Nuova Britannia;
Sparviere della Nuova Britannia

34 *Accipiter brevipes*
Falconiformes - Accipitridae
e Levant Sparrowhawk
f Épervier à pieds courts
d Kurzfangsperber; Kurzfanghabicht
i Sparviero levantino; Sparviero
levantino

35 *Accipiter butleri*
Falconiformes - Accipitridae
e Nicobar Sparrowhawk; Nicobar
Shikra
f Épervier des Nicobar
d Nikobaren-Schikra
i Shikra delle Nicobare; Astore di
Butler

36 *Accipiter castanilius*
Falconiformes - Accipitridae
e Chestnut-flanked Sparrowhawk;
Chestnut-bellied Goshawk; Chestnut-
bellied Sparrowhawk; Chestnut-
flanked Goshawk
f Autour à flancs roux
d Rotflankenhabicht
i Sparviero pettocastano; Astore dai
fianchi castane; Sparviere
pettocastano

37 *Accipiter chilensis*
Falconiformes - Accipitridae
e Chilean Hawk

38 *Accipiter chionogaster*
Falconiformes - Accipitridae
e White-breasted Hawk
f Épervier à poitrine blanche
i Sparviero pettobianco; Sparviere
pettobianco

39 *Accipiter cirrocephalus*
Falconiformes - Accipitridae
e Collared Sparrowhawk; Australian
Collared Sparrowhawk; Australian
Sparrowhawk; Little Hawk
f Épervier à collier roux
d Sydneys Sperber
i Sparviero australiano; Astore dal
collare; Sparviere australiano

40 *Accipiter collaris*
Falconiformes - Accipitridae
e Semicollared Hawk; Semicollared
Sparrowhawk; American Collared
Sparrowhawk; Collared
Sparrowhawk
f Épervier à collier interrompu
d Halsbandsperber
i Sparviero dal collare del Sudamerica;
Astore dal mezzo collare

41 *Accipiter cooperii*
Falconiformes - Accipitridae
e Cooper's Hawk; Chicken Hawk
f Épervier de Cooper
d Rundschwanzsperber
i Sparviero di Cooper

42 *Accipiter erythrauchen*
Falconiformes - Accipitridae
e Rufous-necked Sparrowhawk;
Rufous-necked Collared-
Sparrowhawk; Moluccan
Sparrowhawk; Grey Moluccan
Collared Sparrowhawk; Moluccan
Collared Sparrowhawk; Grey-
throated Sparrowhawk
f Épervier à gorge grise
i Sparviero delle Molucche; Astore
dalla gola grigia; Sparviere delle
Molucche

43 *Accipiter erythronemius*
Falconiformes - Accipitridae
e Rufous-thighed Hawk
f Épervier à cuisses rousses
i Sparviero dai calzoni rossi; Sparviere
dai calzoni rossi

44 *Accipiter erythropus*
Falconiformes - Accipitridae
e Red-thighed Sparrowhawk; Little
Sparrowhawk; Western Little-
Sparrowhawk; West African Little-
Sparrowhawk
f Épervier de Hartlaub
d Waldsperber
i Sparviero dai calzoni rossi; Sparviere
dai calzoni rosssi

45 *Accipiter fasciatus*
Falconiformes - Accipitridae
e Brown Goshawk; Australian
Goshawk; Australasian Goshawk;
Chestnut-collared Goshawk; Western
Goshawk; Grey Goshawk; Lesser
Goshawk
f Autour australien
d Bänderhabicht
i Astore australiano; Sparviere faciato

46 *Accipiter francesiae*
Falconiformes - Accipitridae
e Frances's Goshawk; Frances's
Sparrowhawk; Madagascar Goshawk
f Épervier de Frances
d Eidechsenhabicht
i Sparviero di Frances

47 *Accipiter gentilis*
Falconiformes - Accipitridae
e Northern Goshawk; Goshawk;
Eurasian Goshawk; European
Goshawk
f Busard cendré; Autour des palombes
d Habicht; Hühnerhabicht
i Astore; Sparviero da colombi;
Falcone gentile; Astore comune

48 *Accipiter griseiceps*
Falconiformes - Accipitridae
e Sulawesi Goshawk; Celebes Crested-
Goshawk
f Autour des Célèbes
d Graukopfhabicht
i Astore crestato di Sulawesi;
Sparviere crestato di Celebes

49 *Accipiter gularis*
Falconiformes - Accipitridae
e Japanese Sparrowhawk; Lesser
Sparrowhawk; Japanese Lesser

Sparrowhawk; Asiatic Sparrowhawk

f   Épervier du Japon

d   Trillersperber

i   Sparviero giapponese; Astore minore

50   *Accipiter gundlachi*

Falconiformes - Accipitridae

e   Gundlach's Hawk

f   Épervier de Cuba

d   Gundlachsperber

i   Sparviero di Grundlach; Sparviere di Grundlach

51   *Accipiter haplochrous*

Falconiformes - Accipitridae

e   White-bellied Goshawk; White-bellied Hawk; New Caledonian Goshawk; New Caledonian Sparrowhawk; New Caledonia Goshawk; New Caledonia Sparrowhawk; White-browed Sparrowhawk

f   Autour à ventre blanc

d   Weißbauchhabicht

i   Sparviero della Nuova Caledonia; Sparviere dalla gola nera

52   *Accipiter henicogrammus*

Falconiformes - Accipitridae

e   Moluccan Goshawk; Gray's Goshawk; Moluccan Barred Sparrowhawk; Moluccan Sparrowhawk

f   Autour des Moluques

d   Molukken-Habicht

i   Astore di Gray; Sparviere grigio; Sparviero grigio

53   *Accipiter henstii*

Falconiformes - Accipitridae

e   Henst's Goshawk; Madagascar Goshawk

f   Autour de Henst

d   Madagaskar-Habicht

i   Astore di Henst

54   *Accipiter hiogaster*

Falconiformes - Accipitridae

e   Variable Goshawk

55   *Accipiter imitator*

Falconiformes - Accipitridae

e   Imitator Sparrowhawk; Imitator Hawk; Imitator Goshawk; Little Pied-Sparrowhawk; Little Pied-Goshawk

f   Autour imitateur

d   Trughabicht; Imitatorhabicht

i   Sparviero imitatore; Sparviere imitator

56   *Accipiter luteoschistaceus*

Falconiformes - Accipitridae

e   Slaty-mantled Sparrowhawk; Blue-and-grey Sparrowhawk; Slaty-mantled Goshawk; Slaty-backed Goshawk

f   Autour bleu-et-gris

d   Rabaul-Habicht

i   Sparviero grigio-azzurro; Sparviere blu e grigio; Sparviero blu e grigio

57   *Accipiter madagascariensis*

Falconiformes - Accipitridae

e   Madagascar Sparrowhawk

f   Épervier de Madagascar

d   Madagaskar-Sperber

i   Sparviero di Madagascar; Astore del Madagascar

58   *Accipiter melanochlamys*

Falconiformes - Accipitridae

e   Black-mantled Goshawk; Black-mantled Accipiter; Black-mantled Sparrowhawk

f   Autour à manteau noir

d   Mantelhabicht

i   Astore dal mantello nero; Sparviere nero vestito

59   *Accipiter melanoleucus*

Falconiformes - Accipitridae

e   Black Goshawk; Black Sparrowhawk (CSA); Great Sparrowhawk; Black-and-white Goshawk; Black-and-white Sparrowhawk; Pied Goshawk

f   Autour noir

d   Mohrenhabicht; Trauerhabicht

i   Astore bianco e nero; Sparviero bianco e nero

60 **Accipiter meyerianus**
Falconiformes - Accipitridae
e Meyer's Goshawk; Papuan Goshawk
f Autour de Meyer
d Meyer-Habicht
i Astore di Meyer

61 **Accipiter minullus**
Falconiformes - Accipitridae
e Little Sparrowhawk; African Little-Sparrowhawk
f Épervier minule
d Zwergsperber
i Sparviero minore africano

62 **Accipiter nanus**
Falconiformes - Accipitridae
e Small Sparrowhawk; Dwarf Sparrowhawk; Sulawesi Dwarf Sparrowhawk; Celebes Little-Sparrowhawk
f Épervier des Célèbes
d Archbold-Sperber
i Sparviero nano di Sulawesi; Astore nano di Celebes

63 **Accipiter nisus**
Falconiformes - Accipitridae
e Eurasian Sparrowhawk; Sparrowhawk; European Sparrowhawk; Northern Sparrowhawk
f Épervier d'Europe
d Sperber
i Sparviero eurasiatico; Sparviero dai fringuelli; Smeriglio; Sparviere; Smeriglione m; Sparviero

64 **Accipiter novaehollandiae**
Falconiformes - Accipitridae
e Grey Goshawk; White Goshawk; Variable Goshawk; Varied Goshawk; Lesser White-Goshawk; Rufous-breasted Goshawk; Rufous-breasted Hawk; Grey-backed Goshawk; Papuan Goshawk; Vinous-chested Goshawk; Grey-throated Goshawk
f Autour blanc
d Neuholland-Habicht
i Astore variabile; Sparviere bianco; Sparviere dalla gola grigia; Sparviero bianco

65 **Accipiter ovampensis**
Falconiformes - Accipitridae
e Ovambo Sparrowhawk; Red-thighed Sparrowhawk; Ovampo Sparrowhawk; Sparrowhawk
f Épervier de l'Ovampo
d Ovambo-Sperber
i Sparviero di Ovampo; Astore ovampo

66 **Accipiter poliocephalus**
Falconiformes - Accipitridae
e Grey-headed Goshawk; New Guinea Grey-headed Goshawk; New Guinea Goshawk
f Autour à tête grise
d Aschkopfhabicht
i Astore testagrigia della Nuova Guinea; Astore dalla testa grigia

67 **Accipiter poliogaster**
Falconiformes - Accipitridae
e Grey-bellied Goshawk; Grey-bellied Hawk; Crested Goshawk
f Autour à ventre gris
d Graubauchhabicht
i Sparviero petogrigio; Astore panciuto

68 **Accipiter princeps**
Falconiformes - Accipitridae
e New Britain Goshawk; New Britain Grey-headed Goshawk
f Autour de Mayr
d Prinzenhabicht
i Astore testagrigia della Nuova Britannia

69 **Accipiter rhodogaster**
Falconiformes - Accipitridae
e Vinous-breasted Sparrowhawk
f Épervier à poitrine rousse
d Schlegel-Sperber
i Sparviero pettovinaceo; Astore dal petto rosso

70 **Accipiter rufitorques**
Falconiformes - Accipitridae
e Fiji Goshawk; Grey Goshawk

f Autour des Fidji
d Fidji-Habicht
i Astore delle Isole Figi; Sparviere delle Figi

71 *Accipiter rufiventris*
Falconiformes - Accipitridae
e Rufous-chested Sparrowhawk; Rufous-breasted Sparrowhawk (CSA); Rufous Sparrowhawk; Red-breasted Sparrowhawk
f Épervier menu
d Rotbauchsperber
i Sparviero rufiventre; Astore dal ventre rossiccio

72 *Accipiter soloensis*
Falconiformes - Accipitridae
e Chinese Goshawk; Grey Frog-Hawk; Blue Frog-Hawk; Chinese Sparrowhawk; Grey Goshawk; Horsfield's Sparrowhawk; Horsfield's Goshawk
f Épervier de Horsfield
d Chinesen-Schikra
i Astore cinese; Sparviere cinese

73 *Accipiter striatus*
Falconiformes - Accipitridae
e Sharp-shinned Hawk; Sharpshin
f Épervier brun; Malfini mouche (Ants)
d Eckschwanzsperber; Streifensperber
i Sparviero striato; Astore striato; Falcone gentile

74 *Accipiter superciliosus*
Falconiformes - Accipitridae
e Tiny Hawk; Tiny Sparrowhawk
f Épervier nain
d Däumlingsperber
i Sparviero nano del Sudamerica; Astore minusculo

75 *Accipiter tachiro*
Falconiformes - Accipitridae
e African Goshawk
f Autour tachiro
d Afrika-Habicht; Afrikanischer Sperber; Tachirohabicht
i Astore africano`; Sparviere africano

76 *Accipiter tousseneli*
Falconiformes - Accipitridae
e Vinous-chested Goshawk; Red-chested Goshawk
f Autour de Toussenel
d Guinea-Habicht
i Astore africano occidentale

77 *Accipiter trinotatus*
Falconiformes - Accipitridae
e Spot-tailed Goshawk; Spot-tailed Accipiter; Spot-tailed Sparrowhawk
f Épervier à queue tachetée
d Fleckschwanzsperber
i Sparviero codamacchiata; Astore dalla coda macchiata

78 *Accipiter trivirgatus*
Falconiformes - Accipitridae
e Crested Goshawk; Asian Crested-Goshawk; Celebes Crested-Goshawk
f Autour huppé
d Schopfhabicht
i Astore crestato asiatico; Sparviere crestato

79 *Accipiter ventralis*
Falconiformes - Accipitridae
e Plain-breasted Hawk
f Épervier à gorge rayée
i Sparviero pettochiaro

80 *Accipiter virgatus*
Falconiformes - Accipitridae
e Besra; Besra Sparrowhawk; Philippine Sparrowhawk
f Épervier besra
d Besra-Sperber
i Sparviero besra

81 Accipitridae
Falconiformes
e Hawks; Eagles; Harriers
f Rapaces diurnes; Accipitridés
d Greife; Habichtartige Vögel
i Accipitridi

82 *Aceros cassidix*
Coraciiformes - Bucerotidae
e Knobbed Hornbill; Celebes Hornbill; Buton Hornbill; Sulawesi Wrinkled-

Hornbill; Sulawesi Hornbill; Red-
knobbed Hornbill
f Calao à cimier
d Helmhornvogel
i Bucero di Sulawesi

83 *Aceros comatus*
Coraciiformes - Bucerotidae
e White-crowned Hornbill; Long-
crested Hornbill; Asiatic White-
crested Hornbill; White-crested
Hornbill
f Calao coiffé
d Langschopfhornvogel
i Bucero crestabianca

84 *Aceros corrugatus*
Coraciiformes - Bucerotidae
e Wrinkled Hornbill; Sunda Wrinkled-
Hornbill
f Calao à casque rouge
d Runzelhornvogel
i Bucero corrugato

85 *Aceros everetti*
Coraciiformes - Bucerotidae
e Sumba Hornbill; Sumba Wreathed-
Hornbill; Everett's Hornbill
f Calao de Sumba
d Sumba-Hornvogel
i Bucero di Everett

86 *Aceros leucocephalus*
Coraciiformes - Bucerotidae
e White-headed Hornbill; Writhe-
billed Hornbill; Writhed-billed
Hornbill; Mindanao Wrinkled-
Hornbill; Wrinkled Hornbill; Writhed
Hornbill
f Calao de Vieillot
d Mindanao-Hornvogel
i Bucero testabianca

87 *Aceros narcondami*
Coraciiformes - Bucerotidae
e Narcondam Hornbill; Narcodoman
Hornbill; Narcondom Wreathed
Hornbill
f Calao de Narcondam
d Narcondam-Jahrvogel
i Bucero di Narcondam

88 *Aceros nipalensis*
Coraciiformes - Bucerotidae
e Rufous-necked Hornbill
f Calao à cou roux
d Nepal-Hornvogel
i Bucero collorossiccio

89 *Aceros plicatus*
Coraciiformes - Bucerotidae
e Papuan Hornbill; New Guinea
Wreathed-Hornbill; Plicated
Hornbill; Kokomo
f Calao papou
d Papua-Jahrvogel
i Bucero di Blyth; Riticero; Riticero
dal becco a pieghe

90 *Aceros subruficollis*
Coraciiformes - Bucerotidae
e Plain-pouched Hornbill; Burmese
Hornbill; Blyth's Wreathed-Hornbill;
Blyth's Horbill
f Calao à gorge claire
d Sunda-Jahrvogel
i Bucero birmano

91 *Aceros undulatus*
Coraciiformes - Bucerotidae
e Wreathed Hornbill; Bar-pouched
Wreathed-Hornbill; Bar-throated
Wreathed-Hornbill; Northern Waved
Hornbill
f Calao festonné
d Furchenjahrvogel
i Bucero ondulato

92 *Aceros waldeni*
Coraciiformes - Bucerotidae
e Rufous-headed Hornbill; Writhe-
billed Hornbill; Writhed-billed
Hornbill; Visayan Wrinkled-
Hornbill; Panay Wrinkled-Hornbill
f Calao de Walden
i Bucero di Walden

93 *Acestrura astreans*
Apodiformes - Trochilidae
e Santa Marta Woodstar; Colombian
Woodstar
f Colibri des Santa Marta
i Stella dei boschi della Colombia

94　*Acestrura berlepschi*
　　Apodiformes - Trochilidae
e　Esmeraldas Woodstar; Berlepsch's
　　Woodstar
f　Colibri de Berlepsch
d　Berlepsch-Elfe
i　Stella dei boschi di Esmeralda

95　*Acestrura bombus*
　　Apodiformes - Trochilidae
e　Little Woodstar
f　Colibri bourdon
d　Hummelelfe
i　Stella dei boschi piccola

96　*Acestrura heliodor*
　　Apodiformes - Trochilidae
e　Gorgeted Woodstar; Heliodor's
　　Woodstar
f　Colibri héliodoré
d　Heliodor-Elfe
i　Stella dei boschi dal collare

97　*Acestrura mulsant*
　　Apodiformes - Trochilidae
e　White-bellied Woodstar; Mulsant's
　　Woodstar
f　Colibri de Mulsant
d　Spitzschwanzelfe
i　Stella dei boschi pettobianco

98　*Achaetops pycnopygius*
　　Passeriformes - Sylviidae
e　Damara Rockjumper; Rockrunner
　　(CSA); Damaraland Rockjumper;
　　Rockjumper
f　Achétope à flancs roux; Fauvette
　　des Damaras
d　Damara-Felsenspringer;
　　Klippensänger
i　Corridore delle rocce

99　*Acridotheres albocinctus*
　　Passeriformes - Sturnidae
e　Collared Mynah; White-collared
　　Mynah
f　Martin à collier; Mainate à collier
d　Halsbandmaina
i　Maina comune; Maina dal collare
　　bianco

100　*Acridotheres cristatellus*
　　Passeriformes - Sturnidae
e　Crested Mynah; Chinese Jungle-
　　Mynah; Chinese Crested-Mynah;
　　Tufted Mynah
f　Martin huppé; Mainate huppé
d　Haubenmaina
i　Maina crestata

101　*Acridotheres fuscus*
　　Passeriformes - Sturnidae
e　Jungle Mynah; Indian Jungle Mynah;
　　Buffalo Mynah; Grey Mynah; Field
　　Mynah
f　Martin forestier
d　Braunmaina
i　Maina della giungla

102　*Acridotheres ginginianus*
　　Passeriformes - Sturnidae
e　Bank Mynah
f　Martin des berges; Martin des rivages
d　Ufermaina
i　Maina degli argini

103　*Acridotheres grandis*
　　Passeriformes - Sturnidae
e　White-vented Mynah; Orange-billed
　　Mynah; Jungle Mynah; Great
　　Mynah; Orange-billed Jungle-
　　Mynah; Tufted Mynah; Thai Crested-
　　Mynah
f　Grand Martin; Grand Mainate
d　Langschopfmaina
i　Maina maggiore

104　*Acridotheres javanicus*
　　Passeriformes - Sturnidae
e　Javan Mynah; Javanese Mynah; Pale-
　　bellied Mynah; White-vented Mynah
f　Martin à ventre blanc
d　Graumaina
i　Maina di Giava

105　*Acridotheres tristis*
　　Passeriformes - Sturnidae
e　Common Mynah; Indian Mynah;
　　Mynah; House Mynah; Ceylon
　　Common Mynah; Brown Mynah
f　Martin triste
d　Hirtenmaina; Hirtenstar;

Trauermaina
i   Maina; Maina comune

106   *Acrocephalus aedon*
Passeriformes - Sylviidae
e   Thick-billed Warbler; Thick-billed
Reed-Warbler
f   Rousserolle à gros bec
d   Dickschnabelrohrsänger;
Dickschnabelsänger
i   Cannareccione beccogrosso;
Cannareccione beccoforte

107   *Acrocephalus aequinoctialis*
Passeriformes - Sylviidae
e   Bokikokiko; Polynesian Warbler;
Equinoctial Warbler; Christmas
Warbler; Christmas Island Reed-
Warbler; Polynesian Reed-Warbler;
Christmas Reed-Warbler
f   Rousserolle de la Ligne
d   Fanning-Rohrsänger
i   Cannaiola della Polinesia

108   *Acrocephalus agricola*
Passeriformes - Sylviidae
e   Paddyfield Warbler; Jerdon's Reed-
Warbler
f   Rousserolle isabelle
d   Feldrohrsänger
i   Cannaiola di Jerdon

109   *Acrocephalus arundinaceus*
Passeriformes - Sylviidae
e   Great Reed-Warbler
f   Rousserolle turdoïde
d   Drosselrohrsänger
i   Cannareccione; Cannajuola;
Nonnotto; Guaco; Tarabugino;
Cannareccione comune

110   *Acrocephalus atyphus*
Passeriformes - Sylviidae
e   Tuamotu Reed-Warbler; Tuamotu
Warbler; Atoll Warbler
f   Rousserolle des Touamotou
d   Tuamatu-Rohrsänger
i   Cannaiola di Tuamotu

111   *Acrocephalus australis*
Passeriformes - Sylviidae

e   Australian Reed-Warbler
f   Rousserolle d'Australie
i   Cannareccione australiano

112   *Acrocephalus baeticatus*
Passeriformes - Sylviidae
e   African Reed-Warbler; African
Marsh-Warbler (CSA); South
African Reed-Warbler; Tropical
African Reed-Warbler
f   Rousserolle africaine
d   Gartenrohrsänger
i   Cannaiola africano

**113   *Acrocephalus baeticatus
cinnamomeus***
Passeriformes - Sylviidae
e   Cape Reed-Warbler
f   Rousserolle cannellé

114   *Acrocephalus bistrigiceps*
Passeriformes - Sylviidae
e   Black-browed Reed-Warbler;
Schrenck's Reed-Warbler; Von
Schrenk's Reed-Warbler
f   Rousserolle de Schrenck
d   Brauenrohrsänger; Streifenrohrsänger
i   Cannaiola di Schrenk

115   *Acrocephalus brevipennis*
Passeriformes - Sylviidae
e   Cape Verde Swamp-Warbler; Cape
Verde Warbler; Dohrn's Cane-
Warbler; Cape Verde Islands Cane-
Warbler; Döhrn's Warbler; Cape
Verde Islands Warbler; Cape Verde
Islands Swamp-Warbler
f   Rousserolle du Cap-Vert;
Rousserolle des îles du Cap-Vert
d   Dornrohrsänger
i   Cannaiola del Capo Verde;
Cannaiola di Capo Verde

116   *Acrocephalus caffer*
Passeriformes - Sylviidae
e   Tahiti Reed-Warbler; Tahiti Warbler;
Long-billed Warbler; Long-billed
Reed-Warbler
f   Rousserolle à long bec
d   Langschnabelrohrsänger
i   Cannaiola di Tahiti

117 **Acrocephalus concinens**
Passeriformes - Sylviidae
e Blunt-winged Warbler; Blunt-winged
Paddyfield-Warbler; Blunt-winged
Reed-Warbler
f Rousserolle de Swinhoe
d Strauchrohrsänger
i Cannaiola delle risaie

118 **Acrocephalus dumetorum**
Passeriformes - Sylviidae
e Blyth's Reed-Warbler
f Rousserolle des buissons
d Buschrohrsänger
i Cannaiola di Blyth

119 **Acrocephalus familiaris**
Passeriformes - Sylviidae
e Millerbird; Laysan Millerbird;
Hawaiian Reed-Warbler
f Rousserolle obscure
d Laysan-Rohrsänger
i Cannaiola delle Hawaii

120 **Acrocephalus familiaris kingi**
Passeriformes - Sylviidae
e Nihoa Millerbird; Hawaiian Reed-
Warbler
f Rousserolle de Nihoa

121 **Acrocephalus gracilirostris**
Passeriformes - Sylviidae
e Lesser Swamp-Warbler; Cape Reed-
Warbler (CSA); African Swamp-
Warbler; Swamp Warbler
f Rousserolle à bec fin; Rousserolle à
bec mince
d Kap-Rohrsänger
i Cannaiola beccofine

122 **Acrocephalus griseldis**
Passeriformes - Sylviidae
e Basra Reed-Warbler
f Rousserolle d'Irak; Rousserolle de
Basra
d Basra-Rohrsänger
i Cannareccione di Basra;
Cannareccione di Bassora

123 **Acrocephalus kerearako**
Passeriformes - Sylviidae
e Cook Islands Reed-Warbler; Cook
Islands Warbler; Mangalia Reed-
Warbler
f Rousserolle des Cook
i Cannaiola delle Isole Cook

124 **Acrocephalus luscinia**
Passeriformes - Sylviidae
e Nightingale Reed-Warbler;
Nightingale Warbler; Finsch's Reed-
Warbler
f Rousserolle rossignol
d Sprosserrohrsänger
i Cannaiola usignolo

125 **Acrocephalus melanopogon**
Passeriformes - Sylviidae
e Moustached Warbler, Eurasian
Moustached Warbler
f Lusciniole à moustaches
d Tamariskensänger
i Forapaglie castagnolo

126 **Acrocephalus mendanae**
Passeriformes - Sylviidae
e Marquesan Reed-Warbler;
Marquesas Reed-Warbler; Marquesa
Reed-Warbler
f Rousserolle des Marquises
i Cannaiola delle Isole Marquesas

127 **Acrocephalus newtoni**
Passeriformes - Sylviidae
e Madagascar Swamp-Warbler
f Rousserolle de Newton; Rousserolle
malgache
d Madagaskar-Rohrsänger
i Cannaiola di Madagascar

128 **Acrocephalus orientalis**
Passeriformes - Sylviidae
e Oriental Reed-Warbler; Eastern
Great Reed-Warbler; Eastern Reed-
Warbler
f Rousserolle d'Orient
d China-Rohrsänger

129 **Acrocephalus paludicola**
Passeriformes - Sylviidae
e Aquatic Warbler
f Phragmite aquatique

d   Seggenrohrsänger; Binsenrohrsänger
i   Pagliarolo

130   *Acrocephalus palustris*
      Passeriformes - Sylviidae
e   Marsh Warbler; European Marsh-
      Warbler (CSA)
f   Rousserolle verderolle
d   Sumpfrohrsänger;
      Getreiderohrsänger
i   Cannaiola verdognola

131   *Acrocephalus rehsei*
      Passeriformes - Sylviidae
e   Nauru Reed-Warbler
f   Rousserolle de Nauru
i   Cannaiola di Nauru

132   *Acrocephalus rimatarae*
      Passeriformes - Sylviidae
e   Rimatara Reed-Warbler

133   *Acrocephalus rufescens*
      Passeriformes - Sylviidae
e   Greater Swamp-Warbler; Rufous
      Reed-Warbler (CSA); Rufous
      Swamp-Warbler; Rufous Cane-
      Warbler
f   Rousserolle des cannes
d   Papyrusrohrsänger
i   Cannaiola rossiccia

134   *Acrocephalus schoenobaenus*
      Passeriformes - Sylviidae
e   Sedge Warbler; European Sedge
      Warbler (CSA)
f   Phragmite des joncs
d   Schilfrohrsänger; Uferrohrsänger
i   Forapaglie; Forapaglie comune

135   *Acrocephalus scirpaceus*
      Passeriformes - Sylviidae
e   Eurasian Reed-Warbler; Reed
      Warbler; European Reed-Warbler;
      Common Reed-Warbler
f   Rousserolle effarvatte
d   Teichrohrsänger
i   Cannaiola; Cannaiola comune

136   *Acrocephalus sorghophilus*
      Passeriformes - Sylviidae

e   Streaked Reed-Warbler; Speckled
      Reed-Warbler; Chinese Sedge-
      Warbler; Chinese Reed-Warbler;
      Spectacled Reed-Warbler
f   Rousserolle sorghophile
d   Hirserohrsänger
i   Cannaiola macchiettata

137   *Acrocephalus stentoreus*
      Passeriformes - Sylviidae
e   Clamorous Reed-Warbler; Indian
      Great Reed-Warbler (ISC); Southern
      Great Reed-Warbler; Egyptian Great
      Reed-Warbler; Ceylon Great Reed-
      Warbler; Heinroth's Warbler;
      Heinroth's Reed-Warbler
f   Rousserolle stentor; Rousserolle
      turdoïde d'Égypte
d   Stentorrohrsänger;
      Süddrosselrohrsänger
i   Cannareccione stentoreo

138   *Acrocephalus stentoreus orinus*
      Passeriformes - Sylviidae
e   Large-billed Reed-Warbler; Hume's
      Large-billed Reed-Warbler
f   Rousserolle à grand bec

139   *Acrocephalus syrinx*
      Passeriformes - Sylviidae
e   Caroline Islands Reed-Warbler
f   Rousserolle des Carolines
i   Cannaiola delle Isole Caroline

140   *Acrocephalus taiti*
      Passeriformes - Sylviidae
e   Henderson Island Reed-Warbler;
      Henderson Reed-Warbler

141   *Acrocephalus tangorum*
      Passeriformes - Sylviidae
e   Manchurian Paddyfield Warbler;
      Manchurian Reed-Warbler
f   Rousserolle mandchoue
d   Mandschuren-Rohrsänger
i   Cannaiola della Manciuria

142   *Acrocephalus vaughani*
      Passeriformes - Sylviidae
e   Pitcairn Reed-Warbler; Pitcairn
      Warbler; Pitcairn Island Reed-

Warbler
f Rousserolle des Pitcairn
d Pitcairn-Rohrsänger
i Cannaiola di Pitcairn

143 *Acropternis orthonyx*
Passeriformes - Rhinocryptidae
e Ocellated Tapaculo
f Mérulaxe ocellé
d Perlenmanteltapaculo;
Krallenschlüpfer
i Tapaculo ocellato

144 *Acryllium vulturinum*
Galliformes - Numididae
e Vulturine Guineafowl
f Pintade vulturine
d Geierperlhuhn
i Numidia vulturina; Faraona vulturina

145 *Actenoides bougainvillei*
Coraciiformes - Halcyonidae
e Moustached Kingfisher; Wood
Kingfisher
f Martin-chasseur à moustaches
d Bartliest
i Martin pescatore dai mustacchi

146 *Actenoides concretus*
Coraciiformes - Halcyonidae
e Rufous-collared Kingfisher;
Chestnut-collared Kingfisher
f Martin-chasseur trapu
d Malaien-Liest
i Martin pescatore dal collare rossiccio

147 *Actenoides hombroni*
Coraciiformes - Halcyonidae
e Blue-capped Kingfisher; Hombron's
Kingfisher; Blue-capped Wood-
Kingfisher; Hombron's Wood-
Kingfisher
f Martin-chasseur de Hombron
d Mindanao-Liest
i Martin pescatore di Hombron

148 *Actenoides lindsayi*
Coraciiformes - Halcyonidae
e Spotted Kingfisher; Spotted Wood-
Kingfisher
f Martin-chasseur tacheté

d Tropfenliest
i Martin pescatore di Lindsay

149 *Actenoides monachus*
Coraciiformes - Halcyonidae
e Green-backed Kingfisher; Lonely
Kingfisher; Celebes Lowland
Kingfisher; Hooded Kingfisher;
Blue-headed Kingfisher; Celebes
Kingfisher; Sulawesi Kingfisher;
Celebes Green-Kingfisher
f Martin-chasseur moine
d Einsiedlerliest
i Martin pescatore testablu

150 *Actenoides princeps*
Coraciiformes - Halcyonidae
e Scaly Kingfisher; Princely
Kingfisher; Mountain Kingfisher;
Celebes Mountain-Kingfisher; Scaly-
breasted Kingfisher; Regent
Kingfisher; Bar-headed Kingfisher;
Bar-headed Wood-Kingfisher;
Montain Kingfisher
f Martin-chasseur royal
d Königsliest
i Martin pescatore testabarrata

151 *Actinodura egertoni*
Passeriformes - Sylviidae
e Rusty-fronted Barwing; Spectacled
Barwing; Chestnut Spectacled-
Barwing
f Actinodure d'Egerton
d Rotstirnsibia
i Garrulo alibarrate di Egerton

152 *Actinodura morrisoniana*
Passeriformes - Sylviidae
e Formosan Barwing; Taiwan
Barwing; Nepal Barwing; Himalayan
Barwing
f Actinodure de Taiwan; Actinodure
de Formose; Garulaxe de Mont
Morrison
d Formosa-Sibia
i Garrulo alibarrate di Taiwan

153 *Actinodura nipalensis*
Passeriformes - Sylviidae
e Hoary-throated Barwing; Hoary

Barwing
f Actinodure du Nepal
d Nepal-Sibia
i Garrulo alibarrate del Nepal

154 *Actinodura ramsayi*
Passeriformes - Sylviidae
e Spectacled Barwing; Ramsay's
Spectacled Barwing; Ramsay's
Barwing
f Actinodure de Ramsay
d Brillensibia
i Garrulo alibarrate dagli occhiali

155 *Actinodura souliei*
Passeriformes - Sylviidae
e Streaked Barwing; Soulie's Barwing
f Actinodure de Soulie
d Tonkin-Sibia
i Garrulo alibarrate di Soulie

156 *Actinodura waldeni*
Passeriformes - Sylviidae
e Streak-throated Barwing; Austen's
Barwing; Walden's Barwing
f Actinodure de Walden; Actinodure
grivée
d Yünnan-Sibia
i Garrulo alibarrate di Walden

157 *Actophilornis africanus*
Charadriiformes - Jacanidae
e African Jacana; Lily-Trotter; Greater
African Jacana
f Jacana à poitrine dorée
d Jacana; Blaustirnblatthühnchen;
Blatthühnchen
i Parra africana; Jacana africana

158 *Actophilornis albinucha*
Charadriiformes - Jacanidae
e Madagascar Jacana; Malagasy Jacana
f Jacana malgache; Jacana à nuque
blanche
d Weißnackenblatthühnchen
i Jacana del Madagascar

159 *Adelomyia melanogenys*
Apodiformes - Trochilidae
e Speckled Hummingbird; Pipita
f Colibri moucheté

d Schwarzohrnymphe
i Colibrì guancenere

160 *Aechmophorus clarkii*
Podicipediformes - Podicipedidae
e Clark's Grebe
f Grèbe à face blanche
d Renntaucher
i Svasso cigno di Clark

161 *Aechmophorus occidentalis*
Podicipediformes - Podicipedidae
e Western Grebe
f Grèbe élégant; Grèbe de l'Ouest
d Renntaucher
i Svasso cigno

162 *Aegithalidae*
Passeriformes
e Long-tailed Tits; Long-tailed Titmice
f Mésanges à longue queue;
Égithalidés; Aegithalidés
d Schwanzmeisen
i Egitalidi; Codibugnoli

163 *Aegithalos bonvaloti*
Passeriformes - Aegithalidae
e Black-browed Tit
f Mésange à longue queue

164 *Aegithalos caudatus*
Passeriformes - Aegithalidae
e Long-tailed Tit; Eurasian Long-tailed
Tit
f Mésange à longue queue
d Schwanzmeise
i Codibugnolo eurasiatico;
Codibugnolo; Cincia codona;
Lanciabue; Paglianculo; Codilungo;
Codibugnolo terrestre; Cince col
becco nero

165 *Aegithalos concinnus*
Passeriformes - Aegithalidae
e Black-throated Tit; Red-headed Tit;
Red-headed Long-tailed Tit
f Mésange à tête rousse
d Rostkappenschwanzmeise;
Rotstirnschwanzmeise; Rotkopfmeise
i Codibugnolo testarossa

166   *Aegithalos fuliginosus*
      Passeriformes - Aegithalidae
  e   White-necklaced Tit; Sooty Tit;
      Sooty Long-tailed Tit; Silver-faced
      Tit
  f   Mésange à col blanc; Mésange
      fuligineuse
  d   Graukopfschwanzmeise
  i   Codibugnolo fuligginoso

167   *Aegithalos iouschistos*
      Passeriformes - Aegithalidae
  e   Rufous-fronted Tit; Black-browed Tit
      (ISC); Blyth's Long-tailed Tit;
      Rufous-fronted Long-tailed Tit;
      Black-headed Tit
  f   Mésange de Blyth
  d   Rostwangenschwanzmeise
  i   Codibugnolo di Blyth

168   *Aegithalos laucogenys*
      Passeriformes - Aegithalidae
  e   White-cheeked Tit; White-cheeked
      Long-tailed Tit; Kashmir Tit
  f   Mésange à joues blanches
  d   Schwarzkehlschwanzmeise
  i   Codibugnolo gauancebianche

169   *Aegithalos niveogularis*
      Passeriformes - Aegithalidae
  e   White-throated Tit
  f   Mésange à gorge blanche
  d   Weißkehlschwanzmeise
  i   Codibugnolo golabianca

170   *Aegithina lafresnayei*
      Passeriformes - Corvidae
  e   Great Iora
  f   Iora de Lafresnay
  d   Langschnabelaegithina
  i   Iora maggiore

171   *Aegithina nigrolutea*
      Passeriformes - Corvidae
  e   White-tailed Iora; Marshall's Iora
  f   Iora à queue blanche; Iora jaune-et-
      noir
  d   Schwarzkappenaegithina
  i   Iora di Marshall

172   *Aegithina tiphia*
      Passeriformes - Corvidae
  e   Common Iora; Iora (ISC); Black-
      winged Iora; Small Iora; Caylon Iora
  f   Petit Iora; Iora du Bengale
  d   Schwarzflügelaegithina
  i   Iora comune

173   *Aegithina viridissima*
      Passeriformes - Corvidae
  e   Green-Iora
  f   Iora émeraude
  d   Smaragdaegithina
  i   Iora verde

174   *Aegolius acadicus*
      Strigiformes - Strigidae
  e   Saw-whet Owl; Northern Saw-whet
      Owl
  f   Petite Nyctale
  d   Sägekauz
  i   Civetta acadica

175   *Aegolius funereus*
      Strigiformes - Strigidae
  e   Boreal Owl; Tengmalm's Owl
  f   Nyctale de Tengmalm; Nyctale
      boréale
  d   Rauhfußkauz
  i   Civetta capogrosso

176   *Aegolius harrisii*
      Strigiformes - Strigidae
  e   Buff-fronted Owl
  f   Nyctale de Harris
  d   Blaßstirnkauz
  i   Civetta di Harris

177   *Aegolius ridgwayi*
      Strigiformes - Strigidae
  e   Unspotted Saw-whet Owl; Alfaro's
      Owl
  f   Nyctale immaculée
  d   Ridgway-Kauz
  i   Civetta di Ridgeway

178   *Aegotheles albertisi*
      Caprimulgiformes - Aegothelidae
  e   Mountain Owlet-Nightjar; Albertis's
      Owlet-Nightjar
  f   Égothèle montagnard

d Bergshwalm
i Succiacapre gufo di D'Albertis

179 *Aegotheles archboldi*
Caprimulgiformes - Aegothelidae
e Archbold's Owlet-Nightjar; Eastern
Mountain Owlet-Nightjar
f Égothèle d'Archbold
i Succiacapre gufo di Archbold

180 *Aegotheles bennettii*
Caprimulgiformes - Aegothelidae
e Barred Owlet-Nightjar
f Égothèle de Bennett
d Bennet-Schwalm
i Succiacapre gufo di Bennett

181 *Aegotheles crinifrons*
Caprimulgiformes - Aegothelidae
e Long-whiskered Owlet-Nightjar;
Moluccan Owlet-Nightjar;
Halmahera Owlet-Nightjar
f Égothèle des Moluques
d Molukken-Schwalm
i Succiacapre gufo di Halmahera

182 *Aegotheles cristatus*
Caprimulgiformes - Aegothelidae
e Australian Owlet-Nightjar; Crested
Owlet-Nightjar; Owlet-Nightjar
f Égothèle d'Australie
d Baumschwalm
i Succiacapre gufo australiano;
Egotele; Egotele cristato; Caprimulgo
gufo

183 *Aegotheles cristatus tasmanicus*
Caprimulgiformes - Aegothelidae
e Tasmanian Australian Owlet-
Nightjar

184 *Aegotheles insignis*
Caprimulgiformes - Aegothelidae
e Feline Owlet-Nightjar; Large Owlet-
Nightjar; Rufous Owlet-Nightjar
f Grand Égothèle
d Käuzchenschwalm
i Succiacapre gufo maggiore

185 *Aegotheles savesi*
Caprimulgiformes - Aegothelidae

e New Caledonian Owlet-Nightjar
f Égothèle calédonien
d Schwarzrückenschwalm
i Succiacapre gufo della Nuova
Caledonia

186 *Aegotheles wallacii*
Caprimulgiformes - Aegothelidae
e Wallace's Owlet-Nightjar
f Égothèle de Wallace
d Fleckenschwalm
i Succiacapre gufo di Wallace

187 Aegothelidae
Caprimulgiformes
e Owlet-Nightjars
f Égothelidés
d Höhlenschwalme; Zwergschwalme
i Egotelidi

188 *Aegypius monachus*
Falconiformes - Accipitridae
e Cinereous Vulture; Black Vulture;
Monk Vulture; European Black-
Vulture; Eurasian Black-Vulture
f Vautour moine
d Mönchsgeier; Altweltgeier;
Kuttengeier
i Avvoltoio monaco; Avvoltore

189 *Aenigmatolimnas marginalis*
Gruiformes - Rallidae
e Striped Crake
f Marouette rayée
d Graukehlsumpfhuhn
i Schiribilla striata

190 *Aepypodius arfakianus*
Galliformes - Megapodiidae
e Wattled Brush-Turkey; Wattled
Scrub-Turkey
f Talégalle des Arfak
d Kammtalegalla
i Tacchino di boscaglia del Arfak

191 *Aepypodius bruijnii*
Galliformes - Megapodiidae
e Bruijn's Brush-Turkey; Waigeo
Brush-Turkey; Waigeo Scrub-Turkey
f Talégalle de Bruijn

d   Braunbrusttalegalla
i   Tacchino di boscaglia di Brujin

192   *Aerodramus amelis*
      Apodiformes - Apodidae
e   Grey Swiftlet; Philippine Swiftlet
f   Salangane grise
i   Salangana delle Filippine

193   *Aerodramus bartschi*
      Apodiformes - Apodidae
e   Guam Swiftlet; Guam Cave-Swiftlet;
      Mariana Swiftlet; Micronesian
      Swiftlet; Marianas Swiftlet
f   Salangane de Guam
i   Salangana delle Marianne

194   *Aerodramus brevirostris*
      Apodiformes - Apodidae
e   Himalayan Swiftlet; Himalayan
      Swift; Indian Edible-nest Swiftlet;
      Chinese Swiftlet
f   Salangane de l'Himalaya
d   Himalaja-Salangane
i   Salangana dell'Himalaya

195   *Aerodramus elaphra*
      Apodiformes - Apodidae
e   Seychelles Swiftlet; Seychelles Cave-
      Swiftlet
f   Salangane des Seychelles
i   Salangana delle Seychelle

196   *Aerodramus francisca*
      Apodiformes - Apodidae
e   Mascarene Swiftlet; Grey-rumped
      Swiftlet; Mauritius Swiftlet;
      Mascarene Cave-Swiftlet; Indian
      Ocean Swiftlet
f   Salange des Mascareignes; Salangane
      à croupion gris
d   Mauritius-Salangane
i   Salangana dal groppone grigio

197   *Aerodramus fuciphagus*
      Apodiformes - Apodidae
e   Edible-nest Swiftlet; Thunberg's
      Swiftlet; White-nest Swiftlet; Grey-
      rumped Swiftlet; Brown-rumped
      Swiftlet; Hume's Swiftlet; Andaman
      Grey-rumped Swiftlet

f   Salange à nid blanc
d   Weißnestsalangane; Tafelsalangane
i   Salangana orientale

198   *Aerodramus germani*
      Apodiformes - Apodidae
e   German's Swiftlet; Oustalet's Swiftlet
f   Salangane de German
i   Salangana di Oustalet

199   *Aerodramus hirundinacea*
      Apodiformes - Apodidae
e   Mountain Swiftlet
f   Salangane de montagne
d   Schwalbensalangane
i   Salangana delle montagne

200   *Aerodramus infuscatus*
      Apodiformes - Apodidae
e   Moluccan Swiftlet
f   Salangane des Moluques
i   Salangana delle Molucche

201   *Aerodramus inquietus*
      Apodiformes - Apodidae
e   Micronesian Swiftlet; Carolines
      Swiftlet; Caroline Swiftlet; Caroline
      Islands Swiftlet
f   Salangane des Carolines
i   Salangana delle Caroline

202   *Aerodramus leucophaeus*
      Apodiformes - Apodidae
e   Tahiti Swiftlet; Tahitian Swiftlet;
      Polynesian Swiftlet; South Pacific
      Swiftlet; Marquesan Swiftlet;
      Marquesas Swiftlet; Marquesa
      Swiftlet
f   Salangane de la Société
d   Tahiti-Salangane
i   Salangana di Tahiti

203   *Aerodramus maximus*
      Apodiformes - Apodidae
e   Black-nest Swiftlet; Low's Swiftlet;
      Indomalayan Swiftlet
f   Salangane à nid noir
d   Schwarznestsalangne
i   Salangana di Lowe

204 **Aerodramus mearnsi**
Apodiformes - Apodidae
e Philippine Swiftlet
f Salangane de Mearns
i Salangana dal groppone bruno

205 **Aerodramus nuditarsus**
Apodiformes - Apodidae
e Bare-legged Swiftlet; Naked-legged
Swiftlet; Schrader Mountain Swiftlet;
New Guinea Swiftlet
f Salangane de Salomonsen
i Salangana dei monti

206 **Aerodramus ocista**
Apodiformes - Apodidae
e Marquesan Swiftlet; Marquesa
Swiftlet; Marquesas Swiftlet
f Salangane des Marquises

207 **Aerodramus orientalis**
Apodiformes - Apodidae
e Mayr's Swiftlet; Guadalcanal Swiftlet
f Salangane de Mayr
d Salomonen-Salangane
i Salangana di Guadalcanal

208 **Aerodramus palawanensis**
Apodiformes - Apodidae
e Palawan Swiftlet
f Salangane de Palawan
i Salangana di Palawan

209 **Aerodramus papuensis**
Apodiformes - Apodidae
e Papuan Swiftlet; Idenberg Swiftlet;
Idenberg River Swiftlet; Three-toed
Swiftlet
f Salangane papoue
d Papua-Salangane
i Salangana papua

210 **Aerodramus pelewensis**
Apodiformes - Apodidae
e Palau Swiftlet
f Salangane des Palau
i Salangana di Palauwan

211 **Aerodramus rogersi**
Apodiformes - Apodidae
e Indochinese Swiftlet

f Salangane indochinoise
i Salangana dell indocina

212 **Aerodramus salanganus**
Apodiformes - Apodidae
e Mossy-nest Swiftlet; Mossy Swiftlet;
Sunda Swiftlet
f Salangane de la Sonde
i Salangana muschiata

213 **Aerodramus sawtelli**
Apodiformes - Apodidae
e Atiu Swiftlet; Cook Islands Swift;
Sawtell's Swift; Sawtell's Swiftlet;
Atiu Island Swiftlet
f Salangane de Cook
d Atiu-Salangane
i Salangana delle Cook

214 **Aerodramus spodiopygius**
Apodiformes - Apodidae
e White-rumped Swiftlet; Pacific
White-rumped Swiftlet; Grey-
rumped Swiftlet; Grey Swiftlet
f Salangane à croupion blanc
d Südseesalangane
i Salangana dal groppone bianco

215 **Aerodramus terraereginae**
Apodiformes - Apodidae
e Australian Swiftlet; Grey Swiftlet
f Salangane d'Australie
i Salangana australiana

216 **Aerodramus unicolor**
Apodiformes - Apodidae
e Indian Swiftlet; Indian Edible-nest
Swiftlet (ISC); Indian Edible-nest
Swift
f Salangane de Malabar
d Malabar-Salangane
i Salangana indiana

217 **Aerodramus vanikorensis**
Apodiformes - Apodidae
e Uniform Swiftlet; Vanikoro Swiftlet;
Grey Swiftlet; Lowland Swiftlet;
Island Swiftlet
f Salangane de Vanikoro
d Moosnestsalangane; Vanikoro-

Salangane
i   Salangana unicolore

218   *Aerodramus vulcanorum*
Apodiformes - Apodidae
e   Volcano Swiftlet
f   Salangane des volcans
i   Salangana dei vulcani

219   *Aeronautes andecolus*
Apodiformes - Apodidae
e   Andean Swift
f   Martinet des Andes
d   Anden-Segler
i   Rondone delle Ande

220   *Aeronautes montivagus*
Apodiformes - Apodidae
e   White-tipped Swift; Mountain Swift
f   Martinet montagnard
d   Bergsegler
i   Rondone marginato

221   *Aeronautes saxatalis*
Apodiformes - Apodidae
e   White-throated Swift
f   Martinet à gorge blanche
d   Weißbrustsegler
i   Rondone golabianca

222   *Aethia cristatella*
Charadriiformes - Alcidae
e   Crested Auklet
f   Starique cristatelle; Alque panachée
(Qué)
d   Schopfalk
i   Alca minore crestata; Pulcinella dal
ciuffo

223   *Aethia pusilla*
Charadriiformes - Alcidae
e   Least Auklet; Knob-billed Auklet
f   Starique minuscule; Alque miniscule
(Qué)
d   Zwergalk
i   Alca minima

224   *Aethia pygmaea*
Charadriiformes - Alcidae
e   Whiskered Auklet; Pygmy Auklet
f   Starique pygmée; Alque barbue

(Qué)
d   Bartalk
i   Alca pigmea

225   *Aethopyga boltoni*
Passeriformes - Nectariniidae
e   Apo Sunbird; Mount Apo Sunbird
f   Souïmanga de Bolton
d   Mindanao-Nektarvogel
i   Nettarinia di Apo

226   *Aethopyga christinae*
Passeriformes - Nectariniidae
e   Fork-tailed Sunbird
f   Souïmanga de Christine; Souïmanga
de la reine Christine; Souïmanga à
queue pointue
d   Hainan-Nektarvogel
i   Nettarinia codaforcuta

227   *Aethopyga duyvenbodei*
Passeriformes - Nectariniidae
e   Elegant Sunbird; Sanghir Sunbird;
Sanghir Yellow-backed Sunbird;
Duyvenbode's Sunbird
f   Souïmanga des Sangi
d   Sangihe-Nektarvogel
i   Nettarinia dorsogiallo di Sanghir

228   *Aethopyga eximia*
Passeriformes - Nectariniidae
e   White-flanked Sunbird; Kühl's
Sunbird
f   Souïmanga de Java; Souïmanga de
Kuhl
d   Java-Nektarvogel
i   Nettarinia di Kuhl

229   *Aethopyga flagrans*
Passeriformes - Nectariniidae
e   Flaming Sunbird
f   Souïmanga flamboyant
d   Feuerbrustnektarvogel
i   Nettarinia fiammeggiante

230   *Aethopyga gouldiae*
Passeriformes - Nectariniidae
e   Gould's Sunbird; Mrs.. Gould's
Sunbird; Yellow-breasted Sunbird;
Simla Yellow-backed Sunbird; Blue-
throated Sunbird

f   Souïmanga de Gould; Souïmanga de
Madame Gould
d   Gould-Nektarvogel
i   Nettarinia della signor Gould

231   ***Aethopyga ignicauda***
Passeriformes - Nectariniidae
e   Fire-tailed Sunbird
f   Souïmanga queue-de-feu; Souïmanga
à queue de feu
d   Feuerschwanznektarvogel
i   Nettarinia codadifiamma

232   ***Aethopyga mystacalis***
Passeriformes - Nectariniidae
e   Javan Sunbird; Scarlet Sunbird;
Temminck's Sunbird; Violet-tailed
Sunbird
f   Souïmanga écarlate
d   Temminck-Nektarvogel
i   Nettarinia scarlatta

233   ***Aethopyga nipalensis***
Passeriformes - Nectariniidae
e   Green-tailed Sunbird; Green-throated
Sunbird; Nepal Yellow-backed
Sunbird; Nepal Sunbird
f   Souïmanga à queue verte
d   Grünschwanznektarvogel
i   Nettarinia codaverde

234   ***Aethopyga primigenius***
Passeriformes - Nectariniidae
e   Grey-hooded Sunbird; Hachisuka's
Sunbird
f   Souïmanga d'Hachisuka
d   Graukehlnektarvogel
i   Nettarinia di Hachisuka

235   ***Aethopyga pulcherrima***
Passeriformes - Nectariniidae
e   Metallic-winged Sunbird; Mountain
Sunbird; Loveliest Sunbird
f   Souïmanga montagnard
d   Glanzflügelnektarvogel
i   Nettarinia dei monti

236   ***Aethopyga saturata***
Passeriformes - Nectariniidae
e   Black-throated Sunbird; Black-
breasted Sunbird

f   Souïmanga sombre
d   Schwarzkehlnektarvogel
i   Nettarinia golanera

237   ***Aethopyga shelleyi***
Passeriformes - Nectariniidae
e   Lovely Sunbird; Palawan Sunbird
f   Souïmanga ravissant
d   Goldkehlnektarvogel
i   Nettarinia di Palawan

238   ***Aethopyga siparaja***
Passeriformes - Nectariniidae
e   Crimson Sunbird; Yellow-backed
Sunbird (ISC); Scarlet-breasted
Sunbird; Scarlet-throated Sunbird
f   Souïmanga siparaja
d   Scharlachnektarvogel;
Gelbrückennektarvogel
i   Nettarinia dorsogiallo di Sanghir

239   ***Afropavo congensis***
Galliformes - Phasianidae
e   Congo Peacock; Congo Peafowl
f   Paon du Congo
d   Kongo-Pfau
i   Pavone del Congo

240   ***Agamia agami***
Ciconiiformes - Ardeidae
e   Agami Heron; Chestnut-bellied
Heron; Chestnut-bellied Agami;
Agami
f   Héron agami
d   Speer-Reiher
i   Airone agami

241   ***Agapornis canus***
Psittaciformes - Psittacidae
e   Grey-headed Lovebird; Madagascar
Lovebird
f   Inséparable à tête grise
d   Grauköpfchen
i   Inseparabile testagrigia

242   ***Agapornis fischeri***
Psittaciformes - Psittacidae
e   Fischer's Lovebird
f   Inséparable de Fischer
d   Fischer-Unzertrennlicher;

Pfirsichköpfchen
i Inseparabile di Fischer

243 *Agapornis lilianae*
Psittaciformes - Psittacidae
e Lilian's Lovebird; Nyasa Lovebird
f Inséparable de Lilian
d Rosenpapagei; Pfirsichköpfchen; Nyasa-Unzertrennlicher
i Inseparabile del Niassa

244 *Agapornis nigrigenis*
Psittaciformes - Psittacidae
e Black-cheeked Lovebird
f Inséparable à joues noires
d Rußköpfchen
i Inseparabile guancenere

245 *Agapornis personatus*
Psittaciformes - Psittacidae
e Yellow-collared Lovebird; Masked Lovebird; Zenker's Lovebird; Yellow-collared Parrot; Masked Parrot
f Inséparable masqué
d Schwarzköpfchen; Maskenköpfchen
i Inseparabile mascherato

246 *Agapornis pullarius*
Psittaciformes - Psittacidae
e Red-headed Lovebird; Red-faced Lovebird
f Inséparable à tête rouge
d Orangeköpchen
i Inseparabile facciarossa

247 *Agapornis roseicollis*
Psittaciformes - Psittacidae
e Rosy-faced Lovebird; Peach-faced Lovebird
f Inséparable rosegorge
d Rosenköpfchen; Rosenpapagei
i Inseparabile collorossa

248 *Agapornis swindernianus*
Psittaciformes - Psittacidae
e Black-collared Lovebird
f Inséparable à collier noir
d Grünköpfchen
i Inseparabile dal collare nero

249 *Agapornis taranta*
Psittaciformes - Psittacidae
e Black-winged Lovebird; Abyssinian Lovebird
f Inséparable d'Abyssinie
d Taranta-Unzertrennlicher; Bergpapagei; Tarantiner
i Inseparabile alinere

250 *Agelaius assimilis*
Passeriformes - Icteridae
e Red-shouldered Blackbird; Cuban Red-winged Blackbird

251 *Agelaius cyanopus*
Passeriformes - Icteridae
e Unicoloured Blackbird
f Carouge unicolore
d Einfarbstärling
i Ittero unicolore

252 *Agelaius flavus*
Passeriformes - Icteridae
e Saffron-cowled Blackbird; Saffron-crowned Blackbird
f Carouge safran
d Gilbstärling
i Ittero capozafferano

253 *Agelaius gubernator*
Passeriformes - Icteridae
e Two-coloured Blackbird

254 *Agelaius humeralis*
Passeriformes - Icteridae
e Tawny-shouldered Blackbird
f Merle (Ants); Petit Carouge
d Braunschulterstärling
i Ittero spallefulve

255 *Agelaius icterocephalus*
Passeriformes - Icteridae
e Yellow-hooded Blackbird
f Carouge à capuchon
d Gelbkopfstärling
i Ittero dal cappuccio giallo

256 *Agelaius phoeniceus*
Passeriformes - Icteridae
e Red-winged Blackbird; Redwing; Bicoloured Blackbird

f Carouge à épaulettes
d Rotflügelstärling; Rotschulterstärling
i Ittero dalle ali rosse; Ittero alirosse

257 *Agelaius ruficapillus*
Passeriformes - Icteridae
e Chestnut-capped Blackbird; Rufous-headed Blackbird
f Carouge à calotte rousse
d Braunkopfstärling; Rostkopfstärling
i Ittero capocastano

258 *Agelaius thilius*
Passeriformes - Icteridae
e Yellow-winged Blackbird; Yellow-shouldered Blackbird
f Carouge galonné
d Goldschulterstärling
i Ittero aligialle

259 *Agelaius tricolor*
Passeriformes - Icteridae
e Tricoloured Blackbird
f Carouge de Californie
d Dreifarbenstärling
i Ittero tricolore

260 *Agelaius xanthomus*
Passeriformes - Icteridae
e Yellow-shouldered Blackbird
f Carouge de Porto Rico
d Gelbschulterstärling
i Ittero spallegialle

261 *Agelaius xanthophthalmus*
Passeriformes - Icteridae
e Pale-eyed Blackbird; Yellow-eyed Blackbird
f Carouge à oeil clair
d Gelbaugenstärling
i Ittero occhigialli

262 *Agelastes meleagrides*
Galliformes - Numididae
e White-breasted Guineafowl
f Pintade à poitrine blanche
d Weißbrustperlhuhn
i Faraona pettobianco

263 *Agelastes niger*
Galliformes - Numididae

e Black Guineafowl
f Pintade noire
d Schwarzperlhuhn
i Farona nero

264 *Aglaeactis aliciae*
Apodiformes - Trochilidae
e Purple-backed Sunbeam
f Colibri d'Alice
d Purpurrückenkolibri
i Raggio di sole dorsopurpureo

265 *Aglaeactis castelnaudii*
Apodiformes - Trochilidae
e White-tufted Sunbeam
f Colibri de Castelneau
d Weißbüschelkolibri
i Raggio di sole ciuffibianco

266 *Aglaeactis cupripennis*
Apodiformes - Trochilidae
e Shining Sunbeam; Copper-winged Hummingbird
f Colibri étincelant
d Rosenschillerkolibri
i Raggio di sole splendente

267 *Aglaeactis pamela*
Apodiformes - Trochilidae
e Black-hooded Sunbeam
f Colibri pamela
d Pamela-Kolibri
i Raggio di sole testanera

268 *Aglaiocercus berlepschi*
Apodiformes - Trochilidae
e Venezuelan Sylph

269 *Aglaiocercus coelestis*
Apodiformes - Trochilidae
e Violet-tailed Sylph
f Sylphe à queue violette
d Langschwanzsylphe
i Silfide codaviola

270 *Aglaiocercus kingi*
Apodiformes - Trochilidae
e Long-tailed Sylph; Blue-tailed Sylph; Green-tailed Sylph
f Sylphe à longue queue

d	Himmelssylphe
i	Silfide codalunga

271 *Agriornis andicola*
Passeriformes - Tyrannidae
e White-tailed Shrike-Tyrant
f Gaucho à queue blanche
i Averla tiranna delle Ande

272 *Agriornis livida*
Passeriformes - Tyrannidae
e Great Shrike-Tyrant; Kittlitz's
Shrike-Tyrant
f Grand Gaucho; Gaucho de Kittlitz
d Würgertyrann
i Averla tiranna gigante

273 *Agriornis microptera*
Passeriformes - Tyrannidae
e Grey-bellied Shrike-Tyrant
f Gaucho argentin
d Weißbrauenwürgertyrann
i Averla tiranna ventregrigio

274 *Agriornis montana*
Passeriformes - Tyrannidae
e Black-billed Shrike-Tyrant;
Mountain Shrike-Tyrant; Black-
billed Ground-Tyrant
f Gaucho à bec noir; Gaucho
montagnard
d Bergtyrann
i Averla tiranna becconero

275 *Agriornis murina*
Passeriformes - Tyrannidae
e Lesser Shrike-Tyrant; Brown
Monjita; Mouse-brown Monjita;
Mouse-brown Shrike-Tyrant; Least
Shrike-Tyrant
f Gaucho souris
d Maustyrann
i Averla tiranna grigiotopo

276 *Ailuroedus buccoides*
Passeriformes - Ptilonorhynchidae
e White-eared Catbird; White-throated
Catbird
f Jardinier à joues blanches
d Weißkehlkatzenvogel;

Weißohrlaubenvogel
i Uccello gatto orecchiebianche

277 *Ailuroedus crassirostris*
Passeriformes - Ptilonorhynchidae
e Green Catbird; Spotted Catbird
f Jardinier vert; Oiseau à berceau vert
d Grünlaubenvogel; Grüner
Katzenvogel; Grünkatzenvogel
i Uccello gatto verde

278 *Ailuroedus dentirostris*
Passeriformes - Ptilonorhynchidae
e Tooth-billed Bowerbird (ANZ);
Tooth-billed Catbird; Stagemaker
f Jardinier à bec denté; Dathmocerque
de Winifred
d Zahnlaubenvogel; Zahnkatzenvogel;
Tennenbauer
i Uccello gatto beccodentato;
Ptilonorinco dal becco a sega

279 *Ailuroedus melanotis*
Passeriformes - Ptilonorhynchidae
e Black-eared Catbird
f Jardinier oreillard
i Uccello gatto maculato

280 *Aimophila aestivalis*
Passeriformes - Emberizidae
e Bachman's Sparrow; Pine-woods
Sparrow
f Bruant des pinèdes; Pinson des
pinières
d Hainammer
i Passero di Bachman

281 *Aimophila botterii*
Passeriformes - Emberizidae
e Botteri's Sparrow
f Bruant de Botteri
d Botteri-Ammer
i Passero di Botteri

282 *Aimophila carpalis*
Passeriformes - Emberizidae
e Rufous-winged Sparrow
f Bruant à épaulettes
d Rostflügelammer
i Passero alicastane

283 **Aimophila cassinii**
  Passeriformes - Emberizidae
e Cassin's Sparrow
f Bruant de Cassin
d Cassin-Ammer
i Passero di Cassin

284 **Aimophila humeralis**
  Passeriformes - Emberizidae
e Black-chested Sparrow
f Bruant à plastron
d Schwarzbrustammer
i Passero dalla pettorina nera

285 **Aimophila mystacalis**
  Passeriformes - Emberizidae
e Bridled Sparrow
f Bruant à moustaches
d Zügelammer
i Passero dai mustacchi

286 **Aimophila notosticta**
  Passeriformes - Emberizidae
e Oaxaca Sparrow
f Bruant d'Oaxaca
d Oaxaca-Ammer
i Passero di Oaxaca

287 **Aimophila petenica**
  Passeriformes - Emberizidae
e Peten Sparrow; Yellow-carpalled
  Sparrow

288 **Aimophila rufescens**
  Passeriformes - Emberizidae
e Rusty Sparrow
f Bruant roussâtre
d Rostrückenammer
i Passero rugginoso

289 **Aimophila ruficauda**
  Passeriformes - Emberizidae
e Stripe-headed Sparrow; Russet-tailed
  Sparrow; Striped-headed Sparrow
f Bruant ligné
d Rostschwanzammer
i Passero codarossa

290 **Aimophila ruficeps**
  Passeriformes - Emberizidae
e Rufous-crowned Sparrow

f Bruant à calotte fauve
d Rostscheitelammer
i Passero corona rossa

291 **Aimophila stolzmanni**
  Passeriformes - Emberizidae
e Tumbes Sparrow
f Bruant de Tumbes
d Stolzmann-Ammer
i Passero di Tumbes

292 **Aimophila strigiceps**
  Passeriformes - Emberizidae
e Stripe-capped Sparrow; Striped-
  capped Sparrow
f Bruant à calotte rayée
d Dabbene-Ammer
i Passero capostriato

293 **Aimophila sumichrasti**
  Passeriformes - Emberizidae
e Cinnamon-tailed Sparrow;
  Sumichrast's Sparrow
f Bruant à queue rousse
d Zimtschwanzammer
i Passero codacannella

294 **Aix galericulata**
  Anseriformes - Anatidae
e Mandarin Duck; Mandarin
f Canard mandarin; Aix mandarin; Aix
d Mandarinente; Mandarinenente
i Anatra mandarina

295 **Aix sponsa**
  Anseriformes - Anatidae
e Wood Duck; Carolina Duck
f Canard branchu; Canard carolin; Aix
  carolin
d Brautente
i Anatra sposa; Anatra sposina

296 **Ajaia ajaja**
  Ciconiiformes - Threskiornithidae
e Roseate Spoonbill
f Spatule rosée; Spatule (Ants);
  Spatule rose
d Rosalöffler
i Spatola rosata

297 *Alaemon alaudipes*
Passeriformes - Alaudidae
e Greater Hoopoe-Lark; Hoopoe-Lark;
Bifasciated Lark; Large Desert Lark
f Sirli du désert; Sirli ricoti; Sirli des
déserts
d Wüstenläuferlerche
i Allodola beccocurvo

298 *Alaemon hamertoni*
Passeriformes - Alaudidae
e Lesser Hoopoe-Lark; Witherby's
Lark
f Sirli de Witherby
d Somali-Läuferlerche
i Allodola di Hamerton

299 *Alauda arvensis*
Passeriformes - Alaudidae
e Skylark; Lark; Eurasian Skylark;
European Skylark; Common Skylark;
Western Skylark; Northern Skylark
f Alouette des champs
d Feldlerche
i Allodola; Allodola panterana;
Lodola; Allodola comune; Allodolina

300 *Alauda gulgula*
Passeriformes - Alaudidae
e Oriental Skylark; Lesser Skylark;
Indian Small Skylark (ISC); Small
Skylark; Oriental Lark; Eastern
Skylark; Indian Skylark; Little
Skylark
f Alouette gulgule; Alouette des
champs asiatique; Alouette des
champs sud-asiatique
d Kleine Feldlerche
i Allodola orientale

301 *Alauda japonica*
Passeriformes - Alaudidae
e Japanese Skylark (NA)
f Alouette du Japon
i Allodola giapponese

302 *Alauda razae*
Passeriformes - Alaudidae
e Razo Lark; Razo Island Lark; Razo
Short-toed Lark; Razo Island Short-
toed Lark

f Alouette de Razo; Alouette des îles
du Cap-Vert
d Razolerche
i Allodola dell'Isola Razo

303 **Alaudidae**
Passeriformes
e Larks
f Alouettes; Alaudidés; Calandres
d Lerchen
i Allodole; Allode; Lodole; Alaudidi

304 *Alca torda*
Charadriiformes - Alcidae
e Razorbill; Razor-billed Auk
f Petit Pingouin; Pingouin torda
d Tordalk
i Gazza marina

305 **Alcedinidae**
Coraciiformes
e Alcedinid Kingfishers; Kingfishers
f Martins-pecheurs; Alcédinés
d Eisvögel; Lieste; Fischer
i Alcedinidi; Martin pescatori

306 *Alcedo argentata*
Coraciiformes - Alcedinidae
e Silvery Kingfisher; Forest Kingfisher
f Martin-pêcheur argenté
d Silberfischer
i Martin pescatore argentato

307 *Alcedo atthis*
Coraciiformes - Alcedinidae
e Common Kingfisher; Kingfisher;
Small Blue-Kingfisher (ISC); Little
Kingfisher; River Kingfisher;
European Kingfisher; Eurasian
Kingfisher; Little Blue-Kingfisher;
Ceylon Common Kingfisher
f Martin-pêcheur d'Europe; Martin-
pêcheur
d Eisvogel
i Martin pescatore; Martin pescatore
comune

308 *Alcedo azurea*
Coraciiformes - Alcedinidae
e Azure Kingfisher
f Martin-pêcheur à dos bleu

d   Azurfischer
i   Martin pescatore azzurro

309   *Alcedo azurea diemenensis*
Coraciiformes - Alcedinidae
e   Tasmanian Azure Kingfisher (ANZ)

310   *Alcedo coerulescens*
Coraciiformes - Alcedinidae
e   Cerulean Kingfisher; Small Blue
Kingfisher; Philippine Pectoral
Kingfisher; River Kingfisher
f   Martin-pêcheur aigue-marine
d   Türkis-Fischer
i   Martin pescatore ceruleo

311   *Alcedo cristata*
Coraciiformes - Alcedinidae
e   Malachite Kingfisher
f   Martin-pêcheur huppé
d   Malachiteisvogel;
Zwerghaubenfischer
i   Martin pescatore malachite

312   *Alcedo cyanopecta*
Coraciiformes - Alcedinidae
e   Indigo-banded Kingfisher; Dwarf
River-Kingfisher
f   Martin-pêcheur à poitrine bleue
d   Blaubrustfischer
i   Martin pescatore nano pettazzurro

313   *Alcedo euryzona*
Coraciiformes - Alcedinidae
e   Blue-banded Kingfisher; Broad-
zoned Kingfisher
f   Martin-pêcheur à large bande
d   Brustbandeisvogel
i   Martin pescatore dalla fascia blu

314   *Alcedo hercules*
Coraciiformes - Alcedinidae
e   Blyth's Kingfisher; Great Blue-
Kingfisher
f   Martin-pêcheur de Blyth
d   Herkules-Eisvogel
i   Martin pescatore di Blyth

315   *Alcedo leucogaster*
Coraciiformes - Alcedinidae
e   White-bellied Kingfisher

f   Martin-pêcheur à ventre blanc;
Martin-chasseur à ventre blanc
d   Weißbauchzwergfischer
i   Martin pescatore ventrebianco

316   *Alcedo meninting*
Coraciiformes - Alcedinidae
e   Blue-eared Kingfisher; Deep-blue
Kingfisher
f   Martin-pêcheur meninting
d   Meninting-Eisvogel
i   Martin pescatore dalle orecchie blu

317   *Alcedo nais*
Coraciiformes - Alcedinidae
e   Principe Kingfisher
f   Martin-pêcheur de Principe
d   Prinzen-Zwergfischer
i   Martin pescatore di Principe

318   *Alcedo pusilla*
Coraciiformes - Alcedinidae
e   Little Kingfisher; Mangrove
Kingfisher
f   Martin-pêcheur poucet
d   Papua-Fischer
i   Martin pescatore piccolo

319   *Alcedo pusilla pusilla*
Coraciiformes - Alcedinidae
e   Torres Strait Little Kingfisher (ANZ)

320   *Alcedo quadribrachys*
Coraciiformes - Alcedinidae
e   Shining Blue-Kingfisher
f   Martin-pêcheur azuré
d   Schiller-Eisvogel
i   Martin pescatore splendente

321   *Alcedo semitorquata*
Coraciiformes - Alcedinidae
e   Half-collared Kingfisher
f   Martin-pêcheur à demi-collier
d   Kobalteisvogel
i   Martin pescatore dal semicollare

322   *Alcedo thomensis*
Coraciiformes - Alcedinidae
e   Sao Tomé Kingfisher
f   Martin-pêcheur de Sao Tomé

d   St. Thomas Zwergfischer
i   Martin pescatore di Sao Tomè

323   *Alcedo vintsioides*
      Coraciiformes - Alcedinidae
e   Malagasy Kingfisher; Malachite Kingfisher
f   Martin-pêcheur vintsi
d   Schwarzschwanzzwergfischer
i   Martin pescatore

324   *Alcedo websteri*
      Coraciiformes - Alcedinidae
e   Bismarck Kingfisher; Bismarck Pygmy-Kingfisher
f   Martin-pêcheur des Bismarck
d   Bismarck-Fischer
i   Martin pescatore di Webster

325   **Alcidae**
      Charadriiformes
o   Auks
f   Alcidés
d   Alken; Alkenvögel; Kurzflügler
i   Alcidi

326   *Alcippe brunnea*
      Passeriformes - Sylviidae
e   Dusky Fulvetta; Brown-capped Fulvetta; Gould's Tit-Babbler; Gould's Fulvetta; Rufous Tit-Babbler; Rufous-headed Fulvetta; Rufous Fulvetta; Brown Fulvetta; Brown-eared Fulvetta
f   Alcippe de Gould
d   Rotkopfalcippe
i   Fulvetta di Gould

327   *Alcippe brunneicauda*
      Passeriformes - Sylviidae
e   Brown Fulvetta; Brown Quaker-Babbler; Malaysian Fulvetta; Malaysian Nun-Babbler
f   Alcippe brun
d   Braunschwanzalcippe
i   Fulvetta bruna

328   *Alcippe castaneceps*
      Passeriformes - Sylviidae
e   Rufous-winged Fulvetta; Chestnut-headed Fulvetta; Chestnut-headed

Tit-Babbler; Chestnut-headed Nun-Babbler; Chestnut-headed Wren-Babbler
f   Alcippe à tête marron
d   Kastanienalcippe
i   Fulvetta testacastana

329   *Alcippe chrysotis*
      Passeriformes - Sylviidae
e   Golden-breasted Fulvetta; Golden-breasted Tit-Babbler; Golden Fulvetta
f   Alcippe à poitrine dorée
d   Goldalcippe
i   Fulvetta pettocastano

330   *Alcippe cinerea*
      Passeriformes - Sylviidae
e   Yellow-throated Fulvetta; Dusky Fulvetta; Yellow-throated Tit-Babbler; Gold-fronted Fulvetta; Dusky-green Fulvetta; Dusky-green Tit-Babbler
f   Alcippe à gorge jaune
d   Gelbkehlalcippe
i   Fulvetta golagialla

331   *Alcippe cinereiceps*
      Passeriformes - Sylviidae
e   Streak-throated Fulvetta; Streak-throated Alcippe; Brown-headed Fulvetta; Grey-headed Tit-Babbler; Brown-headed Tit-Babbler; Grey-headed Fulvetta
f   Alcippe à gorge rayée
d   Braunkopfalcippe
i   Fulvetta testagricia

332   *Alcippe dubia*
      Passeriformes - Sylviidae
e   Rusty-capped Fulvetta; Fulvous-capped Fulvetta; Olive-sided Fulvetta; Rufous-capped Tit-Babbler; Rufous-headed Tit-Babbler; Rufous-backed Fulvetta; White-throated Fulvetta
f   Alcippe à calotte rouille
d   Olioveflankenalcippe
i   Fulvetta capirossa

333 *Alcippe ludlowi*
Passeriformes - Sylviidae
e Brown-throated Fulvetta; Ludlow's
Fulvetta

334 *Alcippe morrisonia*
Passeriformes - Sylviidae
e Grey-cheeked Fulvetta; Grey-headed
Quaker-Babbler; Grey-eyed Quaker-
Babbler; Grey-cheeked Tit-Babbler;
Grey-cheeked Quaker-Babbler;
Common Fulvetta; Common Tit-
Babbler; Common Nun-Babbler;
Common Quaker-Babbler; Red-eyed
Fulvetta
f Alcippe à joues grises; Alcippe de
Swinhoe
d Grauwangenalcippe
i Fulvetta guancegrige

335 *Alcippe nipalensis*
Passeriformes - Sylviidae
e Nepal Fulvetta; White-eyed Quaker-
Babbler; Nepal Quaker-Babbler;
Grey-cheeked Tit-Babbler; Nepal
Babbler; White-eyed Fulvetta;
White-eyed Tit-Babbler; Black-
eyebrowed Fulvetta
f Alcippe du Nepal
d Nepal-Alcippe; Weißaugenalcippe
i Fulvetta del Nepal

336 *Alcippe peracensis*
Passeriformes - Sylviidae
e Mountain Fulvetta; Mountain Nun-
Babbler
f Alcippe bridé
d Malaien-Alcippe
i Fulvetta di montagna

337 *Alcippe poioicephala*
Passeriformes - Sylviidae
e Brown-cheeked Fulvetta; Quaker-
Babbler (ISC); Grey-eyed Nun-
Babbler; Grey-eyed Tit-Babbler
f Alcippe à joues brunes
d Graukopfalcippe
i Fulvetta corona grigia

338 *Alcippe pyrrhoptera*
Passeriformes - Sylviidae

e Javan Fulvetta; Javanese Fulvetta;
Javan Nun-Babbler; Javan Quaker-
Babbler
f Alcippe de Java
d Rotrückenalcippe
i Fulvetta di Giava

339 *Alcippe ruficapilla*
Passeriformes - Sylviidae
e Spectacled Fulvetta; Verreaux's
Rufous-headed Tit-Babbler; Rufous-
headed Tit-Babbler; Rufous-headed
Fulvetta; Verreaux's Fulvetta
f Alcippe de Verreaux
d Rotscheitelalcippe
i Fulvetta dagli occhiali

340 *Alcippe rufogularis*
Passeriformes - Sylviidae
e Rufous-throated Fulvetta; Red-
throated Tit-Babbler
f Alcippe à gorge rousse
d Kopfbandalcippe
i Fulvetta golarossa

341 *Alcippe striaticollis*
Passeriformes - Sylviidae
e Chinese Fulvetta; Mountain Tit-
Babbler; Striped Fulvetta; Striped-
neck Fulvetta; Chinese Mountain-
Fulvetta; Chinese Mountain-Tit-
Babbler; Mountain Fulvetta; Streak-
throated Tit-Babbler; Stripe-necked
Fulvetta
f Alcippe montagnard
d Bergalcippe
i Fulvetta montana della Cina

342 *Alcippe variegaticeps*
Passeriformes - Sylviidae
e Golden-fronted Fulvetta; Yellow-
fronted Fulvetta; Yellow-fronted Tit-
Babbler; Gold-fronted Fulvetta;
Gold-fronted Tit-Babbler; Variegated
Fulvetta; Yen's Fulvetta; Gold-
fronted Nun-Babbler
f Alcippe à front jaune
d Buntkopfalcippe
i Fulvetta variegata

343 *Alcippe vinipectus*
Passeriformes - Sylviidae
e White-browed Fulvetta; White-browed Tit-Babbler; Hodgson's Fulvetta
f Alcippe de Hodgson
d Weißbrauenalcippe
i Fulvetta dai sopraccigli bianchi

344 *Aleadryas rufinucha*
Passeriformes - Corvidae
e Rufous-naped Whistler; Rufous-naped Robin-Whistler
f Siffleur à nuque rousse
d Rotnackendickkopf
i Zufolatore nucarossa

345 *Alectoris barbara*
Galliformes - Phasianidae
e Barbary Partridge; Stone Partridge
f Perdrix gambra
d Felsenhuhn; Klippenhuhn
i Pernice sarda

346 *Alectoris chukar*
Galliformes - Phasianidae
e Chukar; Chukar Partridge
f Perdrix chukar; Perdrix choukar
d Chukarhuhn; Chukarsteinhuhn
i Coturnice; Pernice chukar; Pernice ciukar; Ciukar

347 *Alectoris graeca*
Galliformes - Phasianidae
e Rock Partridge; Greek Partridge
f Perdrix bartavelle
d Steinhuhn
i Coturnice orientale; Coturnice

348 *Alectoris magna*
Galliformes - Phasianidae
e Rusty-necklaced Partridge; Przewalski's Rock-Partridge; Przewalski's Partridge
f Perdrix de Przewalski
d Chinesisches Steihnhuhn
i Coturnice di Przewalski

349 *Alectoris melanocephala*
Galliformes - Phasianidae
e Arabian Partridge; Arabian Red-legged Partridge; Arabian Chukar; Rüppell's Partridge
f Perdrix à tête noire
d Schwarzkopfsteinhuhn
i Ciukar d'Arabia

350 *Alectoris philbyi*
Galliformes - Phasianidae
e Philby's Partridge; Philby's Rock-Partridge
f Perdrix de Philby
d Philby-Steinhuhn
i Coturnice di Philby

351 *Alectoris rufa*
Galliformes - Phasianidae
e Red-legged Partridge; European Red-Partridge; French Partridge
f Perdrix rouge
d Rothuhn
i Pernice rossa; Pernice rosa; Pernice comune

352 *Alectroenas madagascariensis*
Columbiformes - Columbidae
e Madagascar Blue-Pigeon; Madagascar Fruit-Dove; Red-tailed Blue-Pigeon
f Founingo bleu
d Madagaskar-Flugtaube; Blaue Madagaskar-Flugtaube
i Colomba azzurra del Madagascar

353 *Alectroenas nitidissima*
Columbiformes - Columbidae
e Mauritius Blue-Pigeon
f Founingo hollandais
d Mauritius-Fruchttaube
i Colomba azzurra di Mauritius

354 *Alectroenas pulcherrima*
Columbiformes - Columbidae
e Seychelles Blue-Pigeon; Seychelles Fruit-Dove; Red-crowned Wart-Pigeon; Red-crowned Blue-Pigeon; Warty-faced Blue-Pigeon
f Founingo rougecap
d Blaue Seychellen-Fruchttaube; Warzenfruchttaube
i Colomba azzurra delle Seychelle

355 *Alectroenas sganzini*
Columbiformes - Columbidae
e Comoro Blue-Pigeon; Comoro Fruit-Dove
f Founingo des Comores
d Komoren-Fruchttaube
i Colomba azzurra delle Comore

356 *Alectrurus risora*
Passeriformes - Tyrannidae
e Strange-tailed Tyrant
f Moucherolle à queue large;
Moucherolle à longue queue
d Wimpeltyrann
i Tiranno codastrana

357 *Alectrurus tricolor*
Passeriformes - Tyrannidae
e Cock-tailed Tyrant
f Moucherolle petit-coq
d Hahnenschwanztyrann
i Tiranno coda di gallo

358 *Alectura lathami*
Galliformes - Megapodiidae
e Australian Brush-Turkey; Brush-Turkey (ANZ); Australian Scrub-Turkey; Barnard's Purple Brush-Turkey; Yellow-pouched Brush-Turkey; Pouched Tallegallus;
Wattled Tallegallus; Wild Turkey (ANZ)
f Talégalle de Latham
d Buschhuhn
i Tacchino di boscaglia australiano

359 *Alethe castanea*
Passeriformes - Muscicapidae
e Fire-crested Alethe
f Alèthe à couronne orangée
d Orangescheitelalethe
i Alete corona rossa

360 *Alethe choloensis*
Passeriformes - Muscicapidae
e Thyolo Alethe; Cholo Alethe; Cholo Mountain Alethe; Mount Cholo Alethe
f Alèthe du Cholo; Alèthe de Mont Cholo

d Cholo-Alethe
i Alete di Monte Cholo

361 *Alethe diademata*
Passeriformes - Muscicapidae
e White-tailed Alethe; Firecrest Alethe; Fire-crested Alethe
f Alèthe à huppe rousse
d Diademalethe
i Alete codabianca

362 *Alethe fuelleborni*
Passeriformes - Muscicapidae
e White-chested Alethe; White-breasted Alethe (ANZ); Fülleborn's Alethe
f Alèthe à poitrine blanche
d Weißbrustalethe
i Alete di Fuelleborn

363 *Alethe poliocephala*
Passeriformes - Muscicapidae
e Brown-chested Alethe; Brown-crowned Alethe
f Alèthe à poitrine brune
d Braunbrustalethe
i Alete pettobruno

364 *Alethe poliophrys*
Passeriformes - Muscicapidae
e Red-throated Alethe
f Alèthe à gorge rousse
d Rotkehlalethe
i Alete golarossa

365 *Alisterus amboinensis*
Psittaciformes - Psittacidae
e Moluccan King-Parrot; Amboina King-Parrot; Amboina King-Parakeet; Ambon King-Parrot
f Perruche tricolore
d Amboina-Königssittich; Amboina-Sittich
i Parrochetto reale di Amboina

366 *Alisterus chloropterus*
Psittaciformes - Psittacidae
e Papuan King-Parrot; Green-winged King-Parrot; Green-winged King-Parakeet; Yellow-winged King-Parakeet

f   Perruche à ailes vertes
d   Gelbflügelkönigssittich;
      Gelbschulterkönigssittich; Papua-
      Sittich
i   Parrochetto reale aliverdi

367   *Alisterus scapularis*
      Psittaciformes - Psittacidae
e   Australian King-Parrot; King Parrot;
      Scarlet-and-green Parrot; King Lory;
      Red Lory; Queensland King-Parrot
f   Perruche royale
d   Königssittich; Australischer
      Königssittich
i   Parrochetto reale australiano

368   *Alle alle*
      Charadriiformes - Alcidae
e   Dovekie; Little Auk
f   Mergule nain
d   Krabbentaucher
i   Gazza marina minore

369   *Alophoixus affinis*
      Passeriformes - Pycnonotidae
e   Golden Bulbul; Moluccan Bulbul
f   Bulbul à queue d'or
d   Goldschwanzbülbül
i   Bulbul delle Molucche

370   *Alophoixus bres*
      Passeriformes - Pycnonotidae
e   Grey-cheeked Bulbul; Grey-cheeked
      Bearded-Bulbul; Olive White-
      throated Bulbul; Scrub Bulbul;
      White-throated Bulbul
f   Bulbul brès
d   Grauwangenbülbül
i   Bulbul guancegrige

371   *Alophoixus finschii*
      Passeriformes - Pycnonotidae
e   Finsch's Bulbul; Finsch's Bearded-
      Bulbul; Dwarf Bearded-Bulbul
f   Bulbul de Finsch
d   Finsch-Bülbül
i   Bulbul di Finsch

372   *Alophoixus flaveolus*
      Passeriformes - Pycnonotidae
e   White-throated Bulbul; Ashy-

      throated Bearded-Bulbul; Ashy-
      fronted Bearded-Bulbul; White-
      throated Bearded-Bulbul; Ashy-
      fronted Bulbul; Yellow-bellied
      Bulbul
f   Bulbul flavéole
d   Weißkehlbülbül
i   Bulbul frontegrigia

373   *Alophoixus ochraceus*
      Passeriformes - Pycnonotidae
e   Ochraceous Bulbul; Ochraceous
      Bearded-Bulbul; Brown Bulbul;
      Brown Bearded-Bulbul; Brown
      White-throated Bulbul
f   Bulbul ocré
d   Rostbauchbülbül
i   Bulbul ocraceo

374   *Alophoixus pallidus*
      Passeriformes - Pycnonotidae
o   Puff-throated Bulbul; Olivaceous
      Bearded-Bulbul; Olivaceous Bulbul;
      White-throated Bulbul
f   Bulbul pâle
d   Blaßbauchbülbül
i   Bulbul golapiumosa

375   *Alophoixus phaeocephalus*
      Passeriformes - Pycnonotidae
e   Yellow-bellied Bulbul; Grey-headed
      Bearded-Bulbul; Crestless Bulbul;
      Crestless White-throated Bulbul;
      White-throated Bulbul
f   Bulbul à calotte grise
d   Gelbbrustbülbül; Schwefelbülbül
i   Bulbul barbuto testagrigia

376   *Alopochen aegyptiacus*
      Anseriformes - Anatidae
e   Egyptian Goose
f   Ouette d'Égypte; Oie d'Égypte
d   Nil-Gans
i   Oca egiziana

377   *Amadina erythrocephala*
      Passeriformes - Passeridae
e   Red-headed Finch; Red-headed
      Weaver-Finch; Paradise Sparrow
f   Amadine à tête rouge; Moineau du
      Paradis

d Rotkopfamadine
i Amadina testarossa

378 *Amadina fasciata*
Passeriformes - Passeridae
e Cut-throat; Cut-throat Finch (CSA);
Ribbon Finch; Cut-throat Weaver
f Amadine cou-coupé; Cou-coupé
d Bandfink; Bandamadine
i Collotagliato; Golatagliata

379 *Amalocichla incerta*
Passeriformes - Petroicidae
e Lesser Ground-Robin; Lesser New
Guinea Thrush
f Petite Pseudobrève
d Pittadrossel
i Tordo minore della Nuova Guinea

380 *Amalocichla sclateriana*
Passeriformes - Petroicidae
e Greater Ground-Robin; Greater New
Guinea Thrush; Sclater's New Guinea
Thrush
f Grande Pseudobrève
d Weichschwanzdrossel
i Tordo maggiore della Nuova Guinea

381 *Amandava amandava*
Passeriformes - Estrildidae
e Red Avadavat; Strawberry Finch;
Red Munia; Avadavat; Tiger Finch;
Avadavade Finch; Strawberry
Waxbill
f Bengali rouge; Bengali moucheté;
Bengali de l'Inde
d Tigerfink; Prachtfink; Tigerastrild;
Kolibrifink
i Bengalino comune; Bengalino

382 *Amandava formosa*
Passeriformes - Estrildidae
e Green Avadavat; Green Munia (ISC);
Green Waxbill
f Bengali vert
d Olivastrild; Olivgrüner Astrild
i Bengalino verde

383 *Amandava subflava*
Passeriformes - Estrildidae
e Zebra Waxbill; Orange-breasted

Waxbill (CSA); Golden-breasted
Waxbill; Goldbreast
f Bengali zébré; Astrild à flancs rayé;
Astrild à ventre orange
d Goldbrütchen
i Bengalino pettodorato

384 *Amaurocichla bocagii*
Passeriformes - Sylviidae
e Bocage's Longbill; Sao Tomé
Shorttail
f Nasique de Bocage; Fauvette de Sao
Tomé
d Bräunling
i Beccolungo di Bocage

385 *Amaurolimnas concolor*
Gruiformes - Rallidae
e Uniform Crake; Red Rail (WI);
Wood Rail (WI); Water Partridge
(WI); Uniform Rail
f Râle concolore
d Einfarbralle
i Rallo unicolore

386 *Amaurornis akool*
Gruiformes - Rallidae
e Brown Crake; Brown Swamphen;
Crimson-legged Crake
f Râle akool; Marouette akool
d Braunbauchkielralle;
Braunbauchkleinralle
i Rallo bruno

387 *Amaurornis bicolor*
Gruiformes - Rallidae
e Black-tailed Crake; Elwes's Crake;
Rufous-backed Crake
f Râle bicolore; Marouette bicolore;
Marouette d' Elwes
d Zweifarbensumpfhuhn
i Rallo di Elwes

388 *Amaurornis flavirostra*
Gruiformes - Rallidae
e Black Crake; African Black-Crake
f Râle à bec jaune; Marouette à bec
jaune
d Negerralle; Mohrenralle
i Gallinella nera; Limnocorace dal
becco giallo

389   *Amaurornis isabellinus*
      Gruiformes - Rallidae
e   Isabelline Waterhen; Celebes
      Swamphen; Isabelline Bush-hen;
      Sulawesi Waterhen; Celebes
      Waterhen
f   Râle isabelle; Marouette isabelle
d   Isabellkielralle; Isabellkleinralle
i   Gallinella di Sulawesi

390   *Amaurornis moluccanus*
      Gruiformes - Rallidae
e   Rufous-tailed Waterhen; Moluccan
      Rufous-tailed Moorhen; Rufous-
      tailed Moorhen; Rufous-tailed Bush-
      hen
d   Molukken-Kielralle
i   Gallinella di Wallace

391   *Amaurornis olivaceus*
      Gruiformes - Rallidae
e   Bush-hen; Plain Bush-hen; Plain
      Swamphen; Common Bush-hen;
      Rufous-tailed Moorhen; Rufous-
      tailed Bush-hen; Rufous-tailed Rail;
      Rufous-tailed Crake; Rufous-vented
      Bush-hen; Rufous-vented Rail;
      Eastern Bush-hen
f   Râle des Philippines; Marouette des
      Philippines
d   Philippinen-Kielralle; Philippinen-
      Kleinralle
i   Gallinella olivacea

392   *Amaurornis olivieri*
      Gruiformes - Rallidae
e   Sakalava Rail; Olivier's Crake;
      Olivier's Rail
f   Râle d'Olivier; Marouette d'Olivier
d   Malegassen-Sumpfhuhn
i   Gallinella di Olivier

393   *Amaurornis phoenicurus*
      Gruiformes - Rallidae
e   White-breasted Waterhen; White-
      breasted Swamphen
f   Râle à poitrine blanche; Marouette à
      poitrine blanche
d   Weißbrustkielralle;
      Weißbrustkleinralle

i   Gallinella d'acqua petto bianco;
      Galliella pettobiancho

394   *Amaurospiza concolor*
      Passeriformes - Emberizidae
e   Blue-Seedeater
f   Sporophile bleu
d   Indigopfäffchen
i   Beccogrosso blu

395   *Amaurospiza moesta*
      Passeriformes - Emberizidae
e   Blackish-blue Seedeater; Blue-black
      Seedeater
f   Sporophile noirâtre
d   Weißachselpfäffchen
i   Beccogrosso blu-nero

396   *Amaurospiza relicta*
      Passeriformes - Emberizidae
e   Slate-blue Seedeater
d   Indigopfäffchen

397   *Amazilia amabilis*
      Apodiformes - Trochilidae
e   Blue-chested Hummingbird; Lovely
      Hummingbird
f   Ariane aimable
d   Blaubrustamazilie
i   Amazilia pettoblu

398   *Amazilia amazilia*
      Apodiformes - Trochilidae
e   Amazilia Hummingbird
f   Ariane de Lesson
d   Lesson-Amazilie
i   Amazilia di Amazili

399   *Amazilia bangsi*
      Passeriformes - Emberizidae
e   Bang's Hummingbird

400   *Amazilia beryllina*
      Apodiformes - Trochilidae
e   Beryline Hummingbird
f   Ariane beryl
d   Beryll-Amazilie
i   Amazilia berillina

401   *Amazilia boucardi*
      Apodiformes - Trochilidae

e Mangrove Hummingbird; Boucard's
Hummingbird
f Ariane de Boucard
d Mangroveamazilie
i Amazilia delle mangrovie

402 *Amazilia candida*
Apodiformes - Trochilidae
e White-bellied Emerald
f Ariane candide
d Bronzekopfamazilie
i Amazilia candida

403 *Amazilia castaneiventris*
Apodiformes - Trochilidae
e Chestnut-bellied Hummingbird
f Ariane à ventre roux
d Braunbauchamazilie
i Amazilia ventrecastano

404 *Amazilia chionogaster*
Apodiformes - Trochilidae
e White-bellied Hummingbird
f Ariane à ventre blanc
d Weißbauchamazilie
i Amazilia ventrebianco

405 *Amazilia chionopectus*
Apodiformes - Trochilidae
e White-chested Emerald
f Ariane à poitrine blanche
d Schneebrustamazilie
i Amazilia pettobianco

406 *Amazilia chlorostephana*
Apodiformes - Trochilidae
e Mosquitia Hummingbird

407 *Amazilia cyanifrons*
Apodiformes - Trochilidae
e Indigo-capped Hummingbird; Blue-
fronted Hummingbird; Blue-capped
Hummingbird
f Ariane à front bleu
d Indigostirnchen
i Amazilia fronteazzurra

408 *Amazilia cyanocephala*
Apodiformes - Trochilidae
e Azure-crowned Hummingbird; Red-
billed Azurecrown

f Ariane à couronne azur
d Blaukopfamazilie
i Amazilia corona blu

409 *Amazilia cyanura*
Apodiformes - Trochilidae
e Blue-tailed Hummingbird
f Ariane à queue bleue
d Blauschwanzamazilie
i Amazilia codablu

410 *Amazilia decora*
Apodiformes - Trochilidae
e Charming Hummingbird
f Ariane charmante
i Amazilia ornata

411 *Amazilia distans*
Apodiformes - Trochilidae
e Tachira Emerald
f Ariane du Tachira
d Tachira-Amazilie
i Amazilia di Tachira

412 *Amazilia edward*
Apodiformes - Trochilidae
e Snowy-breasted Hummingbird;
Snowy-bellied Hummingbird;
Edward's Hummingbird; White-
bellied Hummingbird
f Ariane d'Edward
d Edward-Amazilie
i Amazilia pettodineve

413 *Amazilia fimbriata*
Apodiformes - Trochilidae
e Glittering-throated Emerald
f Ariane de Linne
d Glitzeramazilie
i Amazilia gola scintillante

414 *Amazilia franciae*
Apodiformes - Trochilidae
e Andean Emerald
f Ariane de Francia
d Anden-Amazilie
i Amazilia delle Ande

415 *Amazilia handleyi*
Apodiformes - Trochilidae
e Escudo Hummingbird

416   *Amazilia lactea*
Apodiformes - Trochilidae
e   Sapphire-spangled Emerald
f   Ariane saphirine
d   Saphiramazilie
i   Amazilia di zaffiro

417   *Amazilia leucogaster*
Apodiformes - Trochilidae
e   Plain-bellied Emerald
f   Ariane vert-doré
d   Gmelin-Amazilie
i   Amazilia ventrechiaro

418   *Amazilia luciae*
Apodiformes - Trochilidae
e   Honduran Emerald; Honduras
Emerald; Lucy's Emerald
f   Ariane de Lucy
d   Honduras-Amazilie
i   Amazilia dell'Honduras

419   *Amazilia microrhyncha*
Apodiformes - Trochilidae
e   Small-billed Azurecrown
d   Elliot-Amazilie

420   *Amazilia niveocenter*
Apodiformes - Trochilidae
e   Snowy-bellied Hummingbird

421   *Amazilia ocai*
Passeriformes - Emberizidae
e   d'Oca's Hummiongbird

422   *Amazilia rosenbergi*
Apodiformes - Trochilidae
e   Purple-chested Hummingbird
f   Ariane de Rosenberg
d   Purpurbrustamazilie
i   Amazilia pettopurpureo

423   *Amazilia rutila*
Apodiformes - Trochilidae
e   Cinnamon Hummingbird
f   Ariane cannellé
d   Rostamazilie
i   Amazilia color canella

424   *Amazilia saucerrottei*
Apodiformes - Trochilidae

e   Steely-vented Hummingbird; Blue-
vented Hummingbird
f   Ariane de Sophie
d   Stahlamazilie
i   Amazilia coda d'acciaio

425   *Amazilia tobaci*
Apodiformes - Trochilidae
e   Copper-rumped Hummingbird
f   Ariane de Felicie
d   Tobago-Amazilie
i   Amazilia dal groppone ramato

426   *Amazilia tzacatl*
Apodiformes - Trochilidae
e   Rufous-tailed Hummingbird,
Rieffer's Hummingbird
f   Ariane à ventre gris
d   Braunschwanzamazilie
i   Amazilia coda rossiccia

427   *Amazilia versicolor*
Apodiformes - Trochilidae
e   Versicoloured Emerald; Variegated
Emerald
f   Ariane versicolore
d   Glanzamazilie
i   Amazilia versicolore

428   *Amazilia violiceps*
Apodiformes - Trochilidae
e   Violet-crowned Hummingbird;
Salvin's Hummingbird; Azure-crown
f   Ariane à couronne violette
d   Violetscheitelamazilie
i   Amazilia corona di violetta

429   *Amazilia viridicauda*
Apodiformes - Trochilidae
e   Green-and-white Hummingbird
f   Ariane du Pérou
d   Berlepsch-Amazilie
i   Amazilia verde e b ianco

430   *Amazilia viridifrons*
Apodiformes - Trochilidae
e   Green-fronted Hummingbird
f   Ariane à front vert
d   Grünstirnamazilie
i   Amazilia fronteverde

431 *Amazilia viridigaster*
Apodiformes - Trochilidae
e Green-bellied Hummingbird
f Ariane à ventre vert
d Grünbauchamazilie
i Amazilia ventreverde

432 *Amazilia wagneri*
Apodiformes - Trochilidae
e Cinnamon-sided Hummingbird
i Amazilia

433 *Amazilia yucatanensis*
Apodiformes - Trochilidae
e Buff-bellied Hummingbird; Fawn-
breasted Hummingbird; Yucatan
Hummingbird
f Ariane du Yucatan
d Yucatan-Amazilie
i Amazilia dello Yucatan

434 *Amazona aestiva*
Psittaciformes - Psittacidae
e Blue-fronted Parrot; Turquoise-
fronted Parrot; Blue-fronted Amazon;
Blue-fronted Amazon-Parrot;
Turquoise Parrot; Turquoise-fronted
Parrot
f Amazone à front bleu
d Blaustirnamazone; Rotbugamazone
i Amazzone fronteblu; Amazzonia
dalla fronte azurra

435 *Amazona agilis*
Psittaciformes - Psittacidae
e Black-billed Parrot; Parrot (WI);
Active Amazon; Black-billed
Amzaon; Jamaican Black-billed
Amazon
f Amazone verte
d Rotspiegelamazone
i Amazzone della Giamaica

436 *Amazona albifrons*
Psittaciformes - Psittacidae
e White-fronted Parrot; White-fronted
Amazon; Spectacled Amazon;
Spectacled Amazon-Parrot
f Amazone à front blanc
d Brillenamazone; Weißstirnamazone
i Amazzone frontebianca

437 *Amazona amazonica*
Psittaciformes - Psittacidae
e Orange-winged Parrot; Orange-
winged Amazon
f Amazone aourou
d Amazonen-Papagei; Venezuela-
Amazone
i Amazzone dell'Amazzonia

438 *Amazona arausiaca*
Psittaciformes - Psittacidae
e Red-necked Parrot; Jaco (WI);
Parrot; Red-necked Amazon; Blue-
faced Lesser Dominican Amazon-
Parrot; Bouquet's Lesser Dominican
Amazon-Parrot
f Amazone de Bouquet; Perroquet
(Ants)
d Blaukopfamazone
i Amazzone collorosso

439 *Amazona auropalliata*
Psittaciformes - Psittacidae
e Yellow-naped Parrot
f Amazone à nuque d'or
i Amazzone nucagialla

440 *Amazona autumnalis*
Psittaciformes - Psittacidae
e Red-lored Parrot; Yellow-cheeked
Parrot; Yellow-cheeked Amazon-
Parrot; Red-lored Amazon; Yellow-
cheeked Amazon
f Amazone diadème
d Gelbwangenamazone;
Herbstamazone; Rotstirnamazone
i Amazzone dagli redini rosse

441 *Amazona barbadensis*
Psittaciformes - Psittacidae
e Yellow-shouldered Parrot; Yellow-
shouldered Amazon
f Amazone à épaulettes jaunes
d Gelbflügelamazone;
Gelbschulteramazone; Kleiner
Gelbkopfamazone
i Amazzone delle Barbados

442 *Amazona brasiliensis*
Psittaciformes - Psittacidae
e Red-tailed Parrot; Red-tailed

Amazon; Red-tailed Amazon-Parrot
f   Amazone à joues bleues
d   Rotmaskenamazone;
     Rotschwanzamazone
i   Amazzone del Brasile; Amazzone
     dalla coda rossa

443   *Amazona collaria*
     Psittaciformes - Psittacidae
e   Yellow-billed Parrot; Parrot (WI);
     Red-throated Parrot; Yellow-billed
     Amazon; Red-throated Amazon;
     Red-throated Amazon-Parrot
f   Amazone sasabe
d   Jamaika-Amazone
i   Amazzone golarossa

444   *Amazona dufresniana*
     Psittaciformes - Psittacidae
e   Blue-cheeked Parrot; Blue-cheeked
     Amazone; Dufresne's Amazon
f   Amazone de Dufresne
d   Granada-Amazone
i   Amazzone guanceblu

445   *Amazona farinosa*
     Psittaciformes - Psittacidae
e   Mealy Parrot; Blue-crowned Parrot;
     Mealy Amazon
f   Amazone poudrée
d   Müller-Amazone
i   Amazzone farinosa

446   *Amazona festiva*
     Psittaciformes - Psittacidae
e   Festive Parrot; Festive Amazon
f   Amazone tavoua
d   Blaubartamazone; Blaukinnamazone
i   Amazzone gropponerosso

447   *Amazona finschi*
     Psittaciformes - Psittacidae
e   Lilac-crowned Parrot; Lilac-crowned
     Amazon; Finsch's Amazon-Parrot
f   Amazone à couronne lilas
d   Blaukappenamazone; Finsch-
     Amazone
i   Amazzone corona lilla

448   *Amazona guildingii*
     Psittaciformes - Psittacidae

e   St. Vincent Parrot; Parrot (WI); St.
     Vincent Amazon; Guilding's
     Amazon; Guilding's Amazon-Parrot
f   Amazone de Saint-Vincent
d   Königsamazone; St. Vincent-
     Amazone
i   Amazzone di Guilding; Amazzone di
     San Vincento

449   *Amazona imperialis*
     Psittaciformes - Psittacidae
e   Imperial Parrot; Imperial Amazon;
     August Amazon-Parrot; Dominican
     Amazon-Parrot
f   Amazone impériale
d   Braunschwanzamazone;
     Kaiseramazone
i   Amazzone imperiale

450   *Amazona kawalli*
     Psittaciformes - Psittacidae
o   Kawall's Parrot
f   Amazone de Kawall

451   *Amazona leucocephala*
     Psittaciformes - Psittacidae
e   Cuban Parrot; Rose-throated Parrot
     (WI); Bahama Parrot (WI); Cuban
     Amazon; Cuban Amazon-Parrot
f   Amazone de Cuba
d   Kuba-Amazone
i   Amazzone di Cuba

452   *Amazona mercenaria*
     Psittaciformes - Psittacidae
e   Scaly-naped Parrot; Scaly-naped
     Amazon; Mercenary Amazon-Parrot;
     Tschudi's Amazon-Parrot
f   Amazone mercenaire
d   Soldatenamazone
i   Amazzone loricata

453   *Amazona ochrocephala*
     Psittaciformes - Psittacidae
e   Yellow-crowned Parrot; Yellow-
     headed Amazon; Yellow-headed
     Parrot; Yellow-crowned Amazon;
     Yellow-naped Amazon-Parrot
f   Amazone à front jaune
d   Große Gelbkopfamazone;
     Gelbnackenamazone;

Gelbscheitelamazone; Surinam-
Amazone
i Amazzone frontegialla

454 *Amazona oratrix*
Psittaciformes - Psittacidae
e Yellow-headed Parrot
f Amazone à tête jaune
i Amazzone testagialla

455 *Amazona pretrei*
Psittaciformes - Psittacidae
e Red-spectacled Parrot; Red-
spectacled Amazon; Prêtre's
Amazon-Parrot
f Amazone de Prêtre
d Prachtamazone
i Amazzone dagli occhiali rossi

456 *Amazona rhodocorytha*
Psittaciformes - Psittacidae
e Red-browed Parrot
f Amazone à sourcils rouges
i Amazzone dai corona rossa

457 *Amazona tucumana*
Psittaciformes - Psittacidae
e Tucuman Parrot; Alder Parrot;
Tucuman Amazon; Tucuman
Amazon-Parrot
f Amazone de Tucuman
d Tucuman-Amazone
i Amazzone di Tucuman

458 *Amazona ventralis*
Psittaciformes - Psittacidae
e Hispaniolan Parrot; Hispaniolan
Amazon; San Domingo Amazon;
Salle's Amazon-Parrot
f Amazone d'Hispaniola; Perroquet
(Ants)
d Blaukronenamazone
i Amazzone di Hispaniola

459 *Amazona versicolor*
Psittaciformes - Psittacidae
e St. Lucia Parrot; Jacquot (WI); St.
Lucia Amzon; St. Lucia Amazon-
Parrot; Versicolor Amazon
f Amazone de Sainte-Lucie; Jacquot
(Ants)

d Blaumaskenamazone;
Blaustirnamazone
i Amazzone di Santa Lucia

460 *Amazona vinacea*
Psittaciformes - Psittacidae
e Vinaceous Parrot; Vinaceous
Amazon; Vinaceous Amazon-Parrot
f Amazone vineuse
d Taubenhalsamazone; Weinroter
Amazone
i Amazzone vinacea

461 *Amazona viridigenalis*
Psittaciformes - Psittacidae
e Red-crowned Parrot; Green-cheeked
Parakeet; Green-cheeked Amazon-
Parrot; Green-cheeked Amazon
f Amazone à joues vertes
d Grünwangenamazone
i Amazzone guanceverdi

462 *Amazona vittata*
Psittaciformes - Psittacidae
e Puerto Rican Parrot; Puerto Rican
Amazon (WI); Puerto Rican
Amazon-Parrot; Red-fronted
Amazon
f Amazone de Porto Rico
d Puertorico-Amazone
i Amazzone di Portorico

463 *Amazona xantholora*
Psittaciformes - Psittacidae
e Yellow-lored Parrot; Yucatan Parrot;
Yellow-lored Amazon-Parrot;
Yellow-lored Amazon
f Amazone du Yucatan
d Gelbzügelamazone;
Goldzügelamazone
i Amazzone dalle redini gialle

464 *Amazona xanthops*
Psittaciformes - Psittacidae
e Yellow-faced Parrot; Yellow-faced
Amazon; Yellow-crowned Amazon
f Amazone à face jaune
d Gelbbauchamazone;
Goldbauchamazone
i Amazzone facciagialla

465   *Amazonetta brasiliensis*
       Anseriformes - Anatidae
e   Brazilian Teal; Brazilian Duck
f   Canard amazonette; Amazonette
       maréce
d   Amazonas-Ente
i   Alzavola brasiliana

466   *Amblycercus australis*
       Passeriformes - Icteridae
e   Champman's Cacique

467   *Amblycercus holosericeus*
       Passeriformes - Icteridae
e   Yellow-billed Cacique
d   Gelbschnabelkassike
i   Cacico occhigialli

468   *Amblyornis flavifrons*
       Passeriformes - Ptilonorhynchidae
e   Golden-fronted Bowerbird; Yellow-
       fronted Bowerbird; Yellow-fronted
       Gardener-Bowerbird
f   Jardinier à front d'or
d   Gelbscheitelgärtner
i   Giardiniere frontegialla; Uccello
       giardiniere dalla fronte gialla

469   *Amblyornis inornatus*
       Passeriformes - Ptilonorhynchidae
e   Vogelkop Bowerbird; Gardner
       Bowerbird; Plain Bowerbird; Brown
       Gardnerbird; Vokelkop Gardner-
       Bowerbird
f   Jardinier brun
d   Hüttengärtner; Schopflaubenvogel
i   Giardiniere di Vokelkop

470   *Amblyornis macgregoriae*
       Passeriformes - Ptilonorhynchidae
e   MacGregor's Bowerbird;
       MacGregor's Gardener-Bowerbird;
       MacGregor's Gardner; Crested
       Bowerbird
f   Jardinier de Macgregor
d   Gelbhaubengärtner;
       Goldhaubengärtner
i   Giardiniere di McGregor

471   *Amblyornis subalaris*
       Passeriformes - Ptilonorhynchidae

e   Streaked Bowerbird; Striped
       Bowerbird; Orange-crested Gardner;
       Striped Gardener-Bowerbird;
       Orange-crested Bowerbird; Eastern
       Gardener-Bowerbird; Eastern
       Bowerbird
f   Jardinier à huppe orangé
d   Rothaubengärtner
i   Giardiniere striato

472   *Amblyospiza albifrons*
       Passeriformes - Ploceidae
e   Grosbeak Weaver; Thick-billed
       Weaver (CSA); White-fronted
       Grosbeak
f   Amblyospize à front blanc; Grosbec
       à front blanc; Tisserin à gros bec
d   Weißstirnweber
i   Tessitore beccogrosso

473   *Amblyramphus holosericeus*
       Passeriformes - Fringillidae
e   Scarlet-headed Blackbird; Orange-
       headed Blackbird
f   Troupiale à tête rouge
d   Rotkopfstärling
i   Gracchio testarossa

474   *Ammodramus aurifrons*
       Passeriformes - Emberizidae
e   Yellow-browed Sparrow
f   Bruant à front d'or
d   Gelbwangenammer
i   Passero dai sopraccigli gialli

475   *Ammodramus bairdii*
       Passeriformes - Fringillidae
e   Baird's Sparrow
f   Bruant de Baird
d   Baird-Ammer
i   Passero di Baird

476   *Ammodramus caudacutus*
       Passeriformes - Emberizidae
e   Saltmarsh Sharp-tailed Sparrow;
       Sharp-tailed Sparrow
f   Bruant à queue aiguë; Pinson à queue
       aigue
d   Spitzschwanzammer
i   Passero codacuta

477 **Ammodramus henslowii**
Passeriformes - Emberizidae
e Henslow's Sparrow
f Bruant de Henslow
d Henslow-Ammer
i Passero di Henslow

478 **Ammodramus humeralis**
Passeriformes - Emberizidae
e Grassland Sparrow
f Bruant des savanes
d Wachtelammer
i Passero delle erbe

479 **Ammodramus leconteii**
Passeriformes - Emberizidae
e Le Conte's Sparrow
f Bruant de Le Conte; Pinson de Le Conte
d Le Conte-Ammer
i Passero di LeConte

480 **Ammodramus maritimus**
Passeriformes - Emberizidae
e Seaside Sparrow; Common Seaside-Sparrow
f Bruant maritime; Pinson maritime
d Strandammer; Strandammerfink
i Passero delle coste

481 **Ammodramus mirabilis**
Passeriformes - Emberizidae
e Cape Sable Sparrow; Cape Sable Seaside-Sparrow

482 **Ammodramus nelsoni**
Passeriformes - Emberizidae
e Nelson's Sharp-tailed Sparrow; Sharp-tailed Sparrow
f Bruant de Nelson

483 **Ammodramus nigrescens**
Passeriformes - Emberizidae
e Dusky Seaside-Sparrow

484 **Ammodramus savannarum**
Passeriformes - Emberizidae
e Grasshopper Sparrow; Savannabird (WI); Grass Dodger (WI); Grass Pink (WI)
f Bruant sauterelle; Moineau des

herbes (Ants); Pinson sauterelle
d Heuschreckenammer
i Passero locustella

485 **Ammomanes cincturus**
Passeriformes - Alaudidae
e Bar-tailed Desert Lark; Bar-tailed Lark; Bar-tailed Sandlark; Black-tailed Lark
f Ammomane élégante
d Sandlerche
i Allodola codafaciata; Allodola del deserto codafasciata

486 **Ammomanes deserti**
Passeriformes - Alaudidae
e Desert Lark; Sandlark
f Ammomane isabelline; Ammomane du désert; Alouette du désert
d Steinlerche; Wüstenlerche
i Allodola del deserto

487 **Ammomanes grayi**
Passeriformes - Alaudidae
e Gray's Lark; Gray's Sandlark; Sandlark
f Ammomane de Gray
d Namib-Lerche
i Allodola di Gray

488 **Ammomanes phoenicurus**
Passeriformes - Alaudidae
e Rufous-tailed Lark; Rufous-tailed Finch-Lark (ISC); Rufous-tailed Desert-Lark; Bar-tailed Lark
f Ammomane à queue rouge
d Rotschwanzlerche
i Allodola del deserto codarossa

489 **Ammoperdix griseogularis**
Galliformes - Phasianidae
e See-see Partridge; See-see
f Perdrix si-si
d Persisches Wüstenhuhn
i Pernice del deserto; Pernice di Persia; Pernice di Iran

490 **Ammoperdix heyi**
Galliformes - Phasianidae
e Sand Partridge
f Perdrix de Hey

d   Arabisches Wüstenhuhn
i   Pernice delle sabbie; Pernice d'
      Arabia

491   ***Ampeliceps coronatus***
        Passeriformes - Sturnidae
e   Golden-crested Mynah; Gold-crested
      Mynah (ISC)
f   Martin couronné; Mainate couronnée
d   Kronenatzel
i   Maina crestadorata

492   ***Ampelioides tschudii***
        Passeriformes - Tyrannidae
e   Scaled Fruiteater
f   Cotinga écaillé
d   Schuppenschmuckvogel
i   Cotinga di Tschudi

493   ***Ampelion rubrocristatus***
        Passeriformes - Tyrannidae
e   Red-crested Cotinga
f   Cotinga à huppe rouge
d   Rotkopfzuser
i   Cotinga crestarossa

494   ***Ampelion rufaxilla***
        Passeriformes - Tyrannidae
e   Chestnut-crested Cotinga
f   Cotinga à tête rousse
d   Braunkopfzuser
i   Cotinga crestacastana

495   ***Amphilais seebohmi***
        Passeriformes - Sylviidae
e   Grey Emu-tail; Seebohm's Feather-
      tailed Warbler
f   Amphilais tachetée; Droméocerque
      tacheté
d   Fleckenmausschwanz
i   Forapaglie codaspiumosa di
      Seebohm

496   ***Amphispiza belli***
        Passeriformes - Emberizidae
e   Sage-Sparrow; Bell's Sparrow; Bell's
      Sage-Sparrow
f   Bruant de Bell; Pinson de Bell
d   Beifußammer
i   Passero della salvia

497   ***Amphispiza bilineata***
        Passeriformes - Emberizidae
e   Black-throated Sparrow
d   Schwarzkehlammer
i   Passero pettonero

498   ***Amphispiza nevadensis***
        Passeriformes - Emberizidae
e   Sage-Sparrow

499   ***Amphispiza quinquestriata***
        Passeriformes - Emberizidae
e   Five-striped Sparrow
f   Bruant pentaligne
d   Fünfstreifenammer
i   Passero dalle cinque strie

500   ***Amytornis barbatus***
        Passeriformes - Maluridae
e   Grey Grasswren
f   Amytis gris
d   Brauengrasschlüpfer
i   Scricciolo delle erbe dal collare

501   ***Amytornis barbatus barbatus***
        Passeriformes - Maluridae
e   Bulloo Grey-Grasswren (ANZ)

502   ***Amytornis dorotheae***
        Passeriformes - Maluridae
e   Carpentarian Grasswren; Red-winged
      Grasswren; Carpenter Grasswren;
      Dorothy's Grasswren
f   Amytis de Dorothy
d   Carpentaria-Grasschlüpfer
i   Scricciolo delle erbe di Dorothy

503   ***Amytornis goyderi***
        Passeriformes - Maluridae
e   Eyrean Grasswren
f   Amytis de l'Eyre
d   Eyre-Grasschlüpfer
i   Scricciolo delle erbe del lago

504   ***Amytornis housei***
        Passeriformes - Maluridae
e   Black Grasswren
f   Amytis noir
d   Schwarzkehlgrasschlüpfer
i   Scricciolo delle erbe nero

i Anatra castana australiana; Anatra castana

531 *Anas chlorotis*
Anseriformes - Anatidae
e Brown Teal; Flightless Teal
i Mestolone

532 *Anas clypeata*
Anseriformes - Anatidae
e Northern Shoveler; Shoveler; Shovel-mouth (WI); Spoonbill (WI); Common Shoveler; European Shoveler
f Canard souchet
d Löffelente
i Mestolino; Mestolone; Mestolone comune

533 *Anas couesi*
Anseriformes - Anatidae
e Coue's Gadwall

534 *Anas crecca*
Anseriformes - Anatidae
e Common Teal; Teal; Green-winged Teal (NA); Eurasian Teal; Duck-and-Teal (WI)
f Sarcelle d'hiver; Sarcelle à ailes vertes
d Krickente
i Alzavola; Alzavola comune

535 *Anas crecca carolinensis*
Anseriformes - Anatidae
e Green-winged Teal
f Sarcelle à ailes vertes; Sarcelle de la Caroline
d Amerikanische Krickente
i Alzavole aliverdi; Alzavola aliverdi

536 *Anas cyanoptera*
Anseriformes - Anatidae
e Cinnamon Teal
f Sarcelle cannellée
d Zimtente
i Anatra color cannella; Anatra amaranto

537 *Anas diazi*
Anseriformes - Anatidae

e Mexican Duck
f Canard du Mexique

538 *Anas discors*
Anseriformes - Anatidae
e Blue-winged Teal; Teal (WI); Duck-and-Teal (WI)
f Sarcelle soucrourou; Sarcelle à ailes bleues; Sarcelle (Ants)
d Blauflügelente
i Marzaiola americana

539 *Anas eatoni*
Anseriformes - Anatidae
e Eaton's Pintail; Southern Pintail; Kerguelen Pintail
f Canard d'Eaton
d Kerguelen-Ente
i Anatra di Eaton

540 *Anas erythrorhynchus*
Anseriformes - Anatidae
e Red-billed Duck; Red-billed Teal (CSA); Red-billed Pintail-Teal; Red-billed Pintail
f Canard à bec rouge
d Rotschnabelente
i Anatra beccorosso

541 *Anas falcata*
Anseriformes - Anatidae
e Falcated Teal; Falcated Duck
f Canard à faucilles
d Sichelente
i Anatra falcata

542 *Anas flavirostris*
Anseriformes - Anatidae
e Speckled Teal; Chilean Teal; South American Teal; South American Green-winged Teal; Yellow-billed Teal; Andean Teal; Merida Teal
f Sarcelle tachetée
d Südamerikanische Krickente
i Alzavola macchiettata; Alzavola argentina

543 *Anas flavirostris andinum*
Anseriformes - Anatidae
f Sarcelle des Andes
d Anden-Ente

544 *Anas formosa*
Anseriformes - Anatidae
- e Baikal Teal
- f Sarcelle élégante
- d Gluckente; Baikal-Ente; Prachtente; Zierente
- i Alzavola asiatica

545 *Anas fulvigula*
Anseriformes - Anatidae
- e Mottled Duck; Dusky Duck; American Black-Duck
- f Canard brun; Sarcelle élégante
- d Dunkelente
- i Anatra chiazzata

546 *Anas galapagensis*
Anseriformes - Anatidae
- e Galapagos Pintail

547 *Anas georgica*
Anseriformes - Anatidae
- e Yellow-billed Pintail; Brown Pintail; Georgian Teal; South Georgian Teal; Georgian Pintail; Chilean Pintail; Niceforo's Pintail
- f Canard à queue pointue
- d Spitzschwanzente
- i Anatra georgiana; Xodone della Georgia

548 *Anas gibberifrons*
Anseriformes - Anatidae
- e Indonesian Teal; Sunda Teal (ISC); Grey Teal
- f Sarcelle grise
- d Weißkehlente
- i Alzavola della Sonda

549 *Anas gibberifrons albogularis*
Anseriformes - Anatidae
- f Sarcelle des Andaman

550 *Anas gracilis*
Anseriformes - Anatidae
- e Grey Teal; Australasian Grey-Teal; Slender Teal; Oceanic Teal; Mountain Teal
- f Sarcelle australasienne
- i Alzavola grigia

551 *Anas hottentota*
Anseriformes - Anatidae
- e Hottentot Teal
- f Sarcelle hottentote
- d Hottentotten-Ente; Pünktchenente
- i Anatra ottentotta

552 *Anas laysanensis*
Anseriformes - Anatidae
- e Laysan Duck; Laysan Teal
- f Canard de Laysan
- d Laysan-Ente
- i Anatra di Laysan; Germano di Laysan

553 *Anas luzonica*
Anseriformes - Anatidae
- e Philippine Duck; Luzon Duck; Philippine Mallard
- f Canard des Philippines
- d Philippinen-Ente
- i Alzavola delle Filippine; Germano delle Filippine

554 *Anas melleri*
Anseriformes - Anatidae
- e Meller's Duck
- f Canard de Meller
- d Madagaskar-Ente
- i Anatra di Meller

555 *Anas oxypterum*
Anseriformes - Anatidae
- e Sharp-winged Teal

556 *Anas penelope*
Anseriformes - Anatidae
- e Eurasian Wigeon; Wigeon; European Widgeon
- f Canard siffleur; Canard siffleur d'Europe
- d Pfeifente; Eurpäische Pfeifente
- i Fischione; Fischione europeo

557 *Anas platalea*
Anseriformes - Anatidae
- e Red Shoveler; Argentine Shoveler; South American Shoveler; Blue-winged Shoveler
- f Canard spatule
- d Südamerikanische Löffelente;

Fuchslöffelente
i Mestolone argentino

558 ***Anas platyrhynchos***
Anseriformes - Anatidae
e Mallard; Mallard Duck; Duck-and-Teal (WI); Mexican Duck
f Canard colvert
d Stockente
i Germano reale; Colloverde; Anatra selvatica

559 ***Anas poecilorhyncha***
Anseriformes - Anatidae
e Spot-billed Duck; Spotbill (ISC); Grey Duck (ISC); Spotbill Duck; Spot-billed Grey-Duck; Grey Duck
f Canard à bec tacheté
d Fleckschnabelente
i Anatra beccomacchiato; Germano indiano

560 ***Anas poecilorhyncha harringtoni***
Anseriformes - Anatidae
e Burmese Spottbill (ISC)

561 ***Anas poecilorhyncha zonorhyncha***
Anseriformes - Anatidae
f Canard de Chine

562 ***Anas puna***
Anseriformes - Anatidae
e Puna Teal
f Sarcelle du Puna
d Punaente
i Anatra della Puna

563 ***Anas querquedula***
Anseriformes - Anatidae
e Garganey; Blue-winged Teal (ISC)
f Sarcelle d'été
d Knäkente; Knackente
i Marzaiola

564 ***Anas rhynchotis***
Anseriformes - Anatidae
e Australian Shoveler; Blue-winged Shoveler; Australasian Shoveler; Southern Shoveler; New Zealand Shoveler
f Canard bridé

d Halbmondlöffelente; Australische Löffelente
i Marzaiolo australiano; Mestolone australiano

565 ***Anas rubripes***
Anseriformes - Anatidae
e American Black-Duck; Black Duck
f Canard noir; Canard noirâtre
d Dunkelente; Rotfußente
i Anatra nera americana; Anatra zamperosse

566 ***Anas sibilatrix***
Anseriformes - Anatidae
e Chiloe Wigeon; Southern Wigeon; Chilean Wigeon
f Canard de Chiloe
d Chile-Pfeifente; Prachtpfeifente
i Fischione del Cile; Fischione cileno

567 ***Anas smithii***
Anseriformes - Anatidae
e Cape Shoveler; Smith's Shoveler; South African Shoveler
f Canard de Smith
d Kap-Löffelente
i Mestolone del Capo

568 ***Anas sparsa***
Anseriformes - Anatidae
e African Black-Duck; Black River-Duck
f Canard noirâtre
d Schwarzente; Fleckente
i Anatra nera africana

569 ***Anas specularis***
Anseriformes - Anatidae
e Spectacled Duck; Bronze-winged Duck
f Canard à lunettes
d Kupferspiegelente
i Anatra alibronzate

570 ***Anas specularoides***
Anseriformes - Anatidae
e Crested Duck
f Canard huppé
d Shopfente
i Anatra crestata

571   *Anas strepera*
        Anseriformes - Anatidae
e   Gadwall; Grey Duck
f   Canard chipeau
d   Schnatterente; Mittelente
i   Canapiglia

572   *Anas superciliosa*
        Anseriformes - Anatidae
e   Pacific Black-Duck; Australian Grey-
        Duck; Australian Wild Duck;
        Australasian Wild Duck; Black
        Duck; Grey Duck
f   Canard à sourcils
d   Augenbrauenente
i   Anatra del Pacifico

573   *Anas undulata*
        Anseriformes - Anatidae
e   Yellow-billed Duck; African
        Yellowbill
f   Canard à bec jaune
d   Gelbschnabelente
i   Anatra ondulata; Germano africano

574   *Anas versicolor*
        Anseriformes - Anatidae
e   Silver Teal; Versicolor Teal
f   Sarcelle bariolée
d   Silberente
i   Anatra versicolore

575   *Anas wyvilliana*
        Anseriformes - Anatidae
e   Hawaiian Duck; Koloa
f   Canard des Hawai
d   Zwergstockente
i   Anatra delle Hawaii

576   *Anastomus lamelligerus*
        Ciconiiformes - Ciconiidae
e   African Openbill; African Open-
        billed Stork (CSA); Open-billed
        Stork (CSA)
f   Bec-ouvert africain
d   Klaffschnabel; Schwarzer
        Klaffschnabel; Mohrenklaffschnabel
i   Anastomo d'Africa; Anastomo
        africano; Anatoma

577   *Anastomus oscitans*
        Ciconiiformes - Ciconiidae
e   Asian Openbill; Openbill Stork
        (ISC); Asian Open-billed Stork;
        Asiatic Openbill; Asiatic Open-billed
        Stork; Oriental Openbill
f   Bec-ouvert indien; Bec-ouvert blanc
d   Weißer Klaffschnabel;
        Silberklaffschnabel
i   Anastomo asiatico

578   **Anatidae**
        Anseriformes
e   Ducks; Geese; Swans
f   Anatidés
d   Entenvögel; Entenartige Vögel;
        Gründelenten
i   Anatidi

579   *Ancistrops strigilatus*
        Passeriformes - Furnariidae
e   Chestnut-winged Hookbill
f   Anabate à bec crochu
d   Hakenschnabelblattspäher
i   Ticotico alicastane

580   *Andigena cucullata*
        Piciformes - Ramphastidae
e   Hooded Mountain-Toucan; Hooded
        Toucan
f   Toucan à capuchon
d   Schwarzkopftukan;
        Schwarzkehlblautukan
i   Tucano di monte dl cappuccio

581   *Andigena hypoglauca*
        Piciformes - Ramphastidae
e   Grey-breasted Mountain-Toucan;
        Grey-breasted Toucan
f   Toucan bleu
d   Blautukan; Gelbbürzelblautukan
i   Tucano di monte pettogrigio

582   *Andigena laminirostris*
        Piciformes - Ramphastidae
e   Plate-billed Mountain-Toucan;
        Laminated Toucan; Plain-billed
        Mountain-Toucan; Laminated
        Mountain-Toucan
f   Toucan montagnard

d   Leistenschnabeltukan
i   Tucano di monte beccopiatto

583   ***Andigena nigrirostris***
    Piciformes - Ramphastidae
e   Black-billed Mountain-Toucan;
    Black-billed Toucan
f   Toucan à bec noir
d   Schwarzschnabeltukan
i   Tucano di monte becconero

584   ***Androdon aequatorialis***
    Apodiformes - Trochilidae
e   Tooth-billed Hummingbird; Cadet
    Hummingbird; Equatorial
    Hummingbird
f   Colibri d'Équateur
d   Hakenkolibri
i   Colibrì becco a dente

585   ***Andropadus ansorgei***
    Passeriformes - Pycnonotidae
e   Ansorge's Greenbul; Ansorge's Grey-
    Bulbul; Ansorge's Bulbul
f   Bulbul d'Ansorge
d   Ansorge-Bülbül
i   Bulbul di Ansorge

586   ***Andropadus chlorigula***
    Passeriformes - Pycnonotidae
e   Southern Mountain-Greenbul; Green-
    throated Bulbul; Green-throated
    Greenbul; Greenish-throated
    Greenbul
f   Bulbul à gorge verte
i   Bulbul golaverde

587   ***Andropadus curvirostris***
    Passeriformes - Pycnonotidae
e   Plain Greenbul; Cameroon Sombre
    Greenbul (CSA); Sombre Greenbul;
    Cassin's Greenbul; Cameroon
    Greenbul
f   Bulbul curvirostre
d   Alexander-Bülbül
i   Bulbul del Camerun

588   ***Andropadus fusciceps***
    Passeriformes - Pycnonotidae
e   Northern Mountain-Greenbul

589   ***Andropadus gracilirostris***
    Passeriformes - Pycnonotidae
e   Slender-billed Greenbul; Slender-
    billed Bulbul
f   Bulbul à bec grêle
d   Schmalschnabelbülbül
i   Bulbul beccofine

590   ***Andropadus gracilis***
    Passeriformes - Pycnonotidae
e   Grey Greenbul; Little Grey-
    Greenbul; Little Grey-Bulbul
f   Bulbul gracile
d   Zwergbülbül
i   Bulbul piccolo

591   ***Andropadus hallae***
    Passeriformes - Pycnonotidae
e   Hall's Greenbul; Hall's Green-Bulbul;
    Mrs.. Hall's Greenbul
f   Bulbul de Hall
d   Hall-Bülbül
i   Bulbul della signora Hall

592   ***Andropadus importunus***
    Passeriformes - Pycnonotidae
e   Sombre Greenbul; Zanzibar Sombre
    Greenbul (CSA); Sombre Bulbul
    (CSA); Willie (CSA)
f   Bulbul importun
d   Kap-Grünbülbül; Schlichtbülbül;
    Graugrünbülbül
i   Bulbul di Zanzibar

593   ***Andropadus kakamegae***
    Passeriformes - Pycnonotidae
e   Kakamega Greenbul
f   Bulbul de Kakamega
i   Bulbul di Kakamega

594   ***Andropadus latirostris***
    Passeriformes - Pycnonotidae
e   Yellow-whiskered Greenbul;
    Yellow-whiskered Bulbul
f   Bulbul à moustaches jaunes
d   Gelbbartbülbül
i   Bulbul dai mustacchi gialli

595   ***Andropadus masukuensis***
    Passeriformes - Pycnonotidae
e   Shelley's Greenbul; Shelley's Bulbul

f Bulbul des Masukus; Bulbul des
Monts Masukus
d Shelley-Bülbül
i Bulbul di Masuku

596 *Andropadus milanjensis*
Passeriformes - Pycnonotidae
e Stripe-cheeked Greenbul; Stripe-
cheeked Bulbul (CSA); Striped-
cheek Greenbul; Striped-cheek
Bulbul
f Bulbul montagnard
d Strichelgrünbülbül; Olivbrustbülbül
i Bulbul guancestriate

597 *Andropadus montanus*
Passeriformes - Pycnonotidae
e Cameroon Greenbul; Cameroon
Mountain-Bulbul; Mountain Little-
Greenbul; Cameroon Mountain-
Greenbul; Cameroon Montane
Greenbul; Cameroon Sombre
Greenbul; Sombre Greenbul;
Cameroon Bulbul; Plain Bulbul;
Montane Little-Bulbul
f Bulbul concolore
d Einfarbbülbül
i Bulbul di montagna

598 *Andropadus neumanni*
Passeriformes - Pycnonotidae
e Uluguru Mountain-Greenbul

599 *Andropadus nigriceps*
Passeriformes - Pycnonotidae
e Mountain Greenbul; Mountain
Bulbul; Olive-breasted Mountain-
Bulbul; Kilimanjaro Greenbul
f Bulbul à tête olive

600 *Andropadus olivaceiceps*
Passeriformes - Pycnonotidae
e Olive-headed Greenbul; Olive-
headed Bulbul
i Bulbul testaoliva

601 *Andropadus tephrolaemus*
Passeriformes - Pycnonotidae
e Grey-throated Greenbul; Grey-
throated Bulbul; Grey-throated
Mountain-Greenbul; Olive-breasted

Mountain-Greenbul; Mountain
Greenbul; Olive-breasted Greenbul;
Western Mountain-Greenbul;
Western Mountain-Bulbul
f Bulbul à gorge grise; Bulbul à tête
grise
d Bergwaldbülbül; Graukopfbülbül
i Bulbul petto oliva

602 *Andropadus virens*
Passeriformes - Pycnonotidae
e Little Greenbul; Little Green-Bulbul
f Bulbul verdâtre
d Grünbülbül
i Bulbul verdastro

603 *Androphobus viridis*
Passeriformes - Cinclosomatidae
e Papuan Whipbird; Green-backed
Babbler
f Androphobe vert
i Garrulo dorsoverde

604 *Anhima cornuta*
Anseriformes - Anhimidae
e Horned Screamer
f Kamichi cornu
d Hornwehrvogel
i Kaimichi cornuto; Palamedea cornuta

605 Anhimidae
Anseriformes
e Screamers
f Anhimidés
d Wehrvögel
i Animidi

606 *Anhinga anhinga*
Pelecaniformes - Anhingidae
e Anhinga; Darter; American Darter
(NA); American Anhinga; Snakebird
f Anhinga d'Amérique; Anhinga;
Anhinga noir
d Schlangenhalsvogel; Amerikanischer
Schlangenhalsvogel; Anhinga
i Aninga; Aninga americana; Anhinga
americana

607 *Anhinga melanogaster*
Pelecaniformes - Anhingidae
e Oriental Darter; Darter (CSA);

e   Swan Goose; Chinese Goose
f   Oie cygnoïde
d   Schwanengans
i   Oca cigno; Oca cignoide

634   *Anser domesticus*
Anseriformes - Anatidae
e   Goose
d   Gans

635   *Anser erythropus*
Anseriformes - Anatidae
e   Lesser White-fronted Goose; Lesser-fronted Goose
f   Oie naine
d   Zwerggans; Zwergbläßgans
i   Oca lombardella minore

636   *Anser fabalis*
Anseriformes - Anatidae
e   Bean Goose; Forest Bean-Goose; Taiga Bean-Goose
f   Oie des moissons
d   Saatgans
i   Oca granaiola

637   *Anser flavifrons flavirostris*
Anseriformes - Anatidae
e   Greenland White-fronted Goose
f   Oie de Groenland
d   Grönlandische Bläßgans

638   *Anser hyperborea*
Anseriformes - Anatidae
e   Snow Goose

639   *Anser indicus*
Anseriformes - Anatidae
e   Bar-headed Goose
f   Oie à tête barrée
d   Streifengans
i   Oca indiana

640   *Anser rossii*
Anseriformes - Anatidae
e   Ross's Goose
f   Oie de Ross
d   Zwergschneegans
i   Oca di Ross

641   *Anseranas semipalmata*
Anseriformes - Anseranatidae
e   Magpie Goose; Pied Goose; Black-and-white Goose; Semipalmated Goose
f   Canaroie semipalmée
d   Spaltfußgans
i   Oca gazza

642   Anseranatidae
Anseriformes
e   Magpie Geese
f   Anseranatidés
d   Spaltfußgänse
i   Anseranatidi

643   Anseriformes
e   Waterfowl
f   Ansériformes
d   Entenvögel; Schwimmvögel; Zahnschnäbler
i   Anseriformi

644   *Anthocephala floriceps*
Apodiformes - Trochilidae
e   Blossomcrown
f   Colibri à tête rose
d   Blumenköpchen
i   Colibrì corona fiorita

645   *Anthochaera carunculata*
Passeriformes - Meliphagidae
e   Red Wattlebird; Common Wattlebird
f   Méliphage barbe-rouge
d   Rotlappenhonigfresser; Rotklunkerhonigfresser
i   Mangiamiele dalle caruncole rosse

646   *Anthochaera chrysoptera*
Passeriformes - Meliphagidae
e   Brush Wattlebird; Little Wattlebird
f   Méliphage à gouttelettes
d   Klunkerhonigesser; Zimtflügelhonigfresser
i   Mangiamiele alidorate

647   *Anthochaera lunulata*
Passeriformes - Meliphagidae
e   Little Wattlebird; Western Wattlebird; Brush Wattlebird

f Méliphage mineur
i Mangiamiele piccolo

648 *Anthochaera paradoxa*
Passeriformes - Meliphagidae
e Yellow Wattlebird
f Méliphage à pendeloques
d Gelblappenhonigfresser
i Mangimiele delle caruncole gialle

649 *Anthochaera paradoxa kingi*
Passeriformes - Meliphagidae
e King Island Yellow-Wattlebird
(ANZ)

650 *Anthornis melanura*
Passeriformes - Meliphagidae
e New Zealand Bellbird; Bellbird
f Méliphage carillonneur
d Makomako
i Uccello campanello della Nuova
Zelanda

651 *Anthoscopus caroli*
Passeriformes - Paridae
e African Penduline-Tit; Grey
Penduline-Tit (CSA)
f Rémiz de Carol; Penduline africaine
d Weißstirnbeutelmeise; Schließ-
Beutelmeise
i Pendolino africano

652 *Anthoscopus flavifrons*
Passeriformes - Paridae
e Forest Penduline-Tit; Yellow-fronted
Penduline-Tit
f Rémiz à front jaune
d Goldstirnbeutelmeise
i Pendolino frontegialla

653 *Anthoscopus minutus*
Passeriformes - Paridae
e Southern Penduline-Tit; Cape
Penduline Tit (CSA); Southern
Kapoc Penduline-Tit; African
Penduline-Tit
f Rémiz minute; Rémiz du Cap
d Kapbeutelmeise
i Pendolino del Capo

654 *Anthoscopus musculus*
Passeriformes - Paridae
e Mouse-coloured Penduline-Tit;
Mouse-coloured Tit
f Rémiz souris; Rémiz brune
d Graubeutelmeise
i Pendolino grigiotopo

655 *Anthoscopus parvulus*
Passeriformes - Paridae
e Yellow Penduline-Tit; West African
Penduline-Tit
f Rémiz à ventre jaune
d Senegal-Beutelmeise;
Gelbbauchbeutelmeise
i Pendolino giallo africano

656 *Anthoscopus punctifrons*
Passeriformes - Paridae
e Sennar Penduline-Tit; Sudan
Penduline-Tit; Kapoc Penduline-Tit
f Rémiz du Soudan
d Sudan-Beutelmeise
i Pendolino del Sennar

657 *Anthoscopus sylviella*
Passeriformes - Paridae
e Buff-bellied Penduline-Tit; Rungwe
Penduline-Tit
f Rémiz du roungoué
i Pendolino del Rungwe

658 *Anthracoceros albirostris*
Coraciiformes - Bucerotidae
e Oriental Pied-Hornbill; Indian Pied-
Hornbill; Pied Hornbill; Asian Pied-
Hornbill; Northern Pied-Hornbill;
Sunda Pied-Hornbill; Malaysian
Pied-Hornbill
f Calao malais
i Bucero orientale

659 *Anthracoceros coronatus*
Coraciiformes - Bucerotidae
e Malabar Pied-Hornbill; Oriental
Pied-Hornbill; Indian Pied-Hornbill
(ISC)
f Calao pie; Calao de Malabar
d Malabar-Hornvogel
i Bucero coronato

660 *Anthracoceros malayanus*
Coraciiformes - Bucerotidae
e Black Hornbill; Asian Black
Hornbill; Malaysian Black-Hornbill;
Malay Black-Hornbill
f Calao charbonnier
d Malaien-Hornvogel
i Bucero nero

661 *Anthracoceros marchei*
Coraciiformes - Bucerotidae
e Palawan Hornbill
f Calao de Palawan
d Palawan-Hornvogel
i Bucero di Palawan

662 *Anthracoceros montani*
Coraciiformes - Bucerotidae
e Sulu Hornbill; Montano's Hornbill
f Calao des Sulu
d Sulu-Hornvogel
i Bucero

663 *Anthracothorax dominicus*
Apodiformes - Trochilidae
e Antillean Mango
f Quanga négresse (Ants); Oiseau-
mouche (Ants); Mango doré
d Dominikanermango
i Mango delle Antille

664 *Anthracothorax iridescens*
Apodiformes - Trochilidae
e Ecuadorian Mango

665 *Anthracothorax mango*
Apodiformes - Trochilidae
e Jamaican Mango; Black
Hummingbird (WI); Mango
Hummingbird (WI); Doctor-Bird
(WI)
f Mango de la Jamaïque
d Jamaika-Mango
i Mango della Giamaica

666 *Anthracothorax nigricollis*
Apodiformes - Trochilidae
e Black-throated Mango; Ecuadorian
Mango
f Mango à cravate noire

d Schwarzkehlmango
i Mango collonero

667 *Anthracothorax prevostii*
Apodiformes - Trochilidae
e Green-breasted Mango; Prevost's
Mango; Veraguas Mango;
Ecuadorian Mango
f Mango de Prévost
d Grünbrustmango; Schimmerkolibri
i Mango pettoverde

668 *Anthracothorax veraguensis*
Apodiformes - Trochilidae
e Veraguas Mango; Veraguan Mango

669 *Anthracothorax viridigula*
Apodiformes - Trochilidae
e Green-throated Mango; Black-
breasted Hummingbird; Green-
throated Hummingbird
f Mango à cravate verte
d Grünkehlmango
i Mango golaverde

670 *Anthracothorax viridis*
Apodiformes - Trochilidae
e Green Mango; Puerto Rican Mango
(WI)
f Mango vert
d Smaragdmango
i Mango verde

671 *Anthreptes anchietae*
Passeriformes - Nectariniidae
e Anchieta's Sunbird; Red-and-blue
Sunbird
f Souïmanga d'Anchieta; Souïmanga
malais
d Buntbauchnektarvogel
i Nettarinia di Anchieta

672 *Anthreptes aurantium*
Passeriformes - Nectariniidae
e Violet-tailed Sunbird
f Souïmanga à queue violette
d Violettschwanznektarvogel
i Nettarinia codaviola

673 *Anthreptes axillaris*
Passeriformes - Nectariniidae

e Grey-headed Sunbird
f Souïmanga à tête grise
d Graukopfnektarvogel
i Nettarinia testagrigia

674 *Anthreptes collaris*
Passeriformes - Nectariniidae
e Collared Sunbird
f Souïmanga à collier
d Stahlnektarvogel; Waldnektarvogel
i Nettarinia dal collare nero-blu

675 *Anthreptes fraseri*
Passeriformes - Nectariniidae
e Scarlet tufted Sunbird; Grey-headed Sunbird; Fraser's Sunbird
f Souïmanga de Fraser
d Laubnektarvogel
i Nettarinia di Fraser

676 *Anthreptes gabonicus*
Passeriformes - Nectariniidae
e Mouse-brown Sunbird; Brown Sunbird; Mangrove Sunbird
f Souïmanga brun
d Gabun-Nektarvogel
i Nettarinia bruna del Gabon

677 *Anthreptes longuemarei*
Passeriformes - Nectariniidae
e Western Violet-backed Sunbird; Violet-backed Sunbird (CSA); Honeysucker (CSA); African Violet-backed Sunbird
f Souïmanga violet; Souïmanga à dos violet
d Violettmantelnektarvogel
i Nettarinia dorsoviola

678 *Anthreptes malacensis*
Passeriformes - Nectariniidae
e Plain-throated Sunbird; Brown-throated Sunbird
f Souïmanga à gorge brune
d Braunkehlnektarvogel
i Nettarinia golabruna

679 *Anthreptes metallicus*
Passeriformes - Nectariniidae
e Nile Valley Sunbird; Bronzed Pygmy-Sunbird; Eastern Pygmy-Sunbird; Bronzed Eastern Pygmy-Sunbird; Metallic Sunbird; Northern Pygmy-Sunbird
f Souïmanga du Nil
d Erznektarvogel; Erzhonigsauger
i Nettarinia del Nilo; Nettarinia metallica

680 *Anthreptes neglectus*
Passeriformes - Nectariniidae
e Uluguru Violet-backed Sunbird; Violet-backed Sunbird (CSA)
f Souïmanga des Uluguru; Souïmanga violet des Uluguru
d Uluguru-Nektarvogel
i Nettarinia dorsoviola di Uluguru

681 *Anthreptes orientalis*
Passeriformes - Nectariniidae
e Kenya Violet-backed Sunbird; Eastern Violet-backed Sunbird (CSA), Honeysucker (CSA); Kenya Sunbird
f Souïmanga du Kenya; Souïmanga violet oriental
d Schwalbennektarvogel
i Nettarinia del Kenya

682 *Anthreptes pallidigaster*
Passeriformes - Nectariniidae
e Amani Sunbird; Pale-bellied Sunbird
f Souïmanga d'Amani
d Amani-Nektarvogel
i Nettarinia di Amami

683 *Anthreptes platurus*
Passeriformes - Nectariniidae
e Pygmy Sunbird; Western Pygmy-Sunbird; Pygmy Long-tailed Sunbird; Nile Valley Sunbird; Southern Pygmy-Sunbird
f Souïmanga pygmée; Souïmanga petit à longue queue; Souïmanga pygmé à longue queue; Souïmanga nain
d Grünbrustnektarvogel; Zwerglangschwanznektarvogel
i Nettarinia pigmea; Nettarinia delle spatole; Suimanga

684 *Anthreptes rectirostris*
Passeriformes - Nectariniidae

e   Green Sunbird; Pujoli Sunbird;
    Berlioz's Sunbird; Yellow-chinned
    Sunbird; Grey-chinned Sunbird;
    Green-backed Sunbird; Western
    Sunbird
f   Souïmanga à bec droit; Souïmanga à
    gorge jaune; Souïmanga à gorge grise
d   Goldbandnektarvogel
i   Nettarinia verde

685   *Anthreptes rectirostris tephrolaema*
    Passeriformes - Nectariniidae
f   Souïmanga à gorge grise; Souïmanga
    à menton jaune

686   *Anthreptes reichenowi*
    Passeriformes - Nectariniidae
e   Plain-backed Sunbird; Blue-throated
    Sunbird (CSA); Blue-throated Little-
    Sunbird
f   Souïmanga de Reichenow
d   Blaukehlnektarvogel
i   Nettarinia dorsoverde

687   *Anthreptes rhodolaema*
    Passeriformes - Nectariniidae
e   Red-throated Sunbird; Shelley's
    Sunbird; Rufous-throated Sunbird;
    Red-shouldered Sunbird
f   Souïmanga à gorge rouge
d   Rotkehlnektarvogel
i   Nettarinia golarossa

688   *Anthreptes rubritorques*
    Passeriformes - Nectariniidae
e   Banded Sunbird; Banded Green-
    Sunbird (CSA)
f   Souïmanga à col rouge
i   Nettarinia dal collare rosso

689   *Anthreptes simplex*
    Passeriformes - Nectariniidae
e   Plain Sunbird; Plain-coloured
    Sunbird
f   Souïmanga modeste
d   Schlichtnektarvogel
i   Nettarinia grigioverde

690   *Anthreptes singalensis*
    Passeriformes - Nectariniidae
e   Ruby-cheeked Sunbird; Rubycheek

f   Souïmanga à joues rubis
d   Rubinwangennektarvogel;
    Glanzbürzelnektarvogel
i   Nettarinia guancerosse

691   *Anthus antarcticus*
    Passeriformes - Motacillidae
e   South Georgia Pipit; Sub-antarctic
    Pipit
f   Pipit antarctique
d   Riesenpieper
i   Pispola antartica

692   *Anthus bannermani*
    Passeriformes - Motacillidae
e   Bannerman's Pipit
f   Pipit de Bannerman
i   Calandro di Bannerman

693   *Anthus berthelotii*
    Passeriformes - Motacillidae
e   Berthelot's Pipit; Canarian Pipit;
    Canary Islands Pipit
f   Pipit de Berthelot
d   Kanarien-Pieper
i   Calandro di Berthelot; Pispola di
    Berthelot; Pispola delle Canarie

694   *Anthus bogotensis*
    Passeriformes - Motacillidae
e   Paramo Pipit
f   Pipit du Paramo
i   Pispola del Paramo

695   *Anthus brachyurus*
    Passeriformes - Motacillidae
e   Short-tailed Pipit
f   Pipit à queue courte
d   Kurzschwanzpieper
i   Pispola codacorte

696   *Anthus caffer*
    Passeriformes - Motacillidae
e   Bush Pipit; Little Tawny Pipit
    (CSA); Bushveld Tree Pipit;
    Bushveld Pipit
f   Pipit cafre
d   Buschpieper
i   Pispola cafra

697 *Anthus camaroonensis*
Passeriformes - Motacillidae
e Cameroon Pipit; Bushveld Pipit
(CSA)
f Pipit du Cameroun
i Calandro del Camerun

698 *Anthus campestris*
Passeriformes - Motacillidae
e Tawny Pipit
f Pipit rousseline
d Brachpieper
i Calandro; Calandro comune

699 *Anthus cervinus*
Passeriformes - Motacillidae
e Red-throated Pipit
f Pipit à gorge rousse; Pipit
gorgerousse
d Rotkehlpieper
i Pispola golarossa

700 *Anthus chacoensis*
Passeriformes - Motacillidae
e Chaco Pipit
f Pipit du Chaco
d Chaco-Pieper
i Pispola del Chaco

701 *Anthus chloris*
Passeriformes - Motacillidae
e Yellow-breasted Pipit; Yellow-
bellied Pipit
f Pipit à gorge jaune
d Gelbbrustpieper; Gelbpieper
i Zampagrossa pettogiallo

702 *Anthus cinnamomeus*
Passeriformes - Motacillidae
e African Pipit; Grassveld Pipit (CSA);
Richard's Pipit (CSA); Grassland
Pipit
f Pipit africain
d Spornpieper; Weidelandpieper
i Calandro africano

703 *Anthus correndera*
Passeriformes - Motacillidae
e Correndera Pipit
f Pipit correndera

d Kamp-Pieper
i Pispola corrrendera

704 *Anthus crenatus*
Passeriformes - Motacillidae
e Yellow-tufted Pipit; Rock Pipit
(CSA); Large Yellow-tufted-Pipit;
African Rock-Pipit
f Pipit des rochers
d Klippenpieper
i Pispola ciuffogiallo

705 *Anthus furcatus*
Passeriformes - Motacillidae
e Short-billed Pipit
f Pipit à plastron
d Weißbauchpieper
i Pispola beccocorto

706 *Anthus godlewskii*
Passeriformes - Motacillidae
e Blyth's Pipit; Godlewski's Pipit
f Pipit de Godlewski; Pipit de Blyth
d Steppenpieper; Godlewski-Pieper
i Prispolone di Blyth; Calandro di
Godlewski

707 *Anthus gustavi*
Passeriformes - Motacillidae
e Pechora Pipit; Petchora Pipit;
Siberian Pipit
f Pipit de la Petchora; Pipit de Gustave
d Petschora-Pieper; Petjora-Pieper
i Pispola della Peciora; Pispola
siberiana; Pispola della Pechora

708 *Anthus gutturalis*
Passeriformes - Motacillidae
e Alpine Pipit; New Guinea Pipit;
Mountain Pipit
f Pipit de Nouvelle-Guinée
d Papua-Pieper
i Spioncella della Nuova Guinea

709 *Anthus hellmayri*
Passeriformes - Motacillidae
e Hellmayr's Pipit
f Pipit de Hellmayr
d Hellmayr-Pieper
i Pispola di Hellmayr

710 **Anthus hodgsoni**
Passeriformes - Motacillidae
e Olive-backed Pipit; Ikuve Tree-Pipit;
Olive Tree-Pipit; Indian Tree-Pipit;
Oriental Tree-Pipit; Hodgson's Tree
Pipit; Spotted Pipit; Hodgson's Pipit;
Eastern Tree-Pipit
f Pipit indien; Pipit à dos olive; Pipit
sylvestre; Pipit d'Hodgson
d Waldpieper; Indischer Baumpieper
i Prispolone indiano

711 **Anthus hoeschi**
Passeriformes - Motacillidae
e Mountain Pipit
f Pipit alticole; Pipit du Drakensberg
d Hochlandpieper
i Calandro di montagna

712 **Anthus latistriatus**
Passeriformes - Motacillidae
e Jackson's Pipit
f Pipit à raies larges
i Calandro di Jackson

713 **Anthus leucophrys**
Passeriformes - Motacillidae
e Plain-backed Pipit; Dark Plain-
backed Pipit
f Pipit à dos uni; Pipit à dos roux
d Braunrückenpieper
i Calandro dai sopraccigli bianchi

714 **Anthus lineiventris**
Passeriformes - Motacillidae
e Striped Pipit; Large Striped-Pipit
f Pipit de Sundevall; Pipit strié de
Sundevall
d Streifenpieper
i Spioncello striato

715 **Anthus longicaudatus**
Passeriformes - Motacillidae
e Long-tailed Pipit

716 **Anthus lutescens**
Passeriformes - Motacillidae
e Yellowish Pipit
f Pipit jaunâtre
d Savannenpieper
i Pispola giallastra

717 **Anthus melindae**
Passeriformes - Motacillidae
e Malindi Pipit
f Pipit de Melinda; Pipit de Malindi
d Malindi-Pieper
i Pispola di Malindi

718 **Anthus menzbieri**
Passeriformes - Motacillidae
e Menzbier's Pipit

719 **Anthus nattereri**
Passeriformes - Motacillidae
e Ochre-breasted Pipit
f Pipit ocré
d Ockerbrustpieper
i Pispola di Natterer

720 **Anthus nilghiriensis**
Passeriformes - Motacillidae
e Nilgiri Pipit
f Pipit des Nilgiri
d Nilgiri-Pieper
i Spioncello del Nilgiri

721 **Anthus novaeseelandiae**
Passeriformes - Motacillidae
e Australasian Pipit; Richard's Pipit;
Common Pipit; New Zealand Pipit;
Australian Pipit
f Pipit austral; Pipit de Richard
d Spornpieper
i Calandro maggiore; Calandro
australe

722 **Anthus nyassae**
Passeriformes - Motacillidae
e Woodland Pipit; Wood Pipit (CSA);
Chaplin's Pipit
f Pipit forestier
d Miombo-Pieper
i Pispola del Lago Nyassa

723 **Anthus pallidiventris**
Passeriformes - Motacillidae
e Long-legged Pipit; Long-clawed
Pipit
f Pipit à longues pattes
d Stelzenpieper
i Calandro ventrepallido

724   *Anthus petrosus*
Passeriformes - Motacillidae
e   Rock Pipit; European Rock-Pipit;
British Rock-Pipit; Scandinavian
Pipit; Water Pipit
f   Pipit maritime
d   Strandpieper; Felsenpieper
i   Spioncello; Spioncello di montagna;
Spioncello marino

725   *Anthus pratensis*
Passeriformes - Motacillidae
e   Meadow Pipit
f   Pipit farlouse; Pipit des prés
d   Wiesenpieper
i   Pispola; Pispolino; Prispolino;
Pispolina; Pispoletta

726   *Anthus roseatus*
Passeriformes - Motacillidae
e   Rosy Pipit; Vinaceous-breasted Pipit;
Roseate Pipit; Rose breasted Pipit;
Hodgson's Pipit
f   Pipit rose; Pipit rosé; Pipit à gorge
rosée
d   Rosenpieper
i   Pispola rosata

727   *Anthus rubescens*
Passeriformes - Motacillidae
e   Buff-bellied Pipit; American Pipit;
Water Pipit
f   Pipit d'Amérique; Pipit farlousane
d   Pazifischer Wasserpieper
i   Spioncello ventrerossiccio

728   *Anthus rufulus*
Passeriformes - Motacillidae
e   Paddyfield Pipit; Indian Pipit;
Oriental Pipit
f   Pipit rousset
i   Calandro delle risaie

729   *Anthus sharpei*
Passeriformes - Motacillidae
e   Sharpe's Pipit; Sharpe's Longclaw
f   Pipit de Sharpe; Sentinelle de Sharpe
d   Zitronenpieper
i   Zampagrossa di Sharpe

730   *Anthus similis*
Passeriformes - Motacillidae
e   Long-billed Pipit; Nicholson's Pipit
(CSA); Brown Rock-Pipit (ISC);
Nicholson's Rock-Pipit; Nicholson's
Persian Rock-Pipit; Indian Rock-
Pipit
f   Pipit à long bec
d   Langschnabelpieper
i   Calandro bruno; Pispola beccolungo

731   *Anthus sokokensis*
Passeriformes - Motacillidae
e   Sokoke Pipit; Sokokoe Pipit
f   Pipit de Sokoke
d   Sokoke-Pieper
i   Pispola di Sokoke

732   *Anthus spinoletta*
Passeriformes - Motacillidae
e   Water Pipit; Mountain Pipit
f   Pipit spioncelle; Pipit maritime; Pipit
commun; Pipit aquatique; Pipit
obscur
d   Wasserpieper; Bergpieper;
Strandpieper
i   Spioncello marino; Spioncello

733   *Anthus spragueii*
Passeriformes - Motacillidae
e   Sprague's Pipit
f   Pipit de Sprague; Pipit des prairies
(Qué)
d   Prairiepieper
i   Pispola di Sprague

734   *Anthus sylvanus*
Passeriformes - Motacillidae
e   Upland Pipit; Indian Pipit; Oriental
Pipit; Chinese Pipit
f   Pipit montagnard
d   Schmalschwanzpieper
i   Spioncello delle alture

735   *Anthus trivialis*
Passeriformes - Motacillidae
e   Tree Pipit; Brown Tree-Pipit (NA);
Brown Tree-Pipit
f   Pipit des arbres; Pipit des buissons
d   Baumpieper
i   Prispolone; Prispolone europeo

736 **Anthus vaalensis**
Passeriformes - Motacillidae
e Buffy Pipit; Sandy Pipit; Sandy
Plain-backed Pipit
f Pipit du Vaal
d Vaal-Pieper
i Calandro del Vaal

737 **Antilophia galeata**
Passeriformes - Pipridae
e Helmeted Manakin
f Manakin casqué
d Helmpipra; Rothelmpipra
i Manachino dall'elmo

738 **Anumbius annumbi**
Passeriformes - Furnariidae
e Firewood-gatherer; Woodgatherer
f Annumbi fagoteur
d Anumbi
i Raccoglitore di legna

739 **Anurolimnas castaneiceps**
Gruiformes - Rallidae
e Chestnut-headed Crake; Chestnut-
headed Rail
f Râle à masque rouge
d Rotmaskenralle
i Schiribilla testacastana

740 **Anurolimnas fasciatus**
Gruiformes - Rallidae
e Black-banded Crake; Black-banded
Rail; Hauxwell's Crake
f Râle fascié
d Streifenbauchralle
i Schiribilla fasciata

741 **Anurolimnas viridis**
Gruiformes - Rallidae
e Russet-crowned Crake; Russet-
crowned Rail; Cayenne Crake
f Râle kiolo
d Indigoralle
i Schiribilla rossiccia

742 **Anurophasis monorthonyx**
Galliformes - Phasianidae
e Snow Mountains Quail; Snow
Mountain Quail; New Guinea Quail
f Caille de montagne

d Schneegebirgswachtel
i Quaglia delle nevi

743 **Apalis alticola**
Passeriformes - Cisticolidae
e Brown-headed Apalis
f Apalis à tête brune
d Braunkopffeinsänger
i Apale testabruna

744 **Apalis argentea**
Passeriformes - Cisticolidae
e Kungwe Apalis; Peters's Apalis
f Apalis de Moreau
d Silberfeinsänger
i Apale di Kungwe

745 **Apalis bamendae**
Passeriformes - Cisticolidae
e Bamenda Apalis
f Apalis du Bamenda
d Bamenda-Feinsänger
i Apale di Bamenda

746 **Apalis binotata**
Passeriformes - Cisticolidae
e Masked Apalis
f Apalis masquée
d Maskenfeinsänger
i Apale mascherata

747 **Apalis chapini**
Passeriformes - Cisticolidae
e Chapin's Apalis; Chestnut-headed
Apalis
f Apalis de Chapin; Apalis à tête rouge
i Apale di Chapin

748 **Apalis chariessa**
Passeriformes - Cisticolidae
e White-winged Apalis
f Apalis à ailes blanches
d Spiegelfeinsänger
i Apale alibianche

749 **Apalis chirindensis**
Passeriformes - Cisticolidae
e Chirinda Apalis; Swynnerton's
Apalis; Melsetter Apalis
f Apalis de Chirinda

d Selinda-Feinsänger
i Apale di Chirinda

750 *Apalis cinerea*
Passeriformes - Cisticolidae
e Grey Apalis; Brown-headed Forest-
Warbler
f Fauvette forestière à tête brune;
Apalis grise
d Graurückenfeinsänger
i Apale cenerina

751 *Apalis flavida*
Passeriformes - Cisticolidae
e Yellow-breasted Apalis; Yellow-
chested Apalis
f Apalis à gorge jaune
d Gelbbrustfeinsänger
i Apale pettogiallo

752 *Apalis flavida caniceps*
Passeriformes - Cisticolidae
f Apalis à tête grise

753 *Apalis fuscigularis*
Passeriformes - Cisticolidae
e Gosling's Apalis; Cameroon Apalis
f Apalis de Gosling
i Apale

754 *Apalis goslingi*
Passeriformes - Cisticolidae
e Gosling's Apalis
f Apalis de Gosling
d Gosling-Feinsänger
i Apale di Gosling

755 *Apalis jacksoni*
Passeriformes - Cisticolidae
e Black-throated Apalis
f Apalis à gorge noire; Apalis à
moustaches blancs
d Schwarzkehlfeinsänger
i Apale golanera

756 *Apalis kaboboensis*
Passeriformes - Cisticolidae
e Kabobo Apalis; Karamajoa Warbler;
Prigogine's Apalis
f Apalis du Kabobo

d Kabobo-Feinsänger
i Apale di Kabobo

757 *Apalis karamojae*
Passeriformes - Cisticolidae
e Karamoja Apalis
f Apalis du Karamoja
d Karamoja-Sänger
i Apale di Karamoja

758 *Apalis lynesi*
Passeriformes - Cistocolidae
e Namuli Apalis
i Apale

759 *Apalis melanocephala*
Passeriformes - Cisticolidae
e Black-headed Apalis
f Apalis à tête noire; Apalis
mélanocéphale
d Schwarzkopffeinsänger
i Apale testanera

760 *Apalis nigriceps*
Passeriformes - Cisticolidae
e Black-capped Apalis; Black-capped
Yellow-Warbler
f Apalis à calotte noire; Fauvette
forestière à tête noire; Apalis
masquée des montagnes
d Kappenfeinsänger
i Apale dorsoverde

761 *Apalis personata*
Passeriformes - Cisticolidae
e Black-faced Apalis; Mountain
Masked-Apalis
f Apalis à face noire
i Apale mascherata di montagna

762 *Apalis porphyrolaema*
Passeriformes - Cisticolidae
e Chestnut-throated Apalis
f Apalis à gorge marron
d Bergfeinsänger
i Apale golacastana

763 *Apalis pulchra*
Passeriformes - Cisticolidae
e Black-collared Apalis; Black-
collared Forest-Warbler; Collared

Apalis
f Apalis à col noir; Fauvette forestiêre
à collier noir
d Schmuckfeinsänger
i Apale fianchirossi

764 **Apalis ruddi**
Passeriformes - Cisticolidae
e Rudd's Apalis
f Apalis de Rudd
d Rudds Feinsänger;
Flechtenfeinsänger
i Apale di Rudd

765 **Apalis rufifrons**
Passeriformes - Cisticolidae
e Red-faced Warbler; Red-throated
Warbler; Red-faced Prinia; Red-
faced Apalis; Red-fronted Warbler;
African Red-faced Apalis
f Apalis à front roux; Fauvette à front
roux
d Rotstirnprinie; Schuppenkopfprinie
i Apale fronterosso

766 **Apalis rufogularis**
Passeriformes - Cisticolidae
e Buff-throated Apalis; Black-backed
Apalis
f Apalis à gorge rousse
d Weißbauchfeinsänger
i Apale golafulva

767 **Apalis ruwenzorii**
Passeriformes - Cisticolidae
e Collared Apalis; Ruwenzori Apalis
d Ruwenzori-Feinsänger
i Apale del Ruwenzori

768 **Apalis sharpii**
Passeriformes - Cisticolidae
e Sharpe's Apalis; Bamenda Apalis
f Apalis de Sharpe
d Kurzschwanzfeinsänger
i Apale di Sharpe

769 **Apalis thoracica**
Passeriformes - Cisticolidae
e Bar-throated Apalis
f Apalis à collier; Apalis à collier noir;
Apalis à gorge barrée

d Halsbandfeinsänger
i Apale occhichiari

770 **Apalis viridiceps**
Passeriformes - Cisticolidae
e Brown-tailed Apalis
f Apalis à queue brune
i Apale codabruna

771 **Apaloderma aequatoriale**
Trogoniformes - Trogonidae
e Bare-cheeked Trogon; Yellow-
cheeked Trogon
f Trogon à joues jaunes; Couroucou à
joues jaunes
d Gelbwangentrogon
i Trogone guancenude

772 **Apaloderma narina**
Trogoniformes - Trogonidae
e Narina Trogon
f Trogon narina; Couroucou narina
d Narina-Trogon; Zügeltrogon
i Trogone di Narina

773 **Apaloderma vittatum**
Trogoniformes - Trogonidae
e Bar-tailed Trogon
f Trogon à queue barrée; Couroucou à
queue barrée
d Bergtrogon
i Trogone codabarrata

774 **Apalopteron familiare**
Passeriformes - Meliphagidae
e Bonin White-eye; Bonin Islands
Honeyeater; White-eyed Honeyeater;
Bonin Honeyeater
f Méliphage des Bonin
d Bonin-Honigfresser; Apalopteron
i Mangiamiele dell'Isola Bonin

775 **Aphanotriccus audax**
Passeriformes - Tyrannidae
e Black-billed Flycatcher; Nelson's
Flycatcher
f Moucherolle à bec noir; Moucherolle
de Nelson
d Nelson-Tyrann
i Tiranno becconero

776 **Aphanotriccus capitalis**
Passeriformes - Tyrannidae
e Tawny-chested Flycatcher; Salvin's
Flycatcher
f Moucherolle à poitrine fauve;
Moucherolle de Salvin
d Salvin-Tyrann
i Tiranno pettofulvo

777 **Aphantochroa cirrochloris**
Apodiformes - Trochilidae
e Sombre Hummingbird
f Colibri vert-et-gris
d Erzkolibri
i Colibrì modesto

778 **Aphelocephala leucopsis**
Passeriformes - Pardalotidae
e Southern Whiteface; Common
Whiteface; Whiteface; Eastern
Whiteface
f Gérygone blanchâtre
d Fahlrückenweißstirnchen
i Facciabianca meridionale

779 **Aphelocephala nigricincta**
Passeriformes - Pardalotidae
e Banded Whiteface
f Gérygone à collier noir
d Halsbandweißstirnchen

780 **Aphelocephala pectoralis**
Passeriformes - Pardalotidae
e Chestnut-breasted Whiteface
f Gérygone à collier roux
d Braunbrustweißstirnchen

781 **Aphelocoma californica**
Passeriformes - Corvidae
e California Jay; Western Scrub Jay;
Californian Scrub Jay; Scrub Jay;
Long-tailed Jay; Nicasio Jay;
Belding's Jay; Xantus's Jay; Santa
Cruz Jay
f Geai buissonnier
i Ghiandaia della California

782 **Aphelocoma coerulescens**
Passeriformes - Corvidae
e Florida Jay; Florida Scrub Jay; Scrub
Jay

f Geai à gorge blanche
d Buschhäher
i Ghiandaia della Florida

783 **Aphelocoma couchii**
Passeriformes - Corvidae
e Couch's Jay

784 **Aphelocoma insularis**
Passeriformes - Corvidae
e Island Scrub Jay; Santa Cruz Jay
f Geai de Santa Cruz
i Ghiandaia di Santa Cruz

785 **Aphelocoma sumichrasti**
Passeriformes - Corvidae
e Sumichrast's Scrub Jay

786 **Aphelocoma ultramarina**
Passeriformes - Corvidae
e Grey-breasted Jay; Mexican Jay;
Arizona Jay; Ultramarine Jay;
Couch's Jay
f Geai du Mexique
d Graubrusthäher
i Ghiandaia messicana

787 **Aphelocoma unicolor**
Passeriformes - Corvidae
e Unicoloured Jay; Cerulean Jay;
Griscombe's Jay
f Geai unicolore
d Einfarbhäher
i Ghiandaia unicolore

788 **Aphelocoma woodhouseii**
Passeriformes - Corvidae
e Woodhouse's Scrub Jay;
Woodhouse's Jay; Texas Jay; Blue-
cheeked Jay; Blue-grey Jay

789 **Aphrastura masafuerae**
Passeriformes - Furnariidae
e Masafuera Rayadito; Masafuera
Island Rayadito; Mas Afuera
Rayadito
f Synallaxe de Masafuera
d Masafuera-Schlüpfer
i Rayadito di Masafuera

790 **Aphrastura spinicauda**
Passeriformes - Furnariidae
e Thorn-tailed Rayadito
f Synallaxe Rayadito
d Stachelschwanzschlüpfer
i Rayadito codaspinosa

791 **Aplonis atrifusca**
Passeriformes - Sturnidae
e Samoan Starling
f Stourne de Samoa
d Samoa-Star
i Storno delle Isole Samoa

792 **Aplonis brunneicapilla**
Passeriformes - Sturnidae
e White-eyed Starling; Mimic Starling
f Stourne aux yeux blancs
d Weißaugenstar
i Storno occhibianchi

793 **Aplonis cantoroides**
Passeriformes - Sturnidae
e Singing Starling; Little Starling
f Stourne chanteur
d Singstar
i Storno canoro

794 **Aplonis cinerascens**
Passeriformes - Sturnidae
e Rarotonga Starling; Cook Islands
Starling
f Stourne de Rarotonga
d Rarotonga-Star
i Storno di Rarotonga

795 **Aplonis corvina**
Passeriformes - Sturnidae
e Kosrae Starling; Kittlitz's Starling;
Kusaie Mountain-Starling; Kusaie
Starling; Kittlitz's Mountain-Starling
f Stourne de Kusaie
d Rabenstar
i Storno di Kosrae

796 **Aplonis crassa**
Passeriformes - Sturnidae
e Tanimbar Starling; Little Starling
f Stourne des Tanimbar
d Tenimbar-Star
i Storno di Tanimbar

797 **Aplonis dichroa**
Passeriformes - Sturnidae
e San Cristobal Starling
f Stourne de San Cristobal
i Storno di San Cristobal

798 **Aplonis feadensis**
Passeriformes - Sturnidae
e Atoll Starling; Fead Island Starling
f Stourne des Fead
d Atollstar
i Storno delle Isole Fead

799 **Aplonis fusca**
Passeriformes - Sturnidae
e Norfolk Starling; Tasman Starling;
Norfolk Island Starling; Lord Howe
Island Starling; Tasman Island
Starling
f Stourne de Norfolk
d Norfolk-Star
i Storno dell'Isola Norfolk

800 **Aplonis fusca fusca**
Passeriformes - Sturnidae
e Norfolk Island Tasman-Starling

801 **Aplonis fusca hulliana**
Passeriformes - Sturnidae
e Lord Howe Island Tasman-Starling

802 **Aplonis grandis**
Passeriformes - Sturnidae
e Brown-winged Starling; Large
Glossy-Starling
f Stourne des Salomon
d Kragenstar
i Storno splendente grande

803 **Aplonis insularis**
Passeriformes - Sturnidae
e Rennell Starling; Rennell Island
Starling
f Stourne de Rennell
d Rennel-Star
i Storno dell'Isola Rennell

804 **Aplonis magna**
Passeriformes - Sturnidae
e Long-tailed Starling
f Stourne à longue queue

d Geelvink-Star
i Storno codalunga

805 *Aplonis mavornata*
Passeriformes - Sturnidae
e Mysterious Starling; Raiatea Starling
f Stourne mystérieux
i Storno misterioso

806 *Aplonis metallica*
Passeriformes - Sturnidae
e Metallic Starling; Shining Starling; Colonial Starling
f Stourne luisant; Stourne métallique
d Weberstar
i Storno metallica

807 *Aplonis minor*
Passeriformes - Sturnidae
e Short-tailed Starling; Lesser Glossy-Starling; Short-tailed Glossy-Starling
f Stourne à queue courte
d Sunda-Star
i Storno codabreve

808 *Aplonis mysolensis*
Passeriformes - Sturnidae
e Moluccan Starling; Island Starling
f Stourne des Moluques
d Molukken-Star
i Storno delle Molucche

809 *Aplonis mystacea*
Passeriformes - Sturnidae
e Yellow-eyed Starling; Grant's Starling
f Stourne de Grant
d Mimikstar
i Storno di Grant

810 *Aplonis opaca*
Passeriformes - Sturnidae
e Micronesian Starling
f Stourne de Micronésie
d Karolinen-Star
i Storno della Micronesia

811 *Aplonis panayensis*
Passeriformes - Sturnidae
e Asian Glossy-Starling; Greater Glossy-Starling; Philippine Starling; Indian Glossy-Starling; Philippine Glossy-Starling; Indian Starling; Glossy-Starling
f Stourne bronzé
d Malaien-Star
i Storno delle Filippine

812 *Aplonis pelzelni*
Passeriformes - Sturnidae
e Pohnpei Starling; Ponapé Mountain-Starling; Pohnpei Mountain-Starling
f Stourne de Ponapé
d Petzeln-Star
i Storno di Ponapè

813 *Aplonis santovestris*
Passeriformes - Sturnidae
e Mountain Starling; Watiamassau Starling
f Stourne d'Espiritu Santo
d Rostbürzelstar
i Storno di Spirito Santo

814 *Aplonis striata*
Passeriformes - Sturnidae
e Striated Starling; Striped Glossy-Starling
f Stourne calédonien
d Dickschnabelstar
i Storno striato della Nuova Caledonia

815 *Aplonis tabuensis*
Passeriformes - Sturnidae
e Polynesian Starling; Striped Starling
f Stourne de Polynésie
d Südseestar
i Storno striato del Pacifico

816 *Aplonis zelandica*
Passeriformes - Sturnidae
e Rusty-winged Starling; New Hebrides Starling
f Stourne mélanésien
d Rostflügelstar
i Storno delle Nuove Ebridi

817 *Aplopelia larvata*
Columbiformes - Columbidae
e Lemon Dove; Cinnamon Dove; Forest Dove
f Tourterelle à masque blanc; Pigeon à

masque blanc
d Zimttaube; Weißmaskentaube
i Colomba limone

818 Apodidae
Apodiformes
e Swifts
f Martinets; Apodidés
d Segler; Schwalbensegler
i Apodidi

819 Apodiformes
e Swifts
d Segler; Schwirrflügler
i Apodiformi

820 *Aprosmictus erythropterus*
Psittaciformes - Psittacidae
e Red-winged Parrot; Crimson-winged Parrot; Red-winged Lory; Crimson-winged Lory; Red-winged King-Parrot; Red-winged Parakeet
f Perruche érythroptère
d Rotflügelsittich
i Parrocchetto alirosse

821 *Aprosmictus jonquillaceus*
Psittaciformes - Psittacidae
e Olive-shouldered Parrot; Timor Red-winged Parrot; Timor Crimson-winged Parakeet; Timor Parrot
f Perruche jonquille
d Timor-Rotflügelsittich; Timor-Sittich
i Parrochetto giunchiglia

822 *Aptenodytes forsteri*
Sphenisciformes - Spheniscidae
e Emperor Penguin
f Manchot empéreur
d Kaiserpinguin; Riesenpinguin
i Pinguino imperatore

823 *Aptenodytes patagonicus*
Sphenisciformes - Spheniscidae
e King Penguin
f Manchot royal; Grand Manchot
d Königspinguin
i Pinguino reale

824 Apterygidae
Dinorthiformes

e Kiwis
f Apterygidés
d Kiwis
i Apterigidi

825 Apterygiformes
e Kiwis
f Apterygiformes
d Kiwis; Schnepfenstrauße
i Apterigiformi

826 *Apteryx australis*
Dinorthiformes - Apterygidae
e Brown Kiwi; Common Kiwi; Kiwi
f Kiwi austral; Kiwi
d Streifenkiwi
i Kiwi striato; Kiwi bruno; Kiwi; Chivi; Apterice

827 *Apteryx haastii*
Dinorthiformes - Apterygidae
e Great Spotted-Kiwi; Roaroa Kiwi; Great Grey-Kiwi
f Kiwi roa
d Haast-Kiwi
i Kiwi macchiato maggiore

828 *Apteryx owenii*
Dinorthiformes - Apterygidae
e Little Spotted-Kiwi; Owen's Kiwi; Little Grey-Kiwi
f Kiwi d'Owen
d Zwergkiwi
i Kiwi d'Owen; Kiwi maculato

829 *Apus acuticauda*
Apodiformes - Apodidae
e Dark-rumped Swift; Dark-backed Swift; Khasi Hills Swift
f Martinet de l'Assam
d Glanzrückensegler
i Rondone dorsoscuro

830 *Apus affinis*
Apodiformes - Apodidae
e Little Swift; House Swift
f Martinet à croupion blanc; Martinet des maisons
d Haussegler
i Rondone indiano

831 *Apus alexandri*
Apodiformes - Apodidae
e Alexander's Swift; Cape Verde Swift
f Martinet du Cap-Vert
d Alexander-Segler
i Rondone di Alexander

832 *Apus apus*
Apodiformes - Apodidae
e Common Swift; Swift; European
Swift; Eurasian Swift (CSA);
Northern Swift
f Martinet noir
d Mauersegler; Turmschwalbe;
Turmsegler
i Rondone; Rondone comune;
Rondone eurasiatico

833 *Apus balstoni*
Apodiformes - Apodidae
e Madagascar Swift
f Martinet malgache
d Balston-Segler
i Rondone del Madagascar

834 *Apus barbatus*
Apodiformes - Apodidae
e African Swift; African Black-Swift
(CSA); Black Swift (CSA); Fernando
Po Swift
f Martinet du Cap; Martinet noir
africain
d Kap-Segler
i Rondone nero africano

835 *Apus batesi*
Apodiformes - Apodidae
e Bates's Swift; Bates's Black-Swift
f Martinet de Bates
d Mohrensegler
i Rondone di Bates

836 *Apus berliozi*
Apodiformes - Apodidae
e Forbes-Watson's Swift; Watson's
Swift; Berlioz's Swift
f Martinet de Berlioz
d Sokotra-Segler
i Rondone di Berlioz

837 *Apus bradfieldi*
Apodiformes - Apodidae
e Bradfield's Swift
f Martinet de Bradfield
d Damara-Segler
i Rondone di Bradfield

838 *Apus caffer*
Apodiformes - Apodidae
e White-rumped Swift; Caffer Swift;
African White-rumped Swift
f Martinet cafre
d Weißbürzelsegler; Kaffern-Segler
i Rondone cafro

839 *Apus horus*
Apodiformes - Apodidae
e Horus Swift; Alpine Swift
f Martinet horus
d Horus-Segler; Erdsegler
i Rondone di Horus

840 *Apus niansae*
Apodiformes - Apodidae
e Nyanza Swift; Brown Swift
f Martinet du Nyanza
d Braunsegler
i Rondone di Nyanza

841 *Apus nipalensis*
Apodiformes - Apodidae
e House Swift; Little Swift; Malay
House-Swift
f Martinet malais
d Malaien-Segler
i Rondone delle case

842 *Apus pacificus*
Apodiformes - Apodidae
e Fork-tailed Swift; White-rumped
Swift; Pacific Swift; Pacific Fork-
tailed Swift; Asian White-rumped
Swift; Large White-rumped Swift;
Northern White-rumped Swift
f Martinet de Sibérie
d Pazifik-Segler
i Rondone codaforcuta

843 *Apus pallidus*
Apodiformes - Apodidae
e Pallid Swift; Mouse-coloured Swift

f Martinet pâle
d Fahlsegler
i Rondone pallido

844 *Apus sladeniae*
Apodiformes - Apodidae
e Fernando Po Swift
f Martinet de Fernando Po
d Guinea-Segler

845 *Apus toulsoni*
Apodiformes - Apodidae
e Loanda Swift; Toulson's Swift
f Martinet de Loanda
d Erdsegler
i Rondone di Toulson

846 *Apus unicolor*
Apodiformes - Apodidae
e Plain Swift
f Martinet unicolore
d Einfarbsegler
i Rondone unicolore

847 *Aquila adalberti*
Falconiformes - Accipitridae
e Adalbert's Eagle; Spanish Imperial-Eagle; Adalbert's Imperial-Eagle; White-shouldered Eagle
f Aigle ibérique
i Aquila imperiale spagnola; Aquila imperiale occidentale

848 *Aquila audax*
Falconiformes - Accipitridae
e Wedge-tailed Eagle; Mountain Eagle; Eagle-Hawk
f Aigle d'Australie
d Keilschwanzadler
i Aquila codacuneata; Aquila dalla lunga coda; Aquila audace

849 *Aquila audax fleayi*
Falconiformes - Accipitridae
e Tasmanian Wedge-tailed Eagle (ANZ)

850 *Aquila chrysaetos*
Falconiformes - Accipitridae
e Golden Eagle
f Aigle royal

d Steinadler
i Aquila reale

851 *Aquila clanga*
Falconiformes - Accipitridae
e Greater Spotted-Eagle; Spotted Eagle; Great Spotted-Eagle
f Aigle criard
d Schelladler
i Aquila anatraia maggiore

852 *Aquila gurneyi*
Falconiformes - Accipitridae
e Gurney's Eagle
f Aigle de Gurney
d Molukken-Adler
i Aquila di Gurney

853 *Aquila hastata*
Falconiformes - Accipitridae
e Indian Spotted-Eagle

854 *Aquila heliaca*
Falconiformes - Accipitridae
e Imperial Eagle; Eastern Imperial-Eagle
f Aigle impérial
d Kaiseradler
i Aquila imperiale; Aquila imperiale orientale

855 *Aquila nipalensis*
Falconiformes - Accipitridae
e Steppe Eagle
f Aigle des steppes
d Steppenadler
i Aquila delle steppe

856 *Aquila pomarina*
Falconiformes - Accipitridae
e Lesser Spotted-Eagle
f Aigle pomarin
d Schreiadler
i Aquila anatraia minore

857 *Aquila rapax*
Falconiformes - Accipitridae
e Tawny Eagle; African Tawny-Eagle; Asian Tawny-Eagle
f Aigle ravisseur
d Raubadler; Savannenadler;

Steppenadler
i   Aquila rapace

858   *Aquila verreauxii*
Falconiformes - Accipitridae
e   Verreaux's Eagle; Black Eagle
(CSA); African Black-Eagle
f   Aigle de Verreaux
d   Felsenadler; Kaffern-Adler
i   Aquila del Verreaux; Aquila di
Verreaux

859   *Aquila vindhiana*
Falconiformes - Accipitridae
e   Tawny Eagle (ISC); Eurasian Tawny
Eagle
f   Aigle indien
i   Aquila orientale

860   *Aquila wahlbergi*
Falconiformes - Accipitridae
e   Wahlberg's Eagle
f   Aigle de Wahlberg
d   Wahlberg-Adler; Wahlberg-
Silberadler; Silberadler
i   Aquila di Wahlberg

861   *Ara ambigua*
Psittaciformes - Psittacidae
e   Great Green-Macaw; Grand Military
Macaw; Buffon's Macaw
f   Ara de Buffon
d   Bechsteinara; Bechsteinarara; Großer
Soldatenara; Großer Soldatenarara
i   Ara di Buffon

862   *Ara ararauna*
Psittaciformes - Psittacidae
e   Blue-and-yellow Macaw; Green
Macaw; Blue-and-gold Macaw
f   Ara bleu
d   Arauna; Blaugelber Ara;
Gelbbrustara
i   Ara ararauna; Ara golablu; Arauna

863   *Ara atwoodi*
Psittaciformes - Psittacidae
e   Dominican Macaw

864   *Ara auricollis*
Psittaciformes - Psittacidae

e   Yellow-collared Macaw; Golden-
collared Macaw; Golden-naped
Macaw
f   Ara à collier jaune
d   Gelbnackenara; Gelbnackenarara;
Goldnackenara; Goldnackenarara;
Halsbandara
i   Ara dal collare

865   *Ara chloropterus*
Psittaciformes - Psittacidae
e   Red-and-green Macaw; Green-
winged Macaw; Red-blue-and-green
Macaw; Red-and-blue Macaw
f   Ara chloroptère
d   Dunkelroter Ara; Dunkelroter Arara;
Grünflügelara; Grünflügelarara
i   Ara aliverdi; Ara rossa e verde

866   *Ara couloni*
Psittaciformes - Psittacidae
e   Blue-headed Macaw; Coulon's
Macaw
f   Ara de Coulon
d   Blaukopfara; Blaukopfarara;
Gebirgsara; Gebirgsarara
i   Ara testablu

867   *Ara cubensis*
Psittaciformes - Psittacidae
e   Cuban Macaw
f   Ara de Cuba
d   Dreifarbenara
i   Ara di Cuba

868   *Ara erythrocephalus*
Psittaciformes - Psittacidae
e   Red-headed Green-Macaw

869   *Ara erythrura*
Psittaciformes - Psittacidae
e   Red-tailed Macaw

870   *Ara glaucogularis*
Psittaciformes - Psittacidae
e   Blue-throated Macaw; Caninde
Macaw; Wagler's Macaw
f   Ara canindé
i   Ara golablu

871 **Ara gossei**
Psittaciformes - Psittacidae
e Yellow-headed Macaw; Jamaican
Macaw

872 **Ara guadeloupensis**
Psittaciformes - Psittacidae
e Guadeloupe Macaw

873 **Ara macao**
Psittaciformes - Psittacidae
e Scarlet Macaw; Red-and-yellow
Macaw
f Ara rouge
d Arakanga; Hellroter Ara; Hellroter
Arara
i Ara macao; Ara rossa e verde;
Aracanga

874 **Ara manilata**
Psittaciformes - Psittacidae
e Red-bellied Macaw
f Ara macavouanne
d Macavouanna; Rotbauchara;
Rotbaucharara
i Ara ventre rosso

875 **Ara maracana**
Psittaciformes - Psittacidae
e Blue-winged Macaw; Illiger's Macaw
f Ara d'Illiger
d Marakana; Rotrückenara;
Rotrückenarara
i Ara di Illiger

876 **Ara militaris**
Psittaciformes - Psittacidae
e Military Macaw
f Ara militaire
d Kleiner Soldatenara; Kleiner
Soldatenarara; Soldatenara;
Soldatenarara
i Ara militare

877 **Ara nobilis**
Psittaciformes - Psittacidae
e Red-shouldered Macaw; Hahn's
Macaaw
f Ara noble
d Blaustirnzwergara;
Blaustirnzwergarara; Hahns

Zwergara; Zwergara; Zwergarara
i Ara spallerosse

878 **Ara rubrogenys**
Psittaciformes - Psittacidae
e Red-fronted Macaw; Red-cheeked
Macaw
f Ara de Lafresnaye
d Rotohrara; Rotohrarara
i Ara fronterosso

879 **Ara severa**
Psittaciformes - Psittacidae
e Chestnut-fronted Macaw; Chestnut-
fronted Severe Macaw
f Ara vert
d Rotbugara; Rotbugarara
i Ara severa

880 **Ara tricolor**
Psittaciformes - Psittacidae
e Cuban Macaw; Guacamayo (WI);
Hispaniolan Macaw
f Ara d'Hispaniola
i Ara di Hispaniola

881 **Arachnothera affinis**
Passeriformes - Nectariniidae
e Grey-breasted Spiderhunter
f Arachnothère à poitrine grise
d Graubrustspinnenjäger
i Mangiaragni pettogrigio

882 **Arachnothera chrysogenys**
Passeriformes - Nectariniidae
e Yellow-eared Spiderhunter; Lesser
Yellow-eared Spiderhunter
f Arachnothère à joues jaunes
d Gelbbartspinnenjäger
i Mangiaragni orecchiegialle

883 **Arachnothera clarae**
Passeriformes - Nectariniidae
e Naked-faced Spiderhunter; Bare-
faced Spiderhunter
f Arachnothère à face nue
d Nacktwangenspinnenjäger
i Mangiaragni faccianuda

884 **Arachnothera crassirostris**
Passeriformes - Nectariniidae

e	Thick-billed Spiderhunter
f	Arachnothère à bec epais
d	Dickschnabelspinnenjäger
i	Mangiaragni beccoforte

885   *Arachnothera everetti*
Passeriformes - Nectariniidae
e   Bornean Spiderhunter; Everett's
Spiderhunter; Kinabalu Spiderhunter
f   Arachnothère d'Everett
d   Everett-Spinnenjäger
i   Mangiaragni di Everett

886   *Arachnothera flavigaster*
Passeriformes - Nectariniidae
e   Spectacled Spiderhunter; Greater
Yellow-eared Spiderhunter; Large
Spiderhunter
f   Arachnothère à lunettes
d   Eyton-Spinnenjäger
i   Mangiaragni ventregiallo

887   *Arachnothera juliae*
Passeriformes - Nectariniidae
e   Whitehead's Spiderhunter
f   Arachnothère de Whitehead
d   Bergspinnenjäger
i   Mangiaragni di Whitehead

888   *Arachnothera longirostra*
Passeriformes - Nectariniidae
e   Little Spiderhunter; Small
Spiderhunter
f   Petit Arachnothère; Arachnothère à
long bec
d   Weißkehlspinnenjäger
i   Mangiaragni piccolo

889   *Arachnothera magna*
Passeriformes - Nectariniidae
e   Streaked Spiderhunter
f   Grand Arachnothère
d   Streichelspinnenjäger
i   Mangiaragni striato

890   *Arachnothera robusta*
Passeriformes - Nectariniidae
e   Long-billed Spiderhunter
f   Arachnothère à long bec
d   Langschnabelspinnenjäger
i   Mangiaragni beccolungo

891   Aramidae
Gruiformes
e   Limpkins
f   Aramidés; Courlans
d   Rallenkraniche; Riesenrallen; Echte
Rallenkraniche
i   Aramidi

892   *Aramides axillaris*
Gruiformes - Rallidae
e   Rufous-necked Woodrail; Rufous-
crowned Rail
f   Râle à cou roux
d   Braunkappenralle
i   Rallo collorosso

893   *Aramides cajanea*
Gruiformes - Rallidae
e   Grey-necked Woodrail; Cayenne Rail
f   Râle de Cayenne
d   Cayenne-Ralle
i   Rallo di Cayenna

894   *Aramides calopterus*
Gruiformes - Rallidae
e   Red-winged Woodrail
f   Râle à ailes rouges
d   Rotflügelralle
i   Rallo pettirosso

895   *Aramides mangle*
Gruiformes - Rallidae
e   Little Woodrail; Spix's Woodrail
f   Râle des palétuviers
d   Küstenralle
i   Rallo mangle

896   *Aramides saracura*
Gruiformes - Rallidae
e   Slaty-breasted Woodrail; Slaty-
breasted Rail; Saracura Rail;
Saracura Woodrail
f   Râle saracura
d   Saracura-Ralle
i   Rallo pettoardesia

897   *Aramides wolfi*
Gruiformes - Rallidae
e   Brown Woodrail; Wolf's Woodrail;
Brown-backed Woodrail
f   Râle de Wolf

d   Esmeraldas Ralle
i   Rallo di Wolf

898   ***Aramides ypecaha***
     Gruiformes - Rallidae
e   Giant Woodrail; Giant Rail
f   Râle ypécaha
d   Ypecaha-Ralle
i   Rallo ypecaha

899   ***Aramidopsis plateni***
     Gruiformes - Rallidae
e   Snoring Rail; Platen's Rail; Celebes
     Rail; Platen's Crake
f   Râle de Platen
d   Schnarchralle
i   Rallo di Platen

900   ***Aramus guarauna***
     Gruiformes - Aramidae
e   Limpkin; Clucking Hen (WI)
f   Courlan brun; Poule-à-jolie (Ants);
     Courlan
d   Rallenkranich; Riesenralle
i   Serracura; Aramo

901   ***Aratinga acuticaudata***
     Psittaciformes - Psittacidae
e   Blue-crowned Parakeet; Blue-fronted
     Conure; Blue-crowned Conure;
     Sharp-tailed Conure
f   Conure à tête bleue
d   Blaustirnsittich; Spitzschwanzsittich
i   Conuro testablu

902   ***Aratinga astec***
     Psittaciformes - Psittacidae
e   Aztec Parakeet; Aztek Conure

903   ***Aratinga aurea***
     Psittaciformes - Psittacidae
e   Peach-fronted Parakeet; Halfmoon
     Conure; Peach-fronted Conure;
     Golden-crowned Conure
f   Conure couronnée
d   Goldstirnsittich
i   Conuro frontedorata

904   ***Aratinga auricapilla***
     Psittaciformes - Psittacidae
e   Golden-capped Parakeet; Golden-
     headed Conure; Golden-capped
     Conure; Flame-capped Parakeet
f   Conure à tête d'or
d   Goldkappensittich
i   Conuro testadorata

905   ***Aratinga brevipes***
     Psittaciformes - Psittacidae
e   Socorro Parakeet; Socorro Conure
f   Conure de Socorro

906   ***Aratinga cactorum***
     Psittaciformes - Psittacidae
e   Caatinga Parakeet; Cactus Parakeet;
     Cactus Conure
f   Conure des cactus
d   Kaktussittich
i   Conuro dei cactus

907   ***Aratinga canicularis***
     Psittaciformes - Psittacidae
e   Orange-fronted Parakeet; Halfmoon
     Parakeet; Petz's Parakeet; Petz's
     Conure; Orange-fronted Conure;
     Halfmoon Conure
f   Conure à front rouge
d   Elfenbeinsittich; Halbmondsittich;
     Peltz-Sittich
i   Conuro frontearancio

908   ***Aratinga chloroptera***
     Psittaciformes - Psittacidae
e   Hispaniolan Parakeet; Hispaniolan
     Conure
f   Perruche (Ants); Conure maitresse
d   Grünflügelsittich; Haiti-Sittich
i   Conuro di Hispaniola

909   ***Aratinga erythrogenys***
     Psittaciformes - Psittacidae
e   Red-masked Parakeet; Red-masked
     Conure
f   Conure à tête rouge
d   Guyaquil-Sittich
i   Conuro testarossa

910   ***Aratinga euops***
     Psittaciformes - Psittacidae
e   Cuban Parakeet; Cuban Conure; Red-
     speckled Conure
f   Conure de Cuba

d   Kuba-Sittich
i   Conuro di Cuba

911   *Aratinga finschi*
      Psittaciformes - Psittacidae
e   Crimson-fronted Parakeet; Finsch's
      Parakeet; Finsch's Conure
f   Conure de Finsch
d   Finsch-Sittich; Veragua-Sittich
i   Conuro di Finsch

912   *Aratinga guarouba*
      Psittaciformes - Psittacidae
e   Golden Parakeet; Golden Conure;
      Queen of Bavaria Conure
f   Conure dorée
d   Goldsittich
i   Conuro guaruba; Conuro dorato

913   *Aratinga holochlora*
      Psittaciformes - Psittacidae
e   Green Parakeet; Mexican Green-
      Parakeet; Mexican Green-Conure;
      Green Conure; Salvin's Conure
f   Conure verte
d   Grünsittich; Mexikanischer
      Grünsittich
i   Conuro verde

914   *Aratinga jandaya*
      Psittaciformes - Psittacidae
e   Jandaya Parakeet; Jandaya Conure
f   Conure jandaya
d   Jandaya-Sittich
i   Conuro jandaya

915   *Aratinga leucophthalmus*
      Psittaciformes - Psittacidae
e   White-eyed Parakeet; White-eyed
      Conure; Flaming Parakeet
f   Conure pavouane
d   Pavua-Sittich
i   Conuro occhibianchi

916   *Aratinga mitrata*
      Psittaciformes - Psittacidae
e   Mitred Parakeet; Mitred Comure
f   Conure mitrée
d   Rotmaskensittich
i   Conuro mitrata

917   *Aratinga nana*
      Psittaciformes - Psittacidae
e   Olive-throated Parakeet; Parakeet
      (WI); Jamaican Parakeet; Olive-
      throated Conure; Jamaican Conure
f   Conure aztèque
d   Jamaika-Sittich
i   Conuro nano

918   *Aratinga ocularis*
      Psittaciformes - Psittacidae
e   Veraguas Parakeet

919   *Aratinga pertinax*
      Psittaciformes - Psittacidae
e   Brown-throated Parakeet; Caribbean
      Parakeet (WI); Brown-throated
      Conure; St. Thomas Conure
f   Conure cuivrée
d   Braunwangensittich; St. Thomas-
      Sittich
i   Conuro golabruna

920   *Aratinga rubritorques*
      Psittaciformes - Psittacidae
e   Red-throated Parakeet; Red-throated
      Conure
f   Conure à gorge rouge

921   *Aratinga solstitialis*
      Psittaciformes - Psittacidae
e   Sun Parakeet; Sun Conure
f   Conure soleil
d   Sonnensittich
i   Conuro del sole

922   *Aratinga strenua*
      Psittaciformes - Psittacidae
e   Pacific Parakeet; Pacific Conure
f   Conure de Ridgway
i   Conuro del Pacifico

923   *Aratinga wagleri*
      Psittaciformes - Psittacidae
e   Scarlet-fronted Parakeet; Red-fronted
      Conure; Wagler's Conure
f   Conure de Wagler
d   Kolumbia-Sittich
i   Conuro fronterosso

924 *Aratinga weddellii*
    Psittaciformes - Psittacidae
e Dusky-headed Parakeet; Dusky-
    headed Conure; Weddell's Conure
f Conure de Weddell
d Braunkopfsittich; Wedell-Sitich
i Conuro di Weddell

925 *Arborophila ardens*
    Galliformes - Phasianidae
e Hainan Partridge; Hainan Hill-
    Partridge; Hainan Tree-Partridge;
    White-eared Hill-Partridge; White-
    eared Partridge
d Hainan-Buschwachtel
i Pernice di Hainan

926 *Arborophila atrogularis*
    Galliformes - Phasianidae
e White-cheeked Partridge; White-
    cheeked Hill-Partridge
f Torquéole à joues blanches
d Weißwangenbuschwachtel
i Pernice guancebianche

927 *Arborophila brunneopectus*
    Galliformes - Phasianidae
e Bar-backed Partridge; Brown-
    breasted Hill-Partridge; Bar-backed
    Hill-Partridge; Barred Hill-Partridge;
    Barred Partridge; Brown-breasted
    Partridge
f Torquéole à poitrine brune
d Rotbauchbuschwachtel
i Pernice pettobruno

928 *Arborophila cambodiana*
    Galliformes - Phasianidae
e Chestnut-headed Partridge; Rufous-
    faced Hill-Partridge; Chestnut-
    headed Tree-Partridge; Chestnut-
    headed Hill-Partridge; Cambodian
    Hill-Partridge
f Torquéole du Cambodge
d Kambodscha-Buschwachtel
i Pernice della Cambogia

929 *Arborophila charltonii*
    Galliformes - Phasianidae
e Chestnut-necklaced Partridge;
    Chestnut-breasted Tree-Partridge;

    Scaly-breasted Tree-Partridge; Scaly-
    breasted Hill-Partridge; Eyron's Hill-
    Partridge; Chestnut-necklaced Tree-
    Partridge; Chestnut-necklaced Hill-
    Partridge; Scaly-breasted Partridge
f Torquéole à poitrine châtaine
d Grünfußbuschwachtel
i Pernice di Charlton

930 *Arborophila chloropus*
    Galliformes - Phasianidae
e Scaly-breasted Partridge; Green-
    legged Partridge; Green-legged Hill-
    Partridge; Green-legged Tree-
    Partridge; Woodland Partridge;
    Woodland Hill-Partridge
f Torquéole des bois
i Pernice zampeverdi

931 *Arborophila crudigularis*
    Galliformes - Phasianidae
e Formosan Partridge; Formosan Hill-
    Partridge; Taiwan Partridge; Taiwan
    Hill-Partridge
f Torquéole de Formose
d Formosa-Buschwachtel
i Pernice golabiancha

932 *Arborophila davidi*
    Galliformes - Phasianidae
e Orange-necked Partridge; David's
    Tree-Partridge; Orange-breasted Hill-
    Partridge
f Torquéole de David
d David-Buschwachtel
i Pernice di David

933 *Arborophila diversa*
    Galliformes - Phasianidae
e Siamese Partridge

934 *Arborophila gingica*
    Galliformes - Phasianidae
e White-necklaced Partridge; Rickett's
    Hill-Partridge; Fokien Hill-Partridge;
    Rickett's Partridge; White-browed
    Hill-Partridge; Collared Hill-
    Partridge
f Torquéole de Gingi
d China-Buschwachtel
i Pernice di Rickett

935 *Arborophila hyperythra*
Galliformes - Phasianidae
e   Red-breasted Partridge; Red-breasted
    Hill-Partridge; Red-breasted Tree-
    Partridge; Bornean Partridge;
    Bornean Hill-Partridge
f   Torquéole de Bornéo
d   Borneo-Buschwachtel
i   Pernice pettorosso

936 *Arborophila javanica*
Galliformes - Phasianidae
e   Chestnut-bellied Partridge; Chestnut-
    bellied Tree-Partrdige; Javan Hill-
    Partridge
f   Torquéole de Java; Torquéole de
    Sumatra
d   Rotbauchbuschwachtel
i   Pernice di Giava

937 *Arborophila mandellii*
Galliformes - Phasianidae
e   Chestnut-breasted Partridge; Red-
    breasted Hill-Partridge; Mandell's
    Hill-Partridge; Chestnut-breasted
    Hill-Partridge
f   Torquéole de Mandelli
d   Mandelli-Buschwachtel
i   Pernice di Mandelli

938 *Arborophila merlini*
Galliformes - Phasianidae
e   Annam Partridge; Annam Hill-
    Partrisge; Annam Tree-Partridge;
    Annamese Hill-Partridge
f   Torquéole de Merlin
d   Gelbfußbuschwachtel
i   Pernice dell'Annam

939 *Arborophila orientalis*
Galliformes - Phasianidae
e   Grey-breasted Hill-Partridge; Grey-
    breasted Partridge; Sumatran Hill-
    Partridge; Sumatran Tree-Partridge;
    Sunda Grey-breated Partridge; Grey-
    bellied Partridge; Bare-throated Hill-
    Partridge; Campbell's Hill-Partridge
i   Pernice pettogrigio

940 *Arborophila rubrirostris*
Galliformes - Phasianidae

e   Red-billed Partridge; Red-billed
    Tree-Partridge; Red-billed Hill-
    Partridge
f   Torquéole à bec rouge
d   Rotschnabelbuschwachtel
i   Pernice beccorosso

941 *Arborophila rufipectus*
Galliformes - Phasianidae
e   Sichuan Partridge; Boulton's Hill-
    Partridge; Sichuan Hill-Partridge
f   Torquéole de Boulton
d   Boulton-Buschwachtel
i   Pernice di Boulton

942 *Arborophila rufogularis*
Galliformes - Phasianidae
e   Rufous-throated Partridge; Rufous-
    throated Hill-Partridge
f   Torquéole à gorge rousse
d   Rotkehlbuschwachtel
i   Pernice golarossa

943 *Arborophila torqueola*
Galliformes - Phasianidae
e   HillPartridge; Common Hill-
    Partridge; Indian Hill-Partridge;
    Necklaced Hill-Partridge
f   Torquéole à collier
d   Hügelhuhn
i   Pernice delle alture

944 *Arcanator orostruthus*
Passeriformes - Sylviidae
e   Dapple-throat; Dappled Bulbul;
    Dappled Mountain-Bulbul; Dappled
    Mountain-Robin; Dappled Robin;
    Dappled Greenbul; Dappled
    Illadopsis
f   Modulatrice grivelée
d   Bülbültimalie
i   Cantore maculato delle montagne

945 *Archboldia papuensis*
Passeriformes - Ptilonorhynchidae
e   Archbold's Bowerbird
f   Jardinier d'Archbold
d   Gelbbandgärtner; Archbold-
    Laubenvogel
i   Giardiniere di Archbold

946 *Archboldia sanfordi*
Passeriformes - Ptilonorhynchidae
e Tomba Bowerbird; Sanford's
Bowerbird
f Jardinier de Sanford
i Giardiniere di Sanford; Ptilonorinco
di Sanford

947 *Archilochus alexandri*
Apodiformes - Trochilidae
e Black-chinned Hummingbird
f Colibri à gorge noire
d Schwarzkinnkolibri
i Colibrì golanera; Colibrì di
Alexander

948 *Archilochus colubris*
Apodiformes - Trochilidae
e Ruby-throated Hummingbird
f Colibri à gorge rubis
d Rubinkehlkolibri
i Colibrì golarubino; Colibrì dalla gola
color rubino

949 *Ardea bournei*
Ciconiiformes - Ardeidae
e Cape Verde Purple Heron
f Héron de Bourne
i Airone

950 *Ardea cinerea*
Ciconiiformes - Ardeidae
e Grey Heron; Heron; Common Heron;
Eastern Grey-Heron
f Héron cendré
d Graureiher; Fischreiher
i Airone cenerino; Aghirone; Airone

951 *Ardea cinerea monicae*
Ciconiiformes - Ardeidae
e Mauritian Heron
f Héron pâle
d Blaßreiher

952 *Ardea cocoi*
Ciconiiformes - Ardeidae
e Cocoi Heron; White-necked Heron
f Héron cocoi
d Cokoireiher
i Airone plumbeo

953 *Ardea goliath*
Ciconiiformes - Ardeidae
e Goliath Heron; Giant Heron (ISC)
f Héron goliath
d Goliathreiher
i Airone Golia

954 *Ardea herodias*
Ciconiiformes - Ardeidae
e Great Blue-Heron; Blue Gaulin (WI);
Grey Gaulin (WI); Arsnicker (WI)
f Grand Héron; Crabier bleu (Ants);
Crabier radar (Ants); Crabier noir
(Ants)
d Kanada-Reiher; Amerikanischer
Graureiher
i Airone azzurro maggiore

955 *Ardea humbloti*
Ciconiiformes - Ardeidae
e Humblot's Heron; Madagascar
Heron; Malagasy Heron
f Héron de Humblot
d Madagaskar-Reiher
i Airone malgascio

956 *Ardea insignis*
Ciconiiformes - Ardeidae
e White-bellied Heron; Great White-
bellied Heron (ISC); Imperial Heron
f Héron impérial
d Weißbauchreiher
i Airone imperiale

957 *Ardea intermedia*
Ciconiiformes - Ardeidae
e Intermediate Egret; Yellow-billed
Egret; Smaller Egret (ISC); Median
Egret (ISC); Plumed Egret; Lesser
Egret; Short-billed Egret
f Aigrette intermédiaire; Héron
intermédiaire
d Mittelreiher; Edelreiher
i Garzetta intermedia

958 *Ardea melanocephala*
Ciconiiformes - Ardeidae
e Black-headed Heron
f Héron mélanocéphale
d Schwarzkopfreiher;

Schwarzhalsreiher
i   Airone testanera

959   ***Ardea pacifica***
Ciconiiformes - Ardeidae
e   Pacific Heron; White-necked Heron
f   Héron à tête blanche
d   Weißhalsreiher
i   Airone collobianco

960   ***Ardea picata***
Ciconiiformes - Ardeidae
e   Pied Heron; Allied Heron; White-
headed Egret
f   Héron pie
d   Elsterreiher
i   Airone bianco e nero

961   ***Ardea purpurea***
Ciconiiformes - Ardeidae
e   Purple Heron; Eastern Purple Heron;
Cape Verde Heron
f   Héron pourpré
d   Purpurreiher
i   Airone rosso

962   ***Ardea sumatrana***
Ciconiiformes - Ardeidae
e   Great-billed Heron; Dusky-grey
Heron; Sumatran Heron; Giant Heron
f   Héron typhon
d   Malaien-Reiher
i   Airone di Sumatra

963   **Ardeidae**
Ciconiiformes
e   Herons
f   Hérons; Ardéidés
d   Reiher
i   Ardeidi; Aironi; Tarabusi

964   ***Ardeola bacchus***
Ciconiiformes - Ardeidae
e   Chinese Pond-Heron; Paddybird;
Pond-Heron
f   Crabier chinois
d   Bacchusreiher
i   Sgarza cinese

965   ***Ardeola grayii***
Ciconiiformes - Ardeidae

e   Indian Pond-Heron; Paddybird (ISC);
Pond-Heron (ISC); Indian Paddybird;
Indian Pondbird
f   Crabier de Gray
d   Paddyreiher
i   Sgarza indiana

966   ***Ardeola idae***
Ciconiiformes - Ardeidae
e   Madagascar Pond-Heron; Malagasy
Pond-Heron (CSA); Madagascar
Squacco Heron (CSA); Madagascar
Squacco
f   Crabier blanc; Héron crabier blanc
d   Madagaskar-Rallenreiher;
Dickschnabelreiher
i   Sgarza del Madagascar

967   ***Ardeola ralloides***
Ciconiiformes - Ardeidae
e   Squacco Heron; Common Squacco
Heron
f   Crabier chevelu; Héron crabier
d   Rallenreiher
i   Sgarza ciuffetto

968   ***Ardeola rufiventris***
Ciconiiformes - Ardeidae
e   Rufous-bellied Heron
f   Crabier à ventre roux; Héron à ventre
roux
d   Rotbauchreiher
i   Sgarza rufiventre

969   ***Ardeola speciosa***
Ciconiiformes - Ardeidae
e   Javan Pond-Heron; Javanese Pond-
Heron; Malay Pond-Heron; Pond-
Heron
f   Crabier malais
d   Prachtreiher
i   Sgarza di Giava

970   ***Ardeotis arabs***
Gruiformes - Otididae
e   Arabian Bustard; Great Arabian
Bustard; Sudan Bustard
f   Outarde arabe
d   Arabertrappe
i   Otarda araba; Otarda d'Arabia

971 **Ardeotis australis**
Gruiformes - Otididae
e Australian Bustard
f Outarde d'Australie
d Wammentrappe
i Otarda australiana

972 **Ardeotis kori**
Gruiformes - Otididae
e Kori Bustard; Large Bustard; Kori's
Bustard
d Riesentrappe
i Otarda di Kori; Otarda kori

973 **Ardeotis nigriceps**
Gruiformes - Otididae
e Indian Bustard; Great Indian-Bustard
(ISC)
f Outarde kori; Outarde à tête noire
d Hindu-Trappe
i Otarda dell'India; Otarda maggiore
indiana

974 **Arenaria interpres**
Charadriiformes - Scolopacidae
e Ruddy Turnstone; Turnstone;
Pondbird (WI); European Turnstone
f Tournepierre à collier; Tournepierre
(Ants); Pluvier des salines (Ants);
Pluvier facou (Ants)
d Steinwälzer
i Voltapietre; Voltapietra

975 **Arenaria melanocephala**
Charadriiformes - Scolopacidae
e Black Turnstone
f Tournepierre noir
d Schwarzkopfsteinwältzer
i Voltapietre testanera

976 **Argusianus argus**
Galliformes - Phasianidae
e Great Argus; Argus Pheasant; Great
Argus-Pheasant
f Argus géant
d Argusfasan
i Argo; Grande argo

977 **Argusianus bipunctatus**
Galliformes - Phasianidae
e Double-banded Argus

f Argus bifascié
i Argo bifasciato; Argo

978 **Arremon abeillei**
Passeriformes - Emberizidae
e Black-capped Sparrow
f Tohi d'Abeillé
d Kapuzenruderammer
i Passero capinero

979 **Arremon aurantiirostris**
Passeriformes - Emberizidae
e Orange-billed Sparrow
f Tohi à bec orangé
d Goldschnabelruderammer
i Passero beccoarancio

980 **Arremon flavirostris**
Passeriformes - Emberizidae
e Saffron-billed Sparrow
f Tohi à bec jaune
d Zitronenschnabelruderammer
i Passero beccogiallo

981 **Arremon schlegeli**
Passeriformes - Emberizidae
e Golden-winged Sparrow
f Tohi de Schlegel
d Goldflügelruderammer
i Passero alidorate

982 **Arremon taciturnus**
Passeriformes - Emberizidae
e Pectoral Sparrow
f Tohi silencieux
d Schwarzbrustruderammer
i Passero dal collare nero

983 **Arremonops chloronotus**
Passeriformes - Emberizidae
e Green-backed Sparrow
f Tohi à dos vert
d Grünrückenammer
i Passero dorsoverde

984 **Arremonops conirostris**
Passeriformes - Emberizidae
e Black-striped Sparrow
f Tohi ligné
d Panama-Ammer
i Passero dalle strie nere

985 **Arremonops rufivirgatus**
Passeriformes - Emberizidae
e Olive Sparrow
f Tohi olive
d Olivrückenammer
i Passero olivaceo

986 **Arremonops superciliatus**
Passeriformes - Emberizidae
e Pacific Sparrow

987 **Arremonops tocuyensis**
Passeriformes - Emberizidae
e Tocuyo Sparrow
f Tohi de Tocuyo
d Tocuyo-Ammer
i Passero di Tocuyo

988 **Arses insularis**
Passeriformes - Corvidae
e Rufous-collared Monarch; Rufous-collared Flycatcher
f Monarque à froc roux
i Monarca dal collare rossiccio

989 **Arses kaupi**
Passeriformes - Corvidae
e Pied Monarch; Pied Flycatcher; Australian Pied-Flycatcher; Australian Pied-Monarch; Banded Monarch
f Monarque sanglé
d Elstermonarch
i Monarca dal collare australiano

990 **Arses telescophthalmus**
Passeriformes - Corvidae
e Frilled Monarch; Frilled Flycatcher; Australian Frilled Monarch; Frill-necked Monarch
f Monarque à collerette
d Krausenmonarch; Ringschnäpper
i Monarca arricciato

991 **Arses telescophthalmus harterti**
Passeriformes - Corvidae
e Torres Strait Frilled Monarch (ANZ)

992 **Artamella viridis**
Passeriformes - Corvidae
e White-headed Vanga

f Artamie à tête blanche
d Weißkopfvanga
i Vanga testabianca

993 **Artamidae**
Passeriformes
e Woodswallows; Swallow-Shrikes
f Artamidés
d Schwalbenstare
i Artamidi

994 **Artamus cinereus**
Passeriformes - Artamidae
e Black-faced Woodswallow
f Langrayen gris
d Schwarzgesichtschwalbenstar
i Artamo faccianera

995 **Artamus cinereus normani**
Passeriformes - Artamidae
a Cape York Peninsula Black-faced Woodswallow

996 **Artamus cyanopterus**
Passeriformes - Artamidae
e Dusky Woodswallow
f Langrayen sordide; Langrayen brun; Langrayen asiatique
d Rußschwalbenstar
i Artamo scuro

997 **Artamus fuscus**
Passeriformes - Artamidae
e Ashy Woodswallow; Ashy Swallow-Shrike (ISC); Ashy Woodswallow-Shrike
f Langrayen brun
d Grauschwalbenstar
i Artamo cenerino; Averla rondine dal ventre rosso

998 **Artamus insignis**
Passeriformes - Artamidae
e Bismarck Woodswallow; Insignia Woodswallow; New Britain Woodswallow; Sclater's Woodswallow
f Langrayen des Bismarck
d Bismarck-Schwalbenstar
i Artamo delle Isole Bismarck

999 **Artamus leucorynchus**
Passeriformes - Artamidae
e White-breasted Woodswallow;
Lesser Woodswallow; White-rumped
Woodswallow; Ashy Woodswallow
f Langrayen à ventre blanc
d Weißbauchschwalbenstar
i Artamo pettobianco

1000 **Artamus maximus**
Passeriformes - Artamidae
e Great Woodswallow; Greater
Woodswallow; Black-breasted
Woodswallow; Papuan
Woodswallow; New Guinea
Woodswallow; Giant Woodswallow
f Grand Langrayen
d Riesenschwalbenstar
i Artamo papua

1001 **Artamus mentalis**
Passeriformes - Artamidae
e Fiji Woodswallow; White-breasted
Woodswallow
f Langrayen des Fidji
d Schwalbenstar
i Artamo delle Isole Figi

1002 **Artamus minor**
Passeriformes - Artamidae
e Little Woodswallow
f Petit Langrayen
d Zwergschwalbenstar
i Artamo minore

1003 **Artamus monachus**
Passeriformes - Artamidae
e Ivory-backed Woodswallow; White-
backed Woodswallow; Hooded
Woodswallow; Celebes
Woodswallow
f Langrayen à tête noire
d Weißrückenschwalbenstar
i Artamo dorsobianco

1004 **Artamus personatus**
Passeriformes - Artamidae
e Masked Woodswallow
f Langrayen masqué
d Maskenschwalbenstar
i Artamo mascherato

1005 **Artamus superciliosus**
Passeriformes - Artamidae
e White-browed Woodswallow
f Langrayen bridé; Langrayen à
sourcils blancs
d Weißbrauenschwalbenstar

1006 **Arundinicola leucocephala**
Passeriformes - Tyrannidae
e White-headed Marsh-Tyrant
f Moucherolle à tête blanche
d Weißkopfrohrtyrann; Sumpftyrann
i Tiranno testabianca

1007 **Ashbyia lovensis**
Passeriformes - Meliphagidae
e Gibberbird; Desert Chat; Gibberchat
f Epthianure de Ashby
d Wüstentrugschmätzer;
Trugschmätzer
i Sassicola australiana di Love

1008 **Asio abyssinicus**
Strigiformes - Strigidae
e Abyssinian Owl; Abyssinian Long-
eared Owl; African Owl; African
Long-eared Owl
f Hibou d'Abyssinie
i Gufo dell'Abissinia

1009 **Asio capensis**
Strigiformes - Strigidae
e Marsh Owl; African Marsh-Owl
f Hibou du Cap; Hibou des marais
africain
d Kap-Ohreule
i Gufo di palude del Capo; Gufo di
palude africano

1010 **Asio clamator**
Strigiformes - Strigidae
e Striped Owl
f Hibou strié
d Kreischeule
i Gufo striato

1011 **Asio flammeus**
Strigiformes - Strigidae
e Short-eared Owl
f Hibou des marais; Hibou brachyote;
Chat-huant (Ants)

d Sumpfohreule
i Gufo di palude

1012 *Asio madagascariensis*
Strigiformes - Strigidae
e Madagascar Owl; Madagascar Long-eared Owl; Mascarene Owl
f Hibou malgache
d Malegassen-Eule
i Gufo del Madagascar

1013 *Asio otus*
Strigiformes - Strigidae
e Long-eared Owl; Common Long-eared Owl
f Hibou moyen-duc; Hibou à aigrettes longues (Qué)
d Waldohreule
i Gufo comune; Gufo

1014 *Asio stygius*
Strigiformes - Strigidae
e Stygian Owl
f Maitre-bois (Ants); Chouette (Ants); Hibou maitre-bois
d Styxeule
i Gufo dello Stige

1015 *Aspatha gularis*
Coraciiformes - Momotidae
e Blue-throated Motmot
f Motmot à gorge bleue
d Blaukehlmotmot
i Motmot golablu

1016 *Asthenes anthoides*
Passeriformes - Furnariidae
e Austral Canastero
f Synallaxe austral
d King-Schlüfer
i Canestraio australe

1017 *Asthenes arequipae*
Passeriformes - Furnariidae
e Dark-winged Canastero
i Canestraio

1018 *Asthenes baeri*
Passeriformes - Furnariidae
e Short-billed Canastero
f Synallaxe à bec court

d Kurzschnabelschlüpfer
i Canestraio beccocorto

1019 *Asthenes berlepschi*
Passeriformes - Furnariidae
e Berlepsch's Canastero; Russet-mantled Softtail; Russet-mantled Thornbird; Berlepsch's Softtail
f Synallaxe de Berlepsch; Synallaxe mantelé
d Berlepsch-Schlüpfer; Rostmantelbündelnister
i Canestraio di Berlepsch; Codamorbida di Berlepsch

1020 *Asthenes cactorum*
Passeriformes - Furnariidae
e Cactus Canastero
f Synallaxe des cactus
d Kaktusbuschschlüpfer
i Canestraio dei cactus

1021 *Asthenes dorbignyi*
Passeriformes - Furnariidae
e Creamy-breasted Canastero; Cream-breasted Canastero; D'Orbigny's Canastero
f Synallaxe d'Orbigny
d Fahlbrustschlüpfer; Mantelbuschschlüpfer
i Canestraio di D'Orbigny

1022 *Asthenes flammulata*
Passeriformes - Furnariidae
e Many-striped Canastero
f Synallaxe flammé
d Jardine-Schlüpfer
i Canestraio flammulato

1023 *Asthenes heterura*
Passeriformes - Furnariidae
e Maquis Canastero; Iquico Canastero
f Synallaxe d'Iquico
s Iquico-Schlüpfer
i Canestraio di Iquico

1024 *Asthenes huancavelicae*
Passeriformes - Furnariidae
e Pale-tailed Canastero
f Synallaxe à queue pâle

1025   *Asthenes hudsoni*
       Passeriformes - Furnariidae
  e   Hudson's Canastero
  f   Synallaxe de Hudson
  d   Grundschlüpfer
  i   Canestraio di Hudson

1026   *Asthenes humicola*
       Passeriformes - Furnariidae
  e   Dusky-tailed Canastero; Kittlitz's
      Canastero
  f   Synallaxe à queue noire
  d   Schwarzschwanzschlüpfer
  i   Canestraio codabruna

1027   *Asthenes humilis*
       Passeriformes - Furnariidae
  e   Streak-throated Canastero; Humble
      Canastero
  f   Synallaxe terrestre
  d   Kehlstreifenschlüpfer
  i   Canestraio golastriata

1028   *Asthenes luizae*
       Passeriformes - Furnariidae
  e   Cipo Canastero
  f   Synallaxe du Cipo

1029   *Asthenes maculicauda*
       Passeriformes - Furnariidae
  e   Scribble-tailed Canastero
  f   Synallaxe à queue marbrée
  d   Fleckenschwanzschlüpfer
  i   Canestraio codamaculata

1030   *Asthenes modesta*
       Passeriformes - Furnariidae
  e   Cordilleran Canastero; Cordillera
      Canastero
  f   Synallaxe des rocailles
  d   Eyton-Schlüpfer
  i   Canestraio della Cordigliera

1031   *Asthenes ottonis*
       Passeriformes - Furnariidae
  e   Rusty-fronted Canastero
  f   Synallaxe à front rouille; Synallaxe
      de Garlepp
  d   Garlepp-Schlüpfer
  i   Canestraio fronterossiccia

1032   *Asthenes patagonica*
       Passeriformes - Furnariidae
  e   Patagonian Canastero
  f   Synallaxe de Patagonie
  d   Stutzschnabelschlüfer
  i   Canestraio della Patagonia

1033   *Asthenes pudibunda*
       Passeriformes - Furnariidae
  e   Canyon Canastero
  f   Synallaxe des canyons; Synallaxe des
      cañons
  d   Lima-Schlüpfer
  i   Canestraio dei canyon

1034   *Asthenes punensis*
       Passeriformes - Furnariidae
  e   Puna Canastero
  f   Synallaxe du puna
  d   Punaschlüpfer
  i   Canestraio della Puna

1035   *Asthenes pyrrholeuca*
       Passeriformes - Furnariidae
  e   Sharp-billed Canastero; Lesser
      Canastero
  f   Synallaxe vannier
  d   Ockerkehlschlüpfer
  i   Canestraio minore

1036   *Asthenes sclateri*
       Passeriformes - Furnariidae
  e   Cordoba Canastero
  f   Synallaxe de Cordoba
  d   Cordoba-Schlüpfer
  i   Canestraio di Cordoba

1037   *Asthenes steinbachi*
       Passeriformes - Furnariidae
  e   Chestnut Canastero; Steinbach's
      Canastero
  f   Synallaxe marron
  d   Steinbach-Schlüpfer
  i   Canestraio di Steinbach

1038   *Asthenes urubambensis*
       Passeriformes - Furnariidae
  e   Line-fronted Canastero
  f   Synallaxe inca
  d   Urubamba-Schlüpfer
  i   Canestraio frontelineata

1039   *Asthenes virgata*
        Passeriformes - Furnariidae
e   Junin Canastero
f   Synallaxe de Junin
d   Junin-Schlüpfer
i   Canestraio di Junin

1040   *Asthenes wyatti*
        Passeriformes - Furnariidae
e   Streak-backed Canastero; Wyatt's
        Canastero
f   Synallaxe de Wyatt
d   Wyatt-Schlüpfer
i   Canestraio di Wyatt

1041   *Astrapia mayeri*
        Passeriformes - Corvidae
e   Ribbon-tailed Astrapia; Shaw-
        Mayer's Ribbon-tailed Bird of
        Paradise
f   Paradisier à rubans
d   Seidenbandparadiesvogel;
        Schmalschwanzparadieselster;
        Schmalschwanzastrapia
i   Paradisea coda a nastro; Astrapia di
        Mayer

1042   *Astrapia nigra*
        Passeriformes - Corvidae
e   Arfak Astrapia; Arfak Bird of
        Paradise; Great Astrapia; Black
        Astrapia; Great Bird of Paradise;
        Black Bird of Paradise; Long-tailed
        Astrapia; Long-tailed Bird-of-
        Paradise
f   Paradisier à gorge noire
d   Fächerparadieselster; Fächerastrapia
i   Paradisea dell'Arfak

1043   *Astrapia rothschildi*
        Passeriformes - Corvidae
e   Huon Astrapia; Lord Rothschild's
        Bird of Paradise; Huon Bird of
        Paradise; Rothschild's Astrapia
f   Paradisier de Rothschild
d   Blaubrustparadieselster;
        Blaubrustastrapia
i   Paradisea dell'Huon

1044   *Astrapia splendidissima*
        Passeriformes - Corvidae

e   Splendid Astrapia; Splendid Bird of
        Paradise
f   Paradisier splendide
d   Prachtparadieselster; Pracht Astrapia
i   Paradisea splendida

1045   *Astrapia stephaniae*
        Passeriformes - Corvidae
e   Stephanie's Astrapia; Princess
        Stephanie's Bird of Paradise; Princess
        Stephanie's Astrapia
f   Paradisier de Stéphanie
d   Prinzessin Stephanie-Paradiesvogel;
        Stephanie-Paradieselster; Stephanie-
        Astrapia
i   Paradisea della principessa Stephania

1046   *Asturina nitida*
        Falconiformes - Accipitridae
e   Gray Hawk (NA); Gray-lined Hawk
        (NA); Shining Buzzard-Hawk
f   Buse cendrée
d   Zweibindenbussard
i   Poiana grigia

1047   *Asturina plagiata*
        Falconiformes - Accipitridae
e   Grey Hawk
f   Buse grise
i   Poiana messicana

1048   *Atalotriccus pilaris*
        Passeriformes - Tyrannidae
e   Pale-eyed Pygmy-Tyrant; White-
        eyed Pygmy-Tyrant
f   Microtyran coiffé
d   Isabellwangentyrann
i   Coracia pitta; Tiranno pigmeo
        occhibianchi

1049   *Atelornis crossleyi*
        Coraciiformes - Brachypteraciidae
e   Rufous-headed Ground-Roller;
        Crossley's Ground-Roller
f   Brachyptérolle de Crossley
d   Lätzchenerdracke
i   Coracia pitta di Crossley

1050   *Atelornis pittoides*
        Coraciiformes - Brachypteraciidae
e   Pitta-like Ground-Roller

f Brachyptérolle pittolde
d Blaukopferdracke

1051 *Athene blewitti*
Strigiformes - Strigidae
e Forest Owlet; Forest Spotted-Owlet;
Blewitt's Owlet; Forest Little-Owlet;
Forest Little-Owl
f Chevêche forestière
d Blewitt-Kauz
i Civetta di Blewitt; Civetta delle
foreste

1052 *Athene brama*
Strigiformes - Strigidae
e Spotted Owlet; Spotted Little-Owl;
Tibetan Owl
f Chevêche brame
d Brahmakauz
i Civetta maculata; civetta macchiata

1053 *Athene noctua*
Strigiformes - Strigidae
e Little Owl
f Chevêche commune; Chevêche
d'Athéna; Chouette chevêche;
Chevêche
d Steinkauz
i Civetta; Civetta comune

1054 *Atlantisia rogersi*
Gruiformes - Rallidae
e Inaccessible Island Rail; Inaccesible
Rail; Inaccessible Island Flightless-
Rail
f Râle atlantis
d Atlantis-Ralle
i Rallo dell'Isola Inaccessibile

1055 *Atlapetes albiceps*
Passeriformes - Emberizidae
e White-headed Brush-Finch; White-
headed Finch
f Tohi à tête blanche
d Weißkopfbuschammer
i Fringuello di macchia testabianca

1056 *Atlapetes albinucha*
Passeriformes - Emberizidae
e White-naped Brush-Finch; White-
naped Finch; White-naped Atlapetes

f Tohi à calotte blanche
d Weißnackenbuschammer;
Gelbkehlbuschammer
i Fringuello di macchia nucabianca

1057 *Atlapetes albofrenatus*
Passeriformes - Emberizidae
e Moustached Brush-Finch;
Moustached Finch
f Tohi moustachu
d Weißbartbuschammer
i Fringuello di macchia dai mustacchi

1058 *Atlapetes atricapillus*
Passeriformes - Emberizidae
e Blackheaded Brush-Finch; Black-
headed Finch
f Tohi à tête noire
d Schwarzkopfbuschammer
i Fringuello di macchia capinero

1059 *Atlapetes brunneinucha*
Passeriformes - Emberizidae
e Chestnut-capped Brush-Finch;
Chestnut-capped Finch
f Tohi à nuque brune
d Braunkopfbuschammer
i Fringuello di macchia capocastano

1060 *Atlapetes brunneinucha apertus*
Passeriformes - Emberizidae
f Tohi à gorge claire

1061 *Atlapetes citrinellus*
Passeriformes - Emberizidae
e Yellow-striped Brush-Finch; Yellow-
striped Finch
f Tohi citrin
d Schwarzbartbuschammer
i Fringuello di macchia dalle strie
gialle

1062 *Atlapetes flaviceps*
Passeriformes - Emberizidae
e Olive-headed Brush-Finch; Olive-
headed Finch; Yellow-headed
Brushfinch
f Tohi à tête olive
d Goldkopfbuschammer
i Fringuello di macchia testa oliva

1063   ***Atlapetes fulviceps***
        Passeriformes - Emberizidae
e   Fulvous-headed Brush-Finch;
        Fulvous-headed Finch
f   Tohi à tête rousse
d   Braunbartbuschammer
i   Fringuello di macchia testafulva

1064   ***Atlapetes fuscoolivaceus***
        Passeriformes - Emberizidae
e   Dusky-headed Brush-Finch; Dusky-
        headed Finch
f   Tohi sombre
d   Rußkopfbuschammer
i   Fringuello di macchia testascura

1065   ***Atlapetes gutturalis***
        Passeriformes - Emberizidae
e   Yellow-throated Brush-Finch;
        Yellow-throated Finch
f   Tohi à gorge jaune
d   Gelbkehlbuschammer
i   Fringuello di macchia golagialla

1066   ***Atlapetes leucopis***
        Passeriformes - Emberizidae
e   White-rimmed Brush-Finch; White-
        rimmed Finch
f   Tohi bridé
d   Brillenbuschammer
i   Fringuello di macchia dagli occhiali

1067   ***Atlapetes leucopterus***
        Passeriformes - Emberizidae
e   White-winged Brush-Finch; White-
        winged Finch
f   Tohi leucoptère
d   Spiegelbuschammer
i   Fringuello di macchia alibianchi

1068   ***Atlapetes melanocephalus***
        Passeriformes - Emberizidae
e   Santa Marta Brush-Finch; Santa
        Marta Finch
f   Tohi des Santa Marta
d   Grauohrbuschammer
i   Fringuello di macchia di Santa Marta

1069   ***Atlapetes nationi***
        Passeriformes - Emberizidae
e   Rusty-bellied Brush-Finch; Rusty-

        bellied Finch
f   Tohi à ventre roux
d   Rostbauchbuschammer
i   Fringuello di macchia ventrerosso

1070   ***Atlapetes pallidiceps***
        Passeriformes - Emberizidae
e   Pale-headed Brush-Finch; Pale-
        headed Finch
f   Tohi grisonnant
d   Blaßkopfbuschammer
i   Fringuello di macchia testapallida

1071   ***Atlapetes pallidinucha***
        Passeriformes - Emberizidae
e   Pale-naped Brush-Finch; Pale-naped
        Finch
d   Zimtstirnbuschammer
i   Fringuello di macchia nucapallida

1072   ***Atlapetes personatus***
        Passeriformes - Emberizidae
e   Tepui Brush-Finch; Tepui Finch
f   Tohi des tépuis
d   Tepui-Buschammer
i   Fringuello di macchia del Tepui

1073   ***Atlapetes pileatus***
        Passeriformes - Emberizidae
e   Rufous-capped Brush-Finch; Rufous-
        capped Finch
f   Tohi à calotte rousse
d   Rotkappenbuschammer
i   Fringuello di macchia capirosso

1074   ***Atlapetes rufigenis***
        Passeriformes - Emberizidae
e   Rufous-eared Brush-Finch; Rufous-
        eared Finch
f   Tohi rougeaud
d   Rotohrbuschammer
i   Fringuello di macchia orecchierosse

1075   ***Atlapetes rufinucha***
        Passeriformes - Emberizidae
e   Rufous-naped Brush-Finch; Rufous-
        naped Finch
f   Tohi à nuque rousse
d   Rotnackenbuschammer
i   Fringuello di macchia nucarossa

1076   *Atlapetes schistaceus*
       Passeriformes - Emberizidae
   e   Slaty Brush-Finch; Slaty Finch;
       Sooty Finch
   f   Tohi ardoisé
   d   Graubrustbuschammer
   i   Fringuello di macchia ardesia

1077   *Atlapetes seebohmi*
       Passeriformes - Emberizidae
   e   Bay-crowned Brush-Finch; Bay-
       crowned Finch; Rusty-bellied
       Brushfinch
   f   Tohi de Seebohm
   i   Fringuello di macchia si Seebohm

1078   *Atlapetes semirufus*
       Passeriformes - Emberizidae
   e   Ochre-breasted Brush-Finch; Ochre-
       breasted Finch
   f   Tohi demi-roux
   d   Ockerbrustbuschammer
   i   Fringuello di macchia petto ocra

1079   *Atlapetes torquatus*
       Passeriformes - Emberizidae
   e   Stripe-headed Brush-Finch; Stripe-
       headed Finch
   f   Tohi à tête rayée
   d   Streifenkopfbuschammer;
       Grünscheitelbuschammer
   i   Fringuello di macchia testastriata

1080   *Atlapetes torquatus assimilis*
       Passeriformes - Emberizidae
   f   Tohi à raies grises

1081   *Atlapetes tricolor*
       Passeriformes - Emberizidae
   e   Tricoloured Brush-Finch;
       Tricoloured Finch
   d   Dreifarbenbuschammer
   i   Fringuello di macchia tricolore

1082   *Atlapetes virenticeps*
       Passeriformes - Emberizidae
   e   Green-striped Brush-Finch; Green-
       striped Finch
   f   Tohi à raies vertes
   d   Grünscheitelbuschammer

   i   Fringuello di macchia dalle strie
       verdi

1083   *Atrichornis clamosus*
       Passeriformes - Menuridae
   e   Noisy Scrub-bird; Western Scrub-
       bird
   f   Atrichorne bruyant
   d   Braunbauchdickichtvogel;
       Lärmdickichtvogel; Großer
       Dickichtschlüpfer
   i   Uccello dei cespugli occidentale

1084   *Atrichornis rufescens*
       Passeriformes - Menuridae
   e   Rufous Scrub-bird
   f   Atrichorne roux
   d   Rostbauchdickichtvogel;
       Röteldickichtvogel; Kleiner
       Dickichtschlüpfer
   i   Uccello dei cespugi rossiccio

1085   *Atrichornis rufescens ferrieri*
       Passeriformes - Menuridae
   e   Southern Rufous Scrub-bird (ANZ)

1086   *Atrichornis rufescens rufescens*
       Passeriformes - Menuridae
   e   Northern Rufous Scrub-bird (ANZ)

1087   *Attagis gayi*
       Charadriiformes - Thinocoridae
   e   Rufous-bellied Seedsnipe
   f   Attagis de Gay
   d   Kordillerenläufer
   i   Tinocoride pettorossiccia

1088   *Attagis malouinus*
       Charadriiformes - Thinocoridae
   e   White-bellied Seedsnipe
   f   Attagis de Magellan
   d   Magellan-Läufer
   i   Tinocoride ventrebianco

1089   *Atthis ellioti*
       Apodiformes - Trochilidae
   e   Wine-throated Hummingbird
   f   Colibri d'Elliot
   d   Elliot-Elfe
   i   Colibrì di Elliot

1090 *Atthis heloisa*
Apodiformes - Trochilidae
e Bumblebee Hummingbird; Heloise's
Hummingbird; Morcom's
Hummingbird
f Colibri héloise
d Heloisa-Elfe
i Colibrì di Eloisa

1091 *Atticora fasciata*
Passeriformes - Hirundinidae
e White-banded Swallow; White-
backed Swallow
f Hirondelle à ceinture blanche
d Weißbandschwalbe
i Rondine fasciata

1092 *Atticora melanoleuca*
Passeriformes - Hirundinidae
e Black-collared Swallow
f Hirondelle des torrents
d Halsbandschwalbe
i Rondine dal collare

1093 *Attila bolivianus*
Passeriformes - Tyrannidae
e Dull-capped Attila; White-eyed
Attila; Rufous Attila; White-winged
Attila
f Attila à calotte grise; Attila de
Lafresnaye
d Rostattila
i Attila della Bolivia

1094 *Attila cinnamomeus*
Passeriformes - Tyrannidae
e Cinnamon Attila
f Attila cannellé
d Zimtattila
i Attila color cannella

1095 *Attila citriniventris*
Passeriformes - Tyrannidae
e Citron-bellied Attila
f Attila à ventre jaune
d Gelbbauchattila
i Attila ventregiallo

1096 *Attila phoenicurus*
Passeriformes - Tyrannidae
e Rufous-tailed Attila

f Attila à queue rousse
d Kurzschnabelattila
i Attila codirosso

1097 *Attila rufus*
Passeriformes - Tyrannidae
e Grey-hooded Attila; Grey-throated
Attila
f Attila à tête grise
d Graukopfattila
i Attila testagrigia

1098 *Attila spadiceus*
Passeriformes - Tyrannidae
e Bright-rumped Attila; Streaked
Attila; Polymorphic Attila
f Attila à croupion jaune
d Goldbürzelattila
i Attila polimorfo

1099 *Attila torridus*
Passeriformes - Tyrannidae
e Ochraceous Attila
f Attila ocré
d Ockerattila
i Attila ocraceo

1100 *Augastes geoffroyi*
Apodiformes - Trochilidae
e Wedge-billed Hummingbird
f Colibri de Geoffroy
d Kleinschnabelkolibri
i Colibrì becco a cuneo

1101 *Augastes lumachellus*
Apodiformes - Trochilidae
e Hooded Visorbearer
f Colibri lumachelle
d Kapuzenkolibri
i Colibrì incappucciato

1102 *Augastes scutatus*
Apodiformes - Trochilidae
e Hyacinth Visorbearer
f Colibri superbe
d Schildkolibri
i Colibrì

1103 *Aulacorhynchus calorhynchus*
Piciformes - Ramphastidae
e Yellow-billed Toucanet

1104 **Aulacorhynchus coeruleicinctis**
Piciformes - Ramphastidae
e Blue-banded Toucanet; Blue-throated
Toucanet
f Toucanet à ceinture bleue
d Grauschnabelarassari
i Tucanetto dalla banda cerulea

1105 **Aulacorhynchus derbianus**
Piciformes - Ramphastidae
e Chestnut-tipped Toucanet
f Toucanet de Derby
d Derby-Arassari
i Tucanetto di Derby

1106 **Aulacorhynchus haematopygus**
Piciformes - Ramphastidae
e Crimson-rumped Toucanet
f Toucanet à croupion rouge
d Blutbürzelarassari
i Tucanetto dal groppone rosso

1107 **Aulacorhynchus huallagae**
Piciformes - Ramphastidae
e Yellow-browed Toucanet
f Toucanet à sourcils jaunes
d Gelbbrauenarassari
i Tucanetto dai sopracigglie gialli

1108 **Aulacorhynchus prasinus**
Piciformes - Ramphastidae
e Emerald Toucanet; Blue-throated
Toucanet
f Toucanet émeraude
d Laucharassari
i Tucanetto smeraldo

1109 **Aulacorhynchus sulcatus**
Piciformes - Ramphastidae
e Groove-billed Toucanet; Yellow-
billed Toucanet
f Toucanet à bec silloné
d Blauzügelarassari
i Tucanetto beccoscanalato

1110 **Auriparus flaviceps**
Passeriformes - Certhiidae
e Verdin
f Auripare verdin
d Goldköpfchen; Gelbkopfmeise;

Goldmeise
i Auriparo

1111 **Automolus albogularis**
Passeriformes - Furnariidae
e Neblina Foliagegleaner; Neblina
Leafgleaner
d Neblina-Blattspäher

1112 **Automolus dorsalis**
Passeriformes - Furnariidae
e Crested Foliagegleaner; Crested
Leafgleaner
f Anabate à grands sourcils; Anabate
huppé
d Schopfbaumspäher
i Spigolafoglie crestato

1113 **Automolus infuscatus**
Passeriformes - Furnariidae
e Olive-backed Foliagegleaner;
Olivebacked Leafgleaner
f Anabate olivâtre
d Olivrückenbaumspäher
i Spigolafoglie dorso oliva

1114 **Automolus leucophthalmus**
Passeriformes - Furnariidae
e White-eyed Foliagegleaner; White-
eyed Leafgleaner
f Anabate aux yeux blancs
d Weißzügelbaumspäher
i Spigolafoglie occhibianchi

1115 **Automolus melanopezus**
Passeriformes - Furnariidae
e Brown-rumped Foliagegleaner;
Brown-rumped Leafgleaner
f Anabate brunâtre; Anabate du Napo
d Braunbürzelbaumspäher
i Spigolafoglie gropponebruno

1116 **Automolus ochrolaemus**
Passeriformes - Furnariidae
e Buff-throated Foliagegleaner; Buff-
throated Leafgleaner
f Anabate à gorge fauve
d Braunkehlbaumspäher
i Spigolafoglie golafulva

1117 *Automolus roraimae*
Passeriformes - Furnariidae
e White-throated Foliagegleaner;
White-throated Leafgleaner
f Anabate à gorge blanche
d Weißkehlbaumspäher
i Spigolafoglie golabianca

1118 *Automolus rubiginosus*
Passeriformes - Furnariidae
e Ruddy Foliagegleaner; Ruddy
Leafgleaner
f Anabate rubigineux
d Zimtkehlbaumspäher
i Spigolafoglie rossastro

1119 *Automolus rufipileatus*
Passeriformes - Furnariidae
e Chestnut-crowned Foliagegleaner;
Chestnut-crowned Leafgleaner
f Anabate à couronne rousse
d Rotscheitelbaumspäher
i Spigolafoglie corona rossiccia

1120 *Aviceda cuculoides*
Falconiformes - Accipitridae
e African Baza; African Cuckoo-
Falcon; West African Cuckoo-
Falcon; African Cuckoo-Hawk;
Cuckoo-Hawk
f Baza coucou; Faucon-coucou
d Kuckucksweih
i Baza africana; Falco dal ciuffo
africano

1121 *Aviceda cuculoides verreauxii*
Falconiformes - Accipitridae
e African Cuckoo Hawk (CSA);
Cuckoo Hawk
d Kuckucksweih

1122 *Aviceda jerdoni*
Falconiformes - Accipitridae
e Jerdon's Baza; Blyth's Baza (ISC);
Brown Lizard-Hawk (ISC); Brown-
crested Lizard-Hawk; Brown Baza;
Asian Baza; Legge's Baza; Crested
Lizard-Hawk
f Baza de Jerdon
d Hindu-Weihe
i Baza di Jerdon; Falco crestato

1123 *Aviceda leuphotes*
Falconiformes - Accipitridae
e Black Baza; Indian Black-crested
Baza (ISC); Black-crested Lizard-
Hawk; Black-crested Baza
f Baza huppard
d Dreifarbenweihe
i Baza nero

1124 *Aviceda madagascariensis*
Falconiformes - Accipitridae
e Madagascar Baza; Madagascar
Cuckoo-Falcon; Madagascar Cuckoo
Hawk
f Baza malgache
d Lemurenweihe
i Baza del Madagascar; Falco dal
ciuffo del Madagascar

1125 *Aviceda subcristata*
Falconiformes - Accipitridae
e Pacific Baza; Crested Baza; Crested
Lizard-Hawk; Crested Hawk; Lizard
Hawk; Lizard Baza
f Baza huppé
d Papua-Weihe
i Baza crestato; Baza

1126 *Avocettula recurvirostris*
Apodiformes - Trochilidae
e Fiery-tailed Awlbill; Fiery-tailed
Avocetbill; Swainson's Hummingbird
f Colibri avocette
d Sägeschnabelkolibri; Avosettkolibri
i Colibrì beccoricurvo

1127 *Aythya affinis*
Anseriformes - Anatidae
e Lesser Scaup; Black Duck (WI);
Black-head (WI); Little Bluebill;
Little Broadbill
f Petit Fuligule; Petit Morillon;
Fuligule à bec noire; Canard tête-
noire (Ants)
d Kleine Bergente; Veilchenente
i Moretta grigia minore

1128 *Aythya americana*
Anseriformes - Anatidae
e Redhead; American Redhead;
Pochard; Red-headed Pochard

f   Fuligule à téte rouge; Morillon à tête rouge

d   Rotkopfente; Amerikanische Tafelente

i   Moriglione americano

1129   *Aythya australis*
Anseriformes - Anatidae

e   Hardhead; Australian Pochard; White-eyed Pochard (CSA) (ISC); Australian White-eye; Australian White-eyed Duck; White-eyed Duck; Brownhead; Copperhead; Australasian Pochard

f   Fuligule austral; Fuligule australienne

d   Australische Moorente

i   Moretta australiana

1130   *Aythya baeri*
Anseriformes - Anatidae

e   Baer's Pochard; Siberian White-eye; Asiatic White-eyed Pochard; Eastern White-eyed Pochard

f   Fuligule de Baer

d   Schwarzkopfmoorente

i   Moriglione di Baer

1131   *Aythya collaris*
Anseriformes - Anatidae

e   Ring-necked Duck; Ring-billed Duck; Ringbill

f   Fuligule à collier; Morillon à collier; Fuligule à bec cerclé; Canard tête-noire (Ants); Fuligne (Ants)

d   Ringschnabelente; Halsringente

i   Moretta dal collare

1132   *Aythya ferina*
Anseriformes - Anatidae

e   Common Pochard; Pochard; Dunbird; Northern Pochard; European Pochard; Eurasian Pochard

f   Fuligule milouin; Canard milouin

d   Tafelente

i   Moriglione; Bosco; Moriglione eurasiatico

1133   *Aythya fuligula*
Anseriformes - Anatidae

e   Tufted Duck; Tufted Pochard (ISC)

f   Fuligule morillon

d   Reiherente

i   Moretta; Moretta eurasiatica

1134   *Aythya innotata*
Anseriformes - Anatidae

e   Madagascar Pochard; Madagascar White-eye

f   Fuligule de Madagascar

d   Malagassi-Moorente

i   Moriglione di Madagascar

1135   *Aythya marila*
Anseriformes - Anatidae

e   Greater Scaup; Scaup-Duck; Scaup; Bluebill; Greater Scaup Duck; Broadbill

f   Fuligule milouinan; Grand Morillon; Canard milouinan

d   Bergente

i   Moretta grigia

1136   *Aythya novaeseelandiae*
Anseriformes - Anatidae

e   New Zealand Scaup; Black Scaup; Black Teal

f   Fuligule de Nouvelle-Zélande

d   Maoriente

i   Moretta della Nuova Zelanda

1137   *Aythya nyroca*
Anseriformes - Anatidae

e   Ferruginous Pochard; Ferruginous Duck; White-eyed Pochard (CSA) (ISC); Ferruginous White-eye; White-eye; Pochard; Common White-eye

f   Fuligule nyroca; Canard nyroca; Canard à iris blanc; Sarcelle rousse

d   Moorente

i   Moretta tabaccata

1138   *Aythya valisineria*
Anseriformes - Anatidae

e   Canvasback

f   Fuligule à dos blanc; Morillon à dos blancs

d   Riesentafelente; Kanevasente; Valisneria-Ente

i   Moriglione dorsa di tela; Fuligone a dorso bianco

B

1139 ***Babax koslowi***
Passeriformes - Sylviidae
e Tibetan Babax; Koslow's Babax
f Babaxe de Koslov; Garrulaxe de
Koslow
d Koslow-Babax
i Garrulo di Koslow

1140 ***Babax lanceolatus***
Passeriformes - Sylviidae
e Chinese Babax; Common Babax;
Babax; Streaked Hill-Babbler;
Streaked Hill-Warbler
f Babaxe lancéolé; Garrulaxe lancéolé
d Streifenbabax
i Garrulo lanceolato

1141 ***Babax waddelli***
Passeriformes - Sylviidae
e Giant Babax; Giant Tibetan Babax
f Babaxe de Waddell; Garrulaxe de
Waddell
d Riesenbabax
i Garrulo gigante

1142 ***Baeopogon clamans***
Passeriformes - Pycnonotidae
e White-tailed Greenbul; Sjostedt's
Honeyguide Greenbul; Sjostedt's
White-tailed Bulbul; Sjostedt's
Greenbul; Sjostedt's White-tailed
Greenbul; White-tailed Bulbul
f Bulbul bruyant
d Sjostedt-Bülbül
i Bulbul indicatore

1143 ***Baeopogon indicator***
Passeriformes - Pycnonotidae
e Honeyguide Greenbul; Honeyguide
Bulbul; White-tailed Bulbul
f Bulbul à queue blanche
d Weißschwanzbülbül
i Bulbul indicatore di Sjostedt

1144 ***Baillonius bailloni***
Piciformes - Ramphastidae
e Saffron Toucanet; Baillon's Toucan
f Araçari de Baillon
d Goldtukan; Goldbrusttukan
i Tucanetto zafferano

1145 ***Balaeniceps rex***
Ciconiiformes - Balaenicipitidae
e Shoebill; Shoebill Stork; Whale-
headed Stork; Whalehead
f Bec-en-sabot du Nil; Bec-en-sabot
d Schuhschnabel
i Becco a scarpa

1146 **Balaenicipitidae**
Ciconiiformes
e Shoebills; Shoebill Storks; Whale-
headed Storks; Whalehead Storks
f Balénicipitidés
d Schuhschnäbel
i Balenicipitidi

1147 ***Balearica pavonina***
Gruiformes - Gruidae
e Black Crowned-Crane; Crowned
Crane; Northern Crowned-Crane;
Dark-crowned Crane
f Grue couronnée
d Kronenkranich
i Gru pavonina

1148 ***Balearica regulorum***
Gruiformes - Gruidae
e Grey Crowned-Crane; Crowned
Crane (CSA); Southern Crowned-
Crane; Blue-necked Crane; Royal
Crane
f Grue royale
d Kronenkranich
i Gru coronata

1149 ***Bambusicola fytchii***
Galliformes - Phasianidae
e Mountain Bamboo-Partridge;
Anderson's Bamboo-Partridge;
Bamboo-Partridge
f Bambusicole de Fytch
d Gelbbrauenbambushuhn
i Pernice delle bambù

1150   ***Bambusicola thoracica***
        Galliformes - Phasianidae
   e   Chinese Bamboo-Partridge;
        Formosan Bamboo-Partridge;
        Bamboo-Partridge; Taiwan Bamboo-
        Partridge
   f   Bambusicole de Chine
   d   Graubrauenbambushuhn
   i   Pernice cinese delle bambù

1151   ***Bangsia arcaei***
        Passeriformes - Fringillidae
   e   Blue-and-gold Tanager; Arce's
        Tanager
   f   Tangara jaune-et-bleu
   d   Bangs-Tangare
   i   Tangara blu e dorata

1152   ***Bangsia aureocincta***
        Passeriformes - Fringillidae
   e   Blue-and-gold Tanager; Golden-
        ringed Tanager
   f   Tangara à boucles d'or
   d   Goldringtangare
   i   Tangara dal collare dorato

1153   ***Bangsia edwardsi***
        Passeriformes - Fringillidae
   e   Moss-backed Tanager
   f   Tangara d'Edwards
   d   Edwards-Tangare
   i   Tangara di Milne-Edwards

1154   ***Bangsia melanochlamys***
        Passeriformes - Fringillidae
   e   Black-and-gold Tanager
   f   Tangara à cape noire
   d   Blauschultertangare
   i   Tangara nera e dorata

1155   ***Bangsia rothschildi***
        Passeriformes - Fringillidae
   e   Golden-chested Tanager;
        Rothschild's Tanager
   f   Tangara de Rothschild
   d   Rothschild-Tangare
   i   Tangara pettodorato

1156   ***Barnardius zonarius***
        Psittaciformes - Psittacidae
   e   Australian Ringneck; Port Lincoln

        Ringneck; Banded Parrot; Port
        Lincoln Parrot; Bauer's Parrot;
        Bauer's Parakeet; Port Lincoln
        Parakeet; Ringneck Parrot; Ring-
        necked Parrot; Twenty-eight Parrot
   f   Perruche à collier jaune; Perruche à
        barbe bleue
   d   Ringsittich; Kragensittich
   i   Parrocchetto di Port Lincoln

1157   ***Bartramia longicauda***
        Charadriiformes - Scolopacidae
   e   Upland Sandpiper; Bartram's
        Sandpiper; Upland Plover; Cotton-
        tree Plover (WI)
   f   Maubèche des champs; Bartramie
        des champs; Poule vergenne (Ants)
   d   Prärieläufer; Bartram-Uferläufer
   i   Piro-piro codalunga

1158   ***Baryphthengus martii***
        Coraciiformes - Momotidae
   e   Rufous Motmot; Rufous-capped
        Motmot
   f   Motmot roux; Motmot oranroux
   d   Zimtbrustmotmot
   i   Motmot rugginoso

1159   ***Baryphthengus ruficapillus***
        Coraciiformes - Momotidae
   e   Rufous-capped Motmot; Rufous
        Motmot(ANZ)
   f   Motmot oranroux
   d   Rotkopfmotmot
   i   Motmot testarugginosa; Motmot
        dalla testa rossa

1160   ***Basileuterus auricapillus***
        Passeriformes - Parulidae
   e   Golden-crowned Warbler
   i   Parula reale

1161   ***Basileuterus basilicus***
        Passeriformes - Parulidae
   e   Santa Marta Warbler
   d   Santa Marta-Waldsänger
   i   Parula reale di Santa Marta

1162   ***Basileuterus belli***
        Passeriformes - Parulidae
   e   Golden-browed Warbler; Bell's

Warbler
f Paruline à sourcils dorés
d Goldstreifenwaldsänger
i Parula reale di Bell

1163 *Basileuterus bivittatus*
Passeriformes - Parulidae
e Two-banded Warbler
f Paruline rubanée
d Bindenwaldsänger
i Parula reale dalle due strie

1164 *Basileuterus cabanisi*
Passeriformes - Parulidae
e Cabanis's Warbler

1165 *Basileuterus chlorophrys*
Passeriformes - Parulidae
e Choco Warbler

1166 *Basileuterus chrysogaster*
Passeriformes - Parulidae
e Cuzco Warbler; Golden-bellied
   Warbler
f Paruline à ventre doré
d Goldbauchwaldsänger
i Parula reale ventredorato

1167 *Basileuterus cinereicollis*
Passeriformes - Parulidae
e Grey-throated Warbler; Ashy-
   throated Warbler
f Paruline à cou gris
d Graukehlwaldsänger
i Parula reale dagli occhiali

1168 *Basileuterus conspicillatus*
Passeriformes - Parulidae
e White-lored Warbler; San José
   Warbler
f Paruline à lores blancs

1169 *Basileuterus coronatus*
Passeriformes - Parulidae
e Russet-crowned Warbler; Orange-
   crowned Warbler
f Paruline à diadème
d Goldscheitelwaldsänger
i Parula reale corona rossa

1170 *Basileuterus culicivorus*
Passeriformes - Parulidae
e Golden-crowned Warbler; Stripe-
   crowned Warbler; Cabanis's Warbler
f Paruline à couronne dorée
d Goldhähnchenwaldsänger
i Parula reale corona dorata

1171 *Basileuterus delattrii*
Passeriformes - Parulidae
e Chestnut-capped Warbler

1172 *Basileuterus flaveolus*
Passeriformes - Parulidae
e Flavescent Warbler
f Paruline flavescente
d Gelbwaldsänger
i Parula reale giallastra

1173 *Basileuterus fraseri*
Passeriformes - Parulidae
e Grey-and-gold Warbler; Slate-and-
   gold Warbler; Fraser's Warbler
f Paruline de Fraser
d Feenwaldsänger
i Parula reale di Fraser

1174 *Basileuterus fulvicauda*
Passeriformes - Parulidae
e Buff-rumped Warbler
f Paruline à croupion fauve
d Schmätzerwaldsänger
i Parula reale codafulva

1175 *Basileuterus griseiceps*
Passeriformes - Parulidae
e Grey-headed Warbler
f Paruline à tête grise
d Grauwangenwaldsänger
i Parula reale testagrigia

1176 *Basileuterus hypoleucus*
Passeriformes - Parulidae
e White-bellied Warbler; White-
   rimmed Warbler; White-browed
   Warbler
f Paruline à ventre blanc
d Weißbauchwaldsänger
i Parula reale ventrebianco

1177 **Basileuterus ignotus**
    Passeriformes - Parulidae
e   Pirre Warbler
f   Paruline du Pirre
i   Parula reale di Pirre

1178 **Basileuterus leucoblepharus**
    Passeriformes - Parulidae
e   White-browed Warbler
f   Paruline à paupières blanches
d   Olivflankenwaldsänger
i   Parula reale dai sopraccigli bianchi

1179 **Basileuterus leucophrys**
    Passeriformes - Parulidae
e   White-striped Warbler
f   Paruline bridée
d   Goldflankenwaldsänger
i   Parula reale orecchiebianche

1180 **Basileuterus luteoviridis**
    Passeriformes - Parulidae
e   Citrine Warbler
f   Paruline citrine
d   Bonaparte-Waldsänger
i   Parula reale giallo-verde

1181 **Basileuterus melanogenys**
    Passeriformes - Parulidae
e   Black-cheeked Warbler
f   Paruline sombre
d   Schwarzwangenwaldsänger
i   Parula reale guancenere

1182 **Basileuterus nigrocristatus**
    Passeriformes - Parulidae
e   Black-crested Warbler
f   Paruline à cimier noir
d   Schwarzscheitelwaldsänger
i   Parula reale corona nera

1183 **Basileuterus rivularis**
    Passeriformes - Parulidae
e   Neotropical River-Warbler; River-Warbler; Neotropic River-Warbler; American River-Warbler
f   Paruline des rives
d   Flußwaldsänger
i   Parula reale di fiume

1184 **Basileuterus rufifrons**
    Passeriformes - Parulidae
e   Rufous-capped Warbler; Chestnut-capped Warbler
f   Paruline à calotte rousse
d   Rotkappenwaldsänger
i   Parula reale corona rugginosa

1185 **Basileuterus signatus**
    Passeriformes - Parulidae
e   Pale-legged Warbler; Yellow-green Warbler
f   Paruline à pattes pâles
d   Peru-Waldsänger
i   Parula reale zampechiare

1186 **Basileuterus trifasciatus**
    Passeriformes - Parulidae
e   Three-banded Warbler
f   Paruline trifasciée
d   Cajamarca-Waldsänger
i   Parula reale dalle tre fasce

1187 **Basileuterus tristriatus**
    Passeriformes - Parulidae
e   Three-striped Warbler
f   Paruline triligne
d   Dreistreifenwaldsänger

1188 **Basilornis celebensis**
    Passeriformes - Sturnidae
e   Sulawesi Mynah; Celebes Starling; Celebes Mynah; Sulawesi Crested-Mynah; Celebes Crested-Mynah; Sulawesi King-Starling; Celebes King-Starling; King Starling; Short-crested Mynah
f   Basilorne des Célèbes
d   Celebes-Atzel; Königsatzel
i   Storno reale di Sulawesi

1189 **Basilornis corythaix**
    Passeriformes - Sturnidae
e   Long-crested Mynah; Moluccan Starling; Ceram King-Starling; Ceram Starling; Ceram Mynah
f   Basilorne de Céram
d   Molukken-Atzel
i   Storno reale di Seram

1190   **Basilornis galeatus**
        Passeriformes - Sturnidae
   e   Helmeted Mynah; Crested Starling;
        King Mynah; Greater King-Mynah;
        Greater King-Starling
   f   Basilorne huppé
   d   Helmatzel
   i   Storno reale dall'elmo

1191   **Basilornis miranda**
        Passeriformes - Sturnidae
   e   Apo Mynah; Mount Apo Mynah;
        Mount Apo King-Starling; Mount
        Apo Starling
   f   Basilorne de Mindanao
   d   Prachtatzel
   i   Storno reale di Monte Apo

1192   **Batara cinerea**
        Passeriformes - Formicariidae
   e   Giant Antshrike
   f   Batara géant
   d   Bindenameisenwürger
   i   Averla formichiera gigante

1193   **Bathmocercus cerviniventris**
        Passeriformes - Sylviidae
   e   Black-capped Rufous-Warbler;
        Black-headed Stream-Warbler;
        Stream Warbler; Black-headed
        Rufous-Warbler; Black-faced
        Rufous-Warbler
   f   Bathmocerque à capuchon; Fauvette
        aquatique à capuchon
   d   Fuchssänger; Rostbauchfuchssänger
   i   Forapaglie codagraduata capinero

1194   **Bathmocercus rufus**
        Passeriformes - Sylviidae
   e   Black-faced Rufous-Warbler;
        Rufous-Warbler; Stream Warbler
   f   Bathmocerque à face noire; Fauvettte
        aquatique à face noire
   d   Graubauchfuchssänger
   i   Forapaglie codagraduata rossiccio

1195   **Bathmocercus winifredae**
        Passeriformes - Sylviidae
   e   Mrs.. Moreau's Warbler; Mrs.
        Moreau's Rufous-Warbler
   f   Rousselette de Madame Moreau

   d   Rostkopffuchssänger
   i   Forapaglie codagraduata della
        signora Moreau

1196   **Batis capensis**
        Passeriformes - Platysteiridae
   e   Cape Batis; Cape Puffback-
        Flycatcher; Cape Flycatcher; Cape
        Puffback; Common Batis
   f   Pririt du Cap; Batis du Cap
   d   Kap-Schnäpper
   i   Pigliamosche del Capo

1197   **Batis dimorpha**
        Passeriformes - Platysteiridae
   e   Malawi Batis; Malawi Puffback-
        Flycatcher
   f   Pririt du Malawi
   i   Pigliamosche del Malawi

1198   **Batis diops**
        Passeriformes - Platysteiridae
   e   Ruwenzori Batis; Ruwenzori
        Puffback-Flycatcher; Ruwenzori
        Puffback
   f   Pririt du Ruwenzori; Batis du
        Ruwenzori
   d   Ruwenzori-Schnäpper
   i   Pigliamosche del Ruwenzori

1199   **Batis fratrum**
        Passeriformes - Platysteiridae
   e   Woodward's Batis; Woodward's
        Puffback-Flycatcher; Zululand
        Puffback; Zululand Puffback-
        Flycatcher; Zululand Batis
   f   Pririt de Woodward; Batis des frères
        Woodward
   d   Woodward-Schnäpper; Woodwards
        Schnäpper
   i   Pigliamosche dello Zululand

1200   **Batis ituriensis**
        Passeriformes - Platysteiridae
   e   Ituri Batis; Chapin's Puffback-
        Flycatcher; Chapin's Puffbacked-
        Flycatcher
   f   Pririt de l'Ituri; Batis de l'Ituri
   i   Pigliamosche dell'Ituri

1201 **Batis margaritae**
Passeriformes - Platysteiridae
e Boulton's Batis; Boulton's Puffback-
Flycatcher; Boulton's Puffback;
Margaret's Batis
f Pririt de Boulton; Batis de Margaret
d Margareten-Schnäpper
i Pigliamosche di Boulton

1202 **Batis minima**
Passeriformes - Platysteiridae
e Verreaux's Batis; Verreaux's
Puffback-Flycatcher; Verreaux's
Western Puffback-Flycatcher;
Verreaux's Grey-headed Puffback-
Flycatcher; Forest Puffback;
Verreaux's Puffback; Gabon Batis
f Pririt de Verreaux; Gobemouche
soyeux à tête grise; Gobe-mouches
soyeux à tête grise; Batis à tête grise
d Gabun-Schnäpper
i Pigliamosche di Verreaux

1203 **Batis minor**
Passeriformes - Platysteiridae
e Black-headed Batis; Black-headed
Puffback-Flycatcher; Black-headed
Puffback
f Pririt à joues noires; Gobemouche
soyeux à joues noires; Gobe-
mouches soyeux à joues noires; Batis
à joues noires
d Kongo-Schnäpper
i Pigliamosche testanera

1204 **Batis minulla**
Passeriformes - Platysteiridae
e Angola Batis; Angola Puffback-
Flycatcher; Angola Puffback; West
African Batis
f Pririt de l'Angola; Batis d'Angola
d Angola-Schnäpper
i Pigliamosche dell'Angola

1205 **Batis mixta**
Passeriformes - Platysteiridae
e Short-tailed Batis; Forest Batis
(CSA); Short-tailed Puffback-
Flycatcher; Puffback Flycatcher; East
African Puffback; East African Batis
f Pririt à queue courte; Batis de forêt

d Kurzschwanzschnäpper
i Pigliamosche africano codabreva

1206 **Batis molitor**
Passeriformes - Platysteiridae
e Chinspot Batis; Chin-spot Flycatcher;
Chin-spot Puffback-Flycatcher;
Chin-spot Puffback; White-flanked
Puffback; White-flanked Batis
f Pririt molitor; Batis molitor
d Weißflankenschnäpper;
Weißflankenbuschschnäpper
i Pigliamosche mugnaio

1207 **Batis occulta**
Passeriformes - Platysteiridae
e West African Batis; Lawson's
Puffback-Flycatcher
f Pririt de Lawson
i Pigliamosche di Lawson

1208 **Batis orientalis**
Passeriformes - Platysteiridae
e Grey-headed Batis; Grey-headed
Puffback-Flycatcher; Savanna
Puffback; Savanna Puffback-
Flycatcher; Savannah Puffback;
Savannah Puffback-Flycatcher
f Pririt à tête grise; Batis oriental
d Heuglin-Schnäpper
i Pigliamosche orientale

1209 **Batis perkeo**
Passeriformes - Platysteiridae
e Pygmy Batis; Pygmy Puffback-
Flycatcher; Pygmy Puffback
f Pririt pygmée; Batis nain
d Däumlingsschnäpper
i Pigliamosche pigmeo

1210 **Batis poensis**
Passeriformes - Platysteiridae
e Fernando Po Batis; Fernando Po
Puffback; Fernando Po Puffback-
Flycatcher
f Pririt de Fernando Po; Batis de
Fernando Po
d Alexander-Schnäpper
i Pigliamosche di Fernando Po

1211   *Batis pririt*
Passeriformes - Platysteiridae
e   Pririt Batis; Pririt Puffback-
Flycatcher; Pririt Puffback
f   Pririt de Vieillot; Batis pririt
d   Priritschnäpper
i   Pigliamosche pririt

1212   *Batis reichenowi*
Passeriformes - Platysteiridae
e   Reichenow's Batis; Reichenow's
Puffback; Reichenow's Puffback-
Flycatcher
f   Pririt de Reichenow
i   Pigliamosche di Reichenow

1213   *Batis senegalensis*
Passeriformes - Platysteiridae
e   Senegal Batis; Senegal Puffback-
Flycatcher; Senegal Puffback; Grey-
headed Batis
f   Pririt du Sénégal; Gobemouche
soyeux du Sénégal; Gobe-mouches
soyeux du Sénégal; Batis du Sénégal
d   Senegal-Schnäpper
i   Pigliamosche del Senegal

1214   *Batis soror*
Passeriformes - Platysteiridae
e   Pale Batis; East Coast Batis (CSA);
Mozambique Batis (CSA); Chinspot
Batis; Paler Chinspot Puffback-
Flycatcher; Madagascar Batis
f   Pririt pâle; Batis du Mozambique
d   Sansibar-Schnäpper
i   Pigliamosche africano pallido

1215   Batrachostomidae
Caprimulgiformes
e   Asian Frogmouths
f   Batrachostomidés
d   Froschmäuler
i   Batracostomidi

1216   *Batrachostomus affinis*
Caprimulgiformes -
Batrachostomidae
e   Blyth's Frogmouth
f   Podarge de Blyth
d   Malaien-Froschmaul; Malaien-
Froschschwalm
i   Boccadirana di Blyth

1217   *Batrachostomus auritus*
Caprimulgiformes -
Batrachostomidae
e   Large Frogmouth
f   Podarge oreillard
d   Riesenfroschmaul;
Riesenfroschschwalm
i   Boccadirana maggiore

1218   *Batrachostomus cornutus*
Caprimulgiformes -
Batrachostomidae
e   Sunda Frogmouth; Bornean
Frogmouth; Long-tailed Frogmouth;
Horned Frogmouth
f   Podarge cornu
i   Boccadirana del Borneo

1219   *Batrachostomus harterti*
Caprimulgiformes -
Batrachostomidae
e   Dulit Frogmouth
f   Podarge de Hartert
d   Dulit-Froschmaul; Dulit-
Froschschwalm
i   Boccadirana di Hartert

1220   *Batrachostomus hodgsoni*
Caprimulgiformes -
Batrachostomidae
e   Hodgson's Frogmouth
f   Podarge de Hodgson; Podarge
oriental
d   Langschwanzfroschmaul;
Langschwanzfroschschwalm
i   Boccadirana di Hodgson

1221   *Batrachostomus javensis*
Caprimulgiformes -
Batrachostomidae
e   Javan Frogmouth; Java Frogmouth
f   Podarge de Java
d   Tüpfelfroschmaul;
Tüpfelfroschschwalm
i   Boccadirana di Giava

1222   *Batrachostomus mixtus*
Caprimulgiformes -

Batrachostomidae
e Bornean Frogmouth; Sharpe's
Frogmouth
f Podarge de Bornéo
i Boccadirana di Sharpe

1223 Batrachostomus moniliger
Caprimulgiformes -
Batrachostomidae
e Sri Lanka Frogmouth; Ceylon
Frogmouth (ISC); Sinhalese
Frogmouth
f Podarge de Ceylan
d Ceylon-Froschmaul; Ceylon-
Froschschwalm
i Boccadirana di Sri-Lanka

1224 Batrachostomus poliolophus
Caprimulgiformes -
Batrachostomidae
e Short-tailed Frogmouth; Pale-headed
Frogmouth
f Podarge à tête grise
d Graukopffroschmaul;
Graukopffroschschwalm
i Boccadirana testa pallida

1225 Batrachostomus septimus
Caprimulgiformes -
Batrachostomidae
e Philippine Frogmouth
f Podarge des Philippines
d Philippinen-Froschmaul; Philippinen-
Froschschwalm
i Boccadirana delle Filippine

1226 Batrachostomus stellatus
Caprimulgiformes -
Batrachostomidae
e Gould's Frogmouth
f Podarge étoilé
d Schuppenfroschmaul;
Schuppenfroschschwalm
i Boccadirana di Gould

1227 Bebrornis rodericanus
Passeriformes - Sylviidae
e Rodriguez Brush-Warbler; Rodriguez
Swamp-Warbler; Rodriguez Warbler
f Rousserolle de Rodriguez;
Rousdserolle de l'île Rodriguez

d Rodriguez-Rohrsänger
i Cannaiola dell'Isola Rodriguez

1228 Bebrornis sechellensis
Passeriformes - Sylviidae
e Seychelles Brush-Warbler;
Seychelles Swamp-Warbler;
Seychelles Warbler
f Rousserolle des Seychelles
d Seychellen-Rohrsänger
i Rampichino delle palme; Cannaiola
delle Syechelles

1229 Berlepschia rikeri
Passeriformes - Furnariidae
e Point-tailed Palmcreeper; Pin-tailed
Palmcreeper
f Anabate des palmiers
d Palmsteiger

1230 Bias flammulatus
Passeriformes - Corvidae
e African Shrike-Flycatcher; Shrike
Flycatcher; Common Shrike-
Flycatcher
f Bias écorcheur; Gobemouche
écorcheur
d Seychellen-Rohrsänger;
Schnäpperwürger
i Pigliamosche-averla africano

1231 Bias musicus
Passeriformes - Corvidae
e Black-and-white Shrike-Flycatcher;
Vanga Flycatcher (CSA); Black-and-
white Flycatcher; Black-and-white
Vanga-Flycatcher; Crested Shrike-
Flycatcher
f Bias musicien; Gobemouche
chanteur; Gobe-mouches chanteur
d Vangaschnäpper
i Pigliamosche-averla bianca e nero

1232 Biatas nigropectus
Passeriformes - Formicariidae
e White-bearded Antshrike
f Batara à poitrine noire
d Weißbartameisenwürger
i Averla formichiera pettonero

1233   ***Biziura lobata***
        Anseriformes - Anatidae
e   Musk Duck; Lobbed Duck; Steamer
        Duck
f   Érismature à barbillons
d   Lappenente; Scharbenente
i   Anatra muschiata

1234   ***Bleda canicapilla***
        Passeriformes - Pycnonotidae
e   Grey-headed Bristlebill
f   Bulbul fourmilier; Fourmilier à tête
        grise; Bulbul moustac à tête grise;
        Moustac à tête grise
d   Graukopfbleda
i   Bulbul beccosettoloso testagrigia

1235   ***Bleda eximia***
        Passeriformes - Pycnonotidae
e   Green-tailed Bristlebill
f   Bulbul à queue verte; Moustac à tête
        olive; Bulbul moustac à queue verte
d   Grünschwanzbleda
i   Bulbul beccosetoloso codaverde

1236   ***Bleda notata***
        Passeriformes - Pycnonotidae
e   Lesser Bristlebill
f   Bulbul jaunelore

1237   ***Bleda syndactyla***
        Passeriformes - Pycnonotidae
e   Common Bristlebill; Bristlebill; Red-
        tailed Bristlebill
f   Bulbul moustac; Bulbul moustac à
        queue rousse
d   Rotschwanzbleda
i   Bulbul beccosettoloso

1238   ***Blythipicus pyrrhotis***
        Piciformes - Picidae
e   Bay Woodpecker; Red-eared
        Woodpecker; Greater Bay-
        Woodpecker; Red-eared Rufous
        Woodpecker
f   Pic à oreillons rouges
d   Rotohrspecht; Großer Rindenspalter
i   Picchio orecchierosse

1239   ***Blythipicus rubiginosus***
        Piciformes - Picidae

e   Maroon Woodpecker; Lesser Bay-
        Woodpecker
f   Pic porphyroïde
d   Maronenspecht; Kleiner
        Rindenspalter
i   Picchio bruno-castaneo

1240   ***Boissonneaua flavescens***
        Apodiformes - Trochilidae
e   Buff-tailed Coronet
f   Colibri flavescent
d   Fahlschwanzkolibri
i   Diadema codafulva

1241   ***Boissonneaua Jardini***
        Apodiformes - Trochilidae
e   Velvet-purple Coronet; Jardine's
        Coronet
f   Colibri de Jardine
d   Hyazinthkolibri
i   Diadema purpureo

1242   ***Boissonneaua matthewsii***
        Apodiformes - Trochilidae
e   Chestnut-breasted Coronet
f   Colibri de Matthews
d   Zimtschwanzkolibri
i   Diadema pettocastano

1243   ***Bolbopsittacus lunulatus***
        Psittaciformes - Psittacidae
e   Guaiabero; Luzon Guaiabero
f   Perruche lunulée
d   Luzon Guaiabero; Guaiabero;
        Dickschnabel-
        Stummelschwanzpapagei;
        Stummelschwanzpapagei
i   Guaiabero

1244   ***Bolborhynchus aurifrons***
        Psittaciformes - Psittacidae
e   Mountain Parakeet; Gold-fronted
        Parakeet; Golden-fronted Parakeet
f   Toui à bandeau jaune; Perruche à
        bandeau jaune
d   Zitronensittich
i   Parrocchetto frontedorata

1245   ***Bolborhynchus aymara***
        Psittaciformes - Psittacidae
e   Grey-hooded Parakeeet; Aymara

Parakeet; Sierra Parakeet; Grey-
headed Parakeet
f Toui aymara; Perruche aymara
d Aymara-Sittich
i Parrocchetto della Sierra

1246 *Bolborhynchus ferrugineifrons*
Psittaciformes - Psittacidae
e Rufous-fronted Parakeet
f Toui à front roux
d Rotstirnsittich
i Parrocchetto fronteruggine

1247 *Bolborhynchus lineola*
Psittaciformes - Psittacidae
e Barred Parakeet; Catherine's
Parakeet; Lineolated Parakeet
f Toui catherine; Toui de Catherine;
Perruche rayée
d Katharina-Sittich
i Parrocchetto

1248 *Bolborhynchus orbygnesius*
Psittaciformes - Psittacidae
e Andean Parakeet
f Toui d'Orbigny
d Anden-Sittich
i Parrocchetto delle Ande

1249 *Bombycilla cedrorum*
Passeriformes - Bombycillidae
e Cedar Waxwing
f Jaseur d'Amérique; Jaseur des cèdres
d Zedernseidenschwanz; Zedernvogel
i Beccofrusone dei cedri

1250 *Bombycilla garrulus*
Passeriformes - Bombycillidae
e Bohemian Waxwing; Waxwing;
Greater Waxwing; Common
Waxwing
f Jaseur boréal; Jaseur de Bohème;
Jaseur d'Europe
d Seidenschwanz
i Beccofrusone; Galletto di bosco

1251 *Bombycilla japonica*
Passeriformes - Bombycillidae
e Japanese Waxwing
f Jaseur du Japon

d Blutseidenschwanz
i Beccofrusone giapponese

1252 Bombycillidae
Passeriformes
e Waxwings
f Jaseurs; Bombycillidés
d Seidenschwänze
i Bombicillidi; Beccofrusoni

1253 *Bonasa bonasia*
Galliformes - Phasianidae
e Hazel Grouse; Common Hazelhen;
Hazelhen; Northern Hazelhen
f Gélinotte des bois; Poule des bois
d Haselhuhn
i Francolino di monte; Bonasia

1254 *Bonasa sewerzowi*
Galliformes - Phasianidae
e Chinese Grouse; Chinese Hazelhen;
Severtzov's Grouse; Severtzov's
Hazel Grouse; Black-breasted Hazel
Grouse; Chinese Hazel Grouse
f Gélinotte de Severtzov
d Schwarzbrusthaselhuhn
i Tetraone di Svertzov

1255 *Bonasa umbellus*
Galliformes - Phasianidae
e Ruffed Grouse
f Gélinotte huppée; Gélinotte à fraise
d Rauhußhuhn; Kragenhuhn
i Tetraone dal collare

1256 *Bostrychia bocagei*
Ciconiiformes - Threskiornithidae
e Dwarf Olive-Ibis; Sao Tomé Ibis

1257 *Bostrychia carunculata*
Ciconiiformes - Threskiornithidae
e Wattled Ibis
f Ibis caronculé
d Klunkeribis
i Ibis carunculato

1258 *Bostrychia hagedash*
Ciconiiformes - Threskiornithidae
e Hadada Ibis; Hadada; Hadeda (CSA);
Hadeda Ibis (CSA); Hadedah;
Hagedash

f   Ibis hagedash
d   Hagedasch-Ibis; Hagedash
i   Ibis hadada; Hagedash

1259   ***Bostrychia olivacea***
        Ciconiiformes - Threskiornithidae
e   Olive Ibis; African Green-Ibis
        (CSA); Green Ibis
f   Ibis olive; Ibis olivâtre
d   Guinea-Ibis
i   Ibis olivaceo

1260   ***Bostrychia rara***
        Ciconiiformes - Threskiornithidae
e   Spot-breasted Ibis; Spotted-breasted
        Ibis
f   Ibis vermiculó
d   Fleckenibis
i   Ibis pettomacchiato

1261   ***Botaurus lentiginosus***
        Ciconiiformes - Ardeidae
e   American Bittern; North American
        Bittern
f   Butor d'Amérique
d   Amerikanische Rohrdommel;
        Nordamerikanische Rohrdommel
i   Tarabuso americano

1262   ***Botaurus pinnatus***
        Ciconiiformes - Ardeidae
e   Pinnated Bittern; South American
        Bittern
f   Butor mirasol
d   Südamerikanische Rohrdommel
i   Tarabuso amazzonico

1263   ***Botaurus poiciloptilus***
        Ciconiiformes - Ardeidae
e   Australasian Bittern; Australian
        Bittern; Australian Brown-Bittern;
        Brown Bittern; Black-backed Bittern;
        New Zealand Bittern
f   Butor d'Australie; Butor australien
d   Australische Rohrdommel

1264   ***Botaurus stellaris***
        Ciconiiformes - Ardeidae
e   Great Bittern; European Bittern;
        Bittern; Eurasian Bittern; Bull-of-the-
        bog; Common Bittern

f   Butor étoilé; Grand Butor
d   Rohrdommel; Große Rohrdommel
i   Tarabuso; Tarrabuso; Tarabugio;
        Trabucine; Cappon di palude;
        Capponaccio; Tarabuso australiano

1265   ***Brachycope anomala***
        Passeriformes - Passeridae
e   Bob-tailed Weaver; Anomolous
        Bishop
f   Travailleur à queue courte
d   Kurzschwanzweber
i   Tessitore codabreve

1266   ***Brachygalba albogularis***
        Piciformes - Galbulidae
e   White-throated Jacamar
f   Jacamar à gorge blanche
d   Weißkehlglanzvogel
i   Jacamar golabianca

1267   ***Brachygalba goeringi***
        Piciformes - Galbulidae
e   Pale-headed Jacamar
f   Jacamar à tête pâle
d   Fahlnackenglanzvogel
i   Jacamar testapallida

1268   ***Brachygalba lugubris***
        Piciformes - Galbulidae
e   Brown Jacamar
f   Jacamar brun
d   Braunkehlglanzvogel
i   Jacamar bruno

1269   ***Brachygalba salmoni***
        Piciformes - Galbulidae
e   Dusky-backed Jacamar; Salmon's
        Jacamar
f   Jacamar sombre
d   Salmon-Glanzvogel
i   Jacamar dorsoscuro

1270   ***Brachypteracias leptosomus***
        Coraciiformes - Brachypteraciidae
e   Short-legged Ground-Roller
f   Brachyptérolle leptosome
d   Bindenerdracke
i   Coracia terricola zampecorte

1271 **Brachypteracias squamigera**
Coraciiformes - Brachypteraciidae
e Scaly Ground-Roller; Scaled
Ground-Roller
f Brachyptérolle écaillé;
Brachypterolle écailleux
d Schuppenerdracke
i Coracia terricola squamosa

1272 **Brachypteraciidae**
Coraciiformes
e Ground-Rollers
f Brachypteracidés
d Erdracken
i Braccipteraciidi

1273 **Brachypteryx hyperythra**
Passeriformes - Muscicapidae
e Rusty-bellied Shortwing; Rusty-
breasted Shortwing
f Brachyptère à ventre roux
d Rotbauchkurzflügel
i Alabreve ventreruggine

1274 **Brachypteryx leucophrys**
Passeriformes - Muscicapidae
e Lesser Shortwing; Brown Shortwing
f Petite Brachyptère; Petite Grive à
ailes courtes
d Zwergkurzflügel
i Alabreve minore

1275 **Brachypteryx major**
Passeriformes - Muscicapidae
e White-bellied Shortwing; Rufous-
bellied Shortwing
f Brachyptère à ventre blanc
d Nilgiri-Kurzflügel
i Alabreve ventrebianco

1276 **Brachypteryx montana**
Passeriformes - Muscicapidae
e White-browed Shortwing; Blue
Shortwing; Mountain Bush-Warbler
f Brachyptère bleue; Grive bleue à
ailes courtes; Bouscarle des
montagnes
d Bergkurzflügel
i Alabreve azzurro

1277 **Brachypteryx stellata**
Passeriformes - Muscicapidae
e Gould's Shortwing; Chestnut
Shortwing
f Brachyptère étoilée; Grive à ailes
courtes de Gould
d Braunrückenkurzflügel
i Alabreve di Gould

1278 **Brachyramphus brevirostris**
Charadriiformes - Alcidae
e Kittlitz's Murrelet; Short-billed
Murrelet
f Guillemot de Kittlitz; Alque pâle
(Qué)
d Kurzschnabelalk
i Urietta di Kittlitz

1279 **Brachyramphus hypoleuca**
Charadriiformes - Alcidae
e Xantus's Murrelet
f Guillemot de Xantus
d Lummenalk
i Urietta di Xantus

1280 **Brachyramphus marmoratus**
Charadriiformes - Alcidae
e Marbled Murrelet; Partridge Auk;
Long-billed Murrelet
f Guillemot marbré; Alque marbrée
(Qué)
d Marmelalk
i Urietta marmoreggiata

1281 **Brachyramphus perdix**
Charadriiformes - Alcidae
e Long-billed Murrelet

1282 **Bradornis infuscatus**
Passeriformes - Muscicapidae
e Chat Flycatcher; African Brown-
Flycatcher; Brown Bradornis
f Gobemouche traquet; Gobe-mouches
traquet
d Drosselschnäpper
i Pigliamosche bruno africano

1283 **Bradornis mariquensis**
Passeriformes - Muscicapidae
e Mariqua Flycatcher; Marico
Flycatcher (CSA)

f   Gobemouche du Marico; Gobe-
      mouches du Marico; Gobemouche de
      Marico; Gobe-mouches de Marico
d   Marico-Schnäpper; Marico-
      Blaßschnäpper
i   Pigliamosche di Marqua

1284   Bradornis microrhynchus
        Passeriformes - Muscicapidae
e   Large Flycatcher; African Grey-
      Flycatcher; Grey Flycatcher; Large
      Grey-Flycatcher; Greyish Flycatcher;
      Little Grey-Flycatcher; Small Grey-
      Flycatcher
f   Gobemouche à petit bec; Gobe-
      mouches à petit bec; Gobemouche
      grisâtre; Gobe-mouches grisâtre
d   Strichelkopfschnäpper
i   Pigliamosche grigio

1285   Bradornis pallidus
        Passeriformes - Muscicapidae
e   Pale Flycatcher; Pallid Flycatcher
      (CSA); Mouse-coloured Flycatcher
      (CSA); Pale Bradornis
f   Gobemouche pâle; Gobe-mouches
      pâle
d   Fahlschnäpper; Afrikanischer
      Blaßschnäpper
i   Pigliamosche pallido

1286   Bradornis pumilus
        Passeriformes - Muscicapidae
e   Small Grey-Flycatcher; Little Grey-
      Flycatcher
f   Gobemouche de Sharpe; Gobe-
      mouches de Sharpe
i   Pigliamosche grigio piccolo

1287   Bradypterus accentor
        Passeriformes - Sylviidae
e   Friendly Bush-Warbler; Friendly
      Warbler; Kinabalu Warbler;
      Kinabalu Scrub-Warbler; Kinabalu
      Bush-Warbler; Kinabalu Friendly
      Warbler
f   Bouscarle du Kinabalu
d   Borneo-Buschsänger
i   Forapaglie di Kinabalu

1288   Bradypterus alfredi
        Passeriformes - Sylviidae
e   Stock-Scrub-Warbler; Stock-
      Warbler; Newton's Scrub-Warbler
f   Bouscarle des bambous; Bouscarle
      d'Alfred
d   Graubrustbuschsänger
i   Forapaglie dei bambù

1289   Bradypterus alishaensis
        Passeriformes - Sylviidae
e   Taiwan Bush-Warbler

1290   Bradypterus baboecala
        Passeriformes - Sylviidae
e   African Bush-Warbler; African
      Sedge-Warbler; Little Rush-Warbler;
      Swamp Warbler
f   Bouscarle caqueteuse; Bouscarle des
      marais
d   Sumpfbuschsänger
i   Forapaglie africana

1291   Bradypterus bangwaensis
        Passeriformes - Sylviidae
e   Bangwa Forest-Warbler
f   Bouscarle du Cameroun; Bouscarle
      de Bangwa

1292   Bradypterus barratti
        Passeriformes - Sylviidae
e   African Scrub-Warbler; Barratt's
      Warbler (CSA); Barratt's Scrub-
      Warbler; Scrub-Warbler
f   Bouscarle des fourres; Bouscarle de
      Barrat
d   Barratts Buschsänger
i   Forapaglie di Barrat

1293   Bradypterus carpalis
        Passeriformes - Sylviidae
e   White-winged Scrub-Warbler;
      White-winged Warbler; White-
      winged Bush-Warbler
f   Bouscarle à ailes blanches
d   Bindenbuschsänger
i   Forapaglie alibianche

1294   Bradypterus castaneus
        Passeriformes - Sylviidae
e   Chestnut-backed Bush-Warbler; East

Indies Bush-Warbler; Chestnut
Grass-Warbler; Cinnamon Bracken-
Warbler; Chestnut Bush-Warbler;
Indian Bush-Warbler
f Bouscarle marron
d Molukken-Buschsänger
i Forapaglie dorsocastano

1295 *Bradypterus caudatus*
Passeriformes - Sylviidae
e Long-tailed Bush-Warbler; Long-
tailed Ground-Warbler
f Bouscarle à longue queue
d Langschwanzbuschsänger
i Forapaglie codalunga

1296 *Bradypterus cinnamomeus*
Passeriformes - Sylviidae
e Cinnamon Bracken-Warbler;
Cinnamon Reed-Warbler (CSA);
Cinnamon Scrub-Warbler
f Bouscarle cannellé
d Zimtrohrsänger; Zimtbuschsänger
i Forapaglie dei felceti

1297 *Bradypterus davidi*
Passeriformes - Sylviidae
e David's Bush-Warbler
i Forapaglie

1298 *Bradypterus grandis*
Passeriformes - Sylviidae
e Ja River Scrub-Warbler; Giant
Swamp-Warbler; Dia River-Warbler;
Ju River Warbler; Ja River Warbler;
Dia River Scrub-Warbler; Ju River
Scrub-Warbler; Giant Bush-Warbler
f Bouscarle géante; Bouscarle du Dja
d Gabun-Buschsänger
i Forapaglie del fiume

1299 *Bradypterus graueri*
Passeriformes - Sylviidae
e Grauer's Scrub-Warbler; Grauer's
Warbler; Grauer's Bush-Warbler;
Grauer's Swamp-Warbler; Grauer's
Rush-Warbler
f Bouscarle de Grauer
d Kivu-Buschsänger
i Forapaglie di Grauer

1300 *Bradypterus lopezi*
Passeriformes - Sylviidae
e Evergreen Forest-Warbler;
Cameroon Mountain-Warbler;
Cameroon Scrub-Warbler; White-
tailed Camaroptera; White-tailed
Warbler; Lopez's Warbler
f Bouscarle de Lopes; Bouscarle brune
d Waldbuschsänger
i Forapaglie del Camerun

1301 *Bradypterus luteoventris*
Passeriformes - Sylviidae
e Brown Bush-Warbler; Brown Scrub-
Warbler; Russet Scrub-Warbler
f Bouscarle russule
d Rostbuschsänger
i Forapaglie ventregiallo

1302 *Bradypterus major*
Passeriformes - Sylviidae
e Long-billed Bush-Warbler; Large-
billed Scrub-Warbler; Large-billed
Bush-Warbler; Large-billed Grass-
Warbler
f Bouscarle à long bec
d Langschnabelbuschsänger
i Forapaglie beccogrosso

1303 *Bradypterus mariae*
Passeriformes - Sylviidae
e Evergreen Forest-Warbler
f Bouscarle de forêt
d Waldbuschsänger
i Forapaglie della foresta sempreverde

1304 *Bradypterus montis*
Passeriformes - Sylviidae
e Javan Bush-Warbler
f Bouscarle de Java
i Forapaglie

1305 *Bradypterus palliseri*
Passeriformes - Sylviidae
e Sri Lanka Bush-Warbler; Palliser's
Warbler; Palliser's Bush-Warbler;
Ceylon Bush-Warbler; Ceylon
Warbler
f Bouscarle de Ceylan
d Ceylon-Buschsänger
i Forapaglie di Palliser

1306   ***Bradypterus seebohmi***
        Passeriformes - Sylviidae
    e Russet Bush-Warbler; Russet Scrub-
        Warbler; Mountain Bush-Warbler;
        Mountain Scrub-Warbler
    f Bouscarle de Seebohm
    d Gebirgsbuschsänger
    i Forapaglie di Seebohm

1307   ***Bradypterus sylvaticus***
        Passeriformes - Sylviidae
    e Knysna Scrub-Warbler; Knysna
        Warbler (CSA); Knysna Bush-
        Warbler
    f Bouscarle de Knysna
    d Sujndevals Buschsänger
    i Forapaglie

1308   ***Bradypterus tacsanowskius***
        Passeriformes - Sylviidae
    e Chinese Bush-Warbler; Chinese
        Scrub-Warbler
    f Bouscarle de Taczanowski
    d Taczanowskis Buschsänger
    i Forapaglie cinese

1309   ***Bradypterus thoracicus***
        Passeriformes - Sylviidae
    e Spotted Bush-Warbler; Spotted
        Scrub-Warbler
    f Bouscarle tachetée
    d Fleckenbuschsänger
    i Forapaglie pettomaculato

1310   ***Bradypterus timoriensis***
        Passeriformes - Sylviidae
    e Timor Bush-Warbler
    f Bouscarle de Mayr
    i Forapaglie

1311   ***Bradypterus victorini***
        Passeriformes - Sylviidae
    e Victorin's Scrub-Warbler; Victorin's
        Warbler (CSA); Victorin's Bush-
        Warbler
    f Bouscarle de Victorin
    d Rostbrustbuschsänger
    i Forapaglie di Victorin

1312   ***Branta bernicla***
        Anseriformes - Anatidae

    e Brent Goose; Brant Goose (NA);
        Brant; Brent; Pale-bellied Brent
        Goose
    f Bernache cravant; Barnache cravant;
        Bernache cravant
    d Ringelgans
    i Oca colombaccio

1313   ***Branta bernicla nigricans***
        Anseriformes - Anatidae
    e Black Brant
    f Bernache du Pacifique

1314   ***Branta canadensis***
        Anseriformes - Anatidae
    e Canada Goose
    f Bernache du Canada
    d Kanada-Gans
    i Oca del Canada

1315   ***Branta hutchinsii***
        Anseriformes - Anatidae
    e Tundra Goose; Hutchins's Goose;
        Richardson's Goose

1316   ***Branta leucopsis***
        Anseriformes - Anatidae
    e Barnacle Goose
    f Bernache nonnette; Bernacle
        nonnette
    d Nonnengans; Weißwangengans
    i Oca facciabianca

1317   ***Branta leucoptera***
        Anseriformes - Anatidae
    e Aleutian Goose

1318   ***Branta minima***
        Anseriformes - Anatidae
    e Cackling Goose

1319   ***Branta ruficollis***
        Anseriformes - Anatidae
    e Red-breasted Goose; Siberian Red-
        breasted Goose
    f Bernache à cou roux
    d Rothalsgans
    i Oca collorosso

1320   ***Branta sandvicensis***
        Anseriformes - Anatidae

e   Hawaiian Goose; Néné; Hawaiian
    Néné; Néné Goose
f   Bernache néné
d   Sandwich-Gans
i   Oca delle Hawaii; Nè Nè

1321   *Brotogeris chiriri*
    Psittaciformes - Psittacidae
e   Yellow-chevroned Parakeet; Yellow-
    chevroned Parrotlet; Chiriri Parrotlet
f   Toui à ailes jaunes
d   Kanarienvogelsittich
i   Parrocchetto aligialle

1322   *Brotogeris chrysopterus*
    Psittaciformes - Psittacidae
e   Golden-winged Parakeet; Orange-
    winged Parakeet
f   Toui para; Perruche à menton
    orangée
d   Braunkinnsittich; Goldflügelsittich
i   Parrocchetto alidorate

1323   *Brotogeris cyanoptera*
    Psittaciformes - Psittacidae
e   Cobalt-winged Parakeet; Blue-
    winged Parakeet
f   Toui de Deville
d   Blauflügelsittich; Kobaltflügelsittich
i   Parrocchetto alicobalto

1324   *Brotogeris jugularis*
    Psittaciformes - Psittacidae
e   Orange-chinned Parakeet; Tovi
    Parakeet
f   Toui à menton d'or; Perruche de
    Tovi; Perruche à menton jaune
d   Goldkinnsittich; Tovi-Sittich
i   Parrocchetto tovi; Parrocchetto dal
    mento arancio

1325   *Brotogeris pyrrhopterus*
    Psittaciformes - Psittacidae
e   Grey-cheeked Parakeet; Orange-
    flanked Parakeet
f   Toui flamboyant; Perruche à flancs
    orangés
d   Feuerflügelsittich; Grauwangensittich
i   Parrocchetto alirosse

1326   *Brotogeris sanctithomae*
    Psittaciformes - Psittacidae
e   Tui Parakeet
f   Toui à front d'or; Perruche toui
d   Tui-Sittich; Goldkopfsittich
i   Parrocchetto tui

1327   *Brotogeris tirica*
    Psittaciformes - Psittacidae
e   Plain Parakeet; All-green Parakeet
f   Toui tirica; Perruche tirica
d   Tirika-Sittich
i   Parrocchetto verde

1328   *Brotogeris versicolurus*
    Psittaciformes - Psittacidae
e   Canary-winged Parakeet; White-
    winged Parakeet
f   Toui à ailes variées; Perruche à ailes
    blancs
d   Kanarienflügelsittich;
    Weißflügelsittich
i   Parrocchetto alibianchi

1329   *Bubalornis albirostris*
    Passeriformes - Ploceidae
e   White-billed Buffalo-Weaver;
    Buffalo-Weaver; Black Buffalo-
    Weaver; Common Buffalo-Weaver
f   Alecto à bec blanc; Tisserin alecto
d   Weißschnabelbüffelweber; Alekto-
    Weber
i   Tessitore dei bufali beccobianco

1330   *Bubalornis niger*
    Passeriformes - Passeridae
e   Red-billed Buffalo-Weaver; Buffalo-
    Weaver
f   Alecto à bec rouge
d   Büffelweber;
    Rotschnabelbüffelweber
i   Tessitore dei bufali beccorosso

1331   *Bubo africanus*
    Strigiformes - Strigidae
e   Spotted Eagle-Owl; African Eagle-
    Owl
f   Grand-duc africain
d   Fleckenuhu; Berguhu
i   Gufo reale maculato; Gufo
    picchiettato

1332   ***Bubo ascalaphus***
       Strigiformes - Strigidae
e   Pharaoh Eagle-Owl; Desert Eagle-Owl
f   Grand-duc du désert; Grand-duc ascalaphe
d   Wüstenuhu
i   Gufo reale del deserto

1333   ***Bubo bengalensis***
       Strigiformes - Strigidae
e   Rock Eagle-Owl; Indian Great Horned-Owl; Indian Eagle-Owl
f   Grand-duc indien
i   Gufo reale indiano

1334   ***Bubo bubo***
       Strigiformes - Strigidae
e   Eurasian Eagle-Owl; Eagle-Owl; Indian Great Horned-Owl (ISC); Great Eagle-Owl; Common Eagle-Owl; Northern Eagle-Owl
f   Grand-duc d'Europe; Hibou grand-duc
d   Uhu
i   Gufo reale

1335   ***Bubo capensis***
       Strigiformes - Strigidae
e   Cape Eagle-Owl; Mackinder's Eagle-Owl; Mountain Eagle-Owl
f   Grand-duc du Cap
d   Kap-Uhu
i   Gufo reale del Capo; Gufo del Capo

1336   ***Bubo coromandus***
       Strigiformes - Strigidae
e   Dusky Eagle-Owl; Dusky Horned-Owl
f.   Grand-duc de Coromandel
d   Koromandel-Uhu
i   Gufo reale bruno

1337   ***Bubo lacteus***
       Strigiformes - Strigidae
e   Verreaux's Eagle-Owl; Giant Eagle-Owl (CSA); Milky Eagle-Owl
f   Grand-duc de Verreaux
d   Milchuhu; Blaßuhu
i   Bubo latteo

1338   ***Bubo leucostictus***
       Strigiformes - Strigidae
e   Akun Eagle-Owl; Sooty Eagle-Owl
f   Grand-duc tacheté
d   Schwachschnabeluhu
i   Gufo reale di Akun; Assiolo marmorizzato

1339   ***Bubo nipalensis***
       Strigiformes - Strigidae
e   Spot-bellied Eagle-Owl; Forest Eagle-Owl (ISC)
f   Grand-duc du Népal
d   Nepal-Uhu
i   Gufo reale delle foreste; Gufo delle foreste

1340   ***Bubo philippensis***
       Strigiformes - Strigidae
e   Philippine Eagle-Owl; Philippine Horned-Owl
f   Grand-duc des Philippines
d   Streifenuhu
i   Gufo reale delle Filippine; Gufo delle Filippine

1341   ***Bubo poensis***
       Strigiformes - Strigidae
e   Fraser's Eagle-Owl
f   Grand-duc à aigrettes
d   Guinea-Uhu
i   Gufo reale di Fraser; Gufo di Fraser

1342   ***Bubo shelleyi***
       Strigiformes - Strigidae
e   Shelley's Eagle-Owl; Banded Eagle-Owl
f   Grand-duc de Shelley
d   Bindenuhu
i   Gufo reale di Shelley; Gufo di Shelley

1343   ***Bubo sumatranus***
       Strigiformes - Strigidae
e   Barred Eagle-Owl; Malay Eagle-Owl; Malaysian Eagle-Owl
f   Grand-duc bruyant
d   Malaien-Uhu
i   Gufo reale della Malesia; Guffo malese

1344 **Bubo virginianus**
Strigiformes - Strigidae
e Great Horned-Owl
f Grand-duc d'Amérique; Grand-duc
de Virginie
d Amerikanischer Uhu; Virginia-Uhu
i Bubo virginiano; Gufo della
Virginia; Gufo cornuto

1345 **Bubo vosseleri**
Strigiformes - Strigidae
e Usambara Eagle-Owl; Nduk Eagle-
Owl
f Grand-duc des Usambara

1346 **Bubulcus coromanda**
Ciconiiformes - Ardeidae
e Eastern Cattle-Egret

1347 **Bubulcus ibis**
Ciconiiformes - Ardeidae
e Cattle Egret; Buff-backed Egret;
Common Cattle-Egret; Thickbird
(CSA); Buff-backed Heron; Cattle
Gauldin (WI); Gaulin (WI); Cowbird
(WI); Ibis (WI); Cattle Gaulin (WI)
f Héron garde-boeufs; Gardeboeuf
aigrette; Grue blanche (Ants); Garde-
boeuf (Ants); Pique-boeufs (Ants);
Kio blanc (Ants); Crabier (Ants);
Tiqueur (Ants); Oiseau détiqueur
d Kuhreiher
i Airone guardabuoi

1348 **Buccanodon duchaillui**
Piciformes - Lybiidae
e Yellow-spotted Barbet; Duchaillu's
Yellow-spotted Barbet
f Barbican à taches jaunes
d Gelbfleckbartvogel
i Barbuto macchiegialle

1349 **Bucco capensis**
Piciformes - Bucconidae
e Collared Puffbird
f Tamatia à collier
d Halsbandfaulvogel
i Bucco macuru; Bucco dal collare
nero

1350 **Bucco macrodactylus**
Piciformes - Bucconidae
e Chestnut-capped Puffbird
f Tamatia macrodactyle
d Braunkappenfaulvogel;
Langzehenfaulvogel
i Bucco testacastana

1351 **Bucco noanamae**
Piciformes - Bucconidae
e Sooty-capped Puffbird
f Tamatia de Colombie
d Hellmayrs Faulvogel;
Rußkappenfaulvogel
i Bucco testafuligginosa

1352 **Bucco tamatia**
Piciformes - Bucconidae
e Spotted Puffbird
f Tamatia tacheté
d Tamatia-Faulvogel
i Bucco golacastana

1353 **Bucconidae**
Piciformes
e Puffbirds
f Bucconidés
d Faulvögel
i Bucconidi

1354 **Bucephala albeola**
Anseriformes - Anatidae
e Bufflehead
f Petit Garrot; Garrot albéole
d Büffelkopfente
i Quattrocchi minore

1355 **Bucephala clangula**
Anseriformes - Anatidae
e Common Goldeneye; Goldeneye;
Golden-eye Duck
f Garrot à oeil d'or; Garrot sonneur
d Schellente; Schellenente
i Quattrocchi; Morettone; Cagnola;
Canone; Cagnaccia; Quattrocchi
comune

1356 **Bucephala islandica**
Anseriformes - Anatidae
e Barrow's Goldeneye
f Garrot d'Islande; Garrot de Barrow

d Spatelente
i Quattrocchi d'Islanda

1357 *Buceros bicornis*
Coraciiformes - Bucerotidae
e Great Hornbill; Great Pied-Hornbill
(ISC); Indian Hornbill; Pied
Hornbill; Great Indian Hornbill;
Giant Hornbill
f Calao bicorne
d Doppelhornvogel
i Bucero bicorne; Calao bicorne; Calao
maggiore

1358 *Buceros hydrocorax*
Coraciiformes - Bucerotidae
e Rufous Hornbill; Philippine Rufous
Hornbiull; Philippine Brown-
Hornbill; Great Philippine Hornbill
f Calao à casque plat
d Feuerhornvogel
i Bucero rossiccio

1359 *Buceros rhinoceros*
Coraciiformes - Bucerotidae
e Rhinoceros Hornbill; Great
Rhinoceros-Hornbill
f Calao rhinocéros
d Rhinozerusvogel
i Calao rinoceronte

1360 *Buceros vigil*
Coraciiformes - Bucerotidae
e Helmeted Hornbill; Solid-billed
Hornbill; Great Helmeted-Hornbill
f Calao à casque rond
d Schildschnabel
i Bucero dall'elmo

1361 Bucerotidae
Coraciiformes
e Hornbills
f Bucerotidés
d Hornvögel; Nashornvögel; Tokos
i Bucerotidi; Buceri

1362 Bucorvidae
Coraciiformes
e Ground Hornbills
f Bucorvidés

d Hornraben
i Bucorvidi

1363 *Bucorvus abyssinicus*
Coraciiformes - Bucorvidae
e Northern Ground Hornbill;
Abyssinian Ground Hornbill; African
Ground Hornbill
f Bucorve d'Abyssinie; Calao
d'Abyssinie; Grand Calao
d'Abyssinie
d Sudan-Hornrabe
i Bucorvo dell'Abissinia; Bucorvo
abissino; Bucorvo d'Abissinia

1364 *Bucorvus leadbeateri*
Coraciiformes - Bucorvidae
e Southern Ground Hornbill; Ground
Hornbill (CSA); African Ground
Hornbill
f Bucorve du Sud; Calao terrestre;
Grand Calao terrestre
d Hornrabe; Kaffern-Hornhrabe
i Bucorvo cafro; Bucero

1365 *Buettikoferella bivittata*
Passeriformes - Sylviidae
e Buff-banded Grassbird; Buettikofer's
Warbler; Timor Babbler-Warbler;
Buff-banded Bushbird; Buff-banded
Thicket-Warbler
f Mégalure de Timor
d Timor-Buschsänger
i Cantore bifasciato

1366 *Bulweria bulwerii*
Procellariiformes - Procellariidae
e Bulwer's Petrel; Gadfly-Petrel (CSA)
f Pétrel de Bulwer
d Bulwer-Sturmvogel; Bulwer-
Sturmschwalbe
i Procellaria di Bulwer

1367 *Bulweria fallax*
Procellariiformes - Procellariidae
e Jouanin's Petrel; Gadfly-Petrel
(CSA); Jouanin's Gadfly Petrel
f Pétrel de Jouanin
d Jouanin-Sturmvogel
i Procellaria di Jouanin

1368   **Buphagus africanus**
        Passeriformes - Sturnidae
   e   Yellow-billed Oxpecker; Tickbird
   f   Pique-boeuf à bec jaune
   d   Gelbschnabelmadenhacker
   i   Bufaga africana; Bufaga dal becco
        giallo; Bufaga beccogialla

1369   **Buphagus erythrorhynchus**
        Passeriformes - Sturnidae
   e   Red-billed Oxpecker; Tickbird
   f   Pique-boeuf à bec rouge
   d   Rotschnabelmadenhacker
   i   Bufaga dal becco rosso; Bufaga
        beccorosso

1370   **Burhinidae**
        Charadriiformes
   e   Thick-knees; Stone-Curlews;
        Dikkops (CSA)
   f   Oedicnèmes; Burhinidés
   d   Triele; Dickfüße
   i   Burinidi

1371   **Burhinus bistriatus**
        Charadriiformes - Burhinidae
   e   Double-striped Thick-knee; Double-
        striped Stone-Curlew; Double-striped
        Dikkop; Mexican Thick-knee
   f   Courlis de terrre (Ants); Oedicnème
        bistrié; Oedicnème vocifère
   d   Dominikanertriel
   i   Occhione bistriato

1372   **Burhinus capensis**
        Charadriiformes - Burhinidae
   e   Spotted Thick-knee; Spotted Stone-
        Curlew; Spotted Dikkop; Cape
        Dikkop (CSA); Thick-knees (CSA)
   f   Oedicnème tachard
   d   Kap-Triel
   i   Occhione del Capo

1373   **Burhinus giganteus**
        Charadriiformes - Burhinidae
   e   Beach Thick-knee
   f   Oedicnème des récifs
   d   Riff-Triel
   i   Occhione delle reef

1374   **Burhinus grallarius**
        Charadriiformes - Burhinidae
   e   Bush Thick-knee; Bush Stone-
        Curlew (ANZ); Willaroo; Southern
        Thick-knee; Australian Dikkop;
        Southern Stone-Curlew; Stone
        Curlew
   f   Oedicnème bridé
   d   Langschwanztriel
   i   Occhione australiano

1375   **Burhinus magnirostris**
        Passeriformes - Burhinidae
   e   Great Australian Stone-Curlew;
        Beach Thick-knee; Beach Stone-
        Curlew; Stone Curlew; Reef Stone-
        Curlew; Reef Thick-knee; Beach
        Dikkop; Large-billed Stone-Plover
   d   Langschwanztriel

1376   **Burhinus oedicnemus**
        Charadriiformes - Burhinidae
   e   Eurasian Thick-knee; Eurasian
        Stone-Curlew; Stone Curlew; Thick-
        knee; Stone Thick-knee; Common
        Thick-knee; Northern Thick-knee
   f   Oedicnème criard
   d   Triel; Dickfuß
   i   Occhione; Veregino; Corrione;
        Corrisodo; Tallurino; Brecciolotto;
        Occhione comune

1377   **Burhinus recurvirostris**
        Charadriiformes - Burhinidae
   e   Great Thick-knee; Great Stone-
        Curlew; Great Stone-Plover (ISC);
        Great Dikkop; Oriental Thick-knee
   f   Grand Oedicnème
   d   Krabbentriel
   i   Occhione maggiore indiano;
        Occhione beccogrosso

1378   **Burhinus senegalensis**
        Charadriiformes - Burhinidae
   e   Senegal Thick-knee; Senegal Stone-
        Curlew; Senegal Dikkop
   f   Oedicnème du Sénégal
   d   Senegal-Triel
   i   Occhione del Senegal

1379 ***Burhinus superciliaris***
Charadriiformes - Burhinidae
e Peruvian Thick-knee; Peruvian
Stone-Curlew; Peruvian Dikkop
f Oedicnème du Pérou
d Peruaner Triel; Peru-Triel
i Occhione del Perù

1380 ***Burhinus vermiculatus***
Charadriiformes - Burhinidae
e Water Thick-knee; Water Stone-
Curlew; Water Dikkop
f Oedicnème vermiculé
d Wassertriel; Wellentriel
i Occhione vermicolato

1381 ***Busarellus nigricollis***
Falconiformes - Accipitridae
e Black-collared Hawk; Fishing
Buzzard; Collared Fishing Hawk;
Collared Fishing Buzzard; Chestnut
Hawk
f Busarelle à tête blanche; Buse à tête
blanche
d Fischbussard
i Poiana dal collare nero; Falco
pescatore dal collare nero

1382 ***Butastur indicus***
Falconiformes - Accipitridae
e Grey-faced Buzzard; Grey-faced
Buzzard-Eagle; Frog-Hawk
f Busautour à joues grises
d Kiefernteesa
i Butastore facciagrigia; Poiana
cavaletta dalla faccia grigia

1383 ***Butastur liventer***
Falconiformes - Accipitridae
e Rufous-winged Buzzard; Rufous-
winged Buzzard-Eagle; Cinnamon-
winged Buzzard-Eagle
f Busautour pâle
d Malaien-Teesa; Rotflügelbussard
i Butastore alirosse; Poiana cavaletta
dalle ali rossiccie

1384 ***Butastur rufipennis***
Falconiformes - Accipitridae
e Grasshopper Buzzard; Grasshopper
Buzzard-Eagle

f Busautour des sauterelles; Busard des
sauterelles
d Heuschreckenbussard;
Heuschreckenteesa
i Butastore rufipenne; Poiana cavaletta

1385 ***Butastur teesa***
Falconiformes - Accipitridae
e White-eyed Buzzard; White-eyed
Buzzard-Eagle
f Busautour aux yeux blancs
d Weißaugenteesa
i Butastore occhibianchi; Poiana
cavaletta dagli occhi bianchi

1386 ***Buteo albicaudatus***
Falconiformes - Accipitridae
e White-tailed Hawk
f Buse à queue blanche
d Weißschwanzbussard
i Poiana codabianca americana

1387 ***Buteo albigula***
Falconiformes - Accipitridae
e White-throated Hawk
f Buse à gorge blanche
d Weißkehlbussard
i Poiana golabianca

1388 ***Buteo albonotatus***
Falconiformes - Accipitridae
e Zone-tailed Hawk
f Buse à queue barrée
d Mohrenbussard
i Poiana codafasciata; Poiana dalla
coda fasciata

1389 ***Buteo archeri***
Falconiformes - Accipitridae
e Archer's Buzzard
f Buse d'Archer
i Poiana di Archer

1390 ***Buteo augur***
Falconiformes - Accipitridae
e Augur Buzzard
f Buse augure
d Augurbussard
i Poiana augure

1391 **Buteo auguralis**
Falconiformes - Accipitridae
e   Red-necked Buzzard; Red-tailed
Buzzard; African Red-tailed
Buzzard; African Red-tailed Hawk
f   Buse d'Afrique
d   Salvadoribussard
i   Poiana collonero

1392 **Buteo bannermani**
Falconiformes - Accipitridae
e   Cape Verde Buzzard
i   Poiana

1393 **Buteo brachypterus**
Falconiformes - Accipitridae
e   Madagascar Buzzard
f   Buse de Madagascar
d   Madagaskar-Bussard
i   Poiana del Madagascar

1394 **Buteo brachyurus**
Falconiformes - Accipitridae
e   Short-tailed Hawk
f   Buse à queue courte
d   Kurzschwanzbussard
i   Poiana codacorta; Poiana dalle ali
corte

1395 **Buteo buteo**
Falconiformes - Accipitridae
e   Common Buzzard; Buzzard;
Eurasian Buzzard; Eurasian Buteo;
Western Steppe-Buzzard; Steppe
Buzzzard; Desert Buzzard
f   Buse variable
d   Bussard; Mäusebussard
i   Poiana; Poiana eurasiatica

1396 **Buteo buteo vulpinus**
Falconiformes - Accipitridae
e   Steppe Buzzard (CSA); Buzzard
d   Mäusebussard

1397 **Buteo galapagoensis**
Falconiformes - Accipitridae
e   Galapagos Hawk
f   Buse des Galapagos
d   Galapagos-Bussard
i   Poiana delle Galapagos

1398 **Buteo harlani**
Falconiformes - Accipitridae
e   Harlan's Hawk

1399 **Buteo hemilasius**
Falconiformes - Accipitridae
e   Upland Buzzard; Upland Buteo;
Mongolian Buzzard
f   Buse de Chine
d   Mongolen-Bussard
i   Poiana degli altipiani; Poiana
centroasiatico

1400 **Buteo jamaicensis**
Falconiformes - Accipitridae
e   Red-tailed Hawk; Redtail; Fowl-
Hawk (WI); Chicken-Hawk
f   Buse à queue rousse; Gros Malfini
(Ants)
d   Rotschwanzbussard
i   Poiana dalla coda rossa; Poiana della
Giamaica; Poiana codarossa

1401 **Buteo lagopus**
Falconiformes - Accipitridae
e   Rough-legged Hawk; Rough-legged
Buzzard
f   Buse pattue
d   Rauhfußbussard
i   Poiana calzata

1402 **Buteo leucorrhous**
Falconiformes - Accipitridae
e   White-rumped Hawk; Rufous-
thighed Hawk
f   Buse cul-blanc
d   Weißbürzelbussard
i   Poiana gropponebianco; Poiana dalla
groppa bianca

1403 **Buteo lineatus**
Falconiformes - Accipitridae
e   Red-shouldered Hawk
f   Buse à épaulettes
d   Rotschulterbussard
i   Poiana spallerosse; Poiana dalle
spalle rosse

1404 **Buteo magnirostris**
Falconiformes - Accipitridae
e   Roadside Hawk; Insect Hawk; Large-

billed Hawk; Tropical Broad-winged
Hawk
f   Buse à gros bec
d   Wegebussard
i   Poiana calzata; Poiana beccogrosso;
Poiana delle strade

1405   *Buteo oreophilus*
Falconiformes - Accipitridae
e   Mountain Buzzard; African
Mountain-Buzzard; Forest Buzzard;
Woodland Buzzard
f   Buse montagnarde
d   Bergbussard
i   Poiana dei monti; Poiana di
montagna

1406   *Buteo platypterus*
Falconiformes - Accipitridae
e   Broad-winged Hawk; Chicken Hawk
(WI); Chicken-eater (WI); Gree gree
(WI)
f   Petite Buse; Manger-poulet (Ants);
Malfini (Ants)
d   Breitflügelbussard
i   Poiana alilarghe; Poiana dalle ali
grandi

1407   *Buteo poecilochrous*
Falconiformes - Accipitridae
e   Puna Hawk; Gurney's Hawk;
Variable Hawk
f   Buse du Puna
d   Punabussard
i   Poiana variabile; Poiana di Guerney

1408   *Buteo polyosoma*
Falconiformes - Accipitridae
e   Red-backed Hawk; Variable Hawk
f   Buse tricolore
d   Rotrückenbussard
i   Poiana dorsorosso; Poiana dal dorso
rosso

1409   *Buteo regalis*
Falconiformes - Accipitridae
e   Ferruginous Hawk; Ferruginous
Buzzard; Ferruginous Roughleg
f   Buse rouilleuse
d   Königsbussard

i   Nibbio reale; Poiana ferruginosa;
Poiana reale

1410   *Buteo ridgwayi*
Falconiformes - Accipitridae
e   Ridgway's Hawk
f   Buse de Ridgway
d   Haiti-Bussard
i   Poiana di Ridgway

1411   *Buteo rufinus*
Falconiformes - Accipitridae
e   Long-legged Buzzard; Long-legged
Buteo
f   Buse féroce
d   Adlerbussard
i   Poiana codabianca; Poiana
codabianca eurasiatica

1412   *Buteo rufofuscus*
Falconiformes - Accipitridae
e   Jackal Buzzard; Augur Buzzard
f   Buse rounoir; Buse augure
d   Felsenbussard; Augurbussard;
Schakalbussard
i   Poiana codarossa africana; Poiana
augure; Poiana rufofusca

1413   *Buteo solitarius*
Falconiformes - Accipitridae
e   Hawaiian Hawk; Hawaiian Buzzard;
Io
f   Buse de Hawai; Buse d'Hawal
d   Hawaii-Bussard
i   Poiana delle Hawaii

1414   *Buteo swainsoni*
Falconiformes - Accipitridae
e   Swainson's Hawk; Swainson's
Buzzard
f   Buse de Swainson
d   Präriebussard
i   Poiana di Swainson

1415   *Buteo trizonatus*
Falconiformes - Accipitridae
e   Forest Buzzard; Mountain Buzard
(CSA)
d   Bergbussard

1416   ***Buteo ventralis***
Falconiformes - Accipitridae
e   Rufous-tailed Hawk; Rufous-tailed
Buzzard
f   Buse de Patagonie
d   Magellan-Bussard
i   Poiana codarossa americana; Poiana
dalla coda rossa

1417   ***Buteogallus aequinoctialis***
Falconiformes - Accipitridae
e   Rufous Crab Hawk; Aequinoctal
Hawk; Rufous Hawk
f   Buse buson
d   Rotbauchbussard
i   Poiana dei granchi; Poiana
equinoziale

1418   ***Buteogallus anthracinus***
Falconiformes - Accipitridae
e   Common Black-Hawk; Black Hawk;
Crab Hawk (WI); Lessere Black-
Hawk
f   Crabier (Ants); Buse noire
d   Krabbenbussard
i   Poiana nera

1419   ***Buteogallus gundlachii***
Falconiformes - Accipitridae
e   Cuban Black-Hawk

1420   ***Buteogallus meridionalis***
Falconiformes - Accipitridae
e   Savanna Hawk; Savannah Hawk
f   Buse roussâtre
d   Savannenbussard
i   Poiana delle savane; Falco delle
savane

1421   ***Buteogallus subtilis***
Falconiformes - Accipitridae
e   Mangrove Black-Hawk; Pacific
Black-Hawk
f   Buse des mangroves
i   Poiana delle mangrovie

1422   ***Buteogallus urubitinga***
Falconiformes - Accipitridae
e   Great Black-Hawk
f   Buse urubu

d   Urubitinga
i   Poiana nero maggiore

1423   ***Buthraupis aureodorsalis***
Passeriformes - Fringillidae
e   Golden-backed Mountain-Tanager
f   Tangara à dos d'or
d   Goldrückenbergtangare
i   Tangara dorsodorato

1424   ***Buthraupis eximia***
Passeriformes - Fringillidae
e   Black-chested Mountain-Tanager;
Blue-rumped Mountain-Tanager
f   Tangara à poitrine noire
d   Schwarzbrustbergtangare
i   Tangara dalla pettorina nera

1425   ***Buthraupis montana***
Passeriformes - Fringillidae
e   Hooded Mountain-Tanager
f   Tangara montagnard
d   Blaurückenbergtangare
i   Tangara dal cappuccio

1426   ***Buthraupis wetmorei***
Passeriformes - Fringillidae
e   Masked Mountain-Tanager
f   Tangara de Wetmore
d   Wetmore-Tangare
i   Tangara di Wetmore

1427   ***Butorides striatus***
Ciconiiformes - Ardeidae
e   Striated Heron; Little Heron; Green-
backed Heron; Green Heron; Little
Green Heron (ISC); Thick-billed
Heron; Little Mangrove Heron;
Mangrove Heron; Red Mangrove
Bittern; Red Mangrove Heron
f   Héron strié; Héron vert; Héron à dos
vert
d   Mangrovereiher; Grünreiher
i   Airone dorsoverde

1428   ***Butorides sundevalli***
Ciconiiformes - Ardeidae
e   Galapagos Heron
f   Héron des Galapagos
i   Airone delle Galapagos

1429 ***Butorides virescens***
Ciconiiformes - Ardeidae
- *e* Green Heron; Mary Perk (WI);
Water-witch (WI); Poor Joe (WI);
Little Crabier (WI); Kyallee (WI);
Gaulin (WI); Green-backed Heron
- *f* Héron vert; Valet de Caïman (Ants);
Kio (Ant)
- *d* Grünreiher; Nordamerikanischer
Mangrovereiher
- *i* Airone verde; Airone dorsoverde

C

1430   *Cacatua alba*
Psittaciformes - Cacatuidae
e   White Cockatoo; Great White
Cockatoo; Umbrella Cockatoo;
Greater White-crested Cockatoo;
White-crested Cockatoo
f   Cacatoès blanc
d   Weißhaubenkakadu
i   Cacatua bianca

1431   *Cacatua ducorpsii*
Psittaciformes - Cacatuidae
e   Ducorps's Cockatoo; Solomons
Cockatoo; Solomons Corella; White
Cockatoo
f   Cacatoès de Ducorps
d   Salomonen-Kakadu; Ducorps-
Kakadu
i   Cacatua di Ducorps

1432   *Cacatua galerita*
Psittaciformes - Cacatuidae
e   Sulphur-crested Cockatoo; Greater
Sulphur-crested Cackatoo; Greater
Yellow-crested Cockatoo; White
Cockatoo
f   Cacatoès à huppe jaune; Cacatoés à
crête jaune; Grand Cacatoés à crète
jaune
d   Gelbhaubenkakadu; Großer
Gelbhaubenkakadu
i   Cacatua dal ciuffo giallo; Cacatua
ciuffogiallo maggiore; Grande
cacatua bianco

1433   *Cacatua goffini*
Psittaciformes - Cacatuidae
e   Tanimbar Cockatoo; Goffin's
Cockatoo; Tanimbar Corella
f   Cacatoès de Goffin
d   Goffin-Kakadu
i   Cacatua di Goffin

1434   *Cacatua haematuropygia*
Psittaciformes - Cacatuidae
e   Philippine Cockatoo; Red-vented
Cockatoo
f   Cacatoès des Philippines
d   Rotsteißkakadu
i   Cacatua dal sottocoda rosso

1435   *Cacatua leadbeateri*
Psittaciformes - Cacatuidae
e   Pink Cockatoo; Mitchell's Cockatoo
(ANZ); Major Mitchell's Cockatoo;
Leadbeater's Cockatoo; Pink Desert-
Cockatoo; Wee Juggler Cocklerina;
Chockallot; Kakatoe
f   Cacatoès de Leadbeater; Cacatoès à
huppe tricolore
d   Inkakakadu; Leadbeaters Kakadu
i   Cacatua di Leadbeater

1436   *Cacatua moluccensis*
Psittaciformes - Cacatuidae
e   Salmon-crested Cockatoo; Rose-
crested Cockatoo; Ceram Cockatoo;
Moluccan Cockatoo
f   Cacatoès à huppe rouge
d   Molukken-Kakadu
i   Cacatua delle Molucche; Cacatua
dalla cresta rossa

1437   *Cacatua ophthalmica*
Psittaciformes - Cacatuidae
e   Blue-eyed Cockatoo
f   Cacatoès aux yeux bleus
d   Brillenkakadu
i   Cacatua occhiblu

1438   *Cacatua pastinator*
Psittaciformes - Cacatuidae
e   Western Corella; Western Long-
billed Corella
f   Cacatoès laboureur
i   Cacatua orientale

1439   *Cacatua pastinator pastinator*
Psittaciformes - Cacatuidae
e   Muir's Corella (ANZ)

1440   *Cacatua roseicapilla*
Psittaciformes - Cacatuidae
e   Galah; Rosy Cockatoo; Rose-

breasted Cockatoo; Roseate
Cockatoo
f Cacatoès rosalbin
i Cacatua rosato; Galah; Cacatua
roseicapilla

1441 *Cacatua sanguinea*
Psittaciformes - Cacatuidae
e Little Cockatoo; Little Corella;
Corella; Blood-stained Cockatoo;
Bare-eyed Cockatoo; White
Cockatoo; Bare-eyed Corella; Short-
billed Corella
f Cacatoès corella; Corella à front
rouge; Corella à lunettes rouges;
Corella aux yeux nus
d Nacktaugenkakadu;
Nacktwangenkakadu;
Rotzügelkakadu
i Corella piccola

1442 *Cacatua sulphurea*
Psittaciformes - Cacatuidae
e Yellow-crested Cockatoo; Lesser
Sulphur-crested Cockatoo; Sulphur-
crested Cockatoo; Lemon-crested
Cockatoo; Citron-crested Cockatoo
f Cacatoès soufré
d Kleiner Gelbhaubenkakadu
i Cacatua ciuffogiallo minore; Piccolo
cacatua bianco

1443 *Cacatua tenuirostris*
Psittaciformes - Cacatuidae
e Long-billed Corella; Slender-billed
Corella; Long-billed Cockatoo;
Slender-billed White Cockatoo
f Cacatoès nasique
d Nasenkakadu
i Cacatua beccolungo

1444 Cacatuidae
Psittaciformes
e Cockatoos
f Cacatuidés
d Kakadus

1445 *Cacicus cela*
Passeriformes - Icteridae
e Yellow-rumped Cacique; Cacique
f Cassique cul-jaune

d Gelbbürzelkassike;
Goldbürzelkassike;
Gelbrückenstirnvogel
i Cacico gropponegiallo

1446 *Cacicus chrysonotus*
Passeriformes - Icteridae
e Mountain Cacique
f Cassique montagnard
d Bergkassike
i Cacico di monte

1447 *Cacicus chrysopterus*
Passeriformes - Icteridae
e Golden-winged Cacique; Cacique
f Cassique à épaulettes
d Goldschulterkassike
i Cacico alidorate

1448 *Cacicus haemorrhous*
Passeriformes - Icteridae
e Red-rumped Cacique; Cacique
f Cassique cul-rouge
d Rotbürzelkassike
i Cacico gr500pponerosso

1449 *Cacicus koepckeae*
Passeriformes - Icteridae
e Selva Cacique
f Cassique de Koepcke
d Loreto-Kassike
i Cacico della selva

1450 *Cacicus leucoramphus*
Passeriformes - Icteridae
e Northern Mountain-Cacique
f Cassique à bec blanc
d Bergkassike

1451 *Cacicus melanicterus*
Passeriformes - Icteridae
e Yellow-winged Cacique; Cacique;
Mexican Cacique
f Cassique à ailes jaunes
d Haubenkassike
i Cacico aligialle

1452 *Cacicus microrhynchus*
Passeriformes - Icteridae
e Scarlet-rumped Cacique; Small-
billed Cacique; Flame-rumped

Cacique
f Cassique à bec mince

1453 *Cacicus pacificus*
Passeriformes - Icteridae
e Pacific Cacique

1454 *Cacicus sclateri*
Passeriformes - Icteridae
e Ecuadorian Cacique; Ecuadorian Black-Cacique; Sclater's Black-Cacique
f Cassique d'Équateur
d Trauerkassike
i Cacico dell'Ecuador

1455 *Cacicus solitarius*
Passeriformes - Icteridae
e Solitary Cacique; Solitary Black-Cacique
f Cassique solitaire
d Stahlkassike; Schwarzkassike
i Cacico solitario

1456 *Cacicus uropygialis*
Passeriformes - Icteridae
e Scarlet-rumped Cacique; Cacique; Subtropical Cacique; Curve-billed Cacique; Pacific Cacique
f Cassique à dos rouge
d Scharlachbürzelkassike
i Cacico gropponescarlatto

1457 *Cacicus vitellinus*
Passeriformes - Icteridae
e Saffron-rumped Cacique

1458 *Cacomantis castaneiventris*
Cuculiformes - Cuculidae
e Chestnut-breasted Cuckoo; Chestnut-breasted Brush-Cuckoo
f Coucou à poitrine rousse
d Rostbauchkuckuck

1459 *Cacomantis flabelliformis*
Cuculiformes - Cuculidae
e Fan-tailed Cuckoo; Fan-tailed Brush-Cuckoo
f Coucou à eventail
d Fächerschwanzkuckuck
i Cuculo coda a ventaglio

1460 *Cacomantis heinrichi*
Cuculiformes - Cuculidae
e Moluccan Cuckoo; Heinrich's Cuckoo; Heinrich's Brush-Cuckoo
f Coucou de Heinrich
d Molukken-Kuckuck
i Cuculo di Heinrich

1461 *Cacomantis merulinus*
Cuculiformes - Cuculidae
e Plaintive Cuckoo; Grey-breasted Brush-Cuckoo
f Coucou plaintif
d Klagekuckuck
i Cuculo lamentoso

1462 *Cacomantis passerinus*
Cuculiformes - Cuculidae
e Grey-bellied Cuckoo; Plaintive Cuckoo (ISC)
f Coucou à tête grise
i Cuculo ventregrigio

1463 *Cacomantis sepulcralis*
Cuculiformes - Cuculidae
e Rusty-breasted Cuckoo
f Coucou à ventre roux
i Cuculo indonesiano

1464 *Cacomantis sonneratii*
Cuculiformes - Cuculidae
e Banded Bay-Cuckoo; Indian Bay-banded Cuckoo (ISC)
f Coucou de Sonnerat
d Sonnerat-Kuckuck
i Cuculo di Sonnerat

1465 *Cacomantis variolosus*
Cuculiformes - Cuculidae
e Brush Cuckoo
f Coucou des buissons
d Buschkuckuck
i Cuculo di macchia

1466 *Cairina moschata*
Anseriformes - Anatidae
e Muscovy Duck; Muscovy; Musk Duck
f Canard musqué
d Moschusente
i Anatra muta; Anatra di Barberia

1467   *Cairina scutulata*
        Anseriformes - Anatidae
e   White-winged Duck; White-winged
        Wood-Duck (ISC); Wood Duck
f   Canard à ailes blanches
d   Weißflügelente
i   Anatra alibianche

1468   *Calamanthus campestris*
        Passeriformes - Pardalotidae
e   Rufous Calamanthus; Rufous
        Fieldwren
f   Séricorne roussâtre
i   Sericeo castano

1469   *Calamanthus campestris dorrie*
        Passeriformes - Pardalotidae
e   Dorre Island Rufous Fieldwren
        (ANZ)

1470   *Calamanthus campestris hartogi*
        Passeriformes - Pardalotidae
e   Dirk Hartog Island Rufous Fieldwren
        (ANZ)

1471   *Calamanthus campestris*
        montanellus
        Passeriformes - Pardalotidae
e   Western Wheatbelt Rufous Fieldwren
        (ANZ)

1472   *Calamanthus fuliginosus*
        Passeriformes - Pardalotidae
e   Striated Calamanthus; Striated
        Fieldwren (ANZ); Fieldwren
f   Séricorne strié
d   Feldhuscher
i   Sericeo striato

1473   *Calamonastes fasciolatus*
        Passeriformes - Cisticolidae
e   Barred Wren-Warbler; Barred
        Warbler (CSA); Barred Camaroptera;
        Thornbush Barred Wren-Warbler;
        Common Camaroptera; African
        Barred Warbler; Southern Barred
        Warbler
f   Camaroptère barrée
d   Bindensänger; Damara-Bindensänger
i   Camarottera barrata

1474   *Calamonastes simplex*
        Passeriformes - Cisticolidae
e   Grey Wren-Warbler; Grey
        Camaroptera; Bush-Warbler; Plain
        Bush-Warbler
f   Camaroptère modeste; Camaroptère
        du Miombo
i   Camarottera grigia

1475   *Calamonastes stierlingi*
        Passeriformes - Cisticolidae
e   Stierling's Barred-Warbler; Stierling's
        Barred Wren-Warbler; Stierling's
        Wren-Warbler; Barred Bush-
        Warbler; Eastern Barred Bush-
        Warbler; Woodland Barred Wren-
        Warbler
f   Camaroptère de Stierling
d   Stierling-Bindensänger
i   Camarottera di Stierling

1476   *Calamonastes undosus*
        Passeriformes - Cisticolidae
e   Pale Wren-Warbler; Pale
        Camaroptera; Miombo Bush-
        Warbler; Southern Grey Wren-
        Warbler
f   Caramoptère barrée

1477   *Calamospiza melanocorys*
        Passeriformes - Fringillidae
e   Lark Bunting
f   Bruant noir-et-blanc; Pinson noir-et-
        blanc
d   Prarieammer
i   Zigolo allodola americano

1478   *Calandrella acutirostris*
        Passeriformes - Alaudidae
e   Hume's Lark; Hume's Short-toed
        Lark
f   Alouette de Hume; Alouette
        calandrelle de Hume
d   Tibet-Lerche
i   Calandrella di Hume

1479   *Calandrella athensis*
        Passeriformes - Alaudidae
e   Athi Short-toed Lark; Kenya Short-
        toed Lark
f   Alouette d'Athi

d Athi-Lerche
i Calandrella dell'Athi

1480 *Calandrella blanfordi*
Passeriformes - Alaudidae
e Blanford's Lark; Blanford's Short-toed Lark
f Alouette de Blandford
d Blanford-Lerche
i Calandrella di Blanford

1481 *Calandrella brachydactyla*
Passeriformes - Alaudidae
e Greater Short-toed Lark; Short-toed Lark; Red-capped Lark; European Short-toed Lark
f Alouette calandrelle
d Kurzzehenlerche
i Calandrella; Calandrella araba; Calandrella comune

1482 *Calandrella cheleensis*
Passeriformes - Alaudidae
e Asian Short-toed Lark; Eastern Short-toed Lark; Mongolian Short-toed Lark; Asiatic Short-toed Lark; Grey Short-toed Lark; Lesser Sandlark
f Alouette de Swinhoe
d Tschili-Lerche
i Calandrella della Mongolia

1483 *Calandrella cinerea*
Passeriformes - Alaudidae
e Red-capped Lark; Greater Short-toed Lark; Rufous Short-toed Lark; African Short-toed Lark; Short-toed Lark
f Alouette cendrille
d Rotscheitellerche
i Calandrella capirossa; Calandrella araba

1484 *Calandrella erlangeri*
Passeriformes - Alaudidae
e Erlanger's Lark; Erlanger's Short-toed Lark; Ethiopian Short-toed Lark; Erlanger's Red-capped Lark
f Alouette d'Erlanger
d Erlanger-Lerche
i Calandrella di Erlanger

1485 *Calandrella raytal*
Passeriformes - Alaudidae
e Indian Short-toed Lark; Sandlark; Indian Sandlark; Raytal Lark; Asian Short-toed Lark
f Alouette raytal; Alouette des sables; Calandrelle des sables
d Raytallerche
i Calandrella del Raytal; Calandrella di Raytal; Calandrella del Retal

1486 *Calandrella rufescens*
Passeriformes - Alaudidae
e Lesser Short-toed Lark; Rufous Lark; Common Short-toed Lark; Short-toed Lark; Rufous Short-toed Lark; Rufous Sandlark
f Alouette pispolette
d Stummellerche
i Pispoletta; Calandrina

1487 *Calandrella somalica*
Passeriformes - Alaudidae
e Rufous Short-toed Lark; Somali Short-toed Lark
f Alouette roussâtre
d Benson-Lerche
i Calandrella somala

1488 *Calcarius lapponicus*
Passeriformes - Fringillidae
e Lapland Longspur (NA); Lapland Bunting; Eastern Lapland Bunting
f Bruant lapon; Bruant montain
d Spornammer
i Zigolo di Lapponia

1489 *Calcarius mccownii*
Passeriformes - Fringillidae
e McCown's Longspur
f Bruant de McCown; Bruant à collier gris
d Graubrustammer
i Zigolo di McCown

1490 *Calcarius ornatus*
Passeriformes - Fringillidae
e Chestnut-collared Longspur; Chestnut Longspur
f Bruant à ventre noir

d   Rothalsammer; Schmuckammer
i   Zigolo ornato

1491   *Calcarius pictus*
         Passeriformes - Fringillidae
e   Smith's Longspur
f   Bruant de Smith
d   Smith-Ammer; Gemalte Ammer
i   Zigolo di Smith

1492   *Calicalicus madagascariensis*
         Passeriformes - Corvidae
e   Red-tailed Vanga; Tit-Shrike; Red-
         tailed Shrike
f   Calicalic malgache; Vanga à queue
         rousse
d   Rotschwanzvanga; Meisenvanga
i   Vanga codarossa

1493   *Calicalicus rufocarpalis*
         Passeriformes - Corvidae
e   Red-shouldered Vanga

1494   *Calidris acuminata*
         Charadriiformes - Scolopacidae
e   Sharp-tailed Sandpiper; Sharp-tailed
         Stint; Siberian Pectoral Sandpiper;
         Asian Pectoral Sandpiper; Brown-
         eared Sandpiper; Little Greenshank
f   Bécasseau à queue pointue;
         Bécasseau à queue fine
d   Spitzschwanzstrandläufer
i   Piovanello siberiano; Piro-piro
         siberiano

1495   *Calidris alba*
         Charadriiformes - Scolopacidae
e   Sanderling; Pondbird (WI)
f   Bécasseau sanderling; Bécassine
         blanche (Ants); Gros Maringouin
         blanc (Ants)
d   Sanderling; Strandläufer
i   Piovanello tridattilo; Calidra

1496   *Calidris alpina*
         Charadriiformes - Scolopacidae
e   Dunlin; Oxbird; Red-backed
         Sandpiper (NA)
f   Bécasseau variable
d   Alpenstrandläufer
i   Piovanello pancianera

1497   *Calidris bairdii*
         Charadriiformes - Scolopacidae
e   Baird's Sandpiper
f   Bécasseau de Baird
d   Baird-Strandläufer; Bairds
         Strandläufer
i   Piovanello di Baird; Piro-piro de
         Baird; Gambecchio di Baird

1498   *Calidris canutus*
         Charadriiformes - Scolopacidae
e   Red Knot; Knot; Lesser Knot;
         Common Knot; European Knot;
         Iceland Sandpiper; East Siberian
         Sandpiper
f   Bécasseau maubèche
d   Knutt; Isländischer Strandläufer
i   Piovanello maggiore

1499   *Calidris ferruginea*
         Charadriiformes - Scolopacidae
e   Curlew Sandpiper; Curlew Stint;
         Pygmy Curlew
f   Bécasseau cocorli
d   Sichelstrandläufer
i   Piovanello

1500   *Calidris fuscicollis*
         Charadriiformes - Scolopacidae
e   White-rumped Sandpiper; Pondbird
         (WI); Bonaparte's Sandpiper
f   Bécasseau de Bonaparte; Bécasseau à
         croupion blanc; Bécassine queue-
         blanche (Ants); Bécasse (Ants)
d   Weißbürzelstrandläufer
i   Piovanello dorsobianco; Piro-piro
         dorsobianco; Piro-piro di Bonaparte

1501   *Calidris himantopus*
         Charadriiformes - Scolopacidae
e   Stilt Sandpiper; Pondbird (WI);
         Murky Sandpiper
f   Bécasseau échasse; Bécasseau à
         échasses; Chevalier pied-vert (Ants)
d   Bindenstrandläufer
i   Piovanello zampelunghe; Piro-piro
         zampelunghe

1502   *Calidris maritima*
         Charadriiformes - Scolopacidae
e   Purple Sandpiper

f Bécasseau violet
d Meerstrandläufer;
     Klippenstrandläufer
i Piovanello violetto

1503 *Calidris mauri*
     Charadriiformes - Scolopacidae
e Western Sandpiper; Pondbird (WI);
     Web-footed Sandpiper
f Bécasseau d'Alaska; Maubèche
     (Ants)
d Bergstrandläufer
i Piovanello occidentale; Piro-piro
     occidentale

1504 *Calidris melanotos*
     Charadriiformes - Scolopacidae
e Pectoral Sandpiper; Grassbird (WI);
     Pondbird (WI); American Pectoral
     Sandpiper
f Bécasseau tacheté; Bécasseau à
     poitrine cendrée; Bécasseau pectoral;
     Bécassine à poitrine noire (Ants);
     Dos-rouge (Ants)
d Graubruststrandläufer
i Piro-piro pettorale

1505 *Calidris minuta*
     Charadriiformes - Scolopacidae
e Little Stint; Lesser Stint
f Bécasseau minute
d Zwergstrandläufer
i Gambecchio; Gambecchio comune

1506 *Calidris minutilla*
     Charadriiformes - Scolopacidae
e Least Sandpiper; Pondbird (WI);
     American Stint; American Least-
     Sandpiper
f Bécasseau minuscule; Ricuit (Ants)
d Wiesenstrandläufer; Amerikanischer
     Wiesenstrandläufer
i Gambecchio americano

1507 *Calidris paramelanotos*
     Charadriiformes - Scolopacidae
e Cox's Sandpiper

1508 *Calidris ptilocnemis*
     Charadriiformes - Scolopacidae
e Rock Sandpiper

f Bécasseau des Aléoutiennes
d Bering-Strandläufer

1509 *Calidris pusilla*
     Charadriiformes - Scolopacidae
e Semipalmated Sandpiper; Pondbird
     (WI)
f Bécasseau semipalmé; Bécassine à
     pattes noires (Ants); Alouette (Ants);
     Marangouin (Ants)
d Sandstrandläufer
i Gambecchio semipalmato; Piro-piro
     semipalmato

1510 *Calidris ruficollis*
     Charadriiformes - Scolopacidae
e Red-necked Stint; Rufous-necked
     Stint; Rufous-necked Sandpiper;
     Little Stint; Eastern Little-Stint;
     Least Sandpiper
f Bécasseau à col roux; Bécasseau à
     cou roux
d Rotkehlstrandläufer
i Gambecchio collorosso

1511 *Calidris subminuta*
     Charadriiformes - Scolopacidae
e Long-toed Stint; Least Sandpiper;
     Long-toed Sandpiper; Middendorf's
     Stint
f Bécasseau à longs doigts
d Langzehenstrandläufer
i Gambecchio ditalunghe

1512 *Calidris temminckii*
     Charadriiformes - Scolopacidae
e Temminck's Stint
f Bécasseau de Temminck
d Temminck-Strandläufer
i Gambecchio nano

1513 *Calidris tenuirostris*
     Charadriiformes - Scolopacidae
e Great Knot; Eastern Knot; Greater
     Knot; Asiatic Knot; Great Sandpiper;
     Slender-billed Knot; Snipe-crowned
     Knot
f Bécasseau de l'Anadyr
d Anadyr-Knutt
i Piovanello beccosottile

1514   *Calidris virgata*
           Charadriiformes - Scolopacidae
  e   Surfbird
  f   Bécasseau du Ressac
  d   Gischtläufer
  i   Afriza

1515   *Caliechthrus leucolophus*
           Cuculiformes - Cuculidae
  e   White-crowned Koel
  f   Coucou à calotte blanche
  d   Weißscheitelkoëll
  i   Koel corona bianca

1516   *Callacanthis burtoni*
           Passeriformes - Fringillidae
  e   Spectacled Finch; Burton's Finch;
       Red-browed Finch; Red-browed
       Rose-Finch
  f   Roselin de Burton
  d   Burton-Gimpel; Stieglitz-Gimpel
  i   Fanello rosato di Burton

1517   *Callaeas cinerea*
           Passeriformes - Callaeatidae
  e   Kokako; Wattled Crow
  f   Glaucope cendré
  d   Graulappenvogel; Lappenkrähe
  i   Kokako

1518   **Callaeatidae**
           Passeriformes -
  e   New Zealand Wattlebirds;
       Wattlebirds; Kokakos
  f   Calléatidés
  d   Lappenvögel
  i   Calleidi

1519   *Callipepla californica*
           Galliformes - Odontophoridae
  e   California Quail; Plumed Quail;
       Valley Quail
  f   Colin de Californie
  d   Kalifornische Schopfwachtel;
       Schopfwachtel
  i   Quaglia della California; Colino della
       California

1520   *Callipepla douglasii*
           Galliformes - Odontophoridae
  e   Elegant Quail; Douglas's Quail

  f   Colin élégant
  d   Douglas-Wachtel
  i   Colino di Douglas

1521   *Callipepla gambelii*
           Galliformes - Odontophoridae
  e   Gambel's Quail; Desert Quail
  f   Colin de Gambel
  d   Helmwachtel; Gambel-Wachtel
  i   Quaglia di Gambel; Colino di
       Gambel

1522   *Callipepla squamata*
           Galliformes - Odontophoridae
  e   Scaled Quail; Blue Quail
  f   Colin écaillé
  d   Schuppenwachtel
  i   Quaglia squamata; Callipepla; Colino
       squamato

1523   *Calliphlox amethystina*
           Apodiformes - Trochilidae
  e   Amethyst Woodstar
  f   Colibri améthyste
  d   Amethystkolibri
  i   Colibrì dei boschi ametista

1524   *Calliphlox evelynae*
           Apodiformes - Trochilidae
  e   Bahama Woodstar; Hummingbird
       (WI); Godbird (WI)
  f   Colibri des Bahamas
  d   Bahama-Kolibri
  i   Colibrì delle Bahama

1525   *Callocephalon fimbriatum*
           Psittaciformes - Psittacidae
  e   Gang-Gang Cockatoo; Helmeted
       Cockatoo; Red-crowned Cockatoo;
       Red-headed Parrot; Gang-gang; Red-
       headed Cockatoo
  f   Cacatoès à tête rouge; Cacatoès
       ganga; Cacatoès gang-gang;
       Banksien à tête rouge
  d   Ganga-Kakadu; Helmkakadu;
       Rotkopfkakadu

1526   *Callonetta leucophrys*
           Anseriformes - Anatidae
  e   Ringed Teal; Ring-necked Teal; Red-
       shouldered Teal

f   Canard à collier noir
d   Rotschulterente
i   Alzavola anellata; Alzavola
     brasiliana

1527   ***Calochaetes coccineus***
       Passeriformes - Fringillidae
e   Vermilion Tanager
f   Tangara carmin
d   Mennigtangare
i   Tangara vermiglia

1528   ***Calocitta colliei***
       Passeriformes - Corvidae
e   Black-throated Magpie-Jay; Magpie-
     Jay; Collie's Magpie-Jay
f   Geai à face noire
d   Blauwangenhäher
i   Gazza dal ciuffo golanera

1529   ***Calocitta formosa***
       Passeriformes - Corvidae
e   White-throated Magpie-Jay; Magpie-
     Jay; Plumed Jay
f   Geai à face blanche
d   Langschwanzhäher
i   Gazza dal ciuffo golabianca

1530   ***Caloenas nicobarica***
       Columbiformes - Columbidae
e   Nicobar Pigeon; Vulturine Pigeon;
     Hackled Pigeon; White-tailed Pigeon
f   Nicobar à camail
d   Nikobar-Taube; Kragentaube;
     Mähnentaube
i   Colomba delle Nicobar; Colomba dal
     bavero

1531   ***Calonectris diomedea***
       Procellariiformes - Procellariidae
e   Cory's Shearwater; Mediterranean
     Shearwater; North Atlantic
     Shearwater
f   Puffin cendré
d   Gelbschnabelsturmtaucher
i   Berta maggiore

1532   ***Calonectris edwardsii***
       Procellariiformes - Procellariidae
e   Cape Verde Shearwater; Cory's
     Shearwater; Cape Verde Islands
     Shearwater
f   Puffin du Cap Vert
i   Berta maggiore di Edwards

1533   ***Calonectris leucomelas***
       Procellariiformes - Procellariidae
e   Streaked Shearwater; White-faced
     Shearwater; White-Faced Petrel;
     White-fronted Petrel; White-fronted
     Shearwater; Streak-headed
     Shearwater
f   Puffin leucomèle; Puffin rayé
d   Weißgesichtsturmtaucher;
     Weißgesichtsturmvogel
i   Berta facciabianca

1534   ***Caloperdix oculea***
       Galliformes - Phasianidae
e   Ferruginous Partridge; Ferruginous
     Wood-Partridge
f   Rouloul ocellé
d   Augenwachtel
i   Pernice ferruginosa

1535   ***Calorhamphus fuliginosus***
       Piciformes - Megalaimidae
e   Brown Barbet
f   Barbu fuligineux
d   Braunbartvogel;
     Glattschnabelbartvogel
i   Barbuto bruno

1536   ***Calothorax lucifer***
       Apodiformes - Trochilidae
e   Lucifer Hummingbird
f   Colibri lucifer
d   Luziferkolibri
i   Colibrì luciofero

1537   ***Calothorax pulcher***
       Apodiformes - Trochilidae
e   Beautiful Hummingbird
f   Colibri charmant
d   Schmuckkolibri
i   Colibrì bello

1538   ***Calypte anna***
       Apodiformes - Trochilidae
e   Anna's Hummingbird
f   Colibri d'Anna

d   Annas Kolibri
i   Colibrì Anna; Colibrì di Anna

1539 *Calypte costae*
      Apodiformes - Trochilidae
e   Costa's Hummingbird
f   Colibri de Costa
d   Costas-Kolibri
i   Colibrì di Costa

1540 *Calyptocichla serina*
      Passeriformes - Pycnonotidae
e   Golden Greenbul; Golden Bulbul;
      Serene Greenbul
f   Bulbul doré
d   Goldbülbül
i   Bulbul verdina

1541 *Calyptomena hosii*
      Passeriformes - Eurylaimidae
e   Hose's Broadbill; Magnificent
      Broadbill; Magnificent Green-
      Broadbill; Hose's Magnificent
      Broadbill; Blue-bellied Broadbill
f   Eurylaime de Hose
d   Azurbreitrachen; Azurebreitmaul;
      Blaubauchbreitrachen;
      Blaubauchsmaragdracke
i   Beccolargo verde magnifico;
      Eurilamo verde magnifico

1542 *Calyptomena viridis*
      Passeriformes - Eurylaimidae
e   Green Broadbill; Lesser Green-
      Broadbill
f   Eurylaime vert
d   Smaragdbreitrachen; Smaragdrachen;
      Smaragdbreitmaul
i   Beccolargo verde minore; Eurilamo
      verde minore

1543 *Calyptomena whiteheadi*
      Passeriformes - Eurylaimidae
e   Whitehead's Broadbill; Black-
      throated Green-Broadbill; Black-
      throated Broadbill
f   Eurylaime de Whitehead
d   Schwarzkehlbreitrachen;
      Lauchbreitrachen; Lauchbreitmaul
i   Beccolargo verde di Whitehead;
      Eurilamo verde di Whitehead

1544 *Calyptophilus frugivorus*
      Passeriformes - Thraupidae
e   Chat Tanager; Lowland Chat-
      Tanager
f   Cornichon (Ants); Tangara cornichon
d   Schmätzertangare
i   Tangara tordo

1545 *Calyptophilus tertius*
      Passeriformes - Thraupidae
e   Western Chat-Tanager; Highland
      Chat-Tanager

1546 *Calyptorhynchus banksii*
      Psittaciformes - Psittacidae
e   Red-tailed Black-Cockatoo; Banks's
      Red-tailed Black-Cockatoo;
      Banksian Cockatoo; Red-tailed
      Cockatoo; Banksian Red-tailed
      Cockatoo; Great-billed Black-
      Cockatoo; Great-billed Cockatoo;
      Black Cockatoo; Red-tailed
      Cockatoo
f   Cacatoès banksien
d   Banks-Kakadu; Bartkakadu;
      Rabenkakadu
i   Cacatua nero magnifico; Cacatua di
      Banks; Cacatua coda rossa

1547 *Calyptorhynchus banksii*
      graptogyne
      Psittaciformes - Psittacidae
e   South-Eastern Red-tailed Black-
      Cockatoo (ANZ)

1548 *Calyptorhynchus banksii naso*
      Psittaciformes - Psittacidae
e   South-Western Red-tailed Black-
      Cockatoo (ANZ)

1549 *Calyptorhynchus baudinii*
      Psittaciformes - Psittacidae
e   Long-billed Black-Cockatoo; White-
      tailed Black-Cockatoo; Baudin's
      Black-Cockatoo (ANZ)
f   Cacatoès de Baudin
i   Cacatua nero di Baudin

1550 *Calyptorhynchus funereus*
      Psittaciformes - Psittacidae
e   Yellow-tailed Black-Cockatoo; Black

Cockatoo; White-tailed Cockatoo;
Funereal Cockatoo; Yellow-eared
Cockatoo; Yellow-tailed Cockatoo;
White-tailed Black-Cockatoo; White-
eared Black-Cockatoo; Yellow-tinted
Black-Cockatoo
f Cacatoès funèbre
d Gelbohrrabenkakadu;
Gelbohrkakadu; Rußkakadu;
Weißohrrabenkakadu
i Cacatua nero codagialla; Cacauta
nero

1551 ***Calyptorhynchus lathami***
Psittaciformes - Psittacidae
e Glossy Black-Cockatoo; Glossy
Cockatoo; Solander's Cockatoo;
Leach's Black-Cockatoo; Leach's
Red-tailed Cockatoo; Latham's
Cockatoo
f Cacatoès de Latham
d Braunkopfkakadu; Braunköpfiger
Rabenkakadu; Solanders Kakadu
i Cacatua nero di Latham

1552 ***Calyptorhynchus lathami***
halmaturinus
Psittaciformes - Psittacidae
e Kangaroo Island Glossy Black-
Cockatoo (ANZ)

1553 ***Calyptorhynchus lathami lathami***
Psittaciformes - Psittacidae
e Eastern Glossy Black-Cockatoo
(ANZ)

1554 ***Calyptorhynchus latirostris***
Psittaciformes - Psittacidae
e Short-billed Black-Cockatoo;
Slender-billed Black-Cockatoo;
Carnaby's Black-Cockatoo (ANZ)
f Cacatoès à rectrices blanches
i Cacatua nero beccogiallo

1555 ***Calyptura cristata***
Passeriformes - Tyrannidae
e Kinglet Calyptura; Kinglet Cotinga
f Cotinga roitelet; Cotinga huppé
d Rubinkrönchen; Zwergkotinga
i Caliptura crestata

1556 ***Camarhynchus crassirostris***
Passeriformes - Fringillidae
e Vegetarian Finch; Vegetarian Tree-
Finch; Darwin's Tree-Finch
f Géospize crassirostre
d Dickschnabeldarwinfink
i Fringuello arboricola beccogrosso

1557 ***Camarhynchus heliobates***
Passeriformes - Fringillidae
e Mangrove Finch
f Géospize des mangroves
d Mangrovedarwinfink; Mangrovefink
i Fringuello arboricola delle
mangrovie

1558 ***Camarhynchus pallidus***
Passeriformes - Fringillidae
e Woodpecker Finch
f Géospize pique-bois; Pinson-pic
d Spechtfink; Stocherfink
i Fringuello arboricola picchio

1559 ***Camarhynchus parvulus***
Passeriformes - Fringillidae
e Small Tree-Finch; Small
Insectivorous Tree-Finch
f Géospize minuscule
d Zwergdarwinfink; Kleiner Baumfink
i Fringuello arboricola piccolo

1560 ***Camarhynchus pauper***
Passeriformes - Fringillidae
e Medium Tree-Finch; Santa Maria
Tree-Finch; Charles's Insectivorous
Tree-Finch; Charles's Tree-Finch;
Toreana Tree-Finch
f Géospize modeste
d Kleinschnabeldarwinfink
i Fringuello arboricola di Charles

1561 ***Camarhynchus psittacula***
Passeriformes - Fringillidae
e Large Tree-Finch; Large
Insectivorous Tree-Finch; Parrot
Finch
f Géospize psittacin
d Papageischnabeldarwinfink; Großer
Baumfink; Rindenfink
i Fringuello arboricola grosso

1562   ***Camaroptera brachyura***
        Passeriformes - Cisticolidae
   e   Green-backed Camaroptera; Bleating
        Warbler (CSA); Green-backed
        Bleating-Warbler (CSA); Grey-
        backed Cameroptera; Broad-tailed
        Camaroptera; Bleating Camaroptera;
        Bleating Bush-Warbler
   f   Camaroptère à tête grise
   d   Grünrückencamaroptera;
        Meckergrasmücke
   i   Camarottera dorsoverde

1563   ***Camaroptera brevicauda***
        Passeriformes - Cisticolidae
   e   Grey-backed Bleating-Warbler
        (CSA); Grey-backed Camaroptera;
        Bleeting Bush-Warbler
   f   Camaroptère à dos gris
   d   Meckergrasmücke;
        Graurückencamaroptera
   i   Camarottera dorsogrigio

1564   ***Camaroptera chloronota***
        Passeriformes - Cisticolidae
   e   Olive-green Camaroptera
   f   Camaroptère à dos vert
   d   Olivcamaroptera
   i   Camarottera verde-oliva

1565   ***Camaroptera harterti***
        Passeriformes - Cisticolidae
   e   Hartert's Camaroptera
   f   Camaroptère de Hartert
   i   Camarottera di Hartert

1566   ***Camaroptera superciliaris***
        Passeriformes - Cisticolidae
   e   Yellow-browed Camaroptera
   f   Camaroptère à sourcils jaunes;
        Camaroptère à sourcils
   d   Gelbbrauencamaroptera
   i   Camarottera dai sopraccigli gialli

1567   ***Campephaga flava***
        Passeriformes - Campephagidae
   e   Black Cuckoo-Shrike; African Black
        Cuckoo-Shrike
   f   Échenilleur à épaulettes jaunes
   d   Kuckuckswürger
   i   Mangiabruchi nero

1568   ***Campephaga lobata***
        Passeriformes - Campephagidae
   e   Ghana Cuckoo-Shrike; Wattled
        Cuckoo-Shrike; Western Wattled
        Cuckoo-Shrike; Western Cuckoo-
        Shrike
   f   Échenilleur à barbillons
   d   Lappenraupenfresser
   i   Mangiabruchi caruncolato

1569   ***Campephaga oriolina***
        Passeriformes - Campephagidae
   e   Oriole Cuckoo-Shrike; Eastern
        Wattled Cuckoo-Shrike; Eastern
        Cuckoo-Shrike
   f   Échenilleur loriot
   i   Mangiabruchi rigogolo

1570   ***Campephaga petiti***
        Passeriformes - Campephagidae
   e   Petit's Cuckoo-Shrike
   f   Échenilleur de Petit
   d   Kongo-Raupenfresser; Kongo-
        Kuckuckswürger
   i   Mangiabruchi di Petit

1571   ***Campephaga phoenicea***
        Passeriformes - Campephagidae
   e   Red-shouldered Cuckoo-Shrike;
        Black Cuckoo-Shrike
   f   Échenilleur à épaulettes rouges;
        Échenilleur à épaulettes
   d   Mohrenraupenfresser;
        Kuckuckswürger
   i   Mangiabruchi spallerosse

1572   ***Campephaga quiscalina***
        Passeriformes - Campephagidae
   e   Purple-throated Cuckoo-Shrike
   f   Échenilleur pourpré
   d   Purpurraupenfresser;
        Purpurkuckuckswürger
   i   Mangiabruchi golapurpurea

1573   **Campephagidae**
        Passeriformes
   e   Cuckoo-Shrikes
   f   Campephagidés; Échenilleurs
   d   Stachelbürzler; Kuckuckswürger;
        Raupenfresser
   i   Campefagidi

1574 *Campephilus bairdii*
Piciformes - Picidae
e   Cuban Ivory-billed Woodpecker

1575 *Campephilus gayaquilensis*
Piciformes - Picidae
e   Guayaquil Woodpecker
f   Pic de Guayaquil
d   Guayaquil-Specht
i   Picchio del Guayaquil

1576 *Campephilus guatemalensis*
Piciformes - Picidae
e   Pale-billed Woodpecker; Flint-billed Woodpecker
f   Pic à bec clair; Pic de Lesson
d   Königsspecht
i   Picchio di Guatemala

1577 *Campephilus haematogaster*
Piciformes - Picidae
e   Crimson-bellied Woodpecker
f   Pic superbe
d   Blutbauchspecht
i   Picchio ventrerosso

1578 *Campephilus imperialis*
Piciformes - Picidae
e   Imperial Woodpecker
f   Pic impérial
d   Kaiserspecht
i   Picchio imperiale

1579 *Campephilus leucopogon*
Piciformes - Picidae
e   Cream-backed Woodpecker
f   Pic à dos crème; Pic de Boé
d   Weißmantelspecht; Spitzhaubenspecht
i   Picchio dorsochiaro

1580 *Campephilus magellanicus*
Piciformes - Picidae
e   Magellanic Woodpecker
f   Pic de Magellan
d   Magellan-Specht
i   Picchio di Magellano

1581 *Campephilus melanoleucos*
Piciformes - Picidae
e   Crimson-crested Woodpecker

f   Pic de Malherbe
d   Schwarzkehlspecht; Rotschopfspecht
i   Picchio bianco e nero

1582 *Campephilus pollens*
Piciformes - Picidae
e   Powerful Woodpecker
f   Pic puissant
d   Zimtbindenspecht
i   Picchio poderosa

1583 *Campephilus principalis*
Piciformes - Picidae
e   Ivory-billed Woodpecker; Northern Ivory-billed Woodpecker
f   Pic à bec ivoire
d   Elfenbeinspecht
i   Picchio dal becco d'avorio; Becco d'avorio; Picchio dal becco color avorio

1584 *Campephilus robustus*
Piciformes - Picidae
e   Robust Woodpecker
f   Pic robuste
d   Scharlachkopfspecht
i   Picchio robusto

1585 *Campephilus rubricollis*
Piciformes - Picidae
e   Red-necked Woodpecker
f   Pic à cou rouge
d   Rothalsspecht
i   Picchio collorosso

1586 *Campephilus splendens*
Piciformes - Picidae
e   Splendid Woodpecker
i   Picchio

1587 *Campethera abingoni*
Piciformes - Picidae
e   Golden-tailed Woodpecker
f   Pic à queue dorée
d   Goldschwanzspecht
i   Picchio di Abington

1588 *Campethera bennettii*
Piciformes - Picidae
e   Bennett's Woodpecker
f   Pic de Bennett

d Bennett-Specht
i Picchio di Bennett

1589 *Campethera cailliautii*
     Piciformes - Picidae
e Green-backed Woodpecker; Little
     Spotted-Woodpecker (CSA)
f Pic de Cailliaut; Pic à dos vert
d Tüpfelspecht
i Picchio di Cailliaud

1590 *Campethera caroli*
     Piciformes - Picidae
e Brown-eared Woodpecker
f Pic à oreillons bruns
d Braunohrspecht
i Picchio orecchiebrune

1591 *Campethera maculosa*
     Piciformes - Picidae
e Little Green-Woodpecker; Western
     Golden-backed Woodpecker; African
     Golden-backed Woodpecker;
     Golden-backed Woodpecker
f Pic barré
d Goldmantelspecht
i Picchio maculato

1592 *Campethera mombassica*
     Piciformes - Picidae
e Mombasa Woodpecker
f Pic de Mombassa
i Picchio di Mombasa

1593 *Campethera nivosa*
     Piciformes - Picidae
e Buff-spotted Woodpecker
f Pic tacheté
d Termitenspecht
i Picchio a macchie fulve

1594 *Campethera notata*
     Piciformes - Picidae
e Knysna Woodpecker; Golden-tailed
     Woodpecker
f Pic tigré
d Natal-Specht; Xosa-Specht
i Picchio di Knysna

1595 *Campethera nubica*
     Piciformes - Picidae

e Nubian Woodpecker; Bennett's
     Woodpecker
f Pic de Nubie
d Strichelohrfleckenspecht
i Picchio di Nubia

1596 *Campethera punctuligera*
     Piciformes - Picidae
e Fine-spotted Woodpecker
f Pic à taches noires; Pic ponctué
d Pünktchenspecht
i Picchio punteggiato

1597 *Campethera scriptoricauda*
     Piciformes - Picidae
e Speckle-throated Woodpecker;
     Reichenow's Woodpecker
f Pic de Reichenow
d Schriftschwanzspecht
i Picchio di Reichenow

1598 *Campethera tullbergi*
     Piciformes - Picidae
e Tullberg's Woodpecker; Fine-banded
     Woodpecker (CSA)
f Pic de Tullberg
d Kehlbindenspecht
i Picchio di Tulberg

1599 *Campochaera sloetii*
     Passeriformes - Corvidae
e Golden Cuckoo-Shrike; Orange
     Cuckoo-Shrike; Golden Triller;
     Orange Triller
f Échenilleur doré
d Goldraupenfresser; Goldraupenesser

1600 *Camptorhynchus labradorius*
     Anseriformes - Anatidae
e Labrador Duck
f Eider de Labrador; Canard du
     Labrador
d Labrador-Ente
i Coracina dorata

1601 *Camptostoma imberbe*
     Passeriformes - Tyrannidae
e Northern Beardless-Tyrannulet;
     Northern Beardless-Flycastcher
f Tyranneau imberbe

d Chaparralfliegenstecher
i Tiranno imberbe del Nord

1602 *Camptostoma obsoletum*
Passeriformes - Tyrannidae
e Southern Beardless-Tyrannulet;
Southern Beardless-Flycatcher
f Tyranneau passegris
d Gelbkehlfliegenstecher
i Tiranno imberbe del Sud

1603 *Campylopterus curvipennis*
Apodiformes - Trochilidae
e Wedge-tailed Sabrewing; Curve-
winged Sabrewing
f Campyloptère pampa
d Nachtigallkolibri
i Campilottero codacuneata

1604 *Campylopterus duidae*
Apodiformes - Trochilidae
e Buff-breasted Sabrewing
f Campyloptère montagnard
d Bergdegenflügel
i Campilottero pettofulvo

1605 *Campylopterus ensipennis*
Apodiformes - Trochilidae
e White-tailed Sabrewing
f Campyloptère à queue blanche
d Weißschwanzdegenflügel
i Campilottero codabianca

1606 *Campylopterus excellens*
Apodiformes - Trochilidae
e Long-tailed Sabrewing; Tuxtla
Sabrewing
f Campyloptère de Wetmore
i Campilottero codalunga

1607 *Campylopterus falcatus*
Apodiformes - Trochilidae
e Lazuline Sabrewing
f Campyloptère lazulite
d Lasurdegenflügel
i Campilottero lazulino

1608 *Campylopterus hemileucurus*
Apodiformes - Trochilidae
e Violet Sabrewing; De Lattre's
Sabrewing

f Campyloptère violet
d Purpurdegenflügel
i Campilottero violetto

1609 *Campylopterus hyperythrus*
Apodiformes - Trochilidae
e Rufous-breasted Sabrewing; Tepui
Sabrewing
f Campyloptère rougeâtre
d Rostbauchdegenflügel
i Campilottero pettorossiccio

1610 *Campylopterus largipennis*
Apodiformes - Trochilidae
e Grey-breasted Sabrewing; Bronze-
shafted Sabrewing
f Campyloptère à ventre gris
d Graubrustdegenflügel
i Campilottero pettogrigio

1611 *Campylopterus phainopeplus*
Apodiformes - Trochilidae
e Santa Marta Sabrewing
f Campyloptère des Santa Marta
d Türkis-Degenflügel
i Campilottero di Santa Marta

1612 *Campylopterus rufus*
Apodiformes - Trochilidae
e Rufous Sabrewing
f Campyloptère roux
d Buntschwanzdegenflügel
i Campilottero rossiccio

1613 *Campylopterus villaviscensio*
Apodiformes - Trochilidae
e Napo Sabrewing; Villaviscenio
Sabrewing; Splendid Sabrewing
f Campyloptère du Napo
d Napo-Degenflügel
i Campilottero di Napo

1614 *Campylorhamphus falcularius*
Passeriformes - Dendrocolaptidae
e Black-billed Scythebill
f Grimpar à bec-en-faux
d Trauersensenschnabel
i Becco a scimitarra falcato

1615 *Campylorhamphus procurvoides*
Passeriformes - Dendrocolaptidae

e Curve-billed Scythebill
f Grimpar à bec courbe
d Dunkelsensenschnabel
i Becco a scimitarra ricurvo

1616 *Campylorhamphus pucherani*
Passeriformes - Dendrocolaptidae
e Greater Scythebill
f Grimpar de Pucheran
d Augenstreifsensenschnabel
i Becco a scimitarra maggiore

1617 *Campylorhamphus pusillus*
Passeriformes - Dendrocolaptidae
e Brown-billed Scythebill
f Grimpar à bec brun
d Braunsensenschnabel
i Becco a scimitarra piccolo

1618 *Campylorhamphus trochilirostris*
Passeriformes - Dendrocolaptidae
e Red-billed Scythebill; Common
Scythebill
f Grimpar à bec rouge
d Rotrückensensenschnabel;
Sichelsensenschnabel
i Becco a scimitarra colibri

1619 *Campylorhynchus albobrunneus*
Passeriformes - Certhiidae
e White-headed Wren; White-headed
Cactus-Wren
f Troglodyte à tête blanche
i Scricciolo testabianca

1620 *Campylorhynchus brunneicapillus*
Passeriformes - Certhiidae
e Cactus Wren; Northern Cactus-Wren
f Troglodyte des cactus
d Kaktuszaunkönig
i Scricciolo dei cactus

1621 *Campylorhynchus capistratus*
Passeriformes - Certhiidae
e Rufous-backed Wren

1622 *Campylorhynchus chiapensis*
Passeriformes - Certhiidae
e Giant Wren; Chiapas Wren; Chiapas
Cactus-Wren; Giant Cactus-Wren
f Troglodyte géant

d Riesenzaunkönig
i Scricciolo gigante

1623 *Campylorhynchus fasciatus*
Passeriformes - Certhiidae
e Fasciated Wren; Fasciated Cactus-
Wren
f Troglodyte fascié
d Bindenzaunkönig
i Scricciolo fasciato

1624 *Campylorhynchus griseus*
Passeriformes - Certhiidae
e Bicoloured Wren; Two-coloured
Wren; Bicoloured Cactus-Wren
f Troglodyte bicolore
d Brauenzaunkönig
i Scricciolo bicolore

1625 *Campylorhynchus gularis*
Passeriformes - Certhiidae
e Spotted Wren; Spotted Cactus-Wren;
Mexican Wren; Mexican Spotted-
Wren
f Troglodyte tacheté
d Bartzaunkönig
i Scricciolo macchiato

1626 *Campylorhynchus humilis*
Passeriformes - Certhiidae
e Sclater's Wren

1627 *Campylorhynchus jocosus*
Passeriformes - Certhiidae
e Boucard's Wren; Boucard's Cactus-
Wren; Boucard's Spotted-Wren
f Troglodyte de Boucard
d Harlekinzaunkönig
i Scricciolo di Boucard

1628 *Campylorhynchus megalopterus*
Passeriformes - Certhiidae
e Grey-barred Wren; Gray Cactus-
Wren (NA); Grey Wren; Great-
winged Wren; Grey-barred Cactus-
Wren
f Troglodyte zèbré
d Graubindenzaunkönig
i Scricciolo grigio barrato

1629 **Campylorhynchus nuchalis**
Passeriformes - Certhiidae
e Stripe-backed Wren; Stripe-backed
Cactus-Wren
f Troglodyte rayé
d Pantherzaunkönig
i Scricciolo dorsostriato

1630 **Campylorhynchus rufinucha**
Passeriformes - Certhiidae
e Rufous-naped Wren; Rufous-naped
Cactus-Wren
f Troglodyte à nuque rousse
d Rotnackenzaunkönig
i Scricciolo nucarossiccia

1631 **Campylorhynchus turdinus**
Passeriformes - Certhiidae
e Thrush-like Wren; Thrush-Wren;
Thrush-like Cactus-Wren; Plain-
breasted Wren
f Troglodyte grivelé
d Drosselzaunkönig
i Scricciolo tordo

1632 **Campylorhynchus yucatanicus**
Passeriformes - Certhiidae
e Yucatan Wren; Yucatan Cactus-
Wren; Yucatan Giant-Wren
f Troglodyte du Yucatan
d Yucatan-Zaunkönig
i Scricciolo dello Yucatan

1633 **Campylorhynchus zonatus**
Passeriformes - Certhiidae
e Band-backed Wren; Banded Cactus-
Wren; Barred Wren; Band-backed
Cactus-Wren; Banded-back Wren;
Barred Cactus-Wren
f Troglodyte zoné
d Bänderrückenzaunkönig
i Scricciolo barrato

1634 **Canirallus kioloides**
Gruiformes - Rallidae
e Kioloides Rail; Madagascar Grey-
throated Rail; Madagascar Woodrail;
Grey-throated Woodrail
f Râle à gorge blanche; Râle à front
gris

d Graukehlralle
i Rallo del Madagascar

1635 **Canirallus oculeus**
Gruiformes - Rallidae
e Grey-throated Rail
f Râle à gorge grise
d Augenralle
i Rallo golagrigia

1636 **Capito auratus**
Piciformes - Ramphastidae
e Gilded Barbet

1637 **Capito aurovirens**
Piciformes - Ramphastidae
e Scarlet-crowned Barbet
f Cabézon oranvert
d Olivrückenbartvogel;
Trauerbartvogel
i Barbuto corona scarlatta

1638 **Capito brunneipectus**
Piciformes - Ramphastidae
e Brown-chested Barbet
i Barbuto pettobruno

1639 **Capito dayi**
Piciformes - Ramphastidae
e Black-girdled Barbet
f Cabézon du Brésil
d Kehlbindenbartvogel
i Barbuto di Day

1640 **Capito hypoleucus**
Piciformes - Ramphastidae
e White-mantled Barbet
f Cabézon à manteau blanc
d Weißmantelbartvogel
i Barbuto dorsobianco

1641 **Capito maculicoronatus**
Piciformes - Ramphastidae
e Spot-crowned Barbet
f Cabézon à couronne tachetée
d Tropfenbartvogel
i Barbuto corona macchiata

1642 **Capito niger**
Piciformes - Ramphastidae
e Black-spotted Barbet

f   Cabézon tachteté
d   Tupfenbartvogel; Streifenbartvogel
i   Barbuto macchiato di nero

1643   *Capito quinticolor*
       Piciformes - Ramphastidae
e   Five-coloured Barbet
f   Caézon à cinq couleurs
d   Fünffarbenbartvogel
i   Barbuto cinquecolore

1644   *Capito squamatus*
       Piciformes - Ramphastidae
e   Orange-fronted Barbet
f   Cabézon à nuque blanche
d   Weißnackenbartvogel
i   Barbuto frontearacio

1645   *Capito wallacei*
       Piciformes - Ramphastidae
e   Scarlet-banded Barbet

1646   **Caprimulgidae**
       Caprimulgiformes
e   Nightjars
f   Engoulevents; Caprimulgidés
d   Nachtschwalben; Ziegenmelker
i   Caprimulgidi

1647   **Caprimulgiformes**
e   Goatsuckers
f   Caprimulgiformes
d   Schwalmvögel; Nachtschwalben-
       artige Vögel
i   Caprimulgiformi

1648   *Caprimulgus aegyptius*
       Caprimulgiformes - Caprimulgidae
e   Egyptian Nightjar
f   Engoulevent du Sahara; Engoulevent
       du désert
d   Pharaonenziegenmelker; Ägyptischer
       Ziegenmelker;
       Pharaonennachtschwalbe
i   Succiacapre isabellino

1649   *Caprimulgus affinis*
       Caprimulgiformes - Caprimulgidae
e   Savanna Nightjar; Franklin's Nightjar
       (ISC); Allied Nightjar (ISC);
       Savannah Nightjar

f   Engoulevent affin; Engoulevent de
       Franklin
d   Savannennachtschwalbe
i   Succiacapre della Sonda

1650   *Caprimulgus anthonyi*
       Caprimulgiformes - Caprimulgidae
e   Scrub Nightjar; Anthony's Nightjar
f   Engoulevent d'Anthony
d   Ecuador-Nachtschwalbe
i   Succiacapre di Anthony

1651   *Caprimulgus arizonae*
       Caprimulgiformes - Caprimulgidae
e   Western Whip-poor-will

1652   *Caprimulgus asiaticus*
       Caprimulgiformes - Caprimulgidae
e   Indian Nightjar; Common Indian
       Nightjar (ISC); Southern Common
       Indian Nightjar; Common Nightjar
       (ISC); Little Indian Nightjar
f   Engoulevent indien
d   Hindu-Nachtschwalbe
i   Succiacapre indiano

1653   *Caprimulgus atripennis*
       Caprimulgiformes - Caprimulgidae
e   Jerdon's Nightjar; Long-tailed
       Nightjar; Indian Long-tailed
       Nightjar; Ceylon Nightjar
f   Engoulevent de Jerdon
i   Succiacapre di Jerdon

1654   *Caprimulgus badius*
       Caprimulgiformes - Caprimulgidae
e   Yucatan Nightjar; Yucatan Tawny-
       collared Nightjar
f   Engoulevent maya
i   Succiacapre baio

1655   *Caprimulgus batesi*
       Caprimulgiformes - Caprimulgidae
e   Bates's Nightjar; Bates's Forest-
       Nightjar; Forest Nightjar
f   Engoulevent de Bates
d   Waldnachtschwalbe
i   Succiacapre di Bates

1656   *Caprimulgus binotatus*
       Caprimulgiformes - Caprimulgidae

e Brown Nightjar; Dusky Nightjar
(CSA)
f Engoulevent à deux taches
d Bootschwanznachtschwalbe
i Succiacapre bruno

1657 *Caprimulgus candicans*
Caprimulgiformes - Caprimulgidae
e White-winged Nightjar
f Engoulevent à ailes blanches
d Pelzeln-Nachtschwalbe
i Succiacapre alibianche

1658 *Caprimulgus carolinensis*
Caprimulgiformes - Caprimulgidae
e Chuck-will's-widow; Chuck-will's-
widow Nightjar
f Engoulevent de Caroline; Chouette
(Ants)
d Carolina-Nachtschwalbe
i Succiacapre della Carolina

1659 *Caprimulgus cayennensis*
Caprimulgiformes - Caprimulgidae
e White-tailed Nightjar; Cayenne
Nightjar
f Engoulevent coré; Coré (Ants); Cohé
(Ants)
d Weißschwanznachtschwalbe
i Succiacapre della Cayenna

1660 *Caprimulgus celebensis*
Caprimulgiformes - Caprimulgidae
e Sulawesi Nightjar

1661 *Caprimulgus centralasicus*
Caprimulgiformes - Caprimulgidae
e Vaurie's Nightjar; Xinjiang Nightjar;
Central Asian Nightjar; Chinese
Nightjar
f Engoulevent de Vaurie
d Vaurien-Nachtschwalbe
i Succiacapre di Vaurie

1662 *Caprimulgus clarus*
Caprimulgiformes - Caprimulgidae
e Slender-tailed Nightjar; Reichenow's
Nightjar
f Engoulevent de Reichenow
d Kurzschleppennachtschwalbe
i Succiacapre di Reichenow

1663 *Caprimulgus climacurus*
Caprimulgiformes - Caprimulgidae
e Long-tailed Nightjar
f Engoulevent à longue queue
d Schleppennachtschwalbe
i Succiacapre codalunga

1664 *Caprimulgus concretus*
Caprimulgiformes - Caprimulgidae
e Bonaparte's Nightjar
f Engoulevent de Bonaparte
d Sunda-Nachtschwalbe
i Succiacapre di Bonaparte

1665 *Caprimulgus cubanensis*
Caprimulgiformes - Caprimulgidae
e Greater Antillean Nightjar; Cuban
Nightjar; Antillean Nightjar
f Chouette (Ants); Engoulevent peut-
on-voir
d Kuba-Nachtschwalbe
i Succiacapre delle Grandi Antille

1666 *Caprimulgus donaldsoni*
Caprimulgiformes - Caprimulgidae
e Donaldson-Smith's Nightjar
f Engoulevent des épines
d Dornbuschnachtschwalbe
i Succiacapre di Donaldson

1667 *Caprimulgus ekmani*
Caprimulgiformes - Caprimulgidae
e Hispaniolan Nightjar

1668 *Caprimulgus enarratus*
Caprimulgiformes - Caprimulgidae
e Collared Nightjar
f Engoulevent à nuque rousse;
Engoulevent de Gray; Engoulevent à
collier
d Halsbandnachtschwalbe
i Succiacapre dal collare fulvo

1669 *Caprimulgus europaeus*
Caprimulgiformes - Caprimulgidae
e Eurasian Nightjar; European
Nightjar; Nightjar; Common Nightjar
f Engoulevent d´Europe
d Ziegenmelker; Nachtschwalbe
i Succiacapre europeo; Succiacapre;
Calcabotto; Nottolone

1670   **Caprimulgus eximius**
       Caprimulgiformes - Caprimulgidae
 e   Golden Nightjar
 f   Engoulevent doré
 d   Prachtnachtschwalbe
 i   Succiacapre dorato

1671   **Caprimulgus fossii**
       Caprimulgiformes - Caprimulgidae
 e   Square-tailed Nightjar
 f   Engoulevent du Mozambique
 d   Welwitch-Nachtschwalbe
 i   Succiacapre del Gabon

1672   **Caprimulgus fossii elwitschii**
       Caprimulgiformes - Caprimulgidae
 e   Gabon Nightjar (CSA); Mozambique
       Nightjar (CSA); Gabon Nightjar;
       Gaboon Nightjar
 d   Gabun-Nachtschwalbe

1673   **Caprimulgus fraenatus**
       Caprimulgiformes - Caprimulgidae
 e   Sombre Nightjar; Dusky Nightjar
       (CSA); Dark Nightjar; Northern
       Nightjar
 f   Engoulevent sombre
 d   Zügelnachtschwalbe
 i   Succiacapre modestus

1674   **Caprimulgus hirundinaceus**
       Caprimulgiformes - Caprimulgidae
 e   Pygmy Nightjar
 f   Engoulevent pygmée
 d   Spix-Nachtschwalbe
 i   Succiacapre pigmeo

1675   **Caprimulgus indicus**
       Caprimulgiformes - Caprimulgidae
 e   Grey Nightjar; Jungle Nightjar;
       Indian Jungle Nightjar; Ceylon
       Highland Nightjar; Japanese Nightjar
 f   Engoulevent jotaka
 d   Dschungelnachtschwalbe
 i   Succiacapre della giungla

1676   **Caprimulgus inornatus**
       Caprimulgiformes - Caprimulgidae
 e   Plain Nightjar
 f   Engoulevent terne

 d   Marmornachtschwalbe
 i   Succiacapre disadorno

1677   **Caprimulgus longirostris**
       Caprimulgiformes - Caprimulgidae
 e   Band-winged Nightjar
 f   Engoulevent à miroir
 d   Spiegelnachtschwalbe
 i   Succiacapre beccolungo

1678   **Caprimulgus macrurus**
       Caprimulgiformes - Caprimulgidae
 e   Large-tailed Nightjar; Long-tailed
       Nightjar; White-tailed Nightjar
 f   Engoulevent de Horsfield
 d   Horsfield-Nachtschwalbe
 i   Succiacapre codalarga

1679   **Caprimulgus maculicaudus**
       Caprimulgiformes - Caprimulgidae
 e   Spot-tailed Nightjar; Pitsweet
 f   Engoulevent à queue etoilée
 d   Fleckschwanznachtschwalbe
 i   Succiacapre codamacchiata

1680   **Caprimulgus maculosus**
       Caprimulgiformes - Caprimulgidae
 e   Cayenne Nightjar
 f   Engoulevent de Guyane
 d   Cayenne-Nachtschwalbe
 i   Succiacapre maculato

1681   **Caprimulgus madagascariensis**
       Caprimulgiformes - Caprimulgidae
 e   Madagascar Nightjar
 f   Engoulevent malgache
 d   Madagaskar-Nachtschwalbe
 i   Succiacapre del Madagascar

1682   **Caprimulgus mahrattensis**
       Caprimulgiformes - Caprimulgidae
 e   Sykes's Nightjar; Sind Nightjar
 f   Engoulevent de Sykes
 d   Sind-Nachtschwalbe
 i   Succiacapre di Sykes

1683   **Caprimulgus manillensis**
       Caprimulgiformes - Caprimulgidae
 e   Philippine Nightjar
 f   Engoulevent des Philippines
 i   Succiacapre delle Filippine

1684 **Caprimulgus mininus**
Caprimulgiformes - Caprimulgidae
e Ruddy Nightjar

1685 **Caprimulgus natalensis**
Caprimulgiformes - Caprimulgidae
e Swamp Nightjar; African White-
tailed Nightjar (CSA); Natal Nightjar
(CSA); White-tailed Nightjar
f Engoulevent du Natal; Engoulevent à
queue blanche
d Natal-Nachtschwalbe
i Succiacapre delle paludi

1686 **Caprimulgus nigrescens**
Caprimulgiformes - Caprimulgidae
e Blackish Nightjar
f Engoulevent noirâtre
d Trauernachtschwalbe
i Succiacapre fosco sudamericano

1687 **Caprimulgus nigriscapularis**
Caprimulgiformes - Caprimulgidae
e Dusky Nightjar
f Engoulevent à épaulettes noires
i Succiacapre spallenere

1688 **Caprimulgus noctitherus**
Caprimulgiformes - Caprimulgidae
e Puerto Rican Nightjar; Puerto Rican
Whip-poor-will
f Engoulevent de Porto Rico
d Puertorico-Nachtschwalbe
i Succiacapre di Portorico

1689 **Caprimulgus nubicus**
Caprimulgiformes - Caprimulgidae
e Nubian Nightjar
f Engoulevent de Nubie
d Nubischer Ziegenmelker; Bajuda-
Nachtschwalbe
i Succiacapre della Nubia; Succiacapre
di Numibia

1690 **Caprimulgus otiosus**
Caprimulgiformes - Caprimulgidae
e St. Lucia Nightjar; St. Lucian
Nightjar
f Engoulevent de Sainte-Lucie

1691 **Caprimulgus parvulus**
Caprimulgiformes - Caprimulgidae
e Little Nightjar
f Engoulevent des bois
d Zwergnachtschwalbe
i Succiacapre piccolo

1692 **Caprimulgus pectoralis**
Caprimulgiformes - Caprimulgidae
e Fiery-necked Nightjar; Cuvier's
Nightjar; Black-shouldered Nightjar;
African Dusky Nightjar; Dusky
Nightjar
f Engoulevent musicien
d Rotnackennachtschwalbe;
Pfeifnachtschwalbe
i Succiacapre bruno africano

1693 **Caprimulgus poliocephalus**
Caprimulgiformes - Caprimulgidae
e Montane Nightjar; Abyssinian
Nightjar; Mountain Nightjar
f Engoulevent d'Abyssinie
d Höhennachtschwalbe
i Succiacapre dell'Abissinia

1694 **Caprimulgus prigoginei**
Caprimulgiformes - Caprimulgidae
e Prigogene's Nightjar; Itombwe
Nightjar
f Engoulevent de Prigogine

1695 **Caprimulgus pulchellus**
Caprimulgiformes - Caprimulgidae
e Salvadori's Nightjar
f Engoulevent de Salvadori
d Salvadori-Nachtschwalbe
i Succiacapre di Salvadori

1696 **Caprimulgus ridgwayi**
Caprimulgiformes - Caprimulgidae
e Buff-collared Nightjar; Ridgeway's
Whip-poor-will; Cookacheea;
Tucuchillo
f Engoulevent de Ridgway
d Braunhalsnachtschwalbe
i Succiacapre di Ridgway

1697 **Caprimulgus ruficollis**
Caprimulgiformes - Caprimulgidae
e Red-necked Nightjar

f   Engoulevent à collier roux;
      Engoulevent à cou brun
d   Rothalsziegenmelker;
      Rothalsnachtschwalbe
i   Succiacapre collorosso

1698   ***Caprimulgus rufigena***
        Caprimulgiformes - Caprimulgidae
e   Rufous-cheeked Nightjar
f   Engoulevent à joues rousses
d   Rostwangennachtschwalbe
i   Succiacapre guancerosse

1699   ***Caprimulgus rufus***
        Caprimulgiformes - Caprimulgidae
e   Rufous Nightjar
f   Cent-coups-de couteau (Ants);
      Jacques-pas-papa-pouw (Ants);
      Engoulevent roux
s   Rostnachtschwalbe
i   Succiacapre rossiccio

1700   ***Caprimulgus ruwenzorii***
        Caprimulgiformes - Caprimulgidae
e   Ruwenzori Nightjar; Rwenzori
      Nightjar
f   Engoulevent du Ruwenzori
i   Succiacapre del Ruwenzori

1701   ***Caprimulgus salvini***
        Caprimulgiformes - Caprimulgidae
e   Tawny-collared Nightjar;
      Chipwillow; Salvin's Chuckwill
f   Engoulevent de Salvin
d   Salvin-Nachtschwalbe
i   Succiacapre dal collare fulvo

1702   ***Caprimulgus saturatus***
        Caprimulgiformes - Caprimulgidae
e   Dusky Nightjar; Sooty Nightjar
f   Engoulevent montagnard
d   Bergnachtschwalbe
i   Succiacapre fuligginoso

1703   ***Caprimulgus sericocaudatus***
        Caprimulgiformes - Caprimulgidae
e   Silky-tailed Nightjar
f   Engoulevent à queue de soie
d   Seidennachtschwalbe
i   Succiacappre codadiseta

1704   ***Caprimulgus solala***
        Caprimulgiformes - Caprimulgidae
e   Nechisar Nightjar

1705   ***Caprimulgus stellatus***
        Caprimulgiformes - Caprimulgidae
e   Star-spotted Nightjar
f   Engoulevent étoilé
d   Sternnachtschwalbe
i   Succiacapre stellato

1706   ***Caprimulgus tristigma***
        Caprimulgiformes - Caprimulgidae
e   Freckled Nightjar; West African
      Freckled Nightjar; Rock Nightjar
f   Engoulevent pointillé
d   Fleckennachtschwalbe
i   Succiacapre lentigginoso

1707   ***Caprimulgus vociferus***
        Caprimulgiformes - Caprimulgidae
e   Whip-poor-will; Eastern Whip-poor-
      will; Whip-poor-will Nightjar
f   Engoulevent bois-pourri
d   Whip-poor-will; Amerikanischer
      Ziegenmelker
i   Succiacapre vocifero; Astronomo

1708   ***Caprimulgus whitelyi***
        Caprimulgiformes - Caprimulgidae
e   Roraiman Nightjar
f   Engoulevent du Roraima
d   Roraima-Nachtschwalbe
i   Succiacapre di Roraima

1709   ***Capsiempis flaveola***
        Passeriformes - Tyrannidae
e   Yellow Tyrannulet
f   Tyranneau flavéole
d   Zitronentyrann
i   Tiranno piccolo giallo

1710   ***Cardellina rubrifrons***
        Passeriformes - Fringillidae
e   Red-faced Warbler; American Red-
      faced Warbler
f   Paruline à face rouge
d   Dreifarbenwaldsänger
i   Parula cardelina

1711 Cardinalidae
Passseriformes
e Cardinals; Cardinal-Grosbeaks
f Cardinaux; Cardinalidés
d Kardinäle
i Cardinalidi

1712 *Cardinalis cardinalis*
Passeriformes - Cardinalidae
e Northern Cardinal; Cardinal;
Common Cardinal
f Cardinal rouge; Cardinal de virginie;
Cardinal huppé
d Rotkardinal; Roter Kardinal
i Cardinale rosso; Usignolo della
Virginia

1713 *Cardinalis carneus*
Passeriformes - Cardinalidae
e Long-crested Cardinal

1714 *Cardinalis phoeniceus*
Passeriformes - Cardinalidae
e Vermilion Cardinal; Venezuelan
Cardinal
f Cardinal vermillon
d Purpurkardinal
i Cardinale vermiglio

1715 *Cardinalis sinuatus*
Passeriformes - Cardinalidae
e Pyrrhuloxia
f Cardinal pyrrhuloxia
d Schmalschnabelkardinal
i Cardinale rosa

1716 *Carduelis ambigua*
Passeriformes - Fringillidae
e Black-headed Greenfinch; Yunnan
Greenfinch; Oustalet's Greenfinch;
Oustalet's Black-headed Greenfinch;
Yunnan Black-headed Greenfinch;
Tibetan Greenfinch
f Verdier d'Oustalet
d Schwarzkopfgrünling
i Verdone dello Yunnan

1717 *Carduelis atrata*
Passeriformes - Fringillidae
e Black Siskin
f Chardonneret noir

d Schwarzzeisig
i Negrito della Bolivia

1718 *Carduelis atriceps*
Passeriformes - Fringillidae
e Black-capped Siskin
f Tarin sombre
d Guatemala-Zeisig
i Lucherino capinero

1719 *Carduelis barbata*
Passeriformes - Fringillidae
e Black-chinned Siskin
f Chardonneret à menton noir
d Bartzeisig
i Lucherino mentonero

1720 *Carduelis cannabina*
Passeriformes - Fringillidae
e Eurasian Linnet; Linnet; European
Linnet; Common Linnet; Brown
Linnet
f Linotte commune; Linotte; Linotte
mélodieuse; Linotte ordinaire;
Linotte des vignes
d Bluthänfling; Hänfling
i Fanello; Montanello; Gricciolo;
Fanello eurasiatico

1721 *Carduelis carduelis*
Passeriformes - Fringillidae
e European Goldfinch; Goldfinch;
Eurasian Goldfinch; Eastern
Goldfinch; Grey-crowned Goldfinch
f Chardonneret; Chardonneret élégant;
Chardonneret commun
d Stieglitz; Distelfink;
Graukopfdistelfink
i Cardellino; Cardello; Calderugio;
Carderino; Carderugio; Cardellino
eurasiatico

1722 *Carduelis chloris*
Passeriformes - Fringillidae
e European Greenfinch; Greenfinch;
Common Greenfinch; Western
Greenfinch
f Verdier; Verdier d'Europe; Verdier
commun
d Grünling; Grünfink

i Verdone; Calenzuolo; Verdello;
    Verdone europeo

1723 *Carduelis crassirostris*
    Passeriformes - Fringillidae
e Thick-billed Siskin
f Chardonneret à bec épais
d Dickschnabelzeisig
i Lucherino beccogrosso

1724 *Carduelis cucullata*
    Passeriformes - Fringillidae
e Red Siskin
f Chardonneret rouge
d Feuerzeisig; Kapuzenzeisig
i Cardinalino del Venezuela

1725 *Carduelis dominicensis*
    Passeriformes - Fringillidae
e Antillean Siskin; Hispaniola Siskin;
    Hispaniola Goldfinch
f Petit Serin (Ants); Chardonneret des
    Antilles
d Haiti-Zeisig
i Lucherino delle Antille

1726 *Carduelis exilipes*
    Passeriformes - Fringillidae
e Hoary Redpoll (NA)

1727 *Carduelis flammea*
    Passeriformes - Fringillidae
e Common Redpoll; Redpoll; Holarctic
    Redpoll; Lesser Redpoll; Mealy
    Redpoll
f Sizerin flammé; Sizerin à tête rouge;
    Sizerin boréal
d Birkenzeisig
i Organetto; Fanello eurasiatico

1728 *Carduelis flammea cabaret*
    Passeriformes - Fringillidae
e Lesser Redpoll
f Linotte mélodieuse
d Birkenzeisig
i Organetto minore

1729 *Carduelis flammea rostrata*
    Passeriformes - Fringillidae
e Greenland Redpoll
d Birkenzeisig

1730 *Carduelis flavirostris*
    Passeriformes - Fringillidae
e Twite; Eurasian Twite; Tibetan Twite
f Linotte à bec jaune; Linotte
    montagnarde; Linotte de montagne;
    Grosbec à gorge rousse; Grosbec de
    montagne
d Berghänfling
i Fanello nordico

1731 *Carduelis hornemanni*
    Passeriformes - Fringillidae
e Hoary Redpoll (NA); Arctic Redpoll;
    Hornemann's Redpoll
f Sizerin blanchâtre; Sizerin
    groenlandais
d Polarbirkenzeisig; Schneezeisig
i Organetto artico

1732 *Carduelis johannis*
    Passeriformes - Fringillidae
e Warsangli Linnet
f Linotte de Warsangli
d Somali-Hänfling
i Fanello di Warsangli

1733 *Carduelis lawrencei*
    Passeriformes - Fringillidae
e Lawrence's Goldfinch
f Chardonneret gris
d Maskenzeisig
i Cardellino di Lawrence

1734 *Carduelis magellanica*
    Passeriformes - Fringillidae
e Hooded Siskin
f Chardonneret de Magellan
d Magellan-Zeisig
i Lucherino di Magellano

1735 *Carduelis monguilloti*
    Passeriformes - Fringillidae
e Vietnamese Greenfinch; Vietnam
    Greenfinch
f Verdier du Vietnam
i Verdone del Vietnam

1736 *Carduelis notata*
    Passeriformes - Fringillidae
e Black-headed Siskin; Neotropic
    Black-headed Siskin

f Chardonneret à tête noire
d Schwarzbrustzeisig
i Lucherino testanera

1737 *Carduelis olivacea*
Passeriformes - Fringillidae
e Olivaceous Siskin
f Chardonneret olivâtre
d Olivzeisig
i Lucherino olivaceo

1738 *Carduelis pinus*
Passeriformes - Fringillidae
e Pine Siskin; Siskin
f Tarin des pins; Chardonneret des pins
d Fichtenzeisig
i Lucherino dei pini

1739 *Carduelis psaltria*
Passeriformes - Fringillidae
e Lesser Goldfinch; Dark-backed
Goldfinch; Arkansas Goldfinch
(NA); Green-backed Goldfinch;
Dark-backed Greenfinch
f Chardonneret mineur
d Mexikaner Zeisig; Arkansas-Zeisig
i Cardellino minore americano

1740 *Carduelis siemiradzkii*
Passeriformes - Fringillidae
e Saffron Siskin
f Chardonneret safran
d Safranzeisig
i Lucherino zafferano

1741 *Carduelis sinica*
Passeriformes - Fringillidae
e Grey-capped Greenfinch; Oriental
Greenfinch; Chinese Greenfinch;
Japanese Greenfinch
f Verdier de Chine
d China-Grünling; Chinesen-Grünling;
Bindengrünling; Sibirischer Grünling
i Verdone della Cina

1742 *Carduelis spinescens*
Passeriformes - Fringillidae
e Andean Siskin
f Chardonneret des Andes
d Anden-Zeisig
i Lucherino delle Ande

1743 *Carduelis spinoides*
Passeriformes - Fringillidae
e Yellow-breasted Greenfinch; Black-
headed Greenfinch; Himalayan
Greenfinch; Himalayan Goldfinch
f Verdier de l'Himalaya; Tarin de
l'Himalaya
d Himalaja-Grünling
i Verdone pettogiallo

1744 *Carduelis spinus*
Passeriformes - Fringillidae
e Eurasian Siskin; Siskin; Common
Siskin; Spruce Siskin
f Tarin des aulnes
d Erlenzeisig; Zeisig
i Lucarino; Lucherino; Lucherino
comune

1745 *Carduelis tristis*
Passeriformes - Fringillidae
e American Goldfinch
f Chardonneret jaune
d Trauerzeisig; Goldzeisig
i Cardellino americano

1746 *Carduelis uropygialis*
Passeriformes - Fringillidae
e Yellow-rumped Siskin
f Chardonneret à croupion jaune
d Kordillerenzeisig
i Lucherino gropponegiallo

1747 *Carduelis xanthogastra*
Passeriformes - Fringillidae
e Yellow-bellied Siskin
f Chardonneret à ventre jaune
d Gelbbauchzeisig
i Lucherino ventregiallo

1748 *Carduelis yarrellii*
Passeriformes - Fringillidae
e Yellow-faced Siskin; Yarrell's Siskin
f Chardonneret de Yarrell
d Yarrell-Zeisig
i Lucherino facciagialla

1749 *Carduelis yemenensis*
Passeriformes - Fringillidae
e Yemen Linnet; Linnet; Yemeni
Linnet; Arabian Linnet

f Linotte du Yémen
d Jemen-Hänfling
i Fanello dello Yemen

1750 *Cariama cristata*
Gruiformes - Cariamidae
e Red-legged Seriema; Cariama;
Seriema; Crested Seriema
f Cariama huppé
d Seriema
i Seriema crestato

1751 Cariamidae
Gruiformes
e Seriemas
f Cariamidés
d Seriemas
i Cariamidi

1752 *Caridonax fulgidus*
Coraciiformes - Halcyonidae
e White-rumped Kingfisher; Glittering
Kingfisher; Blue-and-white
Kingfisher
f Martin-chasseur étincelant
d Atlas-Liest
i Martin pescatore dal groppone
bianco

1753 *Carpococcyx radiatus*
Cuculiformes - Cuculidae
e Bornean Ground-Cuckoo; Sunda
Ground-Cuckoo; Malayan Ground-
Cuckoo; Ground-Cuckoo; Malay
Ground-Cuckoo; Malaysian Ground-
Cuckoo; Green-billed Ground-
Cuckoo
f Calobate radieux
d Laufkuckuck
i Cuculo di terra orientale

1754 *Carpococcyx renauldi*
Cuculiformes - Cuculidae
e Coral-billed Ground-Cuckoo;
Renauld's Ground-Cuckoo; Red-
billed Ground-Cuckoo
f Calobate de l' Annam
d Renauld-Kuckuck
i Cuculo di terra beccorallino

1755 *Carpococcyx viridis*
Cuculiformes - Cuculidae
e Sumatran Ground-Cuckoo

1756 *Carpodacus amplus*
Passeriformes - Fringillidae
e Guadalupe House-Finch

1757 *Carpodacus cassinii*
Passeriformes - Fringillidae
e Cassin's Finch; Cassin's Purple Finch
f Roselin de Cassin
d Cassin-Gimpel; Bergpurpurfink
i Ciuffolotto di Cassin

1758 *Carpodacus edwardsii*
Passeriformes - Fringillidae
e Dark-rumped Rosefinch; Large
Rosefinch; Edward's Rosefinch;
Rosefinch; Pink-throated Rosefinch;
Ruddy Rosefinch
f Roselin d'Edwards
d Edwards-Gimpel
i Ciuffolotto di Milne-Edwards

1759 *Carpodacus eos*
Passeriformes - Fringillidae
e Pink-rumped Rosefinch;
Stresemann's Rosefinch; Dawn
Rosefinch
f Roselin de Stresemann
d Aurora-Gimpel
i Ciuffolotto di Stresemann

1760 *Carpodacus erythrinus*
Passeriformes - Fringillidae
e Common Rosefinch; Rosefinch;
Scarlet Rosefinch; Scarlet Grosbeak;
Scarlet Finch; Hodgson's Rosefinch
f Roselin cramoisi
d Karmingimpel
i Ciuffolotto scarlatto

1761 *Carpodacus mcgregori*
Passeriformes - Fringillidae
e McGregor's House-Finch

1762 *Carpodacus mexicanus*
Passeriformes - Fringillidae
e House Finch; Common House-Finch
f Roselin familier

d Mexikanischer Karmingimpel;
Hausgimpel; Hausfink
i Ciuffolotto messicano

1763 *Carpodacus nipalensis*
Passeriformes - Fringillidae
e Dark-breasted Rosefinch; Dark
Rosefinch; Nepal Rosefinch
f Roselin sombre
d Dünnschnabelgimpel
i Ciuffolotto del Nepal

1764 *Carpodacus pulcherrimus*
Passeriformes - Fringillidae
e Beautiful Rosefinch
f Roselin superbe
d Schmuckgimpel
i Ciuffolotto rosato

1765 *Carpodacus puniceus*
Passeriformes - Fringillidae
e Red-fronted Rosefinch; Red-breasted
Rosefinch; Rose-breasted Rosefinch
f Roselin à gorge rouge
d Felsengimpel
i Ciuffolotto pettirosa

1766 *Carpodacus purpureus*
Passeriformes - Fringillidae
e Purple Finch
f Roselin pourpré
d Purpurgimpel; Purpurfink
i Ciuffolotto purpureo

1767 *Carpodacus rhodochlamys*
Passeriformes - Fringillidae
e Red-mantled Rosefinch; Pinkish-
backed Rosefinch
f Roselin à dos rouge
d Rosenmantelgimpel;
Rotmantelgimpel
i Ciuffolotto dal mantello rosso

1768 *Carpodacus rhodochlamys grandis*
Passeriformes - Fringillidae
e Blyth's Rosefinch; Himalayan
Rosefinch; Pink-backed Rosefinch
f Roselin de Blyth

1769 *Carpodacus roborowskii*
Passeriformes - Fringillidae

e Tibetan Rosefinch; Tibet Rosefinch;
Roborowski's Rosefinch; Tsinghai
Rosefinch
f Roselin de Roborowski
d Roborowski-Gimpel
i Ciuffolotto di Roborowski

1770 *Carpodacus rodochrous*
Passeriformes - Fringillidae
e Pink-browed Rosefinch; Pink-
mantled Rosefinch
f Roselin à sourcils roses
d Rosenaugengimpel
i Ciuffolotto dai sopraccigli rosati

1771 *Carpodacus rodopeplus*
Passeriformes - Fringillidae
e Spot-winged Rosefinch; Spotted
Rosefinch; Spotted-wing Rosefinch
f Roselin à ailes tachetées; Roselin
tacheté
d Fleckengimpel; Rosenbauchgimpel
i Ciuffolotto alimaculate

1772 *Carpodacus roseus*
Passeriformes - Fringillidae
e Pallas's Rosefinch; Pallas's Rosy-
Finch
f Roselin rose; Roselin de Pallas
d Rosengimpel
i Ciuffolotto scarlato di Pallas;
Ciuffolotto di Pallas; Ciuffolotto del
Pallas; Ciuffolotto roseo

1773 *Carpodacus rubescens*
Passeriformes - Fringillidae
e Blanford's Rosefinch; Crimson
Rosefinch; Long-tailed Rosefinch
f Roselin de Blanford
d Blanford-Gimpel
i Ciuffolotto di Blanford

1774 *Carpodacus rubicilla*
Passeriformes - Fringillidae
e Great Rosefinch; Caucasian Great-
Rosefinch; Spotted-crowned
Rosefinch; Severtzov's Goldfinch
f Roselin tachteté; Grand Roselin
tacheté
d Berggimpel

i Ciuffolotto scarlatto maggiore;
    Ciuffolotto rosso maggiore

1775 *Carpodacus rubicilloides*
       Passeriformes - Fringillidae
e Streaked Rosefinch; Eastern Great-
    Rosefinch; Straked Great-Rosefinch;
    Crimson-eared Rosefinch
f Roselin strié; Grand Roselin strié
d Alpengimpel
i Ciuffolotto scarlatto maggiore

1776 *Carpodacus synoicus*
       Passeriformes - Fringillidae
e Pale Rosefinch; Sinai Rosefinch
f Roselin du Sinaï
d Sinai-Gimpel; Einödgimpel; Sinai-
    Karmingimpel
i Ciuffolotto del Sinai; Ciuffolotto
    scrlatto del Sinai

1777 *Carpodacus thura*
       Passeriformes - Fringillidae
e White-browed Rosefinch; Thura's
    Rosefinch; Mademoiselle Thura's
    Rosefinch
f Roselin de Thura
d Thura-Gimpel; Waldgimpel
i Ciuffolotto dai sopraccigli bianchi

1778 *Carpodacus trifasciatus*
       Passeriformes - Fringillidae
e Three-banded Rosefinch
f Roselin à trois bandes
d Bindengimpel
i Ciuffolotto trifasciato

1779 *Carpodacus vinaceus*
       Passeriformes - Fringillidae
e Vinaceous Rosefinch
f Roselin vineux
d Burgundergimpel
i Ciuffolotto vinaceo

1780 *Carpodectes antoniae*
       Passeriformes - Cotingidae
e Yellow-billed Cotinga; Antonia's
    Cotinga
f Cotinga à bec jaune; Cotinga
    d'Antonia

d Gelbschnabelschmuckvogel
i Cotinga di Antonia

1781 *Carpodectes hopkei*
       Passeriformes - Cotingidae
e Black-tipped Cotinga; White
    Cotinga; Hopke's Cotinga
f Cotinga blanc
d Silberschmuckvogel
i Cotinga di Hopke

1782 *Carpodectes nitidus*
       Passeriformes - Cotingidae
e Snowy Cotinga
f Cotinga neigeux
d Schneeschmuckvogel
i Cotinga nivea

1783 *Carpornis cucullatus*
       Passeriformes - Tyrannidae
e Hooded Berryeater; Hooded Cotinga
f Cotinga coqueluchon
d Braunmantelbeerenfresser
i Cotinga dal cappuccio

1784 *Carpornis melanocephalus*
       Passeriformes - Tyrannidae
e Black-headed Berryeater; Black-
    headed Cotinga
f Cotinga à tête noire; Cotinga
    melanocéphale
d Schwarzkopfbeerenfresser
i Cotinga testanera

1785 *Carpospiza brachydactyla*
       Passeriformes - Passeridae
e Pale Rock-Finch; Pale Rock-
    Sparrow; Pale Petronia
f Moineau pâle; Moineau soulcie pâle
d Fahlsperling; Arabien-Steinsperling
i Passera lagia del Caucaso; Passera
    lagia chiara

1786 *Caryothraustes canadensis*
       Passeriformes - Cardinalidae
e Yellow-green Grosbeak; Green
    Grosbeak; Black-faced Grosbeak
f Cardinal flavert
d Gelbbauchkardinal
i Beccogrosso giallo-verde

1787   ***Caryothraustes humeralis***
        Passeriformes - Cardinalidae
    e   Yellow-shouldered Grosbeak
    f   Cardinal à épaulettes
    d   Gelbschulterkardinal
    I   Beccogrosso spallegialle

1788   ***Caryothraustes poliogaster***
        Passeriformes - Cardinalidae
    e   Black-faced Grosbeak
    f   Cardinal à ventre blanc
    i   Beccogrosso faccianera

1789   ***Casiornis fusca***
        Passeriformes - Tyrannidae
    e   Ash-throated Casiornis; Dusky
        Casiornis
    f   Casiorne à dos brun
    d   Fahlkehlcasiornis
    i   Piagnone bruno

1790   ***Casiornis rufa***
        Passeriformes - Tyrannidae
    e   Rufous Casiornis
    f   Casiorne roux; Casiorne rouge
    d   Zimtkehlcasiornis
    i   Piagnone rossiccio

1791   **Casuariidae**
        Casuariiformes
    e   Cassowaries
    f   Casuariidés; Casoars
    d   Kasuare; Kasuarvögel
    i   Casuaridi

1792   **Casuariiformes**
    e   Cassowaries; Emus
    f   Casuariformes
    d   Australien Landvögel; Kasuarvögel
    i   Casuariformi

1793   ***Casuarius bennetti***
        Casuariiformes - Casuariidae
    e   Dwarf Cassowary; Australian
        Cassowary; Bennett's Cassowary
    f   Casoar de Bennett
    d   Bennet-Kasuar
    i   Casuario di Bennett; Casuario nano

1794   ***Casuarius casuarius***
        Casuariiformes - Casuariidae

    e   Southern Cassowary; Common
        Cassowary; Cassowary; Two-wattled
        Cassowary; Double-wattled
        Cassowary
    f   Casoar à casque
    d   Helmkasuar
    i   Casuario comune; Casuario

1795   ***Casuarius casuarius johnsonii***
        Casuariiformes - Casuariidae
    e   Australian Southern-Cassowary
        (ANZ)

1796   ***Casuarius unappendiculatus***
        Casuariiformes - Casuariidae
    e   Northern Cassowary; Single-wattled
        Cassowary; One-wattled Cassowary
    f   Casoar unicaronculé
    d   Einlappenkasuar
    i   Casuario unappendicolato

1797   ***Catamblyrhynchus diadema***
        Passeriformes - Fringillidae
    e   Plushcap; Plush-capped Finch
    f   Tete-de-peluche couronné
    d   Plüschkopftangare; Samtkappenfink
    i   Fringuello dal diadema

1798   ***Catamenia analis***
        Passeriformes - Fringillidae
    e   Band-tailed Seedeater; Black-tailed
        Seedeater
    f   Cataménie maculée
    d   Spiegelcatamenie
    i   Beccasemi cadafasciata

1799   ***Catamenia homochroa***
        Passeriformes - Fringillidae
    e   Paramo Seedeater; Santa Marta
        Seedeater; Colombian Seedeater
    f   Cataménie du paramo
    d   Schlankschnabelcatamenie
    i   Beccasemi del Paramo

1800   ***Catamenia inornata***
        Passeriformes - Fringillidae
    e   Plain-coloured Seedeater
    f   Cataménie terne
    d   Schlichtcatamenie
    i   Beccasemi disadorno

1801   *Cataponera turdoides*
       Passeriformes - Muscicapidae
 e   Sulawesi Thrush; Cataponera Thrush;
       Celebes Mountain-Thrush; Sulawesi
       Mountain-Thrush
 f   Grive cataponère
 d   Schwarzbrauendrossel
 i   Tordo cataponera

1802   *Catharacta antarctica*
       Charadriiformes - Stercorariidae
 e   Southern Skua; Falkland Skua;
       Subantarctic Skua (CSA); Southern
       Great-Skua; Brown Skua; Tristan
       Skua; Hamilton's Skua; Lönnberg's
       Skua
 f   Labbe antarctique
 i   Stercorario antartico

1803   *Catharacta chilensis*
       Charadriiformes - Stercorariidae
 e   Chilean Skua
 d   Skua
 i   Stercorario del Cile

1804   *Catharacta maccormicki*
       Charadriiformes - Stercorariidae
 e   South Polar Skua; MacCormick's
       Skua; Antarctic Skua
 f   Labbe de McCormick
 d   Antarktische Raubmöwe;
       Raubmöwe; Antarktik-Skua
 i   Stercorario di McCormick

1805   *Catharacta skua*
       Charadriiformes - Stercorariidae
 e   Great Skua; Bonxie; Northern Skua;
       Subantacrtic Skua; Jaeger; Bonxie
 f   Grand Labbe
 d   Skua; Große Raubmöwe
 i   Stercorario maggiore; Labbo
       maggiore; Skua

1806   *Catharopeza bishopi*
       Passeriformes - Fringillidae
 e   Whistling Warbler; Whistling Bird
       (WI); Lesser Sourfrierebird (WI);
       Black-and-white Soufrierebird (WI)
 f   Paruline de Saint-Vincent
 d   Pfeifwaldsänger
 i   Dendroica di Bishop

1807   *Cathartes aura*
       Falconiformes - Cathartidae
 e   Turkey Vulture; Crow (WI); Carrion
       Crow (WI); John Crow (WI)
 f   Urubu à tête rouge; Vautour (Ants)
 d   Truthahngeier
 i   Avvoltoio collorosso; Avvoltaio dl
       collo rosso

1808   *Cathartes burrovianus*
       Falconiformes - Cathartidae
 e   Lesser Yellow-headed Vulture;
       Savannah Vulture; Savannah Yellow-
       headed Vulture
 f   Urubu à tête jaune
 d   Kleiner Gelbkopfgeier
 i   Avvoltoio testagialla minore;
       Avvoltaio dalla testa gialla

1809   *Cathartes melambrotus*
       Falconiformes - Cathartidae
 e   Greater Yellow-headed Vulture;
       Forest Vulture; Forest Yellow-
       headed Vulture
 f   Grand Urubu
 d   Großer Gelbkopfgeier
 i   Avvoltoio testagialla maggiore

1810   **Cathartidae**
       Falconiformes
 e   American Vultures; New World
       Vultures
 f   Urubus; Cathartidés
 d   Neuwelt Geier
 i   Catartidi

1811   *Catharus aurantiirostris*
       Passeriformes - Turdidae
 e   Orange-billed Nightingale-Thrush;
       Orange-billed Thrush
 f   Grive à bec orangé
 d   Goldschnabelmusendrossel
 i   Tordo usignolo beccoarancio; Cataro
       beccoarancio

1812   *Catharus bicknelli*
       Passeriformes - Turdidae
 e   Bicknell's Thrush
 f   Grive de Bicknell

1813 **Catharus dryas**
Passeriformes - Turdidae
e Spotted Nightingale-Thrush; Spotted
Thrush
f Grive tavelée
d Tropfenbrustmusendrossel
i Tordo usignolo maculato; Cataro
maculato

1814 **Catharus frantzii**
Passeriformes - Turdidae
e Ruddy-capped Nightingale-Thrush;
Highland Nightingale-Thrush;
Highland Thrush; Frantzius's Thrush;
Frantzius's Nightingale-Thrush
f Grive à calotte rousse
d Bergmusendrossel
i Tordo usignolo di Frantzius; Cataro
di Frantzius

1815 **Catharus fuscater**
Passeriformes - Turdidae
e Slaty-backed Nightingale-Thrush;
Slaty-backed Thrush
f Grive ardoisée
d Graurückenmusendrossel
i Tordo usignolo dorsoardesia; Cataro
dorsoardesia

1816 **Catharus fuscescens**
Passeriformes - Turdidae
e Veery; Wilson's Thrush
f Grivette fauve; Grive fauve
d Wiesendrossel; Wilson-Drossel;
Weidendrossel
i Cataro fosco; Tordo usignolo fosco;
Tordo usignolo bruno

1817 **Catharus gracilirostris**
Passeriformes - Turdidae
e Black-billed Nightingale-Thrush;
Black-billed Thrush; Slender-billed
Nightingale-Thrush
f Grive à bec noir
i Tordo usignolo beccofine; Cataro
beccofine

1818 **Catharus griseiceps**
Passeriformes - Turdidae
e Gray-headed Nightingale-Thrush
(NA)

1819 **Catharus guttatus**
Passeriformes - Turdidae
e Hermit Thrush
f Grive solitaire; Grivette solitaire;
Grive éremite
d Einsiedlerdrossel
i Tordo eremita; Cataro eremita

1820 **Catharus hellmayri**
Passeriformes - Turdidae
e Black-backed Nightingale-Thrush

1821 **Catharus mexicanus**
Passeriformes - Turdidae
e Black-headed Nightingale-Thrush;
Black-headed Thrush
f Grive à tête noire
d Schwarzkopfmusendrossel
i Tordo usignolo testanera; Cataro
testanera

1822 **Catharus minimus**
Passeriformes - Turdidae
e Grey-cheeked Thrush
f Grive à joues grises; Grivette à joues
grises
d Grauwangendrossel
i Tordo di Baird; Cataro di Baird;
Tordo nano; Tordo usignolo minimo

1823 **Catharus mustelinus**
Passeriformes - Turdidae
e Wood Thrush
f Grive des bois; Grivette des bois
d Walddrossel; Mäusedrossel
i Tordo dei boschi americano; Cataro
dei boschi americano; Tordo dei
boschi

1824 **Catharus occidentalis**
Passeriformes - Turdidae
e Russet Nightingale-Thrush; Russet
Thrush; Olive Nightingale-Thrush
f Grive roussâtre
d Braunkopfmusendrossel
i Tordo usignolo rossiccio; Cataro
rossiccio

1825 **Catharus ustulatus**
Passeriformes - Turdidae
e Swainson's Thrush; Russet-backed

Thrush; Olive-backed Thrush
f   Grive à dos olive; Grivette à dos
olive; Grivette olive; Merle de
Swainson; Grive de Swainson
d   Zwergdrossel; Swainson-Drossel
i   Tordo di Swainson; Cataro di
Swainson

1826   *Catherpes mexicanus*
Passeriformes - Certhiidae
e   Canyon Wren
f   Troglodyte des canyons
d   Schluchtzaunkönig
i   Scricciolo dei canyon

1827   *Catoptrophorus semipalmatus*
Charadriiformes - Scolopacidae
e   Willet; Laughing Jack (WI); Tell-
billy-willy (WI)
f   Chevalier semipalmé; Bécasse à aile
blanche (Ants); Aile blanche (Ants)
d   Entenschnepfe; Schlammtreter;
Auckland-Schnepfe
i   Totano semipalmato

1828   *Catreus wallichii*
Galliformes - Phasianidae
e   Cheer Pheasant; Chir Pheasant;
Wallace's Pheasant
f   Faisan de Wallich
d   Wallich-Fasan
i   Catreo

1829   *Celeus brachyurus*
Piciformes - Picidae
e   Rufous Woodpecker; Brown
Woodpecker
f   Pic brun
d   Ameisenspecht
i   Picchio rossastro

1830   *Celeus castaneus*
Piciformes - Picidae
e   Chestnut-coloured Woodpecker;
Chestnut Woodpecker
f   Pic roux
d   Kastanienspecht
i   Picchio castano minore

1831   *Celeus elegans*
Piciformes - Picidae

e   Chestnut Woodpecker; Elegant
Woodpecker
f   Pic mordoré
d   Fahlschopfspecht
i   Picchio castano maggiore

1832   *Celeus flavescens*
Piciformes - Picidae
e   Blond-crested Woodpecker
f   Pic ocré
d   Blondschopfspecht
i   Picchio crestabionda

1833   *Celeus flavus*
Piciformes - Picidae
e   Cream-coloured Woodpecker (NA)
f   Pic jaune
d   Strohspecht
i   Picchio giallo

1834   *Celeus grammicus*
Piciformes - Picidae
e   Scaly-breasted Woodpecker; Scale-
breasted Woodpecker
f   Pic de Verreaux
d   Gelbflankenspecht
i   Picchio pettoscaglioso

1835   *Celeus immaculatus*
Piciformes - Picidae
e   Immaculate Woodpecker

1836   *Celeus loricatus*
Piciformes - Picidae
e   Cinnamon Woodpecker
f   Pic cannellé
d   Zimtspecht
i   Picchio loricato

1837   *Celeus lugubris*
Piciformes - Picidae
e   Pale-crested Woodpecker
f   Pic à tête pâle
d   Blaßschopfspecht
i   Picchio crestachiara

1838   *Celeus spectabilis*
Piciformes - Picidae
e   Rufous-headed Woodpecker
f   Pic à tête rousse

d   Zimtkopfspecht
i   Picchio testarossicia

1839   ***Celeus torquatus***
       Piciformes - Picidae
e   Ringed Woodpecker
f   Pic à cravate noire
d   Schwarzbrustspecht
i   Picchio dalla pettorina nera

1840   ***Celeus undatus***
       Piciformes - Picidae
e   Waved Woodpecker
f   Pic ondé
d   Olivbürzelspecht
i   Picchio ondulato

1841   ***Centrocercus minimus***
       Galliformes - Phasianidae
e   Gunnison Sage-Grouse

1842   ***Centrocercus urophasianus***
       Galliformes - Phasianidae
e   Sage Grouse; Greater Sage-Grouse;
       Sage Hen; Sage Chicken
f   Tétras des armoises; Gélinotte des
       armoises
d   Beifußhuhnb
i   Gallo della salvia

1843   ***Centropus andamanensis***
       Cuculiformes - Cuculidae
e   Brown Coucal; Andaman Coucal
f   Coucal des Andaman
d   Andamanen-Kuckuck
i   Cuculo fagiano delle Andamane

1844   ***Centropus anselli***
       Cuculiformes - Cuculidae
e   Gabon Coucal; Gaboon Coucal
f   Coucal du Gabon
d   Ansell-Kuckuck
i   Cuculo fagiano del Gabon

1845   ***Centropus ateralbus***
       Cuculiformes - Cuculidae
e   Pied Coucal; New Britain Coucal;
       White-necked Coucal
f   Coucal atralbin
d   Weißkopfkuckuck
i   Cuculo fagiano bianco e nero

1846   ***Centropus bengalensis***
       Cuculiformes - Cuculidae
e   Lesser Coucal; Black Coucal (CSA)
f   Coucal rutin
d   Grillkuckuck; Bengalen-Kuckuck
i   Cuculo fagiano minore

1847   ***Centropus bernsteini***
       Cuculiformes - Cuculidae
e   Lesser Black-Coucal; Bernstein's
       Coucal; Scrub Coucal
f   Coucal de Bernstein
d   Bernstein-Kuckuck
i   Cuculo fagiano nero minore

1848   ***Centropus burchelli***
       Cuculiformes - Cuculidae
e   Burchell's Coucal; Rainbird (CSA);
       Vlei Lourie (CSA)
f   Coucal de Burchell
d   Tipupit
i   Cuculo fagiano di Burchell

1849   ***Centropus celebensis***
       Cuculiformes - Cuculidae
e   Bay Coucal; Celebes Coucal;
       Celeban Coucal
f   Coucal des Célèbes
d   Celebes-Kuckuck
i   Cuculo fagiano di Sulawesi

1850   ***Centropus chalybeus***
       Cuculiformes - Cuculidae
e   Biak Coucal; Biak Island Coucal;
       Biak Bronze-Coucal
f   Coucal de Biak
d   Atlas-Kuckuck
i   Cuculo fagiano di Biak

1851   ***Centropus chlororhynchus***
       Cuculiformes - Cuculidae
e   Green-billed Coucal; Ceylon Coucal;
       Sri Lanka Green-billed Coucal
f   Coucal de Ceylan
d   Ceylon-Kuckuck
i   Cuculo fagiano di Sri-Lanka

1852   ***Centropus cupreicaudus***
       Cuculiformes - Cuculidae
e   Coppery-tailed Coucal; Copper-tailed
       Coucal; Marsh Coucal

f Coucal des papyrus
d Angola-Mönchkuckuck
i Cuculo fagiano coda di rame

1853 *Centropus goliath*
Cuculiformes - Cuculidae
e Goliath Coucal; Large Coucal; Giant Coucal
f Coucal goliath
d Goliathkuckuck
i Cuculo fagiano gigante

1854 *Centropus grillii*
Cuculiformes - Cuculidae
e Black Coucal; Black-chested Coucal
f Coucal noir
d Schwarzer Spornkuckuck; Grillkuckuck
i Cuculo fagiano pettonero

1855 *Centropus leucogaster*
Cuculiformes - Cuculidae
e Black-throated Coucal
f Coucal à ventre blanc
d Weißbauchkuckuck
i Cuculo fagiano golanera

1856 *Centropus melanops*
Cuculiformes - Cuculidae
e Black-faced Coucal
f Coucal à face noire
d Maskenkuckuck; Brillenkuckuck
i Cuculo fagiano faccianera

1857 *Centropus menbeki*
Cuculiformes - Cuculidae
e Greater Black-Coucal; Great Coucal; Greater Coucal; Jungle Coucal; Menbek's Coucal
f Coucal menébeki
d Mohrenkuckuck
i Cuculo fagiano nero maggiore

1858 *Centropus milo*
Cuculiformes - Cuculidae
e Buff-headed Coucal
f Coucal à tête fauve
d Blaukopfkuckuck
i Cuculo fagiano testafulva

1859 *Centropus monachus*
Cuculiformes - Cuculidae
e Blue-headed Coucal
f Coucal à nuque bleue
d Mönchskuckuck
i Cuculo fagiano testablu

1860 *Centropus neumanni*
Cuculiformes - Cuculidae
e Neumann's Coucal; Smaller Black-throated Coucal
f Coucal de Neumann
i Cuculo fagiano di Neumann

1861 *Centropus nigrorufus*
Cuculiformes - Cuculidae
e Javan Coucal; Sunda Coucal
f Coucal noirou
d Sunda-Kuckuck
i Cuculo fagiano della Sonda

1862 *Centropus phasianinus*
Cuculiformes - Cuculidae
e Pheasant-Coucal; Common Coucal; Timor Coucal
f Coucal faisan
d Fasankuckuck
i Cuculo fagiano comune; Cuculo fagiano

1863 *Centropus rectunguis*
Cuculiformes - Cuculidae
e Short-toed Coucal
f Coucal de Strickland
d Kurzspornkuckuck
i Cuculo fagiano ditabrevi

1864 *Centropus senegalensis*
Cuculiformes - Cuculidae
e Senegal Coucal; Rainbird
f Coucal du Sénégal
d Senegal-Spornkuckuck; Spornkuckuck
i Cuculo fagiano del Senegal; Cuculo del Senegal; Cuculo dallo sperone del Senegal

1865 *Centropus sinensis*
Cuculiformes - Cuculidae
e Greater Coucal; Crow-Pheasant (ISC); Coucal (ISC); Common

Coucal; Large Coucal; Common
Crow-Pheasant
f   Grand Coucal
d   Heckenkuckuck
i   Cuculo fagiano maggiore

1866   *Centropus spilopterus*
Cuculiformes - Cuculidae
e   Kai Coucal; Moluccan Coucal
f   Coucal des Kai
d   Kei-Kuckuck
i   Cuculo fagiano di Kai

1867   *Centropus steerii*
Cuculiformes - Cuculidae
e   Black-hooded Coucal; Steere's
Coucal
f   Coucal de Steere
d   Mindoro-Kuckuck
i   Cuculo fagiano di Steere

1868   *Centropus superciliosus*
Cuculiformes - Cuculidae
e   White-browed Coucal; Rainbird
f   Coucal à sourcils blancs
d   Weißbrauenspornkuckuck
i   Cuculo faggiano dai sopraccigli
bianchi; Cuculo fagiano
sopracciglibianchi

1869   *Centropus toulou*
Cuculiformes - Cuculidae
e   Madagascar Coucal; Black Coucal
f   Coucal toulou; Coucal malgache
d   Tulu-Kuckuck
i   Cuculo fagiano tolù; Cuculo nero del
Madagascar

1870   *Centropus unirufus*
Cuculiformes - Cuculidae
e   Rufous Coucal
f   Coucal roux
d   Bambuskuckuck
i   Cuculo fagiano rossiccio

1871   *Centropus violaceus*
Cuculiformes - Cuculidae
e   Violaceous Coucal; Violet Coucal
f   Coucal violet
d   Purpurkuckuck
i   Cuculo fagiano violetto

1872   *Centropus viridis*
Cuculiformes - Cuculidae
e   Philippine Coucal; Green Coucal
f   Coucal vert
d   Philippinen-Kuckuck
i   Cuculo fagiano delle Filippine

1873   *Cephalopterus glabricollis*
Passeriformes - Cotingidae
e   Bare-necked Umbrellabird
f   Coracine ombrelle
d   Nacktkehlschirmvogel
i   Uccello parasole collonudo

1874   *Cephalopterus ornatus*
Passeriformes - Cotingidae
e   Amazonian Umbrellabird; Ornate
Umbrellabird; Umbrellabird
f   Coracine ornée
d   Schirmvogel;
Kurzlappenschirmvogel
i   Uccello parasole del Amazzonia;
Uccello parasole; Cefalottero adorno

1875   *Cephalopterus penduliger*
Passeriformes - Cotingidae
e   Long-wattled Umbrellabird
f   Coracine casquée
d   Langlappenschirmvogel
i   Uccello parasole caruncolato

1876   *Cephalopyrus flammiceps*
Passeriformes - Paridae
e   Fire-capped Tit; Fire-capped Tit-
Warbler
f   Rémiz tête-de-feu; Mésange à
couronne flammée
d   Flammenstirnchen; Rotscheitelmeise
i   Pendolino testadifiamma

1877   *Cepphus carbo*
Charadriiformes - Alcidae
e   Spectacled Guillemot; Bridled
Guillemot; Sooty Guillemot
f   Guillemot à lunettes
d   Brillenteiste
i   Uria dagli occhiali

1878   *Cepphus columba*
Charadriiformes - Alcidae
e   Pigeon Guillemot; Sea Pigeon

f   Guillemot colombin; Guillemot du
      Pacifique
d   Taubenteiste
i   Uria colomba

1879   *Cepphus grylle*
      Charadriiformes - Alcidae
e   Black Guillemot; Tystie
f   Guillemot à miroir
d   Grylteiste
i   Uria nera

1880   *Cepphus snowi*
      Charadriiformes - Alcidae
e   Kuril Guillemot

1881   *Ceratogymna albotibialis*
      Coraciiformes - Bucerotidae
e   White-thighed Hornbill
f   Calao à cuisses blanches
i   Bucero dai calzoni bianchi

1882   *Ceratogymna atrata*
      Coraciiformes - Bucerotidae
e   Black-casqued Hornbill; Black
      Wattled-Hornbill; Black-casqued
      Wattled-Hornbill; Wattled-Hornbill
f   Calao à casque noir
d   Keulenhornvogel
i   Bucero dal casco nero

1883   *Ceratogymna brevis*
      Coraciiformes - Bucerotidae
e   Silvery-cheeked Hornbill; Silver-
      cheeked Hornbill
f   Calao à joues argent
d   Schopfhornvogel;
      Silberwangenhornvogel
i   Bucero guanceargentata

1884   *Ceratogymna bucinator*
      Coraciiformes - Bucerotidae
e   Trumpeter Hornbill
f   Calao trompette
d   Trompeterhornvogel
i   Bucero trombettiere

1885   *Ceratogymna cylindricus*
      Coraciiformes - Bucerotidae
e   Brown-cheeked Hornbill; White-
      thighed Hornbill

f   Calao à joues brunes
d   Balabi-Hornvogel
i   Bucero guancebrune

1886   *Ceratogymna elata*
      Coraciiformes - Bucerotidae
e   Yellow-casqued Hornbill; Yellow-
      casqued Wattled-Hornbill
f   Calao à casque jaune
d   Goldhelmhornvogel
i   Bucero dal casco giallo

1887   *Ceratogymna fistulator*
      Coraciiformes - Bucerotidae
e   Piping Hornbill; Laughing Hornbill;
      Whistling Hornbill; White-tailed
      Hornbill
f   Calao siffleur
d   Schreihornvogel
i   Bucero zufolatore

1888   *Ceratogymna subcylindricus*
      Coraciiformes - Bucerotidae
e   Black-and-white-casqued Hornbill;
      Grey-cheeked Hornbill; Black-and-
      white Hornbill
f   Calao à joues grises
d   Grauwangenhornvogel
i   Bucero guancegrigge

1889   *Cercibis oxycerca*
      Ciconiiformes - Threskiornithidae
e   Sharp-tailed Ibis; Bare-faced Ibis;
      Long-tailed Ibis
f   Ibis à queue pointue
d   Langschwanzibis
i   Ibis codacuta

1890   *Cercococcyx mechowi*
      Cuculiformes - Cuculidae
e   Dusky Long-tailed Cuckoo;
      Mechow's Long-tailed Cuckoo
f   Coucou de Mechow
d   Schweifkuckuck
i   Cuculo codalunga bruno

1891   *Cercococcyx montanus*
      Cuculiformes - Cuculidae
e   Barred Long-tailed Cuckoo; Barred
      Cuckoo (CSA)
f   Coucou montagnard

d Bergkuckuck
i Cuculo codalunga montano

1892 *Cercococcyx olivinus*
Cuculiformes - Cuculidae
e Olive Long-tailed Cuckoo; Olive Cuckoo
f Coucou olivâtre
d Olivkuckuck
i Cuculo codalunga olivaceo

1893 *Cercomacra brasiliana*
Passeriformes - Formicariidae
e Rio de Janeiro Antbird
f Grisin du Brésil
d Stufenschwanzameisenfänger
i Mangiaformiche brasiliano

1894 *Cercomacra carbonaria*
Passeriformes - Formicariidae
e Rio Branco Antbird; Branco Antbird
f Grisin charbonnier
d Schmalschnabelameisenfänger
i Mangiaformiche di Rio Branca

1895 *Cercomacra cinerascens*
Passeriformes - Formicariidae
e Grey Antbird
f Grisin ardoisé
d Aschkopfameisenfänger
i Mangiaformiche cenerino

1896 *Cercomacra ferdinandi*
Passeriformes - Formicariidae
e Bananal Antbird
f Grisin de Bananal
d Bananal-Ameisenfänger
i Mangiaformiche dei bananal

1897 *Cercomacra manu*
Passeriformes - Formicariidae
e Manu Antbird
f Grisin du Manu

1898 *Cercomacra melanaria*
Passeriformes - Formicariidae
e Mato Grosso Antbird
f Grisin du Mato Grosso
d Jungas-Ameisenfänger
i Mangiaformiche del Matto Grosso

1899 *Cercomacra nigrescens*
Passeriformes - Formicariidae
e Blackish Antbird
f Grisin noirâtre
d Sumpfameisenfänger
i Mangiaformiche nerastro

1900 *Cercomacra nigricans*
Passeriformes - Formicariidae
e Jet Antbird
f Grisin du jais
d Trauerameisenfänger
i Mangiaformiche lucente

1901 *Cercomacra serva*
Passeriformes - Formicariidae
e Black Antbird
f Grisin noir
d Mohrenameisenfänger
i Mangiaformiche nero

1902 *Cercomacra tyrannina*
Passeriformes - Formicariidae
e Dusky Antbird; Tyrannine Antbird
f Grisin sombre
d Tyrannenameisenfänger
i Mangiaformiche tiranno

1903 *Cercomela dubia*
Passeriformes - Muscicapidae
e Sombre Chat; Sombre Rock-Chat; Sooty Rock-Chat; Sooty Chat
f Traquet sombre
d Dunkelschmätzer
i Sassicola modesta

1904 *Cercomela familiaris*
Passeriformes - Muscicapidae
e Familiar Chat; Red-tailed Chat (CSA)
f Traquet familier; Traquet de roche à queue rousse
d Rostschwanzschmätzer; Rotschwanz
i Sassicola familiare

1905 *Cercomela fusca*
Passeriformes - Muscicapidae
e Brown Rock-Chat; Indian Chat; Indian Chat
f Traquet bistré

d   Braunschmätzer
i   Sassicola bruna

1906   *Cercomela melanura*
     Passeriformes - Muscicapidae
e   Blackstart; Black-tailed Rock-Chat
f   Traquet à queue noire; Traquet de roche à queue noire
d   Schwarzschwanz; Schwarzschwanzschmätzer; Grauschmätzer
i   Sassicola codanera; Codinera

1907   *Cercomela schlegelii*
     Passeriformes - Muscicapidae
e   Karoo Chat; Schlegel's Chat; Grey-rumped Sickle-winged Chat; Grey-rumped Chat
f   Traquet du Karoo
d   Bleichschmätzer; Wüstenschmätzer
i   Sassicola di Schlegel

1908   *Cercomela scotocerca*
     Passeriformes - Muscicapidae
e   Brown-tailed Chat
f   Traquet à queue brune
d   Braunschwanz
i   Sassicola codabruna

1909   *Cercomela scotocerca turkana*
     Passeriformes - Muscicapidae
e   Brown-tailed Rock-Chat (CSA); Brown-tailed Chat

1910   *Cercomela sinuata*
     Passeriformes - Muscicapidae
e   Sicklewing Chat; Sickle-winged Chat; Common Sickle-winged Chat; Common Sicklewing Chat
f   Traquet aile-en-faux
d   Namib-Schmätzer; Veldschmätzer
i   Sassicola alifalcate

1911   *Cercomela sordida*
     Passeriformes - Muscicapidae
e   Moorland Chat; Alpine Chat (CSA); Hill Chat; Mountain Chat
f   Traquet afroalpin
d   Almenschmätzer
i   Sassicola di Erlanger

1912   *Cercomela tractrac*
     Passeriformes - Muscicapidae
e   Tractrac Chat; Layard's Chat
f   Traquet tractrac
d   Oranjeschmätzer
i   Sassicola Tractrac

1913   *Cercotrichas barbata*
     Passeriformes - Muscicapidae
e   Miombo Scrub-Robin; Miombo Bearded Scrub-Robin; Western Bearded Scrub-Robin; Central Bearded Scrub-Robin; Bearded Scrub-Robin
f   Agrobate barbu; Agrobate barbu du Miombo
d   Bartheckensänger
i   Usignolo barbuto centrale

1914   *Cercotrichas coryphaeus*
     Passeriformes - Muscicapidae
e   Karoo Scrub-Robin; Karoo Robin (CSA)
f   Agrobate coryphée
d   Karru-Heckensänger
i   Usignolo del Karoo

1915   *Cercotrichas galactotes*
     Passeriformes - Muscicapidae
e   Rufous-tailed Scrub-Robin; Rufous Bush-Chat (CSA); Rufous Scrub-Robin; Rufous-tailed Bush Chat; Rufous Chat; Rufous Bush Chat; Rufous Bush-Robin; Rufous Warbler
f   Agrobate roux; Agrobate rubigineux
d   Afrikanischer Heckensänger; Heckensänger
i   Rusignolo d'Africa; Usignolo d'Africa

1916   *Cercotrichas hartlaubi*
     Passeriformes - Muscicapidae
e   Brown-backed Scrub-Robin
f   Agrobate à dos brun
d   Hartlaub-Heckensänger
i   Usignolo d'Africa di Hartlaub

1917   *Cercotrichas leucophrys*
     Passeriformes - Muscicapidae
e   Red-backed Scrub-Robin; White-browed Scrub-Robin (CSA)

f Agrobate à dos roux
d Weißbrauenheckensänger
i Usignolo d'Africa dai sopraccigli
bianchi

1918 *Cercotrichas leucophrys leucoptera*
Passeriformes - Muscicapidae
f Agrobate à ailes blanches

1919 *Cercotrichas leucosticta*
Passeriformes - Muscicapidae
e Forest Scrub-Robin; Western
Bearded Scrub-Robin; Northern
Bearded Scrub-Robin; Moustached
Scrub-Robin; Gold Coast Scrub-
Robin
f Agrobate du Ghana; Rougequeue de
Ghana; Agrobate de forêt
d Waldheckensänger
i Usignolo barbuto occidentale

1920 *Cercotrichas paena*
Passeriformes - Muscicapidae
e Kalahari Scrub-Robin; Kalahari
Robin (CSA); Kalahari Sandy Scrub-
Robin
f Agrobate du Kalahari
d Kalahari-Heckensänger
i Usignolo del Kalahari

1921 *Cercotrichas podobe*
Passeriformes - Muscicapidae
e Black Scrub-Robin; Black Bush-
Robin
f Agrobate podobé; Merle podobe
d Rußheckensänger
i Merlo podobè

1922 *Cercotrichas quadrivirgata*
Passeriformes - Muscicapidae
e Bearded Scrub-Robin; Bearded
Robin (CSA); Eastern Bearded
Scrub-Robin; Zanzibar Scrub-Robin;
Zanzibar Bearded Scrub-Robin
f Agrobate à moustaches; Agrobate
barbu oriental
d Brauner Bartheckensänger
i Usignolo barbuto orientale

1923 *Cercotrichas signata*
Passeriformes - Muscicapidae

e Brown Scrub-Robin; Brown Robin
(CSA); Brown Robin-Chat
f Agrobate barbu brun
d Natal-Heckensänger
i Usignolo d'Africa bruno

1924 *Cereopsis novaehollandiae*
Anseriformes - Anatidae
e Cape Barren Goose; Cereopsis
Goose; Pigeon Goose
f Céréopse cendré
d Hühnergans
i Oca di Capo Barren; Oca gallina;
Cereopside; Oca incappucciata

1925 *Cereopsis novaehollandiae grisea*
Anseriformes - Anatidae
e South-western Cape Barren Goose
(ANZ)

1926 *Cerorhinca monocerata*
Charadriiformes - Alcidae
e Rhinoceros Auklet; Hornbilled
Puffin; Rhinoceros Auk; Unicorn
Auk
f Macareux rhinocéros
d Nashornalk
i Alca minore rinoceronte

1927 *Certhia americana*
Passeriformes - Certhiidae
e Brown Creeper; Treecreeper;
American Treecreeper; American
Brown-Treecreeper
f Grimpereau brun
d Anden-Baumläufer
i Rampichino americano

1928 *Certhia brachydactyla*
Passeriformes - Certhiidae
e Short-toed Treecreeper
f Grimpereau des jardins; Grimpereau
des arbres; Grimpereau
brachydactyle; Grimpereau à doigts
courts
d Gartenbaumläufer; Hausbaumläufer
i Rampichino; Abriccagnolo;
Scorzaiola; Rampichino comune

1929 *Certhia discolor*
Passeriformes - Certhiidae

e   Brown-throated Treecreeper; Sikkim Treecreeper; Indes Treecreeper
f   Grimpereau discolore
d   Braunkehlbaumläufer; Zweifarbenbaumläufer
i   Rampichino golabruna

1930   *Certhia familiaris*
       Passeriformes - Certhiidae
e   Eurasian Treecreeper; Treecreeper; Common Treecreeper; European Treecreeper; Northern Treecreeper
f   Grimpereau des bois; Grimpereau familier
d   Waldbaumläufer
i   Rampichino alpestre

1931   *Certhia himalayana*
       Passeriformes - Certhiidae
e   Bar-tailed Treecreeper; Himalayan Treecreeper
f   Grimpereau de l'Himalaya
d   Himalaja-Baumläufer
i   Rampichino dell'Himalaya

1932   *Certhia nipalensis*
       Passeriformes - Certhiidae
e   Rusty-flanked Treecreeper; Nepal Treecreeper; Nepalese Treecreeper; Stoliczka's Treecreeper
f   Grimpereau du Nepal
d   Rostbauchbaumläufer; Nepal-Baumläufer
i   Rampichino di Stoliczka

1933   *Certhiaxis cinnamomea*
       Passeriformes - Furnariidae
e   Yellow-chinned Spinetail; Yellow-throated Spinetail
f   Synallaxe à gorge jaune
d   Gelbkehlschlüpfer
i   Codaspinosa color cannella

1934   *Certhiaxis mustelina*
       Passeriformes - Furnariidae
e   Red-and-white Spinetail
f   Synallaxe belette
d   Wieselschlüpfer
i   Codaspinosa rosso e bianco

1935   *Certhidea olivacea*
       Passeriformes - Emberizae
e   Warbler Finch
f   Géospize olive; Pinson fauvette
d   Waldsängerfink; Laubsängerfink
i   Fringuello cantore delle Galapagos

1936   **Certhiidae**
       Passeriformes
e   Northern Creepers; Treecreepers; Creepers
f   Grimperaux; Certhiidés
d   Baumläufer
i   Certidi; Rampichini

1937   *Certhilauda albescens*
       Passeriformes - Alaudidae
e   Karoo Lark; Red-bcked Lark
f   Alouette du Karroo
d   Karru-Lerche
i   Allodola del Karoo

1938   *Certhilauda barlowi*
       Passeriformes - Alaudidae
e   Barlow's Lark
d   Barlows Lerche

1939   *Certhilauda burra*
       Passeriformes - Alaudidae
e   Ferruginous Lark; Red Lark; Red Sandlark; Sandlark
f   Alouette ferrugineuse; Ammomane ferrugineuse
d   Rotscheitellerche; Oranjelerche
i   Allodola ferruginea

1940   *Certhilauda cavei*
       Passeriformes - Alaudidae
e   Cave's Lark

1941   *Certhilauda chuana*
       Passeriformes - Alaudidae
e   Short-clawed Lark; Short-clawed Bush-Lark
f   Alouette à ongles courts
d   Betschuanen-Lerche
i   Allodola del Chuana

1942   *Certhilauda curvirostris*
       Passeriformes - Alaudidae
e   Long-billed Lark

f Alouette à long bec
d Langschnabellerche
i Allodola beccolunga

1943 *Certhilauda erythrochlamys*
Passeriformes - Alaudidae
e Dune Lark; Aristida Lark; Red-backed Lark
f Alouette à dos roux; Alouette des dunes
d Dünenlerche
i Allodola rossa

1944 *Certhionyx niger*
Passeriformes - Meliphagidae
e Black Honeyeater; Dark Honeyeater
f Myzomèle cravaté
d Trauerhonigfresser
i Mangiamiele dorsonero

1945 *Certhionyx pectoralis*
Passeriformes - Meliphagidae
e Banded Honeyeater
f Myzomèle à collier
d Brustbandhonigfresser
i Mangiamiele dal collare

1946 *Certhionyx variegatus*
Passeriformes - Meliphagidae
e Pied Honeyeater
f Myzomèle varié
d Elsterhonigfresser
i Mangiamiele gazza

1947 *Ceryle rudis*
Coraciiformes - Alcedinidae
e Pied Kingfisher; Lesser Pied-Kingfisher; Small Pied-Kingfisher; Indian Pied-Kingfisher
f Alcyon pie; Martin-pêcheur pie
d Graufischer
i Martin pescatore bianco e nero; Cerile

1948 Cerylidae
Coraciiformes
e Cerylid Kingfishers
f Cerylidés
d Fischer
i Cerilidi; Martin pescatori

1949 *Cettia acanthizoides*
Passeriformes - Sylviidae
e Yellowish-bellied Bush-Warbler; Verreaux's Cettia; Verreaux's Bush-Warbler; Yellow-billed Bush-Warbler; Yellow-billed Cettia
f Bouscarle de Verreaux
d Bambusbuschsänger

1950 *Cettia annae*
Passeriformes - Sylviidae
e Palau Bush-Warbler
f Bouscarle des Palau
d Palau-Buschsänger
i Cettia di Palau

1951 *Cettia brunnifrons*
Passeriformes - Sylviidae
e Grey-sided Bush-Warbler; Rufous-capped Bush-Warbler; Rufous-fronted Cettia
f Bouscarle à couronne brune
d Rotkopfbuschsänger
i Cettia frontebruna

1952 *Cettia canturians*
Passeriformes - Sylviidae
e Manchurian Bush-Warbler; Chinese Bush-Warbler
f Bouscarle mandchoue
i Cettia di Manciuria

1953 *Cettia carolinae*
Passeriformes - Sylviidae
e Tanimbar Bush-Warbler; Yamdena Bush-Warbler
f Bouscarle des Tanimbar
i Cettia dell'Isola Tanimbar

1954 *Cettia cetti*
Passeriformes - Sylviidae
e Cetti's Warbler; Cetti's Bush-Warbler
f Bouscarle de Cetti
d Seidensänger; Seidenrohrsänger
i Usignolo di fiume

1955 *Cettia diphone*
Passeriformes - Sylviidae
e Japanese Bush-Warbler; Singing Bush-Warbler; Chinese Cettia; Oriental Bush-Warbler; Bush-

Warbler; Chinese Bush-Warbler
f   Bouscarle chanteuse
d   Japan-Buschsänger
i   Cettia del Giappone

1956   *Cettia flavolivacea*
Passeriformes - Sylviidae
e   Aberrant Bush-Warbler; Aberrant
Cettia
f   Bouscarle jaune-et-vert
d   Olivbuschsänger
i   Cettia ventregiallo

1957   *Cettia fortipes*
Passeriformes - Sylviidae
e   Brownish-flanked Bush-Warbler;
Strong-footed Bush-Warbler; Strong-
footed Cettia; Brown-flanked Bush-
Warbler; Pale Strong-footed Bush-
Warbler; Mountain Bush-Warbler
f   Bouscarle de montagne
d   Bergbuschsänger
i   Cettia flanchibrune

1958   *Cettia major*
Passeriformes - Sylviidae
e   Chestnut-crowned Bush-Warbler;
Large Bush-Warbler; Large Cettia
f   Grande Bouscarle
d   Rhododendronbuschsänger
i   Cettia capocastano

1959   *Cettia pallidipes*
Passeriformes - Sylviidae
e   Pale-footed Bush-Warbler; Pale-
footed Cettia; Pale-footed Stubtail;
Blanford's Bush-Warbler
f   Bouscarle à pattes claires; Bouscarle
de Blanford
d   Weißfußbuschsänger
i   Cettia zampechiare

1960   *Cettia parens*
Passeriformes - Sylviidae
e   Shade Warbler
f   Bouscarle de San Cristobal
d   Schattenhuscher
i   Cettia di San Cristobal

1961   *Cettia robustipes*
Passeriformes - Sylviidae

e   Yellow-bellied Bush-Warbler;
Yellowish-bellied Bush-Warbler;
Verreaux's Bush-Warbler; Verreaux's
Cettia; Verreaux's Bush-Cettia;
Swinhoe's Bush-Warbler; Highland
Bush-Warbler
f   Bouscarle à ventre jaunâtre
d   Bambusbuschsänger
i   Cettia di Swinhoe

1962   *Cettia ruficapilla*
Passeriformes - Sylviidae
e   Fiji Bush-Warbler; Fiji Warbler
f   Bouscarle des Fidji
d   Laubhuscher
i   Cettia delle Figi

1963   *Cettia seebohmi*
Passeriformes - Sylviidae
e   Philippine Bush-Warbler; Seebohm's
Bush-Warbler; Luzon Bush-Warbler
f   Bouscarle de Luçon
d   Luzon-Buschsänger
i   Cettia delle Filippine

1964   *Cettia vulcania*
Passeriformes - Sylviidae
e   Sunda Bush-Warbler; Mountain
Bush-Warbler; Müller's Bush-
Warbler
f   Bouscarle d'Indonesie
d   Sunda-Buschsänger
i   Cettia della Sonda

1965   *Ceuthmochares aereus*
Cuculiformes - Cuculidae
e   Yellowbill; Green Coucal; Yellow-
billed Coucal; Yellowbill Coucal;
Green Malkoha
f   Malcoha à bec jaune; Coucal à bec
jaune
d   Erzkuckuck
i   Beccogiallo

1966   *Ceyx erithacus*
Coraciiformes - Alcedinidae
e   Oriental Dwarf Kingfisher; Black-
backed Kingfisher (ISC); Three-toed
Kingfisher (ISC); Three-toed Forest-
Kingfisher; Indian Three-toed
Kingfisher; Indian Forest Kingfisher

f   Martin-pêcheur pourpre
d   Dschungelfischer
i   Martin pescatore tridattilo

1967   *Ceyx fallax*
Coraciiformes - Alcedinidae
e   Sulawesi Dwarf Kingfisher; Celebes
Pygmy-Kingfisher; Celebes Forest-
Kingfisher; Celebes Dwarf
Kingfisher; Sulawesi Pygmy-
Kingfisher; Sulawesi Kingfisher
f   Martin-pêcheur multicolore
d   Rostfischer
i   Martin pescatore di Sulawesi

1968   *Ceyx goodfellowi*
Coraciiformes - Alcedinidae
e   Goodfellow's Kingfisher; Bluish-
white Kingfisher

1969   *Ceyx lepidus*
Coraciiformes - Alcedinidae
e   Variable Dwarf Kingfisher; Dwarf
Kingfisher; Dwarf Forest-Kingfisher;
Gentian Kingfisher
f   Martin-pêcheur gracieux
d   Waldfischer
i   Martin pescatore variabile

1970   *Ceyx melanurus*
Coraciiformes - Alcedinidae
e   Philippine Dwarf Kingfisher;
Philippine Forest-Kingfisher;
Philippine Kingfisher
f   Martin-pêcheur flamboyant
d   Goldfischer
i   Martin pescatore delle Filippine

1971   *Ceyx rufidorsa*
Coraciiformes - Alcedinidae
e   Rufous-backed Kingfisher; Malay
Forest-Kingfisher; Red-backed
Kingfisher
f   Martin-pêcheur à dos roux
i   Martin pescatore dorsorossiccio

1972   *Chaetocercus jourdanii*
Apodiformes - Trochilidae
e   Rufous-shafted Woodstar; Jourdain's
Woodstar
f   Colibri de Jourdan

d   Rosenelfe
i   Stella dei boschi di Jourdan

1973   *Chaetops aurantius*
Ciconiiformes - Chionidae
e   Orange-breasted Rockjumper;
Rockjumper; Buffy Rockjumper
f   Chétopse doré
d   Natal-Felsenspringer
i   Saltarocce pettoarancio

1974   *Chaetops frenatus*
Ciconiiformes - Chionidae
e   Rufous Rockjumper; Cape
Rockjumper (CSA); South Africa
Jumper; South African Jumper
f   Chétopse bridé; Sauteur des rochers
d   Felsenspringer
i   Saltarocce del Capo

1975   *Chaetoptila angustipluma*
Passeriformes - Meliphagidae
e   Kioea
f   Méliphage kioéa
d   Kioea; Schmalfederhonigesser
i   Kioea

1976   *Chaetorhynchus papuensis*
Passeriformes - Corvidae
e   Pygmy Drongo; Papuan Drongo;
Papuan Mountain-Drongo; Mountain
Drongo
f   Drongo papou
d   Papua-Bergdrongo;
Rundschwanzdrongo

1977   *Chaetornis striatus*
Passeriformes - Sylviidae
e   Bristled Grassbird; Bristled Grass-
Warbler (ISC)
f   Graminicole rayée
d   Nackzügelsänger
i   Forapaglie setoloso

1978   *Chaetura andrei*
Apodiformes - Apodidae
e   Ashy-tailed Swift; Andre's Swift
f   Martinet d'André
d   Buriti-Segler
i   Rondone codadicenere

1979   *Chaetura brachyura*
Apodiformes - Apodidae
e   Short-tailed Swift; Rainbird (WI)
f   Martinet polioure
d   Stutzschwanzsegler
i   Rondone codacorte

1980   *Chaetura chapmani*
Apodiformes - Apodidae
e   Chapman's Swift; Dark-breasted Swift
f   Martinet de Chapman
d   Chapman-Segler
i   Rondone codaspinosa di Chapman

1981   *Chaetura cinereiventris*
Apodiformes - Apodidae
e   Grey-rumped Swift; Rainbird (WI); Ash rumped Swift; Ashy-rumped Swift
f   Oiseau de la pluie (Ants); Martinet à croupion gris
d   Graubürzelsegler
i   Rondone codaspinosa dal groppone grigio

1982   *Chaetura egregia*
Apodiformes - Apodidae
e   Pale-rumped Swift
f   Martinet de Bolivie
d   Blaßbürzelsegler
i   Rondone codaspinosa dal groppone pallido

1983   *Chaetura gaumeri*
Apodiformes - Apodidae
e   Yucatan Swift

1984   *Chaetura martinica*
Apodiformes - Apodidae
e   Lesser Antillean Swift
f   Petit Martinet noir (Ants); Hirondelle (Ants); Martinet chiquesol; Martinet de la Martinique
d   Antillen-Segler
i   Rondone codaspinosa delle Antille

1985   *Chaetura pelagica*
Apodiformes - Apodidae
e   Chimney Swift; American Chimney-Swift

f   Martinet ramoneur
d   Schornsteinsegler; Kaminsegler
i   Rondone dei camini; Rondone codaspinosa dei camini; Rondone dalla coda spinosa

1986   *Chaetura richmondi*
Apodiformes - Apodidae
e   Dusky-backed Swift

1987   *Chaetura sclateri*
Apodiformes - Apodidae
e   Ash-rumped Swift

1988   *Chaetura spinicauda*
Apodiformes - Apodidae
e   Band-rumped Swift
f   Martinet spinicaude
d   Dornensegler
i   Rondone codaspinosa fuligginosa

1989   *Chaetura vauxi*
Apodiformes - Apodidae
e   Vaux's Swift; Dusky-backed Swift
f   Martinet de Vaux
d   Graubauchsegler
i   Rondone codaspinosa di Vaux

1990   *Chaimarrornis leucocephalus*
Passeriformes - Muscicapidae
e   White-capped Water-Redstart; White-capped Redstart; River Chat; White-capped River-Chat; River Redstart
f   Torrentaire à calotte blanche
d   Weißkopfschmätzer
i   Codirossa corona bianca

1991   *Chalcophaps indica*
Columbiformes - Columbidae
e   Emerald Dove; Bronze-winged Dove (ISC); Common Emerald Dove; Green-winged Pigeon; Green-winged Ground-Dove; Green-winged Dove; Green-backed Dove; Ceylon Bronze-winged Pigeon; Long-billed Green-Pigeon; Little Green-Pigeon; Green-and-bronze Pigeon
f   Colombine turvert
d   Glanzkäfertaube; Grünflugeltaube

i   Tortora smeraldina; Colomba verde indiana

1992   *Chalcophaps indica natalis*
Columbiformes - Columbidae
e   Christmas Island Emerald-Dove (ANZ)

1993   *Chalcophaps stephani*
Columbiformes - Columbidae
e   Stephan's Dove; Brown-backed Ground-Pigeon; Brown-backed Emerald Dove; Stephan's Ground-Dove; Stephan's Pigeon; Stephan's Emerald Dove; Brown-backed Green-winged Dove
f   Colombine d'Étienne
d   Stephan-Taube
i   Tortora smeraldina di Jacquinot

1994   *Chalcopsitta atra*
Psittaciformes - Loriidae
e   Black Lory
f   Lori noir
d   Schwarzlori
i   Lori nero

1995   *Chalcopsitta cardinalis*
Psittaciformes - Loriidae
e   Cardinal Lory
f   Lori cardinal
d   Kardinallori
i   Lori cardinale

1996   *Chalcopsitta duivenbodei*
Psittaciformes - Loriidae
e   Brown Lory; Duivenbode's Lory
f   Lori de Duyvenbode
d   Braunlori
i   Lori di Duivenbode

1997   *Chalcopsitta insignis*
Psittaciformes - Loriidae
e   Rajah Lory; Red-quilled Lory

1998   *Chalcopsitta sintillata*
Psittaciformes - Loriidae
e   Yellow-streaked Lory; Red-fronted Lory; Greater Streaked-Lory
f   Lori flamméché

d   Schimmerlori
i   Lori scintillato

1999   *Chalcostigma herrani*
Apodiformes - Trochilidae
e   Rainbow-bearded Thornbill; Herran's Thornbill
f   Métallure arc-en-ciel
d   Rotbrustglanzschänzchen
i   Becco a spina di Herran

2000   *Chalcostigma heteropogon*
Apodiformes - Trochilidae
e   Bronze-tailed Thornbill; Bronzy Thornbill
f   Métallure à queue bronzée
d   Dornschnabelglanzschwänzchen
i   Becco a spina codabronzata

2001   *Chalcostigma olivaceum*
Apodiformes - Trochilidae
e   Olivaceous Thornbill
f   Métallure olivâtre
d   Olivglanzschwänzchen
i   Becco a spina olivaceo

2002   *Chalcostigma ruficeps*
Apodiformes - Trochilidae
e   Rufous-capped Thornbill; Rufous-capped Metaltail; Metaltail
f   Métallure à tête rousse
d   Rotkappenglanzschwänzchen
i   Becco a spina testarossiccia

2003   *Chalcostigma stanleyi*
Apodiformes - Trochilidae
e   Blue-mantled Thornbill
f   Métallure de Stanley
d   Blaurückenglanzschwänzchen
i   Becco a spina di Stanley

2004   *Chalybura buffonii*
Apodiformes - Trochilidae
e   White-vented Plumeleteer; Buffon's Hummingbird
f   Colibri de Buffon
d   Buffon-Kolibri
i   Colibrì di Buffon

2005 *Chalybura melanorrhoa*
Apodiformes - Trochilidae
e Black-vented Plumleteer

2006 *Chalybura urochrysia*
Apodiformes - Trochilidae
e Bronze-tailed Plumeleteer; Gould's
Dusky Plumeleteer
f Colibri à queue bronzée
d Straußkolibri
i Colibrì codaramata di Gould

2007 *Chamaea fasciata*
Passeriformes - Sylviidae
e Wren-Tit
f Cama brune
d Zaunkönigsgrasmücke;
Chaparaltimalie; Zaunkönigsmeise
i Camea americana; Cincia scricciolo

2008 *Chamaepetes goudotii*
Galliformes - Cracidae
e Sickle-winged Guan; Goudot's Guan;
Rufous Sickle-winged Guan
f Pénélope de Goudot
d Sichelguan
i Penelope alifalcate

2009 *Chamaepetes unicolor*
Galliformes - Cracidae
e Black Guan; Black Piping-Guan;
Black Sickle-winged Guan
f Pénélope unicolore
d Mohrenguan
i Penelope nera

2010 *Chamaeza campanisona*
Passeriformes - Formicariidae
e Short-tailed Ant-Thrush
f Tétéma flambé
d Kurzschwanzameisendrossel
i Tordo formichiere codacorta

2011 *Chamaeza meruloides*
Passeriformes - Formicariidae
e Such's Ant-Thrush; Cryptic Ant-
Thrush
f Tétéma de Such

2012 *Chamaeza mollissima*
Passeriformes - Formicariidae

e Barred Ant-Thrush
f Tétéma barré
d Bindenameisendrossel
i Tordo formichiere barrato

2013 *Chamaeza nobilis*
Passeriformes - Formicariidae
e Noble Ant-Thrush; Striated Ant-
Thrush
f Tétéma strié
d Streifenameisendrossel
i Tordo formichiere striato

2014 *Chamaeza ruficauda*
Passeriformes - Formicariidae
e Rufous-tailed Ant-Thrush; Brazilian
Ant-Thrush
f Tétéma à qucuc rousse
d Rotschwanzameisendrossel
i Tordo formichiere codarossiccia

2015 *Chamaeza turdina*
Passeriformes - Formicariidae
e Scalloped Ant-Thrush; Schwartz's
Ant-Thrush; Rufous-tailed Ant-
Thrush
f Tétéma festonné

2016 **Charadriidae**
Charadriiformes
e Plovers; Lapwings
f Pluviers; Charadridés
d Regenpfeifer
i Caradridi

2017 **Charadriiformes**
e Waders; Gulls; Auks
f Charadriiformes
d Watvögel; Alkenvögel; Möwenvögel
i Caradriformi

2018 *Charadrius alexandrinus*
Charadriiformes - Charadriidae
e Kentish Plover; Snowy Plover (NA);
Pondbird (WI); Ceylon Kentish
Plover
f Gravelot à collier interrompu; Pluvier
à collier interrompu; Gravelot patte-
noire; Collier (Ants)

d   Seeregenpfeifer
i   Corriere; Fratino

2019   ***Charadrius alexandrinus nivosus***
Charadriiformes - Charadriidae
e   Snow Plover
f   Pluvier neigeux
d   Schneeregenpfeifer

2020   ***Charadrius alticola***
Charadriiformes - Charadriidae
e   Puna Plover; Puna Two-banded Plover
f   Pluvier du Puna
d   Punaregenpfeifer
i   Corriere della Puna

2021   ***Charadrius asiaticus***
Charadriiformes - Charadriidae
e   Caspian Plover; Lesser Oriental Plover; Caspian Dotterel
f   Pluvier asiatique
d   Wermutregenpfeifer; Kaspischer Regenpfeifer
i   Corriere asiatico

2022   ***Charadrius australis***
Charadriiformes - Charadriidae
e   Inland Dotterel; Australian Dotterel; Australian Courser; Desert Plover; Prairie Plover
f   Pluvier australien
d   Ringrennvogel
i   Piviere tortolino australiano

2023   ***Charadrius bicinctus***
Charadriiformes - Charadriidae
e   Double-banded Plover; Banded Plover; Double-banded Dotterel; Banded Dotterel; Chestnut-breasted Plover
f   Pluvier à double collier
d   Doppelbandregenpfeifer
i   Corriere dai due collari

2024   ***Charadrius collaris***
Charadriiformes - Charadriidae
e   Collared Plover; Little Ploward (WI); Three-banded Plover; Azara's Sandplover
f   Pluvier d'Azara

d   Dreibandregenpfeifer
i   Corriere dal collare

2025   ***Charadrius dubius***
Charadriiformes - Charadriidae
e   Little Ringed-Plover; Little Plover; Little Ring Plover; European Little Ringed-Plover; Indian Little Ringed-Plover
f   Petit Gravelot; Pluvier petit-gravelot
d   Flußregenpfeifer
i   Corriere piccolo; Piviere minore

2026   ***Charadrius falklandicus***
Charadriiformes - Charadriidae
e   Two-banded Plover; Double-banded Plover
f   Pluvier des Falkland
d   Falklands-Regenpfeifer
i   Corriere delle Isole Falkland

2027   ***Charadrius forbesi***
Charadriiformes - Charadriidae
e   Forbes's Plover; Forbes's Banded-Plover; Patagonian Sandplover
f   Pluvier de Forbes
d   Braunstirnregenpfeifer
i   Corriere di Forbes

2028   ***Charadrius hiaticula***
Charadriiformes - Charadriidae
e   Common Ringed-Plover; Ringed Plover; Great Ringed-Plover; Greater Ringed-Plover; Ringed Dotterel
f   Grand Gravelot; Pluvier grand-gravelot
d   Sandregenpfeifer
i   Corriere grosso; Piviere dal collare

2029   ***Charadrius javanicus***
Charadriiformes - Charadriidae
e   Javan Plover; Java Sandplover
f   Pluvier de Java
i   Corriere di Giava

2030   ***Charadrius leschenaultii***
Charadriiformes - Charadriidae
e   Greater Sandplover; Sand Plover (CSA); Great Sandplover (CSA); Geoffrey's Plover; Large Dotterel; Large Sandplover; Large Sand

Dotterel; Geoffroy's Sandplover;
Geoffroy's Dotterel; Large-billed
Dotterel
f Gravelot de Leschenault; Pluvier de
Leschenault; Gravelot du désert;
Pluvier du désert
d Wüstenregenpfeifer
i Corriere di Leschenault; Piviere de
Leschenault; Corriere maggiore

2031 ***Charadrius marginatus***
Charadriiformes - Charadriidae
e White-fronted Plover; White-fronted
Sandplover
f Pluvier à front blanc; Gravelot à front
blanc
d Weißstirnregenpfeifer
i Corriere marginato; Piviere
marginato

2032 ***Charadrius melodus***
Charadriiformes - Charadriidae
e Piping Plover
f Pluvier siffleur
d Gelbfußregenpfeifer
i Corriere canoro

2033 ***Charadrius modestus***
Charadriiformes - Charadriidae
e Rufous-chested Plover; Rufous-
chested Dotterel; Red-breasted
Dotterel; Chilean White Plover;
Winter Plover
f Pluvier d'Urville
d Rotbrustregenpfeifer
i Corriere modesto

2034 ***Charadrius mongolus***
Charadriiformes - Charadriidae
e Mongolian Plover; Mongolian
Sandplover (CSA); Lesser
Sandplover; Mongolian Dotterel;
Mongolian Sand Dotterel; Tibetan
Plover
f Pluvier de Mongolie; Gravelot
mongol
d Mongolen-Regenpfeifer; Mongolei-
Regenpfeifer
i Corriere della Mongolia; Corriere
mongolo

2035 ***Charadrius montanus***
Charadriiformes - Charadriidae
e Mountain Plover; Mountain Dotterel
f Pluvier montagnard
d Bergregenpfeifer; Prärieregenpfeifer
i Corriere montano

2036 ***Charadrius morinellus***
Charadriiformes - Charadriidae
e Eurasian Dotterel (NA); Dotterel;
Common Dotterel
f Pluvier guignard
d Morinell-Regenpfeifer
i Piviere tortolino; Piviere tortolino
eurasiatico

2037 ***Charadrius novaeseelandiae***
Charadriiformes - Charadriidae
e Shore Plover; Tuturuatu; New
Zealand Shore-Plover; Masked New
Zealand Plover; Masked Sandplover;
Shore Dotterel
f Pluvier de Nouvelle-Zélande
d Kappenregenpfeifer
i Corriere dell'Isola Chatham

2038 ***Charadrius obscurus***
Charadriiformes - Charadriidae
e Red-breasted Plover; Red-breasted
Dotterel; New Zealand Red-breasted
Dotterel; New Zealand Dotterel; New
Zealand Plover
f Pluvier roux
d Maoriregenpfeifer
i Piviere della Nuova Zelanda

2039 ***Charadrius occidentalis***
Charadriiformes - Charadriidae
e Peruvian Plover; Snowy Plover (NA)

2040 ***Charadrius pallidus***
Charadriiformes - Charadriidae
e Chestnut-banded Plover; Chestnut-
banded Sandplover
f Pluvier élégant
d Fahlregenpfeifer;
Rotbandregenpfeiffer
i Corriere delle sabbie

2041 ***Charadrius pecuarius***
Charadriiformes - Charadriidae

e Kittlitz's Plover; Kittlitz's Sandplover
f Pluvier pâtre; Gravelot pâtre
d Hirtenregenpfeifer
i Corriere di Kittlitz

2042 *Charadrius peronii*
Charadriiformes - Charadriidae
e Malaysian Plover; Malay Plover;
Malay Sandplover; Malaysian
Sandplover
f Pluvier de Péron
d Sunda-Regenpfeifer
i Corriere del Malesia

2043 *Charadrius placidus*
Charadriiformes - Charadriidae
e Long-billed Plover; Long-billed
Dotterel
f Pluvier à long bec
d Langschnabelregenpfeifer
i Corriere beccolungo

2044 *Charadrius rubricollis*
Charadriiformes - Charadriidae
e Hooded Plover; Hooded Dotterel
f Pluvier à camail
d Mönchsregenpfeifer
i Corriere dal cappuccio

2045 *Charadrius ruficapillus*
Charadriiformes - Charadriidae
e Red-capped Plover; Red-capped
Dotterel; Red-capped Sandplover
f Pluvier à tête rousse; Gravelot de
Madagascar
d Rotkopfregenpfeifer
i Corriere capirosso

2046 *Charadrius sanctaehelenae*
Charadriiformes - Charadriidae
e St. Helena Plover; St. Helena
Sandplover; Wirebird
f Pluvier de Sainte-Hélène
d St. Helena-Regenpfeifer
i Corriere dell'Isola di Santa Elena

2047 *Charadrius semipalmatus*
Charadriiformes - Charadriidae
e Semipalmated Plover; Pondbird
(WI); Ring-neck (WI); Semipalmated
Ringed-Plover

f Pluvier semipalmé; Gravelot
semipalmé; Collier à pattes jaunes
(Ants); Bécasse à collier (Ants);
Collier (Ants)
d Amerikanischer Sandregenpfeifer;
Sandregenpfeifer
i Corriere semipalmato

2048 *Charadrius thoracicus*
Charadriiformes - Charadriidae
e Madagascar Plover; Black-banded
Sandplover; Black-banded Plover
f Pluvier à bandeau noir
d Madagaskar-Regenpfeifer
i Corriere fasciato

2049 *Charadrius tricollaris*
Charadriiformes - Charadriidae
e Three-banded Plover; Treble-banded
Sandplover; Tri-collared Sandplover;
Tri-collared Plover
f Gravelot à triple collier; Pluvier à
triple collier
d Dreibandregenpfeifer
i Corriere dai tre collare

2050 *Charadrius veredus*
Charadriiformes - Charadriidae
e Oriental Plover; Oriental Dotterel;
Eastern Sandplover
f Pluvier oriental
d Hufeisenregenpfeifer
i Corriere orientale

2051 *Charadrius vociferus*
Charadriiformes - Charadriidae
e Killdeer; Killdeer Plover; Tilderee
(WI); Soldierbird (WI); Pondbird
(WI)
f Pluvier kildir; Gravelot kildir; Collier
double (Ants); Chevalier de terre
(Ants); Double Collier (Ants)
d Keilschwanzregenpfeifer;
Schreiregenpfeifer
i Corriere americano; Corriere
vocifero

2052 *Charadrius wilsonia*
Charadriiformes - Charadriidae
e Wilson's Plover; Thick-billed Plover;
Little Ploward (WI); Pondbird (WI);

Sandbird (WI)
f Pluvier de Wilson; Bécassine (Ants);
Collier (Ants)
d Dickschnabelregenpfeifer
i Corriere di Wilson

2053   Charitospiza eucosma
Passeriformes - Fringillidae
e Coal-crested Finch
f Charitospize charbonnier
d Weißohrzwergkardinal
i Fringuella cresta di carbone

2054   Charmosyna amabilis
Psittaciformes - Loriidae
e Red-throated Lorikeet; Golden-
banded Lorikeet
f Lori à gorge rouge
d Goldbandlori; Rothöschen;
Rotkehllori
i Lorichetto golarossa

2055   Charmosyna diadema
Psittaciformes - Loriidae
e New Caledonian Lorikeet; Diademed
Lorikeet
f Lori à diadème
d Diademlori
i Lorichetto della Nuova Caledonia

2056   Charmosyna josefinae
Psittaciformes - Loriidae
e Josephine's Lorikeet
f Lori de Joséphine
d Josephinenlori
i Lorichetto di Josephine

2057   Charmosyna margarethae
Psittaciformes - Loriidae
e Duchess Lorikeet; Duchess
Margaret's Lorikeet
f Lori de Margaret
d Margareten-Lori
i Lorichetto della duchessa

2058   Charmosyna meeki
Psittaciformes - Loriidae
e Meek's Lorikeet
f Lori de Meek
d Meeks Zierlori; Salomonen-Lori
i Lorichetto di Meek

2059   Charmosyna multistriata
Psittaciformes - Loriidae
e Striated Lorikeet; Streaked Lorikeet;
Yellow Streaked- Lorikeet
f Lori strié
d Streifenlori; Vielstrichellori
i Lorichetto striato

2060   Charmosyna palmarum
Psittaciformes - Loriidae
e Palm Lorikeet
f Lori des palmiers
d Palmenlori; Palmzierlori
i Lorichetto delle palme

2061   Charmosyna papou
Psittaciformes - Loriidae
e Papuan Lorikeet; Papuan Lory; Fairy
Lorikeet; Stella's Lorikeet
f Lori papou
d Papua-Lori
i Lorichetto papua

2062   Charmosyna placentis
Psittaciformes - Loriidae
e Red-flanked Lorikeet; Beautiful
Lorikeet; Yellow-fronted Blue-eared
Lorikeet; Yellow-fronted Lorikeet;
Blue-eared Lorikeet; Lowland
Lorikeet
f Lori coquet
d Schönlori
i Lorichetto fianchirossi

2063   Charmosyna pulchella
Psittaciformes - Loriidae
e Fairy Lorikeet; Pectoral Lorikeet;
Little Red-Lorikeet
f Lori féerique
d Goldstrichellori
i Lorichetto grazioso

2064   Charmosyna rubrigularis
Psittaciformes - Loriidae
e Red-chinned Lorikeet; Red-throated
Lorikeet
f Lori à menton rouge
d Rotkehllori; Rotkinnlori
i Lorichetto mentorosso

2065 **Charmosyna rubronotata**
Psittaciformes - Loriidae
e Red-fronted Lorikeet; Red-spotted
Lorikeet; Red-rumped Lorikeet; Red-
marked Lorikeet; Red-fronted Blue-
eared Lorikeet
f Lori à front rouge
d Rotbürzellori; Rotstirnlori
i Lorichetto macchierosso

2066 **Charmosyna toxopei**
Psittaciformes - Loriidae
e Blue-fronted Lorikeet; Buru Lorikeet
f Lori de Buru
d Buru-Lori
i Lorichetto fronteblu

2067 **Charmosyna wilhelminae**
Psittaciformes - Loriidae
e Pygmy Lorikeet; Wilhelmina's
Lorikeet; Pygmy Streaked-Lorkeet
f Lori de Wilhelmina
d Elfenlori; Wilhelminen-Lori
i Lorichetto di Guglielmina

2068 **Chasiempis sandwichensis**
Passeriformes - Corvidae
e Hawaiian Monarch; Elepaio
Monarch; Elapaio
f Monarque élépaio
d Elepaio
i Elepaio

2069 **Chauna chavaria**
Anseriformes - Anhimidae
e Northern Screamer; Black-necked
Screamer
f Kamichi chavaria
d Weißwangentchaja;
Weißwangenchaja
i Kaimichi guancebianche

2070 **Chauna torquata**
Anseriformes - Anhimidae
e Southern Screamer; Crested
Screamer
f Kamichi à collier
d Halsbandtchaja
i Kaimichi dal collare

2071 **Chaunoproctus ferreorostris**
Passeriformes - Fringillidae
e Bonin Grosbeak; Bonin Island
Grosbeak; Bonin Island Finch
f Roselin des Bonin
d Bonin-Gimpel; Bonin-Fink
i Frosone dell'Isola Bonin

2072 **Chelictinia riocourii**
Falconiformes - Accipitridae
e Scissor-tailed Kite; African Swallow-
tailed Kite (CSA); Swallow-tailed
Kite; Fork-tailed Kite
f Élanion naucler; Milan de Riocour
d Schwalbenschwanz
i Nibbio codaforbice; Elani

2073 **Chelidoptera tenebrosa**
Piciformes - Bucconidae
e Swallow-wing; Swallow-wing
Puffbird; Swallow-winged Puffbird
f Barbacou à croupion blanc
d Schwalbenfaulvogel
i Bucco alidirondine; Chelidoptera
tenebrosa

2074 **Chenonetta jubata**
Anseriformes - Anatidae
e Maned Duck; Australian Wood-
Duck; Maned Goose; Maned Wood
Duck; Wood Duck; Blue Duck
f Canard à crinière
d Mähnengans
i Anatra arboricola australiana;
Chenonetta

2075 **Cheramoeca leucosternus**
Passeriformes - Hirundinidae
e White-backed Swallow; Black-and-
white Swallow; White-breasted
Swallow
f Hirondelle à dos blanc
d Weißrückenschwalbe
i Rondine dorsobianco

2076 **Chersomanes albofasciata**
Passeriformes - Alaudidae
e Spike-heeled Lark
f Alouette éperonnée
d Langspornlerche; Zirplerche
i Allodola dagli speroni

2077   ***Chersophilus duponti***
       Passeriformes - Alaudidae
e   Dupont's Lark
f   Sirli de Dupont; Sirli ricoti
d   Dupont-Lerche
i   Allodola di Dupont; Allodola del
       Dupont

2078   ***Chilia melanura***
       Passeriformes - Furnariidae
e   Crag Chilia
f   Chilia des rochers; Cinclode du Chili
d   Chilia
i   Fornaio del Cile

2079   **Chionididae**
       Charadriiformes
e   Sheathbills
f   Chionidés
d   Scheidenschnäbel
i   Chionididi

2080   ***Chionis alba***
       Charadriiformes - Chionididae
e   Snowy Sheathbill; American
       Sheathbill; Snowy Paddy; Greater
       Sheathbill; Greater Paddy; Pale-faced
       Sheathbill; Wattled Sheathbill;
       Yellow-billed Sheathbill; Yellow-
       wattled Sheathbill
f   Chionis blanc
d   Weißgesichtscheidenschnabel
i   Chione bianco; Piccione antartico

2081   ***Chionis minor***
       Charadriiformes - Chionididae
e   Black-faced Sheathbill; Lesser
       Sheathbill (CSA); Kerguelen
       Sheathbill; Hartlaub's Sheathbill
f   Petit Chionis
d   Schwarzgesichtscheidenschnabel
i   Chione becconero; Chione minore

2082   ***Chionis minor nasicornis***
       Charadriiformes - Chionididae
e   Heard Island Black-faced Sheathbill
       (ANZ)

2083   ***Chiroxiphia boliviana***
       Passeriformes - Tyrannidae
e   Yungas Manakin; Blue-backed
       Manakin
f   Manakin des yungas
i   Manachino

2084   ***Chiroxiphia caudata***
       Passeriformes - Tyrannidae
e   Swallow-tailed Manakin; Blue
       Manakin
f   Manakin à longue queue
d   Blaubrustpipra
i   Manachino dalla coda lunga;
       Manachino caudato; Manachino
       codadirondine

2085   ***Chiroxiphia lanceolata***
       Passeriformes - Tyrannidae
e   Lance-tailed Manakin; Sharp-tailed
       Manakin
f   Manakin lancéolé
d   Lanzettschwanzpipra
i   Manachino lanceolato

2086   ***Chiroxiphia linearis***
       Passeriformes - Tyrannidae
e   Long-tailed Manakin
f   Manakin fastueux
d   Langschwanzpipra;
       Spießschnurvogel
i   Manachino codalunga

2087   ***Chiroxiphia pareola***
       Passeriformes - Tyrannidae
e   Blue-backed Manakin
f   Manakin tijé
d   Prachtpipra
i   Manachino dorsoblu

2088   ***Chlamydera cerviniventris***
       Passeriformes - Ptilonorhynchidae
e   Fawn-breasted Bowerbird
f   Jardinier à poitrine fauve
d   Graukopflaubenvogel;
       Braunbauchlaubenvogel
i   Giardiniere ventrefulvo

2089   ***Chlamydera guttata***
       Passeriformes - Ptilonorhynchidae
e   Western Bowerbird
f   Jardinier tacheté
i   Giardiniere occidentale

2116 *Chlorocichla falkensteini*
Passeriformes - Pycnonotidae
e Yellow-necked Greenbul; Yellow-
necked Bulbul; Yellow-throated
Bulbul; Falkenstein's Greenbul
f Bulbul de Falkenstein
d Falkenstein-Bülbül
i Bulbul di Falkenstein

2117 *Chlorocichla flavicollis*
Passeriformes - Pycnonotidae
e Yellow-throated Greenbul; Yellow-
throated Leaflove
f Bulbul à gorge claire; Grand Bulbul;
Bulbul à gorge jaune
d Gelbkehlbülbül
i Bulbul collogiallo

2118 *Chlorocichla flaviventris*
Passeriformes - Pycnonotidae
e Yellow-bellied Greenbul; Yellow-
breasted Bulbul; Yellow-bellied
Bulbul
f Bulbul à poitrine jaune
d Gelbbauchbülbül
i Bulbul ventregiallo

2119 *Chlorocichla laetissima*
Passeriformes - Pycnonotidae
e Joyful Greenbul; Joyful Bulbul
f Bulbul joyeux
d Dotterbülbül
i Bulbul gaio

2120 *Chlorocichla prigoginei*
Passeriformes - Pycnonotidae
e Prigogine's Greenbul; Prigogine's
Bulbul; Butembo Greenbul; Congo
Greenbul
f Bulbul de Prigogine
d Prigogine-Bülbül
i Bulbul di Prigogine

2121 *Chlorocichla simplex*
Passeriformes - Pycnonotidae
e Simple Greenbul; Simple Leaflove;
Simple Bulbul; Leaflove
f Bulbul modeste
d Hartlaub-Bülbül
i Bulbul semplice

2122 *Chloropeta gracilirostris*
Passeriformes - Sylviidae
e Thin-billed Flycatcher-Warbler;
Papyrus Yellow-Warbler (CSA);
Thin-billed Flycatcher; Yellow
Swamp-Warbler; Thin-billed
Yellow-Warbler; Payrus Yellow
Flycatcher-Warbler; Mountain
Yellow Warbler; Yellow Samp
Flycatcher-Warbler
f Chloropète aquatique; Fauvette jaune
aquatique
d Gelbbauchrohrsänger
i Canapino delle paludi

2123 *Chloropeta natalensis*
Passeriformes - Sylviidae
e Yellow Flycatcher-Warbler; Dark-
capped Yellow-Warbler (CSA);
African Yellow-Warbler (CSA);
Yellow Warbler; Natal Yellow-
Warbler; Yellow Flycatcher
f Chloropète jaune; Fauvette jaune
commune
d Schnäpperrohrsänger
i Canapino ventregiallo del Natal

2124 *Chloropeta similis*
Passeriformes - Sylviidae
e Mountain Flycatcher-Warbler;
Mountain Yellow-Warbler (CSA);
Mountain Yellow-Flycatcher;
Mountain Warbler
f Chloropète de montagne; Fauvette
jaune de montagne
d Bambusrohrsänger
i Canapino ventregiallo di montagna

2125 *Chlorophanes spiza*
Passeriformes - Fringillidae
e Green Honeycreeper
f Guit-guit émeraude
d Kappennaschvogel
i Reginetta verde

2126 *Chlorophonia callophrys*
Passeriformes - Fringillidae
e Golden-browed Chlorophonia; Blue-
crowned Chlorophonia
f Organiste à sourcils jaunes
i Tangara verde dai sopraccigli dorati

2127   *Chlorophonia cyanea*
        Passeriformes - Fringillidae
   e   Blue-naped Chlorophonia; Blue
        Chlorophonia
   f   Organiste à nuque bleue
   d   Grünorganist
   i   Tangara verde nucablu

2128   *Chlorophonia flavirostris*
        Passeriformes - Fringillidae
   e   Yellow-collared Chlorophonia
   f   Organiste à col jaune
   d   Halsbandorganist
   i   Tangara verde dal collare

2129   *Chlorophonia occipitalis*
        Passeriformes - Fringillidae
   e   Blue-crowned Chlorophonia; Blue-
        crowned Tanager
   f   Organiste à calotte bleue
   d   Goldbrauenorganist
   i   Tangara verde corona azzurra

2130   *Chlorophonia pyrrhophrys*
        Passeriformes - Fringillidae
   e   Chestnut-breasted Chlorophonia
   f   Organiste à ventre brun
   d   Schwarzbrauenorganist;
        Braunbauchorganist
   i   Tangara verde pettocastano

2131   *Chloropipo flavicapilla*
        Passeriformes - Tyrannidae
   e   Yellow-headed Manakin
   f   Manakin à tête jaune
   d   Olivkehlpipra
   i   Manachino testagialla

2132   *Chloropipo holochlora*
        Passeriformes - Tyrannidae
   e   Green Manakin
   f   Manakin vert
   d   Grünpipra
   i   Manachino verde

2133   *Chloropipo unicolor*
        Passeriformes - Tyrannidae
   e   Jet Manakin
   f   Manakin unicolore; Manakin deuil
   d   Atlas-Pipra
   i   Manachino unicolore

2134   *Chloropipo uniformis*
        Passeriformes - Tyrannidae
   e   Olive Manakin; Uniform Manakin
   f   Manakin olive
   d   Roraima-Pipra
   i   Manachino oliva

2135   *Chloropsis aurifrons*
        Passeriformes - Irenidae
   e   Golden-fronted Leafbird; Golden-
        fronted Chloropsis (ISC); Green
        Bulbul (ISC)
   f   Verdin à front d'or
   d   Goldstirnblattvogel
   i   Verdino frontedorata

2136   *Chloropsis cochinchinensis*
        Passeriformes - Irenidae
   e   Blue-winged Leafbird; Gold-mantled
        Chloropsis (ISC); Jerdon's Leafbird;
        Gold-mantled Leafbird; Jerdon's
        Chloropsis; Golden-headed Leafbird;
        Yellow-headed Leafbird
   f   Verdin à tête jaune
   d   Blauflügelblattvogel;
        Gelbkopfblattvogel
   i   Verdino aliazzurre

2137   *Chloropsis cyanopogon*
        Passeriformes - Irenidae
   e   Lesser Green-Leafbird; Blue-
        whiskered Leafbird
   f   Verdin barbe-bleue
   d   Blaubartblattvogel
   i   Verdino minore

2138   *Chloropsis flavipennis*
        Passeriformes - Irenidae
   e   Philippine Leafbird; Yellow-quilled
        Leafbird; Yellow-billed Leafbird
   f   Verdin à ailes jaunes
   d   Philippinen-Blattvogel
   i   Verdino delle Filippine

2139   *Chloropsis hardwickei*
        Passeriformes - Irenidae
   e   Orange-bellied Leafbird; Harwicke's
        Leafbird
   f   Verdin de Hardwicke; Chloropsis de
        Hardwicke
   d   Blaubartblattvogel;

d   Poortman-Kolibri
i   Colibrì smeraldo di Poortman

2170   *Chlorostilbon ricordii*
    Apodiformes - Trochilidae
e   Cuban Emerald; Emerald
    Hummingbird (WI); Hummingbird
    (WI); Godbird (WI)
f   Émeraude de Ricord
d   Ricord-Kolibri
i   Colibrì smeraldo di Cuba

2171   *Chlorostilbon russatus*
    Apodiformes - Trochilidae
e   Coppery Emerald
f   Émeraude cuivrée
d   Bronzekolibri
i   Colibrì smeraldo ramato

2172   *Chlorostilbon salvini*
    Apodiformes - Trochilidae
e   Salvin's Emerald

2173   *Chlorostilbon stenura*
    Apodiformes - Trochilidae
e   Narrow-tailed Emerald
f   Émeraude à queue étroite
d   Schmalschwanzkolibri
i   Colibrì smeraldo codastretta

2174   *Chlorostilbon swainsonii*
    Apodiformes - Trochilidae
e   Hispaniolan Emerald
f   Quanga négresse (Ants); Colibri
    (Ants); Émeraude d'Hispaniola
d   Swainson-Kolibri
i   Colibrì smeraldo di Hispaniola

2175   *Chlorothraupis carmioli*
    Passeriformes - Thraupidae
e   Carmiol's Tanager; Olive Tanager;
    Yellow-lored Tanager
f   Tangara olive
d   Carmiol-Tangare
i   Tangara di Carmioli

2176   *Chlorothraupis frenata*
    Passeriformes - Thraupidae
e   Olive Tanager; Yellow-lored Tanager

2177   *Chlorothraupis guttata*
    Passeriformes - Thraupidae
e   Yellow-browed Tanager

2178   *Chlorothraupis olivacea*
    Passeriformes - Thraupidae
e   Lemon-spectacled Tanager; Lemon-
    browed Tanager; Yellow-browed
    Tanager; Spectacled Tanager
f   Tangara à lunettes
d   Gelbbrauentangare
i   Tangara dagli occhiali gialli

2179   *Chlorothraupis stolzmanni*
    Passeriformes - Thraupidae
e   Ochre-breasted Tanager
f   Tangara de Stolzmann
d   Ockerbrusttangare
i   Tangara petto ocra

2180   *Chondestes grammacus*
    Passeriformes - Fringillidae
e   Lark Sparrow
f   Bruant à joues marron; Pinson à
    joues marron
d   Rainammer
i   Passero calandra

2181   *Chondrohierax uncinatus*
    Falconiformes - Accipitridae
e   Hook-billed Kite; Mountain Hawk
    (WI)
f   Milan bec-en-croc
d   Langschnabelweihe
i   Nibbio beccouncinato; Nibbio dal
    becco aduco

2182   *Chondrohierax uncinatus wilsonii*
    Falconiformes - Accipitridae
e   Cuban Kite
f   Milan de Cuba

2183   *Chordeiles acutipennis*
    Caprimulgiformes - Caprimulgidae
e   Lesser Nighthawk; Trilling
    Nighthawk
f   Engoulevent ronronneur;
    Engoulevent minime
d   Texas-Nachtschwalbe
i   Succiacapre sparviero minore

2184   *Chordeiles gundlachii*
Caprimulgiformes - Caprimulgidae
e   Antillean Nighthawk; Pirra-ma-dick (WI); Rickery-Dick (WI); Gimme-a-bit (WI); Mosquito Hawk (WI)
f   Peut-on-voir (Ants); Engoulevent piramidig
d   Antillen-Nachtschwalbe
i   Succiacapre sparviero delle Antille

2185   *Chordeiles minor*
Caprimulgiformes - Caprimulgidae
e   Common Nighthawk; Nighthawk; Booming Nighthawk
f   Engoulevent d'Amérique
d   Nachtfalke; Falkennachtschwalbe
i   Succiacapre americano; Caprimulgo sparviero comune; Caprimulgo sparviero; Succciacapre sparviero comune

2186   *Chordeiles pusillus*
Caprimulgiformes - Caprimulgidae
e   Least Nighthawk
f   Engoulevent nain
d   Gnomennachtschwalbe
i   Succicapr

2187   *Chordeiles rupestris*
Caprimulgiformes - Caprimulgidae
e   Sand-coloured Nighthawk; White-throated Nightjar
f   Engoulevent sable
d   Sandnachtschwalbe
i   Succiacapre sparviere color sabbia

2188   *Chordeiles vielliardi*
Caprimulgiformes - Caprimulgidae
e   Caatinga Nighthawk

2189   *Chrysococcyx basalis*
Cuculiformes - Cuculidae
e   Horsfield's Bronze-Cuckoo; Horsfield's Cuckoo; Australian Bronze-Cuckoo; Bronze Cuckoo; Narrow-billed Bronze-Cuckoo
f   Coucou de Horsfield
d   Rotschwanzkuckuck
i   Cuculo dorato di Horsfield

2190   *Chrysococcyx caprius*
Cuculiformes - Cuculidae
e   Diederic Cuckoo; Diderik Cuckoo (CSA); Didric Cuckoo; Dideric Cuckoo
f   Coucou didric
d   Diderikkuckuck; Goldkuckuck
i   Cuculo dorato di Levaillant; Cuculo smeraldino africano

2191   *Chrysococcyx crassirostris*
Cuculiformes - Cuculidae
e   Pied Bronze-Cuckoo; Island Cuckoo
f   Coucou de Salvadori
i   Cuculo bronzato delle Molucche

2192   *Chrysococcyx cupreus*
Cuculiformes - Cuculidae
e   African Emerald-Cuckoo; Emerald Cuckoo (CSA)
f   Coucou foliotocol
d   Smaragdkuckuck
i   Cuculo smeraldino africano

2193   *Chrysococcyx flavigularis*
Cuculiformes - Cuculidae
e   Yellow-throated Cuckoo; Yellow-throated Green-Cuckoo
f   Coucou à gorge jaune
d   Gelbkehlkuckuck
i   Cuculo dorato golagialla

2194   *Chrysococcyx klaas*
Cuculiformes - Cuculidae
e   Klaas's Cuckoo
f   Coucou de Klaas
d   Klaas-Kuckuck
i   Cuculo dorato di Klaas

2195   *Chrysococcyx lucidus*
Cuculiformes - Cuculidae
e   Shining Bronze-Cuckoo; Golden Bronze-Cuckoo; Golden Cuckoo; Shining Cuckoo
f   Coucou éclatant
d   Bronzekuckuck
i   Cuculo bronzato splendente

2196   *Chrysococcyx maculatus*
Cuculiformes - Cuculidae
e   Asian Emerald-Cuckoo; Asiatic

Emerald-Cuckoo; Emerald Cuckoo;
Oriental Emerald-Cuckoo
f   Coucou émeraude
d   Prachtkuckuck
i   Cuculo smeraldino asiatico

2197   *Chrysococcyx malayanus*
Cuculiformes - Cuculidae
e   Malaysian Bronze-Cuckoo
f   Coucou malais
d   Glanzkuckuck

2198   *Chrysococcyx meyeri*
Cuculiformes - Cuculidae
e   White-eared Bronze-Cuckoo;
Meyer's Cuckoo; Meyer's Bronze-
Cuckoo; White-eared Cuckoo
f   Coucou de Meyer
d   Rotschwingenkuckuck
i   Cuculo bronzato di Meyer

2199   *Chrysococcyx minutillus*
Cuculiformes - Cuculidae
e   Little Bronze-Cuckoo; Malay
Cuckoo; Malay Green-Cuckoo;
Malay Bronze-Cuckoo; Malaysian
Bronze-Cuckoo; Australian Bronze-
Cuckoo
f   Coucou menu
d   Glanzkuckuck
i   Cuculo bronzato minore

2200   *Chrysococcyx osculans*
Cuculiformes - Cuculidae
e   Black-eared Cuckoo
f   Coucou oreillard
d   Schwarzohrkuckuck
i   Cuculo orecchienere

2201   *Chrysococcyx ruficollis*
Cuculiformes - Cuculidae
e   Rufous-throated Bronze-Cuckoo;
Reddish-throated Cuckoo; Reddish-
throated Bronze-Cuckoo; Mountain
Bronze-Cuckoo
f   Coucou à gorge rousse
d   Rothalskuckuck
i   Cuculo bronzato golarossa

2202   *Chrysococcyx rufomerus*
Cuculiformes - Cuculidae

e   Green-cheeked Bronze-Cuckoo;
Dark-backed Bronze-Cuckoo
f   Coucou à joues vertes
i   Cuculo bronzato guanceverdi

2203   *Chrysococcyx russatus*
Cuculiformes - Cuculidae
e   Gould's Bronze-Cuckoo; Rufous
Bronze-Cuckoo
f   Coucou roussâtre
i   Cuculo bronzato di Gould

2204   *Chrysococcyx xanthorhynchus*
Cuculiformes - Cuculidae
e   Violet Cuckoo
f   Coucou violet
d   Amethystkuckuck
i   Cuculo violetto

2205   *Chrysocolaptes festivus*
Piciformes - Picidae
e   White-naped Woodpecker; Black-
backed Woodpecker (ISC); Black-
rumped Woodpecker; Black-backed
Yellow Woodpecker
f   Pic de Goa
d   Goldschulterspecht;
Schwarzrückensultanspecht
i   Picchio nucabianca

2206   *Chrysocolaptes lucidus*
Piciformes - Picidae
e   Greater Flameback; Crimson-backed
Woodpecker; Greater Golden-backed
Woodpecker; Greater Goldenback;
Greater Flame-backed Woodpecker;
Larger Gold-backed Woodpecker;
Large Golden-backed Woodpecker;
Larger Golden-backed Woodpecker;
Golden-backed Woodpecker;
Golden-backed Four-toed
Woodpecker
f   Pic sultan
d   Sultanspecht
i   Picchio dorsorosso maggiore

2207   *Chrysolampis mosquitus*
Apodiformes - Trochilidae
e   Ruby-topaz Hummingbird
f   Colibri rubis-topaze

d Moskitokolibri
i Colibrì rubino-topazio

2208 *Chrysolophus amherstiae*
Galliformes - Phasianidae
e Lady Amherst's Pheasant; Chinese
Copper-Pheasant
f Faisan de Lady Amherst; Faisan
d'Amherst
d Diamantfasan
i Fagiano di Lady Amherst

2209 *Chrysolophus pictus*
Galliformes - Phasianidae
e Golden Pheasant
f Faisan doré
d Goldfasan
i Fagiano dorato

2210 *Chrysomma altirostre*
Passeriformes - Sylviidae
e Jerdon's Babbler; Jerdon's Moupinia
f Timalie de Jerdon
d Jerdon-Mupinie
i Garrulo di Jerdon

2211 *Chrysomma poecilotis*
Passeriformes - Sylviidae
e Rufous-tailed Babbler; Rufous-
crowned Babbler; Rufous-crowned
Moupinia; Chestnut-tailed Moupinia;
Chestnut-tailed Babbler; Rufous-
tailed Moupinia; Chestnut-crowned
Moupinia; Chinese Babbler
f Timalie à couronne rousse
d Rotschwanzmupinie
i Garrulo corona rossa

2212 *Chrysomma sinense*
Passeriformes - Sylviidae
e Yellow-eyed Babbler; Oriental
Yellow-eyed Babbler; Ceylon
Yellow-eyed Babbler
f Timalie aux yeux d'or
d Goldaugentimalie
i Garrulo occhigialli orientale

2213 *Chrysothlypis chrysomelas*
Passeriformes - Fringillidae
e Black-and-yellow Tanager
f Tangara loriot

d Zitronentangare
i Tangara nera e gialla

2214 *Chrysothlypis salmoni*
Passeriformes - Fringillidae
e Scarlet-and-white Tanager
f Tangara rouge
d Seidenflankentangare
i Tangara rossa e bianca

2215 *Chrysuronia oenone*
Apodiformes - Trochilidae
e Golden-tailed Sapphire
f Saphir oenome
d Bronzeschwanzsaphir
i Colibrì zafiro codadorato

2216 *Chthonicola sagittatus*
Passeriformes - Pardalotidae
e Speckled Warbler; Little Fieldwren
f Séricorne flèche
d Grundhuscher
i Sericeo macchiettato

2217 *Chunga burmeisteri*
Gruiformes - Cariamidae
e Black-legged Seriema; Cariama;
Seriema; Burmeister's Seriema;
Lesser Seriema
f Cariama de Burmeister
d Tchunja; Tschunga
i Seriema di Burmeister; Piccolo
seriema

2218 *Cichladusa arquata*
Passeriformes - Muscicapidae
e Collared Palm-Thrush; Morning
Warbler; Collared Morning-Warbler;
Collared Morning-Thrush; Scrub
Palm-Thrush
f Cichladuse à collier; Cichladuse de
Peters
d Morgenrötel
i Cantore dal collare

2219 *Cichladusa guttata*
Passeriformes - Muscicapidae
e Spotted Morning-Thrush; Spotted
Palm-Thrush (CSA); Spotted
Morning-Warbler
f Cichladuse tachetée; Cichladuse à

poitrine tachetée
d Tropfenrötel
i Cantore maculato

2220 *Cichladusa ruficauda*
Passeriformes - Muscicapidae
e Rufous-tailed Palm-Thrush; Rufous-tailed Morning-Warbler; Rufous-tailed Morning-Thrush; Red-tailed Palm-Thrush
f Cichladuse à queue rousse
d Graubruströtel
i Cantore codarosso

2221 *Cichlherminla lherminieri*
Passeriformes - Muscicapidae
e Forest Thrush; Yellow-legged Thrush (WI)
f Grive à pattes jaunes (Ants); Mauvis (Ants); Grive à pieds jaunes
d Antillen-Drossel
i Tordo di foresta dei Caraibi

2222 *Cichlocolaptes leucophrus*
Passeriformes - Furnariidae
e Pale-browed Treehunter
f Anabate à sourcils blancs
d Fahlbrauenblattspäher
i Grattafoglie ferrugineo

2223 *Cichlopsis leucogenys*
Passeriformes - Muscicapidae
e Rufous-brown Solitaire
f Solitaire roux
d Rostrückenklarino
i Tordo solitario bruno

2224 *Cicinnurus magnificus*
Passeriformes - Corvidae
e Magnificent Bird of Paradise
f Paradisier magnifique
d Sichelschwanzparadiesvogel
i Paradisea magnifica

2225 *Cicinnurus regius*
Passeriformes - Corvidae
e King Bird of Paradise; Little King Bird of Paradise
f Paradisier royal; Manucaude royal
d Königsparadiesvogel
i Paradisea reale

2226 *Cicinnurus respublica*
Passeriformes - Corvidae
e Wilson's Bird of Paradise; Weigeu Bird of Paradise
f Paradisier républicain
d Wilson-Paradiesvogel; Blauköpfiger Paradiesvogel
i Paradisea della repubblicca; Croce di Cristo; Paradisea repubblicana

2227 *Ciconia abdimii*
Ciconiiformes - Ciconiidae
e Abdim's Stork; White-bellied Stork
f Cigogne d'Abdim
d Abdim-Storch; Regenstorch
i Sfenorinco; Cicogna di Abdim

2228 *Ciconia boyciana*
Ciconiiformes - Ciconiidae
e Oriental Stork; Oriental White-Stork; Eastern White-Stork
f Cigogne orientale
d Schwarzschnabelstorch
i Cicogna bianca orientale

2229 *Ciconia ciconia*
Ciconiiformes - Ciconiidae
e White Stork
f Cigogne blanche
d Weißstorch; Weißer Storch
i Cicogna bianca

2230 *Ciconia episcopus*
Ciconiiformes - Ciconiidae
e Woolly-necked Stork; White-necked Stork (ISC); Storm's Stork; Bishop Stork; Indian White-necked Stork
f Cigogne épiscopale
d Wollhalsstorch
i Cigogna collolanoso

2231 *Ciconia maguari*
Ciconiiformes - Ciconiidae
e Maguari Stork
f Cigogne maguari
d Maguaristorch
i Cicogna maguari

2232 *Ciconia nigra*
Ciconiiformes - Ciconiidae
e Black Stork

f Cigogne noire
d Schwarzstorch
i Cicogna nera

2233 Ciconiformes
e Storks; Herons
f Ciconiformes
d Schreitvögel; Stelzvögel

2234 Ciconiidae
  Ciconiformes
e Storks
f Cigognes; Ciconidés
d Störche

2235 Cinclidae
  Passeriformes
e Dippers
f Cinclidés
d Wasseramseln

2236 *Cinclidium diana*
  Passeriformes - Muscicapidae
e Sunda Robin; Sunda Blue-Robin;
   Indigo Robin
f Notodèle de la Sonde
d Diademschmätzer
i Callene della Sonda

2237 *Cinclidium frontale*
  Passeriformes - Muscicapidae
e Blue-fronted Robin; Blue-fronted
   Callene; Blue-fronted Long-tailed
   Robin
f Notodèle à front bleu
d Callene
i Callene fronteazzura

2238 *Cinclidium leucurum*
  Passeriformes - Muscicapidae
e White-tailed Robin; Blue Robin;
   White-tailed Blue-Robin
f Notodèle à queue blanche
d Schattenschmätzer
i Callene codabianca

2239 *Cinclocerthia gutturalis*
  Passeriformes - Sturnidae
e Grey Trembler
f Trembleur gris
i Mimo grigio delle Antille

2240 *Cinclocerthia ruficauda*
  Passeriformes - Sturnidae
e Brown Trembler; Trembler
f Grive trembleuse (Ants); Trembleur
   brun; Moqueur trembleur; Cocobino
   (Ants)
d Zitterdrossel
i Mimo ruficauda

2241 *Cinclodes antarcticus*
  Passeriformes - Furnariidae
e Blackish Cinclodes
f Cinclode fuligineux
d Einfarbuferwipper
i Batticoda antartico

2242 *Cinclodes aricomae*
  Passeriformes - Furnariidae
e Royal Cinclodes
f Cinclode royal

2243 *Cinclodes atacamensis*
  Passeriformes - Furnariidae
e White-winged Cinclodes
f Cinclode à ailes blanches
d Flügelstreifuferwipper
i Batticoda

2244 *Cinclodes comechingonus*
  Passeriformes - Furnariidae
e Cordoba Cinclodes; Chestnut-winged
   Cinclodes; Comechingones
   Cinclodes; Sierra Cinclodes
f Cinclode gris
d Cordoba-Uferwipper
i Batticoda di Comechingones

2245 *Cinclodes excelsior*
  Passeriformes - Furnariidae
e Stout-billed Cinclodes; Stout-billed
   Miner; Shortbill; Short-billed
   Cinclodes; Paramo Cinclodes
f Cinclode du paramo
d Starkschnabelerdhacker
i Batticoda beccorobusto

2246 *Cinclodes fuscus*
  Passeriformes - Furnariidae
e Bar-winged Cinclodes
f Cinclode brun

d   Bindenuferwipper
i   Batticoda bruno

2247   ***Cinclodes nigrofumosus***
      Passeriformes - Furnariidae
e   Seaside Cinclodes; Chilean Seaside
     Cinclodes
f   Cinclode du Ressac
d   Strandwipper
i   Batticoda nerofumo

2248   ***Cinclodes olrogi***
      Passeriformes - Furnariidae
e   Olrog's Cinclodes
f   Cinclode d'Olrog
d   Olrog-Uferwipper
i   Batticoda di Olrog

2249   ***Cinclodes oustaleti***
      Passeriformes - Furnariidae
e   Grey-flanked Cinclodes
f   Cinclode d'Oustalet
d   Frauflankenuferwipper
i   Batticoda fianchigrigi

2250   ***Cinclodes pabsti***
      Passeriformes - Furnariidae
e   Long-tailed Cinclodes; Pabst's
     Cinclodes
f   Cinclode à longue queue
s   Langschwanzuferwipper
i   Batticoda di Pabst

2251   ***Cinclodes palliatus***
      Passeriformes - Furnariidae
e   White-bellied Cinclodes
f   Cinclode à ventre blanc
d   Weißbauchuferwipper
i   Batticoda ventrebianco

2252   ***Cinclodes patagonicus***
      Passeriformes - Furnariidae
e   Dark-bellied Cinclodes
f   Cinclode à ventre sombre
d   Patagonischer Uferwipper;
     Uferwippschwanz
i   Batticoda della Patagonia

2253   ***Cinclodes taczanowskii***
      Passeriformes - Furnariidae
e   Surf Cinclodes; Taczanowski's

     Cinclodes
f   Cinclode de Taczanowski
d   Küstenwipper
i   Batticoda di Taczanowski

2254   ***Cincloramphus cruralis***
      Passeriformes - Sylviidae
e   Brown Songlark
f   Mégalure brune
d   Schwarzbauchlerchensänger
i   Cantore terricolo bruno

2255   ***Cincloramphus mathewsi***
      Passeriformes - Sylviidae
e   Rufous Songlark
f   Mégalure de Mathews
d   Rostbürzellerchensänger
i   Cantore terricolo rossiccio

2256   ***Cinclosoma ajax***
      Passeriformes - Cinclosomatidae
e   Painted Quail-Thrush; Ajax Quail-
     Thrush; Ajax Scrub-Robin
f   Cinclosome ajax
d   Ajaxflöter
i   Tordo-quaglia variopinto

2257   ***Cinclosoma alisteri***
      Passeriformes - Cinclosomatidae
e   Nullabor Quail-Thrush
d   Donga-Flöter

2258   ***Cinclosoma castaneothorax***
      Passeriformes - Cinclosomatidae
e   Chestnut-breasted Quail-Thrush
f   Cinclosome à poitrine cannellé
i   Tordo-quaglia pettocastano

2259   ***Cinclosoma castanotus***
      Passeriformes - Cinclosomatidae
e   Chestnut Quail-Thrush
f   Cinclosome marron
d   Rotrückenflöter
i   Tordo-quaglia dorsocastano

2260   ***Cinclosoma castanotus castanotus***
      Passeriformes - Cinclosomatidae
e   Eastern Chestnut Quail-Thrush
     (ANZ)

2261   **Cinclosoma cinnamomeum**
       Passeriformes - Cinclosomatidae
e   Cinnamon Quail-Thrush
f   Cinclosome de Nullarbor;
       Cinclosome cannellé
d   Zimtflöter
i   Tordo-quaglia dorsocannella

2262   **Cinclosoma punctatum**
       Passeriformes - Cinclosomatidae
e   Spotted Quail-Thrush
f   Cinclosome pointillé
d   Fleckenflöter
i   Tordo-quaglia macchiettato

2263   **Cinclosoma punctatum anachoreta**
       Passeriformes - Cinclosomatidae
e   Mount Lofty Ranges Spotted Quail-
       Thrush (ANZ)

2264   **Cinclosomatidae**
       Passeriformes
e   Quail-Thrushes
f   Cinclosomatidés
d   Flöter
i   Cinclosomatidi

2265   **Cinclus cinclus**
       Passeriformes - Cinclidae
e   White-throated Dipper; Dipper;
       Common Dipper; Water Ouzel;
       White-breasted Dipper; White-
       bellied Dipper; Eurasian Dipper
f   Cincle plongeur; Merle d'eau
d   Wasseramsel; Wasserschmätzer
i   Merlo acquaiolo

2266   **Cinclus leucocephalus**
       Passeriformes - Cinclidae
e   White-capped Dipper; South
       American Dipper
f   Cincle à tête blanche
d   Weißkopfwasseramsel
i   Merlo acquaiolo testabianca

2267   **Cinclus mexicanus**
       Passeriformes - Cinclidae
e   American Dipper; North American
       Dipper; Mexican Dipper
f   Cincle d'Amérique; Cincle américain
d   Grauwasseramsel; Graue

       Wasseramsel
i   Merlo acquaiolo del Nordamerica

2268   **Cinclus pallasii**
       Passeriformes - Cinclidae
e   Brown Dipper; Pallas's Dipper;
       Asiatic Dipper
f   Cincle de Pallas
d   Flußwasseramsel; Braune
       Wasseramsel
i   Merlo acquaiolo bruno

2269   **Cinclus schulzi**
       Passeriformes - Cinclidae
e   Rufous-throated Dipper; Argentine
       Dipper
f   Cincle à gorge rousse
d   Rostkehlwasseramsel
i   Merlo acquaiolo di Schulz

2270   **Cinnycerthia fulva**
       Passeriformes - Certhiidae
e   Superciliated Wren; Fulvous Wren

2271   **Cinnycerthia olivascens**
       Passeriformes - Certhiidae
e   Sharpe's Wren

2272   **Cinnycerthia peruana**
       Passeriformes - Certhiidae
e   Peruvian Wren; Sepia Wren; Sepia-
       brown Wren
f   Troglodyte brun
d   Sepiazaunkönig
i   Scricciolo bruno-sepia

2273   **Cinnycerthia unirufa**
       Passeriformes - Certhiidae
e   Rufous Wren; Brown Wren
f   Troglodyte roux
d   Einfarbzaunkönig
i   Scricciolo rossiccio

2274   **Cinnyricinclus femoralis**
       Passeriformes - Sturnidae
e   Abbott's Starling
f   Spréo d'Abbott; Étourneau d'Abbott
d   Abbott-Star
i   Storno di Abbott

2275 **Cinnyricinclus leucogaster**
Passeriformes - Sturnidae
e Violet-backed Starling; Plum-coloured Starling (CSA); Amethyst Starling; White-bellied Amethyst-Starling; Violet Starling
f Sprèo améthyste; Merle améthyste; Merle violet à ventre blanc
d Amethystglanzstar
i Storno ametista

2276 **Cinnyricinclus sharpii**
Passeriformes - Sturnidae
e Sharpe's Starling
f Spréo de Sharpe; Étourneau de Sharpe
d Schwalbenglanzstar; Rostbauchstar
i Storno di Sharpe

2277 **Ciorcaetus pectoralis**
Falconiformes - Accipitridae
e Black-chested Snake-Eagle
d Schwarzbrustschlangenadler
i Biancone pettonero

2278 **Circaetus cinerascens**
Falconiformes - Accipitridae
e Banded Snake-Eagle; Western Banded Snake-Eagle (CSA); Smaller Banded-Eagle; Small Snake-Eagle
f Circaète cendré
d Bandschlangenadler; Braunbrustschlangenadler
i Biancone minore; Circeto a strisce

2279 **Circaetus cinereus**
Falconiformes - Accipitridae
e Brown Snake-Eagle; Brown Harrier-Eagle
f Circaète brun
d Brauner Schlangenadler; Einfarbschlangenadler
i Biancone bruno; Biancone cenerino; Circeto cenerino

2280 **Circaetus fasciolatus**
Falconiformes - Accipitridae
e Fasciated Snake-Eagle; Southern Banded Snake-Eagle (CSA); Southern Snake-Eagle; Banded Snake-Eagle

f Circaète barré
d Graubrustschlangenadler
i Biancone fasciato; Circeto fasciolato

2281 **Circaetus gallicus**
Falconiformes - Accipitridae
e Short-toed Snake-Eagle; Short-toed Eagle; Black-chested Snake-Eagle; Black-breasted Snake-Eagle (CSA); Short-toed Serpent-Eagle; European Snake-Eagle
f Vautour fauve; Circaète Jean-le-Blanc; Circaète à poitrine noire
d Schlangenadler; Schwarzbrustschlangenadler
i Biancone; Falco aquilino bianco

2282 **Circaetus gallicus beaudouini**
Falconiformes - Accipitridae
e Beaudoin's Snake-Eagle; Beaudouin's Harrier-Eagle
f Circaète de Beaudouin

2283 **Circus aeruginosus**
Falconiformes - Accipitridae
e Western Marsh-Harrier; Eurasian Marsh-Harrier; Marsh Harrier; European Marsh-Harrier (CSA); Northern Marsh-Harrier
f Busard des roseaux; Busard harpaye; Busard des marais
d Rohrweihe; Europäischer Rohrweihe
i Falco di palude

2284 **Circus approximans**
Falconiformes - Accipitridae
e Swamp Harrier
f Busard de Gould
d Sumpfweihe
i Falco di palude del Pacifico

2285 **Circus assimilis**
Falconiformes - Accipitridae
e Spotted Harrier; Swamp Hawk; Allied Harrier; Jardine's Harrier; Smoke Hawk; Spotted Swamp-Hawk
f Busard tacheté
d Fleckenweihe
i Albanella macchiata

2286   *Circus buffoni*
        Falconiformes - Accipitridae
e   Long-winged Harrier
f   Busard de Buffon
d   Weißbrauenweihe
i   Albanella di Buffon; Albanella
        americana

2287   *Circus cinereus*
        Falconiformes - Accipitridae
e   Cinereous Harrier
f   Busard bariolé
d   Grauweihe
i   Albanella cinerea

2288   *Circus cyaneus*
        Falconiformes - Accipitridae
e   Northern Harrier; Hen-Harrier;
        Marsh Hawk; Northern Hen-Harrier
f   Busard Saint-Martin
d   Kornweihe
i   Albanella reale

2289   *Circus hudsonius*
        Falconiformes - Accipitridae
e   American Harrier; Marsh Hawk;
        American Hen-Harrier
d   Hudson-Weihe

2290   *Circus macrosceles*
        Falconiformes - Accipitridae
e   Madagascar Marsh-Harrier

2291   *Circus macrourus*
        Falconiformes - Accipitridae
e   Pallid Harrier; Pale Harrier (ISC)
f   Busard pâle
d   Steppenweihe
i   Albanella pallida

2292   *Circus maillardi*
        Falconiformes - Accipitridae
e   Réunion Marsh-Harrier; Madagascar
        Marsh-Harrier; Madagascar Harrier;
        Malagasy Marsh-Harrier; Réunion
        Harrier
f   Busard de Maillard

2293   *Circus maurus*
        Falconiformes - Accipitridae
e   Black Harrier

f   Busard maure
d   Mohrenweihe
i   Albanella nera

2294   *Circus melanoleucos*
        Falconiformes - Accipitridae
e   Pied Harrier; Pied Hawk
f   Busard tchoug
d   Elsterweihe
i   Albanella bianca e nera; Albanella
        variapinta

2295   *Circus pygargus*
        Falconiformes - Accipitridae
e   Montagu's Harrier; Montague's
        Harrier; Montagu Harrier
f   Busard pâle; Busard cendré
d   Wiesenweihe
i   Albanella minore

2296   *Circus ranivorus*
        Falconiformes - Accipitridae
e   African Marsh-Harrier
f   Busard grenouillard
d   Afrikanischer Rohrweihe
i   Falco di palude africano; Albanella
        delle paludi

2297   *Circus spilonotus*
        Falconiformes - Accipitridae
e   Eastern Marsh-Harrier; Swamp
        Hawk; Pacific Harrier; Pacific
        Marsh-Harrier; Australasian Harrier;
        Australasian Marsh-Harrier; Marsh
        Harrier; Allied Harrier; Gould's
        Harrier
f   Busard d'Orient
i   Falco di palude orientale

2298   *Ciridops anna*
        Passeriformes - Fringillidae
e   Ula-ai-hawane
f   Ciridopse d'Anna
d   Kohala-Kleidervogel; Ula-ai-hawane;
        Anna-Kleidervogel
i   Ula-ai-hawane

2299   *Cissa chinensis*
        Passeriformes - Fringillidae
e   Common Green-Magpie; Green
        Magpie; Green Jay; Green Pie;

Hunting Cissa; Hunting Crow;
Chinese Green-Magpie; Green
Hunting-Crow
f Pirolle verte
d Jagdelster
i Gazza verde

2300 ***Cissa chinensis margaritae***
Passeriformes - Corvidae
e Yellow-crowned Cissa
i Gazza verde

2301 ***Cissa hypoleuca***
Passeriformes - Corvidae
e Yellow-breasted Magpie; Eastern
Green-Magpie; Yellow-breasted
Green Magpie
f Pirolle à ventre jaune
d Goldbauchelster

2302 ***Cissa thalassina***
Passeriformes - Corvidae
e Short-tailed Magpie; Short-tailed
Green-Jay; Short-tailed Hunting-
Cissa; Short-tailed Hunting-Crow;
Whitehead's Cissa; Short-tailed
Green-Magpie; Short-tailed Green-
Pie; Short-tailed Cissa
f Pirolle à queue courte
d Buschelster; Kinabulu-Elster
i Gazza acquamarina

2303 ***Cissopis leveriana***
Passeriformes - Fringillidae
e Magpie Tanager
f Tangara pillurion
d Elsterling; Elstertangare
i Tangara gazza

2304 ***Cisticola aberdare***
Passeriformes - Cisticolidae
e Aberdare Cisticola
f Cisticole des Aberdare
d Aberdare-Cistensänger
i Beccamoschino dei Monti Aberdare

2305 ***Cisticola aberrans***
Passeriformes - Cisticolidae
e Lazy Cisticola; Rock Cisticola
(CSA); Rock-loving Cisticola (CSA);
Tinktinkie (CSA)

f Cisticole paresseuse; Cisticole des
rochers
d Smiths Cistensänger;
Langschwanzcistensänger
i Beccamoschino pigro

2306 ***Cisticola angolensis***
Passeriformes - Cisticolidae
e Angola Cisticola
f Cisticole angolaise
i Beccamoschino dell' Angola

2307 ***Cisticola angusticauda***
Passeriformes - Cisticolidae
e Tabora Cisticola; Long-tailed
Cisticola (CSA)
f Cisticole à queue fine
i Beccamoschino di Tabora

2308 ***Cisticola anonymus***
Passeriformes - Cisticolidae
e Chattering Cisticola
f Cisticole babillarde
d Waldcistensänger
i Beccamoschino garrulo

2309 ***Cisticola aridulus***
Passeriformes - Cisticolidae
e Desert Cisticola; Tinktinkie (CSA);
Desert Fantail-Warbler
f Cisticole du désert
d Kalahari-Cistensänger
i Beccamoschino del deserto

2310 ***Cisticola ayresii***
Passeriformes - Cisticolidae
e Wing-snapping Cisticola; Ayres's
Cisticola (CSA); Cloudscraper
(CSA); Tinktinkie (CSA); Ayres's
Cloud-Cisticola
f Cisticole gratte-nuage
d Zwergpinkpink
i Beccamoschino di Ayres

2311 ***Cisticola bodessa***
Passeriformes - Cisticolidae
f Cisticole des Borans
d Boran-Cistensänger
i Beccamoschino del Boran

2312   *Cisticola brachypterus*
       Passeriformes - Cisticolidae
   e   Siffling Cisticola; Short-winged
       Cisticola (CSA); Tinktinkie (CSA);
       Shortwing Cisticola
   f   Cisticole à ailes courtes
   d   Kurzflügelcistensänger
   i   Beccamoschino alibreve

2313   *Cisticola brunnescens*
       Passeriformes - Cisticolidae
   e   Pectoral-patch Cisticola; Pale-
       crowned Cisticola (CSA); Tinktinkie
       (CSA); Pale-crowned Cloud-
       Cisticola
   f   Cisticole brune; Cisticole brunâtre
   d   Blaßkopfpinkpink
   i   Beccamoschino pettomarcata

2314   *Cisticola bulliens*
       Passeriformes - Cisticolidae
   e   Bubbling Cisticola
   f   Cisticole murmure
   d   Angola-Cistensänger
   i   Beccamoschino gorgogliante

2315   *Cisticola cantans*
       Passeriformes - Cisticolidae
   e   Singing Cisticola; Tinktinkie (CSA)
   f   Cisticole chanteuse
   d   Grauer Cistensänger;
       Weißbrauencistensänger;
       Graurückencistensänger
   i   Beccamoschino canoro

2316   *Cisticola carruthersi*
       Passeriformes - Cisticolidae
   e   Carruthers's Cisticola
   f   Cisticole de Carruthers
   d   Papyruscistensänger
   i   Beccamoschino di Carruthers

2317   *Cisticola cherinus*
       Passeriformes - Cisticolidae
   e   Madagascar Cisticola
   f   Cisticole malgache
   i   Beccamoschino del Madagascar

2318   *Cisticola chiniana*
       Passeriformes - Cisticolidae
   e   Rattling Cisticola; Tinktinkie (CSA)

   f   Cisticole grinçante; Cisticole à
       crécelles
   d   Rotscheitelcistensänger
   i   Beccamoschino strepitante

2319   *Cisticola chubbi*
       Passeriformes - Cisticolidae
   e   Chubb's Cisticola
   f   Cisticole de Chubb
   d   Farncistensämger
   i   Beccamoschino di Chubb

2320   *Cisticola cinereolus*
       Passeriformes - Cisticolidae
   e   Ashy Cisticola
   f   Cisticole cendrée
   d   Graucistensänger
   i   Beccamoschino cenerino

2321   *Cisticola dambo*
       Passeriformes - Cisticolidae
   e   Cloud-scraping Cisticola; Black-
       tailed Cloud-Cisticola; Cloudscraper
       Cisticola; Black-tailed Cisticola
   f   Cisticole dambo; Cisticole des
       dambos
   d   Dambo-Pinkpink
   i   Beccamoschino grattanuvole

2322   *Cisticola discolor*
       Passeriformes - Cisticolidae
   e   Brown-backed Cisticola
   f   Cisticole à dos brun
   i   Beccamoschino dorsobruno

2323   *Cisticola distinctus*
       Passeriformes - Cisticolidae
   e   Lynes's Cisticola; Kedong Cisticola
   f   Cisticole de Lynes
   d   Kedong-Cistensänger
   i   Beccamoschino di Lynes

2324   *Cisticola dorsti*
       Passeriformes - Cisticolidae
   e   Dorst's Cisticola; Long-tailed
       Cisticola
   f   Cisticole de Dorst

2325   *Cisticola emini*
       Passeriformes - Cisticolidae
   e   Rock-loving Cisticola

f   Cisticole petrophile
i   Beccamoschino delle rocce

2326   ***Cisticola erythrops***
Passeriformes - Cisticolidae
e   Red-faced Cisticola; Tinktinkie
(CSA)
f   Cisticole à face rousse
d   Rotgesichtcistensänger
i   Beccamoschino facciarossa

2327   ***Cisticola exilis***
Passeriformes - Cisticolidae
e   Golden-headed Cisticola; Golden-
capped Cisticola; Red-headed
Cisticola; Yellow-headed Cisticola;
Golden Cisticola; Gold-capped
Cisticola; Bright-headed Cisticola;
Bright-capped Cisticola; Rufous-
headed Fantail Warbler; Fantail-
Warbler; Rufous-headed Cisticola
f   Cisticole à couronne dorée
d   Goldkopfcistensänger
i   Beccamoschino capodorato

2328   ***Cisticola eximius***
Passeriformes - Cisticolidae
e   Black-necked Cisticola; Black-
backed Cisticola (CSA); Tinktinkie
(CSA); Black-backed Cloud-
Cisticola; Black-headed Cloud-
Cisticola
f   Cisticole à dos noir
d   Schwarzrückencistensänger;
Blütenpinkpink; Blütenpinkpink
i   Beccamoschino collonero

2329   ***Cisticola fulvicapillus***
Passeriformes - Cisticolidae
e   Piping Cisticola; Neddicky (CSA);
Tawny-headed Grass-Warbler;
Neddicky Cisticola
f   Cisticole à couronne rousse
d   Brauner Cistensänger; Neddicky
i   Beccamoschino capofulvo

2330   ***Cisticola galactotes***
Passeriformes - Cisticolidae
e   Winding Cisticola; Black-backed
Cisticola (CSA); Rufous Grass-
Warbler; Greater Black-backed

Cisticola
f   Cisticole roussâtre
d   Schwarzrückencistensänger
i   Beccamoschino dorsonero

2331   ***Cisticola haesitatus***
Passeriformes - Cisticolidae
e   Socotra Cisticola; Island Cisticola
f   Cisticole de Socotra
d   Sokotra-Cistensänger
i   Beccamoschino di Scotra

2332   ***Cisticola hunteri***
Passeriformes - Cisticolidae
e   Hunter's Cisticola; Brown-backed
Cisticola
d   Gebirgscistensänger
i   Beccamoschino di Hunter

2333   ***Cisticola incanus***
Passeriformes - Cisticolidae
e   Socotra Cisticola; Socotra Grass-
Warbler; Socotra Warbler; Incana
f   Cisticole pâle; Fauvette de Socotra
d   Sokotra-Sänger
i   Incana di Socotra

2334   ***Cisticola juncidis***
Passeriformes - Cisticolidae
e   Zitting Cisticola; Fan-tailed Warbler
(CSA); Fan-tailed Cisticola (CSA);
Cloudscraper (CSA); Tinktinkie
(CSA); African Warbler; Streaked
Fantail-Warbler (ISC); Common
Fan-tailed Warbler; Rufous Fan-
tailed Warbler; Ceylon Fantail-
Warbler; Streak-headed Fantail-
Warbler
f   Cisticole des joncs
d   Cistensänger
i   Beccamoschino; Beccamoschine
comune

2335   ***Cisticola lais***
Passeriformes - Cisticolidae
e   Wailing Cisticola; Tinktinkie (CSA)
f   Cisticole plaintive
d   Trauercistensänger
i   Beccamoschino lamentoso

2336 *Cisticola lateralis*
       Passeriformes - Cisticolidae
e   Whistling Cisticola
f   Cisticole siffleuse
d   Pfeifcistensänger
i   Beccamoschino fischiante

2337 *Cisticola lepe*
       Passeriformes - Cisticolidae
e   Lepe Cisticola; Angola Cisticola;
       Angola Red-faced Cisticola
f   Cisticole de Lepi
d   Blaßbauchcistensänger
i   Beccamoschino dell'Angola

2338 *Cisticola melanura*
       Passeriformes - Cisticolidae
e   Slender-tailed Cisticola, Angola
       Slender-tailed Cisticola; Pearson's
       Cisticola; Pearson's Warbler; Black-
       tailed Warbler; Black-tailed Cisticola
f   Cisticole à queue noire
d   Schwarzschwanzcistensänger
i   Beccamoschino codanera

2339 *Cisticola mongalla*
       Passeriformes - Cisticolidae
e   Mongalla Cisticola
f   Cisticole de Mongalla
i   Beccamoschino di Mongolia

2340 *Cisticola nanus*
       Passeriformes - Cisticolidae
e   Tiny Cisticola
f   Cisticole naine
d   Dornbuschcistensänger
i   Beccamoschino nano

2341 *Cisticola natalensis*
       Passeriformes - Cisticolidae
e   Croaking Cisticola; Tinktinkie
       (CSA); Striped Cisticola
f   Cisticole striée
d   Strichelcistensänger
i   Beccamoschino del Natal

2342 *Cisticola nigriloris*
       Passeriformes - Cisticolidae
e   Black-lored Cisticola; Black-browed
       Cisticola

f   Cisticole à moustaches
i   Beccamoschino dai sopraccigli neri

2343 *Cisticola njombe*
       Passeriformes - Cisticolidae
e   Churring Cisticola
f   Cisticole njombe; Cisticole des
       alpages
d   Nyika-Cistensänger
i   Beccamoschino del Njombe

2344 *Cisticola pipiens*
       Passeriformes - Cisticolidae
e   Chirping Cisticola; Tinktinkie (CSA)
f   Cisticole pepiante; Cisticole des
       roseaux
d   Sumpfcistensänger
i   Beccamoschino cinguettante

2345 *Cisticola restrictus*
       Passeriformes - Cisticolidae
e   Tana River Cisticola
f   Cisticole du Tana
d   Tana-Cistensänger
i   Beccamoschino del Fiume Tana

2346 *Cisticola robustus*
       Passeriformes - Cisticolidae
e   Stout Cisticola
f   Cisticole robuste
d   Amhara-Cistensänger
i   Beccamoschino robusto

2347 *Cisticola ruficeps*
       Passeriformes - Cisticolidae
e   Red-pate Cisticola
f   Cisticole à tête rousse
d   Rotkopfcistensänger
i   Beccamoschino corona rossa

2348 *Cisticola rufilatus*
       Passeriformes - Cisticolidae
e   Grey Cisticola; Tinkling Cisticola
       (CSA); Tinktinkie (CSA)
f   Cisticole grise; Cisticole des taillis
d   Rotschwanzcistensänger
i   Beccamoschino tintinnante

2349 *Cisticola rufus*
       Passeriformes - Cisticolidae
e   Rufous Cisticola; Rufous Grass-

Warbler
f Cisticole rousse
d Rostcistensänger
i Beccamoschino rossiccio

2350 *Cisticola subruficapillus*
Passeriformes - Cisticolidae
e Red-headed Cisticola; Grey-backed
Cisticola (CSA); Tinktinkie (CSA)
f Cisticole à dos gris
d Bergcistensänger
i Beccamoschino corona castana

2351 *Cisticola textrix*
Passeriformes - Cisticolidae
e Tink-tink Cisticola; Cloud Cisticola
(CSA); Tinktinkie (CSA); Pinc-pinc;
Spotted Cloud-Cisticola; Pinc-pinc
Cisticola
f Cisticole pinc-pinc; Cisticole
ponctuée
d Pinkpink
i Beccamoschino delle nuvole

2352 *Cisticola tinniens*
Passeriformes - Cisticolidae
e Tinkling Cisticola; Levaillant's
Cisticola (CSA); Tinktinkie (CSA);
Lesser Black-backed Cisticola
f Cisticole à sonnette; Petit Cisticole
roussâtre
d Ufercistensänger
i Beccamoschino di Levaillant

2353 *Cisticola troglodytes*
Passeriformes - Cisticolidae
e Foxy Cisticola
f Cisticole russule; Cisticole grisâtre
d Fuchscistensänger
i Beccamoschino volpino

2354 *Cisticola woosnami*
Passeriformes - Cisticolidae
e Trilling Cisticola
f Cisticole de Woosnam
d Miombo-Cistensänger
i Beccamoschino di Woosnam

2355 Cisticolidae
Passeriformes
e African Warblers

f Cisticolidés
d Afrikanische Grasmücken;
Cistensänger
i Cisticolidi

2356 *Cistothorus apolinari*
Passeriformes - Certhiidae
e Apolinar's Wren; Apolinar's Marsh-
Wren; Marsh Wren
f Troglodyte d'Apolinar
d Apolinar-Zaunkönig
i Scricciolo di Apolinar

2357 *Cistothorus meridae*
Passeriformes - Certhiidae
e Merida Wren; Paramo Wren
f Troglodyte du Merida
d Merida-Zaunkönig
i Scricciolo del Paramo

2358 *Cistothorus paludicola*
Passeriformes - Certhiidae
e Western Marsh-Wren

2359 *Cistothorus palustris*
Passeriformes - Certhiidae
e Marsh Wren; Wren; Long-billed
Marsh-Wren; Eastern Marsh-Wren
f Troglodyte des marais
d Sumpfzaunkönig
i Scricciolo di palude

2360 *Cistothorus platensis*
Passeriformes - Certhiidae
e Sedge-Wren; Short-billed Marsh-
Wren; Western Grasswren;
Grasswren
f Troglodyte à bec court
d Seggenzaunkönig
i Scricciolo dei giunchi

2361 *Cistothorus polyglottus*
Passeriformes - Certhiidae
e Eastern Grasswren

2362 *Cistothorus stellaris*
Passeriformes - Certhiidae
e Sedge-Wren

2363 *Cittura cyanotis*
Coraciiformes - Halcyonidae

e Lilac-marked Kingfisher; Masked
Kingfisher; Blue-eared Kingfisher;
Celebes Blue-eared Kingfisher;
Temminck's Kingfisher; Lilac
Kingfisher; Lilac-cheeked
Kingfisher; Sulawesi Blue-eared
Kingfisher
f Martin-chasseur oreillard; Martin-
chasseur oreillon-bleu
d Blauohrliest
i Martin pescatore lilla

2364 *Cladorhynchus leucocephalus*
Charadriiformes - Recurvirostridae
e Banded Stilt
f Échasse à tête blanche
d Schlammstelzer
i Cavaliere fasciato

2365 *Clamator coromandus*
Cuculiformes - Cuculidae
e Chestnut-winged Cuckoo; Red-
winged Crested-Cuckoo (ISC)
f Coucou à collier
d Koromandel-Kuckuck
i Cuculo dal ciuffo alirosse

2366 *Clamator glandarius*
Cuculiformes - Cuculidae
e Great Spotted-Cuckoo
f Coucou-geai
d Häherkuckuck
i Cuculo dal ciuffo

2367 *Clamator levaillantii*
Cuculiformes - Cuculidae
e Levaillant's Cuckoo; Striped Cuckoo
(CSA); Stripe-breasted Cuckoo;
African Striped-Cuckoo; Striped-
crested Cuckoo
f Coucou de Levaillant
d Kap-Kuckuck
i Cuculo di Levaillant

2368 *Clangula hyemalis*
Anseriformes - Anatidae
e Long-tailed Duck; Oldsquaw (NA);
Old Squaw Duck
f Harelde de Miquelon; Harelde
kakawi; Harelde boréale; Canard
kakawi

d Eisente
i Moretta codona

2369 *Claravis godefrida*
Columbiformes - Columbidae
e Purple-winged Ground-Dove;
Purple-barred Ground-Dove;
Godfrey's Dove
f Colombe de Geoffroy
d Purpurbindentäubchen; Geoffrroy-
Täubchen
i Tortora barrata di porpora

2370 *Claravis mondetoura*
Columbiformes - Columbidae
e Maroon-chested Ground-Dove;
Purple-breasted Ground-Dove;
Mondetour's Dove
f Colombe mondetour
d Mondetour-Täubchen;
Purpurbrusttäubchen
i Tortora pettoporpora

2371 *Claravis pretiosa*
Columbiformes - Columbidae
e Blue Ground-Dove; Blue Dove;
Ashy Dove; Cinereous Dove
f Colombe bleutée
d Schmucktäubchen; Graublauers
Täubchen
i Tortora azzurata

2372 *Cleptornis marchei*
Passeriformes - Zosteropidae
e Golden White-eye; Golden
Honeyeater
f Zostérops doré
d Goldhonigfresser
i Occhialino dorato

2373 *Clibanornis dendrocolaptoides*
Passeriformes - Furnariidae
e Canebrake Groundcreeper
f Synallaxe des bambous
d Schilfbündelnister
i Espinero rampichino

2374 **Climacteridae**
Passeriformes
e Australo-Papuan Treecreepers;
Australian Treecreepers; Australian

Creepers
f Climacteridés
d Baumrutscher; Baumsteiger
i Climatteridi

2375 *Climacteris melanura*
Passeriformes - Climacteridae
e Black-tailed Treecreeper
f Échelet à queue noire
d Braunbauchbaumrutscher
i Ramipichino codanera

2376 *Climacteris rufa*
Passeriformes - Climacteridae
e Rufous Treecreeper
f Échelet roux
d Rostbauchbaumrutscher
i Rampichino rugginoso

2377 *Clytoceyx rex*
Coraciiformes - Halcyonidae
e Shovel-billed Kookaburra; Shovel-
billed Kingfisher; Emperor
Kingfisher
f Martin-chasseur bec-en-cuillère
d Froschschnabel
i Martin pescatore becco a pala

2378 *Clytoctantes alixii*
Columbiformes - Cochleariidae
e Recurve-billed Bushbird
f Batara à bec retroussé
d Verkehrtschnabel
i Mangiaformiche beccocurvo

2379 *Clytoctantes atrogularis*
Columbiformes - Cochleariidae
e Rondonia Bushbird
f Batara du Rondonia

2380 *Clytolaema rubricauda*
Apodiformes - Trochilidae
e Brazilian Ruby
f Colibri rubis-émeraude
d Rubinkolibri
i Colibrì rubino del Brasil

2381 *Clytomyias insignis*
Passeriformes - Maluridae
e Orange-crowned Fairywren; Orange-
crowned Wren; Orange-crowned

Wren-Warbler; Rufous Wren-
Warbler; Rufous Fairy-Wren
f Mérion à tête rousse
d Rotkopfstaffelschwanz
i Scricciolo splendente corona
aranciata

2382 *Clytorhynchus hamlini*
Passeriformes - Corvidae
e Rennell Shrikebill
f Monarque de Rennell
d Rennell-Würgermonarch
i Beccouncinato dell'Isola Rennell

2383 *Clytorhynchus nigrogularis*
Passeriformes - Corvidae
e Black-throated Shrikebill; Black-
faced Shrikebill
f Monarque à gorge noire
d Schwarzkehlwürgermonarch
i Beccouncinato golanera

2384 *Clytorhynchus pachycephaloides*
Passeriformes - Corvidae
e Southern Shrikebill
f Monarque brun
d Hebriden-Würgermonarch
i Beccouncinato meridionale

2385 *Clytorhynchus vitiensis*
Passeriformes - Corvidae
e Fiji Shrikebill; Uniform Shrikebill;
Lesser Shrikebill
f Monarque des Fidji
d Fidji-Würgermonarch
i Beccouncinato delle Figi

2386 *Clytospiza monteiri*
Passeriformes - Estrildidae
e Brown Twinspot; Monteiro's
Twinspot
f Sénégali brun; Bengali tacheté à
ventre roux
d Brauner Tropfenastrild; Monteiro-
Astrild
i Amaranto di Monteiro

2387 *Cnemarchus erythropygius*
Passeriformes - Tyrannidae
e Red-rumped Bush-Tyrant; Red-
Rumped Ground-Tyrant

f   Moucherolle à croupion roux
d   Rostbürzeltyrann
i   Tiranno gropponerosso

2388   *Cnemophilus loriae*
Passeriformes - Corvidae
e   Loria's Bird of Paradise
f   Paradisier de Loria
d   Loria-Paradiesvogel
i   Paradisea di Loria

2389   *Cnemophilus macgregorii*
Passeriformes - Corvidae
e   Crested Bird of Paradise; Sickle-
      crested Bird of Paradise;
      Multicrested Bird of Paradise
f   Paradisier huppé
d   Furchenvogel
i   Paradisca crestafalcata

2390   *Cnemoscopus rubrirostris*
Passeriformes - Fringillidae
e   Grey-hooded Bush-Tanager
f   Tangara capucin
d   Graukopfbuschtangare
i   Tangara olivacea testagrigia

2391   *Cnemotriccus fuscatus*
Passeriformes - Tyrannidae
e   Fuscous Flycatcher
f   Moucherolle fuligineux
d   Finkentyrann
i   Tiranno fosco

2392   *Cnipodectes subbrunneus*
Passeriformes - Tyrannidae
e   Brownish Flycatcher; Brownish
      Twistwing; Twistwing; Brownish
      Twisting Flycatcher; Twisting
      Flycatcher
f   Platyrhynque brun; Tyranneau brun
d   Steifschwingentyrann
i   Tiranno brunastro

2393   *Coccothraustes coccothraustes*
Passeriformes - Fringillidae
e   Hawfinch; Grosbeak; Eurasian
      Hawfinch
f   Grosbec; Grosbec casse-noyaux;
      Grosbec commun; Grosbec ordinaire

d   Kernbeißer
i   Frosone; Frosone eurasiatico

2394   *Coccycolius iris*
Passeriformes - Sturnidae
e   Iris Glossy-Starling; Emerald
      Starling; Emerald Glossy-Starling
f   Choucador iris; Merle métallique vert
d   Schiller-Glanzstar
i   Storno splendente iridato

2395   Coccyzidae
Cuculiformes
e   New World Cuckoos
f   Coccyzidés
d   Regenkuckucke

2396   *Coccyzus americanus*
Cuculiformes - Cuculidae
e   Yellow-billed Cuckoo; Maybird
      (WI); Rain Crow (WI); Rainbird
      (WI)
f   Coulicou à bec jaune
d   Gelbschnabelkuckuck
i   Cuculo americano; Cuculo
      occhigialli; Cuculo pettoperlato;
      Cuculo beccogiallo

2397   *Coccyzus cinereus*
Cuculiformes - Cuculidae
e   Ash-coloured Cuckoo; Dwarf
      Cuckoo
f   Coulicou cendré
d   Graukehlkuckuck
i   Cuculo cenerino

2398   *Coccyzus erythropthalmus*
Cuculiformes - Cuculidae
e   Black-billed Cuckoo
f   Coulicou à bec noir
d   Schwarzschnabelkuckuck
i   Cuculo americano occhirossi; Cuculo
      occhirossi; Cuculo becconero

2399   *Coccyzus euleri*
Cuculiformes - Cuculidae
e   Pearly-breasted Cuckoo
f   Coulicou d'Euler
d   Perlbrustkuckuck

2400 **Coccyzus ferrugineus**
Cuculiformes - Cuculidae
e Cocos Cuckoo; Cocos Islands
Cuckoo
f Coulicou de Cocos
i Cuculo di Cocos; Cuculo delle Cocos

2401 **Coccyzus lansbergi**
Cuculiformes - Cuculidae
e Grey-capped Cuckoo
f Coulicou à tête grise
d Lansberg-Kuckuck
i Cuculo di Lansberg

2402 **Coccyzus melacoryphus**
Cuculiformes - Cuculidae
e Dark-billed Cuckoo; Azara Cuckoo
f Coulicou de Vieillot
d Galapagos-Kuckuck
i Cuculo beccoscuro

2403 **Coccyzus minor**
Cuculiformes - Cuculidae
e Mangrove Cuckoo; Sour-sop Bird
(WI); Old Man Crackers (WI);
Catbird (WI); Coffinbird (WI);
Rainbird (WI); Cowbird (WI)
f Coulicou manioc; Coulicou masqué
(Ants); Coucou maicoc; Gangan
(Ants)
d Mangrovenkuckuck
i Cuculo delle mangrovie

2404 **Coccyzus pumilus**
Cuculiformes - Cuculidae
e Dwarf Cuckoo
f Coulicou nain
d Zwergkuckuck
i Cuculo nano

2405 **Cochlearius cochlearia**
Ciconiiformes - Ardeidae
e Boat-billed Heron; Boatbill Heron;
Boatbill
f Savacou huppé
d Kahnschnabel
i Becco a cucchiaio

2406 **Cochlearius zeledoni**
Ciconiiformes - Ardeidae
e Northern Boat-billed Heron

2407 **Cochoa azurea**
Passeriformes - Muscicapidae
e Javan Cochoa; Javan Thrush;
Malaysian Cochoa; Black-and-blue
Cochoa; Sunda Cochoa; Indonesian
Cochoa
f Cochoa azuré
d Sunda-Schnäpperdrossel
i Cocioa di Giava

2408 **Cochoa beccarii**
Passeriformes - Muscicapidae
e Sumatran Cochoa
f Cochoa de Sumatra
i Cocioa di Beccari

2409 **Cochoa purpurea**
Passeriformes - Muscicapidae
e Purple Cochoa; Purple Thrush
f Cochoa pourpré
d Purpurschnäpperdrossel
i Cocioa purpurea

2410 **Cochoa viridis**
Passeriformes - Muscicapidae
e Green Cochoa; Green Thrush
f Cochoa vert
d Smaragdschnäpperdrossel
i Cocioa verde

2411 **Coeligena bonapartei**
Apodiformes - Trochilidae
e Golden-bellied Starfrontlet
f Inca de Bonaparte
d Goldbauchmusketier
i Fronte stellata di Bonaparte

2412 **Coeligena coeligena**
Apodiformes - Trochilidae
e Bronzy Inca
f Inca céleste
d Himmelsmusketier
i Colibrì inca bronzato

2413 **Coeligena helianthea**
Apodiformes - Trochilidae
e Blue-throated Starfrontlet
f Inca porphyre
d Blaukehlmusketier
i Fronte stellata golablu

2414   *Coeligena iris*
            Apodiformes - Trochilidae
e   Rainbow Starfrontlet; King Louis
       Starfrontlet; King Louis XVI
       Starfrontlet
f   Inca iris
d   Aurora-Musketier
i   Fronte stellata arcobaleno

2415   *Coeligena lutetiae*
            Apodiformes - Trochilidae
e   Buff-winged Starfrontlet; Comte de
       Paris's Starfrontlet
f   Inca à gemme bleue
d   Graf von Paris;
       Braunschwingenmusketier
i   Fronte stellata alifulve

2416   *Coeligena orina*
            Apodiformes - Trochilidae
e   Dusky Starfrontlet
f   Inca de Wetmore
d   Grünmusketier
i   Fronte stellata bruno

2417   *Coeligena phalerata*
            Apodiformes - Trochilidae
e   White-tailed Starfrontlet
f   Inca à queue blanche
d   Weißschwanzmusketier
i   Fronte stellata codabianca

2418   *Coeligena prunellei*
            Apodiformes - Trochilidae
e   Black Inca
f   Inca noir
d   Mohrenmusketier
i   Colibrì inca nero

2419   *Coeligena torquata*
            Apodiformes - Trochilidae
e   Collared Inca; White-cravat
       Hummingbird
f   Inca à collier
d   Krawattenmusketier
i   Colibrì inca dal collare

2420   *Coeligena violifer*
            Apodiformes - Trochilidae
e   Violet-throated Starfrontlet
f   Inca violifère

d   Veilchenmusketier
i   Fronte stellata golaviola

2421   *Coeligena wilsoni*
            Apodiformes - Trochilidae
e   Brown Inca
f   Inca brun
d   Königsmusketier
i   Colibrì inca di Wilson

2422   *Coenocorypha aucklandica*
            Charadriiformes - Scolopacidae
e   Subantarctic Snipe; Auckland Snipe;
       New Zealand Snipe; New Zealand
       Semi-woodcock; Auckland Islands
       Snipe, Island Snipe; Southern Island
       Snipe
f   Bécassine d'Auckland
d   Auckland-Schnepfe
i   Beccaccino subantartico

2423   *Coenocorypha pusilla*
            Charadriiformes - Scolopacidae
e   Chatham Islands Snipe; Chatham
       Snipe; New Zealand Snipe; Little
       Snipe; Bush Snipe; Subantarctic
       Snipe
f   Bécassine des Chatham
i   Beccaccino delle Isole Chatham

2424   *Coereba bahamensis*
            Passeriformes - Coerebidae
e   Bahama Bananaquit

2425   *Coereba flaveola*
            Passeriformes - Coerebidae
e   Bananaquit; Common Bananaquit;
       Teasy (WI); Sugarbird (WI);
       Beenybird (WI); Yellowbreast (WI);
       Bananabird (WI); See-see Bird (WI);
       Black See-see (WI)
f   Sucrier (Ants); Falle jaune (Ants);
       Sucrier à poitrine jaune (Ants);
       Sucrier à ventre jaune
d   Bananaquit; Gelbbrustzuckervogel;
       Zuckervogel
i   Cereba gialla

2426   **Coerebidae**
            Passeriformes
e   Bananaquits

f Coerebidés
d Zuckervögel; Gelbbrustzuckervögel

2427 *Colaptes atricollis*
Piciformes - Picidae
e Black-necked Woodpecker
f Pic à cou noir
d Graustirnspecht
i Picchio collonero

2428 *Colaptes auratus*
Piciformes - Picidae
e Northern Flicker; Common Flicker;
Flicker; Yellow-shafted Flicker;
Black-heart
f Pic flamboyant
d Goldspecht; Flicker
i Picchio dorato; Colatte dorat

2429 *Colaptes cafer*
Piciformes - Picidae
e Red-shafted Flicker
i Picchio; Picchio rosato

2430 *Colaptes campestris*
Piciformes - Picidae
e Campo Flicker; Field Flicker;
Campos Flicker
f Pic champêtre
d Feldspecht; Camposspecht
i Picchio di Campo; Colatte campestre

2431 *Colaptes campestris campestroides*
Piciformes - Picidae
f Pic des prés

2432 *Colaptes chrysocaulosus*
Piciformes - Picidae
e Cuban Flicker

2433 *Colaptes chrysoides*
Piciformes - Picidae
e Gilded Flicker
f Pic chrysoïde
i Picchio aureo

2434 *Colaptes fernandinae*
Piciformes - Picidae
e Fernandina's Flicker; Cuban Flicker;
Fernandina's Woodpecker; Cuban
Woodpecker

f Pic de Fernandina
d Kuba-Specht
i Picchio di Fernandina

2435 *Colaptes melanochloros*
Piciformes - Picidae
e Green-barred Woodpecker; Northern
Flicker; Common Flicker; Flicker;
Red-shafted Flicker
f Pic vert-et-noir
d Grünbindenspecht
i Picchio verde e nero

2436 *Colaptes melanochloros melanolainus*
Piciformes - Picidae
e Golden-breasted Woodpecker
f Pic à poitrine d'or

2437 *Colaptes mexicanoides*
Piciformes - Picidae
e Guatemalan Flicker

2438 *Colaptes pitius*
Piciformes - Picidae
e Chilean Flicker
f Pic du Chili
d Bänderspecht
i Picchio del Cile

2439 *Colaptes punctigula*
Piciformes - Picidae
e Spot-breasted Woodpecker
f Pic de Cayenne
d Tüpfelbrustspecht
i Picchio della Cayenna

2440 *Colaptes rupicola*
Piciformes - Picidae
e Andean Flicker
f Pic des rochers
d Anden-Specht
i Picchio delle Ande

2441 *Colibri coruscans*
Apodiformes - Trochilidae
e Sparkling Violet-ear; Chequered
Violet-ear; Colombian Violet-ear;
Gould's Violet-ear
f Colibri anais

d Veilchenohr; Veilchenohrkolibri
i Colibrì orecchieviola scintillante

2442 *Colibri cyanotus*
Apodiformes - Trochilidae
e Mountain Violet-ear

2443 *Colibri delphinae*
Apodiformes - Trochilidae
e Brown Violet-ear
f Colibri de Delphine
d Telesilla-Kolibri
i Colibrì orechieviola bruno; Colibrì
violetto

2444 *Colibri serrirostris*
Apodiformes - Trochilidae
e White-vented Violet-ear; Brazilian
Violet-ear
f Colibri à oreilles mauves
d Amethystohr
i Codabianca

2445 *Colibri thalassinus*
Apodiformes - Trochilidae
e Green Violet-ear; Mexican Violet-
ear; Mountain Violet-ear
f Colibri thalassin
d Zwergveilchenohr
i Colibrì orechieviola talassino; Colibrì
delle guance viola

2446 Coliidae
Coliiformes
e Mousebirds; Colies
f Coliidés
d Mausvögel; Buschkletterer
i Colidi

2447 Coliiformes
e Mousebirds; Colies
f Coliiformes
d Mausvögel; Buschkletterer
i Coliformi

2448 *Colinus cristatus*
Galliformes - Odontophoridae
e Crested Bobwhite; Bobwhite; Quail
(WI)
f Caille (Ants); Colin huppé

d Haubenwachtel
i Colino crestato

2449 *Colinus leucopogon*
Galliformes - Odontophoridae
e Spot-bellied Bobwhite; Spotted-
bellied Bobwhite; Spot-bellied
Crested-Bobwhite; White-breasted
Bobwhite
f Colin à face blanche
d Tapfenwachtel

2450 *Colinus nigrogularis*
Galliformes - Odontophoridae
e Black-throated Bobwhite; Yucatan
Bobwhite; Black-throated Quail;
Yucatan Quail; Yucatan Black-
throated Bobwhite
f Colin à gorge noire
d Schwarzkehlwachtel
i Colino golanera

2451 *Colinus virginianus*
Galliformes - Phasanidae
e Northern Bobwhite; Common
Bobwhite; Bobwhite; Quail (WI);
Virginia Quail
f Colin de Virginie
d Virginia-Wachtel; Baumwachtel
i Colino della Virginia; Quaglia della
Virginia; Colino

2452 *Colius castanotus*
Coliiformes - Coliidae
e Red-backed Mousebird; Coly
f Coliou à dos marron; Coliou à dos
rouge
d Rotrückenmausvogel
i Uccello topo dorsocastano

2453 *Colius colius*
Coliiformes - Coliidae
e White-backed Mousebird; Coly
f Coliou à dos blanc
d Weißrückenmausvogel;
Gartenmausvogel
i Uccello topo dorsobianco

2454 *Colius leucocephalus*
Coliiformes - Coliidae

e White-headed Mousebird
f Coliou à tête blanche
d Weißkopfmausvogel
i Uccello topo testabianca

2455 *Colius striatus*
Coliiformes - Coliidae
e Speckled Mousebird; Bar-breasted Mousebird
f Coliou rayé; Coliou barré
d Braunflügelmausvogel; Streifenmausvogel
i Uccello topo striato; Uccello topo macchiettato

2456 *Collocalia esculenta*
Apodiformes - Apodidae
e Glossy Swiftlet; White-bellied Swiftlet; Grey-rumped Swiftlet
f Salande soyeuse
d Glanzkopfsalagne
i Salangana ventrebianco

2457 *Collocalia esculenta natalis*
Apodiformes - Apodidae
e Christmas Island Glossy Swiftlet (ANZ)

2458 *Collocalia fuciphagus vestita*
Apodiformes - Apodidae
e Philippine Swiftlet
f Salande soyeuse

2459 *Collocalia inexpectata*
Apodiformes - Apodidae
e Edible-nest Swiftlet; White-nest Swiftlet; German's Swiftlet; Grey Swiftlet; Guam Swiftlet; Palu Swiftlet; Mariana Swiftlet
f Salangane des Andaman
d Weißnestsalangane

2460 *Collocalia linchi*
Apodiformes - Apodidae
e Cave Swiftlet; Linchi Swiftlet
f Salangane linchi
i Salangana di Linch

2461 *Collocalia marginata*
Apodiformes - Apodidae
e Grey-rumped Swiftlet; Philippine Swiftlet
f Salangane des Philippines
i Salangana marginata

2462 *Collocalia troglodytes*
Apodiformes - Apodidae
e Pygmy Swiftlet
f Salangane pygmée
d Zwergsalangane
i Salangana pigmea

2463 *Collocalia whiteheadi*
Apodiformes - Apodidae
e Whitehead's Swiftlet; Whitehead's Mountain-Swiftlet
f Salangane de Whitehead
d Philippinen-Salangane
i Salangana di Whitehead

2464 *Colluricincla boweri*
Passeriformes - Corvidae
e Bower's Shrike-Thrush; Stripe-breasted Shrike-Thrush; Stripe-breasted Thrush-Flycatcher
f Pitohui de Bower
d Graurückengudilang
i Tordo-averla pettostriato

2465 *Colluricincla harmonica*
Passeriformes - Corvidae
e Grey Shrike-Thrush; Grey Thrush-Flycatcher; Grey Thrush
f Pitohui gris
d Graubrustgudilang
i Tordo-averla grigio

2466 *Colluricincla megarhyncha*
Passeriformes - Corvidae
e Rufous Shrike-Thrush; Rufous Thrush-Flycatcher; Brown Shrike-Thrush; Brown Thrush-Flycatcher
f Pitohui châtain
d Waldgudilang
i Tordo-averla rossiccio

2467 *Colluricincla megarhyncha parvula*
Passeriformes - Corvidae
e Little Shrike-Thrush; Little Thrush-Flycatcher
f Pitohui menu
d Sumpfgudilang

2468   *Colluricincla tenebrosa*
          Passeriformes - Corvidae
e   Morningbird; Palau Pitohui; Brown
          Pitohui
f   Pitohui des Palau
d   Rußdickkopf; Singpitohui
i   Tordo-averla di Palau

2469   *Colluricincla umbrina*
          Passeriformes - Corvidae
e   Sooty Shrike-Thrush; Sooty
          Whistler; Obscure Shrike-Thrush;
          Obscure Whistler
f   Pitohui ombré
i   Tordo-averla fuligginoso

2470   *Colluricincla woodwardi*
          Passeriformes - Corvidae
e   Sandstone Shrike-Thrush; Brown-
          breasted Shrike-Thrush; Sandstone
          Thrush-Flycatcher
f   Pitohui des rochers
d   Braunbrustgudllang
i   Tordo-averla di Woodward

2471   *Colonia colonus*
          Passeriformes - Tyrannidae
e   Long-tailed Tyrant
f   Moucherolle à raquettes
d   Langschwanztyrann; Stieltyrann
i   Tiranno codalunga

2472   *Colorhamphus parvirostris*
          Passeriformes - Tyrannidae
e   Patagonian Tyrant; Patagonian
          Tyrannulet; Patagonian Chat-Tyrant;
          Small-billed Tyrant; Chat-Tyrant
f   Pitajo de Patagonie
d   Darwin-Tyrann
i   Tiranno di Patagonia

2473   *Columba albilinea*
          Columbiformes - Columbidae
e   White-necked Pigeon

2474   *Columba albinucha*
          Columbiformes - Columbidae
e   White-naped Pigeon
f   Pigeon à nuque blanche
d   Weißgenicktaube
i   Colomba nucabianca

2475   *Columba albitorques*
          Columbiformes - Columbidae
e   White-collared Pigeon; White-
          collared Dove
f   Pigeon à collier blanc
d   Amharen-Taube; Abessinische
          Felsentaube
i   Colomba dal collare bianco

2476   *Columba araucana*
          Columbiformes - Columbidae
e   Chilean Pigeon; Chilean Bandtail
f   Pigeon du Chili
d   Araucaner-Taube; Chile-Taube
i   Colomba del Cile

2477   *Columba argentina*
          Columbiformes - Columbidae
e   Silvery Wood-Pigeon; Silver Pigeon;
          Grey Wood-Pigeon
f   Pigeon argenté
d   Silbertaube
i   Colomba argentata

2478   *Columba arquatrix*
          Columbiformes - Columbidae
e   African Olive-Pigeon; Rameron
          Pigeon (CSA); Olive-Pigeon; African
          Pigeon; Yellow-eyed Pigeon;
          Speckled Wood-Pigeon
f   Pigeon raméron
d   Oliventaube
i   Colomba macchiata africana

2479   *Columba berlepschi*
          Columbiformes - Columbidae
e   Berlepsch's Pigeon

2480   *Columba bollii*
          Columbiformes - Columbidae
e   Bolle's Pigeon; Bolle's Laurel Pigeon
f   Pigeon de Bolle
d   Bolles Lorbeertaube
i   Colomba di Bolle

2481   *Columba caribaea*
          Columbiformes - Columbidae
e   Ring-tailed Pigeon; Ringtail (WI);
          Jamaican Band-tailed Pigeon
f   Pigeon de la Jamaïque; Pigeon à
          queue annelée

d Kariben-Taube
i Colomba della Giamaica

2482 *Columba cayennensis*
Columbiformes - Columbidae
e Pale-vented Pigeon; Rufous Pigeon; Blue-Pigeon; Cayenne Pigeon
f Pigeon rousset
d Rotrückentaube
i Colomba della Cayenna

2483 *Columba corensis*
Columbiformes - Columbidae
e Bare-eyed Pigeon; White-winged Pigeon
f Pigeon jounud
d Nacktaugentaube
i Colomba occhinudi

2484 *Columba delegorguei*
Columbiformes - Columbidae
e Eastern Bronze-naped Pigeon; Delegorgue's Pigeon (CSA); Bronze-naped Pigeon (CSA)
f Pigeon de Délégorgue
d Bronzehalstaube; Glanznackentaube
i Colomba di Delegorgue

2485 *Columba elphinstonii*
Columbiformes - Columbidae
e Nilgiri Wood-Pigeon; Spotted-neck Pigeon; Elphinstone's Pigeon
f Pigeon d'Elphinstone
d Nilgiri-Taube; Halsfleckentaube
i Colomba di Nilgiri

2486 *Columba eversmanni*
Columbiformes - Columbidae
e Pale-backed Pigeon; Yellow-eyed Stock-Dove; Eastern Stock-Dove; Yellow-eyed Pigeon; Yellow-eyed Dove; Pale-backed Dove; Eastern Stock-Pigeon; Pale-backed Eastern Stock-Dove; Eversmann's Stock-Pigeon; Eversmann's Stock-Dove
f Pigeon d'Eversmann
d Gelbaugentaube; Kleine Hohltaube; Ufertaube
i Colombella occhigialli

2487 *Columba fasciata*
Columbiformes - Columbidae
e Band-tailed Pigeon; White-naped Pigeon; White-collared Pigeon; Blue-Pigeon
f Pigeon à queue barrée; Pigeon du Pacifique
d Bindentaube; Schuppenhalstaube
i Colomba fasciata

2488 *Columba flavirostris*
Columbiformes - Columbidae
e Red-billed Pigeon; Blue-Pigeon (WI)
f Pigeon à bec rouge
d Rotschnabeltaube
i Colomba beccogiallo

2489 *Columba goodsoni*
Columbiformes - Columbidae
e Dusky Pigeon; Goodson's Pigeon
f Pigeon de Goodson
d Goodson-Taube
i Colomba di Goodson

2490 *Columba guinea*
Columbiformes - Columbidae
e Speckled Pigeon; Rock Pigeon (CSA)
f Pigeon roussard; Pigeon de Guinée
d Guinea-Taube; Strichelhalstaube
i Colomba di Guinea; Colomba della Guinea

2491 *Columba hodgsonii*
Columbiformes - Columbidae
e Speckled Wood-Pigeon; Hodgson's Pigeon; Jungle Pigeon
f Pigeon de Hodgson
d Schwarzschnabeloliventaube; Dschungeltaube
i Colombaccio di Hodgson

2492 *Columba inornata*
Columbiformes - Columbidae
e Plain Pigeon; Blue-Pigeon (WI)
f Pigeon simple; Ramier (Ants); Ramier ceniza (Ants)
d Rosenschultertaube; Santo Domingo-Taube
i Colomba disadorna

2493 *Columba iriditorques*
Columbiformes - Columbidae
e Western Bronze-naped Pigeon;
Bronze-naped Pigeon
f Pigeon à nuque bronzée
d Glanzkopftaube
i Colomba nucabronzata

2494 *Columba janthina*
Columbiformes - Columbidae
e Japanese Wood-Pigeon; Black
Wood-Pigeon; Black Pigeon
f Pigeon violet
d Veilchentaube; Violettscheiteltaube
i Colomba giapponese

2495 *Columba jouyi*
Columbiformes - Columbidae
e Ryukyu Pigeon; Jouyi's Wood-
Pigeon; Silver-banded Black-Pigeon;
Ryukyu Wood-Pigeon; Silver-
crescented Pigeon
f Pigeon à col d'argent
d Silberbandtaube
i Colomba dal collare argentato

2496 *Columba junoniae*
Columbiformes - Columbidae
e Laurel Pigeon; White-tailed Laurel
Pigeon
f Pigeon des lauriers
d Lorbeertaube
i Colomba dell'alloro; Colomba dei
lauri; Colombaccio minore

2497 *Columba leucocephala*
Columbiformes - Columbidae
e White-crowned Pigeon; Blue-Pigeon
(WI); Baldpate (WI); Whitehead
(WI); White-headed Pigeon
f Pigeon à couronne blanche; Ramier
tête-blanche (Ants); Ramier à tête
blanche (Ants)
d Weißscheiteltaube; Diademtaube
i Colomba corona bianca

2498 *Columba leucomela*
Columbiformes - Columbidae
e White-headed Pigeon; White-headed
Fruit-Pigeon; Cook Pigeon
f Pigeon leucomèle

d Weißbrusttaube; Ostaustralien-Taube
i Colomba testabianca

2499 *Columba leuconota*
Columbiformes - Columbidae
e Snow Pigeon
f Pigeon des neiges
d Schneetaube; Weißrückentaube
i Colomba delle nevi

2500 *Columba livia*
Columbiformes - Columbidae
e Rock Dove; Rock Pigeon; Blue
Rock-Pigeon (ISC); Indian Blue
Rock-Pigeon
f Pigeon biset; Pigeon des champs;
Pigeon de roche; Pigeon des rochers;
Colombe biset
d Felsentaube
i Piccione selvatico; Piccione torraiolo

2501 *Columba livia domestica*
Columbiformes - Columbidae
e Common Pigeon; Pigeon; Domestic
Pigeon; Feral Pigeon; Town Pigeon
(CSA); House Pigeon; Street Pigeon
f Pigeon; Pigeon domestique; Pigeon
de volière; Pigeonne f
d Haustaube; Straßentaube
i Piccione; Piccione torraiola

2502 *Columba maculosa*
Columbiformes - Columbidae
e Spot-winged Pigeon; Spotted Pigeon;
American Spotted- Pigeon
f Pigeon tigré
d Fleckentaube
i Colomba alimacchiate

2503 *Columba malherbii*
Columbiformes - Columbidae
e Sao Tomé Bronze-naped Pigeon; Sao
Tomé Pigeon; Sao Tomé Grey-
Pigeon
f Pigeon de Malherbe
d Sao Tomé-Bronzenackentaube
i Colomba di Malherbe

2504 *Columba mayeri*
Columbiformes - Columbidae
e Pink Pigeon; Mauritius Pink Pigeon;

Chestnut-tailed Pigeon
f Pigeon rose
d Rosentaube
i Colomba rosa

2505 *Columba nigrirostris*
Columbiformes - Columbidae
e Short-billed Pigeon
f Pigeon à bec noir
d Kurzschnabeltaube
i Colomba becconero

2506 *Columba oenas*
Columbiformes - Columbidae
e Stock Pigeon; Stock Dove; Western
Stock-Pigeon
f Pigeon colombin; Pigeon blanc;
Pigeon bleu
d Hohltaube; Kleine Holztaube
i Colombella; Palombella

2507 *Columba oenops*
Columbiformes - Columbidae
e Peruvian Pigeon; Salvin's Pigeon
f Pigeon du Pérou; Pigeon péruvien
d Peru-Taube; Salvin-Taube
i Colomba di Salvin

2508 *Columba oliviae*
Columbiformes - Columbidae
e Somali Pigeon; Somali Rock-Pigeon;
Somali Stock-Dove; Somali Stock-
Pigeon
f Pigeon de Somalie
d Somali-Taube
i Colomba somala

2509 *Columba pallidiceps*
Columbiformes - Columbidae
e Yellow-legged Pigeon; Silver-headed
Pigeon
f Pigeon à tête pâle
d Gelbfußtaube
i Colomba testachiara

2510 *Columba palumboides*
Columbiformes - Columbidae
e Andaman Wood-Pigeon; Andaman
Pigeon
f Pigeon des Andaman

d Andamanen-Taube
i Colomba delle Andamane

2511 *Columba palumbus*
Columbiformes - Columbidae
e Common Wood-Pigeon; Wood-
Pigeon; Ring Dove
f Pigeon ramier; Ramier; Colombe
ramier
d Ringeltaube; Große Holztaube;
Holztaube; Hohltaube; Waldtaube
i Colombaccio

2512 *Columba picazuro*
Columbiformes - Columbidae
e Picazuro Pigeon; Argentine Wood-
Pigeon; Scaly-necked Wood-Pigeon;
Brown Wood-Pigeon
f Pigeon picazuro
d Piccazurtaube
i Colomba picazuro

2513 *Columba plumbea*
Columbiformes - Columbidae
e Plumbeous Pigeon
f Pigeon plombé
d Bleigraue Taube; Weintaube
i Colomba plumbea

2514 *Columba pollenii*
Columbiformes - Columbidae
e Comoro Olive-Pigeon; Comoro
Wood-Pigeon
f Pigeon des Comores
d Komoren-Taube
i Colomba delle Comore

2515 *Columba pulchricollis*
Columbiformes - Columbidae
e Ashy Wood-Pigeon; Ashy Pigeon;
Buff-coloured Pigeon; Nepal Wood-
Pigeon
f Pigeon cendré; Pigeon à col fauve
d Himalaja-Taube; Nepar-Taube
i Colombaccio cenerino

2516 *Columba punicea*
Columbiformes - Columbidae
e Pale-capped Pigeon; Purple Wood-
Pigeon (ISC); Purple Pigeon; Red
Wood-Pigeon; Chestnut Pigeon

f Pigeon marron
d Kupfertaube; Purpurwaldtaube
i Colomba capochiaro

2517 Columba rupestris
Columbiformes - Columbidae
e Hill Pigeon; Blue Hill-Pigeon;
Eastern Rock-Pigeon; Eastern Rock-Dove
f Pigeon des rochers
d Klippentaube
i Piccione torraiolo orientale

2518 Columba sjostedti
Columbiformes - Columbidae
e Cameroon Olive-Pigeon; Cameroon
Rameraon Pigeon; Sjostedt's Pigeon
f Pigeon du Cameroun
d Rotschnabeloliventaube
i Colomba del Camerun

2519 Columba speciosa
Columbiformes - Columbidae
e Scaled Pigeon; Splendid Pigeon;
Scale-necked Pigeon; Scaly-necked
Pigeon; Scallop-necked Pigeon;
Speckle-necked Pigeon; Fair Pigeon;
Red-backed Pigeon; Dominick
Pigeon
f Pigeon ramiret
d Schuppenbauchtaube; Prachttaube
i Colomba squamata

2520 Columba squamosa
Columbiformes - Columbidae
e Scaly-naped Pigeon; Red-necked
Pigeon; Red-head (WI); Blue-Pigeon
(WI); Mountain Pigeon (WI); Scaly-necked Pigeon
f Ramier (Ants); Pigeon à cou rouge;
Ramier à cou rouge (Ants); Pigeons
des mornes
d Antillen-Taube; Portorico-Taube
i Colomba collorosso

2521 Columba subvinacea
Columbiformes - Columbidae
e Ruddy Pigeon; Berlepsch's Pigeon
f Pigeon vineux
d Purpurtaube; Rötliche Taube
i Colomba rossastra

2522 Columba thomensis
Columbiformes - Columbidae
e Sao Tomé Olive-Pigeon; Sao Tomé
Pigeon; Sao Tomé Maroon Pigeon
f Pigeon de Sao Tomé
d Langschwanzoliventaube
i Colomba di Sao Tomè

2523 Columba torringtoni
Columbiformes - Columbidae
e Sri Lanka Wood-Pigeon; Ceylon
Wood-Pigeon
f Pigeon de Ceylan; Pigeon cingalais
d Ceylon-Taube
i Colomba di Sri-Lanka

2524 Columba trocaz
Columbiformes - Columbidae
e Trocaz Pigeon; Long-toed Pigeon
f Pigeon trocaz
d Silberhalstaube
i Colomba trocaz; Colombacci
codabarrata

2525 Columba unicincta
Columbiformes - Columbidae
e Afep Pigeon; Scaly Grey-Pigeon;
African Wood-Pigeon; Congo Wood-Pigeon; Grey Pigeon
f Pigeon gris
d Kongo-Taube; Kongo-Waldtaube
i Colombaccio africano

2526 Columba versicolor
Columbiformes - Columbidae
e Bonin Pigeon; Kittlitz's Wood-Pigeon; Bonin Wood-Pigeon; Bonin
Black-; Bonin Fruit-Pigeon; Shining
Pigeon
f Pigeon de Kittlitz
d Bonin-Taube
i Colomba di Bonin

2527 Columba vitiensis
Columbiformes - Columbidae
e Metallic Pigeon; White-throated
Pigeon (ANZ); Metallic Wood-Pigeon; Chilhi Pigeon
f Pigeon à gorge blanche
d Weißwangentaube; Weißkehltaube
i Colomba golabianca

2528   *Columba vitiensis godmanae*
       Columbiformes - Columbidae
  e Lord Howe Island White-throated
       Pigeon (ANZ)

2529   **Columbidae**
       Columbidae
  e Pigeons; Doves
  f Pigeons; Columbidés
  d Tauben
  i Columbidi; Tortore; Piccioni

2530   **Columbiformes**
  e Pigeons; Doves
  f Columbiformes
  d Taubenvögel
  i Columbiformi

2531   *Columbina buckleyi*
       Columbiformes - Columbidae
  e Ecuadorian Ground-Dove; Buckley's
       Ground-Dove
  f Colombe de Buckley
  d Blaßtäubchen
  i Tortora di Buckley

2532   *Columbina cruziana*
       Columbiformes - Columbidae
  e Croaking Ground-Dove; Golden-
       billed Ground-Dove; Peruvian
       Ground-Dove
  f Colombe à bec jaune; Colombe
       pérouvienne
  d Goldschnabeltäubchen; Peru-
       Täubchen
  i Tortora beccogiallo

2533   *Columbina cyanopis*
       Columbiformes - Columbidae
  e Blue-eyed Ground-Dove
  f Colombe aux yeux bleus
  d Blauaugentäubchen
  i Tortora occhiblu

2534   *Columbina inca*
       Columbiformes - Columbidae
  e Inca Dove
  f Colombe inca
  i Tortora inca

2535   *Columbina minuta*
       Columbiformes - Columbidae
  e Plain-breasted Ground-Dove; Grey
       Ground-Dove; Little Ground-Dove;
       Pygmy Dove
  f Colombe pygmée
  d Zwergtäubchen
  i Tortora minuta

2536   *Columbina passerina*
       Columbiformes - Columbidae
  e Common Ground-Dove; Ground
       Dove; Scaly-breasted Ground-Dove;
       Tobaccobird (WI); Duppybird (WI);
       Grounie (WI); Rosy Ground-Dove;
       Passerine Dove; Scaly Ground-Dove;
       Speckled Ground-Dove
  f Colombe à queue noire; Ortolan
       (Ants); Colombe cocotzin
  d Sperlingstäubchen
  i Tortora passerina

2537   *Columbina picui*
       Columbiformes - Columbidae
  e Picui Ground-Dove; Picui Dove;
       Long-tailed Ground-Dove; White-
       winged Ground-Dove; Steel-barred
       Ground-Dove
  f Colombe picui
  d Picui-Täubchen
  i Tortora picui

2538   *Columbina squammata*
       Columbiformes - Columbidae
  e Scaled Dove; Scaly Dove; Mottled
       Dove; South American Zebra Dove
  f Colombe écaillée
  d Schuppentäubchen
  i Colomba collorosso; Tortora
       squamata

2539   *Columbina talpacoti*
       Columbiformes - Columbidae
  e Ruddy Ground-Dove; Stone Dove;
       Talpacoti Dove; Cinnamon Dove;
       Blue-headed Ground-Dove
  f Colombe rousse; Colombe talpacoti
  d Rosttäubchen
  i Tortora talpacoti

2540 *Compsothraupis loricata*
Passeriformes - Fringillidae
e Scarlet-throated Tanager
f Tangara à gorge écarlate
d Rotbrusttangare

2541 *Compylopterus pampa*
Apodiformes - Trochilidae
e Wedge-tailed Sabrewing

2542 *Conioptilon mcilhennyi*
Passeriformes - Tyrannidae
e Black-faced Cotinga
f Cotinga à face noire
d Schwarzkopfschmuckvogel
i Cotinga di Mclhenny

2543 *Conirostrum albifrons*
Passeriformes - Fringillidae
e Capped Conebill
f Conirostre coiffé
d Kappenspitzschnabel
i Becco a cono blu

2544 *Conirostrum bicolor*
Passeriformes - Fringillidae
e Bicoloured Conebill; Two-coloured Conebill
f Conirostre bicolore
d Zweifarbenspitzschnabel
i Becco a cono bicolore

2545 *Conirostrum cinereum*
Passeriformes - Fringillidae
e Cinereous Conebill
f Conirostre cendré
d Weißstirnspitzschnabel
i Becco a cono cenerino

2546 *Conirostrum ferrugineiventre*
Passeriformes - Fringillidae
e White-browed Conebill
f Conirostre à ventre roux
d Weißbrauenspitzschnabel
i Becco a cono rufiventre

2547 *Conirostrum leucogenys*
Passeriformes - Fringillidae
e White-eared Conebill
f Conirostre oreillard

d Weißohrspitzschnabel
i Becco a cono guancebianche

2548 *Conirostrum margaritae*
Passeriformes - Fringillidae
e Pearly-breasted Conebill
f Conirostre marguerite
d Perlbrustspitzschnabel
i Becco a cono pettoperlato

2549 *Conirostrum rufum*
Passeriformes - Fringillidae
e Rufous-browed Conebill
f Conirostre roux
d Rotbrauenspitzschnabel
i Becco a cono dai sopraccigli rugginosi

2550 *Conirostrum sitticolor*
Passeriformes - Fringillidae
e Blue-backed Conebill
f Conirostre à cape bleue
d Blaurückenspitzschnabel
i Becco a cono dorsoazzurro

2551 *Conirostrum speciosum*
Passeriformes - Fringillidae
e Chestnut-vented Conebill
f Conirostre cul-roux
d Rotsteißspitzschnabel
i Becco a cono dal sottocoda castano

2552 *Conirostrum tamarugense*
Passeriformes - Fringillidae
e Tamarugo Conebill; Chilean Conebill; Johnson's Warbler
f Conirostre de Tamarugo
d Rotstirnspitzschnabel
i Becco a cono di Tamarugo

2553 *Conopias albovittata*
Passeriformes - Tyrannidae
e White-ringed Flycatcher; Yellow-crowned Flycatcher; Yellow-throated Flycatcher
f Tyran diadème
d Kopfbindentyrann
i Tiranno pitango dal collarebianco

2554 *Conopias cinchoneti*
Passeriformes - Tyrannidae

e Lemon-browed Flycatcher; Cinchon
Flycatcher
f Tyran à sourcils jaunes
d Gelbringtyrann
i Tiranno pitango dai sopraccigli gialli

2555 *Conopias parva*
Passeriformes - Tyrannidae
e Yellow-throated Flycatcher; White-ringed Flycatcher
f Tyran diadème
d Kopfbindentyrann
i Tiranno pitango piccolo

2556 *Conopias trivirgata*
Passeriformes - Tyrannidae
e Three-striped Flycatcher
f Tyran à triple bandeau
d Dreistreifentyrann
i Tiranno pitango dalle tre strisce

2557 *Conopophaga ardesiaca*
Passeriformes - Conopophagidae
e Slaty Gnateater
f Conophage ardoisé
d Graubrustmückenfresser
i Mangiamoscerini ardesia

2558 *Conopophaga aurita*
Passeriformes - Conopophagidae
e Chestnut-belted Gnateater
f Conophage à oreilles blanches
d Cayenne-Mückenfresser
i Mangiamoscerini dalle redini

2559 *Conopophaga castaneiceps*
Passeriformes - Conopophagidae
e Chestnut-crowned Gnateater
f Conophage à couronne rousse
d Rotscheitelmückenfresser
i Mangiamoscerini testacastana

2560 *Conopophaga lineata*
Passeriformes - Conopophagidae
e Rufous Gnateater; Silvery-tufted Gnateater
f Conophage roux
d Rotkehlmückenfresser;
Rotkehlmückenesser
i Mangiamoscerini rossiccio

2561 *Conopophaga melanogaster*
Passeriformes - Conopophagidae
e Black-bellied Gnateater
f Conophage à ventre noir
d Schwarzbauchmückenfresser
i Mangiamoscerini guancenere ventrenero

2562 *Conopophaga melanops*
Passeriformes - Conopophagidae
e Black-cheeked Gnateater
f Conophage à joues noires
d Schwarzwangenmückenfresser
i Mangiamoscerini guancenere

2563 *Conopophaga peruviana*
Passeriformes - Conopophagidae
e Ash-throated Gnateater
f Conophage du Pérou
d Graukehlmückenfresser
i Mangiamoscerini del Perù

2564 *Conopophaga roberti*
Passeriformes - Conopophagidae
e Hooded Gnateater
f Conophage capucin
d Schwarzkopfmückenfresser
i Mangiamoscerini di Robert

2565 Conopophagidae
Passeriformes
e Gnateaters
f Conophagidés
d Mückenfresser; Mückenfänger;
Mückenesser
i Conopofagidi

2566 *Conopophila albogularis*
Passeriformes - Meliphagidae
e Rufous-banded Honeyeater; Red-breasted Honeyeater; Rufous-breasted Honeyeater
f Méliphage à gorge blanche
d Rostbandhonigfresser
i Mangiamiele pettoruggino

2567 *Conopophila rufogularis*
Passeriformes - Meliphagidae
e Rufous-throated Honeyeater; Red-throated Honeyeater; Rufous-breasted Honeyeater

f   Méliphage à gorge rousse
d   Rostkehlhonigfresser
i   Mangiamiele golarugginosa

2568   *Conopophila whitei*
         Passeriformes - Meliphagidae
e   Grey Honeyeater; White's Grey-
      Honeyeater; Inconspicuous
      Honeyeater
f   Méliphage de White
d   Grauhonigfressser
i   Mangiamiele grigio

2569   *Conostoma oemodium*
         Passeriformes - Sylviidae
e   Great Parrotbill; Great Crowtit; Great
      Suthora
f   Grande Panure
d   Krähenmeise; Keilschnabel
i   Becco a cono maggiore

2570   *Conothraupis mesoleuca*
         Passeriformes - Fringillidae
e   Cone-billed Tanager
f   Tangara de Berlioz
d   Witwentangare
i   Tangara becco a cono

2571   *Conothraupis speculigera*
         Passeriformes - Fringillidae
e   Black-and-white Tanager
f   Tangara à miroir blanc
d   Spiegeltangare
i   Tangara nera e bianca

2572   *Contopus albogularis*
         Passeriformes - Tyrannidae
e   White-throated Pewee
f   Moucherolle à bavette blanche; Pioui
      à gorge blanche
d   Silberkehltyrann
i   Piuì golabianca

2573   *Contopus borealis*
         Passeriformes - Tyrannidae
e   Olive-sided Flycatcher
f   Moucherolle à côtés olives; Pioui à
      côtés olives
d   Fichtentyrann
i   Piuì fianchi oliva

2574   *Contopus brachytarsus*
         Passeriformes - Tyrannidae
e   Short-legged Peewee

2575   *Contopus caribaeus*
         Passeriformes - Tyrannidae
e   Cuban Pewee; Crescent-eyed Pewee
      (WI); Tity (WI); Flycatcher (WI);
      Greater Antillean Peewee
f   Moucherolle tête-fou; Pioui tête-fou
d   Braunbauchtyrann
i   Piuì maggiore delle Antille

2576   *Contopus cinereus*
         Passeriformes - Tyrannidae
e   Tropical Pewee; Ash-coloured Pewee
f   Moucherolle cendrée; Pioui cendré
d   Spix-Tyrann; Piwih; Pewee
i   Piuì tropicale

2577   *Contopus cooperi*
         Passeriformes - Tyrannidae
e   Olive-sided Flycatcher
f   Moucherolle à côtés olive

2578   *Contopus fumigatus*
         Passeriformes - Tyrannidae
e   Smoke-coloured Pewee; Smoky
      Pewee; Greater Pewee
f   Moucherolle bistrée; Pioui ardoisée
d   Schiefertyrann
i   Piuì fumigato

2579   *Contopus hispaniolensis*
         Passeriformes - Tyrannidae
e   Hispaniolan Pewee

2580   *Contopus latirostris*
         Passeriformes - Tyrannidae
e   Lesser Antillean Pewee
f   Tombé levé (Ants); Loulou fou
      (Ants); Moucherolle (Ants); Gobe-
      mouche (Ants); Moucherolle
      gobemouche; Pioui gobemouches
d   Rostbauchtyrann
i   Piuì minore delle Antille

2581   *Contopus lugubris*
         Passeriformes - Tyrannidae
e   Dark Pewee
f   Moucherolle ombrée

d   Barranea-Tyrann
i   Piuì scuro

2582   *Contopus nigrescens*
Passeriformes - Tyrannidae
e   Blackish Pewee
f   Moucherolle noirâtre
d   Einfarbtyrann
i   Piuì nerastro

2583   *Contopus ochraceus*
Passeriformes - Tyrannidae
e   Ochraceous Pewee
f   Moucherolle ocrée; Pioui ocrè
d   Ockerbrusttyrann
i   Piuì ocraceo

2584   *Contopus pallidus*
Passeriformes - Tyrannidae
e   Jamaican Pewee; Stupid Jimmy
      (WI); Little Tom-fool (WI); Willie
      Pee (WI)

2585   *Contopus pertinax*
Passeriformes - Tyrannidae
e   Greater Pewee; Coue's Flycatcher
f   Moucherolle de Coues; Pioui de
      Coues
d   Coues-Tyrann
i   Piuì di coues

2586   *Contopus portoricensis*
Passeriformes - Tyrannidae
e   Puerto Rican Pewee

2587   *Contopus sordidulus*
Passeriformes - Tyrannidae
e   Western Wood-Pewee; Western
      Pewee; Wood Pewee
f   Pioui de l'Ouest
d   Westlicher Waldtyrann
i   Piuì di bosco occidentale

2588   *Contopus virens*
Passeriformes - Tyrannidae
e   Eastern Wood Pewee; Pewee
f   Pioui de l'Est
d   Östlicher Waldtyrann; Piwih;
      Waldpiwih; Ostwaldtyrann
i   Piuì di bosco orientalis

2589   *Conuropsis carolinensis*
Psittaciformes - Psittacidae
e   Carolina Parakeet; Carolina Conure
f   Conure de Caroline
d   Carolina-Sittich
i   Conuro dellla Carolina

2590   *Conurus labati*
Psittaciformes - Psittacidae
e   Guadeloupe Parakeet

2591   *Copsychus albospecularis*
Passeriformes - Muscicapidae
e   Madagascar Magpie-Robin
f   Shama de Madagascar; Merle dyal
      malgache
d   Malagassen-Dajal
i   Shama del Madagascar

2592   *Copsychus cebuensis*
Passeriformes - Muscicapidae
e   Black Shama; Cebu Black-Shama;
      Cebu Islands Shama
f   Shama de Cebu
d   Cebu-Schama
i   Merlo di Cebu

2593   *Copsychus luzoniensis*
Passeriformes - Muscicapidae
e   White-browed Shama; White-
      eyebrowed Shama
f   Shama bridé
d   Brauenschama
i   Merlo dai sopraccigli bianchi

2594   *Copsychus malabaricus*
Passeriformes - Muscicapidae
e   White-rumped Shama; Shama
      Thrush; Shama (ISC); Common
      Shama; Indian Shama; Ceylon
      Shama
f   Shama à croupion blanc
d   Schama; Schamadrossel
i   Merlo shama

2595   *Copsychus niger*
Passeriformes - Muscicapidae
e   White-vented Shama; Palawan
      Shama; Palawan Black-Shama; Black
      Shama
f   Shama noir

d   Mohrendajal
i   Merlo nero

2596   ***Copsychus saularis***
      Passeriformes - Muscicapidae
e   Oriental Magpie-Robin; Magpie-
      Robin; Dyal-Thrush; Dyal; Asian
      Magpie-Robin; Indian Magpie-
      Robin; Southern Magpie-Robin;
      Robin Dyal
f   Shama dayal; Merle shama
d   Dajal; Dajal-Drossel
i   Merlo dayal

2597   ***Copsychus sechellarum***
      Passeriformes - Muscicapidae
e   Seychelles Magpie-Robin
f   Shama des Seychelles; Merle dyal
      des Seychelles
d   Seychellen-Dajal
i   Shama delle Seychelle

2598   ***Copsychus stricklandii***
      Passeriformes - Muscicapidae
e   White-crowned Shama; Strickland's
      Shama; White-browed Shama
f   Shama de Strickland
d   Weißkappendajal
i   Merlo di Strickland

2599   ***Coracias abyssinica***
      Coraciiformes - Coraciidae
e   Abyssinian Roller
f   Rollier d'Abyssinie
d   Senegal-Racke
i   Ghiandaia marina abissina;
      Ghiandaia marina dell'Abissinia

2600   ***Coracias benghalensis***
      Coraciiformes - Coraciidae
e   Indian Roller; Roller (ISC); Blue Jay
      (ISC)
f   Rollier indien
d   Hindu-Racke; Bengalen-Racke
i   Ghiandaia marina indiana

2601   ***Coracias caudata***
      Coraciiformes - Coraciidae
e   Lilac-breasted Roller; Blue Jay
      (CSA); Mosilikatze's Roller; Lilac-
      throated Roller

f   Rollier à longs brins
d   Gabelracke; Grünscheitelracke
i   Ghiandaia marina pettolilla;
      Ghiandaia marina caudata

2602   ***Coracias cyanogaster***
      Coraciiformes - Coraciidae
e   Blue-bellied Roller
f   Rollier à ventre bleu
d   Blaubrustracke; Blaubauchracke
i   Ghiandaia marina ventreblu

2603   ***Coracias garrulus***
      Coraciiformes - Coraciidae
e   European Roller; Eurasian Roller
      (CSA); Roller; Blue Roller; Common
      Roller
f   Rollier d'Europe
d   Blauracke; Mandelkrähe
i   Ghiandaia marina; Ghiandaia marina
      europea

2604   ***Coracias naevia***
      Coraciiformes - Coraciidae
e   Rufous-crowned Roller; Purple
      Roller (CSA)
f   Rollier varié
d   Strichelracke
i   Ghiandaia marina corona rossiccia

2605   ***Coracias spatulata***
      Coraciiformes - Coraciidae
e   Racket-tailed Roller; Racquet-tailed
      Roller; Weigall's Roller
f   Rollier à raquettes
d   Spatelracke
i   Ghiandaia marina coda a racchetta

2606   ***Coracias temminckii***
      Coraciiformes - Coraciidae
e   Purple-winged Roller; Celebes
      Roller; Temminck's Roller; Sulawesi
      Roller
f   Rollier de Temminck
d   Celebes-Racke; Temmincks Racke
i   Ghiandaia marina di Sulawesi

2607   ***Coraciidae***
      Coraciiformes
e   Rollers
f   Rolliers; Coraciidés

d Rackenvögel; Baumracken;
Eigentliche Racken; Racken

i Coracidi; Ghiandaie marine

2608 Coraciiformes

e Kingfishers

f Coraciaiiformes

d Rackenvögel

i Coraciformi

2609 *Coracina abbotti*
Passeriformes - Campophagidae

e Pygmy Cuckoo-Shrike; Abbott's
Cuckoo-Shrike; Celebes Mountain-
Greybird; Celebes Cuckoo-Shrike;
Celebes Mountain-Greybird;
Moutain Greybird

f Échenilleur d'Abbott

d Abbotts Raupenfänger

i Coracina di Abbott

2610 *Coracina analis*
Passeriformes - Campophagidae

e New Caledonian Cuckoo-Shrike;
Mountain Greybird; New Caledonian
Greybird

f Échenilleur de montagne

d Rotsteißraupenfänger

i Coracina della Nuova Caledonia

2611 *Coracina atriceps*
Passeriformes - Campophagidae

e Moluccan Cuckoo-Shrike; Ceram
Cuckoo-Shrike

f Échenilleur des Moluques;
Échenilleur de Céram

d Molukken-Raupenfänger

i Coracina testanera delle Molucche

2612 *Coracina azurea*
Passeriformes - Campophagidae

e Blue Cuckoo-Shrike; African Blue
Cuckoo-Shrike

f Échenilleur bleu

d Azurraupenfänger

i Coracina azzurra africana

2613 *Coracina bicolor*
Passeriformes - Campophagidae

e Pied Cuckoo-Shrike; Celebes
Cuckoo-Shrike; Muna Greybird;
Bicoloured Cuckoo-Shrike; Munia
Cuckoo-Shrike

f Échenilleur bicolore

d Zweifarbenraupenfänger

i Coracina bicolore

2614 *Coracina boyeri*
Passeriformes - Campophagidae

e Boyer's Cuckoo-Shrike; Boyer's
Greybird; Rufous-underwing
Cuckoo-Shrike; White-lored Cuckoo-
Shrike; Rufous-underwing Greybird

f Échenilleur de Boyer

d Rotachselraupenfänger

i Coracina di Boyer

2615 *Coracina caeruleogrisea*
Passeriformes - Campophagidae

e Stout-billed Cuckoo-Shrike; Stout-
billed Greybird; Blue-grey Cuckoo-
Shrike

f Échenilleur à gros bec

d Dickschnabelraupenfänger

i Coracina beccoforte

2616 *Coracina caesia*
Passeriformes - Campophagidae

e Grey Cuckoo-Shrike; African Grey
Cuckoo-Shrike; Mountain Grey-
Cuckoo-Shrike; African Greybird

f Échenilleur gris

d Grauer Raupenfänger; Graue
Raupendohle

i Coracina grigia africana

2617 *Coracina caledonica*
Passeriformes - Campophagidae

e Melanesian Cuckoo-Shrike;
Melanesian Greybird; Sula Cuckoo-
Shrike; Black-faced Cuckoo-Shrike

f Échenilleur calédonien

d Welchmans Raupenfänger

i Coracina della Melanesia

2618 *Coracina ceramensis*
Passeriformes - Campophagidae

e Pale Cicadabird; Pale-grey Cuckoo-
Shrike; Pale-grey Cicadabird;
Moluccan Cicadabird; Celebes
Cicadabird

f   Échenilleur pâle
i   Coracina di Ceram

2619   *Coracina cinerea*
Passeriformes - Campophagidae
e   Ashy Cuckoo-Shrike; Madagascar
Cuckoo-Shrike
f   Échenilleur malgache
d   Madagaskar-Raupenfänger
i   Coracina del Madagascar

2620   *Coracina coerulescens*
Passeriformes - Campophagidae
e   Blackish Cuckoo-Shrike; Black
Cuckoo-Shrike; Philippine Greybird;
Philippine Cuckoo-Shrike; Black
Greybird; Philippine Black Greybird
f   Échenilleur noir
d   Glanzraupenfänger
i   Coracina nera delle Filippine

2621   *Coracina dispar*
Passeriformes - Campophagidae
e   Kai Cicadabird; Kai Greybird; Kai
Cuckoo-Shrike
f   Échenilleur des Kai
i   Coracina di Kai

2622   *Coracina dohertyi*
Passeriformes - Campophagidae
e   Sumba Cicadabird; Black-barred
Cuckoo-Shrike; Sumba Cuckoo-
Shrike; Doherty's Greybird; Sumba
Greybird; Black-barred Cuckoo-
Shrike
f   Échenilleur de Sumba
d   Sumba-Raupenfänger
i   Coracina di Doherty

2623   *Coracina fimbriata*
Passeriformes - Campophagidae
e   Lesser Cuckoo-Shrike; Lesser
Greybird
f   Échenilleur frangé
d   Zwergraupenfänger
i   Coracina minore

2624   *Coracina fortis*
Passeriformes - Campophagidae
e   Buru Cuckoo-Shrike; Buru Island
Cuckoo-Shrike; Buru Islands

Cuckoo-Shrike; Cheeky Cuckoo-
Shrike
f   Échenilleur de Buru
d   Buru-Raupenfänger
i   Coracina delle Isole Buru

2625   *Coracina graueri*
Passeriformes - Campophagidae
e   Grauer's Cuckoo-Shrike
f   Échenilleur de Grauer
d   Silberraupenfänger
i   Coracina di Grauer

2626   *Coracina holopolia*
Passeriformes - Campophagidae
e   Solomon Islands Cuckoo-Shrike;
Cicada Greybird; Solomon Cuckoo-
Shrike; Cicada Cuckoo-Shrike
f   Échenilleur des Salomon
d   Salomonen-Raupenfänger
i   Coracina ventrenero delle Isole
Salomone

2627   *Coracina incerta*
Passeriformes - Campophagidae
e   Black-shouldered Cicadabird;
Papuan Cicadabird; Papuan Cuckoo-
Shrike
f   Échenilleur à épaulettes noires
i   Coracina spallenere

2628   *Coracina javensis*
Passeriformes - Campophagidae
e   Javan Cuckoo-Shrike; Large Cuckoo-
Shrike; Malaysian Cockooshrike;
White-vented Greybird
f   Échenilleur de Java
i   Coracina della Malesia

2629   *Coracina larvata*
Passeriformes - Campophagidae
e   Sunda Cuckoo-Shrike; Black-faced
Cuckoo-Shrike; Black-eared Cuckoo-
Shrike; Black-faced Greybird
f   Échenilleur de la Sonde
d   Larvenraupenfänger
i   Coracina della Sonda

2630   *Coracina leucopygia*
Passeriformes - Campophagidae
e   White-rumped Cuckoo-Shrike; Muna

Cuckoo-Shrike
f Échenilleur à croupion blanc
d Weißbürzelraupenfänger
i Coracina dal groppone bianco

2631 *Coracina lineata*
Passeriformes - Campophagidae
e Barred Cuckoo-Shrike; Lineated
Cuckoo-Shrike; Lineated Greybird;
Yellow-eyed Cuckoo-Shrike;
Yellow-eyed Greybird; Barred
Greybird
f Échenilleur linéolé
d Gelbaugenraupenfänger
i Coracina occhigialli

2632 *Coracina longicauda*
Passeriformes - Campophagidae
e Hooded Cuckoo-Shrike; Black-
hooded Cuckoo-Shrike; Black-
hooded Greybird; Long-tailed
Cuckoo-Shrike
f Échenilleur à longue queue
d Langschwanzraupenfänger
i Coracina caodalunga

2633 *Coracina macei*
Passeriformes - Campophagidae
e Large Cuckoo-Shrike; Black-throated
Cuckoo-Shrike; Indian Cuckoo-
Shrike
f Échenilleur de Mace
d Maskenraupenfänger
i Coracina maggiore

2634 *Coracina maxima*
Passeriformes - Campophagidae
e Ground Cuckoo-Shrike; Greybird
f Échenilleur terrestre
d Gabelschwanzraupenfänger;
Grundraupenfänger
i Coracina gigante

2635 *Coracina mcgregori*
Passeriformes - Campophagidae
e McGregor's Cuckoo-Shrike; Sharp-
tailed Cuckoo-Shrike; Sharp-tailed
Greybird
f Échenilleur de McGregor
d Spitzschwanzraupenfänger
i Coracina di McGregor

2636 *Coracina melanoptera*
Passeriformes - Campophagidae
e Black-headed Cuckoo-Shrike
f Échenilleur à tête noire
d Schwarzkopfraupenfänger
i Coracina testanera indiana

2637 *Coracina melas*
Passeriformes - Campophagidae
e New Guinea Cuckoo-Shrike; Black
Greybird; Black Cuckoo-Shrike;
New Guinea Black Cuckoo-Shrike
f Échenilleur mélanure
d Raupenfänger
i Coracina nera

2638 *Coracina melaschistos*
Passeriformes - Campophagidae
e Black-winged Cuckoo-Shrike; Large
Grey Cuckoo-Shrike; Grey Cuckoo-
Shrike; Dark-grey Cuckoo-Shrike
f Échenilleur ardoisé; Échenilleur gris
d Trauerraupenfänger;
Schieferraupendohle
i Coracina alinere

2639 *Coracina mindanensis*
Passeriformes - Campophagidae
e Black-bibbed Cicadabird; Black-
bibbed Cuckoo-Shrike; Philippine
Cicadabird
f Échenilleur à menton noir
i Coracina delle Filippine

2640 *Coracina montana*
Passeriformes - Campophagidae
e Black-bellied Cuckoo-Shrike; Black-
bellied Greybird; Mountain Cuckoo-
Shrike
f Échenilleur à ventre noir
d Bergraupenfänger
i Coracina ventrenero della Nuova
Guinea

2641 *Coracina morio*
Passeriformes - Campophagidae
e Sulawesi Cicadabird; Molucccan
Greybird; Müller's Greybird; Müller's
Cuckoo-Shrike; Black-shouldered
Cuckoo-Shrike; Moluccan Cuckoo-
Shrike; Sulawesi Cuckoo-Shrike;

Celebes Cuckoo-Shrike
f   Échenilleur morio
d   Morio-Raupenfänger
i   Coracina di Sulawesi

2642   *Coracina newtoni*
Passeriformes - Campophagidae
e   Réunion Cuckoo-Shrike; Réunion
Greybird; Newton's Cuckoo-Shrike
f   Échenilleur cuisinier; Échenilleur de
la Réunion
d   Newton-Raupenfänger
i   Coracina di Reunion

2643   *Coracina novaehollandiae*
Passeriformes - Campophagidae
e   Black-faced Cuckoo-Shrike; Large
Cuckoo-Shrike; Greater Cuckoo-
Shrike; Black-faced Greybird;
Australian Greybird; White-vented
Cuckoo-Shrike
f   Échenilleur à masque noir; Grand
Échenilleur
d   Schwarzgesichtraupenfänger;
Australischer Raupenfänger
i   Coracina faccianera

2644   *Coracina ostenta*
Passeriformes - Campophagidae
e   White-winged Cuckoo-Shrike;
White-winged Greybird; Philippine
Greybird; Philippine Cuckoo-Shrike
f   Échenilleur à ailes blanches
d   Spiegelraupenfänger
i   Coracina alibianche

2645   *Coracina papuensis*
Passeriformes - Campophagidae
e   White-bellied Cuckoo-Shrike;
Papuan Greybird; Papuan Cuckoo-
Shrike; Little Cuckoo-Shrike; White-
breasted Cuckoo-Shrike; White-
bellied Greybird
f   Échenilleur choucari
d   Weißbauchraupenfänger
i   Coracina ventrebianco

2646   *Coracina parvula*
Passeriformes - Campophagidae
e   Halmahera Cuckoo-Shrike;
Halmahera Greybird

f   Échenilleur d'Halmahera
d   Halmahera-Raupenfänger
i   Coracina di Halmahera

2647   *Coracina pectoralis*
Passeriformes - Campophagidae
e   White-breasted Cuckoo-Shrike;
White-breasted Greybird; Black-
chested Cuckoo-Shrike; Grey-
throated Cuckoo-Shrike
f   Échenilleur à ventre blanc;
Échenilleur à gorge blanche
d   Weißbrustrraupenfänger;
Weißbrustraupendohle
i   Coracina pettobianco

2648   *Coracina personata*
Passeriformes - Campophagidae
e   Wallacean Cuckoo-Shrike; Kai
Kuckooshrike
f   Échenilleur wallacéen
d   Raupenfänger
i   Coracina mascherata

2649   *Coracina polioptera*
Passeriformes - Campophagidae
e   Indochinese Cuckoo-Shrike; Grey
Cuckoo-Shrike
f   Échenilleur indochinois
d   Gartenraupenfänger
i   Coracina indocinese

2650   *Coracina pollens*
Passeriformes - Campophagidae
e   Kai Cuckoo-Shrike; Kai Island
Cuckoo-Shrike; Able Cuckoo-Shrike
f   Échenilleur de Salvadori
d   Salvadoris Raupenfänger

2651   *Coracina schistacea*
Passeriformes - Campophagidae
e   Slaty Cuckoo-Shrike; Sula Cuckoo-
Shrike
f   Échenilleur schistacé
d   Schieferraupenfänger
i   Coracina ardesia

2652   *Coracina schisticeps*
Passeriformes - Campophagidae
e   Grey-headed Cuckoo-Shrike; Gray's
Cuckoo-Shrike; Gray's Greybird;

New Guinea Greybird; Slaty
Cuckoo-Shrike
f   Échenilleur de Gray
d   Grays Raupenfänger
i   Coracina testascura

2653   *Coracina striata*
Passeriformes - Campophagidae
e   Bar-bellied Cuckoo-Shrike; Barred
Cuckoo-Shrike; Barred Greybird
f   Échenilleur barré
d   Bindenraupenfänger
i   Coracina barrata

2654   *Coracina sula*
Passeriformes - Campophagidae
e   Sula Cicadabird; Sula Cuckoo-Shrike
f   Échenilleur des Sula
d   Sula-Raupenfänger
i   Coracina di Sula

2655   *Coracina temminckii*
Passeriformes - Campophagidae
e   Cerulean Cuckoo-Shrike;
Temminck's Cuckoo-Shrike; Celebes
Cuckoo-Shrike; Temmick's Greybird
f   Échenilleur de Temminck
d   Temmincks Raupenfänger
i   Coracina di Temminck

2656   *Coracina tenuirostris*
Passeriformes - Campophagidae
e   Slender-billed Cicadabird; Common
Cicadabird; Cicadabird; Long-billed
Greybird; Long-billed Cuckoo-
Shrike; Slender-billed Greybird;
Jardine Triller
f   Échenilleur cigale
d   Mönchsraupenfänger
i   Coracina beccosottile

2657   *Coracina typica*
Passeriformes - Campophagidae
e   Mauritius Cuckoo-Shrike; Mauritius
Greybird
f   Échenilleur de Maurice; Échenilleur
de l'île Maurice
d   Mauritius-Raupenfänger
i   Coracina di Mauritius

2658   *Coracopsis nigra*
Psittaciformes - Psittacidae
e   Black Parrot; Lesser Vasa-Parrot
f   Perroquet noir
d   Rabenpapagei; Kleiner Vasa; Kleiner
Vasapapagei
i   Vasa minore

2659   *Coracopsis vasa*
Psittaciformes - Psittacidae
e   Vasa Parrot; Greater Vasa-Parrot
f   Perroquet vaza
d   Großer Vasa; Großer Vasapapagei;
Vasapapagei
i   Vasa maggiore

2660   *Coracornis raveni*
Passeriformes - Corvidae
e   Maroon-backed Whistler; Raven's
Whistler; Celebes Whistler; Rano-
Rano Whistler; Maroon-breasted
Whistler
f   Siffleur à dos marron
d   Rotrückendickkopf
i   Zufolatore dorsocastano

2661   *Coragyps atratus*
Falconiformes - Cathartidae
e   Black Vulture; New World Vulture;
American Black-Vulture
f   Urubu noir
d   Rabengeier; Neuweltgeier
i   Avvoltoio nero americano; Urubù;
Avvoltaio nero; Gallinazo

2662   *Corapipo altera*
Passeriformes - Tyrannidae
e   White-bibbed Manakin
f   Manakin à fraise
i   Manachino dalla gorgiera bianca

2663   *Corapipo gutturalis*
Passeriformes - Tyrannidae
e   White-throated Manakin
f   Manakin à gorge blanche
d   Weißkehlpipra
i   Manachino golabianca

2664   *Corapipo leucorrhoa*
Passeriformes - Tyrannidae
e   White-bibbed Manakin; White-ruffed

Manakin
f   Manakin orné; Manakin de Sclater
d   Bäffchenpipra
i   Manachino collobianco

2665   Corcoracidae
Passeriformes
e   Mud-nesters
f   Corcoracidés
d   Schlammnestbauer

2666   *Corcorax melanorhamphos*
Passeriformes - Corvidae
e   White-winged Chough; Australian
Chough
f   Corbicrave leucoptère
d   Drosselhäher; Australische
Bergkrähe; Drosselkrähe
i   Gracchio australiano alibianco

2667   *Cormobates affinis*
Passeriformes - Climacteridae
e   White-browed Treecreeper
f   Échelet à sourcils blancs
d   Weißbrauenbaumrutscher
i   Rampichino dai sopraccigli bianchi

2668   *Cormobates affinis superciliosa*
Passeriformes - Climacteridae
e   Eastern White-browed Treecreeper
(ANZ)

2669   *Cormobates erythrops*
Passeriformes - Climacteridae
e   Red-browed Treecreeper
f   Échelet à sourcils roux
d   Rostbrauenbaumrutscher
i   Rampichino dai sopraccigli rossi

2670   *Cormobates leucophaeus*
Passeriformes - Climacteridae
e   White-throated Treecreeper
f   Échelet leucophée
d   Weißkehlbaumrutscher

2671   *Cormobates picumnus*
Passeriformes - Climacteridae
e   Brown Treecreeper
f   Échelet brun
d   Feldbaumrutscher

i   Rampichino golabianca; Rampichino
bruno australiano

2672   *Cormobates picumnus melanotus*
Passeriformes - Climacteridae
e   Cape York Peninsula Brown-
Treecreeper (ANZ)

2673   *Cormobates picumnus victoriae*
Passeriformes - Climacteridae
e   South-eastern Brown-Treecreeper
(ANZ)

2674   *Cormobates placens*
Passeriformes - Climacteridae
e   Papuan Treecreeper; New Guinea
Treecreeper
f   Échelet papou
d   Papua-Baumrutscher
i   Rampichino papua

2675   Corvidae
Passeriformes
e   Crows
f   Corvidés
d   Rabenvögel
i   Corvidi; Corvi

2676   *Corvinella corvina*
Passeriformes - Laniidae
e   Yellow-billed Shrike; Western Long-
tailed Shrike; Long-tailed Shrike
f   Corvinelle à bec jaune; Corvinelle
d   Gelbschnabelwürger
i   Averla codalunga

2677   *Corvinella melanoleuca*
Passeriformes - Laniidae
e   Magpie Shrike; Long-tailed Shrike
(CSA); Eastern Long-tailed Shrike;
Yellow-billed Shrike; Eastern Long-
tailed Magpie-Shrike
f   Corvinelle noir-et-blanc
d   Elsterwürger
i   Averla gazza

2678   *Corvus albicollis*
Passeriformes - Corvidae
e   White-necked Raven; White-naped
Raven (CSA); Cape Raven; African
Raven; African White-necked Raven;

African White-naped Raven
f Corbeau à nuque blanche
d Geierrabe
i Corvo dal collo bianco; Corvo collobianco africano

2679 *Corvus albus*
Passeriformes - Corvidae
e Pied Crow; African Pied-Crow; White-bellied Crow
f Corbeau pie
d Schildrabe
i Corvo bianco e nero

2680 *Corvus bennetti*
Passeriformes - Corvidae
e Little Crow; Crow (NA); Bennett's Crow; Small-billed Crow
f Corbeau du désert
d Bennett-Krähe
i Corvo di Bennett

2681 *Corvus boreus*
Passeriformes - Corvidae
e Relict Raven; New England Raven
f Corbeau de Rowley
i Corvo relitto

2682 *Corvus brachyrhynchos*
Passeriformes - Corvidae
e American Crow; Crow (NA); Common Crow (NA); Western Crow (NA); Florida Crow (NA)
f Corneille d'Amérique
d Amerikaner-Krähe
i Cornacchia americana

2683 *Corvus capensis*
Passeriformes - Corvidae
e Cape Crow; Cape Rook (CSA); Black Crow (CSA); African Crow; African Rook
f Corneille du Cap; Corbeau du Cap
d Kap-Krähe
i Corvo del Capo

2684 *Corvus caurinus*
Passeriformes - Corvidae
e North-western Crow
f Corneille d'Alaska
i Cornacchia del Nord-Ovest

2685 *Corvus corax*
Passeriformes - Corvidae
e Common Raven; Raven; Northern Raven (NA); Holarctic Raven; Great Raven
f Grand Corbeau
d Kolkrabe
i Corvo imperiale

2686 *Corvus corone*
Passeriformes - Corvidae
e Carrion Crow; Corby Crow; Crow; Common Crow; Hooded Crow; Eurasian Crow; Mesapotamian Crow
f Corneille noire
d Aaskrähe; Rabenkrähe
i Cornacchia; Cornacchia nera; Cornacchia comune

2687 *Corvus corone capellanus*
Passeriformes - Corvidae
e Pied Crow

2688 *Corvus corone cornix*
Passeriformes - Corvidae
e Hooded Crow; Hoodie (Scot); Hoodie Crow (Scotland); Scotch Crow; Danish Crow; Irish Crow; Royston Crow
f Corneille mantelée
d Nebelkrähe
i Cornacchia bigia; Cornacchia grigia

2689 *Corvus coronoides*
Passeriformes - Corvidae
e Australian Raven; Crow
f Corbeau d'Australie
d Neuholland-Krähe
i Corvo australiano

2690 *Corvus crassirostris*
Passeriformes - Corvidae
e Thick-billed Raven; Great-billed Raven; Abyssinian Raven
f Corbeau corbivau
d Erzrabe
i Corvo abissino

2691 *Corvus cryptoleucus*
Passeriformes - Corvidae
e Chihuahuan Raven; White-necked

Raven; American White-necked
Raven
f　Corbeau à cou blanc
d　Weißhalsrabe
i　Corvo di Chihuahua

2692　***Corvus dauuricus***
　　Passeriformes - Corvidae
e　Daurian Jackdaw; Black Jackdaw;
　　Pied Jackdaw; Chinese Jackdaw;
　　Collared Jackdaw
f　Choucas de Daourie; Choucas à
　　collier
d　Weißbauchdohle; Elsterdohle;
　　Elsterndohle
i　Taccola di Dauria; Taccola orientale

2693　***Corvus enca***
　　Passeriformes - Corvidae
e　Slender-billed Crow; Violaceous
　　Crow; Little Crow
f　Corneille à bec fin
d　Sunda Krähe
i　Cornacchia beccofino

2694　***Corvus florensis***
　　Passeriformes - Corvidae
e　Flores Crow
f　Corneille de Florès
d　Flores-Krähe
i　Cornacchia di Flores

2695　***Corvus frugilegus***
　　Passeriformes - Corvidae
e　Rook; Eurasian Rook (NA); Crow
f　Corbeau freux
d　Saatkrähe
i　Corvo; Corvo nero

2696　***Corvus fuscicapillus***
　　Passeriformes - Corvidae
e　Brown-headed Crow; Brown-capped
　　Crow
f　Corneille à tête brune
d　Riesenkrähe; Braunkopfkrähe
i　Corvo testabruna

2697　***Corvus hawaiiensis***
　　Passeriformes - Corvidae
e　Hawaiian Crow; Alala (Hiawai)
f　Corneille d'Hawai

d　Hawaii-Krähe
i　Corvo delle Hawaii

2698　***Corvus imparatus***
　　Passeriformes - Corvidae
e　Mexican Crow; Tamaulipas Crow
f　Corneille du Mexique
d　Mexikaner Krähe
i　Cornacchia messicana

2699　***Corvus jamaicensis***
　　Passeriformes - Corvidae
e　Jamaican Crow; Jabbering Crow
　　(WI); Jamming Crow (WI);
　　Jamicrow (WI)
f　Corneille de la Jamaïque
d　Jamaika-Krähe
i　Cornacchia della Giamaica

2700　***Corvus kubaryi***
　　Passeriformes - Corvidae
e　Mariana Crow; New Caledonian
　　Crow; Guam Crow; Kubary's Crow;
　　Marianas Crow
f　Corneille de Guam
d　Guam-Krähe
i　Cornacchia delle Marianne

2701　***Corvus leucognaphalus***
　　Passeriformes - Corvidae
e　White-necked Crow
f　Corneille d'Hispaniola
d　Antillen-Krähe
i　Cornacchia collobianco

2702　***Corvus levaillantii***
　　Passeriformes - Corvidae
e　Jungle Crow; Large-billed Crow;
　　Japanese Crow
f　Corbeau de Levaillant
i　Cornacchia della giungla

2703　***Corvus macrorhynchos***
　　Passeriformes - Corvidae
e　Large-billed Crow; Jungle Crow;
　　Thick-billed Crow; Indian Crow;
　　Black Crow
f　Corbeau à gros bec; Corneille à gros
　　bec
d　Dickschnabelkrähe

i Cornacchia beccogrosso; Cornacchia
della giungla

2704 *Corvus meeki*
Passeriformes - Corvidae
e Bougainville Crow; Solomons Crow;
Solomon Crow; Solomon Islands
Crow
f Corneille de Meek
d Bougainville-Krähe
i Corvo di Bougainville

2705 *Corvus mellori*
Passeriformes - Corvidae
e Little Raven; South Australian Raven
f Petit Corbeau
d Gesellschafts-Krähe; Kleine
Neuholland-Krähe

2706 *Corvus monedula*
Passeriformes - Corvidae
e Eurasian Jackdaw; Jackdaw;
Common Jackdaw; Western Jackdaw
f Choucas des tours
d Dohle
i Taccola; Monacchia

2707 *Corvus moneduloides*
Passeriformes - Corvidae
e New Caledonian Crow; Wa-wa
f Corbeau calédonien
d Geradschnabelkrähe
i Taccola; Cornacchia della Nuova
Caledonia

2708 *Corvus nasicus*
Passeriformes - Corvidae
e Cuban Crow
f Corneille de Cuba
d Kuba-Krähe
i Cornacchia di Cuba

2709 *Corvus orru*
Passeriformes - Corvidae
e Torresian Crow; Australian Crow;
Crow; Papuan Crow; Tanimbar
Crow
f Corbeau de Torres
d Salvadori-Krähe
i Corvo di Torres

2710 *Corvus orru orru*
Passeriformes - Corvidae
e Torres Straits Torresian crow (ANZ)

2711 *Corvus ossifragus*
Passeriformes - Corvidae
e Fish Crow
f Corneille de rivage
d Fischkrähe
i Cornacchia ossifraga

2712 *Corvus palmarum*
Passeriformes - Corvidae
e Palm Crow
f Corneille palmiste
d Nebelkrähe
i Cornacchia delle palme

2713 *Corvus rhipidurus*
Passeriformes - Corvidae
e Fan-tailed Raven
f Corbeau à queue courte
d Rabenkrähe; Borstenrabe;
Fächerborstenrabe
i Corvo coda a ventaglio; Corvo
codacorta

2714 *Corvus ruficollis*
Passeriformes - Corvidae
e Brown-necked Raven; Dwarf Raven
(CSA); Desert Raven; Desert Crow;
Rufous-necked Raven; Brown Crow;
Somali Raven
f Corbeau brun
d Wüstenrabe; Braunnackenrabe
i Corvo collobruno

2715 *Corvus ruficollis edithae*
Passeriformes - Corvidae
f Corbeau d'Édith

2716 *Corvus sinaloae*
Passeriformes - Corvidae
e Sinaloa Crow; Mexican Crow
f Corneille du Sinaloa
i Cornacchia di Sinaloa

2717 *Corvus splendens*
Passeriformes - Corvidae
e House Crow; Town Crow; Indian
Crow; Ceylon Crow; Grey-necked

Crow; Indian House-Crow
f Corbeau familier; Corbeau d'Indie
d Glanzkrähe; Indischer Hausrabe;
   Hauskrähe
i Cornacchia grigia indiana;
   Cornacchia delle case

2718 *Corvus tasmanicus*
   Passeriformes - Corvidae
e Forest Raven; Tasmanian Raven
f Corbeau de Tasmanie
d Tasman-Krähe
i Corvo della Tasmania

2719 *Corvus tasmanicus boreus*
   Passeriformes - Corvidae
e New England Forest Raven (ANZ)

2720 *Corvus torquatus*
   Passeriformes - Corvidae
e Collared Crow; Ring-necked Crow;
   White-necked Crow
f Corbeau à collier; Corneille à collier
d Halsbandkrähe
i Corvo dal collare

2721 *Corvus tristis*
   Passeriformes - Corvidae
e Grey Crow; Bare-faced Crow; Bare-
   eyed Crow
f Corneille grise
d Greisenkrähe; Neuguinea-Krähe
i Corvo grigio

2722 *Corvus tropicus*
   Passeriformes - Corvidae
e Hawaiian Crow; Alala (Hiawai)
d Hawaii-Krähe

2723 *Corvus typicus*
   Passeriformes - Corvidae
e Piping Crow; Celebes Pied-Crow;
   Celebes Crow
f Corneille des Célèbes
d Celebes-Krähe
i Cornacchia di Sulawesi

2724 *Corvus unicolor*
   Passeriformes - Corvidae
e Banggai Crow

f Corneille des Banggai
i Cornacchia di Banggai

2725 *Corvus validus*
   Passeriformes - Corvidae
e Long-billed Crow; Moluccan Crow
f Corneille des Moluques
d Molukken-Krähe
i Cornacchia delle Molucche

2726 *Corvus woodfordi*
   Passeriformes - Corvidae
e White-billed Crow; Solomon Islands
   Crow; Bougainville Crow
f Corneille à bec blanc
d Buntschnabelkrähe
i Cornacchia delle Isole Salomone

2727 *Corydon sumatranus*
   Passeriformes - Eurylaimidae
e Dusky Broadbill
f Eurylaime corydon
d Riesenbreitrachen
i Beccolargo di Sumatra

2728 *Coryphaspiza melanotis*
   Passeriformes - Fringillidae
e Black-masked Finch
f Coryphaspize à joues noires
d Camposammer
i Fringuello mascherato

2729 *Coryphistera alaudina*
   Passeriformes - Furnariidae
e Lark-like Brushrunner
f Annumbi alouette
d Buschläufer; Lerchenähnlicher
   Furnarius
i Corridore alledola

2730 *Coryphospingus cucullatus*
   Passeriformes - Fringillidae
e Red-crested Finch; Red-pileated
   Finch
f Araguira rougeâtre
d Haubenfink; Roter Kronfink
i Fringuello crestarossa

2731 *Coryphospingus pileatus*
   Passeriformes - Fringillidae
e Pileated Finch; Grey-pileated Finch

f Araguira gris
d Graurückenkronfink; Grauer
   Kronfink
i Fringuello pileato

2732 *Corythaeola cristata*
Musophagiformes - Musophagidae
e Great Blue-Turaco
f Touraco géant
d Riesenturaco
i Turaco azzurro gigante; Turaco
   gigante

2733 *Corythaixoides concolor*
Musophagiformes - Musophagidae
e Grey Go-away-bird; Grey Lourie
   (CSA); Common Go-away-bird; Go-
   away-bird
f Touraco concolore
d Graulärmvogel
i Turaco unicolore

2734 *Corythaixoides leucogaster*
Musophagiformes - Musophagidae
e White-bellied Go-away-bird
f Touraco a ventre blanc
d Weißbauchlärmvogel;
   Langhauenlärmvogel
i Turaco ventrebianco

2735 *Corythaixoides personatus*
Musophagiformes - Musophagidae
e Bare-faced Go-away-bird
f Touraco masqué
d Nacktkehllärmvogel
i Turaco mascherato

2736 *Corythopis delalandi*
Passeriformes - Tyrannidae
e Southern Antpipit; Delalande's
   Antpipit
f Corythopis de Delalande
d Lauftyrann

2737 *Corythopis torquata*
Passeriformes - Tyrannidae
e Ringed Antpipit
f Corythopis à collier
d Brustbandtyran

2738 *Coscoroba coscoroba*
Anseriformes - Anatidae
e Coscoroba Swan; Coscoroba
f Coscoroba blanc
d Coscoroba-Schwan
i Cigno coscoroba

2739 *Cosmopsarus regius*
Passeriformes - Sturnidae
e Golden-breasted Starling; Regal
   Starling
f Spréo royal; Étourneau royal
d Königsglanzstar
i Storno reale africano

2740 *Cosmopsarus unicolor*
Passeriformes - Sturnidae
e Ashy Starling
f Spréo cendré; Étourneau cendré
d Grauglanzstar
i Storno cenerino

2741 *Cossypha albicapilla*
Passeriformes - Muscicapidae
e White-crowned Robin-Chat; White-
   crowned Robin
f Cossyphe à calotte blanche; Grand
   Cossyphe à tête blanche
d Schuppenkopfrötel
i Pettirosso corona bianca

2742 *Cossypha anomala*
Passeriformes - Muscicapidae
e Olive-flanked Robin-Chat; Olive-
   flanked Robin; Olive-flanked Alethe;
   Olive-flanked Ground-Robin
f Cossyphe à flancs olives
d Braunflankenrötel
i Pettirosso fianchioliva

2743 *Cossypha archeri*
Passeriformes - Muscicapidae
e Archer's Robin-Chat; Archer's Chat;
   Archer's Robin; Archer's Ground-
   Robin
f Cossyphe d'Archer
d Ruwenzori-Rötel
i Pettirosso di Archer

2744 *Cossypha caffra*
Passeriformes - Muscicapidae

e Cape Robin-Chat; Cape Robin
    (CSA); Robin-Chat; Common Robin-
    Chat
f Cossyphe du Cap
d Kap-Rötel
i Pettirosso del Capo

2745 *Cossypha cyanocampter*
    Passeriformes - Muscicapidae
e Blue-shouldered Robin-Chat; Blue-
    shouldered Chat; Blue-shouldered
    Robin
f Cossyphe à ailes bleues
d Blauschulterrötel
i Pettirosso spalleblu

2746 *Cossypha dichroa*
    Passeriformes - Muscicapidae
e Chorister Robin-Chat; Chorister
    Robin (CSA); Chorister Chat
f Cossyphe choriste
d Lärmrötel; Spottrötel
i Pettirosso bicolore

2747 *Cossypha heinrichi*
    Passeriformes - Muscicapidae
e White-headed Robin-Chat; Angola
    White-headed Robin; Heinrich's
    Robin-Chat; Rand's Robin-Chat
f Cossyphe à tête blanche
d Weißkopfrötel
i Pettirosso di Heinrich

2748 *Cossypha heuglini*
    Passeriformes - Muscicapidae
e White-browed Robin-Chat; White-
    browed Robin; Heuglin's Chat;
    Heuglin's Robin (CSA); Eye-browed
    Robin-Chat
f Cossyphe de Heuglin
d Weißbrauenrötel
i Pettirosso di Heuglin

2749 *Cossypha humeralis*
    Passeriformes - Muscicapidae
e White-throated Robin-Chat; White-
    throated Robin (CSA); White-
    shouldered Robin-Chat; White-
    shouldered Chat
f Cossyphe à gorge blanche

d Weißkehlrötel
i Pettirosso ventrebianco

2750 *Cossypha isabellae*
    Passeriformes - Muscicapidae
e Mountain Robin-Chat; Cameroon
    Mountain-Chat; Cameroon Robin-
    Chat; Mountain Robin
f Cossyphe d'Isabelle
d Kamerun-Rötel
i Pettirosso di Isabella

2751 *Cossypha natalensis*
    Passeriformes - Muscicapidae
e Red-capped Robin-Chat; Natal Robin
    (CSA); Natal Robin-Chat; Natal
    Chat; Red-capped Robin
f Cossyphe à calotte rousse
d Natal-Rötel
i Pettirosso del Natal

2752 *Cossypha niveicapilla*
    Passeriformes - Muscicapidae
e Snowy-crowned Robin-Chat; Snowy-
    headed Robin-Chat; Snowy-headed
    Robin; Snowy-headed Chat; Snowy-
    crowned Chat
f Cossyphe à calotte neigeuse; Petit
    Cossyphe à tête blanche
d Weißscheitelrötel
i Pettirosso testadineve

2753 *Cossypha polioptera*
    Passeriformes - Muscicapidae
e Grey-winged Robin-Chat; Grey-
    winged Robin (CSA); Grey-winged
    Chat; Grey-winged Ground-Robin;
    Grey-winged Akalat; White-browed
    Akalat; White-browed Robin; White-
    browed Robin-Chat
f Cossyphe à sourcils blancs
d Grauflügelrötel
i Pettirosso aligrige

2754 *Cossypha roberti*
    Passeriformes - Muscicapidae
e White-bellied Akalat; White-bellied
    Robin-Chat; White-bellied Robin;
    White-bellied Chat; Robert's Chat
f Cossyphe à ventre blanc

d   Weißbauchrötel
i   Pettirosso di Robert

2755   *Cossypha semirufa*
      Passeriformes - Muscicapidae
e   Rüppell's Robin-Chat; Rüppell's
      Black-tailed Robin-Chat; Rüppell's
      Chat; Lesser Robin-Chat; Black-
      tailed Robin-Chat
f   Cossyphe de Rüppell
d   Braunrückenrötel
i   Pettirosso codanera

2756   *Cotinga amabilis*
      Passeriformes - Cotingidae
e   Lovely Cotinga
f   Cotinga céleste
d   Azurkotinga
i   Cotinga amabile

2757   *Cotinga cayana*
      Passeriformes - Cotingidae
e   Spangled Cotinga
f   Cotinga de Cayenne
d   Türkis-Kotinga; Halsbandkotinga
i   Cotinga adorna

2758   *Cotinga cotinga*
      Passeriformes - Cotingidae
e   Purple-breasted Cotinga; Blue
      Cotinga
f   Cotinga de Daubenton
d   Purpurbrustkotinga
i   Cotinga pettopurpureo

2759   *Cotinga maculata*
      Passeriformes - Cotingidae
e   Banded Cotinga
f   Cotinga cordonbleu
d   Halsbandkotinga
i   Cotinga maculata

2760   *Cotinga maynana*
      Passeriformes - Cotingidae
e   Plum-throated Cotinga; Maynas
      Cotinga
f   Cotinga des Maynas
d   Veilchenkotinga
i   Cotinga goladiprugna

2761   *Cotinga nattererii*
      Passeriformes - Cotingidae
e   Blue Cotinga; Natterer's Cotinga
f   Cotinga bleu
d   Schwarzbauchkotinga
i   Cotinga di Natterer

2762   *Cotinga ridgwayi*
      Passeriformes - Cotingidae
e   Turquoise Cotinga; Ridgway's
      Cotinga
f   Cotinga turquoise; Cotinga de
      Ridgway
d   Ridgway-Kotinga
i   Cotinga di Ridgway

2763   Cotingidae
      Passeriformes
e   Cotingas
f   Cotingidés
d   Schmuckvögel; Kotingas
i   Cotingidi

2764   *Coturnicops exquisitus*
      Gruiformes - Rallidae
e   Swinhoe's Rail; Swinhoe's Crake;
      Swinhoe's Yellow-Rail; Swinhoe's
      Yellow-Crake
d   Mandschuren-Ralle
i   Rallo di Swinhoe

2765   *Coturnicops notatus*
      Gruiformes - Rallidae
e   Speckled Rail; Speckled Crake;
      Speckled-marked Darwin's Rail;
      Darwin's Rail
d   Merida-Ralle
i   Schiribilla di Darwin

2766   *Coturnicops noveboracensis*
      Gruiformes - Rallidae
e   Yellow Rail; Yellow Crake; Siberian
      Rail
f   Râle jaune
d   Gelbralle
i   Schiribilla gialla

2767   *Coturnix adansonii*
      Galliformes - Phasianidae
e   Blue Quail; African Blue-Quail
f   Caille bleue

d Afrikanische Zwergwachtel;
    Adanson-Wachtel

2768 *Coturnix chinensis*
      Galliformes - Phasianidae
e Blue-breasted Quail; King Quail;
    Blue Quail; Painted Quail; Asian
    Blue-Quail; Indian Blue-Quail;
    Chestnut-bellied Quail; Dwarf Quail;
    Least Quail; Indian Painted Quail
f Caille peinte
d Zwergwachtel

2769 *Coturnix coromandelica*
      Galliformes - Phasianidae
e Rain Quail; Black-breasted Quail
f Caille nattée
d Regenwachtel
i Quaglia dell piogge

2770 *Coturnix coturnix*
      Galliformes - Phasianidae
e Common Quail; Quail; Grey Quail
    (ISC); European Migratory Quail;
    European Quail; Eurasian Migratory
    Quail; Eurasian Quail
f Caille des blés
d Wachtel
i Quaglia; Quaglia comune

2771 *Coturnix delegorguei*
      Galliformes - Phasianidae
e Harlequin Quail
f Caille arlequin
d Harlekinwachtel
i Quaglia arlecchino

2772 *Coturnix japonica*
      Galliformes - Phasianidae
e Japanese Quail; Asian Migratory
    Quail; Asiatic Migratory Quail
f Caille du Japon
d Japan-Wachtel
i Quaglia giapponese

2773 *Coturnix novaezelandiae*
      Galliformes - Phasianidae
e New Zealand Quail
f Caille de Nouvelle-Zélande
d Schwarzbrustwachtel
i Quaglia della Nuova Zelanda

2774 *Coturnix pectoralis*
      Galliformes - Phasianidae
e Stubble Quail; Pectoral Quail; Grey
    Quail (ISC)
f Caille des chaumes
d Australien-Schwarzbrustwachtel
i Quaglia delle stoppie

2775 *Coturnix ypsilophora*
      Galliformes - Phasianidae
e Brown Quail; Swamp Quail;
    Tasmanian Brown-Quail; Silver
    Quail
f Caille tasmane
d Ypsilonwachtel
i Quaglia della Tasmania

2776 *Coturnix ypsilophora australis*
      Galliformes - Phasianidae
f Caille australe

2777 *Coua caerulea*
      Cuculiformes - Cuculidae
e Blue Coua; Blue Madagascar Coua;
    Blue Madagascar Coucal
f Coua bleu
d Blaucoua
i Coua azzurro

2778 *Coua coquereli*
      Cuculiformes - Cuculidae
e Coquerel's Coua
f Coua de Coquerel
d Coquerel-Coua
i Coua di Coquerel

2779 *Coua cristata*
      Cuculiformes - Cuculidae
e Crested Coua; Crested Madagascar
    Coucal
f Coua huppé
d Spitzschopfcoua
i Coua crestato

2780 *Coua cursor*
      Cuculiformes - Cuculidae
e Running Coua; Running Coucal
f Coua coureur
d Gelbkehlcoua; Laufcoua
i Coua corridore

2781 **Coua delalandei**
Cuculiformes - Cuculidae
e Snail-eating Coua; Delalande's Cuckoo
f Coua de Delalande
d Delalande-Coua
i Coua di Delalande

2782 **Coua gigas**
Cuculiformes - Cuculidae
e Giant Coua; Giant Madagascar Coua
f Coua géant
d Riesencoua
i Coua gigante

2783 **Coua reynaudii**
Cuculiformes - Cuculidae
e Red-fronted Coua; Reynaud's Cuckoo
f Coua de Reynaud
d Rotstirncoua
i Coua di Reynaud

2784 **Coua ruficeps**
Cuculiformes - Cuculidae
e Red-capped Coua; Olive-capped Coua; Red-capped Madagascar Coua
f Coua à tête rousse
d Weißkehlcoua
i Coua testarossa

2785 **Coua serriana**
Cuculiformes - Cuculidae
e Red-breasted Coua; Pucheran's Coua; Rufous-breasted Coua
f Coua de Serre
d Rotbrustcoua
i Coua pettorossiccio

2786 **Coua verreauxi**
Cuculiformes - Cuculidae
e Verreaux's Coua; Southern Crested Madagascar Coucal
f Coua de Verreaux
d Breitschopfcoua
i Coua di Verreaux

2787 **Cracidae**
Galliformes
e Curassows; Guans
f Cracidés

d Schackhühner; Hokkohühner
i Cracidi

2788 **Cracticus cassicus**
Passeriformes - Corvidae
e Hooded Butcherbird; Black-headed Butcherbird; Black-and-white Butcherbird; Pied Butcherbird
f Cassican à tête noire
d Papua-Würgatzel
i Gazza chiassosa testanera

2789 **Cracticus louisiadensis**
Passeriformes - Corvidae
e Tagula Butcherbird; Louisiade Butcherbird; Grey Butcherbird; White-rumped Butcherbird; Sudest Butcherbird; Louisiades Butcherbird
f Cassican de Tagula
d Louisiaden-Würgatzel
i Gazza chiassosa gropponebianco; Gazza chiassosa dal groppone bianco

2790 **Cracticus mentalis**
Passeriformes - Corvidae
e Black-backed Butcherbird; White-throated Butcherbird
f Cassican à dos noir
d Schwarzrückenwürgatzel
i Gazza chiassosa dorsonero

2791 **Cracticus nigrogularis**
Passeriformes - Corvidae
e Pied Butcherbird; Black-throated Butcherbird
f Cassican à gorge noire
d Schwarzkehlwürgatzel
i Gazza chiassosa golanera

2792 **Cracticus quoyi**
Passeriformes - Corvidae
e Black Butcherbird
f Cassican des mangroves
d Mangrovenwürgatzel
i Gazza chiassosa di Quoy

2793 **Cracticus quoyi alecto**
Passeriformes - Corvidae
e Torres strait Black Butcherbird (ANZ)

2794   *Cracticus torquatus*
         Passeriformes - Corvidae
e   Grey Butcherbird
f   Cassican à collier
d   Graurückenwürgatzel
i   Gazza chiassosa dal collare

2795   *Cranioleuca albicapilla*
         Passeriformes - Furnariidae
e   Creamy-crested Spinetail
f   Synallaxe à calotte blanche;
         Synallaxe casqué
d   Falbkappenschlüpfer
i   Codaspinosa pettocrema

2796   *Cranioleuca albiceps*
         Passeriformes - Furnariidae
e   Light-crowned Spinetail
f   Synallaxe à bandeaux
d   Kronenschlüpfer
i   Codaspinosa corona bianca

2797   *Cranioleuca antisiensis*
         Passeriformes - Furnariidae
e   Line-cheeked Spinetail; Fraser's
         Spinetail
f   Synallaxe grimpeur
d   Olivrückenschlüpfer
i   Codaspinosa di Fraser

2798   *Cranioleuca baroni*
         Passeriformes - Furnariidae
e   Baron's Spinetail
f   Synallaxe de Baron

2799   *Cranioleuca curtata*
         Passeriformes - Furnariidae
e   Ash-browed Spinetail
f   Synallaxe à sourcils gris
d   Olivstirnschlüpfer
i   Codaspinosa dai sopracigli cenere

2800   *Cranioleuca demissa*
         Passeriformes - Furnariidae
e   Tepui Spinetail
f   Synallaxe de Coiba
d   Tepui-Schlüpfer
i   Codaspinosa del Tepui

2801   *Cranioleuca dissita*
         Passeriformes - Furnariidae

e   Coiba Spinetail
f   Synallaxe à face rouge
i   Codaspinosa di Coiba

2802   *Cranioleuca erythrops*
         Passeriformes - Furnariidae
e   Red-faced Spinetail
f   Synallaxe ponctué
d   Rotwangenschlüpfer
i   Codaspinosa facciarossa

2803   *Cranioleuca gutturata*
         Passeriformes - Furnariidae
e   Speckled Spinetail
f   Synallaxe des broméliades
d   Gelbkinnschlüpfer
i   Codaspinosa macchiettato

2804   *Cranioleuca hellmayri*
         Passeriformes - Furnariidae
e   Streak-capped Spinetail; Streaked-
         capped Spinetail
d   Strichelkopfschlüpfer
i   Codaspinosa di Hellmayr

2805   *Cranioleuca marcapatae*
         Passeriformes - Furnariidae
e   Marcapata Spinetail
f   Synallaxe de Marcapata
d   Marcapata-Schlüpfer
i   Codaspinosa di Marcapata

2806   *Cranioleuca muelleri*
         Passeriformes - Furnariidae
e   Scaled Spinetail
f   Synallaxe écaillé
d   Schuppenschlüpfer
i   Codaspinosa die Müller

2807   *Cranioleuca obsoleta*
         Passeriformes - Furnariidae
e   Olive Spinetail
f   Synallaxe olive
d   Rotschwanzschlüpfer
i   Codaspinosa oliva

2808   *Cranioleuca pallida*
         Passeriformes - Furnariidae
e   Pallid Spinetail
f   Synallaxe pâle

d Fahlschlüpfer
i Codaspinosa pallido

2809 *Cranioleuca pyrrhophia*
Passeriformes - Furnariidae
e Stripe-crowned Spinetail; Striped-
crown Spinetail
f Synallaxe à calotte rayée
d Streifenkopfschlüpfer
i Codaspinosa corona striata

2810 *Cranioleuca semicinerea*
Passeriformes - Furnariidae
e Grey-headed Spinetail
f Synallaxe à tête grise
d Graukopfschlüpfer
i Codaspinosa testagrigia

2811 *Cranioleuca subcristata*
Passeriformes - Furnariidae
e Crested Spinetail
f Synallaxe huppé
d Haubenschlüpfer
i Codaspinosa crestato

2812 *Cranioleuca sulphurifera*
Passeriformes - Furnariidae
e Sulphur-bearded Spinetail; Sulphur-
throated Spinetail
f Synallaxe soufré
d Schwefelkehlschlüpfer
i Codaspinosa golagialla

2813 *Cranioleuca vulpecula*
Passeriformes - Furnariidae
e White-breasted Spinetail

2814 *Cranioleuca vulpina*
Passeriformes - Furnariidae
e Rusty-backed Spinetail
f Synallaxe renard
d Fuchsschlüpfer
i Codaspinosa dorsorossiccio

2815 *Crateroscelis murina*
Passeriformes - Pardalotidae
e Rusty Mouse-Warbler; Lowland
Mouse-Babbler; Lowland Mouse-
Warbler; Rusty Mouse-Babbler
f Séricorne fauve

d Braunrückenwaldhuscher
i Bigia terricola di pianura

2816 *Crateroscelis nigrorufa*
Passeriformes - Pardalotidae
e Bicoloured Mouse-Warbler; Two-
coloured Mouse-Warbler; Mid-
mountain Mouse-Babbler; Mid-
mountain Mouse-Warbler; Black-
backed Mouse-Warbler; Black-
backed Mouse-Babbler; Bicoloured
Mouse-Babbler
f Séricorne noir-et-roux
d Schwarzrückenwaldhuscher
i Bigia terricola di Salvadori

2817 *Crateroscelis robusta*
Passeriformes - Pardalotidae
e Mountain Mouse-Warbler; Mountain
Mouse-Babbler
f Séricorne robuste
d Braunbauchwaldhuscher
i Bigia terricola di montagna

2818 *Crax alberti*
Galliformes - Cracidae
e Blue-knobbed Curassow; Blue-billed
Curassow; Albert's Curassow; Prince
Albert's Curassow; Colombian
Curassow
f Hocco d'Albert
d Blaulappenhokko
i Hocco messicano; Crace beccoblu;
Crace del principe Alberto

2819 *Crax alector*
Galliformes - Cracidae
e Black Curassow; Crested Curassow;
Smooth-billed Curassow
f Hocco alector
d Glatschnabelhokko
i Crace nero; Hocco a caruncolo

2820 *Crax blumenbachii*
Galliformes - Cracidae
e Red-billed Curassow; Blumenbach's
Curassow; Red-wattled Curassow
f Hocco de Blumenbach
d Blumenbach-Hokko
i Crace beccoroso; Crace di
Blumenbach

2821   *Crax daubentoni*
Galliformes - Cracidae
e   Yellow-knobbed Curassow; Yellow-
billed Curassow; Daubenton's
Curassow
f   Hocco de Daubenton
d   Gelblappenhokko
i   Crace dal bernoccolo giallo

2822   *Crax fasciolata*
Galliformes - Cracidae
e   Bare-faced Curassow; Sclater's
Curassow; Banded Curassow;
Fasciated Curassow
f   Hocco à face nue
d   Nacktgesichthokko
i   Crace faccianuda

2823   *Crax globulosa*
Galliformes - Cracidae
e   Wattled Curassow; Yarrell's
Curassow; Globose Curassow
f   Hocco globuleux
d   Karunkelhokko
i   Crace caruncolata

2824   *Crax rubra*
Galliformes - Cracidae
e   Great Curassow; Globose Curassow;
Mexican Curassow; Great Crested-
Curassow
f   Grand Hocco
d   Tuberkelhokko; Roter Hokko
i   Hocco messicano; Hocco

2825   *Creadion carunculatus*
Passeriformes - Callaeatidae
e   Saddleback; Tieko
f   Créadion rounoir
d   Lappenstar; Sattelvogel; Sattelstar
i   Tieko

2826   *Creagrus furcatus*
Charadriiformes - Laridae
e   Swallow-tailed Gull
f   Mouette à queue fourchu
d   Gabelschwanzmöwe
i   Gabbiano codadirondine

2827   *Creatophora cinerea*
Passeriformes - Sturnidae

e   Wattled Starling
f   Étourneau caronculé; Martin
caronculé
d   Lappenstar
i   Storno caruncolato

2828   *Crecopsis egregia*
Gruiformes - Rallidae
e   African Crake
f   Râle des prés
d   Steppenralle
i   Schiribilla africana

2829   *Creurgops dentata*
Passeriformes - Fringillidae
e   Slaty Tanager
f   Tangara ardoisé
d   Schiefertangare
i   Tangara ardesia

2830   *Creurgops verticalis*
Passeriformes - Fringillidae
e   Rufous-crested Tanager
f   Tangara à cimier roux
d   Ockerschopftangare
i   Tangara corona rossa

2831   *Crex crex*
Gruiformes - Rallidae
e   Corncrake; Land Rail
f   Râle des genêts; Râle des prés; Râle
de terre; Râle doré; Râle rouge; Crex
des prés; Roi des cailles; Mère des
cailles; Poule d'eau des genêts
d   Wachtelkönig; Wiesenralle
i   Re di quaglie

2832   *Crinifer piscator*
Musophagiformes - Musophagidae
e   Western Grey-Plantain-eater; Grey
Plantain-eater
f   Touraco gris
d   Schwarzschwanzlärmvogel
i   Turaco grigio occidentale

2833   *Crinifer zonurus*
Musophagiformes - Musophagidae
e   Eastern Grey-Plantain-eater
f   Touraco à queue barrée; Touraco gris
d'Abyssinie

d   Bindenlärmvogel
i   Turaco grigio orentale

2834   ***Criniger barbatus***
Passeriformes - Pycnonotidae
e   Bearded Bulbul; Bearded Greenbul;
Western Bearded-Greenbul
ƒ   Bulbul crinon; Grand Bulbul huppé
d   Haarbülbül
i   Bulbul barbuto

2835   ***Criniger calurus***
Passeriformes - Pycnonotidae
e   Red-tailed Bulbul; Red-tailed
Greenbul; White-bearded Greenbul
ƒ   Bulbul à barbe blanche; Bulbul
huppé à barbe blanche
d   Cassin-Haarbülbül; Swainson-Bülbül
i   Bulbul codirossa

2836   ***Criniger chloronotus***
Passeriformes - Pycnonotidae
e   Green-backed Bulbul; Bulbul;
Eastern Bearded-Bulbul; White-
throated Bulbul; Congo Bulbul;
Congo Bearded-Bulbul
ƒ   Bulbul à dos vert; Bulbul crinon
oriental
i   Bulbul dorsoverde

2837   ***Criniger ndussumensis***
Passeriformes - Pycnonotidae
e   White-bearded Bulbul; White-
bearded Greenbul; Slender-bill
Bearded-Bulbul
ƒ   Bulbul de Reichenow
i   Bulbul barbabianca

2838   ***Criniger olivaceus***
Passeriformes - Pycnonotidae
e   Yellow-bearded Bulbul; Yellow-
bearded Greenbul; Yellow-throated
Olive-Bulbul; Yellow-throated
Olive-Greenbul; Olive-throated
Bulbul
ƒ   Bulbul à barbe jaune
d   Rostbauchbülbül
i   Bulbul golagialla

2839   ***Crocias albonotatus***
Passeriformes - Sylviidae

e   Spotted Crocias; Spotted Sibia
ƒ   Sibia tachetée
d   Fleckenwürgertimalie
i   Sibia maculata

2840   ***Crocias langbianis***
Passeriformes - Sylviidae
e   Grey-crowned Crocias; Grey-
crowned Sibia; Mount Langbian
Sibia; Mount Langbian Crocias
ƒ   Sibia du Langbian
d   Grauscheitelwürgertimalie
i   Sibia del Monte Langbian

2841   ***Crossleyia xanthophrys***
Passeriformes - Sylviidae
e   Yellow-browed Oxylabes; Yellow-
browed Foditany; Yellow-browed
Tetraka; Madagascar Yellowbrow
ƒ   Oxylabe à sourcils jaunes
d   Gelbbrauenfoditany
i   Garrulo orecchiegialle

2842   ***Crossoptilon auritum***
Galliformes - Phasianidae
e   Blue Eared-Pheasant; Pallas's Eared-
Pheasant; Mongolian Eared-Pheasant
ƒ   Hokki bleu
d   Blauer Ohrfasan
i   Fagiano orecchiuto blu

2843   ***Crossoptilon crossoptilon***
Galliformes - Phasianidae
e   White Eared-Pheasant; Tibetan
Eared-Pheasant
ƒ   Hokki blanc
d   Schmalschwanzohrfasan
i   Fagiano orecchiuto bianco

2844   ***Crossoptilon harmani***
Galliformes - Phasianidae
e   Tibetan Eared-Pheasant; Elwe's
Eared-Pheasant
ƒ   Hokki du Tibet
i   Fagiano orecchiuto del Tibet

2845   ***Crossoptilon mantchuricum***
Galliformes - Phasianidae
e   Brown Eared-Pheasant; Manchurian
Eared-Pheasant
ƒ   Hokki brun

d Brauner Ohrfasan
i Fagiano orecchiuto bruno

2846 *Crotophaga ani*
Cuculiformes - Cuculidae
e Smooth-billed Ani; Ani; Crow (WI);
Black Arnold (WI); Old Witch (WI);
Black Parrot (WI); Tickbird (WI);
Savanne Blackbird (WI); Black
Witch (WI); Old Arnold (WI)
f Ani à bec lisse; Bout-de-tabac
(Ants); Merle (Ants); Juif (Ants),
Merle Corbeau (Ants); Bilbitin
(Ants)
d Glattschnabelani
i Ani; Ani beccoliscio

2847 *Crotophaga major*
Cuculiformes - Cuculidae
e Greater Ani; Ani
f Ani des palétuviers
d Riesenani
i Ani maggiore

2848 *Crotophaga sulcirostris*
Cuculiformes - Cuculidae
e Groove-billed Ani; Ani
f Ani à bec cannelé
d Riesenschnabelani
i Ani beccosolcato

2849 *Crypsirina cucullata*
Passeriformes - Corvidae
e Hooded Treepie; Hooded Racket-
tailed Treepie; Hooded Spatulate-
tailed Treepie; Grey Racket-tailed
Treepie; Hooded Racket-tailed
Magpie; Hooded Racquet-tailed
Magpie; Grey Racquet-tailed
Treepie; Hooded Racquet-tailed
Treepie
f Temia à collier; Pie bleue de
l'Himalaya
d Spatelschwanzelster
i Gazza grigia coda a racchetta

2850 *Crypsirina temia*
Passeriformes - Corvidae
e Racket-tailed Treepie; Black Racket-
tailed Treepie; Bronzed Racket-tailed
Treepie; Black Spatulate-tailed

Treepie; Bronzed Spatulate-tailed
Treepie; Black Racket-tailed Magpie;
Bronzed Treepie; Black Racquet-
tailed Magpie; Bronzed Racquet-
tailed Treepie; Black Racquet-tailed
Treepie; Racquet-tailed Treepie;
Black Treepie
f Temia bronzée
d Rackettschwanzelster
i Gazza nera coda a racchetta

2851 *Cryptophaps poecilorrhoa*
Columbiformes - Columbidae
e Sombre Pigeon; Celebes Dusky
Pigeon; Celebes Dusky Fruit-Pigeon;
Rusty-bellied Fruit-Pigeon; Long-
tailed Imperial-Pigeon
f Carpophage des Célèbes
d Fleckenbauchfruchttaube;
Rotsteißfruchttaube
i Piccione bruno di Sulawesi

2852 *Cryptospiza jacksoni*
Passeriformes - Estrildidae
e Dusky Crimsonwing; Jackson's
Crimsonwing; Jackson's Hill-Finch
f Sénégali de Jackson; Bengali de
Jackson
d Jackson-Astrild; Jackson-Bergastrild
i Alarossa di Jackson

2853 *Cryptospiza reichenovii*
Passeriformes - Estrildidae
e Red-faced Crimsonwing; Red-eyed
Crimsonwing; Reichenow's
Crimsonwing; Nyasa Crimsonwing
f Sénégali de Reichenow; Bengali vert
à face rouge; Bengali de Reichenow
d Reichenows Bergastild; Reichenow-
Bergastrild
i Alarossa di Reichenow

2854 *Cryptospiza salvadorii*
Passeriformes - Estrildidae
e Abyssinian Crimsonwing; Ethiopian
Crimsonwing; Salvadori's
Crimsonwing; Crimson-backed
Forest Finch
f Sénégali de Salvadori; Bengali vert;
Bengali de Salvadori
d Salvadori-Astrild; Salvadori-

Bergastrild
i Alarossa di Salvadori

2855 ***Cryptospiza shelleyi***
Passeriformes - Estrildidae
e Shelley's Crimsonwing; Red-billed
Crimsonwing
f Sénégali de Shelley; Bengali vert de
Shelley; Bengali de Shelley
d Rotmantelastrild; Shelley-Bergastrild
i Alarossa di Shelley

2856 ***Cryptosylvicola randrianasoloi***
Passeriformes - Sylviidae
e Cryptic Warbler

2857 ***Crypturellus atrocapillus***
Tinamiformes - Tinamidae
e Black-capped Tinamou; Red-legged
Tinamou
f Tinamou à calotte noire; Tinamou à
pieds rouge
d Rotfußtinamu
i Tinamo capinero

2858 ***Crypturellus bartletti***
Tinamiformes - Tinamidae
e Bartlett's Tinamou
f Tinamou de Bartlett
d Bartlett-Tinamu
i Tinamno di Bartlett

2859 ***Crypturellus berlepschi***
Tinamiformes - Tinamidae
e Berlepsch's Tinamou
f Tinamou de Berlepsch
i Tinamo di Berlepsch

2860 ***Crypturellus boucardi***
Tinamiformes - Tinamidae
e Slaty-breasted Tinamou; Boucard's
Tinamou
f Tinamou de Boucard
d Graukehltinamu
i Tinamo pettoardesia

2861 ***Crypturellus brevirostris***
Tinamiformes - Tinamidae
e Rusty Tinamou; Short-billed
Tinamou
f Tinamou rubigineux

d Rosttinamu
i Tinamo rugginoso

2862 ***Crypturellus casiquiare***
Tinamiformes - Tinamidae
e Barred Tinamou; Cassiquiare
Tinamou
f Tinamou barré
d Bindentinamu
i Tinamo barrato

2863 ***Crypturellus cinereus***
Tinamiformes - Tinamidae
e Cinereous Tinamou; Brushland
Tinamou
f Tinamou cendré
d Grautinamu
i Tinamo cinereo

2864 ***Crypturellus cinnamomeus***
Tinamiformes - Tinamidae
e Thicket Tinamou; Rufescent
Tinamou
f Tinamou cannellé
d Buschtinamu; Zimttinamu
i Tinamo di macchia

2865 ***Crypturellus duidae***
Tinamiformes - Tinamidae
e Grey-legged Tinamou; Duida
Tinamou
f Tinamou de Zimmer
d Graufußtinamu
i Tinamo zampegrige

2866 ***Crypturellus erythropus***
Tinamiformes - Tinamidae
e Red-legged Tinamou; Red-footed
Tinamou
f Tinamou à pieds rouges
e Waldtinamu
i Tinamo piedirossi

2867 ***Crypturellus erythropus***
columbianus
Tinamiformes - Tinamidae
e Colombian Tinamou
f Tinamou de Colombie

2868 *Crypturellus erythropus idoneus*
Tinamiformes - Tinamidae
f Tinamou des Santa Marta

2869 *Crypturellus kerriae*
Tinamiformes - Tinamidae
e Choco Tinamou; Kerr's Tinamou
f Tinamou de Kerr
d Kerr-Tinamu
i Tinamo del Choco

2870 *Crypturellus noctivagus*
Tinamiformes - Tinamidae
e Yellow-legged Tinamou; Weid's
Tinamou
f Tinamou noctivague
d Gelbfußtinamu
i Tinamo zampegialle

2871 *Crypturellus obsoletus*
Tinamiformes - Tinamidae
e Brown Tinamou
f Tinamou brun
d Kastanientinamu
i Tinamo bruno

2872 *Crypturellus parvirostris*
Tinamiformes - Tinamidae
e Small-billed Tinamou
f Tinamou à petit bec
d Kleinschnabeltinamu
i Tinamo beccopiccolo

2873 *Crypturellus ptaritepui*
Tinamiformes - Tinamidae
e Tepui Tinamou; Little Tinamou;
Pileated Tinamou
f Tinamou des tépuis; Tinamou tepui
d Tepui-Tinamu
i Tinamo del Tepui

2874 *Crypturellus saltuarius*
Tinamiformes - Tinamidae
e Magdalena Tinamou
f Tinamou de la Magdalena; Tinamou
forestier
d Waldtinamu

2875 *Crypturellus soui*
Tinamiformes - Tinamidae
e Little Tinamou; Pileated Tinamou

f Tinamou soui
d Brauntinamu
i Tinamo piccolo

2876 *Crypturellus strigulosus*
Tinamiformes - Tinamidae
e Brazilian Tinamou
f Tinamou oariana
d Rotkehltinamu
i Tinamo brasiliano

2877 *Crypturellus tataupa*
Tinamiformes - Tinamidae
e Tataupa Tinamou
f Tinamou tataupa
d Tataupa
i Tinamo tataupa

2878 *Crypturellus transfasciatus*
Tinamiformes - Tinamidae
e Pale-browed Tinamou; Steere's
Tinamou
f Tinamou à grands sourcils
d Brausentinamu
i Tinamo dai sopraccigli chiari

2879 *Crypturellus undulatus*
Tinamiformes - Tinamidae
e Undulated Tinamou; Banded
Tinamou
f Tinamou vermiculé
d Wellentinamu
i Tinamo ondulato

2880 *Crypturellus variegatus*
Tinamiformes - Tinamidae
e Variegated Tinamou
f Tinamou varié
d Rotbrusttinamu
i Tinamo variegato

2881 **Cuculidae**
Cuculiformes
e Old World Cuckoos
f Coucous; Cuculidés
d Kukucke
i Cuculidi; Cuculi

2882 **Cuculiformes**
e Cuckoos
f Cuculiformes

   d   Kukucksvögel
   i   Cuculiformi

2883   ***Cuculus canorus***
      Cuculiformes - Cuculidae
   e   Common Cuckoo; Cuckoo; Eurasian
      Cuckoo (CSA); European Cuckoo
   f   Coucou gris; Coucou
   d   Kuckuck
   i   Cuculo eurasiatico; Cuculo

2884   ***Cuculus clamosus***
      Cuculiformes - Cuculidae
   e   Black Cuckoo
   f   Coucou criard
   d   Schwarzkuckuck
   i   Cuculo nero

2885   ***Cuculus crassirostris***
      Cuculiformes - Cuculidae
   e   Sulawesi Hawk-Cuckoo; Celebes
      Cuckoo
   f   Coucou des Célèbes
   d   Minehassa-Kuckuck
   i   Cuculo sparviero di Sulawesi

2886   ***Cuculus fugax***
      Cuculiformes - Cuculidae
   e   Hodgson's Hawk-Cuckoo; Fugitive
      Hawk-Cuckoo
   f   Coucou fugitif; Coucou de Hodgson
   d   Fluchtkuckuck
   i   Cuculo sparviero di Hodgson

2887   ***Cuculus gularis***
      Cuculiformes - Cuculidae
   e   African Cuckoo
   f   Coucou africain
   d   Afrikanischer Kuckuck
   i   Cuculo africano

2888   ***Cuculus hyperythrus***
      Cuculiformes - Cuculidae
   e   Northern Hawk-Cuckoo

2889   ***Cuculus micropterus***
      Cuculiformes - Cuculidae
   e   Indian Cuckoo; Short-winged
      Cuckoo
   f   Coucou à ailes courtes; Coucou
      indien

   d   Kurzflügelkuckuck
   i   Cuculo indiano

2890   ***Cuculus pallidus***
      Cuculiformes - Cuculidae
   e   Pallid Cuckoo
   f   Coucou pâle
   d   Blaßkuckuck
   i   Cuculo pallido

2891   ***Cuculus pectoralis***
      Cuculiformes - Cuculidae
   e   Philippine Hawk-Cuckoo

2892   ***Cuculus poliocephalus***
      Cuculiformes - Cuculidae
   e   Lesser Cuckoo; Little Cuckoo;
      Eurasian Little-Cuckoo
   f   Petit Coucou; Petit Coucou d'Asie
   d   Gackelkuckuck
   i   Cuculo minore

2893   ***Cuculus rochii***
      Cuculiformes - Cuculidae
   e   Madagascar Cuckoo; Madagascar
      Lesser Cuckoo (CSA)
   f   Coucou de Madagascar; Petit Coucou
      malgache
   d   Madagaskar-Kuckuck
   i   Cuculo del Madagascar

2894   ***Cuculus saturatus***
      Cuculiformes - Cuculidae
   e   Oriental Cuckoo; Sunda Cuckoo;
      Himalayan Cuckoo; Blyth's Cuckoo
   f   Coucou oriental
   d   Hopfkuckuck
   i   Cuculo orientale

2895   ***Cuculus solitarius***
      Cuculiformes - Cuculidae
   e   Red-chested Cuckoo; Piet-my-frou
   f   Coucou solitaire
   d   Einsiedlerkuckuck
   i   Cuculo solitario

2896   ***Cuculus sparverioides***
      Cuculiformes - Cuculidae
   e   Large Hawk-Cuckoo
   f   Coucou épervier

d Sperberkuckuck
i Cuculo sparviero maggiore

2897 ***Cuculus vagans***
Cuculiformes - Cuculidae
e Moustached Hawk-Cuckoo; Lesser Hawk-Cuckoo
f Coucou à moustaches
d Bartkuckuck
i Cuculo sparviero minore

2898 ***Cuculus varius***
Cuculiformes - Cuculidae
e Common Hawk-Cuckoo; Brainfeverbird (ISC); Hawk Cuckoo
f Coucou shikra
d Wechselkuckuck
i Cuculo sparviero comune

2899 ***Culicicapa ceylonensis***
Passeriformes - Muscicapidae
e Grey-headed Canary-Flycatcher; Grey-headed Flycatcher (ISC); Ceylon Grey-headed Flycatcher
f Gobemouche à tête grise; Gobe-mouches à tête grise
d Graukopfkanarienschnäpper
i Pigliamosche giallo testagrigia

2900 ***Culicicapa helianthea***
Passeriformes - Muscicapidae
e Citrine Canary-Flycatcher; Citrine Flycatcher; Sunflower Flycatcher
f Gobemouche canari; Gobe-mouches canari
d Grünkopfkanarienschnäpper
i Pigliamosche giallo

2901 ***Culicivora caudacuta***
Passeriformes - Tyrannidae
e Sharp-tailed Grass-Tyrant; Sharp-tailed Tyrant; Wire-tailed Tyrant
f Tyranneau à queue aiguë; Tyranneau à queue-en-aiguille
d Spitzschwanztyrann
i Tiranno codacuta

2902 ***Curaeus curaeus***
Passeriformes - Fringillidae
e Austral Blackbird
f Quiscale austral

d Stachelkopfstärling
i Gracchio australe

2903 ***Curaeus forbesi***
Passeriformes - Fringillidae
e Forbes's Blackbird
f Quiscale de Forbes
d Forbes-Stärling
i Gracchio di Forbes

2904 ***Cursorius coromandelicus***
Ciconiiformes - Glareolidae
e Indian Courser
f Courvite de Coromandel
d Koromandel-Rennvogel
i Corrione indiano

2905 ***Cursorius cursor***
Ciconiiformes - Glareolidae
e Cream-coloured Courser; Desert Courser
f Courvite isabelle
d Rennvogel; Gewöhnlicher Wüstenläufer; Wüstenläufer
i Corrione biondo

2906 ***Cursorius rufus***
Ciconiiformes - Glareolidae
e Burchell's Courser
f Courvite de Burchell
d Rostrenvogel
i Corrione di Burchell

2907 ***Cursorius somalensis***
Ciconiiformes - Glareolidae
e Somali Courser

2908 ***Cursorius temminckii***
Ciconiiformes - Glareolidae
e Temminck's Courser
f Courvite de Temminck
d Temminck-Rennvogel
i Corrione di Temminck

2909 ***Cutia nipalensis***
Passeriformes - Sylviidae
e Cutia; Nepal Cutia
f Cutie du Nepal
d Cutia; Baumtimalie
i Cutia del Nepal

2910 *Cyanerpes caeruleus*
Passeriformes - Fringillidae
e Purple Honeycreeper; Yellow-legged
Honeycreeper
f Guit-guit ceruléen
d Purpurnaschvogel
i Cianerpe purpurea

2911 *Cyanerpes cyaneus*
Passeriformes - Fringillidae
e Red-legged Honeycreeper; Blue
Honeycreeper
f Guit-guit sai
d Türkis-Naschvogel; Türkisvogel
i Cianerpe zamperosse

2912 *Cyanerpes lucidus*
Passeriformes - Fringillidae
e Shining Honeycreeper
f Guit-guit brillant
d Azurnaschvogel; Gelbfußnaschvogel
i Cianerpe splendente

2913 *Cyanerpes nitidus*
Passeriformes - Fringillidae
e Short-billed Honeycreeper
f Guit-guit à bec court
d Kurzschnabelnaschvogel
i Cianerpe beccobreve

2914 *Cyanicterus cyanicterus*
Passeriformes - Fringillidae
e Blue-backed Tanager
f Tangara cyanictère
d Ziertangare
i Tangara azzurra e gialla

2915 *Cyanochen cyanopterus*
Anseriformes - Anatidae
e Blue-winged Goose; Abyssinian
Blue-winged Goose; Abyssinian
Goose
f Ouette à ailes bleues
d Blauflügelgans
i Oca aliceleste; Oca aliblu

2916 *Cyanocitta cristata*
Passeriformes - Corvidae
e Blue Jay; American Blue-Jay
f Geai bleu; Geai bleu d'Amérique
d Blauhäher

i Ghiandaia azzura; Ghiandaia azzurra
americana

2917 *Cyanocitta stelleri*
Passeriformes - Corvidae
e Steller's Jay
f Geai de Steller
d Diademhäher; Schwarzkopfhäher
i Ghiandaia di Steller

2918 *Cyanocompsa brissonii*
Passeriformes - Fringillidae
e Ultramarine Grosbeak
f Évêque de Brisson
d Ultramarinbischof
i Beccogrosso ultramarino

2919 *Cyanocompsa cyanoides*
Passeriformes - Fringillidae
e Blue-black Grosbeak
f Évêque bleu-noir
d Stahlbischof
i Beccogrosso blu-nero

2920 *Cyanocompsa parellina*
Passeriformes - Fringillidae
e Blue Bunting
f Évêque pare
d Lasurbischof
i Zigolo blu

2921 *Cyanocorax affinis*
Passeriformes - Corvidae
e Black-chested Jay; Talamanca Jay;
Colombian Jay
f Geai à poitrine noire
d Schwarzbrustblaurabe
i Ghiandaia pettonero

2922 *Cyanocorax beecheii*
Passeriformes - Corvidae
e Purplish-backed Jay; Beechey's Jay;
Beechey Jay
f Geai à dos violet
d Trauerblaurabe
i Ghiandaia dorsopurpureo

2923 *Cyanocorax caeruleus*
Passeriformes - Corvidae
e Azure Jay
f Geai azuré

d   Azurblaurabe
i   Ghiandaia cerulea

2924   *Cyanocorax cayanus*
Passeriformes - Corvidae
e   Cayenne Jay; Lavender Jay
f   Geai de Cayenne
d   Cayenne-Blaurabe
i   Ghiandaia della Cayenna

2925   *Cyanocorax chrysops*
Passeriformes - Corvidae
e   Plush-crested Jay; Plush-capped Jay;
Urucca Jay; Pileated Jay; Band-tailed
Jay
f   Geai acahé; Pie akahé
d   Kappenblaurabe; Kappengrünhäher
i   Ghiandaia occhidorati

2926   *Cyanocorax cristatellus*
Passeriformes - Corvidae
e   Curl crested Jay; White tailed Jay
f   Geai à plumet
d   Krauskopfblaurabe;
Weißschwanzblaurabe
i   Ghiandaia crestariccia

2927   *Cyanocorax cyanomelas*
Passeriformes - Corvidae
e   Purplish Jay
f   Geai bleu-noir
d   Purpurblaurabe; Veilchenblaurabe
i   Ghiandaia bruno-purpurea;
Ghiandaia bruno e purpurea

2928   *Cyanocorax cyanopogon*
Passeriformes - Corvidae
e   White-naped Jay; Blue-bearded Jay;
Pileated Jay; White-collared Jay
f   Geai à nuque blanche
d   Weißnackenblaurabe
i   Ghiandaia nucabianca

2929   *Cyanocorax dickeyi*
Passeriformes - Corvidae
e   Tufted Jay; Painted Jay; Dickey's Jay
f   Geai panaché
d   Schopfblaurabe
i   Ghiandaia di Dickey

2930   *Cyanocorax heilprini*
Passeriformes - Corvidae
e   Azure-naped Jay; Black-headed Jay;
Heilprin's Jay
f   Geai à calotte azur
d   Fliederblaurabe
i   Ghiandaia di Heilprin

2931   *Cyanocorax melanocyaneus*
Passeriformes - Corvidae
e   Bushy-crested Jay; Hartlaub's Jay
f   Geai houppé
d   Hartlaub-Blaurabe
i   Ghiandaia azzura e nera

2932   *Cyanocorax mexicanus*
Passeriformes - Corvidae
e   White-tipped Brown-Jay

2933   *Cyanocorax mystacalis*
Passeriformes - Corvidae
e   White tailed Jay; Moustached Jay
f   Geai à moustaches
d   Nacktwangenblaurabe
i   Ghiandaia codabianca

2934   *Cyanocorax sanblasianus*
Passeriformes - Corvidae
e   San Blas Jay; Black-and-blue Jay;
Blue-and-black Jay
f   Geai de San Blas
d   Acapulco-Blaurabe
i   Ghiandaia di San Blas

2935   *Cyanocorax violaceus*
Passeriformes - Corvidae
e   Violaceous Jay
f   Geai violacé
d   Hyanzinthenblaurabe
i   Ghiandaia violacea

2936   *Cyanocorax yncas*
Passeriformes - Corvidae
e   Green Jay; Inca Jay; Blue-headed Jay
f   Geai vert
d   Grünhäher; Peru-Grünhäher
i   Ghiandaia verde

2937   *Cyanocorax yncas cyanodorsalis*
Passeriformes - Corvidae
e   Blue-backed Jay

2938 **Cyanocorax yncas galeatus**
Passeriformes - Corvidae
e Green Jay; Galeated Jay

2939 **Cyanocorax yncas luxuosus**
Passeriformes - Corvidae
e Green Jay; Rio Grande Jay

2940 **Cyanocorax yucatanicus**
Passeriformes - Corvidae
e Yucatan Jay; Blue-and-black Jay
f Geai du Yucatan
i Ghiandaia dello Yucatan

2941 **Cyanolanius madagascarinus**
Passeriformes - Corvidae
e Blue Vanga
f Artamie azurée
d Blauvanga

2942 **Cyanolimnas cerverai**
Gruiformes - Rallidae
e Zapata Rail
f Râle de Zapata
d Kuba-Ralle
i Vanga azzurra

2943 **Cyanoliseus patagonus**
Psittaciformes - Psittacidae
e Burrowing Parakeet; Patagonian
Conure; Lesser Patagonian Conure;
Burrowing Parrot
f Conure de Patagonie
d Felsensittich; Patagonien-Sittich
i Conuro della Patagonia

2944 **Cyanoloxia glaucocaerulea**
Passeriformes - Fringillidae
e Indigo Grosbeak; Glaucous-blue
Grosbeak; Glaucous Grosbeak
f Évêque indigo
d Türkis-Bischof
i Beccogrosso azzurro-indaco

2945 **Cyanolyca argentigula**
Passeriformes - Corvidae
e Silvery-throated Jay; Central
American Jay
f Geai à gorge argentée
d Silberhäher
i Ghiandaia gola argentata

2946 **Cyanolyca armillata**
Passeriformes - Corvidae
e Black-collared Jay; Collared Jay;
Armillated Jay; Angela's Blue-Jay;
Quindio Blue-Jay; Merida Blue-Jay
f Geai à collier
d Halsbandhäher
i Ghiandaia dal collare nero

2947 **Cyanolyca cucullata**
Passeriformes - Corvidae
e Azure-hooded Jay; Hooded Jay
f Geai couronné
d Blaukappenhäher; Blaunackenhäher
i Ghiandaia dal cappuccio azzurro

2948 **Cyanolyca mirabilis**
Passeriformes - Corvidae
e White-throated Jay; Omilteme Jay;
Omilteme
f Geai masqué
d Weißkehlhäher
i Ghiandaia golabianca

2949 **Cyanolyca nana**
Passeriformes - Corvidae
e Dwarf Jay
f Geai nain
d Zwerghäher
i Ghiandaia nana

2950 **Cyanolyca pulchra**
Passeriformes - Corvidae
e Beautiful Jay
f Geai superbe
d Schmuckhäher
i Ghiandaia capobianco

2951 **Cyanolyca pumilo**
Passeriformes - Corvidae
e Black-throated Jay; Strickland's Jay;
Black-throated Dwarf Jay
f Geai à gorge noire
d Schwarzkehlhäher
i Ghiandaia golanera

2952 **Cyanolyca turcosa**
Passeriformes - Corvidae
e Turquoise Jay
f Geai turquoise

d Türkis-Häher
i Ghiandaia turchese

2953 Cyanolyca viridicyana
Passeriformes - Corvidae
e White-collared Jay; Blue-green Jay;
Blue-throated Jay; Joly's Jay;
Collared Jay
f Geai indigo
d Blaukehlhäher
i Ghiandaia dal collare bianco

2954 Cyanophaia bicolor
Apodiformes - Trochilidae
e Blue-headed Hummingbird
f Fou-fou bleu m (Ants); Fou-fou
feuille-blanc f (Ants); Colibri tête-
bleu (Ants)
d Zweifarbenkolibri
i Colibrì testablu

2955 Cyanopica cyana
Passeriformes - Corvidae
e Azure-winged Magpie
f Pie bleue; Pie à calotte noire
d Blauelster
i Gazza azzura; Gazza aliazzurre;
Gazza blu

2956 Cyanopsitta spixii
Psittaciformes - Psittacidae
e Little Blue Macaw; Spix's Macaw
f Ara de Spix
d Spix-Ara; Spix-Blauara
i Ara di Spix

2957 Cyanoptila cyanomelana
Passeriformes - Muscicapidae
e Blue-and-white Flycatcher; Japanese
Blue-Flycatcher
f Gobemouche bleu; Gobe-mouches
bleu du Japon
d Japan-Schnäpper; Blauschnäpper
i Pigliamosche blu e bianco

2958 Cyanoramphus auriceps
Psittaciformes - Psittacidae
e Yellow-fronted Parakeet; New
Zealand Parakeet; Kakariki
f Perruche à tête d'or

d Springsittich
i Kakariki frontegialla

2959 Cyanoramphus cookii
Psittaciformes - Psittacidae
e Norfolk Island Parakeet; Norfolk
Parakeet; Norfolk Island Parrot;
Norfolk Parrot; Norfolk Island
Green-Parrot
f Perruche de Norfolk
i Kakariki di Cook

2960 Cyanoramphus malherbi
Psittaciformes - Psittacidae
e Orange-fronted Parakeet; Alpine
Parakeet
f Perruche alpine
d Alpensittich

2961 Cyanoramphus novaezelandiae
Psittaciformes - Psittacidae
e Red-fronted Parakeet; Red-crowned
Parakeet; New Zealand Parakeet;
Red-fronted Karakiri; Red-fronted
Parrot; Green Parakeet; Green Parrot
f Perruche de Sparrman; Perruche de la
Nouvelle-Zeelande
d Laufsittich; Ziegensittich
i Kakariki fronterossa

2962 Cyanoramphus novaezelandiae
erythrotis
Psittaciformes - Psittacidae
e Macquarie Island Red-crowned
Parakeet (ANZ)

2963 Cyanoramphus novaezelandiae
subflavescens
Psittaciformes - Psittacidae
e Lord Howe Island Red-crowned
Parakeet (ANZ)

2964 Cyanoramphus ulietanus
Psittaciformes - Psittacidae
e Raiatea Parakeet; Society Parakeet
f Perruche de Raiatea
d Braunkopfsittich
i Kakariki delle Societe

2965 Cyanoramphus unicolor
Psittaciformes - Psittacidae

e Antipodes Parakeet; Antipodes
Green-Parakeet; Island Green
Parakeet; Antipodes Island Parakeet
f Perruche des Antipodes
d Einfarblaufsittich; Einfarbsittich
i Kakariki verde

2966 *Cyanoramphus zealandicus*
Psittaciformes - Psittacidae
e Black-fronted Parakeet; Tahiti
Parakeet
f Perruche de Tahiti
d Tahiti-Sittich
i Kakariki frontenera

2967 *Cyclarhis gujanensis*
Passeriformes - Vireonidae
e Rufous-browed Peppershrike
f Sourciroux mélodieux
d Rostbrauenvireo;
Graukopfpapageiwürger;
Großschnabelvireo
i Vireo averla della Guiana

2968 *Cyclarhis nigrirostris*
Passeriformes - Vireonidae
e Black-billed Peppershrike
f Sourciroux à bec noir
d Schwarzschnabelvireo
i Vireo averla becconero

2969 *Cyclarhis ochrocephala*
Passeriformes - Vireonidae
e Ochre-crowned Peppershrike

2970 *Cyclarhis virenticeps*
Passeriformes - Vireonidae
e Yellow-backed Peppershrike

2971 *Cyclarhis viridis*
Passeriformes - Vireonidae
e Chaco Peppershrike

2972 *Cyclopsitta diophthalma*
Psittaciformes - Psittacidae
e Double-eyed Fig-Parrot; Fig-Parrot;
Northern Fig-Parrot; Blue-faced Fig-
Parrot; Ref-faced Fig-Parrot;
Marshall's Fig-Parrot
f Psittacule double-oeil
d Maskenzwergpapagei;

Rotwangenzwergpapagei
i Pappagallo dei ficchi dagli occhiali

2973 *Cyclopsitta diophthalma coxeni*
Psittaciformes - Psittacidae
e Coxen's Fif-Parrot (ANZ)

2974 *Cyclopsitta gulielmitertii*
Psittaciformes - Psittacidae
e Orange-breasted Fig-Parrot;
William's Fig-Parrot; Black-cheeked
Fig-Parrot
f Psittacule à poitrine orangé
d Orangebrustzwergpapagei;
Orangebauchzwergpapagei
i Pappagallo dei ficchi pettoarancia

2975 *Cyclorrhynchus psittacula*
Charadriiformes - Alcidae
e Parakeet Auklet; Parakeet Auk
f Starique perroquet; Alque perroquet
(Qué)
d Rotschnabelalk; Papageischnabelalk
i Alca minore pappagallo

2976 *Cygnus atratus*
Anseriformes - Anatidae
e Black Swan
f Cygne noir
d Trauerschwan; Schwarzschwan
i Cigno nero

2977 *Cygnus bewick*
Anseriformes - Anatidae
e Bewick's Swan; Whistling Swan
f Cygne de Bewick
d Zwergschwan
i Cigno minore

2978 *Cygnus buccinator*
Anseriformes - Anatidae
e Trumpeter Swan
f Cygne trompette
d Trompeterschwan
i Cigno trombettiere

2979 *Cygnus columbianus*
Anseriformes - Anatidae
e Tundra Swan (NA)
f Cygne siffleur

d   Tundraschwan; Zwergschwan
i   Cigno minore

2980   Cygnus cygnus
        Anseriformes - Anatidae
e   Whooper Swan
f   Cygne chanteur; Cygne sauvage
d   Singschwan; Pfeifschwan
i   Cigno selvatico; Cigno canoro

2981   Cygnus melanocorypha
        Anseriformes - Anatidae
e   Black-necked Swan
f   Cygne à cou noir
d   Schwarzhalsschwan
i   Cigno collonero; Cigno dal collo
        nero

2982   Cygnus olor
        Anseriformes - Anatidae
e   Mute Swan; Swan
f   Cygne tubercule
d   Höckerschwan
i   Cigno reale

2983   Cymbilaimus lineatus
        Passeriformes - Thamnophiliidae
e   Fasciated Antshrike; Stock Antshrike
f   Batara fascié
d   Zebraameisenwürger
i   Averla formichiera fasciata

2984   Cymbilaimus sanctaemariae
        Passeriformes - Thamnophiliidae
e   Bamboo Antshrike
i   Averla formichiera di Santa Maria

2985   Cymbirhynchus macrorhynchos
        Passeriformes - Eurylaimidae
e   Black-and-red Broadbill
f   Eurylaime rouge-et-noir
d   Kellenschnabel
i   Beccolargo rosso e nero

2986   Cynanthus doubledayi
        Apodiformes - Trochilidae
e   Doubleday's Hummingbird; Broad-
        billed Hummingbird
d   Kolibri

2987   Cynanthus latirostris
        Apodiformes - Trochilidae
e   Broad-billed Hummingbird
f   Colibri circé
d   Breitschnabelkolibri
i   Colibrì beccolargo

2988   Cynanthus sordidus
        Apodiformes - Trochilidae
e   Dusky Hummingbird
f   Colibri sombre
d   Braunkopfkolibri
i   Colibrì bruno

2989   Cyornis banyumas
        Passeriformes - Muscicapidae
e   Hill Blue-Flycatcher; Hill Blue-
        Niltava
f   Gobemouche des collines; Gobe-
        mouches des collines; Niltava
        banyumas
d   Bergblauschnäpper
i   Niltava delle colline

2990   Cyornis caerulatus
        Passeriformes - Muscicapidae
e   Large-billed Blue-Flycatcher; Long-
        billed Blue-Flycatcher; Large-billed
        Niltava; Sunda Blue-Flycatcher
f   Gobemouche à grand bec; Gobe-
        mouches à grand bec
d   Breitschnabelblauschnäpper
i   Niltava beccogrosso

2991   Cyornis concretus
        Passeriformes - Muscicapidae
e   White-tailed Flycatcher; White-tailed
        Niltava; White-tailed Blue-
        Flycatcher; Dark Blue-Flycatcher;
        Short-tailed Blue-Flycatcher
f   Gobemouche à queue blanche; Gobe-
        mouches à queue blanche
d   Weißschwanzschopfschnäpper;
        Weißschwanzniltava
i   Niltava codabianca

2992   Cyornis hainanus
        Passeriformes - Muscicapidae
e   Hainan Blue-Flycatcher; Hainan
        Niltava; Grant's Niltava; Hainan
        Flycatcher; Grant's Blue-Flycatcher

f　Gobemouche de Hainan; Gobe-
　　mouches de Hainan
d　Hainan-Blauschnäpper
i　Niltava del Hainan

2993　*Cyornis herioti*
　　Passeriformes - Muscicapidae
e　Blue-breasted Flycatcher; Blue-
　　breasted Niltava
f　Gobemouche à poitrine bleue; Gobe-
　　mouches à poitrine bleue
d　Heriot-Blauschnäpper
i　Niltava pettazzurra

2994　*Cyornis hoevelli*
　　Passeriformes - Muscicapidae
e　Blue-fronted Flycatcher; Hoevell's
　　Blue-Flycatcher; Celebes Niltava;
　　Blue-fronted Niltava
f　Gobemouche à front bleu; Gobe-
　　mouches à front bleu
d　Celebes-Blauschnäpper
i　Niltava fronteblu

2995　*Cyornis hyacinthinus*
　　Passeriformes - Muscicapidae
e　Timor Blue-Flycatcher; Hyacinthine
　　Niltava; Blue-backed Niltava;
　　Hyacinthine Flycatcher
f　Gobemouche hyacinthe; Gobe-
　　mouches hyacinthe
d　Hyazynthenblauschnäpper
i　Niltava dorsoblu

2996　*Cyornis lemprieri*
　　Passeriformes - Muscicapidae
e　Palawan Blue-Flycatcher; Palawan
　　Niltava; Palawan Flycatcher; Hill-
　　Blue-Flycatcher; Ramsay's Blue-
　　Flycatcher; Balabac Blue-Flycatcher
f　Gobemouche de Balabac; Gobe-
　　mouches de Balabac
d　Palawan-Blauschnäpper
i　Niltava di Palawan

2997　*Cyornis omissus*
　　Passeriformes - Muscicapidae
e　Sulawesi Blue-Flycatcher; Sulawesi
　　Flycatcher; Sulawesi Niltava; White-
　　throated Blue-Flycatcher
f　Gobemouche des Célèbes; Gobe-

　　mouches des Célèbes
i　Niltava di Sulawesi

2998　*Cyornis pallipes*
　　Passeriformes - Muscicapidae
e　White-bellied Blue-Flycatcher;
　　White-bellied Niltava; White-vented
　　Blue-Flycatcher
f　Gobemouche à ventre blanc; Gobe-
　　mouches à ventre blanc
d　Kerala-Blauschnäpper
i　Niltava pettobianco

2999　*Cyornis poliogenys*
　　Passeriformes - Muscicapidae
e　Pale-chinned Flycatcher; Brooks's
　　Flycatcher (ISC); Brooks's Niltava
f　Gobemouche de Brooks; Gobe-
　　mouches de Brooks
d　Grauwangenschnäpper
i　Niltava di Brook

3000　*Cyornis rubeculoides*
　　Passeriformes - Muscicapidae
e　Blue-throated Flycatcher; Blue-
　　throated Niltava; Blue-throated Blue-
　　Flycatcher
f　Gobemouche à menton bleu; Gobe-
　　mouches à menton bleu; Niltava à
　　gorge bleue
d　Blaukehlschnäpper
i　Niltava golablu

3001　*Cyornis ruckii*
　　Passeriformes - Muscicapidae
e　Rueck's Blue-Flycatcher; Rueck's
　　Niltava; Rueck's Flycatcher
f　Gobemouche de Rueck; Gobe-
　　mouches de Rueck
d　Sumatra-Blauschnäpper
i　Niltava di Rueck

3002　*Cyornis rufigaster*
　　Passeriformes - Muscicapidae
e　Mangrove Blue-Flycatcher;
　　Mangrove Niltava
f　Gobemouche des mangroves; Gobe-
　　mouches des mangroves
d　Mangroveblauschnäpper
i　Niltava delle mangrovie

3003 *Cyornis sanfordi*
Passeriformes - Muscicapidae
e Matinan Flycatcher; Sanford's
Flycatcher; Sanford's Niltava;
Matinan Blue-Flycatcher; Matinan
Niltava; Blue-fronted Niltava
f Gobemouche de Sanford; Gobe-
mouches de Sanford
d Sandford-Schnäpper
i Niltava di Sanford

3004 *Cyornis superbus*
Passeriformes - Muscicapidae
e Bornean Blue-Flycatcher; Bornean
Niltava
f Gobemouche de Bornéo; Gobe-
mouches de Bornéo
d Prachtblauschnäpper
i Niltava del Borneo

3005 *Cyornis tickelliae*
Passeriformes - Muscicapidae
e Tickell's Blue-Flycatcher; Tickell's
Niltava
f Gobemouche de Tickell; Gobe-
mouches de Tickell
d Braunbrustblauschnäpper
i Niltava di Tickell

3006 *Cyornis turcosus*
Passeriformes - Muscicapidae
e Malaysian Blue-Flycatcher;
Malaysian Niltava; Ceylon Orange-
breasted Blue-Flycatcher
f Gobemouche malais; Gobe-mouches
malais
d Malaien-Blauschnäpper
i Niltava della Malesia

3007 *Cyornis unicolor*
Passeriformes - Muscicapidae
e Pale Blue-Flycatcher; Pale
Flycatcher; Pale-blue Niltava; Pale
Niltava
f Gobemouche bleuâtre; Gobe-
mouches bleuâtre; Cyornis à ventre
roux; Niltava à ventre roux
d Blaubrustschnäpper
i Niltava azzura

3008 *Cyphorhinus aradus*
Passeriformes - Certhiidae
e Musician Wren; Southern Musician-
Wren; Song Wren
f Troglodyte arada
d Flageolettzaunkönig
i Scricciolo canoro

3009 *Cyphorhinus phaeocephalus*
Passeriformes - Certhiidae
e Song Wren; Northern Musical Wren;
Northern Musician-Wren
f Troglodyte chanteur
i Scricciolo cantore

3010 *Cyphorhinus thoracicus*
Passeriformes - Certhiidae
e Chestnut-breasted Wren
f Troglodyte ferrugineux
d Braunbrustzaunkönig
i Scricciolo pettocastano

3011 *Cypseloides cherriei*
Apodiformes - Apodidae
e Spot-fronted Swift; Cherrie's Swift
f Martinet à points blancs
d Diademsegler
i Rondone frontemacchiata

3012 *Cypseloides cryptus*
Apodiformes - Apodidae
e White-chinned Swift; Zimmer's Swift
f Martinet à menton blanc; Martinet de
Zimmer
d Weißkinnsegler
i Rondone mentobianco

3013 *Cypseloides fumigatus*
Apodiformes - Apodidae
e Sooty Swift
f Martinet fuligineux
d Rauchsegler
i Rondone fuligginoso

3014 *Cypseloides lemosi*
Apodiformes - Apodidae
e White-chested Swift; Giant Swift
f Martinet à plastron blanc
d Weißbrustsegler
i Rondone pettobianco

3015   ***Cypseloides niger***
Apodiformes - Apodidae
e   Black Swift; Swallow (WI); Black
Swallow (WI); Rainbird (WI)
f   Martinet sombre; Hirondelle de
montagne (Ants); Hirondelle morne
(Ants); Oiseau de la pluie (Ants);
Gros Martinet noir (Ants); Hirondelle
(Ants)
d   Schwarzsegler
i   Rondone nero

3016   ***Cypseloides rothschildi***
Apodiformes - Apodidae
e   Rothschild's Swift; Giant Swift;
Dark-brown Swift; Great Swift
f   Martinet de Rothschild
d   Tucuman-Segler
i   Rondone di Rothschild

3017   ***Cypseloides senex***
Apodiformes - Apodidae
e   Great Dusky Swift
f   Martinet à tête grise; Martinet de
Temminck
d   Rußsegler
i   Rondone maggiore bruno

3018   ***Cypseloides storeri***
Apodiformes - Apodidae
e   White-fronted Swift
f   Martinet de Storer

3019   ***Cypsiurus balasiensis***
Apodiformes - Apodidae
e   Asian Palm-Swift; Palm-Swift
f   Martinet batassia
i   Rondone delle palme asiatico

3020   ***Cypsiurus parvus***
Apodiformes - Apodidae
e   African Palm-Swift; Palm-Swift; Old
World Palm-Swift
f   Martinet des palmes; Martinet des
palmiers
d   Palmensegler; Palmsegler
i   Rondone delle palme africano

3021   ***Cypsnagra hirundinacea***
Passeriformes - Fringillidae
e   White-rumped Tanager

f   Tangara hirundinacé
d   Weißbürzeltangare
i   Tangara gropponebianco

3022   ***Cyrtonyx montezumae***
Galliformes - Odontophoridae
e   Montezuma Quail; Harlequin Quail;
Mearn's Quail; Montezuma's Quail;
Salle's Quail
f   Colin arlequin
d   Massena-Wachtel
i   Colino di Montezuma; Quaglia
arlecchino

3023   ***Cyrtonyx ocellatus***
Galliformes - Odontophoridae
e   Ocellated Quail; Ocellated
Montezuma Quail
f   Colin ocellé
d   Tränenwachtel
i   Colino ocellato

D

3024   *Dacelo gaudichaud*
Coraciiformes - Halcyonidae
e   Rufous-bellied Kookaburra; Rufous-
bellied Giant-Kingfisher;
Gaudichaud's Kingfisher
f   Martin-chasseur de Gaudichaud
d   Rotbauchliest
i   Alcione pettirosso

3025   *Dacelo leachii*
Coraciiformes - Halcyonidae
e   Blue-winged Kookaburra
f   Martin-chasseur à ailes bleues
d   Haubenliest; Lachender Hans;
Kookaburra
i   Alcione aliazzurre

3026   *Dacelo novaeguineae*
Coraciiformes - Halcyonidae
e   Laughing Kookaburra; Laughing
Jackass; Giant Kookaburra
f   Martin-chasseur géant
d   Jägerliest; Lachender Hans
i   Kookaburra; Alcione gigante;
Alcione sghignazzante

3027   *Dacelo tyro*
Coraciiformes - Halcyonidae
e   Spangled Kookaburra; Tyro
Kingfisher; Aru Giant-Kookaburro;
Aru Giant-Kingfisher
f   Martin-chasseur pailleté; Martin-
chasseur tyro
d   Aru-Liest
i   Alcione ornato

3028   *Dacnis albiventris*
Passeriformes - Fringillidae
e   White-bellied Dacnis
f   Dacnis à ventre blanc
d   Weißbauchpitpit
i   Dacne ventrebianco

3029   *Dacnis berlepschi*
Passeriformes - Fringillidae
e   Scarlet-breasted Dacnis; Berlepsch's
Dacnis
f   Dacnis à poitrine rouge
d   Rotbrustpitpit
i   Dacne di Berlepsch

3030   *Dacnis cayana*
Passeriformes - Fringillidae
e   Blue Dacnis
f   Dacnis bleu
d   Blaukopfpitpit
i   Dacne azzurra

3031   *Dacnis egregia*
Passeriformes - Fringillidae
e   Yellow-tufted Dacnis

3032   *Dacnis flaviventer*
Passeriformes - Fringillidae
e   Yellow-bellied Dacnis
f   Dacnis à ventre jaune
d   Gelbbauchpitpit

3033   *Dacnis lineata*
Passeriformes - Fringillidae
e   Black-faced Dacnis
f   Dacnis à coiffe bleue
d   Maskenpitpit
i   Dacne faccianera

3034   *Dacnis nigripes*
Passeriformes - Fringillidae
e   Black-legged Dacnis
f   Dacnis à pattes noires
d   Schwarzfußpitpit
i   Dacne zampenere

3035   *Dacnis venusta*
Passeriformes - Fringillidae
e   Scarlet-thighed Dacnis
f   Dacnis à cuisses rouges
d   Rotschenkelpitpit
i   Dacne dai calzoni scarlatti

3036   *Dacnis viguieri*
Passeriformes - Fringillidae
e   Viridian Dacnis
f   Dacnis vert

d Panama-Pitpit
i Dacne di Viguier

3037 *Dactylortyx thoracicus*
Galliformes - Odontophoridae
e Singing Quail; Long-toed Partridge;
Long-toed Quail
f Colin chanteur
d Singwachtel
i Colino canoro

3038 *Damophila julie*
Apodiformes - Trochilidae
e Violet-bellied Hummingbird; Julie's
Hummingbird
f Colibri julie
d Julia-Kolibri
i Colibrì ventreviola

3039 *Daphoenositta chrysoptera*
Passeriformes - Corvidae
e Varied Sittella; Australian Sitella;
Papuan Sitella
f Néositte variée; Néositte papoue
d Spiegelkleiber; Papua-Kleiber
i Sittella variabile

3040 *Daphoenositta miranda*
Passeriformes - Corvidae
e Black Sittella; Pink-faced Nuthatch;
Pink-faced Sitella; Red-fronted
Creeper
f Néositte noire
d Prachtkleiber; Rotstirnkleiber
i Sitella facciarosa

3041 *Daption capense*
Procellariiformes - Procellariidae
e Cape Petrel; Pintado (CSA); Pintado
Petrel (CSA); Cape Pigeon (CSA);
Pied-Pintado Petrel; Cape Fulmar;
Black-and-white Petrel; Spotted
Petrel
f Damier du Cap; Pétrel du Cap
d Kap-Sturmvogel
i Procellaria del Capo; Piccione del
Capo

3042 *Daption capense capense*
Procellariiformes - Procellariidae
e Southern Cape Petrel (ANZ)

3043 *Daptrius americanus*
Falconiformes - Falconidae
e Red-throated Caracara
f Caracara à gorge rouge; Caracara à
ventre blanc
d Rotkehlwespenfalke
i Caracara golarossa; Caracara dal gola
rossa

3044 *Daptrius ater*
Falconiformes - Falconidae
e Black Caracara; Yellow-throated
Caracara
f Caracara noir
d Gelbkehlwespenfalke
i Caracara golagialla; Caracara dal
gola gialla

3045 *Dasyornis brachypterus*
Passeriformes - Pardalotidae
e Eastern Bristlebird; Brown
Bristlebird; Common Bristlebird
f Dasyorne brun
d Braunkopflackvogel
i Uccello di macchia orientale

3046 *Dasyornis brachypterus monoides*
Passeriformes - Pardalotidae
e Northern Eastern-Bristlebird (ANZ)

3047 *Dasyornis broadbenti*
Passeriformes - Pardalotidae
e Rufous Bristlebird
f Dasyorne roux
d Rotkopflackvogel
i Uccello di macchia castano

3048 *Dasyornis broadbenti caryochrous*
Passeriformes - Pardalotidae
e Otway's Rufous Bristlebird (ANZ)

3049 *Dasyornis broadbenti litoralis*
Passeriformes - Pardalotidae
e Western Rufous Bristlebird (ANZ)

3050 *Dasyornis longirostris*
Passeriformes - Pardalotidae
e Western Bristlebird
f Dasyorne à long bec
d Langschnabellackvogel
i Uccello di macchia occidentale

3051   *Deconychura longicauda*
        Passeriformes - Dendrocolaptidae
e   Long-tailed Woodcreeper; Long-
        tailed Creeper
f   Grimpar à longue queue
d   Langschwanzbaumsteiger
i   Rampichino codalunga

3052   *Deconychura stictolaema*
        Passeriformes - Dendrocolaptidae
e   Spot-throated Woodcreeper; Spot-
        throated Creeper
f   Grimpar à gorge tachetée
d   Kehlfleckenbaumsteiger
i   Rampichino golamacchiata

3053   *Deconychura typica*
        Passeriformes - Dendrocolaptidae
e   Cherrie's Woodcreeper

3054   *Delichon dasypus*
        Passeriformes - Hirundinidae
e   Asian House-Martin; Asiatic House-
        Martin; Asian Martin; Asiatic
        Martin; Kashmir House-Martin
f   Hirondelle de Bonaparte; Hirondelle
        du Cachemire
d   Kaschmir-Schwalbe
i   Balestruccio asiatico

3055   *Delichon nipalensis*
        Passeriformes - Hirundinidae
e   Nepal House-Martin; Nepal Martin
f   Hirondelle du Nepal; Hirondelle de
        Hodgson
d   Nepal-Schwalbe
i   Balestruccio di Nepal

3056   *Delichon urbica*
        Passeriformes - Hirundinidae
e   Northern House-Martin; House
        Martin; Common House-Martin;
        European House-Martin; Western
        House-Martin
f   Hirondelle de fenêtre; Hirondelle
        domestique
d   Mehlschwalbe; Hausschwalbe
i   Balestruccio; Rondine cittadina;
        Rondicchio

3057   *Delothraupis castaneoventris*
        Passeriformes - Fringillidae
e   Chestnut-bellied Mountain-Tanager
f   Tangara à ventre marron
d   Braunbauchbergtangare
i   Tangara di monte ventrecastano

3058   *Deltarhynchus flammulatus*
        Passeriformes - Tyrannidae
e   Flammulated Flycatcher
f   Tyran flammé
d   Deltaschnabel
i   Tiranno flammulato

3059   *Dendragapus canadensis*
        Galliformes - Phasianidae
e   Spruce Grouse; Canada Grouse;
        Franklin's Grouse
f   Tétras du Canada
d   Tannenhuhn
i   Tetraone delle peccete canadese

3060   *Dendragapus falcipennis*
        Galliformes - Phasianidae
e   Siberian Grouse; Sickle-winged
        Spruce-Grouse; Siberian Spruce-
        Grouse; Asian Spruce-Grouse;
        Sickle-winged Grouse; Sharp-winged
        Grouse
f   Tétras de Sibérie
d   Sichelhuhn
i   Tetraone delle peccete sibiriano

3061   *Dendragapus fuliginosus*
        Galliformes - Phasianidae
e   Sooty Grouse
f   Tétras fuligineux

3062   *Dendragapus obscurus*
        Galliformes - Phasianidae
e   Blue Grouse; Dusky Grouse; Grey
        Grouse; Sooty Grouse
f   Tétras sombre
d   Felsengebirgshuhn
i   Tetraone blu

3063   *Dendrexetastes rufigula*
        Passeriformes - Furnariidae
e   Cinnamon-throated Woodcreeper;
        Cinnamon-throated Creeper; Streak-
        throated Woodhewer

f Grimpar à collier; Grimpar à gorge
rousse
d Zimtkehlbaumsteiger
i Rampichino golarossiccia

3064 *Dendrocincla anabatina*
Passeriformes - Dendrocolaptidae
e Tawny-winged Woodcreeper;
Tawny-winged Creeper
f Grimpar à ailes rousses
d Lohschwingenbaumsteiger;
Braunflügelspechtdrossel
i Rampichino alifulve

3065 *Dendrocincla atrirostris*
Passeriformes - Dendrocolaptidae
e D'Orbigny's Woodcreeper

3066 *Dendrocincla fuliginosa*
Passeriformes - Dendrocolaptidae
e Plain-brown Woodcreeper; Line-
throated Woodcreeper; Plain-brown
Creeper; Plain Woodcreeper
f Grimpar enfumé
d Grauwangenbaumsteiger;
Rauchspechtdrossel
i Rampichino fuligginoso

3067 *Dendrocincla homochroa*
Passeriformes - Dendrocolaptidae
e Ruddy Woodcreeper; Ruddy Creeper
f Grimpar roux
d Kappenbaumsteiger
i Rampichino rossiccio

3068 *Dendrocincla macrorhyncha*
Passeriformes - Dendrocolaptidae
e Large Tyranine-Woodcreeper; Large
Tyranine-Creeper; Large-billed
Woodcreeper; Large-billed Creeper
d Starkschnabelbaumsteiger

3069 *Dendrocincla merculoides*
Passeriformes - Dendrocolaptidae
e Plain-brown Woodcreeper

3070 *Dendrocincla merula*
Passeriformes - Dendrocolaptidae
e White-chinned Woodcreeper; White-
chinned Creeper
f Grimpar à menton blanc

d Weißkinnbaumsteiger
i Rampichino mentobianco

3071 *Dendrocincla turdina*
Passeriformes - Dendrocolaptidae
e Thrush-like Woodcreeper; Plain-
winged Woodcreeper
f Grimpar grive
i Rampichino tordo

3072 *Dendrocincla tyrannina*
Passeriformes - Dendrocolaptidae
e Tyrannine Woodcreeper; Tyrannine
Creeper
f Grimpar tyran
d Tyrannenbaumsteiger
i Rampichino tiranno

3073 *Dendrocitta bayleyi*
Passeriformes - Corvidae
e Andaman Treepie
f Temia des Andaman
d Andamanen-Baumelster
i Gazza delle Andamane

3074 *Dendrocitta cinerascens*
Passeriformes - Corvidae
e Bornean Treepie; Malaysian Treepie;
Sunda Treepie
f Temia de Bornéo

3075 *Dendrocitta formosae*
Passeriformes - Corvidae
e Grey-Treepie; Himalayan Treepie
(ISC); Hills Treepie
f Temia de Swinhoe; Pie vagabonde de
Taïwan; Pie vagabonde de Formose
d Graubrustbaumelster; Baumelster
i Gazza dell'Himalaya

3076 *Dendrocitta frontalis*
Passeriformes - Corvidae
e Collared Treepie; Black-faced
Treepie; Black-browed Treepie;
Black-browed Magpie; White-naped
Treepie
f Temia masquée; Pie vagabonde
masquée
d Himalaja-Baumelster;
Maskenbaumelster
i Gazza dal collare

3077   ***Dendrocitta leucogastra***
        Passeriformes - Corvidae
e   White-bellied Treepie; Southern
        Treepie
f   Temia à ventre blanc
d   Weißbauchbaumelster
i   Gazza meridionale

3078   ***Dendrocitta occipitalis***
        Passeifomes - Corvidae
e   Sumatran Treepie; Sunda Treepie;
        Malaysian Treepie
f   Temia coiffée
d   Malaien-Baumelster
i   Gazza della Malesia

3079   ***Dendrocitta vagabunda***
        Passeriformes - Corvidae
e   Rufous Treepie; Treepie (ISC);
        Indian Treepie; Wandering Treepie;
        Common Treepie; Plains Treepie
f   Temia vagabonde
d   Wanderelster
i   Gazza vagabonda; Gazza rossastra

3080   ***Dendrocolaptes certhia***
        Passeriformes - Dendrocolaptidae
e   Barred Woodcreeper; Amazonian
        Barred-Woodcreeper; Barred Creeper
f   Grimpar barré
d   Bindenbaumsteiger
i   Rampichino barrato

3081   ***Dendrocolaptes certhia concolor***
        Passeriformes - Dendrocolaptidae
e   Concolor Woodcreeper; Concolor
        Creeper; Concolored Woodcreeper
f   Grimpar concolore
d   Borbabaumsteiger

3082   ***Dendrocolaptes hoffmannsi***
        Passeriformes - Dendrocolaptidae
e   Hoffmanns's Woodcreeper;
        Hoffmanns's Creeper
f   Grimpar de Hoffmanns
d   Hoffmanns-Baumsteiger
i   Rampichino di Hoffmann

3083   ***Dendrocolaptes multistrigatus***
        Passeriformes - Dendrocolaptidae
e   Cordilleran Woodcreeper

3084   ***Dendrocolaptes pallescens***
        Passeriformes - Dendrocolaptidae
e   Pale-billed Woodcreeper

3085   ***Dendrocolaptes picumnus***
        Passeriformes - Dendrocolaptidae
e   Black-banded Woodcreeper; Black-
        banded Creeper
f   Grimpar varié
d   Blauschnabelbaumsteiger
i   Rampichino bandenere

3086   ***Dendrocolaptes platyrostris***
        Passeriformes - Dendrocolaptidae
e   Planalto Woodcreeper; Planalto
        Creeper
f   Grimpar des plateaux
d   Flachschnabelbaumsteiger
i   Rampichino del Planalto

3087   ***Dendrocolaptes sanctithomae***
        Passeriformes - Dendrocolaptidae
e   Northern Barred-Woodcreeper

3088   ***Dendrocolaptes transfasciatus***
        Passeriformes - Dendrocolaptidae
e   Cross-barred Woodcreeper

3089   **Dendrocolaptidae**
        Passeriformes
e   Woodcreepers
f   Dendrocolaptidés; Grimpereaux du
        Nouveau Monde
d   Eigentliche Baumsteiger;
        Baumkletterer
i   Dendrocolattidi

3090   ***Dendrocopos assimilis***
        Piciformes - Picidae
e   Sind Woodpecker; Sind Pied-
        Woodpecker
f   Pic du Sind
d   Tamariskenspecht
i   Picchio del Sind

3091   ***Dendrocopos atratus***
        Piciformes - Picidae
e   Stripe-breasted Woodpecker; Stripe-
        breasted Pied-Woodpecker
f   Pic à poitrine rayée

d   Streifenbrustspecht
i   Picchio pettostriato

3092   **Dendrocopos auriceps**
Piciformes - Picidae
e   Brown-fronted Woodpecker; Brown-
fronted Pied-Woodpecker
f   Pic à tête jaune
d   Braunstirnspecht
i   Picchio testadorata

3093   **Dendrocopos canicapillus**
Piciformes - Picidae
e   Grey-capped Woodpecker; Grey-
headed Pygmy-Woodpecker; Grey-
headed Woodpecker; Grey-crowned
Pygmy-Woodpecker; Grey-crowned
Woodpecker; Grey-crowned Pied-
Woodpecker; Formosan Woodpecker
f   Pic à coiffe grise
d   Grauscheitelspecht
i   Picchio testacanuta

3094   **Dendrocopos cathpharius**
Piciformes - Picidae
e   Crimson-breasted Woodpecker;
Lesser Pied-Woodpecker; Crimson-
breasted Pied-Woodpecker; Small
Crimson-breasted Pied-Woodpecker
f   Pic à plastron rouge
d   Rotbrustspecht
i   Picchio pettirosso asiatico

3095   **Dendrocopos darjellensis**
Piciformes - Picidae
e   Darjeeling Woodpecker; Darjeeling
Pied-Woodpecker; Brown-throated
Woodpecker
f   Pic de Darjiling
d   Sikkim-Specht
i   Picchio di Darjelling

3096   **Dendrocopos dorae**
Piciformes - Picidae
e   Arabian Woodpecker
f   Pic d'Arabie
d   Araberspecht
i   Picchio d'Arabia

3097   **Dendrocopos himalayensis**
Piciformes - Picidae

e   Himalayan Woodpecker; Himalayan
Pied-Woodpecker
f   Pic de l'Himalaya
d   Himalaja-Specht
i   Picchio dell'Himalaya

3098   **Dendrocopos hyperythrus**
Piciformes - Picidae
e   Rufous-bellied Woodpecker; Rufous-
bellied Pied-Woodpecker; Rufous-
bellied Sapsucker
f   Pic à ventre fauve
d   Braunkehlspecht
i   Picchio ventre rossiccio

3099   **Dendrocopos kizuki**
Piciformes - Picidae
e   Pygmy Woodpecker; Japanese
Pygmy-Woodpecker; Japanese
Spotted-Woodpecker; Japanese
Woodpecker
f   Pic kisuki
d   Kizuki-Specht
i   Picchio kizuki

3100   **Dendrocopos leucopterus**
Piciformes - Picidae
e   White-winged Woodpecker; White-
winged Spotted-Woodpecker; White-
winged Pied-Woodpecker
f   Pic à ailes blanches
d   Weißflügelspecht
i   Picchio alibianche

3101   **Dendrocopos leucotos**
Piciformes - Picidae
e   White-backed Woodpecker
f   Pic à dos blanc
d   Weißrückenspecht
i   Picchio dorsobianco

3102   **Dendrocopos macei**
Piciformes - Picidae
e   Fulvous-breasted Woodpecker;
Fulvous-breasted Pied-Woodpecker;
Streak-bellied Woodpecker; Streak-
breasted Pied-Woodpecker
f   Pic de Mace
d   Isabellbrustspecht
i   Picchio di Mace

3103 ***Dendrocopos maculatus***
Piciformes - Picidae
e Philippine Woodpecker; Philippine Pygmy-Woodpecker; Pygmy-woodpecker
f Pic des Philippines
d Scopoli-Specht
i Picchio pigmeo delle Filippine

3104 ***Dendrocopos mahrattensis***
Piciformes - Picidae
e Yellow-crowned Woodpecker; Yellow-fronted Pied-Woodpecker (ISC); Mahratta Woodpecker (ISC); Yellow-crowned Pied-Woodpecker
f Pic mahratte
d Mahratten-Specht
i Picchio di Mahratte

3105 ***Dendrocopos major***
Piciformes - Picidae
e Great Spotted-Woodpecker; Greater Spotted-Woodpecker; Greater Pied-Woodpecker
f Pic épeiche
d Buntspecht
i Picchio rosso maggiore

3106 ***Dendrocopos medius***
Piciformes - Picidae
e Middle Spotted-Woodpecker
f Pic mar
d Mittelspecht
i Picchio rosso mezzano; Picchio rosso mediano

3107 ***Dendrocopos minor***
Piciformes - Picidae
e Lesser Spotted-Woodpecker; Lesser Pied-Woodpecker
f Pic épeichette
d Kleinspecht; Kleiner Buntspecht
i Picchio rosso minore

3108 ***Dendrocopos moluccensis***
Piciformes - Picidae
e Sunda Woodpecker; Brown-capped Pied-Woodpecker; Malaysian Pygmy-Woodpecker; Sunda Pygmy-Woodpecker; Malaysian Woodpecker; Brown-capped

Woodpecker
f Pic nain
d Braunscheitelspecht
i Picchio testabruna

3109 ***Dendrocopos nanus***
Piciformes - Picidae
e Brown-capped Woodpecker; Pygmy Woodpecker (ISC); Indian Pygmy-Woodpecker; Brown-capped Pygmy-Woodpecker; Brown-crowned Pygmy-Woodpecker; Ceylon Pygmy-Woodpecker
f Pic à calotte brune
d Hindu-Specht
i Picchio capobruno

3110 ***Dendrocopos syriacus***
Piciformes - Picidae
e Syrian Woodpecker
f Pic syriaque
d Blutspecht
i Picchio siriaco; Picchio rosso di Siria

3111 ***Dendrocopos temminckii***
Piciformes - Picidae
e Sulawesi Woodpecker; Sulawesi Pygmy-Woodpecker; Celebes Pygmy-Woodpecker; Temminck's Pygmy-Woodpecker
f Pic de Temminck
d Temminck-Specht
i Picchio di Temminck

3112 ***Dendrocygna arborea***
Anseriformes - Anatidae
e West Indian Whistling-Duck; Whistler (WI); Night Duck (WI); Mangrove Duck (WI); West Indian Tree-Duck; Cuban Whistling-Duck; Black-billed Whistling-Duck; Black-billed Tree-Duck
f Dendrocygne des Antilles; Canard siffleur (Ants); Gingeon (Ants)
d Kuba-Pfeifgans; Kuba-Baumente
i Anatra fischiatrice becconero; Dendrocigna becconero

3113 ***Dendrocygna arcuata***
Anseriformes - Anatidae
e Wandering Whistling-Duck;

Whistling Tree-Duck; Whistling
Duck; Lesser Fulvous Whistling-
Duck; Diving Tree-Duck; Water
Whistling-Duck; Black-legged Tree-
Duck; Black-legged Duck
f Dendrocygne à lunules
d Wanderpfeifgans; Wanderbaumente
i Anatra fischiatrice vagabonda

3114 *Dendrocygna autumnalis*
Anseriformes - Anatidae
e Black-bellied Whistling-Duck;
Black-bellied Tree-Duck; Red-billed
Whistling-Duck; Red-billed Tree-
Duck; Red-billed Whistling Tree-
Duck
f Dendrocygne à ventre noir
d Rotschnabelpfeifgans; Herbstente
i Anatra fischiatrice beccorosso

3115 *Dendrocygna bicolor*
Anseriformes - Anatidae
e Fulvous Whistling-Duck; Fulvous
Tree-Duck; Fulvous Duck (CSA);
Fulvous Whistling-Teal (ISC); Large
Whistling Teal (ISC)
f Dendrocygne fauve; Canard siffleur
(Ants); Dendrocygne fauve (Ants);
Siffleur (Ants)
d Gelbe Baumente; Gelbbrustpfeifgans
i Dendrocigna fulva; Anatra
fischiatrice fulva; Dendrocigna
bicolore

3116 *Dendrocygna eytoni*
Anseriformes - Anatidae
e Plumed Whistling-Duck; Eyton's
Tree-Duck; Grass Whistling-Duck;
Eyton's Plumed Whistling-Duck;
Whistling Tree-Duck; Red-legged
Tree-Duck; Grass Duck
f Dendrocygne d'Eyton
d Gelbfußpfeifgans; Gelbfußbaumente;
Sichelpfeifgans
i Anatra fischiatrice piumata

3117 *Dendrocygna guttata*
Anseriformes - Anatidae
e Spotted Whistling-Duck; Spotted
Tree-Duck
f Dendrocygne tacheté

d Tüpfelpfeifgans; Tüpfelbaumente
i Anatra fischiatrice macchiata;
Dendrocigna maculata

3118 *Dendrocygna javanica*
Anseriformes - Anatidae
e Lesser Whistling-Duck; Indian
Whistling-Duck; Lesser Tree-Duck;
Lesser Whistling Teal; Whistling
Teal
f Dendrocygne siffleur
d Indien-Pfeifganz; Java-Pfeifgans;
Java-Baumente
i Anatra fischiatrice indiana

3119 *Dendrocygna viduata*
Anseriformes - Anatidae
e White-faced Whistling-Duck; White-
faced Tree-Duck; White-faced Duck
(CSA)
f Dendrocygne veuf
d Witwenente; Witwenpfeifgans
i Picchio; Anatra fischiatrice
facciabianca; Dendrocigna
facciabianca

3120 Dendrocygnidae
Anseriformes -
e Whistling-Ducks; Tree-Ducks
f Dendrocygnidés
d Pfeifgänse; Baumenten
i Dendrocignidi

3121 *Dendroica adelaidae*
Passeriformes - Parulidae
e Adelaide's Warbler; Christmasbird
(WI)
f Paruline d'Adelalde
d Antillen-Waldsänger
i Dendroica di Adelaide

3122 *Dendroica aestiva*
Passeriformes - Parulidae
e Yellow Warbler
d Schnäpperrohrsänger

3123 *Dendroica angelae*
Passeriformes - Parulidae
e Elfin-woods Warbler; Puerto Rican
Warbler
f Paruline d'Angela

d Angela-Waldsänger
i Dendroica di Angela

3124 *Dendroica auduboni*
Passeriformes - Parulidae
e Audubon's Warbler

3125 *Dendroica caerulescens*
Passeriformes - Parulidae
e Black-throated Blue-Warbler
f Paruline bleue; Paruline bleue à gorge noire; Fauvette bleu à gorge noire
d Blauer Waldsänger; Blaurückenwaldsänger
i Dendroica blu golanera

3126 *Dendroica castanea*
Passeriformes - Parulidae
c Bay-breasted Warbler
f Paruline à poitrine baie
d Braunbrustwaldsänger
i Dendroica pettocastano

3127 *Dendroica cerulea*
Passeriformes - Parulidae
e Cerulean Warbler
f Paruline azurée; Fauvette azurée
d Lasurwaldsänger; Pappelwaldsänger
i Dendroica cerulea

3128 *Dendroica chrysoparia*
Passeriformes - Parulidae
e Golden-cheeked Warbler
f Paruline à dos noir
d Goldwangenwaldsänger
i Dendroica guancedorate

3129 *Dendroica coronata*
Passeriformes - Parulidae
e Yellow-rumped Warbler; Myrtle Warbler (NA); Audubon's Warbler
f Paruline à croupion jaune; Fauvette à croupion jaune
d Kronwaldsänger; Myrtenwaldsänger; Myrtensänger
i Parula coronata; Dendroica coronata; Dendroica gropponegiallo

3130 *Dendroica delicata*
Passeriformes - Parulidae
e St. Lucia Warbler

3131 *Dendroica discolor*
Passeriformes - Parulidae
e Prairie Warbler
f Paruline des prés; Fauvette des prés
d Rostscheitelwaldsänger
i Dendroica della prateria

3132 *Dendroica dominica*
Passeriformes - Parulidae
e Yellow-throated Warbler
f Paruline à gorge jaune; Fauvette à gorge jaune
d Rotkopflaubsänger; Goldkchlwaldsanger
i Dendroica golagialla

3133 *Dendroica erithachorides*
Passeriformes - Parulidae
e Mangrove Warbler

3134 *Dendroica fusca*
Passeriformes - Parulidae
e Blackburnian Warbler
f Paruline à gorge orangée; Fauvette à gorge orangée
d Gelbbrauner Waldsänger; Fichtenwaldsänger
i Dendroica fosca; Dendroica di Blackburn

3135 *Dendroica graciae*
Passeriformes - Parulidae
e Grace's Warbler
f Paruline de Grace
d Arizona-Waldsänger
i Dendroica di Grace

3136 *Dendroica kirtlandii*
Passeriformes - Parulidae
e Kirtland's Warbler; Jackpine Warbler
f Paruline de Kirtland; Fauvette de Kirtland
d Michigan-Waldsänger; Kirtland-Sänger
i Dendroica di Kirtland

3137   **Dendroica magnolia**
Passeriformes - Parulidae
e   Magnolia Warbler
f   Paruline à tête cendrée; Fauvette à tête cendrée
d   Magnolienwaldsänger; Hemlockwaldsänger
i   Dendroica magnolia; Dendroica delle magnolie

3138   **Dendroica nigrescens**
Passeriformes - Parulidae
e   Black-throated Grey-Warbler
f   Paruline grise; Paruline grise à gorge noire; Fauvette grise à gorge noire
d   Trauerwaldsänger
i   Dendroica grigia

3139   **Dendroica occidentalis**
Passeriformes - Parulidae
e   Hermit Warbler
f   Paruline à tête jaune; Fauvette à tête jaune
d   Einsiedelwaldsänger
i   Dendroica eremita

3140   **Dendroica palmarum**
Passeriformes - Parulidae
e   Palm Warbler
f   Paruline à couronne rousse; Fauvette à couronne rousse
d   Palmenwaldsänger; Sumpfwaldsänger; Palmsänger
i   Dendroica delle palme

3141   **Dendroica pensylvanica**
Passeriformes - Parulidae
e   Chestnut-sided Warbler
f   Paruline à flancs marron; Fauvette à flancs marron
d   Gelbscheitelwaldsänger; Zitronsänger
i   Dendroica fianchicastani

3142   **Dendroica petechia**
Passeriformes - Parulidae
e   Yellow Warbler; Golden Warbler; Mangrove Warbler (WI); Yellowbird (WI); Canary (WI); Bananabird (WI); Goldfinch (WI); American Yellow Warbler

f   Paruline jaune; Petit oiseau mangliers (Ants); Sucrier mangle (Ants); Fauvette jaune; Sucrier mang (Ants)
d   Goldwaldsänger
i   Dendroica gialla

3143   **Dendroica pharetra**
Passeriformes - Parulidae
e   Arrowhead Warbler; Arrow-headed Warbler
f   Paruline de la Jamaïque
d   Strichelwaldsänger
i   Dendroica frecciata

3144   **Dendroica pinus**
Passeriformes - Parulidae
e   Pine Warbler; Chip-chip (WI)
f   Paruline des pins; Petite Chitte des pins (Ants); Fauvette des pins
d   Kiefernwaldsänger
i   Dendroica dei pini

3145   **Dendroica pityophila**
Passeriformes - Parulidae
e   Olive-capped Warbler; Chip-chip (WI)
f   Paruline à calotte verte
d   Kuba-Waldsänger
i   Dendroica capo oliva

3146   **Dendroica plumbea**
Passeriformes - Parulidae
e   Plumbeous Warbler
f   Paruline caféiette; Tic-tic (Ants)
d   Grauwaldsänger
i   Dendroica plumbea

3147   **Dendroica potomac**
Passeriformes - Parulidae
e   Sutton's Warbler

3148   **Dendroica striata**
Passeriformes - Parulidae
e   Blackpoll Warbler
f   Paruline rayée; Petite Chitte rayéee (Ants); Paruline striée; Fauvette rayée
d   Streifenwaldsänger; Kappenwaldsänger
i   Parula di Blackpoll; Dendroica striata

3149 *Dendroica subita*
Passeriformes - Parulidae
e Barbuda Warbler

3150 *Dendroica tigrina*
Passeriformes - Parulidae
e Cape May Warbler
f Paruline tigrée; Fauvette tigrée
d Tigerwaldsänger
i Dendroica di Capo May

3151 *Dendroica townsendi*
Passeriformes - Parulidae
e Townsend's Warbler
f Paruline de Townsend; Fauvette de
Townsend
d Townsend-Waldsänger
i Dendroica di Townsend

3152 *Dendroica virens*
Passeriformes - Parulidae
e Black-throated Green-Warbler
f Paruline à gorge noire; Paruline verte
à gorge noire; Fauvette verte à gorge
noire
d Grünwaldsänger; Grüner Waldsänger
i Dendroica verdastra; Dendroica
golanera

3153 *Dendroica vitellina*
Passeriformes - Parulidae
e Vitelline Warbler; Chip-chip (WI)
f Paruline des Caimans
d Dotterwaldsänger
i Dendroica aranciata

3154 *Dendronanthus indicus*
Passeriformes - Passeridae
e Forest Wagtail; Tree Wagtail
f Bergeronnette de forêt
d Baumstelze; Waldstelze
i Ballerina di foresta

3155 *Dendropicos abyssinicus*
Piciformes - Picidae
e Abyssinian Woodpecker; Golden-
backed Woodpecker; Abyssinian
Golden-backed Woodpecker; African
Golden-backed Woodpecker; Gold-
mantled Woodpecker; Golden-
mantled Woodpecker

f Pic d'Abyssinie
d Wacholderspecht
i Picchio abissino

3156 *Dendropicos elachus*
Piciformes - Picidae
e Little Grey-Woodpecker; Least
Woodpecker
f Pic gris; Petit Pic gris
d Wüstenspecht
i Picchio grigio minore

3157 *Dendropicos elliotii*
Piciformes - Picidae
e Elliot's Woodpecker
f Pic d'Elliot
d Elliot-Specht
i Picchio di Elliot

3158 *Dendropicos fuscescens*
Piciformes - Picidae
e Cardinal Woodpecker
f Pic cardinal
d Kardinalspecht
i Picchio cardinale

3159 *Dendropicos gabonensis*
Piciformes - Picidae
e Gabon Woodpecker
f Pic du Gabon
d Gabun-Specht
i Picchio del Gabon

3160 *Dendropicos goertae*
Piciformes - Picidae
e Grey Woodpecker; African Grey-
Woodpecker
f Pic goertan
d Graubrustspecht
i Picchio di Goerta

3161 *Dendropicos griseocephalus*
Piciformes - Picidae
e Olive Woodpecker; African Grey-
headed Woodpecker
f Pic olive
d Goldrückenspecht
i Picchio oliva

3162 *Dendropicos lugubris*
Piciformes - Picidae

e Melancholy Woodpecker
f Pic à raies noires
d Trauerspecht
i Picchio melancolico

3163 *Dendropicos namaquus*
Piciformes - Picidae
e Bearded Woodpecker
f Pic barbu
d Nama-Specht
i Picchio barbuto

3164 *Dendropicos obsoletus*
Piciformes - Picidae
e Brown-backed Woodpecker; Lesser
White-spotted Woodpecker
f Pic à dos brun
d Braunrückenspecht
i Picchio dorsobruno

3165 *Dendropicos poecilolaemus*
Piciformes - Picidae
e Speckle-breasted Woodpecker;
Uganda Spotted-Woodpecker (CSA)
f Pic à poitrine tachetée
d Tropfenspecht
i Picchio pettochiazzato

3166 *Dendropicos pyrrhogaster*
Piciformes - Picidae
e Fire-bellied Woodpecker
f Pic à ventre de feu; Pic à ventre
rouge
d Rotbauchspecht
i Picchio ventrescarlatto

3167 *Dendropicos spodocephalus*
Piciformes - Picidae
e Grey-headed Woodpecker
f Pic spodocéphale
i Picchio testagrigia

3168 *Dendropicos stierlingi*
Piciformes - Picidae
e Stierling's Woodpecker
f Pic de Stierling
d Stierling-Specht
i Picchio di Stierling

3169 *Dendropicos xantholophus*
Piciformes - Picidae

e Golden-crowned Woodpecker;
Yellow-crested Woodpecker
f Pic à couronne d'or
d Scheitelfleckspecht
i Picchio crestagialla

3170 *Dendrortyx barbatus*
Galliformes - Odontophoridae
e Bearded Wood-Partridge; Bearded
Partridge; Bearded Tree-Partridge;
Bearded Tree-Quail
f Colin barbu
d Bartwachtel
i Colino barbuto

3171 *Dendrortyx leucophrys*
Galliformes - Odontophoridae
e Buffy-crowned Wood-Partridge;
Highland Partridge; Guatemalan
Long-tailed Partridge; Buffy-
crowned Tree-Quail; Buffy-crowned
Tree-Partridge; Buffy-fronted Wood-
Partridge; Highland Wood-Partridge;
Buff-fronted Quail; Costa Rican
Quail
f Colin à sourcils blancs
d Guatemala-Wachtel
i Colino orechiebianche

3172 *Dendrortyx macroura*
Galliformes - Odontophoridae
e Long-tailed Wood-Partridge; Long-
tailed Partridge; Mexican Long-tailed
Partridge; Long-tailed Tree-Quail;
Long-tailed Tree-Partridge
f Colin à longue queue
d Langschwanzwachtel
i Colino codalunga

3173 *Deroptyus accipitrinus*
Psittaciformes - Psittacidae
e Red-fan Parrot; Hawk-headed Parrot
f Papegai maillé; Perroquet maillé
d Fächerpapagei
i Pappagallo accipitrino

3174 *Dicaeum aeneum*
Passeriformes - Nectariniidae
e Midget Flowerpecker; Solomon
Islands Flowerpecker; Solomons
Flowerpecker

f   Dicée des Salomon
d   Brozemistelfresser
i   Beccafiori delle Isole Salomone

3175   *Dicaeum aeruginosum*
     Passeriformes - Nectariniidae
e   Striped Flowerpecker; Fairy
     Flowerpecker
f   Dicée rayé
d   Streifenbrustmistelfresser
i   Beccafiori striato

3176   *Dicaeum agile*
     Passeriformes - Nectariniidae
e   Thick-billed Flowerpecker; Ceylon
     Thick-billed Flowerpecker; Streak-
     breasted Flowerpecker; Striped
     Flowerpecker
f   Dicée à bec epais
d   Dickschnabelmistelfresser
i   Beccafiori beccogrosso

3177   *Dicaeum annae*
     Passeriformes - Nectariniidae
e   Golden-rumped Flowerpecker;
     Anna's Flowerpecker; Flores
     Flowerpecker; Sunda Flowerpecker
f   Dicée de la Sonde
d   Bartmistelfresser
i   Beccafiori della Sonda

3178   *Dicaeum anthonyi*
     Passeriformes - Nectariniidae
e   Flame-crowned Flowerpecker;
     Yellow-crowned Flowerpecker;
     Anthony's Flowerpecker
f   Dicée couronné
d   Goldkronenmistelfresser
i   Beccafiori corona gialla

3179   *Dicaeum aureolimbatum*
     Passeriformes - Nectariniidae
e   Yellow-sided Flowerpecker; Golden-
     edged Flowerpecker; Celebesian
     Flowerpecker; Mynahassa
     Flowerpecker; Golden-flanked
     Flowerpecker; Sulawesi
     Flowerpecker
f   Dicée à flancs jaunes
d   Mynahassa-Mistelfresser
i   Beccafiori fianchigialli

3180   *Dicaeum australe*
     Passeriformes - Nectariniidae
e   Red-striped Flowerpecker; Philippine
     Flowerpecker; Austral Flowerpecker
f   Dicée des Philippines
d   Rotbauchmistelfresser
i   Beccafiori delle Filippine

3181   *Dicaeum bicolor*
     Passeriformes - Nectariniidae
e   Bicoloured Flowerpecker; Two-
     coloured Flowerpecker
f   Dicée bicolore
d   Zweifarbenmistelfresser
i   Beccafiori bicolore

3182   *Dicaeum celebicum*
     Passeriformes - Nectariniidae
e   Grey-sided Flowerpecker; Black-
     sided Flowerpecker
f   Dicée des Célèbes
d   Schwarzwangenmistelfresser;
     Celebes-Mistlefresser
i   Beccafiori fianchineri

3183   *Dicaeum chrysorrheum*
     Passeriformes - Nectariniidae
e   Yellow-vented Flowerpecker
f   Dicée cul-d'or; Dicée à sous-caudales
     jaunes
d   Gelbsteißmistelfresser
i   Beccafiori dal sottocoda giallo

3184   *Dicaeum concolor*
     Passeriformes - Nectariniidae
e   Plain Flowerpecker; Plain-coloured
     Flowerpecker (ISC); Olivaceous
     Flowerpecker; Nilgiri Flowerpecker
f   Dicée Concolore
d   Einfarbmistelfresser
i   Beccafiori unicolore

3185   *Dicaeum cruentatum*
     Passeriformes - Nectariniidae
e   Scarlet-backed Flowerpecker
f   Dicée à dos rouge
d   Scharlachmistelfresser;
     Schrlachblütenpicker
i   Beccafiori dorsorosso

3186 *Dicaeum erythrorhynchos*
Passeriformes - Nectariniidae
e Pale-billed Flowerpecker; Tickell's
Flowerpecker (ISC); Ceylon Small
Flowerpecker; Small Flowerpecker
f Dicée à bec rouge
d Lackschnabelmistelfresser;
Gelbschnabelmistelfresser
i Beccafiori di Tickell

3187 *Dicaeum erythrothorax*
Passeriformes - Nectariniidae
e Flame-breasted Flowerpecker;
Reddish Flowerpecker; Buru
Flowerpecker; White-throated
Flowerpecker; Flame-chested
Flowerpecker; Indonesian
Flowerpecker
f Dicée à gorge blanche
d Buru-Mistelfresser
i Beccafiori golabianca

3188 *Dicaeum everetti*
Passeriformes - Nectariniidae
e Brown-backed Flowerpecker;
Everett's Flowerpecker
f Dicée d'Everett
d Braunrückenmistelfresser
i Beccafiori di Everett

3189 *Dicaeum eximium*
Passeriformes - Nectariniidae
e Red-banded Flowerpecker; Bismarck
Flowerpecker; Beautiful
Flowerpecker; New Ireland
Flowerpecker
f Dicée des Bismarck
d Bismarck-Mistelfresser
i Beccafiori della Nuova Irlanda

3190 *Dicaeum geelvinkianum*
Passeriformes - Nectariniidae
e Red-capped Flowerpecker; Geelvink
Flowerpecker
f Dicée de Geelvink
i Beccafiori papua capirosso

3191 *Dicaeum haematostictum*
Passeriformes - Nectariniidae
e Visayan Flowerpecker

3192 *Dicaeum hirundinaceum*
Passeriformes - Nectariniidae
e Mistletoebird; Mistletoe
Flowerpecker; Australian
Flowerpecker; Fire-breasted
Flowerpecker
f Dicée hirondelle
d Rotsteißmistelfresser;
Schwalbenmistelesser
i Uccello del vischio

3193 *Dicaeum hypoleucum*
Passeriformes - Nectariniidae
e Buzzing Flowerpecker; White-bellied
Flowerpecker
f Dicée à ventre blanc
d Weißbauchmistelfresser
i Beccafiori ventrebiaco

3194 *Dicaeum igniferum*
Passeriformes - Nectariniidae
e Black-fronted Flowerpecker; Rusty
Flowerpecker; Black-banded
Flowerpecker; Lesser Sunda-
Flowerpecker; Fire-tailed
Flowerpecker
f Dicée porte-flamme
d Rotkehlmistelfresser
i Beccafiori ignifero

3195 *Dicaeum ignipectus*
Passeriformes - Nectariniidae
e Fire-breasted Flowerpecker; Green-
backed Flowerpecker; Buff-bellied
Flowerpecker; Fire-throated
Flowerpecker; Scarlet-breasted
Flowerpecker; Bronze-backed
Flowerpecker
f Dicée à gorge feu
d Feuerbrustmistelfresser;
Feuerbrustmistelesser
i Beccafiori dorsoverde

3196 *Dicaeum maugei*
Passeriformes - Nectariniidae
e Red-chested Flowerpecker; Mauge's
Flowerpecker; Timor Flowerpecker;
Blue-cheeked Flowerpecker; Lesser
Sunda Flowerpecker
f Dicée de Mauge

d   Mauge-Mistelfresser
i   Beccafiori guanceblu

3197   *Dicaeum melanoxanthum*
Passeriformes - Nectariniidae
e   Yellow-bellied Flowerpecker
f   Dicée à ventre jaune
d   Gelbbauchmistelfresser;
Schwarzgelbmistelesser
i   Beccafiori ventregiallo

3198   *Dicaeum monticolum*
Passeriformes - Nectariniidae
e   Black-sided Flowerpecker; Bornean
Black-sided Flowerpecker; Bornean
Fire-breasted Flowerpecker; Bornean
Flowerpecker; Mountain
Flowerpecker
f   Dicée de Borneo
d   Borneo-Mistelfresser
i   Beccafiori pettirosso del Borneo

3199   *Dicaeum nehrkorni*
Passeriformes - Nectariniidae
e   Crimson-crowned Flowerpecker;
Red-headed Flowerpecker;
Nehrkorn's Flowerpecker; Celebes
Flowerpecker
f   Dicée à tête rouge
d   Nehrkorn-Mistelfresser
i   Beccafiori testarossa

3200   *Dicaeum nigrilore*
Passeriformes - Nectariniidae
e   Olive-capped Flowerpecker
f   Dicée à calotte olive
d   Olivkopfmistelfresser
i   Beccafiori capo oliva

3201   *Dicaeum nitidum*
Passeriformes - Nectariniidae
e   Louisiade Flowerpecker; Louisiades
Flowerpecker
f   Dicée de la Louisiade
i   Beccafiori nitido

3202   *Dicaeum pectorale*
Passeriformes - Nectariniidae
e   Olive-crowned Flowerpecker;
Papuan Flowerpecker; Pectoral
Flowerpecker; Shining Flowerpecker

f   Dicée à plastron
d   Papua-Mistelfresser
i   Beccafiori pettirosso

3203   *Dicaeum proprium*
Passeriformes - Nectariniidae
e   Whiskered Flowerpecker; Grey-
breasted Flowerpecker
f   Dicée à poitrine grise
d   Graubrustmistelfresser

3204   *Dicaeum pygmaeum*
Passeriformes - Nectariniidae
e   Pygmy Flowerpecker; Palawan
Flowerpecker
f   Dicée pygmée
d   Zwergmistelfresser
i   Beccafiori pigmeo

3205   *Dicaeum quadricolor*
Passeriformes - Nectariniidae
e   Cebu Flowerpecker; Four-coloured
Flowerpecker; Austral Flowerpecker;
Philippine Flowerpecker
f   Dicée quadricolore
d   Vierfarbenmistelfresser
i   Beccafiori quadricolore

3206   *Dicaeum retrocinctum*
Passeriformes - Nectariniidae
e   Scarlet-collared Flowerpecker;
Mindoro Flowerpecker; Red-collared
Flowerpecker
f   Dicée de Mindoro
d   Mindoro-Mistelfresser
i   Beccafiori di Mindoro

3207   *Dicaeum sanguinolentum*
Passeriformes - Nectariniidae
e   Blood-breasted Flowerpecker; Javan
Fire-breasted Flowerpecker; Javan
Flowerpecker; Fire-breasted
Flowerpecker
f   Dicée sanglant
d   Purpurmistelfresser
i   Beccafiori pettopugnalato

3208   *Dicaeum trigonostigma*
Passeriformes - Nectariniidae
e   Orange-bellied Flowerpecker;
Orange-breasted Flowerpecker; Sulu

Flowerpecker
f Dicée à ventre orangé
d Orangebauchmistelfresser;
Orangebrustmistelesser
i Beccafiori ventrearancio

3209 *Dicaeum tristrami*
Passeriformes - Nectariniidae
e Mottled Flowerpecker; Tristram's
Flowerpecker; San Cristobal
Flowerpecker; San Cristobal Midget
f Dicée de San Cristobal
d Tristrams Mistelfresser
i Beccafiori di San Cristobal

3210 *Dicaeum trochileum*
Passeriformes - Nectariniidae
e Scarlet-headed Flowerpecker
f Dicée à tête écarlate
d Feuerkopfmistelfresser
i Beccafiori testarossa

3211 *Dicaeum vincens*
Passeriformes - Nectariniidae
e White-throated Flowerpecker;
Legge's Flowerpecker
f Dicée de Ceylan
d Weißkehlmistelfresser
i Beccafiori di Legge

3212 *Dicaeum vulneratum*
Passeriformes - Nectariniidae
e Ashy Flowerpecker; Moluccan
Flowerpecker; Ceram Flowerpecker;
Ashy-fronted Flowerpecker
f Dicée cendré
d Seran-Mistelfresser
i Beccafiori frontegrigia

3213 *Dichrozona cincta*
Passeriformes - Formicariidae
e Banded Antbird; Banded Antcatcher;
Banded Ant-Wren
f Grisin sanglé
d Buntbürzelameisenfänger
i Mangiaformiche fasciato

3214 Dicruridae
Passeriformes
e Drongos
f Dicruridés

d Drongos
i Dicruridi

3215 *Dicrurus adsimilis*
Passeriformes - Dicruridae
e Fork-tailed Drongo; Common
Drongo (CSA); Drongo (CSA); King
Crow (ISC); African Drongo; Black
Drongo; Glossy-backed Drongo
f Drongo brillant; Drongo à dos
brillant
d Trauerdrongo

3216 *Dicrurus aeneus*
Passeriformes - Dicruridae
e Bronzed Drongo; Bronze Drongo
(CSI); Little Bronze-Drongo
f Drongo bronzé
d Bronzedrongo

3217 *Dicrurus aldabranus*
Passeriformes - Dicruridae
e Aldabra Drongo
f Drongo des Aldabra; Drongo
d'Aldabra
d Aldabra-Drongo

3218 *Dicrurus andamanensis*
Passeriformes - Dicruridae
e Andaman Drongo
f Drongo des Andaman
d Andamanen-Drongo

3219 *Dicrurus annectans*
Passeriformes - Dicruridae
e Crow-billed Drongo
f Drongo à gros bec
d Krähenschnabeldrongo

3220 *Dicrurus atripennis*
Passeriformes - Dicruridae
e Shining Drongo
f Drongo de forêt
d Glanzdrongo

3221 *Dicrurus balicassius*
Passeriformes - Dicruridae
e Philippine Drongo; Balicassiao
Drongo; Balicassiao
f Drongo balicassio
d Philippinen-Drongo

3222   *Dicrurus bracteatus*
Passeriformes - Dicruridae
e   Spangled Drongo
f   Drongo pailleté
d   Drongo

3223   *Dicrurus bracteatus carbonarius*
Passeriformes - Dicruridae
e   Torres Strait Spangled Drongo
      (ANZ)

3224   *Dicrurus caerulescens*
Passeriformes - Dicruridae
e   White-bellied Drongo; Ceylon
      Common Drongo
f   Drongo à ventre blanc
d   Graubrustdrongo

3225   *Dicrurus densus*
Passeriformes - Dicruridae
e   Wallacean Drongo
f   Drongo de la Sonde

3226   *Dicrurus forficatus*
Passeriformes - Dicruridae
e   Crested Drongo; Madagascar
      Crested-Drongo
f   Drongo malgache
d   Gabeldrongo
i   Drongo crestato del Madagascar

3227   *Dicrurus fuscipennis*
Passeriformes - Dicruridae
e   Grand Comoro Drongo; Comoro
      Drongo
f   Drongo de la Grande Comore
d   Braunschwingendrongo

3228   *Dicrurus hottentottus*
Passeriformes - Dicruridae
e   Spangled Drongo; Hair-crested
      Drongo
f   Drongo à crinière
d   Glanzspitzendrongo; Fadendrongo;
      Haarbuschdrongo

3229   *Dicrurus leucophaeus*
Passeriformes - Dicruridae
e   Ashy Drongo; Grey Drongo; Pale
      Drongo; Indian Grey-Drongo

f   Drongo cendré
d   Graudrongo

3230   *Dicrurus ludwigii*
Passeriformes - Dicruridae
e   Square-tailed Drongo
f   Drongo de Ludwig
d   Geradschwanzdrongo

3231   *Dicrurus macrocercus*
Passeriformes - Dicruridae
e   Black Drongo; African Black-
      Drongo; Drongo (CSA); Common
      Drongo (CSA); Himalayan Black-
      Drongo; King Crow; Ceylon Black-
      Drongo
f   Drongo royal
d   Asiatischer Trauerdrongo;
      Königsdrongo
i   Drongo nero; Drongo

3232   *Dicrurus megarhynchus*
Passeriformes - Dicruridae
e   Ribbon-tailed Drongo; New Ireland
      Drongo
f   Drongo de Nouvelle-Irlande
d   Bandschwanzdrongo

3233   *Dicrurus modestus*
Passeriformes - Dicruridae
e   Velvet-mantled Drongo; Principe
      Drongo
f   Drongo modeste

3234   *Dicrurus montanus*
Passeriformes - Dicruridae
e   Sulawesi Drongo; Celebes Mountain-
      Drongo; Celebes Drongo; Mountain
      Drongo
f   Drongo des Célèbes
d   Bergdrongo

3235   *Dicrurus paradiseus*
Passeriformes - Dicruridae
e   Greater Racket-tailed Drongo;
      Racket-tailed Drongo (ISC); Greater
      Racquet-tailed Drongo; Large
      Racquet-tailed Drongo; Ceylon
      Racquet-tailed Drongo; Ceylon
      Crested-Drongo; Ceylon Racket-
      tailed Drongo

f Drongo à raquettes
d Flaggendrongo
i Drongo del Paradiso

3236 *Dicrurus remifer*
Passeriformes - Dicruridae
e Lesser Racket-tailed Drongo; Lesser
Racquet-tailed Drongo; Small
Racquet-tailed Drongo; Small
Racket-tailed Drongo
f Drongo à rames; Petit Drongo à
raquettes
d Ruderdrongo; Spateldrongo

3237 *Dicrurus sumatranus*
Passeriformes - Dicruridae
e Sumatran Drongo
f Drongo de Sumatra

3238 *Dicrurus waldenii*
Passeriformes - Dicruridae
e Mayotte Drongo
f Drongo de Mayotte
d Mayotte-Drongo

3239 *Didunculus strigirostris*
Columbiformes - Columbidae
e Tooth-billed Pigeon
f Diduncule strigirostre
d Zahntaube
i Diduncolo

3240 *Diglossa albilatera*
Passeriformes - Emberizidae
e White-sided Flowerpiercer; White-
sided Honeycreeper; White-sided
Flowerpecker
f Percefleur à flancs blancs
d Schieferhakenschnabel
i Diglossa fianchibianchi

3241 *Diglossa baritula*
Passeriformes - Emberizidae
e Cinnamon-bellied Flowerpiercer;
Cinnamon Flowerpiercer; Slaty
Highland-Honeycreeper; Highland
Honeycreeper; Highland
Flowerpecker; Slaty Flowerpiercer;
Slaty Flowerpecker
f Percefleur cannellé

d Zimtbauchhakenschnabel
i Diglossa ardesia

3242 *Diglossa brunneiventris*
Passeriformes - Emberizidae
e Black-throated Flowerpiercer; Black-
throated Flowerpecker; Vuilleumier's
Flowerpiercer
f Percefleur à gorge noire
i Diglossa golanera

3243 *Diglossa carbonaria*
Passeriformes - Emberizidae
e Grey-bellied Flowerpiercer;
Carbonated Flowerpiercer; Coal-
black Flowerpiercer; Black
Honeycreeper; Grey-bellied
Flowerpecker
f Percefleur charbonnier
d Grauschulterhakenschnabel
i Diglossa carbonaria

3244 *Diglossa duidae*
Passeriformes - Emberizidae
e Scaled Flowerpiercer; Scaled
Honeycreeper; Scaled Flowerpecker
f Percefleur des tépuis
d Schuppenbrusthakenschnabel
i Diglossa squamosa

3245 *Diglossa gloriosa*
Passeriformes - Emberizidae
e Merida Flowerpiercer; Merida
Flowerpecker
f Percefleur de Merida
d Merida-Hakenschnabel
i Diglossa di Merida

3246 *Diglossa gloriosissima*
Passeriformes - Emberizidae
e Chestnut-bellied Flowerpiercer;
Chestnut-bellied Flowerpecker
f Percefleur à ventre marron
d Maronenbauchhakenschnabel
i Diglossa pettocastano

3247 *Diglossa humeralis*
Passeriformes - Emberizidae
e Black Flowerpiercer; Black
Flowerpecker; All-black
Flowerpecker

f   Percefleur noir
d Schwarzbauchhakenschnabel
i   Diglossa nera

3248 *Diglossa lafresnayii*
      Passeriformes - Emberizidae
e Glossy Flowerpiercer; Glossy
      Honeycreeper; Glossy Flowerpecker
f   Percefleur de Lafresnaye
d Stahlhakenschnabel
i   Diglossa di Lafresnaye

3249 *Diglossa major*
      Passeriformes - Emberizidae
e Greater Flowerpiercer; Greater
      Honeycreeper; Greater Flowerpecker
f   Grand Percefleur
d Strichelhakenschnabel
i   Diglossa maggiore

3250 *Diglossa mystacalis*
      Passeriformes - Emberizidae
e Moustached Flowerpiercer;
      Moustached Flowerpecker
f   Percefleur moustachu
d Barthakenschnabel
i   Diglossa dai mustacchi

3251 *Diglossa plumbea*
      Passeriformes - Emberizidae
e Slaty Flowerpiercer; Slaty
      Flowerpecker
f   Percefleur ardoisé
d Einfarbhakenschnabel
i   Diglossa plumbea

3252 *Diglossa sittoides*
      Passeriformes - Emberizidae
e Rusty Flowerpiercer; Rusty
      Flowerpecker; Slaty Flowerpiercer;
      Slaty Flowerpecker
f   Percefleur rouilleux
d Rostbauchhakenschabel
i   Diglossa rossiccia

3253 *Diglossa venezuelensis*
      Passeriformes - Emberizidae
e Venezuelan Flowerpiercer;
      Venezuelan Honeycreeper;
      Venezuelan Flowerpecker
f   Percefleur du Venezuela

d Trauerhakenschnabel
i   Diglossa di Venezuela; Diglossa del
      Venezuela

3254 *Diglossopis caerulescens*
      Passeriformes - Fringillidae
e Bluish Flowerpiercer; Bluish
      Honeycreeper; Bluish Flowerpecker
f   Percefleur bleuté
d Silberhakenschnabel
i   Diglossa azzurra

3255 *Diglossopis cyanea*
      Passeriformes - Fringillidae
e Masked Flowerpiercer; Masked
      Flowerpecker
f   Percefleur masqué
d Maskenhakenschnabel
i   Diglossa mascherata

3256 *Diglossopis glauca*
      Passeriformes - Fringillidae
e Deep-blue Flowerpiercer; Deep-blue
      Honeycreeper; Glaucous
      Honeycreeper; Deep-blue
      Flowerpecker
f   Percefleur glauque
d Ultramarinhakenschnabel
i   Diglossa blu

3257 *Diglossopis indigotica*
      Passeriformes - Fringillidae
e Indigo Flowerpiercer; Indigo
      Honeycreeper; Indigo Flowerpecker
f   Percefleur indigo
d Indigohakenschnabel
i   Diglossa color indica

3258 *Dinemellia dinemelli*
      Passeriformes - Ploceidae
e White-headed Buffalo-Weaver;
      Whitge-faced Buffalo-Weaver
f   Alecto à tête blanche
d Starweber; Weißkopfviehweber
i   Tessitore dei bufali testabianca

3259 *Dinopium benghalense*
      Piciformes - Picidae
e Black-rumped Flameback; Lesser
      Golden-backed Woodpecker (ISC);
      Black-rumped Goldenback; Lesser

Flameback; Ceylon Golden-backed
Woodpecker
f Pic du Bengale
d Orangespecht; Goldrückenspecht
i Picchio dorsodorato di Bengala

3260 *Dinopium javanense*
Piciformes - Picidae
e Common Flameback; Indian Golden-
backed Three-toed-Woodpecker
(ISC); Golden-backed Three-toed
Woodpecker; Golden-backed
Woodpecker; Common Golden-
backed Woodpecker; Common
Goldback; Indian Goldenback
Woodpecker; Indian Flameback;
Common Golden-backed Three-toed
Woodpecker
f Pic à dos rouge
d Stummelspecht
i Picchio dorsodorato di Giava

3261 *Dinopium rafflesii*
Piciformes - Picidae
e Olive-backed Woodpecker; Olive-
backed Three-toed Woodpecker;
Raffles's Woodpecker
f Pic oriflamme
d Olivbauchspecht
i Picchio di Raffles

3262 *Dinopium shorii*
Piciformes - Picidae
e Himalayan Flameback; Himalayan
Three-toed Woodpecker; Himalayan
Golden-backed Woodpecker; Three-
toed Golden-backed Woodpecker
f Pic de Shore
d Braunhalspecht
i Picchio di Shore

3263 *Diomedea albatrus*
Procellariiformes - Diomedeidae
e Short-tailed Albatross; Steller's
Albatross
f Albatros à queue courte
d Kurzschwanzalbatros
i Albatro di Steller; Albatro comune;
Albatro

3264 *Diomedea amsterdamensis*
Procellariiformes - Diomedeidae
e Amsterdam Island Albatross;
Amsterdam Albatross (ANZ)
i Albatro dell'Isola Amsterdam;
Albatro di Amseterdam

3265 *Diomedea antipodensis*
Procellariiformes - Diomedeidae
e Amsterdam Albatross (ANZ)

3266 *Diomedea bulleri*
Procellariiformes - Diomedeidae
e Buller's Albatross; Buller's
Mollymawk
f Albatros de Buller
d Buller-Albatros
i Albatro di Buller

3267 *Diomedea cauta*
Procellariiformes - Diomedeidae
e White-capped Albatross; Shy
Albatross; Shy Mollymawk; Salvin's
Albatross
f Albatros à cape blanche; Albatros des
Chatham
d Weißkappenalbatros; Scheuer
Albatros
i Albatrocauto

3268 *Diomedea chlororhynchos*
Procellariiformes - Diomedeidae
e Yellow-nosed Albatross; Yellow-
nosed Mollymawk; Carter's
Albatross; Atlantic Yellow-nosed
Albatross
f Albatros à nez jaune; Albatros à bec
jaune
d Gelbnasenalbatros
i Albatro beccogiallo

3269 *Diomedea chrysostoma*
Procellariiformes - Diomedeidae
e Grey-headed Albatross; Gould's
Albatross; Grey-headed Mollymawk
f Albatros à tête grise
d Graukopf; Graukopfalbatros
i Albatro testagrigia

3270 **Diomedea dabbena**
Procellariiformes - Diomedeidae
e Tristan Albatross (ANZ)

3271 **Diomedea epomophora**
Procellariiformes - Diomedeidae
e Royal Albatross; Southern Royal-
Albatross (ANZ)
f Albatros royal
d Königsalbatros
i Albatro epomoforo; Albatro reale

3272 **Diomedea exulans**
Procellariiformes - Diomedeidae
e Wandering Albatross; White-winged
Albatross; Snowy Albatross
f Albatros hurleur; Albatros exilé
d Wanderalbatros; Kapschaf; Gemeiner
Albatros; Großer Albatros; Albatros
i Albatro urlatore

3273 **Diomedea gibsoni**
Procellariiformes - Diomedeidae
e Gibson's Albatross (ANZ)

3274 **Diomedea immutabilis**
Procellariiformes - Diomedeidae
e Laysan Albatross
f Albatros de Laysan
d Laysan-Albatros
i Albatro di Laysan

3275 **Diomedea irrorata**
Procellariiformes - Diomedeidae
e Waved Albatross; Galapagos
Albatross
f Albatros des Galapagos
d Galapagos-Albatros
i Albatro marezzato; Albatro delle
Galapagos

3276 **Diomedea melanophris**
Procellariiformes - Diomedeidae
e Black-browed Albatross; Black-
browed Mollymawk
f Albatros à sourcils noirs
d Schwarzbrauenalbatros; Mollymauk;
Schwarzbrauner Albatros
i Albatro dai sopraccigli neri

3277 **Diomedea nigripes**
Procellariiformes - Diomedeidae
e Black-footed Albatross
f Albatros à pieds noirs; Albatros à
pattes noires
d Schwarzfußalbatros
i Albatro piedineri; Albatro dai piedi
neri

3278 **Diomedea salvini**
Procellariiformes - Diomedeidae
e Salvin's Albatross
f Albatros de Salvin

3279 **Diomedea sanfordi**
Procellariiformes - Diomedeidae
e Northern Royal Albatross

3280 **Diomedeidae**
Procellariiformes
e Albatrosses
f Albatros; Diomédéidés
d Albatrosse
i Diomedeidi; Albatri

3281 **Dioptrornis brunneus**
Passeriformes - Muscicapidae
e Angola Slaty-Flycatcher; Angolan
Flycatcher; Angola Chocolate
Flycatcher
f Gobemouche de l'Angola; Gobe-
mouches de l'Angola
d Angola-Drongoschnäpper

3282 **Dioptrornis chocolatinus**
Passeriformes - Muscicapidae
e Abyssinian Slaty-Flycatcher; Slaty
Flycatcher; Chocolate Flycatcher;
Abyssinian Flycatcher
f Gobemouche chocolat; Gobe-
mouches chocolat
d Habesch-Drongoschnäpper
i Pigliamosche ardesia abissino

3283 **Discosura longicauda**
Apodiformes - Trochilidae
e Racket-tailed Coquette; Racquet-
tailed Coquette; Brazilian Racquet-
tail; Racquet-tailed Hummingbird;
Brazilian Racket-tail
f Coquette à raquettes

f	Drome ardéole; Drome
d	Reiherläufer; Meerrenner
i	Droma; Granchiere

3312 **_Dromococcyx pavoninus_**
Coliformes - Neomorphidae
e Pavonine Cuckoo; Peacock Cuckoo
f Géocoucou pavonin
d Pfauenkuckuck
i Cuculo corridore pavonino

3313 **_Dromococcyx phasianellus_**
Coliformes - Neomorphidae
e Pheasant-Cuckoo
f Géocoucou faisan
d Rotschopfkuckuck
i Cuculo corridore fagiano

3314 **_Drymocichla incana_**
Passeriformes - Cisticolidae
e Red-winged Grey-Warbler
f Prinia grise
d Rotschwingensänger
i Cantore alirosse

3315 **_Drymodes brunneopygia_**
Passeriformes - Petroicidae
e Southern Scrub-Robin; Pale Scrub-robin
f Drymode à croupion brun
d Mallee-Scheindrossel
i Usignolo di macchia meridionale

3316 **_Drymodes superciliaris_**
Passeriformes - Petroicidae
e Northern Scrub-Robin; Eastern Scrub-Robin
f Drymode bridé
d Augenstreifscheindrossel
i Usignolo di macchia settentrionale

3317 **_Drymophila caudata_**
Passeriformes - Thamnopilidae
e Long-tailed Antbird
f Grisin à longue queue
d Langschwanzameisenfänger

3318 **_Drymophila devillei_**
Passeriformes - Thamnopilidae
e Striated Antbird
f Grisin de Deville

d Strichelkopfameisenfänger
i Mangiaformiche di Deville

3319 **_Drymophila ferruginea_**
Passeriformes - Thamnopilidae
e Ferruginous Antbird
f Grisin rouilleux
d Weißbraunameisenfänger
i Mangiaformiche ferrugineo

3320 **_Drymophila genei_**
Passeriformes - Thamnopilidae
e Rufous-tailed Antbird
f Grisin à queue rousse
d Rotschwarzameisenfänger
i Mangiaformiche di Genè

3321 **_Drymophila malura_**
Passeriformes - Thamnopilidae
e Dusky-tailed Antbird
f Grisin malure
d Temminck-Ameisenfänger
i Mangiaformiche codabruna

3322 **_Drymophila ochropyga_**
Passeriformes - Thamnopilidae
e Ochre-rumped Antbird
f Grisin à croupion ocré
d Rotbürzelameisenfänger
i Mangiaformiche gropponeocra

3323 **_Drymophila rubricollis_**
Passeriformes - Thamnopilidae
e Bertoni's Antbird
f Grisin de Bertoni
i Mangiaformiche di Bertoni

3324 **_Drymophila squamata_**
Passeriformes - Thamnopilidae
e Scaled Antbird
f Grisin écaillé
d Schuppenameisenfänger
i Mangiaformiche squamoso

3325 **_Drymornis bridgesii_**
Passeriformes - Dedrocolaptidae
e Scimitar-tailed Woodcreeper; Scimitar-billed Woodhewer; Bridges's Woodcreeper
d Degenschnabelbaumsteiger
i Rampichino becco a scimitarra

3326   *Dryocopus galeatus*
         Piciformes - Picidae
  e   Helmeted Woodpecker
  f   Pic casqué
  d   Wellenohrspecht; Helmspecht
  i   Picchio crestato

3327   *Dryocopus hodgei*
         Piciformes - Picidae
  e   Andaman Woodpecker
  f   Pic des Andaman
  i   Picchio delle Andamane

3328   *Dryocopus javensis*
         Piciformes - Picidae
  e   White-bellied Woodpecker; Great
         Black-Woodpecker (ISC); White-
         bellied Black-Woodpecker; Indian
         Great Black-Woodpecker
  f   Pic à ventre blanc
  d   Weißbauchspecht
  i   Picchio nero di Giava

3329   *Dryocopus lineatus*
         Piciformes - Picidae
  e   Lineated Woodpecker; Black-
         mantled Woodpecker
  f   Pic ouentou
  d   Linienspecht; Streifenkehlhelmspecht
  i   Picchio lineato

3330   *Dryocopus martius*
         Piciformes - Picidae
  e   Black Woodpecker; Great Black-
         Woodpecker
  f   Pic noir
  d   Schwarzspecht
  i   Picchio nero

3331   *Dryocopus pileatus*
         Piciformes - Picidae
  e   Pileated Woodpecker
  f   Grand Pic
  d   Schopfspecht; Haubenspecht;
         Helmspecht; Haubenschwarzspecht
  i   Picchio pileato

3332   *Dryocopus schulzi*
         Piciformes - Picidae
  e   Black-bodied Woodpecker
  f   Pic lucifer

  d   Schwarzbauchspecht
  i   Picchio di Schulz

3333   *Dryolimnas cuvieri*
         Gruiformes - Rallidae
  e   White-throated Rail; Madagascar
         White-throated Rail; Cuvier's Rail
  f   Râle de Cuvier
  d   Cuvier-Ralle
  i   Rallo di Cuvier

3334   *Dryoscopus angolensis*
         Passeriformes - Malaconotidae
  e   Pink-footed Puffback; Pink-footed
         Puffback-Shrike
  f   Cubla à pieds roses
  d   Rotfußschneeballwürger
  i   Dorsopiumoso piedirosa

3335   *Dryoscopus cubla*
         Passeriformes - Malaconotidae
  e   Black-backed Puffback; Puffback
         (CSA); Black-backed Puffback-
         Shrike; Southern Puffback
  f   Cubla boule-de-neige; Pie-grièche
         cubla
  d   Schneeballwürger;
         Weißrückenwürger
  i   Dorsopiumoso dorsonero

3336   *Dryoscopus cubla affinis*
         Passeriformes - Malaconotidae
  f   Cubla de Zanzibar

3337   *Dryoscopus gambensis*
         Passeriformes - Malaconotidae
  e   Northern Puffback; Puffback;
         Gambian Puffback-Shrike; Common
         Puffback
  f   Cubla de Gambie
  d   Gambia-Schneeballwürger;
         Waldschneeballwürger
  i   Dorsopiumoso del Nord

3338   *Dryoscopus pringlii*
         Passeriformes - Malaconotidae
  e   Pringle's Puffback; Pringle's
         Puffback-Shrike
  f   Cubla de Pringle
  d   Zwergschneeballwürger
  i   Dorsopiumoso di Pringle

3339 **Dryoscopus sabini**
Passeriformes - Malaconotidae
e Large-billed Puffback; Sabine's
Puffback Shrike; Sabine's Puffback
f Cubla à gros bec
d Dickschnabelschneeballwürger
i Dorsopiumoso di Sabine

3340 **Dryoscopus senegalensis**
Passeriformes - Malaconotidae
e Red-eyed Puffback; Black-
shouldered Puffback; Red-eyed
Puffback-Shrike
f Cubla aux yeux rouges
d Schwarzschulterschneeballwürger
i Dorsopiumoso occhirossi

3341 **Dryotriorchis spectabilis**
Falconiformes - Accipitridae
e Congo Serpent-Eagle; Congo Snake-
Eagle; African Serpent-Eagle
f Serpentaire du Congo
d Schlangenbussard
i Aquila serpentaria del Congo; Aquila
serpentaria della Costa d'Oro

3342 **Dubusia taeniata**
Passeriformes - Fringillidae
e Buff-breasted Mountain-Tanager; Du
Bus's Mountain-Tanager
f Tangara à poitrine fauve
d Silberbrauenbergtangare
i Tangara di monte vermicolata

3343 **Ducula aenea**
Columbiformes - Columbidae
e Green Imperial-Pigeon; Enggano
Imperial-Pigeon
f Carpophage pauline
d Bronzefruchttaube
i Piccione imperiale verde

3344 **Ducula aurorae**
Columbiformes - Columbidae
e Polynesian Imperial-Pigeon; Society
Islands Pigeon; Tahitian Pigeon;
Wilke's Pigeon
f Carpophage de la Société;
Carpophage d'Aurora
d Aurora-Fruchttaube; Tahiti-

Fruchttaube
i Piccione imperiale di Aurora

3345 **Ducula badia**
Columbiformes - Columbidae
e Mountain Imperial-Pigeon; Maroon-
backed Imperial-Pigeon (ISC);
Imperial Pigeon; Hodgson's Imperial-
Pigeon; Jerdon's Imperial-Pigeon;
Grey-headed Imperial-Pigeon; Band-
tailed Imperial-Pigeon; Bronze-
backed Imperial-Pigeon
f Carpophage à manteau brun
d Fahlbauchfruchttaube;
Gebirgsfruchttaube
i Piccione imperiale di montagna

3346 **Ducula bakeri**
Columbiformes - Columbidae
e Baker's Imperial-Pigeon; Vanuatu
Imperial-Pigeon; Baker's Pigeon
f Carpophage de Baker
d Kurzflügelfruchttaube; Baker-
Fruchttaube
i Piccione imperiale di Baker

3347 **Ducula basilica**
Columbiformes - Columbidae
e Cinnamon-bellied Imperial-Pigeon;
Moluccan Rufous-bellied Pigeon;
Moluccan Fruit-Pigeon; Basilica
Pigeon; Moluccan Rufous-bellied
Fruit-Pigeon; Rufous-bellied
Imperial-Pigeon; Pink-headed Fruit-
Pigeon
f Carpophage des Moluques
d Halmahera-Fruchttaube
i Piccione imperiale delle Molucche

3348 **Ducula bicolor**
Columbiformes - Columbidae
e Pied Imperial-Pigeon; Nutmeg
Pigeon; Nutmeg Imperial-Pigeon;
Spice Pigeon; Australian Pied
Imperial-Pigeon
f Carpophage blanc
d Zweifarbenfruchttaube
i Piccione imperiale bicolore

3349 **Ducula brenchleyi**
Columbiformes - Columbidae

e Chestnut-bellied Imperial-Pigeon;
Chestnut-bellied Pigeon
f Carpophage de Brenchley
d Braunbauchfruchttaube
i Piccione imperiale di Benchley

3350 *Ducula carola*
Columbiformes - Columbidae
e Spotted Imperial-Pigeon; Grey-
necked Fruit-Pigeon; Grey-breasted
Fruit-Pigeon
f Carpophage charlotte
d Hufeisenfruchttaube; Gefleckte
Fruchttaube
i Piccione imperiale collogrigio

3351 *Ducula chalconota*
Columbiformes - Columbidae
e Shining Imperial-Pigeon; Red-
breasted Imperial-Pigeon; Mountain
Fruit-Pigeon; Mountain Rufous-
bellied Fruit-Pigeon; Grey-hooded
Fruit-Pigeon; Rufescent Imperial-
Pigeon
f Carpophage brillant
d Rotbauchbergfruchttaube
i Piccione imperiale splendente

3352 *Ducula cineracea*
Columbiformes - Columbidae
e Timor Imperial-Pigeon; Ashy
Imperial-Pigeon
f Carpophage cendrillon
d Timor-Fruchttaube
i Piccione imperiale di Timor

3353 *Ducula concinna*
Columbiformes - Columbidae
e Elegant Imperial-Pigeon; Blue-tailed
Imperial-Pigeon; Yellow-eyed
Imperial-Pigeon; Collared Imperial-
Pigeon; Gold-eyed Pigeon; Gold-
eyed Imperial-Pigeon
f Carpophage à queue bleue
d Blauschwanzfruchttaube; Molukken-
Bronzefruchttaube
i Piccione imperiale codablu

3354 *Ducula constans*
Columbiformes - Columbidae
e Kimberley Imperial-Pigeon

3355 *Ducula finschii*
Columbiformes - Columbidae
e Finsch's Imperial-Pigeon; Finsch's
Rufous-bellied Fruit-Pigeon
f Carpophage de Finsch
d Finsch-Fruchttaube
i Piccione imperiale di Finsch

3356 *Ducula forsteni*
Columbiformes - Columbidae
e White-bellied Imperial-Pigeon;
Green-and-white Zone-tailed Pigeon;
Green-and-white Pigeon; Large
Celebes Zone-tailed Pigeon; Celebes
Zone-tailed Pigeon
f Carpophage de Forsten
d Weißbauchfruchttaube; Große
Celebes-Fruchttaube
i Piccione imperiale verde e bianco

3357 *Ducula galeata*
Columbiformes - Columbidae
e Marquesan Imperial-Pigeon;
Marquesas Pigeon; Marquesan
Pigeon; Nukuhiva Imperial-Pigeon;
Marquesa Imperial-Pigeon;
Marquesas Imperial-Pigeon;
Marquesa Pigeon
f Carpophage des Marquises
d Marquesas-Fruchttaube
i Piccione imperiale delle Marquesas

3358 *Ducula goliath*
Columbiformes - Columbidae
e New Caledonian Imperial-Pigeon;
New Caledonian Pigeon; Giant
Pigeon; Giant Imperial-Pigeon;
Goliath Pigeon
f Carpophage géant
d Riesenfruchttaube
i Piccione imperiale della Nuova
Caledonia

3359 *Ducula lacernulata*
Columbiformes - Columbidae
e Dark-backed Imperial-Pigeon; Black-
backed Imperial-Pigeon; Javanese
Mountain-Pigeon; Javanese Imperial-
Pigeon
f Carpophage mantelé

Fruchttaube
i Piccione imperiale di Wharton

3380 *Ducula zoeae*
Columbiformes - Columbidae
e Banded Imperial-Pigeon; Banded
Fruit-Pigeon; Bar-breasted Fruit-
Pigeon; Zoe Imperial-Pigeon; Zoe's
Imperial-Pigeon
f Carpophage de Zoé
d Gebänderte Fruchttaube;
Halsbandfruchttaube
i Piccione imperiale di Zoe

3381 Dulidae
Passeriformes
e Palmchats
f Dulidés
d Palmenschmätzer; Palmenschwätzer;
Palmschmätzer
i Dulidi

3382 *Dulus dominicus*
Passeriformes - Dulidae
e Palmchat
f Oiseau palmiste (Ants); Esclave
palmiste
d Palmenschmätzer; Palmschmätzer
i Uccello delle palme

3383 *Dumetella carolinensis*
Passeriformes - Sturnidae
e Grey Catbird; Common Catbird;
Catbird; Northern Catbird; Blue
Thrush (WI); Blue Thrasher (WI)
f Moqueur-chat; Oiseau-chat;
Moqueur-chat de la Caroline
d Katzenvogel; Katzendrossel
i Uccello gatto; Uccello gatto
testanero; Mimo della Carolina

3384 *Dumetia hyperythra*
Passeriformes - Sylviidae
e Tawny-bellied Babbler; Rufous-
bellied Babbler (ISC); White-
throated Babbler; Ceylon White-
throated Babbler
f Timalie à ventre roux
d Rotbauchtimalie
i Garrulo ventrerosso

3385 *Dysithamnus leucostictus*
Passeriformes - Thamnophilidae
e White-streaked Antvireo; White-
spotted Antshrike; White-streaked
Antshrike; White-spotted Antvireo
d Weißschulterwürgerling
i Vireo formichiere macchiettato

3386 *Dysithamnus mentalis*
Passeriformes - Thamnophilidae
e Plain Antvireo
d Waldwürgerling
i Vireo formichiere modesto

3387 *Dysithamnus occidentalis*
Passeriformes - Thamnophilidae
e Bicoloured Antvireo; Western
Antshrike; Western Antvireo;
Chapman's Antshrike
f Batara occidental
d Chapman-Würgerling
i Vireo formichiere occidentale

3388 *Dysithamnus plumbeus*
Passeriformes - Thamnophilidae
e Plumbeous Antvireo; Plumbeous
Antshrike
f Batara plombé
d Bleiwürgerling
i Vireo formichiere plumbeo

3389 *Dysithamnus puncticeps*
Passeriformes - Thamnophilidae
e Spot-crowned Antvireo
d Perlkappenwürgerling
i Vireo formichiere corona maculata

3390 *Dysithamnus stictothorax*
Passeriformes - Thamnophilidae
e Spot-breasted Antvireo
d Fleckenbrustwürgerling
i Vireo formichiere pettomacchiato

3391 *Dysithamnus striaticeps*
Passeriformes - Thamnophilidae
e Streak-crowned Antvireo; Streaked
Antvireo
d Streifenkopfwürgerling
i Vireo formichiere corona striata

3392   *Dysithamnus xanthopterus*
Passeriformes - Thamnophilidae
e   Rufous-backed Antvireo
d   Rotbürstelwürgerling
i   Vireo formichere aligialli

3393   *Dysmorodrepanis munroi*
Passeriformes - Fringillidae
e   Lanai Hookbill; Lanai Finch
i   Becco a uncino di Lanai

E

3394 ***Eclectus roratus***
Psittaciformes - Psittacidae
e Eclectus Parrot; Ceram Eclectus
Parrot; Red-sided Parrot; Rocky
River Parrot; Halmahera Hanging
Parrot; Grand Eclectus; Red-sided
Eclectus Parrot; Kalanga
f Grand Eclectus; Perroquet de
Halmahera; Lori de la Nouvelle
Guinée
d Halmahera-Edelpapagei; Ceram
Edelpapagei
i Ecletto

3395 ***Eclectus roratus macgillivrayi***
Psittaciformes - Psittacidae
e Cape York Pensinsula Eclectus
Parrot (ANZ)

3396 ***Eclectus roratus polychloros***
Psittaciformes - Psittacidae
e Torres Strait Pensinsula Eclectus
Parrot (ANZ)

3397 ***Ectopistes migratorius***
Columbiformes - Columbidae
e Passenger Pigeon; Migratory Pigeon;
Wild Pigeon
f Tourte voyageuse; Tourte (Qué);
Pigeon migrateur; Pigeon ectopiste
d Wandertaube
i Colomba migratrice

3398 ***Egretta alba***
Ciconiiformes - Ardeidae
e Great White-Egret; Great White-
Heron; Great Egret; American Egret
(NA); Common Egret; Large Egret
(ISC); Greater Egret; Great American
Egret; Eastern Large Egret; Large
Heron
f Grande Aigrette; Aigrette blanche

d Silberreiher
i Airone bianco maggiore; Egretta

3399 ***Egretta ardesiaca***
Ciconiiformes - Ardeidae
e Black Heron; Black Egret (CSA)
f Aigrette ardoisée
d Glockenreiher
i Garzetta ardesia

3400 ***Egretta caerulea***
Ciconiiformes - Ardeidae
e Little Blue-Heron; Blue Gauldin
(WI); Blue Gaulin (WI); Gaulin (WI)
f Aigrette bleue; Crabier blanc (Ants);
Crabier bleu (Ants); Crabier noir
(Ants); Petit Héron bleu (Ants)
d Kleiner Blaureiher; Blaureiher
i Airone azzurro minore

3401 ***Egretta dimorpha***
Ciconiiformes - Ardeidae
e Dimorphic Egret; Mascarene Reef-
Heron; Madagascar Egret
f Aigrette dimorphe
i Garzetta del Madagascar

3402 ***Egretta eulophotes***
Ciconiiformes - Ardeidae
e Chinese Egret; Little Egret;
Swinhoe's Egret; Yellow-billed
White-Heron
f Aigrette de Chine
d Schneereiher
i Garzetta cinese

3403 ***Egretta garzetta***
Ciconiiformes - Ardeidae
e Little Egret; Lesser Egret; Spotless
Egret; Snowy Egret
f Aigrette garzette; Aigrette blanche;
Petite Aigrette blanche; Petite
Aigrette garzette; Héron garzette
d Seidenreiher
i Garzetta; Garzetta comune

3404 ***Egretta gularis***
Ciconiiformes - Ardeidae
e Western Reef-Egret; Western Reef-
Heron; Indian Reef-Heron (ISC);
Reef Heron; African Reef-Heron;

West African Reef-Heron; Arabian
Reef-Heron
f   Aigrette à gorge blanche; Aigrette
des récifs; Aigrette gorgeblanche
d   Küstenreiher
i   Garzetta ardesia; Garzetta gulare;
Garzetta dai reef occidentale

3405   *Egretta novaehollandiae*
Ciconiiformes - Ardeidae
e   White-faced Heron; White-faced
Egret; White-fronted Heron; Blue
Heron; Blue Crane
f   Aigrette à face blanche
d   Weißwangenreiher
i   Garzetta facciabianca

3406   *Egretta rufescens*
Ciconiiformes - Ardeidae
e   Reddish Egret; Gaulin (WI)
f   Aigrette roussâtre; Aigrette bleue
(Ants); Aigrette rousse
d   Blaufußreiher
i   Garzetta rugginosa

3407   *Egretta sacra*
Ciconiiformes - Ardeidae
e   Pacific Reef-Egret; Eastern Reef-
Egret; Pacific Reef-Heron; Eastern
Reef-Heron; Blue Reef-Egret; Blue
Reef-Heron; White Reef-Egret;
White Reef-Heron; Reef Heron
f   Aigrette sacrée
d   Riffenreiher
i   Garzetta dei reef orientale

3408   *Egretta thula*
Ciconiiformes - Ardeidae
e   Snowy Egret; White-Gauldin (WI);
Golden Slippers (WI); Gaulin (WI);
While Gaulin (WI)
f   Aigrette neigeuse; Aigrette blanche
(Ants); Aigrette (Ants)
d   Schmuckreiher
i   Garzetta nivea

3409   *Egretta tricolor*
Ciconiiformes - Ardeidae
e   Tricoloured Heron; Grey Gauldin
(WI); Switching-neck (WI); Gaulin
(WI); Louisiana Heron

f   Aigrette tricolore; Crabier aux trois
couleurs (Ants); Héron tricolore
(Ants); Crabier (Ants); Aigrette à
ventre blanc
d   Dreifarbenreiher
i   Airone tricolore; Airone della
Louisiana; Airone della Luisiana

3410   *Egretta vinaceigula*
Ciconiiformes - Ardeidae
e   Slaty Egret; Brown-throated Egret;
Red-throated Egret
f   Aigrette vineuse
d   Braunkehlreiher; Schieferreiher
i   Garzetta collobruno

3411   *Elaenia albiceps*
Passeriformes - Tyrannidae
e   White-crested Elaenia
f   Élénie à cimier blanc; Elaène à
couronne blanche
d   Buschelaenie
i   Elenia crestabianca

3412   *Elaenia albiceps modesta*
Passeriformes - Tyrannidae
f   Élénie du Pérou

3413   *Elaenia chinchorroensis*
Passeriformes - Tyrannidae
e   Chinchorro Elaenia

3414   *Elaenia chiriquensis*
Passeriformes - Tyrannidae
e   Lesser Elaenia; Lawrence's Elaenia
f   Élénie menue; Elaène de Lawrence
d   Schlankschnabelelaenie
i   Elenia di chiriqui

3415   *Elaenia cristata*
Passeriformes - Tyrannidae
e   Plain-crested Elaenia; Crested
Elaenia
f   Élénie huppée; Elaène huppée
d   Kappenelaenie
i   Elenia crestata

3416   *Elaenia dayi*
Passeriformes - Tyrannidae
e   Great Elaenia; Giant Elaenia
f   Élénie de Day; Elaène de Day

d Duida-Elaenie
i Elenia di Day

3417 *Elaenia fallax*
Passeriformes - Tyrannidae
e Greater Antillean Elaenia; Antillean Elaenia
f Élénie sara; Elaène sara
d Antillen-Elaenie
i Elenia delle Grandi Antill

3418 *Elaenia flavogaster*
Passeriformes - Tyrannidae
e Yellow-bellied Elaenia
f Élénie à ventre jaune; Elaène à ventre jaune
d Gelbbauchelaenie
i Elenia ventregiallo

3419 *Elaenia frantzii*
Passeriformes - Tyrannidae
e Mountain Elaenia; Frantzius's Elaenia
f Élénie montagnarde; Elaène montagnarde
d Bergelaenie
i Elenia di Frantz

3420 *Elaenia gigas*
Passeriformes - Tyrannidae
e Mottle-backed Elaenia; Giant Elaenia
f Élénie ecaillée; Elaène écaillée
d Schuppenelaenie
i Elenia gigante

3421 *Elaenia martinica*
Passeriformes - Tyrannidae
e Caribbean Elaenia; Whistler (WI); Pea Whistler (WI); Top-knot Judas (WI)
f Élénie siffleuse; Elaène sifleuse; Siffleur (Ants); Siffleur blanc (Ants)
d Weißbauchelaenie
i Elenia dei Caraibi

3422 *Elaenia mesoleuca*
Passeriformes - Tyrannidae
e Olivaceous Elaenia
f Élénie olivâtre; Elaène olivâtre
d Schlichtelaenie
i Elenia olivacea

3423 *Elaenia obscura*
Passeriformes - Tyrannidae
e Highland Elaenia; Dusky Elaenia
f Élénie obscure; Elaène obscure
d Olivkopfelaenie
i Elenia delle alture

3424 *Elaenia pallatangae*
Passeriformes - Tyrannidae
e Sierran Elaenia; Sierra Elaenia; Pallatanga Elaenia
f Élénie de Pallatanga; Elaène de Pallantaga
d Pallantanga-Elaenie
i Elenia della Sierra

3425 *Elaenia parvirostris*
Passeriformes - Tyrannidae
e Small-billed Elaenia
f Élénie à bec court; Elaène à bec court
d Kurzschnabelelaenie
i Elenia beccopiccolo

3426 *Elaenia pelzelni*
Passeriformes - Tyrannidae
e Brownish Elaenia; Pelzeln's Elaenia
f Élénie brune; Elaène brune
d Braunelaenie
i Elenia di Pelzeln

3427 *Elaenia ridleyana*
Passeriformes - Tyrannidae
e Noronha Elaenia
f Élénie de Noronha; Elaène de Noronha
i Elenia dell'Isola Noronha

3428 *Elaenia ruficeps*
Passeriformes - Tyrannidae
e Rufous-crowned Elaenia
f Élénie tête-de-feu; Elaène tête-de-feu
d Rotscheitelelaenie
i Elenia corona rossiccia

3429 *Elaenia spectabilis*
Passeriformes - Tyrannidae
e Large Elaenia
f Élénie remarquable; Elaène remarquable; Elaène de Natterer
d Graubrustelaenie
i Elenia grande

3430   *Elaenia strepera*
           Passeriformes - Tyrannidae
e   Slaty Elaenia
f   Élénie bruyante; Elaène bruyante
d   Schieferelaenie
i   Elenia color lavagna

3431   *Elanoides forficatus*
           Falconiformes - Accipitridae
e   Swallow-tailed Kite; American
      Swallow-tailed Kite
f   Milan à queue fourchue
d   Schwalbenschwanzweihe;
      Schwalbenweihe
i   Nibbio codadirondine; Nibbio a coda
      di rondine-

3432   *Elanus axillaris*
           Falconiformes - Accipitridae
e   Black-shouldered Kite; Australian
      Black-shouldered Kite; Australian
      Kite
f   Élanion d'Australie
d   Gleitaar; Australischer Gleitaar;
      Schwarzschultergleitaar
i   Nibbio bianco australiano

3433   *Elanus caeruleus*
           Falconiformes - Accipitridae
e   Black-winged Kite; Black-
      shouldered Kite; Common Black-
      shouldered Kite; Indonesian Kite
f   Élanion blanc
d   Gleitaar
i   Nibbio bianco

3434   *Elanus leucurus*
           Falconiformes - Accipitridae
e   White-tailed Kite; Black-shouldered
      Kite
f   Élanion à queue blanche
d   Amerikanischer Gleitraar
i   Nibbio codabianca; Nibbio dalla
      coda bianca

3435   *Elanus scriptus*
           Falconiformes - Accipitridae
e   Letter-winged Kite; White-breasted
      Sparrowhawk
f   Élanion lettre
d   Schwarzachselaar

i   Nibbio bianco aliscritte; Nibbio dalle
      ali bordate

3436   *Electron carinatum*
           Coraciiformes - Momotidae
e   Keel-billed Motmot
f   Motmot à bec caréné; Momot caréné
d   Kielschnabelmotmot
i   Motmot beccocarenato

3437   *Electron platyrhynchum*
           Coraciiformes - Momotidae
e   Broad-billed Motmot
f   Motmot à bec large; Momot à bec
      large
d   Plattschnabelmotmot
i   Motmot beccolargo

3438   *Electron pyrrholaemum*
           Coraciiformes - Momotidae
e   Plain-tailed Motmot

3439   *Eleothreptus anomalus*
           Caprimulgiformes - Caprimulgidae
e   Sickle-winged Nightjar
f   Engoulevent à faucilles
d   Sichelschwingennachtschwalbe
i   Succiacapre codafalcata

3440   *Elminia albicauda*
           Passeriformes - Corvidae
e   White-tailed Blue-Flycatcher; White-
      tailed Blue Monarch; Southern Fairy-
      Flycatcher; White-tailed Fairy-
      Flycatcher; White-tailed Elminia
f   Tchitrec à queue blanche
d   Weißschwanzelminie
i   Elminia codabianca

3441   *Elminia albiventris*
           Passeriformes - Corvidae
e   White-bellied Crested-Flycatcher;
      White-bellied Flycatcher; White-
      bellied Crested-Monarch
f   Tchitrec à ventre blanc; Gobemouche
      huppé à tête blanche; Gobe-mouches
      huppé à tête blanche
d   Weißbauchhaubenschnäpper
i   Elminia ventrebianco; Pigliamosche
      crestato ventrebianco

3442 *Elminia albonotata*
Passeriformes - Corvidae
e White-tailed Crested-Flycatcher;
White-tailed Crested-Monarch
ƒ Tchitrec à queue frangée
d Berghaubenschnäpper
i Elminia codabianca; Pigliamosche
crestato codabianca

3443 *Elminia longicauda*
Passeriformes - Corvidae
e African Blue-Flycatcher; Blue-
Flycatcher; Northern Fairy-
Flycatcher; Southern Fairy-
Flycatcher; Fairy Flycatcher; Blue
Fairy Flycatcher
ƒ Tchitrec bleu; Gobemouche bleu;
Gobe-mouches bleu
d Türkis-Elminie
i Elminia africana

3444 *Elminia nigromitrata*
Passeriformes - Corvidae
e Dusky Crested-Flycatcher; Dusky
Crested-Monarch
ƒ Tchitrec à tête noire; Gobemouche
noir huppé; Gobe-mouches noir
huppé
d Schwarzkopfhaubenschnäpper
i Elminia bruno; Pigliamosche crestato
bruno

3445 *Elseyornis melanops*
Charadriiformes - Charadriidae
e Black-fronted Dotterel; Black-
fronted Plover; Australian Black-
fronted Plover
ƒ Pluvier à face noire
d Schwarzstirnregenpfeifer
i Corriere frontenera

3446 *Elvira chionura*
Apodiformes - Trochilidae
e White-tailed Emerald
ƒ Colibri elvire
d Elvira-Kolibri
i Elvira codabianca

3447 *Elvira cupreiceps*
Apodiformes - Trochilidae
e Coppery-headed Emerald

ƒ Colibri à tête cuivrée
d Kupferköpchen
i Elvira testa di rame; Colibrì smeraldo

3448 *Emberiza affinis*
Passeriformes - Emberizidae
e Brown-rumped Bunting; Nigerian
Little-Bunting
ƒ Bruant à ventre jaune
d Braunbürzelammer
i Zigolo gropponebruno

3449 *Emberiza aureola*
Passeriformes - Emberizidae
e Yellow-breasted Bunting; Golden
Bunting; White-shouldered Bunting
ƒ Bruant auréole
d Weidenammer
i Zigolo dal collare

3450 *Emberiza bruniceps*
Passeriformes - Emberizidae
e Red-headed Bunting
ƒ Bruant à tête rousse
d Braunkopfammer
i Zigolo testa aranciata

3451 *Emberiza buchanani*
Passeriformes - Emberizidae
e Grey-necked Bunting; Grey-hooded
Bunting; Buchanan's Bunting
ƒ Bruant à col gris; Bruant à cou gris
d Steinortolan
i Zigolo collogriggio; Zigolo dal
cappuccio grigio

3452 *Emberiza cabanisi*
Passeriformes - Emberizidae
e Cabanis's Bunting; Yellow Bunting;
Cabanis's Yellow Bunting
ƒ Bruant de Cabanis
d Cabanis-Ammer
i Zigolo di Cabanis

3453 *Emberiza caesia*
Passeriformes - Emberizidae
e Cretschmar's Bunting
ƒ Bruant cendrillard
d Grauortolan; Grauer Ortolan;
Rostbartammer
i Ortolano grigio

3454 *Emberiza calandra*
          Passeriformes - Emberizidae
e   Corn Bunting
f   Bruant proyer
d   Grauammer
i   Strillozzo

3455 *Emberiza capensis*
          Passeriformes - Emberizidae
e   Cape Bunting; Southern Rock-
      Bunting
f   Bruant du Cap
d   Kap-Ammer
i   Zigolo dal Capo

3456 *Emberiza chrysophrys*
          Passeriformes - Emberizidae
e   Yellow-browed Bunting
f   Bruant à sourcils jaunes
d   Gelbbrauenammer; Prachtammer
i   Zigolo dai sopraccigli gialli

3457 *Emberiza cia*
          Passeriformes - Emberizidae
e   Rock Bunting; Eurasian Rock-
      Bunting; European Rock-Bunting
f   Bruant fou; Bruant passager
d   Zippammer; Bergammer
i   Zigolo muciatto

3458 *Emberiza cineracea*
          Passeriformes - Emberizidae
e   Cinereous Bunting; Ashy-headed
      Bunting
f   Bruant cendré
d   Türkenammer; Kleinasiatische
      Ammer
i   Zigolo cenerino

3459 *Emberiza cioides*
          Passeriformes - Emberizidae
e   Meadow Bunting; Siberian Meadow
      Bunting; Long-tailed Bunting
f   Bruant à longue queue; Bruant des
      prés
d   Wiesenammer
i   Zigolo muciatto orientale

3460 *Emberiza cirlus*
          Passeriformes - Emberizidae
e   Cirl Bunting

f   Bruant zizi; Bruant des haies
d   Zaunammer
i   Zigolo nero; Zivolo comune

3461 *Emberiza citrinella*
          Passeriformes - Emberizidae
e   Yellowhammer; Yellow Bunting;
      Eastern Yellowhammer
f   Bruant jaune; Bruant des jardins
d   Goldammer
i   Zigolo giallo; Gialletto; Zivolo;
      Nizzola gialla; Settaiuolo

3462 *Emberiza elegans*
          Passeriformes - Emberizidae
e   Yellow-throated Bunting; Yellow-
      headed Bunting; Elegant Bunting
f   Bruant élégant
d   Schmuckammer
i   Zigolo testagialla

3463 *Emberiza flaviventris*
          Passeriformes - Emberizidae
e   African Golden-breasted Bunting;
      Golden-breasted Bunting (CSA);
      Yellow-bellied Bunting; Yellow-
      breasted Bunting
f   Bruant à poitrine dorée
d   Gelbbauchammer
i   Zigolo pettodorato

3464 *Emberiza fucata*
          Passeriformes - Emberizidae
e   Chestnut-eared Bunting; Grey-
      hooded Bunting; Grey-headed
      Bunting
f   Bruant à oreillons; Bruant à oreillons
      marron
d   Graukopfammer; Braunohrammer
i   Zigolo orecchiecastane

3465 *Emberiza godlewskii*
          Passeriformes - Emberizidae
e   Godlewski's Bunting; Godlewski's
      Rock-Bunting
f   Bruant de Godlewski
d   Felsenammer
i   Zigolo di Godlewski

3466 *Emberiza hortulana*
          Passeriformes - Emberizidae

   e   Ortolan Bunting; Ortolan
   f   Bruant ortolan
   d   Ortolan; Gartenammer
   i   Ortolano; Ortolano giallo

3467   ***Emberiza impetuani***
       Passeriformes - Emberizidae
   e   Lark-like Bunting; Pale Rock-Bunting
   f   Bruant des rochers
   d   Lerchenammer
   i   Zigolo allodola africano

3468   ***Emberiza intermedia***
       Passeriformes - Emberizidae
   e   Dark Reed-Bunting
   i   Zigolo

3469   ***Emberiza jankowskii***
       Passeriformes - Emberizidae
   e   Rufous-backed Bunting; Jankowski's Bunting
   f   Bruant de Jankowski
   d   Jankowski-Ammer; Bartammer
   i   Zigolo di Jankowski

3470   ***Emberiza koslowi***
       Passeriformes - Emberizidae
   e   Koslow's Bunting; Tibetan Bunting
   f   Bruant de Koslov
   d   Koslow-Ammer
   i   Zigolo di Koslow

3471   ***Emberiza leucocephalos***
       Passeriformes - Emberizidae
   e   Pine Bunting
   f   Bruant à calotte blanche
   d   Fichtenammer
   i   Zigolo golarossa

3472   ***Emberiza melanocephala***
       Passeriformes - Emberizidae
   e   Black-headed Bunting
   f   Bruant mélanocéphale; Bruant crocote; Bruant à tête noire
   d   Kappenammer
   i   Zigolo capinero; Zigolo testanera

3473   ***Emberiza pallasi***
       Passeriformes - Emberizidae
   e   Pallas's Bunting; Pallas's Reed-Bunting; Mongolian Bunting
   f   Bruant de Pallas
   d   Pallas-Ammer; Grauschulterammer; Grauschulterrohrammer
   i   Migliarino del Pallas; Migliarino polare; Zigolo del Pallas

3474   ***Emberiza poliopleura***
       Passeriformes - Emberizidae
   e   Somali Golden-breasted Bunting; Somali Bunting; Golden-breasted Bunting
   f   Bruant de Somalie
   d   Somali-Ammer
   i   Zigolo pettodorato della Somalia

3475   ***Emberiza pusilla***
       Passeriformes - Emberizidae
   e   Little Bunting
   f   Bruant nain
   d   Zwergammer
   i   Zigolo minore

3476   ***Emberiza pyrrhuloides***
       Passeriformes - Emberizidae
   e   Pale Reed-Bunting
   i   Passera di padule

3477   ***Emberiza rustica***
       Passeriformes - Emberizidae
   e   Rustic Bunting
   f   Bruant rustique
   d   Waldammer
   i   Zigolo boschereccio

3478   ***Emberiza rutila***
       Passeriformes - Emberizidae
   e   Chestnut Bunting; Rufous Bunting
   f   Bruant roux
   d   Rötelammer
   i   Zigolo rutilo

3479   ***Emberiza schoeniclus***
       Passeriformes - Emberizidae
   e   Reed-Bunting; Common Reed-Bunting (NA); Northern Reed-Bunting (NA)
   f   Bruant des roseaux
   d   Rohrammer
   i   Migliarino di palude; Miglarino

3480   *Emberiza socotrana*
Passeriformes - Emberizidae
e   Socotra Bunting; Socotra Mountain-Bunting
f   Bruant de Socotra
d   Sokotra-Ammer
i   Zigolo di Socotra

3481   *Emberiza spodocephala*
Passeriformes - Emberizidae
e   Black-faced Bunting; Masked Bunting; Grey-headed Black-faced Bunting
f   Bruant masqué; Bruant à masque noir
d   Maskenammer; Aschkopfammer
i   Zigolo mascherato

3482   *Emberiza stewarti*
Passeriformes - Emberizidae
e   Chestnut-breasted Bunting; White-capped Bunting; Stewart's Bunting
f   Bruant de Stewart
d   Silberkopfammer
i   Zigolo capobianco

3483   *Emberiza striolata*
Passeriformes - Emberizidae
e   House Bunting; Striolated Bunting (ISC); Striped House-Bunting; Striped Bunting
f   Bruant striolé
d   Hausammer; Streifenammer
i   Zigolo delle case; Zigolo testagrigia

3484   *Emberiza sulphurata*
Passeriformes - Emberizidae
e   Yellow Bunting; Japanese Yellow Bunting; Japanese Bunting
f   Bruant du Japon; Bruant jaune de Japon
d   Schwefelammer
i   Zigolo solforato giapponese

3485   *Emberiza tahapisi*
Passeriformes - Emberizidae
e   Cinnamon-breasted Bunting; Cinnamon-breasted Rock-Bunting (CSA); Rock Bunting (CSA); African Rock-Bunting
f   Bruant cannellé

d   Bergammer; Siebenstreifenammer
i   Zigolo pettocannella

3486   *Emberiza townsendii*
Passeriformes - Emberizidae
e   Townsend's Bunting

3487   *Emberiza tristrami*
Passeriformes - Emberizidae
e   Tristram's Bunting
f   Bruant de Tristram
d   Tristram-Ammer
i   Zigolo di Tristram

3488   *Emberiza variabilis*
Passeriformes - Emberizidae
e   Grey Bunting; Japanese Grey-Bunting
f   Bruant gris
d   Bambusammer
i   Zigolo grigio giapponese

3489   *Emberiza yessoensis*
Passeriformes - Emberizidae
e   Ochre-rumped Bunting; Japanese Reed-Bunting; Far-Eastern Reed-Bunting; Chinese Reed-Bunting; Japanese Bunting; Swinhoe's Bunting
f   Bruant de Yéso; Bruant des roseaux du Japon
d   Mandschuren-Ammer
i   Migliarino di palude del Giappone

3490   **Emberizidae**
Passeriformes
e   Buntings; Emberizids; American Sparrows
f   Bruants; Embérizidés
d   Ammern
i   Emberizidi; Zigoli

3491   *Emberizoides duidae*
Passeriformes - Fringillidae
e   Duida Grass-Finch; Mount Duida Grass-Finch
f   Tardivole du Duida
i   Zigolo delle erbe di Monte Duida

3492   *Emberizoides herbicola*
Passeriformes - Emberizidae
e   Wedge-tailed Grass-Finch; Wedge-

tailed Ground-Finch; Azara's Grass-Finch
d Keilsschwanzammer
i Zigolo; Zigolo delle erbe codacuneata

3493 *Emberizoides ypiranganus*
Passeriformes - Emberizidae
e Grey-cheeked Grass-Finch; Lesser Grass-Finch; Lesser Wedge-tailed Ground-Finch
d Ypiranga-Ammer
i Zigolo delle erbe minore

3494 *Embernagra longicauda*
Passeriformes - Fringillidae
e Pale-throated Pampa-Finch; Pale-throated Sierra-Finch; Buff-throated Pampa-Finch
f Embernagre du Brésil
d Langschwanzammer
i Fringuello golafulva della Pampa

3495 *Embernagra olivascens*
Passeriformes - Fringillidae
e Olivaceous Pampa-Finch; Olive Pampa-Finch

3496 *Embernagra platensis*
Passeriformes - Fringillidae
e Great Pampa-Finch; Red-billed Pampa-Finch; Olive Pampa-Finch
f Embernagre à cinq couleurs
d Pampaammer
i Fringuello maggiore della Pampa

3497 *Emblema pictum*
Passeriformes - Passeridae
e Painted Firetail; Painted Finch; Painted Firetail-Finch; Mountain-Finch
f Diamant peint
d Gemalter Astrild; Spinifexastrild
i Diamante variopinto

3498 *Eminla lepida*
Passeriformes - Cisticolidae
e Grey-capped Warbler
f Éminie à calotte grise
d Eminie
i Oriolino capogrigio

3499 *Empidonax affinis*
Passeriformes - Tyrannidae
e Pine Flycatcher
f Moucherolle des pins
d Kieferntyrann
i Tiranno dei pini

3500 *Empidonax albigularis*
Passeriformes - Tyrannidae
e White-throated Flycatcher
f Moucherolle à gorge blanche
d Fahlkehltyrann
i Tiranno golabianca

3501 *Empidonax alnorum*
Passeriformes - Tyrannidae
e Alder Flycatcher
f Moucherolle des aulnes
d Erlentyrann
i Tiranno degli ontani

3502 *Empidonax atriceps*
Passeriformes - Tyrannidae
e Black-capped Flycatcher
f Moucherolle à tête noire
d Kapuzentyrann
i Tiranno capinero

3503 *Empidonax difficilis*
Passeriformes - Tyrannidae
e Pacific-slope Flycatcher; Western Flycatcher; Coastal Flycatcher; Redwoods Flycatcher
f Moucherolle côtiere; Moucherolle obscure; Moucherolle du Pacifique
d Ufertyrann
i Tiranno del Pacifico

3504 *Empidonax flavescens*
Passeriformes - Tyrannidae
e Yellowish Flycatcher; Ponderosa Flycatcher; Interior Flycatcher
f Moucherolle jaunâtre
i Tiranno giallastro

3505 *Empidonax flaviventris*
Passeriformes - Tyrannidae
e Yellow-bellied Flycatcher
f Moucherolle à ventre jaune
d Birkentyrann
i Tiranno ventregiallo

3506 *Empidonax fulvifrons*
Passeriformes - Tyrannidae
e Buff-breasted Flycatcher
f Moucherolle beige; Moucherolle à
poitrine fauve
d Braunbrusttyrann
i Tiranno frontefulva

3507 *Empidonax griseipectus*
Passeriformes - Tyrannidae
e Grey-breasted Flycatcher
f Moucherolle à poitrine grise
d Graubrusttyrann
i Tiranno pettogrigio

3508 *Empidonax hammondii*
Passeriformes - Tyrannidae
e Hammond's Flycatcher
f Moucherolle de Hammond
d Tannentyrann
i Tiranno di Hammond

3509 *Empidonax minimus*
Passeriformes - Tyrannidae
e Least Flycatcher
f Moucherolle tchébec
d Gartentyrann
i Tiranno minimo

3510 *Empidonax oberholseri*
Passeriformes - Tyrannidae
e Dusky Flycatcher; Wright's
Flycatcher; Oberholser's Flycatcher;
American Dusky Flycatcher
f Moucherolle sombre
d Buschtyrann
i Tiranno di Oberholser

3511 *Empidonax occidentalis*
Passeriformes - Tyrannidae
e Cordilleran Flycatcher; Wetern
Flycatcher
f Moucherolle des ravins
i Tiranno occidentale

3512 *Empidonax traillii*
Passeriformes - Tyrannidae
e Willow Flycatcher; Traill's
Flycatcher
f Moucherolle des saules; Moucherolle
des aulnes

d Weidentyrann
i Tiranno di Traill

3513 *Empidonax virescens*
Passeriformes - Tyrannidae
e Acadian Flycatcher
f Moucherolle verte; Moucherolle
d'Acadie
d Buchentyrann; Grünlicher
Erlentyrann
i Pigliamosche acadico; Tiranno
acadico

3514 *Empidonax wrightii*
Passeriformes - Tyrannidae
e Grey Flycatcher; Wright's Flycatcher
f Moucherolle grise
d Beifußtyrann
i Tiranno di Wright

3515 *Empidonomus varius*
Passeriformes - Tyrannidae
e Variegated Flycatcher; Varied
Flycatcher
f Tyran tacheté
d Fleckentyrann
i Tiranno variegato

3516 *Empidornis semipartitus*
Passeriformes - Muscicapidae
e Silverbird
f Gobemouche argenté; Gobe-mouches
argenté
d Silberschnäpper
i Uccello d'argento

3517 *Enicognathus ferrugineus*
Psittaciformes - Psittacidae
e Austral Parakeet; Austral Conure;
Magellan Conure; Chilean Conure
f Conure magellanique
d Smaragdsittich
i Conuro di Magellano

3518 *Enicognathus leptorhynchus*
Psittaciformes - Psittacidae
e Slender-billed Parakeet; Slender-
billed Conure; Slender-billed
Cockatoo
f Conure à long bec

d Langschnabelsittich
i Canuro beccosottile

3519 *Enicurus immaculatus*
Passeriformes - Muscicapidae
e Black-backed Forktail
f Énicure à dos noir
d Schwarzrückenscherenschwanz
i Codaforcuta dorsonero

3520 *Enicurus leschenaulti*
Passeriformes - Muscicapidae
e White-crowned Forktail;
Leschenault's Forktail
f Énicure de Leschenault
d Weißscheitelscherenschwanz
i Codaforcuta corona bianca

3521 *Enicurus maculatus*
Passeriformes - Muscicapidae
e Spotted Forktail
f Énicure tacheté
d Fleckenscherenschwanz
i Codaforcuta maculato

3522 *Enicurus ruficapillus*
Passeriformes - Muscicapidae
e Chestnut-naped Forktail; Chestnut-
backed Forktail
f Énicure rousse-cape
d Rotkopfscherenschwanz
i Codaforcuta nucarossa

3523 *Enicurus schistaceus*
Passeriformes - Muscicapidae
e Slaty-backed Forktail
f Énicure ardoisé
d Graurückenscherenschwanz
i Codaforcuta dorsoardesia

3524 *Enicurus scouleri*
Passeriformes - Muscicapidae
e Little Forktail
f Énicure nain
d Stummelscherenschwanz
i Codaforcuta minore

3525 *Enicurus velatus*
Passeriformes - Muscicapidae
e Sunda Forktail; Lesser Forktail
f Énicure voile

d Zwergscherenschwanz
i Codaforcuta della Sonda

3526 *Enodes erythrophris*
Passeriformes - Sturnidae
e Fiery-browed Mynah; Celebes
Enodes-Starling; Fiery-browed
Enodes-Starling; Fiery-browed
Starling; Menando Starling; Celebes
Mynah
f Énode à sourcils rouges
d Rotbrauenstar
i Storno orecchierosse

3527 *Ensifera ensifera*
Apodiformes - Trochilidae
e Sword-billed Hummingbird;
Swordbill
f Colibri porte-epée
d Schwertschnabel
i Colibrì becco a spada

3528 *Entomodestes coracinus*
Passeriformes - Muscicapidae
e Black Solitaire
f Solitaire noir
d Mohrenklarino
i Tordo solitario nero

3529 *Entomodestes leucotis*
Passeriformes - Muscicapidae
e White-eared Solitaire
f Solitaire oreillard
d Dreifarbenklarino
i Tordo solitario orecchiebianchi

3530 *Entomyzon cyanotis*
Passeriformes - Meliphagidae
e Blue-faced Honeyeater
f Méliphage à oreillons bleus
d Blauohr; Blauohrhonigesser
i Magiamiele facciazzura

3531 *Eophona migratoria*
Passeriformes - Fringillidae
e Yellow-billed Grosbeak; Black-tailed
Hawfinch; Chinese Grosbeak;
Chinese Hawfinch
f Grosbec migrateur
d Schwarzschwanzkernbeißer;

Weißbandkernbeißer
i   Frosone codanera

3532   ***Eophona personata***
Passeriformes - Fringillidae
e   Japanese Grosbeak; Japanese
Hawfinch; Masked Hawfinch
f   Grosbec masqué
d   Maskenkernbeißer
i   Frosone mascherato

3533   ***Eopsaltria australis***
Passeriformes - Petroicidae
e   Eastern Yellow-Robin; Yellow
Robin; Southern Yellow-Robin
f   Miro à poitrine jaune
d   Goldbauchschnäpper
i   Pigliamosche giallo australiano

3534   ***Eopsaltria flaviventris***
Passeriformes - Petroicidae
e   Yellow-bellied Robin
f   Miro à ventre jaune
d   Gelbbauchschnäpper
i   Pigliamosche ventregiallo della
Nuova Caledonia

3535   ***Eopsaltria georgiana***
Passeriformes - Petroicidae
e   White-breasted Robin; White-
breasted Yellow-Robin; Australian
White-breasted Robin
f   Miro à poitrine blanche
d   Georg-Schnäpper
i   Pigliamosche pettobianco australiano

3536   ***Eopsaltria griseogularis***
Passeriformes - Petroicidae
e   Grey-breasted Yellow-Robin;
Western Yellow-Robin; Grey-
breasted Robin
f   Miro à poitrine grise
d   Graumantelschnäpper
i   Pigliamosche pettogrigio australiano

3537   ***Eopsaltria pulverulenta***
Passeriformes - Petroicidae
e   Mangrove Robin; Mangrove
Flycatcher; Australian Mangrove
Robin
f   Miro des mangroves

d   Mangroveschnäpper
i   Pigliamosche australiano delle
mangrovie

3538   ***Eopsaltriidae***
Passeriformes
e   Australo-Papuan Robins; Scrub-
Robins; Australian Robins
f   Eopsaltriidés
d   Südseeschnäpper; Südseesänger
i   Epsosaltridi

3539   ***Eos bornea***
Psittaciformes - Loriidae
e   Red Lory; Buru Red-Lory;
Goodfellow's Lory
f   Lori écarlate
d   Rotlori
i   Lori rosso

3540   ***Eos cyanogenia***
Psittaciformes - Loriidae
e   Black-winged Lory; Blue-cheeked
Lory; Biak Red-Lory
f   Lori à joues bleues
d   Schwarzschulterlori
i   Lori alinere

3541   ***Eos histrio***
Psittaciformes - Loriidae
e   Red-and-blue Lory; Blue-diademed
Lory; Blue-tailed Lory
f   Lori arlequin
d   Harlekinlori
i   Lori rosso e blu

3542   ***Eos reticulata***
Psittaciformes - Loriidae
e   Blue-streaked Lory; Blue-tailed Lory
f   Lori reticulé
d   Strichellori
i   Lori reticulato

3543   ***Eos semilarvata***
Psittaciformes - Loriidae
e   Blue-eared Lory; Hallf-masked Lory;
Ceram Lory
f   Lori masqué
d   Halbmaskenlori
i   Lori orecchieblu

3544   *Eos squamata*
       Psittaciformes - Loriidae
e   Violet-necked Lory; Violet-naped
       Lory; Moluccan Red-Lory
f   Lori écaillé
d   Kapuzenlori
i   Lori colloviola

3545   *Ephippiorhynchus asiaticus*
       Ciconiiformes - Ciconiidae
e   Black-necked Stork; Green-necked
       Stork; Jabiru
f   Jabiru d'Asie; Jabiru asiatique
d   Riesenstorch
i   Cicogna indiana; Becco a stelo
       asiatico; Jabiru asiatico; Becco a sella
       asiatico

3546   *Ephippiorhynchus senegalensis*
       Ciconiiformes - Ciconiidae
e   Saddle-billed Stork; Saddle-billed
       Jabiru; Saddlebill Stork; African
       Jabiru
f   Jabiru d'Afrique; Jabiru du Sénégal
d   Sattelstorch
i   Mitteria del Senegal; Cigogna dal
       becco a sella; Becco a sella africano;
       Jabirudel Senegal

3547   *Epimachus albertisi*
       Passeriformes - Corvidae
e   Black-billed Sicklebill; Albertis's
       Bird of Paradise; Buff-tailed
       Sicklebill; Black-billed Bird of
       Paradise; Red Sicklebill; Black-billed
       Sicklebill Bird of Paradise
f   Paradisier d'Albertis
d   Gelbschwanzsichelschnabel
i   Paradisea di D'Albertis

3548   *Epimachus bruijnii*
       Passeriformes - Corvidae
e   Pale-billed Sicklebill; Bruijn's Bird
       of Paradise; White-billed Sicklebill
       Bird of Paradise; White-billed
       Sickle-billed Bird of Paradise;
       Lowland Sicklebill
f   Paradisier à bec blanc
d   Weißsichelschnabel;
       Braunschwanzsichelhopf
i   Paradisea di Bruijn

3549   *Epimachus fastuosus*
       Passeriformes - Corvidae
e   Black Sicklebill; Black Sicklebill
       Bird of Paradise; Greater Sicklebill
f   Paradisier fastueux
d   Breitschwanzsichelhopf; Roter
       Sichelschnabel
i   Paradisea fastosa; Epimaco dal becco
       a falce; Epimaco grande

3550   *Epimachus meyeri*
       Passeriformes - Corvidae
e   Brown Sicklebill; Brown Sicklebill
       Bird of Paradise; Meyer's Sicklebill
f   Paradisier de Meyer
d   Meyer-Sichelschnabel;
       Schmalschwanzsichelhopf
i   Paradisea di Meyer

3551   *Epthianura albifrons*
       Passeriformes - Meliphagidae
e   White-fronted Chat; Australian Chat;
       White-faced Chat
f   Epthianure à front blanc
d   Kurzschwanztrugschmätzer;
       Honigfresser
i   Sassicola australiana facciabianca

3552   *Epthianura aurifrons*
       Passeriformes - Meliphagidae
e   Orange Chat
f   Epthianure orangée
d   Goldstirntrugschmätzer
i   Sassicola australiana aranciata

3553   *Epthianura crocea*
       Passeriformes - Meliphagidae
e   Yellow Chat
f   Epthianure à collier
d   Safrantrugschmätzer
i   Sassicola australiana gialla

3554   *Epthianura crocea macgregori*
       Passeriformes - Meliphagidae
e   Dawson Yellow-Chat (ANZ)

3555   *Epthianura crocea tunneyi*
       Passeriformes - Meliphagidae
e   Alligator Rivers Yellow-Chat (ANZ)

3556   *Epthianura tricolor*
Passeriformes - Meliphagidae
e   Crimson Chat
f   Epthianure tricolore
d   Scharlachtrugschmätzer;
Scharlachsänger;
Scharlachepthianura
i   Sassicola australiana ventrerosso

3557   *Eremalauda dunni*
Passeriformes - Alaudidae
e   Dunn's Lark
f   Alouette de Dunn; Ammomane de
Dunn
d   Einödlerche; Kleine
Schwarzschwanzsandlerche
i   Allodola di Dunn

3558   *Eremalauda starki*
Passeriformes - Alaudidae
e   Stark's Lark; Stark's Short-toed Lark
f   Alouette de Stark
d   Starks Kurzhaubenlerche; Stark-
Kurzhaubenlerche; Falblerche
i   Allodola di Stark

3559   *Eremiornis carteri*
Passeriformes - Sylviidae
e   Spinifexbird
f   Mégalure du spinifex
d   Spinifexsänger
i   Uccellino degli spinifex

3560   *Eremobius phoenicurus*
Passeriformes - Furnariidae
e   Band-tailed Earthcreeper
f   Annumbi rougequeue
d   Dornschlüpfer
i   Rampichino terriccola

3561   *Eremomela atricollis*
Passeriformes - Sylviidae
e   Black-necked Eremomela; Black-
collared Eremomela
f   Érémomèle à cou noir; Érémomèle à
collier noir
d   Goldkehleremomela
i   Eremomela collonero

3562   *Eremomela badiceps*
Passeriformes - Sylviidae

e   Rufous-crowned Eremomela; Brown-
crowned Eremomela; Black-crowned
Eremomela
f   Érémomèle à tête brune
d   Rotkopferemomela
i   Eremomela corona bruna

3563   *Eremomela canescens*
Passeriformes - Sylviidae
e   Green-backed Eremomela
f   Érémomèle grisonnante
i   Eremomela dorsoverde

3564   *Eremomela flavicrissalis*
Passeriformes - Sylviidae
e   Yellow-vented Eremomela
f   Érémomèle à ventre jaune
d   Somali-Eremomela
i   Eremomela dl sottocoda giallo

3565   *Eremomela gregalis*
Passeriformes - Sylviidae
e   Yellow-rumped Eremomela; Karoo
Eremomela (CSA); Karoo Green
Warbler; Green-backed Eremomela;
Yellow-rumped Eremomela
f   Érémomèle du Karroo
d   Langschwanzeremomela
i   Eremomela dal groppone giallo

3566   *Eremomela icteropygialis*
Passeriformes - Sylviidae
e   Yellow-bellied Eremomela; Yellow-
backed Eremomela
f   Érémomèle à croupion jaune;
Érémomèle gris-jaune
d   Gelbbaucheremomela
i   Eremomela ventregiallo

3567   *Eremomela pusilla*
Passeriformes - Sylviidae
e   Senegal Eremomela; Green-backed
Eremomela; Smaller Green-backed
Eremomela
f   Érémomèle à dos vert
d   Graukappeneremomela
i   Eremomela piccola

3568   *Eremomela salvadorii*
Passeriformes - Sylviidae
e   Salvadori's Eremomela

f　Érémomèle de Salvadori
i　Eremomela di Salvador

3569　***Eremomela scotops***
　　　Passeriformes - Sylviidae
e　Greencap Eremomela; Green-capped
　　　Eremomela (CSA)
f　Érémomèle à calotte verte;
　　　Érémomèle à tête verte
d　Grünkappeneremomela
i　Eremomela capoverde

3570　***Eremomela turneri***
　　　Passeriformes - Sylviidae
e　Turner's Eremomela; Brown-
　　　crowned Eremomela
f　Érémomèle de Turner
d　Braunstirneremomela
i　Eremomela di Turner

3571　***Eremomela usticollis***
　　　Passeriformes - Sylviidae
e　Burnt-neck Eremomela; Burnt-
　　　necked Eremomela
f　Érémomèle à cou roux
d　Rostkehleremomela;
　　　Rostbanderemomela
i　Eremomela colloscuro

3572　***Eremophila alpestris***
　　　Passeriformes - Alaudidae
e　Horned Lark; Shore Lark;
　　　Przewalski's Lark; Shore Horned-
　　　Lark
f　Alouette haussecol; Alouette cornue;
　　　Alouette de Przewalski; Alouette
　　　oreillard; Alouette de la Sibérie
d　Ohrenlerche
i　Allodola golagialla

3573　***Eremophila bilopha***
　　　Passeriformes - Alaudidae
e　Temminck's Lark; Temminck's
　　　Horned-Lark
f　Alouette bilophe; Alouette hausse-col
　　　du désert; .
d　Sahara-Ohrenlerche; Hornlerche;
　　　Afrikanische Ohrenlerche
i　Allodola di Temminck

3574　***Eremopterix australis***
　　　Passeriformes - Alaudidae
e　Black-eared Sparrow-Lark; Black-
　　　eared Finch-Lark (CSA); Grey-
　　　backed Sparrow-Lark
f　Moinelette à oreillons noirs;
　　　Alouette-moineau à oreillons noirs
d　Schwarzwangenlerche
i　Allodola australe

3575　***Eremopterix grisea***
　　　Passeriformes - Alaudidae
e　Ashy-crowned Sparrow-Lark; Black-
　　　bellied Finch-Lark (ISC); Ceylon
　　　Finch; Ceylon Finch-Lark; Ashy-
　　　crowned Finch-Lark; Black-bellied
　　　Sparrow-Lark
f　Moinelette croisée; Alouette-
　　　moineau croisée
d　Grauscheitellerche
i　Allodola capocenerino

3576　***Eremopterix leucopareia***
　　　Passeriformes - Alaudidae
e　Fischer's Sparrow-Lark; Fischer's
　　　Finch-Lark
f　Moinelette de Fischer; Alouette-
　　　moineau de Fischer
d　Braunscheitelfinkenlerche;
　　　Braunscheitellerche
i　Allodola di Fischer

3577　***Eremopterix leucotis***
　　　Passeriformes - Alaudidae
e　Chestnut-backed Sparrow-Lark;
　　　Chestnut-backed Finch-Lark (CSA);
　　　White-cheeked Sparrow-Lark
f　Moinelette à oreillons blancs;
　　　Alouette-moineau à oreillons blancs
d　Weißwangenlerche
i　Allodola orecchiebianche

3578　***Eremopterix nigriceps***
　　　Passeriformes - Alaudidae
e　Black-crowned Sparrow-Lark;
　　　Black-crowned Finch-Lark; White-
　　　fronted Finch-Lark; White-crested
　　　Sparrow-Lark
f　Moinelette à front blanc; Alouette-
　　　moineau à front blanc
d　Weißstirnlerche;

Weißstirngimpellerche
i Allodola passerina capinera; Allodola
fringuello

3579 *Eremopterix signata*
Passeriformes - Alaudidae
e Chestnut-headed Sparrow-Lark;
Chestnut-headed Finch-Lark
f Moinelette d'Oustalet; Alouette-
moineau d'Oustalet
d Harlekinlerche
i Allodola testacastana

3580 *Eremopterix verticalis*
Passeriformes - Alaudidae
e Grey-backed Sparrow-Lark; Grey-
backed Finch-Lark
f Moinelette à dos gris; Alouette-
moineau à dos gris
d Nonnenlerche
i Allodola dorsogrigio

3581 *Ergaticus ruber*
Passeriformes - Fringillidae
e Red Warbler
f Paruline rouge
d Rotwaldsänger; Karminwaldsänger;
Purpurwaldsänger
i Parula rossa

3582 *Ergaticus versicolor*
Passeriformes - Fringillidae
e Pink-headed Warbler
f Paruline à tête rose
d Rosenwaldsänger; Rosenkopfsänger
i Parula testarossa

3583 *Eriocnemis alinae*
Apodiformes - Trochilidae
e Emerald-bellied Puffleg
f Érione d'Aline
d Smaragdschneehöschen
i Colibrì zampepiumose pettosmeraldo

3584 *Eriocnemis cupreoventris*
Apodiformes - Trochilidae
e Coppery-bellied Puffleg; Copper-
vented Puffleg
f Érione à ventre cuivre
d Kupferbauchschneehöschen
i Colibrì zampepiumose ventreramato

3585 *Eriocnemis derbyi*
Apodiformes - Trochilidae
e Black-thighed Puffleg; Derby's
Puffleg
f Érione de Derby
d Schwarzhöschen
i Colibrì zampepiumose di Derby

3586 *Eriocnemis glaucopoides*
Apodiformes - Trochilidae
e Blue-capped Puffleg; D'Orbigny's
Puffleg
f Érione à front bleu
d Blaukappenschneehöschen
i Colibrì zampepiumose corona blu

3587 *Eriocnemis godini*
Apodiformes - Trochilidae
e Turquoise-throated Puffleg; Godin's
Puffleg
f Érione turquoise
d Türkis Schneehöschen
i Colibrì zampepiumose gola turchese

3588 *Eriocnemis luciani*
Apodiformes - Trochilidae
e Sapphire-vented Puffleg
f Érione catherine
d Blaustirnschneehöschen
i Colibrì zampepiumose coda di
zaffiro

3589 *Eriocnemis mirabilis*
Apodiformes - Trochilidae
e Colourful Puffleg
f Érione multicolore
d Weißohrschneehöschen
i Colibrì zampepiumose multicolore

3590 *Eriocnemis mosquera*
Apodiformes - Trochilidae
e Golden-breasted Puffleg; Mosquera's
Puffleg
f Érione à poitrine d'or
d Goldbrustschneehöschen
i Colibrì zampepiumose pettodorato

3591 *Eriocnemis nigrivestis*
Apodiformes - Trochilidae
e Black-breasted Puffleg
f Érione à robe noire

d Schwarzbrustschneehöschen
i Colibrì zampepiumose pettonero

3592 Eriocnemis vestitus
Apodiformes - Trochilidae
e Glowing Puffleg
f Érione pattue
d Bronzeschneehöschen
i Colibrì zampepiumose smagliante

3593 Erithacus komadori
Passeriformes - Muscicapidae
e Ryukyu Robin; Ryukyu Bush-Robin;
Korean Robin; Temminck's Robin
f Rossignol komadori
d Samtkehlnachtigal; Samtkehlchen
i Pettirosso di Ryukyu

3594 Erithacus rubecula
Passeriformes - Muscicapidae
e European Robin (NA); Robin;
Eurasian Robin; Redbreast-Robin;
Robin Redbreast
f Rougegorge; Rougegorge familier
d Rotkehlchen
i Pettirosso; Pettirosso europeo

3595 Erythrocercus holochlorus
Passeriformes - Corvidae
e Yellow Flycatcher; Little Yellow-
Flycatcher (CSA)
f Érythrocerque jaune; Petit
Gobemouche jaune
d Goldrückenspreizschwanz;
Gelbschnäpper; Zwerggelbschnäpper
i Pigliamosche codirosso verde

3596 Erythrocercus livingstonei
Passeriformes - Corvidae
e Livingstone's Flycatcher;
Livingstone's Yellow-Flycatcher;
Livingstone's Monarch
f Érythrocerque de Livingstone;
Gobemouche de Livingstone
d Livingstones Rotschwanzschnäpper;
Gelbbauchspreizschwanz;
Livingstone-Schnäpper;
Zwergrotschwanzschnäpper
i Pigliamosche codirosso di
Livingstone

3597 Erythrocercus mccallii
Passeriformes - Corvidae
e Chestnut-capped Flycatcher;
Chestnut-capped Monarch
f Érythrocerque à tête rousse;
Gobemouche à tête rousse; Gobe-
mouches à tête rousse
d Braunkappenschnäpper;
Rotkappenspreizschwanz
i Pigliamosche codirosso capocastano

3598 Erythrogonys cinctus
Charadriiformes - Charadriidae
e Red-kneed Dotterel
f Pluvier ceinturé
d Schwarzbrustregenpfeifer
i Corriere dalle ginocchia rossa

3599 Erythrotriorchis buergersi
Falconiformes - Accipitridae
e Chestnut-shouldered Goshawk;
Buerger's Goshawk; Buerger's
Sparrowhawk; Bürger's Goshawk;
Bürger's Sparrowhawk
f Autour de Bürger
d Prachthabicht
i Astore di Bürger

3600 Erythrotriorchis radiatus
Falconiformes - Accipitridae
e Red Goshawk; Red-legged Goshawk;
Rufous-bellied Buzzard; Red
Buzzard
f Autour rouge
d Australhabicht; Fuchshabicht
i Astore rosso; Astore fulvo

3601 Erythrura coloria
Passeriformes - Passeridae
e Red-eared Parrotfinch; Many-
coloured Parrotfinch; Mindanao
Parrotfinch; Red-collared Parrotfinch
f Diamant de Mindanao; Diamant des
montagnes
d Buntkopfpapageiamadine;
Rotohrpapageiamadine; Vielfarbige
Papageiamadine;
Bergpapageiamadine
i Diamante del Monte Katangland

3602   *Erythrura cyaneovirens*
        Passeriformes - Passeridae
  e   Red-headed Parrotfinch; Samoa
        Parrotfinch; Red-capped Parrotfinch;
        Royal Parrotfinch
  f   Diamant vert-bleu; Diamant de
        Samoa
  d   Kurzschwanzpapageiamadine
  i   Diamante testarossa

3603   *Erythrura gouldiae*
        Passeriformes - Passeridae
  e   Gouldian Finch; Rainbow Finch;
        Painted Finch; Purple-breasted Finch
  f   Diamant de Gould
  d   Gould-Amadine
  i   Diamante di Gould; Diamante della
        signora Gould

3604   *Erythrura hyperythra*
        Passeriformes - Passeridae
  e   Tawny-breasted Parrotfinch; Green-
        tailed Parrotfinch; Stock-Parrotfinch;
        Stock-Munia
  f   Diamant à queue verte
  d   Bambuspapageiamadine;
        Grünschwaénzige Papageiamadine
  i   Diamante dei bambù

3605   *Erythrura kleinschmidti*
        Passeriformes - Passeridae
  e   Pink-billed Parrotfinch; Black-faced
        Parrotfinch
  f   Diamant à bec rose; Diamant de
        Kleinschmidt
  d   Schwarzstirnpapageiamadine;
        Kleinschmidts Papageiamadine
  i   Diamante di Kleinschmidt

3606   *Erythrura papuana*
        Passeriformes - Passeridae
  e   Papuan Parrotfinch; Large-tailed
        Parrotfinch
  f   Diamant de Nouvelle-Guinée;
        Diamant de Papua
  d   Papua-Papageiamadine
  i   Diamante papua

3607   *Erythrura pealii*
        Passeriformes - Passeridae
  e   Fiji Parrotfinch; Peale's Parrotfinch

  f   Diamant de Peale
  d   Peale-Papageiamadine
  i   Diamante delle Isole Figi

3608   *Erythrura prasina*
        Passeriformes - Passeridae
  e   Pin-tailed Parrotfinch; Long-tailed
        Munia
  f   Diamant quadricolore; Pape des
        prairies; Quadricolore
  d   Vierfarbenpapageiamadine;
        Lauchgrüne Papageiamadine
  i   Diamante quadricolore

3609   *Erythrura psittacea*
        Passeriformes - Passeridae
  e   Red-throated Parrotfinch; Red-
        headed Parrotfinch; Parrotfinch
  f   Diamant psittaculaire; Diamant à tête
        rouge; Pape de Nouméa
  d   Rotköpfige Papageiamadine;
        Rotkopfpapageiamadine
  i   Diamante papagallo

3610   *Erythrura regia*
        Passeriformes - Passeridae
  e   Royal Parrotfinch; Blue-bellied
        Parrotfinch
  f   Diamant royal
  d   Königspapageiamadine
  i   Diamante reale

3611   *Erythrura trichroa*
        Passeriformes - Passeridae
  e   Blue-faced Parrotfinch; Blue-headed
        Parrotfinch
  f   Diamant de Kittlitz; Diamant
        tricolore de Kittlitz
  d   Dreifarbige Papageiamadine;
        Dreifarbenpapageiamadine
  i   Diamante facciablu

3612   *Erythrura tricolor*
        Passeriformes - Passeridae
  e   Tricoloured Parrotfinch; Three-
        coloured Parrotfinch; Tanimbar
        Parrotfinch; Timor Parrotfinch;
        Forbes's Parrotfinch; Blue-breasted
        Parrotfinch
  f   Diamant azuvert; Diamant de
        Tanimbar

d Forbes-Papageiamadine
i Lemuresthes

3613 ***Erythrura viridifacies***
Passeriformes - Passeridae
e Green-faced Parrotfinch; Green
Parrotfinch; Manilla Parrotfinch
f Diamant de Luçon; Diamant à tête
verte; Pape de Manille
d Manila-Papageiamadine
i Diamante facciaverde

3614 ***Estrilda astrild***
Passeriformes - Estrildidae
e Common Waxbill; Waxbill; Barred
Waxbill; St. Helena Waxbill; Brown
Waxbill; Astrild
f Astrild ondulé; Bec de corail ondulé;
Astrild de Ste.-Hélène; Astrild bec-
de-corail
d Wellenastrild
i Astrilde; Astrilde comune; Estrilda

3615 ***Estrilda atricapilla***
Passeriformes - Estrildidae
e Black-headed Waxbill; Grey-
breasted Waxbill
f Astrild à tête noire; Capucin à tête
noire
d Kappenastrild; Schwarzkappenastrild
i Astrilde testanera

3616 ***Estrilda caerulescens***
Passeriformes - Estrildidae
e Lavender Waxbill; Red-tailed
Lavender-Waxbill; Lavender Fire-
Finch; Bluish Waxbill; Lavender
Finch; Red-tailed Lavender-Finch
f Astrild queue-de-vinaigre; Astrild
lavande; Bengali gris-bleu; Astrild
gris-bleu
d Schönbürzelchen;
Rotschwanzschönbürzel
i Astrilde coda di aceto

3617 ***Estrilda charmosyna***
Passeriformes - Estrildidae
e Red-rumped Waxbill; Pink-bellied
Waxbill; Black-cheeked Waxbill;
Pink-bellied Black-cheeked Waxbill;
Pink-bellied lack-faced Waxbill;

Charmosyn Waxbill
f Astrild des fées
d Feenastrild
i Astrilde ventrerosato

3618 ***Estrilda erythronotos***
Passeriformes - Estrildidae
e Black-cheeked Waxbill; Black-faced
Waxbill (CSA)
f Astrild à moustaches; Astrild à joues
noires; Astrild à moustache noire;
Astrild à face noire
d Elfenastrild
i Astrilde guancenere

3619 ***Estrilda kandti***
Passeriformes - Estrildidae
e Kandt's Waxbill
f Astrild de Kandt
i Astrilde di Kandt

3620 ***Estrilda melanotis***
Passeriformes - Estrildidae
e Swee Waxbill; Yellow-bellied
Waxbill; Dufresnaye's Waxbill
f Astrild à joues noires; Astrild de
Dufresne; Astrild vert; Astrild joue
noire; Astrild à ventre fauve
d Gelbbauchastrild; Angola-
Schwarzbäckchen; Schwarzbäckchen
i Astrilde guancenere

3621 ***Estrilda melpoda***
Passeriformes - Estrildidae
e Orange-cheeked Waxbill; Red-
cheeked Waxbill
f Astrild à joues orangé; Bengali à
joues orangées; Joues-orangées;
Astrild à joues oranges
d Orangebäckchen
i Astrilde guancearancio

3622 ***Estrilda nigriloris***
Passeriformes - Estrildidae
e Black-faced Waxbill; Black-lored
Waxbill; Kiabo Waxbill
f Astrild à masque noir
i Astrilde faccianera

3623 ***Estrilda nonnula***
Passeriformes - Estrildidae

e   Black-crowned Waxbill; Black-
    capped Waxbill; White-breasted
    Waxbill; Blackcap Waxbill
f   Astrild nonnette; Astrild à cape noire
d   Nonnenastrild
i   Astrilde nonnetta

3624   *Estrilda ochrogaster*
    Passeriformes - Estrildidae
e   Abyssinian Waxbill; Abyssinian
    Fawn-breasted Waxbill
f   Astrild abyssinien
d   Ockerastrild
i   Astrilde dell'Abissinia

3625   *Estrilda paludicola*
    Passeriformes - Estrildidae
e   Fawn-breasted Waxbill; Buff-bellied
    Waxbill; Marsh Waxbill
f   Astrild à poitrine fauve; Astrild des
    marais
d   Sumpfastrild
i   Astrilde pettocastano

3626   *Estrilda perreini*
    Passeriformes - Estrildidae
e   Black-tailed Waxbill; Grey Waxbill
    (CSA); Black-tailed Lavender-
    Waxbill; Black-tailed Lavender-
    Finch; Black-tailed Finch; Black-
    tailed Grey-Waxbill
f   Astrild à queue noire; Astrild de
    Perrein; Astrild lavande à queue
    noire
d   Schwarzschwanzschönburstel
i   Astrilde grigio-blu

3627   *Estrilda poliopareia*
    Passeriformes - Estrildidae
e   Anambra Waxbill; Grey-cheeked
    Waxbill
f   Astrild du Niger; astrild de Anambra
d   Anambra-Astrild
i   Astrilde del Anambra

3628   *Estrilda quartinia*
    Passeriformes - Estrildidae
e   Yellow-bellied Waxbill; East African
    Swee (CSA); East African Swee-
    Waxbill; Grey-headed Waxbill
f   Astrild à ventre jaune

d   Grünastrild
i   Astrilde del'Africa Orientale

3629   *Estrilda rhodopyga*
    Passeriformes - Estrildidae
e   Crimson-rumped Waxbill; Rosy-
    rumped Waxbill; Sundevall's
    Waxbill; Rosy-winged Waxbill
f   Astrild à croupion rose; Astrild à dos
    rouge
d   Zügelastrild
i   Astrilde gropponerosso

3630   *Estrilda rufibarba*
    Passeriformes - Estrildidae
e   Arabian Waxbill
f   Astrild barbe-rousse; Astrild du
    Yemen; Astrild d'Arabie
d   Jemen-Astrild
i   Astrilde d'Arabia

3631   *Estrilda thomensis*
    Passeriformes - Estrildidae
e   Cinderella Waxbill; Sao Tomé
    Waxbill; Neumann's Waxbill; Red-
    flanked Lavender Waxbill
f   Astrild de Sao Tomé
d   Cinderella-Schönbürstel
i   Astrilde cenerina

3632   *Estrilda troglodytes*
    Passeriformes - Estrildidae
e   Black-rumped Waxbill; Red-eared
    Waxbill; Grey Waxbill; Pink-
    cheeked Waxbill
f   Astrild cendré; Astrild gris; Astrild
    gris ordinaire; Bec-en-corail cendré;
    Astrild cendrillon
d   Grauastrild
i   Astrilde becco di corallo

3633   *Estrildidae*
    Passeriformes
e   Estrildid Finches; Estrildids
f   Estrildidés; Astrilds
d   Prachtfinken
i   Estrildidi; Astrildi

3634   *Eubucco bourcierii*
    Piciformes - Ramphastidae
e   Red-headed Barbet

f Cabézon de tête rouge
d Anden-Bartvogel; Rotkehlbartvogel
i Barbuto di Bourcier

3635 *Eubucco richardsoni*
Piciformes - Ramphastidae
e Lemon-throated Barbet
f Cabézon à poitrine d'or
d Goldbrustbartvogel
i Barbuto golagialla

3636 *Eubucco tucinkae*
Piciformes - Ramphastidae
e Scarlet-hooded Barbet
f Cabézon de Carabaya
d Carbaya-Bartvogel
i Barbuto di Tucinka

3637 *Eubucco versicolor*
Piciformes - Ramphastidae
e Versicoloured Barbet; Variegated
Barbet
f Cabézon élégant
d Buntbartvogel
i Barbuto versicolore

3638 *Eucometis cristata*
Passeriformes - Thraupidae
e Grey-crested Tanager

3639 *Eucometis penicillata*
Passeriformes - Thraupidae
e Grey-headed Tanager
f Tangara à tête grise
d Pinseltangare; Graukopftangare
i Tangara penicillata

3640 *Eudocimus albus*
Ciconiiformes - Threskiornithidae
e White Ibis; Ibis; American White
Ibis; Curlew (WI)
f Ibis blanc
d Schneesichler; Weißer Ibis; Weißibis
i Ibis bianco

3641 *Eudocimus ruber*
Ciconiiformes - Threskiornithidae
e Scarlet Ibis; Ibis
f Ibis rouge
d Scharlachsichler; Rotibis
i Ibis rosso; Ibis scarlatto

3642 *Eudromia elegans*
Tinamiformes - Tinamidae
e Elegant Crested-Tinamou; Elegant
Tinamou
f Tinamou élégant
d Perlsteißhuhn
i Martinetta dal ciuffo

3643 *Eudromia formosa*
Tinamiformes - Tinamidae
e Quebracho Crested-Tinamou; Lillo's
Tinamou
f Tinamou superbe
d Schmucksteißhuhn
i Martinetta del Quebracho

3644 *Eudynamys cyanocephala*
Cuculiformes - Cuculidae
e Australian Koel; Koel; Australasian
Koel
f Coucou bleuté
i Koel australiano

3645 *Eudynamys melanorhyncha*
Cuculiformes - Cuculidae
e Black-billed Koel; Koel
f Coucou à bec noir
i Koel becconero

3646 *Eudynamys scolopacea*
Cuculiformes - Cuculidae
e Asian Koel; Common Koel; Koel
(ISC); Indian Koel; Koel Cuckoo
f Coucou koël
d Koël
i Koel comune; Koel; Coel

3647 *Eudynamys taitensis*
Cuculiformes - Cuculidae
e Long-tailed Koel; Koel; Long-tailed
Cuckoo (ANZ); Pacific Long-tailed
Cuckoo; Long-tailed New Zealand
Cuckoo
f Coucou de Nouvelle-Zélande
d Langschwanzkoël
i Koel codalunga

3648 *Eudyptes chrysocome*
Sphenisciformes - Spheniscidae
e Rockhopper Penguin; Crested
Rockhopper-Penguin; Drooping-

crested Penguin; Crested Penguin

f   Gorfou sauteur

d   Felsenpinguin

i   Eudipte crestato; Pinguino
saltaroccia; Pinguino crestato;
Pinguino saltatore

3649   *Eudyptes chrysocome filholi*
Sphenisciformes - Spheniscidae

e   Eastern Rockhopper Penguin (ANZ)

3650   *Eudyptes chrysolophus*
Sphenisciformes - Spheniscidae

e   Macaroni Penguin

f   Gorfou doré

d   Goldschopfpinguin

i   Pinguino chiuffodorato; Pinguino
crestagialla

3651   *Eudyptes pachyrhynchus*
Sphenisciformes - Spheniscidae

e   Fiordland Penguin; Fjordland
Crested-Penuin; New Zealand
Crested-Penuin; Victoria Penguin;
Thick-billed Penguin; Drooping-
crested Penguin

f   Gorfou du Fiordland

d   Dickschnabelpinguin

i   Pinguino beccogrosso; Pinguino
beccolargo

3652   *Eudyptes robustus*
Sphenisciformes - Spheniscidae

e   Snares Penguin; Snares Island
Penguin; Snares Crested-Penguin

f   Gorfou des Snares

i   Pinguino dell'Isola Snares; Pinguino
di Snares

3653   *Eudyptes schlegeli*
Sphenisciformes - Spheniscidae

e   Royal Penguin

f   Gorfou de Schlegel

i   Pinguino delle Isole Macquarie;
Pinguino di Schlegel

3654   *Eudyptes sclateri*
Sphenisciformes - Spheniscidae

e   Erect-crested Penguin

f   Gorfou huppé

d   Dickschnabelpinguin

i   Pinguino crestato maggiore; Pinguino
di Sclater

3655   *Eudyptula albosignata*
Sphenisciformes - Spheniscidae

e   White-flippered Penguin

d   Weißflügelpinguin

3656   *Eudyptula minor*
Sphenisciformes - Spheniscidae

e   Little Penguin; Little Blue-Penguin;
Fairy Penguin; Blue Penguin;
Southern Penguin

f   Manchot pygmée; Petit Manchot

d   Zwergpinguin

i   Pinguino minore; Pinguino minore
blu

3657   *Eugenes fulgens*
Apodiformes - Trochilidae

e   Magnificent Hummingbird; Rivoli's
Hummingbird

f   Colibri de Rivoli

d   Dickschnabelkolibri; Rivoli-Kolibri

i   Colibrì magnifico

3658   *Eugenes spectabilis*
Apodiformes - Trochilidae

e   Admirable Hummingbird; Costa Rica
Hummingbird; Panama
Hummingbird; Costa Rican
Hummingbird

3659   *Eugerygone rubra*
Passeriformes - Petroicidae

e   Garnet Robin; Red-backed
Gerygone; Red-backed Robin; Red-
backed Warbler; Garnet Flycatcher;
Garnet Flyeater

f   Miro grenat

i   Pigliamosche dorsorosso

3660   *Eugralla paradoxa*
Passeriformes - Rhinocryptidae

e   Ochre-flanked Tapaculo; Kittlitz's
Tapaculo

f   Mérulaxe à flancs ocre; Mérulaxe de
Kittlitz

d   Rostflankentapaculo;
Rostbauchbürzelstelzer

i   Tapaculo fianchi ocra

3661   *Eulabeornis castaneoventris*
       Gruiformes - Rallidae
  e   Chestnut Rail; Chestnut-bellied Rail;
       Chestnut-breasted Rail; Chestnut-
       breasted Swamphen
  f   Râle à ventre roux
  d   Mangroveralle
  i   Rallo pettocastano

3662   *Eulacestoma nigropectus*
       Passeriformes - Corvidae
  e   Wattled Ploughbill; Wattled Shrike-
       Tit
  f   Écorceur caronculé
  d   Lappendickkopf
  i   Becco a zappa

3663   *Eulampis holosericeus*
       Apodiformes - Trochilidae
  e   Green-throated Carib; Emerald-
       throated Hummingbird; Doctor-Bird
       (WI); Green Doctorbird (WI);
       Doctor-Brushie (WI); Green
       Hummingbird (WI); Green Carib;
       Emerald-throated Carib
  f   Falle vert (Ants); Colibri falle-vert;
       Colibri vert (Ants); Colibri caraïbe
  d   Doktorvogel
  i   Colibrì dei Caraibi sericeo

3664   *Eulampis jugularis*
       Apodiformes - Trochilidae
  e   Purple-throated Carib; Garnet-
       throated Hummingbird; Rubythroat
       (WI); Doctor-Bird (WI); Purple-
       breasted Carib; Garnet
       Hummingbird; Red-breasted
       Hummingbird
  f   Colibri rouge (Ants); Colibri madère;
       Fou-fou d'Espagne (Ants); Falle
       rouge (Ants)
  d   Granatkolibri
  i   Colibrì dei Caraibi golapurpurea

3665   *Eulidia yarrellii*
       Apodiformes - Trochilidae
  e   Chilean Woodstar; Yarrell's
       Woodstar
  f   Colibri d'Arica
  d   Yarrell-Elfe
  i   Stella dei boschi del Cile

3666   *Eumomota superciliosa*
       Coraciiformes - Momotidae
  e   Turquoise-browed Motmot
  f   Motmot à sourcils bleus
  d   Braunmotmot
  i   Motmot dai sopraccigli azzurri

3667   *Eumyias albicaudata*
       Passeriformes - Muscicapidae
  e   Nilgiri Flycatcher; Nilgiri Verditer-
       Flycatcher (ISC)
  f   Gobemouche des Nilgiri; Gobe-
       mouches des Nilgiri
  d   Nilgiri-Schnäpper
  i   Pigliamosche acquamarina del Nilgiri

3668   *Eumyias indigo*
       Passeriformes - Muscicapidae
  e   Indigo Flycatcher
  f   Gobemouche indigo; Gobe-mouches
       indigo
  d   Indigoschnäpper
  i   Pigliamosche indaco

3669   *Eumyias panayensis*
       Passeriformes - Muscicapidae
  e   Island Flycatcher; Panay Flycatcher;
       Philippine Verditer-Flycatcher;
       Island Verditer-Flycatcher; Verditer-
       Flycatcher; Mountain Verditer-
       Flycatcher
  f   Gobemouche des îles; Gobe-mouches
       des îles
  d   Azurschnäpper
  i   Pigliamosche acquamarina delle
       Filippine

3670   *Eumyias sordida*
       Passeriformes - Muscicapidae
  e   Dull Blue-Flycatcher; Dusky Blue-
       Flycatcher; Sri Lankan Dusky-blue
       Flycatcher; Sordid Flycatcher; Sri
       Lankan Dull Blue-Flycatcher; Sri-
       Lankan Blue-Flycatcher
  f   Gobemouche de Ceylan; Gobe-
       mouches de Ceylan
  d   Ceylon-Schnäpper
  i   Pigliamosche blu di Sri-Lanka

3671   *Eumyias thalassina*
       Passeriformes - Muscicapidae

e   Verditer Flycatcher; Indian Verditer-
    Flycatcher
ƒ   Gobemouche vert-de-gris; Gobe-
    mouches vert-de-gris
d   Meerblauer Schnäpper
i   Pigliamosche acquamarina indiano

3672   *Euneornis campestris*
        Passeriformes - Fringillidae
e   Orangequit; Long-mouth Quit (WI);
        Blue Baize (WI); Long-month
        Bluequit (WI); Bluebird (WI); Blue
        Gay (WI)
ƒ   Pique-orange de la Jamaïque
d   Braunlätzchen; Jamaika-Zuckervogel
i   Diglossa della Giamaica

3673   *Eunymphicus cornutus*
        Psittaciformes - Psittacidae
e   Horned Parakeet; Night Parrot;
        Sundown Parrot
ƒ   Perruche cornue
i   Parrochetto cornuto

3674   *Eupetes macrocerus*
        Passeriformes - Corvidae
e   Malaysian Rail-Babbler; Rail-
        Babbler; Malay Rail-Babbler; Malay
        Scrub-Robin; Eupetes
ƒ   Eupète à longue queue
d   Rallenläufer
i   Eupete della Malesia

3675   *Eupetomena macroura*
        Apodiformes - Trochilidae
e   Swallow-tailed Hummingbird;
        Brazilian Hummingbird; Cayenne
        Fork-tailed Hummingbird
ƒ   Colibri hirondelle
d   Breitschwingenkolibri
i   Colibrì codadirondine

3676   *Euphagus carolinus*
        Passeriformes - Fringillidae
e   Rusty Blackbird
ƒ   Quiscale rouilleux; Mainate rouilleux
d   Roststärling
i   Merlo americano

3677   *Euphagus cyanocephalus*
        Passeriformes - Fringillidae

e   Brewer's Blackbird
ƒ   Quiscale de Brewer; Mainate à tête
    pourprée
d   Purpurstärling
i   Merlo di Brewer

3678   *Eupherusa cyanophrys*
        Apodiformes - Trochilidae
e   Blue-capped Hummingbird; Oaxaca
        Hummingbird
ƒ   Colibri d'Oaxaca
d   Blaustirneupherusa
i   Colibrì di Oaxaca

3679   *Eupherusa eximia*
        Apodiformes - Trochilidae
e   Stripe-tailed Hummingbird
ƒ   Colibri à épaulettes
d   Streifenschwanzeupherusa
i   Colibrì codastriata; Colibrì dalla coda
        a strisce

3680   *Eupherusa nigriventris*
        Apodiformes - Trochilidae
e   Black-bellied Hummingbird
ƒ   Colibri à ventre noir
d   Schwarzbauchepherusa
i   Colibrì ventrenero

3681   *Eupherusa poliocerca*
        Apodiformes - Trochilidae
e   White-tailed Hummingbird; Guerrero
        Hummingbird
ƒ   Colibri du Guerrero
d   Weißschwanzeupherusa
i   Colibrì codabianca

3682   *Euphonia affinis*
        Passeriformes - Thraupidae
e   Scrub Euphonia; Lesson's Euphonia;
        Black-throated Euphonia
ƒ   Organiste de brousse
d   Buschorganist
i   Eufonia di macchia

3683   *Euphonia anneae*
        Passeriformes - Thraupidae
e   Tawny-capped Euphonia
ƒ   Organiste à couronne rousse
d   Braunscheitelorganist
i   Eufonia corona fulva

3684 **Euphonia cayennensis**
Passeriformes - Thraupidae
e Golden-sided Euphonia
f Organiste nègre
d Cayenne-Organist
i Eufonia della Cayenna

3685 **Euphonia chalybea**
Passeriformes - Thraupidae
e Green-chinned Euphonia; Green-throated Euphonia
f Organiste chalybée
d Grünkehlorganist; Bronzeorganist
i Eufonia golaverde

3686 **Euphonia chlorotica**
Passeriformes - Thraupidae
e Purple-throated Euphonia; Chlorotic Euphonia
f Organiste chlorotique
d Purpurkehlorganist
i Eufonia golapurpurea

3687 **Euphonia chrysopasta**
Passeriformes - Thraupidae
e White-lored Euphonia; Golden-bellied Euphonia
f Organiste fardé
d Zügelorganist
i Eufonia ventredorato

3688 **Euphonia concinna**
Passeriformes - Thraupidae
e Velvet-fronted Euphonia
f Organiste de la Magdalena
d Samtstirnorganist
i Eufonia fronte di velluto

3689 **Euphonia cyanocephala**
Passeriformes - Thraupidae
e Golden-rumped Euphonia; Blue-hooded Euphonia
f Organiste doré
i Eufonia gropponedorato

3690 **Euphonia elegantissima**
Passeriformes - Thraupidae
e Blue-rumped Euphonia; Blue-hooded Euphonia
f Organiste à capuchon
i Eufonia gropponeblu

3691 **Euphonia finschi**
Passeriformes - Thraupidae
e Finsch's Euphonia
f Organiste de Finsch
d Finschs Organist
i Eufonia di Finsch

3692 **Euphonia fulvicrissa**
Passeriformes - Thraupidae
e Fulvous-vented Euphonia
f Organiste cul-roux
d Rotsteißorganist
i Eufonia sottocoda fulvo; Eufonia dal sottocoda fulvo

3693 **Euphonia goodmani**
Passeriformes - Thraupidae
e Pale-vented Euphonia

3694 **Euphonia gouldi**
Passeriformes - Thraupidae
e Olive-backed Euphonia; Gould's Euphonia; Gouldian Euphonia
f Organiste olive
d Olivrückenorganist
i Eufonia di Gould

3695 **Euphonia hirundinacea**
Passeriformes - Thraupidae
e Yellow-throated Euphonia; Bonaparte's Euphonia; Swallow Euphonia
f Organiste à gorge jaune
d Schwalbenorganist
i Eufonia golagialla

3696 **Euphonia imitans**
Passeriformes - Thraupidae
e Spot-crowned Euphonia; Tawny-bellied Euphonia; Tawny-tailed Euphonia
f Organiste moucheté
s Stirnfleckenorganist
i Eufonia corona maculata

3697 **Euphonia jamaica**
Passeriformes - Thraupidae
e Jamaican Euphonia; Blue Quit (WI); Chocho Quit (WI); Short-mouthed Quit (WI); Short-mouth Blue Quit (WI); Jamaica Euphonia

f Organiste de la Jamaïque
d Gimpelorganist; Jamaika-Organist
i Eufonia della Giamaica

3698 ***Euphonia laniirostris***
Passeriformes - Thraupidae
e Thick-billed Euphonia; Shrike-billed Euphonia
f Organiste à bec épais
d Dickschnabelorganist
i Eufonia beccoforte

3699 ***Euphonia luteicapilla***
Passeriformes - Thraupidae
e Yellow-crowned Euphonia
f Organiste à calotte jaune
d Gelbscheitelorganist
i Eufonia corona gialla

3700 ***Euphonia melanura***
Passeriformes - Thraupidae
e Black-tailed Euphonia

3701 ***Euphonia mesochrysa***
Passeriformes - Thraupidae
e Bronze-green Euphonia; Bronze Euphonia
f Organiste mordoré
d Grünscheitelorganist
i Eufonia verde-bronzata

3702 ***Euphonia minuta***
Passeriformes - Thraupidae
e White-vented Euphonia
f Organiste cul-blanc
d Weißbauchorganist; Kleiner Organist
i Eufonia minuta

3703 ***Euphonia musica***
Passeriformes - Thraupidae
e Antillean Euphonia; Christmasbird (WI); Mistletoebird (WI); Blue-hooded Euphonia; Blue-crowned Euphonia
f Louis d'or (Ants); Avant-Noël (Ants); Oiseau grandpère (Ants); Perruche (Ants); Roi Bois (Ants); Organiste louis-d'or; Carouge (Ants)
d Blauscheitelorganist
i Eufonia delle Antille

3704 ***Euphonia pectoralis***
Passeriformes - Thraupidae
e Chestnut-bellied Euphonia
f Organiste à ventre marron
d Braunbauchorganist
i Eufonia ventrecastano

3705 ***Euphonia plumbea***
Passeriformes - Thraupidae
e Plumbeous Euphonia
f Organiste plombé
d Grauorganist
i Eufonia plumbea

3706 ***Euphonia rufiventris***
Passeriformes - Thraupidae
e Rufous-bellied Euphonia
f Organiste à ventre roux
d Rotbauchorganist
i Eufonia rufiventre

3707 ***Euphonia saturata***
Passeriformes - Thraupidae
e Orange-crowned Euphonia
f Organiste à calotte d'or
d Orangescheitelorganist
i Eufonia corona aranciata

3708 ***Euphonia trinitatis***
Passeriformes - Thraupidae
e Trinidad Euphonia
f Organiste de Trinidad
d Trinidad-Organist
i Eufonia di Trinidad

3709 ***Euphonia violacea***
Passeriformes - Thraupidae
e Violaceous Euphonia; Violet Euphonia
f Organiste téité
d Veilchenorganist; Gutturama; Violettblauer Organist
i Eufonia violacea

3710 ***Euphonia xanthogaster***
Passeriformes - Thraupidae
e Orange-bellied Euphonia
f Organiste à ventre orangé
d Goldbauchorganist
i Eufonia ventrearancio

3711 *Euplectes afer*
Passeriformes - Passeridae
e Yellow-crowned Bishop; Golden Bishop; Napoleon Bishop; Napoleon Weaver; Taha Weaver
f Euplecte vorabé; Euplecte à tête jaune; Tisserin taha; Vorabé
d Tahaweber; Napoleon-Weber
i Vescovo dorato

3712 *Euplectes albonotatus*
Passeriformes - Passeridae
e White-winged Widowbird; White-winged Widow (CSA); Kaffervink (CSA); White-winged Whydah; White-shouldered Whydah; White-fronted Widowbird; Long-tailed Black-Whydah
f Euplecte à épaules blanches
d Spiegelwida
i Vedova alibianche

3713 *Euplectes ardens*
Passeriformes - Passeridae
e Red-collared Widowbird; Red-collared Widow (CSA); Kaffervink (CSA); Red-collared Whydah; Long-tailed Black-Whydah; Red-naped Widowbird
f Euplecte veuve-noire; Veuve noire en feu; Veuve noire
d Schildwida; Weißgezeichnete Wida
i Vedova collare rosso

3714 *Euplectes aureus*
Passeriformes - Passeridae
e Golden-backed Bishop; Golden-backed Weaver
f Euplecte doré; Euplecte à dos doré
d Goldrückenweber
i Vescovo dorsodorato

3715 *Euplectes axillaris*
Passeriformes - Passeridae
e Fan-tailed Widowbird; Red-shouldered Widow (CSA); Kaffervink (CSA); Fan-tailed Whydah; Red-shouldered Whydah; Red-shouldered Widowbird
f Euplecte à épaules orangées; Veuve à épaulettes orangées

d Stummelwida
i Vedova coda a ventaglio

3716 *Euplectes capensis*
Passeriformes - Passeridae
e Yellow Bishop; Yellow-rumped Widow (CSA); Cape Bishop (CSA); Kaffervink (CSA); Black-and-yellow Bishop; Yellow-rumped Bishop
f Euplecte à croupion jaune; Grosbec tacheté du Cap; Grosbec tacheté de Bonne Esperance
d Samtweber; Samtwida
i Vescovo gropponegiallo; Vescovo dal groppone giallo

3717 *Euplectes diadematus*
Passeriformes - Passeridae
e Fire-fronted Bishop; Fire-fronted Weaver
f Euplecte à diadème; Euplecte diadème
d Diademweber
i Vescovo fronterossa

3718 *Euplectes franciscanus*
Passeriformes - Passeridae
e Orange Bishop; Northern Red-Bishop; Kaffervink (CSA); West African Red-Bishop; Red Bishop; Franciscan Bishop
f Grenadier (Ants); Euplecte franciscain; Euplecte ignicolore; Ignicolore
d Feuerweber; Orangeweber
i Vescovo arancio

3719 *Euplectes gierowii*
Passeriformes - Passeridae
e Black Bishop; Gierow's Bishop
f Euplecte de Gierow
d Bischofsweber
i Vescovo di Gierow

3720 *Euplectes hartlaubi*
Passeriformes - Passeridae
e Marsh Widowbird; Hartlaub's Marsh-Widowbird (CSA); Kaffervink (CSA); Marsh Whydah
f Euplecte des marais

d   Hartlaub-Wida
i   Vedova di Hartlaub

3721   *Euplectes hordeaceus*
Passeriformes - Passeridae
e   Black-winged Bishop; Black-winged
Red-Bishop (CSA); Fire-crowned
Bishop (CSA); Red-crowned Bishop;
Fiery-crowned Bishop; Black-
crowned Red-Bishop
f   Euplecte monseigneur; Monseigneur;
Euplecte à couronne de feu
d   Feuerweber; Flammenweber
i   Vescovo corona rossa

3722   *Euplectes jacksoni*
Passeriformes - Passeridae
e   Jackson's Widowbird; Jackson's
Whydah
f   Euplecte de Jackson
d   Leierschwanzwida; Tanzwida
i   Vedova di Jackson

3723   *Euplectes macrourus*
Passeriformes - Passeridae
e   Yellow-shouldered Widowbird;
Yellow-backed Widow (CSA);
Yellow-mantled Whydah; Yellow-
mantled Widowbird; Yellow-
shouldered Whydah
f   Euplecte à dos d'or; Veuve à dos d'or
d   Gelbschulterwida
i   Vedova dorsodorato

3724   *Euplectes nigroventris*
Passeriformes - Passeridae
e   Zanzibar Bishop; Black-vented
Widowbird; Red-crowned Bishop;
Zanzibar Red-Bishop; Black-winged
Bishop
f   Euplecte de Zanzibar
d   Brandweber
i   Vescovo rosso di Zanzibar

3725   *Euplectes orix*
Passeriformes - Passeridae
e   Red Bishop; Grenadier Weaver;
Little Bishop
f   Euplecte ignicolore; Ignicolore;
Monseigneur
d   Oryxweber

i   Vescovo rosso; Tessitore color di
fuoco; Fringuello di fuoco

3726   *Euplectes orix nigrifrons*
Passeriformes - Passeridae
e   Southern Red-Bishop (CSA)

3727   *Euplectes progne*
Passeriformes - Passeridae
e   Long-tailed Widowbird; Long-tailed
Widow (CSA); Flop (CSA);
Sakabula (CSA); Long-tailed
Whydah
f   Euplecte à longue queue
d   Hahnenschweifwida
i   Vedova codalunga

3728   *Euplectes psammocromius*
Passeriformes - Passeridae
e   Buff-shouldered Widowbird;
Mountain Marsh-Whydah; Mountain
Marsh-Widowbird; Highland Marsh-
Widowbird
f   Euplecte montagnard
i   Vedova di palude

3729   *Euplectes taha*
Passeriformes - Passeridae
e   Taha Bishop
f   Tisserin taha

3730   *Eupodotis afra*
Gruiformes - Otididae
e   Black Bustard; Black Korhaan
(CSA); Little Black-Bustard
f   Outarde korhaan
d   Gackeltrappe
i   Otarda korhaan

3731   *Eupodotis afraoides*
Gruiformes - Otididae
e   White-quilled Bustard; White-
winged Black-Korhaan (CSA);
White-quilled Korhaan; Northern
Black-Korhaan
f   Outarde à miroir blanc
d   Gackeltrappe
i   Otarda di Smith

3732   *Eupodotis barrowii*
Gruiformes - Otididae

e   Barrow's Bustard; White-bellied
Korhaan (CSA)
d   Weißbauchtrappe; Senegal-Trappe

3733   *Eupodotis bengalensis*
Gruiformes - Otididae
e   Bengal Florican
f   Outarde du Bengale
d   Barttrappe
i   Otarda del Bengala

3734   *Eupodotis caerulescens*
Gruiformes - Otididae
e   Blue Bustard; Blue Korhaan (CSA)
f   Outarde plombée
d   Blautrappe
i   Otarda azzurra

3735   *Eupodotis gindiana*
Gruiformes - Otididae
e   Buff-crested Bustard; Crested
Bustard (CSA)
f   Outarde d'Oustalet
i   Otarda crestafulva

3736   *Eupodotis hartlaubii*
Gruiformes - Otididae
e   Hartlaub's Bustard; Hartlaub's
Korhaan
f   Outarde de Hartlaub
d   Schwarzbürsteltrappe
i   Otarda di Hartlaub

3737   *Eupodotis humilis*
Gruiformes - Otididae
e   Little Brown-Bustard; Little Brown-
Korhaan; Somali Black-throated
Bustard; Brown Bustard
f   Outarde somalienne
d   Somali-Trappe
i   Otarda umile

3738   *Eupodotis indica*
Gruiformes - Otididae
e   Lesser Florican; Leekh; Likh
f   Outarde passarage
d   Flaggentrappe
i   Otarda minore indiana

3739   *Eupodotis melanogaster*
Gruiformes - Otididae

e   Black-bellied Bustard; Black-bellied
Korhaan (CSA); Long-legged
Korhaahn
f   Outarde à ventre noir
d   Schwarzbauchtrappe
i   Otarda ventrenero

3740   *Eupodotis rueppellii*
Gruiformes - Otididae
e   Rüppell's Bustard; Rüppell's Korhaan
(CSA)
f   Outarde de Rüppell
d   Rüppell-Trappe
i   Otarda di Rüppell

3741   *Eupodotis ruficrista*
Gruiformes - Otididae
e   Red-crested Bustard; Red-crested
Korhaan (CSA); Crested Bustard;
Bush Bustard; Buff-crested Bustard
f   Outarde houppette
d   Rotschopftrappe
i   Otarda crestarossa

3742   *Eupodotis savilei*
Gruiformes - Otididae
e   Savile's Bustard; Lyne's Bustard;
Pygmy Bustard
i   Otarda di Savile

3743   *Eupodotis senegalensis*
Gruiformes - Otididae
e   White-bellied Bustard; Senegal
Bustard; Barrow's Bustard; White-
bellied Korhaan
f   Outarde du Sénégal
d   Senegal-Trappe
i   Otarda del Senegal

3744   *Eupodotis vigorsii*
Gruiformes - Otididae
e   Karoo Bustard; Karoo Korhaan
(CSA); Black-throated Bustard;
Vigors's Bustard; Vigors's Korhaan;
Val Korhaan
f   Outarde de Vigors
d   Knarrtrappe; Nama-Trappe
i   Otarda di Vigors

3745   *Euptilotis neoxenus*
Trogoniformes - Trogonidae

e Eared Trogon; Eared Quetzal
f Trogon oreillard
d Haarbüscheltrogon
i Trogone orecchiuto

3746 *Eurocephalus anguitimens*
Passeriformes - Laniidae
e White-crowned Shrike; Smith's
White-crowned Shrike; Southern
White-crowned Shrike; White-
rumped Helmet-Shrike
f Eurocéphale à couronne blanche
d Weißscheitelwürger;
Schlangenwürger
i Averla corona bianca

3747 *Eurocephalus rueppelli*
Passeriformes - Laniidae
e White-rumped Shrike; Northern
White-crowned Shrike (CSA);
Rüppell's White-crowned Shrike;
White-crowned Shrike
f Eurocéphale de Rüppell
d Rüppell-Würger
i Averla di Rüppell

3748 Eurostopodidae
Caprimulgiformes - Eurostopodidae
e Eared-Nightjars
f Eurostopidés
d Nachtschwalben

3749 *Eurostopodus archboldi*
Caprimulgiformes - Eurostopodidae
e Mountain Eared-Nightjar; Archbold's
Eared-Nightjar; Archbold's Nightjar;
Mountain Nightjar
f Engoulevent d'Archbold
d Archbold-Nachtschwalbe
i Succiacapre di Archbold

3750 *Eurostopodus argus*
Caprimulgiformes - Eurostopodidae
e Spotted Eared-Nightjar; Spotted
Nightjar
f Engoulevent argus
d Argusnachtschwalbe
i Succiacapre maculato

3751 *Eurostopodus diabolicus*
Caprimulgiformes - Eurostopodidae

e Satanic Eared-Nightjar; Diabolical
Nightjar; Kalabat Nightjar; Devilish
Nightjar; Satanic Nightjar; Heinrich's
Eared-Nightjar; Heinrich's Nightjar;
Sulawesi Nightjar; Sulawesi Eared
Nightjar; Celebes Nightjar; Celebes
Eared-Nightjar
f Engoulevent satanique
d Teufelsnachtschwalbe
i Succiacapre diabolico

3752 *Eurostopodus macrotis*
Caprimulgiformes - Eurostopodidae
e Great Eared-Nightjar; Great Nightjar;
Greater Eared-Nightjar; Giant Eared-
Nightjar
f Engoulevent oreillard
d Riesennachtschwalbe
i Succiacapre dai grandi ciuffi

3753 *Eurostopodus mystacalis*
Caprimulgiformes - Eurostopodidae
e White-throated Eared-Nightjar;
White-throated Nightjar
f Engoulevent moustac
d Bartnachtschwalbe
i Succiacapre golabianca

3754 *Eurostopodus papuensis*
Caprimulgiformes - Eurostopodidae
e Papuan Eared-Nightjar; Papuan
Nightjar
f Engoulevent papou
d Papua-Nachtschwalbe
i Succiacapre papua

3755 *Eurostopodus temminckii*
Caprimulgiformes - Eurostopodidae
e Malaysian Eared-Nightjar; Malaysian
Nightjar; Lesser Nightjar; Lesser
Eared-Nightjar
f Engoulevent de Temminck
d Temminck-Nachtschwalbe
i Succiacapre di Temminck

3756 *Euryceros prevostii*
Passeriformes - Corvidae
e Helmetbird; Hellmet Vanga
f Eurycère de Prévost
d Helmvanga
i Vanga dall'elmo

3757 **Eurylaimidae**
Passeriformes
e Broadbills
f Eurylaimidés
d Breitrachen; Breitmäuler;
Rachenvögel
i Eurilaimidi

3758 *Eurylaimus javanicus*
Passeriformes - Eurylaimidae
e Banded Broadbill; Purple-headed
Broadbill
f Eurylaime de Horsfield; Eurylaime
de Java
d Rosenkopfbreitrachen;
Braunkopfbreitrachen
i Beccolargo di Giava; Eurilaimo di
Giava

3759 *Eurylaimus ochromalus*
Passeriformes - Eurylaimidae
e Black-and-yellow Broadbill
f Eurylaime à capuchon
d Halsbandbreitrachen
i Beccolargo giallo e nero

3760 *Eurylaimus samarensis*
Passeriformes - Eurylaimidae
e Visayan Wattled-Broadbill; Samar
Broadbill

3761 *Eurylaimus steerii*
Passeriformes - Eurylaimidae
e Mindanao Wattled-Broadbill;
Wattled Broadbill; Steere's Broadbill
f Eurylaime de Steere
d Philippinen-Breitrachen
i Beccolargo caruncolato

3762 *Eurynorhynchus pygmeus*
Charadriiformes - Scolopacidae
e Spoonbill Sandpiper; Spoon-billed
Sandpiper
f Bécasseau spatule; Bécasseau à
spatule
d Löffelstrandläufer

3763 *Euryptila subcinnamomea*
Passeriformes - Cisticolidae
e Kopje Warbler; Cinnamon-breasted
Warbler (CSA); Cinnamon-breasted
Grey-Warbler
f Camaroptère cannellé; Camaroptère à
ventre cannellé
d Zimtbrustsänger
i Camarottera dei Kopje

3764 *Eurypyga helias*
Gruiformes - Eurypygidae
e Sunbittern
f Caurale soleil
d Sonnenralle
i Euripiga; Airone del sole

3765 **Eurypygidae**
Gruiformes
e Sunbitterns
f Eurypygidés
d Sonnenrallen
i Euripigidi

3766 *Eurystomus azureus*
Coraciiformes - Coraciidae
e Purple Roller; Azure Roller; Purple
Dollarbird
f Rolle azuré
i Euristomo azzurro

3767 *Eurystomus glaucurus*
Coraciiformes - Coraciidae
e Broad-billed Roller; Cinnamon
Roller (CSA); African Broad-billed
Roller
f Rolle violet; Rolle africain
d Zimtroller; Zimtracke
i Euristomo africano

3768 *Eurystomus gularis*
Coraciiformes - Coraciidae
e Blue-throated Roller
f Rolle à gorge bleue
d Blaukehlroller
i Euristomo golablu

3769 *Eurystomus orientalis*
Coraciiformes - Coraciidae
e Dollarbird; Eastern Broad-billed
Roller; Broad-billed Roller (ISC)
f Rolle oriental
d Dollarvogel
i Euristomo orientale

3770   *Euscarthmus meloryphus*
        Passeriformes - Tyrannidae
e   Tawny-crowned Pygmy-Tyrant
f   Tyranneau à huppe fauve
d   Weißbauchtodityrann
i   Tiranno pigmeo corona fulva

3771   *Euscarthmus rufomarginatus*
        Passeriformes - Tyrannidae
e   Rufous-sided Pygmy-Tyrant;
        Rufous-edged Pygmy-Tyrant
f   Tyranneau à flancs roux
d   Gelbbauchtodityrann
i   Tiranno pigmeo fianchiruggine

3772   *Euschistospiza cinereovinacea*
        Passeriformes - Passeridae
e   Dusky Twinspot; Grey Twinspot;
        Dusky Firefinch
f   Sénégali sombre; Astrild grise-
        ardoisé
d   Schieferastrild; Schiefergrauer
        Astrild
i   Amaranto scuro

3773   *Euschistospiza dybowskii*
        Passeriformes - Passeridae
e   Dybowski's Twinspot; Dybowski's
        Dusky Twinspot; Dusky Twinspot
f   Sénégali à ventre noir; Bengali
        tacheté à ventre noir
d   Dybowski-Astrild; Dybowski-
        Tropfenastrild
i   Amaranto di Dybowski

3774   *Euthlypis lachrymosa*
        Passeriformes - Fringillidae
e   Fan-tailed Warbler; Neotropical Fan-
        tailed Warbler
f   Paruline des rochers
d   Fächerwaldsänger
i   Parula coda a ventaglio

3775   *Eutoxeres aquila*
        Apodiformes - Trochilidae
e   White-tipped Sicklebill; Sicklebill;
        Common Sicklebill; Brown-tailed
        Sicklebill
f   Bec-en-faucille aigle
d   Adlerschnabel

i   Colibrì beccodifalce codabronzata;
        Colibrì aquila

3776   *Eutoxeres condamini*
        Apodiformes - Trochilidae
e   Buff-tailed Sicklebill; Condamine's
        Sicklebill; Rufous-tailed Sicklebill
f   Bec-en-faucille de La Condamine
d   Rotschwanzadlerschnabel
i   Colibrì beccodifalce codafulva

3777   *Eutrichomyias rowleyi*
        Passeriformes - Corvidae
e   Cerulean Paradise-Flycatcher;
        Rowley's Paradise-Flycatcher;
        Rowley's Flycatcher
f   Tchitrec de Rowley
d   Silberparadiesschnäpper
i   Pigliamosche crestato di Rowley

3778   *Eutriorchis astur*
        Falconiformes - Accipitridae
e   Madagascar Serpent-Eagle; Long-
        toed Serpent-Eagle
f   Serpentaire de Madagascar
d   Schlangenhabicht
i   Aquila serpentaria del Madagascar

F

3779 Falco alexandri
Falconiformes - Falconidae
e Greater Cape Verde Kestrel
f Faucon

3780 Falco alopex
Falconiformes - Falconidae
e Fox Kestrel
f Crécerelle renard; Faucon renard
d Fuchsfalke
i Gheppio volpino; Gheppio volpe

3781 Falco altaicus
Falconiformes - Falconidae
e Altai Falcon
f Faucon de l'Altai
i Falco dell'Altai

3782 Falco araea
Falconiformes - Falconidae
e Seychelles Kestrel
f Crécerelle des Seychelles; Crécerelle katitie
d Seychellen-Turmfalke
i Falco delle Seychelle; Gheppio delle Sechelle

3783 Falco ardosiaceus
Falconiformes - Falconidae
e Grey Kestrel; White-eyed Kestrel
f Faucon ardoisé
d Graufalke
i Gheppio ardesia; Gheppio grigio

3784 Falco berigora
Falconiformes - Falconidae
e Brown Falcon; Cackling-Falcon; Orange-speckled Falcon; White-breasted Sparrowhawk; Striped Brown-Falcon; Striped Western Falcon
f Faucon bérigora

d Habichtfalke
i Falco bruno; Falchetto bruno

3785 Falco biarmicus
Falconiformes - Falconidae
e Lanner Falcon; Lanner
f Faucon lanier
d Lannerfalke; Lanner; Feldeggsfalke
i Lanario; Falcone lanario

3786 Falco cenchroides
Falconiformes - Falconidae
e Australian Kestrel; Nankeen Kestrel; Mosquito Hawk; Sparrowhawk
f Crécerelle d'Australie; Crécerelle australienne
d Graubartfalke
i Gheppio australiano; Gheppio di Nanchino

3787 Falco cherrug
Falconiformes - Falconidae
e Saker Falcon; Saker; Altai Falcon
f Faucon sacré; Gerfaut sacré
d Sakerfalke; Würgfalke
i Sacro; Falcone sacro; Falco sacro

3788 Falco chicquera
Falconiformes - Falconidae
e Red-necked Falcon; Red-headed Merlin; Red-crowned Bishop; Red-headed Falcon; Turumti
f Faucon chicquera; Faucon à cou roux
d Rothalsfalke; Rotkopffalke
i Falco testarossa; Smeriglio dal testa rossa

3789 Falco columbarius
Falconiformes - Falconidae
e Merlin; Pigeon Hawk; Bird Hawk (WI)
f Faucon émerillon; Faucon des pierres; Émerillon (Ants); Grigri morne; Gligli (Ants); Merlin
d Merlin; Zwergfalke
i Smeriglio

3790 Falco concolor
Falconiformes - Falconidae
e Sooty Falcon
f Faucon concolore

d Schieferfalke; Blaufalke
i Falco unicolore

3791 Falco cuvierii
Falconiformes - Falconidae
e African Hobby; African Hobby-
Falcon
f Faucon de Cuvier; Hobereau africain
d Afrikanischer Baumfalke
i Lodolaio africano; Falcone africano

3792 Falco deiroleucus
Falconiformes - Falconidae
e Orange-breasted Falcon
f Faucon orangé
d Rotbrustfalke
i Falco pettoarancio; Falcone dal petto
arancione

3793 Falco dickinsoni
Falconiformes - Falconidae
e Dickinson's Kestrel; White-rumped
Kestrel
f Faucon de Dickinson
d Schwarzrückenfalke
i Gheppio di Dickinson

3794 Falco eleonorae
Falconiformes - Falconidae
e Eleonora's Falcon
f Faucon d'Éléonore; Faucon saphir
d Eleonorenfalke; Eleonaras Falke
i Falco della regina

3795 Falco fasciinucha
Falconiformes - Falconidae
e Taita Falcon; Teita Falcon
f Faucon taita; Faucon de Taita
d Kurzschwanzfalke; Taitafalke;
Teitafalke
i Falco taita

3796 Falco femoralis
Falconiformes - Falconidae
e Aplomado Falcon
f Faucon aplomado
d Aplomado-Falke
i Falco aplomado

3797 Falco hypoleucos
Falconiformes - Falconidae

e Grey Falcon; Blue Hawk; Smoke
Hawk
f Faucon gris
d Silberfalke
i Falco grigio

3798 Falco jugger
Falconiformes - Falconidae
e Laggar Falcon
f Faucon laggar
d Laggarfalke
i Falco laggar

3799 Falco kreyenborgii
Falconiformes - Falconidae
e Pallid Falcon; Kleinschmidt's Falcon
f Faucon de Kleinschmidt
d Kreyenborg-Falke
i Falcone di Kleinscmidt

3800 Falco longipennis
Falconiformes - Falconidae
e Australian Hobby; Little Falcon;
White-fronted Falcon; Black-faced
Hawk; Little Duck Hawk; Duck
Hawk
f Petit Faucon
d Australischer Baumfalke
i Lodolaio australiano; Falcone
australiano

3801 Falco madens
Falconiformes - Falconidae
e Cape Verde Peregrine-Falcon

3802 Falco mexicanus
Falconiformes - Falconidae
e Prairie Falcon
f Faucon des prairies
d Präriefalke
i Falcone delle praterie; Falco della
prateria; Falco delle praterie

3803 Falco moluccensis
Falconiformes - Falconidae
e Spotted Kestrel; Moluccan Kestrel
f Crécerelle des Moluques
d Molukken-Falke
i Gheppio delle Molucche

3804 **Falco naumanni**
Falconiformes - Falconidae
e Lesser Kestrel
f Faucon crécerellette; Faucon cresserine
d Rötelfalke
i Grillaio; Falco grillaio

3805 **Falco neglectus**
Falconiformes - Falconidae
e Lesser Cape Verde Kestrel

3806 **Falco newtoni**
Falconiformes - Falconidae
e Madagascar Kestrel; Newton's Kestrel; Aldabra's Kestrel
f Crécerelle malgache
d Malagassi-Turmfalke
i Falco del Madagascar; Gheppio di Madagascar

3807 **Falco novaeseelandiae**
Falconiformes - Falconidae
e New Zealand Falcon; New Zealand Bush-Falcon; New Zealand Hobby; New Zealand Quail-Hawk; Bush Hawk; Sparrowhawk
f Faucon de Nouvelle-Zélande
d Maorifalke
i Falco della Nuova Zelanda; Falco australe

3808 **Falco pelegrinoides**
Falconiformes - Falconidae
e Barbary Falcon; Red-capped Falcon; Shahin Falcon; Shaheen Falcon
f Faucon de Barbarie
d Wüstenfalke; Berberfalke
i Falco di Barberia

3809 **Falco peregrinus**
Falconiformes - Falconidae
e Peregrine Falcon; Peregrine; Duck Hawk (NA) (WI); Pigeon Hawk; Eastern Peregrine; Black-cheeked Falcon
f Faucon pélerin; Peregrin (Ants); Malfini; Faucon commun; Faucon
d Wanderfalke
i Pellegrino; Falcone pellegrino; Falcone peregrino; Falco pellegrino

3810 **Falco punctatus**
Falconiformes - Falconidae
e Mauritius Kestrel
f Crécerelle de Maurice; Crécerelle de l'île Maurice
d Mauritius-Turmfalke
i Falco di Mauritius; Gheppio di Mauritius; Gheppio punteggiato

3811 **Falco rufigularis**
Falconiformes - Falconidae
e Bat Falcon
f Faucon des chauves-souris
d Fledermausfalke
i Falco golarossa

3812 **Falco rupicoloides**
Falconiformes - Falconidae
e Greater Kestrel; White-eyed Kestrel
f Crécerelle aux yeux blancs; Faucon aux yeux blancs
d Steppenfalke
i Gheppio maggiore africano; Gheppio dagli occhi bianchi

3813 **Falco rusticolus**
Falconiformes - Falconidae
e Gyrfalcon; Gerfalcon; Jerfalcon
f Faucon gerfaut
d Gerfalke
i Girfalco

3814 **Falco severus**
Falconiformes - Falconidae
e Oriental Hobby; Indian Hobby
f Faucon aldrovandin
d Malaien-Baumfalke
i Falco lodolaio orientale; Falcone orientale

3815 **Falco sparverius**
Falconiformes - Falconidae
e American Kestrel; Kestrel; Sparrowhawk; American Sparrowhawk; Killyhawk (WI); Bastard Hawk (WI); Killy-killy (WI)
f Crécerelle d'Amérique; Faucon des moineaux; Faucon (Ants); Grigri poulet (Ants); Grigri (Ants); Gligli (Ants)

d Buntfalke; Sperlingsfalke
i Gheppio americano

3816 *Falco subbuteo*
Falconiformes - Falconidae
e Eurasian Hobby; Northern Hobby
(NA); Hobby; Hobby Falcon (CSA);
European Hobby
f Faucon hobereau
d Baumfalke; Baumfalk
i Lodolaio

3817 *Falco subniger*
Falconiformes - Falconidae
e Black Falcon
f Faucon noir
d Rußfalke
i Falco nero; Falcone nero

3818 *Falco tinnunculus*
Falconiformes - Falconidae
e Common Kestrel; Kestrel; Rock
Kestrel (CSA); Eurasian Kestrel
(NA); European Kestrel; Old World
Kestrel
f Faucon crécerelle; Crécerelle des
clochers
d Turmfalke
i Gheppio; Gheppio comune

3819 *Falco vespertinus*
Falconiformes - Falconidae
e Red-footed Falcon; Western Red-
footed Kestrel; Western Red-footed
Falcon; Red-legged Falcon (ISC);
Amur Falcon; Amur Kestrel (CSA);
Eastern Red-footed Kestrel (CSA);
Eastern Red-legged Falcon; Amur
Red-footed Falcon; Eastern Red-
footed Falcon; Manchurian Red-
footed Falcon
f Faucon kobez; Faucon à pieds
rouges; Faucon de I'Amour
d Abendfalke; Amur-Falke; Amur-
Rotfußfalke
i Falco cuculo; Falco cuculo
amurensis; Falco cuculo orientale;
Gheppio

3820 *Falco zoniventris*
Falconiformes - Falconidae

e Banded Kestrel; Madagascar
Banded-Kestrel; Madagascar Kestrel;
Barred Kestrel
f Faucon à ventre rayé
d Bindenfalke
i Gheppio fasciato; Gheppio rigato

3821 Falconidae
Falconiformes
e Falcons
f Faucons; Falconidés; Faucons vrais;
Faucons nobles
d Falken; Falkenvögel; Edelfalken
i Falconidi

3822 Falconiformes
e Birds of Prey; Raptors
f Falconiformes
d Greifvögel; Raubvögel
i Falconiformi; Uccelli da preda

3823 *Falculea palliata*
Passeriformes - Corvidae
e Sickle-billed Vanga; Sicklebill
f Falculie mantelée
d Sichelvanga
i Vanga beccodifalce

3824 *Falcunculus frontatus*
Passeriformes - Corvidae
e Crested Shrike-Tit; Shrike-Tit;
Australian Shrike-Tit
f Falconelle à casque
d Meisendickkopf; Meisenwürger
i Cinciaverla crestata

3825 *Falcunculus frontatus leucogaster*
Passeriformes - Corvidae
e Western Crested Shrike-Tit (ANZ)

3826 *Falcunculus frontatus whitei*
Passeriformes - Corvidae
e Northern Crested Shrike-Tit (ANZ)

3827 *Ferminla cerverai*
Passeriformes - Certhiidae
e Zapata Wren; Ferminia; Ferminia
Wren
f Troglodyte de Zapata
d Kuba-Zaunkönig
i Scricciolo di Zapata

3828 **Ficedula albicollis**
Passeriformes - Muscicapidae
e Collared Flycatcher; White-collared
Flycatcher
f Gobemouche à collier occidental;
Gobe-mouches à collier occidental;
Gobe-mouches à collier; Gobe-
mouches à collier
d Halsbandschnäpper
i Balia dal collare

3829 **Ficedula basilanica**
Passeriformes - Muscicapidae
e Little Slaty Flycatcher
f Gobemouche de Basilan; Gobe-
mouches de Basilan
d Schiefergrundschnäpper
i Pigliamosche ardesia piccolo

3830 **Ficedula beijingnica**
Passeriformes - Muscicapidae
e Beijing Flycatcher

3831 **Ficedula bonthaina**
Passeriformes - Muscicapidae
e Lompobattang Flycatcher; Mountain
Flycatcher; Celebes Flycatcher; Blue
Flycatcher; Lisping Flycatcher;
Celebes Blue-Flycatcher
f Gobemouche de Lompobattang;
Gobe-mouches de Lompobattang
d Bothain-Schnäpper
i Balia di Lompobattang

3832 **Ficedula buruensis**
Passeriformes - Muscicapidae
e Cinnamon-chested Flycatcher; Buru
Flycatcher; Moluccan Shrike-Robin
f Gobemouche de Buru; Gobe-
mouches de Buru
d Molukken-Grundschnäpper
i Pigliamosche pettocannella

3833 **Ficedula crypta**
Passeriformes - Muscicapidae
e Vaurie's Flycatcher; Cryptic
Flycatcher; Russet-tailed Flycatcher;
Luzon Flycatcher
f Gobemouche de Vaurie; Gobe-
mouches de Vaurie

d Philippinen-Grundschnäpper
i Pigliamosche di Vaurie

3834 **Ficedula disposita**
Passeriformes - Muscicapidae
e Furtive Flycatcher

3835 **Ficedula dumetoria**
Passeriformes - Muscicapidae
e Rufous-chested Flycatcher; Orange-
breasted Flycatcher; Short-tailed
Flycatcher
f Gobemouche à poitrine rousse;
Gobe-mouches à poitrine rousse
d Spiegelschnäpper
i Balia pettirossa

3836 **Ficedula elisae**
Passeriformes - Muscicapidae
e Chinese Flycatcher

3837 **Ficedula harterti**
Passeriformes - Muscicapidae
e Sumba Flycatcher; Hartert's
Flycatcher; Sumba Blue-Flycatcher
f Gobemouche de Hartert; Gobe-
mouches de Hartert
d Sumba-Grundschnäpper
i Balia pigliamosche di Hartert

3838 **Ficedula henrici**
Passeriformes - Muscicapidae
e Damar Flycatcher; Damar Blue-
Flycatcher
f Gobemouche de Damar; Gobe-
mouches de Damar
d Fleckengrundschnäpper
i Pigliamosche di Damar

3839 **Ficedula hodgsonii**
Passeriformes - Muscicapidae
e Slaty-backed Flycatcher; Rusty-
breasted Flycatcher; Rusty-breasted
Blue-Flycatcher; Slaty-backed Blue-
Flycatcher
f Gobemouche de Hodgson; Gobe-
mouches de Hodgson
d Fichtenschnäpper
i Balia di Hodgson

3840   *Ficedula hyperythra*
       Passeriformes - Muscicapidae
  e   Snowy-browed Flycatcher; White-
       fronted Flycatcher; White-fronted
       Blue-Flycatcher; Thicket Flycatcher;
       Snow-browed Flycatcher; Rufous-
       breasted Blue-Flycatcher
  f   Gobemouche givré; Gobemouche à
       sourcils blancs; Gobe-mouches givré;
       Gobe-mouches à sourcils blancs
  d   Rotbrustgrundschnäpper
  i   Pigliamosche dai sopraciggli di neve

3841   *Ficedula hypoleuca*
       Passeriformes - Muscicapidae
  e   European Pied-Flycatcher (NA);
       Pied-Flycatcher; Western Pied-
       Flycatcher
  f   Gobemouche noir; Gobe-mouches
       noir
  d   Trauerschnäpper
  i   Balia nera

3842   *Ficedula monileger*
       Passeriformes - Muscicapidae
  e   White-gorgetted Flycatcher;
       Gorgetted Flycatcher
  f   Gobemouche à gorge blanche; Gobe-
       mouches à gorge blanche
  d   Diamantschnäpper
  i   Pigliamosche dal collare bianco

3843   *Ficedula mugimaki*
       Passeriformes - Muscicapidae
  e   Mugimaki Flycatcher; Black-and-
       orange Flycatcher; Robin-Flycatcher
  f   Gobemouche mugimaki; Gobe-
       mouches mugimaki
  d   Mugimaki-Schnäpper;
       Tannenschnäpper
  i   Balia di Mugimaki; Balia mugimaki

3844   *Ficedula narcissina*
       Passeriformes - Muscicapidae
  e   Narcissus Flycatcher; Black-and-
       yellow Flycatcher; Narcissina
       Flycatcher
  f   Gobemouche narcisse; Gobe-
       mouches narcisse
  d   Narzissenschnäpper
  i   Balia narcisina

3845   *Ficedula nigrorufa*
       Passeriformes - Muscicapidae
  e   Black-and-rufous Flycatcher; Black-
       and-orange Flycatcher (ISC)
  f   Gobemouche orangé-et-noir; Gobe-
       mouches orangé-et-noir
  d   Orangeschnäpper
  i   Pigliamosche nero e arancio

3846   *Ficedula parva*
       Passeriformes - Muscicapidae
  e   Red-breasted Flycatcher; Red-
       throated Flycatcher
  f   Gobemouche nain; Gobe-mouches
       nain
  d   Zwergschnäpper;
       Zwergfliegenschnäpper
  i   Pigliamosche pettirosso

3847   *Ficedula platenae*
       Passeriformes - Muscicapidae
  e   Palawan Flycatcher
  f   Gobemouche de Palawan; Gobe-
       mouches de Palawan
  d   Palawan-Grundschnäpper
  i   Pigliamosche di Palawan

3848   *Ficedula rufigula*
       Passeriformes - Muscicapidae
  e   Rufous-throated Flycatcher; White-
       vented Flycatcher; Celebes
       Flycatcher; Red-throated Flycatcher;
       Celebes Red-throated Flycatcher;
       Rufous Red-throated Flycatcher
  f   Gobemouche à gorge rousse; Gobe-
       mouches à gorge rousse
  d   Celebes-Grundschnäpper
  i   Pigliamosche golarossa

3849   *Ficedula sapphira*
       Passeriformes - Muscicapidae
  e   Sapphire Flycatcher; Sapphire-
       headed Flycatcher
  f   Gobemouche saphir; Gobe-mouches
       saphir
  d   Saphirschnäpper
  i   Pigliamosche testadizaffiro

3850   *Ficedula semitorquata*
       Passeriformes - Muscicapidae
  e   Semicollared Flycatcher; Half-

collared Flycatcher
f Gobemouche à demi-collier; Gobe-
mouches à demi-collier
d Halbringschnäpper
i Balia caucasica; Balia dal
semicollare

3851 *Ficedula solitaris*
Passeriformes - Muscicapidae
e Rufous-browed Flycatcher; Solitary
Flycatcher; White-throated
Flycatcher; Rufous-browed
Malaysian Flycatcher; White-
throated Flycatcher; White-gorgetted
Flycatcher
f Gobemouche à face rousse; Gobe-
mouches à face rousse
d Rotstirnschnäpper
i Pigliamosche dai sopraccigli rossi

3852 *Ficedula strophiata*
Passeriformes - Muscicapidae
e Rufous-gorgetted Flycatcher;
Orange-gorgetted Flycatcher;
Orange-breasted Flycatcher
f Gobemouche à bavette orangé;
Gobe-mouches à bavette orangé
d Zimtkehlschnäpper;
Orangekehlschnäpper
i Pigliamosche dal collare rossiccio

3853 *Ficedula subrubra*
Passeriformes - Muscicapidae
e Kashmir Flycatcher; Kashmir Red-
breasted Flycatcher; Himalayan
Flycatcher
f Gobemouche du Cachemire; Gobe-
mouches du Cachemire
d Kaschmir-Zwergschnäpper
i Pigliamosche del Kashmir

3854 *Ficedula superciliaris*
Passeriformes - Muscicapidae
e Ultramarine Flycatcher; White-
browed Flycatcher; White-browed
Blue-Flycatcher; Eastern White-
browed Blue-Flycatcher
f Gobemouche ultramarin; Gobe-
mouches bleu à sourcils blancs
d Brauenschnäpper;

Zwergblauschnäpper
i Pigliamosche ultramarino

3855 *Ficedula timorensis*
Passeriformes - Muscicapidae
e Black-banded Flycatcher; Timor
Flycatcher; Timor Blue-Flycatcher;
White-throated Flycatcher
f Gobemouche de Timor; Gobe-
mouches de Timor
d Brustbandgrundschnäpper
i Pigliamosche di Timor

3856 *Ficedula tricolor*
Passeriformes - Muscicapidae
e Slaty Blue-Flycatcher
f Gobemouche bleu ardoisé; Gobe-
mouches bleu ardoisé
d Dreifarbenschnäpper
i Pigliamosche azzurro-lavagna

3857 *Ficedula westermanni*
Passeriformes - Muscicapidae
e Little Pied-Flycatcher; Little Pied
Blue-Flycatcher; Westermann's
Flycatcher
f Gobemouche pie; Gobe-mouches pie
d Elsterschnäpper
i Balia di Westermann

3858 *Ficedula zanthopygia*
Passeriformes - Muscicapidae
e Yellow-rumped Flycatcher; Korean
Flycatcher; Tricoloured Flycatcher
f Gobemouche à croupion jaune;
Gobe-mouches de Corée
d Korea-Goldschnäpper;
Goldschnäpper
i Balia gropponegiallo

3859 *Florisuga mellivora*
Apodiformes - Trochilidae
e White-necked Jacobin; Great Jacobin
Hummingbird; Jacobin
Hummingbird; Collared
Hummingbird; White-bellied
Hummingbird
f Colibri jacobin
d Jakobinerkolibri
i Succhiafiori collobianco

3860   *Fluvicola albiventer*
       Passeriformes - Tyrannidae
e   Black-backed Water-Tyrant
f   Moucherolle à dos noir
i   Tiranno dorsonero

3861   *Fluvicola nengeta*
       Passeriformes - Tyrannidae
e   Masked Water-Tyrant
f   Moucherolle aquatique
d   Wasssertyrann; Schmätzertyrann
i   Tiranno mascherato

3862   *Fluvicola pica*
       Passeriformes - Tyrannidae
e   Pied Water-Tyrant
f   Moucherolle pie
d   Weißschulterwasscrtyrann
i   Tiranno gazza

3863   **Formicarlidae**
       Passeriformes
e   Antbirds
f   Formicariidés
d   Ameisenvögcl
i   Formicaridi

3864   *Formicarius analis*
       Passeriformes - Formicariidae
e   Black-faced Ant-Thrush; Rufous-
       necked Ant-Thrush
f   Tétéma coq-de-bois; Tétéma coq-
       bois
d   Schwarzkehlameisendrossel;
       Schwarzkopfameisenwürger
i   Tordo formichiere faccianera

3865   *Formicarius colma*
       Passeriformes - Formicariidae
e   Rufous-capped Ant-Thrush; Colma
       Ant-Thrush
f   Tétéma colma
d   Colma-Ameisendrossel; Colma-
       Ameisenvogel
i   Tordo formichiere caporossiccio

3866   *Formicarius moniliger*
       Passeriformes - Formicariidae
e   Mexican Ant-Thrush; Black-faced
       Ant-Thrush

3867   *Formicarius nigricapillus*
       Passeriformes - Formicariidae
e   Black-headed Ant-Thrush
f   Tétéma à tête noire
d   Schwarzkopfameisendrossel
i   Tordo formichiere capinero

3868   *Formicarius rufifrons*
       Passeriformes - Formicariidae
e   Rufous-fronted Ant-Thrush
f   Tétéma à front roux
d   Rotstirnameisendrossel
i   Tordo formichiere fronterossiccia

3869   *Formicarius rufipectus*
       Passeriformes - Formicariidae
e   Rufous-breasted Ant-Thrush
f   Tétéma à poitrine roussc
d   Rotbrustameisendrossel
i   Tordo formichicre pellorossiccio

3870   *Formicivora erythronotus*
       Passeriformes - Thamnophilidae
e   Black-hooded Ant-Wren; Hooded
       Ant-Wren
f   Grisin à dos roux; Myrmidon à dos
       roux
d   Rotrückenameisenfänger;
       Rotrückenameisenschlüpfer
i   Mangiaformiche dal cappuccio

3871   *Formicivora grisea*
       Passeriformes - Thamnophilidae
e   White-fringed Ant-Wren; Black-
       breasted Ant-Wren; Bodaert's Ant-
       Wren
f   Grisin de Cayenne
d   Seidenameisenfänger
i   Mangiaformiche grigio

3872   *Formicivora iheringi*
       Passeriformes - Thamnophilidae
e   Narrow-billed Ant-Wren
f   Grisin à bec étroit
d   Dünnschnabelameisenfänger
i   Mangiaformiche di Ihering

3873   *Formicivora littoralis*
       Passeriformes - Thamnophilidae
e   Restinga Ant-Wren

3874 **Formicivora melanogaster**
Passeriformes - Thamnophilidae
e Black-bellied Ant-Wren
f Grisin à ventre noir
d Schwarzbauchameisenfänger
i Mangiaformiche ventrenero

3875 **Formicivora moniliger**
Passeriformes - Thamnophilidae
e Hoffmann's Ant-Thrush

3876 **Formicivora rufa**
Passeriformes - Thamnophilidae
e Rusty-backed Ant-Wren
f Grisin roux
d Zimtrückenameisenfänger
i Mangiaformiche rossiccio

3877 **Formicivora serrana**
Passeriformes - Thamnophilidae
e Serra Ant-Wren; Serra Antbird
f Grisin des montagnes
d Bergameisenfänger
i Mangiaformiche della Serra

3878 **Forpus coelestis**
Psittaciformes - Psittacidae
e Pacific Parrotlet; Lesson's Parrotlet;
Celestial Parrotlet; Western Parrotlet
f Toui céleste
d Himmelspapagei
i Pappagallino del Pacifico

3879 **Forpus conspicillatus**
Psittaciformes - Psittacidae
e Spectacled Parrotlet
f Toui à lunettes
d Brillenpapagei
i Pappagallino dagli occhiali

3880 **Forpus cyanopygius**
Psittaciformes - Psittacidae
e Mexican Parrotlet; Blue-rumped
Parrotlet; Blue-naped Parrotlet
f Toui du Mexique
d Blaubürzelsperlingpapagei
i Pappagallino dal groppone turchese

3881 **Forpus passerinus**
Psittaciformes - Psittacidae
e Green-rumped Parrotlet; Guianan
Parrotlet; Parakeet (WI); Guiana
Parrotlet; Blue-winged Parrotlet;
Common Parrotlet; Parrotlet
f Toui été
d Grünbürzelsperlingpapagei
i Pappagallino dal groppone verde

3882 **Forpus sclateri**
Psittaciformes - Psittacidae
e Dusky-billed Parrotlet; Sclater's
Parrotlet
f Toui de Sclater
d Schwarzschnabelsperlingpapagei
i Pappagallino di Sclater

3883 **Forpus xanthops**
Psittaciformes - Psittacidae
e Yellow-faced Parrotlet
f Toui à tête jaune
d Gelbmaskensperlingpapagei
i Pappagallino facciagialla

3884 **Forpus xanthopterygius**
Psittaciformes - Psittacidae
e Blue-winged Parrotlet
f Toui de Spix
d Blauflügelsperlingpapagei
i Pappagallino aliblu

3885 **Foudia eminentissima**
Passeriformes - Passeridae
e Red-headed Fody; Red Forest-Fody;
Forest Fody; Red-headed Forest
Fody; Comoro Fody; Mascarene
Fody
f Foudi des Comores
d Komoren-Weber
i Tessitore rosso delle Mascarene

3886 **Foudia flavicans**
Passeriformes - Passeridae
e Yellow Fody; Rodriguez Fody
f Foudi de Rodriguez
d Rodriguez-Weber
i Tessitore di Rodriguez

3887 **Foudia madagascariensis**
Passeriformes - Passeridae
e Madagascar Red-Fody; Madagascar
Fody; Madagascar Weaver; Cardinal
Fody; Red Fody

f Foudi rouge
d Madagaskar-Weber
i Tessitore fiammante

3888 *Foudia omissa*
   Passeriformes - Passeridae
e Forest Fody; Red Forest-Fody;
   Rothschild's Fody
f Foudi de forêt
i Tessitore di foresta

3889 *Foudia rubra*
   Passeriformes - Passeridae
e Mauritius Fody; Mascarene Fody;
   Red-headed Fody
f Foudi de Maurice; Foudi de l'ile
   Maurice
d Mauritius-Weber
i Tessitore di Mauritius

3890 *Foudia sechellarum*
   Passeriformes - Passeridae
e Seychelles Fody
f Foudi des Seychelles
d Seychellen-Weber
i Tessitore delle Sychelle

3891 *Foulehaio carunculata*
   Passeriformes - Meliphagidae
e Wattled Honeyeater; Carunculated
   Honeyeater
f Méliphage foulehaio
d Schuppenkopfhonigfresser
i Mangiamiele caruncolato

3892 *Francolinus adspersus*
   Galliformes - Phasianidae
e Red-billed Francolin
f Francolin à bec rouge
d Rotschnabelfrankolin
i Francolino beccorosso

3893 *Francolinus afer*
   Galliformes - Phasianidae
e Red-necked Spurfowl; Red-necked
   Francolin (CSA); Bare-necked
   Francolin; Bare-necked Spurfowl;
   Bare-throated Francolin
d Rotkehlfrankolin
i Francolino collorosso

3894 *Francolinus africanus*
   Galliformes - Phasianidae
e Grey-winged Francolin; Greywing
   Francolin (CSA); Greywing
f Francolin à ailes grises
d Grauflügelgfrankolin
i Francolino africano

3895 *Francolinus ahantensi*
   Galliformes - Phasianidae
e Ahanta Francolin
f Francolin d'Ahanta
d Ahanta-Frankolin;
   Nacktkehlfrankolin
i Francolino di Ahanta

3896 *Francolinus albogularis*
   Galliformes - Phasianidae
e White-throated Francolin
f Francolin à gorge blanche
d Weißkehlfrankolin
i Francolino golabianca

3897 *Francolinus bicalcaratus*
   Galliformes - Phasianidae
e Double-Spurred Francolin; Bush
   Fowl
f Francolin à double éperon
d Doppelspornfrankolin
i Francolino dal doppio sperone

3898 *Francolinus capensis*
   Galliformes - Phasianidae
e Cape Francolin
d Kap-Frankolin
i Francolino del Capo

3899 *Francolinus clappertoni*
   Galliformes - Phasianidae
e Clapperton's Francolin
f Francolin de Clapperton
d Clapperton-Frankolin
i Francolino di Clapperton

3900 *Francolinus coqui*
   Galliformes - Phasianidae
e Coqui Frankolin
f Francolin coqui
d Coquifrankolin
i Francolino coqui

3901 **_Francolinus erckelli_**
Galliformes - Phasianidae
e Erckell's Francolin
f Francolin d'Erckell
d Erckell-Frankolin
i Francolino di Erkel

3902 **_Francolinus finschi_**
Galliformes - Phasianidae
e Finsch's Francolin
f Francolin de Finsch
d Finsch-Frankolin
i Francolino di Finsch

3903 **_Francolinus francolinus_**
Galliformes - Phasianidae
e Black Francolin; Black Partridge
f Francolin noir
d Halsbandfrankolin
i Francolino; Francolino comune

3904 **_Francolinus griseostriatus_**
Galliformes - Phasianidae
e Grey-striped Francolin
f Francolin à bandes grises
d Graustreifenfrankolin
i Francolino striato

3905 **_Francolinus gularis_**
Galliformes - Phasianidae
e Swamp Francolin; Swamp Partridge
   (ISC)
f Francolin multirayé
d Sumpffrankolin
i Francolino delle paludi

3906 **_Francolinus hartlaubi_**
Galliformes - Phasianidae
e Hartlaub's Francolin
f Francolin de Hartlaub
d Hartlaub-Frankolin
i Francolino di Hartlaub

3907 **_Francolinus harwoodi_**
Galliformes - Phasianidae
e Harwood's Francolin
f Francolin de Harwood
d Harwood-Frankolin
i Francolino di Harwood

3908 **_Francolinus hildebrandti_**
Galliformes - Phasianidae
e Hildebrandt's Francolin
f Francolin de Hildebrandt
d Hildebrandt-Frankolin
i Francolino di Hildebrandt

3909 **_Francolinus icterorhynchus_**
Galliformes - Phasianidae
e Heuglin's Francolin; Yellow-billed
   Francolin
f Francolin à bec jaune
d Heuglin-Frankolin
i Francolino beccogiallo

3910 **_Francolinus jacksoni_**
Galliformes - Phasianidae
e Jackson's Francolin
f Francolin de Jackson
d Bambusfrankolin
i Francolino di Jackson

3911 **_Francolinus lathami_**
Galliformes - Phasianidae
e Forest Francolin; Latham's Francolin
f Francolin de Latham
d Waldfrankolin
i Francolino di Latham

3912 **_Francolinus leucoscepus_**
Galliformes - Phasianidae
e Yellow-necked Spurfowl; Yellow-
   necked Francolin
f Francolin à cou jaune
d Gelbkehlfrankolin
i Francolino collogiallo

3913 **_Francolinus levaillantii_**
Galliformes - Phasianidae
e Red-winged Francolin
f Francolin de Levaillant
d Rotflügelfrankolin
i Francolino di Levaillant

3914 **_Francolinus levaillantoides_**
Galliformes - Phasianidae
e Archer's Greywing; Orange River-
   Francolin; Smith's Francolin
f Francolin d'Archer
d Rebhuhnfrankolin
i Francolino del Fiume Orange

3915 *Francolinus nahani*
Galliformes - Phasianidae
e Nahan's Francolin
f Francolin de Nahan
d Ituri-Frankolin
i Francolino di Nahan

3916 *Francolinus natalensis*
Galliformes - Phasianidae
e Natal Francolin
f Francolin du Natal
d Natal-Frankolin
i Francolino del Natal

3917 *Francolinus nobilis*
Galliformes - Phasianidae
e Handsome Francolin; Ruanda
Francolin
f Francolin noble
d Kiwu-Frankolin
i Francolino nobile

3918 *Francolinus ochropectus*
Galliformes - Phasianidae
e Ochre-breasted Francolin; Pale-
bellied Francolin; Dorst's Francolin;
Djibouti Francolin; Tadjourna
Francolin
f Francolin somali
d Wacholderfrankolin
i Francolino di Gibuti

3919 *Francolinus pictus*
Galliformes - Phasianidae
e Painted Francolin; Painted Partridge;
Ceylon Painted Partridge
f Francolin peint
d Tropfenfrankolin
i Francolino indiano

3920 *Francolinus pintadeanus*
Galliformes - Phasianidae
e Chinese Francolin; Burmese
Francolin
f Francolin perlé
d Chinesische Zwergwachtel;
Perlhuhnfrankolin
i Francolino cinese

3921 *Francolinus pondicerianus*
Galliformes - Phasianidae

e Grey Francolin; Indian Grey-
Francolin; Indian Grey-Partridge;
Ceylon Grey-Partridge; Swamp
Partridge; Grey Partridge
f Francolin gris
d Wachtelfrankolin
i Francolino grigio indiano

3922 *Francolinus psilolaemus*
Galliformes - Phasianidae
e Moorland Francolin (CSA); Shoa
Francolin; Montane Francolin
f Francolin montagnard
d Bergheidefrankolin
i Francolino di montagna

3923 *Francolinus rovuma*
Galliformes - Phasianidae
e Kirk's Francolin
d Kirk-Frankolin

3924 *Francolinus rufopictus*
Galliformes - Phasianidae
e Grey-breasted Spurfowl; Painted
Francolin; Grey-breasted Francolin
f Francolin à poitrine grise
d Graubrustfrankolin
i Francolino variopinto

3925 *Francolinus schlegelii*
Galliformes - Phasianidae
e Schlegel's Francolin; Schlegel's
Banded-Francolin
f Francolin de Schlegel
d Schlegel-Frankolin
i Francolino di Schlegel

3926 *Francolinus sephaena*
Galliformes - Phasianidae
e Crested Frankolin
f Francolin huppé
d Schopffrankolin
i Francolino crestato

3927 *Francolinus shelleyi*
Galliformes - Phasianidae
e Shelley's Francolin
f Francolin de Shelley
d Shelley-Frankolin
i Francolino di Shelley

3928   *Francolinus squamatus*
        Galliformes - Phasianidae
e   Scaly Francolin
f   Francolin écaillé
d   Schuppenfrankolin
i   Francolino squamato

3929   *Francolinus streptophorus*
        Galliformes - Phasianidae
e   Ring-necked Francolin
f   Francolin à collier
d   Kragenfrankolin
i   Francolino dal collare

3930   *Francolinus swainsonii*
        Galliformes - Phasianidae
e   Swainson's Spurfowl; Swainson's
        Francolin (CSA)
f   Francolin de Swainson
d   Swainson-Frankolin
i   Francolino di Swainson

3931   *Francolinus swierstrai*
        Galliformes - Phasianidae
e   Swierstra's Francolin
f   Francolin de Swierstra
d   Swierstra-Frankolin
i   Francolino di Swierstra

3932   *Fraseria cinerascens*
        Passeriformes - Muscicapidae
e   White-browed Forest-Flycatcher;
        White-browed Flycatcher
f   Gobemouche à sourcils blancs;
        Gobe-mouches à sourcils blancs
d   Brauenwaldschnäpper
i   Pigliamosche dai sopraccigli bianchi

3933   *Fraseria ocreata*
        Passeriformes - Muscicapidae
e   African Forest-Flycatcher; Common
        Forest-Flycatcher; Forest-Flycatcher;
        Fraser's Forest-Flycatcher
f   Gobemouche forestier; Gobe-
        mouches forestier
d   Waldschnäpper
i   Pigliamosche ocra

3934   *Fratercula arctica*
        Passeriformes - Alcidae
e   Atlantic Puffin; Puffin; Common

        Puffin
f   Macareux moine
d   Papageitaucher
i   Pulcinella di mare

3935   *Fratercula cirrhata*
        Charadriiformes - Alcidae
e   Tufted Puffin
f   Macareux huppé
d   Gelbschopflund
i   Pulcinella dai ciuffi; Lunda dai ciuffi;
        Pulcinella della Kamchatka

3936   *Fratercula corniculata*
        Charadriiformes - Alcidae
e   Horned Puffin
f   Macareux cornu
d   Hornlund
i   Pulcinella dal corno; Pulcinella dai
        cornetti

3937   *Frederickena unduligera*
        Passeriformes - Formicariidae
e   Undulated Antshrike
f   Batara ondé
d   Marmorameisenwürger
i   Averla formichiera golanera

3938   *Frederickena viridis*
        Passeriformes - Formicariidae
e   Black-throated Antshrike
f   Batara à gorge noire
d   Schwarzkehlameisenwürger
i   Averla formichiera ondulata

3939   *Fregata andrewsi*
        Pelecaniformes - Fregatidae
e   Christmas Island Frigatebird;
        Christmas Frigatebird (ISC)
f   Frégate d'Andrews
d   Weißbauchfregattvogel
i   Fregata dell'Isola di Natale

3940   *Fregata aquila*
        Pelecaniformes - Fregatidae
e   Ascension Frigatebird; Ascension
        Island Frigatebird
f   Frégate aigle-de-mer
d   Adlerfregattvogel
i   Fregata aquila

3941   *Fregata ariel*
        Pelecaniformes - Fregatidae
e   Lesser Frigatebird; Least Frigatebird
f   Frégate ariel
d   Zwergfregattvogel; Arielfregattvogel
i   Fregata ariel

3942   *Fregata magnificens*
        Pelecaniformes - Fregatidae
e   Weatherbird (WI); Magnificent
        Frigatebird; Hurricanebird (WI);
        Man-o-war Bird (WI)
f   Frégate superbe; Malfini (Ants);
        Frégate magnifique
d   Prachtfregattvogel
i   Fregata magnifica

3943   *Fregata minor*
        Pelecaniformes - Fregatidae
e   Great Frigatebird; Greater Frigatebird
        (CSA); Man-o-war Bird; Pacific
        Frigatebird
f   Frégate du Pacifique
d   Bindenfregattvogel
i   Fregata minore

3944   *Fregatidae*
        Pelecaniformes
e   Frigatebirds
f   Fregatidés; Frégates
d   Fregattvogel
i   Fregatidi

3945   *Fregetta grallaria*
        Procellariiformes - Hydrobatidae
e   White-bellied Storm-Petrel; Mother
        Carey's Chicken; Mother Carey's
        Chick; White-bellied Petrel; Vieillot's
        Storm-Petrel; Vieillot's Petrel; Broad-
        tailed Petrel
f   Océanite à ventre blanc
d   Weißbauchsturmschwalbe;
        Weißbauchmeerläufer
i   Uccello delle tempeste ventrebianco

3946   *Fregetta grallaria grallaria*
        Procellariiformes - Hydrobatidae
e   Tasman Sea White-bellied Storm-
        Petrel (ANZ)

3947   *Fregetta tropica*
        Procellariiformes - Hydrobatidae
e   Black-bellied Storm-Petrel; Mother
        Carey's Chicken; Mother Carey's
        Chick; Black-bellied Petrel; Gould's
        Storm-Petrel; Striped Storm-Petrel;
        Dusky-vented Storm-Petrel
f   Océanite à ventre noir
d   Schwarzbauchsturmschwalbe;
        Schwarzbauchmeerläufer
i   Uccello delle tempeste ventrenero

3948   *Fregilupus varius*
        Passeriformes - Sturnidae
e   Réunion Starling; Bourbon Crested-
        Starling
f   Étourneau de Bourbon
d   Hopfstar
i   Storno di Reunion

3949   *Fringilla coelebs*
        Passeriformes - Fringillidae
e   Chaffinch; Common Chaffinch;
        European Chaffinch
f   Pinson des arbres; Pinson; Pinson
        ordinaire; Pinson commun
d   Buchfink
i   Fringuello; Fringuello comune

3950   *Fringilla montifringilla*
        Passeriformes - Fringillidae
e   Brambling
f   Pinson du Nord; Pinson des
        Ardennes; Pinson de montagne
d   Bergfink
i   Peppola

3951   *Fringilla teydea*
        Passeriformes - Fringillidae
e   Blue Chaffinch; Canary Islands
        Chaffinch; Teydean Finch; Teyde
        Finch; Teydean Blue-Chaffinch
f   Pinson bleu
d   Teyde-Fink; Blaubuchfink;
        Kanarien-Buchfink
i   Fringuello delle Canarie

3952   **Fringillidae**
        Passeriformes
e   Finches
f   Fringillidés; Pinsons

d   Finken; Finkenvögel
i   Fringillidi; Fringuelli

3953   ***Fulica alai***
Gruiformes - Rallidae
e   Hawaiian Coot
f   Foulque des Hawai
i   Folaga delle Hawaii

3954   ***Fulica americana***
Gruiformes - Rallidae
e   American Coot; White-seal Coot
(WI); Water Fowl (WI)
f   Foulque d'Amérique; Foulque
américaine; Poule d'eau à cachet
blanc (Ants); Judelle (Ants); Poule
d'eau (Ants)
d   Amerikanisches Bläßhuhn; Indianer-
Bläßhuhn
i   Folaga americana

3955   ***Fulica ardesiaca***
Gruiformes - Rallidae
e   Slate-coloured Coot; Andean Coot
f   Foulque ardoisée
d   Schieferbläßhuhn
i   Folaga delle Ande

3956   ***Fulica armillata***
Gruiformes - Rallidae
e   Red-gartered Coot
f   Foulque à jarretières
d   Gelbschnabelbläßhuhn
i   Folaga armillata

3957   ***Fulica atra***
Gruiformes - Rallidae
e   Common Coot; Eurasian Coot (NA);
Coot; European Coot (NA); Bald
Coot; Black Coot
f   Foulque macroule
d   Bläßhuhn; Schwarzes Wasserhuhn
i   Folaga; Folaga comune

3958   ***Fulica caribaea***
Gruiformes - Rallidae
e   Caribbean Coot; White-seal Coot
(WI); Water Fowl (WI)
f   Foulque à cachet blanc (Ants); Poule
d'eau à cachet blanc (Ants)

d   Kariben-Bläßhuhn; Karibe-Bläßhuhn
i   Folaga dei Caraibi

3959   ***Fulica cornuta***
Gruiformes - Rallidae
e   Horned Coot
f   Foulque cornue
d   Rüsselbläßhuhn
i   Folaga cornuta

3960   ***Fulica cristata***
Gruiformes - Rallidae
e   Red-knobbed Coot; Crested Coot
f   Foulque à crête; Foulque caronculée
d   Kammbläßhuhn
i   Folaga crestata

3961   ***Fulica gigantea***
Gruiformes - Rallidae
e   Giant Coot
f   Foulque géante
d   Riesenbläßhuhn
i   Folaga gigante

3962   ***Fulica leucoptera***
Gruiformes - Rallidae
e   White-winged Coot
f   Foulque leucoptère
d   Weißflügelbläßhuhn; Laufhühnchen;
Kampfwachtel; Echte Kampfwachtel
i   Folaga alibianche

3963   ***Fulica newtoni***
Gruiformes - Rallidae
e   Mascarene Coot
d   Mauritius-Bläßhuhn

3964   ***Fulica rufifrons***
Gruiformes - Rallidae
e   Red-fronted Coot
f   Foulque à front rouge
d   Rotstirnbläßhuhn
i   Folaga fronterossa

3965   ***Fulmarus glacialis***
Procellariiformes - Procellariidae
e   Northern Fulmar (NA); Northern
Fulmar-Petrel; Fulmar-Petrel;
Fulmar; Atlantic Fulmar; Arctic
Fulmar
f   Pétrel fulmar; Fulmar boréal; Pétrel

glacial
d Eissturmvogel
i Fulmaro; Fulmaro artico

3966 *Fulmarus glacialoides*
Procellariiformes - Procellariidae
e Southern Fulmar; Antarctic Fulmar
(CSA); Silver-grey Fulmar; Silver
Fulmar; Slender-billed Fulmar;
Silver-grey Fulmar
f Fulmar argenté
d Silbermöwensturmvogel;
Silbersturmvogel
i Fulmaro antartico

3967 Furnariidae
Passeriformes
e Ovenbirds
f Furnariidés
d Ofenvögel; Töpfervögel
i Furnaridi

3968 *Furnarius cristatus*
Passeriformes - Furnariidae
e Crested Hornero; Crested Ovenbird
f Fournier huppé
d Schopftöpfer; Haubentöpfervogel
i Fornaio crestato

3969 *Furnarius figulus*
Passeriformes - Furnariidae
e Band-tailed Hornero; Wing-banded
Hornero; Banded Ovenbird; Wing-
banded Ovenbird; White-banded
Hornero
f Fournier bridé
d Bindentöpfer
i Fornaio bandabianca

3970 *Furnarius leucopus*
Passeriformes - Furnariidae
e Pale-legged Hornero; Pale-legged
Ovenbird
f Fournier variable
d Blaßfußtöpfer
i Fornaio zampechiare

3971 *Furnarius minor*
Passeriformes - Furnariidae
e Lesser Hornero; Lesser Ovenbird
f Petit Fournier

d Kleintöpfer
i Fornaio minore

3972 *Furnarius rufus*
Passeriformes - Furnariidae
e Rufous Hornero; Rufous Ovenbird
f Fournier roux
d Rosttöpfer; Töpfervogel; Ofenvogel;
Hornero
i Fornaio rosso; Hornero

3973 *Furnarius torridus*
Passeriformes - Furnariidae
e Pale-billed Hornero; Bay Hornero
f Fournier à bec clair
i Fornaio beccochiaro

3974 *Furnarius tricolor*
Passeriformes - Furnariidae
e Tricolour Hornero; Giebel's Hornero;
Giebel's Ovenbird
d Dreifarbentöpfer

G

3975   **Galbalcyrhynchus leucotis**
Piciformes - Galbulidae
e White-eared Jacamar; Chestnut Jacamar
d Kurzschwanzglanzvogel
i Jacamar castano

3976   **Galbalcyrhynchus purusianus**
Piciformes - Galbulidae
e Chestnut Jacamar; Purus Jacamar
f Jacamar oreillard
i Jacamar di Purus

3977   **Galbula albirostris**
Piciformes - Galbulidae
e Yellow-billed Jacamar
f Jacamar roux
d Gelbschnabelglanzvogel
i Jacamar beccogiallo; Jacamar dal becco bianco

3978   **Galbula chalcothorax**
Piciformes - Galbulidae
e Purplish Jacamar
f Jacamar à bec jaune
i Jacamar purpureo

3979   **Galbula cyanescens**
Piciformes - Galbulidae
e Bluish-fronted Jacamar; Blue-fronted Jacamar
f Jacamar violacé
d Blaustirnglanzvogel
i Jacamar fronteblu

3980   **Galbula cyanicollis**
Piciformes - Galbulidae
e Blue-necked Jacamar; Purple-necked Jacamar; Blue-cheeked Jacamar
f Jacamar à couronne bleue
i Jacamar colloblu

3981   **Galbula dea**
Piciformes - Galbulidae
e Paradise Jacamar
f Jacamar à joues bleues
d Paradiesglanzvogel
i Jacamar del Paradiso

3982   **Galbula galbula**
Piciformes - Galbulidae
e Green-tailed Jacamar
f Jacamar à longue queue
d Grünschwanzglanzvogel
i Jacamar codaverde; Galbula verde

3983   **Galbula leucogastra**
Piciformes - Galbulidae
e Bronzy Jacamar
f Jacamar vert
d Bronzeglanzvogel
i Jacamar bronzato

3984   **Galbula melanogenia**
Piciformes - Galbulidae
e Black-chinned Jacamar
f Jacamar à ventre blanc

3985   **Galbula pastazae**
Piciformes - Galbulidae
e Coppery-chested Jacamar
f Jacamar des Andes
d Kupferglanzvogel
i Jacamar pettoramato

3986   **Galbula ruficauda**
Piciformes - Galbulidae
e Rufous-tailed Jacamar
f Jacamar à queue rousse
d Rotschwanzjakamar; Rotschwanzglanzvogel
i Jacamar codaruggine

3987   **Galbula tombacea**
Piciformes - Galbulidae
e White-chinned Jacamar
f Jacamar à menton blanc
d Weißkinnglanzvogel
i Jacamar mentobianco

3988   **Galbulidae**
Piciformes
e Jacamars

f Galbulidés
d Glanzvögel
i Galbulidi

3989 *Galerida cristata*
     Passeriformes - Alaudidae
e Crested Lark; Common Crested-Lark
f Cochevis huppé; Cochevis
d Haubenlerche
i Cappellaccia; Allodola cappeluluta;
     Cappellaccia comune

3990 *Galerida deva*
     Passeriformes - Alaudidae
e Sykes's Lark; Tawny Lark; Sykes
     Crested Lark; Tawny Crested-Lark;
     Deccan Lark
f Cochevis de Sykes
d Deva-Lerche
i Cappellaccia di Sykes

3991 *Galerida magnirostris*
     Passeriformes - Alaudidae
e Large-billed Lark; Thick-billed Lark
     (CSA)
f Cochevis à gros bec
d Dickschnabellerche
i Cappellaccia beccogrosso

3992 *Galerida malabarica*
     Passeriformes - Alaudidae
e Malabar Lark; Malabar Crested-Lark
f Cochevis de Malabar
d Malabar-Lerche
i Cappellaccia del Malabar

3993 *Galerida modesta*
     Passeriformes - Alaudidae
e Sun Lark
f Cochevis modeste
d Sonnenlerche
i Cappellaccia modesta

3994 *Galerida theklae*
     Passeriformes - Alaudidae
e Thekla Lark; Short-crested Lark
     (CSA); Thekla's Lark
f Cochevis de Thékla
d Thekla-Lerche; Theklas-
     Haubenlerche; Lorbeerlerche
i Cappellacia di Thekla

3995 *Gallicolumba beccarii*
     Columbiformes - Columbidae
e Bronze Ground-Dove; Grey-throated
     Ground-Dove; Grey-breasted Quail-
     Dove; Grey-breasted Ground-Dove
f Gallicolombe de Beccari
d Graubrusttaube; Graubrusterdtaube
i Colomba terricola di Beccari

3996 *Gallicolumba canifrons*
     Columbiformes - Columbidae
e Palau Ground-Dove; Pelew Ground-
     Dove; Grey-fronted Ground-Dove
f Gallicolombe des Palau
d Graustirntaube; Palau-Erdtaube
i Colomba terricola di Palau

3997 *Gallicolumba criniger*
     Columbiformes - Columbidae
e Mindanao Bleeding-heart; Bartlett's
     Bleeding-heart; Bartlett's Blood-
     breasted Pigeon; Bartlett's Pigeon;
     Bartlett's Punalada; Hair-breasted
     Bleeding-heart; Hair-breasted Pigeon
f Gallicolombe de Bartlett
d Brandtaube
i Colomba pugnalata di Bartlett

3998 *Gallicolumba erythroptera*
     Columbiformes - Columbidae
e Polynesian Ground-Dove; Society
     Islands Ground-Dove; White-
     breasted Ground-Dove; Taumolu
     Ground-Dove
f Gallicolombe érythroptère
d Tahiti-Taube; Gesellschaftsinseln-
     Erdtaube
i Colomba terricola delle Società

3999 *Gallicolumba ferruginea*
     Columbiformes - Columbidae
e Tanna Ground-Dove
f Gallicolombe de Tanna
d Tanna-Taube
i Colomba di Tanna

4000 *Gallicolumba hoedtii*
     Columbiformes - Columbidae
e Wetar Ground-Dove; Wetar Island
     Ground-Dove; Wetar Dove
f Gallicolombe de Wetar

d Wetar-Taube; Wetar-Erdtaube
i Colomba terricola di Wetar

4001 *Gallicolumba jobiensis*
Columbiformes - Columbidae
e White-bibbed Ground-Dove; White-breasted Ground-Dove; White-breasted Ground-Pigeon; Jobi Island Dove; Purple Ground-Dove
f Gallicolombe de Jobi
d Weißbrusttaube; Weißbrusterdtaube
i Colomba terricola di Jobi

4002 *Gallicolumba keayi*
Columbiformes - Columbidae
e Negros Bleeding-heart; Negros Blood-breasted Pigeon; Negros Pigeon; Negros Punalada; Keay's Bleeding-heart
f Gallicolombe de Negros
d Negros-Taube
i Colomba pugnalata di Keay

4003 *Gallicolumba kubaryi*
Columbiformes - Columbidae
e Caroline Islands Ground-Dove; Ponapé Ground-Dove; Truk Island Ground-Dove; White-breasted Ground-Dove; White-fronted Ground-Dove; Purple Ground-Dove
f Gallicolombe de Kubary
d Karolinen-Taube
i Colomba terricola delle Truk

4004 *Gallicolumba luzonica*
Columbiformes - Columbidae
e Luzon Bleeding-heart; Luzon Bleeding-heart Pigeon; Bleeding-heart
f Gallicolombe poignardée
d Dolchstichtaube
i Colomba pugnalata di Luzon; Colomba pugnalata; Colomba cruentata

4005 *Gallicolumba menagei*
Columbiformes - Columbidae
e Sulu Bleeding-heart; Tawitawi Bleeding-heart; Tawitawi Pigeon
f Gallicolombe de Tawi-Tawi

d Tawitawi-Taube
i Colomba pugnalata di Tawitawi

4006 *Gallicolumba norfolciensis*
Columbiformes - Columbidae
e Norfolk Island Ground-Dove (ANZ); Norfolk Island Dove
d Norfolk-Taube

4007 *Gallicolumba platenae*
Columbiformes - Columbidae
e Mindoro Bleeding-heart; Mindoro Pigeon; Mindoro Punalada
f Gallicolombe de Mindoro
d Platen-Taube
i Colomba pugnalata di Mindoro

4008 *Gallicolumba rubescens*
Columbiformes - Columbidae
e Marquesan Ground-Dove; Marquesas Ground-Dove; Grey-hooded Ground-Dove; Grey-hooded Quail-Dove; Grey-bibbed Ground-Dove; Marquesa Ground-Dove; Marquesas Ground-Dove; Marquesan Ground-Dove
f Gallicolombe des Marquises
d Marquesas-Erdtaube; Marquesas-Taube
i Colomba terricola delle Marquesas

4009 *Gallicolumba rufigula*
Columbiformes - Columbidae
e Cinnamon Ground-Dove; Red-throated Ground-Dove; Golden-heart; Chestnut Quail-Dove; Rufous Ground-Dove; Rufous Quail-Dove; Yellow-brown Quail-Dove
f Gallicolombe à poitrine d'or
d Gelbbrusterdtaube
i Colomba pettodorato

4010 *Gallicolumba salamonis*
Columbiformes - Columbidae
e Thick-billed Ground-Dove
f Gallicolombe des Salomon
d San Cristobal-Taube; Dickschnabelerdtaube
i Colomba terricola delle Salomon

4011   *Gallicolumba sanctaecrucis*
Columbiformes - Columbidae
e   Santa Cruz Ground-Dove; Santa
Cruz Ground-Pigeon
f   Gallicolombe de Santa Cruz
d   Santa Crus-Taube
i   Colomba terricola di Santa Cruz

4012   *Gallicolumba stairi*
Columbiformes - Columbidae
e   Friendly Ground-Dove; Friendly
Quail-Dove; Shy Ground-Dove;
Purple-shouldered Quail-Dove;
Samoan Quail-Dove; Tongan Quail-
Dove; Fiji Quail-Dove; West
Polynesian Ground-Dove
f   Gallicolombe de Stair
d   Purpurschultertaube; Samoa-
Erdtaube
i   Colomba terricola di Stair

4013   *Gallicolumba tristigmata*
Columbiformes - Columbidae
e   Sulawesi Ground-Dove; Celebes
Ground-Dove; Sulawesi Quail-Dove;
Celebes Quail-Dove
f   Gallicolombe tristigmate
d   Celebes-Gelbbrusterdtaube;
Hopftaube
i   Colomba terricola di Sulawesi

4014   *Gallicolumba xanthonura*
Columbiformes - Columbidae
e   White-throated Ground-Dove;
White-throated Dove; White-throated
Quail-Dove; White-breasted Quail-
Dove; Marianas Quail-Dove; Yap
Quail-Dove; Mariana Quail-Dove
f   Gallicolombe pampusane
d   Jungferntaube; Marianen-Erdtaube
i   Colomba terricola delle Marianne

4015   *Gallicrex cinerea*
Gruiformes - Rallidae
e   Watercock; Kora
f   Râle à crête
d   Wasserhahn m; Wasserhuhn f
i   Gallo d'acqua

4016   Galliformes
e   Gallinaceous Birds; Fowl-like Birds;
Game Birds
f   Galliformes
d   Hühnervögel
i   Galliformi

4017   *Gallinago andina*
Charadriiformes - Scolopacidae
e   Puna Snipe; Andean Snipe
f   Bécassine du Puna
d   Punabekassine
i   Beccaccino della Puna

4018   *Gallinago delicata*
Charadriiformes - Scolopacidae
e   Wilson's Snipe

4019   *Gallinago gallinago*
Charadriiformes - Scolopacidae
e   Common Snipe; Snipe; European
Snipe; Wilson's Snipe; Fantail Snipe
f   Bécassine des marais; Bécassine
(Ants)
d   Bekassine
i   Beccaccino; Beccaccino comune

4020   *Gallinago hardwickii*
Charadriiformes - Scolopacidae
e   Latham's Snipe; Japanese Snipe;
Australian Snipe; New Holland Snipe
f   Bécassine du Japon
d   Kreisch-Bekassine
i   Beccaccino giapponese

4021   *Gallinago imperialis*
Charadriiformes - Scolopacidae
e   Imperial Snipe; Banded Snipe;
Bogota Snipe
f   Bécassine impériale
d   Kaiserbekassine
i   Beccaccino imperiale

4022   *Gallinago jamesoni*
Charadriiformes - Scolopacidae
e   Andean Snipe; Jameson's Snipe;
Northern Snipe; Cordilleran Snipe
f   Bécassine des paramos
i   Beccaccino di Jameson

4023   *Gallinago macrodactyla*
Charadriiformes - Scolopacidae
e   Madagascar Snipe; Madagascan

Snipe; Malagasy Snipe
  f   Bécassine malgache
  d   Madagaskar-Bekassine
  i   Beccaccino del Madagascar

4024   *Gallinago media*
Charadriiformes - Scolopacidae
  e   Great Snipe; Double Snipe
  f   Bécassine double
  d   Doppelschnepfe
  i   Croccolone

4025   *Gallinago megala*
Charadriiformes - Scolopacidae
  e   Swinhoe's Snipe; Marsh Snipe;
        Forest Snipe; Chinese Snipe
  f   Bécassine de Swinhoe
  d   Waldbekassine
  i   Beccaccino di Swinhoe

4026   *Gallinago minima*
Charadriiformes - Scolopacidae
  e   Jack Snipe; Half Snipe
  f   Bécassine sourde
  d   Zwergschnepfe
  i   Frullino; Frullo; Pinzacchio;
        Beccastrino; Beccaccino sordeo

4027   *Gallinago nemoricola*
Charadriiformes - Scolopacidae
  e   Wood Snipe; Himalayan Snipe
  f   Bécassine des bois
  d   Nepal-Bekassine
  i   Beccaccino dei boschi

4028   *Gallinago nigripennis*
Charadriiformes - Scolopacidae
  e   African Snipe; Ethiopean Snipe
        (CSA)
  f   Bécassine africaine
  d   Afrikanische Bekassine
  i   Beccaccino africano

4029   *Gallinago nobilis*
Charadriiformes - Scolopacidae
  e   Noble Snipe; Paramo Snipe
  f   Bécassine noble
  d   Langschnabelbekassine
  i   Beccaccino nobile

4030   *Gallinago paraguaiae*
Charadriiformes - Scolopacidae
  e   South American Snipe; Magellan
        Snipe; Paraguay Snipe
  f   Bécassine de Magellan
  d   Azara-Bekasssine
  i   Beccaccino del Paraguay

4031   *Gallinago solitaria*
Charadriiformes - Scolopacidae
  e   Solitary Snipe; Hermit Snipe; Tibet
        Snipe; Tibetan Snipe
  f   Bécassine solitaire
  d   Bergbekassine
  i   Beccaccino solitario

4032   *Gallinago stenura*
Charadriiformes - Scolopacidae
  e   Pintail Snipe,; Pin-tailed Snipe
        (ANZ); Asiatic Snipe
  f   Bécassine à queue pointue
  d   Spießbekassine; Stiftbekassine
  i   Beccaccino siberiano; Beccaccino
        stenuro; Beccaccino codappuntita

4033   *Gallinago stricklandii*
Charadriiformes - Scolopacidae
  e   Fuegian Snipe; Cordilleran Snipe;
        Southern Snipe; Strickland's Snipe
  f   Bécassine de Strickland
  d   Cordilleren-Bekassine;
        Kordillerenbekassine
  i   Beccaccino di Strickland

4034   *Gallinago undulata*
Charadriiformes - Scolopacidae
  e   Giant Snipe; Paraguayam Snipe
  f   Bécassine géante
  d   Riesenbekassine
  i   Beccaccino gigante

4035   *Gallinula angulata*
Gruiformes - Rallidae
  e   Lesser Moorhen; Little Moorhen
  f   Gallinule africaine
  d   Zwergteichhuhn
  i   Gallinella d'acqua minore

4036   *Gallinula chloropus*
Gruiformes - Rallidae
  e   Common Moorhen; Moorhen; Grey

Moorhen; Common Gallinule;
Gallinule; Florida Gallinule; Indian
Gallinule; Indian Waterhen;
Gallinule; Waterhen
f   Poule d´eau; Gallinule poule-d'eau;
Gallinule d'eau (Ants); Poule d'eau à
cachet rouge (Ants); Poule d'eau
commune
d   Teichhuhn; Wasserhuhn
i   Gallinella d'acqua

4037   *Gallinula comeri*
Gruiformes - Rallidae
e   Gough Island Moorhen
d   Gough-Teichhuhn

4038   *Gallinula melanops*
Gruiformes - Rallidae
e   Spot-flanked Gallinule; Little
Waterhen
f   Gallinule à face noire
d   Maskenpfuhlhuhn
i   Gallinella fianchi machiati

4039   *Gallinula mortierii*
Gruiformes - Rallidae
e   Tasmanian Native-hen; Tasmanian
Waterhen
f   Gallinule de Tasmanie
d   Grünfußpfulhuhn
i   Gallinella della Tasmania

4040   *Gallinula nesiotis*
Gruiformes - Rallidae
e   Tristan Mooorhen; Gough Moorhen;
Tristan Rail; Tristan Gallinule;
Gough Island Coot; Gough Coot;
Gough Island Moorhen
f   Gallinule de Tristan
d   Tristan-Teichhuhn
i   Gallinella dell'Isola Gough;
Gallinella del Gough

4041   *Gallinula pacifica*
Gruiformes - Rallidae
e   Samoan Moorhen; Samoan Gallinule;
Samoan Woodrail
f   Gallinule punaé
d   Samoa-Pfuhlhuhn
i   Gallinella di Samoa

4042   *Gallinula silvestris*
Gruiformes - Rallidae
e   San Cristobal Moorhen; San
Cristobal Gallinule; San Cristobal
Mountain-Rail; Edith's Gallinule;
Mountain Rail
f   Gallinule d'Édith
d   Blaustirnphuhlhuhn
i   Gallinella San Cristobal

4043   *Gallinula tenebrosa*
Gruiformes - Rallidae
e   Dusky Moorhen; Dusky Gallinule;
Black Moorhen; Black Gallinule;
Sombre Gallinule; Sombre Moorhen
f   Gallinule sombre
d   Papua-Teichhuhn
i   Gallinella tenebrosa

4044   *Gallinula ventralis*
Gruiformes - Rallidae
e   Black-tailed Native-hen; Black-tailed
Waterhen
f   Gallinule aborigène
d   Rotfußpfuhlhuhn
i   Gallinella codanera australiana

4045   *Gallirallus australis*
Gruiformes - Rallidae
e   Weka; New Zealand Woodrail; Weka
Rail; North Island Woodhen; South
Island Woodhen
f   Râle wéka
d   Weka-Ralle
i   Weka

4046   *Gallirallus conditiclus*
Gruiformes - Rallidae
e   Gilbert Rail
f   Râle des Gilbert

4047   *Gallirallus dieffenbachii*
Gruiformes - Rallidae
e   Dieffenbach's Rail
f   Râle de Nouvelle-Bretagne; Râle de
Dieffenbach
d   Dieffenbachralle
i   Rallo di Dieffenbach

4048   *Gallirallus insignis*
Gruiformes - Rallidae

e New Britain Rail; Bismarck Rail;
Pink-legged Rail; Sclater's Rail
d Bartralle
i Rallo della Nuova Britannia

4049 *Gallirallus lafresnayanus*
Gruiformes - Rallidae
e New Caledonian Rail; New
Caledonian Woodrail
f Râle de Lafresnaye
d Pelzralle
i Rallo di Lafrenaye

4050 *Gallirallus modestus*
Gruiformes - Rallidae
e Chatham Islands Rail
f Râle des Chatham
d Chatham-Ralle
i Rallo di Chatham

4051 *Gallirallus okinawae*
Gruiformes - Rallidae
e Okinawa Rail
f Râle d'Okinawa
i Rallo di Okinwa

4052 *Gallirallus owstoni*
Gruiformes - Rallidae
e Guam Rail
f Râle de Guam
d Guam-Ralle
i Rallo di Guam

4053 *Gallirallus pacificus*
Gruiformes - Rallidae
e Tahiti Rail; Red-billed Rail
d Tevea-Ralle
i Rallo di Tahiti

4054 *Gallirallus philippensis*
Gruiformes - Rallidae
e Buff-banded Rail; Banded Land-Rail;
Banded Rail; Pectoral Rail; Painted
Rail; Striped Rail
f Râle tiklin
d Bindenralle
i Rallo delle Filippine

4055 *Gallirallus philippensis andrewsi*
Gruiformes - Rallidae

e Cocos-Keeling Islands Buff-banded
Rail (ANZ)

**4056 *Gallirallus philippensis
macquariensis***
Gruiformes - Rallidae
e Macquarie Island Islands Buff-
banded Rail (ANZ)

4057 *Gallirallus rovianae*
Gruiformes - Rallidae
e Roviana Rail
f Râle roviana

4058 *Gallirallus sharpei*
Gruiformes - Rallidae
e Sharpe's Rail
f Râle de Sharpe
i Rallo di Sharpe

4059 *Gallirallus striatus*
Gruiformes - Rallidae
e Slaty-breasted Rail; Indian Blue-
breasted Banded-Rail (ISC); Blue-
breasted Banded-Rail; Banded Rail
f Râle strié
d Graubrustralle
i Rallo pettoblu

4060 *Gallirallus sylvestris*
Gruiformes - Rallidae
e Lord Howe Island Rail; Lord Howe
Woodhen (ANZ); Lord Howe
Woodrail; Lord Howe Island
Woodrail; Woodhen; Lord Howe's
Rail
f Râle sylvestre
d Waldralle
i Rallo di Lord Howe; Rallo dell'Isola
Lord Howe

4061 *Gallirallus torqutus*
Gruiformes - Rallidae
e Barred Rail
f Râle à collier
d Zebraralle
i Rallo barrato

4062 *Gallirallus wakensis*
Gruiformes - Rallidae
e Wake Island Rail

f   Râle de Wake
d   Wakeralle
i   Rallo di Wake

4063   *Galloperdix bicalcarata*
Galliformes - Phasianidae
e   Ceylon Spurfowl; Sri Lanka
Spurfowl
f   Galloperdix de Ceylon
d   Weißkehlspornhuhn
i   Gallopernice dal doppio sperone

4064   *Galloperdix lunulata*
Galliformes - Phasianidae
e   Painted Spurfowl
d   Perlspornhuhn
i   Gallopernice variopinta

4065   *Galloperdix spadicea*
Galliformes - Phasianidae
e   Red Spurfowl
d   Rotes Spornhuhn; Zwergfasan
i   Gallopernice rossa

4066   *Gallus gallus*
Galliformes - Phasianidae
e   Red Junglefowl; Junglefowl; Wild
Junglefowl; Feral Chicken
f   Coq bankiva
d   Bankiva-Huhn
i   Gallo bankiva; Gallo rosso della
giungla; Gallo dorato

4067   *Gallus lafayetii*
Galliformes - Phasianidae
e   Sri Lanka Junglefowl; Ceylon
Junglefowl; La Fayette's Junglefowl;
Yellow Junglefowl
f   Coq de La Fayette
d   La Fayette-Huhn
i   Gallo di La Fayette

4068   *Gallus sonneratii*
Galliformes - Phasianidae
e   Grey Junglefowl; Sonnerat's
Junglefowl
f   Coq de Sonnerat
d   Sonnerat-Huhn
i   Gallo di Sonnerat; Gallo grigio della
giungla

4069   *Gallus varius*
Galliformes - Phasianidae
e   Green Junglefowl; Javan Junglefowl
f   Coq de Java
d   Gabelschwanzhuhn
i   Gallo selvatico di Giava; Gallo vario;
Gallo variabile; Gallo selvatico

4070   *Gampsonyx swainsonii*
Falconiformes - Accipitridae
e   Pearl Kite
f   Élanion perle
d   Perlaar
i   Nibbio di Swainson

4071   *Gampsorhynchus rufulus*
Passeriformes - Sylviidae
e   White-hooded Babbler; White-
headed Babbler; White-headed
Shrike-Babbler
f   Actinodure à tête blanche; Timalie à
tête blanche
d   Weißkopfwürgertimalie;
Weißkopftimalie
i   Garrulo averla testbianca

4072   *Garrodia nereis*
Procellariiformes - Hydrobatidae
e   Grey-backed Storm-Petrel; Mother
Carey's Chicken (CSA); Grey-backed
Petrel
f   Océanite néréide
d   Graurückensturmschwalbe
i   Uccello delle tempeste dorsogrigio

4073   *Garrulax affinis*
Passeriformes - Sylviidae
e   Black-faced Laughingthrush
f   Garrulaxe à face noire
d   Schwarzscheitelhäherling
i   Garrulo schiamazzante faccianera

4074   *Garrulax albogularis*
Passeriformes - Sylviidae
e   White-throated Laughingthrush;
Collared Laughingthrush
f   Garrulaxe à gorge blanche
d   Weißkehlhäherling
i   Garrulo schiamazzante crestabianca

4075 *Garrulax austeni*
Passeriformes - Sylviidae
e Brown-capped Laughingthrush;
Austen's Laughingthrush; Godwin-
Austen's Laughingthrush
f Garrulaxe d'Austen
d Braunkappenhäherling
i Garrulo schiamazzante capobruno

4076 *Garrulax bieti*
Passeriformes - Sylviidae
e White-Speckled Laughingthrush;
Biet's Laughingthrush
f Garrulaxe de Biet
i Garrulo schiamazzante di Biet

4077 *Garrulax cachinnans*
Passeriformes - Sylviidae
e Rufous-breasted Laughingthrush;
Nilgiri Laughingthrush (ISC)
f Garrulaxe des Nilgiri
d Zimtbrusthäherling
i Garrulo schiamazzante del Nilgiri

4078 *Garrulax caerulatus*
Passeriformes - Sylviidae
e Grey-sided Laughingthrush
f Garrulaxe à flancs gris
d Grauflankenhäherling
i Garrulo schiamazzante fianchigrigi

4079 *Garrulax calvus*
Passeriformes - Sylviidae
e Bare-headed Laughingthrush
f Garrulaxe chauve
d Kahlkopfhäherling
i Garrulo schiamazzante calvo

4080 *Garrulax canorus*
Passeriformes - Sylviidae
e Hwamei; Melodious Laughingthrush;
Hwamei Laughingthrush; Chinese
Thrush
f Garrulaxe hoamy
d Augenbrauenhäherling; Chinesische
Spottdrossel
i Garrulo schiamazzante canoro

4081 *Garrulax chinensis*
Passeriformes - Sylviidae
e Black-throated Laughingthrush;

Chinese Laughingthrush
f Garrulaxe à joues blanches
d Weißohrhäherling
i Garrulo schiamazzante cinese

4082 *Garrulax cineraceus*
Passeriformes - Sylviidae
e Moustached Laughingthrush; Ashy
Laughingthrush; Black-capped
Laughingthrush
f Garrulaxe cendré
d Grauhäherling
i Garrulo schiamazzante cenerino

4083 *Garrulax cinereifrons*
Passeriformes - Sylviidae
e Ashy-headed Laughingthrush
f Garrulaxe à tête cendrée
d Graustirnhäherling
i Garrulo schiamazzante testacenerina

4084 *Garrulax davidi*
Passeriformes - Sylviidae
e Plain Laughingthrush; Père David's
Laughingthrush; David's
Laughingthrush; Peking
Laughingthrush; Beijing
Laughingthrush; Plain Hill-Babbler;
Peking Hill-Babbler
f Garrulaxe de David; Garrulaxe du
père David
d David-Häherling
i Garrulo schiamazzante di David

4085 *Garrulax delesserti*
Passeriformes - Sylviidae
e Wynaad Laughingthrush; Yellow-
throated Laughingthrush (ISC);
Rufous-vented Laughingthrush
f Garrulaxe de Delessert
d Rostflankenhäherling
i Garrulo schiamazzante di Delessert

4086 *Garrulax elliotii*
Passeriformes - Sylviidae
e Elliot's Laughingthrush
f Garrulaxe d'Elliot
d Elliot-Häherling; Braunhäherling
i Garrulo schiamazzante di Elliot

4087   *Garrulax erythrocephalus*
　　　　Passeriformes - Sylviidae
　e   Chestnut-crowned Laughingthrush;
　　　　Red-headed Laughingthrush;
　　　　Chestnut-capped Laughingthrush
　f   Garrulaxe à tête rousse
　d   Rotkopfhäherling
　i   Garrulo schiamazzante capocastano

4088   *Garrulax formosus*
　　　　Passeriformes - Sylviidae
　e   Red-winged Laughingthrush;
　　　　Exquisite Laughingthrush; Crimson-
　　　　winged Laughingthrush
　f   Garrulaxe élégant
　d   Prachthäherling
　i   Garrulo schiamazzante alirosse

4089   *Garrulax galbanus*
　　　　Passeriformes - Sylviidae
　e   Yellow-throated Laughingthrush;
　　　　Austen's Laughingthrush; Courtois's
　　　　Laughingthrush
　f   Garrulaxe à gorge jaune
　d   Gelbbauchhäherling
　i   Garrulo schiamazzante di Austen

4090   *Garrulax gularis*
　　　　Passeriformes - Sylviidae
　e   Rufous-vented Laughingthrush;
　　　　McClelland's Laughingthrush;
　　　　Yellow-browed Laughingthrush
　i   Garrulo schiamazzante pettogiallo

4091   *Garrulax henrici*
　　　　Passeriformes - Sylviidae
　e   Brown-cheeked Laughingthrush;
　　　　Prince Henry's Laughingthrush;
　　　　Prince d'Orleans's Laughingthrush;
　　　　Henry's Laughingthrush
　f   Garrulaxe du prince Henri d'Orléans
　d   Prinzenhäherling
　i   Garrulo schiamazzante del principe
　　　　Enrico

4092   *Garrulax jerdoni*
　　　　Passeriformes - Sylviidae
　e   Grey-breasted Laughingthrush;
　　　　Kerala Laughingthrush (ISC);
　　　　Jerdon's Laughingthrush; White-
　　　　breasted Laughingthrush

　d   Graubrusthäherling
　i   Garrulo schiamazzante di Jerdon

4093   *Garrulax leucolophus*
　　　　Passeriformes - Sylviidae
　e   White-crested Laughingthrush
　f   Garrulaxe à huppe blanche
　d   Weißhaubenhäherling;
　　　　Haubenhäherling
　i   Garrulo schiamazzante crestabianca

4094   *Garrulax lineatus*
　　　　Passeriformes - Sylviidae
　e   Streaked Laughingthrush; Himalayan
　　　　Laughingthrush
　f   Garrulaxe barré
　d   Borstenhäherling
　i   Garrulo schiamazzante dell'Himalaya

4095   *Garrulax lugubris*
　　　　Passeriformes - Sylviidae
　e   Black Laughingthrush
　f   Garrulaxe noir
　d   Trauerhäherling
　i   Garrulo schiamazzante nero

4096   *Garrulax lunulatus*
　　　　Passeriformes - Sylviidae
　e   Barred Laughingthrush; Bar-backed
　　　　Laughingthrush
　f   Garrulaxe à lunules
　d   Wellenhäherling
　i   Garrulo schiamazzante dorsobarrato

4097   *Garrulax maesi*
　　　　Passeriformes - Sylviidae
　e   Grey Laughingthrush; Maes's
　　　　Laughingthrush
　f   Garrulaxe de Maes
　d   Maes-Häherling
　i   Garrulo schiamazzante di Maes

4098   *Garrulax maximus*
　　　　Passeriformes - Sylviidae
　e   Giant Laughingthrush
　f   Garrulaxe géant
　d   Riesenhäherling
　i   Garrulo schiamazzante gigante

4099   *Garrulax merulinus*
　　　　Passeriformes - Sylviidae

e   Spot-breasted Laughingthrush;
    Spotted-breasted Laughingthrush
f   Garrulaxe à poitrine tachetée
d   Fleckenhäherling
i   Garrulo schiamazzante pettomaculato

4100   *Garrulax milleti*
Passeriformes - Sylviidae
e   Black-hooded Laughingthrush;
    Millet's Laughingtrush; Vietnam
    Laughingthrush
f   Garrulaxe de Millet
d   Kapuzenäherling
i   Garrulo schiamazzante dal cappuccio

4101   *Garrulax milnei*
Passeriformes - Sylviidae
e   Red-tailed Laughingthrush
f   Garrulaxe à queue rouge; Garrulaxe à
    queue rousse
d   Rotschwanzhäherling;
    Prachthäherling
i   Garrulo schiamazzante codarossa

4102   *Garrulax mitratus*
Passeriformes - Sylviidae
e   Chestnut-capped Laughingthrush;
    Capped Laughingthrush
f   Garrulaxe mitré
d   Spiegelhäherling
i   Garrulo schiamazzante capocastano

4103   *Garrulax monileger*
Passeriformes - Sylviidae
e   Lesser Necklaced Laughingthrush;
    Necklaced Laughingthrush (ISC);
    Black-necked Laughingthrush
f   Garrulaxe à collier; Petit Garrulaxe à
    collier
d   Lätzchenhäherling
i   Garrulo minore dal collare

4104   *Garrulax morrisonianus*
Passeriformes - Sylviidae
e   White-whiskered Laughingthrush;
    Taiwan Laughingthrush
f   Garrulaxe de Morrison
d   Morrison-Häherling
i   Garrulo schiamazzante di Taiwan

4105   *Garrulax nuchalis*
Passeriformes - Sylviidae
e   Chestnut-backed Laughingthrush;
    Ogle's Laughingthrush
f   Garrulaxe à nuque marron
d   Rotrückenhäherling
i   Garrulo schiamazzante dorsocastano

4106   *Garrulax ocellatus*
Passeriformes - Sylviidae
e   Spotted Laughingthrush; White-
    spotted Laughingthrush; White
    Laughingthrush
f   Garrulaxe ocellé
d   Waldhäherling
i   Garrulo schiamazzante ocellato

4107   *Garrulax palliatus*
Passeriformes - Sylviidae
e   Sunda Laughingthrush; Grey-and-
    brown Laughingthrush; Catbird
    Laughingthrush
f   Garrulaxe mantelé
d   Schieferhäherling
i   Garrulo schiamazzante grigio-bruno

4108   *Garrulax pectoralis*
Passeriformes - Sylviidae
e   Greater Necklaced Laughingthrush;
    Black-gorgeted Laughingthrush;
    Gorgeted Laughingthrush; Necklaced
    Laughingthrush; Large Necklaced
    Laughingthrush
f   Garrulaxe à plastron
d   Brustbandhäherling
i   Garrulo dal collare maggiore;
    Garrulo dal collare; Garrulo
    maggiore dal collare

4109   *Garrulax perspicillatus*
Passeriformes - Sylviidae
e   Masked Laughingthrush; Spectacled
    Laughingthrush; Black-faced
    Laughingthrush
f   Garrulaxe masqué; Garulaxe à
    lunettes
d   Maskenhäherling
i   Garrulo schiamazzante dagli occhiali

4110   *Garrulax poecilorhynchus*
Passeriformes - Sylviidae

e Rusty Laughingthrush; Rufous Laughingthrush; Scaly-headed Laughingthrush
f Garrulaxe à tête rayée
d Rosthalshäherling
i Garrulo schiamazzante rossiccio

4111 Garrulax ruficollis
Passeriformes - Sylviidae
e Rufous-necked Laughingthrush
f Garrulaxe à col roux
d Rothalshäherling
i Garrulo schiamazzante collorosso

4112 Garrulax rufifrons
Passeriformes - Sylviidae
e Rufous-fronted Laughingthrush; Red-fronted Laughingthrush
f Garrulaxe à front roux
d Rotstirnhäherling
i Garrulo schiamazzante fronterossa

4113 Garrulax rufogularis
Passeriformes - Sylviidae
e Rufous-chinned Laughingthrush
f Garrulaxe à gorge rousse
d Rostkinnhäherling
i Garrulo schiamazzante golarossa

4114 Garrulax sannio
Passeriformes - Sylviidae
e White-browed Laughingthrush; White-cheeked Laughingthrush
f Garrulaxe à sourcils blancs
d Weißwangenhäherling
i Garrulo schiamazzante sopracciglibianchi; Garrulo schiamazzante dai sopraciggli bianchi

4115 Garrulax squamatus
Passeriformes - Sylviidae
e Blue-winged Laughingthrush
f Garrulaxe écaillé
d Blauflügelhäherling
i Garrulo schiamazzante aliazzurre

4116 Garrulax strepitans
Passeriformes - Sylviidae
e White-necked Laughingthrush; Tickell's Laughingthrush; Brown-breasted Laughingthrush
f Garrulaxe bruyant
d Weißhalshäherling
i Garrulo schiamazzante di Tickell

4117 Garrulax striatus
Passeriformes - Sylviidae
e Striated Laughingthrush
f Garrulaxe strié
d Streifenhäherling
i Garrulo schiamazzante striato

4118 Garrulax subunicolor
Passeriformes - Sylviidae
e Scaly Laughingthrush; Plain-coloured Laughingthrush
f Garrulaxe modeste
d Goldschwingenhäherling
i Garrulo schiamazzante monocolore

4119 Garrulax sukatschewi
Passeriformes - Sylviidae
e Snowy-cheeked Laughingthrush; Sukatchev's Laughingthrush; Black-fronted Laughingthrush
f Garrulaxe de Sukatschev
d Kansu-Häherling
i Garrulo schiamazzante di Sukatschew

4120 Garrulax variegatus
Passeriformes - Sylviidae
e Variegated Laughingthrush; David's Laughingthrush
f Garrulaxe varié
d Buntflügelhäherling
i Garrulo schiamazzante variegto

4121 Garrulax vassali
Passeriformes - Sylviidae
e White-cheeked Laughingthrush
f Garrulaxe de Vassal
d Schwarzohrhäherling
i Garrulo schiamazzante gauncebianche

4122 Garrulax virgatus
Passeriformes - Sylviidae
e Striped Laughingthrush; Streaked Laughingthrush; Nabipur Laughingthrush

4149 **Geopelia cuneata**
Columbiformes - Columbidae
e Diamond Dove; Little Dove; Little Turtle-Dove; Red-eyed Dove
f Géopélie diamant
d Diamanttäubchen
i Tortora diamantina

4150 **Geopelia humeralis**
Columbiformes - Columbidae
e Bar-shouldered Dove; Bar-shouldered Ground-Dove; Bar-shouldered Mangrove Dove; Bronze-necked Dove; Coppery-necked Dove; Scrub Dove
f Géopélie à nuque rousse
d Kupfernackentaube
i Tortora spallebarrate

4151 **Geopelia maugeus**
Columbiformes - Columbidae
e Barred Dove; Timor Zebra Dove
f Géopélie de Mauge
d Zebratäubchen
i Tortora zebrata di Timor

4152 **Geopelia placida**
Columbiformes - Columbidae
e Peaceful Dove; Gould's Zebra Dove
f Géopélie placide
d Friedenstäubchen
i Tortora placida di Gould

4153 **Geopelia placida papua**
Columbiformes - Columbidae
e Torres Strait Peaceful Dove (ANZ)

4154 **Geopelia striata**
Columbiformes - Columbidae
e Zebra Dove; Barred Dove; Barred Ground-Dove; Peaceful Dove; Placid Dove
f Géopélie zebrée; Colombine zebrée
d Sperbertäubchen
i Tortora zebrata; Tortora striata

4155 **Geophaps lophotes**
Columbiformes - Columbidae
e Crested Pigeon; Crested Dove; Crested Bronzewing; Whistling-winged Pigeon; Topknot Pigeon

f Colombine longup
d Spitzschopftaube; Schopftaube
i Piccione alibronzate crestato

4156 **Geophaps plumifera**
Columbiformes - Columbidae
e Spinifex Pigeon; Plumed Pigeon; Plumed Rock-Pigeon; White-bellied Plumed Pigeon; Plumed Ground-Dove; Crested Ground-Dove; Red-plumed Pigeon
f Colombine plumifère
d Rotschopftaube
i Piccione plumifero

4157 **Geophaps plumifera**
Columbiformes - Columbidae
e Eastern Spinifex-Pigeon (ANZ)

4158 **Geophaps scripta**
Columbiformes - Columbidae
e Squatter Pigeon; Partridge Pigeon; Partridge Bronzewing; Blue-eyed Partridge-Bronzewing
f Colombine marquetée
d Blaubrillentaube; Buchstabentaube
i Piccione ornato

4159 **Geophaps scripta scripta**
Columbiformes - Columbidae
e Southern Squatter-Pigeon (ANZ)

4160 **Geophaps smithii**
Columbiformes - Columbidae
e Partridge Pigeon; Bare-eyed Partridge-Bronzewing; Partridge Bronzewing; Naked-eyed Partridge-Bronzewing; Red-eyed Partridge-Pigeon; Red-eyed Partridge-Bronzewing; Smith's Bare-eyed Pigeon
f Colombine de Smith
d Rotbrillentaube; Schuppenbrusttaube
i Piccione di Smith

4161 **Geophaps smithii blaauwi**
Columbiformes - Columbidae
e Western Partridge-Pigeon (ANZ)

4162   *Geophaps smithii smithii*
　　　　Columbiformes - Columbidae
e　　Eastern Partridge-Pigeon (ANZ)

4163   *Geositta antarctica*
　　　　Passeriformes - Furnariidae
e　　Short-billed Miner
f　　Géositte à bec court
d　　Feuerland-Erdhacker
i　　Minatore beccocorto

4164   *Geositta crassirostris*
　　　　Passeriformes - Furnariidae
e　　Thick-billed Miner
f　　Géositte à bec epais
d　　Dickschnabelerdhacker
i　　Minatore beccogrosso

4165   *Geositta cunicularia*
　　　　Passeriformes - Furnariidae
e　　Common Miner, Miner
f　　Géositte mineuse
d　　Kaninchenerdhacker
i　　Minatore comune; Minatore

4166   *Geositta isabellina*
　　　　Passeriformes - Furnariidae
e　　Creamy-rumped Miner; Cream-
　　　　rumped Miner
f　　Géositte isabelle
d　　Isabellerdhhacker
i　　Minatore isabellino

4167   *Geositta maritima*
　　　　Passeriformes - Furnariidae
e　　Greyish Miner
f　　Géositte grise
d　　Grauerdhacker
i　　Minatore grigiastro

4168   *Geositta peruviana*
　　　　Passeriformes - Furnariidae
e　　Coastal Miner
f　　Géosite du Pérou
d　　Küstenerdhacker; Lerchenerdhacker
i　　Minatore del Perù

4169   *Geositta punensis*
　　　　Passeriformes - Furnariidae
e　　Puna Miner
f　　Géositte du Puna

d　　Punaerdhacker
i　　Minatore della Puna

4170   *Geositta rufipennis*
　　　　Passeriformes - Furnariidae
e　　Rufous-banded Miner
f　　Géositte à ailes rousses
d　　Rotschwanzerdhacker
i　　Minatore rufipenne

4171   *Geositta saxicolina*
　　　　Passeriformes - Furnariidae
e　　Dark-winged Miner
f　　Géositte à ailes sombres
d　　Schwarzflügelerdhacker
i　　Minatore alibrune

4172   *Geositta tenuirostris*
　　　　Passeriformes - Furnariidae
e　　Slender-billed Miner
f　　Géositte à bec grêle
d　　Dünnschnabelerdhacker
i　　Minatore beccosottile

4173   *Geospiza conirostris*
　　　　Passeriformes - Fringillidae
e　　Large Cactus-Finch; Large Cactus
　　　　Ground-Finch
f　　Géospize à bec conique
d　　Opuntiengrundfink; Großer
　　　　Kaktusfink
i　　Fringuello terricolo grosso dei cactus

4174   *Geospiza difficilis*
　　　　Passeriformes - Fringillidae
e　　Sharp-beaked Ground-Finch; Sharp-
　　　　beaked Finch; Sharp-billed Ground-
　　　　Finch; Sharp-billed Finch
f　　Géospize à bec pointu
d　　Spitzschnabelgrundfink
i　　Fringuello terricolo beccotagliente

4175   *Geospiza fortis*
　　　　Passeriformes - Fringillidae
e　　Medium Ground-Finch
f　　Géospize à bec moyen
d　　Mittelgrundfink
i　　Fringuello terricolo medio

4176   *Geospiza fuliginosa*
　　　　Passeriformes - Fringillidae

e   Small Ground-Finch
f   Géospize fuligineux; Petit Pinson
    terrestre
d   Kleingrundfink; Kleiner Grundfink
i   Fringuello terricolo minore;
    Fringuello terricolo piccolo

4177   ***Geospiza magnirostris***
    Passeriformes - Fringillidae
e   Large Ground-Finch
f   Géospize à gros bec; Grand Pinson
    terrestre
d   Großgrundfink; Großer Grundfink
i   Fringuello terricolo grosso

4178   ***Geospiza scandens***
    Passeriformes - Fringillidae
e   Common Cactus-Finch; Cactus-
    Finch; Small Cactus-Finch; Cactus
    Ground-Finch
f   Géospize des cactus; Pinson terrestre
    des cactus
d   Kaktusgrundfink; Kaktusfink
i   Fringuello terricolo dei cactus

4179   ***Geothlypis aequinoctialis***
    Passeriformes - Parulidae
e   Masked Yellowthroat; Black-lored
    Yellowthroat
f   Paruline equatoriale
d   Maskengelbkehlchen
i   Parula dei tropici

4180   ***Geothlypis auricularis***
    Passeriformes - Parulidae
e   Black-lored Yellowthroat

4181   ***Geothlypis beldingi***
    Passeriformes - Parulidae
e   Belding's Yellowthroat; Peninsular
    Yellowthroat
f   Paruline de Belding
d   Belding-Gelbkehlchen
i   Parula di Belding

4182   ***Geothlypis chapalensis***
    Passeriformes - Parulidae
e   Chapala Yellowthroat

4183   ***Geothlypis chiriquensis***
    Passeriformes - Parulidae

e   Chiriqui Yellowthroat
d   Panama-Gelbkehlchen

4184   ***Geothlypis flavovelata***
    Passeriformes - Parulidae
e   Altamira Yellowthroat; Yellow-
    crowned Yellowthroat
d   Goldscheitelgelbkehlchen
i   Parula coronata

4185   ***Geothlypis nelsoni***
    Passeriformes - Parulidae
e   Hooded Yellowthroat; Brush
    Yellowthroat
d   Nelson-Gelbkehlchen
i   Parula di Nelson

4186   ***Geothlypis poliocephala***
    Passeriformes - Parulidae
e   Grey-crowned Yellowthroat; Ground
    Chat; Meadow Warbler; Grey-
    crowned Ground-Chat
d   Wiesengelbkehlchen
i   Parula capogrigio

4187   ***Geothlypis rostrata***
    Passeriformes - Parulidae
e   Bahama Yellowthroat; Black-eyed
    Bird (WI); Sagebird (WI); Bahaman
    Yellowthroat
d   Bahama-Gelbkehlchen
i   Parula delle Bahama

4188   ***Geothlypis semiflava***
    Passeriformes - Parulidae
e   Olive-crowned Yellowthroat
d   Olivscheitelgelbkehlchen
i   Parula corona olivacea

4189   ***Geothlypis speciosa***
    Passeriformes - Parulidae
e   Black-polled Yellowthroat; Orizaba
    Yellowthroat
d   Ockerbrustgelbkehlchen
i   Parula di Orizaba

4190   ***Geothlypis trichas***
    Passeriformes - Parulidae
e   Common Yellowthroat;
    Yellowthroat; Chapala Yellowthroat
f   Paruline masquée; Fauvette masquée

d   Gelbkehlchen; Maryland-
     Waldsänger; Weidenwaldsänger
i   Parula golagialla

4191   Geothlypis velata
     Passeriformes - Parulidae
e   Southern Yellowthroat;
     Yellowthroat; Maryland
     Yellowthroat; Common Yellowthroat
     (NA)

4192   Geotrygon albifacies
     Columbiformes - Columbidae
e   White-faced Quail-Dove
f   Colombe des nuages
i   Tortora quaglia facciabianca

4193   Geotrygon caniceps
     Columbiformes - Columbidae
e   Grey-headed Quail-Dove;
     Moustached Quail-Dove; Grey faced
     Quail-Dove; Grey-headed Ground-
     Pigeon; Hispaniolan Quail-Dove
f   Colombe de Gundlach
d   Graukopferdtaube
i   Tortora quaglia facciagrigia

4194   Geotrygon carrikeri
     Columbiformes - Columbidae
e   Veracruz Quail-Dove

4195   Geotrygon chiriquensis
     Columbiformes - Columbidae
e   Rufous-breasted Quail-Dove;
     Chiriqui Quail-Dove; Chiriqui
     Pigeon; Chiriquiri Dove
f   Colombe du Chiriqui
d   Streifentaube
i   Tortora quaglia pettirosso

4196   Geotrygon chrysia
     Columbiformes - Columbidae
e   Key West Quail-Dove
f   Colombe à joues blanches; Colombe
     chamarée
d   Glanzerdtaube
i   Tortora quaglia dorata

4197   Geotrygon costaricensis
     Columbiformes - Columbidae
e   Buff-fronted Quail-Dove; Costa

     Rican Quail-Dove; Costa Rica Quail-
     Dove
f   Colombe du Costa Rica
d   Costarica-Taube; Costarica-
     Wachteltaube
i   Tortora quaglia della Costarica

4198   Geotrygon frenata
     Columbiformes - Columbidae
e   White-throated Quail-Dove; Pink-
     faced Quail-Dove; Bourcier's Quail-
     Dove; Alamor Quail-Dove; Peruvian
     Quail-Dove
f   Colombe à gorge blanche; Colombe
     de Bourcier
d   Peru-Wachteltaube; Zügeltaube
i   Tortora quaglia golabianca

4199   Geotrygon goldmani
     Columbiformes - Columbidae
o   Russet-crowned Quail-Dove;
     Goldman's Quail-Dove; Mount Pirrie
     Quail Dove
f   Colombe de Goldman
d   Goldmann-Taube; Goldmann-
     Wachteltaube
i   Tortora quaglia Goldman

4200   Geotrygon lawrencii
     Columbiformes - Columbidae
e   Purplish-backed Quail-Dove;
     Lawrence's Quail-Dove
f   Colombe de Lawrence
d   Purpurrückentaube;
     Purpurrückenwachteltaube
i   Tortora quaglia di Lawrence

4201   Geotrygon linearis
     Columbiformes - Columbidae
e   Lined Quail-Dove; White-faced
     Quail-Dove; Chiriquiri Quail-Dove;
     Brown Quail-Dove
f   Colombe bridée
d   Streifentaube;
     Weißgesichtwachteltaube
i   Tortora quaglia lineata

4202   Geotrygon martinica
     Columbiformes - Columbidae
e   Martinique Quail-Dove

4203   *Geotrygon montana*
Columbiformes - Columbidae

e   Ruddy Quail-Dove; Red Partridge
(WI); Partridge (WI); Mountain
Dove (WI); Partridge Dove; Red
Mountain-Dove; Rufous Quail-Dove

f   Colombe rouviolette; Perdrix rouge
(Ants); Perdrix grise (Ants)

d   Bergtaube; Rote Erdtaube

i   Tortora quaglia montana

4204   *Geotrygon mystacea*
Columbiformes - Columbidae

e   Bridled Quail-Dove; Marmy Dove
(WI); Marble Dove (WI); Wood
Dove (WI); Wood Hen (WI);
Partridge (WI)

f   Colombe à croissants; Colombe à
moustaches

d   Schnurrbarttaube

i   Tortora quaglia mistacea

4205   *Geotrygon saphirina*
Columbiformes - Columbidae

e   Sapphire Quail-Dove; Purple Quail-
Dove

f   Colombe saphir

d   Saphirtaube

i   Tortora quaglia zaffiro

4206   *Geotrygon veraguensis*
Columbiformes - Columbidae

e   Olive-backed Quail-Dove; Veragua
Quail-Dove

f   Colombe de Veragua

d   Veragua-Taube; Veragua-
Wachteltaube

i   Tortora quaglia di Veragua

4207   *Geotrygon versicolor*
Columbiformes - Columbidae

e   Crested Quail-Dove; Mountain Witch
(WI)

f   Colombe versicolore

d   Jamaika-Erdtaube; Kurzschopftaube

i   Tortora quaglia crestata

4208   *Geotrygon violacea*
Columbiformes - Columbidae

e   Violaceous Quail-Dove; Blue Dove
(WI); Blue Partridge (WI); Violet

Quail-Dove; White-bellied Quail-
Dove

f   Colombe à nuque violette

d   Bischofstaube; Violette Erdtaube

i   Tortora quaglia violacea

4209   *Geranoaetus melanoleucus*
Falconiformes - Accipitridae

e   Black-chested Buzzard-Eagle; Grey
Eagle-Buzzard; Black Buzzard-
Eagle; Grey Buzzard-Eagle

f   Buse aguia

d   Aguja

i   Poiana-aquila pettonero

4210   *Geranospiza caerulescens*
Falconiformes - Accipitridae

e   Crane Hawk; Grey Crane-Hawk

f   Buse échasse; Serpentaire ardoisé

d   Sperberweihe

i   Sparviero trampoliere; Falco gru

4211   *Geranospiza gracilis*
Falconiformes - Accipitridae

e   Banded Crane-Hawk

4212   *Geranospiza nigra*
Falconiformes - Accipitridae

e   Blackish Crane-Hawk

4213   *Geronticus calvus*
Ciconiiformes - Threskiornithidae

e   Bald Ibis; Southern Bald Ibis

f   Ibis du Cap

d   Glattnackenrapp; Kahlkopfrapp

i   Ibis calvo

4214   *Geronticus eremita*
Ciconiiformes - Threskiornithidae

e   Waldrapp; Northern Bald Ibis;
Hermit Ibis; Bald Ibis; Red-cheeked
Ibis; Red-naped Ibis

f   Ibis chauve

d   Waldrapp

i   Ibis ciuffetto; Ibis eremita

4215   *Gerygone albofrontata*
Passeriformes - Pardalotidae

e   Chatham Islands Gerygone; Chatham
Islands Warbler; Chatham Islands
Gerygone-Warbler; Chatham Islands

Flyeater; Chatham Gerygone
f   Gérygone des Chatham
d   Langschnabelgerygone
i   Gerigone dell'Isola Chatham

4216   *Gerygone chloronotus*
Passeriformes - Pardalotidae
e   Green-backed Gerygone; Green-
backed Flyeater; Grey-headed
Gerygone; Grey-headed Gerygone-
Warbler
f   Gérygone à dos vert
d   Grünrückengerygone
i   Gerigone dorsoverde

4217   *Gerygone chrysogaster*
Passeriformes - Pardalotidae
e   Yellow-bellied Gerygone; Yellow-
bellied Flyeater; Yellow-bellied
Gerygone-Warbler
f   Gérygone à ventre jaune
d   Gelbbauchgerygone
i   Gerigone ventre giallo

4218   *Gerygone cinerea*
Passeriformes - Pardalotidae
e   Mountain Gerygone; Grey Flyeater;
New Guinea Gerygone-Warbler;
Pacific Grey Gerygone-Warbler;
Pacific Grey-Gerygone; Grey
Gerygone; Grey Warbler; Pacific
Gerygone
f   Gérygone grise
d   Weißbürzelgerygone
i   Gerigone grigia

4219   *Gerygone dorsalis*
Passeriformes - Pardalotidae
e   Rufous-sided Gerygone; Rufous-
sided Flyeater; Rufous-sided Fairy
Warbler; Tanimbar Gerygone; Lesser
Sunda Gerygone; Rufous-tailed
Gerygone; Rufous-tailed Flyeater
f   Gérygone à flancs roux
i   Gerigone fianchirossi

4220   *Gerygone flavolateralis*
Passeriformes - Pardalotidae
e   Fan-tailed Gerygone; Fan-tailed
Flyeater; New Caledonian Gerygone-
Warbler; Fan-tailed Gerygone-

Warbler; New Caledonia Gerygone;
South Melanesian Fairy-Warbler
f   Gérygone mélanésienne
d   Fächerschwanzgerygone
i   Gerigone coda a ventaglio

4221   *Gerygone fusca*
Passeriformes - Pardalotidae
e   Western Gerygone; White-tailed
Flyeater; Western Gerygone-
Warbler; White-tailed Gerygone;
Plain Gerygone
f   Gérygone à queue blanche
d   Weißschwanzgerygone
i   Gerigone codabianca

4222   *Gerygone hypoxantha*
Passeriformes - Pardalotidae
e   Biak Gerygone

4223   *Gerygone igata*
Passeriformes - Pardalotidae
e   Grey Gerygone; New Zealand Grey-
Flyeater; New Zealand Grey
Gerygone; Grey Gerygone-Warbler;
Riroriro; Grey Warbler; New
Zealand Flyeater; New Zealand
Gerygone
f   Gérygone de Nouvelle-Zélande
d   Maorigerygone; Graubuschpfeifer
i   Gerigone grigia della Nuova Zelanda

4224   *Gerygone inornata*
Passeriformes - Pardalotidae
e   Plain Gerygone; Plain Flyeater;
Timor Gerygone-Warbler;
Unadorned Gerygone; Plain Fairy
Warbler
f   Gérygone terne
d   Timor-Gerygone
i   Gerigone disadorna

4225   *Gerygone insularis*
Passeriformes - Pardalotidae
e   Lord Howe Island Gerygone; Lord
Howe Gerygone (ANZ); Lord Howe
Gerygone-Warbler; Lord Howe
Island Flyeater
f   Gérygone de Lord Howe
d   Lord Howe-Gerygone
i   Gerigone dell'Isola Lord Howe

4226 *Gerygone levigaster*
Passeriformes - Pardalotidae
e Mangrove Gerygone; Mangrove
Flyeater; Buff-breasted Gerygone-
Warbler; Buff-breasted Gerygone
f Gérygone des mangroves
d Mangrovegerygone
i Gerigone delle mangrovie

4227 *Gerygone magnirostris*
Passeriformes - Pardalotidae
e Large-billed Gerygone; Large-billed
Flyeater; Large-billed Gerygone-
Warbler; Swamp Gerygone-Warbler;
Dusky Gerygone; Swamp Gerygone
f Gérygone à bec fort
d Sumpfgerygone
i Gerigone beccogrosso

4228 *Gerygone magnirostris*
brunneipectus
Passeriformes - Pardalotidae
e Torres Strait Large-billed Gerygone
(ANZ)

4229 *Gerygone modesta*
Passeriformes - Pardalotidae
e Norfolk Island Gerygone; Norfolk
Island Gerygone Warbler; Norfolk
Gerygone; Norfolk Island Flyeater;
Grey Gerygone
f Gérygone de Norfolk
d Norfolk-Gerygone
i Gerigone dell'Isola Norfolk

4230 *Gerygone mouki*
Passeriformes - Pardalotidae
e Brown Gerygone; Brown Flyeater;
Brown Gerygone-Warbler; Northern
Gerygone
f Gérygone brune
d Grauwangengerygone
i Gerigone bruna

4231 *Gerygone olivacea*
Passeriformes - Pardalotidae
e White-throated Gerygone; White-
throated Flyeater; White-throated
Gerygone-Warbler
f Gérygone à gorge blanche

d Weißkehlgerygone
i Gerigone golabianca

4232 *Gerygone palpebrosa*
Passeriformes - Pardalotidae
e Fairy Gerygone; Black-headed
Flyeater; Fairy Gerygone-Warbler;
Black-throated Gerygone; Black-
headed Gerygone
f Gérygone enchanteresse
d Elfengerygone
i Gerigone testanera

4233 *Gerygone ruficollis*
Passeriformes - Pardalotidae
e Brown-breasted Gerygone; Treefern
Gerygone; Treefern Gerygone-
Warbler; Treefern Flyeater
f Gérygone à cou brun
d Baumfarngerygone
i Gerigone dorsobruno

4234 *Gerygone sulphurea*
Passeriformes - Pardalotidae
e Golden-bellied Gerygone; Yellow-
breasted Flyeater; Yellow-breasted
Gerygone-Warbler; Golden-breasted
Flyeater; Yellow-breated Gerygone;
Yellow-breasted Wren-Warbler
f Gérygone soufrée
d Goldbrustgerygone;
Gelbbrustbuschpfeifer
i Gerigone ventresulfureo

4235 *Gerygone tenebrosa*
Passeriformes - Pardalotidae
e Dusky Gerygone; Dusky Gerygone-
Warbler; Dusky Flyeater
f Gérygone blafarde
d Braunrückengerygone
i Gerigone tenebrosa

4236 *Glareola cinerea*
Ciconiiformes - Glareolidae
e Grey Pratincole; Cream-coloured
Pratincole
f Glaréole grise
d Weißachselbrachschwalbe
i Pernice di mare cenerina

4237 *Glareola lactea*
Ciconiiformes - Glareolidae
e Small Pratincole; Little Pratincole;
Small Swallow-Plover (ISC); Small
Indian Pratincole; Indian Little-
Pratincole; Milky Pratincole;
Swallow-Plover
f Glaréole lactée
d Sandbrachchwalbe
i Pernice di mare minore

4238 *Glareola maldivarum*
Ciconiiformes - Glareolidae
e Oriental Pratincole; Eastern Collared-
Pratincole; Large Indian Pratincole;
Indian Pratincole; Collared Swallow-
Plover
f Glaréole orientale
d Orientbrachschwalbe;
Rotbrustbrachschwalbe
i Pernice di mare dal collare

4239 *Glareola nordmanni*
Ciconiiformes - Glareolidae
e Black-winged Pratincole
f Glaréole à ailes noires
d Schwarzflügelbrachschwalbe;
Schwarzflügelige Brachschwalbe
i Pernice di mare orientale

4240 *Glareola nuchalis*
Ciconiiformes - Glareolidae
e Rock Pratincole; White-collared
Pratincole (CSA); Collared
Pratincole
f Glaréole aurcolée
d Weißnackenbrachschwalbe;
Halsbandbrachschwalbe

4241 *Glareola ocularis*
Ciconiiformes - Glareolidae
e Madagascar Pratincole
f Glaréole malgache
d Madagaskar-Brachschwalbe
i Pernice del mare del Madagascar

4242 *Glareola pratincola*
Ciconiiformes - Glareolidae
e Collared Pratincole; Red-winged
Pratincole; Swallow-Plover; Collared
Swallow-Plover (ISC); Large Indian

Pratincole (ISC); Pratincole;
Common Pratincole; European
Pratincole
f Glaréole à collier
d Braunflügelbrachschwalbe;
Braunflügelige Brachschwalbe;
Brachschwalbe;
Rotflügelbrachschwalbe
i Pernice di mare; Pernice di mare
comune

4243 Glareolidae
Ciconiiformes
e Pratincoles; Coursers
d Brachschwalbenartige Vögel

4244 *Glaucidium albertinum*
Strigiformes - Strigidae
e Albertine Owlet; Prigogine's Owlet
i Civetta nana di Prigogine

4245 *Glaucidium bolivianum*
Strigiformes - Strigidae
e Yungas Pygmy-Owl
f Chevêchette de Graben

4246 *Glaucidium brasilianum*
Strigiformes - Strigidae
e Ferruginous Pygmy-Owl;
Ferruginous Owl
f Chevêchette brune
d Strichelkauz
i Civetta nana rossiccia; Gufetto
brasiliano

4247 *Glaucidium brodiei*
Strigiformes - Strigidae
e Collared Owlet; Collared Pygmy-
Owl; Pygmy Owlet; Banded Pygmy-
Owl; Pygmy-Owl
f Chevêchette à collier
d Wachtelkauz
i Civetta nana dal collare; Gufetto dal
collare

4248 *Glaucidium californicum*
Strigiformes - Strigidae
e Northern Pygmy-Owl; Califoronian
Pygmy-Owl
f Chevêchette des rocheuses
i Civetta nana della California

4249 ***Glaucidium capense***
Strigiformes - Strigidae
e African Barred Owlet; Barred Owl
(CSA); Barred Owlet
f Chevêchette du Cap
d Kap-Kauz
i Civetta nana del Capo; Gufetto dalle
stricie

4250 ***Glaucidium castaneum***
Strigiformes - Strigidae
e Chestnut Owlet; Chestnut-barred
Owlet
f Chevêchette châtaine
d Kastanienkauz
i Civetta nana dell'Ituri

4251 ***Glaucidium castanonotum***
Strigiformes - Strigidae
e Chestnut-backed Owlet; Sri Lanka
Chestnut-backed Owlet; Ituri Owlet;
Lake District Owlet
f Chevêchette à dos marron
i Civetta nana di Sri-Lanka

4252 ***Glaucidium castanopterum***
Strigiformes - Strigidae
e Javan Owlet; Chestnut-winged
Owlet; Spadiced Owlet; Cuckoo
Owlet
f Chevêchette spadicée
d Trillerkauz
i Civetta nana alicastane

4253 ***Glaucidium cobanense***
Strigiformes - Strigidae
e Guatemalan Pygmy-Owl

4254 ***Glaucidium cuculoides***
Strigiformes - Strigidae
e Asian Barred Owlet; Cuckoo Owl;
Cuckoo Owlet; Barred Owlet
f Chevêchette cuculolde
d Trillerkauz
i Civetta nana cuculo; Gufetto
cuculoide

4255 ***Glaucidium gnoma***
Strigiformes - Strigidae
e Mountain Pygmy-Owl; Mexican
Pygmy-Owl; Northern Pygmy-Owl;

American Pygmy-Owl; Cape Pygmy-
Owl
f Chevêchette naine
d Gnomenkauz
i Civetta nana del Nordamerica;
Gufetto delle Montagne Rocciose

4256 ***Glaucidium griseiceps***
Strigiformes - Strigidae
e Central American Pygmy-Owl; Least
Pygmy-Owl

4257 ***Glaucidium hardyi***
Strigiformes - Strigidae
e Amzonian Pygmy-Owl; Hardy's
Pygmy-Owl
f Chevêchette d'Amazonie

4258 ***Glaucidium hoskinsii***
Strigiformes - Strigidae
e Cape Pygmy-Owl

4259 ***Glaucidium jardinii***
Strigiformes - Strigidae
e Andean Pygmy-Owl; Mountain
Pygmy-Owl
f Chevêchette des Andes
d Anden-Kauz
i Civetta nana delle Ande; Gufetto
delle Ande

4260 ***Glaucidium minutissimum***
Strigiformes - Strigidae
e Least Pygmy-Owl; Amazonian
Pygmy-Owl; Colima Pygmy-Owl;
Central American Pygmy-Owl
f Chevêchette cabouré
d Zwergkauz
i Civetta pigmeo

4261 ***Glaucidium nanum***
Strigiformes - Strigidae
e Austral Pygmy-Owl; Brazilian
Pygmy-Owl; Least Pygmy-Owl (NA)
f Chevêchette australe
d Araukaner Kauz

4262 ***Glaucidium ngamiense***
Strigiformes - Strigidae
e Ngami Owlet

f Chevêchette de Ngami
i Civetta nana di Ngami

4263 ***Glaucidium palmarum***
         Strigiformes - Strigidae
e Colima Pygmy-Owl

4264 ***Glaucidium parkeri***
         Strigiformes - Strigidae
e Subtropical Pygmy-Owl

4265 ***Glaucidium passerinum***
         Strigiformes - Strigidae
e Eurasian Pygmy-Owl (NA); Pygmy-
     Owl; European Pygmy-Owl;
     Northern Pygmy-Owl; Pygmy Owlet;
     Old World Pygmy-Owl
f Chevêchette d'Europe
d Sperlingskauz
i Civetta nana; Civetta nana eurasiatica

4266 ***Glaucidium perlatum***
         Strigiformes - Strigidae
e Pearl-spotted Owlet; Pearl-spotted
     Owl (CSA)
f Chevêchette perlée
d Perlkauz
i Civetta nana perlata; Gufetto
     perlaceo

4267 ***Glaucidium peruanum***
         Strigiformes - Strigidae
e Pacific Pygmy-Owl; Peruvian
     Pygmy-Owl
f Chevêchette du Pérou

4268 ***Glaucidium radiatum***
         Strigiformes - Strigidae
e Jungle Owlet; Barred Jungle-Owlet
     (ISC)
f Chevêchette du jungle
d Dschungelkauz
i Civetta nana della giungla; Gufetto
     della giungla

4269 ***Glaucidium sanchezi***
         Strigiformes - Strigidae
e Tamaulipas Pygmy-Owl

4270 ***Glaucidium scheffleri***
         Strigiformes - Strigidae

e Scheffler's Owlet; Eastern Barred
     Owlet
f Chevêchette de Scheffler
i Civetta nana di Scheffler

4271 ***Glaucidium siju***
         Strigiformes - Strigidae
e Cuban Pygmy-Owl
f Chevêchette de Cuba
d Kuba-Kauz
i Civetta nana di Cuba

4272 ***Glaucidium sjostedti***
         Strigiformes - Strigidae
e Sjostedt's Owlet; Chestnut-backed
     Owlet; Sjostedt's Barred Owlet
f Chevêchette à queue barrée
d Prachtkauz
i Civetta nana dorsocastano; Gufetto
     dal dorso nocciola

4273 ***Glaucidium tephronotum***
         Strigiformes - Strigidae
e Red-chested Owlet
f Chevêchette à pieds jaunes
d Rotbrustkauz
i Civetta nana pettorossiccio; Gufetto
     dl petto rosso

4274 ***Glaucidium tucumanum***
         Strigiformes - Strigidae
e Tucuman Pygmy-Owl

4275 ***Glaucis aenea***
         Apodiformes - Trochilidae
e Bronzy Hermit; Bronze Hermit;
     Chestnut-coloured Hermit
f Ermite bronzé
d Erzeremit
i Eremita bronzato

4276 ***Glaucis hirsuta***
         Apodiformes - Trochilidae
e Rufous-breasted Hermit; Brown
     Hummingbird (WI); Brown Doctor-
     Bird (WI); Doctor-Bird (WI); Brown
     Breast (WI); Hairy Hermit
f Ermite hirsute; Colibri balisier (Ants)
d Rotschwanzeremit
i Eremita pettorossiccio

4277 *Glossopsitta concinna*
Psittaciformes - Loriidae
e Musk Lorikeet; Musky Lorikeet;
Green Keet; Green Leek; King
Parrot; Red-eared Lorikeet; Musk
Lory
f Lori à bandeau rouge
d Moschuslori
i Lorichetto muschiato

4278 *Glossopsitta porphyrocephala*
Psittaciformes - Loriidae
e Purple-crowned Lorikeet; Blue-
crowned Lorikeet; Purple-capped
Lorikeet
f Lori à couronne pourpre
d Blauscheitellori; Porphyrkopflori
i Lorichetto corona viola

4279 *Glossopsitta pusilla*
Passeriformes - Meliphagidae
e Little Lorikeet; Little Keet; Green
Keet; Green Parakeet; Green Leek;
Jerryang; Dwarf Lory; Little Lory;
Red-faced Lorikeet
f Lori à masque rouge
d Maskenlori; Zwergmoschuslori;
Zwerglori
i Lorichetto nano

4280 *Glycichaera fallax*
Passeriformes - Dendrocolaptidae
e Green-backed Honeyeater; White-
eyed Honeyeater
f Méliphage trompeur
d Grünmantelhonigfresser
i Mangimiele occhialino

4281 *Glyphorynchus spirurus*
Passeriformes - Dendrocolaptidae
e Wedge-billed Woodcreeper; Wedge-
billed Creeper
f Grimpar bec-en-coin
d Rindenpicker
i Rampichino becco a cuneo

4282 *Gnorimopsar chopi*
Apodiformes - Trochilidae
e Chopi Blackbird; Chopi Grackle
f Quiscale chopi

d Chopistärling; Schopi
i Gracchio ciopi

4283 *Goethalsia bella*
Apodiformes - Trochilidae
e Rufous-cheeked Hummingbird; Pirre
Hummingbird; Goethal's
Hummingbird
f Colibri du Pirre
d Rotwangenkolibri
i Colibrì di Goethals

4284 *Goldmania violiceps*
Ciconiiformes - Ardeidae
e Violet-capped Hummingbird;
Goldman's Hummingbird
f Colibri à calotte violette
d Goldman-Kolibri; Goldmans Kolibri
i Colibrì di Goldman

4285 *Gorsachius goisagi*
Ciconiiformes - Ardeidae
e Japanese Night-Heron; Japanese
Bittern; Brown Night-Heron
f Bihoreau goisagi
d Rotscheitelreiher
i Nitticora giapponese

4286 *Gorsachius leuconotus*
Ciconiiformes - Ardeidae
e White-backed Night-Heron
f Bihoreau à dos blanc
d Weißrückennachtreiher
i Nitticora dorsobianco

4287 *Gorsachius magnificus*
Ciconiiformes - Ardeidae
e White-eared Night-Heron; Hainan
Night-Heron; Chinese Night-Heron;
Magnificent Night-Heron
f Bihoreau superbe
d Hainan-Reiher
i Nitticora magnifica

4288 *Gorsachius melanolophus*
Columbiformes - Columbidae
e Malayan Night-Heron; Malay Night-
Heron; Malaysian Night-Heron;
Tiger Bittern
f Bihoreau malais

d   Wellenreiher
i   Nitticora della Malesia

4289   *Goura cristata*
Columbiformes - Columbidae
e   Crowned Pigeon; Common
Crowned-Pigeon; Western Crowned-
Pigeon; Blue Crowned-Pigeon; Grey
Crowned-Pigeon; Blue Crowned-
Goura; Grey Crowned Goura
f   Goura couronné
d   Krontaube
i   Gura coronata; Gura azzura;
Colomba coronata; Goura coronata

4290   *Goura scheepmakeri*
Columbiformes - Columbidae
e   Sheepmaker's Crowned-Pigeon;
Maroon-breasted Crowned-Pigeon;
Southern Crowned-Pigeon; Sclater's
Ground-Pigeon
f   Goura de Sheepmaker
d   Rotbrustkrontaube
i   Gura di Scheepmaker; Goura di
Sheepmaker

4291   *Goura victoria*
Passeriformes - Sturnidae
e   Victoria Crowned-Pigeon; Victoria
Goura; White-tipped Ground-Goura;
Victoria Crowned-Goura; White-
tipped Ground-Pigeon
f   Goura de Victoria
d   Fächertaube; Victoria-Krontaube
i   Gura di Vittoria; Colomba coronata
di Vittoria

4292   *Gracula indica*
Passeriformes - Sturnidae
e   Southern Hill-Mynah
d   Beo

4293   *Gracula ptilogenys*
Passeriformes - Sturnidae
e   Sri Lanka Mynah; Ceylon Mynah;
Ceylon Hill Mynah; Ceylon Grackle
f   Mainate de Ceylan
d   Ceylon-Beo; Dschungelatzel
i   Gracula di Ceylon; Gracula di Sri-
Lanka

4294   *Gracula religiosa*
Passeriformes - Sturnidae
e   Hill Mynah; Talking Mynah; Indian
Hill-Mynah; Grackle; Eastern Hill-
Mynah; Javan Hill-Mynah; Common
Hill-Mynah; Common Grackle (ISC);
Indian Grackle; Hill Grackle
f   Mainate réligieux
d   Beo; Hügelatzel
i   Gracula religiosa; Maina; Merlo
indiano

4295   *Grafisia torquata*
Passeriformes - Sturnidae
e   White-collared Starling
f   Rufipenne à cou blanc; Étourneau à
poitrine blanche
d   Ringstar
i   Storno dal collare bianco

4296   *Grallaria albigula*
Passeriformes - Formicariidae
e   White-throated Ant-Pitta
f   Grallaire à gorge blanche
d   Graubrustameisenpitta
i   Pitta formichiera golabianca

4297   *Grallaria alleni*
Passeriformes - Formicariidae
e   Moustached Ant-Pitta
f   Grallaire à moustaches
d   Bartstreifameisenpitta
i   Pitta formichiera di Allen

4298   *Grallaria andicola*
Passeriformes - Formicariidae
e   Stripe-headed Ant-Pitta
f   Grallaire des Andes
d   Strichelkopfameisenpitta
i   Pitta formichiera delle Ande

4299   *Grallaria bangsi*
Passeriformes - Formicariidae
e   Santa Marta Ant-Pitta; Bangs's Ant-
Pitta
f   Grallaire des Santa Marta
d   Olivrückenameisenpitta
i   Pitta formichiera di Santa Marta

4300   *Grallaria blakei*
Passeriformes - Formicariidae

e	Chestnut Ant-Pitta
f	Grallaire de Blake
i	Pitta formichiera di Blake

4301 *Grallaria capitalis*
Passeriformes - Formicariidae
e   Bay Ant-Pitta
f   Grallaire châtaine
i   Pitta formichiera baia

4302 *Grallaria carrikeri*
Passeriformes - Formicariidae
e   Pale-billed Ant-Pitta
f   Grallaire de Carriker
i   Pitta formichiera di Carriker

4303 *Grallaria chthonia*
Passeriformes - Formicariidae
e   Tachira Ant-Pitta
f   Grallaire du Tachira
d   Tachira-Ameisenpitta
i   Pitta formichiera di Tachira

4304 *Grallaria dignissima*
Passeriformes - Formicariidae
e   Ochre-striped Ant-Pitta; Striped Ant-Pitta; Stripe-sided Ant-Pitta
f   Grallaire flammée
d   Rostkehlameisenpitta
i   Pitta formichiera striata

4305 *Grallaria eludens*
Passeriformes - Formicariidae
e   Elusive Ant-Pitta
f   Grallaire secrète
d   Weißkehlameisenpitta
i   Pitta formichiera elusiva

4306 *Grallaria erythroleuca*
Passeriformes - Formicariidae
e   Red-and-white Ant-Pitta; Chestnut-brown Ant-Pitta
f   Grallaire de Cuzco
i   Pitta formichiera bruna

4307 *Grallaria erythrotis*
Passeriformes - Formicariidae
e   Rufous-faced Ant-Pitta
f   Grallaire masquée
d   Rotohrameisenpitta
i   Pitta formichiera facciarossa

4308 *Grallaria excelsa*
Passeriformes - Formicariidae
e   Great Ant-Pitta
f   Grande Grallaire
d   Großameisenpitta
i   Pitta formichiera grande

4309 *Grallaria flavotincta*
Passeriformes - Formicariidae
e   Yellow-breasted Ant-Pitta
f   Grallaire à poitrine jaune
i   Pitta formichiera pettogiallo

4310 *Grallaria gigantea*
Passeriformes - Formicariidae
e   Giant Ant-Pitta
f   Grallaire géante
d   Riesenameisenpitta; Rostameisenpitta
i   Pitta formichiera gigante

4311 *Grallaria griseonucha*
Passeriformes - Formicariidae
e   Grey-naped Ant-Pitta
f   Grallaire à nuque grise
d   Graunackenameisenpitta
i   Pitta formichiera nucagrigia

4312 *Grallaria guatimalensis*
Passeriformes - Formicariidae
e   Scaled Ant-Pitta
f   Grallaire ecaillée
d   Schuppenkopfameisenpitta
i   Pitta formichiera del Guatemala

4313 *Grallaria haplonota*
Passeriformes - Formicariidae
e   Plain-backed Ant-Pitta; Sclater's Ant-Pitta
f   Grallaire à dos uni
d   Braunrückenameisenpitta
i   Pitta formichiera semplice

4314 *Grallaria hypoleuca*
Passeriformes - Formicariidae
e   White-bellied Ant-Pitta; Bay-backed Ant-Pitta
f   Grallaire à ventre blanc
d   Rotrückenameisenpitta
i   Pitta formichiera dorsobaiao

4315 *Grallaria kaestneri*
Passeriformes - Formicariidae
e Cundinamarca Ant-Pitta
f Grallaire de Kaestner

4316 *Grallaria milleri*
Passeriformes - Formicariidae
e Brown-banded Ant-Pitta; Miller's
Ant-Pitta
f Grallaire ceinturée
d Bandameisenpitta
i Pitta formichiera di Miller

4317 *Grallaria nuchalis*
Passeriformes - Formicariidae
e Chestnut-naped Ant-Pitta
f Grallaire à nuque rousse
d Rotkopfameisenpitta
i Pitta formichiera nucacastana

4318 *Grallaria przewalskii*
Passeriformes - Formicariidae
e Rusty-tinged Ant-Pitta; Rusty-
winged Ant-Pitta; Przewalski's Ant-
Pitta
f Grallaire de Przewalski
i Pitta formichiera di Taczanowski

4319 *Grallaria quitensis*
Passeriformes - Formicariidae
e Tawny Ant-Pitta
f Grallaire de Quito
d Bergameisenpitta
i Pitta formichiera di Quito

4320 *Grallaria ruficapilla*
Passeriformes - Formicariidae
e Chestnut-crowned Ant-Pitta
f Grallaire à tête rousse
d Rostkappenameisenpitta
i Pitta formichiera corona rossiccia

4321 *Grallaria rufocinerea*
Passeriformes - Formicariidae
e Bicoloured Ant-Pitta
f Grallaire bicolore
d Zweifarbenameisenpitta
i Pitta formichiera bicolore

4322 *Grallaria rufula*
Passeriformes - Formicariidae

e Rufous Ant-Pitta
f Grallaire rousse
d Einfarbameisenpitta
i Pitta formichiera rossiccia

4323 *Grallaria squamigera*
Passeriformes - Formicariidae
e Undulated Ant-Pitta
f Grallaire ondée
d Schuppenbauchameisenpitta;
Geschuppter Ameisenstelzer
i Pitta formichiera ondulata

4324 *Grallaria varia*
Passeriformes - Formicariidae
e Variegated Ant-Pitta
f Grallaire roi
d Königsameisenpittta;
Königsameisenstelzer;
Kurzschwänziger Ameisenvogel
i Pitta formichiera variegata

4325 *Grallaria watkinsi*
Passeriformes - Formicariidae
e Scrub Ant-Pitta; Watkins's Ant-Pitta
f Grallaire de Watkins
i Pitta formichiera di Watkins

4326 *Grallaricula cucullata*
Passeriformes - Formicariidae
e Hooded Ant-Pitta
f Grallaire à capuchon
d Rotkopfstelzling
i Pitta formichiera dal cappuccio

4327 *Grallaricula ferrugineipectus*
Passeriformes - Formicariidae
e Rusty-breasted Ant-Pitta; Rusty Ant-
Pitta
f Grallaire à poitrine rousse
d Rotbruststelzling
i Pitta formichiera pettorossiccio

4328 *Grallaricula flavirostris*
Passeriformes - Formicariidae
e Ochre-breasted Ant-Pitta; Ochrceous
Ant-Pitta
f Grallaire ocrée
d Ockerbruststelzling
i Pitta formichiera beccogiallo

4329 **Grallaricula lineifrons**
Passeriformes - Formicariidae
e Crescent-faced Ant-Pitta; Crescented Ant-Pitta
f Grallaire demi-lune
d Halbmondstelzling
i Pitta formichiera frontelineata

4330 **Grallaricula loricata**
Passeriformes - Formicariidae
e Scallop-breasted Ant-Pitta; Scalloped Ant-Pitta
f Grallaire maillée
d Schuppenbruststelzling
i Pitta formichiera loricata

4331 **Grallaricula nana**
Passeriformes - Formicariidae
e Slate-crowned Ant-Pitta; Crowned Ant-Pitta
f Grallaire naine
d Grauscheitelstelzling; Zwergameisenstelzling
i Pitta formichiera nana

4332 **Grallaricula ochraceifrons**
Passeriformes - Formicariidae
e Ochre-fronted Ant-Pitta
f Grallaire à front ocre
i Pitta formichiera fronteocra

4333 **Grallaricula peruviana**
Passeriformes - Formicariidae
e Peruvian Ant-Pitta
f Grallaire du Pérou
d Piura-Stelzling
i Pitta formichiera peruviana

4334 **Grallina bruinji**
Passeriformes - Corvidae
e Torrent Lark; Mudlark (ANZ)
f Gralline papoue
d Papua-Drosselstelze
i Grallina di Bruijn

4335 **Grallina cyanoleuca**
Passeriformes - Corvidae
e Magpie Lark; Mudlark (ANZ); Australian Magpie-Lark
f Gralline pie; Alouette-pie

d Australische Drossel; Drosselstelze
i Grallina australiana

4336 **Graminicola bengalensis**
Passeriformes - Sylviidae
e Rufous-rumped Grassbird; Large Grass-Warbler; Large Grassbird
f Grande Graminicole
d Katzengrassänger
i Forapaglie maggiore del Bengala

4337 **Granatellus pelzelni**
Passeriformes - Fringillidae
e Rose-breasted Chat; Rose-breasted Warbler
f Paruline de Pelzeln
d Rosenbauchgranatellus
i Granatello pettorosa

4338 **Granatellus sallaei**
Passeriformes - Fringillidae
e Grey-throated Chat; Grey-throated Warbler
f Paruline à plastron
d Graukehlgranatellus
i Granatello golagrigia

4339 **Granatellus venustus**
Passeriformes - Fringillidae
e Red-breasted Chat; Red-breasted Warbler
f Paruline multicolore
d Weißkehlgranatellus
i Granatello pettirosso

4340 **Grandala coelicolor**
Passeriformes - Muscicapidae
e Grandala; Hodgson's Grandala; Himalayan Grandala
f Grandala bleu; Grandala bleu-ciel
d Grandala
i Grandala di Hodgson

4341 **Grantiella picta**
Passeriformes - Meliphagidae
e Painted Honeyeater
f Méliphage peint
d Grant-Honigfresser; Mistelhonigesser
i Mangiamiele variopinto

4342   *Graueria vittata*
        Passeriformes - Sylviidae
e   Grauer's Warbler; Grauer's Bush-
        Warbler
f   Grauérie striée; Fauvette de Grauer
d   Sperberbrustsänger
i   Bigia di Grauer

4343   *Graydidascalus brachyurus*
        Psittaciformes - Psittacidae
e   Short-tailed Parrot
f   Caique à queue courte
d   Kurzschwanzpapagei
i   Pappagallo codacorta

4344   *Griseotyrannus
        aurantioatrocristatus*
        Passeriformes - Tyrannidae
e   Crowned Slaty-Flycatcher; Black-
        and-yellow Crested-Flycatcher
f   Tyran oriflamme
d   Inkatyrann
i   Tiranno grigio crestato

4345   **Gruidae**
        Gruiformes
e   Cranes
f   Grues; Gruidés
d   Kraniche
i   Gruidi; Grui

4346   **Gruiformes**
e   Cranes; Rails
f   Gruiformes
d   Kranichvögel; Rallenvögel
i   Gruiformi; Grui

4347   *Grus americana*
        Gruiformes - Gruidae
e   Whooping Crane; Whooper; Big
        White Crane
f   Grue blanche
d   Weißer Schneekranich;
        Schneekranich; Schreikranich
i   Gru americana

4348   *Grus antigone*
        Gruiformes - Gruidae
e   Sarus Crane; Eastern Sarus Crane
f   Grue antigone

d   Saruskranich
i   Gru antigone; Gru di Antigone

4349   *Grus canadensis*
        Gruiformes - Gruidae
e   Sandhill Crane; Grulla (WI); Little
        Brown-Crane; Canadian Crane
f   Grue du Canada
d   Kanada-Kranich; Kanadischer
        Kranich
i   Gru canadese; Gru del Canada

4350   *Grus carunculatus*
        Gruiformes - Gruidae
e   Wattled Crane
f   Grue caronculée
d   Klunkerkranich
i   Gru caruncolata

4351   *Grus grus*
        Gruiformes - Gruidae
e   Common Crane; Crane; Grey Crane-
        Hawk; Eurasian Crane
f   Grue cendrée
d   Kranich; Gemeiner Kranich; Grauer
        Kranich
i   Gru; Gru comune; Gru cenerina

4352   *Grus japonensis*
        Gruiformes - Gruidae
e   Red-crowned Crane; Japanese Crane;
        Manchurian Crane
f   Grue du Japon
d   Mandschuren-Kranich; Japanischer
        Kranich
i   Gru della Manciuria; Gru di
        Mancuria

4353   *Grus leucogeranus*
        Gruiformes - Gruidae
e   Siberian Crane; Siberian White
        Crane; Great White Crane (ISC);
        Asiatic White Crane
f   Grue de Sibérie
d   Nonnenkranich; Weißer Kranich
i   Gru siberiana

4354   *Grus monacha*
        Gruiformes - Gruidae
e   Hooded Crane
f   Grue moine

d Mönchskranich
i Gru monaca

4355 *Grus nigricollis*
Gruiformes - Gruidae
e Black-necked Crane
f Grue à cou noir
d Schwarzhalskranich
i Gru collonero

4356 *Grus paradisea*
Gruiformes - Gruidae
e Blue Crane; Stanley's Crane; Paradise Crane; Stanley Crane
f Grue de paradis
d Paradieskranich
i Gru del Paradiso; Gru di Stanley

4357 *Grus rubicunda*
Gruiformes - Gruidae
e Brolga; Australian Crane
f Grue brolga
d Brolga-Kranich
i Gru brolga

4358 *Grus vipio*
Gruiformes - Gruidae
e White-naped Crane; Japanese White-naped Crane; White-necked Crane
f Grue à cou blanc
d Weißnackenkranich
i Gru collobianco

4359 *Grus virgo*
Gruiformes - Gruidae
e Demoiselle Crane
f Grue demoiselle
d Jungfernkranich
i Damigella di Numidia; Gru damigella

4360 *Guadalcanaria inexpectata*
Passeriformes - Meliphagidae
e Guadalcanal Honeyeater
f Méliphage de Guadalcanal
d Guadalcanal-Honigfresser
i Mangiamiele di Guadalcanal

4361 *Gubernatrix cristata*
Passeriformes - Fringillidae
e Yellow Cardinal

f Commandeur huppé; Bruant commandeur; Cardinal vert
d Grünkardinal
i Cardinale verde

4362 *Gubernetes yetapa*
Passeriformes - Tyrannidae
e Streamer-tailed Tyrant
f Moucherolle yetapa
d Yetapa-Tyrann; Bandtyrann
i Tiranno di Yetapa

4363 *Guira guira*
Cuculiformes - Cuculidae
e Guira Cuckoo
f Guira cantara
d Guira-Kuckuck
i Guira

4364 *Guiraca caerulea*
Passeriformes - Fringillidae
e Blue Grosbeak; Blue Finch
f Guiraca bleu; Passerin bleu; Grosbec bleu
d Azurbischof; Hellblauer Bishof; Blaukardinal
i Beccogrosso azzurro

4365 *Guttera plumifera*
Galliformes - Numididae
e Plumed Guineafowl
f Pintade plumifère
d Schlichthaubenperlhuhn
i Faraona plumifera

4366 *Guttera pucherani*
Galliformes - Numididae
e Crested Guineafowl; Kenya Crested-Guineafowl
f Pintade de Pucheran
d Kräuselhaubenperlhuhn
i Faraona crestata; Numida crestata

4367 *Guttera pucherani edouardi*
Galliformes - Numididae
e Crested Guineafowl
f Pintade huppée
d Kräuselhaubenperlhuhn
i Fraraona crestata

4368   Gygis alba
Charadriiformes - Laridae
e   Common White Tern; Tern; Fairy
Tern; White Noddy; Atlantic Fairy
Tern; Common White Noddy; Cocos
Fairy Tern; Pacific Tern
f   Gygis blanche
d   Feenseeschwalbe
i   Sterna bianca

4369   Gygis candida
Charadriiformes - Laridae
e   Pacific Tern

4370   Gygis microrhyncha
Charadriiformes - Laridae
e   Little Tern; Little White Tern; Little
Fairy Tern
f   Gygis à bec fin
i   Sterna bianca minore

4371   Gymnobucco bonapartei
Piciformes - Lybiidae
e   Grey-throated Barbet; Bonaparte's
Barbet
f   Barbican à gorge grise
d   Trauerbartvogel; Graukopfbartvogel
i   Barbuto testagrigia

4372   Gymnobucco calvus
Piciformes - Lybiidae
e   Naked-faced Barbet
f   Barbican chauve
d   Glatzenbartvogel
i   Barbuto faccianuda

4373   Gymnobucco peli
Piciformes - Lybiidae
e   Bristle-nosed Barbet
f   Barbican à narines emplumées
d   Pel-Bartvogel; Borstenbartvogel
i   Barbuto naso setoloso

4374   Gymnobucco sladeni
Piciformes - Lybiidae
e   Sladen's Barbet; Zaire Barbet
f   Barbican de Sladen
d   Rußbartvogel
i   Barbuto di Sladen

4375   Gymnocichla nudiceps
Passeriformes - Formicariidae
e   Bare-crowned Antbird; Bare-
crowned Antcatcher
f   Alapi à tête nue
d   Nacktkopfameisenvogel;
Nacktkopfameisenwürger

4376   Gymnocrex plumbeiventris
Gruiformes - Rallidae
e   Bare-eyed Rail
f   Râle à ventre gris
d   Rostschwingenralle
i   Rallo ventreplumbeo

4377   Gymnocrex rosenbergii
Gruiformes - Rallidae
e   Bald-faced Rail; Blue-faced Rail;
Rosenberg's Rail; Rosenberg's Bare-
eyed Rail; Bare-faced Rail
f   Râle de Rosenberg
d   Nacktaugenralle
i   Rallo di Rosenberg

4378   Gymnoderus foetidus
Passeriformes - Tyrannidae
e   Bare-necked Fruitcrow
f   Coracine à col nu; Coracine col-nu
d   Nackthalsschmuckvogel
i   Cotinga testanuda

4379   Gymnogyps californianus
Falconiformes - Cathartidae
e   California Condor; Californian
Condor; Condor
f   Condor de Californie
d   Kalifornischer Kondor
i   Condor della California

4380   Gymnomystax mexicanus
Passeriformes - Fringillidae
e   Oriole Blackbird
f   Carouge loriot
d   Nacktaugentrupial
i   Ittero oriolo

4381   Gymnomyza aubryana
Passeriformes - Meliphagidae
e   Crow Honeyeater; Red-faced
Honeyeater
f   Méliphage toulou

d Turu
i Mangiamiele facciarossa

4382 ***Gymnomyza samoensis***
Passeriformes - Meliphagidae
e Mao; Black-breasted Honeyeater;
Mao Honeyeater
f Méliphage mao
d Mao
i Mao

4383 ***Gymnomyza viridis***
Passeriformes - Meliphagidae
e Giant Honeyeater; Giant Forest-
Honeyeater; Green Honeyeater
f Méliphage vert
d Einfarbhonigfresser
i Mangiamiele verde

4384 ***Gymnophaps albertisii***
Columbiformes - Columbidae
e Papuan Mountain-Pigeon; Bare-eyed
Mountain-Pigeon; Bare-eyed Pigeon;
Albertis's Mountain-Pigeon
f Carpophage d'Albertis
d Albertis-Taube
i Piccione di montagna di D'Albertis

4385 ***Gymnophaps mada***
Columbiformes - Columbidae
e Long-tailed Mountain-Pigeon; Mada
Mountain-Pigeon
f Carpophage mada
d Langschwänzige Bergtaube; Mada-
Taube
i Piccione di montagna codaluga

4386 ***Gymnophaps solomonensis***
Columbiformes - Columbidae
e Pale Mountain-Pigeon; Spectacled
Pigeon
f Carpophage des Salomon
d Blaße Bergtaube; Malai-Taube
i Piccione di montagne delle Salomon

4387 ***Gymnopithys bicolor***
Passeriformes - Formicariidae
e Bicoloured Antbird; Two-coloured
Antbird
f Fourmilier bicolore
d Zweifarbenameisenvogel

4388 ***Gymnopithys leucaspis***
Passeriformes - Formicariidae
e White-cheeked Antbird; White-
cheeked Antcatcher; Bicoloured
Antbird
f Fourmilier à joues blanches
d Weißohrameisenvogel

4389 ***Gymnopithys lunulata***
Passeriformes - Formicariidae
e Lunulated Antbird; Lunulated
Antcatcher; White-throated Antbird
f Fourmilier lunule
d Ucayili-Ameisenvogel

4390 ***Gymnopithys rufigula***
Passeriformes - Formicariidae
e Rufous-throated Antbird; Rufous-
throated Antcatcher
f Fourmilier à gorge rousse
d Rotkehlameisenvogel

4391 ***Gymnopithys salvini***
Passeriformes - Formicariidae
e White-throated Antbird; White-
throated Antcatcher
f Fourmilier de Salvin
d Salvin-Ameisenvogel

4392 ***Gymnorhina tibicen***
Passeriformes - Corvidae
e Australian Magpie; Magpie (ANZ);
Bell-Magpie; Black-backed Magpie;
Australasian Magpie
f Cassican flûteur; Corbeau flûteur
d Flötenvogel;
Schwarzrückenflötenvogel
i Gazza australiana

4393 ***Gymnorhinus cyanocephalus***
Passeriformes - Corvidae
e Pinyon Jay; Piñon Jay; Blue Crow;
Piñon Crow; Maximillians's Crow
f Geai des pinèdes
d Nacktschnabelhäher
i Ghiandaia dei pini

4394 ***Gymnostinops bifasciatus***
Passeriformes - Fringillidae
e Amazonian Oropendola; Para
Oropendola

f   Cassique du Para
d   Para-Stirnvogel; Olivkopfstirnvogel
i   Oropendola del Parà

4395   ***Gymnostinops cassini***
        Passeriformes - Fringillidae
e   Baudo Oropendola; Chestnut-
      mantled Oropendola
f   Cassique de Cassin
d   Braunmantelstirnvogel
i   Oropendola di Cassin

4396   ***Gymnostinops montezuma***
        Passeriformes - Fringillidae
e   Montezuma Oropendola; Great
      Oropendola
f   Cassique de Montézuma
d   Montezuma-Stirnvogel
i   Oropendola di Montezuma

4397   ***Gymnostinops yuracares***
        Passeriformes - Fringillidae
e   Olive Oropendola; Amazonian
      Oropendola
f   Cassique bicolore
d   Olivkopfstirnvogel

4398   ***Gypaetus barbatus***
        Falconiformes - Accipitridae
e   Lammergeier; Bearded Vulture;
      Lammergeir; Lammergeyer
f   Gypaète barbu
d   Bartgeier; Lämmergeier
i   Gipeto; Avvoltoio barbuto; Avvoltaio
      delle agnelli

4399   ***Gypohierax angolensis***
        Falconiformes - Accipitridae
e   Palm-nut Vulture; Vulturine Fish-
      Eagle
f   Palmiste africain; Vautour palmiste
d   Palmengeier; Palmgeier
i   Avvoltoio delle palme; Avvoltaio
      delle palme da cocco

4400   ***Gypopsitta vulturina***
        Psittaciformes - Psittacidae
e   Vulturine Parrot
f   Caique vautourin
d   Kahlkopfpapagei
i   Pappagallo vulturino

4401   ***Gyps africanus***
        Falconiformes - Accipitridae
e   White-backed Vulture; Griffon
      (CSA); White-backed Griffon-
      Vulture; African White-backed
      Griffon-Vulture
f   Vautour africain; Gyps africain
d   Weißrückengeier
i   Grifone africano

4402   ***Gyps bengalensis***
        Falconiformes - Accipitridae
e   White-rumped Vulture; Griffon
      (CSA); White-backed Vulture (ISC);
      Bengal Vulture (ISC); Indian White-
      backed Vulture; Indian White-
      rumped Vulture; Oriental White-
      backed Vulture; Asian White-backed
      Vulture
f   Vautour chaugoun
d   Bengalen-Geier
i   Grifone del Bengala; Grifone
      dorsabianca del Bengala; Avvoltaio
      indiano dal dorso bianco

4403   ***Gyps coprotheres***
        Falconiformes - Accipitridae
e   Cape Griffon; Cape Griffon-Vulture;
      Griffon (CSA); Cape Vulture;
      Kolbe's Vulture
f   Vautour chassefiente
d   Kap-Geier; Fahlgeier
i   Grifone del Capo

4404   ***Gyps fulvus***
        Falconiformes - Accipitridae
e   Eurasian Griffon; Eurasian Griffon-
      Vulture; Griffon (CSA); Indian
      Griffon-Vulture (ISC); Griffon-
      Vulture; Gryphon Vulture
f   Percnoptère d´Égypte; Vautour fauve
d   Gänsegeier
i   Grifone; Grifone eurasiatico

4405   ***Gyps himalayensis***
        Falconiformes - Accipitridae
e   Himalayan Griffon; Griffon (CSA);
      Himalayan Vulture; Himalayan
      Griffon-Vulture
f   Vautour de l'Himalaya

d Himalaja-Geier
i Grifone dell'Himalaya

4406 *Gyps indicus*
Falconiformes - Accipitridae
e Long-billed Vulture; Griffon (CSA);
Indian Long-billed Vulture (ISC);
Long-billed Griffon; Indian Griffon;
Indian Vulture
f Vautour indien
d Dünnschnabelgeier
i Grifone indiano

4407 *Gyps rueppellii*
Falconiformes - Accipitridae
e Rüppell's Griffon-Vulture; Rüppell's
Griffon; Griffon (CSA); Griffon
(CSA)
f Vautour de Rüppell
d Sperbergeier
i Grifone di Rüppell

H

4408   Habia atrimaxillaris
Passeriformes - Thraupidae
e   Black-cheeked Ant-Tanager; Black-
billed Ant-Tanager; Black-billed
Tanager; Black-cheeked Tanager
f   Tangara à joues noires
d   Schwarzwangenhabia
i   Tangara formichiera guancenere

4409   Habia cristata
Passeriformes - Thraupidae
e   Crested Ant-Tanager; Crested
Tanager
f   Tangara à crête rouge
d   Scharlachhaubenhabia
i   Tangara formichiera dal ciuffo

4410   Habia fuscicauda
Passeriformes - Thraupidae
e   Red-throated Ant-Tanager; Dusky-
tailed Ant-Tanager; Jungle Tanager
f   Tangara à gorge rouge
d   Schwarzkinnhabia
i   Tangara formichiera golarossa

4411   Habia gutturalis
Passeriformes - Thraupidae
e   Sooty Ant-Tanager; Red-throated
Ant-Tanager; Sooty Tanager; Jungle
Ant-Tanager
f   Tangara fuligineux
d   Graurückenhabia
i   Tangara formichiera fuligginosa

4412   Habia rubica
Passeriformes - Thraupidae
e   Red-crowned Ant-Tanager; Red-
crowned Tanager; Jungle Ant-
Tanager
f   Tangara à couronne rouge
d   Rotkehlhabia
i   Tangara formichiera corona rossa

4413   Habia salvini
Passeriformes - Thraupidae
e   Salvin's Ant-Tanager

4414   Habroptila wallacii
Gruiformes - Rallidae
e   Invisible Rail; Drummer Rail;
Halmahera Rail; Wallace's Rail
f   Râle de Wallace
d   Trommelralle
i   Rallo di Wallace

4415   Haematoderus militaris
Passeriformes - Tyrannidae
e   Crimson Fruitcrow
f   Coracine rouge
d   Blutkotinga
i   Cotinga cremisi

4416   Haematopodidae
Charadriiformes
e   Oystercatchers
f   Huîtriers; Haematopodidés
d   Austernfischer
i   Ematopodidi

4417   Haematopus ater
Charadriiformes - Haematopodidae
e   Blackish Oystercatcher; South
American Black-Oystercatcher;
Black Oystercatcher
f   Huîtrier noir
d   Rußausternfischer
i   Beccaccia di mare nera

4418   Haematopus bachmani
Charadriiformes - Haematopodidae
e   Black Oystercatcher; American
Black-Oystercatcher
f   Huîtrier de Bachman
i   Beccaccia di mare di Bachman

4419   Haematopus chathamensis
Charadriiformes - Haematopodidae
e   Chatham Islands Oystercatcher
f   Huîtrier des Chatham
d   Chatham-Austernfischer
i   Beccaccia di mare

4420   Haematopus finschi
Charadriiformes - Haematopodidae

e South Island Oystercatcher; South Island Pied-Oystercatcher; Pied Oystercatcher
f Huîtrier de Finsch
i Beccaccia di mare dell'Isola Sud

4421 *Haematopus fuliginosus*
Charadriiformes - Haematopodidae
e Sooty Oystercatcher; Black Oystercatcher
f Huîtrier fuligineux
d Klippenausternfischer
i Beccaccia di mare fuligginosa

4422 *Haematopus leucopodus*
Charadriiformes - Haematopodidae
e Magellanic Oystercatcher
f Huîtrier de Garnot
d Magellan-Austernfischer
i Beccaccia di mare di Magellano

4423 *Haematopus longirostris*
Charadriiformes - Haematopodidae
e Pied Oystercatcher; Australian Pied-Oystercatcher; White-breasted Oystercatcher
f Huîtrier à long bec
d Australischer Austernfischer
i Beccaccia di mare orientale

4424 *Haematopus meadewaldoi*
Charadriiformes - Haematopodidae
e Canary Islands Oystercatcher; Canary Oystercatcher; Canary Island Oystercatcher; Canarian Black-Oystercatcher; Meade-Waldo's Oystercatcher
i Beccaccia di mare delle Canarie

4425 *Haematopus moquini*
Charadriiformes - Haematopodidae
e African Oystercatcher; African Black-Oystercatcher; Black Oystercatcher
f Huîtrier de Moquin
d Schwarzer Austernfischer
i Beccaccia di mare di Moquin-Tandon; Beccaccia di mare nera

4426 *Haematopus ostralegus*
Charadriiformes - Haematopodidae

e Eurasian Oystercatcher; Oystercatcher; European Oystercatcher; Common Pied-Oystercatcher; Common Oystercatcher; Palearctic Oystercatcher; Northern Pied-Oystercatcher
f Huîtrier pie
d Austernfischer; Europäischer Austernfischer
i Beccaccia di mare; Beccaccia di mare eurasiatica

4427 *Haematopus palliatus*
Charadriiformes - Haematopodidae
e American Oystercatcher; Whelkcracker (WI); American Pied-Oystercatcher
f Huîtrier d'Amérique; Huîtrier (Ants); Cassseur de Burgau (Ants); Huîtrier américain
d Braunmantelausternfischer
i Beccaccia di mare americana

4428 *Haematopus unicolor*
Charadriiformes - Haematopodidae
e Variable Oystercatcher; North Island Oystercatcher; North Island Pied-Oystercatcher; New Zealand Sooty-Oystercatcher; Black Oystercatcher
f Huîtrier variable
d Chatham-Austernfischer
i Beccaccia di mare variabile

4429 *Haematortyx sanguiniceps*
Galliformes - Phasianidae
e Crimson-headed Partridge; Crimson-headed Wood-Partridge
f Rouloul sanglant
d Rotkopfwachtel
i Pernice testacarminio

4430 *Haematospiza sipahi*
Passeriformes - Fringillidae
e Scarlet Finch
f Cipaye écarlate; Grosbec sipahi
d Blutgimpel; Scharlachgimpel
i Ciuffolotto fiammante

4431 *Halcyon albiventris*
Coraciiformes - Halcyonidae

e	Brown-hooded Kingfisher
f	Martin-chasseur à tête brune
d	Braunkopfliest
i	Martin pescatore capobruno

4432   *Halcyon badia*
Coraciiformes - Halcyonidae
e	Chocolate-backed Kingfisher
f	Martin-chasseur marron
d	Kastanienliest; Prachtliest
i	Martin pescatore dorso baio

4433   *Halcyon chelicuti*
Coraciiformes - Halcyonidae
e	Striped Kingfisher
f	Martin-chasseur strié
d	Streifenliest
i	Martin pescatore striato

4434   *Halcyon coromanda*
Coraciiformes - Halcyonidae
e	Ruddy Kingfisher
f	Martin-chasseur violet
d	Feuerliest
i	Martin pescatore violetto

4435   *Halcyon cyanoventris*
Coraciiformes - Halcyonidae
e	Javan Kingfisher; Java Kingfisher; Blue-bellied Kingfisher
f	Martin-chasseur de Java
d	Java-Liest
i	Martin pescatore di Giava

4436   *Halcyon leucocephala*
Coraciiformes - Halcyonidae
e	Grey-headed Kingfisher; Grey-hooded Kingfisher (CSA); Chestnut-bellied Kingfisher (CSA)
f	Martin-chasseur à tête grise
d	Graukopfliest
i	Martin pescatore testagrigia

4437   *Halcyon malimbica*
Coraciiformes - Halcyonidae
e	Blue-breasted Kingfisher
f	Martin-chasseur à poitrine bleue
d	Zügelliest
i	Martin pescatore pettazzurro

4438   *Halcyon pileata*
Coraciiformes - Halcyonidae
e	Black-capped Kingfisher; Black-capped Purple Kingfisher
f	Martin-chasseur à coiffe noire
d	Kappenliest
i	Martin pescatore testanera

4439   *Halcyon senegalensis*
Coraciiformes - Halcyonidae
e	Woodland Kingfisher; Red-and-black-billed Kingfisher; Senegal Kingfisher
f	Martin-chasseur du Sénégal; Martin-chasseur des mangroves
d	Senegal-Liest; Braunliest
i	Martin pescatore di bosco

4440   *Halcyon senegalensis cyanoleuca*
Coraciiformes - Halcyonidae
e	Angolan Woodland-Kingfisher
f	Martin-chasseur bleu-et-blanc

4441   *Halcyon senegaloides*
Coraciiformes - Halcyonidae
e	Mangrove Kingfisher; African Mangrove Kingfisher
f	Martin-chasseur des mangroves
d	Mangroveliest
i	Martin pescatore delle mangrovie

4442   *Halcyon smyrnensis*
Coraciiformes - Halcyonidae
e	White-throated Kingfisher; Smyrna Kingfisher; White-breasted Kingfisher
f	Martin-chasseur de Smyrne
d	Braunliest
i	Martin pescatore di Smirne; Alcione dalla testa marrone

4443   **Halcyonidae**
Coraciiformes
e	Halyconid Kingfishers
f	Halcyonidés
d	Lieste

4444   *Haliaeetus albicilla*
Falconiformes - Accipitridae
e	White-tailed Eagle; White-tailed Sea-Eagle; Gray Sea-Eagle (NA); Grey

Sea-Eagle; Sea Eagle
f Pygargue à queue blanche; Aigle-
pêcheur pygargue
d Seeadler
i Aquila di mare; Aquila di mare
codabianca

4445 *Haliaeetus leucocephalus*
Falconiformes - Accipitridae
e Bald Eagle; American Eagle; White-
headed Sea-Eagle
f Pygargue à tête blanche
d Weißkopfseeadler
i Aquila di mare dalla testabianca;
Aquila di mare testabianca; Aquila
calva

4446 *Haliaeetus leucogaster*
Falconiformes - Accipitridae
e White-bellied Sea-Eagle; White-
bellied Fish-Eagle (ISC); White-
breasted Sea-Eagle; White-breasted
Fish-Hawk
f Pygargue blagre
d Weißbauchseeadler
i Aquila di mare ventrebianco; Aquila
di mare australiana

4447 *Haliaeetus leucoryphus*
Falconiformes - Accipitridae
e Pallas's Fish-Eagle; Pallas's Sea-
Eagle (ISC); Pallas's Fishing-Eagle
(ISC); Ring-tailed Fishing-Eagle
(ISC); Pallas's Eagle; Band-tailed
Fishing-Eagle
f Pygargue de Pallas
d Bindenseeadler
i Aquila del Pallas; Aquila di mare del
Pallas

4448 *Haliaeetus pelagicus*
Falconiformes - Accipitridae
e Steller's Sea-Eagle; Kamchatkan Sea-
Eagle; White-shouldered Sea-Eagle
f Pygargue empereur; Pygargue de
Steller
d Riesenadler
i Aquila di mare di Steller

4449 *Haliaeetus sanfordi*
Falconiformes - Accipitridae

e Sanford's Fish-Eagle; Sanford's Sea-
Eagle; Brown Sea-Eagle; Solomon
Sea-Eagle
f Pygargue de Sanford
d Salomonen-Seeadler
i Aquila di mare di Sanford

4450 *Haliaeetus vocifer*
Falconiformes - Accipitridae
e African Fish-Eagle; River Eagle;
West African River-Eagle
f Pygargue vocifère
d Schreiseeadler
i Aquila urlatrice; Aquila pescatrice
africana

4451 *Haliaeetus vociferoides*
Falconiformes - Accipitridae
e Madagascar Fish-Eagle; Madagascar
Sea-Eagle
f Pygargue de Madagascar; Pygargue
malgache
d Madagaskar-Seeadler
i Urlatrice di Madagascar; Aquila
pescatirice del Madagascar; Aquila
del Madagascar

4452 *Haliastur indus*
Falconiformes - Accipitridae
e Brahminy Kite; White-headed
Rufous-Eagle; Red-backed Kite;
Rufous-backed Kite; White-headed
Kite; White-and-red Eagle-Kite
f Milan sacré
d Brahminenweihe; Brahminenmilan
i Nibbio brahama; Nibio di Brahama

4453 *Haliastur sphenurus*
Falconiformes - Accipitridae
e Whistling Kite; Whistling Eagle;
Whistling Hawk; Eagle-Hawk;
Carrion Hawk
f Milan siffleur
d Keilschwanzweihe; Pfeifmilan
i Nibbio codacuneata; Nibbio
fisciatore

4454 *Halobaena caerulea*
Procellariiformes - Procellariidae
e Blue Petrel
f Prion bleu; Petrel bleu

d   Blausturmvogel
i   Petrello azzurro

4455   Hamirostra melanosternon
      Falconiformes - Accipitridae
e   Black-breasted Buzzard; Black-
      breasted Kite; Black-breasted
      Buzzard Kite
f   Milan à plastron
d   Bussardmilan; Haubenmilan;
      Schwarzbrustmilan
i   Nibbio pettonero; Nibbio dal petto
      nero

4456   Hapalopsittaca amazonina
      Psittaciformes - Psittacidae
e   Rusty-faced Parrot; Little Amazon
      Parrot
f   Caique à face rouge
d   Zwergamazone
i   Pappagallo facciarugginosa

4457   Hapalopsittaca fuertesi
      Psittaciformes - Psittacidae
e   Indigo-winged Parrot; Fuerte's Parrot
f   Caique de Fuertes
i   Pappagallo aliazzurre

4458   Hapalopsittaca melanotis
      Psittaciformes - Psittacidae
e   Black-winged Parrot; Brown-eared
      Parrot; Black-eared Parrot
f   Caique à ailes noires
d   Schwarzflügelpapagei
i   Pappagallo alinere

4459   Hapalopsittaca pyrrhops
      Psittaciformes - Psittacidae
e   Red-faced Parrot; Ecuadorian Parrot
f   Caique de Salvin
i   Pappagallo dell'Ecuador

4460   Hapaloptila castanea
      Piciformes - Bucconidae
e   White-faced Nunbird
f   Barbacou à face blanche; Barbacou
      de Verreaux
d   Diademfaulvogel;
      Kastanienfaulvogel;
      Braunbauchfaulvogel
i   Monachina facciabianca

4461   Haplophaedia aureliae
      Apodiformes - Trochilidae
e   Greenish Puffleg
f   Érione d'Aurelie
d   Bunthöschen
i   Colibrì zampepiumose verdastro

4462   Haplophaedia lugens
      Apodiformes - Trochilidae
e   Hoary Puffleg
f   Érione givrée
d   Schuppenschneehöschen
i   Colibrì zampepiumose canuto

4463   Haplospiza rustica
      Passeriformes - Fringillidae
e   Slaty Finch
f   Haplospize ardoisé
d   Schieferämmerling
i   Fringuello ardesia

4464   Haplospiza unicolor
      Passeriformes - Fringillidae
e   Uniform Finch
f   Haplospize unicolore
d   Einfarbämmerling
i   Fringuello unicolore

4465   Harpactes ardens
      Trogoniformes - Trogonidae
e   Philippine Trogon
f   Trogon des Philippines; Couroucou
      de Philippines
d   Philippinen-Trogon
i   Trogone delle Filippine

4466   Harpactes diardii
      Trogoniformes - Trogonidae
e   Diard's Trogon
f   Trogon de Diard; Couroucou de
      Diard
d   Halsbandtrogon
i   Trogone di Diard

4467   Harpactes duvaucelii
      Trogoniformes - Trogonidae
e   Scarlet-rumped Trogon; Red-rumped
      Trogon
f   Trogon de Duvaucel; Couroucou de
      Duvaucel

d Rotbürzeltrogon
i Trogone dal groppone scarlatto

4468 *Harpactes erythrocephalus*
Trogoniformes - Trogonidae
e Red-headed Trogon
f Trogon à tête rouge; Couroucou à tête rouge
d Rotkopftrogon
i Trogone testarossa

4469 *Harpactes fasciatus*
Trogoniformes - Trogonidae
e Malabar Trogon; Southern Trogon (ISC); Ceylon Trogon
f Trogon de Malabar; Couroucou de Malabar
d Malabar-Trogon
i Trogone del Malabar

4470 *Harpactes kasumba*
Trogoniformes - Trogonidae
e Red-naped Trogon
f Trogon à nuque rouge; Couroucou à nuque rouge
d Rotnackentrogon
i Trogone nucarossa

4471 *Harpactes oreskios*
Trogoniformes - Trogonidae
e Orange-breasted Trogon
f Trogon à poitrine jaune; Couroucou à poitrine jaune
d Grünkoptrogon
i Trogone pettoarancio

4472 *Harpactes orrhophaeus*
Trogoniformes - Trogonidae
e Cinnamon-rumped Trogon
f Trogon cannellé; Couroucou cannellé
d Zimtbürzeltrogon
i Trogone dal groppone cannella

4473 *Harpactes reinwardtii*
Trogoniformes - Trogonidae
e Blue-tailed Trogon; Reinwardt's Blue-tailed Trogon; Blue-billed Trogon
f Trogon de Reinwardt; Couroucou de Reinwardt

d Reinwardt-Trogon
i Trogone di Reinwardt

4474 *Harpactes wardi*
Trogoniformes - Trogonidae
e Ward's Trogon
f Trogon de Ward; Couroucou à ventre rose
d Rosenschwanztrogon
i Trogone di Ward

4475 *Harpactes whiteheadi*
Trogoniformes - Trogonidae
e Whitehead's Trogon
f Trogon de Whitehead; Couroucou de Whitehead
d Graubrusttrogon
i Trogone di Whitehead

4476 *Harpagus bidentatus*
Falconiformes - Accipitridae
e Double-toothed Kite
f Milan bidenté
d Doppelzahnweihe
i Arpago bidentato

4477 *Harpagus diodon*
Falconiformes - Accipitridae
e Rufous-thighed Kite
f Milan diodon
d Braunschenkelweihe
i Arpago dai calzoni rossi; Arpago dalle strisce rosse

4478 *Harpia harpyja*
Falconiformes - Accipitridae
e Harpy Eagle
f Harpie féroce
d Harpyie
i Arpia

4479 *Harpyhaliaetus coronatus*
Falconiformes - Accipitridae
e Crowned Eagle; Crowned Solitary-Eagle
f Buse couronnée; Aigle couronné
d Kronenadler; Zaunadler; Streitadler
i Aquila solitaria coronata; Aquila coronata

4480   *Harpyhaliaetus solitarius*
Falconiformes - Accipitridae
e   Solitary Eagle; Black Solitary-Eagle
f   Buse solitaire; Aigle solitaire
d   Einsiedleradler
i   Aquila solitaria nera; Aquila solitaria

4481   *Harpyopsis novaeguineae*
Falconiformes - Accipitridae
e   New Guinea Eagle; Harpy-like
Eagle; New Guinea Harpy-Eagle;
New Guinea Harpy; Kapul Eagle
f   Aigle de Nouvelle-Guinée
d   Papua-Adler
i   Arpia della Nuova Guinea

4482   *Heinrichia calligyna*
Passeriformes - Muscicapidae
e   Great Shortwing; Celebes Shortwing;
Sulawesi Shortwing
f   Brachyptère des Célèbes
d   Celebes Kurzflügel
i   Alabreve di Sulawesi

4483   *Heleia crassirostris*
Passeriformes - Zosteropidae
e   Thick-billed White-eye; Large-billed
White-eye; Stripe-headed White-eye;
Lesser Sunda Lowland-Whiteye;
Thick-billed Darkeye; Flores White-
eye
f   Zostérops à bec fort
d   Nacktaugenbrillenvogel
i   Occhialino beccogrosso

4484   *Heleia muelleri*
Passeriformes - Zosteropidae
e   Spot-breasted White-eye; Müller's
White-eye; Timor White-eye; Spot-
breasted Darkeye; Streak-breasted
White-eye
f   Zostérops de Timor
d   Fleckenbrustbrillenvogel
i   Occhialino di Timor

4485   *Heliactin cornuta*
Apodiformes - Trochilidae
e   Horned Sungem; Sungem; Horned
Hummingbird
f   Colibri aux huppes d'or

d   Sonnenstrahlelfe
i   Colibrì gemma del sole

4486   *Heliangelus amethysticollis*
Apodiformes - Trochilidae
e   Amethyst-throated Sunangel;
Amethystine Sunangel
f   Héliange clarisse
d   Sonnennymphe
i   Angelo del sole gola ametista

4487   *Heliangelus exortis*
Apodiformes - Trochilidae
e   Tourmaline Sunangel; Parzudaki's
Sunangel
f   Héliange tourmaline
d   Turmalinnymphe
i   Angelo del sole tormalina

4488   *Heliangelus mavors*
Apodiformes - Trochilidae
e   Orange-throated Sunangel; Mavor's
Sunangel
f   Héliange mars
d   Orangekehlnymphe
i   Angelo del sole gola aranciata

4489   *Heliangelus micraster*
Apodiformes - Trochilidae
e   Little Sunangel

4490   *Heliangelus regalis*
Apodiformes - Trochilidae
e   Royal Sunangel
f   Héliange royale
d   Königsnymphe
i   Angelo del sole regale

4491   *Heliangelus spencei*
Apodiformes - Trochilidae
e   Merida Sunangel; Spence's Sunangel
f   Héliange de Merida
d   Diotima
i   Angelo del sole di Merida

4492   *Heliangelus strophianus*
Apodiformes - Trochilidae
e   Gorgeted Sunangel
f   Héliange à queue bleue
d   Graubauchnymphe
i   Angelo del sole golaornata

4493 Heliangelus viola
Apodiformes - Trochilidae
e Purple-throated Sunangel; Viola
Sunangel
f Héliange violette
d Violettkehlnymphe
i Angelo del sole golapurpurea

4494 Heliangelus zusii
Apodiformes - Trochilidae
e Bogota Sunangel

4495 Heliobletus contaminatus
Passeriformes - Furnariidae
e Sharp-billed Treehunter; Sharp-billed
Xenops
f Sittine à bec fin
d Spitzschnabelbaumspäher
i Xenope beccoacuto

4496 Heliodoxa aurescens
Apodiformes - Trochilidae
e Gould's Jewelfront; Jewelfront
f Brillant à bandeau bleu; Colibri à
bandeau bleu
d Juwellenkrönchen
i Colibrì di Gould

4497 Heliodoxa branickii
Apodiformes - Trochilidae
e Rufous-webbed Brilliant
f Brillant de Branicki
d Rotschwingenbrilliant
i Colibrì di Branick

4498 Heliodoxa gularis
Apodiformes - Trochilidae
e Pink-throated Brilliant; Puce-throated
Brilliant
f Brillant à gorge rose; Brilliant
gorgerette
d Rotkehlbrilliant
i Colibrì diamante golarosa

4499 Heliodoxa imperatrix
Apodiformes - Trochilidae
e Empress Brilliant; Empress
Hummingbird; Empress Eugenie's
Hummingbird
f Brillant impératrice

d Kaiserbrilliant
i Colibrì di Eugenia

4500 Heliodoxa jacula
Apodiformes - Trochilidae
e Green-crowned Brilliant
f Brillant fer-de-lance
d Grünscheitelbrilliant
i Colibrì diamante corona verde

4501 Heliodoxa leadbeateri
Apodiformes - Trochilidae
e Violet-fronted Brilliant; Leadbeater's
Hummingbird
f Brillant à front violet
d Violettstirnbrilliant
i Colibrì di Leadbeater

4502 Heliodoxa rubinoides
Apodiformes - Trochilidae
e Fawn-breasted Brilliant; Lilacthroat;
Rosythroat; Penny-throated
Hummingbird; Lilac-throated
Hummingbird; Lilac-breasted
Brilliant
f Brillant rubinolde
d Braunbauchbrilliant
i Colibrì diamante pettocastano

4503 Heliodoxa schreibersii
Apodiformes - Trochilidae
e Black-throated Brilliant; Black-
throated Hummingbird; Schreiber's
Hummingbird
f Brillant à gorge noire
d Schwarzkehlbrilliant
i Colibrì diamante golanera

4504 Heliodoxa xanthogonys
Apodiformes - Trochilidae
e Velvet-browed Brilliant; Yellow-
cheeked Hummingbird
f Brillant à couronne verte
d Brauenbrilliant
i Colibrì dai sopraccigli di velluto

4505 Heliomaster constantii
Apodiformes - Trochilidae
e Plain-capped Starthroat; Constant's
Starthroat
f Colibri de Constant

d Funkenkehlchen
i Gola stellata di Constant

4506 *Heliomaster furcifer*
Apodiformes - Trochilidae
e Blue-tufted Starthroat; Angela's
Starthroat
f Colibri d'Angèle
d Rotlatzkolibri
i Gola stellata ciuffoblu

4507 *Heliomaster longirostris*
Apodiformes - Trochilidae
e Long-billed Starthroat
f Colibri corinne
d Rosenkehlchen
i Gola stellata beccolungo

4508 *Heliomaster squamosus*
Apodiformes - Trochilidae
e Stripe-breasted Starthroat; White-
striped Hummingbird
f Colibri médiastin
d Temminck-Kolibri
i Gola stellata pettostriato

4509 *Heliopais personata*
Gruiformes - Heliornithidae
e Masked Finfoot; Asian Finfoot
f Grébifoulque d'Asie
d Maskenralle; Indische Binsenralle
i Rallo tuffatoro asiatico

4510 *Heliornis fulica*
Gruiformes - Heliornithidae
e Sungrebe; American Finfoot
f Grébifoulque d'Amérique
d Zwergbinsenralle
i Svasso del sole; Picapare

4511 Heliornithidae
Gruiformes -
e Sungrebes; Finfoots
f Grèbifoulques; Héliornithidés
d Binsenhühner; Binsenrallen
i Eliornitidi

4512 *Heliothryx aurita*
Apodiformes - Trochilidae
e Black-eared Fairy; Green-chinned
Fairy

f Colibri oreillard
d Schwarzohrelfe
i Capello di sole orecchienere

4513 *Heliothryx barroti*
Apodiformes - Trochilidae
e Purple-crowned Fairy; Barrot's Fairy
f Colibri féerique
d Purpurkopfelfe
i Capello di sole corona purpurea

4514 *Hellmayrea gularis*
Passeriformes - Furnariidae
e White-browed Spinetail; Lafresnaye's
Spinetail; Lafresnaye's White-browed
Spinetail
f Synallaxe à sourcils blancs
d Weißbrauenschlüpfer
i Codaspinosa di Lafresnaye

4515 *Helminthophaga cincinnatiensis*
Passeriformes - Fringillidae
e Cincinnati Warbler

4516 *Helmitheros vermivorus*
Passeriformes - Fringillidae
e Worm-eating Warbler
f Paruline vermivore; Fauvette
vermivore
d Haldenwaldsänger
i Mangiavermi

4517 *Hemicircus canente*
Piciformes - Picidae
e Heart-spotted Woodpecker
f Pic canente
d Rundschwanzspecht
i Picchio bicolore

4518 *Hemicircus concretus*
Piciformes - Picidae
e Grey-and-buff Woodpecker; Grey-
breasted Woodpecker; Malaysian
Grey-breasted Woodpecker
f Pic trapu
d Kurzschwanzspecht
i Picchio squamato

4519 *Hemignathus chloris*
Passeriformes - Fringillidae
e Oahu Amakihi

4520 *Hemignathus ellisianus*
Passeriformes - Fringillidae
e Hawaiian Akiola; Greater Akiola
f Hémignathe à long bec

4521 *Hemignathus lucidus*
Passeriformes - Fringillidae
e Nukupuu
f Hémignathe nukupuu
d Nukupuu
i Nukupuu

4522 *Hemignathus obscurus*
Passeriformes - Fringillidae
e Akialoa; Hawaiian Akialoa; Kuai Akiola
f Hémignathe akialoa
d Akiola; Hawaiian Akiola
i Akialoa

4523 *Hemignathus parvus*
Passeriformes - Fringillidae
e Anianiau; Lesser Amakihi; Amakihi
d Anianiau
i Ananiau

4524 *Hemignathus procerus*
Passeriformes - Fringillidae
e Kauai Akiola; Akiola
d Sichelhalbschnäbler

4525 *Hemignathus sagittirostris*
Passeriformes - Fringillidae
e Greater Amakihi; Green Solitaire
d Grünrückenkleidervogel
i Amakihi maggiore

4526 *Hemignathus virens*
Passeriformes - Fringillidae
e Hawaii Amakihi; Common Amikihi; Amikihi
f Amakihi familier
d Amakihi
i Amakihi comune; Amakihi

4527 *Hemignathus wilsoni*
Passeriformes - Fringillidae
e Akiapolaau; Hawaii Nukupuu
f Hémignathe akiapolaau
d Akiopalaau
i Akiapolau

4528 *Hemiphaga novaeseelandiae*
Columbiformes - Columbidae
e New Zealand Pigeon; New Zealand Fruit-Pigeon; Wood-Pigeon (ANZ)
f Carpophage de Nouvelle-Zélande
d Maorifruchttaube; Neuseeland-Fruchttaube
i Piccione della Nuova Zelanda; Kukupa

4529 *Hemiphaga novaeseelandiae spadicea*
Columbiformes - Columbidae
e Norfolk Island New Zealand Pigeon (ANZ)

4530 *Hemiprocne comata*
Apodiformes - Hemiprocnidae
e Whiskered Treeswift; Lesser Treeswift
f Hemiprocné coiffé
d Ohrensegler
i Rondone arboricolo minore

4531 *Hemiprocne coronata*
Apodiformes - Hemiprocnidae
e Crested Treeswift; Crested Swift; Grey-rumped Treeswift; Treeswift
f Hémiprocné couronné
d Kronensegler
i Rondone arboricolo dalla cresta; Rondone arboreo dalla cresta

4532 *Hemiprocne longipennis*
Apodiformes - Hemiprocnidae
e Grey-rumped Treeswift; Crested Treeswift
f Hémiprocné longipenne
d Haubensegler
i Rondone arboricolo dal groppone grigio

4533 *Hemiprocne mystacea*
Apodiformes - Hemiprocnidae
e Moustached Treeswift; Moustached Swift; Whiskered Treeswift
f Hemiprocne à moustaches
d Bartsegler
i Rondone arboricolo dai mustacchi

4534 Hemiprocnidae
Apodiformes -
e Crested Swifts; Treeswifts
f Hemiprocnidés
d Baumsegler
i Emiprocnidi

4535 *Hemipus hirundinaceus*
Passeriformes - Corvidae
e Black-winged Flycatcher-Shrike;
Black-winged Pygmy-Triller; Black-
winged Hemispingus
f Échenilleur véloce
d Schwarzflügelraupenschmätzer
i Averla-pigliamosche alinere

4536 *Hemipus picatus*
Passeriformes - Corvidae
e Bar-winged Flycatcher-Shrike; Pied
Flycatcher-Shrike (ISC); Brown-
backed Flycatcher-Shrike; Bar-
winged Pygmy-Triller; Bar-winged
Hemispingus; Pygmy Triller; Pied
Wood-Shrike; Ceylon Pied-Shrike;
Pygmy Cuckoo-Shrike; Brown-
backed Pied-Shrike
f Échenilleur gobemouche; Échenilleur
pie
d Elsterraupenschmätzer;
Bänderraupenschmätzer
i Averla-pigliamosche alifasciate

4537 *Hemispingus atropileus*
Passeriformes - Fringillidae
e Black-capped Hemispingus; Black-
capped Tanager; Bolivian
Hemispingus; Yungas Hemispingus
f Tangara à calotte noire
d Schwarzkappenhemispingus
i Tangara capinera

4538 *Hemispingus calophrys*
Passeriformes - Fringillidae
e Orange-browed Hemispingus;
Orange-browed Tanager
f Tangara des bambous
d Ockerbrauenhemispingus
i Tangara dai sopraccigli arancio

4539 *Hemispingus frontalis*
Passeriformes - Fringillidae

e Oleaginous Hemispingus;
Oleaginous Tanager; Ochraceous
Hemispingus; Merida Hemispingus
f Tangara ocré
d Olicrückenhemispingus
i Tangara oliva

4540 *Hemispingus goeringi*
Passeriformes - Fringillidae
e Slaty-backed Hemispingus; Slaty-
backed Tanager
f Tangara de Goering
d Graurückenhemispingus
i Tangara dorsoardesia

4541 *Hemispingus melanotis*
Passeriformes - Fringillidae
e Black-eared Hemispingus; Black-
eared Tanager
f Tangara barbouille
d Schwarzwangenhemispingus
i Tangara orecchienere

4542 *Hemispingus parodii*
Passeriformes - Fringillidae
e Parodi's Hemispingus; Parodi's
Tanager
f Tangara de Parodi
d Parodi-Hemispingus
i Tangara di Parodi

4543 *Hemispingus reyi*
Passeriformes - Fringillidae
e Grey-capped Hemispingus; Grey-
capped Tanager
f Tangara à calotte grise
d Graukappenhemispingus
i Tangara capogrigio

4544 *Hemispingus rufosuperciliaris*
Passeriformes - Fringillidae
e Rufous-browed Hemispingus;
Rufous-browed Tanager
f Tangara à sourcils fauves
d Zimtbrauenhemispingus
i Tangara dai sopraccigli rugginosi

4545 *Hemispingus superciliaris*
Passeriformes - Fringillidae
e Superciliaried Hemispingus;
Superciliated Tanager

f Tangara bridé
d Augenbrauenhemispingus
i Tangara canopino

4546 *Hemispingus trifasciatus*
Passeriformes - Fringillidae
e Three-striped Hemispingus; Three-striped Tanager
f Tangara trifascié
d Dreistreifentangare
i Tangara dalle tre strie

4547 *Hemispingus verticalis*
Passeriformes - Fringillidae
e Black-headed Hemispingus; Black-headed Tanager
f Tangara à tête noire
d Schwarzkopfhemispingus
i Tangara testanera

4548 *Hemispingus xanthophthalmus*
Passeriformes - Fringillidae
e Drab Hemispingus; Drab Tanager; Dark Hemispingus
f Tangara aux yeux jaunes
d Gelbaugenhemispingus
i Tangara occhigialli

4549 *Hemitesia neumanni*
Passeriformes - Sylviidae
e Neumann's Warbler; Neumann's Bush-Warbler; Neumann's Short-tailed Warbler
f Crombec de Neumann
d Trugtesia
i Silvietta di Neumann

4550 *Hemithraupis flavicollis*
Passeriformes - Fringillidae
e Yellow-backed Tanager; Yellow-rumped Tanager
f Tangara à dos jaune
d Gelbbürzeltangare

4551 *Hemithraupis guira*
Passeriformes - Fringillidae
e Guira Tanager
f Tangara guira
d Guira-Tangare

4552 *Hemithraupis ruficapilla*
Passeriformes - Fringillidae
e Rufous-headed Tanager
f Tangara à tête rousse
d Rotkappentangare

4553 *Hemitriccus aenigma*
Passeriformes - Tyrannidae
e Zimmer's Tody-Tyrant
f Todirostre de Zimmer
d Weißkinnspateltyrann
i Tiranno todo di Zimmer

4554 *Hemitriccus cinnamomeipectus*
Passeriformes - Tyrannidae
e Cinnamon-breasted Tody-Tyrant; Peruvian Tody-Tyrant
f Todirostre du Pérou
d Zimtbrustspateltyrnn
i Tiranno todo pettocannella

4555 *Hemitriccus diops*
Passeriformes - Tyrannidae
e Drab-breasted Stock-Tyrant; Drab-breasted Pygmy-Tyrant; Temmink's Pygmy-Tyrant
f Todirostre à poitrine ombrée; Todirostre de Temminck
i Tiranno todo pettogrigio

4556 *Hemitriccus flammulatus*
Passeriformes - Tyrannidae
e Flammulated Stock-Tyrant; Flammulated Pygmy-Tyrant; Flammulated Bamboo-Tyrant
f Todirostre flammulé
d Zügelflecktyrann
i Tiranno todo flammulato

4557 *Hemitriccus furcatus*
Passeriformes - Tyrannidae
e Fork-tailed Tody-Tyrant; Fork-tailed Pygmy-Tyrant
f Todirostre à queue fourchue
d Buchschwanztyrann
i Tiranno todo

4558 *Hemitriccus granadensis*
Passeriformes - Tyrannidae
e Black-throated Tody-Tyrant
f Todirostre à gorge noire

d   Schwarzkehlspateltyrann
i   Tiranno todo golanera

4559   *Hemitriccus inornatus*
          Passeriformes - Tyrannidae
e   Pelzeln's Tody-Tyrant
f   Todirostre de Pelzeln
d   Icana-Spateltyrann
i   Tiranno todo di Pelzeln

4560   *Hemitriccus iohannis*
          Passeriformes - Tyrannidae
e   Johannes's Tody-Tyrant
f   Todirostre de Johannes
i   Tiranno todo di Johannes

4561   *Hemitriccus josephinae*
          Passeriformes - Tyrannidae
e   Boat-billed Tody-Tyrant; Josephine's
      Tody-Tyrant
f   Todirostre de Joséphine
d   Kahnschnabeltyrann
i   Tiranno todo di Josephine

4562   *Hemitriccus kaempferi*
          Passeriformes - Tyrannidae
e   Kaempfer's Tody-Tyrant
f   Todirostre de Kaempfer
d   Braunrückenspäteltyrann
i   Tiranno todo di Kaempfer

4563   *Hemitriccus margaritaceiventer*
          Passeriformes - Tyrannidae
e   Pearly-vented Tody-Tyrant
f   Todirostre à ventre perlé
d   Weißbauchspateltyrann
i   Tiranno todo perlato

4564   *Hemitriccus minimus*
          Passeriformes - Tyrannidae
e   Zimmer's Tody-Tyrant

4565   *Hemitriccus minor*
          Passeriformes - Tyrannidae
e   Snethlage's Tody-Tyrant
f   Todirostre de Snethlage
d   Snethlage-Tyrann
i   Tiranno todo di Snethlage

4566   *Hemitriccus mirandae*
          Passeriformes - Tyrannidae

e   Buff-breasted Tody-Tyrant;
      Miranda's Tody-Tyrant
f   Todirostre de Miranda
d   Caatinga-Spateltyrann
i   Tiranno todo pettofulvo

4567   *Hemitriccus nidipendulus*
          Passeriformes - Tyrannidae
e   Hangnest Tody-Tyrant
f   Todirostre de Wied
d   Grünrückenspateltyrann
i   Tiranno todo di Hangnest

4568   *Hemitriccus obsoletus*
          Passeriformes - Tyrannidae
e   Brown-breasted Stock-Tyrant;
      Brown-breasted Pygmy-Tyrant;
      Itatiaya Pygmy-Tyrant
f   Todirostre à poitrine brune;
      Todirostre de Ribeiro
d   Ribeiro-Tyrann
i   Tiranno todo pettobruno

4569   *Hemitriccus orbitatus*
          Passeriformes - Tyrannidae
e   Eye-ringed Tody-Tyrant; Olivaceous
      Tody-Tyrant
f   Todirostre à lunettes
d   Augenringspateltyrann
i   Tiranno todo olivaceo

4570   *Hemitriccus rufigularis*
          Passeriformes - Tyrannidae
e   Buff-throated Tody-Tyrant
f   Todirostre à gorge fauve
d   Braunbrustspateltyrann
i   Tiranno todo golarossiccia

4571   *Hemitriccus spodiops*
          Passeriformes - Tyrannidae
e   Yungas Tody-Tyrant
f   Todirostre de Bolivie
d   Grauzügelspateltyrann
i   Tiranno todo di Yungas

4572   *Hemitriccus striaticollis*
          Passeriformes - Tyrannidae
e   Stripe-necked Tody-Tyrant
f   Todirostre à cou rayé; Todirostre de
      Lafresnaye

d   Streifenbrustspateltyrann
i   Tiranno todo collostriato

4573   *Hemitriccus zosterops*
       Passeriformes - Tyrannidae
e   White-eyed Tody-Tyrant
f   Todirostre zostérops
d   Vireospateltyrann
i   Tiranno todo occhibianche

4574   *Hemixos castanonotus*
       Passeriformes - Pycnonotidae
e   Chestnut Bulbul; Chestnut-backed
       Bulbul
f   Bulbul marron
i   Bulbul castano

4575   *Hemixos flavala*
       Passeriformes - Pycnonotidae
e   Ashy Bulbul; Chestnut-backed
       Bulbul; Brown-eared Bulbul; Brown-
       eared Oriental-Bulbul
f   Bulbul à ailes vertes; Bulbul aux
       ailes vertes
d   Braunohrbülbül
i   Bulbul pettocenerino

4576   *Henicopernis infuscatus*
       Falconiformes - Accipitridae
e   Black Honeybuzzard; New Britain
       Honeybuzzard; New Britain Buzzard
f   Bondrée noire
d   Bismarck-Weihe
i   Pecchiaiolo nero

4577   *Henicopernis longicauda*
       Falconiformes - Accipitridae
e   Long-tailed Honeybuzzard; Long-
       tailed Buzzard
f   Bondrée à longue queue
d   Langschwanzweihe
i   Pecchiaiolo codalunga; Poiana dalla
       lunga coda

4578   *Henicophaps albifrons*
       Columbiformes - Columbidae
e   New Guinea Bronzewing; Black
       Bronzewing; White-capped Ground-
       Pigeon; New Guinea Black-
       Bronzewing; White-capped
       Bronzewing; Long-billed

       Bronzewing
f   Colombine à front blanc
d   Weißstirntaube; Weißscheiteltaube
i   Piccione alibronzate frontebianca

4579   *Henicophaps foersteri*
       Columbiformes - Columbidae
e   New Britain Bronzewing; New
       Britain Ground-Pigeon
f   Colombine de Nouvelle-Bretagne
d   Rotflügeltaube; Neubrittanien-
       Bronzeflügeltaube
i   Piccione alibronzate di Foerster

4580   *Henicorhina leucophrys*
       Passeriformes - Certhiidae
e   Grey-breasted Wood-Wren;
       Highland Wood-Wren
f   Troglodyte à poitrine grise
d   Einsiedlerzaunkönig
i   Scricciolo di bosco pettogrigio

4581   *Henicorhina leucoptera*
       Passeriformes - Certhiidae
e   Bar-winged Wood-Wren
f   Troglodyte à ailes blanches
d   Schwarzschwingenzaunkönig
i   Scricciolo di bosco alibarrate

4582   *Henicorhina leucosticta*
       Passeriformes - Certhiidae
e   White-breasted Wood-Wren; Black-
       capped Wood-Wren; Lowland
       Wood-Wren
f   Troglodyte à poitrine blanche
d   Waldzaunkönig
i   Scricciolo di bosco pettobianco

4583   *Henicorhina pittieri*
       Passeriformes - Certhiidae
e   Cherrie's Wood-Wren

4584   *Henicorhina prostheleuca*
       Passeriformes - Certhiidae
e   Sclater's Wood-Wren

4585   *Herpetotheres cachinnans*
       Falconiformes - Falconidae
e   Laughing Falcon
f   Macagua rieur

d   Lachfalke
i   Falco schignazzante

4586   ***Herpsilochmus atricapillus***
   Passeriformes - Thamnophilidae
e   Black-capped Ant-Wren
f   Grisin mitré
i   Scricciolo formichiere capinero

4587   ***Herpsilochmus axillaris***
   Passeriformes - Thamnophilidae
e   Yellow-breasted Ant-Wren; Yellow-
   winged Ant-Wren
f   Grisin à poitrine jaune
d   Gelbbrustameisenfänger
i   Scricciolo formichiere pettogiallo

4588   ***Herpsilochmus dorsimaculatus***
   Passeriformes - Thamnophilidae
e   Spot-backed Ant-Wren
f   Grisin strié
d   Streifenmantelameisenfanger
i   Scricciolo formichiere
   dorsomacchiato

4589   ***Herpsilochmus dugandi***
   Passeriformes - Thamnophilidae
e   Dugand's Ant-Wren; Colombian Ant-
   Wren
f   Grisin de Dugand
d   Dugand-Ameisenfänger

4590   ***Herpsilochmus longirostris***
   Passeriformes - Thamnophilidae
e   Large-billed Ant-Wren
f   Grisin à grand bec
d   Tupfenbrustameisenfänger
i   Scricciolo formichiere beccolungo

4591   ***Herpsilochmus motacilloides***
   Passeriformes - Thamnophilidae
e   Creamy-bellied Ant-Wren; Yellow-
   bellied Ant-Wren
f   Grisin motacilloïde
i   Scricciolo formichiere ventrecastano

4592   ***Herpsilochmus parkeri***
   Passeriformes - Thamnophilidae
e   Ash-throated Ant-Wren; Parker's
   Ant-Wren

f   Grisin de Parker
i   Scricciolo formichiere di Parker

4593   ***Herpsilochmus pectoralis***
   Passeriformes - Thamnophilidae
e   Pectoral Ant-Wren
f   Grisin à collier
d   Brustbandameisenfänger
i   Scricciolo formichiere pettorale

4594   ***Herpsilochmus pileatus***
   Passeriformes - Thamnophilidae
e   Bahia Ant-Wren; Pileated Ant-Wren;
   Black-capped Ant-Wren; Capped
   Ant-Wren; white-browed Ant-Wren
f   Grisin à coiffe noire
d   Schwarzkopfameisenfänger
i   Scricciolo formichiere di Bahia

4595   ***Herpsilochmus roraimae***
   Passeriformes - Thamnophilidae
e   Roraiman Ant-Wren; Roraima Ant-
   Wren
f   Grisin du Roraraima
d   Roraima-Ameisenfänger
i   Scricciolo formichiere di Roraima

4596   ***Herpsilochmus rufimarginatus***
   Passeriformes - Thamnophilidae
e   Rufous-winged Ant-Wren
f   Grisin à ailes rousses
d   Rotschwanzameisenfänger;
   Rotschwingenameisenfänger
i   Scricciolo formichiere alirosse

4597   ***Herpsilochmus sellowi***
   Passeriformes - Thamnophilidae
e   Caatinga Ant-Wren

4598   ***Herpsilochmus stictocephalus***
   Passeriformes - Thamnophilidae
e   Todd's Ant-Wren
f   Grisin de Todd
d   Cayenne-Ameisenfänger
i   Scricciolo formichiere di Todd

4599   ***Herpsilochmus sticturus***
   Passeriformes - Thamnophilidae
e   Spot-tailed Ant-Wren
f   Grisin givré
d   Fleckschwanzameisenfänger

i   Scricciolo formichiere
     codamacchiata

4600   *Hesperiphona abeillei*
        Passeriformes - Fringillidae
e   Hooded Grosbeak; Abeillé's
     Grosbeak
f   Grosbec à capuchon
d   Abeillé-Kernbeißer
i   Frosone dal cappuccio

4601   *Hesperiphona vespertina*
        Passeriformes - Fringillidae
e   Evening Grosbeak
f   Grosbec errant
d   Abendkernbeißer
i   Frosone vespertino

4602   *Heteralocha acutirostris*
        Passeriformes - Callaeatidae
e   Huia
f   Huia dimorphe
d   Huia; Lappenkopf; Hopflappenvogel;
     Lappenhopf
i   Huia; Uja

4603   *Heterocercus aurantiivertex*
        Passeriformes - Tyrannidae
e   Orange-crested Manakin; Orange-
     crowned Manakin; Yellow-crowned
     Manakin
f   Manakin à bandeau orangé
d   Orangescheitelpipra

4604   *Heterocercus flavivertex*
        Passeriformes - Tyrannidae
e   Yellow-crested Manakin; Yellow-
     crowned Manakin
f   Manakin à bandeau jaune
d   Pelzeln-Pipra

4605   *Heterocercus linteatus*
        Passeriformes - Tyrannidae
e   Flame-crested Manakin; Flame-
     crowned Manakin; Yellow-crested
     Manakin
f   Manakin à moustaches
d   Feuerscheitelpipra

4606   *Heteromirafra archeri*
        Passeriformes - Alaudidae

e   Archer's Lark; Archer's Long-clawed
     Lark; Somali Lark; Somali Long-
     clawed Lark
f   Alouette d'Archer
d   Somali-Spornlerche
i   Allodola di Archer

4607   *Heteromirafra ruddi*
        Passeriformes - Alaudidae
e   Rudd's Lark; Long-clawed Lark;
     Rudd's Long-clawed Lark
f   Alouette de Rudd
d   Spornlerche
i   Allodola di Rudd

4608   *Heteromirafra sidamoensis*
        Passeriformes - Alaudidae
e   Sidamo Lark; Sidamo Bush-Lark;
     Ethiopian Lark; Ethiopean Long-
     clawed Lark; Sidamo Long-clawed
     Lark
f   Alouette d'Erard
d   Sidamo-Spornlerche
i   Allodola del Sidamo

4609   *Heteromunia pectoralis*
        Passeriformes - Passeridae
e   Pictorella Munia; Pictorella
     Mannikin (ANZ); Pictorella Finch;
     Pectoral Munia; Pectoral Finch;
     White-breasted Finch; White-
     breasted Mannikin; White-breasted
     Munia
f   Capucin à poitrine blanche; Donacole
     pectorale; Donacole à poitrine
     blanche
d   Weißbrusthilffink
i   Cappuccino pettobianco

4610   *Heteromyias albispecularis*
        Passeriformes - Petroicidae
e   Black-cheeked Robin; Ashy Robin;
     Grey-headed Robin (ANZ); Ground
     Thicket-Robin; Ground Thicket-
     Flycatcher
f   Miro cendré
d   Farnschnäpper
i   Balia cenerina

4611   *Heteromyias cinereifrons*
        Passeriformes - Petroicidae

e Grey-headed Robin; Grey-headed
Thicket-Robin; Grey-headed
Thicket-Flycatcher
f Miro à tête grise
i Balia frontegrigia

4612 *Heteronetta atricapilla*
Anseriformes - Anatidae
e Black-headed Duck
f Hétéronette à tête noire
d Kuckucksente
i Anatra testanera

4613 *Heterophasia annectens*
Passeriformes - Sylviidae
e Rufous-backed Sibia; Chestnut-
backed Sibia; Chestnut-backed
Minla; Indian Chestnut-backed Sibia
f Sibia à dos marron
d Kletterwürgertimalie
i Sibia dorsorosso

4614 *Heterophasia auricularis*
Passeriformes - Sylviidae
e White-eared Sibia; Formosan Sibia;
Taiwan Sibia
f Sibia de Taïwan; Sibia de Formose
d Weißohrtimalie
i Sibia orecchiebianchi

4615 *Heterophasia capistrata*
Passeriformes - Sylviidae
e Rufous Sibia; Black-capped Sibia
(ISC); Black-headed Sibia
f Sibia casquée
d Schwarzkappentimalie
i Sibia capinera

4616 *Heterophasia desgodinsi*
Passeriformes - Sylviidae
e Black-headed Sibia

4617 *Heterophasia gracilis*
Passeriformes - Sylviidae
e Grey Sibia
f Sibia grise; Sibia gracile
d Grautimalie
i Sibia grigia

4618 *Heterophasia melanoleuca*
Passeriformes - Sylviidae

e Black-backed Sibia; Tickell's Sibia;
Black-headed Sibia; Black-eared
Sibia
f Sibia à tête noire
d Tickell-Timalie
i Sibia ventrebianco

4619 *Heterophasia picaoides*
Passeriformes - Sylviidae
e Long-tailed Sibia
f Sibia à longue queue
d Elstertimalie; Schweiftimalie
i Sibia codalunga

4620 *Heterophasia pulchella*
Passeriformes - Sylviidae
e Beautiful Sibia
f Sibia superbe
d Blaukopftimalie
i Sibia ardesia

4621 *Heterospingus rubrifrons*
Passeriformes - Fringillidae
e Sulphur-rumped Tanager
f Tangara à croupion jaune
i Tangara dai sopraccigli scarlatti

4622 *Heterospingus xanthopygius*
Passeriformes - Fringillidae
e Scarlet-browed Tanager
f Tangara à sourcils roux
d Brauenschopftangare
i Tangara dal gropponesulfureo

4623 *Hieraaetus ayresii*
Falconiformes - Accipitridae
e Ayres's Hawk-Eagle; Ayres's Eagle
f Aigle d'Ayres
d Fleckenadler
i Aquila minore di Ayres

4624 *Hieraaetus fasciatus*
Falconiformes - Accipitridae
e Bonelli's Eagle; Bonelli's Hawk-
Eagle (ISC)
f Aigle de Bonelli
d Habichtsadler
i Poiana a strisce; Aquila del Bonelli

4625 *Hieraaetus kienerii*
Falconiformes - Accipitridae

e Rufous-bellied Eagle; Rufous-bellied
Hawk-Eagle (ISC); Chestnut-bellied
Eagle; Chestnut-bellied Hawk-Eagle;
Rufous-bellied Dwarf-Eagle
f Aigle à ventre roux
d Indien-Zwergadler; Rotbauchadler
i Aquila minore di Kiener; Falco
aquila dalla pancia rossa

4626 *Hieraaetus morphnoides*
Falconiformes - Accipitridae
e Little Eagle; Australian Little-Eagle
f Aigle nain
d Australien-Zwergadler
i Aquila minore australiana; Aquila
piccola

4627 *Hieraaetus pennatus*
Falconiformes - Accipitridae
e Booted Eagle; Booted Hawk-Eagle
(ISC)
f Aigle botté
d Zwergadler
i Aquila minore; Aquila minore
eurasiatica

4628 *Hieraaetus spilogaster*
Falconiformes - Accipitridae
e African Hawk-Eagle; African Eagle
f Aigle fascié
d Habichtsadler
i Aquila minore africana

4629 *Himantopus himantopus*
Charadriiformes - Recurvirostridae
e Black-winged Stilt; Stilt; Common
Stilt; Ceylon Black-winged Stilt;
Pied Stilt
f Échasse blanche
d Stelzenläufer
i Cavaliere d'italia

**4630 *Himantopus himantopus
ceylonensis***
Charadriiformes - Recurvirostridae
e Sri Lanka Stilt
f Échasse de Ceylan

4631 *Himantopus knudseni*
Charadriiformes - Recurvirostridae

e Hawaiian Stilt
f Échasse des Hawai

4632 *Himantopus leucocephalus*
Charadriiformes - Recurvirostridae
e White-headed Stilt; Australian Stilt
f Échasse d'Australie
i Cavaliere testabianca

4633 *Himantopus melanurus*
Charadriiformes - Recurvirostridae
e White-backed Stilt; White-tailed Stilt
f Échasse à queue noire
i Cavaliere dorsobianco

4634 *Himantopus mexicanus*
Charadriiformes - Recurvirostridae
e Black-necked Stilt (NA); South
American Stilt; Redshank (WI);
Cap'n Lewis (WI); Soldier (WI);
Crackpot Soldier (WI); Pète-pète
f Échasse d'Amérique; Échasse (Ants);
Bécasse (Ants); Pète-pète (Ants)
d Amerikanischer Stelzenläufer
i Cavaliere collonero

4635 *Himantopus novaezelandiae*
Charadriiformes - Recurvirostridae
e Black Stilt; New Zealand Stilt
f Échasse noire
i Cavaliere nero

4636 *Himantornis haematopus*
Gruiformes - Rallidae
e Nkulengu Rail; Stripe-browed Crake
f Râle à pieds rouges
d Rotfußralle
i Rallo piedirossi

4637 *Himatione sanguinea*
Passeriformes - Fringillidae
e Apapane
f Picchion cramoisi
d Agapane; Karminkleidervogel
i Agapane

4638 *Hippolais caligata*
Passeriformes - Sylviidae
e Booted Warbler; Booted Tree-
Warbler; Eversmann's Booted
Warbler

f Hupolaïs botté; Hypolaïs russe
d Buschspötter
i Canapino asiatico

4639 *Hippolais icterina*
Passeriformes - Sylviidae
e Icterine Warbler
f Hupolaïs ictérine
d Gelbspötter; Gartenlaubvogel;
Gartenspötter
i Canapino maggiore

4640 *Hippolais languida*
Passeriformes - Sylviidae
e Upcher's Warbler
f Hupolaïs d'Upcher
d Dornspötter
i Canapino di Upcher

4641 *Hippolais olivetorum*
Passeriformes - Sylviidae
e Olive-tree Warbler
f Hupolaïs des oliviers
d Ollvenspötter
i Canapino levantino

4642 *Hippolais pallida*
Passeriformes - Sylviidae
e Olivaceous Warbler
f Hupolaïs pâle
d Blaßspötter
i Canapino pallido

4643 *Hippolais polyglotta*
Passeriformes - Sylviidae
e Melodious Warbler
f Hupolaïs polyglotte
d Orpheus-Spötter
i Canapino; Canapino comune

4644 *Hippolais rama*
Passeriformes - Sylviidae
e Sykes's Warbler
f Hupolaïs rama

4645 *Hirundapus caudacutus*
Apodiformes - Apodidae
e White-throated Needletail; White-
throated Needle-tailed Swift;
Northern Needle-tailed Swift; White-
throated Spinetailed-Swift; White-

throated Spinetail; Spinetailed-Swift;
Northern Spinetailed-Swift; Northern
Needletail
f Martinet épineux
d Stachelschwanzsegler
i Rondone golabianca; Rondone
codaspinosa golabianca

4646 *Hirundapus celebensis*
Apodiformes - Apodidae
e Purple Needletail
f Martinet des Célèbes
i Rondone codaspinosa di Sulawesi;
Rondone di Sulawesi

4647 *Hirundapus cochinchinensis*
Apodiformes - Apodidae
e Silver-backed Needletail; White-
vented Spinetail; White-vented
Spine-tailed Swift; White-vented
Needletail; Silver-backed Spinetail;
Grey-throated Needletail
f Martinet de Cochinchine
d Graukehlsegler
i Rondone codaspinosa della
Cocincina; Rondone della Cocincina

4648 *Hirundapus giganteus*
Apodiformes - Apodidae
e Brown-backed Needletail; Large
Brown-throated Spinetailed-Swift;
Brown Needletail; Brown-throated
Spinetail; Brown-throated
Spinetailed-Swift; Brown Spinetail;
Brown Spinetailed-Swift; Giant
Brown-Needletail
f Martinet géant
d Eilsegler
i Rondone codaspinosa bruno;
Rondone bruno

4649 *Hirundinea bellicosa*
Passeriformes - Tyrannidae
e Swallow Flycatcher
f Moucherolle de Vieillot
i Tiranno rondine

4650 *Hirundinea ferruginea*
Passeriformes - Tyrannidae
e Cliff Flycatcher
f Moucherolle hirondelle

d Schwalbentyrann
i Tiranno dei dirupi

4651 *Hirundinidae*
Passeriformes
e Swallows
f Hirondelles; Hirundinidés
d Schwalben
i Irundinidi; Rondini

4652 *Hirundo abyssinica*
Passeriformes - Hirundinidae
e Lesser Striped-Swallow; Striped
Swallow
f Hirondelle striée; Hirondelle à gorge
striée
d Kleine Streifenschwalbe;
Maidschwalbe
i Rondine dell'Abissinia

4653 *Hirundo aethiopica*
Passeriformes - Hirundinidae
e Ethiopian Swallow
f Hirondelle d'Éthiopie
d Fahlkehlschwalbe
i Rondine etiopica

4654 *Hirundo albigularis*
Passeriformes - Hirundinidae
e White-throated Swallow
f Hirondelle à gorge blanche
d Weißkehlschwalbe
i Rondine golabianca

4655 *Hirundo andecola*
Passeriformes - Hirundinidae
e Andean Swallow; Andean Cliff-
Sparrow
f Hirondelle des Andes
d Anden-Schwalbe
i Rondine delle Ande

4656 *Hirundo angolensis*
Passeriformes - Hirundinidae
e Angola Swallow
f Hirondelle de I'Angola; Hirondelle
d'Angola
d Angola-Schwalbe
i Rondine dell'Angola

4657 *Hirundo ariel*
Passeriformes - Hirundinidae
e Fairy Martin
f Hirondelle ariel
d Arielschwalbe
i Rondine capirossa australiano

4658 *Hirundo atrocaerulea*
Passeriformes - Hirundinidae
e Blue Swallow; Eastern Blue-Swallow
f Hirondelle bleue
d Stahlschwalbe
i Rondine blu

4659 *Hirundo concolor*
Passeriformes - Hirundinidae
e Dusky Crag-Martin; Dusky Martin
f Hirondelle concolore
d Einfarbschwalbe
i Rondine rupestre bruna

4660 *Hirundo cucullata*
Passeriformes - Hirundinidae
e Greater Striped- Swallow
f Hirondelle à tête rousse
d Streifenschwalbe; Große
Strichelschwalbe
i Rondine dal cappuccio

4661 *Hirundo daurica*
Passeriformes - Hirundinidae
e Red-rumped Swallow; Striated
Swallow (ISC); Daurian Swallow;
Lesser Striated-Swallow; Golden-
rumped Swallow
f Hirondelle rousseline
d Rötelschwalbe
i Rondine rossiccia

4662 *Hirundo dimidiata*
Passeriformes - Hirundinidae
e Pearl-breasted Swallow
f Hirondelle à gorge perlée
d Perlbrustschwalbe
i Rondine pettoperlato

4663 *Hirundo domicella*
Passeriformes - Hirundinidae
e West African Swallow; African
Striated-Swallow; Lowland Swallow

f Hirondelle ouest-africaine
i Rondine rossicia del Sahel

4664 Hirundo domicola
Passeriformes - Hirundinidae
e Hill Swallow; Nilgiri House-Swallow
(ISC); House Swallow
f Hirondelle des Nilgiri
d Kurzschnabelschwalbe
i Rondine delle alture

4665 Hirundo fluvicola
Passeriformes - Hirundinidae
e Streak-throated Swallow; Indian
Cliff-Swallow (ISC); Indian Cliff-
Martin
f Hirondelle fluviatile
d Indien-Klippenschwalbe;
Braunscheitelschwalbe
i Rondine rupestre indiano

4666 Hirundo fuliginosa
Passeriformes - Hirundinidae
e Forest Swallow; Dusky Cliff-
Swallow; Forest Cliff-Swallow;
Cameroon Cliff-Swallow
f Hirondelle de forêt
d Kamerun-Schwalbe; Bronzeschwalbe
i Rondine fuligginosa

4667 Hirundo fuligula
Passeriformes - Hirundinidae
e Rock Martin; African Rock-Martin
f Hirondelle isabelline; Hirondelle
brune
d Steinschwalbe
i Rondine rupestre africana

4668 Hirundo fulva
Passeriformes - Hirundinidae
e Cave Swallow; Cinnamon-throated
Swallow; Swallow (WI); Rainbird
(WI)
f Hirondelle à front brun; Hirondelle
fauve (Ants)
d Höhlenschwalbe
i Rondine fulva

4669 Hirundo griseopyga
Passeriformes - Hirundinidae
e Grey-rumped Swallow

f Hirondelle à croupion gris
d Graubürzelschwalbe
i Rondine dal groppone grigio

4670 Hirundo leucosoma
Passeriformes - Hirundinidae
e Pied-winged Swallow; Pied-wing
Swallow
f Hirondelle à ailes tachetées
d Scheckflügelschwalbe
i Rondine alichiazzate

4671 Hirundo lucida
Passeriformes - Hirundinidae
e Red-chested Swallow; Gambia
Swallow; Gambia Barn-Swallow;
Gambian Barn-Swallow
f Hirondelle de Guinée
d Singschwalbe
i Rondine lucente

4672 Hirundo megaensis
Passeriformes - Hirundinidae
e White-tailed Swallow
f Hirondelle à queue blanche
d Benson-Schwalbe
i Rondine codabianca

4673 Hirundo neoxena
Passeriformes - Hirundinidae
e Welcome Swallow
f Hirondelle messagère
d Neuholland-Schwalbe
i Rondine benvenuta

4674 Hirundo nigricans
Passeriformes - Hirundinidae
e Tree Martin; Australian Tree-Martin
f Hirondelle des arbres
d Baumschwalbe;
Australbaumschwalbe
i Rondine arboricola australiano

4675 Hirundo nigrita
Passeriformes - Hirundinidae
e White-throated Blue-Swallow; Little
Blue-Swallow; White-chinned
Swallow
f Hirondelle à bavette; Hirondelle
noire

d Mohrenschwalbe
i Rondine nigrita

4676 ***Hirundo nigrorufa***
Passeriformes - Hirundinidae
e Black-and-rufous Swallow; Black-and-red Swallow; Rufous-breasted Swallow; West African Swallow
f Hirondelle roux-et-noir; Hirondelle rouge-et-noire
d Benguela-Schwalbe
i Rondine nera e rossa

4677 ***Hirundo obsoleta***
Passeriformes - Hirundinidae
e Pale Crag-Martin
f Hirondelle du désert
d Wüstenschwalbe; Wüstenfelsenschwalbe
i Rondine montana pallida

4678 ***Hirundo perdita***
Passeriformes - Hirundinidae
e Red Sea Swallow; Red Sea Cliff-Swallow
f Hirondelle de la Mer Rouge
i Rondine di Fry

4679 ***Hirundo preussi***
Passeriformes - Hirundinidae
e Preuss's Swallow; Preuss's Cliff-Swallow
f Hirondelle de Preuss
d Rotschläfenschwalbe
i Rondine di Preuss

4680 ***Hirundo rufigula***
Passeriformes - Hirundinidae
e Red-throated Swallow; Red-throated Cliff-Swallow; Red-throated Rock-Martin; Angolan Cliff Swallow; Rock Swallow; Angolan Rock-Swallow
f Hirondelle à gorge fauve; Hirondelle de Bocage
d Rotkehlschwalbe
i Rondine golarossa

4681 ***Hirundo rupestris***
Passeriformes - Hirundinidae
e Eurasian Crag-Martin; Crag Martin; European Crag-Martin; Mountain Crag-Martin; Northern Crag-Martin
f Hirondelle des rochers; Hirondelle des roches
d Felsenschwalbe; Gewöhnliche Felsenschwalbe
i Rondine montana

4682 ***Hirundo rustica***
Passeriformes - Hirundinidae
e Barn Swallow; Swallow; European Swallow (NA) (CSA); Christmasbird (WI); Common Swallow; Eurasian Swallow; East Asian Swallow; Chimney Swallow; Rustic Swallow; American Barn-Swallow; Nile Swallow; Egyptian Swallow
f Hirondelle rustique; Hirondelle; Hirondelle de cheminée; Hirondelle des granges; Hirondelle roux (Ants)
d Rauchschwalbe; Schwalbe
i Rondine; Rondine domestica; Rondine comune

4683 ***Hirundo semirufa***
Passeriformes - Hirundinidae
e Rufous-chested Swallow; Red-breasted Swallow (CSA)
f Hirondelle à ventre roux
d Rotbauchschwalbe
i Rondine pettorosso

4684 ***Hirundo senegalensis***
Passeriformes - Hirundinidae
e Mosque Swallow; African Mosque-Swallow
f Hirondelle des mosquées
d Senegal-Schwalbe
i Rondine del Senegal

4685 ***Hirundo smithii***
Passeriformes - Hirundinidae
e Wire-tailed Swallow
f Hirondelle à longs brins
d Rotkappenschwalbe; Rotköpfchenschwalbe
i Rondine di Smith

4686 ***Hirundo spilodera***
Passeriformes - Hirundinidae
e South African Swallow; South

African Cliff Swallow (CSA);
African Cliff-Swallow
f   Hirondelle sud-africaine
d   Klippenschwalbe
i   Rondine rupestre del Sudafrica

4687   *Hirundo striolata*
Passeriformes - Hirundinidae
e   Striated Swallow; Oriental Mosque-
Swallow; Chinese Striated-Swallow;
Larger Striated-Swallow; Greater
Striated-Swallow; Eastern Red-
rumped Swallow
f   Hirondelle striolée; Hirondelle striée
d   Strichelschwalbe
i   Rondine striata maggiore

4688   *Hirundo tahitica*
Passeriformes - Hirundinidae
e   Pacific Swallow; Nilgiri House-
Swallow (ISC); Coast Swallow; Hill-
Swallow; House Swallow; Eastern
House Swallow; Small House-
Swallow
f   Hirondelle de Tahiti
d   Südseeschwalbe
i   Rondine del Pacifico

4689   *Histrionicus histrionicus*
Anseriformes - Anatidae
e   Harlequin Duck; Harlequin
f   Garrot arlequin; Arlequin plongeur;
Canard arlequi
d   Kragenente
i   Moretta arlecchino

4690   *Histurgops ruficauda*
Passeriformes - Ploceidae
e   Rufous-tailed Weaver
f   Histurgopse à queue rouge; Tisserin à
queue rouge
d   Rotschwanzweber
i   Tessitore codarossa

4691   *Hodgsonius phaenicuroides*
Passeriformes - Muscicapidae
e   White-bellied Redstart; Hodgson's
Shortwing; Chinese Shortwing
f   Bradybate à queue rouge;
Rougequeue à ventre blanc

d   Rotschwanzkurzflügel
i   Codirosso blu ventrebianco

4692   *Horizorhinus dohrni*
Passeriformes - Muscicapidae
e   Döhrn's Flycatcher; Döhrn's Thrush-
Babbler; Principe Flycatcher
f   Gobemouche de Dohrn; Gobe-
mouches de Dohrn; Cratérope de
Principe
d   Prinzendrossling
i   Pigliamosche di Dorn

4693   *Humblotia flavirostris*
Passeriformes - Muscicapidae
e   Humblot's Flycatcher; Grand Comoro
Flycatcher
f   Gobemouche des Comores; Gobe-
mouches des Comores
d   Humblot-Schnäpper
i   Pigliamosche di Humblot

4694   *Hydrobates pelagicus*
Procellariiformes - Hydrobatidae
e   European Storm-Petrel; Stormy
Petrel; Storm Petrel; British Storm-
Petrel; Mother Carey's Chicken
(CSA); Mother Carey's Chick
f   Pétrel tempête; Océanite tempête;
Océanite de tempête; Pétrel de
tempête
d   Sturmschwalbe
i   Uccello delle tempeste; Uccello delle
tempeste europeo

4695   Hydrobatidae
Procellariiformes
e   Storm-Petrels
f   Océanites; Hydrobatidés; Pétrels;
Pétrels-tempêtes; Hirondelles de
tempête; Oiseaux de tempête
d   Sturmschwalben
i   Idrobatidi; Uccelli delle tempeste

4696   *Hydrochous gigas*
Apodiformes - Apodidae
e   Waterfall Swift; Giant Swiftlet
f   Salangane géante
d   Riesensalangane
i   Rondone delle cascate

4697 **Hydrophasianus chirurgus**
Charadriiformes - Jacanidae
e Pheasant-tailed Jacana; Water
Pheasant; Chinese Water-Pheasant
f Jacana à longue queue
d Wasserfasan
i Idrofagiano chirurgo; Parra;
Idrofagiano; Fagiano d'acqua

4698 **Hydropsalis brasiliana**
Charadriiformes - Jacanidae
e Scissor-tailed Nightjar
f Engoulevent à queue en ciseaux
d Spießnachtschwalbe
i Succiacapre coda a forbice

4699 **Hydropsalis climacocerca**
Caprimulgiformes - Caprimulgidae
e Ladder-tailed Nightjar
f Engoulevent trifide
d Staffelschwanznachtschwalbe
i Succiacapre codagraduata

4700 **Hyetornis pluvialis**
Cuculiformes - Cocyzidae
e Chestnut-bellied Cuckoo; Old Man's
Bird; Old Man Bird (WI); Hunter
(WI); Rainbird (WI); Maybird (WI);
Jamaican Cuckoo
f Piaye de pluie
d Regenkuckuck
i Cuculo ventrecastano americano

4701 **Hyetornis rufigularis**
Cuculiformes - Cocyzidae
e Bay-breasted Cuckoo; Rufous-
breasted Cuckoo
f Piaye cabrite
d Dominikanerkukuck
i Cuculo pettorugginoso americano

4702 **Hylacola cauta**
Passeriformes - Pardalotidae
e Shy Heathwren; Shy Hylacola; Shy
Groundwren; Mallee Wren; Mallee
Hylacola
f Séricorne timide
d Sandhuscher

4703 **Hylacola cauta spadicea**
Passeriformes - Pardalotidae
e Riverina Shy Heathwren (ANZ)

4704 **Hylacola cauta whitlocki**
Passeriformes - Pardalotidae
e Western Shy Heathwren (ANZ)

4705 **Hylacola pyrrhopygia**
Passeriformes - Pardalotidae
e Chestnut-rumped Hylacola;
Chestnut-rumped Heathwren (ANZ);
Chestnut-tailed Heathwren;
Chestnut-tailed Groundwren;
Chestnut-rumped Groundwren;
Heathwren
f Séricorne à croupion roux
s Heidehuscher

4706 **Hylacola pyrrhopygia parkeri**
Passeriformes - Pardalotidae
e Mt. Loft Ranges Chestnut-rumped
Heathwren (ANZ)

4707 **Hylacola pyrrhopygia pedleri**
Passeriformes - Pardalotidae
e Flinders Ranges Chestnut-rumped
Heathwren (ANZ)

4708 **Hylexetastes brigidai**
Passeriformes - Furnariidae
e Brigida's Woodcreeper

4709 **Hylexetastes perrotii**
Passeriformes - Furnariidae
e Red-billed Woodcreeper; Red-billed
Creeper
f Grimpar de Perrot
d Rotschnabelbaumsteiger
i Rampichino beccorosso

4710 **Hylexetastes stresemanni**
Passeriformes - Furnariidae
e Bar-bellied Woodcreeper; Bar-
bellied Creeper
f Grimpar de Stresemann
d Wellenbauchbaumsteiger
i Rampichino di Stresemann

4711 *Hylexetastes uniformis*
Passeriformes - Furnariidae
e Uniform Woodcreeper

4712 *Hylia prasina*
Passeriformes - Sylviidae
e Green Hylia; Green Hylie; Yellow-
throated Warbler
f Hylia verte
d Hylie
i Ilia verde

4713 *Hyliota australis*
Passeriformes - Sylviidae
e Southern Hyliota; Mashona Hyliota
(CSA); Mashona Flycatcher;
Southern Yellow-bellied Flycatcher;
Southern Yellow-bellied Hyliota;
Southern Yellow-bellied Warbler;
Austral Hyliota
f Hyliote australe
d Maschona-Hyliota
i Ilia meridionale

4714 *Hyliota flavigaster*
Passeriformes - Sylviidae
e Yellow-bellied Hyliota; Yellow-
breasted Hylotia (CSA); Yellow-
bellied Flycatcher; Yellow-bellied
Flycatcher-Warbler
f Hyliote à ventre jaune
d Gelbbauchhyliota
i Ilia ventregiallo

4715 *Hyliota violacea*
Passeriformes - Sylviidae
e Violet-backed Hyliota; Violet-backed
Flycatcher
f Hyliote à dos violet
d Violettmantelhyliota
i Ilia violacea

4716 *Hylocharis chrysura*
Apodiformes - Trochilidae
e Gilded Hummingbird
f Saphir à queue d'or
d Goldsaphir
i Colibrì dorato

4717 *Hylocharis cyanus*
Apodiformes - Trochilidae

e White-chinned Sapphire
f Saphir azuré
d Weißkinnsaphir
i Colibrì zaffiro golabianca

4718 *Hylocharis eliciae*
Apodiformes - Trochilidae
e Blue-throated Goldentail; Elicia's
Sapphire
f Saphir d'Elicia
d Goldschwanzsaphir
i Colibrì di Elisa

4719 *Hylocharis grayi*
Apodiformes - Trochilidae
e Blue-headed Sapphire; Puritan
Sapphire; Gray's Sapphire
f Saphir ulysse
d Blaukopfsaphir
i Colibrì di Gray

4720 *Hylocharis humboldti*
Apodiformes - Trochilidae
e Humboldt's Sapphire

4721 *Hylocharis leucotis*
Apodiformes - Trochilidae
e White-eared Hummingbird
f Saphir à oreilles blanches
d Weißohrsaphir
i Colibrì orecchiebianche

4722 *Hylocharis pyropygia*
Apodiformes - Trochilidae
e Flame-rumped Sapphire
f Saphir embrasé
i Colibrì dal groppone di fiamma

4723 *Hylocharis sapphirina*
Apodiformes - Trochilidae
e Rufous-throated Sapphire; Red-
throated Hummingbird
f Saphir à gorge rousse
d Rotkehlsaphir
i Colibrì zaffiro golarossiccia

4724 *Hylocharis xantusii*
Apodiformes - Trochilidae
e Xantus's Hummingbird; Black-
fronted Hummingbird
f Saphir de Xantus

d Schwarzstirnsaphir
i Colibrì frontenera

4725 *Hylocitrea bonensis*
Passeriformes - Corvidae
e Olive-flanked Whistler; Buff-
throated Thickhead; Hylocitrea;
Celebes Mountain-Whistler; Yellow-
flanked Whistler; Buff-throated
Hylocitrea; Sulawesi Whistler
f Siffleur à flancs jaunes
d Beerenschnäpper
i Capogrosso golafulva

4726 *Hylocryptus erythrocephalus*
Passeriformes - Furnariidae
e Henna-hooded Foliagegleaner;
Henna-hooded Leafgleaner
f Anabate à tête orangée
d Rotkopfbaumspäher
i Spigolafoglie testarossa

4727 *Hylocryptus rectirostris*
Passeriformes - Furnariidae
e Chestnut-capped Foliagegleaner;
Henna-capped Foliagegleaner
f Anabate à bec droit
i Spigolafoglie testacastana

4728 *Hyloctistes subulatus*
Passeriformes - Furnariidae
e Striped Woodhaunter; Striped
Foliagegleaner; Striped Leafgleaner
f Anabate forestier
d Waldspäher
i Ticotico striato

4729 *Hylomanes momotula*
Coraciiformes - Momotidae
e Tody Motmot
f Motmot nain
d Zwergmotmot
i Motmot todo

4730 *Hylonympha macrocerca*
Apodiformes - Trochilidae
e Scissor-tailed Hummingbird
f Colibri à queue-en-ciseaux
d Waldnymphe
i Colibrì codaforcuta

4731 *Hylopezus berlepschi*
Passeriformes - Formicariidae
e Amazonian Ant-Pitta; Berlepsch's
Ant-Pitta
f Grallaire d'Amazonie
d Berlepsch-Ameisenpitta
i Pitta formichiera dell'Amazzonia

4732 *Hylopezus dives*
Passeriformes - Formicariidae
e Fulvous-bellied Ant-Pitta; Thicket
Ant-Pitta

4733 *Hylopezus fulviventris*
Passeriformes - Formicariidae
e White-lored Ant-Pitta; Fulvous-
bellied Ant-Pitta
f Grallaire à ventre fauve
d Schwarzkappenameisenpitta
i Pitta formichiera ventrefulva

4734 *Hylopezus macularius*
Passeriformes - Formicariidae
e Spotted Ant-Pitta
f Grallaire tachetée
d Rostflankenameisenpitta
i Pitta formichiera maculata

4735 *Hylopezus nattereri*
Passeriformes - Formicariidae
e Speckle-breasted Ant-Pitta
f Grallaire de Natterer
i Pitta formichiera pettomacchiettato

4736 *Hylopezus ochroleucus*
Passeriformes - Formicariidae
e White-browed Ant-Pitta; Speckled
Ant-Pitta; Speckle-breasted Ant-Pitta
f Grallaire teguy
d Fleckenbrustameisenpitta
i Pitta formichiera dai copraccigli
bianchi

4737 *Hylopezus perspicillatus*
Passeriformes - Formicariidae
e Spectacled Ant-Pitta; Streak-chested
Ant-Pitta
f Grallaire à lunettes
d Brillenameisenpitta
i Pitta formichiera pettostriato

4738 _Hylophilus amaurocephalus_
Passeriformes - Vireonidae
_e_ Grey-eyed Greenlet
_f_ Viréon aux yeux gris

4739 _Hylophilus aurantiifrons_
Passeriformes - Vireonidae
_e_ Golden-fronted Greenlet; Golden-
fronted Vireo
_f_ Viréon à front d'or
_d_ Goldstirnvireo
_i_ Vireo verdino frontedorata

4740 _Hylophilus brunneiceps_
Passeriformes - Vireonidae
_e_ Brown-headed Greenlet; Brown-
headed Vireo
_f_ Viréon brunâtre
_d_ Braunkopfvireo
_i_ Vireo verdino testabruna

4741 _Hylophilus decurtatus_
Passeriformes - Vireonidae
_e_ Lesser Greenlet; Grey-headed
Greenlet; Grey-headed Vireo
_f_ Viréon menu
_d_ Graukappenvireo;
Graukopfhylophilus
_i_ Vireo verdino minor

4742 _Hylophilus flavipes_
Passeriformes - Vireonidae
_e_ Scrub Greenlet; Scrub Vireo
_f_ Viréon à pattes claires
_d_ Buschvireo
_i_ Vireo verdino di macchia

4743 _Hylophilus griseiventris_
Passeriformes - Vireonidae
_e_ Lemon-chested Greenlet

4744 _Hylophilus hypoxanthus_
Passeriformes - Vireonidae
_e_ Dusky-capped Greenlet; Dusky-
capped Vireo; Yellow-bellied
Greenlet
_f_ Viréon à ventre jaune
_d_ Braunstirnvireo
_i_ Vireo verdino caposcuro

4745 _Hylophilus minor_
Passeriformes - Vireonidae
_e_ Lesser Greenlet; Lesser Vireo
_d_ Darien-Vireo

4746 _Hylophilus muscicapinus_
Passeriformes - Vireonidae
_e_ Buff-cheeked Greenlet; Buff-chested
Vireo; Buff-cheeked Vireo; Buff-
chested Greenlet
_f_ Viréon fardé
_d_ Braunwangenvireo
_i_ Vireo verdino pettochiaro

4747 _Hylophilus ochraceiceps_
Passeriformes - Vireonidae
_e_ Tawny-crowned Greenlet; Tawny-
crowned Vireo
_f_ Viréon à calotte rousse
_d_ Fuchsscheitelvireo
_i_ Vireo verdino corona fulva

4748 _Hylophilus olivaceus_
Passeriformes - Vireonidae
_e_ Olivaceous Greenlet; Olivaceous
Vireo
_f_ Viréon olivâtre
_d_ Olivvireo
_i_ Vireo verdino olivaceo

4749 _Hylophilus pectoralis_
Passeriformes - Vireonidae
_e_ Ashy-headed Greenlet; Ashy-headed
Vireo
_f_ Viréon à tête cendréc
_d_ Aschkopfvireo
_i_ Vireo verdino testagrigia

4750 _Hylophilus poicilotis_
Passeriformes - Vireonidae
_e_ Rufous-crowned Greenlet; Rufous-
crowned Vireo
_f_ Viréon oreillard
_d_ Rostkappenvireo
_i_ Vireo verdino corona rossa

4751 _Hylophilus rubrifrons_
Passeriformes - Vireonidae
_e_ Red-fronted Greenlet

4752 **Hylophilus sclateri**
Passeriformes - Vireonidae
e Tepui Greenlet; Tepui Vireo
f Viréon des tépuis
d Tepui-Vireo
i Vireo verdino del Tepui

4753 **Hylophilus semibrunneus**
Passeriformes - Vireonidae
e Rufous-naped Greenlet; Rufous-naped Vireo
f Viréon à nuque rousse
d Rostnackenvireo
i Vireo verdino nucarossa

4754 **Hylophilus semicinereus**
Passeriformes - Vireonidae
e Grey-chested Greenlet; Grey-chested Vireo
f Viréon à gorge grise
d Graunackenvireo
i Vireo verdino pettogrigio

4755 **Hylophilus thoracicus**
Passeriformes - Vireonidae
e Rio de Janeiro Greenlet; Lemon-chested Greenlet; Lemon-chested Vireo
f Viréon à plastron
d Gelbbrustvireo
i Vireo verdino pettogiallo

4756 **Hylophilus viridiflavus**
Passeriformes - Vireonidae
e Yellow-green Greenlet

4757 **Hylophylax naevia**
Passeriformes - Thamnophilidae
e Spot-backed Antbird
f Fourmilier tacheté
d Braunfleckenwaldwächter
i Mangiaformiche dorsomacchiato

4758 **Hylophylax naevioides**
Passeriformes - Thamnophilidae
e Spotted Antbird
f Fourmilier grivelé
d Fleckenbrustwaldwächter; Fleckenwaldwächter
i Mangiaformiche maculato

4759 **Hylophylax poecilinota**
Passeriformes - Thamnophilidae
e Scale-backed Antbird; Scaly-backed Antbird
f Fourmilier zèbré
d Schuppenwaldwächter; Waldwächter
i Mangiaformiche dorsoscaglioso

4760 **Hylophylax punctulata**
Passeriformes - Thamnophilidae
e Dot-backed Antbird; Des Murs Spotted-Antbird
f Fourmilier perlé
d Silberfleckenwaldwächter
i Mangiaformiche dorsopuntinato

4761 **Hylorchilus navai**
Passeriformes - Certhiidae
e Nava's Wren; Crossin's Wren; Slender-billed Wren

4762 **Hylorchilus sumichrasti**
Passeriformes - Certhiidae
e Sumichrast's Wren; Slender-billed Wren
f Troglodyte de Sumichrast; Troglodyte à bec fin
d Schmalschnabelzaunkönig
i Scricciolo beccosottile

4763 **Hymenolaimus malacorhynchos**
Anseriformes - Anatidae
e Blue Duck; Blue Mountain-Duck; Mountain Duck
f Canard bleu
d Saumschnabelente
i Anatra azzura

4764 **Hymenops perspicillatus**
Passeriformes - Tyrannidae
e Spectacled Tyrant
f Ada; Ada clignot
d Brillentyrann
i Tiranno dagli occhiali

4765 **Hypargos margaritatus**
Passeriformes - Estrildidae
e Pink-throated Twinspot; Rosy Twinspot; Verreaux's Twinspot
f Sénégali de Verreaux; Sénégali à gorge rose

d Perlastrild
i Amaranto rosa

4766 *Hypargos niveoguttatus*
Passeriformes - Estrildidae
e Peters's Twinspot; Red-throated
Twinspot (CSA)
f Sénégali enflammé
d Rotkehltropenastrild
i Amaranto fiammante

4767 *Hypergerus atriceps*
Passeriformes - Cisticolidae
e Oriole Warbler; Oriole Babbler;
Moho
f Noircap loriot; Timalie à tête noire;
Moho à tête noire
d Pirolsänger
i Oriolino caponero

4768 *Hypergerus lepidus*
Passeriformes - Cisticolidae
e Grey-capped Warbler
d Eminie

4769 *Hypnelus ruficollis*
Piciformes - Bucconidae
e Russet-throated Puffbird; Two-
banded Puffbird; Double-banded
Puffbird
f Tamatia à gorge rousse
d Rostkehlfaulvogel
i Bucco collorosso

4770 *Hypocnemis cantator*
Passeriformes - Thamnophilidae
e Warbling Antbird
f Alapi carillonneur
d Singameisenschnäpper
i Mangiaformiche canoro

4771 *Hypocnemis hypoxantha*
Passeriformes - Thamnophilidae
e Yellow-browed Antbird
f Alapi à sourcils jaunes
d Silberbrauenameisenschnäpper;
Gelbbrauenameisenschnäpper
i Mangiaformiche ventregiallo

4772 *Hypocnemoides maculicauda*
Passeriformes - Thamnophilidae

e Band-tailed Antbird; Band-tailed
Antcreeper
f Alapi riverain
d Uferameisenschnäpper
i Rampichino formichiere
codamarginata

4773 *Hypocnemoides melanopogon*
Passeriformes - Thamnophilidae
e Black-chinned Antbird; Black-
chinned Antcreeper
f Alapi à menton noir
d Schwarzkinnameisenschnäpper
i Rampichino formichiere mentonero

4774 *Hypocolius ampelinus*
Passeriformes - Hypocoliidae
e Grey Hypocolius; Hypocolius
f Jaseur boréal; Hypocolius gris
d Seidenwürger; Nachtschattenfresser;
Arabischer Seidenschwanz
i Ipocolio grigio; Ipocolio

4775 *Hypocryptadius cinnamomeus*
Passeriformes - Zosteropidae
e Cinnamon Ibon; Cinnamon White-
eye; Cinnamon Rufouseye
f Zostérops cannellé
d Zimtvogel
i Occhialino color cannella

4776 *Hypoedaleus guttatus*
Passeriformes - Formicariidae
e Spot-backed Antshrike
f Batara moucheté
d Perlenmantelameisenwürger
i Averla formichiera
dorsomacchiettato

4777 *Hypogramma hypogrammicum*
Passeriformes - Nectariniidae
e Purple-naped Sunbird; Blue-naped
Sunbird
f Souïmanga strié
d Streifennektarvogel
i Nettarinia nucablu

4778 *Hypopyrrhus pyrohypogaster*
Passeriformes - Fringillidae
e Red-bellied Grackle
f Quiscale à ventre rouge

d   Rotbauchstärling
i   Gracchio ventrerosso

4779 *Hypositta corallirostris*
     Passeriformes - Corvidae
e   Coral-billed Nuthatch; Madagascar
     Nuthatch; Nuthatch Vanga; Coralbill
f   Hypositte malgache; Vanga-sitelle
     malgache
d   Madagaskar-Kleiber; Kleibervanga
i   Ipositta beccodicorallo

4780 *Hypothymis azurea*
     Passeriformes - Corvidae
e   Black-naped Monarch; Black-naped
     Blue-Flycatcher; Black-naped Blue-
     Monarch; Black-naped Flycatcher;
     Pacific Small-Monarch; Pacific
     Monarch; Moluccan Monarch; Small
     Blue-Monarch; Moluccan Monarch-
     Flycatcher
f   Tchitrec azuré; Gobe-mouches
     monarque azuré
d   Schwarzgenickschnäpper;
     Maidschnepper
i   Monarca azzurro crestabreve

4781 *Hypothymis coelestis*
     Passeriformes - Corvidae
e   Celestial Monarch; Celestial Blue-
     Monarch
f   Tchitrec céleste
d   Himmelschnäpper
i   Monarca celeste

4782 *Hypothymis helenae*
     Passeriformes - Corvidae
e   Short-crested Monarch; Short-crested
     Blue-Monarch
f   Tchitrec d'Hélène
d   Helene-Schnäpper
i   Monarca azzurro crestabreve

4783 *Hypsipetes borbonicus*
     Passeriformes - Pycnonotidae
e   Réunion Black Bulbul; Réunion
     Bulbul; Olivaceous Bulbul; Mauritius
     Bulbul; Mauritius Black-Bulbul
f   Bulbul de Bourbon; Bulbul de l'île
     Maurice

d   Maskarenen-Fluchtvogel
i   Bulbul di Reunion

4784 *Hypsipetes crassirostris*
     Passeriformes - Pycnonotidae
e   Seychelles Bulbul; Thick-billed
     Bulbul; Seychelles Black-Bulbul
f   Bulbul merle
d   Dickschnabelschnäpper
i   Bulbul beccoforte

4785 *Hypsipetes criniger*
     Passeriformes - Pycnonotidae
e   Hairy-backed Bulbul
i   Bulbul dorsopiumoso

4786 *Hypsipetes leucocephalus*
     Passeriformes - Pycnonotidae
e   Black Bulbul; White-headed Bulbul;
     White-headed Black-Bulbul; Ceylon
     Black-Bulbul; Grey Bulbul
d   Borstenmantelbülbül
i   Bulbul nero

4787 *Hypsipetes madagascariensis*
     Passeriformes - Pycnonotidae
e   Black Bulbul; Madagascar Bulbul;
     Madagascar Blackbulbul; Comores
     Bulbul
f   Bulbul de Madagascar; Bulbul noir;
     Bulbul malgache
d   Madagaskar-Fluchtvogel;
     Rotschnabelfluchtvogel
i   Bulbul del Madagascar

4788 *Hypsipetes mcclellandii*
     Passeriformes - Pycnonotidae
e   Mountain Bulbul; McClelland's
     Bulbul; Rufous-bellied Bulbul;
     McLelland's Mountain-Bulbul;
     Mountain Streaked Bulbul; Green-
     winged Bulbul
f   Bulbul de McClelland; Bulbul
     montagnard; Bulbul de l'Himalaya
d   Grünflügelbülbül
i   Bulbul di McClelland

4789 *Hypsipetes nicobariensis*
     Passeriformes - Pycnonotidae
e   Nicobar Bulbul
f   Bulbul des Nicobar

   d   Nicobaren-Fluchtvogel
   i   Bulbul delle Nicobare

4790   ***Hypsipetes olivaceus***
       Passeriformes - Pycnonotidae
   e   Mauritius Black-Bulbul; Mauritius
       Bulbul
   f   Bulbul de Maurice; Bulbul de lîle
       Maurice

4791   ***Hypsipetes parvirostris***
       Passeriformes - Pycnonotidae
   e   Comoro Bulbul; Comoros Black-
       Bulbul; Comoros Bulbul
   f   Bulbul des Comores
   i   Bulbul delle Comore

4792   ***Hypsipetes thompsoni***
       Passeriformes - Pycnonotidae
   e   White-headed Bulbul; Bingham's
       Bulbul; Brown-vented Bulbul
   f   Bulbul à tête blanche
   d   Thompson-Fluchtvogel
   i   Bulbul di Thompson

4793   ***Hypsipetes virescens***
       Passeriformes - Pycnonotidae
   e   Sunda Bulbul; Nicobar Bulbul;
       Green-winged Bulbul; Sumatran
       Bulbul; Sunda Streaked Bulbul;
       Streaked Mountain-Bulbul; Streaked
       Bulbul; Javan Streaked-Bulbul;
       Green Mountain-Bulbul; Green-
       backed Bulbul
   f   Bulbul verdin
   d   Nicobaren-Fluchtvogel
   i   Bulbul di Sumatra

4821 Icterus maculialatus
Passeriformes - Icteridae
e   Bar-winged Oriole
d   Bindentrupial

4822 Icterus mesomelas
Passeriformes - Icteridae
e   Yellow-tailed Oriole
d   Gelbschwanztrupial; Streifentrupial
i   Ittero codagialla

4823 Icterus nigrogularis
Passeriformes - Icteridae
e   Yellow Oriole
d   Orangebrusttrupial
i   Ittero giallo

4824 Icterus oberi
Passeriformes - Icteridae
e   Montserrat Oriole; Tanniabird (WI);
     Blantyrebird (WI)
d   Montserrat-Trupial
i   Ittero di Montserrt

4825 Icterus parisorum
Passeriformes - Icteridae
e   Scott's Oriole
f   Oriole jaune-verdâtre
d   Scotts Trupial; Bonaparte-Trupial

4826 Icterus pectoralis
Passeriformes - Icteridae
e   Spot-breasted Oriole; Spotted-
     breasted Oriole; Spotted Oriole
d   Tropfentrupial; Schwarzbrusttrupial
i   Ittero pettomaculato

4827 Icterus prosthemelas
Passeriformes - Icteridae
e   Black-cowled Oriole

4828 Icterus pustulatus
Passeriformes - Icteridae
e   Streak-backed Oriole; Scarlet-headed
     Oriole; Flame-headed Oriole; Spot-
     breasted Oriole
f   Oriole à dos rayé
d   Piroltrupial
i   Ittero dorsostriato

4829 Icterus spurius
Passeriformes - Icteridae
e   Orchard Oriole; Ochre Oriole;
     Fuete's Oriole
f   Oriole des vergers
d   Gartentrupial
i   Ittero degli orti

4830 Icterus wagleri
Passeriformes - Icteridae
e   Black-vented Oriole; Wagler's Oriole
f   Oriole cul-noir
d   Wagler-Trupial
i   Ittero di Wagler

4831 Icterus xantholaemus
Passeriformes - Icteridae
e   Yellow-throated Oriole

4832 Ictinaetus malayensis
Falconiformes - Accipitridae
e   Black Eagle; Indian Black-Eagle;
     Asian Black-Eagle
f   Aigle noir
d   Malaien-Adler
i   Aquila nera

4833 Ictinia mississippiensis
Falconiformes - Accipitridae
e   Mississippi Kite
f   Milan du Mississippi
d   Mississippi-Weihe
i   Ittinia del Mississippi

4834 Ictinia plumbea
Falconiformes - Accipitridae
e   Plumbeous Kite
f   Milan bleuâtre
d   Schwebeweihe; Grauschwebeweihe
i   Ittinia plumbea

4835 Idiopsar brachyurus
Passeriformes - Fringillidae
e   Short-tailed Finch; Short-tailed
     Diuca-Finch
f   Idiopsar à queue courte
d   Kurzschwanzdiuka
i   Fringuello codabreve delle Ande

4836 Ifrita kowaldi
Passeriformes - Corvidae

e Blue-capped Ifrit; Ilfrit; Blue-capped
   Babbler; Bluecap
f Ifrita de Kowald
d Pittatimalie
i Ifrita

4837 *Ilicura militaris*
   Passeriformes - Tyrannidae
e Pin-tailed Manakin
f Manakin militaire
d Graukehlpipra
i Manachino coda a spillo

4838 *Illadopsis abyssinica*
   Passeriformes - Sylviidae
e Abyssinian Hill-Babbler; African
   Hill-Babbler; Abyssinian Fulvetta;
   Hill-Babbler
f Akalat à tête sombre; Alcippe à tête
   grise
d Mönchsalcippe
i Garrulo tordo di collina

4839 *Illadopsis albipectus*
   Passeriformes - Sylviidae
e Scaly-breasted Illadopsis; White-
   breasted Illadopsis; Mountain
   Illadopsis; Scaly-breasted Thrush-
   Babbler
f Akalat à poitrine ecaillée; Grive-
   akalat à poitrine ecaillée
d Schuppenbrustbuschdrossling
i Garrulo tordo pettoscaglioso

4840 *Illadopsis atriceps*
   Passeriformes - Sylviidae
e Ruwenzori Hill-Babbler; Ruwenzori
   Babbler; Ruwenzori Fulvetta
f Akalat du Ruwenzori
d Grautimalic; Schwarzkopfalcippe
i Garrulo tordo del Ruwenzori

4841 *Illadopsis cleaveri*
   Passeriformes - Sylviidae
e Blackcap Illadopsis; Blackcap
   Akalat; Blackcap Thrush-Babbler;
   Black-capped Illadopsis
f Akalat à tête noire; Grive-akalat à
   tête noire
d Augenbrauenbuschdrossling
i Garrulo tordo capinero

4842 *Illadopsis fulvescens*
   Passeriformes - Sylviidae
e Brown Illadopsis; Brown Akalat;
   Brown Thrush-Babbler; Brown-
   breasted Illadopsis
f Akalat brun; Grive-akalat brune
d Braunbauchbuschdrossling; Braune
   Maustimalie
i Garrulo tordo bruno

4843 *Illadopsis puveli*
   Passeriformes - Sylviidae
e Puvel's Illadopsis; Puvel's Akalat;
   Puvel's Thrush-Babbler
f Akalat de Puvel; Grive-akalat de
   Puvel
d Großfußbuschdrossling; Puvels
   Buschdrossling
i Garrulo tordo di Puvel

4844 *Illadopsis pyrrhoptera*
   Passeriformes - Sylviidae
e Mountain Illadopsis; Mountain
   Thrush-Babbler
f Akalat montagnard; Grive-akalat de
   montagne
d Bergbuschdrossling
i Garrulo tordo di montagna

4845 *Illadopsis rufescens*
   Passeriformes - Sylviidae
e Rufous-winged Illadopsis; Rufous-
   winged Akalat; Rufous-winged
   Thrush-Babbler
f Akalat à ailes rousses; Grive-akalat
   du Libéria
d Rostschwingenbuschdrossling;
   Reichenow-Buschdrossling
i Garrulo tordo rossiccio

4846 *Illadopsis rufipennis*
   Passeriformes - Sylviidae
e Pale-breasted Illadopsis; Pale-
   breasted Thrush-Babbler; White-
   breasted Akalat; White-breasted
   Illadopsis
f Akalat à poitrine blanche; Grive-
   akalat à poitrine blanche
d Grauwangenbuschdrossling
i Garrulo tordo pettochiaro

4847   *Incaspiza laeta*
       Passeriformes - Fringillidae
   e   Buff-bridled Inca-Finch; Buff-bridled
       Inca-Sparrow
   f   Chipiu à moustaches
   d   Goldschnabelammer
   i   Fringuello inca dai mustacchi

4848   *Incaspiza ortizi*
       Passeriformes - Fringillidae
   e   Grey-winged Inca-Finch; Grey-
       winged Inca-Sparrow
   f   Chipiu d'Ortiz
   d   Grauflügelammer
   i   Fringuello inca aligrige

4849   *Incaspiza personata*
       Passeriformes - Fringillidae
   e   Rufous-backed Inca-Finch; Rufous-
       backed Inca-Sparrow
   f   Chipiu costumé
   d   Schwarzstirnammer
   i   Fringuello inca dorsocastano

4850   *Incaspiza pulchra*
       Passeriformes - Fringillidae
   e   Great Inca Finch; Great Inca-
       Sparrow
   f   Chipiu remarquable
   d   Inca-Ammer
   i   Fringuello inca grosso

4851   *Incaspiza watkinsi*
       Passeriformes - Fringillidae
   e   Little Inca-Finch; Little Inca-Sparrow
   f   Chipiu de Watkins
   d   Watkins-Ammer
   i   Fringuello inca piccolo

4852   *Indicator archipelagicus*
       Piciformes - Indicatoridae
   e   Malaysian Honeyguide; Malay
       Honeyguide
   f   Indicateur archipélagique
   d   Schwarzkehlhoniganzeiger
   i   Indicatore malesi

4853   *Indicator conirostris*
       Piciformes - Indicatoridae
   e   Thick-billed Honeyguide

   f   Indicateur à gros bec
   i   Indicatore beccogrosso

4854   *Indicator exilis*
       Piciformes - Indicatoridae
   e   Least Honeyguide; Western Least-
       Honeyguide
   f   Indicateur menu; Indicateur minule
   d   Barthoniganzeiger;
       Zwerghoniganzeiger
   i   Indicatore nano; Indicatore piccolo

4855   *Indicator indicator*
       Piciformes - Indicatoridae
   e   Greater Honeyguide; Black-throated
       Honeyguide
   f   Grand Indicateur; Indicateur mange-
       miel
   d   Großer Honiganzeiger
   i   Indicatore dalla gola nera; Indicatore
       golanera

4856   *Indicator maculatus*
       Piciformes - Indicatoridae
   e   Spotted Honeyguide
   f   Indicateur tacheté
   d   Tropfenbrusthhoniganzeiger
   i   Indicatore maculato

4857   *Indicator meliphilus*
       Piciformes - Indicatoridae
   e   Pallid Honeyguide; Eastern
       Honeyguide (CSA); Eastern Least
       Honeyguide
   f   Indicateur pâle
   d   Olivmantelhoniganzeiger; Taveta-
       Honiganzeiger
   i   Indicatore pallido

4858   *Indicator minor*
       Piciformes - Indicatoridae
   e   Lesser Honeyguide
   f   Petit Indicateur
   d   Kleiner Honiganzeiger;
       Nasenstreifhoniganzeiger
   i   Indicatore minore

4859   *Indicator pumilio*
       Piciformes - Indicatoridae
   e   Dwarf Honeyguide; Pygmy
       Honeyguide

f   Indicateur nain
d   Kurzschnabelhoniganzeiger
i   Indicatore pigmeo; Indicatore
golamacchiata

4860   *Indicator variegatus*
      Piciformes - Indicatoridae
e   Scaly-throated Honeyguide; Scale-
throated Honeyguide; Variegated
Honeyguide
f   Indicateur varié; Indicateur écaillé
d   Gefleckter Honiganzeiger;
Strichelstirnhoniganzeiger;
Schuppenhoniganzeiger
i   Indicatore variegato

4861   *Indicator willcocksi*
      Piciformes - Indicatoridae
e   Willcocks's Honeyguide
f   Indicateur de Willcocks
d   Guinea-Honiganzeiger
i   Indicatore di Willcocks

4862   *Indicator xanthonotus*
      Piciformes - Indicatoridae
e   Yellow-rumped Honeyguide; Indian
Honeyguide; Orange-rumped
Honeyguide; Himalayan Honeyguide
f   Indicateur à dos jaune
d   Gelbbürzelhoniganzeiger
i   Indicatore indiano

4863   **Indicatoridae**
      Piciformes
e   Honeyguides
f   Indicatoridés
d   Honiganzeiger
i   indicatoridi

4864   *Inezia inornata*
      Passeriformes - Tyrannidae
e   Plain Tyrannulet; Plain Inezia
f   Tyranneau terne; Tyranneau de
Salvadori
d   Graukopfinezia
i   Tiranno piccolo disadorno

4865   *Inezia subflava*
      Passeriformes - Tyrannidae
e   Pale-tipped Tyrannulet; Pale-tipped
Inezia

f   Tyranneau givré; Tyranneau inézia
d   Braunkopfinezia
i   Tiranno piccolo dorsoverde

4866   *Inezia tenuirostris*
      Passeriformes - Tyrannidae
e   Slender-billed Tyrannulet; Slender-
billed Inezia
f   Tyranneau à bec fin; Tyranneau à bec
mince
d   Dünnschnabelinezia
i   Tiranno piccolo beccofine

4867   *Iodopleura fusca*
      Passeriformes - Tyrannidae
e   Dusky Purpletuft; Dusky Chatterer
f   Cotinga brun
d   Schwarzkopfseidenfleck
i   Pettoviola bruno

4868   *Iodopleura isabellae*
      Passeriformes - Tyrannidae
e   White-browed Purpletuft; Isabella's
Cotinga; Isabella's Chatterer
f   Cotinga d'Isabelle
d   Jodkotinga; Weißgesichtseidenfleck
i   Pettoviola di Isabella

4869   *Iodopleura pipra*
      Passeriformes - Tyrannidae
e   Buff-throated Purpletuft; Lesson's
Purpletuft; Lesson's Chatterer; Buff-
throated Cotinga
f   Cotinga manakin
d   Zwergseidenfleck; Lachskotinga
i   Pettoviola golafulva

4870   *Iole indica*
      Passeriformes - Pycnonotidae
e   Yellow-browed Bulbul; Golden-
browed Bulbul
f   Bulbul à sourcils d'or
d   Goldbrauenbülbül
i   Bulbul dai sopraccigli dorati

4871   *Iole olivacea*
      Passeriformes - Pycnonotidae
e   Buff-vented Bulbul; Charlotte's
Olive-Bulbul; Finsch's Olive-Bulbul;
Crested Olive-Bulbul; Dull-brown
Bulbul; Charlotte's Bulbul; Olive

Bulbul; Finsch's Olive-brown Bulbul
f Bulbul de Charlotte
d Olivbülbül; Braunbauchbülül
i Bulbul di Carlotta

4872 *Iole propinqua*
Passeriformes - Pycnonotidae
e Grey-eyed Bulbul
f Bulbul aux yeux gris
d Grauaugenbülbül
i Bulbul occhigrigi

4873 *Iole virescens*
Passeriformes - Pycnonotidae
e Olive Bulbul; Viridescent Bulbul;
Blyth's Olive-Bulbul; Blyth's Bulbul;
Green-winged Bulbul
f Bulbul olive; Bulbul birman
d Burma-Bülbül
i Bulbul oliva

4874 *Irania gutturalis*
Passeriformes - Muscicapidae
e White-throated Robin; Persian
White-throated Robin; Irania; Persian
Robin
f Iranie à gorge blanche
d Weißkehlsänger
i Pettirosso dell'Iran; Irania; Pettirosso
golabianca

4875 *Irediparra gallinacea*
Charadriiformes - Jacanidae
e Comb-crested Jacana; Lotusbird;
Comb-crested Parra; Lily-trotter
f Jacana à crête
d Kammblatthühnchen
i Jacana crestata

4876 *Irena cyanogaster*
Passeriformes - Irenidae
e Philippine Fairy-bluebird; Philippine
Bluebird; Black-mantled Fairy-
bluebird
f Irène à ventre bleu
d Kobaltirene; Philippinen-Irene
i Irena ventreazzurro

4877 *Irena puella*
Passeriformes - Irenidae
e Asian Fairy-bluebird; Asian Blue-

backed Fairy-bluebird; Blue-backed
Fairy-bluebird; Fairy-bluebird (ISC);
Blue-mantled Fairy-bluebird;
Palawan Fairy-bluebird
f Irène vierge; Oiseau-bleu des fàes
d Türkis-Irene; Elfenblauvogel
i Irena dorsoazzurro

4878 Irenidae
Passeriformes
e Ioras; Leafbirds
f Irénidés
d Feenvögel
i Irenidi

4879 *Iridosornis analis*
Passeriformes - Fringillidae
e Yellow-throated Tanager
f Tangara à bavette jaune
d Gelbkehltangare
i Tangara golagialla

4880 *Iridosornis jelskii*
Passeriformes - Fringillidae
e Golden-collared Tanager; Jelski's
Tanager
f Tangara à col d'or
d Keslki-Tangare
i Tangara di Jelski

4881 *Iridosornis porphyrocephala*
Passeriformes - Fringillidae
e Purplish-mantled Tanager
f Tangara à cape bleue
d Purpurmanteltangare
i Tangara dal mantello purpureo

4882 *Iridosornis reinhardti*
Passeriformes - Fringillidae
e Yellow-scarfed Tanager
f Tangara de Reinhardt
d Goldbandtangare
i Tangara di Reinhardt

4883 *Iridosornis rufivertex*
Passeriformes - Fringillidae
e Golden-crowned Tanager; Golden-
scarfed Tanager; Yellow-scarfed
Tanager
f Tangara auréolé

d   Goldscheiteltangare
i   Tangara corona dorata

4884   *Ispidina lecontei*
       Coraciiformes - Alcedinidae
e   Dwarf Kingfisher; African Dwarf
       Kingfisher
f   Martin-pêcheur à tête rousse; Martin-
       chasseur à tête rouge; Martin-
       chasseur à tête rousse
d   Braunkopfzwergfisher
i   Martin pescatore nano africano

4885   *Ispidina madagascariensis*
       Coraciiformes - Alcedinidae
e   Madagascar Pygmy-Kingfisher
f   Martin-pêcheur malgache; Martin-
       chasseur roux
d   Madagaskar-Zwergfisher
i   Martin pescatore pigmeo del
       Madagascar

4886   *Ispidina picta*
       Coraciiformes - Alcedinidae
e   African Pygmy-Kingfisher; Pygmy
       Kingfisher (CSA); Miniature
       Kingfisher
f   Martin-pêcheur pygmée; Martin-
       chasseur pygmé
d   Zwergfischer; Natal-Zwergeisvogel;
       Natal-Zwergfischer; Zwergeisvogel
i   Martin pescatore pigmeo africano;
       Martin pescatore pigmeo

4887   *Ithaginis cruentus*
       Galliformes - Phasianidae
e   Blood Pheasant
f   Ithagine ensanglantée
d   Blutfasan
i   Fagiano cruento

4888   *Ixobrychus cinnamomeus*
       Ciconiiformes - Ardeidae
e   Cinnamon Bittern; Chestnut Bittern
       (ISC); Cinnamon Least Bittern
f   Blongios cannellé
d   Zimtdommel
i   Tarabusino cannella

4889   *Ixobrychus eurhythmus*
       Ciconiiformes - Ardeidae

e   Schrenck's Bittern; Schrenck's Little-
       Bittern; Little Bittern; Von Schrenk's
       Bittern; Schrenk's Least-Bittern
f   Blongios mandchou; Blongios de
       Schrenck
d   Mandschuren-Zwergdommel;
       Mandschuren-Dommel
i   Tarabusino orientale

4890   *Ixobrychus exilis*
       Ciconiiformes - Ardeidae
e   Least Bittern; Bittern; Bitlin (WI);
       Gaulin (WI)
f   Petit Blongios; Petit Butor; Kio jaune
       (Ants); Crabier (Ants); Martinet
       (Ants)
d   Amerikanische Zwergdommel;
       Indianer-Dommel
i   Tarabusino americano

4891   *Ixobrychus flavicollis*
       Ciconiiformes - Ardeidae
e   Black Bittern; Mangrove Bittern;
       Yellow-necked Bittern
f   Blongios à cou jaune
d   Schwarzdommel
i   Tarabusino nero

4892   *Ixobrychus involucris*
       Ciconiiformes - Ardeidae
e   Stripe-backed Bittern; Streaked
       Bittern; Azara's Bittern; Pygmy
       Bittern; Little Red-Heron; Variegated
       Heron
f   Blongios varié
d   Streifendommel
i   Tarabusino dorsostriato

4893   *Ixobrychus minutus*
       Ciconiiformes - Ardeidae
e   Little Bittern; Minute Bittern; Leech
       Bittern
f   Blongios nain; Butor blongios;
       Harelde blongios
d   Zwergdommel; Zwergrohrdommel;
       Zwergreiher
i   Tarabusino comune; Tarabusino

4894   *Ixobrychus minutus dubius*
       Ciconiiformes - Ardeidae
e   Australasian Little Bittern (ANZ)

4895   *Ixobrychus novaezelandiae*
  Ciconiiformes - Ardeidae
e Black-backed Bittern; New Zealand
  Little-Bittern
f Blongios à dos noir
d Neuholland-Dommel
i Tarabusino dorsonero

4896   *Ixobrychus sinensis*
  Ciconiiformes - Ardeidae
e Yellow Bittern; Chinese Little-
  Bittern; Chinese Bittern; Chinese
  Least-Bittern; Little Yellow Bittern;
  Long-nosed Bittern
f Blongios de Chine; Blongios chinois
d Chinesendommel
i Tarabusino cinese

4897   *Ixobrychus sturmi*
  Ciconiiformes - Ardeidae
e Dwarf Bittern; Rail Heron (CSA);
  African Dwarf Bittern
f Blongios de Sturm
d Graurückendommel; Sturms
  Zwergrohrdommel; Schieferdommel
i Tarabusino nano africano

4898   *Ixonotus guttatus*
  Passeriformes - Pycnonotidae
e Spotted Greenbul; Spotted Bulbul
f Bulbul tacheté
d Fleckenbülbül
i Bulbul maculato

4899   *Ixos amaurotis*
  Passeriformes - Pycnonotidae
e Brown-eared Bulbul; Chestnut-eared
  Bulbul; Eurasian Chestnut-eared
  Bulbul; Asian Brown-eared Bulbul
f Bulbul à oreillons bruns
d Orpheus-Bülbül
i Bulbul orecchiecastane

4900   *Ixos everetti*
  Passeriformes - Pycnonotidae
e Yellowish Bulbul; Yellow-washed
  Bulbul; Everett's Bulbul; Plain-
  throated Bulbul
f Bulbul d'Everett
d Everett-Bülbül
i Bulbul di Everett

4901   *Ixos malaccensis*
  Passeriformes - Pycnonotidae
e Streaked Bulbul; Green-backed
  Bulbul; Common Streaked-Bulbul
f Bulbul malais
d Strichelbrustbülbül
i Bulbul di Malacca

4902   *Ixos palawanensis*
  Passeriformes - Pycnonotidae
e Sulphur-bellied Bulbul; Golden-eyed
  Bulbul; Olive Bulbul
f Bulbul de Palawan; Bulbul aux yeux
  d'or
d Goldaugenbülbül
i Bulbul di Palawan

4903   *Ixos philippinus*
  Passeriformes - Pycnonotidae
e Philippine Bulbul; Rufous-breasted
  Bulbul
f Bulbul des Philippines
d Rotbrustbülbül
i Bulbul pettirosso

4904   *Ixos rufigularis*
  Passeriformes - Pycnonotidae
e Zamboanga Bulbul; Zamboana
  Bulbul
f Bulbul à gorge rousse
d Zamboana-Bülbül
i Bulbul golarossa

4905   *Ixos siquijorensis*
  Passeriformes - Pycnonotidae
e Streak-breasted Bulbul; Slaty-
  crowned Bulbul; Mottled-breasted
  Bulbul; Siquijor Bulbul
f Bulbul de Siquijor
d Schieferkopfbülbül
i Bulbul corona ardesia

J

4906 *Jabiru mycteria*
Ciconiiformes - Ciconiidae
e Jabiru; Jabiru Stork; American Jabiru
f Jabiru d'Amérique
d Jabiru
i Jabiru; Jabiru americano

4907 *Jabouilleia danjoui*
Passeriformes - Sylviidae
e Short-tailed Scimitar-Babbler;
Danjou's Babbler; Short-tailed
Babbler; Scimitar Babbler
f Pomatorhin à queue courte
d Kurzschwanzsäbler
i Garrulo di Danjou

4908 *Jacamaralcyon tridactyla*
Piciformes - Galbulidae
e Three-toed Jacamar
f Jacamar tridactyle
d Dreizehenglanzvogel
i Jacamar tridattilo

4909 *Jacamerops aureus*
Piciformes - Galbulidae
e Great Jacamar
f Grand Jacamar
d Breitmaulglanzvogel
i Jacamar maggiore

4910 *Jacana jacana*
Charadriiformes - Jacanidae
e Wattled Jacana
f Jacana noir
d Rotstirnjassana
i Jacana dei bargigli

4911 *Jacana spinosa*
Charadriiformes - Jacanidae
e Northern Jacana; River Chink (WI);
Pond Coot (WI); American Jacana;
Middle American Jacana
f Jacana (Ants); Médecin (Ants); Poule
d'eau dorée (Ants)
d Gelbstirnjassana
i Jacana americana; Jacana spinosa;
Parra americana

4912 Jacanidae
Charadriiformes
e Jacanas
f Jacanidés
d Blatthühnchen
i Iacanidi; Jacanidi; Giacanidi

4913 *Jubula lettii*
Strigiformes - Strigidae
e Maned Owl; Akun Scops-Owl;
Collared Scops-Owl
f Duc à crinière; Petit-duc crinière
d Mähneneule
i Gufo dalla criniera; Assiolo crinito

4914 *Junco aikeni*
Passeriformes - Emberizidae
e White-winged Junco

4915 *Junco alticola*
Passeriformes - Emberizidae
e Guatemala Junco

4916 *Junco bairdi*
Passeriformes - Emberizidae
e Baird's Junco

4917 *Junco caniceps*
Passeriformes - Emberizidae
e Grey-headed Junco

4918 *Junco dorsalis*
Passeriformes - Emberizidae
e Red-backed Junco

4919 *Junco fulvescens*
Passeriformes - Emberizidae
e Chiapas Junco

4920 *Junco hyemalis*
Passeriformes - Emberizidae
e Dark-eyed Junco; Slate-colored
Junco (NA); Oregon Junco; Pink-
sided Junco; Red-backed Junco
f Junco ardoisé; Bruant ardoisé; Pinson
niverolle

d   Junko; Winterjunko; Winterammer
i   Junco color lavagna; Junco
   occhiscuri

4921   *Junco insularis*
   Passeriformes - Emberizidae
e   Guadalupe Junco
f   Junco de Guadalupe
i   Junco di Guadelupe

4922   *Junco mearnsi*
   Passeriformes - Emberizidae
e   Pink-sided Junco

4923   *Junco oreganus*
   Passeriformes - Emberizidae
e   Oregon Junco
i   Junco dell'Oregon

4924   *Junco phaeonotus*
   Passeriformes - Emberizidae
e   Yellow-eyed Junco; Mexican Junco;
   Chiapas Junco; Guatemala Junco
f   Junco aux yeux jaunes
d   Rotrückenjunko
i   Junco occhigialli

4925   *Junco vulcani*
   Passeriformes - Emberizidae
e   Volcano Junco; Volcanic Junco
f   Junco des volcans
d   Streifenjunko
i   Junco del Volcano

4926   *Jynx ruficollis*
   Piciformes - Picidae
e   Rufous-necked Wryneck; Red-
   throated Wryneck; Red-breasted
   Wryneck; African Wryneck
f   Torcol à gorge rousse
d   Braunkehlwendehals;
   Rotkehlwendehals
i   Torcicollo africano

4927   *Jynx torquilla*
   Piciformes - Picidae
e   Eurasian Wryneck; Wryneck;
   Northern Wryneck
f   Torcol fourmilier; Torcol
d   Wendehals
i   Torcicollo; Torcicollo eurasiatico

K

4928 Kakamega poliothorax
Passeriformes - Sylviidae
e Grey-chested Illadopsis; Grey-chested Thrush-Babbler
f Akalat à poitrine grise; Grive-akalat à poitrine grise
d Graubrustdrosseltimalie
i Garrulo tordo pettogrigio

4929 Kaupifalco monogrammicus
Falconiformes - Accipitridae
e Lizard Buzzard
f Autour unibande; Buse unibande
d Sperberbussard; Kehlstreifenbussard; Kehlstreifbussard
i Falco monogrammico; Poiana lucertola

4930 Kenopia striata
Passeriformes - Sylviidae
e Striped Wren-Babbler; Falcated Wren-Babbler
f Turdinule striée
d Goldzügeltimalie; Kenopie
i Garrulo scricciolo striato

4931 Ketupa blakistoni
Strigiformes - Strigidae
e Blakiston's Fish-Owl
d Riesenfischuhu
i Gufo pescatore di Blakiston; Gufo pescatore

4932 Ketupa flavipes
Strigiformes - Strigidae
e Tawny Fish-Owl
f Kétoupa roux
d Himalaja-Fischuhu
i Gufo pescatore fulvo

4933 Ketupa ketupu
Strigiformes - Strigidae
e Buffy Fish-Owl; Malaysian Fish-Owl; Malay Fish-Owl
f Kétoupa malais
d Sunda-Fischuhu
i Gufo pescatore della Malesia; Gufo pescatore malese

4934 Ketupa zelonensis
Strigiformes - Strigidae
e Brown Fish-Owl; Ceylon Fish-Owl; Brown Fishing-Owl
d Wellenbrustfischuhu
i Gufo pescatore bruno; Civetta pescatrice di Ceylon; Civetta pescatrice di Sri-Lanka

4935 Klais guimeti
Apodiformes - Trochilidae
e Violet-headed Hummingbird; Violet-crowned Hummingbird
f Colibri à tête violette
d Violettkopfkolibri
i Colibrì testavioletta

4936 Knipolegus aterrimus
Passeriformes - Tyrannidae
e White-winged Black-Tyrant
f Ada à ailes blanches
d Weißspiegelmohrentyrann
i Tiranno nero alibianchi

4937 Knipolegus cabanisi
Passeriformes - Tyrannidae
e Plumbeous Tyrant; Cabanis's Tyrant; Plumbeous Andean-Tyrant
f Ada de Cabanis
d Cabanis-Tyrann

4938 Knipolegus cyanirostris
Passeriformes - Tyrannidae
e Blue-billed Black-Tyrant
f Ada à bec bleu
d Blauschnabeltyrann
i Tiranno nero beccoblu

4939 Knipolegus franciscanus
Passeriformes - Tyrannidae
e Caatinga Black-Tyrant

4940 Knipolegus hudsoni
Passeriformes - Tyrannidae
e Hudson's Black-Tyrant

   f   Ada de Hudson
   d   Hudson-Tyrann
   i   Tiranno nero di Hudson

4941   *Knipolegus lophotes*
       Passeriformes - Tyrannidae
   e   Crested Black-Tyrant
   f   Ada huppé
   d   Langschopfmohrentyrann
   i   Tiranno nero crestato

4942   *Knipolegus nigerrimus*
       Passeriformes - Tyrannidae
   e   Velvety Black-Tyrant; Vieillot's
       Black-Tyrant
   f   Ada noir
   d   Kurzschopfmohrentyrann
   i   Tiranno nero vellutato

4943   *Knipolegus orenocensis*
       Passeriformes - Tyrannidae
   e   Riverside Tyrant; Orinoco Tyrant
   f   Ada de I'Orénoque
   d   Orinoco-Tyrann
   i   Tiranno dei fiumi

4944   *Knipolegus poecilocercus*
       Passeriformes - Tyrannidae
   e   Amazonian Black-Tyrant; Petzeln's
       Black-Tyrant
   f   Ada d'Amazonie; Ada à queue rousse
   d   Kohletyrann
   i   Tiranno nero dell'Amazzonia

4945   *Knipolegus poecilurus*
       Passeriformes - Tyrannidae
   e   Rufous-tailed Tyrant
   f   Ada à queue rousse
   d   Rostschwanztyrann
   i   Tiranno codirosso

4946   *Knipolegus signatus*
       Passeriformes - Tyrannidae
   e   Andean Tyrant; Jelski's Bush-Tyrant;
       Red-rumped Bush-Tyrant;
       Plumbeous Tyrant
   f   Ada de Jelski; Moucherolle de Jelski
   d   Jelski-Tyrann
   i   Tiranno di Jelski

4947   *Knipolegus striaticeps*
       Passeriformes - Tyrannidae
   e   Cinereous Tyrant
   f   Ada cendré
   d   Schmalschwingentyrann
   i   Tiranno testastriata

4948   *Kupeornis chapini*
       Passeriformes - Sylviidae
   e   Chapin's Mountain-Babbler; Chapin's
       Flycatcher-Babbler
   f   Phyllanthe de Chapin; Timalie de
       Chapin
   d   Chapin-Timalie
   i   Garrulo pigliamosche di Chapin

4949   *Kupeornis gilberti*
       Passeriformes - Sylviidae
   e   White-throated Mountain-Babbler;
       Gilbert's Babbler; White-throated
       Flycatcher-Babbler; Gilbert's
       Mountain-Babbler
   f   Phyllanthe à gorge blanche; Timalie
       à gorge blanche
   d   Kupe-Timalie
   i   Garrulo del Monte Kupè

4950   *Kupeornis rufocinctus*
       Passeriformes - Sylviidae
   e   Red-collared Mountain-Babbler;
       Red-collared Flycatcher-Babbler;
       Rufous-collared Mountain-Babbler;
       Red-collared Babbler
   f   Phyllanthe à collier roux; Timalie à
       collier roux
   d   Rostbandtimalie
   i   Garrulo pigliamosche dal collare

L

4951 *Lacedo pulchella*
Coraciiformes - Halcyonidae
e Banded Kingfisher
f Martin-chasseur mignon
d Wellenliest; Schönliest
i Martin pescatore barrato

4952 *Lafresnaya lafresnayi*
Apodiformes - Trochilidae
e Mountain Velvetbreast; Velvetbreast
f Colibri de Lafresnaye
d Fadenschwingenkolibri
i Colibri di Lafresnaye

4953 *Lagonosticta landanae*
Passeriformes - Estrildidae
e Pale-billed Firefinch; Landana Pale-
billed Firefinch; Landana Firefinch
f Amarante de Landana; Amaranthe de
Landana; Sénégali de Landana
d Landana-Amarant
i Amaranto beccochiaro

4954 *Lagonosticta larvata*
Passeriformes - Estrildidae
e Black-throated Firefinch; Black-
faced Firefinch; Masked Firefinch;
Masked Waxbill
f Amarante masqué; Amaranthe
masqué; Bengali à face rouge;
Amaranthe d'Abyssinie; Amarante
d'Abyssinie
d Larvenamarant
i Amaranto mascherato

4955 *Lagonosticta nitidula*
Passeriformes - Estrildidae
e Brown Firefinch
f Amarante nitidule; Amaranthe
nitidule; Amarante barrée
d Braunbürzelamarant; Großer
Pünktchenamarant
i Amaranto bruno

4956 *Lagonosticta rara*
Passeriformes - Estrildidae
e Black-bellied Firefinch; Black-
bellied Waxbill; Black-billed
Firefinch
f Amarante à ventre noir; Amaranthe à
ventre noir; Sénégali à ventre noir
d Seltener Amarant;
Schwarzbauchamarant
i Amaranto ventrenero

4957 *Lagonosticta rhodopareia*
Passeriformes - Estrildidae
e Jameson's Firefinch; Ethipoian
Firefinch
f Amarante de Jameson; Amaranthe de
Jameson
d Jamesons Amarant; Jameson-
Amarant; Rosenamarant
i Amaranto di Jameson

4958 *Lagonosticta rubricata*
Passeriformes - Estrildidae
e African Firefinch; Blue-billed
Firefinch (CSA); Ruddy Waxbill
f Amarante foncé; Amaranthe foncé;
Amaranthe flambée; Amarante
flambé; Bengali à bec bleu; Sénégali
à bec bleu
d Dunkelroter Amarant;
Dunkelamarant
i Amaranto africano

4959 *Lagonosticta rufopicta*
Passeriformes - Estrildidae
e Brown Firefinch; Bar-breasted
Firefinch; Speckled Firefinch
f Amarante pointé; Amaranthe pointé;
Amaranthe pointillé; Amarante
pointillé; Sénégali à poitrine blanche
d Großer Pünktchenamarant
i Amaranto pettobarrato

4960 *Lagonosticta senegala*
Passeriformes - Estrildidae
e Red-billed Firefinch; Senegal
Firefinch; Common Firefinch;
Firefinch; Little Ruddy Waxbill
f Amarante du Sénégal; Amaranthe du
Sénégal; Amaranthe à bec rouge;
Amarante à bec rouge; Amaranthe;

Amaranthe commun; Amaranthe
ordinaire; Amarante; Amarante
ordinaire; Sénégali rouge; Amarante
commun

d Amarant; Kleiner Amarant; Senegal-
Amarant

i Amaranto beccorosso

4961 *Lagonosticta umbrinodorsalis*
Passeriformes - Estrildidae

e Reichenow's Firefinch; Chad
Firefinch; Pink-backed Firefinch

f Amarante de Reichenow; Amaranthe
de Reichenow

i Amaranto di Reichenow

4962 *Lagonosticta vinacea*
Passeriformes - Estrildidae

e Black-faced Firefinch; Vinaceous
Firefinch; Vinaceous Waxbill; Grey
Firefinch; Grey Black-faced
Firefinch

f Amarante vineux; Amaranthe vineux;
Astrild vineux; Bengali vineux

d Schwarzkehlamarant; Weinroter
Amarant

i Amaranto vinaceo

4963 *Lagonosticta virata*
Passeriformes - Estrildidae

e Mali Firefinch; Grey-backed
Firefinch; Blue-billed Firefinch; Kull
Koro Firefinch

f Amarante de Kulikoro; Amaranthe
de Kulikoro

i Amaranto di Kuli Koro

4964 *Lagopus lagopus*
Galliformes - Phasianidae

e Willow Ptarmigan (NA); Willow
Grouse

f Lagopède des saules

d Moorschneehuhn; Waldhuhn

i Pernice bianca nordica; Pernice
bianca

4965 *Lagopus lagopus scoticus*
Galliformes - Phasianidae

e Red Grouse; Moor Fowl; Moor
Game (col); Grouse

f Lagopède d'Écosse

d Schottisches Moorschneehuhn;
Schottisches Moorhuhn; Moorhuhn

i Pernice; Pernice bianca di Scozia;
Pernice bianca

4966 *Lagopus leucurus*
Galliformes - Phasianidae

e White-tailed Ptarmigan; Snow
Grouse

f Lagopède à queue blanche

d Weißschwanzschneehuhn

i Pernice codabianca

4967 *Lagopus mutus*
Galliformes - Phasianidae

e Rock Ptarmigan; Ptarmigan; Rock
Grouse

f Lagopède alpin; Lagopède des
rochers

d Alpenschneehuhn

i Pernice bianca; Pernice di montagna;
Pernice alpestre; Pernice di monte;
Lagopede; Roncaso

4968 *Lalage atrovirens*
Passeriformes - Corvidae

e Black-browed Triller; White-rumped
Triller

f Échenilleur papou

d Papua-Lalage

i Mangiabruchi dal mantello

4969 *Lalage aurea*
Passeriformes - Corvidae

e Rufous-bellied Triller; Moluccan
Triller; Red-bellied Triller; Black-
backed Triller

f Échenilleur orangé

d Rotbauchlalage

i Mangiabruchi ventrerosso

4970 *Lalage leucomela*
Passeriformes - Corvidae

e Varied Triller; White-browed Triller;
Pied Triller

f Échenilleur varié

d Weißbrauenlalage

i Mangiabruchi variabile

4971 *Lalage leucopyga*
Passeriformes - Corvidae

e   Long-tailed Triller
f   Échenilleur pie
d   Langschwanzlalage
i   Mangiabruchi codalunga

4972   *Lalage leucopygialis*
Passeriformes - Corvidae
e   White-rumped Triller; Sulawesi
Triller
f   Échenilleur de Walden
i   Mangiabruchi gropponebianco;
Mangiabruchi dal groppone bianco

4973   *Lalage maculosa*
Passeriformes - Corvidae
e   Polynesian Triller; Spotted Triller
f   Échenilleur de Polynésie
d   Südseelalage
i   Mangiabruchi della Polinesia

4974   *Lalage melanoleuca*
Passeriformes - Corvidae
e   Black-and-white Triller
f   Échenilleur noir-et-blanc;
Échenilleur des Philippines
d   Philippinen-Lalage
i   Mangiabruchi bianco e nero

4975   *Lalage moesta*
Passeriformes - Corvidae
e   White-browed Triller; Tanimbar
Triller
f   Échenilleur des Tanimbar
i   Mangiabruchi dai sopraccigli bianchi

4976   *Lalage nigra*
Passeriformes - Corvidae
e   Pied Triller; Pied Cuckoo-Shrike
f   Échenilleur térat
d   Weißstirnlalage
i   Mangiabruchi gazza

4977   *Lalage sharpei*
Passeriformes - Corvidae
e   Samoan Triller
f   Échenilleur des Samoa
d   Braunrückenlalage
i   Mangiabruchi di Samoa

4978   *Lalage sueurii*
Passeriformes - Corvidae

e   White-shouldered Triller; Sueur's
Triller; White-winged Triller
f   Échenilleur de Lesueur
d   Spiegelraupenschmätzer;
Weißflügellalage
i   Mangiabruchi alibianche

4979   *Lalage tricolor*
Passeriformes - Corvidae
e   White-winged Triller; Australian
Triller
f   Échenilleur tricolore
i   Mangiabruchi australiano

4980   *Lampornis amethystinus*
Apodiformes - Trochilidae
e   Amethyst-throated Hummingbird;
Cazique Hummingbird; Margaret's
Hummingbird
f   Colibri à gorge améthyste
d   Amethystkehlnymphe
i   Colibrì gola ametista

4981   *Lampornis calolaema*
Apodiformes - Trochilidae
e   Purple-throated Mountain-Gem;
Variable Mountain-Gem
f   Colibri à gorge pourprée
i   Colibrì dalla gola purpurea

4982   *Lampornis castaneoventris*
Apodiformes - Trochilidae
e   White-throated Mountain-Gem;
Chestnut-bellied Mountain-Gem;
Variable Mountain-Gem
f   Colibri à ventre châtain
d   Rotbauchnymphe
i   Colibrì di monte golabianca

4983   *Lampornis clemenciae*
Apodiformes - Trochilidae
e   Blue-throated Hummingbird
f   Colibri à gorge bleue
d   Blaukehlnymphe
i   Colibrì golablu

4984   *Lampornis hemileucus*
Apodiformes - Trochilidae
e   White-bellied Mountain-Gem
f   Colibri à gorge lilas

d Weißbauchnymphe
i Colibrì di monte ventrebianco

4985 *Lampornis sybillae*
Apodiformes - Trochilidae
e Green-breasted Mountain-Gem
f Colibri de Sybil
d Sybillennymphe
i Colibrì di monte pettoverde

4986 *Lampornis viridipallens*
Apodiformes - Trochilidae
e Green-throated Mountain-Gem
f Colibri vert-d'eau
d Grünkehlnymphe
i Colibrì di monte golaverde

4987 *Lamprolaima rhami*
Apodiformes - Trochilidae
e Garnet-throated Hummingbird
f Colibri à gorge grenat
d Granatkehlnymphe
i Colibrì gola di granato

4988 *Lamprolia victoriae*
Passeriformes - Corvidae
e Silktail; Satin Flycatcher
f Monarque queue-de-soie
d Lamprolia
i Lamprolia

4989 *Lampropsar tanagrinus*
Passeriformes - Fringillidae
e Velvet-fronted Grackle
f Quiscale velouté
d Samtstirnstärling
i Gracchio fronotevellutata

4990 *Lamprospiza melanoleuca*
Passeriformes - Fringillidae
e Red-billed Pied-Tanager; Red-billed
Tanager; Pied Tanager
f Tangara noir-et-blanc
d Rotschnabeltangare
i Tangara gazza beccorosso

4991 *Lamprotornis acuticaudus*
Passeriformes - Sturnidae
e Sharp-tailed Glossy-Starling; Spreeu
(CSA); Wedge-tailed Glossy-Starling
f Choucador à queue fine; Merle

métallique à queue étroite
d Keilschwanzglanzstar
i Storno splendente codacuta

4992 *Lamprotornis australis*
Passeriformes - Sturnidae
e Burchell's Glossy-Starling; Burchell's
Starling; Spreeu (CSA); Greater
Glossy-Starling
f Choucador de Burchell; Merle
métallique de Burchell
d Glanzelster; Riesenglanzstar
i Storno di Burchell

4993 *Lamprotornis caudatus*
Passeriformes - Sturnidae
e Long-tailed Glossy-Starling;
Northern Long-tailed Glossy-
Starling; Supple-tailed Glossy-
Starling
f Choucador à longue queue; Merle
bronzé à longue queue; Merle
métallique à longue queue
d Langschwanzglanzstar
i Storno splendente codalunga

4994 *Lamprotornis chalcurus*
Passeriformes - Sturnidae
e Bronze-tailed Glossy-Starling;
Bronze-tailed Starling (CSA)
f Choucador à queue violette; Merle
métallique à queue violette
d Erzglanzstar
i Storno splendente codabronzata

4995 *Lamprotornis chalybaeus*
Passeriformes - Sturnidae
e Greater Blue-eared Glossy-Starling;
Greater Blue-eared Starling (CSA);
Spreeu (CSA); Blue-eared Glossy-
Starling; Green Glossy-Starling
f Choucador à oreillons bleus; Merle
métallique commun; Merle
métallique; Merle métallique à
oreilles bleues; Merle métallique à
oreillons bleus
d Grünschwanzglanzstar
i Storno orecchieblu maggiore; Storno
color d'accaio

4996   *Lamprotornis chloropterus*
Passeriformes - Sturnidae
e   Lesser Blue-eared Glossy-Starling;
Lesser Blue-eared Starling (CSA);
Spreeu (CSA); Swainson's Glossy-
Starling; Swainson's Lesser Blue-
eared Glossy-Starling
f   Choucador de Swainson; Merle
métallique de Swainson
d   Messingglazstar
i   Storno orecchieblu minore

4997   *Lamprotornis corruscus*
Passeriformes - Sturnidae
e   Black-bellied Glossy-Starling; Black-
bellied Starling (CSA); Spreeu
(CSA); Black-breasted Glossy-
Starling; Black-breasted Starling
f   Choucador à ventre noir; Merle
métallique à ventre noir
d   Schwarzbauchglanzstar
i   Storno splendente ventrenero

4998   *Lamprotornis cupreocauda*
Passeriformes - Sturnidae
e   Copper-tailed Glossy-Starling;
Coppery-tailed Glossy-Starling
f   Choucador à queue bronzée; Merle
métallique à dos bleu
d   Kupferglanzstar
i   Storno splendente codaramata

4999   *Lamprotornis elisabeth*
Passeriformes - Sturnidae
e   Southern Blue-eared Glossy-Starling;
Southern Lesser Blue-eared Glossy-
Starling
f   Choucador elisabeth
d   Elisabeth-Glanzstar
i   Storno orecchieblu meridionale

5000   *Lamprotornis hildebrandti*
Passeriformes - Sturnidae
e   Hildebrandt's Starling
f   Choucador de Hildebrandt; Merle
métallique de Hildebrandt
d   Hildebrandt-Glanzstar
i   Storno di Hildebrandt

5001   *Lamprotornis mevesii*
Passeriformes - Sturnidae

e   Meves's Glossy-Starling; Long-tailed
Glossy-Starling (CSA); Meve's
Starling; Spreeu (CSA); Long-tailed
Starling; Meves's Long-tailed
Glossy-Starling; Southern Long-
tailed Glossy-Starling
f   Choucador de Meves; Merle
métallique de Meves
d   Meves-Glanzstar
i   Storno di Meves

5002   *Lamprotornis nitens*
Passeriformes - Sturnidae
e   Red-shouldered Glossy-Starling;
Cape Glossy-Starling (CSA); Glossy-
Starling (CSA); Spreeu (CSA); Cape
Red-shouldered Glossy-Starling
f   Choucador à épaulettes rouges;
Merle métallique à épaulettes rouges
d   Rotschulterglanzstar
i   Storno splendente alirosse

5003   *Lamprotornis ornatus*
Passeriformes - Sturnidae
e   Principe Glossy-Starling; Ornate
Starling; Choucador Glossy-Starling
f   Choucador de Principe; Merle
métallique de Principe
d   Prinzen-Glanzstar
i   Splendente di Principe

5004   *Lamprotornis pulcher*
Passeriformes - Sturnidae
e   Chestnut-bellied Starling
f   Choucador à ventre roux; Étourneau
à ventre roux; Merle métallique à
ventre roux
d   Rotbauchglanzstar
i   Storno ventrecastano

5005   *Lamprotornis purpureiceps*
Passeriformes - Sturnidae
e   Purple-headed Glossy-Starling;
Purple-headed Starling; Velvet-
headed Glossy-Starling
f   Choucador à tête pourprée; Merle
métallique à tête pourprée
d   Samtkopfglanzstar; Samtglanzstar
i   Storno splendente purpureo

5006 *Lamprotornis purpureus*
Passeriformes - Sturnidae
e Purple Glossy-Starling; Purple
Starling (CSA)
f Choucador pourpre; Merle bronzé
pourpré; Mcrle métallique pourpré
d Purpurglanzstar
i Storno splendente ventrepurpureo

5007 *Lamprotornis purpuropterus*
Passeriformes - Sturnidae
e Rüppell's Glossy-Starling; Rüppell's
Long-tailed Starling; Rüppell's Long-
tailed Glossy-Starling; Long-tailed
Glossy-Starling; Long-tailed Starling
f Choucador de Rüppell; Merle
métallique de Rüppell
d Langschwanzpurpurglanzstar;
Schweifglanzstar
i Storno di Rüppell

5008 *Lamprotornis shelleyi*
Passeriformes - Sturnidae
e Shelley's Starling
f Choucador de Shelley; Merle
métallique de Shelley
d Shelley-Glanzstar
i Storno di Shelley

5009 *Lamprotornis splendidus*
Passeriformes - Sturnidae
e Splendid Glossy-Starling; Splendid
Starling
f Choucador splendide; Merle
métallique à oeil blanc
d Prachtglanzstar
i Storno splendido

5010 *Lamprotornis superbus*
Passeriformes - Sturnidae
e Superb Starling
f Choucador superbe; Merle métallique
superbe
d Dreifarbenglanzstar
i Storno superbo

5011 *Laniarius aethiopicus*
Passeriformes - Malaconotidae
e Tropical Boubou; Tropical Bush-
Shrike; Ethiopian Bush-Shrike;
Ethiopian Bellshrike

f Gonolek d'Abyssinie; Gonolek des
tropiques
d Flötenwürger; Orgelwürger
i Averla di macchia etiopica; Averla
sibilante

5012 *Laniarius amboimensis*
Passeriformes - Malaconotidae
e Amboim Bush-Shrike; Gabela Bush-
Shrike
f Gonolek de I'Angola
i Averla di macchia di Amboim

5013 *Laniarius atrococcineus*
Passeriformes - Malaconotidae
e Crimson-breasted Gonolek; Crimson-
breasted Boubou (CSA); Crimson-
breasted Shrike (CSA); Burchell's
Gonolek; Black-and-crimson
Gonolek
f Gonolek rouge-et-noir; Gonolek
rouge-noir
d Rotbauchwürger; Reichsvogel
i Averla di macchia pettirossa

5014 *Laniarius atroflavus*
Passeriformes - Malaconotidae
e Yellow-breasted Boubou; Yellow-
breasted Shrike
f Gonolek à ventre jaune
d Gelbwürger
i Averla di macchia pettogiallo

5015 *Laniarius barbarus*
Passeriformes - Malaconotidae
e Common Gonolek; Gonolek; Scarlet-
breasted Shrike; Barbary Shrike;
Yellow-crowned Gonolek;
Abyssinian Gonolek; Sierra Leone
Gonolek
f Gonolek de Barbarie
d Goldscheitelwürger
i Averla di macchia barbara

5016 *Laniarius bicolor*
Passeriformes - Malaconotidae
e Gabon Boubou; Swamp Boubou
(CSA); Ngami Boubou-Shrike;
Ngami Bush-Shrike; Ngami Boubou
f Gonolek à ventre blanc; Gobnolek du
Gabon

d Zweifarbenwürger
i Averla di macchia bicolore

5017 *Laniarius brauni*
Passeriformes - Malaconotidae
e Orange-breasted Bush-Shrike;
   Braun's Bush-Shrike
f Gonolek de Braun
i Averla di macchia di Braun

5018 *Laniarius erythrogaster*
Passeriformes - Malaconotidae
e Black-headed Gonolek; Red-bellied
   Gonolek; Black-headed Bush-Shrike;
   Barbary Shrike
f Gonolek à ventre rouge
d Scharlachwürger
i Averla di macchia ventrerosso

5019 *Laniarius ferrugineus*
Passeriformes - Malaconotidae
e Southern Boubou; Boubou Shrike;
   Bellshrike; Ferruginous Bush-Shrike;
   Ferruginous Boubou; Tropical Bush-
   Shrike; Tropical Boubou; African
   Boubou; African Bush-Shrike
f Gonolek boubou; Gonolek à ventre
   blanc
d Flötenwürger
i Averla di macchia ferruginea

5020 *Laniarius fuelleborni*
Passeriformes - Malaconotidae
e Fuelleborn's Boubou; Fülleborn's
   Black-Boubou (CSA); Fuelleborn's
   Black Boubou-Shrike; Black
   Boubou; Fülleborn's Boubou; Sooty
   Boubou
f Gonolek de Fülleborn; Gonolek noir
   de Fernando Po
d Fuelleborns Würger
i Averla di macchia di Fuelleborn

5021 *Laniarius funebris*
Passeriformes - Malaconotidae
e Slate-coloured Boubou; Slate-
   coloured Boubou-Shrike; Slate-
   coloured Bush-Shrike
f Gonolek ardoisé
d Trauerwürger
i Averla di macchia nera

5022 *Laniarius leucorhynchus*
Passeriformes - Malaconotidae
e Sooty Boubou; Sooty Boubou-
   Shrike; Sooty Bush-Shrike
f Gonolek fuligineux
d Schwarzwürger
i Averla di macchia fuligginosa

5023 *Laniarius liberatus*
Passeriformes - Malaconotidae
e Bulo Burti Boubou
f Gonolek de Bulo Burti
i Averla di macchia

5024 *Laniarius luehderi*
Passeriformes - Malaconotidae
e Lühder's Bush-Shrike; Lühder's
   Bush-Shrike (CSA)
f Gonolek de Lühder
d Braunscheitelwürger
i Averla di macchia di Luehder

5025 *Laniarius mufumbiri*
Passeriformes - Malaconotidae
e Papyrus Gonolek; Mufumbri Shrike;
   Yellow-crowned Gonolek;
   Mufumbiri Gonolek; Papyrus Bush-
   Shrike
f Gonolek des papyrus
d Papyruswürger
i Averla di macchia di Mufumbiri

5026 *Laniarius poensis*
Passeriformes - Malaconotidae
e Mountain Boubou; Mountain Sooty-
   Boubou; Mountain Sooty Boubou-
   Shrike; Mountain Bush-Shrike
f Gonolek de montagne
d Mohrenwürger
i Averla di macchia montana

5027 *Laniarius ruficeps*
Passeriformes - Malaconotidae
e Red-naped Bush-Shrike; Red-
   crowned Bush-Shrike
f Gonolek à nuque rouge
d Rotnackenwürger
i Averla di macchia corona rossa

5028 *Laniarius turatii*
Passeriformes - Malaconotidae

e Turati's Boubou; Tourati Bush-
Shrike; Bellshrike
f Gonolek de Turati; Gonolek de
Verreaux
i Averla di macchia di Turati

5029 Laniidae
Passeriformes
e Shrikes
f Pies-grièches; Laniidés
d Würger
i Lanidi; Averle

5030 *Laniisoma elegans*
Passeriformes - Tyrannidae
e Shrike-like Cotinga; Shrike-Cotinga
f Cotinga élégant
d Schuppenbrustzuser
i Cotinga averla

5031 *Lanio aurantius*
Passeriformes - Fringillidae
e Black-throated Shrike-Tanager
f Tangara à gorge noire
d Schwarzkehlwürgertangare
i Tangara averla golanera

5032 *Lanio fulvus*
Passeriformes - Fringillidae
e Fulvous Shrike-Tanager
f Tangara mordoré
d Braunbrustwürgertangare
i Tangara averla fulva

5033 *Lanio leucothorax*
Passeriformes - Fringillidae
e White-throated Shrike-Tanager;
Black-throated Shrike-Tanager
f Tangara à gorge blanche
d Weißkehlwürgertangare
i Tangara averla golabianca

5034 *Lanio versicolor*
Passeriformes - Fringillidae
e White-winged Shrike-Tanager
f Tangara versicolore
d Gelbstirnwürgertangare
i Tangara averla alibianchi

5035 *Laniocera hypopyrra*
Passeriformes - Tyrannidae

e Cinereous Mourner
f Aulia cendré
d Rotbüschelaulia
i Piagnone cenerino

5036 *Laniocera rufescens*
Passeriformes - Tyrannidae
e Speckled Mourner; Rufescent
Mourner
f Aulia tacheté
d Gelbbüschelaulia
i Piagnone macchiettato

5037 *Lanioturdus torquatus*
Passeriformes - Corvidae
e Chatshrike; White-tailed Shrike
(CSA)
f Lanielle à queue blanche
d Drosselwürger
i Averla-tordo

5038 *Lanius bucephalus*
Passeriformes - Laniidae
e Bull-headed Shrike
f Pie-grièche bucéphale
d Büffelwürger
i Averla giapponese

5039 *Lanius cabanisi*
Passeriformes - Laniidae
e Long-tailed Fiscal; Long-tailed
Fiscal-Shrike
f Pie-grièche à longue queue
d Cabanis-Würger;
Langschwanzwürger;
Graurückenwürger
i Averla dorsonero

5040 *Lanius collaris*
Passeriformes - Laniidae
e Common Fiscal; Fiscal (CSA);
Fiscal-Shrike (CSA); Butcherbird;
Hanger (CSA); Jacky Hangman
(CSA); Common Fiscal-Shrike
f Pie-grièche fiscale
d Fiskalwürger
i Averla fiscale

5041 *Lanius collurio*
Passeriformes - Laniidae
e Red-backed Shrike

f   Pie-grièche écorcheuse
d   Neuntöter; Rotrückenwürger
i   Averla piccola; Averlia piccola

5042   *Lanius collurioides*
     Passeriformes - Laniidae
e   Burmese Shrike; Chestnut-backed
     Shrike; Chestnut-rumped Shrike
f   Pie-grièche à dos marron; Pie-grièche
     birmane
d   Burma-Würger
i   Averla birmana

5043   *Lanius cristatus*
     Passeriformes - Laniidae
e   Brown Shrike; Red-tailed Shrike
f   Pie-grièche brune; Pie-grièche à
     queue rousse
d   Rotschwanzwürger
i   Averla bruna

5044   *Lanius dorsalis*
     Passeriformes - Laniidae
e   Taita Fiscal; Teita Fiscal; Teita
     Fiscal-Shrike; Taita Fiscal-Shrike
f   Pie-grièche des Teita; Pie-grièche des
     Taita; Pie-grièche du Teïta
d   Teitawürger; Taitawürger
i   Averla taita

5045   *Lanius excubitor*
     Passeriformes - Laniidae
e   Great Grey-Shrike; Northern Shrike
     (NA); Grey-backed Fiscal (CSA);
     Great Gray Shrike (NA); Grey Shrike
     (ISC)
f   Pie-grièche grise; Pie-grièche boréale
d   Raubwürger; Graurückenwürger;
     Grauwürger
i   Averla maggiore; Averlia maggiore

5046   *Lanius excubitoroides*
     Passeriformes - Laniidae
e   Grey-backed Fiscal; Grey-backed
     Fiscal-Shrike
f   Pie-grièche à dos gris
d   Graumantelwürger; Afrikanischer
     Bandwürger
i   Averla cenerina africana

5047   *Lanius gubernator*
     Passeriformes - Laniidae
e   Emin's Shrike; Emin's Red-backed
     Shrike
f   Pie-grièche à dos roux; Petite Pie-
     grièche à dos roux
d   Rotbürzelwürger
i   Averla di Emin

5048   *Lanius isabellinus*
     Passeriformes - Laniidae
e   Rufous-tailed Shrike; Isabelline
     Shrike; Central Asian Shrike; Pale-
     brown Shrike; Red-tailed Shrike
f   Pie-grièche isabelle
d   Isabellwürger
i   Averla isabellina

5049   *Lanius leucopygos*
     Passeriformes - Laniidae
e   Saharan Shrike

5050   *Lanius ludovicianus*
     Passeriformes - Laniidae
e   Loggerhead Shrike; Migrant Shrike;
     Loggerhead
f   Pie-grièche migratrice
d   Louisiana-Würger; Amerikanischer
     Raubwürger
i   Averla americana

5051   *Lanius mackinnoni*
     Passeriformes - Laniidae
e   Mackinnon's Shrike; Mackinnon's
     Fiscal (CSA); Mackinnon's Grey-
     Shrike
f   Pie-grièche de Mackinnon
d   Mackinnon-Würger
i   Averla di MacKinnon

5052   *Lanius marwitzi*
     Passeriformes - Laniidae
e   Uhehe Fiscal
f   Pie-grièche de Tanzanie
i   Averla uhehe

5053   *Lanius meridionalis*
     Passeriformes - Laniidae
e   Southern Grey-Shrike; Great Grey-
     Shrike; Indian Grey-Shrike; Steppe
     Grey-Shrike

f   Pie-grièche méridionale
i   Averla maggiore meridinale

5054   ***Lanius minor***
       Passeriformes - Laniidae
e   Lesser Grey-Shrike
f   Pie-grièche à poitrine rose
d   Schwarzstirnwürger
i   Averla cenerina; Averlia cinerina

5055   ***Lanius newtoni***
       Passeriformes - Laniidae
e   Newton's Fiscal; Newton's Fiscal-
      Shrike; Sao Tome Fiscal
f   Pie-grièche de Sao Tomé
d   Büttelwürger
i   Averla di Newton

5056   ***Lanius nubicus***
       Passeriformes - Laniidae
e   Masked Shrike; Nubian Shrike
f   Pie-grièche masquée
d   Maskenwürger
i   Averla mascherata; Averlia
      mascherata

5057   ***Lanius schach***
       Passeriformes - Laniidae
e   Long-tailed Shrike; Black-headed
      Shrike; Rufous-backed Shrike (ISC);
      Black-capped Shrike; Red-backed
      Shrike; Rufous-rumped Shrike;
      Southern Rufous-rumped Shrike;
      Schach Shrike
f   Pie-grièche Schach
d   Königswürger; Schachwürger
i   Averla dorso rossiccio

5058   ***Lanius senator***
       Passeriformes - Laniidae
e   Woodchat-Shrike; Woodchat
f   Pie-grièche à tête rousse
d   Rotkopfwürger
i   Averla capirossa; Capirossa; Averlia
      capirosso

5059   ***Lanius somalicus***
       Passeriformes - Laniidae
e   Somali Fiscal; Somali Shrike; Somali
      Fiscal-Shrike
f   Pie-grièche de Somalie

d   Antinori-Würger
i   Averla somala

5060   ***Lanius souzae***
       Passeriformes - Laniidae
e   Sousa's Shrike
f   Pie-grièche de Souza
d   Souza-Würger; Rostmantelwürger
i   Averla di Souza

5061   ***Lanius sphenocercus***
       Passeriformes - Laniidae
e   Chinese Grey-Shrike; Long-tailed
      Grey-Shrike; Wedge-tailed Grey-
      Shrike; Chinese Shrike; Chinese
      Great Grey-Shrike; Wedge-tailed
      Shrike
f   Pie-grièche géante; Pie-grièche grise
      de Chine
d   Keilschwanzwürger
i   Averla maggiore cinese

5062   ***Lanius tephronotus***
       Passeriformes - Laniidae
e   Grey-backed Shrike; Tibetan Shrike
f   Pie-grièche du Tibet
d   Tibet-Würger
i   Averla dorsogrigio

5063   ***Lanius tigrinus***
       Passeriformes - Laniidae
e   Tiger Shrike; Thick-billed Shrike
f   Pie-grièche tigrine
d   Tigerwürger
i   Averla tigrata

5064   ***Lanius validirostris***
       Passeriformes - Laniidae
e   Mountain Shrike; Strong-billed
      Shrike; Grey-capped Shrike
f   Pie-grièche des Philippines
d   Philippinen-Würger
i   Averla beccoforte

5065   ***Lanius vittatus***
       Passeriformes - Laniidae
e   Bay-backed Shrike
f   Pie-grièche à bandeau; Pie-grièche à
      dos roux
d   Rotschulterwürger
i   Averla dorso castano

5066   Laridae
        Charadriiformes
e   Gulls
f   Laridés
d   Möwen; Seeschwalben; Möwenvögel
i   Laridi; UCC56

5067   *Larosterna inca*
        Charadriiformes - Laridae
e   Inca Tern
f   Sterne inca
d   Inkaseeschwalbe
i   Sterna inca

5068   *Larus argentatus*
        Charadriiformes - Laridae
e   Herring Gull; Ponto-Caspian Gull
f   Goéland argenté; Mauve (Ants);
        Goéland pontique; Goéland à
        manteau bleu
d   Silbermöwe
i   Gabbiano reale; Gabbiano reale
        nordico

5069   *Larus argentatus michahellis*
        Charadriiformes - Laridae
e   Yellow-legged Herring Gull

5070   *Larus armenicus*
        Charadriiformes - Laridae
e   Armenian Gull
f   Goéland d'Arménie
d   Armenien-Möwe
i   Gabbiano d'Armenia; Gabbiano di
        Armenia

5071   *Larus atlanticus*
        Charadriiformes - Laridae
e   Olrog's Gull; Simeon's Gull; Band-
        tailed Gull
f   Goéland d'Olrog
i   Gabbiano di Olrog

5072   *Larus atricilla*
        Charadriiformes - Laridae
e   Laughing Gull; Mauve (WI); Booby
        (WI); Laughingbird (WI); Davy
        (WI); Seagull (WI)
f   Goéland atricille; Mouette marin;
        Mouette aztèque; Mouette atricille;
        Pigeon de mer (Ants); Pigeon la mer

        (Ants); Mauve à tête noire; Mouette
        rieuse (Qué); Mouette à tête noire
d   Aztekenmöwe
i   Gabbiano sghignazzante

5073   *Larus audouinii*
        Charadriiformes - Laridae
e   Audouin's Gull
f   Goéland d'Audouin
d   Korallenmöwe
i   Gabbiano corso

5074   *Larus belcheri*
        Charadriiformes - Laridae
e   Band-tailed Gull; Belcher's Gull;
        Simeon's Gull
f   Goéland siméon; Goeland de Siméon
d   Simeon-Möwe
i   Gabbiano codafasciata

5075   *Larus brunnicephalus*
        Charadriiformes - Laridae
e   Brown-headed Gull; Indian Black-
        headed Gull
f   Mouette du Tibet
d   Braunkopfmöwe; Tibet-Lachmöwe
i   Gabbiano testabruna

5076   *Larus bulleri*
        Charadriiformes - Laridae
e   Black-billed Gull; Buller's Gull
f   Mouette de Buller
d   Maorimöwe
i   Gabbiano di Buller

5077   *Larus cachinnans*
        Charadriiformes - Laridae
e   Yellow-legged Gull; Yellow-legged
        Herring Gulll
f   Goéland leucophée
d   Weißkopfmöwe
i   Gabbiano reale; Gabbiano reale
        zampegialle; Gabbio reale
        mediterraneo

5078   *Larus californicus*
        Charadriiformes - Laridae
e   California Gull
f   Goéland de Californie
d   Indianer-Möwe
i   Gabbiano della California

5079   *Larus canus*
Charadriiformes - Laridae
e   Common Gull; Mew Gull; Seagull;
    Short-billed Gull; Eastern Mew Gull;
    Kamchatka Gull
f   Goéland cendré
d   Sturmmöwe
i   Gavina; Zafferano cenerino;
    Gabbiano; Martinaccio

5080   *Larus cirrocephalus*
Charadriiformes - Laridae
e   Grey-headed Gull; Grey-hooded Gull
f   Mouette à tête grise
d   Graukopfmöwe
i   Gabbiano testagrigia

5081   *Larus crassirostris*
Charadriiformes - Laridae
e   Black-tailed Gull; Japanese Gull;
    Temminck's Gull
f   Goéland à queue noire
d   Japan-Möwe
i   Gabbiano giapponese

5082   *Larus delawarensis*
Charadriiformes - Laridae
e   Ring-billed Gull
f   Goéland à bec cerclé; Goéland (Ants)
d   Ringschnabelmöwe
i   Gabbiano del Delaware; Gavina
    americana

5083   *Larus dominicanus*
Charadriiformes - Laridae
e   Kelp Gull; Southern Black-backed
    Gull; Dominican Gull; Antarctic
    Black-backed Gull
f   Goéland dominicain
d   Dominikanermöwe
i   Gabbiano dominicano

5084   *Larus fuliginosus*
Charadriiformes - Laridae
e   Lava Gull; Dusky Gull
f   Goéland fuligineuse; Mouette
    obscure; Goéland obscur
d   Lavamöwe
i   Gabbiano fuligginoso

5085   *Larus fuscus*
Charadriiformes - Laridae
e   Lesser Black-backed Gull
f   Goéland brun
d   Heringsmöwe
i   Zafferano; Gabbiano zafferano

5086   *Larus genei*
Charadriiformes - Laridae
e   Slender-billed Gull
f   Goéland railleur; Goéland à bec grêle
d   Dünnschnabelmöwe
i   Gabbiano roseo

5087   *Larus glaucescens*
Charadriiformes - Laridae
e   Glaucous-winged Gull
f   Goéland à ailes grises; Goéland à
    ailes glauques; Goéland de Béring
i   Gabbiano glauco del Pacifico

5088   *Larus glaucoides*
Charadriiformes - Laridae
e   Iceland Gull; Greenland Gull;
    Kumlien's Gull
f   Goéland à ailes blanches; Goéland
    arctique; Goéland leucoptère
d   Polarmöwe
i   Gabbiano d'Islanda; Gabbiano
    islandico

5089   *Larus hartlaubii*
Charadriiformes - Laridae
e   King Gull; Hartlaub's Gull (CSA)
f   Mouette de Hartlaub
d   Hartlaubs Möwe;
    Weißkopflachmöwe; Sternmöwe
i   Gabbiano di Hartlaub

5090   *Larus heermanni*
Charadriiformes - Laridae
e   Heermann's Gull
f   Goéland de Heermann; Mouette de
    Heermann
d   Heermann-Möwe
i   Gabbiano di Heermann

5091   *Larus hemprichii*
Charadriiformes - Laridae
e   Sooty Gull; Hemprich's Gull; Aden
    Gull

f Goéland de Hemprich
d Hemprich-Möwe
i Gabbiano di Hemprich; Gabbiano
fuligginoso

5092 *Larus heuglini*
Charadriiformes - Laridae
e Siberian Gull; Heuglin's Gull
f Goéland de Sibérie
d Sibirien-Möwe
i Gabbiano di Heuglin

5093 *Larus hyperboreus*
Charadriiformes - Laridae
e Glaucous Gull
f Goéland bourgmestre
d Eismöwe
i Gabbiano glauco; Gabbiano bianco

5094 *Larus ichthyaetus*
Charadriiformes - Laridae
e Great Black-headed Gull; Pallas's
Gull; Greater Black-headed Gull
f Goéland ichthyaète; Goéland à tête
noire
d Fischmöwe
i Gabbiano del Pallas

5095 *Larus kumlieni*
Charadriiformes - Laridae
e Kumlien's Gull
f Ghoéland de Kumlien

5096 *Larus leucophthalmus*
Charadriiformes - Laridae
e White-eyed Gull; Red-Sea Black-
headed Gull
f Goéland à iris blanc
d Weißaugenmöwe
i Gabbiano occhibianchi; Gabbiano
sopracciglibianchi

5097 *Larus livens*
Charadriiformes - Laridae
e Yellow-footed Gull
f Goéland de Cortez
i Gabbiano zampegialle

5098 *Larus maculipennis*
Charadriiformes - Laridae
e Brown-hooded Gull; Patagonian

Black-headed Gull; Pink-breasted
Gull
f Mouette de Patagonie
d Patagonien-Möwe
i Gabbiano capobruno

5099 *Larus marinus*
Charadriiformes - Laridae
e Great Black-backed Gull; Greater
Black-backed Gull
f Goéland marin; Goéland à manteau
noir
d Mantelmöwe
i Mugnaiaccio

5100 *Larus melanocephalus*
Charadriiformes - Laridae
e Mediterranean Gull; Mediterranean
Black-headed Gull
f Mouette mélanocéphale; Mouette à
tête noire
d Schwarzkopfmöwe
i Gabbiano corallino

5101 *Larus minutus*
Charadriiformes - Laridae
e Little Gull
f Mouette pygmée
d Zwergmöwe
i Gabbianello; Gabbiano minore

5102 *Larus modestus*
Charadriiformes - Laridae
e Grey Gull
f Goéland gris
d Graumöwe
i Gabbiano modesto

5103 *Larus nelsoni*
Charadriiformes - Laridae
e Nelson's Gull

5104 *Larus novaehollandiae*
Charadriiformes - Laridae
e Silver Gull; Red-billed Gull
f Mouette argentée; Mouette
australienne
d Weißkopflachmöwe
i Gabbiano australiano

5105   *Larus occidentalis*
Charadriiformes - Laridae
e   Western Gull
f   Goéland d'Audubon; Goéland occidental
i   Gabbiano occidentale

5106   *Larus pacificus*
Charadriiformes - Laridae
e   Pacific Gull; Large-billed Gull; Australian Gull
f   Goéland austral
d   Dickschnabelmöwe
i   Gabbiano del Pacifico

5107   *Larus philadelphia*
Charadriiformes - Laridae
e   Bonaparte's Gull
f   Mouette de Bonaparte
d   Bonaparte-Möwe
i   Gabbiano di Bonaparte

5108   *Larus pipixcan*
Charadriiformes - Laridae
e   Franklin's Gull
f   Mouette de Franklin
d   Präriemöwe; Franklins Möwe; Franklin-Möwe
i   Gabbiano di Franklin

5109   *Larus relictus*
Charadriiformes - Laridae
e   Relict Gull; Lönnberg's Gull; Mongolian Gull; Central Asian Gull
f   Mouette relique
d   Lönnberg-Möwe
i   Gabbiano relitto

5110   *Larus ridibundus*
Charadriiformes - Laridae
e   Black-headed Gull; Common Black-headed Gull; European Black-headed Gull; Northern Black-headed Gull
f   Mouette rieuse
d   Lachmöwe
i   Gabbiano comune; Gabbiano

5111   *Larus sabini*
Charadriiformes - Laridae
e   Sabine's Gull
f   Mouette de Sabine

d   Schwalbenmöwe
i   Gabbiano di Sabine

5112   *Larus saundersi*
Charadriiformes - Laridae
e   Saunders's Gull; Chinese Black-headed Gull
f   Mouette de Saunders
d   Kappenmöwe
i   Gabbiano di Saunders

5113   *Larus schistisagus*
Charadriiformes - Laridae
e   Slaty-backed Gull; Pacific Gull; Kamchatka Gull
f   Goéland à manteau ardoisé; Goéland du Kamchatka
i   Gabbiano dorsoardesia; Fabbiano della Kamchatka

5114   *Larus scopulinus*
Charadriiformes - Laridae
e   Red-billed Gull
f   Mouette scopuline
i   Gabbiano beccorosso

5115   *Larus scoresbii*
Charadriiformes - Laridae
e   Dolphin Gull; Magellan Gull; Scoresby Gull
f   Goéland de Scoresby
d   Blutschnabelmöwe
i   Gabbiano di Magellano

5116   *Larus serranus*
Charadriiformes - Laridae
e   Andean Gull; Mountain Gull
f   Mouette des Andes
d   Anden-Möwe
i   Gabbiano delle Ande

5117   *Larus thayeri*
Charadriiformes - Laridae
e   Thayer's Gull
f   Goéland de Thayer
d   Thayer-Möwe

5118   *Laterallus albigularis*
Gruiformes - Rallidae
e   White-throated Crake; White-throated Rail

f   Râle à menton blanc
i   Schiribilla golabianca

5119   *Laterallus exilis*
Gruiformes - Rallidae
e   Grey-breasted Crake; Grey-breasted
Rail; Temminck's Crake
f   Râle grêle
d   Amazonas-Ralle
i   Schiribilla pettogrigio

5120   *Laterallus jamaicensis*
Gruiformes - Rallidae
e   Black Rail; Black Crake; Junin Rail;
American Black-Rail
f   Râle noir
d   Schieferralle
i   Schiribilla nera americana

5121   *Laterallus leucopyrrhus*
Gruiformes - Rallidae
e   Red-and-white Crake; Red-and white
Rail
f   Râle blanc-et- roux
d   Weißbrustralle
i   Schiribilla bianca e rossa

5122   *Laterallus levraudi*
Gruiformes - Rallidae
e   Rusty-flanked Crake; Rusty-flanked
Rail; Levraud's Rake
f   Râle de Levraud
d   Venezuela-Ralle
i   Schiribilla di Levraud

5123   *Laterallus melanophaius*
Gruiformes - Rallidae
e   Rufous-sided Crake; Rufous-sided
Rail
f   Râle brunoir
d   Rothalsralle
i   Schiribilla fianchirossi

5124   *Laterallus ruber*
Gruiformes - Rallidae
e   Ruddy Crake; Ruddy Rail; Red Rail;
Red Crake
f   Râle roux
d   Rubinralle
i   Schiribilla rossiccia

5125   *Laterallus spilonotus*
Gruiformes - Rallidae
e   Galapagos Rail; Galapagos Crake
f   Râle des Galapagos
d   Galapagos-Ralle
i   Schiribilla delle Galapagos

5126   *Laterallus tuerosi*
Gruiformes - Rallidae
e   Junin Rail

5127   *Laterallus xenopterus*
Gruiformes - Rallidae
e   Rufous-faced Crake; Rufous-faced
Rail; Horqueta Rufous-faced Crake;
Horqueta Crake
f   Râle de Conover
d   Rotgesichtralle
i   Schiiribilla facciarossa

5128   *Lathamus discolor*
Psittaciformes - Psittacidae
e   Swift Parrot; Swift Lorikeet; Swift
Parakeet; Swift-flying Parakeet; Red-
shouldered Parakeet; Clink Parakeet;
Red-shouldered Parrot; Red-faced
Parrot; Red-faced Parakeet
f   Perruche de Latham
d   Schwalbenlori; Schwalbensittich
i   Parrochetto di Latham

5129   *Lathrotriccus euleri*
Passeriformes - Tyrannidae
e   Euler's Flycatcher
f   Moucherolle d'Euler
d   Euler-Tyrann
i   Tiranno di Euler

5130   *Lathrotriccus flaviventris*
Passeriformes - Tyrannidae
e   Lawrence's Flycatcher

5131   *Latoucheornis siemsseni*
Passeriformes - Fringillidae
e   Slaty Bunting; La Touche's Bunting;
Fokien Blue-Bunting; Chinese Blue-
Bunting; Fokien Slaty Bunting; Blue
Bunting; Chinese Bunting
f   Bruant bleu
d   Blauammer
i   Zigolo blu di Fokien

5132 **Legatus leucophaius**
Passeriformes - Tyrannidae
e Piratic Flycatcher
f Tyran pirate
d Diebstyrann
i Tiranno pirata

5133 **Leiothrix argentauris**
Passeriformes - Sylviidae
e Silver-eared Mesia; Silver-eared
Leiothrix
f Leiothrix à joues argent; Mesia à
oreillons argentés
d Silberohrsonnenvogel
i Usignolo orecchieargentate

5134 **Leiothrix lutea**
Passeriformes - Sylviidae
e Red-billed Mesia; Red-billed
Leiothrix; Pekin Nightingale; Pekin
Robin; Japanese Hill-Robin; Astley's
Leiothrix; Doubtful Leiothrix
f Leiothrix jaune; Rossignol du Japon
d Sonnenvogel; Chinesischer
Sonnenvogel; Chinesische
Nachtigall; China-Nachtigall
i Usignolo di Giappone; Uccello del
sole comune; Uccello del sole

5135 **Leipoa ocellata**
Galliformes - Megapodiidae
e Malleefowl; Malee Hen
f Leipoa ocellé
d Thermometerhuhn
i Fagiano australiano; Megapodia
ocellato

5136 **Leistes militaris**
Passeriformes - Fringillidae
e Red-breasted Blackbird; Northern
Marsh-Meadowlark
f Sturnelle militaire
d Rotbruststärling
i Sturnella pettirossa

5137 **Leistes superciliaris**
Passeriformes - Fringillidae
e White-browed Blackbird;
Bonaparte's Blackbird
f Sturnelle à sourcils blancs

d Weißbrauenstärling
i Sturnella di Bonaparte

5138 **Lemuresthes nana**
Passeriformes - Passeridae
e Madagascar Munia; Bibfinch;
Madagascar Mannikin; Dwarf
Mannikin; African Bibfinch; African
Parsonfinch
f Capucin de Madagascar; Spermète
nain
d Zwergelsterchen
i Diamante del Madagascar

5139 **Lepidocolaptes affinis**
Passeriformes - Dendrocolaptidae
e Spot-crowned Woodcreeper; Spot-
crowned Creeper; Allied Creeper
f Grimpar moucheté; Grimpar
montagnard
d Fleckscheitelbaumsteiger;
Gebirgsstreifenbaumsteiger
i Rampichino corona macchiata

5140 **Lepidocolaptes albolineatus**
Passeriformes - Dendrocolaptidae
e Lineated Woodcreeper; Lineated
Creeper
f Grimpar lancéolé
d Layard-Baumsteiger
i Rampichino lineato

5141 **Lepidocolaptes angustirostris**
Passeriformes - Dendrocolaptidae
e Narrow-billed Woodcreeper;
Narrow-billed Creeper
f Grimpar à bec étroit
d Schmalschnabelbaumsteiger; Kleiner
Baumsteiger; Streifenbaumsteiger
i Rampichino beccosottile

5142 **Lepidocolaptes fuscus**
Passeriformes - Dendrocolaptidae
e Lesser Woodcreeper; Lesser Creeper
f Grimpar brun
d Schlankschnabelbaumsteiger
i Rampichino bruno minore

5143 **Lepidocolaptes lacrymiger**
Passeriformes - Dendrocolaptidae
e Montane Woodcreeper

5144 *Lepidocolaptes leucogaster*
Passeriformes - Dendrocolaptidae
e White-striped Woodcreeper; White-striped Creeper; White-bellied Woodcreeper
f Grimpar givré
d Buntbaumsteiger
i Rampichino ventrebianco

5145 *Lepidocolaptes souleyetii*
Passeriformes - Dendrocolaptidae
e Streak-headed Woodcreeper; Streak-headed Creeper; Souleyet's Creeper; Souleyet's Woodcreeper
f Grimpar de Souleyet
d Souleyet-Baumssteiger
i Rampichino testastriata

5146 *Lepidocolaptes squamatus*
Passeriformes - Dendrocolaptidae
e Scaled Woodcreeper; Scaled Creeper
f Grimpar écaillé
d Fleckenbauchbaumsteiger
i Rampichino squamato

5147 *Lepidopyga coeruleogularis*
Apodiformes - Trochilidae
e Sapphire-throated Hummingbird; Duchassain's Hummingbird
f Colibri faux-saphir
d Blaukehlkolibri
i Colibrì gola di zaffiro

5148 *Lepidopyga goudoti*
Apodiformes - Trochilidae
e Shining-green Hummingbird
f Colibri de Goudot
d Goudot-Kolibri
i Colibrì verde splendente

5149 *Lepidopyga lilliae*
Apodiformes - Trochilidae
e Sapphire-bellied Hummingbird
d Lilli-Kolibri
i Colibrì ventrezaffiro

5150 *Leptasthenura aegithaloides*
Passeriformes - Furnariidae
e Plain-mantled Tit-Spinetail
f Synallaxe mésange
d Schwanzmeisensschlüpfer

5151 *Leptasthenura andicola*
Passeriformes - Furnariidae
e Andean Tit-Spinetail
f Synallaxe des Andes
d Anden-Schlüpfer
i Codaspinosa delle Ande

5152 *Leptasthenura fuliginiceps*
Passeriformes - Furnariidae
e Brown-capped Tit-Spinetail
f Synallaxe à tête brune
d Rotscheitelschlüpfer
i Codaspinosa testabruna

5153 *Leptasthenura pileata*
Passeriformes - Furnariidae
e Rusty-crowned Tit-Spinetail
f Synallaxe couronné
d Peru-Schlüpfer
i Codaspinosa pileato

5154 *Leptasthenura platensis*
Passeriformes - Furnariidae
e Tufted Tit-Spinetail
f Synallaxe de la Plata
d Schopfschlüpfer
i Codaspinosa dal ciuffo

5155 *Leptasthenura setaria*
Passeriformes - Furnariidae
e Araucaria Tit-Spinetail
f Synallaxe à filets
d Araukarien-Schlüpfer
i Codaspinosa di Araucaria

5156 *Leptasthenura striata*
Passeriformes - Furnariidae
e Streaked Tit-Spinetail
f Synallaxe strié
d Streifenschlüpfer
i Codaspinosa striato

5157 *Leptasthenura striolata*
Passeriformes - Furnariidae
e Striolated Tit-Spinetail
f Synallaxe striolé
d Strichelschlüpfer
i Codaspinosa striolato

5158 *Leptasthenura xenothorax*
Passeriformes - Furnariidae
e White-browed Tit-Spinetail
f Synallaxe à gorge rayée
i Codaspinosa dai sopraccigli bianchi

5159 *Leptasthenura yanacensis*
Passeriformes - Furnariidae
e Tawny Tit-Spinetail
f Synallaxe fauve; Synallaxe montagnard
d Yanaca-Schlüpfer
i Codaspinosa fulvo

5160 *Leptodon cayanensis*
Falconiformes - Accipitridae
e Grey-headed Kite; Cayenne Kite
f Milan de Cayenne
d Cayenne-Weihe
i Nibbio testagrigia; Nibbio della Cayenna

5161 *Leptodon forbesi*
Falconiformes - Accipitridae
e White-collared Kite
f Milan de Forbes
d Forbes-Weihe
i Nibio dal collare bianco

5162 *Leptopoecile elegans*
Passeriformes - Sylviidae
e Crested Tit-Warbler
f Pouillot huppé; Roitelet élégant
d Schopfhähnchen
i Magnanina crestata della Cina

5163 *Leptopoecile sophiae*
Passeriformes - Sylviidae
e White-browed Tit-Warbler; Severtzov's Tit-Warbler; Painted Tit-Warbler; Stoliczka's Tit-Warbler
f Pouillot de Sophie; Roitelet de la reine Sophie
d Purpurhähnchen
i Magnanina di Severtzov

5164 *Leptopogon amaurocephalus*
Passeriformes - Tyrannidae
e Sepia-capped Flycatcher; Sepia-capped Leptopogon
f Pipromorphe à tête brune

d Braunkopffliegenstecher
i Tiranno barbuto testabruna

5165 *Leptopogon rufipectus*
Passeriformes - Tyrannidae
e Rufous-breasted Flycatcher; Rufous-breasted Leptopogon
f Pipromorphe à poitrine rousse
d Rotbrustfliegenstecher
i Tiranno barbuto pettirosso

5166 *Leptopogon superciliaris*
Passeriformes - Tyrannidae
e Slaty-capped Flycatcher; White-bellied Leptopogon; Slaty-capped Leptopogon; Slaty-capped Flycatcher
f Pipromorphe à tête grise
d Schieferkopffliegenstecher
i Tiranno barbuto testagrigia

5167 *Leptopogon taczanowskii*
Passeriformes - Tyrannidae
e Inca Flycatcher; Inca Leptopogon
f Pipromorphe inca
d Inca-Fliegenstecher
i Tiranno barbuto degli incas

5168 *Leptopterus chabert*
Passeriformes - Vangidae
e Chabert's Vanga
f Artamie chabert; Artamie de Chabert
d Elstervanga
i Vanga chabert

5169 *Leptoptilos crumeniferus*
Ciconiiformes - Ciconiidae
e Marabou Stork; Marabou
f Marabou d'Afrique
d Marabu
i Marabù africano; Marabù

5170 *Leptoptilos dubius*
Ciconiiformes - Ciconiidae
e Greater Adjutant; Adjutant Stork (ISC); Greater Adjutant-Stork
f Marabout argala
d Argala
i Marabù maggiore; Marabù maggiore asiatico; Marabù indiano

5171   *Leptoptilos javanicus*
Ciconiiformes - Ciconiidae
e   Lesser Adjutant; Lesser Adjutant-
Stork; Hair-crested Stork
f   Marabout chevelu
d   Malaien-Storch
i   Marabù minore; Marabù minore
asiatico

5172   *Leptosittaca branickii*
Psittaciformes - Psittacidae
e   Golden-plumed Parakeet; Golden-
plumed Parrot; Golden-plumed
Conure; Brannick's Conure
f   Conure à pinceaux d'or
d   Hochlandsittich; Pinselsittich
i   Conuro dai ciuffetti dorati

5173   Leptosomidae
Coraciiformes
e   Cuckoo-Rollers; Courols
f   Leptosomidés
d   Kurole
i   Leptosomatidi

5174   *Leptosomus discolor*
Coraciiformes - Leptosomidae
e   Cuckoo-Roller; Courol; Kirombo
Courol; Courol Roller
f   Courol vouroudriou; Courol
d   Kurol
i   Curol; Leptosomo

5175   *Leptotila battyi*
Columbiformes - Columbidae
e   Brown-backed Dove
f   Colombe du Panama
i   Tortora dorsobruno

5176   *Leptotila brasiliensis*
Columbiformes - Columbidae
e   Brazilian Dove
d   Rotringtaube

5177   *Leptotila cassini*
Columbiformes - Columbidae
e   Grey-chested Dove; Cassin's Dove;
Grey-breasted Dove
f   Colombe de Cassin
d   Cassin-Taube; Graubrusttaube
i   Tortora di Cassin

5178   *Leptotila conoveri*
Columbiformes - Columbidae
e   Tolima Dove; Conover's Dove
f   Colombe de Conover
d   Tolima-Taube
i   Tortora di Tolima

5179   *Leptotila jamaicensis*
Columbiformes - Columbidae
e   Caribbean Dove; White-bellied
Dove; White-belly (WI); Ground-
Dove; White-fronted Dove; Violet
Ground-Dove; Jamaican Dove
f   Colombe de la Jamaïque
d   Jamaika-Taube
i   Tortora della Giamaica

5180   *Leptotila megalura*
Columbiformes - Columbidae
e   White-faced Dove; Large-tailed
Dove; Long-tailed Dove
f   Colombe à face blanche
d   Weißgesichtstaube
i   Tortora codalunga

5181   *Leptotila ochraceiventris*
Columbiformes - Columbidae
e   Ochre-bellied Dove; Buff-bellied
Dove
f   Colombe de Chapman
d   Gelbbauchtaube
i   Tortora ventreocra

5182   *Leptotila pallida*
Columbiformes - Columbidae
e   Pallid Dove
f   Colombe pâle
d   Fahltaube
i   Tortora pallida

5183   *Leptotila plumbeiceps*
Columbiformes - Columbidae
e   Grey-headed Dove
f   Colombe à calotte grise
d   Bonaparte-Taube; Graukopftaube
i   Tortora testagrigia

5184   *Leptotila rufaxilla*
Columbiformes - Columbidae
e   Grey-fronted Dove

f Colombe à front gris
d Graustirntaube; Rotachseltaube
i Tortora frontegrigia

5185 *Leptotila verreauxi*
Columbiformes - Columbidae
e White-tipped Dove; White-fronted
Dove; Solitary Pigeon; Pale-fronted
Dove; Brazilian Dove
f Colombe de Verreaux
d Blauringtaube; Weißstirntaube
i Tortora di Verreaux

5186 *Leptotila wellsi*
Columbiformes - Columbidae
e Grenada Dove; Whistling Dove
(WI); Mountain Dove (WI); Wells's
Dove
f Colombe de Grenade
d Grenada-Taube; Wells-Taube
i Tortora di Grenada

5187 *Lerwa lerwa*
Galliformes - Phasianidae
e Snow Partridge
f Lerva des neiges
d Haldenhuhn
i Pernice delle neve

5188 *Lesbia nuna*
Apodiformes - Trochilidae
e Green-tailed Trainbearer
f Porte-traîne nouna; Colibri nouna
d Grünschwanzlesbia
i Lesbia codaverde

5189 *Lesbia victoriae*
Apodiformes - Trochilidae
e Black-tailed Trainbearer; Long-tailed
Trainbearer
f Porte-traîne lesbie
d Schwarzschwanzlesbia
i Lesbia codanera

5190 *Lessonia oreas*
Passeriformes - Tyrannidae
e Andean Negrito; Salvin's Negrito
f Lessonie des Andes
i Negrito di Salvin

5191 *Lessonia rufa*
Passeriformes - Tyrannidae
e Austral Negrito; Patagonian Negrito;
Rufous-backed Negrito; Negrito
Flycatcher
f Lessonie noire
d Sporntyrann
i Negrito dorsorossiccio

5192 *Leucippus baeri*
Apodiformes - Trochilidae
e Tumbes Hummingbird
f Colibri de Tumbes
d Baer-Kolibri
i Colibrì di Tumbes

5193 *Leucippus chlorocercus*
Apodiformes - Trochilidae
e Olive-spotted Hummingbird
f Colibri à queue verte
d Fleckenkolibri
i Colibrì macchiato di oliva

5194 *Leucippus fallax*
Apodiformes - Trochilidae
e Buffy Hummingbird
f Colibri trompeur
d Rostbrüstchen
i Colibrì fulvo

5195 *Leucippus taczanowskii*
Apodiformes - Trochilidae
e Spot-throated Hummingbird
f Colibri de Taczanowski
d Taczanowski-Kolibri
i Colibrì golamacchiata

5196 *Leucochloris albicollis*
Apodiformes - Trochilidae
e White-throated Hummingbird;
Brazilian White-tailed Hummingbird
f Colibri à gorge blanche
d Weißkehlkolibri
i Colibrì golabianca

5197 *Leucopeza semperi*
Passeriformes - Fringillidae
e Semper's Warbler
f Pied-blanc (Ants); Paruline pied-
blanc

d   Blaßfußwaldsänger
i   Parula di Semper

5198   *Leucopsar rothschildi*
        Passeriformes - Sturnidae
e   Bali Starling; Bali Mynah;
        Rothschild's Mynah; White Mynah;
        Rothschild's Grackle
f   Étourneau de Rothschild; Martin de
        Rothschild
d   Bali-Star
i   Storno di Rothschild; Storno di Bali

5199   *Leucopternis albicollis*
        Falconiformes - Accipitridae
e   White Hawk; White Buzzard
f   Buse blanche
d   Schneebussard
i   Poiana bianca; Falco bianco

5200   *Leucopternis kuhli*
        Falconiformes - Accipitridae
e   White-browed Hawk
f   Buse à sourcils blancs
d   Weißbrauenbussard
i   Poiana dai sopraccigli bianchi

5201   *Leucopternis lacernulata*
        Falconiformes - Accipitridae
e   White-necked Hawk
f   Buse lacernulée
d   Weißhalsbussard
i   Poiana collobianco

5202   *Leucopternis melanops*
        Falconiformes - Accipitridae
e   Black-faced Hawk
f   Buse à face noire
d   Zügelbussard
i   Poiana faccianera; Poiana dalla
        faccia nera

5203   *Leucopternis occidentalis*
        Falconiformes - Accipitridae
e   Grey-backed Hawk
f   Buse à dos gris
d   Graurückenbussard
i   Poiana dorsogrigio; Poiano dal dorso
        grigio

5204   *Leucopternis plumbea*
        Falconiformes - Accipitridae
e   Plumbeous Hawk
f   Buse plombée
d   Bleibussard
i   Poiana plumbea

5205   *Leucopternis polionota*
        Falconiformes - Accipitridae
e   Mantled Hawk
f   Buse mantelée
d   Mantelbussard
i   Poiana del mantello; Falco
        mantellato

5206   *Leucopternis princeps*
        Falconiformes - Accipitridae
e   Barred Hawk; Barred Princeps Hawk
f   Buse barrée
d   Prinzenbussard
i   Poiana barrata; Falco a strisce

5207   *Leucopternis schistacea*
        Falconiformes - Accipitridae
e   Slate-coloured Hawk
f   Buse ardoisée
d   Schieferbussard
i   Poiana ardesia

5208   *Leucopternis semiplumbea*
        Falconiformes - Accipitridae
e   Semiplumbeous Hawk
f   Buse semiplombée
d   Möwenbussard
i   Poiana semiplumbea

5209   *Leucosarcia melanoleuca*
        Columbiformes - Columbidae
e   Wonga Pigeon; Wonga Wonga
        Pigeon; White-fleshed Pigeon
f   Colombine wonga
d   Wonga-Taube
i   Colomba uonga

5210   *Leucosticte arctoa*
        Passeriformes - Fringillidae
e   Asian Rosy-Finch; Rosy Finch;
        Grey-crowned Rosy-Finch;
        American Rosy-Finch; Arctic
        Mountain-Finch; Arctic Rosy-Finch;
        Rosy Mountain-Finch; White-winged

Mountain-Finch; Japanese Rosy-
Finch
f   Roselin brun; Roselin rose; Roselin à
tête grise
d   Rosenbauchshneegimpel
i   Fanello rosato di Pallas; Fanello
rosato corona grigia

5211   ***Leucosticte atrata***
Passeriformes - Fringillidae
e   Black Rosy-Finch
f   Roselin noir
i   Fanello rosato nero

5212   ***Leucosticte australis***
Passeriformes - Fringillidae
e   Brown-capped Rosy-Finch
f   Roselin à tête brune
i   Fanello rosato capobruno

5213   ***Leucosticte brandti***
Passeriformes - Fringillidae
e   Black-headed Mountain-Finch;
Brandt's Rosy-Finch; Rosy-rumped
Finch; Brandt's Mountain-Finch;
Rosy-rumped Mountain-Finch
f   Roselin de Brandt
d   Mattenschneegimpel
i   Fanello rosato di Brandt

5214   ***Leucosticte nemoricola***
Passeriformes - Fringillidae
e   Plain Mountain-Finch; Mountain
Finch; Hodgson's Rosy-Finch;
Hodgson's Mountain-Finch
f   Roselin de Hodgson
d   Waldschneegimpel
i   Fanello rosato di Hodgson

5215   ***Leucosticte sillemi***
Passeriformes - Fringillidae
e   Sillem's Mountain-Finch; Sillem's
Rosy-Finch
f   Roselin de Sillem

5216   ***Lewinia mirificus***
Gruiformes - Rallidae
e   Brown-banded Rail; Luzon Rail
f   Râle de Luçon
d   Luzonralle
i   Rallo di Luzon

5217   ***Lewinia muelleri***
Gruiformes - Rallidae
e   Auckland Islands Rail; Aukland Rail
f   Râle d'Auckland
d   Auckland-Ralle
i   Rallo delle Isole Auckland

5218   ***Lewinia pectoralis***
Gruiformes - Rallidae
e   Lewin's Rail; Slate-breasted Rail;
Short-toed Rail; Pectoral Rail
f   Râle à poitrine grise
d   Krickralle
i   Rallo di Lewin

5219   ***Lichenostomus chrysops***
Passeriformes - Meliphagidae
e   Yellow-faced Honeyeater; Yellow-
gaped Honeyeater
f   Méliphage à joues d'or
d   Dreistreifenhonigfresser;
Gelbgesichthonigfresser
i   Mangiamiele facciadorata

5220   ***Lichenostomus cratitius***
Passeriformes - Meliphagidae
e   Purple-gaped Honeyeater; Wattle-
cheeked Honeyeater
f   Méliphage grimé
d   Purpurzügelhonigfresser
i   Mangiamiele boccapurpurea

5221   ***Lichenostomus fasciogularis***
Passeriformes - Meliphagidae
e   Mangrove Honeyeater; Island
Honeyeater; Fasciated Honeyeater
f   Méliphage des mangroves
d   Mangrovehonigfresser;
Mangrovenhonigfresser
i   Mangiamiele delle mangrovie

5222   ***Lichenostomus flavescens***
Passeriformes - Meliphagidae
e   Yellow-tinted Honeyeater; Yellowish
Honeyeater; Pale Yellow Honeyeater
f   Méliphage flavescent
d   Gelbhonigfresser
i   Mangiamiele giallastro

5223   ***Lichenostomus flavicollis***
Passeriformes - Meliphagidae

e   Yellow-throated Honeyeater
f   Méliphage à gorge jaune
d   Gelbkehlhonigfresser
i   Mangiamiele collogiallo

5224   *Lichenostomus flavus*
Passeriformes - Meliphagidae
e   Yellow Honeyeater
f   Méliphage jaune
d   Zitronenhonigfresser
i   Mangiamiele giallo

5225   *Lichenostomus frenatus*
Passeriformes - Meliphagidae
e   Bridled Honeyeater; Mountain
     Honeyeater
f   Méliphage bridé
d   Buntschnabelhonigfresser
i   Mangiamiele dai mustacchi neri

5226   *Lichenostomus fuscus*
Passeriformes - Meliphagidae
e   Fuscous Honeyeater
f   Méliphage grisâtre
d   Olivkehlhonigfresser
i   Mangiamiele fosco

5227   *Lichenostomus hindwoodi*
Passeriformes - Meliphagidae
e   Eungella Honeyeater
f   Méliphage de Hindwood
i   Mangiamiele di Hindwood

5228   *Lichenostomus keartlandi*
Passeriformes - Meliphagidae
e   Grey-headed Honeyeater
f   Méliphage à tête grise
d   Grauscheitelhonigfresser
i   Mangiamiele testagrigia

5229   *Lichenostomus leucotis*
Passeriformes - Meliphagidae
e   White-eared Honeyeater
f   Méliphage leucotique
d   Schwarzkehlhonigfresser
i   Mangiamiele orecchiebianche
     australiano

5230   *Lichenostomus melanops*
Passeriformes - Meliphagidae
e   Yellow-tufted Honeyeater

f   Méliphage cornu
d   Gelbstirnhonigfresser
i   Mangiamiele dai mustacchi gialli

5231   *Lichenostomus melanops cassidix*
Passeriformes - Meliphagidae
e   Helmeted Honeyeater (ANZ)
f   Méliphage casqué
d   Helmhonigfresser

5232   *Lichenostomus obscurus*
Passeriformes - Meliphagidae
e   Obscure Honeyeater; Lemon-
     cheeked Honeyeater
f   Méliphage obscur
d   Laubhonigfresser
i   Mangiamiele guancelimone

5233   *Lichenostomus ornatus*
Passeriformes - Meliphagidae
e   Yellow-plumed Honeyeater; Mallee
     Honeyeater
f   Méliphage orné
d   Gelbscheitelhonigfresser; Kamp-
     Honigfresser
i   Mangiamiele del Mallee

5234   *Lichenostomus penicillatus*
Passeriformes - Meliphagidae
e   White-plumed Honeyeater
f   Méliphage serti
d   Weißbürzelhonigfresser
i   Mangiamiele collomacchiato

5235   *Lichenostomus plumulus*
Passeriformes - Meliphagidae
e   Grey-fronted Honeyeater; Yellow-
     fronted Honeyeater
f   Méliphage à plumet noir
d   Grünscheitelhonigfresser
i   Mangiamiele frontegialla

5236   *Lichenostomus subfrenatus*
Passeriformes - Meliphagidae
e   Black-throated Honeyeater
f   Méliphage à gorge noire
d   Goldstreifenhonigfresser
i   Mangiamiele golanera

5237   *Lichenostomus unicolor*
Passeriformes - Meliphagidae

e   White-gaped Honeyeater; Erect-
    tailed Honeyeater; River Honeyeater
f   Méliphage unicolore
d   Wulst-Honigfresser; Wulst-
    Honigesser
i   Mangiamiele unicolore

5238   *Lichenostomus versicolor*
Passeriformes - Meliphagidae
e   Varied Honeyeater
f   Méliphage versicolore
d   Pirolhonigfresser
i   Mangiamiele variabile

5239   *Lichenostomus virescens*
Passeriformes - Meliphagidae
e   Singing Honeyeater; Black-faced
    Honeyeater; Large-striped
    Honeyeater
f   Méliphage chanteur
d   Pfeifhonigfresser; Sängerhonigesser;
    Zügelhonigfresser
i   Mangiamiele cantore

5240   *Lichmera alboauricularis*
Passeriformes - Meliphagidae
e   Silver-eared Honeyeater; White-
    eared Honeyeater; Eared Honeyeater;
    White-spangled Honeyeater;
    Freckled Honeyeater
f   Méliphage grivelé
d   Ohrfleckenhonigfresser
i   Mangiamiele orecchieargentate

5241   *Lichmera argentauris*
Passeriformes - Meliphagidae
e   Olive Honeyeater; Plain Olive-
    Honeyeater; Silver Honeyeater;
    Silver-spangled Honeyeater
f   Méliphage à joues argent
d   Silberohrhonigfresser
i   Mangiamiele oliva

5242   *Lichmera deningeri*
Passeriformes - Meliphagidae
e   Buru Honeyeater; Deninger's
    Honeyeater
f   Méliphage de Buru
d   Buru-Honigfresser
i   Mangiamiele di Buru

5243   *Lichmera flavicans*
Passeriformes - Meliphagidae
e   Yellow-eared Honeyeater; Timor
    Honeyeater; Vieillot's Honeyeater
f   Méliphage de Timor
d   Timor-Honigfresser
i   Mangiamiele di Timor

5244   *Lichmera incana*
Passeriformes - Meliphagidae
e   Dark-brown Honeyeater; Silver-
    eared Honeyeater; Loyalty Islands
    Honeyeater; Loyalty Honeyeater
f   Méliphage à oreillons gris
d   Grauohrhonigfresser
i   Mangiamiele bruno scuro

5245   *Lichmera indistincta*
Passeriformes - Meliphagidae
e   Brown Honeyeater; Australian
    Brown-Honeyeater
f   Méliphage brunâtre
d   Braunhonigfresser
i   Mangiamiele modesto

5246   *Lichmera limbata*
Passeriformes - Meliphagidae
e   Indonesian Honeyeater
f   Méliphage frangé
i   Mangiamiele indonesiano

5247   *Lichmera lombokia*
Passeriformes - Meliphagidae
e   Scaly-crowned Honeyeater; Lombok
    Honeyeater; Sunda Honeyeater
f   Méliphage de Lombok
d   Lombok-Honigfresser
i   Mangiamiele di Lombok

5248   *Lichmera monticola*
Passeriformes - Meliphagidae
e   Ceram Mountain-Honeyeater;
    Ceram Honeyeater; Speckled
    Honeyeater; Moluccan Honeyeater
f   Méliphage de Ceram
d   Ceram-Honigfresser
i   Mangiamiele di Seram

5249   *Lichmera notabilis*
Passeriformes - Meliphagidae
e   Black-chested Honeyeater; Wetar

Honeyeater; Finsch's Honeyeater;
Black-necklaced Honeyeater
f   Méliphage de Wetar
d   Finsch-Honigfresser
i   Mangiamiele pettonero

5250   *Lichmera squamata*
Passeriformes - Meliphagidae
e   White-tufted Honeyeater; Scaled
Honeyeater; Tanimbar Honeyeater;
Banda Sea Honeyeater; Mottle-
breasted Honeyeater; Scaly
Honeyeater; Scaly-breasted
Honeyeater
f   Méliphage à plumet blanc
d   Salvadori-Honigfresser
i   Mangiamiele squamoso

5251   *Limicola falcinellus*
Charadriiformes - Scolopacidae
e   Broad-billed Sandpiper
f   Bécasseau falcinelle
d   Sumpfläufer
i   Gambecchio frullino

5252   *Limnodromus griseus*
Charadriiformes - Scolopacidae
e   Short-billed Dowitcher; Pondbird
(WI); Common Dowitcher;
Dowitcher; American Dowitcher
f   Bécassin roux; Bécasseau (Ants);
Limnodrome à bec court
d   Kleiner Schlammläufer;
Kurzschnabelschlammläufer
i   Piro-piro pettorossiccio minore;
Limnodromo beccocorto

5253   *Limnodromus scolopaceus*
Charadriiformes - Scolopacidae
e   Long-billed Dowitcher
f   Limnodrome à long bec; Bécasseau à
long bec; Bécassin à long bec
d   Langschnabelschlammläufer; Großer
Schlammläufer
i   Piro-piro pettorossiccio;
Limnodromo beccolungo

5254   *Limnodromus semipalmatus*
Charadriiformes - Scolopacidae
e   Asian Dowitcher; Asiatic Dowitcher;
Oriental Dowitcher; Stripe-billed

Dowitcher; Snipe-billed Godwit
f   Bécassin d'Asie; Limnodrome
semipalmé
d   Steppenschlammläufer
i   Limnodromo semipalmato

5255   *Limnornis curvirostris*
Passeriformes - Furnariidae
e   Curve-billed Reedhaunter
f   Synallaxe à bec courbé; Synallaxe
des roseaux
d   Krummschnabelriedschlüpfer
i   Cannaiola beccocurvo

5256   *Limnornis rectirostris*
Passeriformes - Furnariidae
e   Straight-billed Reedhaunter
f   Synallaxe à bec droit; Synallaxe
rousserole
d   Geradschnabelriedschlüpfer
i   Cannaiola beccodiritto

5257   *Limnothlypis swainsonii*
Passeriformes - Fringillidae
e   Swainson's Warbler
f   Paruline de Swainson
d   Swainson-Waldsänger
i   Parula cannaiola di Swainson

5258   *Limosa fedoa*
Charadriiformes - Scolopacidae
e   Marbled Godwit; Black-tailed
Godwit
f   Barge marbrée
d   Marmorschnepfe
i   Pittima marmoreggiata

5259   *Limosa haemastica*
Charadriiformes - Scolopacidae
e   Hudsonian Godwit; American Black-
tailed Godwit
f   Barge hudsonienne
d   Amerikanische Uferschnepfe;
Hudson-Schnepfe
i   Pittima di Hudson

5260   *Limosa lapponica*
Charadriiformes - Scolopacidae
e   Bar-tailed Godwit; Bar-rumped
Godwit; Pacific Ocean Godwit;
Southern Godwit; Small Godwit

f Barge rousse
d Pfuhlschnepfe; Rostrote
  Pfuhlschnepfe
i Pittima minore

5261 *Limosa limosa*
Charadriiformes - Scolopacidae
e Black-tailed Godwit; Large Godwit
f Barge à queue noire
d Uferschnepfe; Schwarzschänzige
  Uferschnepfe
i Pittima reale; Moschettone

5262 *Linurgus olivaceus*
Passeriformes - Fringillidae
e Oriole Finch
f Linurge loriot; Pinson-loriot
d Pirolgimpel
i Lucherino testanera africano

5263 *Liocichla omeiensis*
Passeriformes - Sylviidae
e Omei Shan Liocichla; Omei
  Liocichla; Mount Omei Shan
  Liocichla; Mount Omei Liocichla;
  Mount Omei Laughingthrush;
  Szechwan Liocichla; Sichaun
  Liocichla; Szechuan Liocichla
f Garrulaxe de l'Omei
d Omei-Häherling
i Liocicla del Monte Omei

5264 *Liocichla phoenicea*
Passeriformes - Sylviidae
e Red-faced Liocichla; Crimson-
  winged Liocichla; Red-faced
  Laughingthrush; Crimson-winged
  Laughingthrush; Crimson-headed
  Liocichla
f Garrulaxe à ailes rouges; Garrulaxe à
  face rouge
d Karminflügelhäherling
i Liocicla facciarossa

5265 *Liocichla steerii*
Passeriformes - Sylviidae
e Steere's Liocichla; Steere's Babbler
f Garrulaxe de Steere
d Formosa-Häherling
i Liocicla di Steere

5266 *Lioptilus nigricapillus*
Passeriformes - Sylviidae
e Blackcap Mountain-Babbler; Bush
  Blackcap (CSA); Blackcap Babbler;
  Black-capped Flycatcher-Babbler
f Lioptile à calotte noire
d Buschschwarzkäppchen
i Garrulo capinero di boscaglia

5267 *Liosceles thoracicus*
Passeriformes - Rhinocryptidae
e Rusty-belted Tapaculo
f Tourco ceinturé
d Brustbandtapaculo; Waldhornist
i Tapaculo cintura rossiccia

5268 *Lipaugus cryptolophus*
Passeriformes - Tyrannidae
e Olivaceous Piha
f Piauhau olivâtre
d Gelbbauchpiha
i Piha olivacea

5269 *Lipaugus fuscocinereus*
Passeriformes - Tyrannidae
e Dusky Piha
f Piauhau sombre
d Langschwanzpiha
i Piha bruna

5270 *Lipaugus lanioides*
Passeriformes - Tyrannidae
e Cinnamon-vented Piha
f Piauhau à tête grise
d Graukopfpiha
i Piha averla

5271 *Lipaugus streptophorus*
Passeriformes - Tyrannidae
e Rose-collared Piha
f Piauhau à collier
d Halsbandpiha; Olivzuser
i Piha rosata

5272 *Lipaugus subalaris*
Passeriformes - Tyrannidae
e Grey-tailed Piha
f Piauhau à queue grise
d Grauschwanzpiha
i Piha codagrigia

5273   *Lipaugus unirufus*
       Passeriformes - Tyrannidae
 e   Rufous Piha
 f   Piauhau roux
 d   Rostpiha
 i   Piha rossiccia

5274   *Lipaugus uropygialis*
       Passeriformes - Tyrannidae
 e   Scimitar-winged Piha
 f   Piauhau à faucilles
 d   Rotbürzelpiha
 i   Piha alifalcate

5275   *Lipaugus vociferans*
       Passeriformes - Tyrannidae
 e   Screaming Piha; Greenheartbird
 f   Piauhau hurleur
 d   Schreipiha
 i   Piha vocifera

5276   *Lipaugus weberi*
       Passeriformes - Tyrannidae
 e   Chestnut-capped Piha; Scimitar-
       winged Piha; Scimitar-winged
       Cotinga

5277   *Loboparadisea sericea*
       Passeriformes - Corvidae
 e   Yellow-breasted Bird of Paradise;
       Wattle-billed Bird of Paradise
 f   Paradisier soyeux
 d   Lappenparadiesvogel
 i   Paradisea pettogiallo

5278   *Lochmias nematura*
       Passeriformes - Furnariidae
 e   Sharp-tailed Streamcreeper;
       Streamside Lochmias; Streamside
       Streamcreeper; Sharp-tailed
       Lochmias
 f   Picerthie de Saint-Hilaire
 d   Bachstachelschwanz;
       Erdhöhlentöpfer
 i   Grattafoglie codarigida

5279   *Locustella amnicola*
       Passeriformes - Sylviidae
 e   Stepanyan's Warbler; Stepanyan's
       Grasshopper-Warbler; Large
       Warbler; Sakhalin Warbler

 f   Locustelle amnicole
 i   Locustella di Stepanyan

5280   *Locustella certhiola*
       Passeriformes - Sylviidae
 e   Pallas's Grasshopper-Warbler;
       Pallas's Warbler; Rusty-rumped
       Warbler
 f   Locustelle de Pallas
 d   Streifenschwirl
 i   Locustella del Pallas; Locustella di
       Pallas

5281   *Locustella fasciolata*
       Passeriformes - Sylviidae
 e   Gray's Grasshopper-Warbler; Gray's
       Warbler; Large-Grasshopper-Warbler
 f   Locustelle fasciée
 d   Riesenschwirl; Taigaschwirl
 i   Locustella di Gray

5282   *Locustella fluviatilis*
       Passeriformes - Sylviidae
 e   Eurasian River-Warbler (NA); River-
       Warbler
 f   Locustelle fluviatile
 d   Schlagschwirl; Flußschwirl
 i   Salciaiola fluviatile; Locustella di
       fiume; Locustella fluviatile

5283   *Locustella lanceolata*
       Passeriformes - Sylviidae
 e   Lanceolated Warbler; Temminck's
       Warbler; Temminck's Grasshopper-
       Warbler; Lanceolated Grasshopper-
       Warbler; Streaked Warbler; Streaked
       Grasshopper-Warbler
 f   Locustelle lancéolée
 d   Strichelschwirl
 i   Locustella lanceolata

5284   *Locustella luscinioides*
       Passeriformes - Sylviidae
 e   Savi's Warbler
 f   Locustelle luscinioïde
 d   Rohrschwirl; Nachtigallrohrsänger;
       Nachtigallschwirl
 i   Salciaiola

5285   *Locustella naevia*
       Passeriformes - Sylviidae

e Common Grasshopper-Warbler;
Grasshopper Warbler; Pale
Grasshopper-Warbler; Western
Grasshopper-Warbler; European
Grasshopper-Warbler
f Locustelle tachetée
d Feldschwirl; Heuschreckensänger
i Forapaglie macchiettato

5286 *Locustella ochotensis*
Passeriformes - Sylviidae
e Middendorf's Grasshopper-Warbler;
Middendorf's Warbler; Asiatic
Grasshopper-Warbler
f Locustelle de Middendorf
d Middendorf-Schirl
i Locustella di Middendorff

5287 *Locustella pleskei*
Passeriformes - Sylviidae
e Pleske's Grasshopper-Warbler;
Pleske's Warbler; Styan's
Grasshopper-Warbler
f Locustelle de Pleske
i Locustella di Styan

5288 *Loddigesia mirabilis*
Apodiformes - Trochilidae
e Marvelous Spatuletail; Peruvian
Racket-tailed Hummingbird;
Marvellous Hummingbird; Peruvian
Racquet-tailed Hummingbird
f Loddigesie admirable
d Wundersylphe
i Colibrì mirabile; Colibrì dalla coda a
racchetta

5289 *Lonchura atricapilla*
Passeriformes - Estrildidae
e Southern Black-headed Munia;
Chestnut Munia; Chestnut Mannikin;
Chestnut Nun; Black-headed Munia;
Black-headed Mannikin; Black-
headed Nun
i Cappuccino

5290 *Lonchura bicolor*
Passeriformes - Estrildidae
e Black-and-white Mannikin (CSA);
Blue-billed Mannikin; Two-coloured
Mannikin; Black-and-white Munia;

Bicoloured Mannikin; Fernando Po
Munia; Fernando Po Mannikin
f Capucin bicolore; Spermète bicolore;
Spermète à bec bleu
d Braunrückenelsterchen;
Graukopfnonne
i Cappuccino bicolore

5291 *Lonchura caniceps*
Passeriformes - Estrildidae
e Grey-headed Munia; Grey-headed
Silverbill; Grey-headed Mannikin
f Capucin gris; Nonne à tète grise;
Spermète à tête grise
d Graukopfnonne
i Cappuccino testagrigia

5292 *Lonchura cantans*
Passeriformes - Estrildidae
e African Silverbill; Black-rumped
Silverbill; Warbling Silverbill;
Silverbill
f Capucin bec-d'argent; Bec d'argent;
Spermète bec d'argent
d Silberschnabelchen
i Becco d'argento africano

5293 *Lonchura castaneothorax*
Passeriformes - Estrildidae
e Chestnut-breasted Munia; Chestnut-
breasted Mannikin (ANZ); Bullybird;
Chestnut Finch
f Capucin donacole; Donacole;
Donacole commun; Tisserin à
poitrine châtaigne; Spermète
donacole
d Braunbrustschilffink
i Cappuccino pettocastano

5294 *Lonchura cucullata*
Passeriformes - Estrildidae
e Bronze Mannikin; Bronze Munia;
Hooded Weaver; Fret (CSA);
Bronze-winged Mannikin;
Bronzewing; Hooded Weaver Finch;
Hooded Mannikin
f Capucin nonnette; Nonne; Nonne
ordinaire; Spermète à capuchon;
Spermète bronzé; Spermète nonnette
d Kleinelsterchen
i Cappuccino bronzato

5295   *Lonchura ferruginosa*
Passeriformes - Estrildidae
e   White-capped Munia; Java
Mannikin; Java Munia; Chestnut
Munia; Chestnut Mannikin; Black-
throated Munia
f   Capucin marron; Nonnette à poitrine
noire; Sspermète à poitrine noire
d   Schwarzkehlnonne
i   Cappuccino castano

5296   *Lonchura flaviprymna*
Passeriformes - Estrildidae
e   Yellow-rumped Munia; Yellow-
rumped Mannikin (ANZ); Yellow-
tailed Finch; Yellow-rumped Finch;
Yellow-tailed Mannikin
f   Capucin à croupion jaune; Donacole
à poitrine jaune; Donacole à tête
grise
d   Gelbbrustschilffink; Gelber
Schilffink; Gelbbauchschilffink
i   Cappuccino codagialla

5297   *Lonchura forbesi*
Passeriformes - Estrildidae
e   New Ireland Munia; New Ireland
Mannikin; Forbes's Mannikin; New
Ireland Finch; Forbes's Munia; Buff-
breasted Mannikin
f   Capucin de Nouvelle-lrlande; Nonne
de Forbes
d   Forbes-Nonne
i   Cappuccino della Nuova Irlanda

5298   *Lonchura fringilloides*
Passeriformes - Estrildidae
e   Magpie Mannikin (CSA); Magpie
Munia; Pied Mannikin (CSA); Giant
Mannikin; Pied Grassfinch; Pied
Weaver Finch
f   Capucin pie; Spermète pie; Grande
Nonne
d   Riesenelsterchen
i   Cappuccino maggiore

5299   *Lonchura fuscans*
Passeriformes - Estrildidae
e   Dusky Munia; Dusky Mannikin;
Borneo Mannikin; Borneo Munia;
Black Mannikin; Black Borneo

Mannikin
f   Capucin sombre; Munie de Bornéo
d   Borneo-Bronzemännchen
i   Cappuccino bruno

5300   *Lonchura grandis*
Passeriformes - Estrildidae
e   Grand Munia; Great-billed Mannikin;
Great-billed Munia; Great Munia;
Great Mannikin; Grand Mannikin
f   Grand Capucin; Nonne à gros bec
d   Dickschnabelnonne
i   Cappuccino beccogrosso

5301   *Lonchura griseicapilla*
Passeriformes - Estrildidae
e   Grey-headed Silverbill (CSA); Pearl-
headed Silverbill; Grey-headed
Mannikin; Pearl-headed Mannikin;
Pearl-winged Mannikin; Pearl-
shouldered Mannikin
f   Capucin à tête grise; Spermète à tête
grise
d   Perlhalsamadine
i   Becco d'argento testagrigia

5302   *Lonchura hunsteini*
Passeriformes - Estrildidae
e   Mottled Munia; Hunstein's
Mannikin; Black-breasted Mannikin;
Hunstein's Munia; White-headed
Finch; Black-breasted Munia; White-
headed Mannikin; Black-breasted
Weaver Finch
f   Capucin de Hunstein; Nonne de
Hunstein
d   Hunstein-Nonne
i   Cappuccino di Huntstein

5303   *Lonchura kelaarti*
Passeriformes - Estrildidae
e   Black-throated Munia; Rufous-
bellied Munia (ISC); Black-throated
Mannikin; Ceylon Hill-Munia;
Jerdon's Mannikin; Hill Munia;
Rufous-breasted Munia
f   Capucin à ventre roux; Capucin des
montagnes
d   Bergbronzemännchen; Jerdon-
Bronzemännchen
i   Cappuccino ventrerosso

5304 *Lonchura leucogastra*
Passeriformes - Estrildidae
e White-bellied Munia; White-breasted
Mannikin; White-breasted Mannikin;
White-breasted Munia
f Capucin à ventre blanc
d Weißbauchbronzemännchen
i Cappuccino ventrebianco

5305 *Lonchura leucogastroides*
Passeriformes - Estrildidae
e Javan Munia; Java Munia; Javan
Mannikin; Javanese Mannikin; Java
White-bellied Munia; Javanese
Munia; Javanese White-bellied
Munia; Black-rumped Munia; Black-
beaked Bronze-Mannikin
f Capucin javanais; Capucin de Java
d Java-Bronzemännchen;
Schwarzbürzelbronzemännchen
i Cappuccino di Giava

5306 *Lonchura leucosticta*
Passeriformes - Estrildidae
e White-spotted Munia; White-spotted
Mannikin
f Capucin tacheté; Capucin à poitrine
écaillée
d Schuppenbrustbronzemännchen;
Perlenbronzemännchen
i Cappuccino maculato

5307 *Lonchura maja*
Passeriformes - Estrildidae
e White-headed Munia; White-headed
Mannikin; Pale-headed Mannikin;
Maya Munia; Cigarbird
f Capucin à tête blanche; Nonnette à
tête blanche
d Weißkopfnonne
i Cappuccino testabianca

5308 *Lonchura malabarica*
Passeriformes - Estrildidae
e White-throated Silverbill; Warbling
Silverbill; Silverbill; White-throated
Munia (ISC); Indian Silverbill;
Common Silverbill; White-rumped
Munia; White-rumped Silverbill
f Capucin bec-de-plomb; Munie à
gorge blanche; Bec-de-plomb; Bec

d'argent
d Malabar-Fasänchen;
Silberschnäbelchen
i Becco d'argento indiano

5309 *Lonchura malacca*
Passeriformes - Estrildidae
e Indian Black-headed Munia;
Chestnut Mannikin; Black-headed
Munia; Black-headed Mannikin;
Chestnut Munia; Black-headed Nun;
Indian Silverbill (WI); Three-
coloured Mannikin; Tricoloured
Munia; Tricoloured Mannikin;
Tricoloured Nun
f Capucin à dos marron; Capucin à tête
noire; Capucin tricolore; Capucin de
l'Inde
d Schwarzkopfnonne;
Schwarzbauchnonne;
Dreifarbennonne
i Cappuccino tricolore

5310 *Lonchura melaena*
Passeriformes - Estrildidae
e Bismarck Munia; Thick-billed
Munia; New Britain Mannikin; New
Britain Finch; Thick-billed
Mannikin; Buff-bellied Mannikin;
Buff-bellied Black-Mannikin
f Capucin de Nouvelle-Bretagne;
Donacole à grosse tête
d Dickkopfschilffink; Dickkopfnonne
i Cappuccino della Nuova Britannia

5311 *Lonchura molucca*
Passeriformes - Estrildidae
e Black-faced Munia; Moluccan
Mannikin; Moluccan Munia
f Capucin jacobin; Capucin des
Moluques
d Wellenbauchbronzemännchen
i Cappuccino delle Molucche

5312 *Lonchura montana*
Passeriformes - Estrildidae
e Snow Mountain Munia; Snow
Mountain Mannikin; Western
Alpine-Mannikin
f Capucin des Snow; Donacole des
hauteurs

d Junge-Schilffink; Höhenschilffink
i Cappuccino dei Monti Nevosi

5313 *Lonchura monticola*
Passeriformes - Estrildidae
e Alpine Munia; Alpine Manikin;
Eastern Alpine-Munia; Eastern
Alpine-Mannikin
f Capucin des montagnes; Donacole
des montagnes
d Bergschilffink
i Cappuccino dei Monti Orientali

5314 *Lonchura nevermanni*
Passeriformes - Estrildidae
e Grey-crowned Munia; White-
crowned Mannikin; White-crowned
Munia; Grey-crowned Mannikin
f Capucin de Nevermann; Nonne à
calotte blanche
d Neevermanns Nonne;
Weißscheitelnonne
i Cappuccino di Nevermann

5315 *Lonchura nigerrima*
Passeriformes - Estrildidae
e New Hanover Munia; New Hanover
Mannikin
f Capucin de Nouvelle-Hanovre;
Nonne de James
d Mohrennonne; Schwarze Nonne
i Cappuccino dell'Isola New Hanover

5316 *Lonchura nigriceps*
Passeriformes - Estrildidae
e Brown-backed Munia; Rufous-
backed Munia; Chestnut-backed
Munia; Red-backed Mannikin
f Capucin à dos brun
i Cappuccino dorsobruno

5317 *Lonchura pallida*
Passeriformes - Estrildidae
e Pale-headed Munia; Pallid Munia;
Celebes Munia; Pale Sunda Munia;
Pallid Mannikin; Pallid Finch; Pale-
headed Nun; Pale-headed Mannikin;
Pale Munia; Sunda Munia; Celebes
Munia
f Capucin pâle; Nonne à tête claire

d Blaßkopfnonne; Gelbbauchnonne
i Cappuccino pallido

5318 *Lonchura punctulata*
Passeriformes - Estrildidae
e Scaly-breasted Munia; Nutmeg
Mannikin; Spice Finch (WI); Spotted
Munia (ISC); Barred Munia; Scaly-
breasted Mannikin; Scaly-breasted
Finch; Nutmeg Finch; Spicebird;
Spice Mannikin; Spotted Mannikin
f Capucin damier; Capucin ponctué
(Ants); Capucin damier muscade;
Damier; Damier commun; Spermète
damier
d Muskatamadine; Muskatfink
i Domino

5319 *Lonchura quinticolor*
Passeriformes - Estrildidae
e Five-colored Munia (NA); Coloured
Finch; Five-coloured Mannikin;
Lesser Sunda Mannikin; Chestnut-
and-white Mannikin; Chestnut-and-
white Munia; Lesser Sunda Munia
f Capucin coloré; Nonne à cinque
couleurs; Nonne quinticolore
d Fünffarbennonne
i Cappuccino dai cinque colore

5320 *Lonchura spectabilis*
Passeriformes - Estrildidae
e Hooded Munia; New Britain
Mannikin; New Guinea Mannikin;
Sclater's Mannikin; Mayr's Munia;
New Britain Munia; Hooded
Mannikin
f Capucin à capuchon; Nonnette à
ventre roux
d Prachtnonne
i Cappuccino dal cappuccio

5321 *Lonchura striata*
Passeriformes - Estrildidae
e White-rumped Munia; White-backed
Munia (ISC); Sharp-tailed Mannikin;
White-rumped Mannikin; Striated
Munia; Striated Mannikin; Striated
Finch; Hodgson's Munia; Sharp-
tailed Munia; Society Finch;
Bengalese Finch

f Capucin domino; Domino à longue queue
d Spitzschwanzbronzemännchen; Weißbürzelbronzemännchen
i Cappuccino gropponebianco

5322 *Lonchura stygia*
Passeriformes - Estrildidae
e Black Munia; Black Mannikin
f Capucin noir; Nonnette noire
d Hadesschilffink; Hadesnonne
i Cappuccino nero

5323 *Lonchura teerinki*
Passeriformes - Estrildidae
e Black-breasted Munia; Grand Valley Mannikin; Teerink's Mannikin; Grand Valley Munia; Teerink's Munia; Black-breasted Mannikin
f Capucin à poitrine noire; Donacole à poitrine noire
d Schwarzbrustschilffink
i Cappuccino pettonero

5324 *Lonchura tristissima*
Passeriformes - Estrildidae
e Streak-headed Munia; Streaked-headed Munia; Streak-headed Mannikin; Streaked-headed Mannikin
f Capucin à tête rayée; Capucin triste
d Trauerbronzemännchen
i Cappuccino striata

5325 *Lonchura vana*
Passeriformes - Estrildidae
e Grey-banded Munia; Arfak Mannikin; Arfak Munia; Grey-banded Mannikin
f Capucin des Arfak; Nonnette des Arfak
d Arfak-Nonne; Weißwangennonne
i Cappuccino dell'Arfak

5326 *Lophaetus occipitalis*
Falconiformes - Accipitridae
e Long-crested Eagle; Long-crested Hawk-Eagle
f Aigle huppard
d Schopfadler

i Aquila dal ciuffo; Aquila dal lungo ciuffo

5327 *Lophodytes cucullatus*
Anseriformes - Anatidae
e Hooded Merganser
f Harle couronné; Bec-scie couronnée
d Kappensäger
i Smergo dal ciuffo; Smergo americano

5328 *Lophoictinia isura*
Falconiformes - Accipitridae
e Square-tailed Kite; Long-winged Kite
f Milan à queue carrée
d Schopfmilan
i Nibbio codasquadrata; Nibbio dalla coda quadra

5329 *Lopholaimus antarcticus*
Columbiformes - Columbidae
e Topknot Pigeon; Topknot; Flock Pigeon; Flock Fruit-Pigeon; Crested Fruit-Pigeon; Chatham Islands Pigeon; Native Pigeon; Wood-Pigeon
f Carpophage à double huppe
d Haubenfruchttaube
i Piccione australe dal ciuffo

5330 *Lophophorus impejanus*
Galliformes - Phasianidae
e Himalayan Monal; Impeyan Monal; Himalayan Monal-Pheasant; Impeyan Monal-Pheasant
f Lophophore resplendissant
d Gelbschwanzglanzfasan
i Lofoforo splendente; Lofoforo dell'Himalaya

5331 *Lophophorus lhuysii*
Galliformes - Phasianidae
e Chinese Monal; Chinese Monal-Pheasant
f Lophophore de Lhuys
d Grünschwanzglanzfasan
i Lofoforo cinese; Lopoforo cinese

5332 *Lophophorus sclateri*
Galliformes - Phasianidae
e Sclater's Monal; Sclater's Monal-

Pheasant; Crestless Pheasant
f Lophophore de Sclater
d Weißchwanzglanzfasan
i Lofoforo di Sclater; Lopoforo di
     Sclater

5333 ***Lophorina superba***
      Passeriformes - Corvidae
e Superb Bird of Paradise
f Paradisier superbe
d Kragenhopf
i Paradisea superba; Lofoforo superba

5334 ***Lophornis adorabilis***
      Apodiformes - Trochilidac
e White-crested Coquette; Adorable
     Coquette
f Coquette adorable
d Weißschopfelfe
i Colibrì dai ciuffi crestabianca

5335 ***Lophornis brachylopha***
      Apodiformes - Trochilidae
e Short-crested Coquette
f Coquette du Guerrero

5336 ***Lophornis chalybeus***
      Apodiformes - Trochilidae
e Festive Coquette
f Coquette chalybée; Coquette de
     Vieillot
d Schmetterlingselfe
i Colibrì dai ciuffi festivo

5337 ***Lophornis delattrei***
      Apodiformcs - Trochilidae
e Rufous-crested Coquette
f Coquette de Delattre
d Rotschopfelfe
i Colibrì dai ciuffi rossiccio

5338 ***Lophornis gouldii***
      Apodiformes - Trochilidae
e Dot-eared Coquette; Gould's
     Coquette
f Coquette de Gould
d Gould-Elfe
i Colibrì dai ciuffi di Gould

5339 ***Lophornis helenae***
      Apodiformes - Trochilidae

e Black-crested Coquette; Princess
     Helena's Coquette
f Coquette d'Hélène
d Schwarzschopfelfe
i Colibrì dai ciuffi crestanera; Colibrì
     dall cresta nera

5340 ***Lophornis magnificus***
      Apodiformes - Trochilidae
e Frilled Coquette
f Coquette magnifique
d Prachtelfe
i Colibrì dai ciuffi magnifico

5341 ***Lophornis ornatus***
      Apodiformes - Trochilidae
e Tufted Coquette; Splendid Coquette
f Coquette huppe-col
d Schmuckelfe
i Colibrì dai ciuffi ornato

5342 ***Lophornis pavoninus***
      Apodiformes - Trochilidae
e Peacock Coquette
f Coquette paon
d Pfauenelfe
i Colibrì dai ciuffi

5343 ***Lophornis stictolophus***
      Apodiformes - Trochilidae
e Spangled Coquette
f Coquette pailletée
d Glanzelfe
i Colibrì dai ciuffi stellato

5344 ***Lophospingus griseocristatus***
      Passeriformes - Fringillidae
e Grey-crested Finch
f Lophospingue gris
d Grauhaubenzwergkardinal
i Fringuello crestagrigia

5345 ***Lophospingus pusillus***
      Passeriformes - Fringillidae
e Black-crested Finch
f Lophospingue à huppe noire
d Schwarzhaubenzwergkardinal
i Fringuello crestanera

5346 ***Lophostrix cristata***
      Strigiformes - Strigidae

e   Crested Owl
f   Duc à aigrettes
d   Haubenkauz
i   Guffo dalla lunga cresta; Assio
    crestato

5347   *Lophotibis cristata*
    Ciconiiformes - Threskiornithidae
e   White-winged Ibis; Crested Wood-
    Ibis; Madagascar Ibis; Madagascar
    Crested-Ibis
f   Ibis huppé
d   Schopfibis
i   Ibis crestato

5348   *Lophotriccus eulophotes*
    Passeriformes - Tyrannidae
e   Long-crested Pygmy-Tyrant; Long-
    crested Tyrant; Todd's Tyrant
f   Microtyran eulophe; Tyranneau
    eulophe
d   Todd-Tyrann
i   Tiranno pigmeo crestalunga

5349   *Lophotriccus galeatus*
    Passeriformes - Tyrannidae
e   Helmeted Pygmy-Tyrant
f   Microtyran casqué; Tyranneau
    casqué
d   Helmtyranm
i   Tiranno pigmeo dall'elmo

5350   *Lophotriccus pileatus*
    Passeriformes - Tyrannidae
e   Scale-crested Pygmy-Tyrant; Scale-
    crested Tyrant
f   Microtyran chevelu; Tyranneau
    chevelu
d   Schuppenkopftyrann
i   Tiranno pigmeo crestascagliosa

5351   *Lophotriccus vitiosus*
    Passeriformes - Tyrannidae
e   Double-banded Pygmy-Tyrant;
    Double-banded Tyrant
f   Microtyran bifascié; Tyranneau
    coiffé
d   Ährenschopftyrann
i   Tiranno pigmeo bifasciato

5352   *Lophozosterops dohertyi*
    Passeriformes - Zosteropidae
e   Crested White-eye; Doherty's White-
    eye; Crested Darkeye; Dark-crowned
    White-eye
f   Zostérops de Doherty
d   Schopfbrillenvogel;
    Haubenbrillenvogel
i   Occhialino crestato

5353   *Lophozosterops goodfellowi*
    Passeriformes - Zosteropidae
e   Black-masked White-eye;
    Goodfellow's White-eye; Apo White-
    eye; Mindanao White-eye
f   Zostérops de Goodfellow
d   Mindanao-Brillenvogel
i   Occhialino di Goodfellow

5354   *Lophozosterops javanicus*
    Passeriformes - Zosteropidae
e   Javan Grey-throated White-eye;
    Grey-throated White-eye; Grey-
    throated Darkeye; Javan White-eye
f   Zostérops javanais
d   Java-Brillenvogel
i   Occhialino golagrigia di Giava

5355   *Lophozosterops pinaiae*
    Passeriformes - Zosteropidae
e   Grey-hooded White-eye; Ceram
    White-eye; Brown-breasted White-
    eye; Pinaia White-eye; Grey-hooded
    Dark-eye
f   Zostérops à froc gris
d   Pinaia-Brillenvogel
i   Occhialino dal cappuccio grigio

5356   *Lophozosterops squamiceps*
    Passeriformes - Zosteropidae
e   Streaky-headed White-eye; Pygmy
    Grey White-eye; Celebes Mountain-
    White-eye; Streak-headed White-eye;
    Streak-headed Darkeye; Celebes
    Grey-throated White-eye
f   Zostérops à tête rayée
d   Schuppenkopfbrillenvogel
i   Occhialino testagrigia

5357   *Lophozosterops superciliaris*
    Passeriformes - Zosteropidae

e Yellow-browed White-eye; White-
browed White-eye; Sunda White-eye;
Lesser Sunda Mountain White-eye;
Yellow-browed Darkeye; Eyebrowed
White-eye
f Zostérops à sourcils
d Gelbbrauenbrillenvogel
i Occhialino dai sopraccigli bianchi

5358 *Lophura bulweri*
Galliformes - Phasianidae
e Bulwer's Pheasant; Wattled Pheasant;
White-tailed Pheasant
f Faisan de Bulwer
d Bulwer-Fasan
i Fagiano di Bulwer

5359 *Lophura diardi*
Galliformes - Phasianidae
e Siamese Fireback; Siamese Fireback-
Pheasant
f Faisan prélat
d Prälatfasan
i Fagiano siamese; Fagiano prelato

5360 *Lophura edwardsi*
Galliformes - Phasianidae
e Edwards's Pheasant; Annam Pheasant
f Faisan d'Edwards
d Edwards-Fasan
i Fagiano di Edwards

5361 *Lophura erythrophthalma*
Galliformes - Phasianidae
e Crestless Fireback; Crestless
Fireback-Pheasant
f Faisan à queue rousse
d Gabelschwanzfasan
i Fagiano occhirossi

5362 *Lophura hatinhensis*
Galliformes - Phasianidae
e Vietnamese Pheasant; Vietnamese
Fireback; Vo Quy's Pheasant
f Faisan du Vietnam
i Fagiano di Vo Quy

5363 *Lophura hoogerwerfi*
Galliformes - Phasianidae
e Hoogerwerf's Pheasant; Sumatran
Fireback; Sumatran Fireback

Pheasant; Atjeh Pheasant
f Faisan de Sumatra
i Fagiano di Sumatra

5364 *Lophura ignita*
Galliformes - Phasianidae
e Crested Fireback; Crested Fireback-
Pheasant
f Faisan noble
d Feuerrückenfasan
i Fagiano della Malesia

5365 *Lophura imperialis*
Galliformes - Phasianidae
e Imperial Pheasant
f Faisan impérial
d Kaiserfasan
i Fagiano imperiale

5366 *Lophura inornata*
Galliformes - Phasianidae
e Salvadori's Pheasant
f Faisan de Salvadori
d Salvadori-Fasan

5367 *Lophura leucomelanos*
Galliformes - Phasianidae
e Kalij Pheasant; Kalij; Kallej
Pheasant; Horsfield's Pheasant;
Lineated Pheasant
f Faisan leucomèle
d Schwarzfasan; Kalij
i Fagiano di Kalij

5368 *Lophura nycthemera*
Galliformes - Phasianidae
e Silver Pheasant
f Faisan argenté
d Silberfasan
i Fagiano argentato

5369 *Lophura swinhoii*
Galliformes - Phasianidae
e Swinhoe's Pheasant; Taiwan Blue-
Pheasant; Formosan Blue-Pheasant
f Faisan de Swinhoe
d Swinhoe-Fasan
i Fagiano di Swinhoe

5370 *Loriculus amabilis*
Psittaciformes - Psittacidae

e Moluccan Hanging-Parrot
f Coryllis des Moluques
d Zierfledermauspapageichen;
  Zierpapageichen; Lieblicher Lori
i Loricolo delle Molucche

5371 *Loriculus aurantiifrons*
Psittaciformes - Psittacidae
e Orange-fronted Hanging-Parrot;
  Golden-fronted Hanging Parrot; Bat
  Lorikeet; Papuan Hanginging-Parrot
f Coryllis à front orangé; Loricule à
  tête d'or
d Goldstirnpapageichen
i Loricolo frontedorata

5372 *Loriculus beryllinus*
Psittaciformes - Psittacidae
e Sri Lanka Hanging-Parrot; Ceylon
  Hanging-Parrot; Sinhalese Hanging-
  Parrot; Ceylon Hanging-Parakeet;
  Ceylon Lorikeet
f Coryllis de Ceylan; Loricule de
  Ceylan
d Blumenpapageichen; Ceylon-
  Papageichen
i Loricolo di Sri-Lanka

5373 *Loriculus catamene*
Psittaciformes - Psittacidae
e Sangihe Hanging-Parrot
f Coryllis des Sangi
i Loricolo di Sangihe

5374 *Loriculus exilis*
Psittaciformes - Psittacidae
e Red-billed Hanging-Parrot; Pygmy
  Hanging-Parrot; Green Hanging-
  Parrot; Green Hanging-Parakeet
f Coryllis vert
d Däumlingspapageichen; Celebes-
  Fledermauspapageichen
i Loricolo beccorosso

5375 *Loriculus flosculus*
Psittaciformes - Psittacidae
e Wallace's Hanging-Parrot; Flores
  Hanging-Parakeet
f Coryllis de Wallace
d Blütenpapageichen; Flores-

Papageichen
i Loricolo di Wallace

5376 *Loriculus galgulus*
Psittaciformes - Psittacidae
e Blue-crowned Hanging-Parrot;
  Malay Hanging Lorkeet; Malay
  Lorikeet; Malaysian Hanging
  Lorikeet; Blue-crowned Hanging
  Lorikeet
f Coryllis à tête bleue
d Blaukrönchen
i Loricolo corona blu; Corilli dalla
  gola rossa

5377 *Loriculus philippensis*
Psittaciformes - Psittacidae
e Colasisi; Philippine Hanging-Parrot;
  Golden-backed Hanging-Parrot;
  Luzon Hanging-Parrot; Philippine
  Hanging-Parakeet; Colazizi
f Coryllis des Philippines
d Philippinen-Papageichen;
  Philippinen-Fledermauspapageichen;
  Goldrückenfledermauspapagei
i Loricolo delle Filippine

5378 *Loriculus pusillus*
Psittaciformes - Psittacidae
e Yellow-throated Hanging-Parrot;
  Javan Hanging-Parrot; Little
  Hanging-Parrot; Javanese Hanging
  Parakeet
f Coryllis à gorge jaune
d Elfenpapageichen
i Loricolo golagrigia

5379 *Loriculus stigmatus*
Psittaciformes - Psittacidae
e Sulawesi Hanging-Parrot; Celebes
  Hanging-Parrot; Red-crowned
  Hanging-Parrot; Large Sulawesi
  Hanging-Parrot; Black-bellied
  Hanging-Parrot
f Coryllis des Célèbes
d Rotplättchen
i Loricolo di Sulawesi

5380 *Loriculus tener*
Psittaciformes - Psittacidae
e Green-fronted Hanging-Parrot

f Coryllis des Bismarck
d Grünstirnpapagei
i Loricolo fronteverde

5381 *Loriculus vernalis*
Psittaciformes - Psittacidae
e Vernal Hanging-Parrot; Lorikeet
(ISC); Vernal Hanging-Parakeet;
Indian Lorikeet; Hanging Lorikeet
f Coryllis vernal
d Frühlingspapageichen
i Loricolo primaverile

5382 Loriidae
Psittaciformes
e Lories; Lorikeets
f Loriidés
d Loris

5383 *Lorius albidinuchus*
Psittaciformes - Loriidae
e White-naped Lory
f Lori à nuque blanche
d Weißnackenlori
i Lori nucabianca

5384 *Lorius amabilis*
Psittaciformes - Loriidae
e Stresemann's Lory; Halmahera
Hanging-Parrot
d Zierfledermauspapagei

5385 *Lorius chlorocercus*
Psittaciformes - Loriidae
e Yellow-bibbed Lory; Green-tailed
Lory
f Lori à collier jaune
d Grünschwanzlori
i Lori dal collare giallo

5386 *Lorius domicella*
Psittaciformes - Loriidae
e Purple-naped Lory
f Lori des dames
d Erzlori; Schwarzkappenlori
i Lori nucaviola

5387 *Lorius domicella tibialis*
Psittaciformes - Loriidae
e Blue-thighed Lory

5388 *Lorius garrulus*
Psittaciformes - Loriidae
e Chattering Lory; Yellow-backed
Lory
f Lori noira
d Prachtlori; Gelbmantellori
i Lori garrulo

5389 *Lorius hypoinochrous*
Psittaciformes - Loriidae
e Purple-bellied Lory; Louisiade Lory;
Louisiades Lory
f Lori à ventre violet
d Louisiaden-Lori; Schwarzsteißlori
i Lori ventreviola

5390 *Lorius lory*
Psittaciformes - Loriidae
e Black-capped Lory; Western Black-
capped Lory; Tricoloured Lory
f Lori tricolore; Lori à calotte noire
d Frauenlori
i Lori testanera

5391 *Loxia curvirostra*
Passeriformes - Fringillidae
e Red Crossbill; Crossbill; Common
Crossbill
f Bec-croisé des sapins; Bec-croisé
rouge; Bec-croisé commun; Bec-
croisé ordinaire; Bec-croisé
d Fichtenkreuzschnabel;
Kreuzschnabel
i Crociere; Crociero; Becco storto;
Crocione; Becc'a forbice

5392 *Loxia leucoptera*
Passeriformes - Fringillidae
e White-winged Crossbill; Two-barred
Crossbill
f Bec-croisé bifascié; Bec-croisé à
ailes blanches; Bec-croisé (Ants);
Grosbec (Ants)
d Bindenkreuzschnabel
i Crociere fasciato

5393 *Loxia pytyopsittacus*
Passeriformes - Fringillidae
e Parrot Crossbill
f Bec-croisé perroquet; Bec-croisé des
pins

d Kiefernkreuzschnabel
i Crociere delle pinete

5394 *Loxia scotica*
Passeriformes - Fringillidae
e Scottish Crossbill
f Bec-croisé d'Écosse
d Schottischer Kreuzschnabel;
Schottenkreuzschnabel
i Crociere di Scozia; Crociere scozzese

5395 *Loxigilla noctis*
Passeriformes - Fringillidae
e Lesser Antillean Bullfinch; West
Indian Robin (WI); Red-throat See-
see (WI); Robin (WI); Sparrow (WI);
Cheecheebird (WI)
f Grosbec rouge-gorge (Ants);
Sporophile rougegorge
d Bartgimpelfink
i Ciuffolotto delle Piccole Antille

5396 *Loxigilla portoricensis*
Passeriformes - Fringillidae
e Puerto Rican Bullfinch; Mountain
Blacksmith (WI)
f Sporophile de Porto Rico
d Rotkopfgimpelfink
i Ciuffolotto di Portorico

5397 *Loxigilla violacea*
Passeriformes - Fringillidae
e Greater Antillean Bullfinch; Black
Sparrow (WI); Jack Sparrow (WI);
Cotton-tree Sparrow (WI)
f Petit-coq (Ants); Père-noir (Ants);
Grosbec père-noir; Sporophile petit-
coq; Petit-coq
d Rotsteißgimpelfink
i Ciuffolotto delle Grande Antille

5398 *Loxioides bailleui*
Passeriformes - Fringillidae
e Palila
f Psittirostre palila
d Safranpapageischnäbler
i Palila

5399 *Loxipasser anoxanthus*
Passeriformes - Fringillidae
e Yellow-shouldered Grassquit;

Yellow-backed Finch; Yellow-back
(WI); Yellow-backed Grassbird
(WI); Yellow-shouldered Finch
(WI); Yellow-backed Grassquit
f Sporophile mantelé
d Goldbuggimpelfink
i Fringuello spallegialle

5400 *Loxops caeruleirostris*
Passeriformes - Fringillidae
e Akekee; Kauai Akepa
f Loxopse de Kauai
d Akekee; Kauai Akekee
i Akekee

5401 *Loxops coccineus*
Passeriformes - Fringillidae
e Akepa; Common Akepa; Hawaii
Akepa
f Loxopse des Hawai
d Akepa; Kreuzschnabelkleidervogel
i Akepa

5402 *Loxops ochraceus*
Passeriformes - Fringillidae
e Maui Akepa

5403 *Loxops wolstenholmei*
Passeriformes - Fringillidae
e Oahu Akepa

5404 *Lugensa brevirostris*
Procellariiformes - Procellariidae
e Kerguelen Petrel; Little Black-Petrel;
Short-billed Petrel
f Pétrel de Kerguélen
d Kerguelen-Sturmvogel
i Petrello delle Isole Kerguelen

5405 *Lullula arborea*
Passeriformes - Alaudidae
e Wood Lark
f Lulu des bois; Petite Alouette
d Heidelerche
i Tottavilla; Tottovilla; Allodola
mattolina; Mattolina; Covillelo

5406 *Lurocalis nattereri*
Caprimulgiformes - Caprimulgidae
e Chestnut-banded Nighthawk

5407   *Lurocalis rufiventris*
Caprimulgiformes - Caprimulgidae
e   Rufous-bellied Nighthawk;
Taczanowski's Nightjar

5408   *Lurocalis semitorquatus*
Caprimulgiformes - Caprimulgidae
e   Short-tailed Nighthawk;
Semicollared Nighthawk
f   Engoulevent à queue courte
d   Bändernachtschwalbe
i   Succiacapre sparviero dal semicollare

5409   *Luscinia akahige*
Passeriformes - Muscicapidae
e   Japanese Robin
f   Rossignol akahigé; Rouge-gorge
japonais
d   Rostkehlnachtigall; Japanisches
Rotkehlchen
i   Pettirosso giapponese

5410   *Luscinia brunnea*
Passeriformes - Muscicapidae
e   Indian Blue-Robin; Blue Chat (ISC);
Indian Blue-Chat
f   Rossignol indien; Rossignol bleu de
l'Inde
d   Orangenachtigall
i   Usignolo azzurro indiano

5411   *Luscinia calliope*
Passeriformes - Muscicapidae
e   Siberian Rubythroat; Rubythroat;
Common Rubythroat; Eurasian
Rubythroat
f   Rossignol calliope; Calliope
sibérienne; Calliope de Sibérie
d   Rubinkehlchen
i   Calliope; Calliope siberiana

5412   *Luscinia cyane*
Passeriformes - Muscicapidae
e   Siberian Blue-Robin; Siberian Blue-
Chat
f   Rossignol bleu; Rossignol bleu du
Japon
d   Blaunachtigall
i   Usignolo azzurro siberiano

5413   *Luscinia luscinia*
Passeriformes - Muscicapidae
e   Thrush-Nightingale; Sprosser
Nightingale (CSA); Eastern
Nightingale
f   Rossignol progné
d   Sprosser
i   Rusignolo maggiore; Usignolo
maggiore; Rosignolo maggiore;
Usignuolo maggiore

5414   *Luscinia megarhynchos*
Passeriformes - Muscicapidae
e   Common Nightingale; Nightingale;
Rufous Nighjingale (NA); European
Nightingale
f   Rossignol philomèle
d   Nachtigall
i   Rusignolo; Usignolo comune;
Rosignolo; Usignuolo

5415   *Luscinia obscura*
Passeriformes - Muscicapidae
e   Black-throated Blue-Robin; Black-
throated Robin; Blackthroat; La
Touche's Shortwing
f   Rossignol à gorge noire
d   Schwarzkehlnachtigall
i   Usignolo golanera

5416   *Luscinia pectardens*
Passeriformes - Muscicapidae
e   Firethroat; David's Firethroat; Père
David's Firethroat; David's
Orangethroat; Père David's
Orangethroat; David's Rubythroat;
Père David's Rubythroat; Golden-
breasted Rubythroat
f   Rossignol de David; Rossignol de
père David
d   David-Nachtigall
i   Calliopedi David

5417   *Luscinia pectoralis*
Passeriformes - Muscicapidae
e   White-tailed Rubythroat; Rubythroat
(ISC); Himalayan White-tailed
Rubythroat; Himalayan Rubythroat;
Black-breasted Rubythroat
f   Rossignol à gorge rubis

d   Bergrubinkehlchen
i   Calliope dell'Himalaya

5418 ***Luscinia ruficeps***
Passeriformes - Muscicapidae
e   Rufous-headed Robin; Red-headed
Robin
f   Rossignol à tête rousse
d   Rotkopfnachtigall
i   Usignolo testarugginosa

5419 ***Luscinia sibilans***
Passeriformes - Muscicapidae
e   Rufous-tailed Robin; Swinhoe's
Pseudo-Robin; Swinhoe's Robin;
Swinhoes Red-tailed Robin;
Swinhoe's Rufous-tailed Robin; Red-
tailed Robin; Pseudorobin; Whistling
Nightingale
f   Rossignol siffleur; Rouge-gorge
siffleur
d   Schwirrnachtigall
i   Usignolo di Swinhoe; Rusignolo di
Swinhoe

5420 ***Luscinia svecica***
Passeriformes - Muscicapidae
e   Bluethroat; Red-spotted Bluethroat
f   Gorgebleue; Gorgebleue à mirroir
d   Blaukehlchen
i   Pettazzurro

5421 **Lybiidae**
Piciformes
e   African Barbets
f   Lybiidés
d   Afrikanische Bartvögel

5422 ***Lybius bidentatus***
Piciformes - Lybiidae
e   Double-toothed Barbet; Tooth-billed
Barbet
f   Barbican bidenté
d   Doppelzahnbartvogel
i   Capitone; Barbuto bidentato

5423 ***Lybius chaplini***
Piciformes - Lybiidae
e   Chaplin's Barbet
f   Barbican de Chaplin

d   Feigenbartvogel
i   Barbuto di Chaplin

5424 ***Lybius dubius***
Piciformes - Lybiidae
e   Bearded Barbet
f   Barbican à poitrine rouge
d   Senegal-Furchenschnabel
i   Barbuto pettirosso

5425 ***Lybius guifsobalito***
Piciformes - Lybiidae
e   Black-billed Barbet
f   Barbican guifsobalito; Barbican à bec
noir
d   Purpurmaskenbartvogel
i   Barbuto becconero

5426 ***Lybius leucocephalus***
Piciformes - Lybiidae
e   White-headed Barbet
f   Barbican à tête blanche
d   Weißkopfbartvogel
i   Barbuto testabianca

5427 ***Lybius melanopterus***
Piciformes - Lybiidae
e   Brown-breasted Barbet; Black-
winged Barbet
f   Barbican à poitrine brune
d   Braunbrustbartvogel
i   Barbuto alinere

5428 ***Lybius minor***
Piciformes - Lybiidae
e   Black-backed Barbet; Levaillant's
Barbet
f   Barbican de Levaillant
d   Rosenbauchbartvogel
i   Barbuto dorsonero

5429 ***Lybius minor macclounii***
Piciformes - Lybiidae
e   MacClounie's Barbet
f   Barbican de MacClounie

5430 ***Lybius rolleti***
Piciformes - Lybiidae
e   Black-breasted Barbet
f   Barbican à poitrine noire

d Schwarzbrustfurchenschnabel
i Barbuto pettonero

5431 ***Lybius rubrifacies***
  Piciformes - Lybiidae
e Red-faced Barbet
f Barbican à face rouge
d Rotgesichtbartvogel
i Barbuto facciarossa

5432 ***Lybius torquatus***
  Piciformes - Lybiidae
e Black-collared Barbet
f Barbican à collier
d Halsbandbartvogel
i Capitone dal collare; Barbuto dal collare

5433 ***Lybius undatus***
  Piciformes - Lybiidae
e Banded Barbet
f Barbican barré
d Wellenbartvogel
i Barbuto barrato

5434 ***Lybius vieilloti***
  Piciformes - Lybiidae
e Vieillot's Barbet
f Barbican de Vieillot
d Blutbrustbartvogel
i Barbuto di Vieillot

5435 ***Lycocorax pyrrhopterus***
  Passeriformes - Corvidae
e Paradise Crow; Silky Crow; Brown-winged Bird of Paradise
f Paradisier corvin
d Krähenparadiesvogel; Paradieskrähe
i Cornacchia del Paradiso

5436 ***Lysurus castaneiceps***
  Passeriformes - Fringillidae
e Olive Finch; Olive Brush-Finch
f Tohi lysure
d Olivbuschammer
i Fringuello olivaceo

5437 ***Lysurus crassirostris***
  Passeriformes - Fringillidae
e Sooty-faced Finch; Sooty-faced Brushfinch; Barranca Finch;

  Barranca Brushfinch
f Tohi masqué
d Dickschnabelbuschammer
i Fringuello facciafuligginosa

M

5438 *Macgregoria pulchra*
Passeriformes - Corvidae
e MacGregor's Bird of Paradise
f Paradisier de Macgregor
d Brillenparadiesvogel
i Paradisea di MacGregor

5439 *Machaerirhynchus flaviventer*
Passeriformes - Corvidae
e Yellow-breasted Boatbill; Yellow-breasted Flatbill-Flycatcher; Yellow-breasted Boat-billed Flycatcher; Yellow-breasted Flatbill; Yellow-breasted Flycatcher; Boat-billed Monarch-Flycatcher; Boat-billed Flycatcher
f Monarque à poitrine jaune
d Flachschnabel; Gelbbauchflachschnabel
i Beccotagliente pettogiallo

5440 *Machaerirhynchus nigripectus*
Passeriformes - Corvidae
e Black-breasted Boatbill; Black-breasted Boat-billed Flycatcher; Black-breasted Flatbill; Black-breasted Flatbill-Flycatcher
f Monarque à poitrine noire
d Brustfleckflachschnabel
i Beccotagliente pettonero

5441 *Machaeropterus deliciosus*
Passeriformes - Tyrannidae
e Club-winged Manakin
f Manakin à ailes blanches
d Keulenpipra
i Manachino delizioso

5442 *Machaeropterus pyrocephalus*
Passeriformes - Tyrannidae
e Fiery-capped Manakin
f Manakin tête-de-feu

d Buntpipra; Klingelpipra
i Manachino testadifiamma

5443 *Machaeropterus regulus*
Passeriformes - Tyrannidae
e Striped Manakin
f Manakin rubis
d Streifenpipra
i Manachino regolo

5444 *Macheiramphus alcinus*
Falconiformes - Accipitridae
e Bat Hawk; Bat Falcon; Bat Kite; Bat-eating Buzzard; Bat-eating Hawk
f Milan des chauves-souris
d Fledermausaar
i Nibbio dei pipistrelli; Poiana dei pipistrelli

5445 *Machetornis rixosus*
Passeriformes - Tyrannidae
e Cattle Tyrant; Cattle Flycatcher; Fire-crowned Tyrant
f Moucherolle querelleuse
d Stelzentyrann; Streifentyrann
i Tiranno guardabuoi

5446 *Mackenziaena leachii*
Passeriformes - Formicariidae
e Large-tailed Antshrike
f Batara de Leach
d Langschwanzameisenwürger; Tüpfelparanawürger
i Averla formichiera di Leach

5447 *Mackenziaena severa*
Passeriformes - Formicariidae
e Tufted Antshrike
f Batara othello
d Schwarzmaskenameisenwürger
i Averla formichiera dal ciuffo

5448 *Macroagelaius imthurni*
Passeriformes - Fringillidae
e Golden-tufted Grackle; Tepuis Mountain-Grackle
f Quiscale des tépuis
d Goldachselstärling
i Gracchio del Tepui

5449   *Macroagelaius subalaris*
          Passeriformes - Fringillidae
  e   Mountain Grackle; Colombian
          Mountain-Grackle
  f   Quiscale montagnard
  d   Braunachselstärling
  i   Gracchio colombiano

5450   *Macrocephalon maleo*
          Galliformes - Megapodiidae
  e   Maleo; Maleo Fowl; Grey Brush-
          Turkey
  f   Mégapode maléo
  d   Hammerhuhn
  i   Maleo

5451   *Macrodipteryx longipennis*
          Caprimulgiformes - Caprimulgidae
  e   Standard-winged Nightjar
  f   Engoulevent à balanciers
  d   Flaggenflügel; Fahnennachtschwalbe
  i   Succiacapre dal bilanciere;
          Macroditterice

5452   *Macrodipteryx vexillarius*
          Caprimulgiformes - Caprimulgidae
  e   Pennant-winged Nightjar
  f   Engoulevent porte-étendard
  d   Ruderflügel
  i   Succiacapre dal vessilo; Cosmetorno

5453   *Macronectes giganteus*
          Procellariiformes - Procellariidae
  e   Antarctic Giant-Petrel; Southern
          Giant-Petrel; Giant Petrel; Stinker
          Petrel; Nellie (CSA); Giant Fulmar;
          Southern Giant-Fulmar
  f   Fulmar géant; Pétrel géant
  d   Riesensturmvogel; Südlicher
          Sturmvogel
  i   Ossifraga del Sud; Ossifraga;
          Fulmaro gigante

5454   *Macronectes halli*
          Procellariiformes - Procellariidae
  e   Hall's Giant-Petrel; Northern Giant-
          Petrel (ANZ)
  f   Pétrel de Hall
  d   Nördlicher Riesensturmvogel;
          Riesensturmvogel

  i   Ossifraga del Nord; Ossifraga;
          Ossifraga di Hall

5455   *Macronous flavicollis*
          Passeriformes - Sylviidae
  e   Grey-cheeked Tit-Babbler; Grey-
          faced Tit-Babbler; Yellow-collared
          Tit-Babbler; Javan Tit-Babbler
  f   Timalie à face grise
  d   Graunackentimalie
  i   Garrulo golagialla

5456   *Macronous gularis*
          Passeriformes - Sylviidae
  e   Striped Tit-Babbler; Yellow-breasted
          Babbler (ISC); Striated Tit-Babbler;
          Stripe-throated Tit-Babbler
  f   Timalie à gorge striée; Timalie à
          gorge jaune
  d   Gelbbrustbaumtimalie
  i   Garrulo pettogiallo

5457   *Macronous kelleyi*
          Passeriformes - Sylviidae
  e   Grey-faced Tit-Babbler; Kelley's Tit-
          Babbler
  f   Timalie de Kelley
  d   Kelley-Timalie
  i   Garrulo facciagrigia

5458   *Macronous ptilosus*
          Passeriformes - Sylviidae
  e   Fluffy-backed Tit-Babbler; Plume-
          backed Tit-Babbler
  f   Timalie chamasa
  d   Stachelrückentimalie
  i   Garrulo dorsopiumoso

5459   *Macronous striaticeps*
          Passeriformes - Sylviidae
  e   Brown Tit-Babbler
  f   Timalie brune
  d   Streifenkopftimalie
  i   Garrulo testastriata

5460   *Macronyx ameliae*
          Passeriformes - Passeridae
  e   Rosy-throated Longclaw; Pink-
          throated Longclaw (CSA); Rosy-
          breasted Longclaw
  f   Sentinelle à gorge rose

d   Rotkehlgroßsporn; Rubinkehlpieper
i   Zampagrossa pettorosa

5461   *Macronyx aurantiigula*
Passeriformes - Passeridae
e   Pangani Longclaw
f   Sentinelle dorée
d   Pangani-Pieper
i   Zampagrossa del Pangani

5462   *Macronyx capensis*
Passeriformes - Passeridae
e   Cape Longclaw; Orange-throated
Longclaw (CSA)
f   Sentinelle du Cap
d   Kap-Großsporn
i   Zampagrossa del Capo

5463   *Macronyx croceus*
Passeriformes - Passeridae
e   Yellow-throated Longclaw
f   Sentinelle à gorge jaune; Alouette
sentinelle
d   Gelbkehlgroßsporn;
Gelbkehlsaffrangroßsporn;
Gelbkehlpieper
i   Zampagrossa golagialla; Macronice

5464   *Macronyx flavicollis*
Passeriformes - Passeridae
e   Abyssinian Longclaw; Ethiopian
Longclaw
f   Sentinelle d'Abyssinie
d   Goldhalspieper
i   Zampagrossa dell'Abissinia

5465   *Macronyx fuellebornii*
Passeriformes - Passeridae
e   Fuelleborn's Longclaw; Fülleborn's
Longclaw (CSA)
f   Sentinelle de Fülleborn
d   Fülleborn-Pieper
i   Zampagrossa di Fuelleborn

5466   *Macronyx grimwoodi*
Passeriformes - Passeridae
e   Grimwood's Longclaw
f   Sentinelle de Grimwood
d   Dambo-Pieper
i   Zampagrossa di Grimwood

5467   *Macropsalis creagra*
Caprimulgiformes - Caprimulgidae
e   Long-trained Nightjar; Long-tailed
Nightjar
f   Engoulevent à traine
d   Scherenachtschwalbe
i   Succiacapre coda a strascico

5468   *Macropygia amboinensis*
Columbiformes - Columbidae
e   Slender-billed Cuckoo-Dove; Brown
Cuckoo-Dove (ANZ); Amboina
Cuckoo-Dove; Pink-breasted
Cuckoo-Dove; Brown Pigeon;
Pheasant-tailed Pigeon; Large
Pigeon; Philippine Cuckoo-dove;
Ruddy Cuckoo-Dove
f   Phasianelle d'Amboine
d   Kuckuckstaube;
Rosabrustkuckuckstaube
i   Colomba cuculo bruna

5469   *Macropygia emiliana*
Columbiformes - Columbidae
e   Ruddy Cuckoo-Dove; Indonesian
Cuckoo-Dove
f   Phasianelle rousse
i   Colomba cuculo dell'indonesia

5470   *Macropygia mackinlayi*
Columbiformes - Columbidae
e   Mackinlay's Cuckoo-Dove; Rufous
Cuckoo-Dove; Rufous-brown
Cuckoo-Dove; Rufous Pheasant-
Pigeon; Rufous-brown Pheasant-
Dove; Spot-breasted Cuckoo-Dove;
Black-spotted Cuckoo-Dove
f   Phasianelle de Mackinlay
d   Fuchsrote Kuckuckstaube;
Mackinlay-Taube
i   Colomba cuculo di Mackinlay

5471   *Macropygia magna*
Columbiformes - Columbidae
e   Dusky Cuckoo-Dove; Large Cuckoo-
Dove
f   Grande Phasianelle
d   Große Kuckuckstaube; Schweiftaube
i   Colomba cuculo maggiore

5472 *Macropygia nigrirostris*
Columbiformes - Columbidae
e Black-billed Cuckoo-Dove; Lesser Bar-tailed Cuckoo-Dove
f Phasianelle barrée
d Kastanientaube; Schwarzschnabelkuckuckstaube
i Colomba cuculo becconero

5473 *Macropygia phasianella*
Columbiformes - Columbidae
e Brown Cuckoo-Dove; Large Brown Cuckoo-Dove; Red Cuckoo-Dove; Slender-billed Cuckoo-Dove; Dark Cuckoo-Dove
f Phasianelle brune
d Dunkle Kuckuckstaube; Maronentaube
i Colomba cuculo fagianella

5474 *Macropygia ruficeps*
Columbiformes - Columbidae
e Little Cuckoo-Dove; Lesser Red Cuckoo-Dove; Red-headed Cuckoo-Dove; Red-faced Cuckoo-Dove
f Phasianelle à tête rousse
d Kleine Kuckuckstaube; Rotmanteltaube
i Colomba cuculo minore

5475 *Macropygia rufipennis*
Columbiformes - Columbidae
e Andaman Cuckoo-Dove; Nicobar Cuckoo-Dove
f Phasianelle des Nicobar
d Andamanen-Kuckuckstaube; Rotsteißtaube
i Colomba cuculo delle Andamane

5476 *Macropygia tenuirostris*
Columbiformes - Columbidae
e Philippine Cuckoo-Dove
f Phasianelle des Philippines
i Colomba cuculo delle Filippine

5477 *Macropygia unchall*
Columbiformes - Columbidae
e Barred Cuckoo-Dove; Bar-tailed Cuckoo-Dove (ISC); Long-tailed Cuckoo-Dove; Larger Indian Cuckoo-Dove; Larger Malay Cuckoo-Dove
f Phasianelle onchall; Phasianelle unchall
d Bindenschwanztaube
i Colomba cuculo codabarrata

5478 *Macrosphenus concolor*
Passeriformes - Sylviidae
e Grey Longbill; Olive Longbill; Olive Bushcreeper
f Nasique grise; Fauvette nasique grise
d Einfarbbülbügrasmücke
i Beccolungo grigio

5479 *Macrosphenus flavicans*
Passeriformes - Sylviidae
e Yellow Longbill; Long-billed Bushcreeper
f Nasique jaune; Fauvette nasique jaune
d Gelbbauchbülbügrasmücke
i Beccolungo giallastro

5480 *Macrosphenus kempi*
Passeriformes - Sylviidae
e Kemp's Longbill; Kemp's Bushcreeper
f Nasique de Kemp
d Rostflankenbülbügrasmücke
i Beccolungo di Kemp

5481 *Macrosphenus kretschmeri*
Passeriformes - Sylviidae
e Kretschmer's Longbill; Kretschmer's Greenbul; Kretschmer's Bulbul
f Nasique de Kretschmer
d Suaheli-Bülbügrasmücke
i Beccolungo di Kretschmer

5482 *Macrosphenus pulitzeri*
Passeriformes - Sylviidae
e Pulitzer's Longbill; Pullitzer's Greenbul
f Nasique de Pulitzer
d Angola-Bülbügrasmücke
i Beccolungo di Pulitzer

5483 *Madanga ruficollis*
Passeriformes - Zosteropidae
e Rufous-throated White-eye; Madanga White-eye; Buru Mountain-

White-eye; Rufous-throated Darkeye;
Rufous-collared White-eye
f Zostérops à gorge rousse
d Orangekehlbrillenvogel
i Occhialino golarossa

5484 *Malacocincla abbotti*
Passeriformes - Sylviidae
e Abbott's Babbler; Abbot's Jungle-
Babbler
f Akalat d'Abbott
d Rotschwanzmaustimalie;
Rotschwanzdschungeltimalie
i Garrulo di Abbott

5485 *Malacocincla cinereiceps*
Passeriformes - Sylviidae
e Ashy-headed Babbler; Ashy-headed
Jungle-Babbler; Ashy-headed
Ground-Babbler
f Akalat à tête cendrée
d Graukopfmaustimalie
i Garrulo testagrigia

5486 *Malacocincla malaccensis*
Passeriformes - Sylviidae
e Short-tailed Babbler; Short-tailed
Jungle-Babbler; Sarawak Babbler;
Ochre-throated Babbler
f Akalat à queue courte
d Kurzschwanzmaustimalie
i Garrulo codabreve

5487 *Malacocincla perspicillata*
Passeriformes - Sylviidae
e Black-browed Babbler; Black-
browed Jungle-Babbler; Büttikofer's
Babbler
f Akalat à sourcils noirs
d Schwarzbrauenmaustimalie;
Großdschungeltimalie
i Garrulo dai sopraccigli nere

5488 *Malacocincla sepiarium*
Passeriformes - Sylviidae
e Horsfield's Babbler; Horsfield's
Jungle-Babbler
f Akalat de Horsfield
d Horsfield-Maustimalie
i Garrulo di Horsfield

5489 *Malacocincla vanderbilti*
Passeriformes - Sylviidae
e Vanderbilt's Babbler; Vanderbildt's
Jungle-Babbler; Koengke Babbler;
Koengke Jungle-Babbler
f Akalat de Vanderbilt
d Atjeh-Maustimalie
i Garrulo di Vanderbilt

5490 **Malaconotidae**
Passeriformes
e Bush-Shrikes
f Malaconotidés
d Buschwürger
i Malaconotidi

5491 *Malaconotus alius*
Passeriformes - Malaconotidae
e Uluguru Bush-Shrike; Black-cap
Bush-Shrike; Uluguru Shrike
f Gladiateur des Uluguru; Gladiateur à
tête noire
i Averla di macchia capinera; Averla
capinera

5492 *Malaconotus blanchoti*
Passeriformes - Malaconotidae
e Grey-headed Bush-Shrike
f Gladiateur de Blanchot; Pie-grièche
de Blanchot
d Graukopfwürger;
Graukopfbuschwürger
i Averla di macchia di Blanchot;
Averla di Blanchot

5493 *Malaconotus bocagei jacksoni*
Passeriformes - Malaconotidae
e Grey-green Bush-Shrike (CSA)

5494 *Malaconotus cruentus*
Passeriformes - Malaconotidae
e Fiery-breasted Bush-Shrike; Rosy-
patched Shrike; Fiery-breasted Shrike
f Gladiateur ensanglanté; Pie-grièche
verte ensanglantée
d Blutbrustwürger
i Averla petto di flamma; Averla di
macchia pettodifiamma

5495 *Malaconotus dohertyi*
Passeriformes - Malaconotidae

e   Doherty's Bush-Shrike; Doherty's
    Shrike
f   Gladiateur de Doherty
d   Rotstirnwürger
i   Averla di macchia di Doherty; Averla
    di Doherty

5496   *Malaconotus gladiator*
    Passeriformes - Malaconotidae
e   Green-breasted Bush-Shrike;
    Gladiator Bush-Shrike; Cameroon
    Mountain-Bush-shrike
f   Gladiateur à poitrine verte
d   Grünkehlwürger
i   Averla di macchia pettoverde; Averla
    pettoverde

5497   *Malaconotus lagdeni*
    Passeriformes - Malaconotidae
e   Lagden's Bush-Shrike; Lagden's
    Shrike
f   Gladiateur de Lagden
d   Lagden-Würger
i   Averla di macchia di Lagden; Averla
    di Langden

5498   *Malaconotus monteiri*
    Passeriformes - Malaconotidae
e   Monteiro's Bush-Shrike
f   Gladiateur de Monteiro
d   Monteiro-Würger
i   Averla di macchia di Monteiro;
    Averla di Monteiro

**5499   *Malaconotus quadricolor
          nigricauda***
    Passeriformes - Malaconotidae
e   Four-coloured Bush-Shrike (CSA)

5500   *Malaconotus sulfureopectus similis*
    Passeriformes - Malaconotidae
e   Sulphur-breasted Bush-Shrike (CSA)

5501   *Malacopteron affine*
    Passeriformes - Sylviidae
e   Sooty-capped Babbler; Plain
    Babbler; Sooty-capped Tree-Babbler;
    Sooty-headed Babbler
f   Akalat affin
d   Schwarzscheitelzweigtimalie;

    Kappenzweigtimalie
i   Garrulo fuligginoso

5502   *Malacopteron albogulare*
    Passeriformes - Sylviidae
e   Grey-breasted Babbler; White-
    throated Babbler
f   Akalat à gorge blanche
d   Graubrustzweigtimalie
i   Garrulo pettogrigio

5503   *Malacopteron cinereum*
    Passeriformes - Sylviidae
e   Scaly-crowned Babbler; Lesser Red-
    headed Babbler; Scaly-crowned
    Tree-Babbler; Scaly-capped Babbler
f   Akalat à calotte maillée
d   Rotstirnzweigtimalie
i   Garrulo scaglioso

5504   *Malacopteron magnirostre*
    Passeriformes - Sylviidae
e   Moustached Babbler; Brown-headed
    Babbler; Moustached Tree-Babbler
f   Akalat moustachu
d   Bartstreifzweigtimalie
i   Garrulo dai mustacchi

5505   *Malacopteron magnum*
    Passeriformes - Sylviidae
e   Rufous-crowned Babbler; Greater
    Red-headed Babbler; Red-headed
    Babbler; Rufous-crowned Tree-
    Babbler
f   Akalat géant
d   Rotscheitelzweigtimalie; Große
    Zweigtimalie
i   Garrulo corona rossicia

5506   *Malacopteron palawanense*
    Passeriformes - Sylviidae
e   Melodious Babbler; Palawan Tree-
    Babbler; Palawan Babbler; Red-
    headed Babbler; Red-headed Tree-
    Babbler
f   Akalat de Palawan
d   Palawan-Zweigtimalie
i   Garrulo di Palawan

5507   *Malacoptila fulvogularis*
    Piciformes - Bucconidae

e   Black-streaked Puffbird
f   Tamatia à gorge fauve
d   Ockerkehlfaulvogel;
    Braunkehlfaulvogel
i   Bucco golafulava

5508   *Malacoptila fusca*
    Piciformes - Bucconidae
e   White-chested Puffbird
f   Tamatia brun
d   Weißbrustfaulvogel
i   Bucco pettobianco; Monasa bruna

5509   *Malacoptila mystacalis*
    Piciformes - Bucconidae
e   Moustached Puffbird
f   Tamatia à moustaches
d   Schnurrbartfaulvogel
i   Bucco dai mustacchi; Monasa baffuta

5510   *Malacoptila panamensis*
    Piciformes - Bucconidae
e   White-whiskered Puffbird; Brown
    Puffbird
f   Tamatia de Lafresnaye
d   Weißzügelfaulvogel; Weißfaulvogel
i   Bucco di Panama

5511   *Malacoptila rufa*
    Piciformes - Bucconidae
e   Rufous-necked Puffbird
f   Tamatia à col roux; Tamatia
    soussâtre
d   Goldstirnfaulvogel; Rostroter
    Faulvogel
i   Bucco rugginoso

5512   *Malacoptila semicincta*
    Piciformes - Bucconidae
e   Semicollared Puffbird
f   Tamatia à semi-collier
d   Beni-Faulvogel
i   Bucco dal semicollare

5513   *Malacoptila striata*
    Piciformes - Bucconidae
e   Crescent-chested Puffbird; Striated
    Puffbird
f   Tamatia rayé
d   Halbmondfaulvogel;
    Weißkopffaulvogel;

    Schwarzweißfaulvogel
i   Bucco dal mezzaluna

5514   *Malacorhynchus membranaceus*
    Anseriformes - Anatidae
e   Pink-eared Duck; Pink-eyed Duck;
    Zebra Duck; Zebra Teal
f   Canard à oreilles roses
d   Rosenohrenente; Spatelschnabelente
i   Anatra orecchierosa

5515   *Malcorus pectoralis*
    Passeriformes - Cisticolidae
e   Rufous-eared Warbler; Rufous-eared
    Prinia
f   Prinia à joues rousses; Prinia à
    oreilles rousses
d   Rotbackensänger
i   Prinia orecchierosse

5516   *Malia grata*
    Passeriformes - Pycnonotidae
e   Malia; Celebes Malia; Celebes
    Babbler
f   Malia des Célèbes
d   Mooswaldtimalie; Mooshacker
i   Malia

5517   *Malimbus ballmanni*
    Passeriformes - Ploceidae
e   Ballmann's Malimbe; Gola Malimbe;
    Tai Malimbe
f   Malimbe de Ballmann; Malimbe de
    Gola
d   Ballmann-Weber
i   Malimbo di Tai

5518   *Malimbus cassini*
    Passeriformes - Ploceidae
e   Black-throated Malimbe; Cassin's
    Malimbe
f   Malimbe de Cassin
d   Cassin-Weber
i   Malimbo golanera

5519   *Malimbus coronatus*
    Passeriformes - Ploceidae
e   Red-crowned Malimbe; Scarlete-
    crowned Malimbe
f   Malimbe couronné

d   Kronenweber
i   Malimbo corona rossa

5520   *Malimbus erythrogaster*
Passeriformes - Ploceidae
e   Red-bellied Malimbe
f   Malimbe à ventre rouge
d   Rotbauchweber
i   Malimbo ventrerosso

5521   *Malimbus flavipes*
Passeriformes - Ploceidae
e   Yellow-legged Malimbe; Yellow-
footed Weaver; Yellow-legged
Weaver; Yellow-fronted Malimbe
f   Malimbe à pieds jaunes; Tisserin à
pieds jaunes
d   Gelbfußweber
i   Malimbo piedigialli

5522   *Malimbus ibadanensis*
Passeriformes   Ploceidae
e   Ibadan Malimbe
f   Malimbe d'Ibadan
d   Ibadan-Weber
i   Malimbo di Ibadan

5523   *Malimbus malimbicus*
Passeriformes - Ploceidae
e   Crested Malimbe
f   Malimbe huppé
d   Haubenweber
i   Malimbo dalla cresta

5524   *Malimbus nitens*
Passeriformes - Ploceidae
e   Gray's Malimbe; Blue-billed
Malimbe
f   Malimbe à bec bleu
d   Rotkehlweber
i   Malimbo di Gray

5525   *Malimbus racheliae*
Passeriformes - Ploceidae
e   Rachel's Malimbe
f   Malimbe de Rachel
d   Rachel-Weber
i   Malimbo di Rachel

5526   *Malimbus rubricollis*
Passeriformes - Ploceidae

e   Red-headed Malimbe; Red-headed
Quelea; Red-collared Malimbe
f   Malimbe à tête rouge
d   Kletterweber
i   Malimbo testarossa

5527   *Malimbus scutatus*
Passeriformes - Ploceidae
e   Red-vented Malimbe
f   Malimbe à queue rouge
d   Schildweber
i   Malimbo dal sottocoda rosso

5528   Maluridae
Passeriformes
e   Australasian Wrens; Wren-Warblers
f   Maluridés
d   Staffelschwänze; Australische Sänger
i   Maluridi

5529   *Malurus alboscapulatus*
Passeriformes - Maluridae
e   White-shouldered Fairywren; Black-
and-white Wren; Black-and-white
Wren-Warbler; Black-and-white
Fairywren
f   Mérion à épaulettes
d   Weißschulterstaffelschwanz
i   Scricciolo splendente spallebianche

5530   *Malurus amabilis*
Passeriformes - Maluridae
e   Lovely Fairywren; Lovely Wren;
Lovely Wren-Warbler
f   Mérion ravissant
i   Scricciolo splendente amabile

5531   *Malurus callainus*
Passeriformes - Maluridae
e   Turquoise Wren; Turquoise Wren-
Warbler
f   Mérion turquoise

5532   *Malurus campbelli*
Passeriformes - Maluridae
e   Campbell's Fairywren
f   Mérion de Campbell
i   Scricciolo splendente di Campbell

5533   *Malurus coronatus*
Passeriformes - Maluridae

e Purple-crowned Fairywren; Lilac-
crowned Wren; Purple-crowned
Wren; Purple-crowned Wren-
Warbler
f Mérion couronné
d Purpurstaffelschwanz
i Scricciolo splendente coronato

5534 *Malurus coronatus coronatus*
Passeriformes - Maluridae
e Western Purple-crowned Fairywren
(ANZ)

5535 *Malurus cyaneus*
Passeriformes - Maluridae
e Superb Fairywren; Superb Blue-
Wren; Superb Wren-Warbler; Superb
Blue-Fairywren; Blue Wren-Warbler;
Australian Blue-Wren; Blue
Fairywren; Imperial Fairywren; New
Guinea Wren; New Guinea Blue-
Wren
f Mérion superbe
d Prachtstaffelschwanz;
Lasurstaffelschwanz
i Scricciolo azzurro superbo

5536 *Malurus cyanocephalus*
Passeriformes - Maluridae
e Emperor Fairywren; Imperial Wren;
New Guinea Blue-Wren; Blue Wren-
Warbler
f Mérion empéreur
d Blaukopfstaffelschwanz
i Scricciolo splendente imperatore

5537 *Malurus elegans*
Passeriformes - Maluridae
e Red-winged Fairywren; Red-winged
Wren; Red-winged Wren-Warbler
f Mérion élégant
d Silberkopfstaffelschwanz
i Scricciolo splendente di palude

5538 *Malurus grayi*
Passeriformes - Maluridae
e Broad-billed Fairywren; Broad-billed
Wren; Broad-billed Wren-Warbler
f Mérion à bec large
d Breitschnabelstaffelschwanz
i Scricciolo splendente di Gray

5539 *Malurus lamberti*
Passeriformes - Maluridae
e Variegated Fairywren; Variegated
Wren; Variegated Wren-Warbler;
Purple-backed Wren; Purple-backed
Wren-Warbler
f Mérion de Lambert
d Weißbauchstaffelschwanz;
Vielfarbenstachelschwanz
i Scricciolo splendente variegato

5540 *Malurus lamberti bernieri*
Passeriformes - Maluridae
e Shark Bay Variegated-Fairywren

5541 *Malurus leucopterus*
Passeriformes - Maluridae
e White-winged Fairywren; White-
winged Wren; Blue-and-white Wren-
Warbler; Blue-and-white Flycatcher
f Mérion leucoptère
d Weißflügelstaffelschwanz
i Scricciolo splendente alibianche

5542 *Malurus leucopterus edouardi*
Passeriformes - Maluridae
e Barrow Island White-winged
Fairywren (ANZ)

5543 *Malurus leucopterus leucopterus*
Passeriformes - Maluridae
e Dirk Hartog Island White-winged
Fairywren (ANZ)

5544 *Malurus melanocephalus*
Passeriformes - Maluridae
e Red-backed Fairywren; Red-backed
Wren; Red-backed Wren-Warbler
f Mérion à dos rouge
d Rotrückenstaffelschwanz
i Scricciolo splendente dorsorosso

5545 *Malurus melanotus*
Passeriformes - Maluridae
e Black-backed Wren; Black-backed
Blue-Wren

5546 *Malurus pulcherrimus*
Passeriformes - Maluridae
e Blue-breasted Fairywren; Blue-
breasted Wren; Blue-breasted Wren-

Warbler
f Mérion à gorge bleue
d Blaubruststaffelschwanz
i Scricciolo splendente pettoblu

5547 *Malurus splendens*
Passeriformes - Maluridae
e Splendid Fairywren; Banded Wren;
Splendid Wren; Splendid Wren-
Warbler; Splendid Blue-Wren;
Splendid Blue Wren-Warbler;
Banded Fairywren; Banded Wren-
Warbler
f Mérion splendide
d Türkis-Staffelschwanz
i Scricciolo splendente azzurro;
Scricciolo azzurro splendente

5548 *Manacus aurantiacus*
Passeriformes - Pipridae
e Orange-collared Manakin; Salvin's
Manakin
f Manakin à col orangé
i Manachino dal collare arancio

5549 *Manacus candei*
Passeriformes - Pipridae
e White-collared Manakin; Cande's
Manakin
f Manakin à col blanc
i Manachino dal collare bianco

5550 *Manacus manacus*
Passeriformes - Pipridae
e White-bearded Manakin; Collared
Manakin; Bearded Manakin
f Manakin casse-noisette
d Weißsäbelpipra
i Manachino barbato; Manachino
monaco

5551 *Manacus viridiventris*
Passeriformes - Pipridae
e Greenish-bellied Manakin

5552 *Manacus vitellinus*
Passeriformes - Pipridae
e Golden-collared Manakin; Gould's
Manakin; Almirante Manakin
f Manakin à col d'or
i Manachino dal collare dorato

5553 *Mandingoa nitidula*
Passeriformes - Estrildidae
e Green-backed Twinspot; Green
Twinspot (ANZ); Green-eyed
Twinspot; Schlegel's Twinspot
f Sénégali vert; Bengali vert tacheté;
Bengali vert pointillé; Astrild tacheté
à dos vert; Sénégali vert tacheté
d Grüner Tropenastrild
i Mandingo bimaculato

5554 *Manorina flavigula*
Passeriformes - Meliphagidae
e Yellow-throated Miner; White-
rumped Miner
f Méliphage à cou jaune
d Gelbstirnschwatzvogel
i Manorina golagialla

5555 *Manorina melanocephala*
Passeriformes - Meliphagidae
e Noisy Miner
f Méliphage bruyant
d Weißstirnschwatzvogel
i Manorina chiassoso

5556 *Manorina melanophrys*
Passeriformes - Meliphagidae
e Bell Miner
f Méliphage à sourcils noirs
d Glockenschwatzvogel;
Klingelschwatzvogel
i Uccello campanello australiano

5557 *Manorina melanotis*
Passeriformes - Meliphagidae
e Black-eared Miner; Dusky Miner
d Mallee-Schwatzvogel

5558 *Manorina obscura*
Passeriformes - Meliphagidae
e Dusky Miner

5559 *Manucodia atra*
Passeriformes - Corvidae
e Glossy-mantled Manucode
f Paradisier noir
d Glanzmanukode
i Manucodia dal mantello rilucente

5560 **Manucodia chalybata**
Passeriformes - Corvidae
e Crinkle-collared Manucode; Crinkle-
breasted Manucode; Green-breasted
Manucode
f Paradisier vert
i Manucodia collare riccio

5561 **Manucodia comrii**
Passeriformes - Corvidae
e Curl-crested Manucode; Curl-
breasted Manucode; Curly-crested
Manucode
f Paradisier d'Entrecasteaux
d Kräuselmanukode
i Manucodia crestariccia

5562 **Manucodia jobiensis**
Passeriformes - Corvidae
e Jobi Manucode; Trumpet Manucode;
Australian Trumpet Manucode
f Paradisier de Jobi
d Jobi-Manukode
i Manucodia di Jobi

5563 **Manucodia keraudrenii**
Passeriformes - Corvidae
e Trumpet Manucode; Trumpetbird
f Paradisier de Keraudren
d Trompeterparadiesvogel;
Schallmanucodia; Schallmanukode;
Schalldrossel
i Manucodia trombettiere

5564 **Manucodia keraudrenii jamesii**
Passeriformes - Corvidae
e Torres Strait Trumpet Manucode
(ANZ)

5565 **Margaroperdix madagarensis**
Galliformes - Phasianidae
e Madagascar Partridge
f Perdrix de Madagascar; Caille de
Madagascar
d Perlwachtel
i Pernice del Madagascar

5566 **Margarops fuscatus**
Passeriformes - Sturnidae
e Pearly-eyed Thrasher; Jackbird (WI);
Paw-paw Bird (WI); Black Thrasher

(WI); Mangobird (WI); Wall-eyed
Thrush (WI); Soursopbird (WI)
f Pie voleuse (Ants); Louis Jo (Ants);
Grive-corossol (Ants); Grosse Grive
(Ants); Grive-corosol (Ants);
Moqueur-corossol (Ants); Truche
(Ants)
d Perlaugenspottdrossel
i Mimo occhidiperla

5567 **Margarops fuscus**
Passeriformes - Sturnidae
e Scaly-breasted Thrasher; Black-
billed Thrush (WI); Grieve (WI);
Spotted Grieve (WI)
f Grivotte (Ants); Moqueur grivotte;
Grive fine (Ants)
d Schuppenspottdrossel
i Mimo pettoscaglioso

5568 **Margarornis bellulus**
Passeriformes - Furnariidae
e Beautiful Treerunner
f Anabasitte superbe
d Schmuckstachelschwanz
i Corridore arboricolo elegante

5569 **Margarornis rubiginosus**
Passeriformes - Furnariidae
e Ruddy Treerunner
f Anabasitte rousse
d Bergstachelschwanz
i Corridore arboricolo rossiccio

5570 **Margarornis squamiger**
Passeriformes - Furnariidae
e Pearled Treerunner
f Anabasitte perlée
d Perlenstachelschwanz
i Corridore arboricolo perlato

5571 **Margarornis stellatus**
Passeriformes - Furnariidae
e Fulvous-dotted Treerunner; Dotted
Treerunner
f Anabasitte etoilée
d Tropfenstachelschwanz
i Corridore arboricolo stellato

5572 **Marmaronetta angustirostris**
Anseriformes - Anatidae

e   Marbled Teal; Marbled Duck
f   Sarcelle marbrée; Marmaronette
      marbrée; Sarcelle angustirostre;
      Canard marbré
d   Marmelente
i   Anatra marmorizzata

5573   *Mascarinus mascarinus*
         Psittaciformes - Psittacidae
e   Mascarene Parrot
f   Mascarin de la Réunion
d   Maskarenen-Papagei
i   Pappagallo delle Mascarene

5574   *Masius chrysopterus*
         Passeriformes - Tyrannidae
e   Golden-winged Manakin
f   Manakin aux ailes d'or
d   Goldschwingenpipra
i   Manachino alidorate

5575   *Mayrornis lessoni*
         Passeriformes - Corvidae
e   Slaty Monarch; White-tipped Slaty-
      Flycatcher; Fiji Slaty-Flycatcher;
      Cinereous Flycatcher; Slaty
      Flycatcher; Large Monarch;
      Cinereous Monarch
f   Monarque de Lesson
d   Schwarzschwanzmayrornis
i   Pigliamosche ardesia delle Figi

5576   *Mayrornis schistaceus*
         Passeriformes - Corvidae
e   Vanikoro Monarch; Slaty Flycatcher;
      Small Slaty Flycatcher; Small Slaty
      Monarch
f   Monarque schistacé
d   Vanikoro-Mayrornis
i   Pigliamosche ardesia di Vanikoro

5577   *Mayrornis versicolor*
         Passeriformes - Corvidae
e   Ogea Monarch; Fiji Versicoloured
      Flycatcher; Versicoloured
      Flycatcher; Mayr's Flycatcher;
      Mayr's Versicoloured Flycatcher;
      Ogea Monarch-Flycatcher; Ogea
      Flycatcher; Versicoloured Monarch;
      Versicolour Flycatcher; Versicolour
      Monarch

f   Monarque versicolore
d   Rotbrustmayrornis
i   Pigliamosche di Ogea

5578   *Mearnsia novaeguineae*
         Apodiformes - Apodidae
e   Papuan Needletail; New Guinea
      Spinetailed-Swift; New Guinea
      Needletail; New Guinea Spinetail;
      Papuan Spinetail; Papuan
      Spinetailed-Swift; Grey-bellied
      Needletail
f   Martinet papou
d   Papua-Segler
i   Rondone codaspinosa della Nuova
      Guinea

5579   *Mearnsia picina*
         Apodiformes - Apodidae
e   Philippine Needletail; Philippine
      Spinetailed-Swift; Philippine
      Spinetail
f   Martinet des Philippines
d   Philippinen-Segler
i   Rondone codaspinosa delle Filippine

5580   *Mecocerculus calopterus*
         Passeriformes - Tyrannidae
e   Rufous-winged Tyrannulet
f   Tyranneau à ailes rousses
d   Rotschwingentachuri
i   Tiranno piccolo alirosse

5581   *Mecocerculus hellmayri*
         Passeriformes - Tyrannidae
e   Buff-banded Tyrannulet; Hellmayr's
      Tyrannulet
f   Tyranneau de Hellmayr
d   Braunschwanztachuri
i   Tiranno piccolo di Hellmayr

5582   *Mecocerculus leucophrys*
         Passeriformes - Tyrannidae
e   White-throated Tyrannulet
f   Tyranneau à gorge blanche
d   Weißkehltachuri
i   Tiranno piccolo golabianca

5583   *Mecocerculus minor*
         Passeriformes - Tyrannidae
e   Sulphur-bellied Tyrannulet

f   Tyranneau soufré
d   Gelbbauchtachuri
i   Tiranno piccolo pettosulfureo

5584   *Mecocerculus poecilocercus*
    Passeriformes - Tyrannidae
e   White-tailed Tyrannulet
f   Tyranneau à queue blanche
d   Weißschwanztachuri
i   Tiranno piccolo codabianca

5585   *Mecocerculus stictopterus*
    Passeriformes - Tyrannidae
e   White-banded Tyrannulet
f   Tyranneau à sourcils blancs
d   Weißbindentachuri
i   Tiranno piccolo alifasciate

5586   *Megaceryle alcyon*
    Coraciifo - Cerylidae
e   Belted Kingfisher; Sally Benjamin
    (WI); Kingfisherman (WI);
    Kingfisher (WI)
f   Martin-pêcheur d'Amérique; Martin-
    pêcheur (Ants); Pie (Ants); Alcyon
    ceinturé
d   Gürtelfischer; Halsbandfischer
i   Martin pescatore americano; Martin
    pescatore fasciato

5587   *Megaceryle lugubris*
    Coraciiformes - Cerylidae
e   Crested Kingfisher; Greater Pied-
    Kingfisher; Great Pied-Kingfisher;
    Large Pied-Kingfisher
f   Martin-pêcheur tacheté
d   Trauerfischer
i   Martin pescatore dalla cresta

5588   *Megaceryle maxima*
    Coraciiformes - Cerylidae
e   Giant Kingfisher; African Giant-
    Kingfisher; Alucan Giant-Kingfisher
f   Martin-pêcheur géant
d   Riesenfischer
i   Martin pescatore gigante

5589   *Megaceryle torquata*
    Coraciiformes - Cerylidae
e   Ringed Kingfisher
f   Martin-pêcheur à ventre roux; Cloche

    (Ants); Pie (Ants); Cracra (Ants)
d   Rotbrustfischer
i   Martin pescatore dal collare

5590   *Megacrex inepta*
    Gruiformes - Rallidae
e   New Guinea Flightless-Rail; Papuan
    Flightless-Rail; Grey-faced Rail
f   Râle géant
d   Baumralle
i   Rallo di D'Albertis

5591   *Megadyptes antipodes*
    Sphenisciformes - Spheniscidae
e   Yellow-eyed Penguin
f   Manchot antipode
d   Gelbaugenpinguin
i   Pinguino degli Antipodi; Pinguino
    occhigialli

5592   *Megalaima armillaris*
    Piciformes - Megalaimidae
e   Flame-fronted Barbet; Blue-crowned
    Barbet; Orange-fronted Barbet
f   Barbu souci-col
d   Temminck-Bartvogel
i   Barbuto corona azzurra

5593   *Megalaima asiatica*
    Piciformes - Megalaimidae
e   Blue-throated Barbet; Blue-breasted
    Barbet
f   Barbu à gorge bleue
d   Blauwangenbartvogel
i   Barbuto gola azzurra; Barubto dalla
    gola azzurra

5594   *Megalaima australis*
    Piciformes - Megalaimidae
e   Blue-eared Barbet; Little Barbet;
    Crimson-browed Barbet; Large-
    billed Barbet
f   Barbu à calotte bleue; Barbu de
    Duvaucel
d   Blauohrbartvogel
i   Barbuto orecchieazzurre

5595   *Megalaima chrysopogon*
    Piciformes - Megalaimidae
e   Gold-whiskered Barbet
f   Barbu à joues jaunes; Barbu à

moustaches
d Goldwangenbartvogel
i Barbuto dai mustacchi dorati

5596 *Megalaima corvina*
Piciformes - Megalaimidae
e Brown-throated Barbet; Javan
Barbet; Javan Brown-throated Barbet
f Barbu corbin
d Braunkehlbartvogel
i Barbuto gollabruna

5597 *Megalaima eximia*
Piciformes - Megalaimidae
e Bornean Barbet; Black-throated
Barbet
f Barbu à gorge noire
d Schwarzkehlbartvogel
i Barbuto golanera

5598 *Megalaima faiostricta*
Piciformes - Megalaimidae
e Green-eared Barbet; Lineated Barbet
f Barbu grivelé
d Grünohrbartvogel
i Barbuto orecchieverdi

5599 *Megalaima flavifrons*
Piciformes - Megalaimidae
e Yellow-fronted Barbet; Sri Lanka
Yellow-fronted Barbet; Yellow-faced
Barbet
f Barbu à front d'or
d Gelbstirnbartvogel
i Barbuto frontegialla

5600 *Megalaima franklinii*
Piciformes - Megalaimidae
e Golden-throated Barbet
f Barbu de Franklin; Barbu à gorge
dorée
d Goldkehlbartvogel
i Barbuto goladorata

5601 *Megalaima haemacephala*
Piciformes - Megalaimidae
e Coppersmith Barbet; Coppersmith
(ISC); Crimson-breasted Barbet
(ISC)
f Barbu à plastron rouge
d Goldbartvogel; Kupferschmied;

Rotscheitelbartvogel
i Barbuto fabbro; Barbuto dorato

5602 *Megalaima henricii*
Piciformes - Megalaimidae
e Yellow-crowned Barbet
f Barbu à sourcils jaunes
d Gelbscheitelbartvogel
i Barbuto coronoa gialla

5603 *Megalaima incognita*
Piciformes - Megalaimidae
e Moustached Barbet; Hume's Blue-
throated Barbet
f Barbu de Hume
d Grünscheitelbartvogel
i Barbuto gola azzurra di Hume

5604 *Megalaima javensis*
Piciformes - Megalaimidae
e Black-banded Barbet; Kortorea
Barbet; Javan Barbet
f Barbu de Java
d Java-Bartvogel
i Barbuto di Giava

5605 *Megalaima lagrandieri*
Piciformes - Megalaimidae
e Red-vented Barbet; Lagrandier's
Barbet
f Barbu à ventre rouge; Barbu de
Lagrandière
d Rotsteißbartvogel
i Barbuto di Lagrandier

5606 *Megalaima lineata*
Piciformes - Megalaimidae
e Lineated Barbet; Grey-headed
Barbet; Green Barbet
f Barbu rayé
d Streifenbartvogel
i Barbuto lineato

5607 *Megalaima malabarica*
Piciformes - Megalaimidae
e Crimson-fronted Barbet

5608 *Megalaima monticola*
Piciformes - Megalaimidae
e Mountain Barbet
f Barbu montagnard

d   Borneo-Bartvogel
i   Barbuto di monte

5609   *Megalaima mystacophanos*
Piciformes - Megalaimidae
e   Red-throated Barbet; Gaudy Barbet
f   Barbu arlequin
d   Harlekinbartvogel;
Buntkopfbartvogel
i   Barbuto golarossa

5610   *Megalaima oorti*
Piciformes - Megalaimidae
e   Black-browed Barbet; Müller's
Barbet; Embroidered Barbet;
Mueller's Barbet; Malayan Barbet
f   Barbu malais
d   Schwarzbrauenbartvogel;
Schwarzstirnbartvogel
i   Barbuto dai sopraccigli neri

5611   *Megalaima oorti faber*
Piciformes - Megalaimidae
f   Barbu forgeron

5612   *Megalaima oorti nuchalis*
Piciformes - Megalaimidae
e   Formosan Barbet
f   Barbu de Formose
d   Formosa-Bartvogel

5613   *Megalaima pulcherrima*
Piciformes - Megalaimidae
e   Golden-naped Barbet; Kinabalu
Barbet
f   Barbu élégant
d   Prachtbartvogel
i   Barbuto nucadorata

5614   *Megalaima rafflesii*
Piciformes - Megalaimidae
e   Red-crowned Barbet; Golden-
rumped Barbet; Many-coloured
Barbet
f   Barbu bigarré
d   Vielfarbenbartvogel
i   Barbuto multicolore

5615   *Megalaima rubricapilla*
Piciformes - Megalaimidae
e   Crimson-throated Barbet; Crimson-
fronted Barbet (ISC); Small Barbet;
Ceylon Small Barbet
f   Barbu à couronne rouge
d   Malabar-Schmied
i   Barbuto capirosso

5616   *Megalaima virens*
Piciformes - Megalaimidae
e   Great Barbet; Himalayan Great
Barbet (ISC); Great Himalayan
Barbet; Great Hill-Barbet
f   Barbu géant; Grand Barbu
d   Blaukopfbartvogel
i   Barbuto grosso

5617   *Megalaima viridis*
Piciformes - Megalaimidae
e   White-cheeked Barbet; Small Green-
Barbet (ISC); Little Green Barbet;
Small Brown-headed Barbet; Indian
Small Green Barbet; Indian Little
Green-Barbet; Brown-headed Barbet
f   Barbu vert
d   Grünbartvogel
i   Barbuto piccolo verde

5618   *Megalaima zeylanica*
Piciformes - Megalaimidae
e   Brown-headed Barbet; Large Green-
Barbet (ISC); Oriental Green Barbet;
Great Green Barbet; Green Barbet
f   Barbu à tête brune
d   Braunkopfbartvogel; Ceylon-
Bartvogel
i   Barbuto verde

5619   Megalaimidae
Piciformes
e   Asian Barbets
f   Megalaimidés
d   Asiatische Bartvögel
i   Megalaimidi

5620   *Megalurulus grosvenori*
Passeriformes - Sylviidae
e   Bismarck Thicketbird; Wightman
Mountains Warbler; New Britain
Thicket-Warbler; Bismarck Thicket
Warbler
f   Mégalure de Gilliard

d   Maskenbuschsänger
i   Cantore di macchia di Grosvenor

5621   *Megalurulus llaneae*
    Passeriformes - Sylviidae
e   Bougainville Thicketbird;
    Bougainville Thicket Warbler
f   Mégalure de Bougainville
i   Cantore di macchia di Bougainville

5622   *Megalurulus mariei*
    Passeriformes - Sylviidae
e   New Caledonian Grassbird; New
    Caledonian Grass-Warbler
f   Mégalure calédonienne
d   Verreaux-Buschsänger
i   Cantore della Nuova Caledonia

5623   *Megalurulus rubiginosus*
    Passeriformes - Sylviidae
e   Rusty Thicketbird; New Britain
    Babbler-Warbler; Rufous-faced
    Thicket-Warbler
f   Mégalure rubigineuse
d   Rotbrustbuschsänger
i   Cantore di macchia facciarossa

5624   *Megalurulus whitneyi*
    Passeriformes - Sylviidae
e   Guadalcanal Thicketbird; Thicket
    Warbler; Solomons Thicketbird
f   Mégalure de Whitney
d   Whitney-Buschsänger
i   Cantore di macchia di Whitney

5625   *Megalurulus whitneyi turipave*
    Passeriformes - Sylviidae
f   Mégalure de Guadalcanal

5626   *Megalurus albolimbatus*
    Passeriformes - Sylviidae
e   Fly River Grassbird; Fly River Grass-
    Warbler; Albertis's Grassbird
f   Mégalure à ventre blanc
d   Weißflügelschilfsteiger
i   Forapaglie codone di D'Albertis

5627   *Megalurus gramineus*
    Passeriformes - Sylviidae
e   Little Grassbird; Little Marshbird;
    Little Grass-Warbler

f   Mégalure menue
d   Zwergschilfsteiger
i   Forapaglie codone piccolo

5628   *Megalurus palustris*
    Passeriformes - Sylviidae
e   Striated Grassbird; Striated
    Canegrass-Warbler; Striated Marsh-
    Warbler (ISC); Striated Warbler
f   Mégalure des marais; Grande
    Fauvette des marais
d   Strichelkopfschilfsteiger
i   Forapaglie codone striato

5629   *Megalurus pryeri*
    Passeriformes - Sylviidae
e   Japanese Marsh-Warbler
f   Mégalure du Japon
d   Riedsänger
i   Forapaglie codone cinese

5630   *Megalurus punctatus*
    Passeriformes - Sylviidae
e   New Zealand Fernbird; Fernbird;
    Mata
f   Mégalure matata
d   Farnsteiger
i   Forapaglie codone putenggiato

5631   *Megalurus rufescens*
    Passeriformes - Sylviidae
e   Chatham Islands Fernbird
f   Mégalure des Chatham

5632   *Megalurus timoriensis*
    Passeriformes - Sylviidae
e   Tawny Grassbird; Tawny Marshbird;
    Rufous-capped Marsh-Warbler;
    Tawny Grass-Warbler; Rufous-
    capped Grassbird; Rufous-capped
    Canegrass-Warbler
f   Mégalure fauve
d   Rostkopfschilfsteiger
i   Forapaglie codone fulvo

5633   Megapodiidae
    Galliformes
e   Megapodes; Moundbuilders
f   Megapodiidés
d   Großfußhühner
i   Megapodidi

5634   *Megapodius affinis*
Galliformes - Megapodiidae
e   New Guinea Scrubfowl
f   Mégapode de Nouvelle-Guinée
i   Megapodio della Nuova Guinea

5635   *Megapodius bernsteinii*
Galliformes - Megapodiidae
e   Sula Scrubfowl; Sula Megapode
f   Mégapode de Bernstein
i   Megapodio di Bernstein

5636   *Megapodius cumingii*
Galliformes - Megapodiidae
e   Tabon Scrubfowl; Tabon; Philippine
Scrubfowl; Philippine Megapode;
Megapode; Cuming's Scrubfowl
f   Mégapode des Philippines
i   Megapodio delle Filippine

5637   *Megapodius eremita*
Galliformes - Megapodiidae
e   Melanesian Scrubfowl; Bismarck
Scrubfowl
f   Mégapode mélanésien
i   Megapodio eremita

5638   *Megapodius forstenii*
Galliformes - Megapodiidae
e   Forsten's Scrubfowl; Forsten's
Megapode

5639   *Megapodius freycinet*
Galliformes - Megapodiidae
e   Dusky Scrubfowl; Common
Scrubhen; Common Scrubfowl;
Scrubfowl; Scrubhen; Megapode;
Common Megapode; Dusky
Megapode
f   Mégapode de Freycinet
d   Großfußhuhn
i   Megapodio di Freycinet

5640   *Megapodius geelvinkianus*
Galliformes - Megapodiidae
e   Geelvink Scrubfowl

5641   *Megapodius laperouse*
Galliformes - Megapodiidae
e   Micronesian Scrubfowl; Marianas
Scrubhen; Marianas Scrubfowl;

Micronesian Megapode; Palau
Scrubfowl; Mariana Scrubhen;
Mariana Scrubfowl
f   Mégapode de La Pérouse
d   Lapérousehuhn
i   Megapodio di Laperouse

5642   *Megapodius layardi*
Galliformes - Megapodiidae
e   Vanuatu Scrubfowl; Banks Island
Scrubfowl; New Hebrides Scrubfowl;
Tristram's Scrubfowl
f   Mégapode de Layard
i   Megapodio di Layard

5643   *Megapodius nicobariensis*
Galliformes - Megapodiidae
e   Nicobar Scrubfowl; Nicobar
Megapode
f   Mégapode des Nicobar
i   Megapodio delle Nicobare

5644   *Megapodius pritchardii*
Galliformes - Megapodiidae
e   Niuafou Scrubfowl; Pritchard's
Scrubfowl; Polynesian Scrubhen;
Niuafou Megapode; Malay Fowl;
Polynesian Scrubfowl; Tongan
Megapode; Tongan Scrubfowl
f   Mégapode de Pritchard
d   Pritchard-Huhn
i   Megapodio della Polinesia

5645   *Megapodius reinwardt*
Galliformes - Megapodiidae
e   Orange-footed Scrubfowl;
Reinwardt's Scrubfowl; Orange-
footed Megapode
f   Mégapode de Reinwardt
i   Megapodio zampearancio

5646   *Megapodius tenimberensis*
Galliformes - Megapodiidae
e   Tanimbar Megapode; Tanimbar
Scrubfowl

5647   *Megapodius wallacei*
Galliformes - Megapodiidae
e   Moluccan Scrubfowl; Moluccan
Scrubhen; Moluccan Megapode;
Wallace's Scrubfowl; Moluccas

Scrubhen; Moluccas Scrubfowl
f Mégapode de Wallace
d Molukken-Huhn
i Megapodio di Wallace

5648 *Megarynchus pitangua*
Passeriformes - Tyrannidae
e Boat-billed Flycatcher
f Tyran pitangua
d Bauchschnabel;
Bauchschnabeltyrann
i Pitango beccogrosso

5649 *Megastictus margaritatus*
Passeriformes - Formicariidae
e Pearly Antshrike
f Batara perlé
d Perlenwollrücken
i Averla formichiera perlata

5650 *Megatriorchis doriae*
Falconiformes - Accipitridae
e Doria's Goshawk; Doria's Hawk
f Autour de Doria
d Neuguinea-Habicht
i Astore di Doria

5651 *Megaxenops parnaguae*
Passeriformes - Furnariidae
e Great Xenops; Reiser's Recurvebill
f Megasittine du Brésil
d Riesensteigschnabel
i Xenope maggiore

5652 *Megazosterops palauensis*
Passeriformes - Zosteropidae
e Giant White-eye; Large Palau White-eye; Palau White-eye
f Zostérops des Palau
d Bronzebrillenvogel
i Occhialino di Palau

5653 *Meiglyptes jugularis*
Piciformes - Picidae
e Black-and-buff Woodpecker
f Pic jugulaire
d Dommelspecht
i Picchio nero maculato

5654 *Meiglyptes tristis*
Piciformes - Picidae

e Buff-rumped Woodpecker; Fulvous-rumped Woodpecker; Fulvous-rumped Barred Woodpecker
f Pic strihup
d Braunbürzelspecht
i Picchio barrato chiaro

5655 *Meiglyptes tukki*
Piciformes - Picidae
e Buff-necked Woodpecker; Barred Woodpecker; Buff-necked Barred Woodpecker
f Pic tukki
d Tukkispecht
i Picchio barrato scuro

5656 *Melaenornis annamarulae*
Passeriformes - Muscicapidae
e West African Black-Flycatcher; Liberian Flycatcher; Mrs.. Forbes-Watson's Black-Flycatcher; Nimba Flycatcher; Liberian Black-Flycatcher
f Gobemouche du Libéria; Gobe-mouches du Liberia
d Marula-Schnäpper
i Pigliamosche nero liberiano

5657 *Melaenornis ardesiacus*
Passeriformes - Muscicapidae
e Yellow-eyed Black-Flycatcher; Western Mountain-Flycatcher; Berlioz's Black-Flycatcher; Berlioz's Flycatcher
f Gobemouche de Berlioz; Gobe-mouches de Berlioz
d Gelbaugendrongoschnäpper
i Pigliamosche nero di Berlioz

5658 *Melaenornis edolioides*
Passeriformes - Muscicapidae
e Northern Black-Flycatcher; Black Flycatcher; Eidolon Flycatcher; Western Black Flycatcher; African Black-Flycatcher
f Gobemouche drongo; Gobe-mouches drongo
d Senegal-Drongoschnäpper; Swainson-Schnäpper; Schieferschwarzer Drongoschnäpper
i Pigliamosche nero

5659 **Melaenornis fischeri**
Passeriformes - Muscicapidae
e White-eyed Slaty Flycatcher (CSA);
White-eyed Slaty Flycatcher
f Gobemouche de Fischer; Gobe-
mouches de Fischer
d Bergdrongoschnäpper
i Pigliamosche ardesia occhibianci

5660 **Melaenornis pammelaina**
Passeriformes - Muscicapidae
e Southern Black-Flycatcher; Black
Flycatcher (CSA); South African
Black-Flycatcher; African Black-
Flycatcher
f Gobemouche sud-africain; Gobe-
mouches sud-africain
d Drongoschnäpper
i Pigliamosche nero sudafricano

5661 **Melampitta gigantea**
Passeriformes - Corvidae
e Greater Melampitta
f Grande Mélampitte
d Rußflöter
i Melampitta gigante

5662 **Melampitta lugubris**
Passeriformes - Corvidae
e Lesser Melampitta
f Petite Mélampitte
d Glanzflöter
i Melampitta minore

5663 **Melamprosops phaeosoma**
Passeriformes - Fringillidae
e Poo-uli; Black-faced Honeycreeper;
Po'ouli
f Po-o-uli masqué
d Maui-Gimpel
i Po'ouli

5664 **Melanerpes aurifrons**
Piciformes - Picidae
e Golden-fronted Woodpecker; Gold-
fronted Woodpecker
f Pic à front doré; Pic à front d'or
d Goldstirnspecht
i Picchio frontedorata

5665 **Melanerpes cactorum**
Piciformes - Picidae
e White-fronted Woodpecker
f Pic des cactus
d Kaktusspecht
i Picchio dei cactus

5666 **Melanerpes candidus**
Piciformes - Picidae
e White Woodpecker
f Pic dominicain
d Weißspecht
i Picchio bianco

5667 **Melanerpes carolinus**
Piciformes - Picidae
e Red-bellied Woodpecker
f Pic à ventre roux; Pic de Caroline
d Carolina-Specht
i Picchio della Carolina

5668 **Melanerpes chrysauchen**
Piciformes - Picidae
e Golden-naped Woodpecker; Gold-
naped Woodpecker
f Pic masqué
d Buntkopfspecht; Goldnackenspecht
i Picchio nucadorata

5669 **Melanerpes chrysogenys**
Piciformes - Picidae
e Golden-cheeked Woodpecker; Gold-
cheeked Woodpecker
f Pic élégant
d Goldwangenspecht
i Picchio guancedorate

5670 **Melanerpes cruentatus**
Piciformes - Picidae
e Yellow-tufted Woodpecker; Red-
fronted Woodpecker
f Pic à chevron d'or
d Goldschopfspecht
i Picchio fronterossa

5671 **Melanerpes erythrocephalus**
Piciformes - Picidae
e Red-headed Woodpecker
f Pic à tête rouge
d Rotkopfspecht
i Picchio testarossa; Picchio capirosso

5672   *Melanerpes flavifrons*
Piciformes - Picidae
e   Yellow-fronted Woodpecker
f   Pic à front jaune
d   Goldmaskenspecht
i   Picchio frontegialla; Picchio dalla
     fronte gialla

5673   *Melanerpes formicivorus*
Piciformes - Picidae
e   Acorn Woodpecker
f   Pic glandivore
d   Eichelspecht
i   Picchio delle ghiande

5674   *Melanerpes herminieri*
Piciformes - Picidae
e   Guadeloupe Woodpecker
f   Pic de la Guadeloupe; Tapeur (Ants);
     Pic de l'Herminier
d   Guadeloupe-Specht
i   Picchio della Guadalupa

5675   *Melanerpes hoffmannii*
Piciformes - Picidae
e   Hoffmann's Woodpecker
f   Pic de Hoffman
i   Picchio di Hoffmann

5676   *Melanerpes hypopolius*
Piciformes - Picidae
e   Grey-breasted Woodpecker; Balsas
     Woodpecker
f   Pic alezan
d   Graukehlspecht
i   Picchio pettogrigio

5677   *Melanerpes lewis*
Piciformes - Picidae
e   Lewis's Woodpecker
f   Pic de Lewis
d   Blutgesichtsspecht; Seidenspecht
i   Picchio di Lewis

5678   *Melanerpes portoricensis*
Piciformes - Picidae
e   Puerto Rican Woodpecker
f   Pic de Porto Rico
d   Scharlachbrustspecht
i   Picchio di Portorico

5679   *Melanerpes pucherani*
Piciformes - Picidae
e   Black-cheeked Woodpecker;
     Pucheran's Woodpecker
f   Pic de Pucheran
d   Schläfenfleckspecht
i   Picchio guancenere

5680   *Melanerpes pulcher*
Piciformes - Picidae
e   Beautiful Woodpecker
f   Pic splendide

5681   *Melanerpes pygmaeus*
Piciformes - Picidae
e   Yucatan Woodpecker; Red-vented
     Woodpecker
f   Pic du Yucatan
d   Yucatan-Specht
i   Picchio dello Yucatan

5682   *Melanerpes radiolatus*
Piciformes - Picidae
e   Jamaican Woodpecker; Woodpecker
     (WI)
f   Pic de la Jamaïque
d   Jamaika-Specht
i   Picchio della Giamaica

5683   *Melanerpes rubricapillus*
Piciformes - Picidae
e   Red-crowned Woodpecker
f   Pic à couronne rouge
d   Rotkappenspecht
i   Picchio corona rossa

5684   *Melanerpes striatus*
Piciformes - Picidae
e   Hispaniolan Woodpecker
f   Charpentier (Ants); Pic d'Hispaniola
d   Haiti-Specht
i   Picchio di Hispaniola

5685   *Melanerpes superciliaris*
Piciformes - Picidae
e   West Indian Woodpecker; Great
     Red-bellied Woodpecker; West
     Indian Red-bellied Woodpecker;
     Bahama Woodpecker; Readhead
     (WI); Cuban Red-headed
     Woodpecker

f   Pic à sourcils noirs
d   Bahama-Specht
i   Picchio panciarossa

5686   *Melanerpes uropygialis*
       Piciformes - Picidae
e   Gila Woodpecker
f   Pic des saguaros
d   Gila-Specht
i   Picchio di Gila

5687   *Melanitta americana*
       Anseriformes - Anatidae
e   American Scoter; Black Scoter
f   Macreuse à bec jaune
d   Amerikanische Trauerente

5688   *Melanitta deglandi*
       Anseriformes - Anatidae
e   White-winged Scoter (NA);
       Degland's Scoter; Degland's White-
       winged Scoter
f   Macreuse à ailes blanches
d   Höckerschnabelente

5689   *Melanitta fusca*
       Anseriformes - Anatidae
e   Velvet Scoter
f   Macreuse brune; Macreuse à ailes
       blanches
d   Samtente
i   Orco marino

5690   *Melanitta nigra*
       Anseriformes - Anatidae
e   Black Scoter (NA); Common Scoter;
       Scoter; American Scoter
f   Macreuse noire; Macreuse à bec
       jaune; Canard macreuse
d   Trauerente
i   Orchetto marino; Orco marino

5691   *Melanitta perspicillata*
       Anseriformes - Anatidae
e   Surf Scoter
f   Macreuse à lunettes; Macreuse à
       front blanc
d   Brillenente
i   Orco marino dagli occhiali; Anatra
       dal becco largo

5692   *Melanitta stejnegeri*
       Anseriformes - Anatidae
e   Asiatic Scoter

5693   *Melanocharis arfakiana*
       Passeriformes - Melanocharitidae
e   Obscure Berrypecker
f   Piquebaie obscur
d   Arfak-Beerenpicker
i   Beccafrutti dell'Arfak

5694   *Melanocharis crassirostris*
       Passeriformes - Melanocharitidae
e   Spotted Berrypecker; Thick-billed
       Berrypecker
f   Piquebaie tacheté
d   Beerenpicker;
       Schlankschnabelbeerenpicker;
       Dickschnabelmistelesser
i   Beccafrutti maculato

5695   *Melanocharis longicauda*
       Passeriformes - Melanocharitidae
e   Lemon-breasted Berrypecker; Mid-
       mountain Berrypecker; Yellow-billed
       Berrypecker; Long-tailed Black-
       Berrypecker
f   Piquebaie à longue queue
d   Gelbbüschelbeerenpicker
i   Beccafrutti di collina

5696   *Melanocharis nigra*
       Passeriformes - Melanocharitidae
e   Black Berrypecker
f   Piquebaie noir
d   Weißbüschelbeerenpicker
i   Beccafrutti nero

5697   *Melanocharis striativentris*
       Passeriformes - Melanocharitidae
e   Streaked Berrypecker; Striated
       Berrypecker; Green Berrypecker
f   Piquebaie strié
d   Streifenbauchbeerenpicker
i   Beccafrutti striato

5698   *Melanocharis versteri*
       Passeriformes - Melanocharitidae
e   Fan-tailed Berrypecker; Verster's
       Berrypecker
f   Piquebaie éventail

d   Fächerschwanzbeerenpicker
i   Beccafrutti coda a ventaglio

5699   Melanocharitidae
Passeriformes
e   Solitary Berrypeckers; Berrypeckers
f   Melanocharitidés
d   Beerenpicker
i   Melanocaritidi

5700   *Melanochlora sultanea*
Passeriformes - Paridae
e   Sultan Tit
f   Mésange sultane
d   Sultansmeise
i   Cincia sultna

5701   *Melanocorypha bimaculata*
Passeriformes - Alaudidae
e   Bimaculated Lark; Eastern Calandra;
Eastern Calandra-Lark
f   Alouette monticole; Calandre
orientale
d   Bergkalanderlerche;
Ostkalanderlerche
i   Calandra asiatica; Calandra
bimaculata

5702   *Melanocorypha calandra*
Passeriformes - Alaudidae
e   Calandra Lark; European Calandra-
Lark
f   Alouette calandre
d   Kalanderlerche
i   Allodola calandra; Calandra;
Calandra comune

5703   *Melanocorypha leucoptera*
Passeriformes - Alaudidae
e   White-winged Lark
f   Alouette leucoptère; Alouette à ailes
blanches; Calandre leucoptère
d   Weißflügellerche
i   Calandra siberiana

5704   *Melanocorypha maxima*
Passeriformes - Alaudidae
e   Tibetan Lark; Long-billed Calandra-
Lark; Long-billed Lark; Asiatic Lark;
Asiatic Long-billed Lark
f   Alouette du Tibet; Grande Calandre

d   Sumpflerche; Riesensumpflerche
i   Calandra maggiore

5705   *Melanocorypha mongolica*
Passeriformes - Alaudidae
e   Mongolian Lark; Mongolian Skylark
f   Alouette de Mongolie; Alouette
mongole; Calandre de Mongolie
d   Mongolen-Lerche
i   Calandra della Mongolia

5706   *Melanocorypha yeltoniensis*
Passeriformes - Alaudidae
e   Black Lark
f   Alouette nègre
d   Mohrenlerche
i   Calandra nera

5707   *Melanodera melanodera*
Passeriformes - Fringillidae
e   Canary-winged Finch; Black-
throated Finch
f   Mélanodère à sourcils blancs
d   Schwarzkehlammerfink
i   Fringuello golanera

5708   *Melanodera xanthogramma*
Passeriformes - Fringillidae
e   Yellow-bridled Finch
f   Mélanodère à sourcils jaunes
d   Zügelammerfink
i   Fringuello dai mustacchi gialli

5709   *Melanodryas cucullata*
Passeriformes - Petroicidae
e   Hooded Robin; Hooded Robin-
Flycatcher
f   Miro à capuchon
d   Scheckenschnäpper
i   Petroica dal cappuccio

5710   *Melanodryas cucullata cucullata*
Passeriformes - Petroicidae
e   South-eastern Hooded Robin (ANZ)
i   Petroica di Tasmania

5711   *Melanodryas cucullata melvillensis*
Passeriformes - Petroicidae
e   Tiwi Islands Hooded Robin (ANZ)

5712 **Melanodryas vittata**
Passeriformes - Petroicidae
e Dusky Robin; Dusky Robin-
Flycatcher
f Miro de Tasmanie
d Tasman-Schnäpper
i Pigliamosche

5713 **Melanopareia elegans**
Passeriformes - Rhinocryptidae
e Elegant Crescentchest
f Cordon-noir élégant
d Schmuckbandvogel
i Pettolunato elegante

5714 **Melanopareia maranonica**
Passeriformes - Rhinocryptidae
e Marañon Crescentchest
f Cordon-noir du Marañon
d Marañon-Bandvogel

5715 **Melanopareia maximiliani**
Passeriformes - Rhinocryptidae
e Olive-crowned Crescentchest;
Maximilian's Crescentchest
f Cordon-noir à dos olive; Cordon-noir
d'Orbigny
d Olivkappenbandvogel
i Pettolunato del principe
Massimiliano

5716 **Melanopareia torquata**
Passeriformes - Rhinocryptidae
e Collared Crescentchest
f Cordon-noir à col roux; Cordon-noir
à collier
d Zimtbandvogel
i Pettolunato dal collare

5717 **Melanoperdix nigra**
Galliformes - Phasianidae
e Black Partridge; Black Wood-
Partridge
f Perdrix noir
d Schwarzwachtel
i Pernice nera

5718 **Melanoptila glabrirostris**
Passeriformes - Sturnidae
e Black Catbird
f Moqueur noir

d Glanzkatzendrossel;
Glanzspottdrossel
i Uccello gatto nero

5719 **Melanospiza richardsoni**
Passeriformes - Fringillidae
e St. Lucia Black-Finch
f Moisson pied-blanc
d Schwarzammer; Blaßfußgimpelfink
i Fringuello nero di Santa Lucia

5720 **Melanotis caerulescens**
Passeriformes - Sturnidae
e Blue Mockingbird
f Moqueur bleu
d Blaukopfspottdrossel;
Blauspottdrossel
i Mimo azzurro; Mimo blu del
Messico

5721 **Melanotis hypoleucus**
Passeriformes - Sturnidae
e Blue-and-white Mockingbird
f Moqueur bleu-et-blanc
d Lasurspottdrossel
i Mimo bianco e azzurro

5722 **Melanotrochilus fuscus**
Apodiformes - Trochilidae
e Black Jacobin; Pied Jacobin; Dusky
Jacobin; White-tailed Jacobin
f Colibri demi-deuil
d Trauerkolibri
i Colibrì nero

5723 **Meleagrididae**
Galliformes
e Turkeys
f Dindons sauvages
d Truthüner
i Meleagrididi

5724 **Meleagris gallopavo**
Galliformes - Meleagrididae
e Wild Turkey; Turkey; Common
Turkey; Gang (col)
f Dindon; Dindon bronzé; Dindon
commun; Dindon sauvage
d Truthuhn; Wildes Truthuhn;
Gemeines Truthuhn; Truthahn m;
Wilder Truthahn m; Gemeiner

Truthahn m
i Tacchino; Tacchino selvatico; Tacchino comune

5725 *Meleagris ocellata*
Galliformes - Meleagrididae
e Ocellated Turkey
f Dindon ocellée
d Pfauentruthuhn
i Tacchino ocellato

5726 *Melichneutes robustus*
Piciformes - Indicatoridae
e Lyre-tailed Honeyguide
f Indicateur à queue-en-lyre
d Leierschwanzhoniganzeiger
i Indicatore coda a lira

5727 *Melidectes belfordi*
Passeriformes - Meliphagidae
e Belford's Honeyeater; Belford's Melidectes
f Méliphage de Belford
d Belford-Honigfresser
i Mangiamiele di Belford

5728 *Melidectes foersteri*
Passeriformes - Meliphagidae
e Huon Wattled-Honeyeater; Förster's Melidectes; Huon Melidectes; Huon Wattled-Melidectes; Huon Honeyeater; Förster's Honeyeater
f Méliphage de Foerster
i Mangiamiele di Foerster

5729 *Melidectes fuscus*
Passeriformes - Meliphagidae
e Sooty Melidectes; Sooty Honeyeater
f Méliphage fuligineux
d Buntwarzenhonigfresser
i Mangiamiele fuligginoso

5730 *Melidectes leucostephes*
Passeriformes - Meliphagidae
e Vogelkop Honeyeater; White-fronted Melidectes; Arfak White-fronted Honeyeater; Vogelkop Melidectes; White-fronted Honeyeater; Arfak Melidectes
f Méliphage à face blanche

d Diademhonigfresser
i Mangiamiele di Vogelkop

5731 *Melidectes nouhuysi*
Passeriformes - Meliphagidae
e Short-bearded Honeyeater; Short-bearded Melidectes; Bearded Melidectes; Houhuysi
f Méliphage à barbe courte
d Kurzbarthonigfresser
i Mangiamiele barbacorta

5732 *Melidectes ochromelas*
Passeriformes - Meliphagidae
e Cinnamon-browed Honeyeater; Mid-mountain Honeyeater; Mid-mountain Melidectes; Dark-mantled Honeyeater; Cinnamon-browed Melidectes
f Méliphage à sourcils roux
d Rostohrhonigfresser
i Mangiamiele delle colline

5733 *Melidectes princeps*
Passeriformes - Meliphagidae
e Long-bearded Honeyeater; Long-bearded Melidectes
f Méliphage à barbe longue
d Langbarthonigfresser
i Mangiamiele barbalunga

5734 *Melidectes rufocrissalis*
Passeriformes - Meliphagidae
e Yellow-browed Honeyeater; Yellow-browed Melidectes; Reichenow's Honeyeater
f Méliphage de Reichenow
i Mangiamiele di Reichenow

5735 *Melidectes sclateri*
Passeriformes - Meliphagidae
e San Cristobal Honeyeater; San Cristobal Melidectes
f Méliphage de San Cristobal
d Rostschwanzhonigfresser
i Mangiamiele di San Cristobal

5736 *Melidectes torquatus*
Passeriformes - Meliphagidae
e Ornate Honeyeater; Cinnamon-breasted Melidectes; Cinnamon-

breasted Wattlebird; Ornamental
Melidectes
f   Méliphage maquille
d   Zimtbrusthonigfresser;
Nacktaugenhonigesser
i   Mangiamiele ornato

5737   *Melidectes whitemanensis*
Passeriformes - Meliphagidae
e   Bismarck Honeyeater; Whiteman
Honeyeater; Whiteman Mountain
Honeyeater; Gilliard's Honeyeater;
Vose Melidectes; Whiteman
Melidectes; New Britain Melidectes;
Bismarck Melidectes
f   Méliphage de Whiteman
d   Gilliard-Honigfresser
i   Mangiamiele di Gilliard

5738   *Melidora macrorrhina*
Coraciiformes - Halcyonidae
e   Hook-billed Kingfisher
f   Martin-chasseur d'Euphrosine
d   Hakenliest
i   Martin pescatore beccouncinato

5739   *Melierax canorus*
Falconiformes - Accipitridae
e   Pale Chanting-Goshawk
f   Autour chanteur; Autour-chanteur
pâle
d   Weißbürzelsinghabicht; Weißbürzel-
Heller Singhabicht; Weißbürzel
Großer Singhabicht; Heller
Singhabicht; Großer Singhabicht
i   Astore canoro pallido; Astore
cantante pallido

5740   *Melierax gabar*
Falconiformes - Accipitridae
e   Gabar Goshawk
f   Autour gabar
d   Gabarhabicht
i   Astore gabar

5741   *Melierax metabates*
Falconiformes - Accipitridae
e   Dark Chanting-Goshawk; Chanting
Goshawk; Southern Pale Chanting-
Goshawk; Eastern Chanting-
Goshawk

f   Autour sombre
d   Graubürzel-Dunkler Singhabicht;
Graubürzelsinghabicht; Dunkler
Singhabicht; Kleiner Singhabicht
i   Astore cantante scuro; Astore
cantante; Astore canore scuro; Astore
cantante bruno

5742   *Melierax poliopterus*
Falconiformes - Accipitridae
e   Eastern Chanting-Goshawk; Eastern
Pale Chanting-Goshawk (CSA); Grey
Chanting-Goshawk; Somali
Chanting-Goshawk
f   Autour à ailes grises
i   Astore canoro pallido

5743   *Melignomon eisentrauti*
Piciformes - Indicatoridae
e   Yellow-footed Honeyguide;
Eisentraut's Honeyguide; Coe's
Honeyguide
f   Indicateur d'Eisentraut
i   Indicatore zampegialle

5744   *Melignomon zenkeri*
Piciformes - Indicatoridae
e   Zenker's Honeyguide
f   Indicateur de Zenker
d   Weißschwanzhoniganzeiger; Zenker-
Honiganzeiger
i   Indicatore di Zenker

5745   *Melilestes megarhynchus*
Passeriformes - Meliphagidae
e   Long-billed Honeyeater
f   Méliphage à long bec
d   Langschnabelhonigfresser
i   Mangiamiele beccolungo

5746   *Meliphaga albilineata*
Passeriformes - Meliphagidae
e   White-lined Honeyeater; White-
striped Honeyeater
f   Méliphage à boucle blanche
d   Weißbarthonigfresser
i   Mangiamiele dai mustacchi bianchi

5747   *Meliphaga albonotata*
Passeriformes - Meliphagidae
e   Scrub Honeyeater; White-marked

Meliphaga; White-marked
Honeyeater; Scrub White-eared
Honeyeater; White-eyed Honeyeater;
White-eyed Mountain-Honeyeater;
Diamond Honeyeater; Southern
Honeyeater; Mountain Meliphaga;
Mountain Honeyeater
f   Méliphage buissonnier
d   Schneeohrhonigfresser
i   Mangiamiele di macchia

5748   *Meliphaga analoga*
Passeriformes - Meliphagidae
e   Mimetic Honeyeater; Mimic
Honeyeater; Mimic Meliphaga;
Allied Honeyeater; Yellow-spotted
Honeyeater
f   Méliphage sosie
d   Papua-Honigfresser
i   Mangiamiele mimo

5749   *Meliphaga aruensis*
Passeriformes - Meliphagidae
e   Puff-backed Honeyeater; Puff-
backed Meliphaga; Large Tufted
Honeyeater
f   Méliphage bouffant
d   Aru-Honigfresser
i   Mangiamiele di Aru

5750   *Meliphaga auga*
Passeriformes - Meliphagidae
e   Southern White-eared Honeyeater;
Southern White-eared Mountain-
Honeyeater
d   Rand-Honigfresser

5751   *Meliphaga flavirictus*
Passeriformes - Meliphagidae
e   Yellow-gaped Honeyeater; Yellow-
gaped Meliphaga
f   Méliphage souriant
d   Gelbkinnhonigfresser
i   Mangiamiele boccagialla

5752   *Meliphaga gracilis*
Passeriformes - Meliphagidae
e   Graceful Honeyeater; Slender-billed
Meliphaga
f   Méliphage gracile

d   Feenhonigfresser
i   Mangiamiele gracile

5753   *Meliphaga lewinii*
Passeriformes - Meliphagidae
e   Lewin's Honeyeater; Greater Lewin's
Honeyeater
f   Méliphage de Lewin
d   Goldohrhonigfresser
i   Mangiamiele maggiore di Lewin

5754   *Meliphaga mimikae*
Passeriformes - Meliphagidae
e   Mottle-breasted Honeyeater; Large
Spot-breasted Meliphaga; Large
Spot-breasted Honeyeater; Spot-
breasted Honeyeater
f   Méliphage de Mimika
d   Mimika-Honigfresser
i   Mangiamiele pettomacchiato
maggiore

5755   *Meliphaga montana*
Passeriformes - Meliphagidae
e   Forest Honeyeater; White-eared
Mountain-Meliphaga; White-eared
Mountain-Honeyeater; Mountain
Meliphaga; Mountain Honeyeater;
Forest White-eared Honeyeater
f   Méliphage forestier
d   Bergwaldhonigfresser
i   Mangiamiele delle foreste

5756   *Meliphaga notata*
Passeriformes - Meliphagidae
e   Yellow-spotted Honeyeater; Lesser
Lewin's Honeyeater; Lesser Lewin
Honeyeater
f   Méliphage marqué
d   Torres-Honigfresser
i   Mangiamiele minore di Lewin

5757   *Meliphaga orientalis*
Passeriformes - Meliphagidae
e   Hill-forest Honeyeater; Small Spot-
breasted Meliphaga; Small Spot-
breasted Honeyeater; Mountain
Honeyeater; Mountain Meliphaga;
Mountain Yellow-eared Honeyeater
f   Méliphage montagnard

d   Schlankschnabelhonigfresser
i   Mangiamiele pettomacchiato minore

5758   *Meliphaga reticulata*
    Passeriformes - Meliphagidae
e   Streaky-breasted Honeyeater;
    Reticulated Honeyeater; Temmick's
    Honeyeater; Streak-breasted
    Honeyeater; Streak-breasted
    Meliphaga; Timor Meliphaga
f   Méliphage reticulé
d   Temminck-Honigfresser
i   Mangiamiele reticolato

5759   *Meliphaga vicina*
    Passeriformes - Meliphagidae
e   Tagula Honeyeater; Louisiades
    Honeyeater; Louisiades Meliphaga;
    Louisiade Meliphaga; Louisiade
    Honeyeater
f   Méliphage de Tagula
d   Tagula-Honigfresser
i   Mangiamiele delle Luisiade

5760   **Meliphagidae**
    Passeriformes
e   Honeyeaters
f   Meliphagidés
d   Honigfresser; Honigesser
i   Melifagidi

5761   *Melipotes ater*
    Passeriformes - Meliphagidae
e   Spangled Honeyeater; Huon
    Melipotes; Huon Honeyeater; Black
    Honeyeater
f   Méliphage pailleté
d   Huon-Honigfresser
i   Succhiamiele ornato di Huon

5762   *Melipotes fumigatus*
    Passeriformes - Meliphagidae
e   Smoky Honeyeater; Common
    Melipotes; Melipotes; Smoky
    Melipotes; Common Honeyeater;
    Eastern Smoky Honeyeater
f   Méliphage enfumé
d   Aschbrusthonigfresser
i   Succhiamiele fuligginoso

5763   *Melipotes gymnops*
    Passeriformes - Meliphagidae
e   Arfak Honeyeater; Arfak Melipotes;
    Western Smoky Honeyeater; Smoky
    Honeyeater; Bare-eyed Honeyeater
f   Méliphage à ventre tacheté
d   Fleckenbrusthonigfresser
i   Succhiamiele di D'Albertis

5764   *Melithreptus affinis*
    Passeriformes - Meliphagidae
e   Black-headed Honeyeater
f   Méliphage à tête noire
d   Schwarzkopfhonigschmecker
i   Mangiamiele testanera

5765   *Melithreptus albogularis*
    Passeriformes - Meliphagidae
e   White-throated Honeyeater
f   Méliphage à menton blanc
d   Weißkinnhonigschmecker
i   Mangiamiele dai sopraccigli bianchi

5766   *Melithreptus brevirostris*
    Passeriformes - Meliphagidae
e   Brown-headed Honeyeater; Short-
    billed Honeycreeper
f   Méliphage à tête brune
d   Braunkopfhonigschmecker
i   Mangiamiele capobruno

5767   *Melithreptus gularis*
    Passeriformes - Meliphagidae
e   Black-chinned Honeyeater; Black-
    throated Honeyeater
f   Méliphage à menton noir
d   Schwarzkinnhonigschmecker
i   Mangiamiele mentonero

5768   *Melithreptus gularis gularis*
    Passeriformes - Meliphagidae
e   Eastern Black-chinned Honeyeater
    (ANZ)

5769   *Melithreptus laetior*
    Passeriformes - Meliphagidae
e   Golden-backed Honeyeater
d   Goldrückenhonigschmecker
i   Mangiamiele dorsodorato

5770   *Melithreptus lunatus*
       Passeriformes - Meliphagidae
e   White-naped Honeyeater; Lunulated
       Honeycreeper
f   Méliphage à lunule
d   Mondstreifhonigschmecker;
       Weißnackenhonigschmecker
i   Mangiamiele dai sopraccigli rossi

5771   *Melithreptus validirostris*
       Passeriformes - Meliphagidae
e   Strong-billed Honeyeater
f   Méliphage à bec fort
d   Starkschnabelhonigschmecker;
       Kletterhonigfresser
i   Mangiamiele beccoforte

5772   *Melitograis gilolensis*
       Passeriformes - Meliphagidae
e   White-streaked Friarbird; Striated
       Friarbird; Bonaparte's Friarbird;
       Gilolo Friarbird
f   Polochion strié
d   Halmahera-Lederkopf
i   Uccello frate striato

5773   *Mellisuga helenae*
       Apodiformes - Trochilidae
e   Bee Hummingbird
f   Colibri d'Helen
d   Bienenelfe
i   Colibrì di Elena; Colibrì ape; Colibrì
       di Cuba

5774   *Mellisuga minima*
       Apodiformes - Trochilidae
e   Vervain Hummingbird; Little
       Doctor-Bird (WI); Bee Hummingbird
       (WI); Little Bee-Hummingbird (WI)
f   Ouanga négresse (Ants); Suce-fleur
       (Ants); Colibri nain
d   Zwergelfe
i   Colibrì minimo

5775   *Melocichla mentalis*
       Passeriformes - Sylviidae
e   Moustached Grass-Warbler; African
       Moustached-Warbler (CSA);
       Moustached Warbler (CSA)
f   Mélocichle à moustaches; Fauvette à
       moustaches

d   Bartgrassänger
i   Melocicla dai mustacchi

5776   *Melophus lathami*
       Passeriformes - Fringillidae
e   Crested Bunting
f   Bruant huppé
d   Haubenammer
i   Zigolo blu dal ciuffo

5777   *Melopsittacus undulatus*
       Psittaciformes - Psittacidae
e   Australian Budgerigar; Parakeet;
       Shell Parakeet; Budgerygah; Shell
       Parrot; Undulated Grass-Parakeet;
       Budgerigar
f   Perruche ondulée
d   Wellensittich
i   Pappagallino ondulato; Parrocchetto
       ondulato; Parrocchetto canoro

5778   *Melopyrrha nigra*
       Passeriformes - Fringillidae
e   Cuban Bullfinch; Black Sparrow
       (WI); Black Bullfinch
f   Sporophile négrito
d   Schwarzgimpelfink
i   Ciuffolotto di Cuba

5779   *Melospiza georgiana*
       Passeriformes - Fringillidae
e   Swamp Sparrow
f   Bruant des marais; Pinson des marais
d   Sumpfammer
i   Passero delle paludi

5780   *Melospiza lincolnii*
       Passeriformes - Fringillidae
e   Lincoln's Sparrow
f   Bruant de Lincoln; Pinson de Lincoln
d   Lincoln-Ammer
i   Passero di Lincoln

5781   *Melospiza melodia*
       Passeriformes - Fringillidae
e   Song Sparrow
f   Bruant chanteur; Pinson chanteur
d   Singammer
i   Passero cantore

5782 *Melozone biarcuatum*
Passeriformes - Emberizidae
e Prevost's Ground-Sparrow; Chiapas
Ground-Sparrow; White-faced
Ground-Sparrow; White-faced
Sparrow; Rusty-crowned Ground-
Sparrow; Cabanis's Ground-Sparrow
f Tohi à face blanche
d Weißwangenammer
i Passero terricolo di Prevost

5783 *Melozone kieneri*
Passeriformes - Emberizidae
e Rusty-crowned Ground-Sparrow;
Rusty-crowned Sparrow; Ground-
Sparrow
f Tohi de Kiener
d Rostnackenammer
i Passero terricolo corona rossa

5784 *Melozone leucotis*
Passeriformes - Emberizidae
e White-eared Ground-Sparrow;
White-eared Sparrow
f Tohi oreillard
d Weißohrammer
i Passero terricolo orecchiebianchi

5785 *Menura alberti*
Passeriformes - Menuridae
e Albert's Lyrebird; Prince Albert's
Lyrebird
f Ménure d'Albert; Ménure du prince
Albert; Oiseau-lyre du prince Albert
d Braunrückenleierschwanz;
Schwarzleierschwanz
i Uccello lira del principe Alberto

5786 *Menura novaehollandiae*
Passeriformes - Menuridae
e Superb Lyrebird
f Ménure superbe; Fauvette à
moustaches
d Prachtleierschwanz;
Graurückenleierschwanz
i Uccello lira comune; Uccello lira
superbo; Ucello lira

5787 **Menuridae**
Passeriformes
e Lyrebirds

f Menuridés; Oiseaux-lyres
d Leierschwänze
i Menuridi

5788 *Merganetta armata*
Anseriformes - Anatidae
e Torrent Duck
f Merganette des torrents
d Sturzbachente
i Anatra di torrente

5789 *Mergus albellus*
Anseriformes - Anatidae
e Smew
f Harle piette; Petite harle huppé
d Zwergsäger
i Pesciaiola

5790 *Mergus australis*
Anseriformes - Anatidae
e Auckland Islands Merganser;
Auckland Merganser; Auckland
Island Merganser
f Harle austral
d Auckland-Säger
i Smergo di Auckland

5791 *Mergus merganser*
Anseriformes - Anatidae
e Common Merganser; Goosander;
Merganser; Eastern Merganser (ISC)
f Grand Harle; Harle bièvre; Grand
Bec-scie; Bec-scie commun; Bec-scie
d Gänsesäger
i Smergo maggiore

5792 *Mergus octosetaceus*
Anseriformes - Anatidae
e Brazilian Merganser
f Harle huppard; Harle du Brésil
d Dunkelsäger
i Smergo del Brasile; Smergo
brasiliano

5793 *Mergus serrator*
Anseriformes - Anatidae
e Red-breasted Merganser
f Harle huppé; Bec-scie à poitrine
rousse
d Mittelsäger
i Smergo minore

5794   ***Mergus squamatus***
        Anseriformes - Anatidae
   e   Scaly-sided Merganser; Chinese
        Merganser
   f   Harle écaillé
   d   Schuppensäger
   i   Smergo della Cina

5795   **Meropidae**
        Coraciiformes
   e   Bee-eaters
   f   Guêpiers; Méropidés
   d   Spinte; Bienenfresser; Immenvögel
   i   Meropidi; Gruccioni

5796   ***Meropogon forsteni***
        Coraciiformes - Meropidae
   e   Purple-bearded Bee-eater; Bearded
        Bee-eater; Celebes Bearded Bee-
        eater; Sulawesi Bearded Bee-eater;
        Sulawesi Bee-eater; Celebes Bee-
        eater
   f   Guêpier des Célèbes
   d   Celebes-Spint
   i   Gruccione di Sulawesi

5797   ***Merops albicollis***
        Coraciiformes - Meropidae
   e   White-throated Bee-eater
   f   Guêpier à gorge blanche
   d   Weißkehlspint
   i   Gruccione golabianca

5798   ***Merops apiaster***
        Coraciiformes - Meropidae
   e   European Bee-eater; Eurasian Bee-
        eater (CSA); Bee-eater; Golden-
        backed Bee-eater (CSA); Common
        Bee-eater
   f   Guêpier d'Europe
   d   Bienenfresser; Europäischer
        Bienenfresser
   i   Gruccione europeo; Tordo marino;
        Gorgoglione; Grottaione;
        Barbiglione; Merope; Gruccione

5799   ***Merops boehmi***
        Coraciiformes - Meropidae
   e   Böhm's Bee-eater
   f   Guêpier de Böhm

   d   Böhm-Spint
   i   Gruccione di Boehm

5800   ***Merops breweri***
        Coraciiformes - Meropidae
   e   Black-headed Bee-eater
   f   Guêpier à tête noire
   d   Schwarzkopfspint
   i   Gruccione testanera

5801   ***Merops bullockoides***
        Coraciiformes - Meropidae
   e   White-fronted Bee-eater
   f   Guêpier à front blanc
   d   Weißstirnbienenfresser;
        Weißstirnspint
   i   Gruccione frontebianca

5802   ***Merops bulocki***
        Coraciiformes - Meropidae
   e   Red-throated Bee-eater; Yellow-
        throated Bee-eater
   f   Guêpier à gorge rouge
   d   Grünstirnspint; Rotkehlspint
   i   Gruccione golarossa; Melittofago di
        Bulock

5803   ***Merops gularis***
        Coraciiformes - Meropidae
   e   Black Bee-eater
   f   Guêpier noir
   d   Purpurspint
   i   Gruccione nero

5804   ***Merops hirundinaceus***
        Coraciiformes - Meropidae
   e   Swallow-tailed Bee-eater
   f   Guêpier à queue d'aronde; Guêpier à
        queue d'hirondelle
   d   Schwalbenschwanzbienenfresser;
        Gabelschwanzspint;
        Schwalbenschwanzspint
   i   Gruccione codadirondine

5805   ***Merops lafresnayii***
        Coraciiformes - Meropidae
   e   Cinnamon-chested Bee-eater;
        Lafresnaye's Bee-eater; Ethiopean
        Bee-eater
   f   Guêpier de Lafresnaye
   d   Schwarzbrustspint

5806 *Merops leschenaulti*
Coraciiformes - Meropidae
e Chestnut-headed Bee-eater; Bay-headed Bee-eater
f Guêpier de Leschenault
d Braunkopfspint
i Gruccione di Leschenault

5807 *Merops malimbicus*
Coraciiformes - Meropidae
e Rosy Bee-eater
f Guêpier gris-rose
d Rosenspint
i Gruccione grigio e carmino

5808 *Merops muelleri*
Coraciiformes - Meropidae
e Blue-headed Bee-eater
f Guêpier à tête bleue
d Samtspint
i Gruccione testablu; Gruccione di Mülller

5809 *Merops nubicus*
Coraciiformes - Meropidae
e Northern Carmine Bee-eater; Carmine Bee-eater (CSA); Nubian Carmine Bee-eater
f Guêpier écarlate
d Karminspint
i Gruccione carmino; Gruccione carmino settentrionale; Gruccione rosea dalla gola azzurra

5810 *Merops oreobates*
Coraciiformes - Meropidae
e Cinnamon-chested Bee-eater; Cinnamon-breasted Bee-eater
f Guêpier montagnard
d Großer Blaubrustspint
i Gruccione pettocannella

5811 *Merops orientalis*
Coraciiformes - Meropidae
e Green Bee-eater; Little Green Bee-eater; Small Green Bee-eater (ISC); Ceylon Green Bee-eater
f Guêpier d'Orient
d Schmaragdspint
i Gruccione verde piccolo; Gruccione orientale

5812 *Merops ornatus*
Coraciiformes - Meropidae
e Rainbow Bee-eater; Rainbowbird
f Guêpier arc-en-ciel
d Regenbogenspint
i Gruccione ornato; Gruccione adorno

5813 *Merops persicus*
Coraciiformes - Meropidae
e Blue-cheeked Bee-eater
f Guêpier de Perse
d Blauwangenspint; Blauwangenbienenfresser
i Gruccione persiano; Grucione di Persia; Gruccione guanceblu; Gruccione egiziano; Gruccione d'Iran

5814 *Merops philippinus*
Coraciiformes - Meropidae
e Blue-tailed Bee-eater; Brown-breasted Bee-eater
f Guêpier à queue d'azur
d Blauschwanzspint
i Gruccione cadablu

5815 *Merops pusillus*
Coraciiformes - Meropidae
e Little Bee-eater
f Guêpier nain
d Zwergbienenfresser; Zwergspint
i Gruccione pettocannella minore

5816 *Merops revoilii*
Coraciiformes - Meropidae
e Somali Bee-eater
f Guêpier de Revoil
d Blaßspint
i Gruccione della Somalia

5817 *Merops rubicoides*
Coraciiformes - Meropidae
e Southern Carmine Bee-eater; Carmine Bee-eater (CSA)
f Guêpier carmin
d Scharlachspint; Karminspint
i Gruccione carmino meridionale

5818 *Merops superciliosus*
Coraciiformes - Meropidae
e Madagascar Bee-eater; Olive Bee-eater (CSA); Blue-cheeked Bee-

eater; Brown-breasted Bee-eater;
Blue-tailed Bee-eater
f Guêpier de Madagascar
d Madagaskar-Bienenfresser;
Blauwangenspint;
Blauwangenbienenfresser; Persischer
Bienenfresser
i Gruccione egiziano; Gruccione dai
sopraccigli chiari

5819 *Merops variegatus*
Coraciiformes - Meropidae
e Blue-breasted Bee-eater
f Guêpier à collier bleu
d Blaubrustspint
i Gruccione pettazzurro

5820 *Merops viridis*
Coraciiformes - Meropidae
e Blue-throated Bee-eater; Chestnut-
headed Bee-eater
f Guêpier à gorge bleue
d Hindu-Spint
i Gruccione golablu

5821 *Merulaxis ater*
Passeriformes - Rhinocryptidae
e Slaty Bristlefront
f Mérulaxe noir
d Bürstentapaculo
i Frontesetolosa color lavagna

5822 *Merulaxis stresemanni*
Passeriformes - Rhinocryptidae
e Stresemann's Bristlefront
f Mérulaxe de Stresemann
d Stresemann-Tapaculo
i Frontesetoloso di Stresemann

5823 *Mesembrinibis cayennensis*
Ciconiiformes - Threskiornithidae
e Green Ibis; Cayenne Ibis
f Ibis vert
d Cayenne-Ibis
i Ibis verde

5824 *Mesitornis unicolor*
Gruiformes - Mesitornithidae
e Brown Mesite; Roatelo; Brown
Roatelo
f Mésite unicolore

d Einfarbstelzenralle
i Mesena bruna

5825 *Mesitornis variegata*
Gruiformes - Mesitornithidae
e White-breasted Mesite; White-
breasted Roatelo
f Mésite variée
d Kurzfußstelzenralle
i Mesena pettobianco

5826 **Mesitornithidae**
Gruiformes
e Mesites
f Mesitornithidés
d Stelzenrallen; Madagaskar-Rallen
i Mesitornitidi

5827 *Metabolus rugensis*
Passeriformes - Corvidae
e Truk Monarch
f Monarque de Truk
d Truk-Monarch
i Monarca di Truk

5828 *Metallura aeneocauda*
Apodiformes - Trochilidae
e Scaled Metaltail; Brassy Metaltail;
Andean Metaltail; Reddish Metaltail
f Métallure à queue d'airain
d Schuppenglanzschwänzchen
i Coda metallica scaglioso

5829 *Metallura baroni*
Apodiformes - Trochilidae
e Violet-throated Metaltail
f Métallure de Baron
d Purpurkehlglanzschwänzchen
i Coda metallica golaviola

5830 *Metallura eupogon*
Apodiformes - Trochilidae
e Fire-throated Metaltail
f Métallure à gorge feu
d Rotkehlglanzschwänzchen
i Coda metallica goladifiamma

5831 *Metallura iracunda*
Apodiformes - Trochilidae
e Perija Metaltail
f Métallure dorée

d Goldglanzschwänzchen
i Coda metallica di Perija

5832 *Metallura odomae*
Apodiformes - Trochilidae
e Neblina Metaltail
f Métallure du Chinguela
d Neblina-Glanzschwänzchen
i Coda metallica di Neblina

5833 *Metallura phoebe*
Apodiformes - Trochilidae
e Black Metaltail
f Métallure phébé
d Rußglanzschwänzchen
i Coda metallica nero

5834 *Metallura theresiae*
Apodiformes - Trochilidae
e Coppery Metaltail
f Métallure de Thérèse
d Kupferglanzschwänzchen
i Coda metallica ramato

5835 *Metallura tyrianthina*
Apodiformes - Trochilidae
e Tyrian Metaltail
f Métallure émeraude
d Smaragdkehlglanzschwänzchen
i Coda metallica rosso tirio

5836 *Metallura williami*
Apodiformes - Trochilidae
e Viridian Metaltail; Ecuadorian
   Metaltail; Colombian Metaltail;
   Black-throated Metaltail
f Métallure verte
d Grünglanzschwänzchen
i Coda metallica di William

5837 *Metopidius indicus*
Charadriiformes - Jacanidae
e Bronze-winged Jacana
f Jacana bronzé
d Hindu-Blatthühnchen
i Jacana alibronzate

5838 *Metopothrix aurantiacus*
Passeriformes - Furnariidae
e Orange-fronted Plushcrown; Orange-
   fronted Softtail

f Tête-de-feu pelucheux
d Goldschlüpfer
i Frontearancio coronato

5839 *Metriopelia aymara*
Columbiformes - Columbidae
e Golden-spotted Ground-Dove;
   Bronze-winged Ground-Dove
f Colombe aymara
d Aymara-Täubchen
i Tortora alibronzate

5840 *Metriopelia ceciliae*
Columbiformes - Columbidae
e Bare-faced Ground-Dove; Bare-eyed
   Ground-Dove; Yellow-eyed Dove;
   Spectacled Dove; Cecilia's Dove
f Colombe de Cécile
d Brillentäubchen;
   Nacktgesichttäubchen
i Tortora faccianuda

5841 *Metriopelia melanoptera*
Columbiformes - Columbidae
e Black-winged Ground-Dove
f Colombe à ailes noires
d Kordillerentaube; Weißbugtäubchen
i Tortora alinere

5842 *Metriopelia morenoi*
Columbiformes - Columbidae
e Bare-eyed Ground-Dove; Moreno's
   Bare-faced Ground-Dove
f Colombe de Moreno
d Moreno-Täubchen
i Tortora occhinudi

5843 *Micrastur buckleyi*
Falconiformes - Falconidae
e Buckley's Forest-Falcon; Taylor's
   Forest-Falcon
f Carnifex de Buckley
d Taylor-Waldfalke
i Falco di Buckley; Falcone di Taylor

5844 *Micrastur gilvicollis*
Falconiformes - Falconidae
e Lined Forest-Falcon
f Carnifex à gorge cendrée
d Zweibindenwaldfalke
i Falco collogrigio

5845 *Micrastur mirandollei*
          Falconiformes - Falconidae
    e   Slaty-backed Forest-Falcon
    f   Carnifex ardoisé
    d   Graurückenwaldfalke
    i   Falco di Mirandolle; Falcone dal
          dorso ardesia

5846 *Micrastur plumbeus*
          Falconiformes - Falconidae
    e   Plumbeous Forest-Falcon; Sclater's
          Forest-Falcon
    f   Carnifex plombé
    d   Einbindenwaldfalke
    i   Falco plumbeo; Falcone plumbeo

5847 *Micrastur ruficollis*
          Falconiformes - Falconidae
    e   Barred Forest-Falcon
    f   Carnifex barré
    d   Sperberwaldfalke
    i   Falco golarossa; Falcone a striscie

5848 *Micrastur semitorquatus*
          Falconiformes - Falconidae
    e   Collared Forest-Falcon
    f   Carnifex à collier
    d   Kappenwaldfalke
    i   Falco dal collare; Falcone dal collare

5849 *Micrathene whitneyi*
          Strigiformes - Strigidae
    e   Elf-Owl
    f   Chevêchette des saguaros;
          Chevêchette elfe
    d   Elfenkauz; Kaktuskauz
    i   Elfo dei cactus; Elfo

5850 *Microbates cinereiventris*
          Passeriformes - Muscicapidae
    e   Tawny-faced Gnatwren; Half-
          collared Gnatwren
    f   Microbate cendré
    d   Graubauchdegenschnäbler
    i   Zanzariere dal semicollare

5851 *Microbates collaris*
          Passeriformes - Muscicapidae
    e   Collared Gnatwren
    f   Microbate à collier

    d   Halsbanddegenschnäbler
    i   Zanzariere dal collare

5852 *Microcerculus bambla*
          Passeriformes - Troglotidae
    e   Wing-banded Wren
    f   Troglodyte bambla
    d   Weißbindenzaunkönig
    i   Scricciolo bambla

5853 *Microcerculus luscinia*
          Passeriformes - Troglotidae
    e   Whistling Wren
    f   Troglodyte rossignol

5854 *Microcerculus marginatus*
          Passeriformes - Troglotidae
    e   Southern Nightingale-Wren; Scaly-
          breasted Wren; Southern Nightingale
          Whistler-Wren
    f   Troglodyte siffleur
    d   Nachtigalzaunkönig
    i   Scricciolo pettoscaglioso

5855 *Microcerculus philomela*
          Passeriformes - Troglotidae
    e   Northern Nightingale-Wren;
          Nightingale Wren; Dark-throated
          Nightingale-Wren; Nightingale
          Whistler-Wren
    f   Troglodyte philomè
    i   Scricciolo usignolo

5856 *Microcerculus taeniatus*
          Passeriformes - Troglotidae
    e   Scaly Nightingale-Wren

5857 *Microcerculus ustulatus*
          Passeriformes - Troglotidae
    e   Flutist Wren
    f   Troglodyte flûtiste
    d   Flötenzaunkönig
    i   Scricciolo flautista

5858 *Microchera albocoronata*
          Apodiformes - Trochilidae
    e   Snowcap; White-crowned
          Hummingbird
    f   Colibri à coiffe blanche
    d   Schneekrönchen
    i   Colibrì corona di neve

5859 **Microdynamis parva**
Cuculiformes - Cuculidae
e Dwarf Koel; Black-capped Cuckoo;
Little Koel
f Coucou à tête noire
d Schwarzkappenkuckuck
i Koel minore

5860 **Microeca fascinans**
Passeriformes - Petroicidae
e Jacky-winter; Australian Brown-
Flycatcher; Brown Flycatcher;
White-tailed Microeca; Brown
Flyrobin; White-tailed Flyrobin;
Flycatcher Robin; Flycatcher
Microeca
f Miro enchanteur
d Weißschwanzschnäpper
i Pigliamosche bruno australiano

5861 **Microeca flavigaster**
Passeriformes - Petroicidae
e Lemon-bellied Flyrobin; Lemon-
bellied Flycatcher (ANZ); Lemon-
breasted Flycatcher; Lemon-breasted
Microeca; Lemon-breased Flyrobin
f Miro à ventre citron
d Gelbbrustschnäpper
i Pigliamosche ventregiallo australiano

5862 **Microeca flavigaster tormenti**
Passeriformes - Petroicidae
e Brown-tailed Flycatcher; Brown-
tailed Microeca; Kimberley
Flycatcher; Kimberley Robin
f Miro à queue brune
d Braunschwanzschnäpper

5863 **Microeca flavovirescens**
Passeriformes - Petroicidae
e Olive Flyrobin; Olive Microeca;
Olive Flycatcher; Orange-chinned
Flyrobin
f Miro olive
d Aru-Schnäpper
i Pigliamosche giallo oliva

5864 **Microeca griseoceps**
Passeriformes - Petroicidae
e Yellow-legged Flyrobin; Yellow-
footed Flycatcher; Yellow-legged

Flycatcher (ANZ); Yellow-footed
Microeca; Little Yellow Flyrobin;
Grey-headed Flyrobin
f Miro à pattes jaunes
d Gelbfußschnäpper
i Pigliamosche piedigialli

5865 **Microeca hemixantha**
Passeriformes - Petroicidae
e Golden-bellied Flyrobin; Tanimbar
Flycatcher; Tanimbar Microeca;
Tanimbar Microeca-Flycatcher;
Golden-bellied Flycatcher
f Miro des Tanimbar
d Tanimbar-Schnäpper
i Pigliamosche di Tanimbar

5866 **Microeca papuana**
Passeriformes - Petroicidae
e Canary Flyrobin; Papuan Flycatcher;
Papuan Microeca; Canary Flycatcher
f Miro papou
d Papua-Schnäpper
i Pigliamosche papua

5867 **Microgoura meeki**
Columbiformes - Columbidae
e Choiseul Pigeon; Meek's Pigeon;
Solomon Islands Crowned-Pigeon;
Solomon Islands Ground-Pigeon
f Microgoura de Choiseul
d Große Salomonen-Erdtaube;
Salomonen-Taube
i Microgura di Meek

5868 **Microhierax caerulescens**
Falconiformes - Falconidae
e Collared Falconet; Red-thighed
Falconet; Red-breasted Falconet
(ISC); Red-legged Falconet; Red-
collared Falconet
f Fauconnet à collier
d Rotschenkelzwergfalke;
Rotkehlfälkchen; Finkenfälkchen
i Falchetto dal collare; Falchetto

5869 **Microhierax erythrogenys**
Falconiformes - Falconidae
e Philippine Falconet
f Fauconnet des Philippines

d Zweifarbenfälkchen
i Falchetto delle Filippine

5870 *Microhierax fringillarius*
Falconiformes - Falconidae
e Black-thighed Falconet; Black-
legged Falconet; Black-tailed
Falconet; Black-sided Falconet;
Malay Falconet
f Fauconnet de Bornéo
d Finkenfälkchen
i Falchetto dai calzoni neri; Falchetto
dalle zampei nere

5871 *Microhierax latifrons*
Falconiformes - Falconidae
e White-fronted Falconet; Bornean
Falconet
d Weißscheitelfälkchen
i Falchetto del Borneo; Falchetto dalla
testa bianca

5872 *Microhierax melanoleucus*
Falconiformes - Falconidae
e Pied Falconet; Pied Pygmy (ISC);
Pied Pygmy-Falconet (ISC); White-
legged Falconet (ISC)
f Fauconnet noir-et-blanc
d Elsterfälkchen
i Falchetto bianco e nero; Falchetto
variopinta

5873 *Microligea palustris*
Passeriformes - Fringillidae
e Green-tailed Ground-Warbler;
Green-tailed Warbler; Grey-breasted
Ground-Warbler; Grey-breasted
Warbler
f Paruline aux yeux rouges
d Graubrustwaldsänger
i Parula codaverde

5874 *Micromacronus leytensis*
Passeriformes - Sylviidae
e Miniature Tit-Babbler; Leyte Tit-
Babbler
f Timalie miniature
d Laubsängertimalie
i Garrulo di Leyte

5875 *Micromonacha lanceolata*
Piciformes - Bucconidae
e Lanceolated Monklet
f Barbacou lancéolé
d Streifenfaulvogel
i Monachina lanceolata; Monasa
lanceolata

5876 *Microparra capensis*
Charadriiformes - Jacanidae
e Lesser Jacana; Smaller Jacana;
Lesser African Jacana; Lesser Lily-
trotter
f Jacana nain
d Zwergblatthühnchen
i Jacana minore

5877 *Micropsitta bruijnii*
Psittaciformes - Psittacidae
e Red-breasted Pygmy-Parrot; Rosy-
breasted Pygmy-Parrot; Bruijn's
Pygmy-Parrot; Rose-breasted
Pygmy-Parrot; Rose-breasted
Pygmy-Lorilet; Mountain Pygmy-
Parrot
f Micropsitte de Bruijn
d Rotbrustspechtpapagei; Bruijn-
Spechtpapagei
i Pappagallo pigmeo di Brujin

5878 *Micropsitta finschii*
Psittaciformes - Psittacidae
e Finsch's Pygmy-Parrot; Emerald Fig-
Parrot
f Micropsitte de Finsch
d Finsch-Spechtpapagei; Salomonen-
Spechtpapagei
i Pappagallo pigmeo verde

5879 *Micropsitta geelvinkiana*
Psittaciformes - Psittacidae
e Geelvink Pygmy-Parrot; Mafor
Pygmy-Parrot
f Micropsitte de Geelvink; Perruche de
Mafor
d Geelvink-Spechtpapagei; Schlegel-
Spechtpapagei
i Pappagallo pigmeo testabruna

5880 *Micropsitta keiensis*
Psittaciformes - Psittacidae

e Yellow-capped Pygmy-Parrot; Kai
Island Pygmy-Parrot
f Micropsitte pygmée
d Gelbkappenspechtpapagei;
Salvadori-Spechtpapagei
i Pappagallo pigmeo delle Kai

5881 *Micropsitta meeki*
Psittaciformes - Psittacidae
e Meek's Pygmy-Parrot
f Micropsitte de Meek
d Meek-Papagei; Meek-Spechtpapagei
i Pappagallo pigmeo pettogiallo

5882 *Micropsitta pusio*
Psittaciformes - Psittacidae
e Buff-faced Pygmy-Parrot; Sclater's
Pygmy-Parrot; Solomon Islands
Pygmy-Parrot
f Micropsitte à tête fauve
d Sclater-Spechtpapagei;
Braunstirnspechtpapagei
i Pappagallo pigmeo facciafulva

5883 *Micropygia schomburgkii*
Gruiformes - Rallidae
e Ocellated Crake; Ocellated Rail;
Dotted Rail; Dotted Crake
f Râle ocellé
d Schomburgk-Ralle; Schomburgks
Ralle
i Schiribilla ocellata

5884 *Microrhopias boucardi*
Passeriformes - Formicariidae
e Boucard's Ant-Wren

5885 *Microrhopias quixensis*
Passeriformes - Formicariidae
e Dot-winged Ant-Wren; Amazonian
Ant-Wren; Velvety Ant-Wren;
Lower Amazonian Antwren
f Grisin étoilé
d Tropfenflügelameisenfänger
i Scricciolo formichiere di Quixos

5886 *Microstilbon burmeisteri*
Apodiformes - Trochilidae
e Slender-tailed Woodstar;
Burmeister's Woodstar
f Colibri de Burmeister

d Burmeister-Kolibri
i Stella dei boschi codasottile

5887 *Milvago chimachima*
Falconiformes - Falconidae
e Yellow-headed Caracara;
Chimachima Caracara
f Caracara à tête jaune
d Chimachima
i Caracara testagialla; Caracara dalla
testa gialla

5888 *Milvago chimango*
Falconiformes - Falconidae
e Chimango Caracara; Chimango;
Caracara; Chimango Hawk
f Caracara chimango
d Chimango
i Caracara cimango; Caracara
chimangio

5889 *Milvus aegypteus*
Falconiformes - Accipitridae
e Yellow-billed Kite; Egyptian Kite
f Milan d'Égypte
d Schmarotzermilan

5890 *Milvus fasciicauda*
Falconiformes - Accipitridae
e Cape Verde Kite

5891 *Milvus lineatus*
Falconiformes - Accipitridae
e Black-eared Kite; Large Indian Kite
f Milan brun
i Nibbio orecchienere

5892 *Milvus migrans*
Falconiformes - Accipitridae
e Black Kite; Common Pariah-Kite
(ISC); Pariah Kite; Dark Kite; Allied
Kite; Black-eared Kite; Fork-tailed
Kite
f Élanion noir; Milan noir
d Schwarzmilan; Schwarzer Milan
i Nibbio bruno

5893 *Milvus milvus*
Falconiformes - Accipitridae
e Red Kite; Kite; Cape Verde Kite
f Milan royal

d   Rotmilan; Roter Milan; Gabelweihe
i   Nibbio reale

5894   *Mimodes graysoni*
      Passeriformes - Sturnidae
e   Socorro Mockingbird; Socorro
      Thrasher
f   Moqueur de Socorro
d   Socorro-Spottdrossel
i   Mimo di Socorro

5895   *Mimus dorsalis*
      Passeriformes - Sturnidae
e   Brown-backed Mockingbird
f   Moqueur à dos brun
d   Braunrückenspottdrossel
i   Mimo dorsobruno

5896   *Mimus gilvus*
      Passeriformes - Sturnidae
e   Tropical Mockingbird; Nightingale
      (WI); Southern Mockingbird
f   Grive blanche (Ants); Pierre-fouillé
      (Ants); Pied carreau (Ants); Grive
      des savannes (Ants); Moqueur des
      savannes
i   Mimo tropiccale

5897   *Mimus gundlachii*
      Passeriformes - Sturnidae
e   Bahama Mockingbird; Spanish
      Nightingale (WI); Salt Island
      Nightingale (WI); Hill's Mockingbird
      (WI); Grundlach's Mockingbird;
      Bahaman Mockingbird
f   Moqueur des Bahamas
d   Grundlach-Spottdrossel
i   Mimo delle Bahama

5898   *Mimus longicaudatus*
      Passeriformes - Sturnidae
e   Long-tailed Mockingbird
f   Moqueur à longue queue
d   Langschwanzspottdrossel
i   Mimo codalunga

5899   *Mimus magnirostris*
      Passeriformes - Sturnidae
e   St. Andrew Mockingbird
f   Moqueur à gros bec

5900   *Mimus patagonicus*
      Passeriformes - Sturnidae
e   Patagonian Mockingbird
f   Moqueur de Patagonie
d   Rostflankenspottdrosssel
i   Mimo della Patagonia

5901   *Mimus polyglottos*
      Passeriformes - Sturnidae
e   Northern Mockingbird; Common
      Mockingbird; Mockingbird; English
      Thrasher (WI); Nightingale (WI)
f   Moqueur polyglotte; Rossignol
      (Ants)
d   Spottdrossel
i   Mimo poliglotto

5902   *Mimus saturninus*
      Passeriformes - Sturnidae
e   Chalk-browed Mockingbird;
      Saturnine Mockingbird
f   Moqueur plombé
d   Camposspottdrossel
i   Mimo dei campi

5903   *Mimus thenca*
      Passeriformes - Sturnidae
e   Chilean Mockingbird
f   Moqueur du Chili
d   Chile-Spottdrossel
i   Mimo del Cile

5904   *Mimus triurus*
      Passeriformes - Sturnidae
e   White-banded Mockingbird
f   Moqueur à ailes blanches
d   Weißbindenspottdrossel
i   Mimo alifasciate

5905   *Minla cyanouroptera*
      Passeriformes - Sylviidae
e   Blue-winged Minla; Blue-winged
      Siva
f   Minla à ailes bleues; Minla à aile
      bleue
d   Blauflügelsiva;
      Blauflügelsonnenvogel
i   Minla aliazzurre

5906   *Minla ignotincta*
      Passeriformes - Sylviidae

e Red-tailed Minla; Fire-tailed Minla;
   Red-tailed Siva
f Minla à queue rousse
d Rotschwanzsiva
i Minla codarossa

5907 *Minla strigula*
   Passeriformes - Sylviidae
e Chestnut-tailed Minla; Bar-throated
   Minla; Stripe-throated Minla; Bar-
   throated Siva; Stripe-throated Siva;
   Chestnut-tailed Siva
f Minla à gorge striée
d Bändersiva
i Minla codacastana

5908 *Mino anais*
   Passeriformes - Sturnidae
e Golden Mynah; Golden-breasted
   Mynah
f Mino anais
d Orangeatzel
i Maina pettodorato

5909 *Mino dumontii*
   Passeriformes - Sturnidae
e Yellow-faced Mynah; Papuan
   Mynah; Orange-faced Mynah;
   Orange-faced Grackle; Long-tailed
   Mynah
f Mino de Dumont
d Papua-Atzel
i Maina facciagialla

5910 *Mionectes macconnelli*
   Passeriformes - Tyrannidae
e MacConnell's Flycatcher
f Pipromorphe de McConnel
d Olivschwingenpipratyrann
i Tiranno di McConnell

5911 *Mionectes oleagineus*
   Passeriformes - Tyrannidae
e Ochre-bellied Flycatcher; Oleaginous
   Flycatcher; Oleaginous Pipromorpha
f Pipromorphe roussâtre
d Ockerbauchpipratyrann
i Tiranno petto ocra

5912 *Mionectes olivaceus*
   Passeriformes - Tyrannidae

e Olive-striped Flycatcher
f Pipromorphe olive
d Rundschwingenstricheltyrann
i Tiranno olivaceo

5913 *Mionectes rufiventris*
   Passeriformes - Tyrannidae
e Grey-hooded Flycatcher
f Pipromorphe à ventre roux
d Graukopfpipratyrann
i Tiranno rufiventre

5914 *Mionectes striaticollis*
   Passeriformes - Tyrannidae
e Streak-necked Flycatcher; Streaked-
   necked Flycatcher
f Pipromorphe strié
d Spitzschwingenstricheltyrann
i Tiranno collostriato

5915 *Mirafra africana*
   Passeriformes - Alaudidae
e Rufous-naped Lark; Rufous-naped
   Bush-Lark
f Alouette à nuque rousse
d Rotnackenlerche
i Allodola nucarossicia

5916 *Mirafra africanoides*
   Passeriformes - Alaudidae
e Fawn-coloured Lark; Fawn-coloured
   Bush-Lark; Fawn Lark; Fawn Bush-
   Lark
f Alouette fauve
d Steppenlerche
i Allodola castana

5917 *Mirafra albicauda*
   Passeriformes - Alaudidae
e White-tailed Lark; White-tailed
   Bush-Lark; Northern White-tailed
   Bush-Lark; Northern White-tailed
   Lark
f Alouette à queue blanche
d Weißschwanzlerche
i Allodola dimacchia settentrionale

5918 *Mirafra alopex*
   Passeriformes - Alaudidae
e Abyssinian Lark; Abyssinian Bush-
   Lark

f Alouette abyssinienne
d Fuchslerche
i Allodola abissina

5919 *Mirafra angolensis*
Passeriformes - Alaudidae
e Angola Lark; Angolan Lark;
Angolan Bush-Lark
f Alouette de I'Angola
d Angola-Lerche
i Allodola dell'Angola

5920 *Mirafra apiata*
Passeriformes - Alaudidae
e Clapper-Lark; Cape Clapper-Lark
(CSA); Clapper Bush-Lark; Bar-ailed
Lark
f Alouette bateleuse
d Grasklapperlerche
i Allodola delle nuvole

5921 *Mirafra apiata adendorffii*
Passeriformes - Alaudidae
e Adendorff's Clapper-Lark
f Alouette bateleuse

5922 *Mirafra apiata hewitti*
Passeriformes - Alaudidae
e Highveld Clapper-Lark

5923 *Mirafra ashi*
Passeriformes - Alaudidae
e Ash's Lark; Ash's Bush-Lark
f Alouette de Ash
i Allodola di Ash

5924 *Mirafra assamica*
Passeriformes - Alaudidae
e Rufous-winged Lark; Bush Lark
(ISC); Rufous-winged Bush-Lark;
Bengal Bush-Lark; Bengal Lark;
Ceylon Bush-Lark
f Alouette du Siam
d Bengalen-Lerche
i Allodola dell'Assam

5925 *Mirafra cantillans*
Passeriformes - Alaudidae
e Singing Bush-Lark; Singing Lark;
Western Singing Bush-Lark; African
Singing Bush-Lark

f Alouette chanteuse
d Buschlerche
i Allodola di Giava; Allodola canora

5926 *Mirafra cheniana*
Passeriformes - Alaudidae
e Latakoo Lark; Melodious Lark;
Singing Bush-Lark (CSA); Southern
Singing Bush-Lark; Latakoo Bush-
Lark; Southern Singing Lark
f Alouette mélodieuse
d Spottlerche
i Allodola di macchia meridionale

5927 *Mirafra collaris*
Passeriformes - Alaudidae
e Collared Lark; Collared Bush-Lark
f Alouette à collier
d Halsbandlerche
i Allodola dal collare

5928 *Mirafra cordofanica*
Passeriformes - Alaudidae
e Kordofan Lark; Kordofan Bush-Lark;
Golden Lark
f Alouette du Kordofan
d Kordofan-Lerche
i Allodola del Kordofan

5929 *Mirafra degodiensis*
Passeriformes - Alaudidae
e Degodi Lark; Ergard's Lark
f Alouette du Degodi
d Degodi-Lerche
i Allodola di Erard

5930 *Mirafra erythroptera*
Passeriformes - Alaudidae
e Indian Bush-Lark; Indian Lark; Red-
winged Bush-Lark (ISC); Indian
Red-winged Bush-Lark; Indian Red-
winged Lark; Red-winged Lark;
Rusty-winged Lark
f Alouette à ailes rousses
d Rotflügellerche
i Allodola alirosse

5931 *Mirafra gilletti*
Passeriformes - Alaudidae
e Gillett's Lark
f Alouette de Gillett

d Ogaden-Lerche
i Allodola di Gillett

5932 *Mirafra hova*
Passeriformes - Alaudidae
e Madagascar Lark; Madagascar Bush-Lark; Hova Lark
f Alouette malgache
d Hova-Lerche
i Allodola hova

5933 *Mirafra hypermetra*
Passeriformes - Alaudidae
e Red-winged Lark; Red-winged Bush-Lark; Rufous-winged Bush-Lark
f Alouette polyglotte
d Riesenlerche
i Allodola di macchia alirosse

5934 *Mirafra javanica*
Passeriformes - Alaudidae
e Australasian Lark; Singing Bush-Lark; Horsfield's Bush-Lark; Cinnamon Bush-Lark; Eastern Singing Bush-Lark; Javan Lark; Australasian Bush-Lark; Eastern Bush-Lark; Bush-Lark
f Alouette de Java
d Horsfield-Lerche; Buschlerche
i Allodola di Giava

5935 *Mirafra naevia*
Passeriformes - Alaudidae
e Bradfield's Lark
f Alouette tachée
d Somali-Riesenlerche
i Allodola di Bradfield

5936 *Mirafra passerina*
Passeriformes - Alaudidae
e Monotonous Lark; Monotonous Bush-Lark
f Alouette monotone
d Sperlingslerche
i Allodola passerina

5937 *Mirafra poecilosterna*
Passeriformes - Alaudidae
e Pink-breasted Lark
f Alouette à poitrine rose

d Fahlbrustlerche
i Allodola Erard

5938 *Mirafra pulpa*
Passeriformes - Alaudidae
e Friedmann's Lark; Friedmann's Bush-Lark; Sagon Lark; Sagon Bush-Lark
f Alouette de Friedmann
d Friedmann-Lerche
i Allodola di Friedmann

5939 *Mirafra rufa*
Passeriformes - Alaudidae
e Rusty Lark; Rusty Bush-Lark
f Alouette rousse
d Rostlerche
i Allodola rugginosa

5940 *Mirafra rufocinnamomea*
Passeriformes - Alaudidae
e Flappet Lark; Flappet Bush-Lark; Cinnamon Lark; Cinnamon Bush-Lark
f Alouette bourdonnante
d Baumklapperlerche
i Allodola di Salvadori

5941 *Mirafra sabota*
Passeriformes - Alaudidae
e Sabota Lark
f Alouette sabota; Alouette de Sabota
d Sabotalerche
i Allodola sabota

5942 *Mirafra sharpii*
Passeriformes - Alaudidae
e Somali Lark; Somali Bush-Lark; Red Somali Lark
f Alouette de Sharpe
d Elliott-Lerche
i Allodola di Sharpe

5943 *Mirafra somalica*
Passeriformes - Alaudidae
e Somali Long-billed Lark; Somali Long-billed Bush-Lark; Somali Lark
f Alouette de Somalie
d Somali-Riesenlerche
i Allodola somala

5944   *Mirafra williamsi*
           Passeriformes - Alaudidae
  e   Williams's Lark; William's Bush-
        Lark; Marsabit Lark
  f   Alouette de Williams
  d   Williams-Lerche
  i   Allodola del Marsabit

5945   *Mitrephanes olivaceus*
           Passeriformes - Tyrannidae
  e   Olive Flycatcher; Olive Tufted
        Flycatcher
  f   Moucherolle olive
  d   Yungas-Tyrann
  i   Tiranno dal ciuffo olivaceo

5946   *Mitrephanes phaeocercus*
           Passeriformes - Tyrannidae
  e   Tufted Flycatcher; Common Tufted
        Flycatcher
  f   Moucherolle huppé
  d   Gelbbauchtyrann
  i   Tiranno dal ciuffo

5947   *Mitrospingus cassinii*
           Passeriformes - Fringillidae
  e   Dusky-faced Tanager
  f   Tangara obscur
  d   Rußgesichttangare
  i   Tangara dal ciuffo facciascura

5948   *Mitrospingus oleagineus*
           Passeriformes - Fringillidae
  e   Olive-backed Tanager
  f   Tangara à dos olive
  d   Olivmanteltangare
  i   Tangara dal ciuffo dorso oliva

5949   *Mitu mitu*
           Galliformes - Cracidae
  e   Alagoas Curassow; Razor-billed
        Curassow
  f   Hocco mitou
  d   Mitu
  i   Mitu

5950   *Mitu salvini*
           Galliformes - Cracidae
  e   Salvin's Curassow; Salvin's Razor-
        billed Curassow
  f   Hocco de Salvin

  d   Salvin-Hokko
  i   Crace di Salvin

5951   *Mitu tomentosa*
           Galliformes - Cracidae
  e   Crestless Curassow; Lesser Razor-
        billed Curassow
  f   Hocco de Spix
  d   Samthokko
  i   Crace senza cresta

5952   *Mitu tuberosa*
           Galliformes - Cracidae
  e   Razor-billed Curassow; Greater
        Amazonian Razor-billed Curassow
  f   Hocco nocturne
  i   Crace becco a rasaio

5953   *Mniotilta varia*
           Passeriformes - Fringillidae
  e   Black-and-white Warbler, Antsbird
        (WI); Ants Picker (WI)
  f   Paruline noire-et-blanc; Demi-deuil
        (Ants); Mi-deuil (Ants); Madras
        (Ants); Fauvette noir-et-blanc
  d   Kletterwaldsänger;
        Baumläuferwaldsänger
  i   Parula variegata; Mniotilta bianca e
        nera; Parula bianca e nera

5954   *Modulatrix stictigula*
           Passeriformes - Sylviidae
  e   Spot-throat
  f   Modulatrice à lunettes
  d   Fleckenkehlchen
  i   Cantore golamacchiata

5955   *Moho apicalis*
           Passeriformes - Meliphagidae
  e   Oahu Oo
  f   Moho d'Oahu
  d   Krausschwanzmoho
  i   Moho di Oahu

5956   *Moho bishopi*
           Passeriformes - Meliphagidae
  e   Bishop's Oo; Molokai Oo
  f   Moho de Bishop
  d   Molokai-Krausschwanz;
        Ohrbüschelmoho
  i   Moho di Bishop

5957 **Moho braccatus**
Passeriformes - Meliphagidae
e Kauai Oo; Ooaa
f Moho de Kauai
d Braunkrausschwanz;
Schuppenkehlmoho
i Moho di Kauai

5958 **Moho nobilis**
Passeriformes - Meliphagidae
e Hawaii Oo; Great Oo
f Moho d'Hawai
d Edelkrausschwanz; Prachtmoho
i Moho delle Hawaii

5959 **Mohoua albicilla**
Passeriformes - Corvidae
e Whitehead; Popokatea
f Mohoua à tête blanche
d Weißköpchen
i Testabianca

5960 **Mohoua novaeseelandiae**
Passeriformes - Corvidae
e Pipipi; New Zealand Creeper; Brown
Creeper; New Zealand Brown-
Creeper
f Mohoua pipipi
d Finschia
i Rampichino della Nuova Zelanda

5961 **Mohoua ochrocephala**
Passeriformes - Corvidae
e Yellowhead; Mohua
f Mohoua à tête jaune
d Gelbköpchen; Maorigrasmücke
i Testagialla

5962 **Molothrus aeneus**
Passeriformes - Icteridae
e Bronzed Cowbird; Red-eyed
Cowbird; Brown Cowbird
f Vacher bronzé
d Rotaugenkuhstärling
i Molotro bronzato

5963 **Molothrus armenti**
Passeriformes - Icteridae
e Bronze-brown Cowbird; Red-eyed
Cowbird
f Vacher brun

d Amazonas-Kuhstärling
i Molotro testabruna

5964 **Molothrus ater**
Passeriformes - Icteridae
e Brown-headed Cowbird; Eastern
Cowbird; Nevada Cowbird; Common
Cowbird
f Vacher à tête brune; Vacher;
Molothre noir
d Maskenkuhstärling; Schwarzer
Kuhstärling; Braunkopfkuhstärling;
Kuhstärling; Nordamerikanischer
Kuhstärling
i Molotro nero; Molotro testabruna

5965 **Molothrus badius**
Passeriformes - Icteridae
e Bay-winged Cowbird
f Vacher à ailes baies
d Braunkuhstärling;
Braunflügelkuhstärling
i Molotro badio; Molotro alicastane

5966 **Molothrus bonariensis**
Passeriformes - Icteridae
e Shiny Cowbird; Glossy Cowbird;
Blackbird's Cousin (WI); Cornbird
(WI); Common Cowbird; Cowbird
f Vacher luisant; Merle Ste.-Lucie
(Ants); Merle de Barbade (Ants)
d Seidenkuhstärling; Glanzkuhstärling
i Molotro bonariense; Molotra
splendente

5967 **Molothrus rufoaxillaris**
Passeriformes - Icteridae
e Screaming Cowbird
f Vacher criard
d Rotachselkuhstärling;
Lärmkuhstärling
i Molotro ascelle castane

5968 **Momotidae**
Coraciiformes
e Motmots
f Momotidés
d Sägeracken
i Momotidi

5969   *Momotus aequatorialis*
        Coraciiformes - Momotidae
e   Highland Motmot
f   Motmot d'Équateur

5970   *Momotus coeruliceps*
        Coraciiformes - Momotidae
e   Blue-crowned Motmot
f   Motmot à tête bleue

5971   *Momotus lessoni*
        Coraciiformes - Momotidae
e   Lesson's Motmot
f   Motmot de Lesson

5972   *Momotus mexicanus*
        Coraciiformes - Momotidae
e   Russet-crowned Motmot
f   Motmot à tête rousse
d   Braunscheitelmotmot
i   Motmot corona rossiccia

5973   *Momotus momota*
        Coraciiformes - Momotidae
e   Blue-crowned Motmot; Blue-
        diademed Motmot; Highland
        Motmot; Equatorial Motmot
f   Motmot houtouc
d   Blauscheitelmotmot
i   Prionite; Motmot corona azzura;
        Motmot

5974   *Momotus subrufescens*
        Coraciiformes - Momotidae
e   Tawny-bellied Motmot
f   Motmot caraibe

5975   *Monachella muelleriana*
        Passeriformes - Petroicidae
e   Torrent Robin; River Flycatcher;
        River Robin
f   Miro des torrents
d   Uferschnäpper
i   Pigliamosche di fiume

5976   *Monarcha axillaris*
        Passeriformes - Corvidae
e   Black Monarch; Black Monarch-
        Flycatcher; Black Monarch; Black
        Fantail
f   Monarque noir

d   Fächerschwanzmonarch
i   Monarca nero

5977   *Monarcha barbatus*
        Passeriformes - Corvidae
e   Black-and-white Monarch; Pied
        Monarch; Pied Monarch-Flycatcher;
        Black-throated Monarch; Solomons
        Pied-Monarch; Solomon Islands
        Pied-Monarch; Fluttering Monarch
f   Monarque pie
d   Weißbartmonarch
i   Monarca gazza

5978   *Monarcha boanensis*
        Passeriformes - Corvidae
e   Black-chinned Monarch; Boana
        Monarch
f   Monarque dc Boano
i   Monarca mentenero

5979   *Monarcha brehmii*
        Passeriformes - Corvidae
e   Biak Monarch; Biak Monarch-
        Flycatcher
f   Monarque de Brehm
d   Falbschwanzmonarch
i   Monarca di Biak

5980   *Monarcha browni*
        Passeriformes - Corvidae
e   Kulambangra Monarch; Brown's
        Monarch
f   Monarque de Brown
d   Salomonen-Monarch
i   Monarca di Kulambangra

5981   *Monarcha castaneiventris*
        Passeriformes - Corvidae
e   Chestnut-bellied Monarch; Chestnut-
        bellied Monarch-Flycatcher
f   Monarque à ventre marron
d   Schwarzrückenmonarch
i   Monarca ventrecastano

5982   *Monarcha castus*
        Passeriformes - Corvidae
e   Loetoe Monarch
f   Monarque des Tanimbar
i   Monarca di Loetoe

5983 *Monarcha chrysomela*
Passeriformes - Corvidae
e Golden Monarch; Black-and-yellow
Monarch; Black-and-yellow
Monarch-Flycatcher
f Monarque doré
d Goldmonarch
i Monarca nero e oro

5984 *Monarcha cinerascens*
Passeriformes - Corvidae
e Island Monarch; Ashy Monarch;
Grey-headed Monarch; Islet
Monarch; Island Grey-headed
Monarch; Islet Flycatcher
f Monarque des îles
d Mangrovemonarch;
Graukopfmonarch
i Monarca cenerino

5985 *Monarcha erythrostictus*
Passeriformes - Corvidae
e Bougainville Monarch
f Monarque de Bougainville
i Monarca di Bougainville

5986 *Monarcha everetti*
Passeriformes - Corvidae
e White-tipped Monarch; Everett's
Monarch; Djampea Monarch
f Monarque d'Everett
d Everett-Monarch
i Monarca di Everett

5987 *Monarcha frater*
Passeriformes - Corvidae
e Black-winged Monarch; Black-
chinned Monarch; Pearly Monarch
f Monarque à ailes noires
d Schwarzflügelmonarch
i Monarca alinere

5988 *Monarcha godeffroyi*
Passeriformes - Corvidae
e Yap Monarch; Yap Island Monarch
f Monarque de Yap
d Karolinen-Monarch; Yap-Monarch
i Monarca dell'Isola Yap

5989 *Monarcha guttulus*
Passeriformes - Corvidae

e Spot-winged Monarch; Spot-wing
Monarch; Spot-winged Monarch-
Flycatcher; Spot-wing Monarch-
Flycatcher
f Monarque à ailes tachetées
d Perlenflügelmonarch
i Monarca aligocciolate

5990 *Monarcha infelix*
Passeriformes - Corvidae
e Manus Monarch; Admiralty Islands
Monarch-Flycatcher; Unhappy
Monarch; Sombre Monarch;
Admiralty Islands Monarch;
Admiralty Monarch; Admiralty
Monarch-Flycatcher
f Monarque triste
d Silberschwanzmonarch
i Monarca infelice

5991 *Monarcha julianae*
Passeriformes - Corvidae
e Black-backed Monarch; Kofiau
Monarch; Kofiau Monarch-
Flycatcher; Black-backed Monarch
f Monarque de Kofiau
d Juliana-Monarch
i Monarca di Kofiau

5992 *Monarcha leucotis*
Passeriformes - Corvidae
e White-eared Monarch; White-eared
Monarch-Flycatcher; White-eared
Flycatcher
f Monarque oreillard
d Weißohrmonarch
i Monarca orecchiebianche

5993 *Monarcha leucurus*
Passeriformes - Corvidae
e White-tailed Monarch; Kai Monarch;
Kai Monarch-Flycatcher; White-
tailed Flycatcher
f Monarque des Kai
d Weißschwanzmonarch
i Monarca di Kai

5994 *Monarcha loricatus*
Passeriformes - Corvidae
e Black-tipped Monarch; Buru
Monarch

f   Monarque de Buru
d   Buru-Monarch
i   Monarca loricato

5995   ***Monarcha manadensis***
        Passeriformes - Corvidae
e   Hooded Monarch; Black-and-white
      Monarch; Black-and-white Monarch-
      Flycatcher; White-bellied Monarch
f   Monarque à capuchon
d   Zweifarbenmonarch
i   Monarca bianco e nero

5996   ***Monarcha melanopsis***
        Passeriformes - Corvidae
e   Black-faced Monarch; Grey-winged
      Monarch; Pearly-winged Monarch;
      Grey-winged Monarch-Flycatcher
f   Monarque à face noire
d   Brillenmonarch; Maskenmonarch
i   Monarca faccianera

5997   ***Monarcha menckei***
        Passeriformes - Corvidae
e   White-breasted Monarch; Mencke's
      Monarch; Mencke's Monarch-
      Flycatcher; St. Matthias Monarch;
      Musao Island Monarch; Monarch-
      Flycatcher
f   Monarque des Saint-Matthias
d   Weißrückenmonarch
i   Monarca di San Matthias

5998   ***Monarcha mundus***
        Passeriformes - Corvidae
e   Black-bibbed Monarch; Tanimbar
      Monarch; Tanimbar Monarch-
      Flycatcher; Mundane Monarch
f   Monarque à menton noir
d   Tanimbar-Monarch
i   Monarca di Tanimbar

5999   ***Monarcha pileatus***
        Passeriformes - Corvidae
e   White-naped Monarch; Tufted
      Monarch-Flycatcher; Pileated
      Monarch; Moluccan Monarch
f   Monarque à nuque blanche
d   Halmahera-Monarch
i   Monarca nucabianca

6000   ***Monarcha richardsii***
        Passeriformes - Corvidae
e   White-capped Monarch; Richard's
      Monarch
f   Monarque de Richards
i   Monarca di Richards

6001   ***Monarcha rubiensis***
        Passeriformes - Corvidae
e   Rufous Monarch; Rufous Monarch-
      Flycatcher
f   Monarque roux
d   Fuchsmonarch
i   Monarca rossiccio

6002   ***Monarcha sacerdotum***
        Passeriformes - Corvidae
e   Flores Monarch; Mees's Monarch;
      Mees's Monarch-Flycatcher; Priestly
      Monarch; Flores Mountain-Monarch
f   Monarque de Florès
d   Flores-Monarch
i   Monarca di Flores

6003   ***Monarcha takatsukasae***
        Passeriformes - Corvidae
e   Tinian Monarch; Tinian Island
      Monarch
f   Monarque de Tinian
d   Tinian-Monarch
i   Monarca dell'Isola Tinian

6004   ***Monarcha trivirgatus***
        Passeriformes - Corvidae
e   Spectacled Monarch; Spectacled
      Monarch-Flycatcher
f   Monarque à lunettes
d   Brillenmonarch
i   Monarcha dagli occhiali

6005   ***Monarcha verticalis***
        Passeriformes - Corvidae
e   Black-tailed Monarch; Bismarck
      Monarch; New Britain Pied-
      Monarch; Duke of York's Monarch;
      New Britain Monarch; York
      Monarch
f   Monarque des Bismarck
d   Stirnschopfmonarch
i   Monarca delle Isole Bismarck

6006 **Monarcha viduus**
Passeriformes - Corvidae
e White-collared Monarch; Scaled
Monarch; San Cristobal Monarch;
White-collared Flycatcher
f Monarque à col blanc
d Schuppenmonarch
i Monarca di San Cristobal

6007 **Monasa atra**
Piciformes - Bucconidae
e Black Nunbird; Black Nunlet
f Barbacou noir
d Schwarztrappist; Mohrentrappist
i Monaca nera

6008 **Monasa flavirostris**
Piciformes - Bucconidae
e Yellow-billed Nunbird; Yellow-
billed Nunlet
f Barbacou à bec jaune
d Gelbschnabeltrappist;
Gelbschnabelfaulvogel
i Monaca beccogiallo

6009 **Monasa morphoeus**
Piciformes - Bucconidae
e White-fronted Nunbird; White-
fronted Nunlet
f Barbacou à front blanc
d Weißstirntrappist;
Weißgesichtfaulvogel
i Monaca frontebrianca

6010 **Monasa nigrifrons**
Piciformes - Bucconidae
e Black-fronted Nunbird; Black-
fronted Nunlet
f Barbacou unicolore
d Schwarzstirntrappist
i Monaca frontenera

6011 **Monias benschi**
Gruiformes - Mesitornithidae
e Sub-desert Mesite; Monia; Bensch's
Rail; Bensch's Monia; Monias;
Bensch's Mesite
f Mésite monias; Monias de Bensch
d Naka; Moniasralle
i Monias di Bensch; Monia

6012 **Monticola angolensis**
Passeriformes - Muscicapidae
e Miombo Rock-Thrush; Mottled
Rock-Thrush; Angola-Thrush;
Angola Rock-Thrush
f Monticole angolais; Merle de roche
d'Angola
d Miombo-Rötel; Angola-Rötel;
Großes Waldrötel
i Codirossone del Miombo

6013 **Monticola brevipes**
Passeriformes - Muscicapidae
e Short-toed Rock-Thrush
f Monticole à doigts courts; Merle de
roche brachydactyle
d Kurzzehenrötel
i Codirossone corona chiara

6014 **Monticola cinclorhynchus**
Passeriformes - Muscicapidae
e Blue-capped Rock-Thrush; Blue-
headed Rock-Thrush
f Monticole à croupion roux
d Bergrötel
i Codirossone capoblu

6015 **Monticola explorator**
Passeriformes - Muscicapidae
e Sentinel Rock-Thrush
f Monticole espion; Merle de roche
montagnard
d Langzehenrötel
i Codirossone sentinella

6016 **Monticola gularis**
Passeriformes - Muscicapidae
e White-throated Rock-Thrush;
Swinhoe's White-throated Rock-
Thrush; Swinhoe's Rock-Thrush;
White-breasted Rock-Thrush
f Monticole à gorge blanche
d Amur-Rötel
i Codirossone golabianca

6017 **Monticola pretoriae**
Passeriformes - Muscicapidae
e Transvaal Rock-Thrush
f Monticole du Transvaal
i Codirossone del Transvaal

6018 *Monticola rufiventris*
Passeriformes - Muscicapidae
- e Chestnut-bellied Rock-Thrush;
  Chestnut-breasted Rock-Thrush
- f Monticole à ventre marron; Merle de
  roche à ventre marron
- d Rötelmerle
- i Passero solitario rufiventre

6019 *Monticola rufocinereus*
Passeriformes - Muscicapidae
- e Little Rock-Thrush
- f Monticole rougequeue; Petit Merle
  de roche
- d Schluchtenrötel
- i Codirossone piccolo

6020 *Monticola rupestris*
Passeriformes - Muscicapidae
- e Cape Rock-Thrush
- f Monticole rocar; Merle de roche du
  Cap
- d Klippenrötel; Kap-Rötel
- i Codirossone del Capo

6021 *Monticola saxatilis*
Passeriformes - Muscicapidae
- e Rufous-tailed Rock-Thrush;
  Common Rock-Thrush (CSA); Rock-
  Thrush; Mountain Rock-Thrush
  (CSA); Chestnut-tailed Rock-Thrush;
  White-backed Rock-Thrush;
  European Rock-Thrush
- f Merle de roche; Monticole de roche;
  Monticole merle-de-roche; Merle de
  roche commun; Merle solitaire
  rouge; Petrocincle de roche
- d Steinrötel
- i Codirossone europeo; Codirossone

6022 *Monticola solitarius*
Passeriformes - Muscicapidae
- e Blue Rock-Thrush; Red-bellied
  Rock-Thrush; Indian Blue Rock-
  Thrush
- f Merle bleu; Monticole bleue;
  Monticole merle-bleu; Merle de
  roche bleu; Petrocincle bleu;
  Petrocincle bleu solidaire
- d Blaumerle; Blaudrossel
- i Passero solitario

6023 *Montifringilla adamsi*
Passeriformes - Passeridae
- e Tibetan Snowfinch; Black-winged
  Snowfinch; Adams's Snowfinch;
  Tibet Snowfinch
- f Niverolle du Tibet
- d Adams-Schneefink; Adams-
  Schneesperling
- i Fringuello alpino di Adams

6024 *Montifringilla blanfordi*
Passeriformes - Passeridae
- e Plain-backed Snowfinch; Blanford's
  Snowfinch; Blanford's Finch
- f Niverolle de Blanford
- d Blanford-Schneefink; Blanford-
  Sperling
- i Fringuello alpino di Blanford

6025 *Montifringilla davidiana*
Passeriformes - Passeridae
- e Small Snowfinch; Père David's
  Snowfinch; Snowfinch; David's
  Snowfinch; Père David's Finch;
  Mongolian Snowfinch
- f Niverolle de David; Niverolle du
  père David
- d David-Schneefink; Erdsperling
- i Fringuello alpino di padre David

6026 *Montifringilla nivalis*
Passeriformes - Passeridae
- e White-winged Snowfinch;
  Snowfinch; Eurasian Snowfinch;
  Common Snowfinch
- f Niverolle alpine; Niverolle des
  neiges; Pinson des neiges
- d Schneefink; Schneesperling
- i Fringuello alpino; Fringuello alpino
  europeo

6027 *Montifringilla ruficollis*
Passeriformes - Passeridae
- e Rufous-necked Snowfinch; Red-
  necked Snowfinch
- f Niverolle à cou roux
- d Rothalsschneefink; Rothalssperling
- i Fringuello alpino collorosso

6028 *Montifringilla taczanowskii*
Passeriformes - Passeridae

e   White-rumped Snowfinch;
    Taczanowski's Snowfinch;
    Taczanowski's Finch
f   Niverolle de Taczanowski
d   Taczanowski-Schneefink;
    Weißbürzelsperling
i   Fringuello alpino di Mandelli

6029   *Montifringilla theresae*
         Passeriformes - Passeridae
e   Afghan Snowfinch; Theresa's
    Snowfinch; Bar-tailed Snowfinch;
    Meinertzhagen's Snowfinch
f   Niverolle d'Afghanistan
d   Afghanen-Schneefink; Afghanen-
    Sperling
i   Fringuello alpino di Meinertzhagen

6030   *Morococcyx erythropygus*
         Cuculiformes - Neomorphidae
e   Lesser Ground-Cuckoo
f   Géocoucou de Lesson
d   Drosselkuckuck
i   Cuculo di terra minore

6031   *Morphnus guianensis*
         Falconiformes - Accipitridae
e   Crested Eagle; Guiana Crested-Eagle
f   Harpie huppée; Aigle huppé
d   Würgadler
i   Aquila crestata della Guiana; Arpia
    crestata

6032   *Morus bassanus*
         Pelecaniformes - Sulidae
e   Northern Gannet (NA); Gannet;
    Atlantic Gannet; North Atlantic
    Gannet
f   Fou de Bassan
d   Basstölpel; Bassantölpel
i   Sula; Sula bassana

6033   *Morus capensis*
         Pelecaniformes - Sulidae
e   Cape Gannet; African Gannet; South
    African Gannet; Gannet
f   Fou du Cap
d   Kap-Tölpel
i   Sula del Capo

6034   *Morus serrator*
         Pelecaniformes - Sulidae
e   Australasian Gannet; Australian
    Gannet (CSA); Malgas; Gannet
f   Fou austral; Flamant austral
d   Australtölpel
i   Sula australiana

6035   *Motacilla aguimp*
         Passeriformes - Motacillidae
e   African Pied-Wagtail; African
    Wagtail; Pied Wagtail; Moroccan
    Wagtail
f   Bergeronnette pie
d   Witwenstelze
i   Ballerina nera africana

6036   *Motacilla alba*
         Passeriformes - Motacillidae
e   White Wagtail; Pied Wagtail;
    Wagtail; Common Pied-Wagtail;
    Indian Pied-Wagtail; Masked
    Wagtail; Black-eared Wagtail;
    White-faced Wagtail; British Pied-
    Wagtail; Moroccan Wagtail
f   Bergeronnette grise; Lavandière grise
d   Bachstelze; Weiße Bachstelze
i   Ballerina bianca; Batticoda;
    Coditremola

6037   *Motacilla alba yarelli*
         Passeriformes - Motacillidae
e   Yarrell's White-Wagtail; Pied
    Wagtail; British Pied-Wagtail (NA)
f   Bergeronnette de Yarrell
d   Trauerbachstelze
i   Ballerina nera

6038   *Motacilla capensis*
         Passeriformes - Motacillidae
e   Cape Wagtail; Willie-Wagtail (CSA)
f   Bergeronnette du Cap
d   Kap-Stelze
i   Ballerina del Capo

6039   *Motacilla cinerea*
         Passeriformes - Motacillidae
e   Grey Wagtail
f   Bergeronnette des ruisseaux
d   Gebirgsstelze; Bergstelze
i   Ballerina gialla

6040   *Motacilla citreola*
Passeriformes - Motacillidae
e   Citrine Wagtail; Yellow-hooded
Wagtail (ISC); Yellow-headed
Wagtail (ISC)
f   Bergeronnette citrine
d   Zitronenstelze
i   Cutrettola testagialla orientale

6041   *Motacilla clara*
Passeriformes - Motacillidae
e   Mountain Wagtail; Long-tailed
Wagtail (CSA)
f   Bergeronnette à longue queue
d   Langschwanzstelze
i   Ballerina codalunga

6042   *Motacilla feldegg*
Passeriformes - Motacillidae
e   Black-headed Wagtail
f   Bergeronnette à tête noire
d   Maskenstelze

6043   *Motacilla flava*
Passeriformes - Motacillidae
e   Blue-headed Wagtail; Yellow
Wagtail; Central European Yellow-
Wagtail
f   Bergeronnette printanière;
Bergeronnette flavéole; Bergeronette
jaune
d   Schafstelze; Wiesenstelze; Kuhstelze
i   Cutrettola; Ballerina gialla;
Codinzinzola; Cutrettola di
primavere

6044   *Motacilla flava flavissima*
Passeriformes - Motacillidae
e   Yellow-Wagtail; British Yellow-
Wagtail; Yellowish-crowned Wagtail
f   Bergeronnette flavéole
d   Englische Schafstelze; Grünköpfige
Schafstelze

6045   *Motacilla flava thunbergi*
Passeriformes - Motacillidae
e   Grey-headed Wagtail; Fenno-
Scandinavian Yellow-Wagtail
f   Bergeronnette printanière nordique;
Bergeronnette à tête grise

6046   *Motacilla flaviventris*
Passeriformes - Motacillidae
e   Madagascar Wagtail
f   Bergeronnette malgache
d   Madagaskar-Stelze
i   Ballerina del Madagascar

6047   *Motacilla grandis*
Passeriformes - Motacillidae
e   Japanese Wagtail; Japanese Pied-
Wagtail
f   Bergeronnette du Japon;
Bergeronnette grise du Japon
d   Schieferstelze; Japan-Stelze
i   Ballerina del Giappone

6048   *Motacilla lugens*
Passeriformes - Motacillidae
e   Black-backed Wagtail; Kamchatka
Pied-Wagtail; Kamchatkan Pied-
Wagtail; Japanese Pied-Wagtail
f   Bergeronnette lugubre
i   Ballerina dorsonero

6049   *Motacilla madaraspatensis*
Passeriformes - Motacillidae
e   White-browed Wagtail; Large Pied-
Wagtail
f   Bergeronnette indienne
d   Mamula-Stelze; Brauenstelze
i   Ballerina nera indiana

6050   *Motacilla subpersonata*
Passeriformes - Motacillidae
e   Masked Wagtail

6051   *Motacilla tschutschensis*
Passeriformes - Passeridae
e   Alaska Yellow-Wagtail

6052   **Motacillidae**
Passeriformes
e   Pipits; Wagtails
f   Pipits; Motacillidés
d   Stelzen
i   Motacillidi; Pispole; Cutrettole

6053   *Mulleripicus fulvus*
Piciformes - Picidae
e   Ashy Woodpecker; Fulvous
Woodpecker; Celebes Woodpecker

f Pic fauve
d Celebes-Specht
i Picchio fulvo

6054 ***Mulleripicus funebris***
Piciformes - Picidae
e Sooty Woodpecker
f Pic en deuil
d Philippinen-Specht
i Picchio fuligginoso

6055 ***Mulleripicus pulverulentus***
Piciformes - Picidae
e Great Slaty Woodpecker; Himalayan Great Slaty Woodpecker
f Pic meunier
d Puderspecht
i Picchio polveroso

6056 ***Muscicapa adusta***
Passeriformes - Muscicapidae
e African Dusky Flycatcher; Dusky Flycatcher (CSA); Dusky Alseonax; Chapin's Flycatcher
f Gobemouche sombre; Gobe-mouches sombre
d Dunkelschnäpper
i Pigliamosche scuro

6057 ***Muscicapa aquatica***
Passeriformes - Muscicapidae
e Swamp Flycatcher; Swamp Alseonax
f Gobemouche des marais; Gobe-mouches des marais
d Sumpfschnäpper
i Pigliamosche delle paludi

6058 ***Muscicapa boehmi***
Passeriformes - Muscicapidae
e Böhm's Flycatcher
f Gobemouche de Böhm; Gobe-mouches de Böhm
d Böhm-Schnäpper
i Pigliamosche di Boehm

6059 ***Muscicapa caerulescens***
Passeriformes - Muscicapidae
e Ashy Flycatcher; Blue-grey Flycatcher (CSA); White-eyed Flycatcher; Ashy Alseonax; Blue-grey Alseonax; Cinereous Flycatcher

f Gobemouche à lunettes; Gobe-mouches à lunettes; Gobemouche à lunettes blanc; Gobe-mouches à lunettes blanc
d Schieferschnäpper
i Pigliamosche cenerino

6060 ***Muscicapa cassini***
Passeriformes - Muscicapidae
e Cassin's Flycatcher; Cassin's Grey-Flycatcher; Cassin's Alseonax; Cassin's Grey-Alseonax
f Gobemouche de Cassin; Gobe-mouches de Cassin
d Cassin-Schnäpper
i Pigliamosche grigio di Cassin

6061 ***Muscicapa comitata***
Passeriformes - Muscicapidae
e Dusky-blue Flycatcher
f Gobemouche ardoisé; Gobe-mouches ardoisé
d Stuhlmann-Schnäpper
i Pigliamosche blu scuro

6062 ***Muscicapa daurica***
Passeriformes - Muscicapidae
e Asian Brown Flycatcher; Brown Flycatcher; Grey-breasted Flycatcher
f Gobemouche brun; Gobe-mouches brun; Gobemouche de Daourie; Gobe-mouches de Daourie
d Braunschnäpper; Brauner Fliegenschnäpper
i Pigliamosche bruno; Pigliamosche beccolargo; Pigliamosche di Dauria

6063 ***Muscicapa epulata***
Passeriformes - Muscicapidae
e Little Grey-Flycatcher; Little Grey-Alseonax; Grey Flycatcher
f Gobemouche cendré; Gobe-mouches cendré
d Fanti-Schnäpper
i Pigliamosche grigio piccolo

6064 ***Muscicapa ferruginea***
Passeriformes - Muscicapidae
e Ferruginous Flycatcher
f Gobemouche ferrugineux; Gobe-mouches ferrugineux

d Rostchnäpper
i Pigliamosche ferrugineo

6065 *Muscicapa gambagae*
Passeriformes - Muscicapidae
e Gambaga Flycatcher; Gambaga
Dusky-Flycatcher; Gambaga
Spotted-Flycatcher
f Gobemouche de Gambaga; Gobe-
mouches de Gambaga
d Gambaga-Schnäpper
i Pigliamosche del Gambaga

6066 *Musclcapu griseisticta*
Passeriformes - Muscicapidae
e Grey-streaked Flycatcher; Grey-
spotted Flycatcher; Spot-breasted
Flycatcher
f Gobemouche à taches grises; Gobe-
mouches à taches grises
d Fleckenschnäpper
i Pigliamosche striata

6067 *Muscicapa infuscata*
Passeriformes - Muscicapidae
e Sooty Flycatcher; African Sooty-
Flycatcher
f Gobemouche enfumé; Gobe-
mouches enfumé
d Schieferbrustschnäpper
i Pigliamosche fuligginoso africano

6068 *Muscicapa itombwensis*
Passeriformes - Muscicapidae
e Itombwe Flycatcher; Itombwe
Alseonax; Prigogine's Alseonax
f Gobemouche de l'Itombwe; Gobe-
mouches de l'Itombwe
d Itombwe-Schnäpper
i Pigliamosche di Itombwe

6069 *Muscicapa lendu*
Passeriformes - Muscicapidae
e Chapin's Flycatcher; Lendu
Flycatcher; Chapin's Alseonax;
Lendu Alseonax
f Gobemouche de Chapin; Gobe-
mouches de Chapin; Gobemouche de
Lendu; Gobe-mouches de Lendu
d Lendu-Schnäpper
i Pigliamosche di Chapin

6070 *Muscicapa muttui*
Passeriformes - Muscicapidae
e Brown-breasted Flycatcher; Layard's
Flycatcher
f Gobemouche muttui; Gobe-mouches
muttui
d Bambusschnäpper
i Pigliamosche pettobruno

6071 *Muscicapa olivascens*
Passeriformes - Muscicapidae
e Olivaceous Flycatcher; Dusky-
capped Flycatcher; Olivaceous
Alseonax; African Olivaceous
Alseonax
f Gobemouche olivâtre; Gobe-
mouches olivâtre
d Olivschnäpper
i Pigliamosche oliva africano

6072 *Muscicapa randi*
Passeriformes - Muscicapidae
e Ashy-breasted Flycatcher; Ash-
breasted Flycatcher
f Gobemouche à poitrine grise; Gobe-
mouches à poitrine grise
i Pigliamosche pettogrigio

6073 *Muscicapa ruficauda*
Passeriformes - Muscicapidae
e Rusty-tailed Flycatcher; Rufous-
tailed Flycatcher
f Gobemouche à queue rousse; Gobe-
mouches à queue rousse
d Rostschwanzschnäpper
i Pigliamosche codaruggine

6074 *Muscicapa segregata*
Passeriformes - Muscicapidae
e Sumba Brown-Flycatcher; Sumba
Flycatcher
f Gobemouche de Sumba; Gobe-
mouches de Sumba
d Sumba-Braunschnäpper
i Pigliamosche bruno di Sumba

6075 *Muscicapa sethsmithi*
Passeriformes - Muscicapidae
e Yellow-footed Flycatcher; Yellow-
footed Alseonax
f Gobemouche à pattes jaunes; Gobe-

mouches à pattes jaunes;
Gobemouche de Seth-Smith; Gobe-
mouches de Seth-Smith
d   Gelblaufschnäpper
i   Pigliamosche piedigialli

6076   *Muscicapa sibirica*
Passeriformes - Muscicapidae
e   Dark-sided Flycatcher; Sooty
Flycatcher; Siberian Flycatcher;
Siberian Sooty-Flycatcher;
Himalayan Flycatcher
f   Gobemouche de Sibérie;
Gobemouche fuligineux; Gobe-
mouches de Sibérie; Gobe-mouches
fuligineux
d   Rußschnäpper
i   Pigliamosche siberiano

6077   *Muscicapa striata*
Passeriformes - Muscicapidae
e   Spotted Flycatcher
f   Gobemouche gris; Gobe-mouches
gris
d   Grauschnäpper
i   Pigliamosche europeo

6078   *Muscicapa tessmanni*
Passeriformes - Muscicapidae
e   Tessmann's Flycatcher
f   Gobemouche de Tessmann; Gobe-
mouches de Tessmann
d   Tessmann-Schnäpper
i   Pigliamosche di Tessmann

6079   *Muscicapa ussheri*
Passeriformes - Muscicapidae
e   Ussher's Flycatcher; Ussher's Sooty-
Flycatcher; Ussher's Dusky
Flycatcher
f   Gobemouche d'Ussher; Gobe-
mouches d'Ussher
d   Schwalbenschnäpper
i   Pigliamosche di Ussher

6080   *Muscicapa williamsoni*
Passeriformes - Muscicapidae
e   Brown-streaked Flycatcher;
Chocolate Flycatcher; Williamson's
Flycatcher
f   Gobemouche de Williamson; Gobe-

mouches de Williamson
d   Fuchsschnäpper

6081   *Muscicapella hodgsoni*
Passeriformes - Muscicapidae
e   Pygmy Blue-Flycatcher; Pygmy
Blue-Niltava
f   Gobemouche pygmée; Gobe-
mouches pygmée
d   Goldhähnchenblauschnäpper
i   Niltava di Hodgson

6082   Muscicapidae
Passeriformes
e   Old World Flycatchers
f   Gobemouches; Muscicapidés; Gobe-
mouches
d   Fliegenschnäpper
i   Muscicapidi; Pigliamoschi

6083   *Muscigralla brevicauda*
Passeriformes - Tyrannidae
e   Short-tailed Field-Tyrant
f   Dormilon à queue courte;
Moucherolle à queue courte
d   Stummeltyrann
i   Tiranno codacorta

6084   *Muscipipra vetula*
Passeriformes - Tyrannidae
e   Shear-tailed Grey-Tyrant
f   Moucherolle à queue-de-pie;
Moucherolle de Lichtenstein
d   Kerbschwanztyrann
i   Tiranno grigio codaforcuta

6085   *Muscisaxicola albifrons*
Passeriformes - Tyrannidae
e   White-fronted Ground-Tyrant
f   Dormilon à front blanc
d   Weißstirngrundtyrann
i   Tiranno terricolo frontebianca

6086   *Muscisaxicola albilora*
Passeriformes - Tyrannidae
e   White-browed Ground-Tyrant;
White-lored Ground-Tyrant
f   Dormilon à sourcils blancs
d   Rotkäppchentyrann
i   Tiranno terricolo sopracciglibianchi

6087   *Muscisaxicola alpina*
Passeriformes - Tyrannidae
e   Plain-capped Ground-Tyrant; Alpine
    Ground-Tyrant
f   Dormilon à grands sourcils;
    Dormilon andin
d   Felsentyrann
i   Tiranno terricolo delle Ande

6088   *Muscisaxicola capistrata*
Passeriformes - Tyrannidae
e   Cinnamon-bellied Ground-Tyrant;
    Burmeister's Ground-Tyrant
f   Dormilon à ventre roux
d   Zügeltyrann
i   Tiranno terricolo ventrecannella

6089   *Muscisaxicola cinerea*
Passeriformes - Tyrannidae
e   Cinereous Ground-Tyrant
f   Dormilon cendré
d   Graunackentyrann
i   Tiranno terricolo cenerino

6090   *Muscisaxicola flavinucha*
Passeriformes - Tyrannidae
e   Ochre-naped Ground-Tyrant
f   Dormilon à nuque jaune
d   Gelbnackentyrann
i   Tiranno terricolo nucaocra

6091   *Muscisaxicola fluviatilis*
Passeriformes - Tyrannidae
e   Little Ground-Tyrant
f   Dormilon fluviatile
d   Piepertyrann
i   Tiranno terricolo minore

6092   *Muscisaxicola frontalis*
Passeriformes - Tyrannidae
e   Black-fronted Ground-Tyrant
f   Dormilon à front noir
d   Schwarzstirntyrann
i   Tiranno terricolo frontenera

6093   *Muscisaxicola juninensis*
Passeriformes - Tyrannidae
e   Puna Ground-Tyrant
f   Dormilon de Junin
d   Punatyrann
i   Tiranno terricolo della Puna

6094   *Muscisaxicola macloviana*
Passeriformes - Tyrannidae
e   Dark-faced Ground-Tyrant
f   Dormilon bistré; Dormilon des
    Maloines
d   Maskentyrann
i   Tiranno terricolo facciascura

6095   *Muscisaxicola maculirostris*
Passeriformes - Tyrannidae
e   Spot-billed Ground-Tyrant
f   Dormilon à bec maculé; Dormilon à
    bec jaune
d   Lerchentyrann
i   Tiranno terricolo beccomacchiato

6096   *Muscisaxicola rufivertex*
Passeriformes - Tyrannidae
e   Rufous-naped Ground-Tyrant
f   Dormilon à calotte rousse; Dormilon
    à tête rousse
d   Rotnackentyrann
i   Tiranno terricolo nucarossiccia

6097   *Musophaga johnstoni*
Musophagiformes - Musophagidae
e   Ruwenzori Turaco; Johnstone's
    Touraco
f   Touraco du Ruwenzori
d   Kammschnabelturako;
    Kammschnabelturaco
i   Turaco del Ruwenzori

6098   *Musophaga porphyreolopha*
Musophagiformes - Musophagidae
e   Purple-crested Turaco; Purple-
    crested Lourie (CSA); Violet-crested
    Turaco; Purple-crested Touraco;
    Violet-crested Touraco
f   Touraco à huppe splendide
d   Glanzhaubenturako;
    Glanzhaubenturaco
i   Turaco crestavioletta

6099   *Musophaga rossae*
Musophagiformes - Musophagidae
e   Ross's Turaco; Ross's Lourie (CSA);
    Lady Ross's Touraco; Lady Ross's
    Turaco; Ross's Touraco
f   Touraco de Lady Ross

d Ross-Turako; Ross-Turaco
i turaco di Lady Ross

6100 *Musophaga violacea*
Musophagiformes - Musophagidae
e Violet Turaco; Violet Touraco
f Touraco violet
d Schildturaco
i Turaco violetto

6101 Musophagidae
Musophagiformes
e Turacos
f Musophagidés
d Turakos; Lärmvögel; Turacos
i Musofagidi

6102 *Myadestes coloratus*
Passeriformes - Muscicapidae
e Varied Solitaire
f Solitaire varié
i Tordo solitario variegato

6103 *Myadestes elisabeth*
Passeriformes - Muscicapidae
e Cuban Solitaire
f Solitaire de Cuba
d Kuba-Klarino
i Tordo solitario di Cuba

6104 *Myadestes genibarbis*
Passeriformes - Muscicapidae
e Rufous-throated Solitaire; Solitaire
(WI); Mountain Whistler (WI);
Fiddler (WI); Soufrierebird (WI);
Rufous-brown Solitaire
f Oiseau musicien (Ants); Siffleur
morne (Ants); Solitaire à gorge rouge
(Ants); Siffleur de montagne (Ants);
Solitaire siffleur
d Bartklarino
i Tordo solitario golarossa

6105 *Myadestes lanaiensis*
Passeriformes - Muscicapidae
e Olomao; Lanai Thrush
f Solitaire de Lanai
i Tordo di Lanai

6106 *Myadestes melanops*
Passeriformes - Muscicapidae

e Black-faced Solitaire
f Solitaire masqué
i Tordo solitario faccianera

6107 *Myadestes myadestinus*
Passeriformes - Muscicapidae
e Kamao; Hawaiian Thrush; Large
Kaual; Large Kaual Thrush
f Solitaire kamao
i Tordo grande di Kauai

6108 *Myadestes oahensis*
Passeriformes - Muscicapidae
e Amaui; Oahui Thrush
f Solitaire d'Oahu
i Tordo amaul

6109 *Myadestes obscurus*
Passeriformes - Muscicapidae
e Omao; Brown-backed Solitaire;
Hawaiian Thrush; Hawaii Thrush
f Solitaire d'Hawai
d Hawaii-Drossel
i Tordo delle Hawaii

6110 *Myadestes occidentalis*
Passeriformes - Muscicapidae
e Brown-backed Solitaire; Omao
f Solitaire à dos brun
d Braunrückenklarino
i Tordo solitario dorsobruno

6111 *Myadestes palmeri*
Passeriformes - Muscicapidae
e Puaiohi; Small Kauai Thrush
f Solitaire pualohi
d Palmerdrossel
i Tordo piccolo di Kauai

6112 *Myadestes ralloides*
Passeriformes - Muscicapidae
e Andean Solitaire
f Solitaire des Andes
d Rallenklarino
i Tordo solitario delle Ande

6113 *Myadestes sibilans*
Passeriformes - Muscicapidae
e St. Vincent Solitaire

6114   *Myadestes townsendi*
           Passeriformes - Muscicapidae
e   Townsend's Solitaire
f   Solitaire de Townsend
d   Bergklarino; Klarinettenvogel
i   Tordo solitario di Townsend

6115   *Myadestes unicolor*
           Passeriformes - Muscicapidae
e   Slaty Solitaire; Slate-coloured
     Solitaire
f   Solitaire ardoisé
d   Schieferklarino
i   Tordo solitario color ardesia

6116   *Mycerobas affinis*
           Passeriformes - Fringillidae
e   Collared Grosbeak; Yellow-collared
     Grosbeak; Allied Grosbeak
f   Grosbec voisin
d   Gelbschenkelkernbeißer;
     Halsbandkernbeißer
i   Frosone eurasiatico; Frosone dal
     collare

6117   *Mycerobas carnipes*
           Passeriformes - Fringillidae
e   White-winged Grosbeak
f   Grosbec à ailes blanches
d   Wacholderkernbeißer
i   Frosone alibianche; Beccogrosso
     alibianche

6118   *Mycerobas icterioides*
           Passeriformes - Fringillidae
e   Black-and-yellow Grosbeak; Black-
     and-yellow Hawfinch
f   Grosbec noir-et-jaune
d   Goldkernbeißer;
     Schwarzschenkelkernbeißer
i   Frosone giallo e nero

6119   *Mycerobas melanozanthos*
           Passeriformes - Fringillidae
e   Spot-winged Grosbeak; Spotted-wing
     Grosbeak
f   Grosbec à ailes tachetées
d   Fleckenkernbeißer;
     Zahnschnabelkernbeißer
i   Frosone alimaculate

6120   *Mycteria americana*
           Ciconiiformes - Ciconiidae
e   Wood Stork; Wood Ibis; American
     Wood-Stork; American Wood-Ibis
f   Tantale d'Amérique
d   Waldstorch; Waldibis
i   Mitteria americana; Tantalo
     americano; Tantalo

6121   *Mycteria cinerea*
           Ciconiiformes - Ciconiidae
e   Milky Stork; Wood Stork; Southern
     Painted Stork
f   Tantale blanc
d   Milchstorch
i   Mitteria cinerea; Tantalo cinereo

6122   *Mycteria ibis*
           Ciconiiformes - Ciconiidae
e   Yellow-billed Stork; Wood-Ibis
     (CSA)
f   Tantale ibis; Tantale africain
d   Nimmersatt; Afrikanischer
     Nimmersatt
i   Tantalo; Tantalo africano; Mitteria
     africana

6123   *Mycteria leucocephala*
           Ciconiiformes - Ciconiidae
e   Painted Stork
f   Tantale indien
d   Buntstorch
i   Tantalo indiano; Mitteria indiana;
     Cigogna variopinta

6124   *Mylagra albiventris*
           Passeriformes - Corvidae
e   Samoan Flycatcher; White-vented
     Flycatcher; Samoan White-vented
     Flyctcher; Samoan Broadbill;
     Samoan Myiagra
f   Monarque des Samoa
d   Samoa-Myiagra
i   Miagra dal sottocoda bianco

6125   *Myiagra alecto*
           Passeriformes - Corvidae
e   Shining Flycatcher; Shining
     Monarch-Flycatcher; Shining
     Myiagra-Flycatcher; Shining
     Monarch

f   Monarque luisant
d   Glanzmyiagra
i   Miagra splendente

6126   *Myiagra atra*
Passeriformes - Corvidae
e   Biak Black-Flycatcher; Biak
Flycatcher; Black Myiagra-
Flycatcher; Black Monarch; Black
Myiagra; Black Flycatcher
f   Monarque de Biak
d   Stahlmyiagra
i   Miagra nera

6127   *Myiagra azureocapilla*
Passeriformes - Corvidae
e   Blue-crested Flycatcher; Blue-crested
Broadbill; Blue-crested Myiagra-
Flycatcher; Blue-headed Flycatcher
f   Monarque à crête bleue
d   Schmuckmyiagra
i   Miagra capoazzurro

6128   *Myiagra caledonica*
Passeriformes - Corvidae
e   Melanesian Flycatcher; New
Caledonian Myiagra-Flycatcher;
New Hebrides Broadbill; Melanesian
Myiagra; Caledonian Myiagra;
Melanesian Broadbill
f   Monarque mélanésien
d   Hebriden-Myiagra
i   Miagra della Nuova Caledonia

6129   *Myiagra cervinicauda*
Passeriformes - Corvidae
e   Ochre-headed Flycatcher; San
Cristobal Myiagra; San Cristobal
Myiagra-Flycatcher; Red-tailed
Flycatcher
f   Monarque de San Cristobal
i   Miagra di San Cristobal

6130   *Myiagra cyanoleuca*
Passeriformes - Corvidae
e   Satin Flycatcher; Satin Myiagra;
Satin Myiagra-Flycatcher
f   Monarque satiné
d   Seidenmyiagra
i   Miagra di seta

6131   *Myiagra erythrops*
Passeriformes - Corvidae
e   Mangrove Flycatcher; Palau
Broadbill; Micronesian Flycatcher
f   Monarque des Palau
d   Rotstirnmyiagra
i   Miagra di Palau

6132   *Myiagra ferrocyanea*
Passeriformes - Corvidae
e   Steel-blue Flycatcher; Steel-blue
Myiagra; Solomons Broadbill-
Flycatcher; Solomons Flycatcher;
Solomons Broadbill; Solomons Satin
Flycatcher
f   Monarque acier
d   Salomonen-Myiagra
i   Miagra blu-acciaio

6133   *Myiagra freycineti*
Passeriformes - Corvidae
e   Guam Flycatcher; Guam Myiagra-
Flycatcher; Guam Myiagra; Guam
Broadbill
f   Monarque de Guam
d   Marianen-Myiagra
i   Miagra di Guam

6134   *Myiagra galeata*
Passeriformes - Corvidae
e   Dark-grey Flycatcher; Helmet
Flycatcher; Moluccan Myiagra-
Flycatcher; Moluccan Flycatcher;
Helmeted Flycatcher; Slaty Monarch;
Slaty Flycatcher; Helmeted Broadbill
f   Monarque des Moluques
d   Molukken-Myiagra
i   Miagra dall'elmo

6135   *Myiagra hebetior*
Passeriformes - Corvidae
e   Dull Flycatcher; Dull Monarch; Dull
Monarch-Flycatcher; Hartert's
Monarch; Island Flycatcher
f   Monarque terne
d   Eichhorn-Myiagra
i   Miagra modesta

6136   *Myiagra inquieta*
Passeriformes - Corvidae
e   Restless Flycatcher

f   Monarque infatiguable
d   Weißkehlmyiagra
i   Miagra maggiore

6137   *Myiagra inquieta nana*
          Passeriformes - Corvidae
f   Monarque menu

6138   *Myiagra oceanica*
          Passeriformes - Corvidae
e   Oceanic Flycatcher; Melanesian
      Myiagra-Flycatcher; Truk Myiagra-
      Flycatcher; Truk Island Myiagra-
      Flycatcher; Micronesian Myiagra;
      Micronesian Broadbill; Micronesian
      Flycatcher; Truk Flycatcher
f   Monarque océanite
d   Microantipoden-Schnäpper
i   Miagra della Micronesia

6139   *Myiagra pluto*
          Passeriformes - Corvidae
e   Ponapé Flycatcher; Ponapé Broadbill
f   Monarque de Ponapé
d   Ponapé-Myiagra
i   Miagra di Ponapè

6140   *Myiagra rubecula*
          Passeriformes - Corvidae
e   Leaden Flycatcher; Leaden Myiagra
f   Monarque rougegorge
d   Silbermyiagra
i   Myiagra plumbea

6141   *Myiagra rubecula papuana*
          Passeriformes - Corvidae
e   Torres Strait Leaden Flycatcher
      (ANZ)

6142   *Myiagra ruficollis*
          Passeriformes - Corvidae
e   Broad-billed Flycatcher; Broad-billed
      Myiagra; Broad-billed Myiagra-
      Flycatcher; Broad-billed Monarch
f   Monarque à bec large
d   Breitschnabelmyiagra
i   Miagra pettirossa

6143   *Myiagra vanikorensis*
          Passeriformes - Corvidae
e   Vanikoro Flycatcher; Red-bellied

      Flycatcher; Vanikoro Myiagra;
      Vanikoro Broadbill
f   Monarque de Vanikoro
d   Rotbauchmyiagra
i   Miagra ventrerosso

6144   *Myiarchus antillarum*
          Passeriformes - Tyrannidae
e   Puerto Rican Flycatcher; Stolid
      Flycatcher (WI)
f   Tyran de Porto Rico
d   Antillen-Tyrann
i   Tiranno di Portorico

6145   *Myiarchus apicalis*
          Passeriformes - Tyrannidae
e   Apical Flycatcher
f   Tyran à queue givrée; Tyran de
      Bogota
d   Saumschwanztyrann
i   Tiranno apicale

6146   *Myiarchus atriceps*
          Passeriformes - Tyrannidae
e   Dark-capped Flycatcher

6147   *Myiarchus barbirostris*
          Passeriformes - Tyrannidae
e   Sad Flycatcher; Little Tom-fool
      (WI); Dusky-capped Flycatcher
      (WI); Jamaican Flycatcher
f   Tyran triste
d   Jamaika-Tyrann
i   Tiranno barbuto

6148   *Myiarchus brachyurus*
          Passeriformes - Tyrannidae
e   Ometepe Flycatcher
d   Ometepe-Tyrann

6149   *Myiarchus cephalotes*
          Passeriformes - Tyrannidae
e   Pale-edged Flycatcher;
      Taczanowski's Flycatcher
f   Tyran givré; Tyran de Taczanowski
d   Taczanowski-Tyrann
i   Tiranno testagrossa

6150   *Myiarchus cinerascens*
          Passeriformes - Tyrannidae
e   Ash-throated Flycatcher

f Tyran à gorge cendrée
d Graukehltyrann
i Tiranno golacenerina

6151 *Myiarchus crinitus*
Passeriformes - Tyrannidae
e Great Crested-Flycatcher; Crested
Flycatcher
f Tyran huppé
d Gelbbrusttyrann; Schnäppertyrann
i Tiranno crestato maggiore

6152 *Myiarchus ferox*
Passeriformes - Tyrannidae
e Short-crested Flycatcher
f Tyran féroce
d Kurzschopftyrann
i Tiranno crestacorta

6153 *Myiarchus magister*
Passeriformes - Tyrannidae
e Wied's Flycatcher; Brown-crested
Flycatcher
d Braunschopftyrann

6154 *Myiarchus magnirostris*
Passeriformes - Tyrannidae
e Large-billed Flycatcher; Galapagos
Flycatcher
f Tyran des Galapagos
d Galapagos-Tyrann
i Tiranno delle Galapagos

6155 *Myiarchus nugator*
Passeriformes - Tyrannidae
e Grenada Flycatcher; Loggerhead
(WI); Sunsetbird (WI)
f Tyran bavard
d Grenada-Tyrann
i Tiranno di Grenada

6156 *Myiarchus nuttingi*
Passeriformes - Tyrannidae
e Nutting's Flycatcher; Pale-throated
Flycatcher
f Tyran de Nutting
d Blaßkehltyrann
i Tiranno di Nutting

6157 *Myiarchus oberi*
Passeriformes - Tyrannidae

e Lesser Antillean Flycatcher; St.
Lucia Pewee (WI); Ober's
Flycatcher; Loggerhead (WI);
Guadeloupe Flycatcher
f Gobemouche (Ants); Gros-tête
(Ants); Arbitre (Ants); Gobe-mouche
huppée (Ants); Tyran Janneau;
Siffleur (Ants); Siffleur huppé
(Ants); Pipiri gros-tête (Ants)
d Zimtflügeltyrann
i Tiranno di Wied

6158 *Myiarchus panamensis*
Passeriformes - Tyrannidae
e Panama Flycatcher
f Tyran du Panama
d Panama-Tyrann
i Tiranno di Panama

6159 *Myiarchus phaeocephalus*
Passeriformes - Tyrannidae
e Sooty-crowned Flycatcher; Ash-
fronted Flycatcher; Ashy-fronted
Flycatcher
f Tyran à front gris
d Graustirntyrann
i Tiranno corona fuligginosa

6160 *Myiarchus sagrae*
Passeriformes - Tyrannidae
e La Sagra's Flycatcher; Tom-fool
(WI); Bahama Flycatcher; Cuban
Crested-Flycatcher
f Tyran de La Sagra
d Sagratyrann
i Tiranno di La Sagra

6161 *Myiarchus semirufus*
Passeriformes - Tyrannidae
e Rufous Flycatcher; Seaboard
Flyctcher
f Tyran roux
d Küstentyrann
i Tiranno rossiccio

6162 *Myiarchus stolidus*
Passeriformes - Tyrannidae
e Stolid Flycatcher; Tom-fool (WI)
f Louis (Ants); Pipirite gros-tête
(Ants); Alouette huppée (Ants);
Tyran grosse-tete

d Dickkopftyrann
i Tiranno della Giamaica

6163 *Myiarchus swainsoni*
Passeriformes - Tyrannidae
e Swainson's Flycatcher; Whitley's
Flycatcher; Pelzeln's Flycatcher
f Tyran de Swainson
d Swainson-Tyrann
i Tiranno di Swainson

6164 *Myiarchus tuberculifer*
Passeriformes - Tyrannidae
e Olivaceous Flycatcher; Dusky-
capped Flycatcher (NA)
f Tyran olivâtre
d Kappentyran
i Tiranno capobruno

6165 *Myiarchus tyrannulus*
Passeriformes - Tyrannidae
e Brown-crested Flycatcher; Wied's
Crested-Flycatcher; Wied's
Flycatcher; Guiananan Flycatcher;
Rusty-tailed Flycatcher
f Tyran de Wied
d Cayenne-Tyrann
i Tiranno crestato bruno

6166 *Myiarchus validus*
Passeriformes - Tyrannidae
e Rufous-tailed Flycatcher; Big Tom-
fool (WI); Big-head Bob (WI)
f Tyran à queue rousse
d Rotschwanztyrann
i Tiranno codasossiccia

6167 *Myiarchus venezuelensis*
Passeriformes - Tyrannidae
e Venezuelan Flycatcher
f Tyran du Venezuela
d Venezuela-Tyrann
i Tiranno del Venezuela; Tiranno di
Venezuela

6168 *Myiarchus yucatanensis*
Passeriformes - Tyrannidae
e Yucatan Flycatcher
f Tyran du Yucatan
d Yucatan-Tyrann
i Tiranno dello Yucatan

6169 *Myiobius atricaudus*
Passeriformes - Tyrannidae
e Black-tailed Flycatcher
f Moucherolle à queue noire
d Schwarzschwanztyrann
i Tiranno codanera

6170 *Myiobius barbatus*
Passeriformes - Tyrannidae
e Sulphur-rumped Flycatcher; Bearded
Flycatcher
f Moucherolle barbichon
d Gelbbürzeltyrann
i Tiranno gropponegiallo

6171 *Myiobius barbatus sulphureipygius*
Passeriformes - Tyrannidae
e Sulphur-rumped Flycatcher
f Moucherolle à croupion jaune

6172 *Myiobius erythrurus*
Passeriformes - Tyrannidae
e Ruddy-tailed Flycatcher
f Moucherolle rougequeue
d Zimtschwanztyrann
i Tiranno codirosso

6173 *Myiobius villosus*
Passeriformes - Tyrannidae
e Tawny-breasted Flycatcher
f Moucherolle hérissée
d Borstentyrann
i Tiranno villoso

6174 *Myioborus albifacies*
Passeriformes - Fringillidae
e White-faced Redstart; White-faced
Whitestart
f Paruline à face blanche
d Weißwangenwaldsänger
i Parula pigliamosche facciabianca

6175 *Myioborus albifrons*
Passeriformes - Fringillidae
e White-fronted Redstart; White-
fronted Whitestart
f Paruline à front blanc
d Weißstirnwaldsänger
i Parula pigliamosche frontebianca

6176 *Myioborus brunniceps*
Passeriformes - Fringillidae
e Brown-capped Redstart; Brown-capped Whitestart
f Paruline basanée
d Braunkappenwaldsänger
i Parula pigliamosche capobruno

6177 *Myioborus cardonai*
Passeriformes - Fringillidae
e Saffron-breasted Redstart; Saffron-breasted Whitestart; Guaiquinima Whitestart
f Paruline de Cardona
d Cardona-Waldsänger
i Parula pigliamosche pettozafferano

6178 *Myioborus castaneocapillus*
Passeriformes - Fringillidae
e Tepui Redstart; Tepui Whitestart
f Paruline des tépuis
i Parula pigliamosche del Tepui

6179 *Myioborus flavivertex*
Passeriformes - Fringillidae
e Yellow-crowned Redstart; Yellow-crowned Whitestart; Santa Marta Redstart; Santa Marta Whitestart; Paragua Redstart
f Paruline à cimier jaune
d Salvin-Waldsänger
i Parula pigliamosche corona gialla

6180 *Myioborus melanocephalus*
Passeriformes - Fringillidae
e Spectacled Redstart; Spectacled Warbler; Spectacled Whitestart
f Paruline à lunettes
d Brillenwaldsänger
i Parula pigliamosche dagli occhiali

6181 *Myioborus miniatus*
Passeriformes - Fringillidae
e Slate-throated Redstart; Slate-throated Whitestart
f Paruline ardoisée
d Larvenwaldsänger
i Parula pigliamosche gola ardesia

6182 *Myioborus ornatus*
Passeriformes - Fringillidae

e Golden-fronted Redstart; Golden-fronted Warbler; Golden-fronted Whitestart; Ornate Redstart
f Paruline dorée
d Schwarzohrwaldsänger
i Parula pigliamosche frontedorata

6183 *Myioborus pariae*
Passeriformes - Fringillidae
e Yellow-faced Redstart; Paria Redstart; Yellow-faced Whitestart; Paria Whitestart
f Paruline de Paria
d Goldaugenwaldsänger
i Parula pigliamosche facciagialla

6184 *Myioborus pictus*
Passeriformes - Fringillidae
e Painted Redstart; Painted Whitestart
f Paruline à ailes blanches
d Rotbrustwaldsänger; Bunter Waldsänger
i Parula pigliamosche variopinta

6185 *Myioborus torquatus*
Passeriformes - Fringillidae
e Collared Redstart; Collared Whitestart
f Paruline ceinturée
d Halsbandwaldsänger; Halsbandmyioborus
i Parula pigliamosche dal collare

6186 *Myiodynastes bairdii*
Passeriformes - Tyrannidae
e Baird's Flycatcher
f Tyran de Baird
d Samtstirntyrann
i Tiranno pitango

6187 *Myiodynastes chrysocephalus*
Passeriformes - Tyrannidae
e Golden-crowned Flycatcher
f Tyran à casque d'or
d Goldkrontyrann
i Tiranno pitango corona dorata

6188 *Myiodynastes hemichrysus*
Passeriformes - Tyrannidae
e Golden-bellied Flycatcher
f Tyran à ventre d'or

d Goldbauchtyrann
i Tiranno pitango ventredorato

6189 *Myiodynastes luteiventris*
Passeriformes - Tyrannidae
e Sulphur-bellied Flycatcher
f Tyran tigré; Tyran à bec court
d Weißstirntyrann
i Tiranno pitango ventresulfureo

6190 *Myiodynastes maculatus*
Passeriformes - Tyrannidae
e Streaked Flycatcher
f Tyran audacieux
d Streifentyrann
i Tiranno pitango striato

6191 *Myiodynastes solitarius*
Passeriformes - Tyrannidae
e Solitary Flycatcher
f Tyran solitaire

6192 *Myiopagis caniceps*
Passeriformes - Tyrannidae
e Grey Elaenia; Bananal Tyrannulet
f Élénie grise; Elaène grise; Tyranneau
de l'Araguaya
d Graukopfelaenie; Olivbürzeltachuri
i Elenia grigia

6193 *Myiopagis cotta*
Passeriformes - Tyrannidae
e Jamaican Elaenia; Jamaican Yellow-
crowned Elaenia; Yellow-crowned
Elaenia (WI); Sarahbird (WI); Cotta's
Elaenia; Yellow Elaenia
f Élénie de la Jamaïque; Elaène de
Cotta
d Cotta-Elaenie
i Elenia gialla

6194 *Myiopagis flavivertex*
Passeriformes - Tyrannidae
e Yellow-crowned Elaenia
f Élénie à couronne d'or; Elaène à
couronne d'or
d Goldscheitelelaenie
i Elenia corona gialla

6195 *Myiopagis gaimardii*
Passeriformes - Tyrannidae

e Forest Elaenia; Gaimard's Elaenia
f Élénie de Gaimard; Elaène de
Gaimard
d Waldelaenie
i Elenia di foresta

6196 *Myiopagis subplacens*
Passeriformes - Tyrannidae
e Pacific Elaenia; Fraser's Elaenia
f Élénie striée; Elaène striée
d Strichelbrustelaenie
i Elenia pacifica

6197 *Myiopagis viridicata*
Passeriformes - Tyrannidae
e Greenish Elaenia; Azara's Elaenia
f Élénie verdâtre; Elaène verdâtre
d Grünelaenie
i Elenia verdastra

6198 *Myioparus griseigularis*
Passeriformes - Muscicapidae
e Grey-throated Tit-Flycatcher; Grey-
throated Flycatcher
f Gobemouche à gorge grise; Gobe-
mouches à gorge grise
d Graukehlschnäpper
i Pigliamosche golagrigia

6199 *Myioparus plumbeus*
Passeriformes - Muscicapidae
e Grey Tit-Flycatcher; Lead-coloured
Flycatcher (CSA); Fan-tailed
Flycatcher (CSA); Grey Tit-Babbler
f Gobemouche mésange; Gobe-
mouches mésange
d Meisenschnäpper
i Pigliamosche plumbeo

6200 *Myiophobus cryptoxanthus*
Passeriformes - Tyrannidae
e Olive-chested Flycatcher; Olive-
crested Flycatcher
f Moucherolle à poitrine olive;
Moucherolle de Zamora
d Zamora-Tyrann
i Tiranno cresta oliva

6201 *Myiophobus fasciatus*
Passeriformes - Tyrannidae
e Bran-coloured Flycatcher; Brown-

coloured Flycatcher; Rufescent
Flycatcher; Lima Flycatcher
f Moucherolle fascié
d Rosttyrann
i Tiranno fasciato

6202 *Myiophobus flavicans*
Passeriformes - Tyrannidae
e Flavescent Flycatcher
f Moucherolle flavescente;
Moucherolle jaunâtre
d Gelbtyrann
i Tiranno gialliccio

6203 *Myiophobus inornatus*
Passeriformes - Tyrannidae
e Unadorned Flycatcher
f Moucherolle simple; Moucherolle de
Carriker
d Carriker-Tyrann
i Tiranno disadorno

6204 *Myiophobus lintoni*
Passeriformes - Tyrannidae
e Orange-banded Flycatcher
f Moucherolle de Linton; Moucherolle
de Schauensee
d Loja-Tyrann
i Tiranno di Linton

6205 *Myiophobus ochraceiventris*
Passeriformes - Tyrannidae
e Ochraceous-breasted Flycatcher;
Ochreous Flycatcher
f Moucherolle à poitrine ocré;
Moucherolle ocrée
d Goldkehltyrann
i Tiranno ventre ocraceo

6206 *Myiophobus phoenicomitra*
Passeriformes - Tyrannidae
e Orange-crested Flycatcher
f Moucherolle à cimier orangé;
Moucherolle à couronne orangée
d Goldscheiteltyrann
i Tiranno cresta aranciata

6207 *Myiophobus pulcher*
Passeriformes - Tyrannidae
e Handsome Flycatcher
f Moucherolle superbe

d Goldbrusttyrann
i Tiranno bello

6208 *Myiophobus roraimae*
Passeriformes - Tyrannidae
e Roraiman Flycatcher; Roraima
Flycatcher
f Moucherolle du Roraraima
d Roraima-Tyrann
i Tiranno di Roraima

6209 *Myiophobus rufescens*
Passeriformes - Tyrannidae
e Rufescent Flycatcher

6210 *Myiopsitta monachus*
Psittaciformes - Psittacidae
e Monk Parakeet; Grey-breasted
Parakeet; Quaker Parakeet
f Perruche moine; Perruche jaune-
veuve; Conure veuve; Perruche
souris (Ants)
d Mönchssittich; Mönchsittich
i Parrocchetto monaco; Monaco;
Pappagallo dal petto giallo

6211 *Myiornis albiventris*
Passeriformes - Tyrannidae
e White-bellied Pygmy-Tyrant; White-
breasted Pygmy-Tyrant
f Microtyran à ventre blanc;
Tyranneau à ventre blanc
d Weißbauchzwergtyrann
i Tiranno pigmeo pettobianco

6212 *Myiornis atricapillus*
Passeriformes - Tyrannidae
e Black-capped Pygmy-Tyrant; Short-
tailed Pygmy-Tyrant
f Microtyran à calotte noire
i Tiranno pigmeo capinero

6213 *Myiornis auricularis*
Passeriformes - Tyrannidae
e Eared Pygmy-Tyrant
f Microtyran oreillard; Tyranneau à
oreillons noirs
d Zwergtyrann
i Tiranno pigmeo orecchiuto

6214   *Myiornis ecaudatus*
Passeriformes - Tyrannidae
e   Short-tailed Pygmy-Tyrant
f   Microtyran à queue courte;
Tyranneau à queue courte
d   Stummelschwanzzwergtyrann
i   Tiranno pigmeo codacorta

6215   *Myiotheretes fumigatus*
Passeriformes - Tyrannidae
e   Smoky Bush-Tyrant
f   Moucherolle enfumée; Moucherolle
de Boissonneau
d   Rußtyrann
i   Tiranno fumigato

6216   *Myiotheretes fuscorufus*
Passeriformes - Tyrannidae
e   Rufous-bellied Bush-Tyrant
f   Moucherolle à ventre fauve;
Moucherolle des buissons
d   Rostbindentyrann
i   Tiranno ventrerosso

6217   *Myiotheretes pernix*
Passeriformes - Tyrannidae
e   Santa Marta Bush-Tyrant
f   Moucherolle des Santa Marta;
Moucherolle de Bangs
d   Bangs-Tyrann
i   Tiranno di Santa Marta

6218   *Myiotheretes striaticollis*
Passeriformes - Tyrannidae
e   Streak-throated Bush-Tyrant
f   Moucherolle à gorge rayée;
Moucherolle ornée
d   Strauchtyrann
i   Tiranno golastriata

6219   *Myiotriccus ornatus*
Passeriformes - Tyrannidae
e   Ornate Flycatcher
f   Moucherolle orné
d   Schmucktyrann
i   Tiranno ornato

6220   *Myiozetetes cayanensis*
Passeriformes - Tyrannidae
e   Rusty-margined Flycatcher
f   Tyran de Cayenne

d   Rotschwingentyrann
i   Pintago della Cayenna

6221   *Myiozetetes granadensis*
Passeriformes - Tyrannidae
e   Grey-capped Flycatcher; Gray-
capped Flycatcher (NA)
f   Tyran à tête grise
d   Grauscheiteltyrann
i   Pitango capogrigio

6222   *Myiozetetes luteiventris*
Passeriformes - Tyrannidae
e   Dusky-chested Flycatcher; Orange-
vented Flycatcher
f   Tyran à gorge rayée
d   Kurzschnabeltyrann
i   Pitango pettobruno

6223   *Myiozetetes similis*
Passeriformes - Tyrannidae
e   Social Flycatcher; Vermillion-
crowned Flycatcher
f   Tyran sociable
d   Rotkronkyrann
i   Pitango di Giraud

6224   *Myiozetetes texensis*
Passeriformes - Tyrannidae
e   Vermillion-crowned Flycatcher

6225   *Myophonus blighi*
Passeriformes - Muscicapidae
e   Sri Lanka Whistling-Thrush; Ceylon
Whistling-Thrush; Bligh's Whistling-
Thrush
f   Arrenga de Ceylan
d   Ceylon-Pfeifdrossel
i   Tordo zufolatore di Sri-Lanka

6226   *Myophonus caeruleus*
Passeriformes - Muscicapidae
e   Blue Whistling-Thrush; Himalayan
Whistling-Thrush (ISC); Whistling-
Thrush; Violet Whistling-Thrush
f   Arrenga siffleur; Merle bleu siffleur
d   Purpurpfeifdrossel
i   Tordo zufolatore blu

6227   *Myophonus castaneus*
Passeriformes - Muscicapidae

e Brown-winged Whistling-Thrush;
Sumatran Whistling-Thrush

6228 *Myophonus glaucinus*
Passeriformes - Muscicapidae
e Sunda Whistling-Thrush
f Arrenga bleuet
d Sunda-Pfeifdrossel
i Tordo zufolatore della Sonda

6229 *Myophonus horsfieldii*
Passeriformes - Muscicapidae
e Malabar Whistling-Thrush
f Arrenga de Malabar
d Horsfield-Pfeiffdrossel
i Tordo zufolatore del Malabar

6230 *Myophonus insularis*
Passeriformes - Muscicapidae
e Formosan Whistling-Thrush; Taiwan
Whistling-Thrush
f Arrenga de Taïwan; Merle bleu de
Taïwan; Merle bleu siffleur de
Taïwan
d Formosa-Pfeifdrossel
i Tordo zufolatore di Taiwan

6231 *Myophonus melanurus*
Passeriformes - Muscicapidae
e Shiny Whistling-Thrush; Sumatran
Whistling-Thrush
f Arrenga de Sumatra
d Glanzpfeifdrossel
i Tordo zufolatore smagliante

6232 *Myophonus robinsoni*
Passeriformes - Muscicapidae
e Malayan Whistling-Thrush;
Robinson's Whistling-Thrush
f Arrenga de Robinson
d Malaien-Pfeifdrossel
i Tordo zufolatore della Malesia

6233 *Myornis senilis*
Passeriformes - Rhinocryptidae
e Ashy Tapaculo; Ash-coloured
Tapaculo
f Mérulaxe cendré
d Grautapaculo
i Tapaculo cenerino

6234 *Myrmeciza atrothorax*
Passeriformes - Formicariidae
e Black-throated Antbird
f Alapi de Buffon
d Pechbrustameisenvogel

6235 *Myrmeciza berlepschi*
Passeriformes - Formicariidae
e Stub-tailed Antbird; Berlepsch's
Antbird
d Stutzschwanzsipia

6236 *Myrmeciza disjuncta*
Passeriformes - Formicariidae
e Yapacana Antbird
f Alapi du Yapacana
d Yapacana-Ameisenvogel

6237 *Myrmeciza exsul*
Passeriformes - Formicariidae
e Chestnut-backed Antbird
f Alapi à dos roux
d Braunrückenameisenvogel
i Mangiaformiche dorsocastano

6238 *Myrmeciza ferruginea*
Passeriformes - Formicariidae
e Ferruginous-backed Antbird;
Ferruginous Antbird
f Alapi à cravate noire
d Rostrückenameisenvogel

6239 *Myrmeciza fortis*
Passeriformes - Formicariidae
e Sooty Antbird
f Alapi fuligineux
d Dunkelameisenvogel

6240 *Myrmeciza goeldii*
Passeriformes - Formicariidae
e Goeldi's Antbird; Göldi's Antbird
f Alapi de Goeldi
d Goeldi-Ameisenvogel

6241 *Myrmeciza griseiceps*
Passeriformes - Formicariidae
e Grey-headed Antbird
f Alapi à tête grise
d Palambla-Ameisenvogel

6242 **Myrmeciza hemimelaena**
Passeriformes - Formicariidae
e Chestnut-tailed Antbird
f Alapi rougequeue
d Rotmantelameisenvogel

6243 **Myrmeciza hyperythra**
Passeriformes - Formicariidae
e Plumbeous Antbird
f Alapi plombé
d Chamicuros-Ameisenvogel

6244 **Myrmeciza immaculata**
Passeriformes - Formicariidae
e Immaculate Antbird
f Alapi immaculé
d Weißbugameisenvogel

6245 **Myrmeciza laemosticta**
Passeriformes - Formicariidae
e Dull-mantled Antbird; Salvin's
Antbird
f Alapi tabac
d Grauscheitelameisenvogel

6246 **Myrmeciza longipes**
Passeriformes - Formicariidae
e White-bellied Antbird; Swainson's
Antcatcher
f Alapi à ventre blanc
d Rotsteißameisenvogel;
Feuerameisenläufer

6247 **Myrmeciza loricata**
Passeriformes - Formicariidae
e White-bibbed Antbird
f Alapi cuirassé
d Schmuckbrustameisenvogel

6248 **Myrmeciza melanoceps**
Passeriformes - Formicariidae
e White-shouldered Antbird
f Alapi à épaules blanches; Alapi de
Spinx
d Spix-Ameisenvogel

6249 **Myrmeciza nigricauda**
Passeriformes - Formicariidae
e Esmeraldas Antbird; Rosenberg's
Antbird
d Schiefersipia

6250 **Myrmeciza pelzelni**
Passeriformes - Formicariidae
e Grey-bellied Antbird
f Alapi à ventre gris; Alapi à joues
blanches
d Weißwangenameisenvogel

6251 **Myrmeciza ruficauda**
Passeriformes - Formicariidae
e Scalloped Antbird; Rufous-tailed
Antbird
f Alapi barbu
d Braunscheitelameisenvogel

6252 **Myrmeciza squamosa**
Passeriformes - Formicariidae
e Squamate Antbird
f Alapi écaillé
d Schwarzmantelameisenvogel

6253 **Myrmeciza stictothorax**
Passeriformes - Formicariidae
e Spot-breasted Antbird
f Alapi strié
d Strichelbrustameisenvogel

6254 **Myrmecocichla aethiops**
Passeriformes - Muscicapidae
e Northern Anteater Chat; Anteater
Chat; Ant-Chat; Northern Ant-eating
Chat
f Traquet brun; Traquet-fourmilier
brun; Traquet-fourmilier brun du
nord
d Ameisenschmätzer;
Termitenschmätzer; Rußschmätzer
i Sassicola mangiaformiche

6255 **Myrmecocichla albifrons**
Passeriformes - Muscicapidae
e White-fronted Black-Chat; White-
foreheaded Chat
f Traquet à front blanc; Traquet noir à
front blanc
d Weißstirnschmätzer
i Sassicola nera frontebianca

6256 **Myrmecocichla arnotti**
Passeriformes - Muscicapidae
e White-headed Black-Chat; Arnot's
Chat; Arnot's White-headed Chat;

Arnot's Black-Chat
f Traquet d'Arnott; Traquet noir
d'Arnott
d Arnott-Schmätzer
i Sassicola di Arnott

6257 *Myrmecocichla formicivora*
Passeriformes - Muscicapidae
e Southern Anteater Chat; Ant-eating
Chat (CSA); Ant-Chat; Anteater
Chat; Southern Ant-eating Chat
f Traquet fourmilier; Traquet-
fourmilier brun du sud
d Ameisenschmätzer;
Termitenschmätzer; Südafrika-
Termitenschmätzer
i Sassicola mangiaformiche
meridionale

6258 *Myrmecocichla melaena*
Passeriformes - Muscicapidae
e Rüppell's Chat; Rüppell's Black-
Chat; Black Chat
f Traquet de Rüppell; Traquet noir
d'Abyssinie
d Einfarbschmätzer
i Sassicola di Rüppell

6259 *Myrmecocichla nigra*
Passeriformes - Muscicapidae
e Sooty Chat; Anteater Chat; Sooty
Anteater Chat
f Traquet commandeur; Traquet-
fourmilier noir
d Hadesschmätzer
i Sassicola nera

6260 *Myrmecocichla tholloni*
Passeriformes - Muscicapidae
e Congo Moorchat; Thollon's
Moorchat
f Traquet du Congo; Traquet-
fourmilier du Congo
d Kongo-Schmätzer
i Sassicola del Congo

6261 *Myrmia micrura*
Apodiformes - Trochilidae
e Short-tailed Woodstar
f Colibri à queue courte
d Kurzschwanzelfe

i Stella dei boschi codacorta

6262 *Myrmoborus leucophrys*
Passeriformes - Formicariidae
e White-browed Antbird; White-
browed Antcreeper
f Alapi à sourcils blancs
d Augenbrauenameisenschnäpper
i Rampichino formichiere dai
sopraccigli bianchi

6263 *Myrmoborus lugubris*
Passeriformes - Formicariidae
e Ash-breasted Antbird; Ash-breasted
Antcreeper
f Alapi lugubre
d Fahlstirnameisenschnäpper
i Rampichino formichiere
pettocenerino

6264 *Myrmoborus melanurus*
Passeriformes - Formicariidae
e Black-tailed Antbird; Black-tailed
Antcreeper
f Alapi à queue noire
d Schwarzschwanzameisenschnäpper
i Rampichino formichiere codanera

6265 *Myrmoborus myotherinus*
Passeriformes - Formicariidae
e Black-faced Antbird; Black-faced
Antcreeper
f Alapi masqué
d Schuppenflügelameisenschnäpper
i Rampichino formichiere faccianera

6266 *Myrmochanes hemileucus*
Passeriformes - Formicariidae
e Black-and-white Antbird; Black-and-
white Antcatcher
f Alapi noir-et-blanc
d Weißbauchameisenschnäpper

6267 *Myrmorchilus strigilatus*
Passeriformes - Formicariidae
e Stripe-backed Antbird
f Grisin à dos rayé
d Weißbartameisenfänger
i Mangiaformiche dorsostriato

6268   *Myrmornis torquata*
            Passeriformes - Formicariidae
    e   Wing-banded Antbird; Wing-banded
            Ant-Thrush
    f   Palicour de Cayenne
    d   Graubauchameisenpitta
    i   Tordo formichiere alifasciate

6269   *Myrmothera campanisona*
            Passeriformes - Formicariidae
    e   Thrush-like Ant-Pitta; Thrush Ant-
            Pitta
    f   Grallaire grand-beffroi
    d   Fleckenbrustameisenjäger
    i   Pitta formichiera tordina

6270   *Myrmothera simplex*
            Passeriformes - Formicariidae
    e   Brown-breasted Tepui Ant-Pitta;
            Brown Ant-Pitta; Tepui Ant-Pitta
    f   Grallaire sobre
    d   Graubrustameisenjäger
    i   Pitta formichiere pettobruno

6271   *Myrmotherula ambigua*
            Passeriformes - Formicariidae
    e   Yellow-throated Ant-Wren
    f   Myrmidon à gorge jaune
    d   Gelbkehlameisenschlüpfer
    i   Scricciolo formichiere golagialla

6272   *Myrmotherula assimilis*
            Passeriformes - Formicariidae
    e   Leaden Ant-Wren; White-backed
            Ant-Wren
    f   Myrmidon plombé
    d   Pelzeln-Ameisenschlüpfer
    i   Scricciolo formichiere

6273   *Myrmotherula axillaris*
            Passeriformes - Formicariidae
    e   White-flanked Ant-Wren; Black Ant-
            Wren
    f   Myrmidon à flancs blancs
    d   Weißflankenameisenschlüpfer
    i   Scricciolo formichiere fianchibianchi

6274   *Myrmotherula behni*
            Passeriformes - Formicariidae
    e   Plain-winged Ant-Wren; Pale-
            winged Ant-Wren

    f   Myrmidon de Behn
    d   Schwarzbrustameisenschlüpfer

6275   *Myrmotherula brachyura*
            Passeriformes - Formicariidae
    e   Pygmy Ant-Wren
    f   Myrmidon pygmée
    d   Zwergameisenschlüpfer;
            Kurzschwänziger
            Zwergameisenwürger
    i   Scricciolo formichiere pigmeo

6276   *Myrmotherula cherriei*
            Passeriformes - Formicariidae
    e   Cherrie's Ant-Wren
    f   Myrmidon de Cherrie
    d   Kehlstreifenameisenschlüpfer
    i   Scricciolo formichiere di Cherrie

6277   *Myrmotherula erythrura*
            Passeriformes - Formicariidae
    e   Rufous-tailed Ant-Wren
    f   Myrmidon à queue rousse
    d   Rotschwarzameisenschlüpfer
    i   Scricciolo formichiere codarossiccia

6278   *Myrmotherula fluminensis*
            Passeriformes - Formicariidae
    e   Rio de Janeiro Ant-Wren; Rio de
            Janeiro Antbird
    f   Myrmidon de Rio de Janeiro
    i   Scricciolo formichiere di Rio de
            Janeiro

6279   *Myrmotherula fulviventris*
            Passeriformes - Formicariidae
    e   Chequer-throated Ant-Wren;
            Chequered Ant-Wren; Fulvous-
            bellied Ant-Wren; Fulvous Ant-Wren
    f   Myrmidon fauve
    d   Marmorkehlameisenschlüpfer
    i   Scricciolo formichiere ventrefulvo

6280   *Myrmotherula grisea*
            Passeriformes - Formicariidae
    e   Ashy Ant-Wren; Yungas Ant-Wren
    f   Myrmidon cendré
    d   Carriker-Ameisenschlüpfer
    i   Scricciolo formichiere grigio

6281   *Myrmotherula gularis*
       Passeriformes - Formicariidae
  e   Star-throated Ant-Wren
  f   Myrmidon à gorge etoilée
  d   Perlenkehlameisenschlüpfer
  i   Scricciolo formichiere golastellata

6282   *Myrmotherula guttata*
       Passeriformes - Formicariidae
  e   Rufous-bellied Ant-Wren
  f   Myrmidon moucheté
  d   Rotbauchameisenschlüpfer
  i   Scricciolo formichiere macchiettato

6283   *Myrmotherula gutturalis*
       Passeriformes - Formicariidae
  e   Brown-bellied Ant-Wren
  f   Myrmidon à ventre brun
  d   Braunbauchameisenschlüpfer
  i   Scricciolo formichiere ventrebruno

6284   *Myrmotherula haematonota*
       Passeriformes - Formicariidae
  e   Stipple-throated Ant-Wren; Stippled
        Ant-Wren
  f   Myrmidon cravaté
  d   Graubrustameisenschlüpfer
  i   Scricciolo formichiere golamacchiata

6285   *Myrmotherula hauxwelli*
       Passeriformes - Formicariidae
  e   Plain-throated Ant-Wren; Hauxwell's
        Ant-Wren
  f   Myrmidon de Hauxwell
  d   Graubauchameisenschlüpfer
  i   Scricciolo formichiere di Hauxwell

6286   *Myrmotherula ignota*
       Passeriformes - Formicariidae
  e   Griscombe's Ant-Wren; Colombian
        Ant-Wren; Pygmy Ant-Wren
  f   Myrmidon de Griscom
  d   Panama-Ameisenschlüpfer

6287   *Myrmotherula iheringi*
       Passeriformes - Formicariidae
  e   Ihering's Ant-Wren
  f   Myrmidon d'Ihering
  d   Ithering-Ameisenschlüpfer
  i   Scricciolo formichiere di Ihering

6288   *Myrmotherula klagesi*
       Passeriformes - Formicariidae
  e   Klages's Ant-Wren
  f   Myrmidon de Klages
  d   Obidos-Ameisenschlüpfer
  i   Scricciolo formichiere di Klages

6289   *Myrmotherula leucophthalma*
       Passeriformes - Formicariidae
  e   White-eyed Ant-Wren
  f   Myrmidon aux yeux blancs
  d   Weißaugenameisenschlüpfer
  i   Scricciolo formichiere occhibianchi

6290   *Myrmotherula longicauda*
       Passeriformes - Formicariidae
  e   Stripe-chested Ant-Wren; Striped-
        chested Ant-Wren
  f   Myrmidon à ventre blanc; Myrmidon
        à longue queue
  d   Langschwanzameisenschlüpfer
  i   Scricciolo formichiere codalunga

6291   *Myrmotherula longipennis*
       Passeriformes - Formicariidae
  e   Long-winged Ant-Wren
  f   Myrmidon longipenne; Myrmidon
        argenté
  d   Silberameisenschlüpfer
  i   Scricciolo formichiere alilunghe

6292   *Myrmotherula menetriesii*
       Passeriformes - Formicariidae
  e   Grey Ant-Wren; Ménétries's Ant-
        Wren
  f   Myrmidon gris
  d   Buntflügelameisenschlüpfer
  i   Scricciolo formichiere di Menetries

6293   *Myrmotherula minor*
       Passeriformes - Formicariidae
  e   Salvadori's Ant-Wren
  f   Myrmidon de Salvadori
  d   Salvadori-Ameisenschlüpfer
  i   Scricciolo formichiere di Salvadori

6294   *Myrmotherula obscura*
       Passeriformes - Formicariidae
  e   Short-billed Ant-Wren
  f   Myrmidon à bec court

d   Kurzschnabelameisenschlüpfer
i   Scricciolo formichiere beccocorto

6295   *Myrmotherula ornata*
Passeriformes - Formicariidae
e   Ornate Ant-Wren; Black-throated
Ant-Wren
f   Myrmidon orné
d   Schwarzkehlameisenschlüpfer
i   Scricciolo formichiere ornato

6296   *Myrmotherula schisticolor*
Passeriformes - Formicariidae
e   Slaty Ant-Wren
f   Myrmidon ardoisé
d   Schieferameisenschlüpfer
i   Scricciolo formichiere lavagna

6297   *Myrmotherula sclateri*
Passeriformes - Formicariidae
e   Sclater's Ant-Wren
f   Myrmidon de Sclater
d   Gelbstreifenameisenschlüpfer
i   Scricciolo formichiere di Sclater

6298   *Myrmotherula snowi*
Passeriformes - Formicariidae
e   Alagoas Ant-Wren

6299   *Myrmotherula spodionota*
Passeriformes - Formicariidae
e   Foothill Ant-Wren; Foothills Ant-
Wren; Ecuadorian Ant-Wren
f   Myrmidon des contreforts
i   Scricciolo formichiere delle colline

6300   *Myrmotherula sunensis*
Passeriformes - Formicariidae
e   Rio Suno Ant-Wren; Suno Ant-Wren
f   Myrmidon du Suno
d   Suno-Ameisenschlüpfer
i   Scricciolo formichiere di Rio Suna

6301   *Myrmotherula surinamensis*
Passeriformes - Formicariidae
e   Streaked Ant-Wren; Amazonian
Streaked Ant-Wren (NA); Guianan
Ant-Wren
f   Myrmidon du Surinam
d   Surinam-Ameisenschlüpfer
i   Scricciolo formichiere striato

6302   *Myrmotherula unicolor*
Passeriformes - Formicariidae
e   Unicoloured Ant-Wren
f   Myrmidon unicolore
d   Einfarbameisenschlüpfer
i   Scricciolo formichiere unicolore

6303   *Myrmotherula urosticta*
Passeriformes - Formicariidae
e   Band-tailed Ant-Wren
f   Myrmidon à queue blanche
d   Weißschwarzameisenschlüpfer
i   Scricciolo formichiere codafasciata

6304   *Myrtis fanny*
Apodiformes - Trochilidae
e   Purple-collared Woodstar; Fanny's
Woodstar
f   Colibri fanny
d   Bandelfe
i   Stella dei boschi di Fanny

6305   *Mystacornis crossleyi*
Passeriformes - Sylviidae
e   Crossley's Babbler; Yellow-browed
Oxylabes
f   Mystacorne de Crossley; Mystacorne
d   Mystacornis
i   Garrulo di Crossley

6306   *Myza celebensis*
Passeriformes - Meliphagidae
e   Dark-eared Myza; Meyer's Myza;
Celebes Honeyeater; Brown
Honeysucker; Lesser Streaked-
Honeyeater; Brown Honeyeater;
Lesser Sulawesi Honeyeater
f   Méliphage des Célèbes
d   Celebes Honigfresser
i   Succhiamiele di Sulawesi

6307   *Myza sarasinorum*
Passeriformes - Meliphagidae
e   White-eared Myza; Wiglesworth's
Myza; Spot-headed Honeysucker;
Mengkoka Honeyeater; Greater
Streaked-Honeyeater; Greater
Sulawesi Honeyeater; White-eared
Honeyeater
f   Méliphage à points
d   Sarasinhonigfresser;

Schattenhonigfresser

i   Succhiamiele testamaculata

6308   *Myzomela adolphinae*
Passeriformes - Meliphagidae
e   Mountain Myzomela; Red-headed
Honeyeater; Adolphina's Myzomela;
Mountain Red-headed Myzomela;
Mountain Red-headed Honeyeater;
Elfin Myzomela
f   Myzomèle montagnard
d   Arfak-Honigfresser
i   Mangiamiele di Adolfina

6309   *Myzomela albigula*
Passeriformes - Meliphagidae
e   White-chinned Myzomela; White-
chinned Honeyeater
f   Myzomèle à menton blanc
d   Weißkinnhonigfresser
i   Mangiamiele golabianca

6310   *Myzomela blasii*
Passeriformes - Meliphagidae
e   Drab Myzomela; Blas's Honeyeater;
Amboina Honeyeater; Ambon
Honeyeater; Amboina Myzomela;
Drab Honeyeater; Ceram Honeyeater
f   Myzomèle sobre
d   Amboina-Honigfresser
i   Mangiamiele di Amboina

6311   *Myzomela boiei*
Passeriformes - Meliphagidae
e   Banda Myzomela; Banda
Honeyeater; Boie's Honeyeater
f   Myzomèle de Banda
d   Honigfresser
i   Mangiamiele di Banda

6312   *Myzomela caledonica*
Passeriformes - Meliphagidae
e   New Caledonian Myzomela; New
Caledonian Honeyeater
f   Myzomèle calédonien
d   Honigfresser
i   Mangiamiele della Nuova Caledonia

6313   *Myzomela cardinalis*
Passeriformes - Meliphagidae
e   Cardinal Myzomela; Cardinal

Honeyeater
f   Myzomèle cardinal
d   Kardinalhonigfresser;
Kardinalhonigesser
i   Mangiamiele cardinale

6314   *Myzomela chermesina*
Passeriformes - Meliphagidae
e   Rotuma Myzomela; Rotuma
Honeyeater
d   Karmesinhonigfresser
i   Mangiamiele di Rotuma

6315   *Myzomela chloroptera*
Passeriformes - Meliphagidae
e   Sulawesi Myzomela; Celebes
Myzomela; Sulawesi Honeyeater
f   Myzomèle des Célèbes
i   Mangiamiele di Sulawesi

6316   *Myzomela cineracea*
Passeriformes - Meliphagidae
e   Ashy Myzomela; Bismarck
Honeyeater; Umboi Honeyeater;
Sclater's Honeyeater; Bismarck
Myzomela
f   Myzomèle cendré
d   Schlichthonigfresser

6317   *Myzomela cruentata*
Passeriformes - Meliphagidae
e   Red Myzomela; Red Honeyeater;
Red-tinted Myzomela
f   Myzomèle vermillon
d   Bluthonigfresser
i   Mangiamiele rosso

6318   *Myzomela dammermani*
Passeriformes - Meliphagidae
e   Sumba Myzomela; Sumba
Honeyeater
f   Myzomèle de Sumba
i   Mangiamiele dell'Isola Sumba

6319   *Myzomela eichhorni*
Passeriformes - Meliphagidae
e   Yellow-vented Myzomela; Yellow-
vented Honeyeater; Eichhorn's
Honeyeater; Kulambangra
Honeyeater; Eichhorn's Myzomela
f   Myzomèle à ventre jaune

d   Rotbürzelhonigfresser
i   Mangiamiele di Eichhorn

6320   *Myzomela eques*
     Passeriformes - Meliphagidae
e   Red-throated Myzomela; Red-spotted
     Myzomela; Red-spotted Honeyeater;
     Red-spot Honeyeater; Red-chinned
     Honeyeater; Ruby-throated
     Myzomela
f   Myzomèle à menton rouge
d   Dolchstichhonigfresser
i   Mangiamiele golarossa

6321   *Myzomela erythrocephala*
     Passeriformes - Meliphagidae
e   Red-headed Myzomela; Red-headed
     Honeyeater (ANZ); Mangrove Red-
     headed Honeyeater
f   Myzomèle à tête rouge
d   Rotkopfhonigfresser
i   Mangiamiele testarossa

6322   *Myzomela erythrocephala infuscata*
     Passeriformes - Meliphagidae
e   Torres Strait Red-headed Honeyeater
     (ANZ)

6323   *Myzomela erythromelas*
     Passeriformes - Meliphagidae
e   Black-bellied Myzomela; Black-
     bellied Honeyeater; New Britain
     Myzomela; New Britain Honeyeater;
     New Britain Red-headed Honeyeater
f   Myzomèle à ventre noir
d   Flammenkopfhonigfresser
i   Mangiamiele ventrenero

6324   *Myzomela jugularis*
     Passeriformes - Meliphagidae
e   Orange-breasted Myzomela; Orange-
     breasted Honeyeater
f   Myzomèle desFidji
d   Orangebrusthonigfresser
i   Mangiamiele pettoarancio

6325   *Myzomela kuehni*
     Passeriformes - Meliphagidae
e   Crimson-hooded Myzomela; Wetar
     Honeyeater; Rothschild's
     Honeyeater; Kühn's Myzomela;

     Crimson-hooded Honeyeater; Banda
     Myzomela
f   Myzomèle de Wetar
d   Wetar-Honigfresser
i   Mangiamiele cappucciorosso

6326   *Myzomela lafargei*
     Passeriformes - Meliphagidae
e   Scarlet-naped Myzomela; Small
     Bougainville-Honeyeater; Red-naped
     Honeyeater; Solomon Islands
     Honeyeater; Red-crowned
     Honeyeater; Islet Myzomela
f   Myzomèle à nuque rouge
d   Scharlachnackenhonigfresser
i   Mangiamiele nucarossa

6327   *Myzomela malaitae*
     Passeriformes - Meliphagidae
e   Red-bellied Myzomela; Malaita
     Honeyeater; Malaita Myzomela
f   Myzomèle de Malaita
d   Malaita-Honigfresser
i   Mangiamiele di Malaita

6328   *Myzomela melanocephala*
     Passeriformes - Meliphagidae
e   Black-headed Myzomela; Black-
     headed Honeyeater; Savo
     Honeyeater; Guadalcanal Honeyeater
f   Myzomèle à tête noire
d   Savo-Honigfresser
i   Mangiamiele testanera

6329   *Myzomela nigrita*
     Passeriformes - Meliphagidae
e   Black Myzomela; Black Honeyeater;
     Gray's Honeyeater; Papuan Black-
     Myzomela; Papuan Black-
     Honeyeater
f   Myzomèle noir
d   Mohrenhonigfresser
i   Mangiatore di miele nero

6330   *Myzomela obscura*
     Passeriformes - Meliphagidae
e   Dusky Myzomela; Dusky Honeyeater
     (ANZ)
f   Myzomèle ombré
d   Rußhonigfresser
i   Mangiamiele scuro

6331 *Myzomela obscura fumata*
Passeriformes - Meliphagidae
e Torres Strait Dusky-Honeyeater
(ANZ)

6332 *Myzomela pammelaena*
Passeriformes - Meliphagidae
e Ebony Myzomela; Bismarck Black-
Honeyeater; Bismarck Black-
Myzomela; Ebony Honeyeater;
Admiralty Myzomela; Islet
Myzomela
f Myzomèle ébène
i Mangiamiele delle Isole Bismarck

6333 *Myzomela pulchella*
Passeriformes - Meliphagidae
e Olive-yellow Myzomela; Beautiful
Myzomela; New Ireland Honeyeater;
Dainty Honeyeater
f Myzomèle de Nouvelle-Irlande
d Neuirland-Honigfresser
i Mangiamiele della Nuova Irlanda

6334 *Myzomela rosenbergii*
Passeriformes - Meliphagidae
e Red-collared Myzomela; Black-and-
red Honeyeater; Red-collared
Honeyeater; Red-capped Myzomela;
Rosenberg's Myzomela
f Myzomèle de Rosenberg
d Rosenberg-Honigfresser
i Mangiamiele di Rosenberg

6335 *Myzomela rubratra*
Passeriformes - Meliphagidae
e Micronesian Myzomela; Micronesian
Cardinal-Honeyeater; Micronesian
Honeyeater
f Myzomèle de Micronésie
d Feuerhonigfresser
i Mangiamiele della Micronesia

6336 *Myzomela sanguinolenta*
Passeriformes - Meliphagidae
e Scarlet Myzomela; Scarlet
Honeyeater (ANZ); Crimson
Myzomela; Crimson Honeyeater
f Myzomèle écarlate
d Scharlachhonigfresser
i Mangiamiele scarlatto

6337 *Myzomela sclateri*
Passeriformes - Meliphagidae
e Scarlet-bibbed Myzomela; Sclater's
Honeyeater; Sclater's Myzomela;
Scarlet-throated Honeyeater
f Myzomèle de Sclater
d Palakuru-Honigfresser; Kokos-
Honigschmecker
i Mangiamiele di Sclater

6338 *Myzomela tristrami*
Passeriformes - Meliphagidae
e Sooty Myzomela; Tristram's
Honeyeater; Santa Ana Honeyeater;
Tristram's Myzomela; St. Ana
Myzomela
f Myzomèle de Tristram
d Tristram-Honigfresser
i Mangiamiele di Tristram

6339 *Myzomela vulnerata*
Passeriformes - Meliphagidae
e Red-rumped Myzomela; Sunda
Honeyeater; Red-crowned
Honeyeater; Timor Honeyeater;
Timor Myzomela; Red-rumped
Honeyeater; Sunda Myzomela;
Black-breasted Myzomela
f Myzomèle de Timor
d Dreifarbenhonigfresser
i Mangiamiele dal groppone rosso

6340 *Myzomela wakoloensis*
Passeriformes - Meliphagidae
e Wakolo Myzomela; Wakolo
Honeyeater
f Myzomèle de Forbes
i Mangiamiele di Wakolo

6341 *Myzornis pyrrhoura*
Passeriformes - Sylviidae
e Fire-tailed Myzornis
f Myzorne queue-de-feu; Myzonrne à
queue de feu
d Feuerschwanz; Feuerschwänzchen;
Feuerschwanztimalie
i Succhiafiori codirosso

N

6342   Namibornis herero
Passeriformes - Muscicapidae
e   Herero Chat; Herero Thrush-
Flycatcher
f   Namiorne herero; Traquet des
Héréros
d   Namib-Schnäpper
i   Saltimpalo dell'Herero

6343   Nandayus nenday
Psittaciformes - Psittacidae
e   Nanday Parakeet; Black-hooded
Parakeet; Black-headed Parakeet;
Nanday Conure; Black-headed
Conure; Black-masked Conure
f   Conure nanday
d   Nanday-Sittich; Schwarzkopfsittich
i   Conuro nanday; Pappagallino dalla
testa nera

6344   Nannopsittaca dachilleae
Psittaciformes - Psittacidae
e   Amazonian Parrotlet
f   Toui de D'Achille
i   Pappagallino della signora D'Achille

6345   Nannopsittaca panychlora
Psittaciformes - Psittacidae
e   Tepui Parrotlet; Mount Roraima
Parrotlet
f   Toui des tépuis
d   Grünsperlingspapagei; Tepui-Sittich
i   Pappagallino del Tepui

6346   Napothera atrigularis
Passeriformes - Sylviidae
e   Black-throated Wren-Babbler
f   Turdinule à gorge noire
d   Fahlbauchtimalie
i   Garrulo scricciolo golanera

6347   Napothera brevicaudata
Passeriformes - Sylviidae
e   Streaked Wren-Babbler; Short-tailed
Wren-Babbler; Intermediate Wren-
Babbler; Streaked Wren-Warbler
f   Turdinule à queue courte
d   Stutzschwanztimalie
i   Garrulo scricciolo golastriata

6348   Napothera crassa
Passeriformes - Sylviidae
e   Mountain Wren-Babbler
f   Turdinule des montagnes
d   Blaßkehltimalie
i   Garrulo scricciolo montano

6349   Napothera crispifrons
Passeriformes - Sylviidae
e   Limestone Wren-Babbler; Limestone
Babbler
f   Turdinule des rochers
d   Kalkstein-Timalie
i   Garrulo scricciolo variabile

6350   Napothera epilepidota
Passeriformes - Sylviidae
e   Eyebrowed Wren-Babbler; Small
Wren-Babbler; Eyebrowed Wren-
Warbler; Lesser Wren-Babbler;
Streak-breasted Wren-Babbler
f   Petite Turdinule
d   Brustfleckentimalie
i   Garrulo scricciolo minore

6351   Napothera macrodactyla
Passeriformes - Sylviidae
e   Large Wren-Babbler; Large-footed
Wren-Babbler
f   Grande Turdinule
d   Graubauchtimalie
i   Garrulo scricciolo zampegrosse

6352   Napothera marmorata
Passeriformes - Sylviidae
e   Marbled Wren-Babbler; Müller's
Wren-Babbler; Large Wren-Warbler
f   Turdinule marbrée
d   Marmortimalie
i   Garrulo scricciolo marmoreggiato

6353 **Napothera rabori**
Passeriformes - Sylviidae
e Rabor's Wren-Babbler; Luzon Wren-
Babbler; Rabor's Babbler
f Turdinule de Luçon
d Rabors Timalie
i Garrulo scricciolo di Luzon

6354 **Napothera rufipectus**
Passeriformes - Sylviidae
e Rusty-breasted Wren-Babbler;
Sumatra Wren-Babbler; Large
Sumatran Wren-Babbler; Rufous-
chested Wren-Babbler; Sumatran
Wren-Babbler; Sumatran Large
Wren-Babbler
f Turdinule de Sumatra
d Kastanienbauchtimalie
i Garrulo scricciolo di Sumatra

6355 **Nasica longirostris**
Passeriformes - Furnariidae
e Long-billed Woodcreeper; Long-
billed Creeper
f Grimpar nasican; Grimpar nasican
d Langschnabelbaumsteiger
i Rampichino beccolungo

6356 **Neafrapus boehmi**
Apodiformes - Apodidae
e Bat-like Spinetail; Böhm's Spinetail
(CSA); Bat-like Spinetailed-Swift;
Böhm's Spinetailed-Swift
f Martinet de Böhm
d Fledermaussegler
i Rondone codaspinosa di Boehm

6357 **Neafrapus cassini**
Apodiformes - Apodidae
e Cassin's Spinetail; Cassin's
Spinetailed-Swift
f Martinet de Cassin
d Cassin-Segler
i Rondone codaspinosa di Cassin

6358 **Necropsar rodericanus**
Passeriformes - Sturnidae
e Rodriguez Starling; Leguat's Starling
f Étourneau de Rodriguez
d Leguat-Star
i Storno di Rodriguez

6359 **Necrosyrtes monachus**
Falconiformes - Accipitridae
e Hooded Vulture
f Vautour charognard; Percnoptère
brun
d Kappengeier
i Capovaccaio pileato

6360 **Nectarinia adelberti**
Passeriformes - Nectariniidae
e Buff-throated Sunbird
f Souïmanga à gorge rousse;
Souïmanga à gorge jaune;
Souïmanga à gorge fauve
d Fahlkehlglanzköpchen
i Nettarinia golafulva

6361 **Nectarinia afra**
Passeriformes - Nectariniidae
e Greater Double-collared Sunbird
f Souïmanga à plastron rouge
d Großer Halsbandnektarvogel
i Nettarinia maggiore dal doppio
collare; Nettarinia maggiore

6362 **Nectarinia alinae**
Passeriformes - Nectariniidae
e Blue-headed Sunbird; Ruwenzori
Blue-headed Sunbird
f Souïmanga d'Aline
d Ruwenzori-Nektarvogel
i Nettarinia testablu

6363 **Nectarinia amethystina**
Passeriformes - Nectariniidae
e Amethyst Sunbird; Black Sunbird
(CSA); Honeysucker (CSA);
Amethist Black-Sunbird
f Souïmanga améthyste
d Amethystglanzköpchen;
Amethystnektarvogel
i Nettarinia ametista

6364 **Nectarinia asiatica**
Passeriformes - Nectariniidae
e Purple Sunbird
f Souïmanga asiatique
d Purpurnektarvogel
i Nettarinia purpurea

6365   *Nectarinia aspasia*
        Passeriformes - Nectariniidae
e   Black Sunbird
f   Souïmanga satiné; Souïmanga
        soyeux
d   Seidennektarvogel
i   Nettarinia sericea

6366   *Nectarinia balfouri*
        Passeriformes - Nectariniidae
e   Socotra Sunbird
f   Souïmanga de Socotra
d   Sokotra-Nektarvogel
i   Nettarinia di Socotra

6367   *Nectarinia bannermani*
        Passeriformes - Nectariniidae
e   Bannerman's Sunbird
f   Souïmanga de Bannerman
d   Bannermann-Nektarvogel
i   Nettarinia di Bannerman

6368   *Nectarinia batesi*
        Passeriformes - Nectariniidae
e   Bates's Sunbird; Bate's Olive-Sunbird
f   Souïmanga de Bates
d   Einfarbnektarvogel
i   Nettarinia di Bates

6369   *Nectarinia bifasciata*
        Passeriformes - Nectariniidae
e   Purple-banded Sunbird; Tsavo
        Purple-banded Sunbird; Little Purple-
        banded Sunbird
f   Souïmanga bifascié; Souïmanga à
        double bande; Souïmanga vert et
        brun
d   Kleiner Bindennektarvogel
i   Nettarinia bifasciata

6370   *Nectarinia bocagii*
        Passeriformes - Nectariniidae
e   Bocage's Sunbird
f   Souïmanga de Bocage
d   Bocage-Nektarvogel
i   Nettarinia di Bocage

6371   *Nectarinia bouvieri*
        Passeriformes - Nectariniidae
e   Orange-tufted Sunbird; Southern
        Orange-tufted Sunbird; Bouvier's

        Sunbird
f   Souïmanga de Bouvier
d   Bouvier-Nektarvogel
i   Nettarinia di Bouvier

6372   *Nectarinia buettikoferi*
        Passeriformes - Nectariniidae
e   Apricot-breasted Sunbird; Sumba
        Sunbird; Sumba Island Sunbird
f   Souïmanga de Sumba
d   Sumba-Nektarvogel
i   Nettarinia di Sumba

6373   *Nectarinia calcostetha*
        Passeriformes - Nectariniidae
e   Copper-throated Sunbird; Macklot's
        Sunbird
f   Souïmanga de Macklot
d   Kupferkehlnektarvogel
i   Nettarinia di Macklot

6374   *Nectarinia chalcomelas*
        Passeriformes - Nectariniidae
e   Violet-breasted Sunbird
f   Souïmanga à poitrine violette

6375   *Nectarinia chalybea*
        Passeriformes - Nectariniidae
e   Southern Double-collared Sunbird;
        Lesser Double-collared Sunbird
        (CSA); Honeysucker (CSA)
f   Souïmanga chalybée; Souïmanga à
        double collier
d   Halsbandnektarvogel; Kleiner
        Halsbandnektarvogel;
        Blaubandnektarvogel
i   Nettarinia minore dal doppio collare;
        Nettarinia minore

6376   *Nectarinia chloropygia*
        Passeriformes - Nectariniidae
e   Olive-bellied Sunbird
f   Souïmanga à ventre olive
d   Olivbauchnektarvogel
i   Nettarinia ventreoliva

6377   *Nectarinia coccinigaster*
        Passeriformes - Nectariniidae
e   Splendid Sunbird
f   Souïmanga éclatant

d   Rotbauchnektarvogel
i   Nettarinia splendido

6378   *Nectarinia comorensis*
Passeriformes - Nectariniidae
e   Anjouan Sunbird
f   Souïmanga d'Anjouan
d   Anjouan-Nektarvogel
i   Nettarinia di Anjouan

6379   *Nectarinia congensis*
Passeriformes - Nectariniidae
e   Congo Sunbird; Congo Black-bellied
Sunbird; Black-bellied Sunbird
f   Souïmanga du Congo; Souïmanga à
longue queue du Congo
d   Kongo-Nektarvogel
i   Nettarinia ventrenero del Congo;
Nettarinia del Congo; Nettarinia
ventrenero

6380   *Nectarinia coquerellii*
Passeriformes - Nectariniidae
e   Mayotte Sunbird; Mayotte Yellow-
bellied Sunbird
f   Souïmanga de Mayotte
d   Mayotte-Nektarvogel
i   Nettarinia di Mayotte

6381   *Nectarinia cuprea*
Passeriformes - Nectariniidae
e   Copper Sunbird; Coppery Sunbird
(CSA)
f   Souïmanga cuivré
d   Kupfernektarvogel
i   Nettarinia cuprea

6382   *Nectarinia cyanolaema*
Passeriformes - Nectariniidae
e   Blue-throated Brown-Sunbird; Blue-
throated Sunbird; Blue-headed
Brown-Sunbird
f   Souïmanga à gorge bleue
d   Braunrückennektarvogel
i   Nettarinia golablu

6383   *Nectarinia dussumieri*
Passeriformes - Nectariniidae
e   Seychelles Sunbird
f   Souïmanga des Seychelles

d   Seychellen-Nektarvogel
i   Nettarinia delle Seychelle

6384   *Nectarinia erythrocerca*
Passeriformes - Nectariniidae
e   Red-chested Sunbird
f   Souïmanga à ceinture rouge
d   Schmucknektarvogel
i   Nettarinia codarossa

6385   *Nectarinia famosa*
Passeriformes - Nectariniidae
e   Malachite Sunbird; Yellow-tufted
Malachite-Sunbird; Common
Malachite-Sunbird
f   Souïmanga malachite
d   Malachitnektarvogel
i   Nettarinia malachite

6386   *Nectarinia fuliginosa*
Passeriformes - Nectariniidae
e   Carmelite Sunbird
f   Souïmanga carmelite
d   Karmelglanzköpfchen
i   Nettarinia fuligginosa

6387   *Nectarinia fusca*
Passeriformes - Nectariniidae
e   Dusky Sunbird; White-vented
Sunbird
f   Souïmanga fuligineux
d   Rußnektarvogel; Rußbrauner
Nektarvogel
i   Nettarinia fosca

6388   *Nectarinia habessinica*
Passeriformes - Nectariniidae
e   Shining Sunbird
f   Souïmanga brillant; Souïmanga
luisant
d   Glanznektarvogel; Abessinien-
Nektarvogel
i   Nettarinia abissinica; Nettarinia
abissina

6389   *Nectarinia hartlaubii*
Passeriformes - Nectariniidae
e   Principe Sunbird; Principe Island
Sunbird; Hartlaub's Sunbird
f   Souïmanga de Hartlaub

d   Hartlaub-Nektarvogel
i   Nettarinia dell'Isola Principe

6390   *Nectarinia humbloti*
      Passeriformes - Nectariniidae
e   Humblot's Sunbird
f   Souïmanga de Humblot
d   Schlichtmantelnektarvogel
i   Nettarinia di Humblot

6391   *Nectarinia hunteri*
      Passeriformes - Nectariniidae
e   Hunter's Sunbird
f   Souïmanga de Hunter
d   Violettbürzelnektarvogel;
      Purpurbürzelglanzköpchen
i   Nettarinia di Hunter

6392   *Nectarinia johannae*
      Passeriformes - Nectariniidae
e   Johanna's Sunbird; Madame
      Verreaux's Sunbird
f   Souïmanga de Johanna; Souïmanga
      de Jeanne
d   Grünscheitelnektarvogel
i   Nettarinia della signora Verreaux

6393   *Nectarinia johnstoni*
      Passeriformes - Nectariniidae
e   Red-tufted Sunbird; Scarlet-tufted
      Malachite-Sunbird (CSA);
      Honeysucker (CSA); Johnston's Red-
      tufted Sunbird; Red-tufted Malachite-
      Sunbird; Scarlet-tufted Sunbird;
      Long-tailed Sunbird
f   Souïmanga de Johnston
d   Lobeliennektarvogel
i   Nettarinia malachita di Johnston

6394   *Nectarinia jugularis*
      Passeriformes - Nectariniidae
e   Olive-backed Sunbird; Yellow-
      bellied Sunbird (ANZ); Yellow-
      breasted Sunbird; Black-breasted
      Sunbird; Black-throated Sunbird
f   Souïmanga à dos vert; Souïmanga à
      gorge bleue; Souïmanga à front bleu
d   Grünrückennektarvogel
i   Nettarinia ventregiallo

6395   *Nectarinia kilimensis*
      Passeriformes - Nectariniidae
e   Bronze Sunbird; Bronzy Sunbird
f   Souïmanga bronzé
d   Bronzenektarvogel
i   Nettarinia bronzata

6396   *Nectarinia lotenia*
      Passeriformes - Nectariniidae
e   Loten's Sunbird; Long-billed
      Sunbird; Maroon-breasted Sunbird
      (ISC)
f   Souïmanga de Loten
d   Lotens Nektarvogel
i   Nettarinia di Loten

6397   *Nectarinia loveridgei*
      Passeriformes - Nectariniidae
e   Loveridge's Sunbird
f   Souïmanga de Loveridge
d   Orangebauchnektarvogel
i   Nettarinia di Loveridge

6398   *Nectarinia ludovicensis*
      Passeriformes - Nectariniidae
e   Montane Double-collared Sunbird;
      Preuss's Double-collared Sunbird
f   Souïmanga de l'Angola
d   Hochlandnektarvogel
i   Nettarinia di monte dal doppio
      collare; Nettarinia di monte;
      Nettarinia dal doppio collare

6399   *Nectarinia manoensis*
      Passeriformes - Nectariniidae
e   Miombo Double-collared Sunbird;
      Miombo Sunbird; Collared Sunbird
f   Souïmanga du Miombo
d   Miombo-Nektarvogel
i   Nettarinia del Miombo dal doppio
      collare; Nettarinia del Miombo

6400   *Nectarinia mariquensis*
      Passeriformes - Nectariniidae
e   Mariqua Sunbird; Marico Sunbird;
      Southern Bifasciated Sunbird
f   Souïmanga de Mariqua; Souïmanga
      de Marico
d   Bindennektarvogel
i   Nettarinia di Mariqua

6401 *Nectarinia mediocris*
Passeriformes - Nectariniidae
e Eastern Double-collared Sunbird
f Souïmanga du Kilimandjaro
d Fülleborn-Nektarvogel
i Nettarinia orientale dal doppio
collare; Nettarinia orientale

6402 *Nectarinia minima*
Passeriformes - Nectariniidae
e Crimson-backed Sunbird; Small
Sunbird (ISC)
f Souïmanga menu
d Däumlingsnektarvogel
i Nettarinia piccola

6403 *Nectarinia minulla*
Passeriformes - Nectariniidae
e Tiny Sunbird
f Souïmanga minule
d Zwergnektarvogel
i Nettarinia minuta

6404 *Nectarinia moreaui*
Passeriformes - Nectariniidae
e Moreau's Sunbird
d Moreau-Nektarvogel

6405 *Nectarinia nectarinioides*
Passeriformes - Nectariniidae
e Black-bellied Sunbird; Smaller
Black-bellied Sunbird
f Souïmanga nectarin; Souïmanga à
ventre noir
d Mennigbrustnektarvogel;
Orangebrustnektarvogel
i Nettarinia piccola pettonero

6406 *Nectarinia neergaardi*
Passeriformes - Nectariniidae
e Neergaard's Sunbird; Neergard's
Double-collared Sunbird
f Souïmanga de Neergaard
d Neergards Nektarvogel; Neergard-
Nektarvogel
i Nettarinia di Neergard

6407 *Nectarinia newtonii*
Passeriformes - Nectariniidae
e Newton's Sunbird; Sao Tomé

Yellow-breasted Sunbird; Newton's
Yellow-breasted Sunbird; Yellow-
breasted Sunbird
f Souïmanga de Newton
d Gelbbrustnektarvogel
i Nettarinia di Newton

6408 *Nectarinia notata*
Passeriformes - Nectariniidae
e Long-billed Green-Sunbird; Noted
Sunbird; Pemba Violet-breasted
Sunbird; Green Sunbird; Madagascar
Sunbird; Madagascar Green-Sunbird;
Violet-breasted Sunbird; Violet-
green Sunbird
f Souïmanga angaladian
d Stahlnektarvogel
i Nettarinia marcata

6409 *Nectarinia olivacea*
Passeriformes - Nectariniidae
e Olive Sunbird
f Souïmanga olivâtre
d Olivnektarvogel
i Nettarinia olivacea

6410 *Nectarinia oritis*
Passeriformes - Nectariniidae
e Cameroon Sunbird; Cameroon Blue-
headed Sunbird; Green-headed
Sunbird; Blue-headed Sunbird
f Souïmanga à tête bleue
d Blaukopfnektarvogel
i Nettarinia testablu del Camerun

6411 *Nectarinia osea*
Passeriformes - Nectariniidae
e Palestine Sunbird; Orange-tufted
Sunbird; Northern Orange-tufted
Sunbird
f Souïmanga de Palestine
d Jeriko-Nektarvogel;
Orangebüschelnektarvogel
i Nettarinia della Palestina; Nettarinia
violetta

6412 *Nectarinia oustaleti*
Passeriformes - Nectariniidae
e Oustalet's Sunbird; Angola White-
bellied Sunbird; Angola Sunbird;
Oustalet's White-bellied Sunbird

f Souïmanga d'Oustalet
d Angola-Nektarvogel
i Nettarinia di Oustalet

6413 *Nectarinia pembae*
Passeriformes - Nectariniidae
e Pemba Sunbird; Violet-breasted
Sunbird; Pemba Purple-banded
Sunbird
f Souïmanga de Pemba
d Veilchenbrustnektarvogel
i Nettarinia pettoviola

6414 *Nectarinia preussi*
Passeriformes - Nectariniidae
e Northern Double-collared Sunbird
f Souïmanga de Preuss; Souïmanga
d'Angola
d Preuss-Nektarvogel
i Nettarinia settentrionale dal doppio
collare; Nettarinia settentrionale

6415 *Nectarinia prigoginei*
Passeriformes - Nectariniidae
e Prigogine's Double-collared Sunbird;
Prigogine's Sunbird
f Souïmanga de Prigogine

i Nettarinia di Prigogine

6416 *Nectarinia pulchella*
Passeriformes - Nectariniidae
e Beautiful Sunbird; Beautiful Long-
tailed Sunbird
f Souïmanga à longue queue;
Souïmanga élégant; Souïmanga vert-
doré à longue queue
d Elfennektarvogel
i Nettarinia multicolore

6417 *Nectarinia purpureiventris*
Passeriformes - Nectariniidae
e Purple-breasted Sunbird; Rainbow
Sunbird
f Souïmanga à ventre pourpre
d Purpurbauchnektarvogel
i Nettarinia ventrepurpureo

6418 *Nectarinia regia*
Passeriformes - Nectariniidae
e Regal Sunbird

f Souïmanga royal
d Königsnektarvogel
i Nettarinia reale

6419 *Nectarinia reichenbachii*
Passeriformes - Nectariniidae
e Reichenbach's Sunbird
f Souïmanga de Reichenbach
d Kamerun-Nektarvogel
i Nettarinia di Reichenbach

6420 *Nectarinia reichenowi*
Passeriformes - Nectariniidae
e Golden-winged Sunbird; Northern
Double-collared Sunbird
f Souïmanga à ailes dorées;
Souïmanga de Reichenow;
Souïmanga à ailes d'or
d Sichelnektarvogel
i Nettarinia alidorate

6421 *Nectarinia rockefelleri*
Passeriformes - Nectariniidae
e Rockefeller's Sunbird
f Souïmanga de Rockefeller
d Blutbrustnektarvogel
i Nettarinia di Rockerfeller

6422 *Nectarinia rubescens*
Passeriformes - Nectariniidae
e Green-throated Sunbird
f Souïmanga à gorge verte
d Grünkehlglanzköpchen
i Nettarinia golaverde

6423 *Nectarinia rufipennis*
Passeriformes - Nectariniidae
e Rufous-winged Sunbird
f Souïmanga à ailes rousses;
Souïmanga à ailes rouges
i Nettarinia alirugginose

6424 *Nectarinia seimundi*
Passeriformes - Nectariniidae
e Little Green-Sunbird; Little Olive-
Sunbird
f Souïmanga de Seimund
d Stutzschwanznektarvogel
i Nettarinia verde piccola

6425 *Nectarinia senegalensis*
Passeriformes - Nectariniidae
e Scarlet-chested Sunbird; Scarlet-breasted Sunbird
f Souïmanga à poitrine rouge; Souïmanga à poitrine écarlate
d Rotbrustnektarvogel; Rotbrustglanzköpchen; Natal-Glanzköpchen
i Nettarinia pettirossa

6426 *Nectarinia shelleyi*
Passeriformes - Nectariniidae
e Shelley's Sunbird; Shelley's Double-collared Sunbird; Black-bellied Sunbird
f Souïmanga de Shelley
d Shelleys Nektarvogel; Shelley-Nektarvogel; Scharlachbrustnektarvogel
i Nettarinia di Shelley

6427 *Nectarinia solaris*
Passeriformes - Nectariniidae
e Flame-breasted Sunbird; Timor Sunbird; Sunda Sunbird
f Souïmanga de Timor
d Sonnennektarvogel
i Nettarinia di Timor

6428 *Nectarinia sovimanga*
Passeriformes - Nectariniidae
e Souïmanga Sunbird
f Souïmanga malgache
d Malegassen-Nektarvogel
i Nettarinia souimanga

6429 *Nectarinia sperata*
Passeriformes - Nectariniidae
e Purple-throated Sunbird; Van Hasselt's Sunbird
f Souïmanga de Hasselt
d Purpurkehlnektarvogel; Hassel-Nektarvogel
i Nettarinia di Van Hasselt

6430 *Nectarinia stuhlmanni*
Passeriformes - Nectariniidae
e Stuhlmann's Double-collared Sunbird; Stuhlmann's Sunbird
f Souïmanga de Stuhlmann
d Stuhlmann-Nektarvogel
i Nettarinia di Stuhlmann

6431 *Nectarinia superba*
Passeriformes - Nectariniidae
e Superb Sunbird
f Souïmanga superbe
d Prachtnektarvogel
i Nettarinia superba

6432 *Nectarinia tacazze*
Passeriformes - Nectariniidae
e Tacazze Sunbird
f Souïmanga tacazze
d Tacazze-Nektarvogel
i Nettarinia di Tacazze

6433 *Nectarinia talatala*
Passeriformes - Nectariniidae
e White-breasted Sunbird; White-bellied Sunbird (CSA); Honeysucker (CSA); Southern White-bellied Sunbird
f Souïmanga à ventre blanc
d Weißbauchnektarvogel
i Nettarinia ventrebianco

6434 *Nectarinia thomensis*
Passeriformes - Nectariniidae
e Sao Tomé Sunbird; Sao Tomé Giant-Sunbird; Giant Sunbird
f Souïmanga de Sao Tomé; Souïmanga géant de Sao Tomé
d Riesennektarvogel
i Nettarinia gigante di Sao Tomè

6435 *Nectarinia ursulae*
Passeriformes - Nectariniidae
e Ursula's Sunbird; Fernando Po Sunbird; Ursula's Mouse-coloured Sunbird
f Souïmanga d'Ursula
d Graubrustnektarvogel
i Nettarinia di Ursula

6436 *Nectarinia venusta*
Passeriformes - Nectariniidae
e Variable Sunbird; Yellow-bellied Sunbird (CSA); Honeysucker (CSA); Yellow-breasted Sunbird
f Souïmanga à ventre jaune;

Souïmanga élégant
d   Gelbbauchnektarvogel
i   Nettarinia variabile

6437   *Nectarinia veroxii*
Passeriformes - Nectariniidae
e   Grey Sunbird (CSA); Mouse-
coloured Sunbird
f   Souïmanga murin; Souïmanga gris
d   Graunektarvogel
i   Nettarinia di Verreaux

6438   *Nectarinia verticalis*
Passeriformes - Nectariniidae
e   Green-headed Sunbird; Olive-backed
Sunbird
f   Souïmanga à tête verte; Souïmanga
olive à tête bleue; Souïmanga olive à
tête brune
d   Grünkopfnektarvogel
i   Nettarinia testaverde

6439   *Nectarinia violacea*
Passeriformes - Nectariniidae
e   Orange-breasted Sunbird
f   Souïmanga orangé; Souïmanga à
gorge violette
d   Goldbrustnektarvogel
i   Nettarinia ventrearancio

6440   *Nectarinia zeylonica*
Passeriformes - Nectariniidae
e   Purple-rumped Sunbird
f   Souïmanga à croupion pourpre;
Souïmanga à ceinture marron
d   Ceylon-Nektarvogel
i   Nettarinia gropponepurpureo

6441   **Nectariniidae**
Passeriformes
e   Sunbirds
f   Souïmangas
d   Nektarvögel
i   Nettarinidi

6442   *Nemosia pileata*
Passeriformes - Fringillidae
e   Hooded Tanager; Pileated Tanager
f   Tangara coiffe-noire
d   Nemosia
i   Tangara dal cappuccio

6443   *Nemosia rourei*
Passeriformes - Fringillidae
e   Cherry-throated Tanager
f   Tangara rougegorge
d   Rubinkehltangare; Rotkehltangare
i   Tangara gola amaranto

6444   *Neochelidon tibialis*
Passeriformes - Hirundinidae
e   White-thighed Swallow
f   Hirondelle à cuisses blanches
d   Zwergschwalbe
i   Rondine dai calzoni bianchi

6445   *Neochen jubata*
Anseriformes - Anatidae
e   Orinoco Goose
f   Ouette de l'Orénoque
d   Orinoco-Gans
i   Oca dell'Orinoco

6446   *Neochmia modesta*
Passeriformes - Passeridae
e   Plum-headed Finch; Plum-headed
Cherryfinch; Modest Grassfinch;
Plum-capped Finch; Cherry Finch;
Diadem Finch
f   Diamant modeste; Modeste
d   Zeres-Amadine; Zeres-Astrild;
Zeresfink
i   Diamante modesto

6447   *Neochmia phaeton*
Passeriformes - Passeridae
e   Crimson Finch; Red-tailed Finch;
Red-faced Finch; Blood Finch;
Australian Firefinch
f   Diamant phaéton; Rubin d'Australie
d   Sonnenastrild
i   Diamante rosso

6448   *Neochmia phaeton evangelinae*
Passeriformes - Passeridae
e   White-bellied Crimson Finch (ANZ)

6449   *Neochmia ruficauda*
Passeriformes - Passeridae
e   Star Finch; Red-faced Grassfinch;
Rufous-tailed Grassfinch
f   Diamant à queue rousse; Diamant
ruficaude

d Binsenastrild
i Diamante codarossa

6450 *Neochmia ruficauda clarescens*
Passeriformes - Passeridae
e Cape York Peninsula Star Finch
(ANZ)

6451 *Neochmia ruficauda ruficauda*
Passeriformes - Passeridae
e Southern Star Finch (ANZ)

6452 *Neochmia ruficauda subclarescens*
Passeriformes - Passeridae
e Western Star Finch (ANZ)

6453 *Neochmia temporalis*
Passeriformes - Passeridae
e Red-browed Firetail; Red-browed
Finch (ANZ); Red-browed Waxbill;
Sydney Waxbill; Australian Waxbill;
Red-browed Firetail-Finch; Sydney
Firetail
f Diamant à cinq couleurs; Astrild à
cinq couleurs; Diamant à sourcils
rouges
d Dornastrild
i Diamante dai sopraccigli rossi

6454 *Neocichla gutturalis*
Passeriformes - Sturnidae
e Babbling Starling; White-winged
Babbling-Starling; White-winged
Starling
f Spréo à gorge noire; Étourneau pâle
d Weißflügelstar
i Storno alibianchi

6455 *Neocossyphus finschii*
Passeriformes - Muscicapidae
e Finsch's Flycatcher-Thrush; Finsch's
Rusty Flycatcher; Finsch's Rusty
Flycatcher-Thrush; Finsch's Ant-
Thrush; White-tailed Flycatcher-
Thrush; White-tailed Flycatcher
f Stizorhin de Finsch; Néocossyphe de
Finsch
d Finsch-Drossel
i Tordo di Finsch

6456 *Neocossyphus fraseri*
Passeriformes - Muscicapidae
e Rufous Flycatcher-Thrush; Rufous
Broad-billed Ant-Thrush; Fraser's
Rusty Flycatcher; Fraser's
Flycatcher-Thrush; Rufous
Flycatcher; Fraser's Rusty
Flycatcher-Thrush; Fraser's Ant-
Thrush; Rufous Ant-Thrush
f Stizorhin de Fraser; Néocossyphe de
Fraser; Gobemouche roux; Gobe-
mouches roux; Grive fourmilière
rousse
d Kurzlaufdrossel
i Tordo di Fraser

6457 *Neocossyphus poensis*
Passeriformes - Muscicapidae
e White-tailed Ant-Thrush; White-
tailed Thrush
f Néocossyphe à queue blanche; Grive
fourmilière à queue blanche
d Weißschwanzfuchsdrossel
i Tordo codabianca

6458 *Neocossyphus rufus*
Passeriformes - Muscicapidae
e Red-tailed Ant-Thrush; Red-tailed
Thrush
f Néocossyphe à queue rousse; Grive
fourmilière à queue rousse
d Rotschwanzfuchsdrossel
i Tordo codarossa

6459 *Neocrex colombianus*
Gruiformes - Rallidae
e Colombian Crake
f Râle de Colombie
i Schiribilla colombiana

6460 *Neocrex erythrops*
Gruiformes - Rallidae
e Paint-billed Crake
f Râle à bec peint
d Goldschnabelralle
i Schiribilla beccorosso

6461 *Neoctantes niger*
Passeriformes - Formicariidae
e Black Bushbird
f Batara des fourres

d   Mohrenwürgerling
i   Mangiaformiche nero

6462   *Neodrepanis coruscans*
      Passeriformes - Philepittidae
e   Sunbird Asity; Wattled False-
      Sunbird; False Sunbird; Wattled
      Asity; Wattled Sunbird
f   Philépitte souimanga; Faux-
      Souïmanga caronculé
d   Langschnabelnektarjala; Nektarpitta;
      Pseudonektarvogel
i   Asity caruncolato

6463   *Neodrepanis hypoxantha*
      Passeriformes - Philepittidae
e   Yellow-bellied Asity; Small-billed
      False-Sunbird; Small-billed
      Neodrepanis; Small-billed Asity;
      Yellow-bellied Sunbird Asity
f   Philépitte de Salomonsen; Faux-
      Souïmanga à ventre jaune
d   Kurzschnabelnektarjala
i   Asity ventregiallo

6464   *Neolalage banksiana*
      Passeriformes - Corvidae
e   Buff-bellied Monarch; Buff-bellied
      Flycatcher; New Hebrides
      Flycatcher; Silky Flycatcher;
      Banksian Flycatcher; New Hebrides
      Monarch; Pacific Monarch; Banksian
      Monarch
f   Monarque des Banks
d   Harlekinmonarch
i   Monarca dell'Isola Banks

6465   *Neolestes torquatus*
      Passeriformes - Pycnonotidae
e   Black-collared Bulbul
f   Bulbul à collier noir
d   Rüttelbülbül
i   Bulbul dal collare

6466   *Neomixis flavoviridis*
      Passeriformes - Sylviidae
e   Wedge-tailed Jery
f   Éréonesse à queue étagée
d   Keilschwanzeroessa
i   Jery codacuneata

6467   *Neomixis striatigula*
      Passeriformes - Sylviidae
e   Stripe-throated Jery
f   Grande Éréonesse
d   Streifenkehleroessa
i   Jery golastriata

6468   *Neomixis tenella*
      Passeriformes - Sylviidae
e   Common Jery; Northern Jery;
      Madagascar Jery; Jery
f   Petite Éréonesse
d   Graunackeneroessa
i   Jery del Nord

6469   *Neomixis viridis*
      Passeriformes - Sylviidae
e   Green Jery; Southern Green-Jery
f   Éréonesse verte
d   Grüneroessa; Grünjery
i   Jery del Sud

6470   Neomorphidae
      Cuculiformes
e   Ground-Cuckoos
f   Neomorphidés
e   Erdkuckucke
i   Neomorfidi

6471   *Neomorphus geoffroyi*
      Cuculiformes - Neomorphidae
e   Rufous-vented Ground-Cuckoo;
      Rufous-vented Cuckoo
f   Géocoucou de Geottroy
d   Tajazuira
i   Cuculo di terra ventrerosso

6472   *Neomorphus pucheranii*
      Cuculiformes - Neomorphidae
e   Red-billed Ground-Cuckoo
f   Géocoucou de Pucheran
d   Rotschnabelgrundkuckuck
i   Cuculo di terra beccorosso

6473   *Neomorphus radiolosus*
      Cuculiformes - Neomorphidae
e   Banded Ground-Cuckoo
f   Géocoucou barré
d   Bindengrundkuckuck
i   Cuculo di terra fasciato

6474 *Neomorphus rufipennis*
Cuculiformes - Neomorphidae
e   Rufous-winged Ground-Cuckoo;
Chapman's Ground-Cuckoo
f   Géocoucou à ailes rousses
d   Rotschwingengrundkuckuck
i   Cuculo di terra alirossicce

6475 *Neomorphus squamiger*
Cuculiformes - Neomorphidae
e   Scaled Ground-Cuckoo
f   Géocoucou écaillé
d   Schuppengrundkuckuck
i   Cuculo di terra squamoso

6476 *Neopelma aurifrons*
Passeriformes - Tyrannidae
e   Wied's Tyrant-Manakin; Wied's
Manakin
f   Manakin tyran
d   Schnäpperpipra; Gelbscheitelpipra

6477 *Neopelma chrysocephalum*
Passeriformes - Tyrannidae
e   Saffron-crested Tyrant-Manakin;
Orange Manakin; Saffron-crested
Manakin
f   Manakin à panache doré
d   Gelbhaubenpipra

6478 *Neopelma chrysolophum*
Passeriformes - Tyrannidae
e   Serra do Mar Tyrant-Manakin
f   Manakin à ventre blanc

6479 *Neopelma pallescens*
Passeriformes - Tyrannidae
e   Pale-bellied Tyrant-Manakin;
Yellow-crested Manakin; Pale-
bellied Manakin
f   Manakin à ventre blanc
d   Weißbauchpipra;
Graukehlschnurrvogel

6480 *Neopelma sulphureiventer*
Passeriformes - Tyrannidae
e   Sulphur-bellied Tyrant-Manakin;
Sulphur-bellied Manakin
f   Manakin à ventre jaune
d   Gelbbauchpipta

6481 *Neophema chrysogaster*
Psittaciformes - Psittacidae
e   Orange-bellied Parrot; Orange-
breasted Parrot; Orange-breasted
Grass-Parakeet; Orange-bellied
Grass-Parakeet
f   Perruche à ventre orangé; Perruche à
lunettes vertes
d   Goldbauchsittich;
Orangebauchsittich; Grünzügeligers-
Schönsittich; Grünzügelschönsittich
i   Parrochetto ventrearancio

6482 *Neophema chrysostoma*
Psittaciformes - Psittacidae
e   Blue-winged Parrot; Blue-banded
Grass-Parakeet; Blue-winged Grass-
Parakeet; Blue-banded Grass-Parrot
f   Perruche à bouche d'or; Perruche
vénuste
d   Blauflügeliger Schönsittich;
Blauflügeligerziersittich
i   Parrochetto aliazzurre

6483 *Neophema elegans*
Psittaciformes - Psittacidae
e   Elegant Parrot; Elegant Grass-Parrot;
Elegant Grass-Parakeet
f   Perruche élégante
d   Schmucksittich; Ziersittich
i   Parrochetto elegante

6484 *Neophema petrophila*
Psittaciformes - Psittacidae
e   Rock Parrot; Rock Elegant-Parrot;
Rock Grass-Parakeet; Rock Grass-
Parrot
f   Perruche des rochers; Perruche
petrophile
d   Felsensittich; Klippensittich;
Felsengrassittich
i   Parrochetto delle rocce

6485 *Neophema pulchella*
Psittaciformes - Psittacidae
e   Turquoise Parrot; Turquoisine Parrot;
Turquoisine Grass-Parakeet;
Beautiful Grass-Parakeet; Turquoise
Grass-Parrot; Beautiful Parrot;
Chestnut-winged Parrot; Chestnut-
shouldered Parrot

f   Perruche turquoisine; Perruche
      d'Edwards
d   Schönsittich; Turquoisinesittich
i   Parracchetto turchese

6486   *Neophema splendida*
           Psittaciformes - Psittacidae
e   Scarlet-chested Parrot; Scarlet-
      breasted Parrot; Splendid Grass-
      Parrot; Splendid Grass-Parakeet;
      Scarlet-chested Grass-Parakeet;
      Orange-bellied Parrot; Orange-
      bellied Grass-Parakeet
f   Perruche splendide
d   Glanzsittich; Rotbrüstiger
      Schönsittich
i   Parrachetto splendido

6487   *Neophron percnopterus*
           Falconiformes - Accipitridae
e   Egyptian Vulture; White-Scavenger-
      Vulture (ISC); Pharaoh's Chicken
      (ISC); Small Scavenger-Vulture
f   Pygargue à queue blanche; Vautour
      percnoptère; Percnoptère d'Égypte;
      Percnoptère stercoraire
d   Schmutzgeier
i   Capovaccaio

6488   *Neopipo cinnamomea*
           Passeriformes - Tyrannidae
e   Cinnamon Tyrant; Cinnamon
      Manakin; Cinnamon Tyrant-Manakin
f   Manakin cannellé
d   Zimtpipra

6489   *Neopsephotus bourkii*
           Psittaciformes - Psittacidae
e   Bourke's Parrot; Night Parrot;
      Sundown Parrot; Bourke's Parakeet;
      Bourke's Grass-Parakeet; Pink-
      bellied Parakeet
f   Perruche de Bourke
d   Bourke-Sittich; Rosenbauchsittich
i   Parrocchetto di Bourke

6490   *Neopsittacus musschenbroekii*
           Psittaciformes - Loriidae
e   Yellow-billed Lorikeet;
      Muschenbroek's Lorikeet
f   Lori de Musschenbroek

d   Gelbschnabelberglori
i   Lorichetto di Musschenbroek

6491   *Neopsittacus pullicauda*
           Psittaciformes - Loriidae
e   Orange-billed Lorikeet; Emerald
      Lorikeet; Hartert's Lorikeet; Alpine
      Lorikeet
f   Lori émeraude
d   Orangeschnabelberglori
i   Lorichetto smeraldo

6492   *Neospiza concolor*
           Passeriformes - Fringillidae
e   Sao Tomé Grosbeak; Sao Tomé
      Seedeater; Neospiza; Sao Tomé
      Canary; St. Thomas Canary;
      Grosbeak Bunting
f   Néospize de Sao Tomé; Grosbec de
      Sao Tomé
d   Einfarbgirlitz
i   Neospiza di Sao Tomè

6493   *Neothraupis fasciata*
           Passeriformes - Fringillidae
e   White-banded Tanager; Shrike-like
      Tanager
f   Tangara unifascié
d   Flügelbindentangare
i   Tangara fasciata

6494   *Neotis denhami*
           Gruiformes - Otididae
e   Denham's Bustard; Jackson's Bustard
      (CSA)
f   Outarde de Denham
d   Denham-Trappe; Kaffern-Trappe
i   Otarda di Denham

6495   *Neotis denhami stanleyi*
           Gruiformes - Otididae
e   Stanley's Bustard
f   Outarde de Stanley
d   Stanley-Trappe
i   Otarda di Stanley

6496   *Neotis heuglinii*
           Gruiformes - Otididae
e   Heuglin's Bustard
f   Outarde de Heuglin

d Heuglin-Trappe
i Otarda di Heuglin

6497 *Neotis ludwigii*
Gruiformes - Otididae
e Ludwig's Bustard
f Outarde de Ludwig
d Ludwigs Trappe
i Otarda di Ludwig

6498 *Neotis nuba*
Gruiformes - Otididae
e Nubian Bustard
f Outarde nubienne
d Nubier-Trappe; Nubien-Trappe
i Otarda nubiana

6499 *Neoxolmis rufiventris*
Passeriformes - Tyrannidae
e Chocolate Tyrant; Chocolate-vented Tyrant
f Pépoaza à ventre rougeâtre
d Weißschultermonjita
i Monjita rufiventre

6500 *Nephelornis oneilli*
Passeriformes - Fringillidae
e Pardusco
f Pardusco d'O'Neill
d Pardusco
i Pardusco

6501 *Nesasio solomonensis*
Strigiformes - Strigidae
e Fearful Owl
f Hibou redoutable
d Salomonen-Eule
i Gufo delle Salomon; Gufo terribile

6502 *Nesillas aldabrana*
Passeriformes - Sylviidae
e Aldabra Brush-Warbler; Aldabra Warbler; Aldabra Tsikirity
f Nésille d'Aldabra; Fauvette d'Aldabra
d Aldabra-Buschsänger
i Cantore di Aldabra

6503 *Nesillas brevicaudata*
Passeriformes - Sylviidae
e Grand Comoro Brush-Warbler; Comoro Warbler; Comoro Brush-Warbler; Comoros Brush-Warbler
f Nésille de la Grande Comore; Fauvette des Comores

6504 *Nesillas longicaudata*
Passeriformes - Sylviidae
e Anjouan Brush-Warbler

6505 *Nesillas mariae*
Passeriformes - Sylviidae
e Mrs.. Benson's Brush-Warbler; Moheli Brush-Warbler; Moheli Tsikirity; Comoro Tsikirity; Mrs.. Benson's Warbler
f Nésille de Moheli; Fauvette de Madame Benson
d Moheli-Buschsänger
i Cantore delle Comore

6506 *Nesillas typica*
Passeriformes - Sylviidae
e Malagasy Brush-Warbler; Tsikirity Warbler; Tsikirity Brush-Warbler; Tsikirity; Common Tsikirity; Madagascar Brush-Warbler; Madagascar Tsikirity
f Nésille malgache; Fauvette malgache
d Madagaskar-Buschsänger
i Cantore di Tsikirity

6507 *Nesocharis ansorgei*
Passeriformes - Passeridae
e White-collared Oliveback; White-collared Olive-Weaver; Olive Weaver Finch
f Dos-vert à collier; Bengali vert d'Ansorge
d Halsbandastrild
i Dorso oliva dal collare

6508 *Nesocharis capistrata*
Passeriformes - Passeridae
e Grey-headed Oliveback; White-cheeked Oliveback; White-cheeked Olive-Weaver; White-cheeked Waxbill
f Dos-vert à joues blanches; Bengali vert à joues blanches; Sénégali vert à joues blanches
d Weißwangenastrild
i Dorso oliva testagrigia

6509   *Nesocharis shelleyi*
   Passeriformes - Passeridae
   e   Fernando Po Oliveback; Shelley's
       Oliveback; Little Olive-Waxbill;
       Little Olive-Weaver; Little
       Oliveback; Little Olive-Waxback
   f   Dos-vert à tête noire; Dos-olive de
       Shelley; Petit Sénégali vert
   d   Meisenastrild
   i   Dorso oliva di Shelley

6510   *Nesocichla eremita*
   Passeriformes - Muscicapidae
   e   Tristan Thrush; Starchy Thrush
   f   Grive de Tristan da Cunha
   d   Tristan-Dommel
   i   Tordo di Tristan da Cunha; Tordo
       dell'Isola Tristan da Cunha

6511   *Nesoclopeus poecilopterus*
   Gruiformes - Rallidae
   e   Bar-winged Rail; Barred-winged
       Rail; Fiji Rail
   f   Râle des Fidji
   d   Fidji-Ralle
   i   Rallo alibarrate

6512   *Nesoclopeus woodfordi*
   Gruiformes - Rallidae
   e   Woodford's Rail; Solomons Rail
   f   Râle de Woodford
   d   Salomonen-Ralle
   i   Rallo delle Salomon; Rallo delle
       Isole Salomon

6513   *Nesoctites micromegas*
   Piciformes - Picidae
   e   Antillean Piculet
   f   Charpentier (Ants); Charpentier-bois
       (Ants); Charpentier camelle (Ants);
       Picumne des Antilles
   d   Fidji-Ralle
   i   Picchio nano delle Antille

6514   *Nesofregetta fuligunosa*
   Procellariiformes - Hydrobatidae
   e   Polynesian Storm-Petrel; White-
       throated Storm-Petrel; White-
       throated Petrel; Samoan Storm-Petrel
   f   Océanite à gorge blanche

   d   Weißkehlmeerläufer
   i   Uccello delle tempeste golabianca

6515   *Nesomimus macdonaldi*
   Passeriformes - Sturnidae
   e   Hood Mockingbird; Española
       Mockingbird
   f   Moqueur d'Espanola
   i   Mimo di McDonald

6516   *Nesomimus melanotis*
   Passeriformes - Sturnidae
   e   San Cristobal Mockingbird; Chatham
       Mockingbird; Chatham Islands
       Mockingbird
   f   Moqueur de San Cristobal
   i   Mimo di San Cristobal

6517   *Nesomimus parvulus*
   Passeriformes - Sturnidae
   e   Galapagos Mockingbird
   f   Moqueur des Galapagos
   d   Galapagos-Spottdrossel;
       Erdspottdrossel
   i   Mimo delle Galapagos

6518   *Nesomimus trifasciatus*
   Passeriformes - Sturnidae
   e   Floreana Mockingbird; Charles
       Mockingbird; Three-banded
       Mockingbird; Santa Maria
       Mockingbird
   f   Moqueur de Floreana; Moqueur de
       Charles
   i   Mimo do Darwin

6519   *Nesopsar nigerrimus*
   Passeriformes - Fringillidae
   e   Jamaican Blackbird; Black Banana-
       bird (WI); Wild Pine Sargeant (WI);
       Corporalbird (W.I)
   f   Carouge de la Jamaïque
   d   Bromelienstärling
   i   Ittero nero della Giamaica

6520   *Nesospingus speculiferus*
   Passeriformes - Fringillidae
   e   Puerto Rican Tanager
   f   Tangara de Porto Rico
   d   Brustfleckentangare
   i   Tangara di Portorico

6521 **Nesospiza acunhae**
Passeriformes - Fringillidae
e Nightingale Finch; Tristan Bunting;
Tristan Finch
f Nésospize acunha
d Tristan-Ammerfink
i Fringuello dell'Isola Tristan da
Cunha

6522 **Nesospiza wilkinsi**
Passeriformes - Fringillidae
e Wilkins's Finch; Grosbeak Bunting;
Wilkins's Bunting (CSA); Big-billed
Tristan-Finch; Tristan Grosbeak;
Tristan Finch
f Nésospize de Wilkins
d Wilkins-Ammerfink
i Fringuello di Wilkins

6523 **Nesotriccus ridgwayi**
Passeriformes - Tyrannidae
e Cocos Flycatcher; Cocos Islands
Flycatcher
f Tyranneau des Cocos
d Cocos-Tyrann
i Tiranno delle Isole Cocos

6524 **Nestor meridionalis**
Psittaciformes - Psittacidae
e New Zealand Kaka; Kaka (ANZ);
Brown Parrot; Common Kaka
f Nestor superbe
d Kaka; Kakanestor
i Nestore meridionale; Kaka

6525 **Nestor notabilis**
Psittaciformes - Psittacidae
e Kea; Mountain Kea; Mountain Parrot
f Nestor kea
d Kea
i Nestore notabile; Kea

6526 **Nestor productus**
Psittaciformes - Psittacidae
e Norfolk Island Kaka; Norfolk Kaka;
Philip Island Kaka
f Nestor de Norfolk
d Norfolk-Kaka
i Kaka delle Norfolk

6527 **Netta erythrophthalma**
Anseriformes - Anatidae
e Southern Pochard; Red-eyed Pochard
f Nette brune; Canard plongeur austral
d Rotaugenente
i Fistione meridonale

6528 **Netta peposaca**
Anseriformes - Anatidae
e Rosy-billed Pochard; Rosybill
f Nette demi-deuil
d Peposak-Ente
i Fistione beccorosa

6529 **Netta rufina**
Anseriformes - Anatidae
e Red-crested Pochard; Red-crested
Duck
f Nette rousse
d Kolbenente
i Fistione turco; Fischione turco

6530 **Nettapus auritus**
Anseriformes - Anatidae
e African Pygmy-Goose; Pygmy
Goose (CSA); Dwarf Pugmy Goose
f Anserelle naine
d Afrikanische Zwerggans;
Rotbrustzwerggans
i Oca pigmea africana

6531 **Nettapus coromandelianus**
Anseriformes - Anatidae
e Cotton Pygmy-Goose; Cotton Teal
(ISC); White-Pygmy Goose; Indian
Pygmy-Goose; White-quilled Dwarf
Goose
f Anserelle de Coromandel
d Koromandel-Zwergente;
Weißbauchzwerggans
i Oca pigmea indiana

6532 **Nettapus coromandelianus
albipennis**
Anseriformes - Anatidae
e Australian Cotton Pygmy-Goose

6533 **Nettapus pulchellus**
Anseriformes - Anatidae
e Green Pygmy-Goose; Green Dwarf
Goose; Goose-Teal

f Anserelle élégante
d Halsbandzwerggans; Grüne
     Zwerggans
i Oca pigmea verde; Oca pigmea
     australiana

6534 *Newtonia amphichroa*
     Passeriformes - Sylviidae
e Dark Newtonia; Dark Tulear
     Newtonia; Tulear Newtonia
f Newtonie sombre
d Olivbauchnewtonie
i Newtonia di Tulear

6535 *Newtonia archboldi*
     Passeriformes - Sylviidae
e Archbold's Newtonia; Tabity
     Newtonia
f Newtonie d'Archbold
d Braunstirnnewtonie
i Newtonia di Tabity

6536 *Newtonia brunneicauda*
     Passeriformes - Sylviidae
e Common Newtonia; Newtonia
f Newtonie commune
d Rostbauchnewtonie
i Newtonia; Newtonia comune

6537 *Newtonia fanovanae*
     Passeriformes - Sylviidae
e Red-tailed Newtonia; Fanovana
     Newtonia
f Newtonie de Fanovana; Newtonie à
     queue rouge
d Fanovana-Newtonie
i Newtonia di Fanovana

6538 *Nicator chloris*
     Passeriformes - Malaconotidae
e Yellow-spotted Nicator; West
     African Nicator; African Nicator
f Bulbul nicator; Pie-grièche nicator;
     Nicator à gorge blanche; Nicator vert
d Bülbülwürger; Graukehlnicator
i Nicator occidentale

6539 *Nicator gularis*
     Passeriformes - Malaconotidae
e Eastern Nicator; Yellow-spotted
     Nicator (CSA)

f Bulbul à tête brune; Nicator à tête
     brune
d Braunkopfnicator
i Nicator orientale

6540 *Nicator vireo*
     Passeriformes - Malaconotidae
e Yellow-throated Nicator
f Bulbul à gorge jaune; Nicator à gorge
     jaune
d Gelbkehlnicator
i Nicator golagialla

6541 *Nigrita bicolor*
     Passeriformes - Estrildidae
e Chestnut-breasted Negrofinch
f Nigrette à ventre roux; Nègrette à
     dessous châtain; Bengali brun à
     ventre roux; Sénégali brun à ventre
     roux
d Zweifarbenschwärzling
i Nigrita pettocastano

6542 *Nigrita canicapilla*
     Passeriformes - Estrildidae
e Grey-headed Negrofinch; Grey-
     crowned Negrofinch; Grey-crowned
     Blackfinch; Western Negrofinch
f Nigrette à calotte grise; Nègrette à
     front noir; Bengali nègre; Sénégali
     nègre
d Graunackenschwärzling
i Nigrita corona grigia

6543 *Nigrita fusconota*
     Passeriformes - Estrildidae
e White-breasted Negrofinch; White-
     breasted Blackfinch
f Nigrette à ventre blanc; Nègrette à
     dessous blanc; Bengali brun à ventre
     blanc; Sénégali brun à ventre blanc
d Mantelschwärzling;
     Weißbrustschwärzling
i Nigrita pettobianco

6544 *Nigrita luteifrons*
     Passeriformes - Estrildidae
e Pale-fronted Negrofinch
f Nigrette à front jaune; Nègrette à
     dessous gris; Bengali nègre à front
     jaune; Sénégali nègre à front jaune

d   Blaßstirnschwärzling
i   Nigrita frontechiara

6545 *Nilaus afer*
Passeriformes - Malaconotidae
e   Brubru; Brubru Shrike
f   Brubru africain; Pie-grièche brubru;
Brubru
d   Brubru-Würger; Brubru
i   Brubru

6546 *Niltava davidi*
Passeriformes - Muscicapidae
e   Fujian Niltava; Fukien Niltava;
David's Niltava
f   Gobemouche de David; Gobe-
mouches de David
d   David-Niltava
i   Niltava del Fukien

6547 *Niltava grandis*
Passeriformes - Muscicapidae
e   Large Niltava; Great Niltava; Greater
Niltava; Great Blue-Flycatcher
f   Grand Gobemouche; Grand Gobe-
mouches; Grand Niltava
d   Kobaltniltava; Großniltava
i   Niltava grande

6548 *Niltava macgrigoriae*
Passeriformes - Muscicapidae
e   Small Niltava
f   Gobemouche de McGrigor; Gobe-
mouches de McGrigor; Petit Niltava
d   Feenniltava
i   Niltava piccola

6549 *Niltava sumatrana*
Passeriformes - Muscicapidae
e   Rufous-vented Niltava; Rufous-
bellied Niltava; Malaysian Niltava;
Sumatran Niltava
f   Gobemouche de Sumatra; Gobe-
mouches de Sumatra
i   Niltava di Sumatra

6550 *Niltava sundara*
Passeriformes - Muscicapidae
e   Rufous-bellied Niltava; Blue-and-
orange Niltava; Orange-bellied
Niltava; Beautiful Niltava; Sundara

Niltava
f   Gobemouche sundara
d   Rotbauchniltava; Schwarzkehlniltava
i   Niltava ventrerosso

6551 *Niltava vivida*
Passeriformes - Muscicapidae
e   Vivid Niltava; Rufous-bellied Blue-
Flycatcher; Rufous-bellied Blue-
Niltava; Rufous-vented Blue-Niltava;
Vivid Flycatcher; Vivid Blue-
Flycatcher
f   Gobemouche à ventre roux; Gobe-
mouches à ventre roux; Niltava à
ventre roux; Cyornis à ventre roux
d   Swinhoe-Niltava
i   Niltava vivida

6552 *Ninox affinis*
Strigiformes - Strigidae
e   Andaman Hawk-Owl; Andaman
Boobook; Andaman Brown Hawk-
Owl
f   Ninoxe des Andaman
d   Andamanen-Kauz
i   Civetta sparviero delle Andamane;
Ulula delle Andamane

6553 *Ninox boobook*
Strigiformes - Strigidae
e   Southern Boobook; Boobook (ANZ);
Boobook Owl; Spotted Boobook;
Red Boobook; Dark Boobook
f   Ninoxe d'Australie
i   Civetta sparviero meridionale

6554 *Ninox connivens*
Strigiformes - Strigidae
e   Barking Owl; Barking Hawk-Owl;
Winking Owl
f   Ninoxe aboyeuse
d   Kläfferkauz
i   Civetta ululante; Ulula ammicante

6555 *Ninox connivens connivens*
Strigiformes - Strigidae
e   Southern Barking Owl (ANZ)

6556 *Ninox jacquinoti*
Strigiformes - Strigidae
e   Solomon Islands Boobook; Solomon

Hawk-Owl; Solomon Islands Hawk-
Owl
f   Ninoxe de Jacquinot
d   Salomonen-Kauz
i   Civetta sparviero di Jacquinot; Ulula
delle Salomone

6557   *Ninox meeki*
Strigiformes - Strigidae
e   Manus Boobook; Meek's Hawk-Owl;
Admiralty Islands Hawk-Owl
f   Ninoxe de l'Amiraute
d   Manus-Kauz
i   Ulula delle Isole dell'Ammiragliato

6558   *Ninox natalis*
Strigiformes - Strigidae
e   Christmas Boobook; Christmas
Island Hawk-Owl (ANZ)

6559   *Ninox novaeseelandiae*
Strigiformes - Strigidae
e   Morepork; Book-book Owl; Mopoke;
Morepork Owl; Boobook; Lord
Howe Boobook; Lord Howe Island
Boobook; Norfolk Boobook; Norfolk
Island Boobook
f   Ninoxe boubouk
d   Langflügelkauz; Kuckuckskauz
i   Civetta bubuk; Ulula della Nuova
Zelanda

6560   *Ninox novaeseelandiae albaria*
Strigiformes - Strigidae
e   Lord Howe Island Southern Boobook
(ANZ)

6561   *Ninox novaeseelandiae undulata*
Strigiformes - Strigidae
e   Norfolk Island Southern Boobook
(ANZ)

6562   *Ninox ochracea*
Strigiformes - Strigidae
e   Ochre-bellied Boobook; Ochre-
bellied Hawk-Owl
f   Ninoxe ocrée; Ninoxe perverse
d   Ockerbauchkauz
i   Civetta sparviero ocracea; Ulula dal
becco ocra

6563   *Ninox odiosa*
Strigiformes - Strigidae
e   Russet Boobook; New Britain Hawk-
Owl
f   Ninoxe odieuse
d   Bismarck-Kauz
i   Civetta sparviero della Nuova
Britannia

6564   *Ninox philippensis*
Strigiformes - Strigidae
e   Philippine Boobook; Philippine
Hawk-Owl; Philippine Boobook-Owl
f   Ninoxe des Philippines
d   Philippinen-Kauz
i   Civetta sparviero delle Filippine;
Ulula delle Filippine

6565   *Ninox punctulata*
Strigiformes - Strigidae
e   Speckled Boobook; Speckled Hawk
Owl
f   Ninoxe pointillée
d   Pünktchenkauz
i   Civetta sparviero punteggiata; Ulula
picchiettata

6566   *Ninox rudolfi*
Strigiformes - Strigidae
e   Sumba Boobook; Sumba Boobook-
Owl; Sumba Hawk-Owl; Boobook
Owl; Boobook
f   Ninoxe de Sumba
i   Civetta del principe Rodolfo

6567   *Ninox rufa*
Strigiformes - Strigidae
e   Rufous Owl; Rufous Hawk-Owl
f   Ninoxe rousse
d   Rostkauz
i   Civetta sparviero rossiccio; Ulula
rufa

6568   *Ninox rufa meesi*
Strigiformes - Strigidae
e   Cape York Peninsula Rufous Owl
(ANZ)

6569   *Ninox rufa queenslandica*
Strigiformes - Strigidae
e   Eastern Rufous Owl (ANZ)

6570 *Ninox scutulata*
Strigiformes - Strigidae
e Brown Hawk-Owl; Brown Boobook;
Oriental Hawk-Owl; Philippine
Hawk-Owl; Boobook; Boobook Owl
f Ninoxe hirsute
d Falkenkauz; Zugkauz
i Civetta sparviero bruno; Ulula bruno

6571 *Ninox squamipila*
Strigiformes - Strigidae
e Moluccan Boobook; Indonesian
Hawk-Owl; Moluccan Hawk-Owl
f Ninoxe des Moluques
d Molukken-Kauz
i Civetta sparviero delle Molucche;
Ulula delle Molucche

6572 *Ninox strenua*
Strigiformes - Strigidae
e Powerful Owl
f Ninoxe puissante
d Riesenkauz
i Civetta reale australiana; Gufo
potente

6573 *Ninox superciliaris*
Strigiformes - Strigidae
e White-browed Boobook; White-
browed Owl; Madagascar Hawk-
Owl; White-browed Hawk-Owl
f Ninoxe à sourcils blancs
d Madagaskar-Kauz
i Civetta sparviero del Madagascar;
Ulula del Madagascar

6574 *Ninox theomacha*
Strigiformes - Strigidae
e Jungle Boobook; Brown Owl; Sooty-
backed Owl; Sooty-backed Hawk-
Owl
f Ninoxe brune
d Einfarbkauz
i Civetta sparviero di Bonaparte; Ulula
dl dorsofuligginosa

6575 *Ninox variegata*
Strigiformes - Strigidae
e Bismarck Boobook; New Ireland
Hawk-Owl; Carteret's Hawk-Owl;
Bismarck Hawk-Owl

f Ninoxe bariolée
d Neuirland-Kauz
i Civetta sparviero delle Bismarck

6576 *Nipponia nippon*
Ciconiiformes - Threskiornithidae
e Crested Ibis; Ibis; Japanese Ibis;
Japanese Crested-Ibis; Oriental
Crested-Ibis
f Ibis nippon
d Nippon-Ibis
i Ibis giapponese

6577 *Nonnula amaurocephala*
Piciformes - Bucconidae
e Chestnut-headed Nunlet
f Barbacou à face rousse
d Rotkopffaulvogel
i Monachina testacastana

6578 *Nonnula brunnea*
Piciformes - Bucconidae
e Brown Nunlet
f Barbacou brun
d Einfarbfaulvogel; Brauner Faulvogel
i Monachina bruna

6579 *Nonnula frontalis*
Piciformes - Bucconidae
e Grey-cheeked Nunlet
f Barbacou a joues grises
i Monachina guancegrige

6580 *Nonnula rubecula*
Piciformes - Bucconidae
e Rusty-breasted Nunlet
f Barbacou rufalbin
d Rotkehlfaulvogel
i Monachina pettirossa

6581 *Nonnula ruficapilla*
Piciformes - Bucconidae
e Rufous-capped Nunlet; Grey-
cheeked Nunlet
f Barbacou à couronne rousse
d Rotscheitelfaulvogel;
Grauwangenfaulvogel
i Monachina corona ruggine

6582 *Nonnula sclateri*
Piciformes - Bucconidae

e   Fulvous-chinned Nunlet
f   Barbacou de Sclater
d   Gelbkinnfaulvogel; Sclaters Faulvogel
i   Monachina di Sclater

6583 *Northiella haematogaster*
       Psittaciformes - Psittacidae
e   Bluebonnet; Bulloak Parrot; Bluebonnet Parrot; Crimson-bellied Parrot; Yellow-vented Bluebonnet Parakeet
f   Perruche à bonnet bleu; Perruche à ventre rouge
d   Blutbauchsittich; Gelbsteißsittich
i   Parrochetto facciablu

6584 *Notharchus macrorhynchos*
       Piciformes - Bucconidae
e   White-necked Puffbird
f   Tamatia à gros bec
d   Weißhalsfaulvogel; Großer Fleckenfaulvogel
i   Bucco beccogrosso

6585 *Notharchus ordii*
       Piciformes - Bucconidae
e   Brown-banded Puffbird
f   Tamatia de Cassin
d   Braunbindenfaulvogel; Ordes Faulvogel
i   Bucco fasciato

6586 *Notharchus pectoralis*
       Piciformes - Bucconidae
e   Black-breasted Puffbird
f   Tamatia à plastron
d   Gürtelfaulvogel; Schwarzbrustfaulvogel
i   Bucco pettonero

6587 *Notharchus swainsoni*
       Piciformes - Bucconidae
e   Buff-bellied Puffbird

6588 *Notharchus tectus*
       Piciformes - Bucconidae
e   Pied Puffbird
f   Tamatia pie
d   Elsterfaulvogel; Bänderfaulvogel
i   Bucco bianco e nero

6589 *Nothocercus bonapartei*
       Tinamiformes - Tinamidae
e   Highland Tinamou; Bonaparte's Tinamou
f   Tinamou de Bonaparte
d   Bergtinamu
i   Tinamo di Bonaparte

6590 *Nothocercus julius*
       Tinamiformes - Tinamidae
e   Tawny-breasted Tinamou; Verreaux's Tinamou
f   Tinamou à tête rousse
d   Gelbbrusttinamu
i   Tinamo pettofulvo

6591 *Nothocercus nigrocapillus*
       Tinamiformes - Tinamidae
e   Hooded Tinamou
f   Tinamou à capuchon
d   Schwarzkappentinamu
i   Tinamo dal cappuccio

6592 *Nothocrax urumutum*
       Galliformes - Cracidae
e   Nocturnal Curassow; Rufous Curassow; Flat-chested Curassow; Flat-chested Urumutum
f   Hocco nocturne
d   Rothokko
i   Crace notturno; Hocco dal ciuffo

6593 *Nothoprocta cinerascens*
       Tinamiformes - Tinamidae
e   Brushland Tinamou
f   Tinamou sauvageon
d   Cordoba-Steißhuhn
i   Tinamo delle erbe

6594 *Nothoprocta curvirostris*
       Tinamiformes - Tinamidae
e   Curve-billed Tinamou
f   Tinamou curvirostre
d   Krumschnabelsteißhuhn
i   Tinamo beccocurvo

6595 *Nothoprocta kalinowskii*
       Tinamiformes - Tinamidae
e   Kalinowski's Tinamou
f   Tinamou de Kalinowski

d   Kalinowski-Steißhuhn
i   Tinamo di Kalinowski

6596   *Nothoprocta ornata*
Tinamiformes - Tinamidae
e   Ornate Tinamou
f   Tinamou orné
d   Pissaca
i   Tinamo ornato; Nottoprotta

6597   *Nothoprocta pentlandii*
Tinamiformes - Tinamidae
e   Andean Tinamou
f   Tinamou des Andes
d   Anden-Steißhuhn
i   Tinamo delle Ande

6598   *Nothoprocta perdicaria*
Tinamiformes - Tinamidae
e   Chilean Tinamou
f   Tinamou perdrix
d   Chile-Steißhuhn
i   Tinamo del Cile

6599   *Nothoprocta taczanowskii*
Tinamiformes - Tinamidae
e   Taczanowski's Tinamou
f   Tinamou de Taczanowski
d   Taczanowski-Steißhuhn
i   Tinamo di Taczanowski

6600   *Nothura boraquira*
Tinamiformes - Tinamidae
e   White-bellied Nothura; Marbled
Nothura; Yellow-legged Nothura
f   Tinamou boraquira
d   Weißbauchsteißhuhn
i   Cotorna pettobianco

6601   *Nothura chacoensis*
Tinamiformes - Tinamidae
e   Chaco Nothura
f   Tinamou du Chaco
d   Chaco-Steißhuhn
i   Cotorna del Ciaco

6602   *Nothura darwinii*
Tinamiformes - Tinamidae
e   Darwin's Nothura
f   Tinamou de Darwin

d   Darwin-Steißhuhn
i   Cotorna di Darwin

6603   *Nothura maculosa*
Tinamiformes - Tinamidae
e   Spotted Nothura
f   Tinamou tacheté
d   Fleckensteißhuhn
i   Cotorna macchiata

6604   *Nothura minor*
Tinamiformes - Tinamidae
e   Lesser Nothura; Least Nothura
f   Petit Tinamou
d   Wachtelsteißhuhn
i   Cotorna minore

6605   *Notiochelidon flavipes*
Passeriformes - Hirundinidae
e   Pale-footed Swallow; Coban
Swallow; Cloud Forest Swallow
f   Hirondelle de Chapman
d   Blaufußschwalbe
i   Rondine zampechiare

6606   *Notiochelidon murina*
Passeriformes - Hirundinidae
e   Brown-bellied Swallow
f   Hirondelle à ventre brun
d   Mausschwalbe
i   Rondine ventrebruno

6607   *Notiochelidon pileata*
Passeriformes - Hirundinidae
e   Black-capped Swallow; Cuban
Swallow
f   Hirondelle à tête noire
d   Kappenschwalbe
i   Rondine testanera

6608   *Notiomystis cincta*
Passeriformes - Meliphagidae
e   Stitchbird; Hihi
f   Méliphage hihi
d   Hihi; Gelbbandhonigesser
i   Hihi della Nuova Zelanda

6609   *Nucifraga caryocatactes*
Passeriformes - Corvidae
e   Spotted Nutcracker; Eurasian
Nutcracker; Nutcracker; Thick-billed

Nutcracker; European Nutcracker;
Indian Nutcracker; Larger Spotted-
Nutcracker
f   Cassenoix moucheté
d   Tannenhäher; Nußknacker
i   Nocciolaia; Nocciolaia eurasiatica

6610   *Nucifraga columbiana*
Passeriformes - Corvidae
e   Clark's Nutcracker; Clark's Crow;
Woodpecker Crow
f   Cassenoix d'Amérique
d   Amerikanischer Tannenhäher;
Kiefernhäher
i   Nocciolaia di Clark

6611   *Numenius americanus*
Charadriiformes - Scolopacidae
e   Long-billed Curlew
f   Courlis à long bec
d   Rostbrachvogel
i   Chiurlo americano

6612   *Numenius arquata*
Charadriiformes - Scolopacidae
e   Eurasian Curlew; Curlew; European
Curlew; Common Curlew; Western
Curlew
f   Courlis cendré
d   Großer Brachvogel
i   Chiurlo; Chiurlo maggiore; Fischione
maggiore

6613   *Numenius borealis*
Charadriiformes - Scolopacidae
e   Eskimo Curlew
f   Courlis esquimau
d   Eskimobrachvogel
i   Chiurlo boreale

6614   *Numenius hudsonicus*
Charadriiformes - Scolopacidae
e   Hudsonian Curlew

6615   *Numenius madagascariensis*
Charadriiformes - Scolopacidae
e   Far Eastern Curlew; Eastern Curlew;
Australian Curlew; Long-billed
Curlew; Red-rumped Curlew; Sea
Curlew
f   Courlis de Sibérie

d   Isabellbrachvogel
i   Chiurlo orientale

6616   *Numenius minutus*
Charadriiformes - Scolopacidae
e   Little Curlew; Little Whimbrel;
Pygmy Curlew; Siberian Bay Curlew
f   Courlis nain
d   Zwergbrachvogel
i   Chiurlo minore

6617   *Numenius phaeopus*
Charadriiformes - Scolopacidae
e   Whimbrel; Curlew (WI); Eurasian
Whimbrel; Little Curlew; Asiatic
Whimbrel; Hudsonian Curlew
f   Courlis corlieu; Courlis (Ants); Bec-
crochu (Ants)
d   Regenbrachvogel
i   Chiurlo piccolo; Chiurletto; Chirolo
piccolo di Hudson

6618   *Numenius tahitiensis*
Charadriiformes - Scolopacidae
e   Bristle-thighed Curlew; Otaheite
Curlew
f   Courlis de Tahiti; Courlis d'Alaska
d   Borstenbrachvogel
i   Chiurlo di Tahiti

6619   *Numenius tenuirostris*
Charadriiformes - Scolopacidae
e   Slender-billed Curlew
f   Courlis à bec grêle
d   Dünnschnabelbrachvogel
i   Chiurilottello

6620   *Numida galatea*
Galliformes - Numididae
e   West African Guineafowl

6621   *Numida meleagris*
Galliformes - Numididae
e   Helmeted Guineafowl; Guineafowl;
Grey-breasted Helmet Guineafowl;
Grey-throated Guineafowl;
Reicheow's Guineafowl; Tufted
Guineafowl
f   Pintade de Numidie; Pintade
sauvage; Pintade huppée; Pintade
vulgaire; Pintade; Pintade commune

d   Haubenperlhuhn; Helmperlhuhn
i   Gallina di Numidia; Faraona;
     Faraona mitrata; Faraona comune

6622   *Numida mitrata*
       Galliformes - Numididae
e   Tufted Guineafowl

6623   **Numididae**
       Galliformes
e   Guineafowl
f   Numididés
d   Perlhühner
i   Numididi

6624   *Nyctanassa violacea*
       Ciconiiformes - Ardeidae
e   Yellow-crowned Night-Heron; Quok
     (WI); Crabcracker (WI); Crabcatcher
     (WI)
f   Bihoreau violacé; Crabier gris
     (Ants); Coq de nuit (Ants); Crabier
     (Ants)
d   Krabbenreiher; Cayenne-Nachtreiher
i   Nitticora violacea

6625   *Nyctea scandiaca*
       Strigiformes - Strigidae
e   Snowy Owl; Snow Owl
f   Harfang des neiges; Harfang;
     Chouette harfang
d   Schneeeule
i   Gufo delle nevi; Civetta delle nevi

6626   **Nyctibiidae**
       Caprimulgiformes
e   Potoos
f   Nyctibiidés
d   Tagschläfer; Schwalke
i   Nittibidi

6627   *Nyctibius aethereus*
       Caprimulgiformes - Nyctibiidae
e   Long-tailed Potoo
f   Ibijau à longue queue
d   Langschwanzschwalk
i   Nittibio codalunga

6628   *Nyctibius bracteatus*
       Caprimulgiformes - Nyctibiidae
e   Rufous Potoo

f   Ibijau roux
d   Tropfenschwalk
i   Nittibio rossiccio

6629   *Nyctibius grandis*
       Caprimulgiformes - Nyctibiidae
e   Great Potoo; Grand Potoo
f   Grand Ibijau
d   Riesenschwalk
i   Nittibio maggiore

6630   *Nyctibius griseus*
       Caprimulgiformes - Nyctibiidae
e   Common Potoo; Potoo; Poor-me-
     one; Antillean Potoo
f   Ibijau gris
d   Uratau; Urutau
i   Nittibio comune; Nittibio

6631   *Nyctibius jamaicensis*
       Caprimulgiformes - Nyctibiidae
e   Northern Potoo; Patoo (WI)
f   Chat-huant (Ants); Ibijau jamaïcain
i   Nittibio della Giamaica

6632   *Nyctibius leucopterus*
       Caprimulgiformes - Nyctibiidae
e   White-winged Potoo
f   Ibijau à ailes blanches
d   Weißflügelschwalk
i   Nittibio alibianche

6633   *Nyctibius maculosus*
       Caprimulgiformes - Nyctibiidae
e   Andean Potoo; Potoo
f   Ibijau des Andes
i   Nittibio delle Ande

6634   *Nycticorax caledonicus*
       Ciconiiformes - Ardeidae
e   Rufous Night-Heron; Nankeen
     Night-Heron; Nankeen Heron
f   Bihoreau cannellé
d   Rotrückennachtreiher
i   Nitticora rossiccia

6635   *Nycticorax nycticorax*
       Ciconiiformes - Ardeidae
e   Black-crowned Night-Heron; Night-
     Heron; Quok (WI); Crabcracker
     (WI); Black-capped Night-Heron;

Common Night-Heron
f Bihoreau gris; Bihoreau à couronne
noire; Coq-d'eau (Ants); Coq de nuit
(Ants); Crabier de nuit (Ants);
Crabier bois (Ants); Crabier grosse-
tête (Ants); Héron bihoreau
d Nachtreiher
i Nitticora; Nonna col ciuffo;
Pavoncella di palude; Nitticora
comune

6636 *Nyctidromus albicollis*
Caprimulgiformes - Caprimulgidae
e Pauraque; Common Pauraque;
White-necked Nightjar; White-
necked Cuejo
f Engoulevent pauraqué
d Pauraque
i Succiacapre collobianco

6637 *Nyctiphrynus mcleodii*
Caprimulgiformes - Caprimulgidae
e Eared Poorwill
f Engoulevent aztèque
d Ohrennachtschwalbe
i Succiacapre di McLeod

6638 *Nyctiphrynus ocellatus*
Caprimulgiformes - Caprimulgidae
e Ocellated Poorwill
f Engoulevent ocellé
d Augennachtschwalbe
i Succiacapre ocellato

6639 *Nyctiphrynus rosenbergi*
Caprimulgiformes - Caprimulgidae
e Choco Poorwill

6640 *Nyctiphrynus yucatanicus*
Caprimulgiformes - Caprimulgidae
e Yucatan Poorwill
f Engoulevent du Yucatan
d Yucatan-Nachtschwalbe
i Succiacapre dello Yucatan

6641 *Nyctiprogne leucopyga*
Caprimulgiformes - Caprimulgidae
e Band-tailed Nighthawk
f Engoulevent leucopyge; Engoulevent
minute
d Bindenschwanznachtschwalbe

i Succiacapre sparviero codafasciata;
Succiacapre sparviere codafasciata

6642 *Nyctyornis amictus*
Coraciiformes - Meropidae
e Red-bearded Bee-eater; Red-breasted
Bee-eater
f Guêpier à fraise
d Rotbartspint
i Gruccione barbarossa; Gruccione a
barba rossa della Malesia

6643 *Nyctyornis athertoni*
Coraciiformes - Meropidae
e Blue-bearded Bee-eater
f Guêpier à barbe bleue
d Blaubartspint
i Gruccione barbablu; Gruccione di
Atherton

6644 *Nymphicus hollandicus*
Psittaciformes - Psittacidae
e Cockatiel; Cockatoo Parrot; Crested
Parrot; Weero; Quarrion
f Calopsitte élégante
d Nymphensittich
i Calopsitta

6645 *Nystalus chacuru*
Piciformes - Bucconidae
e White-eared Puffbird
f Tamatia chacuru
d Cerrado-Faulvogel;
Weißohrfaulvogel
i Bucco ciacuru

6646 *Nystalus maculatus*
Piciformes - Bucconidae
e Spot-backed Puffbird; Spotted
Puffbird
f Tamatia tamajac
d Fleckenfaulvogel
i Bucco maculato

6647 *Nystalus radiatus*
Piciformes - Bucconidae
e Barred Puffbird
f Tamatia barré
d Grünschnabelfaulvogel
i Bucco barrato

6648 *Nystalus striolatus*
Piciformes - Bucconidae
e Striolated Puffbird
f Tamatia striolé
d Strichelfaulvogel
i Bucco striato

O

6649   *Oceanites gracilis*
Procellariiformes - Hydrobatidae
e   White-vented Storm-Petrel; Eliot's
       Storm-Petrel; Graceful Storm-Petrel;
       Elliot's Petrel
f   Océanite d'Elliot; Pétrel d'Elliot
d   Elliot-Sturmschwalbe
i   Uccello delle tempeste di Elliot

6650   *Oceanites oceanicus*
Procellariiformes - Hydrobatidae
e   Wilson's Storm-Petrel; Mother
       Carey's Chicken (CSA); Mother
       Carey's Chick; Wilson's Petrel; Flat-
       clawed Petrel
f   Océanite de Wilson; Pétrel de
       Wilson; Pétrel océanite
d   Buntfußsturmschwalbe; Buntfüssige
       Sturmschwalbe
i   Uccello delle tempeste di Wilson

6651   *Oceanites oceanicus oceanicus*
Procellariiformes - Hydrobatidae
e   Subantarctic Wilson's Storm-Petrel
       (ANZ)

6652   *Oceanodroma castro*
Procellariiformes - Hydrobatidae
e   Band-rumped Storm-Petrel; Madeira
       Storm-Petrel; Harcourt's Storm-
       Petrel; Madeiran Storm-Petrel;
       Madeiran Fork-tailed Petrel;
       Madeiran Petrel
f   Océanite de Castro; Pétrel de Castro;
       Pétrel océanique
d   Madeira-Wellenläufer
i   Uccello delle tempeste di Castro

6653   *Oceanodroma furcata*
Procellariiformes - Hydrobatidae
e   Fork-tailed Storm-Petrel; Grey Fork-
       tailed Storm-Petrel; Grey Fork-tailed
       Petrel; Grey Storm-Petrel

f   Océanite à queue fourchue; Pétrel à
       queue fourchue
d   Gabelschwanzwellenläufer
i   Uccello delle tempeste codafurcata
       del Pacifico

6654   *Oceanodroma homochroa*
Procellariiformes - Hydrobatidae
e   Ashy Storm-Petrel; Ashy Petrel
f   Océanite cendré; Pétrel cendré
d   Einfarbwellenläufer; Aschgrauer
       Wellenläufer
i   Uccello delle tempeste cinereo

6655   *Oceanodroma hornbyi*
Procellariiformes - Hydrobatidae
e   Ringed Storm-Petrel; Hornby's
       Petrel; Hornby's Storm-Petrel
f   Océanite de Hornby
d   Anden-Wellenläufer
i   Uccello delle tempeste di Hornby

6656   *Oceanodroma leucorhoa*
Procellariiformes - Hydrobatidae
e   Leach's Storm-Petrel; Leach's Petrel;
       Mother Carey's Chicken (CSA);
       Mother Carey's Chick; Leach's
       Forktailed Storm-Petrel; Leach's
       Fork-tailed Petrel
f   Pétrel cul-blanc; Océanite cul-blanc
d   Wellenläufer
i   Uccello delle tempeste codaforcuta

6657   *Oceanodroma macrodactyla*
Procellariiformes - Hydrobatidae
e   Guadaloupe Storm-Petrel
f   Océanite de Guadeloupe; Pétrel de
       Guadeloupe
d   Guadalupe-Wellenläufer
i   Uccello delle tempeste di Guadalupa

6658   *Oceanodroma markhami*
Procellariiformes - Hydrobatidae
e   Markham's Storm-Petrel; Sooty
       Storm-Petrel; Markham's Petrel
f   Océanite de Markham; Pétrel de
       Markham
d   Rußwellenläufer
i   Uccello delle tempeste di Markham

6659 *Oceanodroma matsudairae*
Procellariiformes - Hydrobatidae
e Matsudaira's Storm-Petrel; Mother
Carey's Chicken (CSA); Mother
Carey's Chick; Matsudaira's Petrel;
Sooty Storm-Petrel
f Océanite de Matsudaira
d Matsudaira-Wellenläufer
i Uccello delle tempeste di Matsudaira

6660 *Oceanodroma melania*
Procellariiformes - Hydrobatidae
e Black Storm-Petrel; Black Fork-
tailed Petrel
f Océanite noir; Pétrel noir
d Schwarzwellenläufer
i Uccello delle tempeste nero

6661 *Oceanodroma microsoma*
Procellariiformes - Hydrobatidae
e Least Storm-Petrel; Least Petrel
f Océanite minute; Pétrel minute;
Océanite nain
d Sturmschwalbe
i Uccello delle tempeste minore

6662 *Oceanodroma monorhis*
Procellariiformes - Hydrobatidae
e Swinhoe's Storm-Petrel; Swinhoe's
Petrel
f Océanite de Swinhoe; Pétrel de
Swinhoe
d Swinhoe-Wellenläufer
i Uccello delle tempeste di Swinhoe

6663 *Oceanodroma tethys*
Procellariiformes - Hydrobatidae
e Wedge-rumped Storm-Petrel;
Galapagos Storm-Petrel; Galapagos
Petrel
f Océanite tethys; Pétrel tethys
d Galapagos-Wellenläufer
i Uccello delle tempeste delle
Galapagos

6664 *Oceanodroma tristrami*
Procellariiformes - Hydrobatidae
e Tristram's Storm-Petrel; Sooty
Storm-Petrel; Stejneger's Storm-
Petrel
f Océanite de Tristram; Pétrel de

Tristram
d Tristram-Wellenläufer
i Uccello delle tempeste di Tristram

6665 *Ochthoeca cinnamomeiventris*
Passeriformes - Tyrannidae
e Slaty-backed Chat-Tyrant
f Pitajo noir
d Graumanteltyrann
i Tiranno dorsoardesia

6666 *Ochthoeca frontalis albidiadema*
Passeriformes - Tyrannidae
e Crowned Chat-Tyrant

6667 *Ochthoeca fumicolor*
Passeriformes - Tyrannidae
e Brown-backed Chat-Tyrant
f Pitajo à dos brun
d Rauchtyrann
i Tiranno dorsobruno

6668 *Ochthoeca leucophrys*
Passeriformes - Tyrannidae
e White-browed Chat-Tyrant
f Pitajo à sourcils blancs
d Schluchttyrann
i Tiranno sopracciglibianchi

6669 *Ochthoeca oenanthoides*
Passeriformes - Tyrannidae
e D'Orbigny's Chat-Tyrant
f Pitajo d'Orbigny
d Klufttyrann; Schmätzertyrann
i Tiranno di D'Orbigny

6670 *Ochthoeca piurae*
Passeriformes - Tyrannidae
e Piura Chat-Tyrant
f Pitajo de Piura
d Piura-Tyrann
i Tiranno di Piura

6671 *Ochthoeca rufipectoralis*
Passeriformes - Tyrannidae
e Rufous-breasted Chat-Tyrant
f Pitajo à poitrine rousse
d Röteltyrann
i Tiranno pettorossiccio

6672   *Ochthoeca salvini*
Passeriformes - Tyrannidae
e   Tumbes Tyrant; Salvin's Tyrant
f   Pitajo de Tumbes; Moucherolle de
Tumbes
d   Gelbstirntyrann
i   Tiranno di Salvin

6673   *Ochthoeca spodionota*
Passeriformes - Tyrannidae
e   Peruvian Chat-Tyrant

6674   *Ochthoeca thoracica*
Passeriformes - Tyrannidae
e   Maroon-chested Chat-Tyrant

6675   *Ochthornis littoralis*
Passeriformes - Tyrannidae
e   Drab Water-Tyrant
f   Pitajo riverain; Moucherolle riveraine
d   Fahltyrann
i   Tiranno dei litorali

6676   *Ocreatus underwoodii*
Apodiformes - Trochilidae
e   Booted Racket-tail; Booted Racquet-
tail; Racquet-tailed Hummingbird
f   Haut-de-chausses à palettes
d   Flaggensylphe
i   Coda di racchetta di Underwood;
Coda a rachetta di Underwood

6677   *Oculocincta squamifrons*
Passeriformes - Zosteropidae
e   Pygmy White-eye; Pygmy Darkeye;
Pygmy Grey White-eye
f   Zostérops pygmée
d   Zwergbrillenvogel
i   Occhialino pigmeo

6678   *Ocyalus latirostris*
Passeriformes - Fringillidae
e   Band-tailed Oropendola
f   Cassique à queue frangée
d   Breitschnabelstirnvogel
i   Oropendola codafasciata

6679   *Ocyceros birostris*
Coraciiformes - Bucerotidae
e   Indian Grey-Hornbill; Common
Grey-Hornbill (ISC); Grey Hornbill

d   Keilschwanztoko
i   Bucero grigio indiano

6680   *Ocyceros gingalensis*
Coraciiformes - Bucerotidae
e   Sri Lanka Grey-Hornbill; Sri Lankan
Grey-Hornbill; Ceylon Grey-Hornbill
f   Calao de Gingi
i   Bucero grigio di Sri-Lanka

6681   *Ocyceros griseus*
Coraciiformes - Bucerotidae
e   Malabar Grey-Hornbill
f   Calao gris
d   Malabar-Toko
i   Bucero grigio del Malabar

6682   **Odontophoridae**
Galliformes
e   New World Quails
f   Odontophoridés
d   Wachteln
i   Odontoforidi

6683   *Odontophorus atrifrons*
Galliformes - Odontophoridae
e   Black-fronted Wood-Quail; Black-
fronted Quail
f   Tocro à front noir
d   Schwarzstirnwachtel
i   Colino frontenera

6684   *Odontophorus balliviani*
Galliformes - Odontophoridae
e   Stripe-faced Wood-Quail; Stripe-
faced Quail; Ballivan's Quail;
Ballivan's Wood-Quail
f   Tocro de Ballivian
d   Streifengesichtwachtel
i   Colino fasciastriata

6685   *Odontophorus capueira*
Galliformes - Odontophoridae
e   Spot-winged Wood-Quail; Capueira
Wood-Quail
f   Tocro uru
d   Capueirawachtel
i   Colino alimacchiate

6686   *Odontophorus columbianus*
Galliformes - Odontophoridae

e   Venezuelan Wood-Quail;
      Venezuelan Quail
f   Tocro du Venezuela
d   Venezuela-Wachtel
i   Colino del Venezuela

6687   *Odontophorus dialeucos*
      Galliformes - Odontophoridae
e   Tacarcuna Wood-Quail; Tacarcuna
      Quail; Black-crowned Wood-Quail
f   Tocro du Panama
d   Tacarcuna-Wachtel
i   Colino di Tacarcuana

6688   *Odontophorus erythrops*
      Galliformes - Odontophoridae
e   Rufous-fronted Wood-Quail;
      Paramba Quail; Rufous-breasted
      Wood-Quail
f   Tocro à front roux
d   Rotstirnwachtel
i   Colino fronterosso

6689   *Odontophorus gujanensis*
      Galliformes - Odontophoridae
e   Marbled Wood-Quail; Marbled Quail
f   Tocro de Guyane
d   Guyana-Wachtel
i   Colino marmorizzato

6690   *Odontophorus guttatus*
      Galliformes - Odontophoridae
e   Spotted Wood-Quail; Spotted Quail
f   Tocro tacheté
d   Tropfenwachtel
i   Colino macchiato

6691   *Odontophorus hyperythrus*
      Galliformes - Odontophoridae
e   Chestnut Wood-Quail; Chestnut-
      throated Quail; Chestnut-throated
      Wood-Quail
f   Tocro marron
d   Kastanienwachtel
i   Colino castano

6692   *Odontophorus leucolaemus*
      Galliformes - Odontophoridae
e   Black-breasted Wood-Quail; White-
      throated Wood-Quail; Black-breasted
      Quail

f   Tocro à poitrine noire
d   Weißkehlwachtel
i   Colino golabianca

6693   *Odontophorus melanonotus*
      Galliformes - Odontophoridae
e   Dark-backed Wood-Quail; Dark-
      backed Quail
f   Tocro à dos noir
d   Schwarzrückenwachtel
i   Colino dorsonero

6694   *Odontophorus melanotis*
      Galliformes - Odontophoridae
e   Black-eared Wood-Quail
f   Tocro à face noire
d   Rotstirnwachtel
i   Colino orecchienere

6695   *Odontophorus speciosus*
      Galliformes - Odontophoridae
e   Rufous-breasted Wood-Quail;
      Rufous-breasted Quail
f   Tocro à poitrine rousse
d   Rotbrustwachtel
i   Colino pettorosso

6696   *Odontophorus stellatus*
      Galliformes - Odontophoridae
e   Starred Wood-Quail; Starred Quail
f   Tocro étoilé
d   Sternwachtel
i   Colino stellato

6697   *Odontophorus strophium*
      Galliformes - Odontophoridae
e   Gorgeted Wood-Quail; Gorgeted
      Quail
f   Tocro à miroir
d   Kragenwachtel
i   Colino dal collare

6698   *Odontorchilus branickii*
      Passeriformes - Certhiidae
e   Grey-mantled Wren; Tooth-billed
      Wren
f   Troglodyte de Branicki
d   Graumantelzaunkönig
i   Scricciolo di Branick

6699   *Odontorchilus cinereus*
Passeriformes - Certhiidae
e   Tooth-billed Wren
f   Troglodyte denté
d   Zahnschnabelzaunkönig
i   Scricciolo beccodentato

6700   *Oedistoma iliolophum*
Passeriformes - Meliphagidae
e   Dwarf Honeyeater; Plumed Longbill;
Grey-bellied Longbill; Grey-bellied
Honeyeater; Dark-crested
Honeyeater; Long-plumed False-
Sunbird
f   Toxoramphe à ventre gris
d   Graubauchpfriemschnabel
i   Beccolungo nano

6701   *Oedistoma pygmaeum*
Passeriformes - Meliphagidae
a   Pygmy Longbill; Pygmy Honeyeater
f   Toxoramphe pygmée
d   Zwergpfriemschnabel
i   Beccolungo pigmeo

6702   *Oena capensis*
Columbiformes - Columbidae
e   Namaqua Dove; Masked Dove
f   Tourterelle masquée; Tourterelle à
masque de fer
d   Kap-Täubchen
i   Tortora maschero di ferro

6703   *Oenanthe alboniger*
Passeriformes - Muscicapidae
e   Hume's Wheatear; Hume's Chat;
Black-headed Wheatear
f   Traquet de Hume
d   Schwarzkopfsteinschmätzer
i   Monachella di Hume

6704   *Oenanthe bifasciata*
Passeriformes - Muscicapidae
e   Buff-streaked Wheatear; Buff-
streaked Chat (CSA)
d   Fahlschulterschmätzer
i   Sassicola fulva; Monachella fulva

6705   *Oenanthe bottae*
Passeriformes - Muscicapidae
e   Botta's Wheatear; Red-breasted

Wheatear; Red-breasted Chat
f   Traquet à poitrine rousse
d   Braunbrustschmätzer
i   Monachella pettorosso

6706   *Oenanthe cypriaca*
Passeriformes - Muscicapidae
e   Cyprus Wheatear; Cyprus Pied-
Wheatear; Pied Wheatear
f   Traquet de Chypre
d   Zypernsteinschmätzer
i   Monachella di Cipro

6707   *Oenanthe deserti*
Passeriformes - Muscicapidae
e   Desert Wheatear
f   Traquet du désert
d   Wüstensteinschmätzer
i   Monachella del deserto

6708   *Oenanthe finschii*
Passeriformes - Muscicapidae
e   Finsch's Wheatear; Barnes's
Wheatear; White-backed Wheatear;
Arabian Wheatear
f   Traquet de Finsch
d   Felsensteinschmätzer
i   Monachella di Finsch

6709   *Oenanthe heuglini*
Passeriformes - Muscicapidae
e   Heuglin's Wheatear
f   Traquet de Heuglin
i   Monachella di Heuglin

6710   *Oenanthe hispanica*
Passeriformes - Muscicapidae
e   Black-eared Wheatear; Spanish
Wheatear
f   Traquet oreillard; Traquet stapazin
d   Mittelmeer-Steinschmätzer
i   Monachella; Monachella comune

6711   *Oenanthe isabellina*
Passeriformes - Muscicapidae
e   Isabelline Wheatear; Isabelline Chat
(ISC)
f   Traquet isabelle
d   Isabellsteinschmätzer;
Isabellschmätzer

i Culbianco isabellino; Monachella isabellina

6712 Oenanthe leucopyga
Passeriformes - Muscicapidae
e White-tailed Wheatear; White-crowned Black-Wheatear; White-rumped Black-Chat; White-rumped Black-Wheatear; White-crowned Wheatear
f Traquet à tête blanche
d Sahara-Steinschmätzer
i Monachella nera codabianca; Monachella bella testabianca

6713 Oenanthe leucura
Passeriformes - Muscicapidae
e Black Wheatear; Common Black-Wheatear
f Traquet rieur; Traquet noir
d Trauersteinschmätzer; Trauerschmätzer
i Monachella nera

6714 Oenanthe lugens
Passeriformes - Muscicapidae
e Mourning Wheatear; White-underwinged Wheatear
f Traquet deuil
d Schwarzrückensteinschmätzer; Schwarzrückenschmätzer; Weißachselnonnensteinschmätzer
i Monachella lamentosa; Monachella di roccia

6715 Oenanthe lugentoides
Passeriformes - Muscicapidae
e Arabian Wheatear; South Arabian Wheatear
f Traquet d'Arabie
i Monachella del Sud Arabia; Monachella araba

6716 Oenanthe lugubris
Passeriformes - Muscicapidae
e Schalow's Wheatear; Abyssinian Black-Wheatear (CSA); Eastern Black-Wheatear; East African Wheatear
d Rüppell-Steinschmätzer
i Monachella di Schalow

6717 Oenanthe moesta
Passeriformes - Muscicapidae
e Red-rumped Wheatear; Grey-headed Wheatear; Tristram's Wheatear
f Traquet à tête grise
d Fahlbürzelsteinschmätzer; Fahlbürzelschmätzer
i Monachella testagrigia; Monachella mesta

6718 Oenanthe monacha
Passeriformes - Muscicapidae
e Hooded Wheatear; White-rumped Wheatear; Hooded Chat
f Traquet à capuchon
d Kappensteinschmätzer; Kappenschmätzer; Haubensteinschmätzer
i Monachella dal cappuccio; Monachella pettonero

6719 Oenanthe monticola
Passeriformes - Muscicapidae
e Mountain Wheatear; Mountain Chat (CSA)
f Traquet montagnard
d Bergschmätzer
i Monachella variabile di montagna

6720 Oenanthe oenanthe
Passeriformes - Muscicapidae
e Northern Wheatear; Wheatear; European Wheatear (CSA); Common Wheatear; Eurasian Whetear
f Traquet motteux; Traquet traîne-charrue
d Steinschmätzer
i Culbianco; Culbianco comune

6721 Oenanthe phillipsi
Passeriformes - Muscicapidae
e Somali Wheatear; Phillips's Wheatear
f Traquet de Somalie
d Somali-Schmätzer
i Monachella somala

6722 Oenanthe picata
Passeriformes - Muscicapidae
e Variable Wheatear; Eastern Pied-Wheatear; Pied Wheatear; Pied Chat

f   Traquet variable; Traquet pie
      d'Orient
d   Elsternsteinschmätzer;
      Elsterschmätzer
i   Monachella variabile

6723   *Oenanthe pileata*
      Passeriformes - Muscicapidae
e   Capped Wheatear
f   Traquet du Cap
d   Erdschmätzer;
      Brustschildsteinschmätzer
i   Monachella testanera

6724   *Oenanthe pleschanka*
      Passeriformes - Muscicapidae
e   Pied Wheatear; Pleschanka's
      Wheatear; Common Pied-Wheatear;
      Pleschanka's Pied-Chat
f   Traquet pie; Traquet leucomèle
d   Nonnensteinschmätzer;
      Nonnenschmätzer
i   Monachella dorsonero

6725   *Oenanthe seebohmi*
      Passeriformes - Muscicapidae
e   Black-throated Wheatear
i   Culbianco africano

6726   *Oenanthe xanthoprymna*
      Passeriformes - Muscicapidae
e   Rufous-tailed Wheatear; Red-tailed
      Wheatear; Red-rumped Wheatear;
      Red-tailed Chat
f   Traquet à queue rousse
d   Rostbürzelsteinschmätzer;
      Rotbürzelsteinschmätzer;
      Rostbürzelschmätzer;
      Rotschwanzsteinschmätzer
i   Monachella codarossa; Monachella
      codirossa

6727   *Ognorhynchus icterotis*
      Psittaciformes - Psittacidae
e   Yellow-eared Parrot; Yellow-eared
      Parakeet; Yellow-eared Conure;
      Yellow-eared Macaw
f   Conure à joues d'or
d   Gelbohrsittich
i   Conuro orecchiegialle

6728   *Oncostoma cinereigulare*
      Passeriformes - Tyrannidae
e   Northern Bentbill; Bentbill
f   Tyranneau à bec courbé
d   Aschkehlkrummschnabel
i   Beccoricurvo del Nord

6729   *Oncostoma olivaceum*
      Passeriformes - Tyrannidae
e   Southern Bentbill; Bentbill
f   Tyranneau de Lawrence
d   Gelbkehlkrummschnabel
i   Beccoricurvo del Sud

6730   *Onychognathus albirostris*
      Passeriformes - Sturnidae
e   White-billed Starling
f   Rufipenne à bec blanc; Étourneau à
      bec blanc
d   Weißschnabelstar
i   Storno beccobianco

6731   *Onychognathus blythii*
      Passeriformes - Sturnidae
e   Somali Starling; Somali Chestnut-
      winged Starling
f   Rufipenne de Blyth; Étourneau de
      Blyth
d   Somali-Star
i   Storno di Blyth

6732   *Onychognathus frater*
      Passeriformes - Sturnidae
e   Socotra Starling; Socotra Chestnut-
      winged Starling
f   Rufipenne de Socotra; Étourneau de
      Socotra
d   Socotra-Star
i   Storno di Socotra

6733   *Onychognathus fulgidus*
      Passeriformes - Sturnidae
e   Chestnut-winged Starling; Forest
      Chestnut-winged Starling; Common
      Chestnut-winged Starling
f   Rufipenne de forêt; Étourneau
      rufipenne d'Alexander; Étourneau
      roupenne
d   Kastanienflügelstar
i   Storno fulgido

6734 *Onychognathus morio*
Passeriformes - Sturnidae
e Red-winged Starling; African Red-winged Starling; Crag Chestnut-winged Starling
f Rufipenne morio; Étourneau morio
d Rotschwingenstar
i Storno alicastane

6735 *Onychognathus nabouroup*
Passeriformes - Sturnidae
e Pale-winged Starling
f Rufipenne nabouroup; Étourneau à ailes pâles
d Bergstar; Fahlflügelstar
i Storno alichiare

6736 *Onychognathus salvadorii*
Passeriformes - Sturnidae
e Bristle-crowned Starling
f Rufipenne de Salvadori; Étourneau de Salvadori
d Helmstar
i Storno di Salvadori

6737 *Onychognathus tenuirostris*
Passeriformes - Sturnidae
e Slender-billed Starling; Chestnut-winged Starling; Thin-billed Starling; Slender-billed Chestnut-winged Starling; Slender-billed Red-winged Starling; Red-winged Starling
f Rufipenne à bec fin; Étournea à bec fin
d Star; Zimtflügelstar
i Storno beccofino

6738 *Onychognathus tristramii*
Passeriformes - Sturnidae
e Tristram's Starling; Tristram's Grackle; Arabian Chestnut-winged Starling
f Rufipenne de Tristram; Étourneau de Tristram
d Tristram-Star
i Storno di Tristram; Gracula di Tristram

6739 *Onychognathus walleri*
Passeriformes - Sturnidae
e Waller's Starling; Waller's Chestnut-winged Starling; Waller's Red-winged Starling; Mountain Red-winged Starling
f Rufipenne de Waller; Étourneau à bec court
d Waller-Star
i Storno di Waller

6740 *Onychorhynchus coronatus*
Passeriformes - Tyrannidae
e Royal Flycatcher; Amazonian Royal-Flycatcher; Crowned Flycatcher; Pacific Royal-Flycatcher
f Moucherolle couronnée
d Kronentyrann; Königstyrann
i Tiranno reale

6741 *Onychorhynchus mexicanus*
Passeriformes - Tyrannidae
e Northern Royal-Flycatcher
f Moucherolle royale
d Mexiko-Königstyrann; Mexikanischer Königstyrann

6742 *Onychorhynchus occidentalis*
Passeriformes - Tyrannidae
e Pacific Royal-Flycatcher; Western Royal-Flycatcher

6743 *Onychorhynchus swainsoni*
Passeriformes - Tyrannidae
e Atlantic Royal-Flycatcher; Swainson's Royal-Flycatcher
f Moucherolle de Swaison

6744 Ophisthocomidae
Gruiformes
e Hoatzins; Hoactzins
f Ophistocomidés; Hoazins
d Hoatzine

6745 *Ophisthocomus hoazin*
Gruiformes - Ophistocomidae
e Hoatzin; Hoactzin
f Hoazin huppé
d Zigeunerhuhn; Hoatzin
i Hoazin

6746 *Ophrysia superciliosa*
Galliformes - Phasianidae
e Himalayan Quail; Mountain Quail

(ISC); Indian Mountain-Quail;
Himalayan Mountain Quail
f   Ophrysie de I.Himalaya
d   Hangwachtel
i   Quaglia dell'Himalaya

6747   *Opisthoprora euryptera*
Apodiformes - Trochilidae
e   Mountain Avocetbill; Loddige's
Thornbill; Avocetbill
f   Colibri avocettin
d   Degenschnabelkolibri
i   Colibrì di monte becco di avocetta

6748   *Oporornis agilis*
Passeriformes - Parulidae
e   Connecticut Warbler
f   Paruline à gorge grise; Fauvette à
gorge grise
d   Augenringwaldsänger
i   Parula del Connecticut

6749   *Oporornis formosus*
Passeriformes - Parulidae
e   Kentucky Warbler
f   Paruline du Kentucky; Fauvette de
Kentucky
d   Kentucky-Waldsänger
i   Parula del Kentucky

6750   *Oporornis philadelphia*
Passeriformes - Parulidae
e   Mourning Warbler
f   Geothlypis; Fauvette triste; Paruline
triste
d   Graukopfwaldsänger
i   Parula lamentosa

6751   *Oporornis tolmiei*
Passeriformes - Parulidae
e   MacGillivray's Warbler
f   Paruline des buissons; Fauvette des
buissons
d   Dickichtwaldsänger
i   Parula di MacGillivray

6752   *Orchesticus abeillei*
Passeriformes - Fringillidae
e   Brown Tanager
f   Tangara brun

d   Fuchstangare
i   Tangara bruna

6753   *Oreocharis arfaki*
Passeriformes - Paramythidae
e   Tit Berrypecker; Arkak
Flowerpecker; Arfak Berrypecker;
Painted Berrypecker; New Guinea
Berrypecker
f   Oréochare des Arfak
d   Gelbbauchbeerenpicker; Trugmeise;
Arfak-Gelbwangenvogel
i   Beccafrutti cincia

6754   *Oreoica gutturalis*
Passeriformes - Corvidae
e   Crested Bellbird
f   Carillonneur huppé
d   Haubengudilang
i   Campanero crestato

6755   *Oreoica gutturalis gutturalis*
Passeriformes - Corvidae
e   Southern Crested Bellbird (ANZ)

6756   *Oreomanes fraseri*
Passeriformes - Fringillidae
e   Giant Conebill
f   Conirostre géant
d   Riesenspitzschnabel
i   Becco di storno di Fraser

6757   *Oreomystis bairdi*
Passeriformes - Fringillidae
e   Kauai Creeper; Akikiki Creeper;
Akikiki; Baird's Creeper
f   Grimpeur de Kauai
i   Rampichino di Kauai

6758   *Oreomystis mana*
Passeriformes - Fringillidae
e   Hawaii Creeper
f   Grimpeur d'Hawai
i   Rampichino delle Hawaii

6759   *Oreonympha nobilis*
Apodiformes - Trochilidae
e   Bearded Mountaineer
f   Colibri noble
d   Bergnymphe
i   Ninfa dei monti

6760 **Oreophasis derbianus**
Galliformes - Cracidae
e Horned Guan; Derby's Guan; Lord
Derby's Mountain-Pheasant
f Oréophase cornu
d Zapfenguan; Bergguan
i Crace di Derby

6761 **Oreopholus ruficollis**
Charadriiformes - Charadriidae
e Tawny-throated Dotterel; Slender-
billed Plover
f Pluvier oréophile
d Klippenregenpfeifer
i Piviere tortolino collorosso

6762 **Oreopsar bolivianus**
Passeriformes - Fringillidae
e Bolivian Blackbird
f Quiscale de Bolivie
d Anden-Stärling
i Gracchio della Bolivia

6763 **Oreopsittacus arfaki**
Psittaciformes - Loriidae
e Plum-faced Lorikeet; Whiskered
Lorikeet; Whiskered Arfak-Lorikeet;
Arfak Alpine-Lorikeet; Blue-cheeked
Lorikeet; Mountain Lory
f Lori bridé; Loriquet des Monts
Arfak; Lorilet à joues blanches
d Arfak-Lori
i Lorichetto dell'Arfak

6764 **Oreornis chrysogenys**
Passeriformes - Meliphagidae
e Orange-cheeked Honeyeater; Small
Mountain-Honeyeater
f Méliphage des Snow
d Goldwangenhonigfresser
i Mangiamiele guancearancio

6765 **Oreortyx pictus**
Galliformes - Odontophoridae
e Mountain Quail; Plumed Quail
f Colin des montagnes
d Bindenwachtel; Bergwachtel
i Colino plumifero; Quaglia di
montagna

6766 **Oreoscoptes montanus**
Passeriformes - Sturnidae
e Sage Thrasher
f Moqueur des armoises
d Salbeisichelspötter; Bergspottdrossel
i Mimo della salvia

6767 **Oreoscopus gutturalis**
Passeriformes - Pardalotidae
e Fernwren; Fernbird; Australian
Fernwren
f Séricorne des fougères
d Farnhuscher

6768 **Oreostruthus fuliginosus**
Passeriformes - Passeridae
e Mountain Firetail; Crimson-sided
Mountain-Finch; Red-sided
Mountain-Finch; Crimson-sided
Weaver Finch
f Diamant des montagnes
d Bergamadine
i Codadifuoca di monte

6769 **Oreothraupis arremonops**
Passeriformes - Fringillidae
e Tanager Finch
f Oréotangara élégant
d Tangarenbuschammer
i Fringuello tangre

6770 **Oreotrochilus adela**
Apodiformes - Trochilidae
e Wedge-tailed Hillstar; Adela's
Hillstar
f Colibri adèle
d Adela-Kolibri
i Stella dei monti codacuneata

6771 **Oreotrochilus chimborazo**
Apodiformes - Trochilidae
e Ecuadorian Hillstar; Chimborazo
Hillstar; Violet-hooded Hillstar
f Colibri du Chimborazo
i Stella dei monti dell'Ecuador

6772 **Oreotrochilus estella**
Apodiformes - Trochilidae
e Andean Hillstar; Estella's Hillstar
f Colibri estelle
d Anden-Kolibri

i   Stella dei monti andina; Colibrì delle
      Ande

6773   Oreotrochilus leucopleurus
      Apodiformes - Trochilidae
e   White-sided Hillstar
f   Colibri à flancs blancs
d   Weißflankenkolibri
i   Stella dei monti fianchibianchi

6774   Oreotrochilus melanogaster
      Apodiformes - Trochilidae
e   Black-breasted Hillstar
f   Colibri à plastron noir
d   Samtkolibri
i   Stella dei monti pettonero

6775   Origma solitaria
      Passeriformes - Pardalotidae
e   Origma; Rock-Warbler
f   Origma des rochers
d   Steinhuscher
i   Origma

6776   Oriolia bernieri
      Passeriformes - Corvidae
e   Bernier's Vanga
f   Oriolie de Bernier
d   Schwarzvanga; Stahlvanga
i   Vanga di Bernier

6777   Oriolidae
      Passeriformes
e   Old World Orioles; Golden Orioles
f   Loriots; Oriolidés
d   Pirole
i   Oriolidi; Rigogoli

6778   Oriolus albiloris
      Passeriformes - Oriolidae
e   White-lored Oriole
f   Loriot à face blanche
i   Rigogolo dalle redini

6779   Oriolus auratus
      Passeriformes - Oriolidae
e   African Golden-Oriole; African
      Oriole
f   Loriot doré
d   Schwarzohrpirol
i   Rigogolo giallo africano

6780   Oriolus bouroensis
      Passeriformes - Oriolidae
e   Black-eared Oriole; Buru Oriole;
      Tanimbar Oriole; Black-fced Oriole
f   Loriot de Buru
d   Buru-Pirol
i   Rigogolo di Buru

6781   Oriolus brachyrhynchus
      Passeriformes - Oriolidae
e   Western Black-headed Oriole;
      Greenish-backed Oriole
f   Loriot à tête noire; Loriot à tête noire
      occidental
d   Blauflügclpirol
i   Rigogolo testanera occidentale

6782   Oriolus chinensis
      Passeriformes - Oriolidae
e   Black-naped Oriole
f   Loriot de Chine
d   Schwarznackenpirol
i   Rigogolo giallo orientale

6783   Oriolus chlorocephalus
      Passeriformes - Oriolidae
e   Green-headed Oriole
f   Loriot à tête verte
d   Grünkopfpirol
i   Rigogolo testaverde

6784   Oriolus crassirostris
      Passeriformes - Oriolidae
e   Sao Tomé Oriole; Great-billed Sao
      Tomé Oriole
f   Loriot de Sao Tomé
d   Fahlbrustpirol
i   Rigogolo di Sao Tomè

6785   Oriolus cruentus
      Passeriformes - Oriolidae
e   Black-and-crimson Oriole; Crimson-
      breasted Oriole
f   Loriot ensanglanté
d   Rotbrustpirol
i   Rigogolo rosso e nero

6786   Oriolus flavocinctus
      Passeriformes - Oriolidae
e   Yellow Oriole; Green Oriole;
      Australian Yellow Oriole; Yellow-

bellied Oriole
f Loriot verdâtre
d Mangrovepirol
i Rigogolo giallo australiano

6787 *Oriolus forsteni*
Passeriformes - Oriolidae
e Grey-collared Oriole; Forsten's
Oriole; Ceram Oriole
f Loriot de Ceram
d Forsten-Pirol
i Rigogolo di Ceram

6788 *Oriolus hosii*
Passeriformes - Oriolidae
e Black Oriole
f Loriot noir
d Mohrenpirol
i Rigogolo nero

6789 *Oriolus isabellae*
Passeriformes - Oriolidae
e Isabella Oriole; Green-lored Oriole;
Olive-lored Oriole
f Loriot d'Isabella
d Isabella-Pirol
i Rigogolo di Isabella

6790 *Oriolus larvatus*
Passeriformes - Oriolidae
e African Black-headed Oriole; Black-
headed Oriole (CSA); Eastern Black-
headed Oriole
f Loriot masqué; Loriot à tête noire
oriental
d Maskenpirol
i Rigogolo testanera africano

6791 *Oriolus melanotis*
Passeriformes - Oriolidae
e Olive-brown Oriole; Dark Oriole;
Timor Oriole; Sunda Oriole
f Loriot de Timor
i Rigogolo orecchienere

6792 *Oriolus mellianus*
Passeriformes - Oriolidae
e Silver Oriole; Mell's Maroon Oriole;
Mell's Oriole; Stresemann's Maroon
Oriole
f Loriot argenté

d Seidenpirol
i Rigogolo argentato

6793 *Oriolus monacha*
Passeriformes - Oriolidae
e Dark-headed Oriole; Black-headed
Forest Oriole; Ethiopian Oriole;
Abyssinian Black-headed Oriole;
Forest Oriole
f Loriot moine; Loriot d'Abyssinie
d Mönchspirol
i Rigogolo testascura

6794 *Oriolus nigripennis*
Passeriformes - Oriolidae
e Black-winged Oriole
f Loriot à ailes noires
d Gabun-Pirol
i Rigogolo alinere

6795 *Oriolus oriolus*
Passeriformes - Oriolidae
e Eurasian Golden-Oriole; Golden
Oriole; European Golden-Oriole
(CSA); Indian Golden-Oriole
f Loriot d'Europe
d Pirol
i Rigogolo eurasiatico; Golo;
Rigogolo; Oriolo

6796 *Oriolus percivali*
Passeriformes - Oriolidae
e Black-tailed Oriole; Montane Oriole
(CSA); Percival's Oriole
f Loriot de Percival
d Bergpirol
i Rigogolo di Percival

6797 *Oriolus phaeochromus*
Passeriformes - Oriolidae
e Dusky-brown Oriole; Ruddy Oriole;
Halmahera Oriole; Dusky Oriole;
Moluccan Oriole; Gray's Oriole
f Loriot d'Halmahera
d Halmahera-Pirol
i Rigogolo delle Molucche

6798 *Oriolus sagittatus*
Passeriformes - Oriolidae
e Olive-backed Oriole; White-bellied
Oriole; Green Oriole

f Loriot sagittal
d Streifenpirol
i Rigogolo dorso oliva

6799 *Oriolus steerii*
Passeriformes - Oriolidae
e Philippine Oriole; Grey-throated
Oriole
f Loriot des Philippines
i Rigogolo delle Filippine

6800 *Oriolus szalayi*
Passeriformes - Oriolidae
e Brown Oriole; Striated Oriole; New
Guinea Oriole
f Loriot papou
d Grant-Pirol
i Rigogolo bruno

6801 *Oriolus tenuirostris*
Passeriformes - Oriolidae
e Slender-billed Oriole
f Loriot à bec effilé
i Rigogolo beccofine

6802 *Oriolus traillii*
Passeriformes - Oriolidae
e Maroon Oriole
f Loriot pourpre
d Blutpirol; Indischer Blutpirol
i Rigogolo castano

6803 *Oriolus xanthonotus*
Passeriformes - Oriolidae
e Dark-throated Oriole; Black-headed
Oriole; Malaysian Oriole; Black-
throated Oriole
f Loriot à gorge noire
d Weißbauchpirol; Gelbmantelpirol
i Rigogolo golascura

6804 *Oriolus xanthornus*
Passeriformes - Oriolidae
e Black-hooded Oriole; Black-headed
Oriole (ISC); Oriental Black-headed
Oriole; Asian Black-headed Oriole;
Indian Black-headed Oriole; Ceylon
Black-headed Oriole
f Loriot à capuchon noir; Loriot à tête
noire
d Schwarzkopfpirol

i Rigogolo testanera asiatico; Rigogola
testascura

6805 *Oriturus superciliosus*
Passeriformes - Fringillidae
e Striped Sparrow
f Bruant rayéé
d Streifenammer
i Passero striato

6806 *Ornithion brunneicapillum*
Passeriformes - Tyrannidae
e Brown-capped Tyrannulet
f Tyranneau à tête brune
i Tiranno piccolo capobruno

6807 *Ornithion inerme*
Passeriformes - Tyrannidae
e White-lored Tyrannulet; Hartlaub's
Tyrannulet
f Tyranneau minute
d Weißflügelfliegenstecher
i Tiranno piccolo inerme

6808 *Ornithion semiflavum*
Passeriformes - Tyrannidae
e Yellow-bellied Tyrannulet
f Tyranneau à ventre jaune
d Gelbbauchfliegenstecher
i Tiranno piccolo pettogiallo

6809 *Oroaetus isidori*
Falconiformes - Accipitridae
e Black-and-chestnut Eagle; Isidore's
Eagle
f Aigle d'Isidore
d Isidor-Adler; Glanzhaubenadler
i Spizaeto bianco e nero

6810 *Ortalis canicollis*
Galliformes - Cracidae
e Chaco Chachalaca; Grey-headed
Chachalaca
f Ortalide du Chaco
d Chaco-Guan
i Ciacialaca del Chaco

6811 *Ortalis cinereiceps*
Galliformes - Cracidae
e Grey-headed Chachalaca; Chestnut-
winged Chachalaca

f Ortalide à tête grise
d Graukopfguan
i Ciacialaca testagrigia

6812 *Ortalis erythroptera*
Galliformes - Cracidae
e Rufous-headed Chachalaca;
Ecuadorian Chachalaca
f Ortalide à tête rousse
d Rotkopfguan
i Ciacialaca testarossa

6813 *Ortalis garrula*
Galliformes - Cracidae
e Chestnut-winged Chachalaca
f Ortalide babillarde
d Rotflügelguan
i Ciacialaca alicastane

6814 *Ortalis guttata*
Galliformes - Cracidae
e Speckled Chachalaca; Spotted
Chachalaca
f Ortalide maillée
i Ciacialaca macchiettato

6815 *Ortalis leucogastra*
Galliformes - Cracidae
e White-bellied Chachalaca
f Ortalide à ventre blanc
d Weißbauchguan
i Ciacialaca ventrebianco

6816 *Ortalis motmot*
Galliformes - Cracidae
e Little Chachalaca; Variable
Chachalaca; Guiana Chachalaca;
Rufous-headed Chachalaca
f Ortalide motmot
d Paraka
i Ciacialaca minore

6817 *Ortalis poliocephala*
Galliformes - Cracidae
e West Mexican Chachalaca
f Ortalide de Wagler
d Graubrustguan
i Ciacialaca occidentale

6818 *Ortalis ruficauda*
Galliformes - Cracidae

e Rufous-vented Chachalaca; Rufous-
tailed Chachalaca; Rufous-tipped
Chachalaca; Red-tailed Chachalaca
f Ortalide à ventre roux
d Rotschwanzguan
i Ciacialaca codarossa

6819 *Ortalis superciliaris*
Galliformes - Cracidae
e Buff-browed Chachalaca;
Superciliated Chachalaca
f Ortalide à sourcils
i Ciacialaca di Gray

6820 *Ortalis vetula*
Galliformes - Cracidae
e Plain Chachalaca; Chachalaca;
Common Chachalaca; Mexican
Chachalaca; Eastern Chachalaca
f Ortalide chacamel
d Braunflügelguan
i Ciacialaca comune; Ciacialaca

6821 *Ortalis wagleri*
Galliformes - Cracidae
e Rufous-bellied Chachalaca; Waglers
Chachalaca
f Ortalide à ventre marron
i Ciacialaca di Wagler

6822 *Orthogonys chloricterus*
Passeriformes - Fringillidae
e Olive-green Tanager
f Tangara viréon
d Olivtangare
i Tangara verde-oliva

6823 Orthonychidae
Passeriformes
e Logrunners; Chowchillas (ANZ)
f Orthonychidés
d Laufflöter; Erdtimalien

6824 *Orthonyx spaldingii*
Passeriformes - Orthonychidae
e Chowchilla; Northern Chowchilla;
Northern Logrunner; Spalding's
Logrunner; Black-headed Logrunner
f Orthonyx de Spalding
d Schwarzkopfflöter
i Corridore dei tronchi di Spalding

6825   *Orthonyx temminckii*
Passeriformes - Orthonychidae
e   Logrunner; Spine-tailed Chowchilla;
Southern Spinetailed Chowchila;
Southern Logrunner; Brown
Logrunner; Spine-tailed Logrunner
f   Orthonyx de Temminck
d   Stachelschwanzflöter
i   corridore dei tronchi di Temminck

6826   *Orthorhyncus cristatus*
Apodiformes - Trochilidae
e   Antillean Crested-Hummingbird;
Doctor-Brushie (WI); Doctor-Bird
(WI); Little Doctor-Bird (WI);
Coulibri (WI); Crested Hummingbird
f   Colibri huppé; Fou-fou (Ants)
d   Haubenkolibri
i   Colibrì crestato delle Antille

6827   *Orthotomus atrogularis*
Passeriformes - Sylviidae
e   Dark-necked Tailorbird; Black-
necked Tailorbird; Dark-cheeked
Tailorbird
f   Couturière à col noir; Fauvette
couturière à col noir
d   Strichelschneidervogel
i   Uccello sarto collonero

6828   *Orthotomus castaneiceps*
Passeriformes - Sylviidae
e   Philippine Tailorbird
f   Couturière à calotte rousse; Fauvette
couturière à calotte rousse
i   Uccello sarto delle Filippine

6829   *Orthotomus cinereiceps*
Passeriformes - Sylviidae
e   White-eared Tailorbird
f   Couturière oreillarde; Fauvette
couturière oreillarde
d   Graukopfschneidervogel
i   Uccello sarto testagrigia

6830   *Orthotomus cuculatus*
Passeriformes - Sylviidae
e   Mountain Tailorbird; Golden-headed
Tailorbird; Golden-crowned
Tailorbird
f   Couturière montagnarde; Fauvette

couturière montagnarde; Couturière
de montagne
d   Bergschneidervogel
i   Uccello sarto di montagna

6831   *Orthotomus derbianus*
Passeriformes - Sylviidae
e   Grey-backed Tailorbird; Luzon
Tailorbid
f   Couturière de Luçon; Fauvette
couturière de Luçon
d   Luzon-Schneidervogel
i   Uccello sarto di Luzon

6832   *Orthotomus frontalis*
Passeriformes - Sylviidae
e   Rufous-fronted Tailorbird
f   Couturière à front roux; Fauvette
couturière à front roux
i   Uccello sarto fronterossa

6833   *Orthotomus heterolaemus*
Passeriformes - Sylviidae
e   Rufous-headed Tailorbird
f   Couturière à capuchon; Fauvette
couturière à capuchon
i   Uccello sarto testarossa

6834   *Orthotomus metopias*
Passeriformes - Sylviidae
e   African Tailorbird; Red-capped
Forest-Warbler (CSA)
f   Couturière d'Afrique; Fauvette
couturière d'Afrique; Couturière des
Usambaras
d   Rotkappensänger
i   Uccello sarto capirosso

6835   *Orthotomus moreaui*
Passeriformes - Sylviidae
e   Long-billed Tailorbird; Long-billed
Apalis; Long-billed Forest-Warbler;
Moreau's Tailorbird (CSA)
f   Couturière de Moreau; Fauvette
couturière de Moreau; Apalis à long
bec
d   Langschnabelsänger
i   Uccello sarto di Moreau

6836   *Orthotomus nigriceps*
Passeriformes - Sylviidae

   e   Black-headed Tailorbird; White-browed Tailorbird; White-faced Tailorbird

   f   Couturière à tête noire; Fauvette couturière à tête noire

   d   Schwarzkopfschneidervogel

   i   Uccello sarto testanera

6837 ***Orthotomus ruficeps***
     Passeriformes - Sylviidae

   e   Ashy Tailorbird; Grey Tailorbird; Red-headed Tailorbird; Rufous-crowned Tailorbird

   f   Couturière à tête rousse; Fauvette couturière à tête rousse

   i   Uccello sarto cenerino

6838 ***Orthotomus samarensis***
     Passeriformes - Sylviidae

   e   Yellow-breasted Tailorbird; Samar Tailorbird

   f   Couturière de Samar; Fauvette couturière de Samar

   i   Uccello sarto di Samar

6839 ***Orthotomus sepium***
     Passeriformes - Sylviidae

   e   Olive-backed Tailorbird; Ashy Tailorbird; Red-backed Tailorbird; Bali Tailorbird; Javan Tailorbird

   f   Couturière à dos vert; Fauvette couturière à dos vert

   d   Rostwangenschneidervogel

   i   Uccello sarto dorso oliva

6840 ***Orthotomus sericeus***
     Passeriformes - Sylviidae

   e   Rufous-tailed Tailorbird; Red-headed Tailorbird; Rufous-headed Tailorbird; Silky Rufous-tailed Tailorbird; Red-tailed Tailorbird; Rufous-crowned Tailorbird; Rufous-backed Tailorbird

   f   Couturière à queue rousse; Fauvette couturière à queue rousse

   d   Rotschwanzschneidervogel

   i   Uccello sarto sericeo

6841 ***Orthotomus sutorius***
     Passeriformes - Sylviidae

   e   Common Tailorbird; Tailorbird;

     Long-tailed Tailorbird; Ceylon Tailorbird

   f   Couturière à longue queue; Couturière; Fauvette couturière

   d   Rotstirnschneidervogel

   i   Uccello sarto codalunga; Uccello sarto dalla codalunga

6842 ***Ortygospiza atricollis***
     Passeriformes - Estrildidae

   e   African Quailfinch; Quailfinch (CSA); Common Quailfinch; Ground Finch; Partridge Finch

   f   Astrild-caille à lunettes

   d   Wachtelastrild; Rebhuhnastrild

   i   Astro quaglia africano

6843 ***Ortygospiza gabonensis***
     Passeriformes - Estrildidae

   e   Red-billed Quailfinch; Black-chinned Quailfinch; Gabon Quailfinch; Black-faced Quailfinch

   f   Astrild-caille à gorge noire

   d   Schwarzkinnwachtelastrild

   i   Astro quaglia del Gabon

6844 ***Ortygospiza locustella***
     Passeriformes - Estrildidae

   e   Locust Finch; Marsh Finch; Red Quailfinch; Common Locust-Finch; Locust Quailfinch; Red Marsh-Finch

   f   Astrild-caille à gorge rouge; Astrild locustelle

   d   Heuschreckenastrild; Rotflügelwachtelastrild

   i   Astro quaglia locustella

6845 ***Ortyxelos meiffrenii***
     Gruiformes - Turnicidae

   e   Lark Button-Quail; Quail-Plover; Lark-Quail

   f   Turnix à ailes blanches

   d   Lerchenlaufhühnchen

6846 ***Oryzoborus angolensis***
     Passeriformes - Emberizidae

   e   Lesser Seed-Finch; Chestnut-bellied Seed-Finch

   f   Sporophile curio

   d   Schwarzkopfreisknacker

   I   Beccogrosso minore

6847 *Oryzoborus atrirostris*
Passeriformes - Emberizidae
e   Black-billed Seed-Finch; Peruvian
Seed-Finch
f   Sporophile à bec noir
I   Beccogrosso becconero

6848 *Oryzoborus crassirostris*
Passeriformes - Emberizidae
e   Large-billed Seed-Finch; Great Seed-
Finch
f   Sporophile crassirostre
d   Mohrenreisknacker
I   Beccogrosso nero del Sudamerica

6849 *Oryzoborus funereus*
Passeriformes - Emberizidae
e   Thick-billed Seed-Finch
f   Sporophile à bec fort

6850 *Oryzoborus maximiliani*
Passeriformes - Emberizidae
e   Great-billed Seed-Finch; Greater
Large-billed Seed-Finch; Greater
Seed-Finch
f   Sporophile de Maximilian
d   Dickschnabelreisknacker
I   Beccogrosso nero di Wied

6851 *Oryzoborus nuttingi*
Passeriformes - Emberizidae
e   Nicaraguan Seed-Finch; Pink-billed
Seed-Finch
f   Sporophile de Nutting
I   Beccogrosso del Nicaragua

6852 **Otididae**
Gruiformes
e   Bustards
f   Outardes; Otididés
d   Trappen
i   Otididi; Otarde

6853 *Otidiphaps nobilis*
Columbiformes - Columbidae
e   Pheasant-Pigeon; Green-naped
Pheasant-Pigeon; Bronze-naped
Pheasant-Pigeon; Green-collared
Pigeon; Bronze-collared Pigeon
f   Otidiphaps noble

d   Fasantaube
i   Colomba fagiano

6854 *Otis tarda*
Gruiformes - Otididae
e   Great Bustard
f   Grande Outarde; Outarde barbue
d   Großtrappe; Trappe
i   Otarda; Otarda maggiore

6855 *Otus albogularis*
Strigiformes - Strigidae
e   White-throated Screech-Owl
f   Petit-duc à gorge blanche
d   Weißkehleule
i   Assiolo golabianca; Assiolo dal petto
bianco

6856 *Otus alfredi*
Strigiformes - Strigidae
e   Flores Scops-Owl; Everett's Owl
f   Petit-duc de Florès
d   Everett-Eule
i   Assiolo di Flores

6857 *Otus angelinae*
Strigiformes - Strigidae
e   Javan Scops-Owl; Angeline's Scops-
Owl
f   Petit-duc de Java
d   Angelina-Eule
i   Assiolo di Giava

6858 *Otus asio*
Strigiformes - Strigidae
e   Eastern Screech-Owl; Screech-Owl;
Common Screech-Owl
f   Petit-duc maculé
d   Kreisch-Eule; Schreieule
i   Assiolo americano orientale; Assiolo
urlatore

6859 *Otus atricapillus*
Strigiformes - Strigidae
e   Black-capped Screech-Owl; Variable
Screech-Owl
f   Petit-duc à mèches noires; Petit-duc
chaperonné
d   Kappeneule
i   Assiolo testanera; Assiolo dalla
ciuffa

6860 **Otus bakkamoena**
Strigiformes - Strigidae
e Indian Scops-Owl; Collared Scops-Owl (ISC)
f Petit-duc à collier
d Halsbandeule
i Assiolo indiano; Assiolo dal collare

6861 **Otus balli**
Strigiformes - Strigidae
e Andaman Scops-Owl
f Petit-duc des Andaman
d Andamanen-Eule
i Assiolo delle Andamane

6862 **Otus barbarus**
Strigiformes - Strigidae
e Bearded Screech-Owl; Santa Barbara Screech-Owl; Bridled Screech-Owl
f Petit-duc bridé
d Tropfeneule
i Assiolo dalle redini; Assiolo barbuto

6863 **Otus beccarii**
Strigiformes - Strigidae
e Beccari's Scops-Owl; Biak Island Scops-Owl
f Petit-duc de Beccari
d Beccari-Eule

6864 **Otus brookii**
Strigiformes - Strigidae
e Rajah Scops-Owl; Brooke's Scops-Owl
f Petit-duc radjah
d Radjaheule
i Assiolo del Rajah; Assiolo ragià

6865 **Otus brucei**
Strigiformes - Strigidae
e Pallid Scops-Owl; Striated Scops-Owl; Bruce's Scops-Owl
f Petit Duc de Bruce
d Streifenohreule; Wüstenzwergohreule
i Assiolo di Bruce

6866 **Otus capnodes**
Strigiformes - Strigidae
e Anjouan Scops-Owl
f Petit-duc d'Anjouan

6867 **Otus choliba**
Strigiformes - Strigidae
e Tropical Screech-Owl; Choliba Scops-Owl
f Petit-duc choliba
d Choliba-Eule
i Assiolo dei tropici

6868 **Otus clarkii**
Strigiformes - Strigidae
e Bare-shanked Screech-Owl; Bare-legged Screech-Owl
f Petit-duc de Clark; Petit-duc nudipède
d Nacktbeineule
i Assiolo di Clark

6869 **Otus cooperi**
Strigiformes - Strigidae
e Pacific Screech-Owl; Cooper's Screech-Owl
f Petit-duc de Cooper
d Mangroveeule
i Assiolo di Cooper; Assiolo del Pacifico

6870 **Otus elegans**
Strigiformes - Strigidae
e Elegant Scops-Owl; Ryukyu Scops-Owl
f Petit-duc élégant
d Schmuckeule
i Assiolo di Ryukyu

6871 **Otus enganensis**
Strigiformes - Strigidae
e Enggano Scops-Owl
f Petit-duc d'Enggano
i Assiolo di Enggano

6872 **Otus flammeolus**
Strigiformes - Strigidae
e Flammulated Owl; Flammulated Screech-Owl; Screech-Owl
f Petit-duc nain
d Ponderosa-Eule
i Assiolo flammulato; Assiolo color di fiamma

6873 **Otus fuliginosus**
Strigiformes - Strigidae

e   Palawan Scops-Owl
f   Petit-duc de Palawan
i   Assiolo di Palawan

6874   *Otus guatemalae*
Strigiformes - Strigidae
e   Middle American Screech-Owl;
Vermiculated Screech-Owl; Long-
tufted Screech-Owl
f   Petit-duc guatémaltèque; Petit-duc
vermiculé
d   Rotgesichteule
i   Assiolo di Guatemala

6875   *Otus gurneyi*
Strigiformes - Strigidae
e   Giant Scops-Owl; Mindanao Eagle-
Owl; Lesser Eagle-Owl
f   Petit-duc de Gurney
d   Rotohreule
i   Assiolo gigante di Gurney; Assiolo
gigante

6876   *Otus hartlaubi*
Strigiformes - Strigidae
e   Sao Tomé Scops-Owl
f   Petit-duc de Sao Tomé
d   Hartlaub-Eule
i   Assiolo di Sao Tomè

6877   *Otus hoyi*
Strigiformes - Strigidae
e   Hoy's Screech-Owl
f   Petit-duc de Hoy

6878   *Otus huberi*
Strigiformes - Strigidae
e   Cloud-forest Screech-Owl

6879   *Otus icterorhynchus*
Strigiformes - Strigidae
e   Sandy Scops-Owl; Cinnamon Scops-
Owl
f   Petit-duc à bec jaune
d   Gelbschnabeleule
i   Assiolo beccogiallo; Assiolo
cinnamomo

6880   *Otus ingens*
Strigiformes - Strigidae
e   Rufescent Screech-Owl

f   Petit-duc de Salvin
d   Salvin-Eule
i   Assiolo rossiccio americano

6881   *Otus ingens colombianus*
Strigiformes - Strigidae
f   Petit-duc de Colombie

6882   *Otus insularis*
Strigiformes - Strigidae
e   Seychelles Scops-Owl; Bare-legged
Scops-Owl
f   Petit-duc scieur
d   Seychellen-Eule
u   Assiolo delle Seychelle

6883   *Otus ireneae*
Strigiformes - Strigidae
e   Sokoke Scops-Owl; Mrs.. Morden's
Owlet; Morden's Scops-Owl
f   Petit-duc d'Irène
d   Sokoke-Eule
i   Assiolo di Sokoke

6884   *Otus kennicottii*
Strigiformes - Strigidae
e   Western Screech-Owl; Kennicott's
Screech-Owl
f   Petit-duc des montagnes
d   Mangroveeule
i   Assiolo di Kennicott

6885   *Otus koepckeae*
Strigiformes - Strigidae
e   Koepcke's Screech-Owl; Maria
Koepke's Screech-Owl
f   Petit-duc de Koepcke
i   Assiolo di Maria Koepcke

6886   *Otus lambi*
Strigiformes - Strigidae
e   Oaxaca Screech-Owl

6887   *Otus lawrencii*
Strigiformes - Strigidae
e   Bare-legged Owl; Cuban Screech-
Owl; Bare-legged Screech-Owl
f   Petit-duc de Cuba
d   Kuba-Eule
i   Assiolo di Cuba

6888   *Otus lempiji*
Strigiformes - Strigidae
e   Sunda Scops-Owl
f   Petit-duc de Horsfield
i   Assiolo dal collare

6889   *Otus leucotis*
Strigiformes - Strigidae
e   White-faced Scops-Owl; White-faced
Owl (CSA); White-faced Screech-
Owl
f   Petit-duc à face blanche
d   Weißgesichtohreule;
Weißgesichteule; Büscheleule;
Schwarzbüscheleule
i   Assiolo facciabianca

6890   *Otus longicornis*
Strigiformes - Strigidae
e   Luzon Scops-Owl
f   Petit-duc longicorne
i   Assiolo di Luzon

6891   *Otus madagascariensis*
Strigiformes - Strigidae
e   Torotoroka Scops-Owl

6892   *Otus magicus*
Strigiformes - Strigidae
e   Moluccan Scops-Owl; Mysterious
Scops-Owl; Seychelles Scops-Owl;
Insular Scops-Owl; Engano Scops-
Owl; Flores Scops-Owl; Papuan
Scops-Owl
f   Petit-duc mystérieux; Hibou petit-duc
scieur
e   Everett-Eule; Beccari-Eule;
Seychellen-Eule
i   Assiolo delle Molucche

6893   *Otus manadensis*
Strigiformes - Strigidae
e   Sulawesi Scops-Owl; Celebes Scops-
Owl
f   Petit-duc de Manado
d   Manado-Eule
i   Assiolo di Sulawesi; Assiolo di
Celebes

6894   *Otus mantananensis*
Strigiformes - Strigidae

e   Mantanani Scops-Owl; South
Philippines Scops-Owl
f   Petit-duc de Mantanani
i   Assiolo di Mantanani

6895   *Otus marshalli*
Strigiformes - Strigidae
e   Cloud-forest Screech-Owl;
Cinnamon Screech-Owl
f   Petit-duc de Marshall; Petit-duc de
Peterson
i   Assiolo di Marshall; Assiolo di
Peterson

6896   *Otus mayottensis*
Strigiformes - Strigidae
e   Malagasy Scops Owl

6897   *Otus megalotis*
Strigiformes - Strigidae
e   Philippine Scops-Owl; Whitehead's
Scops-Owl
f   Petit-duc de Luçon
i   Assiolo delle Filippine

6898   *Otus mentawi*
Strigiformes - Strigidae
e   Mentawai Scops-Owl; Mentawi
Scops-Owl; Sipora Scops-Owl
f   Petit-duc des Mentawei
i   Assiolo di Sipora

6899   *Otus mindorensis*
Strigiformes - Strigidae
e   Mindoro Scops-Owl
f   Petit-duc de Mindoro
i   Assiolo di Mindoro

6900   *Otus mirus*
Strigiformes - Strigidae
e   Mindanao Scops-Owl; Bornean
Scops-Owl
f   Petit-duc de Mindanao
i   Assiolo di Mindanao

6901   *Otus nudipes*
Strigiformes - Strigidae
e   Puerto Rican Screech-Owl; Puerto
Rican Bare-legged Owl; Cuckoobird
(WI)
f   Petit-duc de Porto Rico

d Nacktfußeule
i Assiolo di Portorico

6902 *Otus pauliani*
Strigiformes - Strigidae
e Comoro Scops-Owl; Grand Comoro
Scops-Owl; Kathala Scops-Owl
f Petit-duc du Karthala

6903 *Otus pembaensis*
Strigiformes - Strigidae
e Pemba Scops-Owl
f Petit-duc de Pemba

6904 *Otus podarginus*
Strigiformes - Strigidae
e Palau Owl; Palau Scops-Owl
f Petit-duc des Palau
d Palau-Eule
i Assiolo di Palau

6905 *Otus roboratus*
Strigiformes - Strigidae
e West Peruvian Screech-Owl;
Roborate Screech-Owl; Peruvian
Scops-Owl; Coastal Screech-Owl
f Petit-duc du Pérou
d Buscheule
i Assiolo del Perù; Assiolo dalla
corona bruna

6906 *Otus rufescens*
Strigiformes - Strigidae
e Reddish Scops-Owl; Rufous Scops-
Owl; Rufescent Scops-Owl
f Petit-duc roussâtre
d Röteleule
i Assiolo rossiccio asiatico; Assiolo
rossiccio

6907 *Otus rutilus*
Strigiformes - Strigidae
e Malagasy Scops-Owl; Madagascar
Scops-Owl
f Petit-duc malgache
d Inseleule
i Assiolo del Madagascar; Assiolo
color ruggine

6908 *Otus sagittatus*
Strigiformes - Strigidae

e White-fronted Scops-Owl; Malayan
Scops-Owl
f Petit-duc à front blanc
d Weißstirneule
i Assiolo frontebianca; Assiolo dalla
fronte bianca

6909 *Otus sanctaecatarinae*
Strigiformes - Strigidae
e Long-tufted Screech-Owl
f Petit-duc à aigrettes longues

6910 *Otus scops*
Strigiformes - Strigidae
e Scops-Owl; Common Scops-Owl;
Eurasian Scops-Owl
f Petit-duc scops; Hibou petit-duc
d Zwergohreule; Afrikanische
Zwergohreule
i Assiolo; Assiuolo; Chiù; Usciolo;
Tassolo; Assiolo eurasiatico

6911 *Otus seductus*
Strigiformes - Strigidae
e Balsas Screech-Owl
f Petit-duc du Balsas
i Assiolo di Balsas

6912 *Otus semitorques*
Strigiformes - Strigidae
e Japanese Scops-Owl

6913 *Otus senegalensis*
Strigiformes - Strigidae
e African Scops-Owl; Senegal Scops-
Owl
f Petit-duc africain
d Zwergohreule
i Assiolo africano

6914 *Otus silvicola*
Strigiformes - Strigidae
e Wallace's Scops-Owl; Lesser Sunda
Scops-Owl; Flores Scops-Owl;
Sunda Scops-Owl
f Petit-duc de Wallace
d Wallace-Eule
i Assiolo di Wallace; Assiolo minore
della Sonda

6915   ***Otus spilocephalus***
        Strigiformes - Strigidae
   e   Mountain Scops-Owl; Spotted Scops-
        Owl; Spotted-Mountain Scops-Owl;
        Stresemann's Scop's Owl;
        Vanderwater's Scops-Owl
   f   Petit-duc tacheté
   d   Fuchseule
   i   Assiolo macchiato; Assiolo di
        montgna

6916   ***Otus spilocephalus stresemanni***
        Strigiformes - Strigidae
   f   Petit-duc de Stresemann

6917   ***Otus sunia***
        Strigiformes - Strigidae
   e   Oriental Scops-Owl; Little Scops-
        Owl; Eastern Scops-Owl; Indian
        Scops-Owl
   f   Petit-duc d'Orient
   d   Streifenohreule
   i   Assiolo orientale

6918   ***Otus trichopsis***
        Strigiformes - Strigidae
   e   Whiskered Screech-Owl; Whiskered
        Owl; Spotted-Screech-Owl
   f   Petit-duc à moustaches; Petit-duc à
        favoris
   d   Fleckeneule
   i   Assiolo dai mustacchi; Assiolo
        baffuto

6919   ***Otus umbra***
        Strigiformes - Strigidae
   e   Simeulue Scops-Owl; Mentaur
        Screech-Owl; Simalur Scops-Owl;
        Mentaur Scops-Owl
   f   Petit-duc de Simalur
   i   Assiolo di Simalur

6920   ***Otus usta***
        Strigiformes - Strigidae
   e   Austral Screech-Owl; Southern
        Screech-Owl
   f   Petit-duc austral
   i   Assiolo australe

6921   ***Otus vermiculatus***
        Strigiformes - Strigidae

   e   Vermiculated Screech-Owl
   f   Petit-duc vermiculé
   i   Assiolo vermicolato

6922   ***Otus vinaceus***
        Strigiformes - Strigidae
   e   Vinaceous Screech-Owl

6923   ***Otus watsonii***
        Strigiformes - Strigidae
   e   Tawny-bellied Screech-Owl
   f   Petit-duc de Watson
   d   Watson-Eule
   i   Assiolo di Watson; Assiolo dal becco
        bruno

6924   ***Oxylabes madagascariensis***
        Passeriformes - Sylviidae
   e   White-throated Oxylabes; White-
        throated Foditany; Foditany
   f   Oxylabe à gorge blanche
   d   Weißkehlfoditany
   i   Garrulo di Madagascar

6925   ***Oxylophus jacobinus***
        Passeriformes - Cuculidae
   e   Jacobin Cuckoo; Pied Crested-
        Cuckoo; Pied Cuckoo; Black-and-
        white Cuckoo
   f   Coucou jacobin
   d   Jakobinerkuckuck; Elsterkuckuck
   i   Cuculo bianco e nero

6926   ***Oxypogon guerinii***
        Apodiformes - Trochilidae
   e   Bearded Helmetcrest; Guerin's
        Helmetcrest
   f   Colibri casqué
   d   Helmkolibri
   i   Colibrì dall'elmo di Guerin

6927   ***Oxyruncus cristatus***
        Passeriformes - Cotingidae
   e   Sharpbill
   f   Oxyrhynque huppé; Oxyrhynque en
        feu
   d   Feuerkopf; Flammenkopf
   i   Testadifiamma

6928   ***Oxyura australis***
        Anseriformes - Anatidae

e   Blue-billed Duck; Australian Blue-
     billed Duck; Bluebill; Stiff-tailed
     Duck; Spring-tailed Duck
f   Érismature australe
d   Australische Ruderente;
     Schwarzkinnruderente
i   Gobbo australiano

6929 *Oxyura dominica*
      Anseriformes - Anatidae
e   Masked Duck; White-winged Lake-
     Duck; Squat Duck (WI); Duck-and-
     Teal (WI)
f   Érismature routoutou; Canard
     masqué (Ants); Canard routoutou
     (Ants); Canard zombie (Ants)
d   Maskenente
i   Gobbo mascherato

6930 *Oxyura ferruginea*
      Anseriformes - Anatidae
e   Andean Duck; Andean Ruddy Duck;
     Peruvian Duck
f   Érismature des Andes
i   Gobbo delle Ande

6931 *Oxyura jamaicensis*
      Anseriformes - Anatidae
e   Ruddy Duck; Blue-bill (WI); Rubber
     Duck (WI); Red Diver (WI); Diving
     Teal (WI)
f   Érismature rousse; Canard roux;
     Canard plongeur (Ants); Canard
     plongeon (Ants)
d   Schwarzkopfruderente
i   Gobbo rugginoso americano; Gobbo
     della Giamaica; Gobbo americano;
     Gobbo rugginoso giamaicano

6932 *Oxyura leucocephala*
      Anseriformes - Anatidae
e   White-headed Duck; Stifftail; White-
     faced Duck; White-headed Stifftail
     Duck; White-headed Stiff-tailed
     Duck
f   Érismature à tête blanche
d   Weißkopfruderente; Ruderente
i   Gobbo rugginoso

6933 *Oxyura maccoa*
      Anseriformes - Anatidae

e   Maccoa Duck
f   Érismature maccoa
d   Maccoa-Ente; Afrika-Ruderente
i   Gobbo maccoa

6934 *Oxyura vittata*
      Anseriformes - Anatidae
e   Lake Duck; Argentine Ruddy Duck;
     Argentine Lake Duck; Argentine
     Bluebill Duck; Argentine Blue-billed
     Duck
f   Érismature ornée
d   Bindenruderente; Argentinische
     Ruderente
i   Gobbo dell'Argentina

P

6935 **Pachycare flavogrisea**
Passeriformes - Corvidae
e Goldenface; Dwarf Whistler;
Golden-faced Whistler; Golden-faced
Pachycare
f Pachycare nain
d Goldstirndickkopf
i Zufolatore giallo e grigio

6936 **Pachycephala albiventris**
Passeriformes - Corvidae
e Green-backed Whistler
f Siffleur à dos vert
i Zufolatore dorsoverde

6937 **Pachycephala arctitorquis**
Passeriformes - Corvidae
e Wallacean Whistler
f Siffleur de Wallace
i Zufolatore di Wallace

6938 **Pachycephala aurea**
Passeriformes - Corvidae
e Golden-backed Whistler; Yellow-
backed Whistler
f Siffleur à cape jaune
d Gelbrückendickkopf
i Zufolatore dorsogiallo

6939 **Pachycephala caledonica**
Passeriformes - Corvidae
e New Caledonian Whistler; Gmelin's
Whistler
f Siffleur calédonien
d Ockerbauchdickkopf
i Zufolatore della Nuova Caledonia

6940 **Pachycephala flavifrons**
Passeriformes - Corvidae
e Samoan Whistler; Yellow-fronted
Whistler
f Siffleur des Samoa

d Diademdickkopf
i Zufolatore frontegialla

6941 **Pachycephala griseiceps**
Passeriformes - Corvidae
e Grey-headed Whistler; Grey Whistler
f Siffleur à tête grise
d Papua-Dickkopf
i Zufolatore testagrigia

6942 **Pachycephala griseonota**
Passeriformes - Corvidae
e Drab Whistler
f Siffleur terne
d Schnäpperdickkopf
i Zufolatore disadorno

6943 **Pachycephala grisola**
Passeriformes - Corvidae
e Mangrove Whistler; Grey Thickhead;
White-bellied Whistler
f Siffleur cendré
i Zufolatore delle mangrovie

6944 **Pachycephala homeyeri**
Passeriformes - Corvidae
e White-vented Whistler; White-
bellied Whistler
f Siffleur de Blasius
i Zufolatore di Homeyer

6945 **Pachycephala hyperythra**
Passeriformes - Corvidae
e Rusty Whistler; Rufous-breasted
Whistler; Brownish Whistler
f Siffleur rouilleux
d Rostbauchdickkopf
i Zufolatore pettirosso

6946 **Pachycephala hypoxantha**
Passeriformes - Corvidae
e Bornean Whistler; Bornean
Mountain-Whistler
f Siffleur de Bornéo
d Borneo-Dickkopf
i Zufolatore del Borneo

6947 **Pachycephala implicata**
Passeriformes - Corvidae
e Hooded Whistler; Mountain
Whistler; Solomons Mountain-

Whistler; Solomons Whistler; Black-
headed Whistler
f Siffleur des Salomon
d Olivbauchdickkopf
i Zufolatore dei monti

6948   *Pachycephala inornata*
Passeriformes - Corvidae
e Gilbert's Whistler; Black-lored
Whistler
f Siffleur de Gilbert
d Schwarzflügeldickkopf
i Zufolatore di Gilbert

6949   *Pachycephala jacquinoti*
Passeriformes - Corvidae
e Tongan Whistler
f Siffleur des Tonga
i Zufolatore di Tonga

6950   *Pachycephala lanioides*
Passeriformes - Corvidae
e White-breasted Whistler; White-
bellied Whistler
f Siffleur à bavette blanche
d Weißbrustdickckopf
i Zufolatore pettobianco

6951   *Pachycephala leucogastra*
Passeriformes - Corvidae
e White-bellied Whistler
f Siffleur à ventre blanc
i Zufolatore ventrebianco

6952   *Pachycephala lorentzi*
Passeriformes - Corvidae
e Lorentz's Whistler
f Siffleur de Lorentz
d Lorentz-Dickkopf
i Zufolatore di Lorentz

6953   *Pachycephala melanura*
Passeriformes - Corvidae
e Black-tailed Whistler; Golden
Mangrove-Whistler; Mangrove
Golden-Whistler
f Siffleur à queue noire
d Mangrovedickkopf
i Zufolatore codanera

6954   *Pachycephala meyeri*
Passeriformes - Corvidae
e Vogelkop Whistler; Meyer's
Whistler; Grey-crowned Whistler
f Siffleur du Vogelkop
d Braunohrdickkopf
i Zufolatore di Meyer

6955   *Pachycephala modesta*
Passeriformes - Corvidae
e Brown-backed Whistler
f Siffleur modeste
d Braunrückendickkopf
i Zufolatore dorsobruno

6956   *Pachycephala monacha*
Passeriformes - Corvidae
e Black-headed Whistler; Aru Whistler
f Siffleur moine
i Zufolatore di Aru

6957   *Pachycephala nudigula*
Passeriformes - Corvidae
e Bare-throated Whistler; Lesser
Sunda-Whistler; Sunda Whistler
f Siffleur à gorge nue
d Nacktkehldickkopf
i Zufolatore golanuda

6958   *Pachycephala olivacea*
Passeriformes - Corvidae
e Olive Whistler; Olive-flanked
Whistler; Olivaceous Whistler
f Siffleur olivâtre
d Buchendickkopf
i Zufolatore olivaceo

6959   *Pachycephala olivacea hesperus*
Passeriformes - Corvidae
e Glenelg Olive Whistler (ANZ)

6960   *Pachycephala orpheus*
Passeriformes - Corvidae
e Fawn-breasted Whistler; Timor
Whistler; Sunda Whistler
f Siffleur orphée
d Orpheus-Dickopf
i Zufolatore pettocastano

6961   *Pachycephala pectoralis*
Passeriformes - Corvidae

e   Golden Whistler; Common Golden-
    Whistler
f   Siffleur doré
d   Goldbauchschnäpper;
    Großraumdickopfschnäpper
i   Zufolatore dorato

6962   *Pachycephala pectoralis contempta*
    Passeriformes - Corvidae
e   Lord Howe Island Golden Whistler
    (ANZ)

**6963   *Pachycephala pectoralis
    xanthoprocta***
    Passeriformes - Corvidae
e   Norfolk Island Golden Whistler

6964   *Pachycephala phaionotus*
    Passeriformes - Corvidae
e   Island Whistler
f   Siffleur des Moluques
d   Küstendickkopf
i   Zufolatore isolano

6965   *Pachycephala philippinensis*
    Passeriformes - Corvidae
e   Yellow-bellied Whistler
f   Siffleur des Philippines
d   Philippinen-Dickkopf
i   Zufolatore ventregiallo

6966   *Pachycephala rufiventris*
    Passeriformes - Corvidae
e   Rufous Whistler
f   Siffleur itchong
d   Schlichtmanteldickkopf
i   Zufolatore ventrerossiccio

6967   *Pachycephala rufogularis*
    Passeriformes - Corvidae
e   Red-lored Whistler; Red-throated
    Whistler; Buff-breasted Whistler
f   Siffleur à face rousse
d   Rotzügeldickkopf
i   Zufolatore golarossa

6968   *Pachycephala schlegelii*
    Passeriformes - Corvidae
e   Regent Whistler; Schlegel's Whistler
f   Siffleur de Schlegel

d   Schlegel-Dickkopf
i   Zufolatore di Schlegel

6969   *Pachycephala simplex*
    Passeriformes - Corvidae
e   Grey Whistler; Brown Whistler
f   Siffleur sobre
d   Weißbauchdickkopf
i   Zufolatore grigio

6970   *Pachycephala soror*
    Passeriformes - Corvidae
e   Sclater's Whistler
f   Siffleur de Sclater
d   Grünnackendickkopf
i   Zufolatore di Sclater

6971   *Pachycephala sulfuriventer*
    Passeriformes - Corvidae
e   Sulphur-bellied Whistler; Sulphur-
    vented Whistler; Yellow-bellied
    Whistler; Yellow-vented Whistler;
    Celebes Mountain-Whistler; Celebes
    Whistler
f   Siffleur à ventre jaune
d   Celebes Dickkopf
i   Zufolatore dal sottocoda sulfureo

6972   *Pachycephalopsis hattamensis*
    Passeriformes - Petroicidae
e   Green-backed Robin; Green Thicket-
    Flycatcher; Green Whistler-Robin;
    Green White-eyed Robin
f   Miro à dos vert
d   Grünrückenpachycephalopsis
i   Pigliamosche dorsoverde

6973   *Pachycephalopsis poliosoma*
    Passeriformes - Petroicidae
e   White-eyed Robin; White-throated
    Thicket-Flycatcher; Eastern White-
    eyed Robin
f   Miro aux yeux blancs
d   Graurückenpachycephalopsis
i   Pigliamosche grigio dal cappuccio

6974   *Pachycoccyx audeberti*
    Cuculiformes - Cuculidae
e   Thick-billed Cuckoo
f   Coucou d'Audebert

d   Dickschnabelkuckuck
i   Cuculo di Audebert

6975   *Pachyptila belcheri*
    Procellariiformes - Procellariidae
e   Slender-billed Prion; Narrow-billed
    Prion; Thin-billed Prion
f   Prion de Belcher
d   Dünnschnabelsturmvogel;
    Belchersturmvogel
i   Prione beccosottile

6976   *Pachyptila crassirostris*
    Procellariiformes - Procellariidae
e   Fulmar Prion; Thick-billed Prion
f   Prion à bec epais
d   Feensturmvogel
i   Prione beccogrosso

6977   *Pachyptila crassirostris eatoni*
    Procellariiformes - Procellariidae
e   Southern Fulmar-Prion (ANZ)

6978   *Pachyptila desolata*
    Procellariiformes - Procellariidae
e   Antarctic Prion; Dove-Prion (CSA);
    Dove-Petrel; Blue-Dove-Petrel;
    Banks's Dove-Petrel
f   Prion de la Désolation
d   Taubensturmvogel
i   Prione antartico

6979   *Pachyptila salvini*
    Procellariiformes - Procellariidae
e   Medium-billed Prion; Salvin's Prion;
    Lesser Broad-billed Petrel; Marion
    Island Petrel
f   Prion de Salvin
d   Kleiner Entensturmvogel
i   Prione di Salvin

6980   *Pachyptila turtur*
    Procellariiformes - Procellariidae
e   Fairy Prion; Gould Petrel; Short-
    billed Petrel
f   Prion colombe
d   Feensturmvogel
i   Prione tortora

6981   *Pachyptila turtur subantarctica*
    Procellariiformes - Procellariidae
e   Southern Fairy-Prion (ANZ)

6982   *Pachyptila vittata*
    Procellariiformes - Procellariidae
e   Broad-billed Prion; Blue Petrel;
    Broadbilled Dove-Petrel; Long-billed
    Prion; Common Prion; Prion
f   Prion de Forster
d   Entensturmvogel; Großer
    Entensturmvogel
i   Prione beccolargo

6983   *Pachyramphus aglaiae*
    Passeriformes - Tyrannidae
e   Rose-throated Becard; Larger Becard
f   Bécarde à gorge rose
d   Dickkopfbekarde
i   Beccaio golarosata

6984   *Pachyramphus albogriseus*
    Passeriformes - Tyrannidae
e   Black-and-white Becard
f   Bécarde pie
d   Elsterbekarde
i   Beccaio bianco e grigio

6985   *Pachyramphus castaneus*
    Passeriformes - Tyrannidae
e   Chestnut-crowned Becard; Rufous
    Becard; Cinnamon Becard
f   Bécarde à calotte rousse; Bécarde à
    couronne rousse
d   Kastanienbekarde
i   Beccaio capocastano

6986   *Pachyramphus cinnamomeus*
    Passeriformes - Tyrannidae
e   Cinnamon Becard
f   Bécarde cannellée
d   Zimtbekarde
i   Beccaio cannella

6987   *Pachyramphus homochrous*
    Passeriformes - Tyrannidae
e   One-coloured Becard
f   Bécarde unicolore
i   Beccaio unicolor

6988   *Pachyramphus major*
Passeriformes - Tyrannidae
e   Grey-collared Becard; Mexican
Becard
f   Bécarde du Mexique
d   Glanzkopfbekarde
i   Beccaio messicano

6989   *Pachyramphus marginatus*
Passeriformes - Tyrannidae
e   Black-capped Becard
f   Bécarde à calotte noire; Bécarde à
courone noire
d   Kappenbekarde
i   Beccaio marginato

6990   *Pachyramphus minor*
Passeriformes - Tyrannidae
e   Pink-throated Becard; Lesser Becard;
Rose-throated Becard
f   Bécarde de Lesson
d   Rotkehlbekarde
i   Beccaio minore

6991   *Pachyramphus niger*
Passeriformes - Tyrannidae
e   Jamaican Becard; Judy f (WI);
Mountain Dick m (WI); Kissidy
(WI); London City (WI); Weaverbird
(WI)
f   Bécarde de la Jamaïque
d   Jamaika-Bekarde
i   Beccaio della Giamaica

6992   *Pachyramphus polychopterus*
Passeriformes - Tyrannidae
e   White-winged Becard; Vieillot's
Becard
f   Bécarde à ailes blanches; Bécarde
noire
d   Weißflügelbekarde
i   Beccaio alibianche

6993   *Pachyramphus rufus*
Passeriformes - Tyrannidae
e   Cinereous Becard
f   Bécarde cendrée
d   Graubekarde
i   Beccaio rossiccio

6994   *Pachyramphus spodiurus*
Passeriformes - Tyrannidae
e   Slaty Becard
f   Bécarde ardoisée
d   Schieferbekarde
i   Beccaio lavagna

6995   *Pachyramphus surinamus*
Passeriformes - Tyrannidae
e   Glossy-backed Becard
f   Bécarde du Surinam
d   Surinam-Bekarde
i   Beccaio dorsosplendente

6996   *Pachyramphus validus*
Passeriformes - Tyrannidae
e   Crested Becard; Plain Becard
f   Bécarde huppée
d   Schopfbekarde
i   Beccaio modesto

6997   *Pachyramphus versicolor*
Passeriformes - Tyrannidae
e   Barred Becard
f   Bécarde barrée
d   Wellenbekarde
i   Beccaio barrato

6998   *Pachyramphus viridis*
Passeriformes - Tyrannidae
e   Green-backed Becard
f   Bécarde verte
d   Grünrückenbekarde
i   Beccaio dorsoverde

6999   *Pachyramphus xanthogenys*
Passeriformes - Tyrannidae
e   Yellow-cheeked Becard
f   Bécarde à joues jaunes

7000   *Padda fuscata*
Passeriformes - Tyrannidae
e   Timor Finch; Timor Sparrow; Timor
Dusky Sparrow; Brown Ricebird
f   Padda brun
d   Timor-Reisfink; Brauner Reisfink
i   Padda bruno di Timor

7001   *Padda oryzivora*
Passeriformes - Estrildidae
e   Java Sparrow; Java Finch; Ricebird;

Rice Munia; Paddybird; Templebird
f Padda de Java; Calfat; Padda; Padda
de riz; Spermète de Java
d Reisfink; Reisvogel
i Padda; Uccello delle risaie

7002 *Pagodroma nivea*
Procellariiformes - Procellariidae
e Snow Petrel; Lesser Snow Petrel;
Snowy Petrel
f Pétrel des neiges
d Schneesturmvogel
i Procellaria delle nevi; Petrello delle
neve minore

7003 *Pagodroma nivea confusa*
Procellariiformes - Procellariidae
e Greater Snow-Petrel
f Pétrel blanc
i Peterello delle neve maggiore

7004 *Pagophila eburnea*
Charadriiformes - Laridae
e Ivory Gull
f Mouette blanche; Mouette ivoire;
Mouette sénateur; Goéland sénateur;
Pagophile blanche; Goeland blanc
d Elfenbeinmöwe
i Gabbiano eburneo; Gabbiano
d'avorio

7005 *Palmeria dolei*
Passeriformes - Fringillidae
e Akohekohe; Crested Honeycreeper
f Palmérie huppée
d Schopfkleidervogel
i Akohekoe

7006 *Pandion haliaetus*
Falconiformes - Accipitridae
e Osprey; Fish Hawk; Sea Eagle(WI);
Fish Eagle (WI); White-headed
Osprey
f Balbuzard; Balbuzard pêcheur;
Malfini de mer (Ants); Malfini la mer
(Ants); Aiglon (Ants); Gligli
montagne (WI); Malfini de la mer
(Ants)
d Fischadler
i Falco pescatore

7007 *Panterpe insignis*
Apodiformes - Trochilidae
e Fiery-throated Hummingbird; Irazu
Hummingbird
f Colibri insigne
d Feuerkehlkolibri
i Colibrì goladifiamma

7008 *Panurus biarmicus*
Passeriformes - Sylviidae
e Bearded Parrotbill; Bearded Tit;
Bearded Reedling; Reedling;
Bearded Tit-Babbler; Eastern
Bearded Tit
f Mésange à moustaches; Panure à
moustaches; Mézette à moustaches;
Mésange barbue
d Bartmeise
i Basettino

7009 *Panyptila cayennensis*
Apodiformes - Apodidae
e Lesser Swallow-tailed Swift;
Cayenne Swift
f Martinet de Cayenne
d Steigrohrsegler
i Rondone codadirondine minore

7010 *Panyptila sanctihieronymi*
Apodiformes - Apodidae
e Great Swallow-tailed Swift;
Geronimo Swift; Geronimo Swiftlet
f Martinet de San Geronimo
d San Geronimo-Segler
i Rondone codadirondine maggiore

7011 *Papasula abbotti*
Pelecaniformes - Sulidae
e Abbott's Booby
f Fou d'Abbott
d Abbott-Tölpel
i Sula di Abbott

7012 *Parabuteo unicinctus*
Falconiformes - Accipitridae
e Harris's Hawk; Bay-winged Hawk;
Harris's Bay-winged Hawk
f Buse de Harris
d Wüstenbussard
i Poiana di Harris

7013   *Paradigalla brevicauda*
        Passeriformes - Corvidae
e   Short-tailed Paradigalla; Short-tailed
        Bird of Paradise
f   Paradisier à queue courte
d   Kurzschwanzparadigalla
i   Paradigalla codalunga

7014   *Paradigalla carunculata*
        Passeriformes - Corvidae
e   Long-tailed Paradigalla; Wattled Bird
        of Paradise
f   Paradisier caronculé
d   Schwarzkehlparadieselster;
        Langschschwanzparadigalla
i   Paradigalla codabreve

7015   *Paradisaea apoda*
        Passeriformes - Paradisaeidae
e   Greater Bird of Paradise
f   Paradisier grand-émeraude; Grand
        Paradisier
d   Großer Paradiesvogel
i   Paradisea apoda; Paradisea maggiore

7016   *Paradisaea decora*
        Passeriformes - Paradisaeidae
e   Goldie's Bird of Paradise
f   Paradisier de Goldie
d   Lavendelparadiesvogel
i   Paradisea di Goldie

7017   *Paradisaea guilielmi*
        Passeriformes - Paradisaeidae
e   Emperor Bird of Paradise; Emperor
        of Germany; Emperor of Germany's
        Bird of Paradise; White-plumed Bird
        of Paradise
f   Paradisier de Guillaume
d   Kaiserparadiesvogel
i   Paradisea di Guglielmo di Germania;
        Paradisea dell'imperatore Guglielmo

7018   *Paradisaea minor*
        Passeriformes - Paradisaeidae
e   Lesser Bird of Paradise
f   Paradisier petit-émeraude
d   Kleiner Paradiesvogel
i   Paradisea minore

7019   *Paradisaea raggiana*
        Passeriformes - Paradisaeidae
e   Raggiana Bird of Paradise; Count
        Raggi's Bird of Paradise; Red-
        plumed Bird of Paradise
f   Paradisier de Raggi
d   Göttervogel
i   Paradisea di Raggi

7020   *Paradisaea rubra*
        Passeriformes - Paradisaeidae
e   Red Bird of Paradise
f   Paradisier rouge
d   Roter Paradiesvogel
i   Paradisea rossa

7021   *Paradisaea rudolphi*
        Passeriformes - Paradisaeidae
e   Blue Bird of Paradise; Archduke
        Rudolph's Blue Bird of Paradise
f   Paradisier bleu; Paradisier de
        Rudolph; Paradisier de Rodolphe
d   Blauer Paradiesvogel
i   Paradisea dell'arciduca Rodolfo;
        Paradisea azzurra

7022   **Paradisaeidae**
        Passeriformes
e   Birds-of-Paradise
f   Paradiséidé; Paradisiers
d   Paradiesvögel
i   Paradiseidi

7023   *Paradoxornis alphonsianus*
        Passeriformes - Sylviidae
e   Ashy-throated Parrotbill
f   Paradoxornis à gorge cendrée
i   Becco a cono golacenerina

7024   *Paradoxornis atrosuperciliaris*
        Passeriformes - Sylviidae
e   Black-browed Parrotbill; Black-
        browed Crowtit; Black-throated
        Parrotbill; Lesser Rufous-headed
        Parrotbill; Lesser Red-headed
        Parrotbill; Lesser Red-headed
        Suthora
f   Paradoxornis à sourcils noirs
d   Schopfpapageischnabel
i   Becco a cono testarossa minore

7025   *Paradoxornis brunneus*
Passeriformes - Sylviidae
e   Brown-winged Parrotbill; Rickett's
       Parrotbill; Yunnan Parrotbill
f   Paradoxornis à ailes brunes
i   Becco a cono bruno

7026   *Paradoxornis conspicillatus*
Passeriformes - Sylviidae
e   Spectacled Parrotbill; Sspectacled
       Crowtit
f   Paradoxornis à lunettes
d   Brillenpapageischnabel
i   Becco a cono dagli occhiali

7027   *Paradoxornis davidianus*
Passeriformes - Sylviidae
e   Short-tailed Parrotbill; David's
       Parrotbill; David's Crowtit
f   Paradoxornis de David; Paradoxornis
       du père David
d   Kurzschwanzpapageischnabel
i   Becco a cono di David

7028   *Paradoxornis flavirostris*
Passeriformes - Sylviidae
e   Black-breasted Parrotbill; Gould's
       Black-breasted Parrotbill; Gould's
       Yellow-billed Parrotbill; Gould's
       Parrotbill
f   Paradoxornis de Gould
d   Schwarzkehlpapageischnabel;
       Dschungelpapageimeise
i   Becco a cono di Gould

7029   *Paradoxornis fulvifrons*
Passeriformes - Sylviidae
e   Fulvous Parrotbill; Fulvous-fronted
       Parrotbill; Fulvous-fronted Suthora;
       Fulvous-fronted Crowtit
f   Paradoxornis à front fauve
d   Gelbstirnpapageischnabel
i   Becco a cono frontefulvo

7030   *Paradoxornis gularis*
Passeriformes - Sylviidae
e   Grey-headed Parrotbill; Grey-headed
       Crowtit; Hoary-headed Parrotbill
f   Paradoxornis à tête grise
d   Graukopfpapageischnabel;

Kernbeißertimalie
i   Becco a cono testagrigia

7031   *Paradoxornis guttaticollis*
Passeriformes - Sylviidae
e   Spot-breasted Parrotbill; Rufous-
       headed Parrotbill; Spotted-breasted
       Parrotbill; White-throated Parrotbill;
       Spot-necked Parrotbill
f   Paradoxornis flèché
d   Brustfleckenpapageischnabel
i   Becco a cono pettomaculato

7032   *Paradoxornis heudei*
Passeriformes - Sylviidae
e   Reed Parrotbill; Yangtse Parrotbill;
       Yangtse Crowtit; Heude's Parrotbill;
       Chinese Parrotbill; Chinese Crowtit;
       Eyebrowed Parrotbill
f   Paradoxornis du Yangtsé
d   Jangtse-Papageischnabel;
       Rosenpapageimeise
i   Becco a cono di Heude

7033   *Paradoxornis nipalensis*
Passeriformes - Sylviidae
e   Black-throated Parrotbill; Ashy-eared
       Parrotbill; Grey-cheeked Parrotbill;
       Nepal Parrotbill; Black-fronted
       Parrotbill; Blyth's Parrotbill; Orange
       Suthora; Orange Parrotbill
f   Paradoxornis à menton noir;
       Paradoxornis à oreillons cendrés
d   Grauohrpapageischnabel
i   Becco a cono del Nepal

7034   *Paradoxornis paradoxus*
Passeriformes - Sylviidae
e   Three-toed Parrotbill; Three-toed
       Crowtit
f   Paradoxornis tridactyle
d   Dreizehenpapageischnabel;
       Dreizehen-Timalie
i   Becco a cono tridattilo

7035   *Paradoxornis przewalskii*
Passeriformes - Sylviidae
e   Rusty-throated Parrotbill; Grey-
       crowned Parrotbill; Przewalski's
       Parrotbill; Grey-crowned Crowtit

f   Paradoxornis de Przewalski
d   Przewalski-Papageischnabel
i   Becco a cono di Przewalski

7036   *Paradoxornis ruficeps*
Passeriformes - Sylviidae
e   Rufous-headed Parrotbill; Greater
Red-headed Parrotbill; Greater
Rufous-headed Parrotbill; Red-
headed Parrotbill; Red-headed
Crowtit
f   Paradoxornis à tête rousse
d   Rotkopfpapageischnabel
i   Becco a cono testarossa maggiore

7037   *Paradoxornis unicolor*
Passeriformes - Sylviidae
e   Brown Parrotbill; Brown Suthora;
Himalayan Brown-Crowtit
f   Paradoxornis unicolore
d   Einfarbpapageischnabel
i   Becco a cono bruno

7038   *Paradoxornis verreauxi*
Passeriformes - Sylviidae
e   Golden Parrotbill
f   Paradoxornis de Verreaux
d   Papageischnabel
i   Becco a cono dorato

7039   *Paradoxornis webbianus*
Passeriformes - Sylviidae
e   Vinous-throated Parrotbill; Webb's
Parrotbill; Rufous-headed Crowtit
f   Paradoxornis de Webb
d   Braunkopfpapageischnabel;
Braunkopfpapageimeise;
Rundschwanztimalie
i   Becco a cono di Webb

7040   *Paradoxornis zappeyi*
Passeriformes - Sylviidae
e   Grey-hooded Parrotbill; Dusky
Parrotbill; Zappey's Parrotbill;
Crested Parrotbill; Dusky Crowtit
f   Paradoxornis de Zappey
d   Zappey-Papageischnabel
i   Becco a cono di Zappa

7041   *Paramythia montium*
Passeriformes - Paramythiidae

e   Crested Berrypecker; Mountain
Berrypecker
f   Paramythie huppée
d   Schopbeerenfresser;
Haubenmistelfresser;
Schwarzhaubenblauvogel
i   Beccafrutti dal ciuffo

7042   Pardalotidae
Passeriformes
e   Pardalotes
f   Pardalotidés
d   Schopbeerenfresser;
Haubenmistelfresser;
Schwarzhaubenblauvögel
i   Pardalotidi

7043   *Pardalotus punctatus*
Passeriformes - Pardalotidae
e   Spotted Pardalote; Fernbird; Matata;
Tataki Thrush; Spotted Diamondbird
f   Pardalote pointillé
d   Farnsteiger; Fleckenpanthervogel
i   Pardaloto macchiato

7044   *Pardalotus punctatus xanthopygus*
Passeriformes - Pardalotidae
e   Yellow-tailed Pardalote; Yellow-
rumped Pardalote; Yellow-tailed
Diamondbird
f   Pardalote à croupion jaune
d   Gelbürzelpanthervogel

7045   *Pardalotus quadragintus*
Passeriformes - Pardalotidae
e   Forty-spotted Pardalote; Forty-
spotted Diamondbird
f   Pardalote de Tasmanie
d   Tasman-Panthervogel
i   Pardoloto dalle quaranta macchie

7046   *Pardalotus rubricatus*
Passeriformes - Pardalotidae
e   Red-browed Pardalote; Red-browed
Diamondbird
f   Pardalote à sourcils rouges
d   Rotbrauenpanthervogel
i   Pardoloto dai spraciggli rossi

7047   *Pardalotus striatus*
Passeriformes - Pardalotidae

e   Striated Pardalote; Striped Pardalote
    (ANZ); Yellow-tipped Pardalote;
    Yellow-tipped Diamondbird
f   Pardalote à point jaune
d   Streifenpanthervogel
i   Pardaloto striato

7048   *Pardalotus striatus melanocephalus*
    Passeriformes - Pardalotidae
e   Black-headed Pardalote; Black-
    headed Diamondbird
f   Pardalote à calotte noire

7049   *Pardalotus striatus ornatus*
    Passeriformes - Pardalotidae
e   Red-tipped Pardalote; Red-tipped
    Diamondbird
f   Pardalote orné

7050   *Pardalotus striatus substriatus*
    Passeriformes - Pardalotidae
e   Striated Pardalote; Striated
    Diamondbird
f   Pardalote strié

7051   *Pardirallus maculatus*
    Gruiformes - Rallidae
e   Spotted Rail
f   Râle tacheté
d   Pardelralle
i   Rallo macchiato

7052   *Pardirallus nigricans*
    Gruiformes - Rallidae
e   Blackish Rail
f   Râle noirâtre
d   Trauerralle
i   Rallo nerastro

7053   *Pardirallus sanguinolentus*
    Gruiformes - Rallidae
e   Plumbeous Rail
f   Râle à bec ensanglanté
d   Grauralle
i   Rallo plumbeo

7054   Paridae
    Passeriformes
e   Titmice; Tits; Chickadees
f   Mésanges; Paridés

d   Meisen; Eigentliche Meisen
i   Paridi; Cince

7055   *Parmoptila rubrifrons*
    Passeriformes - Passeridae
e   Red-fronted Antpecker; Red-fronted
    Flowerpecker Weaver-Finch;
    Jameson's Antpecker
f   Parmoptile à front rouge
i   Beccaformiche fronterossa

7056   *Parmoptila woodhousei*
    Passeriformes - Passeridae
e   Woodhouse's Antpecker; Antpecker;
    Flowerpecker Weaver-Finch
f   Parmoptile à gorge rousse; Astrild
    fourmilier; Parmoptile à tête rouge
d   Woodhouse-Ameisenpicker
i   Beccaformiche

7057   *Paroaria baeri*
    Passeriformes - Fringillidae
e   Crimson-fronted Cardinal
f   Paroare de Baer
d   Blutstirnkardinal
i   Cardinale fronterossa

7058   *Paroaria capitata*
    Passeriformes - Fringillidae
e   Yellow-billed Cardinal
f   Paroare à bec jaune
d   Mantelkardinal
i   Cardinale beccogiallo

7059   *Paroaria coronata*
    Passeriformes - Fringillidae
e   Red-crested Cardinal; Brazilian
    Cardinal; Crested Cardinal
f   Paroare huppé; Cardinal gris
d   Graukardinal; Rothaubenkardinal
i   Cardinale ciufforosso

7060   *Paroaria dominicana*
    Passeriformes - Fringillidae
e   Red-cowled Cardinal; Dominican
    Cardinal; Pope Cardinal; Brazilian
    Cardinal
f   Paroare dominicain
d   Dominikanerkardinal
i   Cardinale dal cappuccio rosso

7061 *Paroaria gularis*
Passeriformes - Fringillidae
e Red-capped Cardinal
f Paroare rougecap
d Braunkehlkardinal
i Cardinale capirosso

7062 *Parophasma galinieri*
Passeriformes - Sylviidae
e Abyssinian Catbird
f Phyllanthe de Galinier; Timalie d'Abyssinie
d Singtimalie
i Uccello gatto dell'Abissinia

7063 *Paroreomyza flammea*
Passeriformes - Fringillidae
e Molokai Creeper; Kakawahie Creeper; Kakawahie
f Grimpeur de Molokai
i Rampichino di Molokai

7064 *Paroreomyza maculata*
Passeriformes - Fringillidae
e Oahu Alauahio; Oahu Creeper; Alauahio; Hawaiian Creeper; Lanai Creeper; Hawaii Creeper
f Grimpeur d'Oahu
d Alauwahio; Kletterkleidervogel; Hawaii-Baumläufer
i Rampichino di Oahu

7065 *Paroreomyza montana*
Passeriformes - Fringillidae
e Maui Alauahio; Maui Creeper; Alauahio Creeper; Alauahio
f Grimpeur de Maui
i Rampichino di Maui

7066 *Parotia carolae*
Passeriformes - Corvidae
e Carola's Parotia; Queen Carola's Parotia; Queen Carola's Bird of Paradise; Queen Carola's Six-wired Bird of Paradise
f Paradisier de Carola
d Carola-Strahlenparadiesvogel; Carola-Paradiesvogel
i Paradisea della regina Carola

7067 *Parotia helenae*
Passeriformes - Corvidae
e Eastern Parotia
f Paradisier d'Helena
d Helena-Paradiesvogel; Helena-Strahlenparadiesvogel
i Paradisea di Elena

7068 *Parotia lawesii*
Passeriformes - Corvidae
e Lawes's Parotia; Lawes's Six-wired Parotia; Six-plumed Bird of Paradise; Lawes's Six-plumed Bird of Paradise; Lawe's Six-wired Bird of Paradise
f Paradisier de Lawes; Sifilet de Lawes
d Blaunackenstrahlenparadiesvogel; Blaunackenparadiesvogel
i Paradisea di Lawes

7069 *Parotia sefilata*
Passeriformes - Corvidae
e Western Parotia; Arfak Six-wired Parotia; Arfak Parotia; Arfak Six-plumed Bird of Paradise
f Paradisier sifilet
d Arfak-Strahlenparadiesvogel; Arfak-Paradiesvogel
i Paradisea dai sei fili; Parozia; Paradisea delle sei penne

7070 *Parotia wahnesi*
Passeriformes - Corvidae
e Wahnes's Parotia; Wahnes's Six-wired Parotia; Wahnes's Six-plumed Bird of Paradise; Huon Parotia; Wahnes's Six-wired Bird of Paradise
f Paradisier de Wahnes
d Wahnes-Paradiesvogel; Wahnes-Strahlenparadiesvogel
i Paradisea di Wahnes

7071 *Parula americana*
Passeriformes - Parulidae
e Northern Parula; Parula Warbler; American Warbler
f Paruline à collier; Mésange bleue (Ants); Fauvette parula; Fauvette d'Amérique
d Meisenwaldsänger
i Parula americana

7072 *Parula graysoni*
Passeriformes - Parulidae
e Socorro Warbler

7073 *Parula gutturalis*
Passeriformes - Parulidae
e Flame-throated Warbler
f Paruline embrasée
d Feuerwaldsänger
i Parula goladifiamma

7074 *Parula pitiayumi*
Passeriformes - Parulidae
e Tropical Parula; Olive-backcd
Warbler; Socorro Warbler
f Paruline à joues noires
d Elfenwaldsänger
i Parula tropicale

7075 *Parula superciliosa*
Passeriformes - Parulidae
e Crescent-chested Warbler; Spot-
breasted Warbler; Hartlaub's Warbler
f Paruline à croissant
d Schmuckwaldsänger
i Parula dalla pettorina rossa

7076 **Parulidae**
Passeriformes
e Wood-Warblers; Warblers; New
World Warblers
f Parulines; Parulidés
d Waldsänger
i Parulidi

7077 *Parus afer*
Passeriformes - Paridae
e Grey Tit; Southern Grey-Tit (CSA);
African Grey-Tit; Acacia Grey-Tit;
Acacia Tit
f Mésange petit-deuil; Mésange grise
australe
d Kapmeise
i Cincia grigia africana

7078 *Parus albiventris*
Passeriformes - Paridae
e White-bellied Tit; White-breasted
Tit; White-bellied Titmouse; White-
breasted Titmouse; White-bellied
Black-Tit

f Mésange à ventre blanc
d Weißbauchmeise
i Cincia nera ventrebianco

7079 *Parus amabilis*
Passeriformes - Paridae
e Palawan Tit; Palawan Titmouse;
Black-headed Titmouse
f Mésange de Palawan
d Kapuzenmeise
i Cincia di Palawan

7080 *Parus ater*
Passeriformes - Paridae
e Coal Tit; Coal Titmouse
f Mésange noire; Petite Charbonnière
d Tannenmeise
i Cincia mora

7081 *Parus atricapillus*
Passeriformes - Paridae
e Black-capped Chickadee; Black-
capped Tit
f Mésange à tête noire
d Schwarzkopfmeise; Chickadeemeise
i Cincia bigia americana; Cincia bigia
alpestre

7082 *Parus atricristatus*
Passeriformes - Paridae
e Black-crested Titmouse; Black-
crested Tit
f Mésange à plumet noir
i Cincia crestanera

7083 *Parus bicolor*
Passeriformes - Paridae
e Tufted Titmouse; Tufted Tit
f Mésange bicolore; Mésange huppée
d Indianer-Meise; Zweifarbenmeise
i Cincia bicolore

7084 *Parus bokharensis*
Passeriformes - Paridae
e Turkestan Tit; Turkestan Great Tit;
Turkestan Grey-Tit
f Mésange du Turkestan
d Turkestan-Meise
i Cinccia del Turkestan

7085 *Parus caeruleus*
Passeriformes - Paridae
e European Blue-Tit (NA); Blue Tit;
Blue Titmouse
f Mésange bleue; Mésange à tête bleue
d Blaumeise
i Cinciarella; Cincia; Cincia piccola;
Perlonza piccola; Cincia putichia;
Potazzina

7086 *Parus carolinensis*
Passeriformes - Paridae
e Carolina Chickadee; Carolina Tit
f Mésange de Caroline; Mésange
minime
d Carolina-Meise
i Cincia della Carolina

7087 *Parus carpi*
Passeriformes - Paridae
e Carp's Tit; Carp's Black-Tit (CSA)
f Mésange de Carp

7088 *Parus cinctus*
Passeriformes - Paridae
e Siberian Tit; Grey-headed Chickadee
(NA); Siberian Chickadee (NA);
Taita Tit; Grey-headed Tit
f Mésange lapone; Mésange
sibérienne; Mésange à plastron
d Lappland-Meise
i Cincia siberiana

7089 *Parus cinerascens*
Passeriformes - Paridae
e Ashy Tit; Ashy Grey-Tit; Acacia Tit;
Acacia Grey-Tit; Grey Tit
f Mésange cendrée
d Aschenmeise
i Cincia delle acacie

7090 *Parus cristatus*
Passeriformes - Paridae
e Crested Tit
f Mésange huppée
d Haubenmeise
i Cincia dal ciuffo

7091 *Parus cyanus*
Passeriformes - Paridae
e Azure Tit

f Mésange azurée; Mésange bleue de
Pleske
d Lasurmeise
i Cinciarella azzurra; Cincia azzura

7092 *Parus davidi*
Passeriformes - Paridae
e Rusty-breasted Tit; Père David's Tit;
Red-bellied Tit
f Mésange de David; Mésange du père
David
d David-Meise
i Cincia di padre David

7093 *Parus degener*
Passeriformes - Paridae
e Fuerteventura Blue-Tit
f Mésange de Fuerteventura

7094 *Parus dichrous*
Passeriformes - Paridae
e Grey-crested Tit; Brown-crested Tit;
Northern Brown-Tit
f Mésange des bouleaux; Mésange
d'Asie
d Birkenmeise
i Cincia bruna

7095 *Parus elegans*
Passeriformes - Paridae
e Elegant Tit; Elegant Titmouse
f Mésange élégante
d Panthermeise
i Cincia elegante

7096 *Parus fasciiventer*
Passeriformes - Paridae
e Stripe-breasted Tit
f Mésange à ventre strié
d Schwarzbrustmeise
i Cincia pettostriato

7097 *Parus flavipectus*
Passeriformes - Paridae
e Yellow-breasted Tit; Yellow-
breasted Azure Tit; Turkestan Azure
Tit
f Mésange à poitrine jaune
i Cincia azzurra pettogiallo

7098 *Parus fringillinus*
Passeriformes - Paridae
e Red-throated Tit
f Mésange à gorge rousse
d Rostkehlmeise
i Cincia golarossa

7099 *Parus funereus*
Passeriformes - Paridae
e Dusky Tit
f Mésange enfumée; Mésange ardoisée
d Einfarbmeise; Dunkelmeise
i Cincia tenebrosa

7100 *Parus gambeli*
Passeriformes - Paridae
e Mountain Chickadee; Gambel's Chickadee; Gambel's Tit
f Mésange de Gambel
d Gambel-Meise; Felsengebirgsmeise
i Cincia delle Montagne Rocciose

7101 *Parus griseiventris*
Passeriformes - Paridae
e Miombo Tit; Northern Grey-Tit (CSA); Tabora Grey-Tit; Miombo Grey-Tit
f Mésange à ventre gris
d Miombo-Meise; Afrikanische Graumeise; Graubauchmeise
i Cincia grigia del Miombo

7102 *Parus guineensis*
Passeriformes - Paridae
e Northern Black-Tit (CSA); White-shouldered Black-Tit; White-shouldered Tit
f Mésange galonnée
i Cincia nera spallebianche

7103 *Parus holsti*
Passeriformes - Paridae
e Yellow Tit; Formosan Yellow-Tit
f Mésange de Taïwan; Mésange de Formose
d Formosa-Meise
i Cincia di Taiwan

7104 *Parus hudsonicus*
Passeriformes - Paridae
e Boreal Chickadee; Brown-capped Chickadee (NA); Brown-headed Chickadee; Hudsonian Chickadee; Hudsonian Tit; Brown-capped Tit
f Mésange à tête brune
d Hudson-Meise; Braunkappenmeise
i Cincia boreale

7105 *Parus hypermelaena*
Passeriformes - Paridae
e Black-bibbed Marsh-Tit; Black-bibbed Tit
f Mésange à bavette

7106 *Parus hyrcanus*
Passeriformes - Paridae
e Caspian Tit; Elburz Tit; Iranian Sombre Tit; Hyrcanian Tit

7107 *Parus inornatus*
Passeriformes - Paridae
e Plain Titmouse; Oak Titmouse; Plain Tit
f Mésange unicolore
d Schlichtmeise
i Cincia grigiotopo

7108 *Parus leucomelas*
Passeriformes - Paridae
e White-winged Tit; White-winged Black-Tit; Common Black-Tit; White-shouldered Black-Tit; Black Tit; Northern Black Tit
f Mésange à épaulettes; Mésange noire à épaulettes blanches; Mésange à épaulettes blanches
d Schwarzmeise; Rüppells Meise
i Cincia nera alibianche

7109 *Parus leuconotus*
Passeriformes - Paridae
e White-backed Tit; White-backed Black-Tit
f Mésange à dos blanc
d Weißrückenmeise
i Cincia nera dorsobianco

7110 *Parus lugubris*
Passeriformes - Paridae
e Sombre Tit
f Mésange lugubre

d Trauermeise
i Cincia dalmatina

7111 *Parus major*
Passeriformes - Paridae
e Great Tit; Grey Tit (ISC); Great
Titmouse; Ceylon Great Tit;
Cinereous Tit; Japanese Tit
f Mésange charbonnière; Grande
Mésange charbonnière; Mésange
grosse
d Kohlmeise
i Cinciallegra; Cincia; Cincia allegra;
Cincia grossa; Cincera; Cingallina;
Cingallegra

7112 *Parus melanolophus*
Passeriformes - Paridae
e Black-crested Tit; Spot-winged Tit;
Vigors's Black-crested Tit; Vigors's
Crested-Tit; Spot-winged Black-Tit;
Crested Black Tit
f Mésange de Vigors
d Schwarzschopfmeise;
Graubrustmeise
i Cincia di Vigors

7113 *Parus montanus*
Passeriformes - Paridae
e Willow Tit
f Mésange boréale; Mésange à calotte
matte
d Weidenmeise; Mönchsmeise;
Alpenmeise; Mattkopfmeise
i Cincia bigia alpestre; Cincia boreale

7114 *Parus monticolus*
Passeriformes - Paridae
e Green-backed Tit
f Mésange montagnarde
d Bergmeise; Bergkohlmeise
i Cincia dorsoverde

7115 *Parus niger*
Passeriformes - Paridae
e Black Tit; Southern Black-Tit
f Mésange nègre; Mésange noire
australe
d Mohrenmeise
i Cincia nera

7116 *Parus nuchalis*
Passeriformes - Paridae
e White-naped Tit; White-winged
Black-Tit (ISC); White-winged Tit;
Collared Tit
f Mésange à ailes blanches
d Weißflügelmeise
i Cincia nucabianca

7117 *Parus ombriosus*
Passeriformes - Paridae
e Hierro Blue-Tit; Palma Blue-Tit
f Mésange de Palma

7118 *Parus pallidiventris*
Passeriformes - Paridae
e Cinnamon-breasted Tit
f Mésange à oeil jaunâtre
i Cincia pettocannella

7119 *Parus palustris*
Passeriformes - Paridae
e Marsh Tit; Marsh Titmouse; Chinese
Marsh-Tit; Burmese Marsh-Tit
f Mésange nonnette
d Sumpfmeise; Nonnenmeise;
Glanzkopfmeise
i Cincia bigia

7120 *Parus ridgwayi*
Passeriformes - Paridae
e Ridgway's Titmouse; Juniper
Titmouse

7121 *Parus rubidiventris*
Passeriformes - Paridae
e Rufous-vented Tit; Rufous-bellied
Tit; Black-crested Tit; Rufous-bellied
Crested-Tit; Sikkim Rufous-vented
Tit; Rufous-bellied Black Tit;
Rufous-breasted Black-Tit; Sikkim
Black-Tit
f Mésange cul-roux; Mésange huppée
de Blyth
d Sikkim-Meise; Rotbrustmeise
i Cincia crestanera

7122 *Parus rufescens*
Passeriformes - Paridae
e Chestnut-backed Chickadee;
Chestnut-backed Tit
f Mésange à dos marron

d   Rotrückenmeise; Braunrückenmeise
i   Cincia dorsocastano

7123   *Parus rufiventris*
       Passeriformes - Paridae
e   Rufous-bellied Tit; Rufous Tit;
      Cinnamon-breasted Tit
f   Mésange à ventre cannellé
d   Rotbauchmeise
i   Cincia ventreruggine

7124   *Parus rufonuchalis*
       Passeriformes - Paridae
e   Dark-grey Tit; Rufous-naped Tit;
      Rufous-vented Tit; Simla Tit; Simla
      Black-Tit; Black-crested Tit; Black-
      breasted Tit
f   Mésange à nuque rousse
d   Fichtenmeise
i   Cincia nucafulva

7125   *Parus sclateri*
       Passeriformes - Paridae
e   Mexican Chickadee; Grey-sided Tit;
      Grey-sided Chickadee
f   Mésange grise
d   Grauflankenmeise; Mexico-Meise
i   Cincia messicana

7126   *Parus semilarvatus*
       Passeriformes - Paridae
e   White-fronted Tit; White-fronted
      Titmouse
f   Mésange à front blanc
d   Weißstirnmeise
i   Cincia frontebianca

7127   *Parus songarus*
       Passeriformes - Paridae
e   Songar Tit; Chinese Willow Tit
d   Weißgoldmeise

7128   *Parus spilonotus*
       Passeriformes - Paridae
e   Yellow-cheeked Tit; Chinese
      Yellow-Tit; Chinese Black-spotted
      Yellow-Tit; Black-spotted Yellow-
      Tit
f   Mésange à dos tacheté
d   Königsmeise
i   Cincia gialla cinese

7129   *Parus superciliosus*
       Passeriformes - Paridae
e   White-browed Tit
f   Mésange à sourcils blancs
d   Weißbrauenmeise; Brauenmeise
i   Cincia dai sopraccigli bianchi

7130   *Parus teneriffae*
       Passeriformes - Paridae
e   Tenerife Blue-Tit
f   Mésange de Ténériffe

7131   *Parus thruppi*
       Passeriformes - Paridae
e   Somali Tit; Somali Grey-Tit;
      Northern Grey-Tit; Grey Tit; Acacia
      Tit
f   Mésange somalienne
d   Somali-Meise
i   Cincia somala

7132   *Parus thruppi barakae*
       Passeriformes - Paridae
e   Tit (CSA)
f   Mésange maghrébine

7133   *Parus ultramarinus*
       Passeriformes - Paridae
e   African Blue-Tit; North African
      Blue-Tit

7134   *Parus varius*
       Passeriformes - Paridae
e   Varied Tit
f   Mésange variée; Mésange du Japon
d   Buntmeise
i   Cincia varia

7135   *Parus venustulus*
       Passeriformes - Paridae
e   Yellow-bellied Tit
f   Mésange gracieuse; Mésange à
      ventre jaune
d   Schmuckmeise
i   Cincia ventregiallo

7136   *Parus wollweberi*
       Passeriformes - Paridae
e   Bridled Titmouse; Bridled Tit;
      Bridled Chickadee (NA)
f   Mésange arlequin

7161 **Passer swainsonii**
Passeriformes - Passeridae
e Swainson's Sparrow
f Moineau de Swainson
i Passero di Swainson

7162 **Passer zarudnyi**
Passeriformes - Passeridae
e Asian Desert Sparrow

7163 **Passerculus princeps**
Passeriformes - Emberizidae
e Ipswich Sparrow

7164 **Passerculus rostratus**
Passeriformes - Emberizidae
e Large-billed Sparrow

7165 **Passerculus sandwichensis**
Passeriformes - Emberizidae
e Savannah Sparrow; Savanna Sparrow
f Bruant des prés; Bruant des savanes;
Pinson des prés
d Grasammer; Savannenammer
i Passero delle praterie

7166 **Passerella iliaca**
Passeriformes - Emberizidae
e Red Fox-Sparrow; Fox Sparrow
d Fuchsammer
i Passerella variabile

7167 **Passerella schistacea**
Passeriformes - Emberizidae
e Slate-coloured Fox-Sparrow

7168 **Passerella unalschensis**
Passeriformes - Emberizidae
e Sooty Fox-Sparrow; Fox Sparrow

7169 **Passeridae**
Passeriformes
e Sparrows; Old World Sparrows
f Moineaux; Passeridés
d Sperlinge
i Passeri

7170 **Passeriformes**
e Perching Birds; Passerines; Passerine
Birds
f Passereaux; Passériformes

d Sperlingsvögel
i Passeriformi

7171 **Passerina amoena**
Passeriformes - Cardinelidae
e Lazuli Bunting
f Passerin azuré; Bruant azuré
d Lazulifink; Lasurfink
i Papa lazuli

7172 **Passerina ciris**
Passeriformes - Cardinelidae
e Painted Bunting
f Passerin nonpareil
d Papstfink
i Settecolori; Papa della Louisiana;
Papa della Luisiana

7173 **Passerina cyanea**
Passeriformes - Cardinelidae
e Indigo Bunting
f Passerin indigo
d Indigofink
i Ministro

7174 **Passerina leclancherii**
Passeriformes - Cardinelidae
e Orange-breasted Bunting;
Leclancher's Buntinmg
f Passerin arc-en-ciel
d Orangeblaufink
i Papa di Leclancher

7175 **Passerina rositae**
Passeriformes - Cardinelidae
e Rose-bellied Bunting; Rosita's
Bunting
f Passerin à ventre rose
d Rosenbauchfink
i Papa pettorosa

7176 **Passerina versicolor**
Passeriformes - Cardinelidae
e Varied Bunting
f Passerin varié
d Vielfarbenfink
i Papa variabile

7177 **Patagona gigas**
Apodiformes - Trochilidae
e Giant Hummingbird

f Colibri géant
d Riesenkolibri
i Colibrì gigante

7178 *Pauxi pauxi*
Galliformes - Cracidae
e Helmeted Curassow; Northern
Helmeted-Curassow; Galeated
Curassow
f Hocco à pierre
d Helmhokko
i Crace dall'elmo

7179 *Pauxi unicornis*
Galliformes - Cracidae
e Horned Curassow; Southern
Helmeted-Curassow; Bolivian
Helmeted-Curassow; Helmeted
Curassow
f Hocco unicorne
d Bolivianischer Helmhokko
i Crace unicorno

7180 *Pavo cristatus*
Galliformes - Phasianidae
e Indian Peafowl; Common Peafowl;
Peafowl; Peahen f; Peacock m;
Indian Peacock; Blue Peafowl
f Paon bleu
d Pfau; Blauer Pfau
i Pavone comune; Pavone

7181 *Pavo muticus*
Galliformes - Phasianidae
e Green Peafowl; Green Peahen f;
Peacock m; Peafowl; Congo
Peafowl; Congo Peacock; African
Peafowl
f Paon spicifère
d Ährenpfau
i Pavone verde; Pavone spicifera

7182 Pedionomidae
Charadriiformes
e Plains Wanderers
f Pedionomidés
d Trappenlaufhühnchen; Steppenläufer
i Pedionomidi

7183 *Pedionomus torquatus*
Charadriiformes - Pedionomidae

e Plains Wanderer; Collared
Hemipode; Collared Plains
Wanderer; Turkey Quail; Plain
Wanderer
f Pédionome errant
d Trappenlaufhühnchen; Steppenläufer
i Vagabondo delle pianure; Emipode
dal collare

7184 *Pelagodroma marina*
Procellariiformes - Hydrobatidae
e White-faced Storm-Petrel; Mother
Carey's Chicken (CSA); Mother
Carey's Chick; Frigate Petrel; White-
browed Storm-Petrel
f Océanite frégate; Pétrel frégate
d Weißgesichtsturmschwalbe;
Fregattensturmschwalbe
i Uccello delle tempeste fregata

7185 *Pelargopsis amauropterus*
Coraciiformes - Halcyonidae
e Brown-winged Kingfisher; Brown-
winged Stork-billed Kingfisher
f Martin-chasseur à ailes brunes
d Braunflügelgurial
i Martin pescatore alibrune

7186 *Pelargopsis capensis*
Coraciiformes - Halcyonidae
e Stork-billed Kingfisher; Brown-
headed Stork-billed Kingfisher (ISC)
f Martin-chasseur gurial
d Gurial; Storchschnabelgurial
i Martin pescatore dal becco di
cicogna

7187 *Pelargopsis melanorhyncha*
Coraciiformes - Halcyonidae
e Black-billed Kingfisher; Celebes
Stork-billed Kingfisher; Great-billed
Kingfisher
f Martin-chasseur à bec noir
d Schwarzschnabelgurial
i Martin pescatore becconero

7188 Pelecanidae
Pelecaniformes
e Pelicans
f Pélicans; Pélécanidés

d Pelikane
i Pelecanidi; Pelecani

7189 Pelecaniformes
e Pelicans
f Pélécaniformes
d Ruderfüßler; Ruderfüsser
i Pelecaniformi; Pelecani

7190 *Pelecanoides garnotii*
Procellariiformes - Pelecanoididae
e Peruvian Diving-Petrel
f Puffinure de Garnot
d Garnot-Sturmvogel
i Petrello tuffatore del Perù

7191 *Pelecanoides georgicus*
Procellariiformes - Pelecanoididae
e South Georgia Diving-Petrel;
Georgian Diving-Petrel; South
Georgian Diving-Petrel
f Puffinure de Géorgie du Sud
d Breitschnabelsturmvogel
i Petrello tuffatore georgiano

7192 *Pelecanoides magellani*
Procellariiformes - Pelecanoididae
e Magellanic Diving-Petrel; Magellan's
Diving-Petrel; Magellan Diving-
Petrel
f Puffinure de Magellan
d Magellan-Sturmvogel
i Petrello tuffatore di Magellano

7193 *Pelecanoides urinatrix*
Procellariiformes - Pelecanoididae
e Common Diving-Petrel; Tristan
Diving-Petrel; Subarctic Diving-
Petrel; Diving Petrel; Smaller
Diving-Petrel; Berard's Diving-Petrel
f Puffinure plongeur
d Lummensturmvogel
i Petrello tuffatore comune; Petrello
tuffatore

7194 Pelecanoididae
Procellariiformes
e Diving-Petrels
f Pelecanoididés
d Tauchersturmvögel;

Lummensturmvögel
i Pelecanoididi

7195 *Pelecanus conspicillatus*
Pelecaniformes - Pelecanidae
e Australian Pelican; Spectacled
Pelican
f Pélican à lunettes
d Brillenpelikan
i Pellicano australiano

7196 *Pelecanus crispus*
Pelecaniformes - Pelecanidae
e Dalmatian Pelican
f Pélican frisé
d Krauskopfpelikan
i Pellicano riccio

7197 *Pelecanus erythrorhynchos*
Pelecaniformes - Pelecanidae
e American White Pelican; Rough-
billed Pelican
f Pélican d'Amérique; Pélican blanc
d'Amérique; Pélican à bec rouge
d Nashornpelikan
i Pellicano bianco americano;
Pellicano comune d'America;
Pelicano d'America

7198 *Pelecanus occidentalis*
Pelecaniformes - Pelecanidae
e Brown Pelican; Old Joe (WI);
Ganuche (WI)
f Pélican brun; Grand Gosier (Ants);
Pélican (Ants); Blague-à-diable
(Ants)
d Braunpelikan
i Pellicano bruno

7199 *Pelecanus onocrotalus*
Pelecaniformes - Pelecanidae
e Great White-Pelican; White Pelican;
Eastern White-Pelican (CSA); Rosy
Pelican (ISC); European White-
Pelican
f Pélican blanc; Pélican commun;
Pélican
d Rosapelikan; Gemeiner Pelikan;
Pelikan
i Tambau; Pellicano; Pellicano
comune; Pellicano bianco

7200 *Pelecanus philippensis*
Pelecaniformes - Pelecanidae
e Spot-billed Pelican; Grey Pelican
(ISC); Spotted-billed Pelican;
Philippine Pelican
f Pélican à bec tacheté
d Grauppelikan
i Pellicano grigio

7201 *Pelecanus rufescens*
Pelecaniformes - Pelecanidae
e Pink-backed Pelican
f Pélican blanc; Pélican gris; Pélican
roussâtre
d Rötelpelikan
i Pellicano rossiccio; Pelicano
rossiccio africano

7202 *Pelecanus thagus*
Pelecaniformes - Pelecanidae
e Peruvian Pelican; Chilean Pelican
f Pélican thage
i Pellicano peruviano

7203 *Pellorneum albiventre*
Passeriformes - Sylviidae
e Spot-throated Babbler; Brown
Babbler; Plain Babbler; White-
bellied Jungle-Babbler; Plain Brown-
Babbler; White-bellied Babbler
f Akalat à gorge tachetée
d Weißbaucherdtimalie
i Garrulo golamacchiata

7204 *Pellorneum capistratum*
Passeriformes - Sylviidae
e Black-capped Babbler; Black-capped
Jungle-Babbler
f Akalat à calotte noire
d Schwarzkappenerdtimalie
i Garrulo capinero

7205 *Pellorneum fuscocapillum*
Passeriformes - Sylviidae
e Brown-capped Babbler; Brown-
capped Jungle-Babbler; Sri Lanka
Babbler
f Akalat à calotte brune
d Braunkappenerdtimalie
i Garrulo capobruno

7206 *Pellorneum palustre*
Passeriformes - Sylviidae
e Marsh Babbler; Marsh Spotted-
Babbler
f Akalat des marais
d Sumpftimalie
i Garrulo dei paludi

7207 *Pellorneum pyrrogenys*
Passeriformes - Sylviidae
e Temminck's Babbler; Temminck's
Jungle-Babbler
f Akalat de Temminck
d Rostwangenmaustimalie
i Garrulo di Temminck

7208 *Pellorneum ruficeps*
Passeriformes - Sylviidae
e Puff-throated Babbler; Spotted
Babbler (ISC); Spotted-Jungle-
Babbler; Puff-throated Jungle-
Babbler; Streak-breasted Jungle-
Babbler; Striped Babbler; Striped
Jungle-Babbler
f Akalat à poitrine tachetée; Timalie à
poitrine tachetée
d Streifenbrusttimalie
i Garrulo maculato; Timalide a petto
macchiettato

7209 *Pellorneum tickelli*
Passeriformes - Sylviidae
e Buff-breasted Babbler; Tickell's
Babbler; Tickell's Jungle-Babbler;
Mountain Brown-Babbler;
Büttikofer's Babbler
f Akalat de Tickell; Timalie de Tickell
d Rotbrustmaustimalie
i Garrulo di Tickell

7210 *Peltops blainvillii*
Passeriformes - Corvidae
e Lowland Peltops; Lowland Peltops-
Flycatcher; Clicking Peltops;
Clicking Peltops-Flycatcher
f Peltopse des plaines
d Waldpeltops
i Uccello scudato di pianura

7211 *Peltops montanus*
Passeriformes - Corvidae

e Mountain Peltops; Mountain Peltops-
  Flycatcher; Singing Peltops
f Peltopse des montagnes
d Bergpeltops
i Uccello scudato di montagna

7212 *Penelope albipennis*
  Galliformes - Cracidae
e White-winged Guan
f Pénélope à ailes blanches
d Weißschwingenguan
i Penelope alibianche

7213 *Penelope argyrotis*
  Galliformes - Cracidae
e Band-tailed Guan; Banded Guan
f Pénélope à queue barrée
d Bindenschwanzguan
i Penelope codafasciata

7214 *Penelope barbata*
  Galliformes - Cracidae
e Bearded Guan
f Pénélope barbue
d Bartguan
i Penelope barbuta

7215 *Penelope dabbenei*
  Galliformes - Cracidae
e Red-faced Guan; Dabbene's Guan
f Pénélope de Dabbene
d Rotgesichtguan
i Penelope facciarossa

7216 *Penelope jacquacu*
  Galliformes - Cracidae
e Spix's Guan; Green-backed Guan
f Pénélope de Spix
d Spix-Guan
i Penelope di Spix

7217 *Penelope jacucaca*
  Galliformes - Cracidae
e White-browed Guan; Brown Guan
f Pénélope à front blanc
d Schakukaka
i Penelope dai sopraccigli bianchi

7218 *Penelope marail*
  Galliformes - Cracidae
e Marail Guan; Cayenne Guan

f Pénélope marail
d Cayenne-Guan
i Penelope di foresta

7219 *Penelope montagnii*
  Galliformes - Cracidae
e Andean Guan
f Pénélope des Andes; Pénélope
  andine
d Anden-Guan
i Penelope delle Ande

7220 *Penelope obscura*
  Galliformes - Cracidae
e Dusky-legged Guan; Dusky Guan;
  Common Guan; Guan
f Pénélope yacouhou
d Bronzeguan
i Penelope zampescure

7221 *Penelope ochrogaster*
  Galliformes - Cracidae
e Chestnut-bellied Guan
f Pénélope à ventre roux
d Rotbrustguan
i Penelope pettocastano

7222 *Penelope ortoni*
  Galliformes - Cracidae
e Baudo Guan; Orton's Guan
f Pénélope d'Orton
d Orton-Guan
i Penelope di Baudo

7223 *Penelope perspicax*
  Galliformes - Cracidae
e Cauca Guan
f Pénélope de Cauca
d Cauca-Guan
i Penelope di Cauca

7224 *Penelope pileata*
  Galliformes - Cracidae
e White-crested Guan; White-headed
  Guan; Pileated Guan; Green-backed
  Guan
f Pénélope à poitrine rousse
d Weißschopfguan
i Penelope crestabianca

7225   *Penelope purpurascens*
Galliformes - Cracidae
e   Crested Guan; Purple Guan; Purplish
Guan
f   Pénélope panachée; Pénélope huppée
d   Rostbauchguan
i   Penelope crestata

7226   *Penelope superciliaris*
Galliformes - Cracidae
e   Rusty-margined Guan; Superciliated
Guan; White-eyebrowed Guan
f   Pénélope péoa
d   Schakupemba
i   Penelope jacupemba; Penelope

7227   *Penelopides affinis*
Coraciiformes - Bucerotidae
e   Mindanao Hornbill; Tarictic
Hornbill; Visayan Hornbill; Rufous-
tailed Hornbill; Mindanao Tarictic-
Hornbill
f   Calao de Mindanao; Calao tarictic
d   Tariktik-Hornvogel
i   Bucero di Mindanao

7228   *Penelopides exarhatus*
Coraciiformes - Bucerotidae
e   Sulawesi Hornbill; Celebes Hornbill;
Temminck's Hornbill; Sulawesi
Tarictic-Hornbill; Sulawesi Dwarf
Hornbill; Celebes Tarictic-Hornbill
f   Calao des Célèbes; Calao à
cannelures
d   Celebes-Hornvogel
i   Bucero di Sulawesi

7229   *Penelopides manillae*
Coraciiformes - Bucerotidae
e   Luzon Hornbill; Luzon Tarictic-
Hornbill; Tarictic Hornbill
f   Calao de Manille
i   Bucero di Luzon

7230   *Penelopides mindorensis*
Coraciiformes - Bucerotidae
e   Mindoro Hornbill; Mindoro Tarictic-
Hornbill
f   Calao de Mindoro
i   Bucero di Mindoro

7231   *Penelopides panini*
Coraciiformes - Bucerotidae
e   Tarictic Hornbill; Visayan Tarictic-
Hornbill; Rufous-tailed Hornbill;
Panay Tarictic-Hornbill
d   Tariktik-Hornvogel
i   Bucero codarossiccia

7232   *Penelopides samarensis*
Coraciiformes - Bucerotidae
e   Samar Hornbill
f   Calao de Samar
i   Bucero di Samar

7233   *Penelopina nigra*
Galliformes - Cracidae
e   Highland Guan; Black Penelopina;
Black Chachalaca; Black Pajuil;
Black Guan; Little Guan; Little
Penelopina
f   Pénélope pajuil
d   Schluchtenguan
i   Penelopina nera

7234   *Peneothello bimaculatus*
Passeriformes - Petroicidae
e   White-rumped Robin; White-rumped
Thicket-Flycatcher; White-winged
Robin; White-winged Flyrobin;
White-winged Thicket-Flycatcher;
White-winged Thicket-Robin
f   Miro à croupion blanc
d   Weißbürzeldickichtschnäpper
i   Balia bimaculata

7235   *Peneothello cryptoleucus*
Passeriformes - Petroicidae
e   Smoky Robin; Grey Thicket-
Flycatcher
f   Miro ombré
d   Fahlbauchdickichtschnäpper
i   Balia fuligginosa

7236   *Peneothello cyanus*
Passeriformes - Petroicidae
e   Blue-grey Robin; Slaty Thicket-
Flycatcher; Slaty Blue-Robin;
Salvadori's Blue-Robin; Slaty Robin-
Flycatcher
f   Miro gris-bleu

d Graubauchdickichtschnäpper
i Balia blu-antracite

7237 ***Peneothello sigillatus***
Passeriformes - Petroicidae
e White-winged Robin; White-winged Thicket-Flycatcher
f Miro à ailes blanches
d Spiegeldickichtschnäpper
i Balia alibianche

7238 ***Percnostola caurensis***
Passeriformes - Formicariidae
e Caura Antbird
f Alapi du Caura
d Caura-Ameisenvogel

7239 ***Percnostola leucostigma***
Passeriformes - Formicariidae
e Spot-winged Antbird
f Alapi ponctué
d Sternflügelameisenvogel

7240 ***Percnostola lophotes***
Passeriformes - Formicariidae
e White-lined Antbird; Rufous-crested Antbird
f Alapi à huppe rousse; Alapi de Berlioz
d Rotschopfameisenvogel; Langschopfameisenvogel

7241 ***Percnostola rufifrons***
Passeriformes - Formicariidae
e Black-headed Antbird
f Alapi à tête noire
d Mohrenkopfameisenvogel

7242 ***Percnostola schistacea***
Passeriformes - Formicariidae
e Slate-coloured Antbird
f Alapi ardoisé
d Einfarbameisenvogel

7243 ***Perdicula argoondah***
Galliformes - Phasianidae
e Rock Bush-Quail; Rock Quail
f Perdicule argoondah
d Madras-Wachtel
i Quaglia delle rocce

7244 ***Perdicula asiatica***
Galliformes - Phasianidae
e Jungle Bush-Quail; Jungle Quail; Ceylon Jungle Bush-Quail
f Perdicule rousse-gorge
d Frankolinwachtel
i Quaglia della giungla

7245 ***Perdicula erythrorhyncha***
Galliformes - Phasianidae
e Painted Bush-Quail; Bush-Quail
f Perdicule à bec rouge
d Buntwachtel
i Quaglia beccorosso

7246 ***Perdicula manipurensis***
Galliformes - Phasianidae
e Manipur Bush-Quail; Manipur Quail; Assam Quail; Assam Bush-Quail
f Perdicule du Manipur
d Manipur-Wachtel
i Quaglia di Manipur

7247 ***Perdix daurica***
Galliformes - Phasianidae
e Daurian Partridge; Bearded Partridge
f Perdrix de Daourie
d Bartrebhuhn
i Starna daurica

7248 ***Perdix hodgsoniae***
Galliformes - Phasianidae
e Tibetan Partridge; Hodgson's Partridge
f Perdrix de Hodgson
d Tibet-Rebhuhn
i Starna tibetana

7249 ***Perdix perdix***
Galliformes - Phasianidae
e Grey Partridge; Partridge; Hungarian Partridge (NA); Common Partridge; English Partridge
f Perdrix grise
d Rebhuhn
i Starna; Pernice griggia

7250 ***Pericrocotus brevirostris***
Passeriformes - Corvidae
e Short-billed Minivet
f Minivet à bec court

d   Kurzschnabelmennigvogel;
    Bergmennigvogel
i   Uccello di fiamma beccocorto

7251   *Pericrocotus cantonensis*
         Passeriformes - Corvidae
e   Brown-rumped Minivet; Canton
    Minivet; Swinhoe's Minivet
f   Minivet de Swinhoe
i   Uccello di fiamma dal groppone
    bruno

7252   *Pericrocotus cinnamomeus*
         Passeriformes - Corvidae
e   Small Minivet; Little Minivet; Lesser
    Minivet
f   Minivet oranor
d   Zwergmennigvogel
i   Uccello di fiamma piccolo

7253   *Pericrocotus divaricatus*
         Passeriformes - Corvidae
e   Ashy Minivet
f   Minivet cendré
d   Graumennigvogel
i   Uccello di fiamma di cenere

7254   *Pericrocotus erythropygius*
         Passeriformes - Corvidae
e   White-bellied Minivet; Jerdon's
    Minivet
f   Minivet à ventre blanc
d   Weißbauchmennigvogel
i   Uccello di fiamma ventrebianco

7255   *Pericrocotus ethologus*
         Passeriformes - Corvidae
e   Long-tailed Minivet; Flame-coloured
    Minivet
f   Minivet rouge
d   Langschwanzmennigvogel
i   Uccello di fiamma codalunga

7256   *Pericrocotus flammeus*
         Passeriformes - Corvidae
e   Scarlet Minivet; Orange Minivet;
    Indioan Minivet; Flame Minivet
f   Grand Minivet
d   Scharlach Mennigvogel
i   Uccello di fiamma scarlatto; Uccello
    fiamma; Uccello vermiglio

7257   *Pericrocotus igneus*
         Passeriformes - Corvidae
e   Fiery Minivet
f   Minivet flamboyant
i   Uccello di fiamma; Uccello di fuoco

7258   *Pericrocotus lansbergei*
         Passeriformes - Corvidae
e   Flores Minivet; Sumbawa Minivet;
    Little Minivet; Orange-breasted
    Minivet
f   Minivet de Sumbawa
d   Sumbawa-Mennigvogel
i   Uccello di fiamma di Flores

7259   *Pericrocotus minlatus*
         Passeriformes - Corvidae
e   Sunda Minivet
f   Minivet vermillon
d   Sunda-Mennigvogel
i   Uccello di fiamma della Sonda

7260   *Pericrocotus roseus*
         Passeriformes - Corvidae
e   Rosy Minivet
f   Minivet rose
d   RosenMennigvogel
i   Uccello di fiamma rosato

7261   *Pericrocotus solaris*
         Passeriformes - Corvidae
e   Grey-chinned Minivet; Yellow-
    throated Minivet; Mountain Minivet;
    Grey-throated Minivet
f   Minivet mandarin; Minivet
    montagnard
d   Graukinnmennigvogel
i   Uccello di fiamma golagrigia

7262   *Pericrocotus tegimae*
         Passeriformes - Corvidae
e   Ryukyu Minivet
f   Minivet des Ryukyu
i   Uccello di fiamma di Ryukyu

7263   *Periporphyrus erythromelas*
         Passeriformes - Fringillidae
e   Red-and-black Grosbeak
f   Cardinal érythromèle
d   Schwarzkopfkardinal
i   Beccogrosso rosso e nero

7264 **Perisoreus canadensis**
Passeriformes - Corvidae
e Grey Jay; Canada Jay; Gray Jay
(NA); Whisky Jack; Moosebird;
Meatbird; Hudson Bay Bird; Camp
Robber; Cariboubird; Oregon Jay;
Labrador Jay; Rocky Mountain Jay
f Mésangeai du Canada; Geai du
Canada
d Meisenhäher; Kanadischer
Unglückshäher
i Ghiandaia canadese

7265 **Perisoreus infaustus**
Passeriformes - Corvidae
e Siberian Jay; Boreal Jay; Northern
Jay; Red-tailed Jay; Grey Jay
f Mésangeai imitateur; Geai de Sibérie
d Unglückshäher
i Ghiandaia siberiana

7266 **Perisoreus internigrans**
Passeriformes - Corvidae
e Sichuan Jay; Szechwan Grey-Jay;
Sichuan Grey-Jay; Szechuan Grey-
Jay; Sooty Jay; Black-headed Grey-
Jay; Black-headed Jay; Chinese Jay
f Mésangeai du Sitchouan; Mesangeai
du Setchouan
d Szetschwan-Häher
i Ghiandaia del Szechwan

7267 **Perisoreus obscurus**
Passeriformes - Corvidae
e Oregon Jay

7268 **Perissocephalus tricolor**
Passeriformes - Tyrannidae
e Capuchinbird; Calfbird
f Coracine chauve
d Kapuzinervogel
i Uccello cappuccino

7269 **Pernis apivorus**
Falconiformes - Accipitridae
e European Honeybuzzard (NA);
Honeybuzzard; Western
Honeybuzzard; Common
Honeybuzzard; Eurasian
Honeybuzzard
f Bondrée apivore

d Wespenbussard
i Falco pecchiaiolo; Pecchiaiolo;
Pecchiaiolo occidentale

7270 **Pernis celebensis**
Falconiformes - Accipitridae
e Barred Honeybuzzard
f Bondrée des Célèbes
d Celebes-Wespenbussard
i Pecchiaiolo barrato; Pecchiaiolo
striato

7271 **Pernis ptilorhynchus**
Falconiformes - Accipitridae
e Oriental Honeybuzzard; Crested
Honeybuzzard; Honey Buzzard
(ISC); Eurasian Honeybuzzard;
Honey Kite
f Bondré orientale
d Schopfwespenbussard
i Pecchiaiolo orientale

7272 **Petrochelidon pyrrhonota**
Passeriformes - Hirundinidae
e Cliff Swallow; American Cliff-
Swallow; North American Cliff-
Swallow
f Hirondelle à front blanc; Hirondelle
blanc (Ants)
d Fahlstirnschwalbe
i Rondine rupestre americana

7273 **Petrochelidon rufocollaris**
Passeriformes - Hirundinidae
e Chestnut-collared Swallow; Peruvian
Swallow
f Hirondelle à bande rousse
i Rondine dal collare castano

7274 **Petroica archboldi**
Passeriformes - Petroicidae
e Snow Mountain-Robin; Rock Robin-
Flycatcher; Rock Robin; Alpine
Robin
f Miro des rochers
d Felsenschnäpper
i Petroica della Snow Mountain

7275 **Petroica australis**
Passeriformes - Petroicidae
e New Zealand Robin; Robin-

Flycatcher; New Zealand Robin-
Flycatcher; Toutouwai
f Miro rubisole
d Langbeinschnäpper
i Petroica bruna neozelandese

7276 *Petroica bivittata*
Passeriformes - Petroicidae
e Alpine Robin; Forest Robin; Forest
Robin-Flycatcher; Mountain Robin-
Flycatcher
f Miro montagnard
d Bergwaldschnäpper
i Petroica alpina

7277 *Petroica goodenovii*
Passeriformes - Petroicidae
e Red-capped Robin; Red-capped
Robin-Flycatcher
f Miro à front rouge
d Rotstirnschnäpper
i Petroica fronterossa

7278 *Petroica macrocephala*
Passeriformes - Petroicidae
e Tomtit; Pied Tit; New Zealand Tit
f Miro mésange
d Gelbbrustpetroica; Maorischnäpper;
Scheckmeisenschnäpper
i Petroica della Nuova Zelanda

7279 *Petroica multicolor*
Passeriformes - Petroicidae
e Scarlet Robin; Scarlet Robin-
Flycatcher; New Zealand Tomtit
f Miro écarlate
d Scharlachschnäpper;
Vielfarbenscnäpper
i Petroica pettirossa

7280 *Petroica multicolor multicolor*
Passeriformes - Petroicidae
e Norfolk Island Scarlet-Robin (ANZ)

7281 *Petroica phoenicea*
Passeriformes - Petroicidae
e Flame Robin; Flame Robin-
Flycatcher
f Miro embrasé
d Flammenbrustschnäpper
i Petroica di fiamma

7282 *Petroica rodinogaster*
Passeriformes - Petroicidae
e Pink Robin; Pink Robin-Flycatcher
f Miro incarnat
d Rosenbrustschnäpper
i Petroica ventrerosa

7283 *Petroica rosea*
Passeriformes - Petroicidae
e Rose Robin; Rose Robin-Flycatcher
f Miro rose
d Rosenschnäpper
i Petroica pettorosa

7284 *Petroica traversi*
Passeriformes - Petroicidae
e Chatham Islands Robin; Black
Robin; Chatham Islands Robin-
Flycatcher; Chatham Black-Robin
f Miro des Chatham
d Chatham-Schnäpper
i Petroica dell'Isola Chatham

7285 Petroicidae
Passeriformes
e Australo-Papuan Robins;
Australasian Robins
f Petroicidés
d Südseeschnäpper
i Petroicidi

7286 *Petronia dentata*
Passeriformes - Passeridae
e Bush Petronia; Lesser Rock-Sparrow;
Bush Sparrow; Lesser Petronia
f Petit Moineau; Petit Moineau
soulcie; Moineau soulcie pâle
d Buschsperling
i Passera lagia minore

7287 *Petronia petronia*
Passeriformes - Passeridae
e Rock-Sparrow; Rock Petronia;
Streaked Rock-Sparrow; Eurasian
Rock-Sparrow; European Rock-
Sparrow
f Moineau soulcie
d Steinsperling
i Passera lagia; Passera lagia europea

7288 **Petronia pyrgita**
Passeriformes - Passeridae
e Yellow-spotted Petronia; Abyssinian
Yellow-throated Sparrow; Yellow-
spotted Rock-Sparrow; Abyssinian
Yellow-throated Petronia; Sudan
Petronia; Sudan Rock-Sparrow
f Moineau à point jaune; Moineau
soulcie à point jaune
d Kehlflecksperling
i Passera lagia golagialla

7289 **Petronia superciliaris**
Passeriformes - Passeridae
e Southern Yellow-throated Sparrow;
Yellow-throated Sparrow; Yellow-
throated Petronia; South African
Rock-Sparrow; South African
Petronia
f Moineau à gorge jaune; Moineau
bridé; Moineau à sourcils; Moineau
soulcie austral
d Gelbkehlsperling;
Augenbrauensperling
i Passera lagia del Sudafrica

7290 **Petronia xanthocollis**
Passeriformes - Passeridae
e Chestnut-shouldered Petronia;
Yellow-throated Petronia; Yellow-
throated Sparrow; Yellow-throated
Rock-Sparrow; Indian Yellow-
throated Sparrow
d Gelbhalssperling; Gelbkehlsperling
i Passera lagia maculata; Passera lagia
orientale

7291 **Petrophassa albipennis**
Columbiformes - Columbidae
e White-quilled Rock-Pigeon; White-
winged Rock-Pigeon; White-winged
Bronzewing
f Colombine des rochers
d Weißflügelsteintaube;
Weißspiegeltaube
i Piccione albipenne

7292 **Petrophassa rufipennis**
Columbiformes - Columbidae
e Chestnut-quilled Rock-Pigeon; Red-
quilled Rock-Pigeon; Rufous-quilled

Rock-Pigeon; Red-winged Rock-
Pigeon
f Colombine rufipenne
d Rotspiegelftaube;
Schopfwachteltaube
i Piccione rufipenne

7293 **Peucedramus taeniatus**
Passeriformes - Fringillidae
e Olive Warbler
f Fauvine des pins
d Trugwaldsänger
i Pseudoparula olivacea

7294 **Pezopetes capitalis**
Passeriformes - Fringillidae
e Large-footed Finch; Large-footed
Brush-Finch; Big-footed Sparrow
f Tohi à grands pieds
d Großfußbuschammer
i Fringuello piedigrossi

7295 **Pezophaps solitaria**
Passeriformes - Fringillidae
e Rodriguez Solitaire
f Dronte de Rodriguez
d Einsiedlerlori; Einsidler
i Solitario di Rodriguez; Solitario

7296 **Pezoporus occidentalis**
Psittaciformes - Psittacidae
e Night-Parrot; Night Parakeet;
Western Ground Parrot; Spinifex
Parrot
f Perruche nocturne
d Höhlensittich; Nachtsittich
i Parrochetto nocturno

7297 **Pezoporus wallicus**
Psittaciformes - Psittacidae
e Ground Parrot; Swamp Parrot;
Button-grass Parrot; Green Parrot;
Ground Parakeet
f Perruche terrestre
d Erdsittich
i Parrochetto terragnolo

7298 **Pezoporus wallicus flaviventris**
Psittaciformes - Psittacidae
e Western Ground-Parrot (ANZ)

7299   ***Pezoporus wallicus wallicus***
        Psittaciformes - Psittacidae
e   Eastern Ground-Parrot (ANZ)

7300   ***Phacellodomus dorsalis***
        Passeriformes - Furnariidae
e   Chestnut-backed Thornbird
f   Synallaxe à dos marron
d   Rotrückenbündelnister
i   Espinero dorsocastano

7301   ***Phacellodomus erythrophthalmus***
        Passeriformes - Furnariidae
e   Red-eyed Thornbird
f   Synallaxe aux yeux rouges
d   Rotaugenbündelnister
i   Espinero occhirossi

7302   ***Phacellodomus maculipectus***
        Passeriformes - Furnariidae
e   Spot-breasted Thornbird
f   Synallaxe maculé

7303   ***Phacellodomus ruber***
        Passeriformes - Furnariidae
e   Greater Thornbird; Yellow-eyed
        Thornbird
f   Synallaxe rouge
d   Rotschwingenbündelnister
i   Espinero maggiore

7304   ***Phacellodomus rufifrons***
        Passeriformes - Furnariidae
e   Rufous-fronted Thornbird; Common
        Thornbird; Thornbird; Rufous-
        throated Thornbird; Plain-fronted
        Thornbird
f   Synallaxe à front roux
d   Rotstirnbündelnister
i   Espinero fronterossiccia

7305   ***Phacellodomus sibilatrix***
        Passeriformes - Furnariidae
e   Little Thornbird
f   Synallaxe siffleur
d   Zwergbündelnister
i   Espinero piccolo

7306   ***Phacellodomus striaticeps***
        Passeriformes - Furnariidae
e   Streak-fronted Thornbird

f   Synallaxe de Lafresnaye
d   Schuppenkopfbündelnister
i   Espinero frontestriata

7307   ***Phacellodomus striaticollis***
        Passeriformes - Furnariidae
e   Freckle-breasted Thornbird; Freckled
        Thornbird
f   Synallaxe rousselé; Synallaxe des
        épines
d   Fleckenbrustbündelnister
i   Espinero collostriato

7308   ***Phaenicophaeus calyorhynchus***
        Cuculiformes - Cuculidae
e   Yellow-billed Malkoha; Celebes
        Malkoha; Fiery-billed Malkoha
f   Malcoha à bec peint
d   Buntschnabelkuckuck
i   Malcoha di Sulawesi

7309   ***Phaenicophaeus chlorophaeus***
        Cuculiformes - Cuculidae
e   Raffles's Malkoha
f   Malcoha de Raffles
d   Bubu
i   Malcoha di Raffles

7310   ***Phaenicophaeus cumingi***
        Cuculiformes - Cuculidae
e   Scale-feathered Malkoha; Scale-
        feathered Cuckoo
f   Malcoha frisé
d   Schuppenhalskuckuck
i   Malcoha squamosa

7311   ***Phaenicophaeus curvirostris***
        Cuculiformes - Cuculidae
e   Chestnut-breasted Malkoha; Palawan
        Malkoha
f   Malcoha rouverdin
d   Schimmerkuckuck
i   Malcoha pettocastano

7312   ***Phaenicophaeus diardi***
        Cuculiformes - Cuculidae
e   Black-bellied Malkoha; Lesser
        Green-billed Malkoha
f   Malcoha de Diard
d   Diard-Kuckuck
i   Malcoha minore

7313   *Phaenicophaeus javanicus*
Cuculiformes - Cuculidae
e   Red-billed Malkoha
f   Malcoha javanais
d   Kastanienbauchkuckuck
i   Malcoha beccorosso

7314   *Phaenicophaeus leschenaultii*
Cuculiformes - Cuculidae
e   Sirkeer Malkoha; Sirkeer Cuckoo
(ISC); Sirkeer; Southern Serkeer
f   Malcoha sirkir
d   Sirkir
i   Malcoha di Sirkeer

7315   *Phaenicophaeus pyrrhocephalus*
Cuculiformes - Cuculidae
e   Red-faced Malkoha
f   Malcoha à face rouge
d   Malcoha
i   Malcoha facciarossa

7316   *Phaenicophaeus sumatranus*
Cuculiformes - Cuculidae
e   Chestnut-bellied Malkoha; Rufous-
bellied Malkoha
f   Malcoha à ventre roux
i   Malcoha ventrerossiccio

7317   *Phaenicophaeus superciliosus*
Cuculiformes - Cuculidae
e   Red-crested Malkoha; Rough-crested
Cuckoo
f   Malcoha à sourcils rouges
i   Malcoha dai sopraccigli

7318   *Phaenicophaeus tristis*
Cuculiformes - Cuculidae
e   Green-billed Malkoha; Large Green-
billed Malkoha (ISC); Greater Green-
billed Malkoha
f   Malcoha sombre
d   Kokil
i   Malcoha beccoverde grosso

7319   *Phaenicophaeus viridirostris*
Cuculiformes - Cuculidae
e   Blue-faced Malkoha; Small Green-
billed Malkoha (ISC)
f   Malcoha à bec vert

d   Gabelkuckuck
i   Malcoha beccoverde minore

7320   *Phaenicophilus palmarum*
Passeriformes - Fringillidae
e   Black-crowned Palm-Tanager;
Hispaniolan Palm-Tanager
f   Quatre-yeux (Ants); Tangara à
couronne noire
d   Schwarzscheitelpalmist
i   Tangara delle palme capinera

7321   *Phaenicophilus poliocephalus*
Passeriformes - Fringillidae
e   Grey-crowned Palm-Tanager
f   Quatre-yeux (Ants); Tangara quâtre-
yeux
d   Grauscheitelpalmist
i   Tangara delle palme capogrigio

7322   *Phaenostictus mcleannani*
Passeriformes - Thamnophilidae
e   Ocellated Antbird
f   Fourmilier ocellé
d   Halsbandameisenvogel
i   Mangiaformiche ocellato

7323   *Phaeochroa cuivieri*
Apodiformes - Trochilidae
e   Scaly-breasted Hummingbird;
Cuvier's Hummingbird; Cuvier's
Scaly-breasted Hummingbird;
Cuvier's Sabrewing
f   Colibri de Cuivier
d   Schuppenbrustkolibri
i   Colibrì di Cuvier

7324   *Phaeochroa roberti*
Apodiformes - Trochilidae
e   Robert's Hummingbird
f   Colibri de Robert

7325   *Phaeomyias murina*
Passeriformes - Tyrannidae
e   Mouse-coloured Tyrannulet
f   Tyranneau souris
d   Brauenfliegenstecher
i   Tiranno piccolo grigiotopo

7326   *Phaeoprogne tapera*
Passeriformes - Hirundinidae

e   Brown-chested Martin
f   Hirondelle tapère
d   Braunbrustschwalbe
i   Rondine pettobruno

7327   *Phaethon aethereus*
Pelecaniformes - Phaethontidae
e   Red-billed Tropicbird; Red-billed
Bosunbird; Lesser Red-billed
Tropicbird
f   Phaéton à bec rouge; Paille en queue
(Ants); Paille en queue à bec rouge
(Ants); Cibérou (Ants); Couac
(Ants); Grand Phaéton
d   Rotschnabeltropikvogel
i   Fetonte dal becco rosso; Fetonte
beccorosso

7328   *Phaethon lepturus*
Pelecaniformes - Phaethontidae
e   White-tailed Tropicbird; Yellow-
billed Tropicbird; White-tailed
Bosunbird; Yellow-billed Bosunbird;
Scissor-tail (WI); Boatswainbird
(WI); Bosunbird (WI); Long-tailed
Tropicbird; Golden Tropicbird
f   Phaéton à bec jaune; Flèche en queue
(Ants); Phaéton à bec jaune (Ants);
Petit Paille-en-queue (Qué)
d   Weißschwanztropikvogel
i   Fetonte beccogiallo; Fetonte a becco
giallo

7329   *Phaethon lepturus dorotheae*
Pelecaniformes - Phaethontidae
e   Indian Ocean White-tailed Tropicbird
(ANZ

7330   *Phaethon lepturus fulvus*
Pelecaniformes - Phaethontidae
e   Christmas Island White-tailed
Tropicbird (ANZ)

7331   *Phaethon rubricauda*
Pelecaniformes - Phaethontidae
e   Red-tailed Tropicbird; Red-tailed
Bosunbird; Silver Bosunbird
f   Phaéton à brins rouges
d   Rotschwanztropikvogel
i   Fetonte codarossa

7332   **Phaethontidae**
Pelecaniformes
e   Tropicbirds
f   Phaéthontidés
d   Tropikvögel
i   Fetoneidi; Fetontidi

7333   *Phaethornis adolphi*
Apodiformes - Trochilidae
e   Dusky Hermit; Boucard's Hermit
i   Colibrì del sole

7334   *Phaethornis anthophilus*
Apodiformes - Trochilidae
e   Pale-bellied Hermit
f   Ermite anthophile
d   Fahlbaucheremit
i   Colibrì del sole beccochiaro

7335   *Phaethornis augusti*
Apodiformes - Trochilidae
e   Sooty-capped Hermit; Salle's Hermet
f   Ermite d'Auguste
d   Kappeneremit
i   Colibrì del sole capofuligginoso

7336   *Phaethornis baroni*
Apodiformes - Trochilidae
e   Baron's Hermit
i   Colibrì del sole

7337   *Phaethornis bourcieri*
Apodiformes - Trochilidae
e   Straight-billed Hermit; Bourcier's
Hermit
f   Ermite de Bourcier
d   Geradschnabeleremit
i   Colibrì del sole di Bourcier

7338   *Phaethornis eurynome*
Apodiformes - Trochilidae
e   Scale-throated Hermit
f   Ermite eurynome
d   Schuppenkehleremit
i   Colibrì del sole golascaliosa

7339   *Phaethornis gounellei*
Apodiformes - Trochilidae
e   Broad-tipped Hermit; Gounell's
Hermit
f   Ermite de Gounelle

d Rostbarteremit
i Colibrì del sole di Gounelle

7340 *Phaethornis griseogularis*
Apodiformes - Trochilidae
e Grey-chinned Hermit; Grey-throated
Hermit; Gould's Hermit
f Ermite à gorge grise
d Graukinneremit
i Colibrì del sole golagrigia

7341 *Phaethornis griseoventer*
Apodiformes - Trochilidae
e Jalisco Hermit

7342 *Phaethornis guy*
Apodiformes - Trochilidae
e Green Hermit; Guy's Hermit; Guy's
White-tailed Hermit
f Ermite vert; Ermite de Guy
d Graubrusteremit
i Colibrì del sole verde

7343 *Phaethornis hispidus*
Apodiformes - Trochilidae
e White-bearded Hermit
f Ermite d'Osery
d Weißbarteremit
i Colibrì del sole barbabianca

7344 *Phaethornis idaliae*
Apodiformes - Trochilidae
e Minute Hermit; Obscure Hermit
f Ermite d'Idalie
d Braunkehleremit
i Colibrì del sole minuta

7345 *Phaethornis koepckeae*
Apodiformes - Trochilidae
e Koepcke's Hermit
f Ermite de Koepcke
d Koepke-Eremit
i Colibrì del sole di Koepcke

7346 *Phaethornis longirostris*
Apodiformes - Trochilidae
e Long-tailed Hermit

7347 *Phaethornis longuemareus*
Apodiformes - Trochilidae
e Little Hermit; Longemare's Hermit

f Ermite nain; Ermite de Longuemare
d Zwergeremit
i Colibrì del sole piccolo

7348 *Phaethornis malaris*
Apodiformes - Trochilidae
e Great-billed Hermit; Nordmann's
Hermit
f Ermite à long bec
d Langschnabeleremit
i Colibrì del sole beccogrande

7349 *Phaethornis malaris margarettae*
Apodiformes - Trochilidae
e Margaretta's Hermit
f Ermite de Margaretta
d Margareten-Eremit

7350 *Phaethornis maranhaoensis*
Apodiformes - Trochilidae
e Maranhao Hermit
f Ermite de Maranhao
d Maranhao-Eremit

7351 *Phaethornis mexicanus*
Apodiformes - Trochilidae
e Hartert's Hermit; Mexican Hermit

7352 *Phaethornis nattereri*
Apodiformes - Trochilidae
e Cinnamon-throated Hermit;
Natterer's Hermit
f Ermite de Natterer
d Zimtkehleremit
i Colibrì del sole Natterer

7353 *Phaethornis philippii*
Apodiformes - Trochilidae
e Needle-billed Hermit; De Philippi's
Hermit
f Ermite de De Filippi
d Dünnschnabeleremit
i Colibrì del sole beccoacuto

7354 *Phaethornis pretrei*
Apodiformes - Trochilidae
e Planalto Hermit; Prêtre's Hermit
f Ermite de Prêtre
d Planalto-Eremit
i Colibrì del sole del Planalto

7355   *Phaethornis ruber*
        Apodiformes - Trochilidae
   e   Reddish Hermit; Pygmy Hermit
   f   Ermite roussâtre; Ermite pygmée
   d   Rotbaucheremit
   i   Colibrì del sole rossastro

7356   *Phaethornis rupurumii*
        Apodiformes - Trochilidae
   e   Streak-throated Hermit

7357   *Phaethornis squalidus*
        Apodiformes - Trochilidae
   e   Dusky-throated Hermit; Medium
        Hermit; Sooty-throated Hermit
   f   Ermite terne
   d   Schwarzkehleremit
   l   Colibrì del sole golabruna

7358   *Phaethornis striigularis*
        Apodiformes - Trochilidae
   e   Stripe-throated Hermit

7359   *Phaethornis stuarti*
        Apodiformes - Trochilidae
   e   White-browed Hermit; Stuart's
        Hermit
   f   Ermite de Stuart
   d   Weißkehleremit
   i   Colibrì del sole di Stuart

7360   *Phaethornis subochraceus*
        Apodiformes - Trochilidae
   e   Buff-bellied Hermit
   f   Ermite ocré
   d   Ockerbaucheremit
   i   Colibrì del sole ventrefulvo

7361   *Phaethornis superciliosus*
        Apodiformes - Trochilidae
   e   Rusty-breasted Hermit; Long-tailed
        Hermit; Guiana Hermit; Cayenne
        Hermit; Allied Hermit; Buff-browed
        Hermit; Mexican Hermit
   f   Ermite à brins blancs
   d   Langschwanzeremit
   i   Colibrì del sole codalunga

7362   *Phaethornis syrmatophorus*
        Apodiformes - Trochilidae
   e   Tawny-bellied Hermit; Train-bearing

        Hermit; Baron Rothschild's Hermit
   f   Ermite à ventre fauve
   d   Braunbaucheremit
   i   Colibrì del sole ventrefulvo

7363   *Phaethornis yaruqui*
        Apodiformes - Trochilidae
   e   White-whiskered Hermit; Yaruquian
        Hermit; Black-winged Hermit
   f   Ermite yaruqui
   d   Blauschwanzeremit
   i   Colibrì del sole dai mustacchi bianchi

7364   *Phaetusa simplex*
        Charadriiformes - Laridae
   e   Gull-billed Tern; Large-billed Tern
   f   Sterne à gros bec
   d   Großschnabelseeschwalbe
   i   Sterna beccogrosso

7365   *Phainopepla nitens*
        Passeriformes - Bombycillidae
   e   Phainopepla
   f   Phénopèple luisante; Phénopèple
   d   Trauerseidenschnäpper;
        Glanzseidenschnäpper
   i   Fainopepla; Fainopepla nera

7366   *Phainoptila melanoxantha*
        Passeriformes - Bombycillidae
   e   Black-and-yellow Silky-Flycatcher;
        Phainoptila; Salvin's Silky Flycatcher
   f   Phénoptile noir-et-jaune; Phénoptile
        jaune-et-noir
   d   Gelbflankenseidenschnäpper
   i   Uccello di seta giallo e nero

7367   **Phalacrocoracidae**
        Pelecaniformes
   e   Cormorants
   f   Cormorans; Phalacrocoracidés
   d   Kormorane; Scharben
   i   Falacrocoracidi; Cormorani

7368   *Phalacrocorax africanus*
        Pelecaniformes - Phalacrocoracidae
   e   Long-tailed Cormorant; Reed
        Cormorant; Reed Duiker
   f   Cormoran africain
   d   Riedscharbe
   i   Marangone minore dal becco giallo;

Cormorano africano; Cormorano codalunga

7369 *Phalacrocorax albiventer*
Pelecaniformes - Phalacrocoracidae
e King Cormorant; White-necked Shag
f Cormoran à ventre blanc
d Königsscharbe

7370 *Phalacrocorax aristotelis*
Pelecaniformes - Phalacrocoracidae
e European Shag (NA); Shag;
Common Shag; Green Cormorant;
Green Shag
f Cormoran huppé; Cormoran largup
d Krähenscharbe; Mittelmeerkormoran
i Marangone dal ciuffo; Marangone
col ciuffo

7371 *Phalacrocorax atriceps*
Pelecaniformes - Phalacrocoracidae
e Imperial Shag; King Cormorant;
Blue-eyed Cormorant; Imperial
Cormorant (CSA); Magellan Blue-
eyed Shag
f Cormoran impérial
d Blauaugenscharbe; Königsscharbe
i Cormorano imperiale

7372 *Phalacrocorax atriceps nivalis*
Pelecaniformes - Phalacrocoracidae
e Heard Shag; Heard Island Shag;
Blue-eyed Shag; Heard Island
Imperial-Shag (ANZ)
f Cormoran de Heard

7373 *Phalacrocorax atriceps
purpurascens*
Pelecaniformes - Phalacrocoracidae
e Macquaric Island Imperial-Shag
(ANZ)
f Cormoran de Macquarie

7374 *Phalacrocorax auritus*
Pelecaniformes - Phalacrocoracidae
e Double-crested Cormorant; White-
crested Cormorant
f Cormoran à aigrettes; Cormoran
(Ants)
d Ohrenscharbe
i Marangone dalla doppia cresta

7375 *Phalacrocorax bougainvillii*
Pelecaniformes - Phalacrocoracidae
e Guanay Cormorant
f Cormoran de Bougainville
d Guanoscharbe; Guanokormoran
i Marangone di Bougainville;
Cormorano del guano

7376 *Phalacrocorax bransfieldensis*
Pelecaniformes - Phalacrocoracidae
e Antarctic Shag
f Cormoran antarctique
i Cormorano antartico

7377 *Phalacrocorax brasilianus*
Pelecaniformes - Phalacrocoracidae
e Neotropic Cormorant; Olivaceous
Cormorant
f Cormoran vigua
d Bigua-Scharbe
i Cormorano olivaceo

7378 *Phalacrocorax campbelli*
Pelecaniformes - Phalacrocoracidae
e Campbell Island Shag; Campbell
Island Cormorant; Campbell Shag;
Campbell Cormorant
f Cormoran de Campbell
d Campbell-Scharbe
i Marangone dell'Isola Campbell

7379 *Phalacrocorax capensis*
Pelecaniformes - Phalacrocoracidae
e Cape Cormorant; Duiker (CSA)
f Cormoran du Cap
d Kap-Scharbe; Kap-Kormoran
i Cormorano del Capo

7380 *Phalacrocorax capillatus*
Pelecaniformes - Phalacrocoracidae
e Japanese Cormorant; Temninck's
Cormorant
f Cormoran de Temminck
d Japan-Kormoran
i Cormorano giapponese

7381 *Phalacrocorax carbo*
Pelecaniformes - Phalacrocoracidae
e Great Cormorant; Cormorant; Black
Cormorant; Common Cormorant;
White-breasted Cormorant; Large

Cormorant (ISC); White-necked
Cormorant; European Cormorant;
Large Black-Cormorant; Large Shag;
Large Black-Shag; Black Shag
f   Grand Cormoran; Cormoran
ordinaire; Cormoran noir; Cormoran
d   Kormoran; Weißbrustkormoran
ı   Marangone; Cormorano comune;
Cormorano; Marangone comune

7382   *Phalacrocorax carbo sinensis*
Pelecaniformes - Phalacrocoracidae
f   Grand Cormoran continental

7383   *Phalacrocorax carunculatus*
Pelecaniformes - Phalacrocoracidae
e   Rough-faced Shag; New Zealand
Shag; Cook Straights Cormorant;
New Zealand King-Cormorant; King
Shag; New Zealand King-Shag;
Carrunculated Shag; Malborough
Sound Shag
f   Cormoran caronculé
d   Warzenscharbe
i   Marangone caruncolato

7384   *Phalacrocorax chalconotus*
Pelecaniformes - Phalacrocoracidae
e   Bronze Shag; Stewart Island Shag;
Bronzed Shag; Stewart Shag; Stewart
Cormorant; Stuart Island Cormorant;
Gray's Shag
f   Cormoran bronzé
i   Marangone bronzato

7385   *Phalacrocorax colensoi*
Pelecaniformes - Phalacrocoracidae
e   Auckland Islands Shag; Auckland
Shag; Auckland Islands Cormorant;
Auckland Cormorant
f   Cormoran des Auckland
i   Marangone delle Isole Auckland

7386   *Phalacrocorax coronatus*
Pelecaniformes - Phalacrocoracidae
e   Crowned Cormorant
f   Cormoran couronné
d   Wahlberg-Scharbe
i   Cormorano coronato

7387   *Phalacrocorax featherstoni*
Pelecaniformes - Phalacrocoracidae
e   Pitt Island Shag; Pitt Shag; Pitt Island
Cormorant
f   Cormoran de Featherston
i   Cormorano delle Isole Chatham

7388   *Phalacrocorax fuscescens*
Pelecaniformes - Phalacrocoracidae
e   Black-faced Cormorant; Black-faced
Shag; Black-and-white Shag; White-
browed Cormorant; White-breasted
Cormorant
f   Cormoran de Tasmanie
d   Braunwangenscharbe
i   Cormorano minore nero

7389   *Phalacrocorax fuscicollis*
Pelecaniformes - Phalacrocoracidae
e   Indian Cormorant; Indian Shag
f   Cormoran à cou brun
d   Gould-Scharbe
i   Cormorano indiano

7390   *Phalacrocorax gaimardi*
Pelecaniformes - Phalacrocoracidae
e   Red-legged Cormorant; Red-footed
Shag; Red-legged Shag; Red-footed
Cormorant; Gaimard's Cormorant
f   Cormoran de Gaimard
d   Buntscharbe; Rotfußkormoran
i   Cormorano zamperosse; cormorano
di Gaimard

7391   *Phalacrocorax georgianus*
Pelecaniformes - Phalacrocoracidae
e   South Georgia Shag; Georgia Shag;
South Georgia Cormorant
f   Cormoran géorgien
i   Marangone della Georgia australe

7392   *Phalacrocorax harrisi*
Pelecaniformes - Phalacrocoracidae
e   Flightless Cormorant; Galapagos
Cormorant; Galapagos Flightless-
Cormorant
f   Cormoran aptère
d   Galapagos-Sharbe;
Stummelkormoran
i   Cormorano delle Galapagos

7393 **Phalacrocorax kenyoni**
Pelecaniformes - Phalacrocoracidae
e Amchitka Cormorant; Kenyon's Shag
f Cormoran de Kenyon

7394 **Phalacrocorax lucidus**
Pelecaniformes - Phalacrocoracidae
e White-breasted Cormorant
f Cormoran à poitrine blanche
d Weißbrustkormoran
i Cormorano pettobianco

7395 **Phalacrocorax magellanicus**
Pelecaniformes - Phalacrocoracidae
e Rock Shag; Rock Cormorant;
Magellan Cormorant; Magellan Shag
f Cormoran de Magellan; Cormoran
magellanique
d Felsenscharbe
i Cormorano di Magellano

7396 **Phalacrocorax melanogenis**
Pelecaniformes - Phalacrocoracidae
e Crozet Shag; Marion Island Shag;
Marion Shag
f Cormoran de Crozet

7397 **Phalacrocorax melanoleucos**
Pelecaniformes - Phalacrocoracidae
e Little Pied-Cormorant; Little Black-
and-white Cormorant; White-
throated Shag; Little Shag; Little
Cormorant; Frilled Shag; Little
River-Shag; White-throated
Cormorant
f Cormoran pie
d Kräuselscharbe
i Marangone minore australiano;
Marangone minore bianco e nero

7398 **Phalacrocorax neglectus**
Pelecaniformes - Phalacrocoracidae
e Bank Cormorant; Duiker (CSA)
f Cormoran des bancs
d Küstenscharbe
i Cormorano di riva africano

7399 **Phalacrocorax niger**
Pelecaniformes - Phalacrocoracidae
e Little Cormorant; Javan Cormorant;
Javanese Cormorant

f Cormoran de Vieillot
d Mohrenscharbe
i Cormorano di Giava

7400 **Phalacrocorax nigrogularis**
Pelecaniformes - Phalacrocoracidae
e Socotra Cormorant
f Cormoran de Socotra
d Sokotra-Kormoran;
Schwarzgesichtscharbe
i Cormorano di Socotra

7401 **Phalacrocorax olivaceus**
Pelecaniformes - Phalacrocoracidae
e Olivaceous Cormorant; Neotropic
Cormorant; Bigua Cormorant
f Cormoran Bigua; Cormoran de Bigua
d Bigua-Scharbe
i Cormorano olivaceo

7402 **Phalacrocorax onslowi**
Pelecaniformes - Phalacrocoracidae
e Chatham Islands Shag; Chatham
Shag; Chatham Cormorant; Chatham
Islands Cormorant
f Cormoran des Chatham
i Marangone di Onslow

7403 **Phalacrocorax pelagicus**
Pelecaniformes - Phalacrocoracidae
e Pelagic Cormorant; Pelagic Shag;
Baird's Cormorant
f Cormoran pélagique
d Meerscharbe
i Cormorano pelagico

7404 **Phalacrocorax penicillatus**
Pelecaniformes - Phalacrocoracidae
e Brandt's Cormorant; Brown
Cormorant; Pencilled Cormorant;
Townsend's Cormorant
f Cormoran de Brandt
d Pinselscharbe
i Cormorano di Brandt

7405 **Phalacrocorax perspicillatus**
Pelecaniformes - Phalacrocoracidae
e Pallas's Cormorant; Spectacled
Cormorant; Steller's Cormorant
f Cormoran de Pallas

d Brillenscharbe
i Cormorano dagliocchiali

7406   *Phalacrocorax punctatus*
Pelecaniformes - Phalacrocoracidae
e Spotted Shag; Spotted Cormorant;
Blue Shag
f Cormoran moucheté
d Tüpfelscharbe
i Cormorano macchiato

7407   *Phalacrocorax pygmeus*
Pelecaniformes - Phalacrocoracidae
e Pygmy Cormorant; African Pygmy-
Cormorant; Little Cormorant
f Cormoran pygmée
d Zwergscharbe
i Marangore minore; Marangone nano;
Cormorano pigmeo; Marangone
pigmeo

7408   *Phalacrocorax ranfurlyi*
Pelecaniformes - Phalacrocoracidae
e Bounty Islands Shag; Bounty Shag;
Bounty Cormorant; Bounty Islands
Cormorant
f Cormoran de Bounty
i Marangone delle Isole Bounty

7409   *Phalacrocorax sulcirostris*
Pelecaniformes - Phalacrocoracidae
e Little Black-Cormorant; Little Black-
Shag; Large Cormorant; Large
Black-Shag
f Cormoran noir
d Schwarzscharbe
i Cormorano nero

7410   *Phalacrocorax urile*
Pelecaniformes - Phalacrocoracidae
e Red-faced Cormorant; Red-faced
Shag
f Cormoran à face rouge
d Rotgesichtscharbe
i Cormorano facciarossa

7411   *Phalacrocorax varius*
Pelecaniformes - Phalacrocoracidae
e Pied Cormorant; Yellow-faced
Cormorant; Pied Shag; Larger Pied-
Cormorant; Greater Pied-Cormorant;

Large Black-and-white Cormorant;
Yellow-faced Shag; Black-and-white
Cormorant; Black-and-white Shag
f Cormoran varié
d Elsterscharbe
i Cormorano bianco e nero

7412   *Phalacrocorax verrucosus*
Pelecaniformes - Phalacrocoracidae
e Kerguelen Shag; Kerguelen
Cormorant
f Cormoran de Kerguélen
d Kerguelen-Scharbe
i Cormorano delle Isole Kerguelen

7413   *Phalaenoptilus nuttallii*
Caprimulgiformes - Caprimulgidae
e Common Poorwill; Poorwill
f Engoulevent de Nuttall
d Poorwill; Winternachtschwalbe
i Succiacapre di Nuttall

7414   *Phalaropus fulicaria*
Charadriiformes - Scolopacidae
e Red Phalarope; Grey Phalarope; Flat-
nosed Phalarope
f Phalarope à bec large; Phalarope
roux
d Thorshühnchen; Breitschnabeliger
Wassertreter; Thorswassertreter
i Falaropo beccolargo; Falarope a
becco largo

7415   *Phalaropus lobatus*
Charadriiformes - Scolopacidae
e Red-necked Phalarope; Northern
Phalarope (NA); Round-nosed
Phalarope
f Phalarope à bec étroit; Phalarope
hyperboréen
d Odinshühnchen; Schmalschnabeliger
Wassertreter; Odinswassertreter
i Falaropo beccosottile

7416   *Phalaropus tricolor*
Charadriiformes - Scolopacidae
e Wilson's Phalarope
f Phalarope de Wilson; Phalarope à
bec mince
d Wilson-Wassertreter; Willsons
Wassertreter; Amerikanisches

Odinshühnchen; Wilson-Hühnchen
i Falaropo di Wilson

7417 Phalcoboenus albogularis
Falconiformes - Falconidae
e White-throated Caracara; Darwin's
Caracara
f Caracara à gorge blanche
d Weißkehlkarakara
i Caracara golabianca; Caracara dal
petto bianco

7418 Phalcoboenus australis
Falconiformes - Falconidae
e Striated Caracara; Forster's Caracara
f Caracara austral
d Falkland-Karakara
i Caracara di Forster

7419 Phalcoboenus carunculatus
Falconiformes - Falconidae
e Carunculated Caracara
f Caracara caronculé
d Streifenkarakara
i Caracara caruncolato

7420 Phalcoboenus megalopterus
Falconiformes - Falconidae
e Mountain-Caracara
f Caracara montagnard
d Bergkarakara
i Caracara aligrandi; Caracara dei
montagne

7421 Phapitreron amethystina
Columbiformes - Columbidae
e Amethyst Brown-Dove; Amethyst
Dove; Greater Brown Fruit-Pigeon;
Greater Brown Fruit-Dove; Dark-
eared Brown-Dove; Dark-eared
Dove; Amethyst Fruit-Dove;
Amethyst Brown Fruit-Dove;
Amethystine Brown-Pigeon; Greater
White-eared Fruit-Pigeon
f Phapitréron améthyste
d Amethysttaube; Cebu-Amethysttaube
i Colomba frugivora ametistina

7422 Phapitreron cinereiceps
Columbiformes - Columbidae
e Dark-eared Brown-Dove; Dark-eared

Dove; Southern Brown Fruit-Dove
f Phapitréron à oreillons bruns
i Colomba frugivora orecchiescure

7423 Phapitreron leucotis
Columbiformes - Columbidae
e White-eared Brown-Dove; White-
eared Dove; Lesser Brown Fruit-
Pigeon; Brown Fruit-Pigeon; Lesser
Brown Fruit-Dove
f Phapitréron à oreillons blancs
d Ohrstreiftaube
i Colomba frugivora guancebianche

7424 Phaps chalcoptera
Columbiformes - Columbidae
e Common Bronzewing; Bronzewing;
Common Bronzewing Pigeon; Scrub
Bronzewing; Forest Bronzewing
f Colombine lumachelle
d Bronzeflügeltaube; Gewöhnliche
Bronzeflügeltaube
i Piccione alibronzate comune

7425 Phaps elegans
Columbiformes - Columbidae
e Brush Bronzewing; Little Bronze-
Pigeon; Brown Bronzewing Pigeon
f Colombine élégante
d Buschtaube; Kleine
Bronzeflügeltaube
i Piccione alibronzate elegante;
Piccione alibronzate

7426 Phaps histrionica
Columbiformes - Columbidae
e Flock Bronzewing; Flock Pigeon;
Harlequin Bronzewing; Harlequin
Pigeon
f Colombine arlequin
d Harlekintaube
i Piccione alibronzate arlecchino

7427 Pharomachrus antisianus
Trogoniformes - Trogonidae
e Crested Quetzal; Crested Trogon
f Quetzal antisien
d Kammtrogon
i Quetzal crestato; Trogone crestato

7428   *Pharomachrus auriceps*
Trogoniformes - Trogonidae
e   Golden-headed Quetzal; Golden-
headed Trogon
f   Quetzal doré
d   Goldkopftrogon
i   Quetzal testadorata; Trogone
testadorato

7429   *Pharomachrus fulgidus*
Trogoniformes - Trogonidae
e   White-tipped Quetzal
f   Quetzal brillant
d   Glanztrogon
i   Quetzal fulgido; Trogone fulgido

7430   *Pharomachrus mocinno*
Trogoniformes - Trogonidae
e   Resplendent Quetzal; White-tipped
Trogon; Magnificent Quetzal;
Northern Quetzal
f   Quetzal resplendissant
d   Quesal; Quetzal
i   Trogone splendido; Quetzal; Quetzal
splendente; Caluro splendente

7431   *Pharomachrus pavoninus*
Trogoniformes - Trogonidae
e   Pavonine Quetzal; Pavonine Trogon;
Golden-headed Quetzal
f   Quetzal pavonin
d   Pfauentrogon
i   Quetzal pavonino; Trogone pavonino

7432   Phasianidae
Galliformes
e   Pheasants; Quails; Partridges
f   Faisans; Phasianidés
d   Glattfußhühner; Fasanenartige
Vögel; Fasanvögel; Fasanartige
Vögel; Fasane
i   Fasianidi; Fagiani; Quaglie

7433   *Phasianus colchicus*
Galliformes - Phasianidae
e   Common Pheasant; Pheasant; Ring-
necked Pheasant (NA); English
Pheasant; Chinese Pheasant
f   Faisan de Colchide; Faisan de chasse
d   Fasan; Ringfasan; Jagdfasan

i   Fagiano comune; Fagiano; Gaiano
nobile

7434   *Phasianus colchicus mongolicus*
Galliformes - Phasianidae
e   Mongolian Pheasant

7435   *Phasianus versicolor*
Galliformes - Phasianidae
e   Green Pheasant; Japanese Pheasant
f   Faisan versicolore
d   Buntfasan
i   Fagiano giapponese

7436   *Phedina borbonica*
Passeriformes - Hirundinidae
e   Mascarene Martin; Madagascar
Martin
f   Hirondelle des Mascareignes
d   Maskarenen-Schwalbe;
Maskarenschwalbe
i   Topino delle Mascarene

7437   *Phedina brazzae*
Passeriformes - Hirundinidae
e   Brazza's Martin; Congo Martin;
Brazza's Swallow
f   Hirondelle de Brazza
d   Brazza-Schwalbe
i   Topino di Brazza

7438   *Phegornis mitchellii*
Charadriiformes - Charadriidae
e   Diademed Sandpiper-Plover;
Diademed Plover
f   Pluvier des Andes
d   Bänderegenpfeifer
i   Piviere di Mitchell

7439   *Phelpsia inornata*
Passeriformes - Tyrannidae
e   White-bearded Flycatcher
f   Tyran des llanos
d   Weißbarttyrann
i   Tiranno barbabianca

7440   *Pheucticus aureoventris*
Passeriformes - Fringillidae
e   Black-backed Grosbeak
f   Cardinal à dos noir

d   Goldbauchkernknacker
i   Beccogrosso dorsonero

7441   *Pheucticus chrysogaster*
Passeriformes - Fringillidae
e   Golden-bellied Grosbeak; Southern
Yellow Grosbeak; Yellow Grosbeak;
Yellow-bellied Grosbeak
f   Cardinal à tête jaune
i   Beccogrosso ventregiallo

7442   *Pheucticus chrysopeplus*
Passeriformes - Fringillidae
e   Yellow Grosbeak; Mexican Yellow
Grosbeak
f   Cardinal jaune
d   Gelbkopfkernknacker;
Gelbkopfkardinal; Gelbkardinal
i   Beccogrosso giallo

7443   *Pheucticus ludovicianus*
Passeriformes - Fringillidae
e   Rose-breasted Grosbeak; Common
Grosbeak; Grosbeak; Black-headed
Grosbeak
f   Cardinal à poitrine rouge; Cardinal à
poitrine rose
d   Rosenbrustkernknacker; Bischof;
Schwarzkopfkernknacker;
Rosenbrustknacker
i   Beccogrosso pettorosso; Beccogrosso
pettorosa

7444   *Pheucticus melanocephalus*
Passeriformes - Fringillidae
e   Black-headed Grosbeak
f   Cardinal à tête noire
d   Schwarzkopfkernknacker;
Schwarzkopfknacker
i   Beccogrosso testanera

7445   *Pheucticus tibialis*
Passeriformes - Fringillidae
e   Black-thighed Grosbeak
f   Cardinal à cuisses noires
i   Beccogrosso dai calzoni neri

7446   *Phibalura flavirostris*
Passeriformes - Tyrannidae
e   Swallow-tailed Cotinga
f   Cotinga à queue fourchue; Cotinga à

bec jaune
d   Gabelschwanzzuser;
Gabelschwanzkotinga
i   Cotinga codadirondine

7447   *Phigys solitarius*
Psittaciformes - Loriidae
e   Collared Lory; Solitary Lory; Ruffed
Lory
f   Lori des Fidji
d   Einsiedlerlori; Einsiedler
i   Lori solitario

7448   *Philemon albitorques*
Passeriformes - Meliphagidae
e   Manus Friarbird; White-naped
Friarbird; Admiralty Friarbird;
Admiralty Island Friarbird
f   Polochion à nuque blanche
d   Manus-Lederkopf
i   Uccello frate nucabianca

7449   *Philemon argenticeps*
Passeriformes - Meliphagidae
e   Silver-crowned Friarbird
f   Polochion couronné
d   Weißscheitellederkopf
i   Uccello frate testacanuta

7450   *Philemon brassi*
Passeriformes - Meliphagidae
e   Brass's Friarbird
f   Polochion de Brass
d   Mamberano-Lederkopf
i   Uccello frate di Brass

7451   *Philemon buceroides*
Passeriformes - Meliphagidae
e   Helmeted Friarbird; Timor
Helmeted-Friarbird; Noisy Friarbird;
Sandstone Friarbird; Mansoor
Friarbird
f   Polochion casqué
d   Helmlederkopf
i   Uccello frate bucero

7452   *Philemon citreogularis*
Passeriformes - Meliphagidae
e   Little Friarbird; Yellow-throated
Friarbird; Yellow-throated Friarbird
f   Polochion à menton jaune

d   Glattstirnlederkopf
i   Uccello frate piccolo

7453   *Philemon cockerelli*
Passeriformes - Meliphagidae
e   New Britain Friarbird; Bismarck
    Friarbird
f   Polochion de Nouvelle-Bretagne
d   Cockerell-Lederkopf
i   Uccello frate della Nuova Britannia

7454   *Philemon corniculatus*
Passeriformes - Meliphagidae
e   Noisy Friarbird; Leatherhead; Bald
    Friarbird
f   Polochion criard
d   Lärmlederkopf;
    Kahlscheitellederkopf
i   Uccello frate schiamazzante

7455   *Philemon diemenensis*
Passeriformes - Meliphagidae
e   New Caledonian Friarbird; New
    Caledonia Friarbird
f   Polochion moine
d   Lesson-Lederkopf
i   Uccello frate della Nuova Caledonia

7456   *Philemon eichhorni*
Passeriformes - Meliphagidae
e   New Ireland Friarbird; Eichhorn's
    Friarbird; Helmeted Friarbird
f   Polochion de Nouvelle-Irlande
d   Eichhorn-Lederkopf
i   Uccello frate della Nuova Irlanda

7457   *Philemon fuscicapillus*
Passeriformes - Meliphagidae
e   Dusky Friarbird; Morotai Friarbird;
    Moluccan Friarbird
f   Polochion sombre
d   Morotai-Lederkopf
i   Uccello frate caposcuro

7458   *Philemon inornatus*
Passeriformes - Meliphagidae
e   Timor Friarbird; Plain Friarbird
f   Polochion sobre
d   Timor-Lederkopf
i   Uccello frate disadorno

7459   *Philemon kisserensis*
Passeriformes - Meliphagidae
e   Grey Friarbird
f   Polochion de Kisar
d   Banda-Lederkopf
i   Uccello frate grigio

7460   *Philemon meyeri*
Passeriformes - Meliphagidae
e   Meyer's Friarbird
f   Polochion de Meyer
d   Zwerglederkopf
i   Uccello frate di Meyer

7461   *Philemon moluccensis*
Passeriformes - Meliphagidae
e   Black-faced Friarbird; Moluccan
    Friarbird; Moluccas Friarbird
f   Polochion des Moluques
d   Buru-Lederkopf
i   Uccello frate delle Molucche

7462   *Philemon novaeguineae*
Passeriformes - Meliphagidae
e   New Guinea Friarbird; Melville
    Island Firebird; Leatherhead
f   Polochion de Nouvelle-Guinée
i   Uccello frate della Nuova Guinea

7463   *Philemon subcorniculatus*
Passeriformes - Meliphagidae
e   Grey-necked Friarbird; Ceram
    Friarbird
f   Polochion de Céram
d   Ceram-Lederkopf
i   Uccello frate collogrigio

7464   *Philentoma pyrhopterum*
Passeriformes - Corvidae
e   Rufous-winged Philentoma;
    Chestnut-winged Monarch
f   Philentome à ailes rousses
d   Kastanienflügelschnäpper
i   Filentoma alirosse

7465   *Philentoma velatum*
Passeriformes - Corvidae
e   Maroon-breasted Philentoma;
    Maroon-breasted Monarch; Maroon-
    breasted Flycatcher
f   Philentome à poitrine marron

d   Kastanienbrustschnäpper
i   Filentoma pettomarrone

7466   *Philepitta castanea*
Passeriformes - Philepittidae
e   Velvet Asity; Velvety Asity
f   Philépitte veloutée
d   Seidenjala; Schwarzlappenpitta;
Seidenlappenpitta
i   Asity velluto

7467   *Philepitta schlegeli*
Passeriformes - Philepittidae
e   Schlegel's Asity
f   Philépitte de Schlegel
d   Gelbbauchjala
i   Asity di Schlegel

7468   **Philepittidae**
Passeriformes
e   Asities
f   Phillépittidés
d   Lappenpittas; Jalas
i   Filepittidi

7469   *Philetairus socius*
Passeriformes - Passeridae
e   Social Weaver; Sociable Weaver
(CSA); Common Social-Weaver
f   Républicain social; Tisserin social
d   Siedelweber; Siedelsperling
i   Passero repubblicano

7470   *Philodice bryantae*
Apodiformes - Trochilidae
e   Magenta-throated Woodstar; Costa
Rican Woodstar; Costa Rica
Woodstar
f   Colibri magenta
d   Magentakolibri
i   Stella dei boschi golamagenta;
Colibrì dall gola magenta

7471   *Philodice mitchellii*
Apodiformes - Trochilidae
e   Purple-throated Woodstar; Mitchell's
Woodstar
f   Colibri de Mitchell
d   Mitchell-Kolibri
i   Stella dei boschi di Mitchell

7472   *Philohydor lictor*
Passeriformes - Tyrannidae
e   Lesser Kiskadee
f   Tyran licteur
d   Liktor
i   Pitango solforato minore

7473   *Philomachus pugnax*
Charadriiformes - Scolopacidae
e   Ruff m; Reeve f
f   Chevalier combattant; Combattant
varié; Bécasseau combattant
d   Kampfläufer
i   Combattente

7474   *Philortyx fasciatus*
Galliformes - Odontophoridae
e   Banded Quail; Barred Quail
f   Colin barré
d   Bergwachtel; Bindenwachtel
i   Colino barrato

7475   *Philydor amaurotis*
Passeriformes - Furnariidae
e   White-browed Foliagegleaner;
White-browed Leafgleaner
f   Anabate bridé; Anabate oreillon-brun
d   Weißbrauenblattspäher
i   Ticotico dai sopraccigli bianchi

7476   *Philydor atricapillus*
Passeriformes - Furnariidae
e   Black-capped Foliagegleaner; Black-
capped Leafgleaner
f   Anabate à tête noire
d   Kappenblattspäher; Klettertöpfer
i   Ticotico capinero

7477   *Philydor dimidiatus*
Passeriformes - Furnariidae
e   Russet-mantled Foliagegleaner;
Planalto Foliagegleaner; Russet-
mantled Leafgleaner
f   Anabate mantelé; Anabate cannellé
d   Rostmantelblattspäher
i   Ticotico dal mantello rosso

7478   *Philydor erythrocercus*
Passeriformes - Furnariidae
e   Rufous-rumped Foliagegleaner;
Rufous-rumped Leafgleaner

f   Anabate à croupion rouge
d   Rotbürzelblattspäher
i   Ticotico gropponerossiccio

7479   *Philydor erythronotus*
Passeriformes - Furnariidae
e   Rufous-backed Foliagegleaner;
Santander Foliagegleaner
f   Anabate à dos roux
d   Rotrückenblattspäher

7480   *Philydor erythropterus*
Passeriformes - Furnariidae
e   Chestnut-winged Foliagegleaner;
Chestnut-winged Leafgleaner
f   Amanate à ailes rousses
d   Rotschwingenblattspäher
i   Ticotico alirosse

7481   *Philydor fuscipennis*
Passeriformes - Furnariidae
e   Slaty-winged Foliagegleaner; Dusky-
winged Foliagegleaner; Dusky-
backed Foliagegleaner; Rufous-
bellied Foliagegleaner
f   Anabate à ailes sombres
i   Ticotico aliardesia

7482   *Philydor lichtensteini*
Passeriformes - Furnariidae
e   Ochre-breasted Foliagegleaner;
Lichtenstein's Foliagegleaner; Ochre-
breasted Leafgleaner; Lichtenstein's
Leafgleaner
f   Anabate de Lichtenstein
d   Ockerbrustblattspäher
i   Ticotico di Lichtenstein

7483   *Philydor novaesi*
Passeriformes - Furnariidae
e   Alagoas Foliagegleaner; Novaes's
Foliagegleaner; Greater Black-
capped Foliagegleaner
f   Anabate d'Alagoas
i   Ticotico di Novaes

7484   *Philydor ochrogaster*
Passeriformes - Furnariidae
e   Ochre-bellied Foliagegleaner
f   Anabate ocré
i   Ticotico ventreocra

7485   *Philydor pyrrhodes*
Passeriformes - Furnariidae
e   Cinnamon-rumped Foliagegleaner;
Cinnamon-rumped Leafgleaner
f   Anabate flamboyant
d   Zimtbürstelblattspäher
i   Ticotico gropponecannella

7486   *Philydor ruficaudatus*
Passeriformes - Furnariidae
e   Rufous-tailed Foliagegleaner;
Rufous-tailed Leafgleaner
f   Anabate rougequeue
d   Rotschwanzblattspäher
i   Ticotico codarossiccio

7487   *Philydor rufus*
Passeriformes - Furnariidae
e   Buff-fronted Foliagegleaner; Buff-
fronted Leafgleaner
f   Anabate roux
d   Goldstirnblattspäher
i   Ticotico frontefulva

7488   *Phimosus infuscatus*
Ciconiiformes - Threskiornithidae
e   Whispering Ibis; Bare-faced Ibis
f   Ibis à face nue
d   Mohrenibis
i   Ibis faccianuda

7489   *Phlegopsis barringeri*
Passeriformes - Thamnophilidae
e   Argus Bare-eye
f   Fourmilier argus
d   Glanzkopfameisenvogel
i   Mangiaformiche di Barringer

7490   *Phlegopsis erythroptera*
Passeriformes - Thamnophilidae
e   Reddish-winged Bare-eye
f   Fourmilier à miroir roux
d   Rotspiegelameisenvogel
i   Mangiaformiche alirosse

7491   *Phlegopsis nigromaculata*
Passeriformes - Thamnophilidae
e   Black-spotted Bare-eye
f   Fourmilier maculé
d   Rotaugenameisenvogel;

Brillenameisenvogel
i Mangiaformiche neromaculato

7492 *Phleocryptes melanops*
Passeriformes - Furnariidae
e Wren-like Rushbird; Rushbird
f Synallaxe des joncs
d Rohrschlüpfer; Kappenschlüpfer
i Lavoratore scricciolo

7493 *Phlogophilus harterti*
Apodiformes - Trochilidae
e Peruvian Piedtail
f Colibri de Hartert
d Hartert-Kolibri
i Colibrì di Hartert

7494 *Phlogophilus hemileucurus*
Apodiformes - Trochilidae
e Ecuadorian Piedtail
f Colibri à queue mi-blanche
d Elsterschwänzchen
i Colibrì coda bianca e nero

7495 *Phodilus badius*
Strigiformes - Tytonidae
e Oriental Bay-Owl; Bay-Owl (ISC);
Asian Bay-Owl; Ceylon Bay-Owl
f Phodile calong
d Fratzeneule; Maskeneule
i Gufo baio orientale; Gufo baio

7496 *Phodilus prigoginei*
Strigiformes - Tytonidae
e Congo Bay-Owl; Bay Owl;
Tanzanian Bay-Owl; African Bay-
Owl; Congo Owl; Itombe Owl;
Prigogine's Owl
f Phodile de Prigogine
d Prigogine-Eule
i Gufo baio di Prigogine; Gufo baio
del Congo

7497 *Phoebetria fusca*
Procellariiformes - Diomedeidae
e Sooty Albatross; Dark-mantled
Sooty-Albatross (CSA)
f Albatros brun
d Dunkler Rußalbatros
i Albatro fuligginoso; Albatro
fuligginoso del Nord

7498 *Phoebetria palpebrata*
Procellariiformes - Diomedeidae
e Light-mantled Albatross; Light-
mantled Sooty-Albatross; Grey
Albatross; Grey-mantled Albatross
f Albatros fuligineux
d Heller Rußalbatros
i Albatro palpebrata; Albatro
fuligginoso del Sud

7499 *Phoenicircus carnifex*
Passeriformes - Tyrannidae
e Guianan Red-Cotinga; Guianian Red-
Cotinga; Red Cotinga
f Cotinga ouette
d Blutkotinga
i Cotinga rossa della Guiana

7500 *Phoenicircus nigricollis*
Passeriformes - Tyrannidae
e Black-necked Red-Cotinga; Black-
necked Cotinga
f Cotinga à col noir
d Samtkotinga
i Cotinga rossa collonero

7501 *Phoeniconaias minor*
Ciconiiformes - Phoenicopteridae
e Lesser Flamingo
f Flamant nain; Petit Flamant
d Zwergflamingo
i Fenicottero minore; Fenicottero nano

7502 *Phoenicopteridae*
Ciconiiformes
e Flamingos
f Flamants; Phoenicoptéridés
d Flamingos
i Fenicotteridi; Fenicotteri

7503 *Phoenicopterus andinus*
Ciconiiformes - Phoenicopteridae
e Andean Flamingo; Greater Andean
Flamingo
f Flamant des Andes
d Anden-Flamingo
i Fenicottero delle Ande

7504 *Phoenicopterus chilensis*
Ciconiiformes - Phoenicopteridae
e Chilean Flamingo

f   Flamant du Chili
d   Chile-Flamingo
i   Fenicottero del Cile

7505   *Phoenicopterus jamesi*
Ciconiiformes - Phoenicopteridae
e   Puna Flamingo; James's Flamingo;
Lesser Andean Flamingo
f   Flamant de James
d   James-Flamingo
i   Fenicottero di James

7506   *Phoenicopterus roseus*
Ciconiiformes - Phoenicopteridae
e   Greater Flamingo (NA); Fillymingo
(WI); Flamingo (ISC)
d   Flamingo
i   Fenicottero americano

7507   *Phoenicopterus ruber*
Ciconiiformes - Phoenicopteridae
e   Greater Flamingo; American
Flamingo; Flamingo
f   Flamant rose
d   Rosaflamingo; Flamingo; Großer
Flamingo; Roter Flamingo; Kuba-
Flamingo
i   Fenicottero; Fenicottero rosa;
Fiammingo; Fiammante

7508   Phoeniculidae
Coraciiformes
e   Wood-Hoopoes
f   Pheoniculidés
d   Baumhöpfe
i   Feniculidi

7509   *Phoeniculus bollei*
Coraciiformes - Phoeniculidae
e   White-headed Wood-Hoopoe
f   Irrisor à tête blanche; Moqueur à tête
blanche
d   Weißmaskenhopf
i   Upupa arboricola testabianca

7510   *Phoeniculus castaneiceps*
Coraciiformes - Phoeniculidae
e   Forest Wood-Hoopoe
f   Irrisor à tête brune
d   Waldhopf
i   Upupa arboricola delle foreste

7511   *Phoeniculus damarensis*
Coraciiformes - Phoeniculidae
e   Violet Wood-Hoopoe; Southern
Violet Wood-Hoopoe
f   Irrisor damara
d   Steppenbaumhopf
i   Upupa arboricola violetta

7512   *Phoeniculus granti*
Coraciiformes - Phoeniculidae
e   Grant's Wood-Hoopoe
f   Irrisor de Grant

7513   *Phoeniculus purpureus*
Coraciiformes - Phoeniculidae
e   Green Wood-Hoopoe; Red-billed
Wood-Hoopoe (CSA); Kakelaar
(CSA)
f   Irrisor moqueur
d   Damara-Baumhopf; Baumhopf
i   Upupa arboricola purpurea; Upupa
arborea dal becco rosso

7514   *Phoeniculus somaliensis*
Coraciiformes - Phoeniculidae
e   Black-billed Wood-Hoopoe
f   Irrisor à bec noir
i   Upupa arboricola della Somalia

7515   *Phoenicurus alaschanicus*
Passeriformes - Muscicapidae
e   Ala Shan Redstart; Asian Redstart;
Asian Scrub-Robin; Tibetan Scrub-
Robin; Przewalski's Redstart
f   Rougequeue de Przewalski
d   Rostkehlrotschwanz
i   Codirosso di Przewalski

7516   *Phoenicurus auroreus*
Passeriformes - Muscicapidae
e   Daurian Redstart
f   Rougequeue de Daourie;
Rougequeue aurore
d   Spiegelrotschwanz
i   Codirosso daurico

7517   *Phoenicurus caeruleocephalus*
Passeriformes - Muscicapidae
e   Blue-capped Redstart; Blue-headed
Redstart
f   Rougequeue à tête bleue

d Blaukkopfrötel
i Codirosso testablu

7518 *Phoenicurus erythrogaster*
Passeriformes - Muscicapidae
e White-winged Redstart;
Güldenstädt's Redstart; Large White-
winged Redstart; Afghan Redstart
f Rougequeue de Güldenstädt
d Riesenrotschwanz;
Weißkappenrotschwanz
i Codirosso di Güldenstädt

7519 *Phoenicurus erythronota*
Passeriformes - Muscicapidae
e Rufous-backed Redstart;
Eversmann's Redstart
f Rougequeue d'Eversmann
d Sprosserrotschwanz
i Codirosso di Eversmann

7520 *Phoenicurus frontalis*
Passeriformes - Muscicapidae
e Blue-fronted Redstart; Chinese Blue-
fronted Redstart; Blue-breasted
Redstart
f Rougequeue à front bleu
d Alpenrotschwanz
i Codirosso fronteazzurro

7521 *Phoenicurus hodgsoni*
Passeriformes - Muscicapidae
e Hodgson's Redstart
f Rougequeue de Hodgson
d Feldrotschwanz
i Codirosso di Hodgson

7522 *Phoenicurus moussieri*
Passeriformes - Muscicapidae
e Moussier's Redstart; Coroneted
Redstart
f Rougequeue diadème; Rougequeue
de Moussier; Rubiette de Moussier
d Diademrotschwanz
i Codirosso algerino

7523 *Phoenicurus ochruros*
Passeriformes - Muscicapidae
e Black Redstart
f Roguequeue noir; Rubiette rougequeue

d Hausrotschwanz
i Codirosso spazzacamino

7524 *Phoenicurus phoenicurus*
Passeriformes - Muscicapidae
e Common Redstart; Redstart;
European Redstart (NA); Eurasian
Redstart (CSA); White-fronted
Redstart
f Rougequeue à front blanc; Rubiette à
front blanc; Rougequeue des
murailles
d Gartenrotschwanz
i Codirosso; Codirosso comune

7525 *Phoenicurus schisticeps*
Passeriformes - Muscicapidae
e White-throated Redstart
f Rougequeue à gorge blanche
d Baumrötel
i Codirosso golabianca

7526 *Pholidornis rushiae*
Passeriformes - Paridae
e Tit-Hylia; Tit-Warbler; Tiny Tit-
Warbler
f Mésangette rayée; Astrild-mésange
d Strichelköpfchen
i Ilia scagliosa

7527 *Phrygilus alaudinus*
Passeriformes - Fringillidae
e Band-tailed Sierra-Finch
f Phrygile à queue barrée
d Schwanzfleckenämmerling
i Fringuello della Sierra codafasciata

7528 *Phrygilus atriceps*
Passeriformes - Fringillidae
e Black-hooded Sierra-Finch; Northern
Sierra-Finch
f Phrygile à tête noire
d Kapuzenämmerling
i Fringuello della Sierra testanera

7529 *Phrygilus carbonarius*
Passeriformes - Fringillidae
e Carbonated Sierra-Finch
f Phrygile charbonnier
d Schwarzbrustämmerling
i Fringuello della Sierra fuligginoso

7530 **Phrygilus dorsalis**
Passeriformes - Fringillidae
e Red-backed Sierra-Finch
f Phrygile à dos roux
d Braunmantelämmerling
i Fringuello della Sierra dorsocastano

7531 **Phrygilus erythronotus**
Passeriformes - Fringillidae
e White-throated Sierra-Finch
f Phrygile bicolore
d Weißkehlämmerling
i Fringuello della Sierra golabianca

7532 **Phrygilus fruticeti**
Passeriformes - Fringillidae
e Mourning Sierra-Finch
f Phrygile petit-deuil
d Strauchämmerling
i Fringuello della Sierra lamentoso

7533 **Phrygilus gayi**
Passeriformes - Fringillidae
e Grey-hooded Sierra-Finch; Gay's
Sierra-Finch; Southern Hooded
Sierra-Finch
f Phrygile à tête grise
d Cordilleraämmerling
i Fringuello della Sierra testagrigia

7534 **Phrygilus patagonicus**
Passeriformes - Fringillidae
e Patagonian Sierra-Finch; Patagonian
Hooded Sierra-Finch
f Phrygile de Patagonie
d Magellan-Ämmerling
i Fringuello della Sierra della
Patagonia; Fringuello della Patagonia

7535 **Phrygilus plebejus**
Passeriformes - Fringillidae
e Ash-breasted Sierra-Finch
f Phrygile plébéien
d Aschbrustämmerling
i Fringuello della Sierra pettocenerino

7536 **Phrygilus punensis**
Passeriformes - Fringillidae
e Peruvian Sierra-Finch; Hooded
Sierra-Finch

f Phrygile du Pérou
i Fringuello della Sierra del Perù

7537 **Phrygilus unicolor**
Passeriformes - Fringillidae
e Plumbeous Sierra-Finch
f Phrygile gris-de-plomb
d Bleiämmerling
i Fringuello della Sierra plumbeo

7538 **Phsophodes olivaceus**
Passeriformes - Paridae
e Eastern Whipbird
f Psophode à tête noire
d Schwarzschopfwippflöter

7539 **Phylidonyris albifrons**
Passeriformes - Meliphagidae
e White-fronted Honeyeater
f Méliphage à front blanc
d Weißstirnhonigfresser
i Mangiamiele frontebianca
australiano

7540 **Phylidonyris melanops**
Passeriformes - Meliphagidae
e Tawny-crowned Honeyeater;
Fulvous-crowned Honeyeater
f Méliphage à calotte fauve
d Goldscheitelhonigfresser;
Braunscheitelhonigesser
i Mangiamiele corona fulva

7541 **Phylidonyris nigra**
Passeriformes - Meliphagidae
e White-cheeked Honeyeater
f Méliphage fardé
d Weißohrhonigfresser
i Mangiamiele guancebianche

7542 **Phylidonyris notabilis**
Passeriformes - Meliphagidae
e New Hebrides Honeyeater; White-
bellied Honeyeater; White-breasted
Honeyeater
f Méliphage des Nouvelles-Hébrides
d Weißbauchhonigfresser
i Mangiamiele ventrebianco

7543 **Phylidonyris novaehollandiae**
Passeriformes - Meliphagidae

e New Holland Honeyeater; Yellow-
 winged Honeyeater; White-bearded
 Honeyeater
f Méliphage de Nouvelle-Hollande
d Weißaugenhonigfresser;
 Gelbflügelhonigesser
i Mangiamiele della Nuova Olanda

7544 *Phylidonyris pyrrhoptera*
Passeriformes - Meliphagidae
e Crescent Honeyeater; Tasmanian
 Honeyeater; Horseshoe Honeyeater
f Méliphage à croissants
d Goldflügelhonigfresser;
 Halbmondhonigesser
i Mangiamiele ferrodicavallo

7545 *Phylidonyris undulata*
Passeriformes - Meliphagidae
e Barred Honeyeater
f Méliphage barré
d Sperberhonigfresser; Gestreifter
 Honigesser
i Mangiamiele barrato

7546 *Phyllanthus atripennis*
Passeriformes - Sylviidae
e Capuchin Babbler; Black-winged
 Babbler; Chestnut Babbler
f Phyllanthe capucin; Cratérope
 capucin
d Schwarzflügeltimalie
i Garrulo cappuccino

7547 *Phyllastrephus albigularis*
Passeriformes - Pycnonotidae
e White-throated Greenbul; African
 White-throated Bulbul; White-
 throated Bulbul
f Bulbul à gorge blanche
d Shuppenstirnbülbül; Weißstirnbülbül
i Bulbul golabianca

7548 *Phyllastrephus alfredi*
Passeriformes - Pycnonotidae
e Sharpe's Greenbul; Malawi Greenbul
f Bulbul d'Alfred
i Bulbul di Sharpe

7549 *Phyllastrephus apperti*
Passeriformes - Pycnonotidae

e Appert's Greenbul; Colston's Bulbul;
 Apert's Tetraka
f Bulbul d'Appert; Bernière d'Appert
d Appert-Bülbül
i Bulbul di Appert

7550 *Phyllastrephus baumanni*
Passeriformes - Pycnonotidae
e Baumann's Olive-Greenbul;
 Baumann's Bulbul; Baumann's Olive-
 Bulbul; Baumann's Greenbul
f Bulbul de Baumann
d Baumann-Bülbül
i Bulbul di Baumann

7551 *Phyllastrephus cabanisi*
Passeriformes - Pycnonotidae
e Cabanis's Greenbul; Olive Greenbul
f Bulbul de Cabanis
d Cabanis-Bülbül
i Bulbul di Cabanis

7552 *Phyllastrephus cerviniventris*
Passeriformes - Pycnonotidae
e Grey-olive Greenbul; Grey-olive
 Bulbul
f Bulbul vert-olive
d Fahlbauchbülbül
i Bulbul grigio-oliva

7553 *Phyllastrephus cinereiceps*
Passeriformes - Pycnonotidae
e Grey-crowned Greenbul; Grey-
 crowned Foditany; Grey-crowned
 Tetraka; Grey-crowned Oxylabes
f Bulbul à tête grise; Bernière à
 courone grise
d Grauscheitelbülbül
i Bulbul corona grigia

7554 *Phyllastrephus debilis*
Passeriformes - Pycnonotidae
e Tiny Greenbul; Slender Bulbul
 (CSA); Slender Greenbul; Miniature
 Greenbul; Yellow-streaked Greenbul;
 Small Yellow-streaked Greenbul
f Bulbul minute
d Kleiner Gelbstreifenlaubbülbül;
 Schlankbülbül
i Bulbul snello

7555   *Phyllastrephus fischeri*
Passeriformes - Pycnonotidae
e   Fischer's Greenbul; Fischer's Bulbul
f   Bulbul de Fischer
d   Fischer-Bülbül
i   Bulbul di Fischer

7556   *Phyllastrephus flavostriatus*
Passeriformes - Pycnonotidae
e   Yellow-streaked Greenbul; Yellow-
streaked Bulbul (CSA); Yellow-
bellied Bulbul; Yellow-striped
Greenbul; Yellow-streaked Bulbul
f   Bulbul à striés jaunes; Bulbul à
ventre jaune
d   Gelbstreifenlaubbülbül;
Gelbstreifenbülbül
i   Bulbul striato di giallo

7557   *Phyllastrephus fulviventris*
Passeriformes Pycnonotidae
e   Pale-olive Greenbul; Pale-olive
Bulbul
f   Bulbul à ventre roux
d   Angola-Bülbül
i   Bulbul ventrefulvo

7558   *Phyllastrephus hypochloris*
Passeriformes - Pycnonotidae
e   Toro Olive-Greenbul; Toro
Greenbul; Toro Olive-Bulbul
f   Bulbul du Toro
d   Toro-Bülbül
i   Bulbul oliva di Toro

7559   *Phyllastrephus icterinus*
Passeriformes - Pycnonotidae
e   Icterine Greenbul; Lesser Icterine
Bulbul; Icterine Bulbul
f   Bulbul ictérin
d   Zeisigbülbül
i   Bulbul itterino

7560   *Phyllastrephus leucolepis*
Passeriformes - Pycnonotidae
e   Liberian Greenbul; White-winged
Greenbul
f   Bulbul du Libéria; Bulbul ictérin
tacheté
i   Bulbul alimacchiate

7561   *Phyllastrephus lorenzi*
Passeriformes - Pycnonotidae
e   Sassi's Greenbul; Sassi's Bulbul;
Sassi's Olive-Greenbul; Lorenz's
Bulbul
f   Bulbul de Lorenz
d   Lorenz-Bülbül
i   Bulbul oliva di Sassi

7562   *Phyllastrephus madagascariensis*
Passeriformes - Pycnonotidae
e   Long-billed Greenbul; Tetraka;
Common Tetraka; Common
Greenbul; Madagascar Greenbul;
Madagascar Tetraka
f   Bulbul tétraka; Bernière tétraka
d   Gmelin-Bülbül
i   Bulbul del Madagascar

7563   *Phyllastrephus placidus*
Passeriformes - Pycnonotidae
e   Placid Greenbul; Kenya Highlands
Olive-Greenbul; Kenya Highlands
Greenbul; Olive Mountain-Greenbul;
Shelley's Greenbul
f   Bulbul placide
d   Kenia-Bülbül
i   Bulbul oliva di montagna

7564   *Phyllastrephus poensis*
Passeriformes - Pycnonotidae
e   Cameroon Olive-Greenbul; Olive
Bulbul; Cameroon Bulbul; Cameroon
Greenbul
f   Bulbul olivâtre
d   Bamenda-Bülbül
i   Bulbul oliva del Camerun

7565   *Phyllastrephus poliocephalus*
Passeriformes - Pycnonotidae
e   Grey-headed Greenbul; Yellow-
bellied Greenbul; Yellow-bellied
Bulbul
f   Bulbul à ventre jaune
d   Goldbauchbülbül
i   Bulbul testagrigia africano

7566   *Phyllastrephus scandens*
Passeriformes - Pycnonotidae
e   Leaflove; Common Leaflove;
African Leaflove

 f Bulbul à queue rousse
 d Uferbülbül
 i Bulbul voltafoglie

7567 ***Phyllastrephus strepitans***
   Passeriformes - Pycnonotidae
 e Northern Brownbul; Northern
   Brown-Bulbul
 f Bulbul brun
 d Schillings Bülbül
 i Bulbul chiassoso

7568 ***Phyllastrephus tenebrosus***
   Passeriformes - Pycnonotidae
 e Dusky Greenbul; Dusky Tetraka
 f Bulbul obscur; Bernière obscure
 d Sianka Bülbül
 i Bulbul bruno

7569 ***Phyllastrephus terrestris***
   Passeriformes - Pycnonotidae
 e Terrestrial Brownbul; Terrestial
   Bulbul (CSA); Brownbul; Bristle-
   necked Brownbul
 f Bulbul jaboteur
 d Laubbülbül
 i Bulbul collosetoloso

7570 ***Phyllastrephus xavieri***
   Passeriformes - Pycnonotidae
 e Xavier's Greenbul; Greater icterine
   Greenbul; Greater Icterine Bulbul
 f Bulbul de Xavier
 d Xavier-Bülbül
 i Bulbul di Xavier

7571 ***Phyllastrephus zosterops***
   Passeriformes - Pycnonotidae
 e Spectacled Greenbul; Short-billed
   Tetraka; Short-billed Greenbul
 f Bulbul à bec court; Bernière à bnec
   court
 d Kurzschnabelbülbül
 i Bulbul beccocorte

7572 ***Phyllolais pulchella***
   Passeriformes - Sylviidae
 e Buff-bellied Warbler; Acacia
   Warbler
 f Phyllolais à ventre fauve; Apalis à
   ventre fauve; Apalis à ventre jaune

 d Akaziensänger
 i Donzella delle foglie

7573 ***Phyllomyias burmeisteri***
   Passeriformes - Tyrannidae
 e Rough-legged Tyrannulet
 f Tyranneau pattu; Tyranneau de
   Burmeister
 d Höckerfußfliegenstecher
 i Tiranno piccolo di Burmeister

7574 ***Phyllomyias cinereiceps***
   Passeriformes - Tyrannidae
 e Ashy-headed Tyrannulet
 f Tyranneau à tête cendrée
 d Graukopffliegenstecher
 i Tiranno piccolo testacenerina

7575 ***Phyllomyias fasciatus***
   Passeriformes - Tyrannidae
 e Planalto Tyrannulet
 f Tyranneau fascié; Tyranneau gris-et-
   jaune
 d Schuppenkopffliegenstecher
 i Tiranno piccolo del Planalto

7576 ***Phyllomyias griseiceps***
   Passeriformes - Tyrannidae
 e Sooty-headed Tyrannulet; Crested
   Tyrannulet
 f Tyranneau nain
 d Zwergfliegenstecher
 i Tiranno piccolo testafulligginosa

7577 ***Phyllomyias griseocapilla***
   Passeriformes - Tyrannidae
 e Grey-capped Tyrannulet
 f Tyranneau à tête grise
 d Graukappenfliegenstecher
 i Tiranno piccolo capogrigio

7578 ***Phyllomyias nigrocapillus***
   Passeriformes - Tyrannidae
 e Black-capped Tyrannulet
 f Tyranneau à tête noire
 d Mönchsfliegenstecher
 i Tiranno piccolo capinero

7579 ***Phyllomyias plumbeiceps***
   Passeriformes - Tyrannidae
 e Plumbeous-crowned Tyrannulet

f	Tyranneau plombé; Tyranneau montagnard		

d Bleikopffliegenstecher
i Tiranno piccolo corona plumbea

7580 *Phyllomyias reiseri*
Passeriformes - Tyrannidae
e Reiser's Tyrannulet
f Tyranneau de Reiser
d Reiser-Grünrückenfliegenstecher
i Tiranno piccolo di Reiser

7581 *Phyllomyias sclateri*
Passeriformes - Tyrannidae
e Sclater's Tyrannulet
f Tyranneau de Sclater
d Streifenbauchfliegenstecher
i Tiranno piccolo di Sclater

7582 *Phyllomyias urichi*
Passeriformes - Tyrannidae
e Urich's Tyrannulet

7583 *Phyllomyias uropygialis*
Passeriformes - Tyrannidae
e Tawny-rumped Tyrannulet
f Tyranneau à croupion fauve
d Goldbürzelfliegenstecher
i Tiranno piccolo grapponefulvo; Tiranno piccolo dal groppone fulvo

7584 *Phyllomyias virescens*
Passeriformes - Tyrannidae
e Greenish Tyrannulet
f Tyranneau verdin
d Grünrückenfliegenstecher
i Tiranno piccolo verdastro

7585 *Phyllomyias zeledoni*
Passeriformes - Tyrannidae
e White-fronted Tyrannulet; Zeledoni's Tyrannulet; Rough-legged Tyrannulet
f Tyranneau de Zeledon
i Tiranno piccolo frontebianca

7586 *Phylloscartes beckeri*
Passeriformes - Tyrannidae
e Bahia Tyrannulet

7587 *Phylloscartes ceciliae*
Passeriformes - Tyrannidae
e Alagoas Tyrannulet; Long-tailed Tyrannulet
f Tyranneau de Cecilia
i Tiranno piccolo di Cecilia

7588 *Phylloscartes chapmani*
Passeriformes - Tyrannidae
e Chapman's Tyrannulet; Chapman's Bristle-Tyrant
f Tyranneau de Chapman
d Weißbrauenlaubtyrann
i Tiranno piccolo di Chapman

7589 *Phylloscartes difficilis*
Passeriformes - Tyrannidae
e Serra do Mar Tyrannulet; Ihering's Tyrannulet; Difficult Tyrannulet
f Tyranneau d'Ihering
d Weißaugenlaubtyrann
i Tiranno piccolo di Ihering

7590 *Phylloscartes eximius*
Passeriformes - Tyrannidae
e Southern Bristle-Tyrant; Natterer's Tyrant
f Tyranneau distingué
d Weißaugenborstentyrann
i Tiranno piccolo meridionale

7591 *Phylloscartes flaviventris*
Passeriformes - Tyrannidae
e Rufous-lored Tyrannulet; Yellow-bellied Bristle-Tyrant
f Tyranneau masqué
d Rotzügelborstentyrann
i Tiranno piccolo ventregiallo

7592 *Phylloscartes flavovirens*
Passeriformes - Tyrannidae
e Yellow-green Tyrannulet
f Tyranneau jaune-vert; Tyranneau de Panama
d Panama-Laubtyrann
i Tiranno piccolo verde-giallo

7593 *Phylloscartes gualaquizae*
Passeriformes - Tyrannidae
e Ecuadorian Tyrannulet; Gualaquiza Bristle-Tyrant; Bristle-Tyrant

e Pale-rumped Warbler; Lemon-
 rumped Warbler
d Zilpzalp

7620 *Phylloscopus collybita*
 Passeriformes - Sylviidae
e European Chiffchaff; Eurasian
 Chiffchaff; Chiffchaff; Common
 Chiffchaff; Brown Leaf-Warbler
f Pouillot véloce; Pouillot roux
d Zilpzalp; Weidenlaubsänger
i Luí piccolo; Luicchio

7621 *Phylloscopus coronatus*
 Passeriformes - Sylviidae
e Eastern Crowned-Warbler;
 Temminck's Crowned Willow-
 Warbler; Crowned Willow-Warbler;
 Eastern Crowned Leaf-Warbler;
 Crowned Leaf-Warbler; Crowned
 Warbler
f Pouillot de Temminck; Pouillot
 couronné
d Kronenlaubsänger
i Luì coronato di Temminck

7622 *Phylloscopus davisoni*
 Passeriformes - Sylviidae
e White-tailed Leaf-Warbler; White-
 tailed Willow-Warbler; White-tailed
 Warbler; Oates's Crowned Leaf-
 Warbler; Oates's Leaf-Warbler
f Pouillot de Davison
d Weißschwanzlaubsänger
i Luì codabianca

7623 *Phylloscopus emeiensis*
 Passeriformes - Sylviidae
e Emei Leaf-Warbler

7624 *Phylloscopus fuligiventer*
 Passeriformes - Sylviidae
e Smoky Warbler; Smoky Leaf-
 Warbler; Smoky Willow-Warbler
f Pouillot enfumé
d Dunkellaubsänger; Rußlaubsänger
i Luì fuligginoso

7625 *Phylloscopus fuscatus*
 Passeriformes - Sylviidae
e Dusky Warbler; Dusky Leaf-

 Warbler; Dusky Willow-Warbler
f Pouillot brun
d Dunkellaubsänger; Dunkler
 Laubsänger; Braunlaubsänger
i Luí scuro

7626 *Phylloscopus griseolus*
 Passeriformes - Sylviidae
e Sulphur-bellied Warbler; Olivaceous
 Willow-Warbler; Jerdon's Willow-
 Warbler; Sulphur Leaf-Warbler;
 Olivaceous Leaf-Warbler; Sulphur-
 bellied Willow-Warbler; Sulphur
 Willow-Warbler; Jerdon's Leaf-
 Warbler; Greyish Willow-Warbler
f Pouillot griséole
d Pamir-Laubsänger; Felsenlaubsänger
i Luì di Jerdon

7627 *Phylloscopus hainanus*
 Passeriformes - Sylviidae
e Hainan Leaf-Warbler
f Pouillot de Hainan

7628 *Phylloscopus herberti*
 Passeriformes - Sylviidae
e Black-capped Woodland-Warbler;
 Herbert's Leaf-warbler; Herbert's
 Woodland-Warbler
f Pouillot à tête noire
d Tienschan-Laubsänger;
 Schwarzscheitellaubsänger
i Luì capinero

7629 *Phylloscopus humei*
 Passeriformes - Sylviidae
e Buff-browed Warbler; Hume's
 Warbler; Hume's Yellow-browed
 Warbler; Hume's Leaf-Warbler
f Pouillot de Hume
i Luì di Hume

7630 *Phylloscopus ijimae*
 Passeriformes - Sylviidae
e Ijima's Leaf-Warbler; Ijima's
 Willow-Warbler; Ijima's Warbler
f Pouillot d'Ijima
d Ijima-Laubsänger
i Luì di Ijima

7631   *Phylloscopus inornatus*
Passeriformes - Sylviidae
e   Yellow-browed Warbler; Inornate
Warbler; Yellow-browed Leaf-
Warbler
f   Pouillot à grands sourcils
d   Gelbbrauenlaubsänger
i   Luí forestiero

7632   *Phylloscopus kansuensis*
Passeriformes - Sylviidae
e   Gansu Leaf-Warbler

7633   *Phylloscopus laetus*
Passeriformes - Sylviidae
e   Red-faced Woodland-Warbler; Red-
faced Leaf-Warbler; Rusty-faced
Warbler
f   Pouillot à face rousse
d   Braunwangenlaubsänger
i   Luì facciarossa

7634   *Phylloscopus laurae*
Passeriformes - Sylviidae
e   Laura's Woodland-Warbler; Mrs.
Boulton's Warbler; Mrs. Boulton's
Woodland-Warbler; Laura's Leaf-
Warbler; Laura's Warbler
f   Pouillot de Laura
d   Boultons Laubsänger
i   Luì della signora Boulton

7635   *Phylloscopus lorenzii*
Passeriformes - Sylvidae
e   Caucasian Chiffchaff
i   Luì orientale

7636   *Phylloscopus maculipennis*
Passeriformes - Sylviidae
e   Ash-throated Warbler; Grey-faced
Warbler; Grey-faced Willow-
Warbler; Grey-faced Leaf-Warbler;
Ashy-throated Leaf-Warbler; Ashy-
throated Warbler
f   Pouillot à face grise
d   Graukehllaubsänger
i   Luì facciagrigia

7637   *Phylloscopus magnirostris*
Passeriformes - Sylviidae
e   Large-billed Leaf-Warbler; Large-

billed Willow-Warbler; Large-billed
Tree-Warbler
f   Pouillot à gros bec
d   Schluchtenlaubsänger
i   Luì beccogrosso

7638   *Phylloscopus makirensis*
Passeriformes - Sylviidae
e   San Cristobal Leaf-Warbler
f   Pouillot de San Cristobal
i   Luì di San Cristobal

7639   *Phylloscopus neglectus*
Passeriformes - Sylviidae
e   Plain Leaf-Warbler; Plain Willow-
Warbler
f   Pouillot modeste
d   Eichenlaubsänger
i   Luì grosso orientale

7640   *Phylloscopus nitidus*
Passeriformes - Sylviidae
e   Bright-green Warbler; Green
Warbler; Green Willow-Warbler;
Green Leaf-Warbler; Yellowish-
breasted Warbler
f   Pouillot du Caucase
d   Wacholderlaubsänger
i   Luì pallido; Luì nitido

7641   *Phylloscopus occipitalis*
Passeriformes - Sylviidae
e   Western Crowned-Warbler; Large
Crowned Willow-Warbler; Great
Crowned Willow-Warbler; Western
Crowned Leaf-Warbler; Large
Crowned-Warbler
f   Pouillot couronné
d   Dachskopflaubsänger
i   Luì coronato grosso

7642   *Phylloscopus olivaceus*
Passeriformes - Sylviidae
e   Philippine Leaf-Warbler
f   Pouillot des Philippines
d   Philippinen-Laubsänger
i   Luì delle Filippine

7643   *Phylloscopus orientalis*
Passeriformes - Sylviidae
e   Eastern Bonelli's Warbler; Bonelli's

Warbler
f Pouillot oriental
i Luì bianco orientale

7644 *Phylloscopus plumbeitarsus*
Passeriformes - Sylviidae
e Two-barred Warbler; Two-barred
Greenish Warbler; Grey-legged
Warbler
f Pouillot à pattes sombres; Pouillot à
deux barres
d Middendorf-Laubsänger
i Luì verdastro barrato

7645 *Phylloscopus poliocephalus*
Passeriformes - Sylviidae
e Island Leaf-Warbler; New Guinea
Leaf-Warbler; Numfor Leaf-Warbler
f Pouillot des îles
d Numfort-Laubsänger
i Luì testagrigia

7646 *Phylloscopus presbytes*
Passeriformes - Sylviidae
e Timor Leaf-Warbler
f Pouillot de Timor
i Luì di Timor

7647 *Phylloscopus proregulus*
Passeriformes - Sylviidae
e Pallas's Leaf-Warbler; Pallas's
Warbler; Pallas's Willow-Warbler;
Lemon-rumped Warbler; Yellow-
rumped Warbler; Yellow-rumped
Leaf-Warbler
f Pouillot de Pallas; Pouillot roitelet
d Goldhähnchenlaubsänger
i Luí di Pallas; Luì del Pallas

7648 *Phylloscopus pulcher*
Passeriformes - Sylviidae
e Buff-barred Warbler; Orange-barred
Warbler; Orange-barred Leaf-
Warbler; Orange-barred Willow-
Warbler
f Pouillot élégant
d Goldbindenlaubsänger
i Luì fulvo

7649 *Phylloscopus reguloides*
Passeriformes - Sylviidae

e Blyth's Leaf-Warbler; Blyth's
Crowned Willow-Warbler; Blyth's
Crowned Leaf-Warbler; Greater
White-tailed Leaf-Warbler; Crowned
Leaf-Warbler; Crowned Warbler
f Pouillot de Blyth; Pouillot roitelet
d Streifenkopflaubsänger
i Luì di Blyth

7650 *Phylloscopus ricketti*
Passeriformes - Sylviidae
e Sulphur-breasted Warbler; Rickett's
Willow-Warbler; Slater's Leaf-
Warbler; Yellow-breasted Willow-
Warbler; Black-browed Leaf-
Warbler
f Pouillot de Rickett
d Goldscheitellaubsänger
i Luì pettosullfureo

7651 *Phylloscopus ruficapillus*
Passeriformes - Sylviidae
e Yellow-throated Woodland-Warbler;
Yellow-throated Warbler (CSA);
Yellow-throated Leaf-Warbler;
Yellow-throated Wood-Warbler;
African Yellow-throated Warbler
f Pouillot à gorge jaune
d Rostscheitellaubsänger
i Luì capirosso

7652 *Phylloscopus sarasinorum*
Passeriformes - Sylviidae
e Sulawesi Leaf-Warbler
f Pouillot des Célèbes
d Bartlaubsänger
i Luì di Sulawesi

7653 *Phylloscopus schwarzi*
Passeriformes - Sylviidae
e Radde's Warbler; Radde's Willow-
Warbler; Radde's Leaf-warbler;
Radde's Bush-Warbler; Thick-billed
Willow-Warbler
f Pouillot de Schwarz; Pouillot roitelet
d Waldlaubsänger
i Luí di Radde; Luì de Schwarz

7654 *Phylloscopus sibilatrix*
Passeriformes - Sylviidae
e Wood Warbler

f   Pouillot siffleur
d   Waldlaubsänger; Waldschwirrvogel
i   Luí verde

7655   *Phylloscopus sichuanensis*
       Passeriformes - Sylviidae
e   Sichuan Warbler
f   Pouillot du Sitchouan
d   Bergzipzalp

7656   *Phylloscopus sindianus*
       Passeriformes - Sylviidae
e   Mountain-Chiffchaff; Eastern
       Chiffchaff; Sind Chiffchaff
f   Pouillot montagnard
i   Luí orientale

7657   *Phylloscopus subaffinis*
       Passeriformes - Sylviidae
e   Buff-throated Warbler; Buff-throated
       Willow-Warbler; Chinese Willow-
       Warbler; Grant's Leaf-Warbler;
       Chinese Leaf-Warbler; Ogilvie-
       Grant's Warbler; Chinese Yellow-
       bellied Leaf-Warbler; Yellow-bellied
       Leaf-Warbler
f   Pouillot subaffin; Pouillot fitis de
       Chine
d   Dornlaubsänger
i   Luí golafulva

7658   *Phylloscopus subviridis*
       Passeriformes - Sylviidae
e   Brooks's Leaf-Warbler; Brooks's
       Willow-Warbler; Brooks's Warbler
f   Pouillot de Brooks
d   Brooks-Laubsänger
i   Luí di Brooks

7659   *Phylloscopus tenellipes*
       Passeriformes - Sylviidae
e   Pale-legged Leaf-Warbler; Pale-
       legged Willow-Warbler
f   Pouillot à pattes claires
d   Ussuri-Laubsänger
i   Luí grosso zampechiare

7660   *Phylloscopus tristis*
       Passeriformes - Sylviidae
e   Siberian Chiffchaff
f   Pouillot du Sibérie

7661   *Phylloscopus trivirgatus*
       Passeriformes - Sylviidae
e   Mountain-Leaf-Warbler; Island Leaf-
       Warbler; Green Flycatcher-Warbler
f   Pouillot à triple bandeau
d   Südseelaubsänger
i   Luì isolano

7662   *Phylloscopus trochiloides*
       Passeriformes - Sylviidae
e   Greenish Warbler; Dull-green Leaf-
       Warbler (ISC); Greenish Leaf-
       Warbler; Dull-green Warbler;
       Greenish Willow-Warbler
f   Pouillot verdâtre; Pouillot terne
d   Grünlaubsänger; Grüner Laubsänger
i   Luí verdastro; Luì giallo

7663   *Phylloscopus trochilus*
       Passeriformes - Sylviidae
e   Willow Warbler
f   Pouillot fitis; Pouillot chantre
d   Fitis; Fitis-Laubsänger
i   Luí grosso

7664   *Phylloscopus tytleri*
       Passeriformes - Sylviidae
e   Tytler's Leaf-Warbler; Tytler's
       Willow-Warbler; Slender-billed
       Leaf-Warbler; Slender-billed
       Warbler
f   Pouillot de Tytler
d   Dünnschnabellaubsänger
i   Luì di Tytler

7665   *Phylloscopus umbrovirens*
       Passeriformes - Sylviidae
e   Brown Woodland-Warbler; Brown
       Warbler; Brown Leaf-Warbler
f   Pouillot ombré
d   Umberlaubsänger
i   Luì bruno

7666   *Phytotoma raimondii*
       Passeriformes - Phytotomidae
e   Peruvian Plantcutter
f   Rara du Pérou
d   Graubrustpflanzenmäher
i   Tagliafoglie di Raimondi

7667 **Phytotoma rara**
Passeriformes - Phytotomidae
e Rufous-tailed Plantcutter; Chilean
Plantcutter
f Rara à queue rousse
d Rotschwanzpflanzenmäher; Rara;
Rarita
i Tagliafoglie codirosso

7668 **Phytotoma rutila**
Passeriformes - Phytotomidae
e White-tipped Plantcutter; Reddish
Plantcutter; Red-breasted Plantcutter
f Rara du Paraguay
d Bandpflanzenmäher
i Tagliafoglie pettirosso

7669 **Phytotomidae**
Passeriformes
e Plantcutters
f Phytotomidés
d Pflanzenmäher
i Fitotomidi

7670 **Piaya cayana**
Coliiformes - Coccyzidae
e Squirrel Cuckoo
f Piaye ecureuil
d Cayenne-Kuckuck
i Cuculo scoiattolo della Cayenna

7671 **Piaya melanogaster**
Coliiformes - Coccyzidae
e Black-bellied Cuckoo
f Piaye à ventre noir
d Schwarzbauchkuckuck
i Cuculo scoiattolo ventrenero

7672 **Piaya minuta**
Coliiformes - Coccyzidae
e Little Cuckoo; Minute Cuckoo
f Petit Piaye
d Rötelkuckuck
i Cuculo scoiattolo nano

7673 **Pica hudsonia**
Passeriformes - Corvidae
e Black-billed Magpie (NA); American
Black-billed Magpie

7674 **Pica nuttalli**
Passeriformes - Corvidae
e Yellow-billed Magpie; Blue-winged
Magpie; Blue Magpie; Blue-tailed
Magpie
f Pie à bec jaune
d Gelbschnabelelster
i Gazza beccogiallo

7675 **Pica pica**
Passeriformes - Corvidae
e Black-billed Magpie; Magpie;
Eurasian Magpie; Pie; Pyet;
Common Magpie; White-rumped
Magpie; Pied Magpie
f Pie bavarde; Pie
d Elster
i Gazza; Pica; Agassa; Gazza comune

7676 **Pica pica asirensis**
Passeriformes - Corvidae
e Arabian Magpie

7677 **Picathartes gymnocephalus**
Passeriformes - Callaeatidae
e White-necked Rockfowl; White-
necked Picathartes; White-necked
Bald-Crow; Yellow-headed
Picathartes; Guinea Bare-headed
Rockfowl; Guinea Rockfowl; Bare-
headed Rockfowl
f Picatharte à cou blanc; Picatharte de
Guinée
d Gelbkopffeldhüpfer;
Gelbkopfstelzenkrähe;
Weißhalsstelzenkrähe
i Picatarte dalla testa nuda; Picatarte
collonudo; Picatarte dal collo nudo

7678 **Picathartes oreas**
Passeriformes - Callaeatidae
e Grey-necked Rockfowl; Grey-necked
Bald-Crow; Grey-necked Picathartes;
Red-headed Rockfowl; Red-headed
Picathartes; Cameroon Bare-headed
Rockfowl; Cameroon Rockfowl
f Picatharte à coup gris; Picatharte du
Cameroun
d Buntkopffeldhüpfer; Kamerun-
Feldhüpfer; Stelzenkrähe;

      Bergstelzenkrähe
i   Picatarte collogrigio

7679 Picathartidae
      Passeriformes
e   Rockfowl
f   Picarthartidés
d   Felshüpfer

7680 Picidae
      Piciformes
e   Woodpeckers
f   Pics; Picidés
d   Spechte
i   Picidi; Picchi

7681 Piciformes
e   Woodpeckers
f   Piciformes
d   Spechtvögel
i   Piciformi

7682 *Picoides albolarvatus*
      Piciformes - Picidae
e   White-headed Woodpecker
f   Pic à tête blanche
d   Nonnenspecht
i   Picchio testabianca

7683 *Picoides arcticus*
      Piciformes - Picidae
e   Black-backed Woodpecker; Black-backed Three-toed Woodpecker; Arctic Three-toed Woodpecker; Arctic Woodpecker
f   Pic à dos noir; Pic à poitrine rayée
d   Schwarzrückenspecht
i   Picchio artico

7684 *Picoides arizonae*
      Piciformes - Picidae
e   Arizona Woodpecker

7685 *Picoides borealis*
      Piciformes - Picidae
e   Red-cockaded Woodpecker
f   Pic à face blanche
d   Kokardenspecht
i   Picchio boreale

7686 *Picoides dorsalis*
      Piciformes - Picidae
e   American Three-toed Woodpecker

7687 *Picoides lignarius*
      Piciformes - Picidae
e   Striped Woodpecker
f   Pic bûchéron
d   Strichelkopfspecht
i   Picchio boscaiolo

7688 *Picoides mixtus*
      Piciformes - Picidae
e   Chequered Woodpecker
f   Pic varié
d   Streifenschwanzspecht
i   Picchio a scacchi

7689 *Picoides nuttallii*
      Piciformes - Picidae
e   Nuttall's Woodpecker
f   Pic de Nuttall
d   Nuttalspecht
i   Picchio di Nuttall

7690 *Picoides pubescens*
      Piciformes - Picidae
e   Downy Woodpecker
f   Pic mineur
d   Dünenspecht; Flaumspecht
i   Picchio vellutato

7691 *Picoides scalaris*
      Piciformes - Picidae
e   Ladder-backed Woodpecker
f   Pic arlequin
d   Texas-Specht
i   Picchio arlecchino

7692 *Picoides stricklandi*
      Piciformes - Picidae
e   Strickland's Woodpecker; Arizona Woodpecker; Brown-backed Woodpecker
f   Pic de Strickland
d   Strickland-Specht
i   Picchio di Strickland; Picchio dell'Arizona

7693 *Picoides tridactylus*
      Piciformes - Picidae

e   Eurasian Three-toed Woodpecker
(NA); Three-toed Woodpecker;
Northern Three-toed Woodpecker
f   Pic tridactyle
d   Dreizehenspecht
i   Picchio tridattilo

7694   ***Picoides villosus***
Piciformes - Picidae
e   Hairy Woodpecker; Spanish
Woodpecker (WI); Sook (WI)
f   Pic chevelu
d   Haarspecht
i   Picchio villoso

7695   ***Piculus aeruginosus***
Piciformes - Picidae
e   Bronze-winged Woodpecker
f   Pic à ailes bronzées
d   Goldflügelspecht
i   Picchio

7696   ***Piculus auricularis***
Piciformes - Picidae
e   Grey-crowned Woodpecker
f   Pic à tête grise
d   Weißbrauenspecht
i   Picchio capocenerino

7697   ***Piculus aurulentus***
Piciformes - Picidae
e   Yellow-browed Woodpecker; White-
browed Woodpecker
f   Pic à bandeaux; Pic d'Azara
d   Graukappenspecht
i   Picchio ciliato

7698   ***Piculus callopterus***
Piciformes - Picidae
e   Stripe-cheeked Woodpecker
f   Pic bridé
d   Bronzespecht
i   Picchio guancestriate

7699   ***Piculus chrysochloros***
Piciformes - Picidae
e   Golden-green Woodpecker
f   Pic vert-doré
d   Gelbkehlspecht
i   Picchio verde-dorato

7700   ***Piculus flavigula***
Piciformes - Picidae
e   Yellow-throated Woodpecker
f   Pic à gorge jaune
d   Gelbkehlspecht
i   Picchio golagialla

7701   ***Piculus leucolaemus***
Piciformes - Picidae
e   White-throated Woodpecker;
Rufous-winged Woodpecker
f   Pic à gorge blanche
d   Weißkehlspecht
i   Picchio golabianca

7702   ***Piculus litae***
Piciformes - Picidae
e   Lita Woodpecker
f   Pic de Lita
i   Picchio di Lita

7703   ***Piculus rivolii***
Piciformes - Picidae
e   Crimson-mantled Woodpecker
f   Pic de Rivoli
d   Rotmantelspecht
i   Picchio cremisi

7704   ***Piculus rubiginosus***
Piciformes - Picidae
e   Golden-olive Woodpecker
f   Pic or-olive
d   Grauscheitelspecht;
Olivmantelspecht
i   Picchio oliva-dorato

7705   ***Piculus simplex***
Piciformes - Picidae
e   Rufous-winged Woodpecker
f   Pic à ailes rousses
d   Braunschwingenspecht
i   Picchio alirosse

7706   ***Picumnus albosquamatus***
Piciformes - Picidae
e   White-wedged Piculet; White-scaled
Piculet; Blackish Piculet
f   Picumne noir-et-blanc
i   Picchio nano squamebianche

7707   *Picumnus aurifrons*
     Piciformes - Picidae
e   Bar-breasted Piculet; Gold-fronted
     Piculet; Golden-fronted Piculet
f   Picumne barré
d   Goldstirnzwergspecht;
     Wellenbrustzwergspecht
i   Picchio nano barrato

7708   *Picumnus castelnau*
     Piciformes - Picidae
e   Plain-breasted Piculet
f   Picumne de Castelnau
d   Gelbbauchzwergspecht; Castelnau-
     Zwergspecht
i   Picchio nano di Castelnaud

7709   *Picumnus cinnamomeus*
     Piciformes - Picidae
e   Chestnut Piculet
f   Picumne cannellé
d   Zimtzwergspecht
i   Picchio nano castano

7710   *Picumnus cirratus*
     Piciformes - Picidae
e   White-barred Piculet
f   Picumne frangé
d   Zebrazwergspecht;
     Haubenzwergspecht;
     Drosselzwergspecht
i   Picchio nano cirrato

7711   *Picumnus dorbygnianus*
     Piciformes - Picidae
e   Ocellated Piculet
f   Picumne d'Orbigny
d   Tüpfelzwergspecht
i   Picchio nano ocellato

7712   *Picumnus fulvescens*
     Piciformes - Picidae
e   Tawny Piculet
f   Picumne fauve
i   Picchio nano fulvo

7713   *Picumnus fuscus*
     Piciformes - Picidae
e   Rusty-necked Piculet; Natterer's
     Piculet
f   Picumne à nuque rousse

d   Rostnackenzwergspecht
i   Picchio nano bruno

7714   *Picumnus granadensis*
     Piciformes - Picidae
e   Greyish Piculet
f   Picumne gris
d   Grauer Zwergspecht;
     Braunrückenzwergspecht
l   Picchio nano grigiastro

7715   *Picumnus innominatus*
     Piciformes - Picidae
e   Speckled Piculet; Spotted Piculet
f   Picumne tacheté
d   Indien-Zwergspecht
i   Picchio nano maculato

7716   *Picumnus lafresnayi*
     Piciformes - Picidae
e   Lafresnaye's Piculet
f   Picumne de Lafresnaye
i   Picchio nano di Lafresnaye

7717   *Picumnus limae*
     Piciformes - Picidae
e   Ochraceous Piculet
f   Picumne ocré
d   Lima-Zwergspecht;
     Ockerzwergspecht
i   Picchio nano ocraceo

7718   *Picumnus minutissimus*
     Piciformes - Picidae
e   Guianan Piculet; Arrowhead Piculet
f   Picumne de Cayenne
d   Däumlingsspecht;
     Schuppenohrzwergspecht
i   Picchio nano della Guiana

7719   *Picumnus nebulosus*
     Piciformes - Picidae
e   Mottled Piculet
f   Picumne strié
d   Sundevall-Zwergspecht;
     Braunrückenzwergspecht
i   Picchio nano nebuloso

7720   *Picumnus olivaceus*
     Piciformes - Picidae
e   Olivaceous Piculet

f Picumne olivâtre
d Olivrückenzwergspecht
i Picchio nano polivaceo

7721 *Picumnus pumilus*
Piciformes - Picidae
e Orinoco Piculet
f Picumne de I'Orénoque
d Orinoco-Zwergspecht; Stella-
Zwergspecht
i Picchio nano dell'Orinoco

7722 *Picumnus pygmaeus*
Piciformes - Picidae
e Spotted Piculet
f Picumne ocellé
d Kleinster Zwergspecht;
Fleckenzwergspecht
i Picchio nano ocellato

7723 *Picumnus rufiventris*
Piciformes - Picidae
e Rufous-breasted Piculet
f Picumne à ventre roux
d Zimtzwergspecht;
Rotbauchzwergspecht
i Picchio nano rufiventre

7724 *Picumnus sclateri*
Piciformes - Picidae
e Ecuadorian Piculet
f Picumne de Sclater
d Braunohrzwergspecht; Sclater-
Zwergspecht
i Picchio nano dell'Ecuador

7725 *Picumnus spilogaster*
Piciformes - Picidae
e White-bellied Piculet
f Picumne à ventre blanc
d Weißbauchzwergspecht
i Picchio nano ventrebianco

7726 *Picumnus squamulatus*
Piciformes - Picidae
e Scaled Piculet
f Picumne squamulé
d Shuppenzwergspecht
i Picchio nano squamato

7727 *Picumnus steindachneri*
Piciformes - Picidae
e Speckle-chested Piculet
f Picumne perlé
d Perlenbrustzwergspecht
i Picchio nano perlato

7728 *Picumnus subtilis*
Piciformes - Picidae
e Fine-barred Piculet; Stager's Piculet;
Marcapata Piculet
f Picumne de Cuzco
d Cusco-Zwergspecht
i Picchio nano di Cuzco

7729 *Picumnus temminckii*
Piciformes - Picidae
e Ochre-collared Piculet
f Picumne de Temminck
d Bänderzwergspecht
i Picchio nano di Temminck

7730 *Picumnus varzeae*
Piciformes - Picidae
e Varzea Piculet
f Picumne des Varzeas
d Varzea-Zwergspecht
i Picchio nano di Verzea

7731 *Picus awokera*
Piciformes - Picidae
e Japanese Woodpecker; Japanese
Green-Woodpecker; Wavy-bellied
Woodpecker
f Pic awokera
d Japan-Specht
i Picchio awokera

7732 *Picus canus*
Piciformes - Picidae
e Grey-faced Woodpecker; Grey-
headed Woodpecker; Grey-headed
Green Woodpecker; Ashy
Woodpecker; Black-naped
Woodpecker
f Pic cendré
d Grauspecht
i Picchio cenerino

7733 *Picus chlorolophus*
Piciformes - Picidae

e   Lesser Yellownape; Small Yellow-
    naped Woodpecker (ISC); Lesser
    Yellow-naped Woodpecker; Ceylon
    Yellow-naped Woodpecker
f   Pic à huppe jaune
d   Gelbhaubenspecht
i   Picchio nucagialla minore

7734   *Picus erythropygius*
Piciformes - Picidae
e   Black-headed Woodpecker; Red-
    rumped Green-Woodpecker
f   Pic à tête noire
d   Rotbürzelspecht
i   Picchio testanera

7735   *Picus flavinucha*
Piciformes - Picidae
e   Greater Yellownape; Greater Yellow-
    naped Woodpecker; Large Yellow-
    naped Woodpecker
f   Pic à nuque jaune
d   Gelbnackenspecht
i   Picchio nucagialla maggiore

7736   *Picus mentalis*
Piciformes - Picidae
e   Checker-throated Woodpecker
f   Pic gorgeret
d   Tropfenkehlspecht
i   Picchio gola a scacchi

7737   *Picus mineaceus*
Piciformes - Picidae
e   Banded Woodpecker; Banded Red-
    Woodpecker
f   Pic minium
d   Mennigspecht
i   Picchio color minio

7738   *Picus puniceus*
Piciformes - Picidae
e   Crimson-winged Woodpecker;
    Crimson-winged Yellownape
f   Pic grenadin
d   Rotflügelspecht
i   Picchio alirosse

7739   *Picus rabieri*
Piciformes - Picidae
e   Red-collared Woodpecker; Rabiere's

Woodpecker
f   Pic de Rabier
d   Halsbandspecht
i   Picchio di Rabier

7740   *Picus squamatus*
Piciformes - Picidae
e   Scaly-bellied Woodpecker; Common
    Scaly-Woodpecker; Scaly
    Woodpecker
f   Pic écaillé
d   Almora-Specht
i   Picchio pettosquamato

7741   *Picus vaillantii*
Piciformes - Picidae
e   Levaillant's Woodpecker; Levaillant's
    Green-Woodpecker; North African
    Green Woodpecker; Algerian Green
    Woodpecker
f   Pic de Levaillant
d   Vaillant-Specht; Atlas-Grünspecht
i   Picchio verde africano

7742   *Picus viridanus*
Piciformes - Picidae
e   Streak-breasted Woodpecker;
    Burmese Scaly-bellied Woodpecker;
    Streaked-breasted Woodpecker;
    Small Scaly-breasted Woodpecker;
    Small Scaly-breasted Green
    Woodpecker; Little Scaly-bellied
    Green-Woodpecker; Little Scaly-
    bellied Woodpecker
f   Pic verdâtre
d   Brauenspecht
i   Picchio pettostriato

7743   *Picus viridis*
Piciformes - Picidae
e   Eurasian Green-Woodpecker (NA);
    Green Woodpecker; European Green
    Woodpecker
f   Pic vert
d   Grünspecht
i   Picchio verde; Picchio gallinaccio;
    Picchio grosso; Picchio galletto

7744   *Picus vittatus*
Piciformes - Picidae
e   Laced Woodpecker; Laced Green-

Woodpecker
f Pic médiastin
d Netzbauchspecht
i Picchio lanceolato

7745 *Picus xanthopygaeus*
Piciformes - Picidae
e Streak-throated Woodpecker; Little Scaly-bellied Green-Woodpecker (ISC); Little Scaly-bellied Woodpecker
d Schuppenbauchspecht
i Picchio dal groppone giallo

7746 *Piezorhina cinerea*
Passeriformes - Fringillidae
e Cinereous Finch
f Piézorhin cendré
d Grauämmerling
i Fringuello cenerino del Perù

7747 *Pilherodius pileatus*
Ciconiiformes - Ardeidae
e Capped Heron
f Héron coiffé; Bihoreau blanc
d Kappenreiher
i Airone dal cappuccio

7748 *Pinarocorys erythropygia*
Passeriformes - Alaudidae
e Rufous-rumped Lark; Red-tailed Lark; Rufous-rumped Bush-Lark; Red-rumped Bush-Lark
f Alouette à queue rousse
d Rotbürzellerche
i Allodola di macchia codarossa

7749 *Pinarocorys nigricans*
Passeriformes - Alaudidae
e Dusky Lark; Dusky Bush-Lark
f Alouette brune
d Drossellerche
i Allodola di macchia bruna

7750 *Pinaroloxias inornata*
Passeriformes - Fringillidae
e Cocos Finch; Cocos Islands Finch
f Spizin des Cocos
d Kokos-Fink
I Fringuello delle Isole Cocos

7751 *Pinarornis plumosus*
Passeriformes - Muscicapidae
e Boulder Chat; Sooty Rock-Chat; Sooty Babbler; Sooty Chat
f Rochassier des éboulis; Merle des rochers
d Steindrossling
i Merlo dei macigni

7752 *Pinguinus impennis*
Charadriiformes - Alcidae
e Great Auk
f Grand Pingouin
d Riesenalk
i Alca impenne; Alca maggiore

7753 *Pinicola enucleator*
Passeriformes - Fringillidae
e Pine Grosbeak; Pine Rosefinch
f Durbec des sapins; Durbec des pins; Grosbec des pins; Grosbec des sapins; Durbec vulgaire; Durbec; Bouvreuil durbec
d Hakengimpel; Fichtengimpel
i Ciuffolotto delle pinete

7754 *Pinicola subhimachalus*
Passeriformes - Fringillidae
e Crimson-browed Finch; Red-headed Finch; Red-headed Rosefinch; Juniper Finch
f Durbec à tête rouge
d Rhododendrongimpel
i Ciuffoletto dei ginepri

7755 *Pionites leucogaster*
Psittaciformes - Psittacidae
e White-bellied Parrot; White-breasted Caique; White-bellied Caique
f Caique à ventre blanc
d Weißbauchpapagei; Rostkappenpapagei
i Caicco ventrebianco

7756 *Pionites melanocephala*
Psittaciformes - Psittacidae
e Black-headed Parrot; Black-headed Caique; Caique
f Caique malpourri; Caïque à ventre noir

d   Grünzügelpapagei
i   Caicco testanera

7757   *Pionopsitta barrabandi*
Psittaciformes - Psittacidae
e   Orange-cheeked Parrot; Barraband's
Parrot
f   Caique de Barraband
d   Goldwangenpapagei
i   Pappagallo di Barraband

7758   *Pionopsitta caica*
Psittaciformes - Psittacidae
e   Caica Parrot; Hooded Parrot
f   Caique à tête noire; Perroquet caïca
d   Kappenpapagei
i   Pappagallo dei Caica

7759   *Pionopsitta haematotis*
Psittaciformes - Psittacidae
e   Brown-hooded Parrot
f   Caique à capuchon
d   Blutohrpapagei; Grauwangenpapagei
i   Pappagallo capobruno

7760   *Pionopsitta pileata*
Psittaciformes - Psittacidae
e   Pileated Parrot; Red-capped Parrot
f   Caique mitré; Perroquet mitré
d   Scharlachkopfpapagei
i   Pappagallo caporosso

7761   *Pionopsitta pulchra*
Psittaciformes - Psittacidae
e   Rose-faced Parrot; Beautiful Parrot
f   Caique à joues roses
d   Rosenwangenpapagei
i   Pappagallo facciarossa

7762   *Pionopsitta pyrilia*
Psittaciformes - Psittacidae
e   Saffron-headed Parrot
f   Caique de Bonaparte
d   Goldkopfpapagei; Feuerauge
i   Pappagallo testazafferano

7763   *Pionus chalcopterus*
Psittaciformes - Psittacidae
e   Bronze-winged Parrot
f   Pione noire; Perroquet aux ailes
brunes

d   Glanzflügelpapagei
i   Pappagallo alibronzate

7764   *Pionus fuscus*
Psittaciformes - Psittacidae
e   Dusky Parrot; Violaceous Parrot;
Violet Parrot
f   Pione violette; Perroquet violet
d   Veilchenpapagei
i   Pappagallo fosco

7765   *Pionus maximiliani*
Psittaciformes - Psittacidae
e   Scaly-headed Parrot; Maximilian's
Parrot
f   Pione de Maximilan
d   Maximilians Papagei; Maximilians
Langflügelpapagei
i   Pappagallo di Massimiliano

7766   *Pionus menstruus*
Psittaciformes - Psittacidae
e   Blue-headed Parrot
f   Pione à tête bleue
d   Schwarzohrpapagei
i   Pappagallo testablu

7767   *Pionus senilis*
Psittaciformes - Psittacidae
e   White-crowned Parrot
f   Pione à couronne blanche; Perroquet
à tête blanche
d   Glatzkopfpapagei;
Weißkappenpapagei;
Weißkopfpapagei
i   Pappagallo corona bianca

7768   *Pionus seniloides*
Psittaciformes - Psittacidae
e   White-capped Parrot; White-headed
Parrot; Grey-headed Parrot
f   Pione givrée; Perroquet de Masséna
d   Greisenkopf

7769   *Pionus sordidus*
Psittaciformes - Psittacidae
e   Red-billed Parrot
f   Pione à bec rouge
d   Dünenkopf
i   Pappagallo becco di corallo

7770   ***Pionus tumultuosus***
     Psittaciformes - Psittacidae
e   Speckle-faced Parrot; Restless Parrot
f   Pione pailletée; Pione pourprée;
     Perroquet à tête rose
d   Purpurstirnpapagei;
     Rosenkopfpapagei
i   Pappagallo corona di prugna

7771   ***Pipile cujubi***
     Galliformes - Cracidae
e   Red-throated Piping-Guan; Common
     Piping-Guan; Piping-Guan
f   Pénélope cujubi
i   Penelope cujubi

7772   ***Pipile cumanensis***
     Galliformes - Cracidae
e   Blue-throated Piping-Guan
f   Pénélope à gorge bleue
i   Penelope golablu

7773   ***Pipile jacutinga***
     Galliformes - Cracidae
e   Black-fronted Piping-Guan; Black-
     fronted Guan; White-crested Guan
f   Pénélope à front noir
d   Schakutinga
i   Penelope jacutinga

7774   ***Pipile pipile***
     Galliformes - Cracidae
e   Trinidad Piping-Guan; Blue-throated
     Piping-Guan
f   Pénélope siffleuse
d   Blaukehlguan
i   Penelope di Trinidad

7775   ***Pipilo aberti***
     Passeriformes - Emberizidae
e   Abert's Towhee
f   Tohi d'Abert
d   Schwarzkinngrundammer
i   Toui di Abert

7776   ***Pipilo albicollis***
     Passeriformes - Emberizidae
e   White-throated Towhee
f   Tohi à gorge blanche
d   Weißkehlgrundammer
i   Toui golabianca

7777   ***Pipilo chlorurus***
     Passeriformes - Emberizidae
e   Green-tailed Towhee
f   Tohi à queue verte
d   Grünschwanzgrundammer
i   Toui codaverde

7778   ***Pipilo crissalis***
     Passeriformes - Emberizidae
e   California Towhee; Brown Towhee
f   Tohi de Californie
i   Toui della California

7779   ***Pipilo erythrophthalmus***
     Passeriformes - Emberizidae
e   Eastern Towhee; Rufous-sided
     Towhee; Red-eyed Towhee; Spotted
     Towhee
f   Tohi à flancs roux; Tohi aux yeux
     rouges; Tohi commun; Tohi
d   Rötelgrundammer; Grundrötel;
     Grundammer
i   Toui fianchirossi

7780   ***Pipilo fuscus***
     Passeriformes - Emberizidae
e   Canyon Towhee; Brown Towhee
f   Tohi des canyons
d   Braunrückengrundammer
i   Toui bruno

7781   ***Pipilo maculatus***
     Passeriformes - Emberizidae
e   Spotted Towhee; Rufous-sided
     Towhee
f   Tohi tacheté

7782   ***Pipilo ocai***
     Passeriformes - Emberizidae
e   Collared Towhee
f   Tohi à collier
d   Halsbandgrundammer
i   Toui dal collare

7783   ***Pipilo soccoroensis***
     Passeriformes - Emberizidae
e   Socorro Towhee; Olive-backed
     Towhee

7784   ***Pipra anthracina***
     Passeriformes - Tyrannidae

e   Zeledon's Manakin
d   Pipra
i   Manachino

7785   Pipra aureola
Passeriformes - Tyrannidae
e   Crimson-hooded Manakin
f   Manakin auréole
d   Goldstirnpipra; Rotbauchpipra
i   Manachino cappuccio rosso

7786   Pipra caeruleocapilla
Passeriformes - Tyrannidae
e   Cerulean-capped Manakin
f   Manakin céruléen
d   Blaukappenpipra
i   Manachino capoceruleo

7787   Pipra chloromeros
Passeriformes - Tyrannidae
e   Round-tailed Manakin
f   Manakin à queue ronde
d   Breitschwanzpipra
i   Manachino coda a ventaglio

7788   Pipra cornuta
Passeriformes - Tyrannidae
e   Scarlet-horned Manakin
f   Manakin à cornes rouges
d   Schopfpipra
i   Manachino cornuto

7789   Pipra coronata
Passeriformes - Tyrannidae
e   Blue-crowned Manakin; Crowned
    Manakin; Exquisite Manakin
f   Manakin à tête bleue
d   Blauscheitelpipra
i   Manachino corona azzurra

7790   Pipra erythrocephala
Passeriformes - Tyrannidae
e   Golden-headed Manakin; Yellow-
    thighed Manakin
f   Manakin à tête d'or
d   Goldkopfpipra; Uirapuru
i   Manachino testadorata

7791   Pipra exquisita
Passeriformes - Tyrannidae
e   Exquisite Manikin

7792   Pipra fasciicauda
Passeriformes - Tyrannidae
e   Band-tailed Manakin
f   Manakin à queue barrée
d   Schwanzbindenpipra
i   Manachino codafasciata

7793   Pipra filicauda
Passeriformes - Tyrannidae
e   Wire-tailed Manakin
f   Manakin filifère
d   Fadenpipra
i   Manachino coda a fili

7794   Pipra iris
Passeriformes - Tyrannidae
e   Opal-crowned Manakin; Opal
    Manakin; Pearl-headed Manakin;
    Sick's Manakin
f   Manakin à tête d'opale
d   Opalpipra; Opalscheitelpipra
i   Manachino corona opale

7795   Pipra isidorei
Passeriformes - Tyrannidae
e   Blue-rumped Manakin; Isidore's
    Manakin; Isidore's Rufous Babbler;
    Rufous Babbler
f   Manakin à dos bleu
d   Blaubürzelpipra; Blausäbler;
    Beuteljahoo
i   Manachino dal groppone blu

7796   Pipra mentalis
Passeriformes - Tyrannidae
e   Red-capped Manakin; Yellow-
    thighed Manakin
f   Manakin à cuisses jaunes
d   Gelbhosenpipra
i   Manachino capirosso

7797   Pipra nattereri
Passeriformes - Tyrannidae
e   Snow-capped Manakin; Natterer's
    Manakin
f   Manakin neigeux; Manakin à tête
    blanche
d   Weißkappenpipra
i   Manachino di Natterer

7798		***Pipra pipra***
		Passeriformes - Tyrannidae
	e	White-crowned Manakin
	f	Manakin à tête blanche
	d	Weißscheitelpipra
	i	Manachino corona bianca

7799 ***Pipra rubrocapilla***
Passeriformes - Tyrannidae
e Red-headed Manakin
f Manakin à tête rouge
d Rotkopfpipra
i Manachino testarossa

7800 ***Pipra serena***
Passeriformes - Tyrannidae
e White-fronted Manakin
f Manakin à front blanc
d Weißstirnpipra
i Manachino frontebianca

7801 ***Pipra suavissima***
Passeriformes - Tyrannidae
e Tepui Manakin; Orange-breasted
Manakin

7802 ***Pipra velutina***
Passeriformes - Tyrannidae
e Velvety Manakin
d Samtpipra

7803 ***Pipra vilasboasi***
Passeriformes - Tyrannidae
e Golden-crowned Manakin
f Manakin doré
d Gelbscheitelpipra
i Manachino di Vilasboas

7804 ***Pipraeidea melanonota***
Passeriformes - Fringillidae
e Fawn-breasted Tanager
f Tangara à dos noir
d Schwarzrückentangare
i Tangara mascherata

7805 ***Pipreola arcuata***
Passeriformes - Tyrannidae
e Barred Fruiteater
f Cotinga barré
d Grünkotinga; Bindenschmuckvogel
i Pipreola barrata

7806 ***Pipreola aureopectus***
Passeriformes - Tyrannidae
e Golden-breasted Fruiteater
f Cotinga à poitrine d'or
d Goldbauchschmuckvogel
i Pipreola pettodorato

7807 ***Pipreola chlorolepidota***
Passeriformes - Tyrannidae
e Fiery-throated Fruiteater
f Cotinga à gorge rouge
d Rotkehlschmuckvogel
i Pipreola goladifiamma

7808 ***Pipreola formosa***
Passeriformes - Tyrannidae
e Handsome Fruiteater
f Cotinga magnifique
d Hartlaub-Schmuckvogel
i Pipreola magnifica

7809 ***Pipreola frontalis***
Passeriformes - Tyrannidae
e Scarlet-breasted Fruiteater
f Cotinga chevalier
d Scharlachbrustschmuckvogel
i Pipreola pettorosso

7810 ***Pipreola intermedia***
Passeriformes - Tyrannidae
e Band-tailed Fruiteater
f Cotinga à queue rayée; Cotinga à
collier
d Buntschwanzschmuckvogel
i Pipreola intermedia

7811 ***Pipreola jucunda***
Passeriformes - Tyrannidae
e Orange-breasted Fruiteater
f Cotinga jucunda
i Pipreola pettoarancio

7812 ***Pipreola lubomirskii***
Passeriformes - Tyrannidae
e Black-chested Fruiteater
f Cotinga de Lubomirsk
i Pipreola pettonero

7813 ***Pipreola pulchra***
Passeriformes - Tyrannidae
e Masked Fruiteater

f   Cotinga masqué
i   Pipreola mascherata

7814   *Pipreola riefferii*
        Passeriformes - Tyrannidae
e   Green-and-black Fruiteater;
        Tallman's Fruiteater; Black-headed
        Fruiteater
f   Cotinga vert-et-noir
d   Grünrückenschmuckvogel;
        Gelbbrustkotinga
i   Pipreola verde e nera

7815   *Pipreola whitelyi*
        Passeriformes - Tyrannidae
e   Red-banded Fruiteater; Whitely's
        Fruiteater
f   Cotinga cordon-rouge
d   Goldbauchschmuckvogel
i   Piprcola di Whiteley

7816   **Pipridae**
        Passeriformes
e   Manakins
f   Pipridés
d   Schnurvögel; Pipras
i   Pipridi

7817   *Piprites chloris*
        Passeriformes - Pipridae
e   Wing-barred Piprites; Wing-barred
        Manakin; Temminck's Manakin;
        White-winged Piprites
f   Piprite verdin
d   Graunackenpiprites

7818   *Piprites griseiceps*
        Passeriformes - Pipridae
e   Grey-headed Piprites; Grey-headed
        Manakin; Grey-hooded Mannakin
f   Piprite à tête grise; Manakin à tête
        grise
d   Graukopfpiprites

7819   *Piprites pileatus*
        Passeriformes - Pipridae
e   Black-capped Piprites; Black-capped
        Manakin; Pileated Manakin
f   Piprite chaperon
d   Zimtpiprites

7820   *Piranga bidentata*
        Passeriformes - Thraupidae
e   Flame-coloured Tanager; Streak-
        backed Tanager; Striped Tanager;
        Swainson's Tanager
f   Tangara à dos rayé
d   Bluttangare
i   Piranga color di fiamma

7821   *Piranga erythrocephala*
        Passeriformes - Thraupidae
e   Red-headed Tanager
f   Tangara érythrocéphale
d   Rotkopftangare
i   Piranga testarossa

7822   *Piranga flava*
        Passeriformes - Thraupidae
e   Red Tanager; Hepatic Tanager;
        Lowland Hepatic Tanager
f   Tangara orangé
d   Zinnobert-Tangare
i   Piranga epatica

7823   *Piranga hepatica*
        Passeriformes - Thraupidae
e   Hepatic Tanager; Northern Hepatic
        Tanager

7824   *Piranga leucoptera*
        Passeriformes - Thraupidae
e   White-winged Tanager
f   Tangara bifascié
d   Weißbindentangare
i   Piranga alibianchi

7825   *Piranga ludoviciana*
        Passeriformes - Thraupidae
e   Western Tanager
f   Tangara à tête rouge
d   Louisiana-Tangare; Kieferntangare
i   Tanagra occidentale

7826   *Piranga lutea*
        Passeriformes - Thraupidae
e   Tooth-billed Tanager; Highland
        Hepatic Tanager

7827   *Piranga olivacea*
        Passeriformes - Thraupidae
e   Scarlet Tanager

f Tanagra écarlate
d Scharlachtangare
i Tanagra scarlatta; Piranga scarlatta

7828 Piranga roseogularis
Passeriformes - Thraupidae
e Rose-throated Tanager
f Tangara à gorge rose
d Rosenkehltangare
i Piranga golarossa

7829 Piranga rubra
Passeriformes - Thraupidae
e Summer Tanager
f Tangara vermillon
d Sommertangare; Feuertangare
i Tanagra rossa; Piranga estiva

7830 Piranga rubriceps
Passeriformes - Thraupidae
e Red-hooded Tanager; Gray's Tanager
f Tangara à capuchon
d Scharlachkopftangare
i Piranga dal cappuccio rosso

7831 Pitangus sulphuratus
Passeriformes - Thraupidae
e Great Kiskadee; Kiskadee
Flycatcher; Greater Kiskadee
f Tyran quiquivi; Tyran quesquildit
s Schwefeltyrann; Bentevi
i Pitango solforato

7832 Pithecophaga jefferyi
Falconiformes - Accipitridae
e Great Philippine Eagle; Monkey-
eating Eagle; Philippine Monkey-
eating Eagle; Philippine Eagle
f Aigle des singes; Aigle pithécophage
d Affenadler
i Aquila delle scimmie

7833 Pithys albifrons
Passeriformes - Thamnophilidae
e White-plumed Antbird; White-faced
Antbird; White-faced Antcatcher
f Fourmilier manikup
d Weißbartameisenvogel;
Bartameisenvogel
i Mangiaformiche facciabianca

7834 Pithys castanea
Passeriformes - Thamnophilidae
e White-masked Antbird; White-
masked Antcatcher
f Fourmilier à masque blanc
d Weißmaskenameisenvogel

7835 Pitohui cristatus
Passeriformes - Corvidae
e Crested Pitohui
f Pitohui huppé
d Schopfpitohui
i Pitoi crestato

7836 Pitohui dichrous
Passeriformes - Corvidae
e Hooded Pitohui; Black-headed
Pitohui; Lesser Pitohui
f Pitohui bicolore
d Zweifarbenpitohui
i Pitoi testanera

7837 Pitohui ferrugineus
Passeriformes - Corvidae
e Rusty Pitohui; Ferruginous Pitohui
f Pitohui rouilleux
d Einfarbpitohui
i Pitoi rugginoso

7838 Pitohui incertus
Passeriformes - Corvidae
e White-bellied Pitohui; Mottle-breast
Pitohui; Mottle-breasted Pitohui;
Mottled Pitohui; Brown Pitohui
f Pitohui à ventre clair
d Fleckenbrustpitohui
i Pitoi pettochiazzato

7839 Pitohui kirhocephalus
Passeriformes - Corvidae
e Variable Pitohui; Greater Wood-
Shrike; Greater Pitohui
f Pitohui variable
d Ockerpitohui
i Pitoi variabile

7840 Pitohui nigrescens
Passeriformes - Corvidae
e Black Pitohui; Dusky Pitohui
f Pitohui noir

d   Mohrenpitohui
i   Pitoi nero

7841   *Pitta anerythra*
       Passeriformes - Pittidae
e   Black-faced Pitta; Masked Pitta;
       Black-faced Jewel Thrush
f   Brève masquée
d   Salomonen-Pitta
i   Pitta faccianera

7842   *Pitta angolensis*
       Passeriformes - Pittidae
e   African Pitta; Angola Pitta (CSA);
       Angolan Pitta
f   Brève de l'Angola; Brève d'Angola
d   Angola-Pitta; Afrikanische Pitta
i   Pitta africana

7843   *Pitta arquata*
       Passeriformes - Pittidae
e   Blue-banded Pitta
f   Brève à bandeau
d   Rotrückenpitta; Rotkopfpitta
i   Pitta del Borneo

7844   *Pitta baudii*
       Passeriformes - Pittidae
e   Blue-headed Pitta
f   Brève à tête bleue
i   Pitta testablu

7845   *Pitta brachyura*
       Passeriformes - Pittidae
e   Indian Pitta; Blue-winged Pitta;
       Green-winged Pitta; Bengal Pitta
f   Brève du Bengale; Brève indienne
d   Neunfarbenpitta; Kronpitta
i   Pitta indiana; Pitta del Bengala

7846   *Pitta caerulea*
       Passeriformes - Pittidae
e   Giant Pitta; Great Blue-Pitta
f   Brève géante
d   Riesenpitta; Bengalen-Pitta; Nurang
i   Pitta gigante

7847   *Pitta cyanea*
       Passeriformes - Pittidae
e   Blue Pitta; Lesser Blue-Pitta
f   Brève bleue

d   Blaupitta; Große Blaupitta
i   Pitta blu

7848   *Pitta dohertyi*
       Passeriformes - Pittidae
e   Sula Pitta; Red-breasted Pitta; Blue-
       breasted Pitta
f   Brève des Sula
i   Pitta dell'Isola Sula

7849   *Pitta elegans*
       Passeriformes - Pittidae
e   Elegant Pitta; Noisy Pitta; Irene Pitta
f   Brève élégante
d   Schmuckpitta; Goldbauchpitta
i   Pitta elegante

7850   *Pitta elegans vigorsii*
       Passeriformes - Pittidae
e   Two-striped Pitta
f   Brève de Vigors

7851   *Pitta elliotii*
       Passeriformes - Pittidae
e   Bar-bellied Pitta; Elliot's Pitta
f   Brève d'Elliot
d   Elliot-Pitta
i   Pitta di Elliot

7852   *Pitta erythrogaster*
       Passeriformes - Pittidae
e   Red-bellied Pitta; Red-breasted Pitta;
       Blue-breasted Pitta
f   Brève à ventre rouge
d   Rotbauchpitta
i   Pitta ventrerosso; Pitta dal ventre
       rosso

7853   *Pitta granatina*
       Passeriformes - Pittidae
e   Garnet Pitta; Red-headed Scarlet
       Pitta
f   Brève grenadine
d   Granatpitta; Rotbrustpitta
i   Pitta granatina

7854   *Pitta guajana*
       Passeriformes - Pittidae
e   Banded Pitta; Blue-tailed Pitta
f   Brève azurine
d   Blauschwanzpitta

i   Pitta dalla coda azzurra; Pitta
      codablu; Pitta striata

7855   *Pitta gurneyi*
         Passeriformes - Pittidae
e   Gurney's Pitta; Black-breasted Pitta
f   Brève de Gurney
d   Goldkehlpitta; Flammenkopfpitta
i   Pitta di Gurney

7856   *Pitta iris*
         Passeriformes - Pittidae
e   Rainbow Pitta; Black-breasted Pitta
f   Brève iris
d   Regenbogenpitta
i   Pitta arcobaleno

7857   *Pitta kochi*
         Passeriformes - Pittidae
e   Whiskered Pitta; Koch's Pitta
f   Brève de Koch
d   Luzon-Pitta
i   Pitta di Koch

7858   *Pitta maxima*
         Passeriformes - Pittidae
e   Ivory-breasted Pitta; Great Pitta;
      Halmahera Pitta; Moluccan Pitta
f   Brève d'Halmahera
d   Halmahera-Pitta
i   Pitta grande

7859   *Pitta megarhyncha*
         Passeriformes - Pittidae
e   Mangrove Pitta; Large Blue-winged
      Pitta; Larger Blue-winged Pitta;
      Malay Pitta
f   Brève des palétuviers
d   Mangrovepitta; Riesenpitta
i   Pitta delle mangrovie

7860   *Pitta moluccensis*
         Passeriformes - Pittidae
e   Blue-winged Pitta; Lesser Blue-
      winged Pitta; Moluccan Pitta; Little
      Blue-winged Pitta
f   Brève à ailes bleues
d   Blauflügelpitta
i   Pitta aliazzurre

7861   *Pitta nipalensis*
         Passeriformes - Pittidae
e   Blue-naped Pitta
f   Brève à nuque bleue
d   Blaunackenpitta; Spiegelpitta
i   Pitta nucablu

7862   *Pitta nympha*
         Passeriformes - Pittidae
e   Fairy Pitta; Lesser Blue-winged Pitta;
      Chinese Pitta; Swinhoe's Pitta
f   Brève migratrice
d   Nymphenpitta; Nepal-Pitta
i   Pitta ninfa

7863   *Pitta oatesi*
         Passeriformes - Pittidae
e   Rusty-naped Pitta; Fulvous Pitta
f   Brève à nuque fauve
d   Braunkopfpitta
i   Pitta fulva

7864   *Pitta phayrei*
         Passeriformes - Pittidae
e   Eared Pitta; Phayre's Pitta
f   Brève ornée; Brève à oreillons
d   Sichelpitta
i   Pitta di Phayre

7865   *Pitta reichenowi*
         Passeriformes - Pittidae
e   Green-breasted Pitta
f   Brève à poitrine verte
i   Pitta di Reichenow

7866   *Pitta schneideri*
         Passeriformes - Pittidae
e   Schneider's Pitta
f   Brève de Schneider
d   Schneider-Pitta
i   Pitta di Schneider

7867   *Pitta sordida*
         Passeriformes - Pittidae
e   Hooded Pitta; Green-breasted Pitta
      (ISC); Black-headed Pitta
f   Brève à capuchon
d   Kappenpitta
i   Pitta dal cappuccio

7868   **Pitta soror**
Passeriformes - Pittidae
  e   Blue-rumped Pitta; Blue-backed
      Pitta; Blue-headed Pitta
  f   Brève à dos bleu
  d   Blaubürzelpitta; Schwarzkopfpitta
  i   Pitta dorsoblu

7869   **Pitta steerii**
Passeriformes - Pittidae
  e   Azure-breasted Pitta; Steere's Pitta
  f   Brève de Steere
  d   Babaqua-Pitta; Blaubauchpitta
  i   Pitta di Steere

7870   **Pitta superba**
Passeriformes - Pittidae
  e   Superb Pitta; Black-backed Pitta;
      Black-headed Pitta
  f   Brève superbe
  d   Mohrenpitta; Philippinen-Pitta
  i   Pitta superba

7871   **Pitta ussheri**
Passeriformes - Pittidae
  e   Black-and-crimson Pitta; Black-
      headed Pitta; Black-and-scarlet Pitta

7872   **Pitta venusta**
Passeriformes - Pittidae
  e   Black-crowned Pitta; Graceful Pitta;
      Black-crowned Garnet Pitta
  f   Brève gracieuse
  i   Pitta corona nera

7873   **Pitta versicolor**
Passeriformes - Pittidae
  e   Noisy Pitta; Buff-breasted Pitta
  f   Brève versicolore
  d   Lärmpitta
  i   Pitta rumorosa

7874   **Pittasoma michleri**
Passeriformes - Formicariidae
  e   Black-crowned Ant-Pitta
  f   Grallaire à tête noire
  d   Schwarzscheitelameisenpitta
  i   Pitta formichiera corona nera

7875   **Pittasoma rufopileatum**
Passeriformes - Formicariidae

  e   Rufous-crowned Ant-Pitta
  f   Grallaire à sourcils noirs
  d   Rotscheitelameisenpitta
  i   Pitta formichiera corona rossiccia

7876   **Pittidae**
Passeriformes
  e   Pittas
  f   Pittidés; Brèves
  d   Pittas
  i   Pittidi

7877   **Pityriasis gymnocephala**
Passeriformes - Corvidae
  e   Bornean Bristlehead; Bristled Shrike;
      Bald-headed Wood-Shrike; Bristled
      Shrike-Starling
  f   Barite chauve; Pityorase
      gymnocéphale
  d   Warzenkopf; Kahlkopfwürger
  i   Testasetolosa del Borneo

7878   **Platalea alba**
Ciconiiformes - Threskiornithidae
  e   African Spoonbill
  f   Spatule d'Afrique
  d   Afrikanischer Löffler;
      Rosenfußlöffler
  i   Spatola africana

7879   **Platalea flavipes**
Ciconiiformes - Threskiornithidae
  e   Yellow-billed Spoonbill; Yellow-
      legged Spoonbill
  f   Spatule à bec jaune
  d   Gelbschnabellöffler
  i   Spatola beccogiallo

7880   **Platalea leucorodia**
Ciconiiformes - Threskiornithidae
  e   Eurasian Spoonbill; Spoonbill;
      European Spoonbill; White
      Spoonbill; Common Spoonbill
  f   Spatule blanche
  d   Löffler
  i   Spatola; Spatola bianca

7881   **Platalea minor**
Ciconiiformes - Threskiornithidae
  e   Black-faced Spoonbill; Lesser
      Spoonbill

f Petite Spatule
d Schwarzstirnlöffler
i Spatola faccianera

7882 *Platalea regia*
Ciconiiformes - Threskiornithidae
e Royal Spoonbill; Black-billed
Spoonbill; Australian White
Spoonbill
f Spatule royale
d Königslöffler
i Spatola reale

7883 *Platycercus adelaidae*
Psittaciformes - Psittacidae
e Adelaide Rosella; Adelaide Parrot;
Pheasant-Parrot; Adelaide's Parakeet;
Pennant's Rosella; Blue-cheeked
Rosella
f Perruche d'Adélaide
d Adelaide-Sittich; Fasansittich;
Hyazinthsittich; Adelaide-Rosella

7884 *Platycercus adscitus*
Psittaciformes - Psittacidae
e Pale-headed Rosella; Mealy Rosella;
Moreton Bay Rosella; Blue-cheeked
Rosella; Blue Rosella; White-headed
Rosella
f Perruche à tête pâle; Perruche à tête
blanchâtre; Perruche palliceps à joues
blanches
d Blaßkopfrosella; Blaßköpfiger
Buntsittich; Blaßkopfsittich;
Blauwangensittich;
Blauwangenrosella
i Rosella pallida

7885 *Platycercus barnardi*
Psittaciformes - Psittacidae
e Barnard's Parakeet; Barnard's
Broadtail; Mallee Ring-necked
Parrot; Ring-necked Parrot; Mallee
Parrot; Mallee Ringneck; Scrub
Parrot
f Perruche de Barnard; Perruche de
Bullabulla
d Barnard-Sittich; Bullabulla-Sittich;
Gelbnackensittich
i Parrochetto di Barnard

7886 *Platycercus caledonicus*
Psittaciformes - Psittacidae
e Green Rosella; Tasmanian Rosella;
Yellow-bellied Rosella; Yellow-
bellied Parakeet; Mountain Parrot;
Green Parrot; Yellow-billed Parrot;
Tasman Parrot
f Perruche à ventre jaune
d Gelbbauchsittich; Port Lincoln-
Sittich
i Rosella ventregiallo

7887 *Platycercus caledonicus brownii*
Psittaciformes - Psittacidae
e King Island Green Rosella (ANZ)

7888 *Platycercus elegans*
Psittaciformes - Psittacidae
e Crimson Rosella; Pennant's Rosella;
Crimson Parrot; Mountain Lory; Red
Lory; Pennant's Parakeet; Mountain
Parrot
f Perruche de Pennant
d Pennant-Sittich; Buschwaldsittich
i Rosella di Pennant; Pappagallo di
Pennant

7889 *Platycercus eximius*
Psittaciformes - Psittacidae
e Eastern Rosella; Red Rosella; Red-
headed Rosella; White-cheeked
Rosella; Golden-mantled Rosella
f Perruche omnicolore
d Buntsittich; Rosella; Rosellasittich
i Rosella comune; Rosella

7890 *Platycercus eximius diemenensis*
Psittaciformes - Psittacidae
e Tasmanian Eastern Rosella (ANZ)

7891 *Platycercus flaveolus*
Psittaciformes - Psittacidae
e Yellow Rosella; Murumbidgee
Rosella; Yellow-rumped Parakeet;
Murumbidgee Lory; Murumbidgee
Swamp-Lory; Murray Smoker;
Yellow Parrot
f Perruche flavéole; Perruche à couleur
de paille
d Strohsittich; Strichsittich
i Rosella gialla

7892   *Platycercus icterotis*
        Psittaciformes - Psittacidae
e   Western Rosella; Stanley's Rosella;
        Yellow-cheeked Rosella; Western
        Parakeet; Yellow-cheeked Parakeet;
        Stanley's Parakeet; Earl of Derby's
        Parakeet
f   Perruche à oreilles jaunes; Perruche
        de Stanley
d   Gelbwangenrosella; Stanley-Sittich;
        Scharlachsittich
i   Rosella di Stanley

7893   *Platycercus icterotis xanthogenys*
        Psittaciformes - Psittacidae
e   Wheatbelt Western Rosella (ANZ)

7894   *Platycercus venustus*
        Psittaciformes - Psittacidae
e   Northern Rosella; Brown's Rosella;
        Smutty Rosella; Brown's Parakeet;
        Smutty Parakeet
f   Perruche gracieuse; Perruche de
        Brown
d   Brown-Sittich; Brown-Rosella
i   Rosella di Brown

7895   *Platycichla flavipes*
        Passeriformes - Muscicapidae
e   Yellow-legged Thrush
f   Merle à pattes jaunes
d   Köhler-Drossel
i   Merlo zampegialle

7896   *Platycichla leucops*
        Passeriformes - Muscicapidae
e   Pale-eyed Thrush
f   Merle à oeil clair
d   Taczanowski-Drossel
i   Merlo occhibianchi

7897   *Platylophus galericulatus*
        Passeriformes - Corvidae
e   Crested Jay; Crested Shrike-Jay;
        Crested Maylayan Jay; Malay
        Crested-Jay
f   Geai longup
d   Haubenhäher
i   Ghiandaia dal ciffo

7898   *Platyrinchus cancrominus*
        Passeriformes - Tyrannidae
e   Stub-tailed Spadebill; Mexican
        Spadebill
f   Platyrhynque à queue courte
i   Beccopiatto codamozza

7899   *Platyrinchus coronatus*
        Passeriformes - Tyrannidae
e   Golden-crowned Spadebill
f   Platyrhynque à tête d'or
d   Goldkopfbreitschnabel
i   Beccopiatto corona dorata

7900   *Platyrinchus flavigularis*
        Passeriformes - Tyrannidae
e   Yellow-throated Spadebill
f   Platyrhynque à gorge jaune
d   Gelbkehlbreitschnabel
i   Beccopiatto golagialla

7901   *Platyrinchus leucoryphus*
        Passeriformes - Tyrannidae
e   Russet-winged Spadebill
f   Platyrhynque à ailes rousses
d   Rotflügelbreitschnabel
i   Beccopiatto alirosse

7902   *Platyrinchus mystaceus*
        Passeriformes - Tyrannidae
e   White-throated Spadebill
f   Platyrhynque à moustaches
d   Weißkehlbreitschnabel
i   Beccopiatto golabianca

7903   *Platyrinchus platyrhynchos*
        Passeriformes - Tyrannidae
e   White-crested Spadebill; Yellow-
        crested Spadebill
f   Platyrhynque à cimier blanc;
        Platyrhynque brun
d   Silberkopfbreitschnabel
i   Beccopiatto crestabianca

7904   *Platyrinchus saturatus*
        Passeriformes - Tyrannidae
e   Cinnamon-crested Spadebill
f   Platyrhynque à cimier orangé;
        Platyrhinque à tache orangé
d   Zimtkopfbreitschnabel
i   Beccopiatto crestacannella

7905 **Platysmurus leucopterus**
Passeriformes - Corvidae
e Black Magpie; White-winged
Magpie; White-winged Jay; Black
Jay; White-winged Black-Magpie;
Black-crested Magpie
f Geai à ailes blanches
d Trauerelster
i Gazza alibianche

7906 **Platysmurus leucopterus aterimus**
Passeriformes - Corvidae
e Black Crested-Magpie

7907 **Platysteira albifrons**
Passeriformes - Platysteiridae
e White-fronted Wattle-eye
f Pririt à front blanc; Gobemouche
caronculé à front blanc
d Lappenschnäpper
i Occhiorosso frontebianca

7908 **Platysteira blissetti**
Passeriformes - Platysteiridae
e Red-cheeked Wattle-eye; Blissett's
Wattle-eye
f Pririt de Blisset; Gobemouche
caronculé de Blisset; Gobe-mouches
caronculé de Blisset; Gobemouche
caronculé joues rouges
d Weißstirnlappenschnäpper
i Occhiorosso guancerosse

7909 **Platysteira castanea**
Passeriformes - Platysteiridae
e Chestnut Wattle-eye
f Pririt châtain; Gobemouche
caronculé châtain; Gobe-mouches
caronculé châtain
d Glanzlappenschnäpper
i Occhiorosso castano

7910 **Platysteira chalybea**
Passeriformes - Platysteiridae
e Black-necked Wattle-eye;
Reichenow's Wattle-eye; Cameroon
Wattle-eye
f Pririt chalybée; Cobemouche
caronculé à cou noir
d Weißbürzellappenschnäpper
i Occhiorosso di Reichenow

7911 **Platysteira concreta**
Passeriformes - Platysteiridae
e Yellow-bellied Wattle-eye; Golden-
bellied Wattle-eye; Chestnut-bellied
Wattle-eye
f Pririt à ventre dorée; Gobemouche
caronculé à ventre doré; Gobe-
mouches caronculé à ventre doré;
Gobemouche caronculé à collier
d Lappenschnäpper
i Occhiorosso ventregiallo

7912 **Platysteira cyanea**
Passeriformes - Platysteiridae
e Common Wattle-eye (CSA); Wattle-
eye; Scarlet-spectacled Wattle-eye;
Brown-throated Wattle-eye; Wattle-
eyed Flycatcher; Banded Wattle-eye
f Pririt à collier; Gobemouche
caronculé à collier; Gobe-mouches
caronculé à collier
d Gelbbauchlappenschnäpper;
Braunkehllappenschnäpper
i Occhiorosso golabruna

7913 **Platysteira jamesoni**
Passeriformes - Platysteiridae
e Jameson's Wattle-eye
f Pririt de Jameson; Gobemouche
caronculé de Jameson
d Lappenschnäpper
i Occhiorosso di Jameson

7914 **Platysteira laticincta**
Passeriformes - Platysteiridae
e Banded Wattle-eye; Bamenda
Wattle-eye
f Pririt du Bamenda
i Occhiorosso di Bamenda

7915 **Platysteira peltata**
Passeriformes - Platysteiridae
e Black-throated Wattle-eye; Wattle-
eyed Flycatcher (CSA)
f Pririt à gorge noire; Gobemouche
caronculé à gorge noire
d Schwarzkehllappenschnäpper
i Occhiorosso golanera

7916 **Platysteira tonsa**
Passeriformes - Platysteiridae

e   White-spotted Wattle-eye
f   Pririt à taches blanches; Gobemouche
      caronculé à taches blanches; Gobe-
      mouches caronculé à taches blanches
d   Weißbrauenschnäpperwürger
i   Occhiorosso maculato

7917   Platysteiridae
      Passeriformes -
e   Wattle-eyes; Puffback-Flycatchers
f   Platystéiridés
d   Kleinschnäpper; Schnäpperwürger
i   Platısteiridi

7918   *Plectorhyncha lanceolata*
      Passeriformes - Meliphagidae
e   Striped Honeyeater; Lanceolated
      Honeyeater
f   Méliphage lancéolé
d   Strichelhonigfresser
i   Mangiamiele striato piccolo

7919   *Plectrophenax hyperboreus*
      Passeriformes - Fringillidae
e   McKay's Bunting
f   Bruant blanc; Plectrophane blanc
i   Zigolo delle neve di McKay

7920   *Plectrophenax nivalis*
      Passeriformes - Fringillidae
e   Snow Bunting
f   Bruant des neiges; Plectrophane des
      neiges
d   Schneeammer
i   Zigolo delle nevi

7921   *Plectropterus gambensis*
      Anseriformes - Anatidae
e   Spur-winged Goose; Spur-winged
      Duck
f   Oie-armée de Gambie
d   Sporngans; Sporengans
i   Oca delle sperone; Oca dallo sperone

7922   *Plegadis chihi*
      Ciconiiformes - Threskiornithidae
e   White-faced Ibis; White-faced
      Glossy Ibis
f   Ibis à face blanche; Ibis guarana
d   Brillensichler
i   Ibis facciabianca

7923   *Plegadis falcinellus*
      Ciconiiformes - Threskiornithidae
e   Glossy Ibis; Curlew (WI)
f   Ibis falcinelle; Pêcheur (Ants); Ibis
      peché (Ants); Ibis noir (Ants)
d   Braunsichler; Brauner Sichler;
      Sichler; Brauner Ibis
i   Mignattaio

7924   *Plegadis ridgwayi*
      Ciconiiformes - Threskiornithidae
e   Puna Ibis
f   Ibis de Ridgeway
d   Schmalschnabellöffler
i   Ibis della Puna

7925   Ploceidae
      Passeriformes
e   Weavers; Cuckoo-Finches (CSA)
f   Plocéidés
d   Webervögel; Weber; Webertinken
i   Ploceidi

7926   *Plocepasser donaldsoni*
      Passeriformes - Passeridae
e   Donaldson-Smith's Sparrow-Weaver
f   Mahali de Donaldson; Moineau-
      tisserin de Donaldson
d   Dornbuschweber; Dornbuschmahali
i   Passero tessitore di Donaldson

7927   *Plocepasser mahali*
      Passeriformes - Passeridae
e   White-browed Sparrow-Weaver;
      Mahali Sparrow-Weaver; Stripe-
      breasted Sparrow-Weaver; Stripe-
      breasted Weaver
f   Mahali à sourcils blancs; Mahali;
      Moineau-tisserin à sourcils blancs
d   Mahaliweber; Augenbrauenmahali;
      Mahali
i   Passero tessitore mahali

7928   *Plocepasser rufoscapulatus*
      Passeriformes - Passeridae
e   Chestnut-backed Sparrow-Weaver;
      Rufous-backed Sparrow-Weaver;
      Red-mantled Sparrow-Weaver;
      Chestnut-mantled Sparrow-Weaver
f   Mahali à dos roux; Moineau-tisserin
      à dos roux

<table>
<tr><td>d</td><td>Rotrückenmahali</td></tr>
<tr><td>i</td><td>Passero tessitore mantello castano</td></tr>
</table>

7929 *Plocepasser superciliosus*
Passeriformes - Passeridae
e Chestnut-crowned Sparrow-Weaver;
Sparrow-Weaver
f Mahali à calotte marron; Moineau
tisserin; Moineau-tisserin à calotte
maronne
d Braunwangenweber;
Braunwangenmahali
i Passero tessitore corona castana

7930 *Ploceus albinucha*
Passeriformes - Ploceidae
e Maxwell's Black-Weaver; White-
naped Black-Weaver; White-eyed
Widow
f Tisserin de Maxwell; Tisserin noir de
Maxwell
d Trauerweber
i Tessitore nero di Maxwell

7931 *Ploceus alienus*
Passeriformes - Ploceidae
e Strange Weaver
f Tisserin de montagne; Tisserin
bizarre
d Meisenweber
i Tessitore straniero

7932 *Ploceus angolensis*
Passeriformes - Ploceidae
e Bar-winged Weaver
f Tisserin malimbe; Tisserin du
Miombo
d Miombo-Weber
i Tessitore alibarrate

7933 *Ploceus aurantius*
Passeriformes - Ploceidae
e Orange Weaver
f Tisserin orangé
d Königsweber
i Tessitore arancio

7934 *Ploceus aureonucha*
Passeriformes - Ploceidae
e Golden-naped Weaver; Gold-naped
Weaver

f Tisserin à nuque d'or
d Goldnackenweber
i Tessitore nucadorata

7935 *Ploceus badius*
Passeriformes - Ploceidae
e Cinnamon Weaver
f Tisserin cannellé
d Schulterfleckenweber
i Tessitore color cannella

7936 *Ploceus baglafecht*
Passeriformes - Ploceidae
e Baglafecht Weaver; Reichenow's
Weaver (CSA)
f Tisserin baglafecht
d Reichenow-Weber; Baglafecht-
Weber
i Tessitore baglafecht

7937 *Ploceus bannermani*
Passeriformes - Ploceidae
e Bannerman's Weaver
f Tisserin de Bannerman
d Bannerman-Weber
i Tessitore di Bannerman

7938 *Ploceus batesi*
Passeriformes - Ploceidae
e Bates's Weaver
f Tisserin de Bates
d Bates-Weber
i Tessitore di Bates

7939 *Ploceus benghalensis*
Passeriformes - Ploceidae
e Black-breasted Weaver; Black-
throated Weaverbird (ISC); Black-
throated Weaver; Bengal Weaver
f Tisserin du Bengale
d Bengalen-Weber
i Tessitore del Bengala

7940 *Ploceus bertrandi*
Passeriformes - Ploceidae
e Bertrand's Weaver; Xanthoploceus;
Bertram's Weaver
f Tisserin de Bertrand
d Bertrand-Weber
i Tessitore di Bertrand

7941 *Ploceus bicolor*
Passeriformes - Ploceidae
e Forest Weaver; Dark-backed Weaver
(CSA)
f Tisserin bicolore
d Waldweber
i Tessitore di foresta

7942 *Ploceus bojeri*
Passeriformes - Ploceidae
e Golden Palm-Weaver
f Tisserin palmiste; Tisserin des
palmiers
d Palmenweber; Bojers Weber
i Tessitore delle palme

7943 *Ploceus burnieri*
Passeriformes - Ploccidae
e Kilombero Weaver
f Tisserin de Burnier; Tisserin de
Kilombéro

7944 *Ploceus cupensis*
Passeriformes - Ploceidae
e Cape Weaver
f Tisserin du Cap
d Kapweber
i Tessitore del Capo

7945 *Ploceus castaneiceps*
Passeriformes - Ploceidae
e Taveta Golden-Weaver; Taveta
Weaver
f Tisserin de Taveta; Tisserin doré de
Taveta
d Genickbandgoldweber;
Genickbandweber; Fuchsweber
i Tessitore dorato di Taveta

7946 *Ploceus castanops*
Passeriformes - Ploceidae
e Northern Brown-throated Weaver
f Tisserin à gorge noire
d Riedweber
i Tessitore golabruna settentrionale

7947 *Ploceus collaris*
Passeriformes - Ploceidae
e Mottled Weaver
i Tessitore

7948 *Ploceus cucullatus*
Passeriformes - Ploceidae
e Black-headed Weaver; Village
Weaver
f Tisserin gendarme
d Textor; Textorweber; Dorfweber
i Gendarme

7949 *Ploceus cucullatus nigriceps*
Passeriformes - Ploceidae
e Layard's Black-headed Weaver;
Spotted-backed Weaver (CSA);
Yellow Finch (CSA); Layard's
Weaver
f Madame Sara (Ants), Tisserin Cap
Moor; Tisserin de Layard
i Gendarme

7950 *Ploceus dicrocephalus*
Passeriformes - Ploceidae
e Salvadori's Weaver; Juba Weaver;
Jubaland Weaver; Yellow-backed
Weaver
f Tisserin de Salvadori; Tisserin de
Somalie
d Gelbrückenweber
i Tessitore di Salvadori

7951 *Ploceus dorsomaculatus*
Passeriformes - Ploceidae
e Yellow-capped Weaver
f Tisserin à cape jaune
d Gelbkappenweber
i Tessitore capogiallo

7952 *Ploceus galbula*
Passeriformes - Ploceidae
e Rüppell's Weaver
f Tisserin de Rüppell; Tisserin doré de
Rüppell
d Gilbweber; Pirolweber
i Tessitore di Rüppell; Tessitore giallo

7953 *Ploceus golandi*
Passeriformes - Ploceidae
e Clarke's Weaver; Goland's Weaver
f Tisserin de Clarke
d Goland-Weber
i Tessitore di Clarke

7954	***Ploceus grandis***
	Passeriformes - Ploceidae
e	Giant Weaver
f	Tisserin géant
d	Riesenweber
i	Tessitore gigante

7955 ***Ploceus heuglini***
Passeriformes - Ploceidae
e Heuglin's Masked-Weaver; Heuglin's Weaver
f Tisserin masqué; Tisserin masqué de Heuglin
d Heuglin-Weber
i Tessitore mascherto di Heuglin

7956 ***Ploceus hypoxanthus***
Passeriformes - Ploceidae
e Asian Golden-Weaver; Asiatic Golden-Weaver; Golden Weaver
f Tisserin doré
d Kernbeißerweber
i Tessitore dorato asiatico

7957 ***Ploceus insignis***
Passeriformes - Ploceidae
e Brown-capped Weaver
f Tisserin à cape brune
d Braunkappenweber
i Tessitore capobruno

7958 ***Ploceus intermedius***
Passeriformes - Ploceidae
e Lesser Masked-Weaver; Masked Weaver
f Tisserin intermédiaire; Tisserin masqué austral
d Cabanis-Weber
i Tessitore mascherato minore

7959 ***Ploceus jacksoni***
Passeriformes - Ploceidae
e Golden-backed Weaver; Jackson's Weaver; Jackson's Golden-backed Weaver; Eastern Golden-Weaver
f Tisserin à dos d'or; Tisserin de Jackson
d Jackson-Weber
i Tessitore di Jackson

7960 ***Ploceus katangae***
Passeriformes - Ploceidae
e Katanga Masked-Weaver
f Tisserin du Katanga
i Tessitore mascherato del Katanga

7961 ***Ploceus luteolus***
Passeriformes - Ploceidae
e Little Weaver; Slender-billed Weaver; Little Masked-Weaver
f Tisserin minule
d Zwergweber; Zwergmaskenweber
i Tessitore mascherato

7962 ***Ploceus manyar***
Passeriformes - Ploceidae
e Streaked Weaver; Streaked Weaverbird (ISC); Manyar Weaver; Striated Weaverbird; Striated Weaver
f Tisserin manyar
d Manyar-Weber
i Tessitore striato

7963 ***Ploceus megarhynchus***
Passeriformes - Ploceidae
e Finn's Weaver; Finn's Baya; Yellow Weaver (ISC); Himalayan Weaver
f Tisserin de Finn
d Großschnabelweber
i Tessitore di Finn

7964 ***Ploceus melanocephalus***
Passeriformes - Ploceidae
e Black-headed Weaver; Yellow-backed Weaver; Yellow-collared Weaver
f Tisserin à tête noire
d Schwarzkopfweber; Kleiner Textor
i Tessitore testanera

7965 ***Ploceus melanogaster***
Passeriformes - Ploceidae
e Black-bellied Weaver; Black-Mountain-Weaver
f Tisserin à tête jaune
d Schwarzbauchweber
i Tessitore nero facciagialla

7966 ***Ploceus nelicourvi***
Passeriformes - Ploceidae
e Nelicourvi Weaver

f Tisserin nélicourvi
d Grünweber
i Tessitore nelicourvi

7967 *Ploceus nicolli*
Passeriformes - Ploceidae
e Usambara Weaver; Nicoll's Weaver;
Tanzanian Mountain-Weaver
f Tisserin des Usambara
d Nicoll-Weber
i Tessitore dell'Usambara

7968 *Ploceus nigerrimus*
Passeriformes - Ploceidae
e Vieillot's Black-Weaver; Vieillot's
Weaver; Black Weaver
f Tisserin noir; Tisserin noir de
Vieillot
d Mohrenweber
i Tessitore nero di Vieillot

7969 *Ploceus nigricollis*
Passeriformes - Ploceidae
e Black-necked Weaver; Spectacled
Weaver
f Tisserin à cou noir
d Kurzflügelweber
i Tessitore giallo e nero

7970 *Ploceus nigrimentum*
Passeriformes - Ploceidae
e Black-chinned Weaver; Angola
Weaver
f Tisserin à menton noir
d Schwarzkinnweber
i Tessitore mentorosso

7971 *Ploceus ocularis*
Passeriformes - Ploceidae
e Spectacled Weaver; Bottle Weaver
f Tisserin à lunettes
d Brillenweber
i Tessitore dagli occhiali

7972 *Ploceus olivaceiceps*
Passeriformes - Ploceidae
e Olive-headed Weaver; Olive-headed
Golden-Weaver (CSA)
f Tisserin à tête olive
d Olivenkopfweber; Olivkopfweber
i Tessitore dorato testaoliva

7973 *Ploceus pelzelni*
Passeriformes - Ploceidae
e Slender-billed Weaver; Common
Slender-billed Weaver
f Tisserin de Pelzeln; Tisserin nain
d Mönchsweber
i Tessitore di Pelzeln

7974 *Ploceus philippinus*
Passeriformes - Ploceidae
e Baya Weaver; Baya Weaverbird
(ISC); Common Weaver; Indian
Baya; Baya
f Tisserin baya
d Baja-Weber
i Tessitore baya

7975 *Ploceus preussi*
Passeriformes - Ploceidae
e Preuss's Weaver; Golden-backed
Weaver; Western Golden-backed
Weaver; Preuss's Golden-backed
Weaver
f Tisserin de Preuss
d Preuss-Weber
i Tessitore dorsodorato occidentale

7976 *Ploceus princeps*
Passeriformes - Ploceidae
e Principe Golden-Weaver; Principe
Weaver
f Tisserin de Principe; Tisserin doré de
Principe
d Prinzen-Weber
i Tessitore dorato dell'Isola Principe

7977 *Ploceus reichardi*
Passeriformes - Ploceidae
e Tanzania Masked-Weaver;
Tanganyika Masked-Weaver; Swamp
Masked-Weaver; Lake Lufira
Weaver
f Tisserin de Reichard
d Reichard-Weber
i Tessitore mascherato della Tanzania

7978 *Ploceus rubiginosus*
Passeriformes - Ploceidae
e Chestnut Weaver
f Tisserin roux; Tisserin rubigineux

d　Rotbrauner Weber
i　Tessitore castano

7979　*Ploceus ruweti*
　　　Passeriformes - Ploceidae
e　Ruwet's Masked-Weaver; Lufira
　　　Masked-Weaver; Lake Lufira
　　　Weaver
f　Tisserin de Ruwet
i　Tessitore mascherato di Lufira

7980　*Ploceus sakalava*
　　　Passeriformes - Ploceidae
e　Sakalava Weaver; Sakalava Fody
f　Tisserin sakalave
d　Sakalaven-Weber
i　Tessitore sakalava

7981　*Ploceus sanctaethomae*
　　　Passeriformes - Ploceidae
e　Sao Tomé Weaver
f　Tisserin de Sao Tomé
d　St. Thomas-Weber
i　Tessitore di Sao Tomè

7982　*Ploceus spekei*
　　　Passeriformes - Ploceidae
e　Speke's Weaver
f　Tisserin de Speke
d　Spekes-Weber; Somali-Weber
i　Tessitore di Speke; Ploceo di Speke

7983　*Ploceus spekeoides*
　　　Passeriformes - Ploceidae
e　Fox's Weaver
f　Tisserin de Fox
d　Foxs Weber
i　Tessitore di Fox

7984　*Ploceus subaureus*
　　　Passeriformes - Ploceidae
e　African Golden-Weaver; Golden
　　　Weaver; Eastern Golden-Weaver;
　　　African Yellow Weaver; Yellow
　　　Weaver
f　Tisserin jaune; Petit Tisserin doré
d　Goldweber
i　Tessitore dorato

7985　*Ploceus subpersonatus*
　　　Passeriformes - Ploceidae

e　Loango Weaver; Western Golden-
　　　Weaver; Loango Slender-billed
　　　Weaver
f　Tisserin à bec grêle; Tisserin de
　　　Cabinda
d　Loango-Weber
i　Tessitore di Loanga

7986　*Ploceus superciliosus*
　　　Passeriformes - Plocidae
e　Compact Weaver
f　Tisserin à gros bec; Tisserin gros-bec
d　Braunbürzelweber;
　　　Augenbrauenweber
i　Tessitore compatto

7987　*Ploceus taeniopterus*
　　　Passeriformes - Ploceidae
e　Northern Masked-Weaver; Sudan
　　　Masked-Weaver
f　Tisserin du Nil; Tisserin masqué du
　　　Nil
d　Goldmantelweber
i　Tessitore mascherato settentrionale

7988　*Ploceus temporalis*
　　　Passeriformes - Ploceidae
e　Bocage's Weaver; Bocage's Golden-
　　　Weaver; Angolan Weaver; Cape
　　　Weaver
f　Tisserin de Bocage; Tisserin à joues
　　　olives
d　Angola-Weber; Bocage-Weber
i　Tessitore di Bocage

7989　*Ploceus tricolor*
　　　Passeriformes - Ploceidae
e　Yellow-mantled Weaver
f　Tisserin tricolore
d　Dreifarbenweber
i　Tessitore tricolore

7990　*Ploceus velatus*
　　　Passeriformes - Ploceidae
e　Southern Masked-Weaver; Masked
　　　Weaver (CSA); Black-fronted
　　　Weaver; Vitelline Masked-Weaver;
　　　Greater Masked-Weaver; Black-
　　　faced Weaver
f　Tisserin à tête rousse; Tisserin à front
　　　noir

d Maskenweber; Dotterweber;
　　Schwarzstirnweber
i Tessitore mascherato africano;
　　Tessitore mascherato vitellino

7991 *Ploceus velatus uluensis*
Passeriformes - Ploceidae
e Vitelline Masked-Weaver (CSA)
f Tisserin vitellin

7992 *Ploceus victoriae*
Passeriformes - Ploceidae
e Victoria Masked-Weaver
f Tisserin du Victoria
i Tessitore del Lago Victoria

7993 *Ploceus weynsi*
Passeriformes - Ploceidae
e Weyns's Weaver
f Tisserin de Weyns
d Weyns-Weber
i Tessitore di Weyns

7994 *Ploceus xanthops*
Passeriformes - Ploceidae
e Holub's Golden-Weaver; Golden
　　Weaver (CSA); Monteiro's Golden-
　　Weaver; Large Golden-Weaver;
　　Larger Golden-Weaver
f Tisserin safran
d Großer Goldweber; Safranweber;
　　Holub-Weber
i Tessitore dorato di Holub

7995 *Ploceus xanthopterus*
Passeriformes - Ploceidae
e Southern Brown-throated Weaver;
　　Brown-throated Weaver (CSA);
　　Brown-throated Golden-Weaver
f Tisserin à gorge brune
d Braunkehlweber;
　　Braunkehlgoldweber
i Tessitore dorato golabruna

7996 *Pluvialis apricaria*
Charadriiformes - Charadriidae
e European Golden-Plover; Greater
　　Golden-Plover; Golden Plover;
　　Eurasian Golden-Plover
f Pluvier doré; Pluvier bronzé; Pluvier
　　doré d'Eurasie (Qué)

d Goldregenpfeifer
i Piviere dorato

7997 *Pluvialis dominica*
Charadriiformes - Charadriidae
e American Golden-Plover; Lesser
　　Golden-Plover; Golden Plover
f Pluvier dominicain; Pluvier doré
　　d'Amérique; Pluvier bronzé (Ants);
　　Pluvier doré (Ants); Pluvier doré
　　américain
d Kleiner Goldregenpfeifer;
　　Amerikanischer Goldregenpfeifer;
　　Wanderregenpfeifer; Kleiner
　　Amerikanischer Goldregenpfeifer
i Piviere dorato americano; Piviere
　　americano; Piviere dorato asiatico
　　americano; Piviere orientale; Piviere
　　dorato minore

7998 *Pluvialis fulva*
Charadriiformes - Charadriidae
e Pacific Golden-Plover; Asiatic
　　Golden-Plover; Asian Golden-
　　Plover; Lesser Golden-Plover;
　　Eastern Golden-Plover; Least
　　Golden-Plover
f Pluvier sibérien; Pluvier fauve
d Pazifischer Goldregenpfeifer;
　　Sibirischer Goldregenpfeifer
i Piviere dorato orientale; Piviere
　　orientale; Piviere del Pacifico;
　　Piviere dorato del Pacifico

7999 *Pluvialis squatarola*
Charadriiformes - Charadriidae
e Grey Plover; Black-bellied Plover
　　(NA); Lapwing (WI); Soldierbird
　　(WI); Silver Plover
f Pluvier argenté; Pluvier (Ants);
　　Pluvier gris (Ants); Pluvier grosse
　　tête (Ants); Bécasseau (Ants)
d Kiebitzregenpfeifer
i Pivieressa

8000 *Pluvianellus socialis*
Charadriiformes - Charadriidae
e Magellanic Plover
f Pluvianelle magéllanique
d Magellan-Läufer; Magellan-

Regenpfeifer
i Piviere di Magellano

8001 *Pluvianus aegyptius*
Ciconiiformes - Glareolidae
e Egyptian Plover; Crocodile Bird;
Egyptian Courser
f Pluvian fluviatile; Pluvian d'Égypte
d Krokodilwächter
i Guardiano dei coccodrilli; Guardiano
del coccodrillo

8002 *Pnoepyga albiventer*
Passeriformes - Sylviidae
e Scaly-breasted Wren-Babbler;
Greater Scaly Wren-Babbler; White-
vented Wren-Babbler; Scaly-breasted
Wren-Warbler
f Turdinule à ventre blanc; Pnoepyga à
ventre blanc
d Schuppentimalie
i Garrulo scricciolo pettoscaglioso

8003 *Pnoepyga immaculata*
Passeriformes - Sylviidae
e Nepal Wren-Babbler; Immaculate
Wren-Babbler
f Turdinule immaculée

8004 *Pnoepyga pusilla*
Passeriformes - Sylviidae
e Pygmy Wren-Babbler; Lesser Scaly
Wren-Babbler; Brown Wren-
Babbler; Lesser Scaly-breasted
Wren-Babbler
f Turdinule maillée; Pnoepyga à
poitrine maillée
d Moostimalie
i Garrulo scricciolo pigmeo

8005 *Podager nacunda*
Caprimulgiformes - Caprimulgidae
e Nacunda Nighthawk
f Engoulevent nacunda
d Weißbauchnachtschwalbe
i Succiacapre sparviero nacunda;
Succiacapre sparviere nacunda

8006 **Podargidae**
Caprimulgiformes
e Frogmouths

f Podargidé
d Eulenschwalme
i Podargidi

8007 *Podargus ocellatus*
Caprimulgiformes - Podargidae
e Marbled Frogmouth; Little Papuan
Frogmouth
f Podarge ocellé
d Marmorschwalm
i Podargo ocellato

8008 *Podargus ocellatus plumiferus*
Caprimulgiformes - Podargidae
e Plumed Frogmouth (ANZ)

8009 *Podargus papuensis*
Caprimulgiformes - Podargidae
e Papuan Frogmouth; Giant
Frogmouth; Great Frogmouth
f Podarge papou
d Papua-Schwalm
i Podargo papua

8010 *Podargus strigoides*
Caprimulgiformes - Podargidae
e Tawny Frogmouth
f Podarge gris
d Eulenschwalm
i Podargo strigoide; Podargo maggiore

8011 *Podica senegalensis*
Gruiformes - Heliornithidae
e African Finfoot; Finfoot; Peters's
Finfoot
f Grébifoulque du Sénégal;
Grébifoulque d'Afrique
d Afrikanische Binsenralle; Binsenralle
i Rallo tuffatore africano

8012 *Podiceps andinus*
Podicipediformes - Podicipedidae
e Colombian Grebe
f Grèbe des Andes
d Anden-Taucher
i Svasso colombiano

8013 *Podiceps auritus*
Podicipediformes - Podicipedidae
e Horned Grebe; Slavonian Grebe
f Grèbe esclavon; Grèbe cornu; Grèbe

à cou brun; Grèbe à crête; Grèbe
oreillard
d   Ohrentaucher
i   Svasso cornuto

8014   *Podiceps cristatus*
Podicipediformes - Podicipedidae
e   Great Crested-Grebe; Crested Grebe;
Southern Crested-Grebe
f   Grèbe huppé
d   Haubentaucher
i   Svasso maggiore

8015   *Podiceps gallardoi*
Podicipediformes - Podicipedidae
e   Hooded Grebe; Mitred Grebe
f   Grèbe mitré
d   Goldscheiteltaucher
i   Svasso dal cappuccio

8016   *Podiceps grisegena*
Podicipediformes - Podicipedidae
e   Red-necked Grebe; Holböll's Grebe
f   Grèbe jougris; Grèbe à joues grises
d   Rothalstaucher
i   Svasso collorosso

8017   *Podiceps major*
Podicipediformes - Podicipedidae
e   Great Grebe
f   Grand Grèbe
d   Magellan-Taucher
i   Svasso maggiore del Sudamerica

8018   *Podiceps nigricollis*
Podicipediformes - Podicipedidae
e   Black-necked Grebe; Eared Grebe
(NA)
f   Grèbe à cou noir
d   Schwarzhalstaucher
i   Svasso piccolo

8019   *Podiceps occipitalis*
Podicipediformes - Podicipedidae
e   Silvery Grebe; Silver Grebe
f   Grèbe aux belles joues
d   Inkataucher
i   Svasso argentato

8020   *Podiceps taczanowskii*
Podicipediformes - Podicipedidae

e   Puna Grebe; Junin Grebe; Crested
Grebe; Taczanowski's Grebe; Silver
Grebe
f   Grèbe de Taczanowski
d   Punataucher
i   Svasso della Puna

8021   Podicipedidae
Podicipediformes
e   Grebes
f   Grèbes; Podicipédidés
d   Lappentaucher; Steißfüßer
i   Podicipitidi

8022   Podicipediformes
e   Grebes
f   Grèbes; Podicipediformes
d   Lappentaucher; Steißtaucher
i   Podicipediformi

8023   *Podilymbus gigas*
Podicipediformes - Podicipedidae
e   Atitlan Grebe
f   Grèbe de l'Atitlan; Grèbe du Lac
Atitlan
d   Atita-aucher
i   Podilimbo del Lago Atitlan

8024   *Podilymbus podiceps*
Podicipediformes - Podicipedidae
e   Pied-billed Grebe; Duck-and-Teal
(WI); Dapper (WI); Diver (WI)
f   Grèbe à bec bigarré; Grèbe à bec
cerclé; Grand Plongeon; Plongeon
(Ants)
d   Bindentaucher; Fleckschnabeltaucher
i   Podilimbo; Podilimbo comune

8025   *Podoces biddulphi*
Passeriformes - Corvidae
e   Xinjiang Ground-Jay; Biddulph's
Ground-Jay; Biddulphs Ground-
Chough; Biddulph's Desert-Chough;
Biddulphs Desert-Jay; White-tailed
Ground-Chough; White-tailed
Ground-Jay; Sinkiang Ground-Jay
f   Podoce de Biddulph; Geai terrestre
de Biddulph
d   Weißschwanzhäher
i   Podoce di Biddulph

8026   *Podoces hendersoni*
       Passeriformes - Corvidae
e   Mongolian Ground-Jay; Henderson's
       Ground-Jay; Henderson's Chough;
       Mongolian Desert Jay
f   Podoce de Henderson; Geai terrestre
       de Henderson
d   Mongolen-Häher; Hendersons
       Wüstenhäher
i   Podoce di Henderson

8027   *Podoces panderi*
       Passeriformes - Corvidae
e   Turkestan Ground-Jay; Grey Ground-
       Jay; Grey Ground-Chough; Pander's
       Ground-Chough; Pander's Ground-
       Jay; Saxaul Desert-Jay
f   Podoce de Pander
d   Saxaul-Häher
i   Podoce di Pander

8028   *Podoces pleskei*
       Passeriformes - Corvidae
e   Iranian Ground-Jay; Pleske's Ground-
       Jay; Persian Ground-Jay; Persian
       Ground-Chough; Pleske's Ground-
       Chough; Fawn-coloured Chough
f   Podoce de Pleske
d   Pleske-Häher
i   Podoce di Pleske

8029   *Poecilodryas albonotata*
       Passeriformes - Petroicidae
e   Black-throated Robin; Black-throated
       Flycatcher
f   Miro à gorge noire
d   Halsfleckenschnäpper
i   Balia golanera

8030   *Poecilodryas brachyura*
       Passeriformes - Petroicidae
e   Black-chinned Robin; White-
       breasted Robin; Short-tailed Robin;
       Short-tailed Flycatcher; White-
       breasted Flyrobin; New Guinea
       Robin; New Guinea Flyrobin
f   Miro à menton noir
d   Kurzschwanzschnäpper
i   Balia mentobianco

8031   *Poecilodryas hypoleuca*
       Passeriformes - Petroicidae
e   Black-sided Robin; Black-and-white
       Robin; Blacksided Flyrobin
f   Miro à flancs noirs
d   Scheckenschnäpper
i   Balia fianchineri

8032   *Poecilodryas placens*
       Passeriformes - Petroicidae
e   Olive-yellow Robin; Olive-yellow
       Flycatcher; Olive-yellow Flyrobin;
       Banded Yellow Robin; Yellow
       Thicket-Flycatcher
f   Miro ceinturé
d   Olivbandschnäpper
i   Balia ventregiallo dal collare

8033   *Poecilodryas superciliosa*
       Passeriformes - Petroicidae
e   White-browed Robin; White-browed
       Flycatcher
f   Miro bridé
d   Augenstreifschnäpper
i   Balia dai sopraccigli bianchi

8034   *Poecilotriccus albifacies*
       Passeriformes - Tyrannidae
e   White-cheeked Tody-Tyrant; White-
       chested Tody-Tyrant; Blake's Tody-
       Flycatcher; Blake's Tody-Tyrant
f   Microtyran à face blanche
d   Weißwangenspateltyrann
i   Tiranno todo guancebianche

8035   *Poecilotriccus capitalis*
       Passeriformes - Tyrannidae
e   Black-and-white Tody-Tyrant;
       Black-and-white Tody-Flycatcher
f   Microtyran noir-et-blanc; Tyranneau
       noir-et-blanc
d   Dreifarbenspateltyrann
i   Tiranno todo bianco e nero

8036   *Poecilotriccus capitalis tricolor*
       Passeriformes - Tyrannidae
e   White-cheeked Tody-Tyrant
f   Microtyran tricolore
d   Weißwangenspateltyrann

8037 *Poecilotriccus ruficeps*
Passeriformes - Tyrannidae
e Rufous-crowned Tody-Tyrant;
Rufous-headed Tody-Tyrant
f Microtyran bariolé; Tyranneau
bariolé
d Rotscheitelspateltyrann
i Tiranno todo testarossiccia

8038 *Poeoptera kenricki*
Passeriformes - Sturnidae
e Kenrick's Starling
f Rufipenne de Kenrick; Étourneau de
Kenrick
d Kenrick-Star
i Storno di Kenrick

8039 *Poeoptera lugubris*
Passeriformes - Sturnidae
e Narrow-tailed Starling
f Rufipenne à queue étroite; Étourneau
à queue étroite
d Waldstar; Spitzschwanzstar
i Storno codagraduata

8040 *Poeoptera stuhlmanni*
Passeriformes - Sturnidae
e Stuhlmann's Starling
f Rufipenne de Stuhlmann; Étournea
de Stuhlmann
d Stuhlmann-Star
i Storno di Stuhlmann

8041 *Poephila acuticauda*
Passeriformes - Passeridae
e Long-tailed Finch; Long-tailed
Grassfinch; Black-hearted Finch;
Blackheart Finch
f Diamant à longue queue
d Spitzschwanzamadine;
Spitzschwanzgürtelgrasfink
i Diamante codalunga

8042 *Poephila cincta*
Passeriformes - Passeridae
e Black-throated Finch; Black-throated
Grassfinch; Parson Finch
f Diamant à bavette
d Gürtelamadine; Gürtelgrasfink
i Diamante bavetta

8043 *Poephila cincta cincta*
Passeriformes - Passeridae
e Southern Black-throated Finch
(ANZ)

8044 *Poephila personata*
Passeriformes - Passeridae
e Masked Finch; Masked Grassfinch
f Diamant masqué
d Maskenamadine
i Diamante mascherato

8045 *Pogoniulus atroflavus*
Piciformes - Lybiidae
e Red-rumped Tinkerbird
f Barbion à croupion rouge
d Rotbürzelbartvogel
i Barbatula dal groppone rosso

8046 *Pogoniulus bilineatus*
Piciformes - Lybiidae
e Yellow-rumped Tinkerbird; Golden-
rumped Tinkerbird; Golden-rumped
Tinker-Barbet (CSA); Tinkerbird
(CSA); Lemon-rumped Tinkerbird
f Barbion à croupion jaune
d Goldbürzelbartvogel;
Bindenbartvogel; Zwergbartvogfel;
Gelbbüschelbartvogel
i Barbatula dal groppone gialla

8047 *Pogoniulus bilineatus leucolaima*
Piciformes - Lybiidae
f Barbion à gorge blanche

8048 *Pogoniulus chrysoconus*
Piciformes - Lybiidae
e Yellow-fronted Tinkerbird; Yellow-
fronted Tinker-Barbet (CSA);
Tinkerbird (CSA); Yellow-fronted
Barbet
f Barbion à front jaune
d Gelbstirnbartvogel
i Barbatula frontegialla

8049 *Pogoniulus coryphaeus*
Piciformes - Lybiidae
e Western Tinkerbird; Western Green-
Tinkerbird; Western Green Tinker-
Barbet; Tinkerbird (CSA); Green
Tinkerbird

f   Barbion montagnard
d   Gelbrückenbartvogel
i   Barbatula verde occidentale

8050   ***Pogoniulus leucomystax***
      Piciformes - Lybiidae
e   Moustached Tinkerbird; Moustached
      Green-Tinkerbird (CSA); Whiskered
      Tinker-Barbet
f   Barbion à moustaches
d   Bergbartvogel
i   Barbatula verde dai mustacchi

8051   ***Pogoniulus makawai***
      Piciformes - Lybiidae
e   Black-chinned Tinkerbird; White-
      chested Tinkerbird; White-chested
      Tinker-Barbet
f   Barbion à potrine blanche
d   Makawa-Bartvogel

8052   ***Pogoniulus pusillus***
      Piciformes - Lybiidae
e   Red-fronted Tinkerbird; Red-fronted
      Tinker-Barbet (CSA); Tinkerbird
      (CSA)
f   Barbion à front rouge
d   Feuerstirnbartvogel
i   Barbatula fronterossa

8053   ***Pogoniulus scolopaceus***
      Piciformes - Lybiidae
e   Speckled Tinkerbird; Speckled
      Tinker-Barbet; Tinker-Barbet;
      Spectacled Tinkerbird
f   Barbion grivelé
d   Schuppenbartvogel
i   Barbatula macchiettata

8054   ***Pogoniulus simplex***
      Piciformes - Lybiidae
e   Green Tinkerbird; Eastern Green-
      Tinkerbird (CSA); Green Tinker-
      Barbet (CSA); Tinkerbird (CSA);
      African Green Tinkerbird
f   Barbion vert
d   Schlichtbartvogel
i   Barbatula verde

8055   ***Pogoniulus subsulphureus***
      Piciformes - Lybiidae

e   Yellow-throated Tinkerbird; Yellow-
      throated Tinker-Barbet
f   Barbion à gorge jaune
d   Gelbkehlbartvogel
i   Barbatula golagialla

8056   ***Pogonocichla stellata***
      Passeriformes - Muscicapidae
e   White-starred Robin; Starred Robin
      (CSA); Starred Bush-Robin; White-
      starred Bush-Robin; Bush-Robin;
      White-starred Forest Robin
f   Rougegorge étoilé
d   Sternrötel
i   Pettirosso stellato

8057   ***Poicephalus crassus***
      Psittaciformes - Psittacidae
e   Niam-niam Parrot
f   Perroquet des Niam-Niam
d   Niamniam-Papagei
i   Papagallo del Niam-Niam

8058   ***Poicephalus cryptoxanthus***
      Psittaciformes - Psittacidae
e   Brown-headed Parrot
f   Perroquet à tête brune
d   Braunkopfpapagei
i   Papagallo testabruna

8059   ***Poicephalus flavifrons***
      Psittaciformes - Psittacidae
e   Yellow-fronted Parrot; Yellow-faced
      Parrot; African Yellow-faced Parrot
f   Perroquet à face jaune
d   Schoa-Papagei;
      Gelbstirnmohrenkopfpapagei
i   Papagallo facciagialla

8060   ***Poicephalus gulielmi***
      Psittaciformes - Psittacidae
e   Red-fronted Parrot; Jardine's Parrot;
      Red-headed Parrot
f   Perroquet à calotte rouge; Perroquet à
      front rouge; Perroquet à tête orangée
d   Kongo-Papagei; Gulielmis
      Rotstirnpapagei
i   Papagallo di Jardine

8061   ***Poicephalus meyeri***
      Psittaciformes - Psittacidae

e Meyer's Parrot; Brown Parrot
f Perroquet de Meyer
d Goldbugpapagei
i Papagallo di Meyer

8062 *Poicephalus robustus*
Psittaciformes - Psittacidae
e Brown-necked Parrot; Cape Parrot (CSA)
f Perroquet robuste; Perroquet de Levaillant
d Kap-Papagei
i Pappagallo del Congo; Papagallo robusto

8063 *Poicephalus rueppellii*
Psittaciformes - Psittacidae
e Rüppell's Parrot
f Perroquet de Rüppell
d Rüppel-Papagei; Rüppels Blausteißpapagei
i Pappagallo di Ruppell

8064 *Poicephalus rufiventris*
Psittaciformes - Psittacidae
e Red-bellied Parrot; African Orange-bellied Parrot (CSA); Orange-bellied Parrot; Red-breasted Parrot; Abyssinian Parrot
f Perroquet à ventre rouge
d Rotbauchmohrenkopf; Rotbauchpapagei
i Papagallo rufiventre

8065 *Poicephalus senegalus*
Psittaciformes - Psittacidae
e Senegal Parrot
f Perroquet youyou; Perroquet du Sénégal
d Mohrenkopfpapagei; Mohrenkopf
i Jou-jou; Papagallo del Senegal

8066 *Polemaetus bellicosus*
Psittaciformes - Psittacidae
e Martial Eagle; Martial Hawk-Eagle
f Aigle martial
d Kampfadler
i Aquila marziale

8067 *Polihierax insignis*
Falconiformes - Falconidae

e White-rumped Falcon; Fielden's Falconet; Fielden's Pygmy-Falcon; White-rumped Pygmy-Falcon
f Fauconnet à pattes jaunes
d Langschwanzzwergfalke
i Falco pigmeo dal groppone bianco; Falchetto di Fielden

8068 *Polihierax semitorquatus*
Falconiformes - Falconidae
e Pygmy Falcon; African Pygmy-Falcon
f Fauconnet d'Afrique; Faucon d'Afrique
d Zwergfalke; Halsbandzwergfalke
i Falchetto africano

8069 *Poliocephalus poliocephalus*
Podicipediformes - Podicipedidae
e Hoary-headed Grebe; Hoary-headed Dabchick
f Grèbe argenté
d Haarschopftaucher; Silbertaucher
i Falco pigmeo africano; Svasso testacanuta

8070 *Poliocephalus rufopectus*
Podicipediformes - Podicipedidae
e New Zealand Grebe; New Zealand Dabchick
f Grèbe de Nouvelle-Zélande
d Maoritaucher
i Svasso della Nuova Zelanda

8071 *Poliolais lopesi*
Passeriformes - Sylviidae
e White-tailed Warbler; Cameroon Warbler
f Poliolais à queue blanche; Camaroptère à queue blanche
d Weißschwanzsänger
i Bigia di Lopez

8072 *Polioptila albiloris*
Passeriformes - Muscicapidae
e White-lored Gnatcatcher
f Gobemoucheron à face blanche
d Weißzügelmückenfänger
i Zanzariere guancebianche

8073 *Polioptila bilineata*
Passeriformes - Muscicapidae
e White-faced Gnatcatcher; White-browed Gnatcatcher

8074 *Polioptila caerulea*
Passeriformes - Muscicapidae
e Blue-grey Gnatcatcher; Catbird (WI); Chewbird (WI); Spain-Spain (WI); Cottonbird (WI)
f Gobemoucheron gris-bleu; Gobemouches gris-bleu
d Blaumückenschnäpper; Blaumückenfänger
i Zanzariere grigio-blu

8075 *Polioptila californica*
Passeriformes - Muscicapidae
e California Gnatcatcher
f Gobemoucheron de Californie
i Zanzariere della California

8076 *Polioptila dumicola*
Passeriformes - Muscicapidae
e Masked Gnatcatcher; Berlepsch's Gnatcatcher
f Gobemoucheron masqué
d Maskenmückenfänger
i Zanzariere mascherato

8077 *Polioptila guianensis*
Passeriformes - Muscicapidae
e Guianan Gnatcatcher
f Gobemoucheron guyanais
d Cayenne-Mückenfänger
i Zanzariere della Guiana

8078 *Polioptila lactea*
Passeriformes - Muscicapidae
e Creamy-bellied Gnatcatcher; Cream-bellied Gnatcatcher; Creamy Gnatcatcher; Cream-coloured Gnatcatcher
f Gobemoucheron lacté
d Rahmbauchmückenfänger
i Zanzariere ventrelatteo

8079 *Polioptila lembeyei*
Passeriformes - Muscicapidae
e Cuban Gnatcatcher
f Gobemoucheron de Cuba

d Kuba-Mückenfänger
i Zanzariere di Cuba

8080 *Polioptila maior*
Passeriformes - Muscicapidae
e Marañon Gnatcatcher

8081 *Polioptila maranonica*
Passeriformes - Muscicapidae
e Marañon Gnatcatcher

8082 *Polioptila melanura*
Passeriformes - Muscicapidae
e Black-tailed Gnatcatcher; Plumbeous Gnatcatcher; Long-billed Gnatcatcher
f Gobemoucheron à queue noire
d Schwarzschwanzmückenfänger
i Zanzariere codanera

8083 *Polioptila nigriceps*
Passeriformes - Muscicapidae
e Black-capped Gnatcatcher
f Gobemoucheron à coiffe noire
i Zanzariere testanera

8084 *Polioptila plumbea*
Passeriformes - Muscicapidae
e Tropical Gnatcatcher; Marañon Gnatcatcher; Mountain Gnatcatcher; White-browed Gnatcatcher
f Gobemoucheron tropical
d Schwarzkappenmückenfänger
i Zanzariere tropicale

8085 *Polioptila schistaceigula*
Passeriformes - Muscicapidae
e Slate-throated Gnatcatcher
f Gobemoucheron ardoisé
d Graukehlmückenfänger
i Zanzariere gola ardesia

8086 *Polioxolmis rufipennis*
Passeriformes - Tyrannidae
e Rufous-webbed Bush-Tyrant; Rufous-webbed Tyrant; Rufous-webbed Ground-Tyrant; Rufous-webbed Monjita
f Pépoaza à ailes rousses; Pépoaza à ventre rougeâtre

d   Rotspiegeltyrann
i   Tiranno rufipenne

8087   Polyboroides radiatus
Falconiformes - Accipitridae
e   Madagascar Harrier-Hawk; Harrier-
Hawk (CSA); African Harrier-Hawk;
Gymnogene; Madagascar
Gymnogene; Banded Gymnogene
f   Gymnogène de Madagascar;
Gymnogène malgache
d   Höhlenweihe; Höhenweihe
i   Sparviero serpentaria del
Madagascar; Caracara del
Madagascar

8088   Polyboroides typus
Falconiformes - Accipitridae
e   African Harrier-Hawk; Gymnogene;
Banded Harrier-Hawk; Madagascar
Gymnogene
f   Gymnogène d'Afrique
d   Schlangensperber; Höhlenweihe
i   Sparviero serpentario africano;
Serpentario minore

8089   Polyborus lutosus
Falconiformes - Falconidae
e   Guadalupe Caracara
f   Caracara de Guadalupe
i   Caracara della Guadelupa

8090   Polyborus plancus
Falconiformes - Falconidae
e   Southern Caracara; Crested Caracara;
Common Caracara; Caracara
f   Caracara huppé
d   Karakara
i   Caracara comune; Caracara crestato;
Caracara

8091   Polyonymus caroli
Apodiformes - Trochilidae
e   Bronze-tailed Comet
f   Colibri de Bourcier
d   Breitschwanzsylphe
i   Colibrì cometa codabronzata

8092   Polyplectron bicalcaratum
Galliformes - Phasianidae
e   Grey Peacock-Pheasant; Peacock-

Pheasant; Iris Peacock-Pheasant;
Chinquis Peacock-Pheasant;
Burmese Peacock-Pheasant
f   Éperonnier chinquis
d   Grauer Fasan
i   Speroniere della Birmania

8093   Polyplectron chalcurum
Galliformes - Phasianidae
e   Bronze-tailed Peacock-Pheasant;
Sumatran Peacock-Pheasant; Bronze-
tailed Pheasant; Lesson's Peacock-
Pheasant
f   Éperonnier à queue bronzée
d   Bronzeschwanzfasan
i   Speroniere di Sumatra

8094   Polyplectron emphanum
Galliformes - Phasianidae
e   Palawan Peacock-Pheasant;
Napoleon's Peacock-Pheasant
f   Éperonnier napoléon
d   Napoleon-Fasan
i   Speroniere di Palawan

8095   Polyplectron germaini
Galliformes - Phasianidae
e   Germain's Peacock-Pheasant
f   Éperonnier de Germain
d   Brauner Pfaufasan
i   Speroniere di Germain

8096   Polyplectron inopinatum
Galliformes - Phasianidae
e   Mountain-Peacock-Pheasant;
Bronze-tailed Peacock-Pheasant;
Rothschild's Peacock-Pheasant;
Bronze-tailed Pheasant; Rothschild's
Pheasant; Mirror Peacock Pheasant;
Malayan Peacock-Pheasant;
Malaysian Peacock-Pheasant
f   Éperonnier de Rothschild
d   Rothschild-Pfaufasan
i   Speroniere di Rothschild

8097   Polyplectron katsumatae
Galliformes - Phasianidae
e   Hainan Peacock-Pheasant

8098   Polyplectron malacense
Galliformes - Phasianidae

e Malayan Peacock-Pheasant; Malay
Peacock-Pheasant; Malaysian
Peacock-Pheasant; Malaysian
Crested Peacock-Pheasant; Malayan
Crested Peacock-Pheasant
f Éperonnier malais
d Malaien-Pfaufasan
i Speroniere di Malacca

8099 *Polyplectron schleiermacheri*
Galliformes - Phasianidae
e Bornean Peacock-Pheasant
f Éperonnier de Bornéo
i Speroniere del Borneo

8100 *Polysticta stelleri*
Anseriformes - Anatidae
e Steller's Eider
f Eider de Steller
d Scheckente
i Edredone di Steller

8101 *Polystictus pectoralis*
Passeriformes - Tyrannidae
e Bearded Tachuri; Narrow-tailed
Tyrant
f Tyranneau barbu; Tyranneau du
Paraguay
d Schmalschwanztyrann
i Tachuri barbuto

8102 *Polystictus superciliaris*
Passeriformes - Tyrannidae
e Grey-backed Tachuri; Superciliated
Tachuri
f Tyranneau bridé; Tyranneau des
campos
d Silberbrauentyrann
i Tachuri dorsogrigio

8103 *Polytelis alexandrae*
Psittaciformes - Psittacidae
e Alexandra's Parrot; Princess Parrot;
Queen Alexandra's Parakeet; Queen
Parakeet; Princess Parakeet;
Alexandra's Parakeet; Princess of
Wales Parakeet; Princess Alexandra's
Parakeet
f Perruche d'Alexandra; Perruche à
calotte bleue
d Alexandra-Sittich; Princess-of-Wales

Sittich; Blaukappensittich
i Parrocchetto di Alessandra;
Parrochetta dellla principessa
Alexandra

8104 *Polytelis anthopeplus*
Psittaciformes - Psittacidae
e Regent Parrot; Smoker Parrot; Regal
Parrot; Mountain Parrot; Rock-
pebbler Parakeet; Marlock Parakeet;
Black-tailed Parakeet; Royal Parrot;
Black-tailed Parrot
f Perruche mélanure; Perruche à queue
noire
d Bergsittich
i Parrochetto codanera

8105 *Polytelis anthopeplus monarchoides*
Psittaciformes - Psittacidae
e Eastern Regent Parrot (ANZ)

8106 *Polytelis swainsonii*
Psittaciformes - Psittacidae
e Superb Parrot; Scarlet-breasted
Parrot; Superb Parakeet; Barraband
Parakeet; Green Parakeet; Green
Leek; Baraband Parrot
f Perruche de Barraband
d Barraband-Sittich; Schildsittich
i Parrochetto di Swainson

8107 *Polytmus guainumbi*
Apodiformes - Trochilidae
e White-tailed Goldenthroat; White-
tailed Hummingbird
f Colibri guainumbi
d Weißschwanzgoldkehlchen
i Colibrì guainumbi

8108 *Polytmus milleri*
Apodiformes - Trochilidae
e Tepui Goldenthroat
f Colibri des tépuis
d Tepui-Goldkehlchen
i Colibrì di Miller

8109 *Polytmus theresiae*
Apodiformes - Trochilidae
e Green-tailed Goldenthroat
f Colibri tout-vert

d   Grünschwanzgoldkehlchen
i   Colibrì di Teresa

8110   *Pomarea dimidiata*
Passeriformes - Corvidae
e   Rarotonga Monarch; Rarotonga
Flycatcher; Cook Islands Flycatcher;
Rarotonga Monarch-Flycatcher;
Cook Islands Monarch-Flycatcher;
Rarotongan Flycatcher
f   Monarque de Rurotonga
d   Rarotonga-Monarch
i   Pomarea di Raratonga

8111   *Pomarea iphis*
Passeriformes - Corvidae
e   Iphis Monarch; Huahuna Flycatcher;
Allied Flycatcher
f   Monarque iphis
d   Fleckenmonarch
i   Pomarea di Huahuna

8112   *Pomarea mendozae*
Passeriformes - Corvidae
e   Marquesan Monarch; Marquesas
Flycatcher; Marquesas Monarch;
Mendoza Flycatcher; Marquesa
Monarch; Marquesa Flycatcher;
Marquesan Flycatcher
f   Monarque des Marquises
d   Marquesas-Monarch; Marquesas-
Fliegenschnäpper
i   Pomarea delle Isole Marquesas

8113   *Pomarea nigra*
Passeriformes - Corvidae
e   Tahiti Monarch; Tahiti Flycatcher;
Society Islands Flycatcher; Tahiti
Monarch-Flycatcher
f   Monarque de Tahiti
d   Tahiti-Monarch
i   Pomarea delle Isole Società

8114   *Pomarea whitneyi*
Passeriformes - Corvidae
e   Fatuhiva Monarch; Fatuhiva
Flycatcher; Large Flycatcher;
Fatuhiva Monarch-Flycatcher; Large
Monarch
f   Monarque de Fatuhiva

d   Fatuhiva-Monarch
i   Pomarea di Whitney

8115   *Pomatorhinus erythrocnemis*
Passeriformes - Sylviidae
e   Spot-breasted Scimitar-Babbler;
Chinese Spot-breasted Scimitar-
Babbler; Chinese Rusty-cheeked
Scimitar-Babbler
f   Pomatorhin tacheté
d   Drosselsäbler
i   Garrulo scimitarra della Cina

8116   *Pomatorhinus erythrogenys*
Passeriformes - Sylviidae
e   Ferruginous-cheeked Scimitar-
Babbler
f   Pomatorhin à joues rousses;
Pomatorhin à gorge rousse
d   Rotwangensäbler
i   Garrulo tordo guancerosse

8117   *Pomatorhinus ferruginosus*
Passeriformes - Sylviidae
e   Coral-billed Scimitar-Babbler
f   Pomatorhin à bec corail
d   Korallschnabelsäbler
i   Garrulo scimitarra beccodicorallo

8118   *Pomatorhinus horsfieldii*
Passeriformes - Sylviidae
e   Indian Scimitar-Babbler; Slaty-
headed Scimitar-Babbler (ISC);
Horsfield's Scimitar-Babbler;
Scimitar-Babbler; Travancore
Scimitar-Babbler; Ceylon Scimitar-
Babbler; Deccan Scimitar-Babbler;
Peninsular Scimitar-Babbler; White-
browed Scimitar-Babbler
f   Pomatorhin de Horsfield
d   Horsfield-Säbler
i   Garrulo scimitarra di Travancore

8119   *Pomatorhinus hypoleucos*
Passeriformes - Sylviidae
e   Large Scimitar-Babbler; Greater
Scimitar-Babbler; Long-billed
Scimitar-Babbler
f   Pomatorhin à long bec
d   Langschnabelsäbler; Riesensäbler
i   Garrulo scimitarra beccolungo

8120 **Pomatorhinus montanus**
Passeriformes - Sylviidae
e Chestnut-backed Scimitar-Babbler;
Yellow-billed Scimitar-Babbler
f Pomatorhin à dos marron
d Rotrückensäbler; Weißbrauensäbler
i Garrulo scimitarra dorsocastano

8121 **Pomatorhinus ochraceiceps**
Passeriformes - Sylviidae
e Red-billed Scimitar-Babbler; Lloyd's
Scimitar-Babbler; Ochraceous-
headed Scimitar-Babbler
f Pomatorhin à bec rouge
d Rotschnabelsäbler
i Garrulo scimitarra beccorosso

8122 **Pomatorhinus ruficollis**
Passeriformes - Sylviidae
e Streak-breasted Scimitar-Babbler;
Rufous-necked Scimitar-Babbler
f Pomatorhin à col roux; Pomatorhin à
cou roux
d Rothalssäbler; Rotkehlsäbler
i Garrulo scimitarra pettostriato

8123 **Pomatorhinus schisticeps**
Passeriformes - Sylviidae
e Slaty-headed Scimitar-Babbler;
White-browed Scimitar-Babbler
(ISC)
f Pomatorhin à tête ardoisé
d Himalaja-Säbler
i Garrulo scimitarra testa ardesia

8124 **Pomatostomidae**
Passeriformes
e Australo-Papuan Babblers
f Pomatostomidés
d Säbler; Weißbrauensäbler
i Pomatostomidi

8125 **Pomatostomus halli**
Passeriformes - Pomatostomidae
e Hall's Babbler; Major Hall's Babbler;
Hall's White-throated Babbler
f Pomatostome de Hall
d Rußbauchsäbler
i Garrulo di Hall

8126 **Pomatostomus isidorei**
Passeriformes - Pomatostomidae
e New Guinea Babbler; Isidore's
Rufous Babbler
f Pomatostome isidore
d Beutelsäbler
i Garrulo di Saint-Hilaire; Timalide
dal becco a falce

8127 **Pomatostomus ruficeps**
Passeriformes - Pomatostomidae
e Chestnut-crowned Babbler; Red-
capped Babbler
f Pomatostome à calotte marron
d Rotscheitelsäbler
i Garrulo corona castana

8128 **Pomatostomus superciliosus**
Passeriformes - Pomatostomidae
e White-browed Babbler
f Pomatostome bridé
d Brauensäbler
i Garrulo dai sopraccigli bianchi

8129 **Pomatostomus superciliosus ashbyi**
Passeriformes - Pomatostomidae
e Western Wheatbelt White-browed
Babbler (ANZ)

8130 **Pomatostomus temporalis**
Passeriformes - Pomatostomidae
e Grey-crowned Babbler
f Pomatostome à calotte grise
d Grauscheitelsäbler
i Garrulo corona grigia

8131 **Pomatostomus temporalis
temporalis**
Passeriformes - Pomatostomidae
e Western Wheatbelt Grey-crowned
Babbler (ANZ)

8132 **Pooecetes gramineus**
Passeriformes - Fringillidae
e Vesper Sparrow
f Bruant vespéral
d Abendammer
i Passero del vespro

8133 **Poospiza alticola**
Passeriformes - Fringillidae

e	Plain-tailed Warbling-Finch	*d*	Garlepp-Ammerfink
f	Chipiu alticole	*i*	Fringuello cantore di Cochamba
d	Grauschwanzammerfink		
i	Fringuello cantore del Perù	**8140**	***Poospiza hispaniolensis***

8134 ***Poospiza baeri***
Passeriformes - Fringillidae
e Tucuman Mountain-Finch
f Chipiu de Tucuman
d Baer-Ammerfink
i Fringuello cantore tucumano

8135 ***Poospiza boliviana***
Passeriformes - Fringillidae
e Bolivian Warbling-Finch
f Chipiu de Bolivie
d Zimtbrustammerfink
i Fringuello cantore della Bolivia

8136 ***Poospiza caesar***
Passeriformes - Fringillidae
e Chestnut-breasted Mountain-Finch;
Chestnut-bellied Mountain-Finch;
Chestnut Warbling-Finch; Cuzco
Mountain-Finch
f Chipiu césar
d Kaiserammerfink
i Fringuello cantore pettocastano

8137 ***Poospiza cinerea***
Passeriformes - Fringillidae
e Cinereous Warbling-Finch; Black-
capped Warbling-Finch; Grey-and-
white Warbling-Finch
f Chipiu à tête cendrée
i Fringuello cantore grigio e bianco

8138 ***Poospiza erythrophrys***
Passeriformes - Fringillidae
e Rusty-browed Warbling-Finch
f Chipiu à sourcils roux
d Rostbrauenammerfink
i Fringuello cantore dai sopraccigli
rossi

8139 ***Poospiza garleppi***
Passeriformes - Fringillidae
e Cochabamba Mountain-Finch; Slate-
and-rufous Warbling-Finch
f Chipiu de Cochabamba

8140 ***Poospiza hispaniolensis***
Passeriformes - Fringillidae
e Collared Warbling-Finch
f Chipiu à col noir
d Schwarzbrustammerfink
i Fringuello cantore dal collare

8141 ***Poospiza hypochondria***
Passeriformes - Fringillidae
e Rufous-sided Warbling-Finch
f Chipiu à flancs roux
d Rotflankenammerfink
i Fringuello cantore fianchirossi

8142 ***Poospiza lateralis***
Passeriformes - Fringillidae
e Red-rumped Warbling-Finch; Grey-
throated Warbling-Finch
f Chipiu à croupion roux
d Rotbürzelammerfink
i Fringuello cantore gropponerosso

8143 ***Poospiza melanoleuca***
Passeriformes - Fringillidae
e Black-capped Warbling-Finch;
White-and-grey Warbling-Finch;
Grey-and-white Warbling-Finch
f Chipiu à capuchon
d Schwarzwangenammerfink
i Fringuello cantore capinero

8144 ***Poospiza nigrorufa***
Passeriformes - Fringillidae
e Black-and-rufous Warbling-Finch
f Chipiu noiroux
d Rotbrustammerfink
i Fringuello cantore nero e ruggine

8145 ***Poospiza ornata***
Passeriformes - Fringillidae
e Cinnamon Warbling-Finch
f Chipiu cannellé
d Schmuckammerfink
i Fringuello cantore ornato

8146 ***Poospiza rubecula***
Passeriformes - Fringillidae
e Rufous-breasted Warbling-Finch

f	Chipiu rougegorge
d	Rostbauchammerfink
i	Fringuello cantore pettirosso

8147 *Poospiza thoracica*
Passeriformes - Fringillidae
e Bay-chested Warbling-Finch; Bay-breasted Warbling-Finch
f Chipiu à poitrine baie
d Maronenbrustsammerfink
i Fringuello cantore pettocannella

8148 *Poospiza torquata*
Passeriformes - Fringillidae
e Ringed Warbling-Finch
f Chipiu sanglé
d Bandammerfink
i Fringuello cantore grigio dal collare

8149 *Poospiza whitii*
Passeriformes - Fringillidae
e Black-and-chestnut Warbling-Finch
f Chipiu noiron
i Fringuello cantore nero e castano

8150 *Popelairia conversii*
Apodiformes - Trochilidae
e Green Thorntail
f Coquette à queue fine
d Dornschwanzelfe
i Colibrì codalunga verde

8151 *Popelairia langsdorffi*
Apodiformes - Trochilidae
e Black-bellied Thorntail; Langsdorf's Thorntail; Black-breasted Thorntail
f Coquette de Langsdorff
d Langsdorff-Elfe
i Colibrì codalunga di Langsdorff

8152 *Popelairia letitiae*
Apodiformes - Trochilidae
e Coppery Thorntail; Letitia's Thorntail
f Coquette de Létitia
d Kupferelfe
i Colibrì codalunga di Letizia

8153 *Popelairia popelairii*
Apodiformes - Trochilidae
e Wire-crested Thorntail; Popelair's Thorntail

f	Coquette de Popelaire
d	Fadenschopfelfe
i	Colibrì codalunga di Popelaire

8154 *Porphyrio albus*
Gruiformes - Rallidae
e Lord Howe Island Swamphen; White Gallinule (ANZ); Lord Howe Swamphen; New Britain Gallinule
f Talève de Lord Howe
i Pollo sultano dell'Isola Lord Howe

8155 *Porphyrio alleni*
Gruiformes - Rallidae
e Allen's Gallinule; Lesser Gallinule (CSA); Allen's Reedhen
f Talève d'Allen; Poule-sultane d'Allen
d Afrikanisches Sultanshuhn; Bronzesultanshuhn; Afrika-Sultanshuhn
i Pollo sultano di Allen

8156 *Porphyrio caerulescens*
Gruiformes - Rallidae
e Giant Gallinule
d Maskarenen-Purpurhuhn

8157 *Porphyrio flavirostris*
Gruiformes - Rallidae
e Azure Gallinule; Little Gallinule
f Talève favorite; Poule-sultane favorite
d Azursultanshuhn
i Gallinella azzurra

8158 *Porphyrio mantelli*
Gruiformes - Rallidae
e Takahe; Notornis; Moho
f Talève takahé
d Takahe
i Takahe

8159 *Porphyrio martinicus*
Gruiformes - Rallidae
e Purple Gallinule; American Purple Gallinule; Blue-pate Coot (WI); American Gallinule
f Talève violacée; Poule-sultane violacée; Gallinule violacée; Poule d'eau à cachet vert (Ants); Talève pourprée; Poule-sultane d'Amérique

d   Amerikanisches Sultanshuhn;
      Amerikanischer Uferläufer;
      Zwergsultanshuhn
i   Pollo sultano americano; Pollo
      sultano della Martinica

8160   *Porphyrio porphyrio*
        Gruiformes - Rallidae
e   Purple Swamphen; Purple Gallinule
      (CSA); Purple Coot; Purple
      Moorhen; Indian Swamphen; African
      Swamphen; Western Swamphen;
      Black-breasted Swamphen
f   Talève sultane; Poule-sultane; Talève
      poule-sultane
d   Purpurhuhn
i   Pollo sultano

8161   *Porphyrolaema porphyrolaema*
        Passeriformes - Tyrannidae
e   Purple-throated Cotinga
f   Cotinga à gorge mauve
d   Purpurkehlkotinga
i   Cotinga golapurpurea

8162   *Porphyrospiza caerulescens*
        Passeriformes - Fringillidae
e   Yellow-billed Blue-Finch; Blue
      Finch
f   Porphyrin à bec jaune
d   Kobaltämmerling
i   Fringuello blu del Brasile

8163   *Porzana albicollis*
        Gruiformes - Rallidae
e   Ash-throated Crake; White-throated
      Crake; White-necked Rail
f   Marouette plombée
d   Wieselsumpfhuhn
i   Schiribilla collobianco

8164   *Porzana atra*
        Gruiformes - Rallidae
e   Henderson Island Crake; Henderson
      Crake; Henderson Island Rail;
      North's Crake; Henderson Crake
f   Marouette de Henderson
d   Tuamotu-Sumpfhuhn
i   Schiribilla dell'Isola Henderson

8165   *Porzana carolina*
        Gruiformes - Rallidae
e   Sora; Sora Crake; Carolina Crake;
      Sora Rail; Carolina Rail
f   Marouette de Caroline; Marouette
      caroline; Marouette de la Caroline;
      Râle de Caroline
d   Carolina-Sumpfhuhn; Karolinen-
      Sumphhuhn
i   Voltolino americano

8166   *Porzana cinerea*
        Gruiformes - Rallidae
e   White-browed Crake; White-browed
      Water Crake; White-browed Rail;
      Ashy Crake; Grey-bellied Crake
f   Marouette grise
d   Weißbrauenralle
i   Schiribilla cinerea

8167   *Porzana fluviventer*
        Gruiformes - Rallidae
e   Yellow-breasted Crake; Twopenny
      Chick (WI)
f   Marouette à sourcils blancs
d   Gelbbrustsumpfhuhn
i   Schiribilla pettogiallo

8168   *Porzana fluminea*
        Gruiformes - Rallidae
e   Australian Crake; Australian Spotted-
      Crake (ANZ); Water Crake
f   Marouette d'Australie; Marouette
      australienne
d   Flußsumpfhuhn
i   Voltolino australiano

8169   *Porzana fusca*
        Gruiformes - Rallidae
e   Ruddy-breasted Crake; Ruddy Crake
f   Marouette brune
d   Zimtsumpfhuhn
i   Schiribilla fusca

8170   *Porzana monasa*
        Gruiformes - Rallidae
e   Kosrae Crake; Kusaie Crake; Ponapé
      Crake; Kittlitz's Crake
f   Marouette de Kusaie
d   Karolinen-Sumpfhuhn
i   Schiribilla di Kosrae

8171   *Porzana palmeri*
        Gruiformes - Rallidae
    e   Laysan Crake; Laysan Rail
    f   Marouette de Laysan
    d   Laysan-Sumpfhuhn
    i   Schiribilla di Laysan

8172   *Porzana parva*
        Gruiformes - Rallidae
    e   Little Crake
    f   Marouette poussin; Râle poussin;
        Poule poussin; Poule d'eau poussin
    d   Kleines Sumpfhuhn;
        Kleinsumpfhuhn
    i   Schiribilla grigiata; Schiribilla;
        Schiribilla comune

8173   *Porzana paykullii*
        Gruiformes - Rallidae
    e   Band-bellied Crake; Chinese
        Banded-Crake; Siberian Ruddy
        Crake; Chestnut-breasted Crake
    f   Marouette mandarin
    d   Mandarinsumpfhuhn
    i   Schiribilla fasciata

8174   *Porzana porzana*
        Gruiformes - Rallidae
    e   Spotted Crake; Eurasian Spotted-
        Crake
    f   Marouette ponctuée
    d   Tüpfelsumpfhuhn
    i   Voltolino; Sutro; Teccola; Voltoline
        eurasiatico

8175   *Porzana pusilla*
        Gruiformes - Rallidae
    e   Baillon's Crake; Lesser Spotted-
        Crake; Dwarf Rail; Little Crake; Tiny
        Crake; Marsh Crake; Pallas's Crake
    f   Marouette de Baillon
    d   Zwergsumpfhuhn
    i   Schiribilla grigiata

8176   *Porzana sandwichensis*
        Gruiformes - Rallidae
    e   Hawaiian Crake; Hawaiian Rail
    f   Marouette des Hawai
    d   Hawaii-Sumpfhuhn
    i   Schiribilla delle Hawaii

8177   *Porzana spiloptera*
        Gruiformes - Rallidae
    e   Dot-winged Crake; Dot-winged Rail
    f   Marouette maillée; Rale maillé
    d   Fleckenralle
    i   Schiribilla alipunteggiate

8178   *Porzana tabuensis*
        Gruiformes - Rallidae
    e   Spotless Crake; Sooty Crake; Sooty
        Rail; Tabuan Crake; Leaden Crake;
        Blue Rail; Little Swamphen
    f   Marouette fuligineuse
    d   Südseesumpfhuhn
    i   Schiribilla fuligginosa

8179   *Premnoplex brunnescens*
        Passeriformes - Furnariidae
    e   Spotted Barbtail
    f   Anabasitte tachetée
    d   Fleckbruststachelschwanz
    i   Corridore arboricolo maculato

8180   *Premnoplex tatei*
        Passeriformes - Furnariidae
    e   White-throated Barbtail; Tate's
        Barbtail
    f   Anabasitte à gorge blanche
    d   Weißkehlstachelschwanz
    i   Corridore arboricolo golabianca

8181   *Premnornis guttuligera*
        Passeriformes - Furnariidae
    e   Rusty-winged Barbtail
    f   Anabasitte à gouttelettes
    d   Rostschwingenstachelschwanz
    i   Corridore arboricolo alirosse

8182   *Prinia atrogularis*
        Passeriformes - Cisticolidae
    e   HillPrinia; Black-throated Hill-
        Prinia; White-browed Prinia; White-
        browed Hill-Prinia; White-breasted
        Hill-Warbler; Black-throated Prinia
    f   Prinia à gorge noire
    d   Weißbrustprinie; Schwarzkehlprinie
    i   Prinia pettostriato

8183   *Prinia bairdii*
        Passeriformes - Cisticolidae
    e   Banded Prinia

f   Prinia rayée
d   Zebraprinie
i   Prinia barrata

8184   *Prinia buchanani*
      Passeriformes - Cisticolidae
e   Rufous-fronted Prinia; Rufous-
      fronted Wren-Warbler (ISC);
      Rufous-fronted Longtail-Warbler
f   Prinia à front roux
d   Rotscheitelprinie
i   Prinia di Buchanan

8185   *Prinia burnesii*
      Passeriformes - Cisticolidae
e   Rufous-vented Prinia; Long-tailed
      Prinia; Long-tailed Grass-Warbler
f   Prinia de Burnes
d   Schweifprinie
i   Prinia codalunga

8186   *Prinia cinerascens*
      Passeriformes - Cistocolidae
e   Swamp Prinia
f   Prinia des marais

8187   *Prinia cinereocapilla*
      Passeriformes - Cisticolidae
e   Grey-crowned Prinia; Hodgson's
      Grey-crowned Prinia; Hodgson's
      Long-tailed Warbler; Grey-capped
      Prinia; Yellow-breasted Prinia;
      Hodgson's Longtail-Warbler;
      Hodgson's Prinia
f   Prinia à calotte grise
d   Graukopfprinie
i   Prinia capocenerino

8188   *Prinia criniger*
      Passeriformes - Cisticolidae
e   Striated Prinia; Hill Prinia; Brown
      Prinia; Brown Hill-Prinia; Hill-
      Warbler; Brown Hill-Warbler;
      David's Hill-Warbler
f   Prinia crinigère
d   Bergprinie
i   Prinia striata

8189   *Prinia erythroptera*
      Passeriformes - Cisticolidae
e   Red-winged Prinia; Red-winged

      Warbler; Redwing Prinia
f   Prinia à ailes rousses; Fauvette à ailes
      rousses
d   Sonnenprinie; Sonnensänger
i   Prinia alirosse

8190   *Prinia familiaris*
      Passeriformes - Cisticolidae
e   Bar-winged Prinia; Bar-winged
      Wren-Warbler
f   Prinia bifasciée
d   Sunda-Prinie
i   Prinia alibarrate

8191   *Prinia flavicans*
      Passeriformes - Cisticolidae
e   Black-chested Prinia
f   Prinia à plastron
d   Brustbandprinie
i   Prinia dal collare

8192   *Prinia flaviventris*
      Passeriformes - Cisticolidae
e   Yellow-bellied Prinia; Yellow-
      bellied Warbler; Yellow-bellied
      Wren-Warbler
f   Prinia à ventre jaune
d   Gelbbauchprinie
i   Prinia ventregiallo

8193   *Prinia fluviatilis*
      Passeriformes - Cisticolidae
e   River Prinia; Lake Chad Prinia
f   Prinia aquatique
d   Flußprinie
i   Prinia del lago

8194   *Prinia gracilis*
      Passeriformes - Cisticolidae
e   Graceful Prinia; Graceful Warbler;
      Streaked Wren-Warbler (ISC);
      Striped-backed Prinia; Streaked
      Prinia; Streaked Long-tail Warbler;
      Stripe-backed Prinia; Fulvous-
      streaked Prinia
f   Prinia gracile
d   Streifenprinie
i   Prinia gracile; Beccomoschino
      codalunga

8195   *Prinia hodgsonii*
      Passeriformes - Cisticolidae
  e   Grey-breasted Prinia; Franklin's
      Wren-Warbler (ISC); Franklin's
      Prinia; Ashy-breasted Prinia;
      Hodgson's Prinia; Ashy-grey Prinia;
      Ashy-grey Wren-Warbler
  f   Prinia de Hodgson
  d   Fleckenprinie; Graubrustprinie
  i   Prinia pettogrgio

8196   *Prinia hypoxantha*
      Passeriformes - Cistocolidae
  e   Saffron-breasted Prinia; Saffron
      Prinia (CSA); Spotted-Prinia (CSA);
      Drakensberg Prinia
  f   Prinia du Drakensberg; Prinia à
      poitrine canelle

8197   *Prinia inornata*
      Passeriformes - Cisticolidae
  e   Plain Prinia; Greater Brown Wren-
      Warbler; Plain-coloured Prinia; Plain
      Wren-Warbler; Ceylon White-
      browed Prinia; White-browed Prinia
  f   Prinia simple
  d   Braunkopfprinie
  i   Prinia disadorna

8198   *Prinia leontica*
      Passeriformes - Cisticolidae
  e   White-eyed Prinia; Sierra Leone
      Prinia
  f   Prinia du Sierra Leone
  d   Weißaugenprinie
  i   Prinia della Sierra Leone

8199   *Prinia leucopogon*
      Passeriformes - Cisticolidae
  e   White-chinned Prinia; White-chinned
      Longtail
  f   Prinia à gorge blanche; Fauvette-
      roitelet à gorge blanche
  d   Weißkehlprinie
  i   Prinia mentobianco

8200   *Prinia maculosa*
      Passeriformes - Cisticolidae
  e   Karoo Prinia; Spotted Prinia
  f   Prinia du Karroo

  d   Fleckenprinie
  i   Prinia del Karoo

8201   *Prinia melanops*
      Passeriformes - Cisticolidae
  e   Black-faced Prinia
  f   Prinia à face noire
  i   Prinia faccianera

8202   *Prinia molleri*
      Passeriformes - Cisticolidae
  e   Sao Tomé Prinia; Sao Tomé Long-
      tailed Prinia
  f   Prinia de Sao Tomé
  d   St. Thomas-Prinia
  i   Prinia di Sao Tomè

8203   *Prinia polychroa*
      Passeriformes - Cisticolidae
  e   Brown Prinia; Brown Hill-Prinia;
      Brown Hill-Warbler; Javan Brown-
      Prinia; Javan Brown Hill-Warbler
  f   Prinia des montagnes
  d   Malaien-Prinie

8204   *Prinia robertsi*
      Passeriformes - Cisticolidae
  e   Briar Warbler; Brier Warbler (CSA);
      Roberts's Prinia (CSA); Robert's
      Forest-Prinia; Forest Prinia
  f   Prinia de Roberts
  d   Roberts-Prinie
  i   Prinia di Roberts

8205   *Prinia rufescens*
      Passeriformes - Cisticolidae
  e   Rufescent Prinia; Rufous Prinia;
      Lesser Brown-Prinia; Dark-crowned
      Wren-Warbler; Lesser Brown Wren-
      Warbler; Dark-crowned Prinia;
      Beavan's Wren-Warbler
  f   Prinia roussâtre
  d   Rostprinie
  i   Prinia rossiccia

8206   *Prinia socialis*
      Passeriformes - Cisticolidae
  e   Ashy Prinia; Ashy Wren-Warbler
      (ISC); Ashy Long-tailed Warbler;
      Ashy Longtail Warbler; Ceylon Ashy
      Prinia

f Prinia cendrée
d Rostbauchprinie
i Prinia cenerina

8207 *Prinia somalica*
Passeriformes - Cisticolidae
e Pale Prinia; Pale-coloured Prinia
f Prinia pâle
d Somali-Pirinie
i Prinia della Somalia

8208 *Prinia subflava*
Passeriformes - Cisticolidae
e Tawny-flanked Prinia; Plain Wren-
Warbler (ISC); Plain Prinia (ISC);
Tawny-flanked Prinia; West African
Prinia
f Prinia modeste; Fauvette-roitelet
d Rahmbrustprinie
i Prinia fianchifulve

8209 *Prinia substriata*
Passeriformes - Cisticolidae
e White-breasted Prinia; Namaqua
Warbler; Namaqua Prinia
f Prinia du Namaqua
d Nama-Prinie
i Prinia di Namaqua

8210 *Prinia sylvatica*
Passeriformes - Cisticolidae
e Jungle Prinia; Jungle Wren-Warbler
(ISC); Jungle Long-tailed Warbler;
Large Prinia; Ceylon Large Prinia;
White-tailed Prinia
f Prinia forestière
d Dschungelprinie
i Prinia della giungla

8211 *Prioniturus discurus*
Psittaciformes - Psittacidae
e Blue-crowned Racquet-tail; Blue-
crowned Racquet-tailed Parrot;
Philippine Racquet-tailed Parakeet;
Blue-crowned Racket-tail; Blue-
crowned Racket-tailed Parrot;
Philippine Racket-tailed Parakeet;
Blue-headed Racquet-tail; Blue-
headed Rackettail
f Palette à couronne bleue; Perroquet à
palettes

d Blauköpfiger Spatelschwanzpapagei
i Pappagallo coda a racchetta corona
blu

8212 *Prioniturus flavicans*
Psittaciformes - Psittacidae
e Yellow-breasted Racquet-tail;
Yellow-breasted Racket-tail; Red-
spotted Racket-tailed Parrot;
Crimson-spotted Racket-tailed
Parakeet; Red-spotted Racquet-tailed
Parrot; Yellowish-breasted Racquet-
tail; Yellowish-breasted Racket-tail
f Palette de Cassin; Perroquet
couronné
d Flaggenschwanzpapagei;
Spatelschwanzpapagei
i Pappagallo coda a racchetta corona
rossa

8213 *Prioniturus luconensis*
Psittaciformes - Psittacidae
e Green Racquet-tail; Green Racket-
tail; Green Racket-tailed Parrot;
Luzon Racket-tailed Parakeet; Green
Racquet-tailed Parrot; Luzon
Racquet-tailed Parrot
f Palette verte
d Luzon-Papagei;
Raketenschwanzpapagei
i Pappagallo coda a racchetta verde

8214 *Prioniturus mada*
Psittaciformes - Psittacidae
e Buru Racquet-tail; Buru Racket-tail;
Mount Mada Racket-tailed Parrot;
Buru Racket-tailed Parakeet; Mount
Mada Racquet-tailed Parrot; Buru
Racquet-tailed Parakeet; Buru
Racket-tailed Parrot; Buru Racquet-
tailed Parrot
f Palette de Buru; Perroquet à raquettes
de Buru
d Mada-Papagei; Grüner
Spatelschwanzpapagei; Grünköpfiger
Spatelschwanzpapagei
i Pappagallo coda a racchetta di Buru

8215 *Prioniturus montanus*
Psittaciformes - Psittacidae
e Montane Racquet-tail; Mountain

Racket-tailed Parrot; Montane
Racket-tail; Mountain Racquet-tailed
Parrot; Luzon Racquet-tail; Crimson-
spotted Racquet-tail; Red-crowned
Racquet-tail; Racket Parrot; Racket-
tailed Parrot; Racquet-tailed Parrot;
Luzon Racket-tail; Crimson-spotted
Racket-tail

f   Palette momot
d   Motmotpapagei; Mada-
Spatelschwanzpapagei
i   Pappagallo coda a racchetta di
montagna

8216   *Prioniturus platenae*
Psittaciformes - Psittacidae
e   Blue-headed Racquet-tail; Blue-
headed Racket-tail; Palawan
Racquet-tail; Palawan Racket-tailed
Parrot; Palawan Racket-tail; Palawan
Racquet-tailed Parrot
f   Palette de Palawan
i   Pappagallo coda a racchetta di
Palawan

8217   *Prioniturus platurus*
Psittaciformes - Psittacidae
e   Golden-mantled Racquet-tail;
Golden-mantled Racket-tail; Golden-
mantled Racket-tailed Parrot;
Celebes Racket-tailed Parrot;
Golden-mantled Racquet-tailed
Parrot; Celebes Racquet-tailed
Parakeet; Gold-backed Racquet-tail;
Celebes Racket-tailed Parakeet;
Gold-backed Racket-tail
f   Palette à manteau d'or
d   Rotscheitelpapagei
i   Pappagallo coda a racchetta
dorsodorato

8218   *Prioniturus verticalis*
Psittaciformes - Psittacidae
e   Blue-winged Racquet-tail; Blue-
winged Racket-tail; Sulu Racket-
tailed Parrot; Sulu Racquet-tailed
Parrot
f   Palette des Sulu
i   Pappagallo coda a racchetta di Sulu

8219   *Prioniturus waterstradti*
Psittaciformes - Psittacidae
e   Mindanao Racquet-tail; Mindanao
Racket-tail; Mindanao Racket-tailed
Parrot; Mindanao Racquet-tailed
Parrot
f   Palette de Mindanao
i   Pappagallo coda a racchetta di
Mindanao

8220   *Prionochilus maculatus*
Passeriformes - Nectariniidae
e   Yellow-breasted Flowerpecker;
Yellow-throated Flowerpecker
f   Dicée tacheté
d   Goldbrustmistelfresser
i   Beccafiori maculato

8221   *Prionochilus olivaceus*
Passeriformes - Nectariniidae
e   Olive-backed Flowerpecker
f   Dicée olive
d   Grünmantelmistelfresser
i   Beccafiori dorso oliva

8222   *Prionochilus percussus*
Passeriformes - Nectariniidae
e   Crimson-breasted Flowerpecker
f   Dicée poignarde
d   Mennigbrustmistelfresser
i   Beccafiori pettocremisi

8223   *Prionochilus plateni*
Passeriformes - Nectariniidae
e   Palawan Flowerpecker; Yellow-
rumped Flowerpecker; Platen's
Flowerpecker; Palawan Yellow-
rumped Berrypecker
f   Dicée de Palawan
d   Palawan-Mistelfresser
i   Beccafiori di Palawan

8224   *Prionochilus thoracicus*
Passeriformes - Nectariniidae
e   Scarlet-breasted Flowerpecker
f   Dicée à poitrine écarlate
d   Rubinkehlmistelfresser
i   Beccafiori pettoscarlatto

8225   *Prionochilus xanthopygius*
Passeriformes - Nectariniidae

e Yellow-rumped Flowerpecker;
   Borneo Yellow-rumped
   Flowerpecker; Borneo Flowerpecker;
   Bornean Flowerpecker
f Dicée à croupion jaune
d Gelbbürzelmistelfresser;
   Gelbbürzelblütenpicker
i Beccafirori gropponegiallo

8226 *Prionodura newtoniana*
   Passeriformes - Ptilonorhynchidae
e Golden Bowerbird; Newton's
   Golden-Bowerbird; Newton's
   Bowerbird; Newton's Gardner-
   Bowerbird
f Jardinier de Newton
d Säulengärtner
i Giardiniere dorato; Uccello
   giardiniere di Newton

8227 Prionopidae
   Passerformes
e Helmet Shrikes
f Prionopidés
d Brillenwürger
i Prionipidi

8228 *Prionops alberti*
   Passeriformes - Prionopidae
e Yellow-crested Helmet-Shrike;
   Chestnut-bellied Helmet-Shrike
f Bagadais d'Albert; Bagadais du roi
   Albert
d Gelbschopfbrillenwürger;
   Kronbrillenwürger
i Prionope crestagialla

8229 *Prionops caniceps*
   Passeriformes - Prionopidae
e Chestnut-bellied Helmet-Shrike;
   Red-billed Helmet-Shrike; Red-billed
   Shrike; Western Helmet-Shrike
f Bagadais à bec rouge
d Rostbauchwürger
i Prionope beccorosso

8230 *Prionops gabela*
   Passeriformes - Prionopidae
e Gabela Helmet-Shrike; Angola
   Helmet-Shrike; Angola Red-billed
   Shrike; Angola Red-billed Helmet-

   Shrike; Rand's Red-billed Shrike;
   Rand's Red-billed Helmet-Shrike
f Bagadais de Gabela
d Gabela-Brillenwürger; Gabela-
   Würger
i Prionope beccorosso di Rand

8231 *Prionops plumatus*
   Passeriformes - Prionopidae
e White Helmet-Shrike; White-crested
   Helmet-Shrike (CSA); Long-crested
   Helmet-Shrike; Straight-crested
   Helmet-Shrike; Common Helmet-
   Shrike; Helmet-Shrike; Southern
   Helmet-Shrike; Smith's Helmet-
   Shrike
f Bagadais casqué
d Brillenwürger
i Prionope crestalunga

8232 *Prionops plumatus cristatus*
   Passeriformes - Prionopidae
f Bagadais bouclé
d Helmbrillenwürger;
   Schopfbrillenwürger

8233 *Prionops poliolophus*
   Passeriformes - Prionopidae
e Grey-crested Helmet-Shrike
f Bagadais à huppe grise
d Grauschopfbrillenwürger
i Prionope crestagrigia

8234 *Prionops retzii*
   Passeriformes - Prionopidae
e Red-billed Helmet-Shrike; Retz's
   Helmet-Shrike; Retz's Red-billed
   Helmet-Shrike; Retz's Red-billed
   Shrike
f Bagadais de Retz
d Dreifarbenwürger;
   Dreifarbenbrillenwürger
i Prionope beccorosso di Retz

8235 *Prionops rufiventris*
   Passeriformes - Prionopidae
e Gabon Helmet-Shrike; Gabon
   Chestnut-breasted Helkmetshrike
f Bagadais à ventre roux
i Prionope del Gabon

8236 **Prionops scopifrons**
Passeriformes - Prionopidae
e Chestnut-fronted Helmet-Shrike;
Chestnut-fronted Shrike
f Bagadais à front roux
d Braunstirnwürger;
Braunstirnbrillenwürger
i Prionope frontecastano

8237 **Priotelus roseigaster**
Trogoniformes - Trogonidae
e Hispaniolan Trogon
f Caleçon rouge (Ants); Pic de
montagne (Ants); Dame anglaise
(Ants); Demoiselle anglaise (Ants);
Trogon damoiseau
d Rosentrogon
i Trogone di Hispaniola

8238 **Priotelus temnurus**
Trogoniformes - Trogonidae
e Cuban Trogon
f Trogon de Cuba
d Kuba-Trogon
i Trogone di Cuba

8239 **Probosciger aterrimus**
Psittaciformes - Psittacidae
e Palm Cockatoo; Black Macaw; Great
Palm-Cockatoo; Great Black-
Cockatoo; Goliath Cockatoo
f Cacatoès noir
d Ara-Kakadu; Arara-Kakadu;
Palmkakadu
i Cacatua dalla proboscide;
Caliptorinco nero; Cacatua delle
palme

8240 **Probosciger aterrimus macgillivrayi**
Psittaciformes - Psittacidae
e Southern Palm Cockatoo (ANZ)

8241 **Procellaria aequinoctialis**
Procellariiformes - Procellariidae
e White-chinned Petrel; Cape Hen
(CSA); Shoemaker Hen; Spectacled
Petrel
f Puffin à menton blanc; Puffin brun
d Weißkinnsturmvogel
i Procellaria mentobianco

8242 **Procellaria cinerea**
Procellariiformes - Procellariidae
e Grey Petrel; Great Grey-Sheerwater;
Grey Shearwater; Black-tailed Petrel;
Black-tailed Shearwater; Brown
Petrel; Brown Bulky Petrel;
Pediunker
f Puffin gris
d Grausturmvogel
i Procellaria cenerina

8243 **Procellaria conspicillata**
Procellariiformes - Procellariidae
e Spectacled Petrel
d Brillensturmvogel

8244 **Procellaria parkinsoni**
Procellariiformes - Procellariidae
e Parkinson's Petrel; Black Petrel;
Black Fulmar
f Puffin de Parkinson
d Parkinson-Sturmvogel
i Procellaria nera

8245 **Procellaria westlandica**
Procellariiformes - Procellariidae
e Westland Petrel; Westland Black
Petrel
f Puffin du Westland
d Westland-Sturmvogel
i Procellaria di Westland

8246 **Procellariidae**
Procellariiformes
e Petrels; Shearwaters
f Procéllaridés; Oiseaux des tempêtes
d Sturmvögel; Röhrennasen
i Procellariidi; Procellaridi

8247 **Procellariiformes**
e Albatrosses; Petrels; Tube-nosed
Swimmers; Tubinares
f Procellariformes
d Röhrennasen
i Procellariformi

8248 **Procelsterna albivitta**
Charadriiformes - Laridae
e Grey Noddy; Grey Ternlet
f Noddi gris

d   Graunoddi
i   Sterna stolida grigia

8249   ***Procelsterna cerulea***
Charadriiformes - Laridae
e   Blue-grey Noddy; Blue Noddy; Blue
Ternlet; Little Grey-Tern; Little
Blue-Petrel; f
f   Noddi bleu
d   Blaunoddi
i   Sterna stolida cerulea

8250   ***Procnias alba***
Passeriformes - Cotingidae
e   White-Bellbird
f   Araponga blanc
d   Zapfenglöckner; Zapfenglockenvogel
i   Campanaro bianco

8251   ***Procnias averano***
Passeriformes - Cotingidae
e   Bearded Bellbird
f   Araponga barbu; Araponga avérano
d   Araponga; Flechtenglöckner;
Flechtenglockenvogel
i   Campanaro barbuto

8252   ***Procnias nudicollis***
Passeriformes - Cotingidae
e   Bare-throated Bellbird; Bare-necked
Belkbird
f   Araponga à gorge nue
d   Nacktkehlglockenvogel;
Nacktkehlglöckner
i   Campanaro collonudo

8253   ***Procnias tricarunculata***
Passeriformes - Tyrannidae
e   Three-wattled Bellbird
f   Araponga tricaronculé; Araponga
caronculé
d   Hämmerling
i   Campanaro collonudo dalle tre
caruncole

8254   ***Prodotiscus insignis***
Piciformes - Indicatoridae
e   Cassin's Honeyguide; Cassin's
Honeybird; Cassin's Sharp-billed
Honeyguide; Western Green-backed
Honeybird

f   Indicateur pygmé
d   Liliputhoniganzeiger
i   Indicatore di Cassin

8255   ***Prodotiscus regulus***
Piciformes - Indicatoridae
e   Wahlberg's Honeyguide; Sharp-billed
Honeyguide (CSA); Wahlberg's
Honeybird; Brown-backed
Honeybird
f   Indicateur de Wahlberg
d   Zwerghoniganzeiger; Wahlberg-
Laubpicker
i   Indicatore di Wahlberg

8256   ***Prodotiscus zambesiae***
Piciformes - Indicatoridae
e   Green-backed Honeyguide; Eastern
Honeyguide (CSA); Slender-billed
Honeyguide (CSA); Eastern Green-
backed Honeybird; Eastern
Honeybird; Eastern Slender-billed
Honeyguide
f   Indicateur gris; Indicateur verdâtre
d   Graubauchlaubpicker
i   Indicatore dorsoverde

8257   ***Progne chalybea***
Passeriformes - Hirundinidae
e   Grey-breasted Martin
f   Hirondelle chalybée
d   Graubrustschwalbe
i   Rondine pettogrigio

8258   ***Progne cryptoleuca***
Passeriformes - Hirundinidae
e   Cuban Martin
f   Hirondelle de Cuba
i   Rondine di Cuba

8259   ***Progne dominicensis***
Passeriformes - Hirundinidae
e   Caribbean Martin; Purple Swallow
(WI); Swallow (WI); Galebird (WI);
White-bellied Martin; Snowy-bellied
Martin
f   Hirondelle à ventre blanc; Hirondelle
(Ants)
d   Dominikanerschwalbe
i   Rondine dei Caraibi

8260 **Progne elegans**
Passeriformes - Hirundinidae
e Southern Martin
f Hirondelle gracieuse; Hirondelle
sombre; Hirondelle des Galapagos
d Gabelschwalbe
i Rondine modesta

8261 **Progne sinaloae**
Passeriformes - Hirundinidae
e Sinaloa Martin; Snowy-bellied
Martin; White-bellied Martin
f Hirondelle du Sinaloa
i Rondine di Sinaloa

8262 **Progne subis**
Passeriformes - Hirundinidae
e Purple Martin
f Hirondelle noire; Hirondelle
pourprée
d Purpurschwalbe
i Rondine pupurea

8263 **Promerops cafer**
Passeriformes - Nectariniidae
e Cape Sugarbird; Sugarbird; Long-
tailed Sugarbird
f Promérops du Cap
d Kap-Honigfresser; Kap-Honigesser
i Mangianettare del Capo

8264 **Promerops gurneyi**
Passeriformes - Nectariniidae
e Gurney's Sugarbird; Natal Honeybird
f Promérops de Gurney
d Guerneys Honigfresser; Natal-
Honigfresser
i Mangianettare di Gurney

8265 **Prosobonia cancellata**
Charadriiformes - Scolopacidae
e Tuamotu Sandpiper
f Chevalier des Touamotou
d Südseeläufer
i Piro-piro di Tuamotu

8266 **Prosobonia leucoptera**
Charadriiformes - Scolopacidae
e Tahitian Sandpiper; White-winged
Sandpiper
f Chevalier à ailes blanches

d Gesellschafts-Läufer
i Piro-piro di Tahiti

8267 **Prosopeia personata**
Psittaciformes - Psittacidae
e Masked Shining-Parrot; Masked
Parakeet; Yellow-breasted Musk-
Parakeet; Yellow-breasted Shining
Parrot; Sulphur-breasted Musk Parrot
f Perruche masquée
d Maskensittich
i Parrocchetto splendente mascherato

8268 **Prosopeia splendens**
Psittaciformes - Psittacidae
e Crimson Shining-Parrot; Kadavu
Shining Parrot
f Perruche écarlate
d Pompadoursittich
i Parrocchetto splendente di Kandavu

8269 **Prosopeia tabuensis**
Psittaciformes - Psittacidae
e Red Shining-Parrot; Shining
Parakeet; Tabuan Parakeet; Red-
breasted Shining Parrot; Red-
breasted Musk Parrot
f Perruche pompadour
d Pompadoursittich
i Parrocchetto splendente rosso

8270 **Prosthemadera novaeseelandiae**
Passeriformes - Meliphagidae
e Tui; Parsonbird
f Tui cravate-frisée; Tui
d Pastorvogel; Priestervogel; Tui
i Tui; Poe

8271 **Protonotaria citrea**
Passeriformes - Fringillidae
e Prothonotary Warbler
f Paruline orangée; Fauvette orangée
d Zitronenenwaldsänger;
Protonotarsänger;
Protonotarwaldsänger
i Parula citrina

8272 **Prunella atrogularis**
Passeriformes - Passeridae
e Black-throated Accentor
f Accenteur à gorge noire; Accenteur

gorgenoire
d Schwarzkehlbraunelle
i Sordone golanera; Passera scopaiola
golanera

8273 *Prunella collaris*
Passeriformes - Passeridae
e Alpine Accentor; Collared Accentor;
Rock Accentor
f Accenteur alpin
d Alpenbraunelle; Flühvogel
i Sordone

8274 *Prunella fagani*
Passeriformes - Passeridae
e Yemen Accentor; Arabian Accentor;
Yemeni Accentor
f Accenteur d'Arabie; Accenteur arabe
i Sordone arabo; Sordone dello Yemen

8275 *Prunella fulvescens*
Passeriformes - Passeridae
e Brown Accentor; Pale Accentor
f Accenteur brun
d Fahlbraunelle
i Sordone bruno

8276 *Prunella himalayana*
Passeriformes - Passeridae
e Altai Accentor; Rufous-streaked
Accentor; Himalayan Accentor;
Rufous-breasted Accentor
f Accenteur de I'Himalaya; Accenteur
de l'Altaï
d Himalaja-Braunelle; Steinbraunelle
i Sordone dell'Himalaya

8277 *Prunella immaculata*
Passeriformes - Passeridae
e Maroon-backed Accentor
f Accenteur immaculé; Accenteur à
dos roux
d Waldbraunelle
i Sordone dorsobruno

8278 *Prunella koslowi*
Passeriformes - Passeridae
e Mongolian Accentor; Koslow's
Accentor
f Accenteur de Koslov

d Steppenbraunelle
i Sordone di Koslow

8279 *Prunella modularis*
Passeriformes - Passeridae
e Hedge Accentor; Dunnock; Hedge
Sparrow; European Dunnock
f Accenteur mouchet; Passereau
balayeur; Traîne-buisson
d Heckenbraunelle
i Passera scopaiola

8280 *Prunella montanella*
Passeriformes - Passeridae
e Siberian Accentor; Mountain
Accentor
f Accenteur montanelle; Accenteur de
Sibérie; Accenteur montagnarde
d Bergbraunelle
i Passera scopaiola asiatica; Sordone
siberiano

8281 *Prunella ocularis*
Passeriformes - Passeridae
e Radde's Accentor; Spoot-throated
Accentor; Spotted-Accentor
f Accenteur de Rade
d Steinbraunelle; Felsenbraunelle
i Passera scopaiola di Radde; Sordone
di Radde

8282 *Prunella rubeculoides*
Passeriformes - Passeridae
e Robin Accentor; Robin Hedge-
Sparrow
f Accenteur rougegorge
d Rotkehlbraunelle; Rotbrustbraunelle
i Sordone pettirosso

8283 *Prunella rubida*
Passeriformes - Passeridae
e Japanese Accentor; Japanese Hedge
Sparrow
f Accenteur du Japon
d Rötelbraunelle
i Passera scopaiola del Giappone

8284 *Prunella strophiata*
Passeriformes - Passeridae
e Rufous-breasted Accentor
f Accenteur à poitrine rousse

d   Streichelbraunelle
i   Sordone pettocastano

8285   *Psalidoprocne albiceps*
Passeriformes - Hirundinidae
e   White-headed Saw-wing; White-headed Roughwing (CSA); White-headed Saw-wing Swallow (CSA); White-headed Rough-winged Swallow; White-headed Roughwing-Swallow
f   Hirondelle à tête blanche
d   Weißkopfschwalbe
i   Rondine testabianca

8286   *Psalidoprocne antinorii*
Passeriformes - Hirundinidae
e   Brown Saw-wing; Roughwing Swallow
f   Hirondelle de Salvadori
i   Rondine di Antinori

8287   *Psalidoprocne antinorii blanfordi*
Passeriformes - Hirundinidae
f   Hirondelle de Blanford

8288   *Psalidoprocne chalybea*
Passeriformes - Hirundinidae
e   Shari Saw-wing; Western Saw-wing
f   Hirondelle à queue fourchue
i   Rondine shari

8289   *Psalidoprocne fuliginosa*
Passeriformes - Hirundinidae
e   Mountain Saw-wing; Cameroon Saw-wing; Mountain Saw-winged Swallow; Cameroon Rough-winged Swallow; Cameroon Roughwing; Mountain Roughwing; Mountain Rough-winged Swallow; Cameroon Mountain Saw-wing; Mountain Roughwing-Swallow
f   Hirondelle brune
d   Kamerun-Schwalbe
i   Rondine del Camerun

8290   *Psalidoprocne holomelas*
Passeriformes - Hirundinidae
e   Black Saw-wing; Black Saw-wing Swallow (CSA); Black Roughwing Swallow (CSA); Black Saw-winged Swallow; Black Roughwing; Black Roughwinged Swallow
f   Hirondelle du Ruwenzori
d   Sägeflügelschwalbe
i   Rondine nera

8291   *Psalidoprocne mangbettorum*
Passeriformes - Hirundinidae
e   Mangbettu Saw-wing; Mangbetu Swallow; Congo Saw-wing
f   Hirondelle des Mangbetu
i   Rondine di Mangbettu

8292   *Psalidoprocne nitens*
Passeriformes - Hirundinidae
e   Square-tailed Saw-wing; Square-tailed Saw-winged Swallow; Square-tailed Roughwing; Square-tailed Rough-winged Swallow
f   Hirondelle à queue courte; Hirondelle hérissée à queue courte
d   Glanzschwalbe
i   Rondine codasquadrata

8293   *Psalidoprocne obscura*
Passeriformes - Hirundinidae
e   Fanti Saw-wing; Fantee Saw-wing; Fantee-Rough-winged Swallow; Fanti Rough-winged Swallow; Fanti Saw-winged Swallow; Fantee Saw-winged Swallow; Fanti Roughwing-Swallow
f   Hirondelle fanti; Hirondelle fanti hérissée; Hirondelle de Fanti
d   Scherenschwanzschwalbe
i   Rondine scura

8294   *Psalidoprocne oleaginea*
Passeriformes - Hirundinidae
e   Ethiopian Saw-wing; Eastern Saw-wing Swallow (CSA); Eastern Roughwing Swallow (CSA); Kaffra Saw-wing
f   Hirondelle de Kafa
i   Rondine di Kaffa

8295   *Psalidoprocne orientalis*
Passeriformes - Hirundinidae
e   Eastern Saw-wing
f   Hirondelle de Reichenow
i   Rondine spallebianche

8296 *Psalidoprocne petiti*
      Passeriformes - Hirundinidae
e   Petit's Saw-wing; Petit's Swallow;
      Roughwing Swallow
f   Hirondelle de Petit
i   Rondine di Petit

8297 *Psalidoprocne pristoptera*
      Passeriformes - Hirundinidae
e   Blue Saw-wing; Blue Saw-winged
      Swallow; Blue Roughwing; Blue
      Rough-winged Swallow; African
      Blue Saw-wing; Black Saw-wing;
      Black Roughwing-Swallow; Black
      Roughwing; Fork-tailed Swallow
f   Hirondelle hérissée
d   Erzschwalbe
i   Rondine blu

8298 *Psaltria exilis*
      Passeriformes - Aegithalidae
e   Pygmy Tit
f   Mésange pygmée
d   Zwergmeise; Java-Schwanzmeise
i   Codibugnolo pigmeo

8299 *Psaltriparus melanotis*
      Passeriformes - Aegithalidae
e   Black-eared Bush-Tit
f   Mésange masquée
d   Streifenbuschmeise
i   Codibugnolo americano

8300 *Psaltriparus minimus*
      Passeriformes - Aegithalidae
e   Bush Tit; Common Bush-Tit (NA);
      Coast Bush-Tit; Plain Bush-Tit;
      American Bush-Tit
f   Mésange buissonnière
d   Buschmeise; Schwanzmeise;
      Kappenbuschmeise
i   Codibugnolo americano

8301 *Psarisomus dalhousiae*
      Passeriformes - Eurylaimidae
e   Long-tailed Broadbill
f   Eurylaime psittacin
d   Papageibreitrachen
i   Beccolargo pappagallo

8302 *Psarocolius angustifrons*
      Passeriformes - Icteridae
e   Russet-backed Oropendola
f   Cassique roussâtre
d   Breithaubenstirnvogel;
      Braunhaubenstärling; Conoto
i   Oropendola dorsocastano

8303 *Psarocolius atrovirens*
      Passeriformes - Icteridae
e   Dusky-green Oropendola
f   Cassique olivâtre
d   Grünschnabelstirnvogel
i   Oropendola di Wagner; Oropendola
      verde scuro

8304 *Psarocolius bifasciatus*
      Passeriformes - Icteridae
e   Olive Oropendola; Para Oropendola
d   Para-Stirnvogel
i   Oropendola del Parà

8305 *Psarocolius cassini*
      Passeriformes - Icteridae
e   Chestnut-mantled Oropendola;
      Cassin's Oropendola
d   Braunmantelstirnvogel
I   Oropendola

8306 *Psarocolius decumanus*
      Passeriformes - Icteridae
e   Crested Oropendola
f   Cassique huppé
d   Krähenstirnvogel;
      Schwarzhaubenstärling; Schapu
I   Oropendola dal ciuffo

8307 *Psarocolius guatimozinus*
      Passeriformes - Icteridae
e   Black Oropendola
f   Cassique noir .
d   Mohrenstirnvogel
i   Oropendola nera

8308 *Psarocolius oseryi*
      Passeriformes - Icteridae
e   Casqued Oropendola
f   Cassique casqué
d   Helmstirnvogel
I   Oropendola dal casco

8309 **Psarocolius viridis**
Passeriformes - Icteridae
e Green Oropendola
f Cassique vert
d Grünschopfstirnvogel; Olivgrüner Stirnvogel
I Oropendola verde

8310 **Psarocolius wagleri**
Passeriformes - Icteridae
e Chestnut-headed Oropendola; Wagler's Oropendola
f Cassique à tête brune
d Rotkopfstirnvogel; Erzflügelstirnvogel; Waglers Stirnvogel
I Oropendola di Wagler

8311 **Pselliophorus luteoviridis**
Passeriformes - Fringillidae
e Yellow-green Finch; Yellow-green Brush-Finch
f Tohi jaune-vert
d Griscom-Buschammer
i Passero giallo-verde

8312 **Pselliophorus tibialis**
Passeriformes - Fringillidae
e Yellow-thighed Finch; Yellow-thighed Brush-Finch
f Tohi à cuisses jaunes
d Gelbschenkelbuschammer
i Passero dai calzoni gialli

8313 **Psephotus chrysopterygius**
Psittaciformes - Psittacidae
e Golden-shouldered Parrot; Hooded Parrot; Golden-shouldered Parakeet; Golden-winged Parakeet; Chestnut-crowned Parakeet; Golden-winged Parrot; Anthill Parrot; Antbed Parrot
f Perruche à ailes d'or; Perruche aux ailes d'or
d Gelbschultersittich; Goldschultersittich
i Parrochetto aligialli; Parrocchetto alidorate; Parroccheto dalle ali dorate

8314 **Psephotus dissimilis**
Psittaciformes - Psittacidae
e Hooded Parrot; Anthill Parrot;

Antbed Parrot
f Perruche à capuchon noir
i Parrochetto dal cappuccio

8315 **Psephotus haematonotus**
Psittaciformes - Psittacidae
e Red-rumped Parrot; Red-backed Parrot; Red-rumped Parakeet; Blood-rumped Parakeet; Redrump
f Perruche à croupion rouge
d Blutrumpfsittich
i Parrochetto dal groppone rosso

8316 **Psephotus pulcherrimus**
Psittaciformes - Psittacidae
e Paradise Parrot; Soldier Parrot; Ground Parrot; Grass-Parrot; Elegant Parrot; Beautiful Parakeet; Paradise Parakeet; Red-shouldered Parakeet; Scarlet-shouldered Parrot
f Perruche de paradis
d Paradiessittich; Singsittich; Prachtsittich; Soldatensittich; Ameisenhügelsittich
i Parrochetto del Paradiso

8317 **Psephotus varius**
Psittaciformes - Psittacidae
e Mulga Parrot; Many-coloured Parrot; Many-coloured Parakeet; Varied Parrot
f Perruche multicolore
d Vielfarbensittich
i Parrochetto multicolore

8318 **Pseudalaemon fremantlii**
Passeriformes - Alaudidae
e Short-tailed Lark
f Cochevis à queue courte
d Bartlerche
i Allodola di Fremantle

8319 **Pseudelaenia leucospodia**
Passeriformes - Tyrannidae
e Grey-and-white Tyrannulet; Ash-coloured Tyrannulet
f Élénie gris-et-blanc; Elaène grise-et-blanche
d Graustirnelaenie

8320   **Pseudeos fuscata**
        Psittaciformes - Loriidae
e   Dusky Lory; White-rumped Lory;
        Dusky Orange Lory
f   Lori sombre
d   Weißbürzellori
ı   Lori bruno

8321   **Pseudibis davisoni**
        Ciconiiformes - Threskiornithidae
e   White-shouldered Ibis; Davison's Ibis
f   Ibis de Davison
i   Ibis spallebianche

8322   **Pseudibis gigantea**
        Ciconiiformes - Threskiornithidae
e   Giant Ibis
f   Ibis géant
d   Riesenibis
i   Ibis gigante

8323   **Pseudibis papillosa**
        Ciconiiformes - Threskiornithidae
e   Red-naped Ibis; Black Ibis; Indian
        Black-Ibis
f   Ibis noir
d   Warzenibis
i   Ibis nero

8324   **Pseudobias wardi**
        Passeriformes - Muscicapidae
e   Ward's Shrike-Flycatcher; Ward's
        Flycatcher
f   Bias de Ward; Gobemouche de ward
d   Ward-Schnäpper
i   Pigliamosche-averla di Ward

8325   **Pseudocalyptomena graueri**
        Passeriformes - Eurylaimidae
e   Grauer's Broadbill; Green Broadbill;
        African Green Broadbill
f   Eurylaime de Grauer
d   Grauer-Breitmaul;
        Blaukehlbreitrachen; Grauer
        Breitrachen
i   Beccolargo di Grauer

8326   **Pseudochelidon eurystomina**
        Passeriformes - Hirundinidae
e   African River-Martin; River-Martin
f   Pseudolangrayen d'Afrique;

        Pseudolangrayen de rivière;
        Hirondelle de rivière
d   Trugschwalbe
i   Rondine di fiume africano

8327   **Pseudochelidon sirintarae**
        Passeriformes - Hirundinidae
e   White-eyed River-Martin
f   Pseudolangrayen d'Asie;
        Pseudolangrayen à lunettes
d   Sirintara-Schwalbe;
        Stachelschwanzschwalbe
i   Rondine di fiume occhibianchi

8328   **Pseudocolaptes boissonneautii**
        Passeriformes - Furnariidae
e   Streaked Tuftedcheek; Tuftedcheek
f   Anabate de Boissonneau
d   Anden-Schopfohr
i   Pseudopicchio di Boissoneau

8329   **Pseudocolaptes johnsoni**
        Passeriformes - Furnariidae
e   Pacific Tuftedcheek

8330   **Pseudocolaptes lawrencii**
        Passeriformes - Furnariidae
e   Buffy Tuftedcheek; Lawrence's
        Tuftedcheek
f   Anabate chamois
d   Panama-Shopfohr
i   Pseudopicchio di Lawrence

8331   **Pseudocolopteryx acutipennis**
        Passeriformes - Tyrannidae
e   Subtropical Doradito; Sharp-winged
        Doradito
f   Doradite à ailes pointues
d   Schmalschwingendoradito
i   Doradito subtropicale

8332   **Pseudocolopteryx dinellianus**
        Passeriformes - Tyrannidae
e   Dinelli's Doradito; Rufous-edged
        Pygmy-Tyrant
f   Doradite de Dinelli
d   Dreibindendoradito
i   Doradito di Dinelli

8333   **Pseudocolopteryx flaviventris**
        Passeriformes - Tyrannidae

e   Warbling Doradito
f   Doradite babillarde
d   Rohrdoradito
i   Doradito canoro

8334   *Pseudocolopteryx sclateri*
       Passeriformes - Tyrannidae
e   Crested Doradito; Sclater's Doradito
f   Doradite de Sclater
d   Schopfdoradito
i   Doradito di Sclater

8335   *Pseudocossyphus bensoni*
       Passeriformes - Muscicapidae
e   Benson's Rock-Thrush; Benson's
    Robin-Chat; Farke's Robin-Chat
f   Monticole de Benson; Merle de roche
    de Benson
d   Benson-Rötel
i   Tordo pettirosso di Benson

8336   *Pseudocossyphus imerinus*
       Passeriformes - Muscicapidae
e   Littoral Rock-Thrush; Madagascar
    Robin-Chat; Madagascar Rock-
    Thrush
f   Monticole du littoral; Merle de roche
    du sub-désert
d   Dünenrötel
i   Tordo pettirosso del Madagascar

8337   *Pseudocossyphus sharpei*
       Passeriformes - Muscicapidae
e   Forest Rock-Thrush; Forest Robin-
    Chat; Sharpe's Robin-Chat; Eastern
    Robin-Chat
f   Monticole de forêt; Merle de roche
    de forêt
d   Laubrötel
i   Tordo pettirosso orientale

8338   *Pseudodacnis hartlaubi*
       Passeriformes - Fringillidae
e   Turquoise Dacnis-Tanager;
    Turquoise Dacnis
f   Dacnis de Hartlaub
d   Türkis-Pipit
i   Dacne turchese

8339   *Pseudoleistes guirahuro*
       Passeriformes - Fringillidae

e   Yellow-rumped Marshbird
f   Troupiale guirahu
d   Gelbbürzelstärling
i   Ittero di palude gropponegiallo

8340   *Pseudoleistes virescens*
       Passeriformes - Fringillidae
e   Brown-and-yellow Marshbird;
    Brown-yellow Marshbird; Yellow-
    breasted Marshbird
f   Troupiale dragon
d   Drachenstärling
i   Ittero di palude bruno e giallo

8341   *Pseudonestor xanthophrys*
       Passeriformes - Fringillidae
e   Maui Parrotbill; Pseudonestor;
    Parrot-billed Koa-Finch
f   Psittirostre de Maui
d   Papageischnabelgimpel; Pseudokoa;
    Koakleidervogel
i   Becco di pappagallo di Maui

8342   *Pseudonigrita arnaudi*
       Passeriformes - Ploceidae
e   Grey-headed Social-Weaver; Grey-
    capped Social-Weaver
f   Républicain d'Arnaud; Moineau
    sociale à tête grise
d   Marmorweber; Marmorspätzling
i   Tessitore sociale capogrigio

8343   *Pseudonigrita cabanisi*
       Passeriformes - Ploceidae
e   Black-capped Social-Weaver
f   Républicain de Cabanis; Moineau
    sociale à tête noire
d   Schwarzkopfspätzling
i   Tessitore sociale caponero

8344   *Pseudopodoces humilis*
       Passeriformes - Corvidae
e   Tibetan Ground-Jay; Hume's
    Ground-Jay; Humes Ground-
    Chough; Little Ground-Jay; Little
    Ground-Chough; Tibetan Ground-
    Chough; Hume's Groundpecker;
    Ground Chough
f   Podoce de Hume; Geai terrestre de
    Hume

d Höhlenhäher
i Podoce di Hume

8345 *Pseudoscops grammicus*
Strigiformes - Strigidae
e Jamaican Owl; Brown Owl (WI);
Patoo (WI); Potoo (WI)
f Hibou de la Jamaïque
d Jamaika-Eule
i Gufo della Giamaica

8346 *Pseudoseisura cristata*
Passeriformes - Furnariidae
e Rufous Cacholote; Caatinga
Cachalote; Rufous Catchalote;
Caatinga Catchalote; Rufous-
breasted Catchalote
f Cacholote roux
d Haubencachalote
i Cacialote rossiccio

8347 *Pseudoseisura gutturalis*
Passeriformes - Furnariidae
e White-throated Cacholote; White-
throated Cachalote; White-throated
Catchalote
f Cacholote à gorge blanche
d Weißkehlcachalote;
Stachelburgnister
i Cacialote golabianca

8348 *Pseudoseisura lophotes*
Passeriformes - Furnariidae
e Brown Cacholote
f Cacholote brun
d Brauncachalote
i Cacialote bruno

8349 *Pseudoseisura unirufa*
Passeriformes - Furnariidae
e Grey-crested Cachalote

8350 *Pseudotriccus pelzelni*
Passeriformes - Tyrannidae
e Bronze-olive Pygmy-Tyrant; Bronze
Pygmy-Tyrant; Petzeln's Pygmy-
Tyrant; Streak-crowned Pygmy-
Tyrant
f Tyranneau bronzé
d Bronzetyrann
i Tiranno pigmeo di Pelzeln

8351 *Pseudotriccus ruficeps*
Passeriformes - Tyrannidae
e Rufous-headed Pygmy-Tyrant
f Tyranneau à tête rousse
d Rotkopftyrann
i Tiranno pigmeo testarossiccia

8352 *Pseudotriccus simplex*
Passeriformes - Tyrannidae
e Hazel-fronted Pygmy-Tyrant;
Bolivian Pygmy-Tyrant
f Tyranneau à front brun
d Braunstirntyrann
i Tiranno pigmeo frontenocciola

8353 *Psilopogon pyrolophus*
Piciformes - Megalaimidae
e Fire-tufted Barbet
f Barbu à collier
d Rotbüschelbartvogel
i Barbuto ciufforosso

8354 *Psilorhamphus guttatus*
Passeriformes - Rhinocryptidae
e Spotted Bamboo-Wren; Bamboo-
Wren
f Mérulaxe des bambous
d Trugzaunkönig
i Scricciolo dei bambù maculato

8355 *Psilorhinus morio*
Passeriformes - Corvidae
e Brown Jay; Plain-tipped Brown-Jay;
Plain-tailed Brown-Jay; White-tipped
Brown-Jay
f Geai enfumé
d Trugbürzelstelzer; Braunhäher
i Ghiandaia bruna

8356 *Psittacella brehmii*
Psittaciformes - Psittacidae
e Brehm's Tiger-Parrot; Brehm's Parrot
f Perruche de Brehm
d Brehms Bindensittich; Großer
Bindensittich
i Pappagallo di Brehm

8357 *Psittacella madaraszi*
Psittaciformes - Psittacidae
e Madarasz's Tiger-Parrot; Madarasz's
Parrot; Plain-breasted Little-Tiger-

Parrot
f Perruche de Madarasz
d Schuppenkopfpapagei
i Pappagallo di Madarasz

8358 *Psittacella modesta*
Psittaciformes - Psittacidae
e Modest Tiger-Parrot; Modest Parrot;
Barred Little-Parrot; Barred Little
Tiger-Parrot
f Perruche modeste
d Kleiner Bindensittich; Olivpapagei
i Pappagallo modesto

8359 *Psittacella picta*
Psittaciformes - Psittacidae
e Painted Tiger-Parrot; Painted Parrot;
Timberline Parrot; Timbereline
Tiger-Parrot
f Perruche peinte
d Braunscheitelpapagei; Gemalter
Bindensittich
i Pappagallo variopinto

8360 Psittacidae
Psittaciformes
e Parrots
f Perroquets; Psittacidés
d Papageien; Eigentliche Papageien
i Psittacidi; Parrocchetti

8361 Psittaciformes
e Parrots
f Psittaciformes
d Papageien; Handfüßler
i Psittaciformi

8362 *Psittacula alexandri*
Psittaciformes - Psittacidae
e Red-breasted Parakeet; Indian Red-
breasted Parakeet (ISC); Moustached
Parakeet; Javan Parakeet
f Perruche à moustaches; Perruche à
poitrine rose
d Bartsittich; Rosenbrustpapagei
i Parrocchetto dai mustacchi

8363 *Psittacula calthropae*
Psittaciformes - Psittacidae
e Layard's Parakeet; Emerald-collared
Parakeet; Sri Lanka Layard's

Parakeet
f Perruche de Layard
d Blauschwanzsittich;
Blauschwanzedelsittich
i Parrocchetto dal collare smeraldo

8364 *Psittacula caniceps*
Psittaciformes - Psittacidae
e Nicobar Parakeet; Grey-headed
Parakeet; Blyth's Nicobar Parakeet;
Blyth's Parakeet
f Perruche des Nicobar; Perrruche de
Blyth; Perruche à tête grise
d Graukopfsittich; Graukopfedelsittich
i Parrocchetto testagrigia

8365 *Psittacula columboides*
Psittaciformes - Psittacidae
e Malabar Parakeet; Blue-winged
Parakeet (ISC); Malabar Blue-
winged Parakeet
f Perruche de Malabar
d Taubensittich
i Parrocchetto del Malabar

8366 *Psittacula cyanocephala*
Psittaciformes - Psittacidae
e Plum-headed Parakeet; Blossom-
headed Parakeet (ISC)
f Perruche à tête prune
d Pflaumenkopfsittich
i Parrocchetto testa di prugna;
Parrocchetto dalla testa viola

8367 *Psittacula derbiana*
Psittaciformes - Psittacidae
e Derbyan Parakeet; Lord Derby's
Parakeet; Derbian Parakeet
f Perruche de Derby; Perruche de
Salvadori
d China-Sittich
i Parrocchetto Lord Derby

8368 *Psittacula echo*
Psittaciformes - Psittacidae
e Mauritius Parakeet; Echo Parakeet
f Perruche de Maurice; Perruche de
l'île Maurice
d Mauritius-Halsbandsittich
i Parrocchetto di Mauritius

8369   *Psittacula eupatria*
        Psittaciformes - Psittacidae
  e   Alexandrine Parakeet; Large Indian
        Parakeet (ISC); Large Parakeet;
        Ceylon Large Parakeet
  f   Perruche alexandre; Perruche
        d'Alexandre
  d   Alexander-Sittich; Großer
        Alexandersittich
  l   Parrocchetto alessandrino

8370   *Psittacula exsul*
        Psittaciformes - Psittacidae
  e   Newton's Parakeet
  f   Perruche de Newton
  d   Rodriguez-Sittich
  i   Parrocchetto di Newton

8371   *Psittacula finschii*
        Psittaciformes - Psittacidae
  e   Grey-headed Parakeet; Eastern Slaty-
        headed Parakeet; Hume's Parakeet;
        Finsch's Parrot
  f   Perruche de Finsch
  d   Finsch-Sittich
  i   Parrocchetto testagrigia

8372   *Psittacula himalayana*
        Psittaciformes - Psittacidae
  e   Slaty-headed Parakeet; Himalayan
        Slaty-headed Parakeet (ISC)
  f   Perruche de I'Himalaya; Perruche à
        tête noire
  d   Himalaja-Sittich;
        Schwarzkopfedelsittich
  i   Parrocchetto dell'Himalaya

8373   *Psittacula intermedia*
        Psittaciformes - Psittacidae
  e   Intermediate Parakeet; Intermediate
        Parrot; Rothschild's Parakeet
  f   Perruche intermediaire
  i   Parrocchetto intermedio

8374   *Psittacula krameri*
        Psittaciformes - Psittacidae
  e   Rose-ringed Parakeet; Ring-necked
        Parakeet; Indian Ring-necked
        Parakeet; Senegal Long-tailed
        Parakeet
  f   Perruche à collier; Perruche rose de

        l'Inde
  d   Halsbandsittich; Afrikanischer
        Halsbandsittich; Kleiner
        Alexandersittich
  i   Parrocchetto dal collare

8375   *Psittacula longicauda*
        Psittaciformes - Psittacidae
  e   Pink-cheeked Parakeet; Long-tailed
        Parakeet; Malacca Parakeet; Red-
        cheeked Parakeet
  f   Perruche à longs brins; Perruche de
        Malacca
  d   Langschwanzedelsittich;
        Langschwanzsittich
  i   Parrocchetto codalunga

8376   *Psittacula roseata*
        Psittaciformes - Psittacidae
  e   Blossom-headed Parakeet; Rose-
        headed Parakeet; Eastern Blossom-
        headed Parakeet
  f   Perruche à tête rose
  d   Rosenkopfsittich;
        Pflaumenkopfsittich
  i   Parrocchetto testarosa

8377   *Psittacula wardi*
        Psittaciformes - Psittacidae
  e   Seychelles Parakeet
  f   Perruche des Seychelles
  d   Seychellen-Sittich
  i   Parrocchetto delle Seychelle

8378   **Psittaculidae**
        Psittaciformes
  e   Parakeets
  f   Pstittaculidés
  d   Edelpapageien
  i   Psittaculidi

8379   *Psittaculirostris desmarestii*
        Psittaciformes - Psittacidae
  e   Large Fig-Parrot; Golden-headed
        Fig-Parrot; Desmarest's Fig-Parrot
  f   Psittacule de Desmarest
  d   Keilschwanzzwergpapagei;
        Buntbrustzwergpapagei
  i   Pappagallo dei fichi di Desmarest

8380 **Psittaculirostris edwardsii**
Psittaciformes - Psittacidae
e Edwards's Fig-Parrot
f Psittacule d'Edwards
d Edwards-Zwergpapagei; Edwards-Feigenpapagei
i Pappagallo dei fichi di Edwards

8381 **Psittaculirostris salvadorii**
Psittaciformes - Psittacidae
e Salvadori's Fig-Parrot; Whiskered Fig-Parrot
f Psittacule de Salvadori
d Salvadori-Zwergpapagei; Bartzwergpapagei; Schmuckohrpapagei
i Pappagallo dei fichi di Salvadori

8382 **Psittacus erithacus**
Psittaciformes - Psittacidae
e Grey Parrot; African Grey-Parrot
f Perroquet jaco
d Graupapagei; Jako
i Pappagallo cenerino; Giaco

8383 **Psitteuteles goldiei**
Psittaciformes - Loriidae
e Goldie's Lorikeet; Red-capped Streaked-Lorikeet
f Loriquet de Goldie
d Veilchenlori
i Lorichetto di Goldie

8384 **Psitteuteles iris**
Psittaciformes - Loriidae
e Iris Lorikeet
f Loriquet iris
d Irislori
i Lorichetto iride

8385 **Psitteuteles versicolor**
Psittaciformes - Loriidae
e Varied Lorikeet; Red-crowned Lorikeet; Red-caped Lorikeet
f Loriquet versicolore
d Warol; Buntlori
i Lorichetto multicolore

8386 **Psittinus cyanurus**
Psittaciformes - Psittacidae
e Blue-rumped Parrot; Little Malay Parrot; Blue-rumped Parakeet
f Perruche à croupion bleu; Psitacule de Malacca
d Rotachselpapagei; Zwergedelpapagei
i Pappagallo gropponeblu

8387 **Psittirostra psittacea**
Passeriformes - Fringillidae
e Ou; Yellow Laysan-Finch; Yellow Hawaii Finch
f Psittirostre psittacin
d Laysan-Gimpel; Ou; Grünpapageischnäbler
i Ou

8388 **Psittrichas fulgidus**
Psittaciformes - Psittacidae
e Pesquet's Parrot; Vulturine Parrot; Bristlehead
f Psittrichas de Pesquet
d Borstenkopf
i Pappagallo di Pesquet; Pappagallo vulturino

8389 **Psophia crepitans**
Gruiformes - Psophiidae
e Grey-winged Trumpeter; Common Trumpeter; Trumpeter
f Agami trompette
d Graurückentrompetervogel
i Agami aligrige; Agami

8390 **Psophia leucoptera**
Gruiformes - Psophiidae
e Pale-winged Trumpeter; White-winged Trumpeter
f Agami à ailes blanches
d Weißflügeltrompetervogel
i Agami alibianche

8391 **Psophia viridis**
Gruiformes - Psophiidae
e Dark-winged Trumpeter; Green-winged Trumpeter
f Agami vert
d Grünflügeltrompetervogel
i Agami aliverdi

8392 **Psophiidae**
Gruiformes
e Trumpeters

f   Psophidés
d   Trompetervögel
i   Psofidi

8393   ***Psophocichla litsipsirupa***
Passeriformes - Muscicapidae
e   Groundscraper Thrush
f   Merle litsipsirupa
d   Wipflöter
i   Tordo grattaterra

8394   ***Psophodes cristatus***
Passeriformes - Cinclosomatidae
e   Chirruping Wedgebill; Wedgebill
f   Psophode babillard
d   Buschflöter
i   Becco a cuneo crestato

8395   ***Psophodes nigrogularis***
Passeriformes - Cinclosomatidae
e   Western Whipbird; Black-throated
      Whipbird
f   Psophode à menton noir
d   Graukopfwippflöter
i   Uccello frustino occidentale

8396   ***Psophodes nigrogularis lashmari***
Passeriformes - Cinclosomatidae
e   Kangaroo Island Western-Whipbird
      (ANZ)

8397   ***Psophodes nigrogularis leucogaster***
Passeriformes - Cinclosomatidae
e   Eastern Western-Whipbird (ANZ)

8398   ***Psophodes nigrogularis nigrogularis***
Passeriformes - Cinclosomatidae
e   Western Heath Western-Whipbird
      (ANZ)

8399   ***Psophodes nigrogularis oberon***
Passeriformes - Cinclosomatidae
e   Western Mallee Western Whipbird
      (ANZ)

8400   ***Psophodes occidentalis***
Passeriformes - Cinclosomatidae
e   Chiming Wedgebill
f   Psophode carillonneur
d   Glockenflöter
i   Becco a cuneo occidentale

8401   ***Psophodes olivaceus***
Passeriformes - Cinclosomatidae
e   Eastern Whipbird
d   Schwarzkopfwippflöter
i   Uccello frustino orientale

8402   ***Pteridophora alberti***
Passeriformes - Corvidae
e   King-of-Saxony Bird of Paradise
f   Paradisier du prince Albert
d   Albert-Paradiesvogel; Wimpelträger
i   Paradisea del re di Sassonia

8403   ***Pternistis camerunensis***
Galliformes - Phasianidae
e   Cameroon Francolin; Cameroon
      Mountain-Francolin; Mount
      Cameroon Francolin
f   Francolin du Cameroun; Francolin du
      Mont Cameroun
d   Kamerun-Frankolin
i   Francolino del Monte Camerun

8404   ***Pternistis castaneicollis***
Galliformes - Phasianidae
e   Chestnut-naped Francolin
d   Braunnackenfrankolin
i   Francolino collocastano

8405   ***Pterocles alchata***
Pteroclidiformes - Pteroclididae
e   Pin-tailed Sandgrouse; Large Pintail
      Sandgrouse; Large Pin-tailed
      Sandgrouse
f   Ganga cata
d   Spießflughuhn
i   Grandule; Grandule comune;
      Grandule mediterranea

8406   ***Pterocles bicinctus***
Pteroclidiformes - Pteroclididae
e   Double-banded Sandgrouse
f   Ganga bibande
d   Nachtflughuhn
i   Grandule doppio collare

8407   ***Pterocles burchelli***
Pteroclidiformes - Pteroclididae
e   Burchell's Sandgrouse; Spotted
      Sandgrouse (CSA); Variegated
      Sandgrouse

f	Ganga de Burchell		*f*	Ganga de Lichtenstein
d	Fleckenflughuhn		*d*	Wellenflughuhn
i	Grandule di Burchell		*i*	Grandule di Lichtenstein; Pterocle di Lichtenstein

8408 **Pterocles coronatus**
Pteroclidiformes - Pteroclididae
e Crowned Sandgrouse; Coroneted Sandgrouse
f Ganga couronné
d Kronenflughuhn
i Grandule coronato; Pterocle coronato

8409 **Pterocles decoratus**
Pteroclidiformes - Pteroclididae
e Black-faced Sandgrouse
f Ganga à face noire
d Schmuckflughuhn
i Grandule faccianera

8410 **Pterocles exustus**
Pteroclidiformes - Pteroclididae
e Chestnut-bellied Sandgrouse; Small Pin-tailed Sandgrouse; Indian Sandgrouse (ISC)
f Ganga à ventre brun; Ganga à ventre châtain; Ganga tacheté
d Braunbauchflughuhn
i Grandule ventrecastano; Grandule dal ventre castano; Pterocle dal ventre castano

8411 **Pterocles gutturalis**
Pteroclidiformes - Pteroclididae
e Yellow-throated Sandgrouse
f Ganga à gorge jaune
d Gelbkehlflughuhn; Senegal-Spießflughuhn; Spießflughuhn
i Grandule golagialla

8412 **Pterocles indicus**
Pteroclidiformes - Pteroclididae
e Painted Sandgrouse; Close-barred Sandgrouse
f Ganga indien; Ganga des Indes
d Bindenflughuhn
i Grandule indiana

8413 **Pterocles lichtensteinii**
Pteroclidiformes - Pteroclididae
e Lichtenstein's Sandgrouse; Close-barred Sandgrouse

8414 **Pterocles namaqua**
Pteroclidiformes - Pteroclididae
e Namaqua Sandgrouse
f Ganga namaqua
d Nama-Flughuhn
i Grandule di Namaqua

8415 **Pterocles orientalis**
Pteroclidiformes - Pteroclididae
e Black-bellied Sandgrouse; Imperial Sandgrouse (ISC)
f Ganga unibande
d Sandflughuhn
i Ganga unibande; Ganga

8416 **Pterocles personatus**
Pteroclidiformes - Pteroclididae
e Madagascar Sandgrouse; Gould's Masked Sandgrouse
f Ganga masqué
d Maskenflughuhn
i Grandule del Madagascar

8417 **Pterocles quadricinctus**
Pteroclidiformes - Pteroclididae
e Four-banded Sandgrouse
f Ganga quadribande
d Buschflughuhn
i Grandule quadricincta

8418 **Pterocles senegallus**
Pteroclidiformes - Pteroclididae
e Spotted Sandgrouse
f Ganga tacheté
d Tropfenflughuhn
i Grandule di Senegal; Grandule maculata

8419 **Pteroclididae**
Ciconiiformes
e Sandgrouse
f Gangas; Ptéroclidés; Ptéroclididés
d Flughühner
i Pteroclidi; Pteroclididi

8420 *Pterodroma alba*
Procellariiformes - Procellariidae
e   Phoenix Petrel
f   Pétrel à poitrine blanche
d   Phoenixsturmvogel
i   Petrello di Phoenix

8421 *Pterodroma arminjoniana*
Procellariiformes - Procellariidae
e   Trinidade Petrel; Trinidade Island
    Petrel; South Trinidad Petrel; Herald
    Petrel; Round Island Petrel
f   Pétrel de la Trinité du Sud; Pétrel des
    Hawaii
d   Südtrinidad-Sturmvogel
i   Petrello di Arminjon

8422 *Pterodroma aterrima*
Procellariiformes - Procellariidae
e   Mascarene Petrel, Mascarene Black
    Petrel
f   Pétrel de Bourbon
d   Maskarenen-Sturmvogel
i   Petrello nero delle Mascarene

8423 *Pterodroma atrata*
Procellariiformes - Procellariidae
e   Henderson Petrel

8424 *Pterodroma axillaris*
Procellariiformes - Procellariidae
e   Chatham Islands Petrel; Chatham
    Petrel
f   Pétrel des Chatham
d   Runguru-Sturmvogel
i   Petrello delle Isole Chatham

8425 *Pterodroma baraui*
Procellariiformes - Procellariidae
e   Barau's Petrel; Réunion Petrel
f   Pétrel de Barau; Pétrel taillevent
d   Barau-Sturmvogel
i   Petrello di Barau

8426 *Pterodroma becki*
Procellariiformes - Procellariidae
e   Beck's Petrel; Solomon Islands Petrel
f   Pétrel de Beck
d   Maskarenen-Sturmvogel
i   Petrello delle Salomon

8427 *Pterodroma brevipes*
Procellariiformes - Procellariidae
e   Collared Petrel
f   Pétrel à collier
i   Petrello dal collare

8428 *Pterodroma cahow*
Procellariiformes - Procellariidae
e   Bermuda Petrel; Cahow; Gadfly-
    Petrel (USA); Cahow Petrel
f   Pétrel des Bermudes; Pétrel cahow
d   Bermuda-Sturmvogel
i   Cahow

8429 *Pterodroma caribbaea*
Procellariiformes - Procellariidae
e   Jamaica Petrel; Blue Mountain Duck;
    Diablotin (WI); Jamaican Petrel;
    Capped Petrel; Capped Jamaican
    Petrel; Black-capped Petrel
f   Pétrel de la Jamaïque

8430 *Pterodroma cervicalis*
Procellariiformes - Procellariidae
e   White-necked Petrel; Gadfly-Petrel;
    Kermadec White-necked Petrel;
    Kermadec Petrel; White-naped
    Petrel; Black-capped Petrel; Sundy
    Island Petrel
f   Pétrel à col blanc
i   Petrello collobianco dell'Isola
    Kermadec

8431 *Pterodroma cookii*
Procellariiformes - Procellariidae
e   Cook's Petrel; Blue-footed Petrel
f   Pétrel de Cook
d   Cook-Sturmvogel
i   Petrello di Cook

8432 *Pterodroma defilippiana*
Procellariiformes - Procellariidae
e   Defillipi's Petrel; Mas Atierra Petrel;
    Mas a Tierra Petrel
f   Pétrel de Defilippi
d   Masatierra-Sturmvogel
i   Petrello di Defilippi

8433 *Pterodroma externa*
Procellariiformes - Procellariidae
e   Juan Fernandez Petrel; White-necked

Petrel
f Pétrel de Juan Fernandez
d Salvin-Sturmvogel
i Petrello di Juan Fernandez

8434 *Pterodroma feae*
Procellariiformes - Procellariidae
e Fea's Petrel; Cape Verde Petrel
f Pétrel gongon
d Kap Verde-Sturmvogel
i Petrello di Fea

8435 *Pterodroma hasitata*
Procellariiformes - Procellariidae
e Black-capped Petrel; Diablotin;
Capped Petrel
f Pétrel diablotin; Diablotin (Ants);
Pétrel (Ant); Canard de montagne
(Ants); Diablotin errant (Qué)
d Teufelsturmvogel
i Petrello dal cappuccio

8436 *Pterodroma heraldica*
Procellariiformes - Procellariidae
e Herald Petrel
f Pétrel herault
d Wappensturmvogel
i Petrello di Tonga

8437 *Pterodroma hypoleuca*
Procellariiformes - Procellariidae
e Bonin Petrel; Stout-billed Petrel
f Pétrel des Bonin
d Bonin-Sturmvogel
i Petrello delle Isole Bonin

8438 *Pterodroma incerta*
Procellariiformes - Procellariidae
e Atlantic Petrel; Hooded Petrel;
Schlegel's Petrel
f Pétrel de Schlegel
d Schlegels Sturmvogel; Schlegel-
Sturmvogel

8439 *Pterodroma inexpectata*
Procellariiformes - Procellariidae
e Mottled Petrel; Scaled Petrel
f Pétrel maculé; Pétrel de Peale;
Diablotin maculé (Qué)
d Regensturmvogel
i Petrello di Peale

8440 *Pterodroma lessonii*
Procellariiformes - Procellariidae
e White-headed Petrel; White-headed
Fulmar
f Pétrel de Lesson
d Weißkopfsturmvogel
i Petrello testabianca

8441 *Pterodroma leucoptera*
Procellariiformes - Procellariidae
e White-winged Petrel; Gould's Petrel;
Collared Petrel; Stout-billed Gadfly;
Stout-billed Gadfly-Petrel; White-
throated Petrel; Sooty-capped Petrel
f Pétrel de Gould; Pétrel à ailes
blanches
d Brustbandsturmvogel; Gould-
Sturmschwalbe;
Brustbandsturmschwalbe;
Taubensturmschwalbe
i Petrello alibianche

8442 *Pterodroma leucoptera leucoptera*
Procellariiformes - Procellariidae
e White-winged Petrel; Australian
Gould's Petrel

8443 *Pterodroma longirostris*
Procellariiformes - Procellariidae
e Stejneger's Petrel
f Pétrel de Stejneger
d Stejneger-Sturmvogel
i Petrello di Stejneger

8444 *Pterodroma macgillivrayi*
Procellariiformes - Procellariidae
e Fiji Petrel; Gadfly-Petrel (CSA);
McGillivray's Petrel
f Pétrel des Fidji
d Macgillivray-Sturmvogel
i Petrello di Macgillivray

8445 *Pterodroma macroptera*
Procellariiformes - Procellariidae
e Great-winged Petrel; Grey-faced
Petrel; Long-winged Petrel; Long-
winged Fulmar
f Pétrel noir
d Langflügelsturmvogel
i Petrello aligrandi

8446   *Pterodroma madeira*
        Procellariiformes - Procellariidae
e   Zino's Petrel; Madeira Petrel; Soft-
        plumed Petrel; Zino's Petrel
f   Pétrel de Madère
d   Madeira-Sturmvogel
i   Petrello di Madeira

8447   *Pterodroma magentae*
        Procellariiformes - Procellariidae
e   Magenta Petrel; Taiko; Chatham
        Islands Taiko
f   Pétrel de Magenta
d   Madeira-Sturmvogel
i   Petrello della Magenta

8448   *Pterodroma mollis*
        Procellariiformes - Procellariidae
e   Soft-plumaged Petrel; Soft-plumaged
        Fulmar; Gon-gon
f   Pétrel soyeux
d   Weichfedersturmvogel;
        Weichfedernsturmvogel
i   Petrello soffice

8449   *Pterodroma mollis deceptornis*
        Procellariiformes - Procellariidae
e   Northern Soft-plumaged Petrel
        (ANZ)

8450   *Pterodroma neglecta*
        Procellariiformes - Procellariidae
e   Kermadec Petrel; Kermadek Petrel
f   Pétrel des Kermadec
d   Kermadek-Sturmvogel; Kermadek-
        Sturmschwalbe
i   Petrello dell'Isola Kermadec

8451   *Pterodroma neglecta neglecta*
        Procellariiformes - Procellariidae
e   Western Kermadec-Petrel (ANZ)

8452   *Pterodroma nigripennis*
        Procellariiformes - Procellariidae
e   Black-winged Petrel
f   Pétrel à ailes noires
d   Schwarzflügelsturmvogel
i   Petrello alinere

8453   *Pterodroma phaeopygia*
        Procellariiformes - Procellariidae

e   Galapagos Petrel; Dark-rumped
        Petrel; Hawaiian Petrel; White-
        necked Petrel
f   Pétrel des Galapagos; Pétrel de
        Hawaii
d   Hawaii-Sturmvogel
i   Petrello delle Hawaii

8454   *Pterodroma pycrofti*
        Procellariiformes - Procellariidae
e   Pycroft's Petrel
f   Pétrel de Pycroft
d   Pycroft-Sturmvogel
i   Petrello di Pycroft

8455   *Pterodroma rostrata*
        Procellariiformes - Procellariidae
e   Tahiti Petrel; Peale's Petrel
f   Pétrel de Tahiti
d   Tahiti-Sturmvogel
i   Petrello di Tahiti

8456   *Pterodroma sandwichensis*
        Procellariiformes - Procellariidae
e   Hawaiian Petrel

8457   *Pterodroma solandri*
        Procellariiformes - Procellariidae
e   Providence Petrel; Solander's Petrel;
        Bird-of-Providence; Brown-headed
        Petrel
f   Pétrel de Solander; Pétrel de
        Providence
d   Solander-Sturmvogel
i   Petrello di Solander

8458   *Pterodroma ultima*
        Procellariiformes - Procellariidae
e   Murphy's Petrel
f   Pétrel de Murphy
d   Murphy-Sturmvogel
i   Petrello di Murphy

8459   *Pteroglossus aracari*
        Piciformes - Ramphastidae
e   Black-necked Aracari; Aracari
        Toucan
f   Araçari grigri
d   Arassari; Schwarzkehlarassari
i   Aracari collonero; Pteroglosso dal
        collare nero

8460 **Pteroglossus azara**
Piciformes - Ramphastidae
e Ivory-billed Aracari; Ivory-billed
Toucan; Azara Aracari
f Araçari d'Azara
d Rotkopfarassari
i Aracari becco d'avorio

8461 **Pteroglossus beauharnaesii**
Piciformes - Ramphastidae
e Curl-crested Aracari; Curl-crested
Toucan
f Araçari de Beauharnais
d Krauskopfarassari; Plättchentukan
i Aracari crestariccia

8462 **Pteroglossus bitorquatus**
Piciformes - Ramphastidae
e Red-necked Aracari; Red-necked
Toucan
f Araçari à double collier
d Rotnackenarassari;
Doppelbandarassari
i Aracari collorosso

8463 **Pteroglossus castanotis**
Piciformes - Ramphastidae
e Chestnut-eared Aracari; Chestnut-
eared Toucan
f Araçari à oreillons roux
d Braunohrarassari
i Aracari orrecchie castane

8464 **Pteroglossus erythropygius**
Piciformes - Ramphastidae
e Pale-mandibled Aracari; Pale-
mandibled Toucan
f Araçari à bec clair
i Aracari beccochiaro

8465 **Pteroglossus frantzii**
Piciformes - Ramphastidae
e Fiery-billed Aracari; Fiery-billed
Toucan
f Araçari de Frantzius
i Aracari beccorosso

8466 **Pteroglossus inscriptus**
Piciformes - Ramphastidae
e Lettered Aracari; Lettered Toucan
f Araçari de Humboldt

d Schriftarassari
i Aracari scritto

8467 **Pteroglossus mariae**
Piciformes - Ramphastidae
e Brown-mandibled Aracari; Maria's
Touccan; Brown-mandibled Toucan
f Araçari de Maria
i Aracari becco bruno

8468 **Pteroglossus pluricinctus**
Piciformes - Ramphastidae
e Many-banded Aracari; Many-banded
Toucan
f Araçari multibande; Araçari à double
ceinture
d Doppelbindenarassari
i Aracari ventre a bande

8469 **Pteroglossus sanguineus**
Piciformes - Ramphastidae
e Stripe-billed Aracari; Stripe-billed
Toucan
f Araçari à bec maculé
i Aracari becco striato

8470 **Pteroglossus torquatus**
Piciformes - Ramphastidae
e Collared Aracari; Spot-chested
Aracari; Collared Toucan
f Araçari à collier
d Halsbandarassari
i Aracari dal collare

8471 **Pteroglossus viridis**
Piciformes - Ramphastidae
e Green Aracari; Green Toucan
f Araçari vert
d Grünarassari
i Aracari verde

8472 **Pteronetta hartlaubii**
Anseriformes - Anatidae
e Hartlaub's Duck; Hartlaub's Goose;
Hartlaub's Teal
f Canard de Hartlaub
d Hartlaub-Ente
i Anatra di Hartlaub

8473 **Pterophanes cyanopterus**
Apodiformes - Trochilidae

e Great Sapphirewing; Sapphirewing;
    Paramo Sapphirewing; Temminck's
    Sapphirewing
f Colibri à ailes saphir
d Blauflügelkolibri
i Colibrì ali di zaffiro

8474 *Pteroptochos castaneus*
    Passeriformes - Rhinocryptidae
e Chestnut-throated Huet-huet;
    Chestnut-breasted Huet-huet; Huet-
    Huet
f Tourco à gorge marron
d Braunkehl Huëthuët; Turkenvogel;
    Huëthuët; Braunes Hähnchen

8475 *Pteroptochos megapodius*
    Passeriformes - Rhinocryptidae
e Moustached Turca; Turca;
    Moustached Turco; Turco
f Tourco à moustaches
d Turkenvogel; Turco;
    Großfußrallenschlüpfer
i Turka dai mustacchi

8476 *Pteroptochos tarnii*
    Passeriformes - Rhinocryptidae
e Black-throated Huet-huet; Huet-huet
f Tourco huet-huet
d Schwarzkehlhuëthuët
i Uet-uet golanera

8477 *Pteruthius aenobarbus*
    Passeriformes - Sylviidae
e Chestnut-fronted Shrike-Babbler;
    Lesser Shrike-Babbler
f Allotrie à front marron
d Rotstirnwürgertimalie
i Garrulo averla frontecastano

8478 *Pteruthius flaviscapis*
    Passeriformes - Sylviidae
e White-browed Shrike-Babbler;
    Greater Shrike-Babbler; Red-winged
    Shrike-Babbler; Black-crowned
    Shrike-Babbler
f Allotrie à sourcils blancs; Allotrie à
    ailes rouges
d Schwarzkappenwürgertimalie;
    Weißbrauenwürgertimalie
i Garrulo averla alirosse

8479 *Pteruthius melanotis*
    Passeriformes - Sylviidae
e Black-eared Shrike-Babbler;
    Chestnut-throated Shrike-Babbler;
    Yellow-throated Shrike-Babbler
f Allotrie à gorge marron
d Zimtkehlwürgertimalie
i Garrulo averla orecchienere

8480 *Pteruthius rufiventer*
    Passeriformes - Sylviidae
e Black-headed Shrike-Babbler;
    Rufous-bellied Shrike-Babbler
f Allotrie à ventre roux
d Rotbauchwürgertimalie;
    Kappenwürgertimalie
i Garrulo averla ventrerosso

8481 *Pteruthius xanthochlorus*
    Passeriformes - Sylviidae
e Green Shrike-Babbler
f Allotrie vert
d Vireotimalie
i Garrulo averla verde

8482 *Ptilinopus alligator*
    Columbiformes - Columbidae
e Black-banded Fruit-Dove; Black-
    banded Pigeon; Banded Pigeon
f Ptilope de l'Alligator
d Graubauchfruchttaube
i Colomba frugivora collare nero

8483 *Ptilinopus arcanus*
    Columbiformes - Columbidae
e Negros Fruit-Dove; Ripley's Fruit-
    Dove; Ripley'sDove
f Ptilope de Ripley
d Negros-Fruchttaube; Ripley-
    Fruchttaube
i Colomba frugivora di Ripley

8484 *Ptilinopus aurantiifrons*
    Columbiformes - Columbidae
e Orange-fronted Fruit-Dove; Golden-
    fronted Fruit-Dove; Yellow-fronted
    Fruit-Dove
f Ptilope à front d'or
d Goldstirnfruchttaube
i Colomba frugivora frontearancio

8485 *Ptilinopus bernsteinii*
Columbiformes - Columbidae
e   Scarlet-breasted Fruit-Dove
f   Ptilope à poitrine écarlate
d   Scharlachbrustfruchttaube
i   Colomba frugivora di Bernstein

8486 *Ptilinopus chalcurus*
Columbiformes - Columbidae
e   Mahatea Fruit-Dove; Makatea Fruit-Dove
f   Ptilope de Makatea
i   Colomba frugivora di Mahatea

8487 *Ptilinopus cinctus*
Columbiformes - Columbidae
e   Black-backed Fruit-Dove; White-headed Pigeon; White-headed Fruit-Dove
f   Ptilope à ceinture
d   Weißkopfflaumfußtaube
i   Colomba frugivora dorsonero

8488 *Ptilinopus cinctus alligator*
Columbiformes - Columbidae
e   Australian Banded Fruit-Dove (ANZ)

8489 *Ptilinopus coralensis*
Columbiformes - Columbidae
e   Atoll Fruit-Dove; Coral Fruit-Dove; Tuamotu Fruit-Dove
f   Ptilope des Touamotou
i   Colomba frugivora di Tuamoto

8490 *Ptilinopus coronulatus*
Columbiformes - Columbidae
e   Coroneted Fruit-Dove; Little Coroneted Fruit-Dove; Little Capped Fruit-Dove; Lilac-capped Fruit-Dove; Lilac-crowned Fruit-Dove; Lilac-crowned Pigeon; Diadem Fruit-Dove; Diademed Fruit-Dove
f   Ptilope à couronne lilas
d   Gemalte Fruchttaube; Veilchenkappenfruchttaube
i   Colomba frugivora corona lilla

8491 *Ptilinopus dohertyi*
Columbiformes - Columbidae
e   Red-naped Fruit-Dove; Purple-tiled Fruit-Dove; Doherty's Fruit-Dove

f   Ptilope de Sumba
d   Rahmkopffruchttaube; Rotnackenflaumfußtaube
i   Colomba frugivora nucarossa

8492 *Ptilinopus dupetithouarsii*
Columbiformes - Columbidae
e   White-capped Fruit-Dove; Marquesa Fruit-Dove; Marquesas Fruit-Dove; Marquesan Fruit-Dove
f   Ptilope de Dupetit-Thouars
d   Weißkappenfruchttaube
i   Colomba frugivora corona bianca

8493 *Ptilinopus eugeniae*
Columbiformes - Columbidae
e   White-headed Fruit-Dove; Eugenie's Fruit-Dove; Eugenie's Fruit-Pigeon
f   Ptilope d'Eugénie
d   Schneekopffruchttaube
i   Colomba frugivora testabianca

8494 *Ptilinopus fischeri*
Columbiformes - Columbidae
e   Red-eared Fruit-Dove; Fischer's Fruit-Dove; Celebes Fruit-Dove
f   Ptilope de Fischer
d   Fischer-Fruchttaube; Rotohrfruchttaube
i   Colomba frugivora di Fischer

8495 *Ptilinopus granulifrons*
Columbiformes - Columbidae
e   Carunculated Fruit-Dove; Carunculated Dove; Wattled Fruit-Dove; Obi Fruit-Dove
f   Ptilope caronculé
d   Karunkelfruchttaube; Warzenfruchttaube
i   Colomba frugivora caruncolata

8496 *Ptilinopus greyii*
Columbiformes - Columbidae
e   Red-bellied Fruit-Dove; Grey's Fruit-Dove
f   Ptilope de Grey
d   Grey-Fruchttaube
i   Colomba frugivora di Grey

8497 *Ptilinopus huttoni*
Columbiformes - Columbidae

e   Rapa Fruit-Dove; Rapa Island Fruit-
Dove; Long-billed Fruit-Dove;
Hutton's Fruit-Dove
f   Ptilope de Hutton
d   Langschnabelfruchttaube; Rapa-
Fruchttaube
i   Colomba frugivora di Hutton

8498   *Ptilinopus hyogastra*
Columbiformes - Columbidae
e   Grey-headed Fruit-Dove; Halmahera
Fruit-Dove; Purple-bellied Fruit-
Dove
f   Ptilope hyogastre
d   Blaukopffruchttaube;
Graukopffruchttaube
i   Colomba frugivora ventrepurpureo

8499   *Ptilinopus insolitus*
Columbiformes - Columbidae
e   Knob-billed Fruit-Dove; Knob-billed
Orange-bellied Fruit-Dove
f   Ptilope casqué
d   Knopffruchttaube
i   Colomba frugivora dal tubercolo

8500   *Ptilinopus insularis*
Columbiformes - Columbidae
e   Henderson Island Fruit-Dove;
Henderson Fruit-Dove; Henderson
Island Fruit-Pigeon; Henderson fruit-
Pigeon
f   Ptilope de Henderson
d   Silberwangenfruchttaube
i   Colomba frugivora di Henderson

8501   *Ptilinopus iozonus*
Columbiformes - Columbidae
e   Orange-bellied Fruit-Dove
f   Ptilope à ventre orangé
d   Orangebauchfruchttaube
i   Colomba frugivora ventrearancio

8502   *Ptilinopus jambu*
Columbiformes - Columbidae
e   Jambu Fruit-Dove; Jambu Dove;
Jambu Pigeon; Jambu Fruit-Pigeon;
Pink-headed Fruit-Dove; Crimson-
headed Fruit-Dove
f   Ptilope jambou

d   Jambu-Fruchttaube
i   Colomba frugivora jambu

8503   *Ptilinopus layardi*
Columbiformes - Columbidae
e   Whistling Dove; Velvet Fruit-Dove;
Yellow-headed Fruit-Dove; Yellow-
headed Dove; Velvet Dove; Kadavu
Dove; Green Dove
f   Ptilope de Layard
d   Gelbkopffruchttaube
i   Colomba frugivora di Layard

8504   *Ptilinopus leclancheri*
Columbiformes - Columbidae
e   Black-chinned Fruit-Dove;
Leclancher's Fruit-Dove
f   Ptilope de Leclancher
d   Schwarzkinnfruchttaube
i   Colomba frugivora di Leclancher

8505   *Ptilinopus luteovirens*
Columbiformes - Columbidae
e   Golden Dove; Yellow Dove; Lemon
Dove
f   Ptilope jaune
d   Gelbe Fidschi-Flaumfußtaube;
Goldtaube
i   Colomba frugivora dorata

8506   *Ptilinopus magnificus*
Columbiformes - Columbidae
e   Wompoo Fruit-Dove; Wompoo
Pigeon; Magnificent Dove;
Magnificent Fruit-Dove; Magnificent
Fruit-Pigeon; Allied Fruit-Pigeon;
Purple-breasted Fruit-Dove; Purple-
breasted Pigeon; Purple-bellied
Pigeon; Painted Pigeon
f   Ptilope magnifique
d   Langschwanzfruchttaube;
Purpurbrustfruchttaube
i   Colomba frugivora magnifica

8507   *Ptilinopus marchei*
Columbiformes - Columbidae
e   Flame-breasted Fruit-Dove; Marché's
Fruit-Dove; Marché's Dove;
Marché's Fruit-Pigeon; Marché's
Pigeon; Black-eared Fruit-Pigeon
f   Ptilope de Marché

d   Blutschwingenfruchttaube;
Schwarzohrfruchttaube
i   Colomba frugivora di Marché

8508   *Ptilinopus melanospila*
Columbiformes - Columbidae
e   Black-naped Fruit-Dove; Black-naped Pigeon; Black-headed Fruit-Dove
f   Ptilope turgris
d   Schwarznackenfruchttaube
i   Colomba frugivora nucanera

8509   *Ptilinopus mercierii*
Columbiformes - Columbidae
e   Red-moustached Fruit-Dove;
Moustached Dove; Moustached
Fruit-Dove; Red-capped Fruit-Dove;
Yellow-bellied Fruit-Dove;
Marquesa Fruit-Dove; Marquesas
Fruit-Dove; Marquesan Fruit-Dove
f   Ptilope de Mercier
d   Rotbartfruchttaube; Rotkappen
Marquesas-Fruchttaube
i   Colomba frugivora dai mustacchi
rossi

8510   *Ptilinopus merrilli*
Columbiformes - Columbidae
e   Cream-bellied Fruit-Dove; Merrill's
Fruit-Dove; Cream-breasted Fruit-Dove
f   Ptilope de Merrill
d   Merrill-Fruchttaube
i   colomba frugivora di Merrill

8511   *Ptilinopus monacha*
Columbiformes - Columbidae
e   Blue-capped Fruit-Dove; Blue-capped Dove; Blue-crowned Fruit-Dove; Molucca Fruit-Dove;
Moluccan Fruit-Dove
f   Ptilope moine
d   Blaukappenfruchttaube
i   Colomba frugivora corona blu

8512   *Ptilinopus naina*
Columbiformes - Columbidae
e   Dwarf Fruit-Dove; Small Green
Fruit-Dove; Small Fruit-Dove; Little
Fruit-Dove

f   Ptilope nain
d   Zwergfruchttaube
i   Colomba frugivora nana

8513   *Ptilinopus occipitalis*
Columbiformes - Columbidae
e   Yellow-breasted Fruit-Dove; Yellow-breasted Fruit-Pigeon; Sulphur-breasted Fruit-Dove
f   Ptilope batilde
d   Gelbbrustfruchttaube
i   Colomba frugivora pettogiallo

8514   *Ptilinopus ornatus*
Columbiformes - Columbidae
e   Ornate Fruit-Dove; Ornate Dove;
Gestroi's Fruit-Dove
f   Ptilope orné
d   Schmuckfruchttaube
i   Colomba frugivora ornata

8515   *Ptilinopus pelewensis*
Columbiformes - Columbidae
e   Palau Fruit-Dove; Pellew Fruit-Dove;
Pallu Fruit-Dove
f   Ptilope des Palau
d   Palau-Fruchttaube
i   Colomba frugivora di Palau

8516   *Ptilinopus perlatus*
Columbiformes - Columbidae
e   Pink-spotted Fruit-Dove
f   Ptilope perlé
d   Rosafleckenfruchttaube
i   Colomba frugivora perlata

8517   *Ptilinopus perousii*
Columbiformes - Columbidae
e   Rainbow Dove; Many-coloured
Fruit-Dove; Rainbow Fruit-Dove;
Nutmeg Dove; Painted Dove
f   Ptilope de La Pérouse
d   Perouse-Fruchttaube; Samoa-Fruchttaube
i   Colomba frugivora di Laperouse

8518   *Ptilinopus porphyraceus*
Columbiformes - Columbidae
e   Crimson-crowned Fruit-Dove;
Purple-capped Fruit-Dove; Ponapé
Dove

f   Ptilope de Clementine
d   Purpurscheitelfruchttaube
i   Colomba frugivora corona purpurea

8519   *Ptilinopus porphyreus*
Columbiformes - Columbidae
e   Pink-headed Fruit-Dove; Pink-
necked Fruit-Dove
f   Ptilope porphyre
d   Rothalsfruchttaube
i   Colomba frugivora collorosso

8520   *Ptilinopus pulchellus*
Columbiformes - Columbidae
e   Beautiful Fruit-Dove; Crimson-
capped Fruit-Dove; Rose-fronted
Pigeon; Grey-breasted Fruit-Dove
f   Ptilope mignon
d   Rotkappenfruchttaube; Schöne
Flaumfußtaube
i   Colomba frugivora bella

8521   *Ptilinopus purpuratus*
Columbiformes - Columbidae
e   Grey-green Fruit-Dove; Purple-
capped Fruit-Dove; Tahiti Fruit-
Dove; Tahitian Fruit-Dove; Tuamotu
Fruit-Dove
f   Ptilope de la Societé
d   Purpurkappenfruchttaube; Tahiti-
Fruchttaube
i   Colomba frugivora corona purpurea

8522   *Ptilinopus rarotongensis*
Columbiformes - Columbidae
e   Cook Islands Fruit-Dove; Rarotongan
Fruit-Dove
f   Ptilope de Rarotonga
d   Rarotonga-Fruchttaube
i   Colomba frugivora di Rarotanga

8523   *Ptilinopus regina*
Columbiformes - Columbidae
e   Rose-crowned Fruit-Dove; Red-
crowned Fruit-Dove; Swainson's
Fruit-Dove; Pink-capped Fruit-Dove;
Pink-capped Dove; Pink-headed
Fruit-Dove; Red-capped Fruit-Dove;
Ewing's Fruit-Pigeon; Swainson's
Fruit-Pigeon; Blue-spotted Fruit-
Dove; Grey-spotted Fruit-Dove;

Grey-capped Fruit-Dove
f   Ptilope à diadème
d   Königsfruchttaube;
Rosakappenfruchttaube
i   Colomba frugivora corona rosa

8524   *Ptilinopus richardsii*
Columbiformes - Columbidae
e   Silver-capped Fruit-Dove; Silver-
capped Dove; Pink-spotted Fruit-
Dove; Richard's Fruit-Dove
f   Ptilope de Richards
d   Salomonen-Flaumfußtaube;
Silberkappenfruchttaube
i   Colomba frugivora corona argentata

8525   *Ptilinopus rivoli*
Columbiformes - Columbidae
e   White-bibbed Fruit-Dove; White-
breasted Fruit-Dove; Moon Dove;
Beautiful Fruit-Dove; Moon Fruit-
Dove; Moon-breasted Fruit-Dove;
High Mountain Fruit-Dove
f   Ptilope de Rivoli
d   Korallenfruchttaube
i   Colomba frugivora di Rivoli

8526   *Ptilinopus roseicapilla*
Columbiformes - Columbidae
e   Mariana Fruit-Dove; Rose-capped
Dove; Marianas Fruit-Dove
f   Ptilope des Mariannes
d   Marianen-Fruchttaube;
Rosenkopffruchttaube
i   Colomba frugivora delle Marianne

8527   *Ptilinopus solomonensis*
Columbiformes - Columbidae
e   Yellow-bibbed Fruit-Dove; Splendid
Fruit-Dove; Yellow-breasted Fruit-
Dove
f   Ptilope des Salomon
d   Gelbbauchfruchttaube;
Gelblatzfruchttaube
i   Colomba frugivora delle Salomon

8528   *Ptilinopus subgularis*
Columbiformes - Columbidae
e   Maroon-chinned Fruit-Dove; Dark-
chinned Fruit-Dove
f   Ptilopeà mentonnière

 d Braunkinnfruchttaube;
    Dunkelkinnfruchttaube
 i Colomba frugivora mentoscuro

8529 *Ptilinopus superbus*
    Columbiformes - Columbidae
 e Superb Fruit-Dove; Purple-crowned
    Fruit-Dove; Superb Dove; Superb
    Fruit-Pigeon; Purple-capped Fruit-
    Dove; Purple-crowned Pigeon
 f Ptilope superbe
 d Prachtfruchttaube
 i Colomba frutivora superba

8530 *Ptilinopus tannensis*
    Columbiformes - Columbidae
 e Tanna Fruit-Dove; Silver-shouldered
    Fruit-Dove; Yellow-headed Fruit-
    Dove; New Hebrides Fruit-Dove
 f Ptilope de Tanna
 d Silberfleckfruchttaube
 i Colomba frugivora spallearegentate

8531 *Ptilinopus victor*
    Columbiformes - Columbidae
 e Orange Dove; Flame Dove
 f Ptilope orangé
 d Orangetaube; Rote Fidschi-
    Flaumfußtaube
 i Colomba frugivora arancio

8532 *Ptilinopus viridis*
    Columbiformes - Columbidae
 e Claret-breasted Fruit-Dove; Red-
    breasted Fruit-Dove; Red-bibbed
    Fruit-Dove; Red-bibbed Dove; Red-
    throated Fruit-Dove
 f Ptilope turvert
 d Frühlingsfruchttaube;
    Rotlatzfruchttaube
 i Colomba frugivora dalla pettorina
    rossa

8533 *Ptilinopus wallacii*
    Columbiformes - Columbidae
 e Wallace's Fruit-Dove; Wallace's
    Green Fruit-Dove; Gold-shouldered
    Fruit-Dove; Golden-shouldered
    Fruit-Dove; Crimson-capped Fruit-
    Dove
 f Ptilope de Wallace

 d Goldschulterfruchttaube;
    Weißkehlfruchttaube
 i Colomba frugivora di Wallace

8534 *Ptilocichla falcata*
    Passeriformes - Sylviidae
 e Falcated Wren-Babbler; Falcated
    Ground-Babbler; Palawan Wren-
    Babbler
 f Turdinule de Palawan
 d Palawan-Wolltimalie
 i Garrulo scricciolo di Palawan

8535 *Ptilocichla leucogrammica*
    Passeriformes - Sylviidae
 e Bornean Wren-Babbler; Borneo
    Wren-Babbler; Bornean Ground-
    Warbler
 f Turdinule de Bornéo
 d Borneo-Wolltimalie
 i Garrulo scricciolo del Borneo

8536 *Ptilocichla mindanensis*
    Passeriformes - Sylviidae
 e Striated Wren-Babbler; Streaked
    Ground-Babbler; Streaked Wren-
    Babbler; Blasius's Wren-Babbler;
    Striated Ground-Babbler
 f Turdinule des Philippines
 d Philippinen-Wolltimalie
 i Garrulo scricciolo di Mindano

8537 *Ptilogonys caudatus*
    Passeriformes - Bombycillidae
 e Long-tailed Silky-Flycatcher
 f Ptilogon à longue queue; Ptilogon à
    longue queue
 d Langschwanzseidenschnäpper
 i Uccello di seta codalunga

8538 *Ptilogonys cinereus*
    Passeriformes - Bombycillidae
 e Grey Silky-Flycatcher
 f Ptilogon cendré
 d Grauseidenschnäpper
 i Uccello di seta cenerino

8539 Ptilonorhynchidae
 e Bowerbirds
 f Ptilonorhynchidés

d   Laubenvögel
i   Ptilonrincidi

8540   *Ptilonorhynchus violaceus*
Passeriformes - Ptilonorhynchidae
e   Satin Bowerbird
f   Jardinier satiné; Ptilonorhynque
satiné; Oiseau à berceau satiné
d   Seidenlaubenvogel
i   Ptilonorinco violaceo; Uccello di
raso

8541   *Ptilopachus petrosus*
Galliformes - Phasianidae
e   Stone Partridge
f   Poulette de roche; Poule de roche
i   Pernice delle rocce

8542   *Ptiloprora erythropleura*
Passeriformes - Meliphagidae
e   Rufous-sided Honeyeater; Red-sided
Honeyeater; Red-sided Streaked-
Honeyeater
f   Méliphage à flancs roux
d   Rotflankenhonigfresser
i   Mangiamiele striato fianchirossi

8543   *Ptiloprora guisei*
Passeriformes - Meliphagidae
e   Rufous-backed Honeyeater; Red-
backed Honeyeater; Red-backed
Streaked-Honeyeater
f   Méliphage à dos roux
d   Rotrückenhonigfresser
i   Mangiamiele dorsocastano

8544   *Ptiloprora mayri*
Passeriformes - Meliphagidae
e   Mayr's Honeyeater
f   Méliphage de Mayr
i   Mangiamiele di Mayr

8545   *Ptiloprora meekiana*
Passeriformes - Meliphagidae
e   Olive-streaked Honeyeater; Meek's
Honeyeater; Meek's Streaked-
Honeyeater; Yellowish Honeyeater
f   Méliphage de Meek
d   Meek-Honigfresser
i   Mangiamiele striato di Meek

8546   *Ptiloprora perstriata*
Passeriformes - Meliphagidae
e   Black-backed Honeyeater; Black-
backed Streaked-Honeyeater; Grey-
streaked Honeyeater; Many-streaked
Honeyeater
f   Méliphage strié
d   Streifenhonigfresser
i   Mangiamiele dorsoscuro

8547   *Ptiloprora plumbea*
Passeriformes - Meliphagidae
e   Leaden Honeyeater; Leaden
Streaked-Honeyeater
f   Méliphage gris-de-plomb
d   Bleikehlhonigfresser
i   Mangiamiele striato plumbeo

8548   *Ptiloris intercedens*
Passeriformes - Corvidae
e   Eastern Riflebird

8549   *Ptiloris magnificus*
Passeriformes - Corvidae
e   Magnificent Riflebird
f   Paradisier gorge-d'acier
d   Prachtparadiesvogel
i   Uccello fucile magnifico

8550   *Ptiloris paradiseus*
Passeriformes - Corvidae
e   Paradise Riflebird
f   Paradisier festonné
d   Schildparadiesvogel
i   Uccello fucile del Paradiso

8551   *Ptiloris victoriae*
Passeriformes - Corvidae
e   Victoria's Riflebird; Queen Victoria's
Riflebird
f   Paradisier de Victoria
d   Victoria-Paradiesvogel
i   Uccello fucile di Vittoria

8552   *Ptilorrhoa caerulescens*
Passeriformes - Corvidae
e   Blue Jewel-Babbler; Lowland Rail-
Babbler; Lowland Quail-Thrush;
Lowland Jewel-Babbler; Lowland
Eupates
f   Ptilorrhoa bleu

d   Blauflöter

i   Eupete azzurro

8553   ***Ptilorrhoa castanonota***
Passeriformes - Corvidae

e   Chestnut-backed Jewel-Babbler;
Mid-mountain Rail-Babbler; Mid-
mountain Scrub-Robin; Mid-
mountain Jewel-Babbler; Mid-
mountain Eupates

f   Ptilorrhoa à dos roux

d   Buntflöter

i   Eupete dorsocastano

8554   ***Ptilorrhoa leucosticta***
Passeriformes - Corvidae

e   Spotted Jewel-Babbler; High-
mountain Rail-Babbler; High-
mountain Scrub-Robin; Mountain
Scrub-Robin; High Mountain
Eupates

f   Ptilorrhoa tacheté

d   Bergwaldflöter

i   Eupete alimacchiate

8555   ***Ptilostomus afer***
Passeriformes - Corvidae

e   Piapiac; Black Magpie; African
Magpie

f   Piapiac africain; Piacpiac

d   Spitzschwanzelster; Piapia

i   Piaciac

8556   ***Ptychoramphus aleuticus***
Charadriiformes - Alcidae

e   Cassin's Auklet; Aleutian Auk

f   Starique de Cassin; Alque de Cassin
(Qué)

d   Aleuten-Alk

i   Alca minore di Cassin

8557   ***Ptyrticus turdinus***
Passeriformes - Sylviidae

e   White-bellied Thrush-Babbler;
White-headed Thrush-Babbler;
Thrush-Babbler; African Thrush-
Babbler; Spotted Thrush-Babbler

f   Akalat à dos roux; Grive-akalat à dos
rouge

d   Weißbauchdrosseltimalie;

Buschdrossling

i   Garrulo tordo africano

8558   ***Pucrasia macrolopha***
Galliformes - Phasianidae

e   Koklass Pheasant; Koklass

f   Eulophe koklass

d   Koklas-Fasan

i   Fagiano koklass

8559   ***Puffinus assimilis***
Procellariiformes - Procellariidae

e   Little Shearwater; Allied Shearwater;
Muttonbird (CSA); Allied Petrel;
Dusky Petrel; Dusky Shearwater

f   Petit Puffin; Puffin semblable; Puffin
obscur (Qué)

d   Kleiner Sturmtaucher; Sturmtaucher

i   Berta minore fosca

8560   ***Puffinus assimilis assimilis***
Procellariiformes - Procellariidae

e   Tasman Sea Little Shearwater (ANZ)

8561   ***Puffinus atrodorsalis***
Procellariiformes - Procellariidae

e   Mascarene Shearwater

8562   ***Puffinus auricularis***
Procellariiformes - Procellariidae

e   Townsend's Shearwater

f   Puffin de Townsend

d   Schuppensturmvogel

i   Berta di Townsend

8563   ***Puffinus bannermani***
Procellariiformes - Procellariidae

e   Bannerman's Shearwater

f   Puffin de Bannerman

i   Berta di Bannerman

8564   ***Puffinus boydi***
Procellariiformes - Procellariidae

e   Cape Verde Little-Shearwater

f   Puffin de Buller

8565   ***Puffinus bulleri***
Procellariiformes - Procellariidae

e   Buller's Shearwater; Grey-backed
Shearwater; New Zealand
Shearwater; Ashy-backed Shearwater

f Puffin de Buller
d Graunackensturmtaucher
i Berta di Buller

8566 ***Puffinus carneipes***
      Procellariiformes - Procellariidae
e Flesh-footed Shearwater; Pale-footed
      Shearwater; Muttonbird (CSA);
      Fleshy-fronted Shearwater; Flesh-
      footed Petrel; Fleshy-fronted Petrel;
      Lord Howe Island Muttonbird
f Puffin à pieds pâles; Puffin à pattes
      pâles
d Blaßfußsturmtaucher;
      Barfußsturmtaucher
i Berta piedicarnicini

8567 ***Puffinus creatopus***
      Procellariiformes - Procellariidae
e Pink-footed Shearwater
f Puffin à pieds roses; Puffin à pattes
      roses
i Berta piedirosa

8568 ***Puffinus gavia***
      Procellariiformes - Procellariidae
e Fluttering Shearwater; Forster's
      Petrel; Forster's Shearwater; Brown-
      beaked Petrel; Brown-beaked
      Shearwater
f Puffin volage
i Berta vagabonda

8569 ***Puffinus gravis***
      Procellariiformes - Procellariidae
e Great Shearwater; Greater
      Shearwater (NA); Muttonbird (CSA)
f Puffin majeur; Grand Puffin
d Kappensturmtaucher; Großer
      Sturmtaucher
i Berta dell'Atlantico

8570 ***Puffinus griseus***
      Procellariiformes - Procellariidae
e Sooty Shearwater; Muttonbird (CSA)
f Puffin fuligineux
d Dunkelsturmtaucher; Dunkler
      Sturmtaucher
i Berta grigia

8571 ***Puffinus heinrothi***
      Procellariiformes - Procellariidae
e Heinroth's Shearwater
f Puffin de Heinroth
d Heinroth-Sturmtaucher
i Berta di Heinroth

8572 ***Puffinus huttoni***
      Procellariiformes - Procellariidae
e Hutton's Shearwater
f Puffin de Hutton
i Berta di Hutton

8573 ***Puffinus lherminieri***
      Procellariiformes - Procellariidae
e Audubon's Shearwater; Muttonbird
      (CSA); Plimico (WI); Little Devil
      (WI); Diablotin (WI); Dusky-backed
      Shearwater
f Puffin d'Audubon; Puffin (Ants);
      Puffin sombre
d Audubon-Sturmtaucher
i Berta di Audubon

8574 ***Puffinus mauretanicus***
      Procellariiformes - Procellariidae
e Balearic Shearwater; Mediterranean
      Shearwater
f Puffin des Baléares
d Mittelmeer-Sturmtaucher;
      Mauretanischer Sturmtaucher
i Berta minore delle Baleari

8575 ***Puffinus nativitalis***
      Procellariiformes - Procellariidae
e Christmas Island Shearwater;
      Christmas Shearwater; Black
      Shearwater
f Puffin de la Nativité
d Weihnachts-Sturmtaucher
i Berta di Natale; Berta dell'Isola di
      Natale

8576 ***Puffinus newelli***
      Procellariiformes - Procellariidae
e Newell's Shearwater
f Puffin de Newell

8577 ***Puffinus opisthomelas***
      Procellariiformes - Procellariidae
e Black-vented Shearwater

f   Puffin cul-noir
i   Berta codanera

8578   *Puffinus pacificus*
Procellariiformes - Procellariidae
e   Wedge-tailed Shearwater;
Muttonbird (CSA); Wedge-tailed
Petrel; Green-billed Shearwater;
Little Petrel
f   Puffin fouquet; Puffin du Pacifique
d   Keilschwanzsturmtaucher;
Keilschwanzsturmvogel
i   Berta del Pacifico

8579   *Puffinus persicus*
Procellariiformes - Procellariidae
e   Persian Shearwater
f   Puffin persique; Puffin obscur
d   Persischer Sturmtaucher
i   Berta dal Golfo Persico

8580   *Puffinus puffinus*
Procellariiformes - Procellariidae
e   Manx Shearwater; Muttonbird
(CSA); Levantine Shearwater
f   Puffin des Anglais; Puffin de Manx
d   Schwarzschnabelsturmtaucher
i   Berta minore; Berta minore atlantica

8581   *Puffinus tenuirostris*
Procellariiformes - Procellariidae
e   Short-tailed Shearwater; Slender-
billed Shearwater; Muttonbird;
Slender-billed Petrel; Tasmanian
Muttonbird; Short-tailed Petrel
f   Puffin à bec grêle; Puffin à bec mince
(Qué)
d   Dünnschnabeliger Sturmvogel;
Kurzschwanzsturmtaucher
i   Berta codacorta

8582   *Puffinus yelkouan*
Procellariiformes - Procellariidae
e   Mediterranean Shearwater; Yelkouan
Shearwater; Levantine Shearwater
f   Puffin yelkouan
d   Mittelmeer-Sturmtaucher
i   Berta minore mediterranea

8583   *Pulsatrix koeniswaldiana*
Strigiformes - Strigidae

e   Tawny-browed Owl; White-chinned
Owl
f   Chouette à sourcils jaunes
d   Gelbrauenkauz; Weißkinnkauz
i   Gufo dai sopraccigli fulvi; Gufo a
righe bianche

8584   *Pulsatrix melanota*
Strigiformes - Strigidae
e   Band-bellied Owl; Rusty-barred Owl
f   Chouette à collier
d   Bindenkauz
i   Gufo ventrefasciato; Gufo a righe
ruggine

8585   *Pulsatrix perspicillata*
Strigiformes - Strigidae
e   Spectacled Owl
f   Chouette à lunettes
d   Brillenkauz
i   Gufo dagli occhiali; Civetta dagli
occhiali

8586   *Purpureicephalus spurius*
Psittaciformes - Psittacidae
e   Red-capped Parrot; King Parrot;
Red-capped Parakeet; Pileated
Parrot; Pileated Parakeet; Western
King-Parrot
f   Perruche à tête pourpre
d   Rotkappensittich
i   Parrochetto capirosso

8587   Pycnonotidae
Passeriformes
e   Bulbuls
f   Bulbuls; Pycnonotidés
d   Bülbüls
i   Picnonotidi

8588   *Pycnonotus atriceps*
Passeriformes - Pycnonotidae
e   Black-headed Bulbul
f   Bulbul cap-nègre
d   Schwarzkopfbülbül
i   Bulbul testanera

8589   *Pycnonotus aurigaster*
Passeriformes - Pycnonotidae
e   Sooty-headed Bulbul; White-eared
Bulbul; Swinhoe's Bulbul; Black-

capped Bulbul; Yellow-vented
Bulbul; Golden-vented Bulbul; Red-
vented Bulbul
f Bulbul cul-d'or; Bulbul de Swinhoe
d Kotilang-Bülbül
i Bulbul ventredorato

8590 *Pycnonotus barbatus*
Passeriformes - Pycnonotidae
e Garden Bulbul; Common Bulbul
(CSA); Dulbul (CSA), Dark-capped
Bulbul (CSA); Black-eyed Bulbul
(CSA); Toppie (CSA); White-vented
Bulbul; Common Brown-Bulbul;
Black-eyed Brown-Bulbul; Common
Black-eyed Bulbul; Yellow-vented
Bulbul
f Bulbul des jardins; Bulbul commun;
Bulbul
d Graubülbül
i Bulbul golanera; Bulbul grigio

8591 *Pycnonotus bimaculatus*
Passeriformes - Pycnonotidae
e Orange-spotted Bulbul
f Bulbul bimaculé
d Goldzügelbülbül
i Bulbul bimaculato

8592 *Pycnonotus blanfordi*
Passeriformes - Pycnonotidae
e Streak-eared Bulbul; Blanford's
Bulbul; Blanford's Olive-Bulbul
f Bulbul de Blanford
d Blanford-Bülbül
i Bulbul di Blanford

8593 *Pycnonotus brunneus*
Passeriformes - Pycnonotidae
e Red-eyed Bulbul; Red-eyed Brown-
Bulbul; Brown-Bulbul
f Bulbul aux yeux rouges
d Maskenbülbül; Rotaugenbülbül
i Bulbul bruno occhirossi

8594 *Pycnonotus cafer*
Passeriformes - Pycnonotidae
e Red-vented Bulbul; Common Bulbul
(CSA)
f Bulbu à ventre rouge; Bulbul cul-
rouge; Bulbul indien

d Rußbülbül; Tonki-Bülbül
i Bulbul dal sottocoda rosso

8595 *Pycnonotus capensis*
Passeriformes - Pycnonotidae
e Cape Bulbul; Toppie (CSA); Dark-
capped Bulbul
f Bulbul du Cap
d Kapbülbül; Graubülbül
i Bulbul del Capo

8596 *Pycnonotus cyaniventris*
Passeriformes - Pycnonotidae
e Grey-bellied Bulbul
f Bulbul à ventre gris
d Graubauchbülbül
i Bulbul ventreardesia

8597 *Pycnonotus dodsoni*
Passeriformes - Pycnonotidae
e Dodson's Bulbul; African White-
eared Bulbul; White-eared Bulbul
f Bulbul de Dodson
i Bulbul di Dodson

8598 *Pycnonotus erythropthalmus*
Passeriformes - Pycnonotidae
e Spectacled Bulbul; Lesser Brown-
Bulbul
f Bulbul oeil-de-feu
d Salvadori-Bülbül
i Bulbul bruno minore

8599 *Pycnonotus eutilotus*
Passeriformes - Pycnonotidae
e Puff-backed Bulbul; Puff-backed
Brown-Bulbul; Crested Brown-
Bulbul
f Bulbul laineux
d Wollrückenbülbül
i Bulbul dorsosetoloso

8600 *Pycnonotus finlaysoni*
Passeriformes - Pycnonotidae
e Stripe-throated Bulbul; Streak-
throated Bulbul; Striped-throated
Bulbul
f Bulbul de Finlayson
d Streifenkehlbülbül
i Bulbul di Finlayson

8601 **Pycnonotus flavescens**
Passeriformes - Pycnonotidae
e Flavescent Bulbul; Pale-faced
Bulbul; Blyth's Bulbul; Round-tailed
Bulbul
f Bulbul flavescent; Bulbul jaunâtre
d Gelbwangenbülbül;
Blaßwangenbülbül
i Bulbul giallo-verde

8602 **Pycnonotus goiavier**
Passeriformes - Pycnonotidae
e Yellow-vented Bulbul
f Bulbul goiavier
d Gelbbauchbülbül
i Bulbul corona nera

8603 **Pycnonotus jocosus**
Passeriformes - Pycnonotidae
e Red-whiskered Bulbul; Crested
Bulbul
f Bulbul orphée
d Rotohrbülbül
i Bulbul dai mustacchi rossi; Bulbul
dai mustacchi rossi; Bulbul orfeo

8604 **Pycnonotus leucogenys**
Passeriformes - Pycnonotidae
e Himalayan Bulbul; White-cheeked
Bulbul
f Bulbul à joues blanches
d Weißohrbülbül
i Bulbul guancebianche

8605 **Pycnonotus leucogrammicus**
Passeriformes - Pycnonotidae
e Cream-striped Bulbul; Striated
Bulbul; Streaked Bulbul; Striated
Green Bulbul
f Bulbul rayé
d Strichelbülbül
i Bulbul striato

8606 **Pycnonotus leucotis**
Passeriformes - Pycnonotidae
e White-eared Bulbul; White-cheeked
Bulbul
f Bulbul à oreillons blancs
i Bulbul orecchiebianche

8607 **Pycnonotus luteolus**
Passeriformes - Pycnonotidae
e White-browed Bulbul; Ceylon
White-browed Bulbul
f Bulbul à sourcils blancs
d Brauenbülbül
i Bulbul dai sopraccigli bianchi

8608 **Pycnonotus melanicterus**
Passeriformes - Pycnonotidae
e Black-crested Bulbul; Black-headed
Yellow-Bulbul (ISC); Black-capped
Yellow-Bulbul; Black-crested
Yellow Bulbul; Black-capped
Bulbul; Yellow Bulbul
f Bulbul à tête noire
d Goldbrustbülbül
i Bulbul crestanera

8609 **Pycnonotus melanicterus gularis**
Passeriformes - Pycnonotidae
e Ruby-throated Bulbul (ISC)

8610 **Pycnonotus melanoleucus**
Passeriformes - Pycnonotidae
e Black-and-white Bulbul
f Bulbul demi-deuil
d Trauerbülbül; Schwarzweißbülbül
i Bulbul nero e bianco

8611 **Pycnonotus nieuwenhuisii**
Passeriformes - Pycnonotidae
e Blue-wattled Bulbul; Malaysian
Wattled-Bulbul; Malaysian Blue-
wattled Bulbul; Nieuwenhuis's
Bulbul
f Bulbul à lunettes bleues
d Blaubrillenbülbül; Nieuwenhuis-
Bülbül
i Bulbul di Nieuwenhuis

8612 **Pycnonotus nigricans**
Passeriformes - Pycnonotidae
e Red-eyed Bulbul; Black-fronted
Bulbul; African Red-eyed Bulbul
f Bulbul brunoir
d Maskenbülbül
i Bulbul occhirossi

8613 **Pycnonotus penicillatus**
Passeriformes - Pycnonotidae

e Yellow-eared Bulbul; Yellow-Tufted
    Bulbul; Ceylon Bulbul; Sri Lanka
    Bulbul; Sri Lanka Yellow-eared
    Bulbul
f Bulbul oreillard
d Schmuckbülbül
i Bulbul ciuffogiallo

8614 *Pycnonotus plumosus*
    Passeriformes - Pycnonotidae
e Olive-winged Bulbul; Olive-brown
    Bulbul; Olive Bulbul; Large Olive-
    Bulbul
f Bulbul aux ailes olives
d Malaien-Bülbül
i Bulbul bruno-oliva

8615 *Pycnonotus priocephalus*
    Passeriformes - Pycnonotidae
e Grey-headed Bulbul
f Bulbul colombar
d Graukopfbülbül
i Bulbul testagrigia indiana

8616 *Pycnonotus simplex*
    Passeriformes - Pycnonotidae
e White-eyed Bulbul; White-eyed
    Brown-Bulbul; Cream-vented Bulbul
f Bulbul aux yeux blancs
d Weißaugenbülbül
i Bulbul bruno occhibianchi

8617 *Pycnonotus sinensis*
    Passeriformes - Pycnonotidae
e Light-vented Bulbul; Chinese Bulbul;
    White-vented Bulbul
f Bulbul de Chine
d Gartenbülbül; Chinesen-Bülbül
i Bulbul cinese

8618 *Pycnonotus somaliensis*
    Passeriformes - Pycnonotidae
e Somali Bulbul
f Bulbul somalien
i Bulbul somalo

8619 *Pycnonotus squamatus*
    Passeriformes - Pycnonotidae
e Scaly-breasted Bulbul
f Bulbul écaillé

d Schuppenbülbül
i Bulbul pettosquamato

8620 *Pycnonotus striatus*
    Passeriformes - Pycnonotidae
e Striated Bulbul; Striped Bulbul;
    Striated Green-Bulbul (ISC)
f Bulbul strié
d Streifenbülbül
i Bulbul verde striato

8621 *Pycnonotus taivanus*
    Passeriformes - Pycnonotidae
e Styan's Bulbul; Formosan Bulbul;
    Taiwan Bulbul
f Bulbul de Taïwan; Bulbul de
    Formose
d Formosa-Bülbül
i Bulbul di Taiwan

8622 *Pycnonotus tricolor*
    Passeriformes - Pycnonotidae
e Dark-capped Bulbul; Yellow-vented
    Bulbul; Tricoloured Bulbul
f Bulbul tricolore
i Bulbul capinero

8623 *Pycnonotus tympanistrigus*
    Passeriformes - Pycnonotidae
e Spot-necked Bulbul; Olive-crowned
    Bulbul
f Bulbul à cou tacheté; Bulbul à tête
    olive
d Goldbürzelbülbül
i Bulbul corona oliva

8624 *Pycnonotus urostictus*
    Passeriformes - Pycnonotidae
e Yellow-wattled Bulbul; Wattled
    Bulbul
f Bulbul à lunettes jaunes
d Gelbbrillenbülbül
i Bulbul dalle caruncole gialle

8625 *Pycnonotus xantholaemus*
    Passeriformes - Pycnonotidae
e Yellow-throated Bulbul; Yellow-
    eared Bulbul
f Bulbul à menton jaune; Bulbul à
    gorge jaune

d Goldkehlbülbül
i Bulbul golagialla

8626 Pycnonotus xanthopygos
Passeriformes - Pycnonotidae
e White-spectacled Bulbul; Yellow-vented Bulbul; Black-capped Bulbul; White-eyed Bulbul; Black-headed Bulbul; Dark-capped Bulbul
f Bulbul d'Arabie; Bulbul des jardins
d Vallombrosa-Bülbül
i Bulbul dagli occhiali bianchi; Bulbul capinero

8627 Pycnonotus xanthorrhous
Passeriformes - Pycnonotidae
e Brown-breasted Bulbul; Anderson's Bulbul; Yellow-vented Bulbul
f Bulbul à poitrine brune; Bulbul à ventre jaune; Bulbul d'Anderson
d Gelbsteißbülbül; Braunbrustbülbül
i Bulbul di Anderson

8628 Pycnonotus zeylanicus
Passeriformes - Pycnonotidae
e Straw-headed Bulbul; Straw-crowned Bulbul; Yellow-crested Bulbul; Yellow-crowned Bulbul
f Bulbul à tête jaune
d Gelbscheitelbülbül
i Bulbul corona di paglia

8629 Pycnoptilus floccosus
Passeriformes - Pardalotidae
e Pilotbird
f Pycnoptile compagnon
d Leierschwanzlakai
i Uccello pilota

8630 Pycnopygius cinereus
Passeriformes - Meliphagidae
e Marbled Honeyeater; Grey-fronted Honeyeater; Grey Honeyeater; Greyish-brown Honeyeater
f Méliphage marbré
d Marmorhonigfresser
i Mangiamiele marmoreggiato

8631 Pycnopygius ixoides
Passeriformes - Meliphagidae
e Plain Honeyeater; New Guinea Brown-Honeyeater; New Guinea Olive-brown Honeyeater; New Guinean Brown-Honeyeater; Nondescript Honeyeater; Brown Honeyeater
f Méliphage ocré
d Bülbülhonigfresser
i Mangiamiele di Salvadori

8632 Pycnopygius stictocephalus
Passeriformes - Meliphagidae
e Streak-headed Honeyeater; Streaked-capped Honeyeater
f Méliphage à tête rayée
d Strichelkopfhonigfresser
i Mangiamiele capomaculato

8633 Pygarrhichas albogularis
Passeriformes - Furnariidae
e Rufous-breasted Leaftosser; White-throated Treerunner
f Picotelle à gorge blanche
d Spechttöpfer
i Coridore arboricolo golabianca

8634 Pygiptila stellaris
Passeriformes - Formicariidae
e Spot-winged Antshrike; Starred Antshrike
f Batara étoilé
d Fleckflügelwollrücken
i Averla formichiera alimacchiattate

8635 Pygochelidon cyanoleuca
Passeriformes - Hirundinidae
e Blue-and-white Swallow
f Hirondelle bleue-et-blanche
d Schwarzsteißschwalbe
i Rondine bianca e blu

8636 Pygochelidon patagonica
Passeriformes - Hirundinidae
e Patagonian Swallow

8637 Pygoscelis adeliae
Sphenisciformes - Spheniscidae
e Adelie's Penguin
f Manchot d'Adélie
d Adelie-Pinguin
i Pigoscelide di Adelia; Pinguino di Adelia

8638   *Pygoscelis antarctica*
Sphenisciformes - Spheniscidae
e   Chinstrap Penguin; Bearded Penguin; Ringed Penguin; Antarctic Penguin
f   Manchot à jugulaire; Manchot jugulaire
d   Kehlstreifpinguin; Zügelpinguin
i   Pigoscelide dell'Antartide; Pinguino dell'Antartide; Pinguino antartico

8639   *Pygoscelis papua*
Sphenisciformes - Spheniscidae
e   Gentoo Penguin
f   Manchot papou
d   Eselspinguin
i   Pigoscelide papua; Pinguino papua

8640   *Pygoscelis papua papua*
Sphenisciformes - Spheniscidae
e   Subantarctic Gentoo Penguin (ANZ)

8641   *Pyrenestes minor*
Passeriformes - Estrildidae
e   Lesser Seedcracker; Nyasa Seedcracker (CSA); Nyasaland Seedcracker
f   Petit Pyréneste; Petit bec-poinceau; Petit Grosbec à ponceau
d   Kleiner Purpurastrild
i   Pireneste minore

8642   *Pyrenestes ostrinus*
Passeriformes - Estrildidae
e   Black-bellied Seedcracker; Rothschild's Seedcracker; Large Seedcracker; Notch-billed Seedcracker
f   Pyréneste ponceau; Grosbec poinceau; Grosbec ponceau à ventre noir
d   Purpurastrild
i   Pireneste ventrenero

8643   *Pyrenestes sanguineus*
Passeriformes - Estrildidae
e   Crimson Seedcracker
f   Pyréneste gros-bec; Grosbec poinceau à ventre brun
d   Karmesinastrild
i   Pireneste rosso

8644   *Pyriglena atra*
Passeriformes - Thamnophilidae
e   Fringe-backed Fire-eye; Black Fire-eye; Swainson's Fire-eye
f   Alapi noir
d   Fleckenmantelfeuerauge
i   Occhiodifuoco di Swainson

8645   *Pyriglena leuconota*
Passeriformes - Thamnophilidae
e   White-backed Fire-eye
f   Alapi à dos blanc
d   Weißrückenfeuerauge
i   Occhiodifuoco dorsobianco

8646   *Pyriglena leucoptera*
Passeriformes - Thamnophilidae
e   White-shouldered Fire-eye
f   Alapi demoiselle
d   Weißschulterfeuerauge; Feuerauge
l   Occhiodifuoco alibianche

8647   *Pyrocephalus nanus*
Passeriformes - Tyrannidae
e   Galapagos Flycatcher
f   Moucherolle des Galapagos

8648   *Pyrocephalus rubinus*
Passeriformes - Tyrannidae
e   Vermilion Flycatcher
f   Moucherolle vermillon
d   Purpurtyrann
i   Tiranno vermiglio

8649   *Pyroderus scutatus*
Passeriformes - Tyrannidae
e   Red-ruffed Fruitcrow
f   Coracine ignite
d   Pavao; Schildvogel; Krähenschmuckvogel
i   Cotinga pettirossa

8650   *Pyrrhocoma ruficeps*
Passeriformes - Fringillidae
e   Chestnut-headed Tanager
f   Tangara à tête marron
d   Zimtkopftangare
i   Tangara testacastana

8651   *Pyrrhocorax graculus*
Passeriformes - Corvidae

e   Yellow-billed Chough; Yellow-bellied Chough; Alpine Chough; Mountain Chough
f   Chocard à bec jaune
d   Alpendohle
i   Gracchio alpino; Gracchio; Gracculo

8652   *Pyrrhocorax pyrrhocorax*
Passeriformes - Corvidae
e   Red-billed Chough; Chough; Cornish Chough; Red-legged Crow; Fire Crow; Keg; Keeog; Common Chough
f   Crave à bec rouge
d   Alpenkrähe
i   Gracchio corallino

8653   *Pyrrholaemus brunneus*
Passeriformes - Pardalotidae
e   Redthroat
f   Séricorne rougegorge
d   Dornhuscher; Australrotkehlchen
i   Sericeo golarossa

8654   *Pyrrhomyias cinnamomea*
Passeriformes - Tyrannidae
e   Cinnamon Flycatcher
f   Moucherolle cannellée; Moucherolle fauve
d   Zimttyrann
i   Tiranno cannella

8655   *Pyrrhoplectes epauletta*
Passeriformes - Fringillidae
e   Gold-naped Finch; Gold-headed Finch; Epauletted Finch; Gold-headed Black-Finch; Gold-crowned Black-Finch; Epaulet Finch
f   Pyrrhoplecte à nuque d'or; Bouvreuil à couronne d'or
d   Mohrengimpel
i   Frosone nucadorata

8656   *Pyrrhula aurantiaca*
Passeriformes - Fringillidae
e   Orange Bullfinch
f   Bouvreuil orangé
d   Goldrückengimpel; Orangegimpel
i   Ciuffolotto arancio

8657   *Pyrrhula cineracea*
Passeriformes - Fringillidae
e   Baikal Bulfinch

8658   *Pyrrhula erythaca*
Passeriformes - Fringillidae
e   Grey-headed Bullfinch; Beavan's Bullfinch
f   Bouvreuil à tête grise
d   Maskengimpel; China-Maskengimpel
i   Ciuffolotto di Beavan

8659   *Pyrrhula erythrocephala*
Passeriformes - Fringillidae
e   Red-headed Bullfinch
f   Bouvreuil à tête rouge
d   Rotkopfgimpel
i   Ciuffolotto testarossa

8660   *Pyrrhula leucogenis*
Passeriformes - Fringillidae
e   White-cheeked Bullfinch; Philippine Bullfinch
f   Bouvreuil des Philippines
d   Weißwangengimpel
i   Ciuffolotto delle Filippine

8661   *Pyrrhula murina*
Passeriformes - Fringillidae
e   Azores Bullfinch
f   Bouvreuil des Açores
d   Azoren-Gimpel
i   Ciuffolotto delle Azzorre

8662   *Pyrrhula nipalensis*
Passeriformes - Fringillidae
e   Brown Bullfinch
f   Bouvreuil brun
d   Schuppenkopfgimpel
i   Ciuffolotto bruno

8663   *Pyrrhula pyrrhula*
Passeriformes - Fringillidae
e   Eurasian Bullfinch (NA); Bullfinch; Common Bullfinch; Northern Bullfinch; Baikal Bullfinch; Grey Bullfinch; Grey-bellied Bullfinch; Oriental Bullfinch; Japanese Bullfinch
f   Bouvreuil pivoine; Bouvreuil commun; Bouvreuil; Bouvreuil

ponceau; Bouvreuil vulgaire
d   Gimpel; Dompfaff; Blutfink
i   Ciuffolotto; Fringuello marino; Monachine; Borgognone; Ciuffolotto comune

8664   *Pyrrhura albipectus*
      Psittaciformes - Psittacidae
e   White-necked Parakeet; White-breasted Parakeet; White-throated Conure; White-necked Conure; White-breasted Conure
f   Conure à col blanc; Perruche à col blanc
d   Weißhalssittich; Weißbrustsittich
i   Conuro collobianco

8665   *Pyrrhura calliptera*
      Psittaciformes - Psittacidae
e   Brown-breasted Parakeet; Brown-breasted Conure; Flame-winged Parakeet; Flame-winged Conure; Beautiful Conure; Brown-backed Parakeet
f   Conure à poitrine brune; Perruche à poitrine brune
d   Braunbrustsittich; Prachtflügelsittich
i   Conuro pettobruno

8666   *Pyrrhura cruentata*
      Psittaciformes - Psittacidae
e   Blue-throated Parakeet; Ochre-marked Parakeet; Red-rumped Parakeet; Blue-throated Conure; Black-tailed Conure; Red-rumped Conure
f   Conure tiriba; Perruche tiriba
d   Blaulatzsittich; Rohrsittich
i   Conuro golablu

8667   *Pyrrhura devillei*
      Psittaciformes - Psittacidae
e   Blaze-winged Parakeet; Blaze-winged Conure; Bolivia Conure; Deville's Conure
f   Conure de Deville; Perruche de Deville
d   Deville-Sittich; Bolivia-Rotschwanzsittich; Bolivien-Rotschwanzsittich
i   Conuro di Deville

8668   *Pyrrhura egregia*
      Psittaciformes - Psittacidae
e   Fiery-shouldered Parakeet; Fiery-shouldered Conure; Demerara Conure
f   Conure aile-de-feu; Perruche aile-de-feu; Perruche aile-de-feu de Demara
d   Demerara-Sittich; Feuerbugsittich
i   Conuro spallerosse

8669   *Pyrrhura frontalis*
      Psittaciformes - Psittacidae
e   Maroon-bellied Parakeet; Reddish-bellied Parakeet; Red-bellied Conure; Maroon-bellied Conure
f   Conure de Vieillot; Perruche de Vieillot; Perruche à bandeau
d   Braunohrsittich
i   Conuro ventrecastano

8670   *Pyrrhura haematotis*
      Psittaciformes - Psittacidae
e   Red-eared Parakeet; Red-eared Conure; Blood-eared Parakeet
f   Conure à oreillons; Perruche à oreillons
d   Blutohrsittich; Blutohrrotschwanzsittich
i   Conuro orecchierosse

8671   *Pyrrhura hoffmanni*
      Psittaciformes - Psittacidae
e   Sulphur-winged Parakeet; Hoffmann's Conure; Hoffmann's Parakeet; Sulphur-winged Conure
f   Conure de Hoffmann; Perruche de Hoffmann
d   Gelbgrüner Rotschwanzsittich; Hoffmann-Sittich
i   Conuro di Hoffmann

8672   *Pyrrhura lepida*
      Psittaciformes - Psittacidae
e   Pearly Parakeet

8673   *Pyrrhura leucotis*
      Psittaciformes - Psittacidae
e   White-eared Parakeet; Maroon-faced Parakeet; White-eared Conure; Maroon-faced Conure
f   Conure emma; Perruche emma;

Perruche à oreillons blancs
d Weißohrsittich
i Conuro orecchiebianche

8674 ***Pyrrhura melanura***
Psittaciformes - Psittacidae
e Maroon-tailed Parakeet; Berlepsch's Parakeet; Maroon-tailed Conure; Berlepsch's Conure; Black-tailed Conure; Magdalena Parakeet
f Conure de Souancé; Perruche de Souancé; Perruche de Berlepsch
d Braunschwanzsittich; Schwarzschwanzsittich
i Conuro codabruna

8675 ***Pyrrhura molinae***
Psittaciformes - Psittacidae
e Green-cheeked Parakeet; Molina's Conure; Green-cheeked Conure; Yellow-sided Parrot
f Conure de Molina; Perruche de Molina
d Grünwangenrotschwanzsittich; Molina-Sittich
i Conuro guanceverdi

8676 ***Pyrrhura orcesi***
Psittaciformes - Psittacidae
e El Oro Parakeet; El Oro Conure
f Conure d'Orces
i Conuro di El Oro

8677 ***Pyrrhura perlata***
Psittaciformes - Psittacidae
e Crimson-bellied Parakeet; Pearly Parakeet; Pearly Conure; Crimson-bellied Conure
f Conure perlée; Perruche perlée
d Blauwangensittich; Blausteißsittich
i Conuro perlato

8678 ***Pyrrhura picta***
Psittaciformes - Psittacidae
e Painted Parakeet; Painted Conure; Blue-winged Conure
f Conure versicolore; Perruche versicolore; Perruche à gorge variée
d Rotzügelsittich; Blaustirnrotschwanzsittich;

Blaunackensittich
i Conuro variopinto

8679 ***Pyrrhura rhodocephala***
Psittaciformes - Psittacidae
e Rose-headed Parakeet; Rose-headed Conure; Rose-crowned Conure; Rose-crowned Parakeet
f Conure tête-de-feu; Perruche tête-de-feu; Perruche à tête rouge
d Rotkopfsittich
i Conuro corona rosa

8680 ***Pyrrhura rhodogaster***
Psittaciformes - Psittacidae
e Crimson-bellied Parakeet; Crimson-breasted Conure; Crimson-bellied Conure
f Conure à ventre rouge; Perruche à ventre rouge; Perruche à poitrine cramoisie
d Rotbauchsittich
i Conuro ventrerosso

8681 ***Pyrrhura rupicola***
Psittaciformes - Psittacidae
e Black-capped Parakeet; Black-capped Conure; Rock Conure; Rock Parakeet
f Conure à cape noire; Perruche à cape noire
d Schwarzkappensittich; Steinsittich
i Conuro capinero

8682 ***Pyrrhura viridicata***
Psittaciformes - Psittacidae
e Santa Marta Parakeet; Santa Marta Conure
f Conure des Santa Marta; Perruche des Santa Marta; Perruche de Santa Marta; Conure de Santa Marta
d Santa Marta-Sittich; Grünlicher Rotschwanzsittich
i Conuro di Santa Marta

8683 ***Pytilia afra***
Passeriformes - Estrildidae
e Orange-winged Pytilia; Golden-backed Pytilia (CSA); Yellow-backed Pyrtilia; Red-faced Finch; Yellow-winged Pytilia

f   Beaumarquet à dos jaune; Pytilie à
     dos jaune; Astrild de Wiener
d   Wiener-Astrild; Wieners Astrild
i   Astro aliarancio

8684   *Pytilia hypogrammica*
     Passeriformes - Estrildidae
e   Red-faced Pytilia; Yellow-winged
     Pytilia; Golden-winged Pytilia; Red-
     faced Aurora-Waxbill; Aurora Finch;
     Red-faced Aurora-Finch
f   Beaumarquet à ailes jaunes; Pytilie à
     ailes jaunes
d   Rotmaskenastrild
i   Astro facciarossa

8685   *Pytilia lineata*
     Passeriformes - Estrildidae
e   Lineated Pytilia; Aurora Finch; Red-
     billed Pytilia
f   Beaumarquet à bec rouge
i   Astro lineato

8686   *Pytilia melba*
     Passeriformes - Estrildidae
e   Green-winged Pytilia; Melba Finch
     (CSA); Grey-naped Pytilia; Melba
     Waxbill; Crimson-faced Waxbill
f   Beaumarquet melba; Pytilie melba;
     Ptylie élégante
d   Buntastrild
i   Melba

8687   *Pytilia phoenicoptera*
     Passeriformes - Estrildidae
e   Red-winged Pytilia; Crimson-winged
     Pytilia; Aurora Waxbill; Aurora
     Finch; Crimson-winged Waxbill
f   Beaumarquet aurore; Pytilie à ailes
     rouges; Astrild aurore; Diamant
     aurore
d   Aurora-Astrild
i   Astro aurora

Q

8688 **Quelea cardinalis**
Passeriformes - Ploceidae
e Cardinal Quelea
f Travailleur cardinal
d Kardinalweber
i Quelea cardinale

8689 **Quelea erythrops**
Passeriformes - Ploceidae
e Red-headed Quelea; Red-headed
Dioch
f Travailleur à tête rouge; Quéléa à tête
rouge
d Rotkopfweber
i Quelea testarossa

8690 **Quelea quelea**
Passeriformes - Ploceidae
e Red-billed Quelea; Sudan Dioch;
Common Quelea; Quelea; Black-
faced Dioch; Red-billed Weaver;
Pink-billed Quelea; Sudan Quelea;
Black-faced Quelea
f Travailleur à bec rouge
d Blutschnabelweber
i Quelea; Lavatore dal becco rosso;
Quelea beccorosso

8691 **Querula purpurata**
Passeriformes - Cotingidae
e Purple-throated Fruitcrow
f Coracine noire
d Pioho; Schildschmuckvogel
i Cotinga golapurpurea

8692 **Quiscalus lugubris**
Passeriformes - Icteridae
e Carib Grackle; Lesser Antillean
Grackle; Blackbird (WI); Merle
(WI); Bequia Sweet (WI)
f Quiscale merle; Bilbitin (Ants);
Crédit (Ants); Merle (Ants);
Cancangnan (Ants)

d Trauergrackel
i Gracchio delle Antille

8693 **Quiscalus major**
Passeriformes - Icteridae
e Boat-tailed Grackle
f Quiscale des marais
d Bootschwanzgrackel; Großer
Bootschwanz
i Gracchio maggiore americano

8694 **Quiscalus mexicanus**
Passeriformes - Icteridae
e Great-tailed Grackle
f Quiscale à longue queue; Grand
Quiscale
d Dohlengrackel
i Gracchio messicano

8695 **Quiscalus nicaraguensis**
Passeriformes - Icteridae
e Nicaraguan Grackle; Nicaragua
Grackle
f Quiscale du Nicaragua
d Nicaragua-Grackel
i Gracchio del Nicaragua

8696 **Quiscalus niger**
Passeriformes - Icteridae
e Greater Antillean Grackle; Ching-
ching (WI); Cling-cling (WI);
Antillean Grackle
f Merle diable; Quiscale noir
d Antillen-Grackel; Weißaugengrackel
i Gracchio delle Antille

8697 **Quiscalus palustris**
Passeriformes - Icteridae
e Slender-billed Grackle
f Quiscale de Mexico
d Sumpfgrackel
i Gracchio beccosottile

8698 **Quiscalus quiscula**
Passeriformes - Icteridae
e Common Grackle; Purple Grackle;
Grackle; Bronzed Grackle
f Quiscale bronzé; Mainate bronzé
d Purpurgrackel
i Gracchio comune americano

8699   *Quiscalus versicolor*
           Passeriformes - Icteridae
  e   Bronzed Grackle

R

8700 Rallidae
Gruiformes
e Rails; Crakes
f Rallidés
d Rallen; Schilfschlüpfer
i Rallidi; Ralli

8701 Rallina canningi
Gruiformes - Rallidae
e Andaman Crake; Andaman Banded-Rail; Andaman Banded-Crake
f Râle des Andaman
d Andamanen-Ralle
i Rallina delle Andamane

8702 Rallina eurizonoides
Gruiformes - Rallidae
e Slaty-legged Crake; Slaty-legged Banded-Crake (ISC); Slaty-legged Banded-Rail; Philippine Banded-Crake; Banded Crake; Ryukyu Banded-Crake
f Râle de forêt
d Hindu-Ralle
i Rallina zampeardesia

8703 Rallina fasciata
Gruiformes - Rallidae
e Red-legged Crake; Malay Banded-Rail; Malay Banded-Crake; Malaysian Banded-Crake; Red-legged Banded-Crake
f Râle barré
d Malaien-Ralle
i Rallina fasciata

8704 Rallina forbesi
Gruiformes - Rallidae
e Forbes's Forest-Rail; Forbes's Chestnut Rail; Red-backed Forest-Rail
f Râle de Forbes
d Nymphenralle
i Rallina di Forbes

8705 Rallina leucospila
Gruiformes - Rallidae
e White-striped Forest-Rail; White-striped Chestnut Rail
f Râle vergeté
d Strichelralle
i Rallina striata

8706 Rallina mayri
Gruiformes - Rallidae
e Mayr's Forest-Rail; Mayr's Chestnut Rail
f Râle de Mayr
d Zyklopenralle
i Rallina di Mayr

8707 Rallina rubra
Gruiformes - Rallidae
e Chestnut Forest-Rail; New Guinea Chestnut Rail
f Râle marron; Râle roux
d Kastanienralle
i Rallina castana

8708 Rallina tricolor
Gruiformes - Rallidae
e Red-necked Crake; Red-necked Rail
f Râle tricolore
d Dreifarbenralle
i Rallina collorosso

8709 Rallus antarcticus
Gruiformes - Rallidae
e Austral Rail
f Râle austral
d Magellan-Ralle
i Porciglione australe

8710 Rallus aquaticus
Gruiformes - Rallidae
e Water Rail; European Water-Rail; Indian Water-Rail
f Râle d'eau
d Wasserralle
i Porciglione; Gallinella; Porciglione comune

8711   *Rallus caerulescens*
         Gruiformes - Rallidae
e   Kaffir Rail; African Rail (CSA);
         African Water-Rail; Cape Rail
f   Râle bleuâtre
d   Kap-Ralle
i   Porciglione africano

8712   *Rallus elegans*
         Gruiformes - Rallidae
e   King Rail
f   Râle élégant
d   Königsralle
i   Porciglione reale

8713   *Rallus limicola*
         Gruiformes - Rallidae
e   Virginia Rail
f   Râle de Virginie
d   Virginia-Ralle
i   Porciglione della Virginia

8714   *Rallus longirostris*
         Gruiformes - Rallidae
e   Clapper Rail; Marsh Hen (WI);
         Mangrove Hen (WI); Pond Shakee
         (WI); Mexican Rail
f   Râle gris; Rateau (Ants); Râle d'eau
         (Ants); Pintade (Ants)
d   Klapperralle
i   Porciglione americano

8715   *Rallus madagascariensis*
         Gruiformes - Rallidae
e   Madagascar Rail
f   Râle de Madagascar
d   Madagaskar-Ralle
i   Porciglione del Madagascar

8716   *Rallus pectoralis clellandi*
         Gruiformes - Rallidae
e   Western Australian Lewins-Rail
         (ANZ)

8717   *Rallus pectoralis pectoralis*
         Gruiformes - Rallidae
e   Eastern Lewin's-Rail (ANZ)

8718   *Rallus semiplumbeus*
         Gruiformes - Rallidae
e   Bogota Rail

f   Râle de Bogota
d   Bogota-Ralle
i   Porciglione di Bogotà

8719   *Rallus wetmorei*
         Gruiformes - Rallidae
e   Plain-flanked Rail; Wetmore's Rail
f   Râle de Wetmore
d   Wetmore-Ralle
i   Porciglione di Wetmore

8720   **Ramphastidae**
         Piciformes
e   Toucans
f   Rhamphastidés
d   Tukane; Pfefferfresser
i   Ramfastidi

8721   *Ramphastos ambiguus*
         Piciformes - Ramphastidae
e   Black-mandibled Toucan; Yellow-
         breasted Toucan
f   Toucan tocard
d   Goldkehltukan
i   Tucano ambiguo

8722   *Ramphastos brevis*
         Piciformes - Ramphastidae
e   Choco Toucan
f   Toucan du Choco
d   Choko-Tyrann
i   Tucano cioco

8723   *Ramphastos citreolaemus*
         Piciformes - Ramphastidae
e   Citron-throated Toucan
f   Toucan à gorge citron
i   Tucano golagialla

8724   *Ramphastos culminatus*
         Piciformes - Ramphastidae
e   Yellow-ridged Toucan
f   Toucan à culmen jaune
i   Tucano crestagialla

8725   *Ramphastos cuvieri*
         Piciformes - Ramphastidae
e   Cuvier's Toucan; Red-billed Toucan;
         White-throated Toucan; Inca Toucan
f   Toucan de Cuvier; Toucan à bec
         rouge

d   Cuvier-Tukan; Weißbrusttukan
i   Tucano di Cuvier

8726   *Ramphastos dicolorus*
        Piciformes - Ramphastidae
e   Red-breasted Toucan; Green-billed
        Toucan; Keel-billed Toucan
f   Toucan à ventre rouge
d   Bunttukan
i   Tucano bicolore

8727   *Ramphastos sulfuratus*
        Piciformes - Ramphastidae
e   Keel-billed Toucan
f   Toucan à caréne
d   Regenbogen-Tukan; Fischer-Tukan
i   Tucano sulfureo; Tucano solforato

8728   *Ramphastos swainsonii*
        Piciformes - Ramphastidae
e   Chestnut-mandibled Toucan;
        Yellow-breasted Toucan; Swainson's
        Toucan
f   Toucan de Swainson
d   Braunrückentukan
i   Tucano di Swainson

8729   *Ramphastos toco*
        Piciformes - Ramphastidae
e   Toco Toucan
f   Toucan toco
d   Riesentukan
i   Toco; Tucano toco

8730   *Ramphastos tucanus*
        Piciformes - Ramphastidae
e   Red-billed Toucan; White-throated
        Toucan
d   Weißbrusttukan
i   Tucano

8731   *Ramphastos vitellinus*
        Piciformes - Ramphastidae
e   Channel-billed Toucan
f   Toucan ariel
d   Dottertukan
i   Tucano beccoscanalato

8732   *Ramphastos vitellinus*
        aurantiirostris
        Piciformes - Ramphastidae

e   Orange-billed Toucan
f   Toucan à bec orangé

8733   *Ramphocaenus melanurus*
        Passeriformes - Muscicapidae
e   Long-billed Gnatwren; Black-tailed
        Gnatwren; Straight-billed Gnatwren
f   Microbate à long bec
d   Schwarzschwanzdegensäbler
i   Zanzariere beccolungo

8734   *Ramphocaenus rufiventris*
        Passeriformes - Muscicapidae
e   Long-billed Gnatwren

8735   *Ramphocelus bresilius*
        Passeriformes - Thraupidae
e   Brazilian Tanager
f   Tangara du Brésil
i   Tangara brasiliana

8736   *Ramphocelus carbo*
        Passeriformes - Thraupidae
e   Silver-beaked Tanager; Common
        Silverbeak-Tanager; Common
        Maroon Tanager
f   Tangara à bec d'argent
d   Silberschnabeltangare; Purpurtangare
i   Tangara beccoargentato

8737   *Ramphocelus dimidiatus*
        Passeriformes - Thraupidae
e   Crimson-backed Tanager
f   Tangara à dos rouge
d   Scharlachbauchtangare
i   Tangara dorsocremisi

8738   *Ramphocelus dominicensis*
        Passeriformes - Thraupidae
e   Hispaniolan Tanager; Hispaniolan
        Spindalis; Hispaniolan Stripe-headed
        Tanager

8739   *Ramphocelus flammigerus*
        Passeriformes - Thraupidae
e   Flame-rumped Tanager; Lemon-
        rumped Tanager; Yellow-rumped
        Tanager
f   Tangara flamboyant
d   Feuerbürzeltangare
i   Tangara dal groppone di fiamma

8740   *Ramphocelus icteronotus*
        Passeriformes - Thraupidae
e   Yellow-rumped Tanager
f   Tangara à dos citron
d   Gelbbürzeltangare

8741   *Ramphocelus melanogaster*
        Passeriformes - Thraupidae
e   Huallaga Tanager; Black-bellied
     Tanager
f   Tangara du Huallaga
i   Tangara ventrenero

8742   *Ramphocelus nigricephalus*
        Passeriformes - Thraupidae
e   Jamaican Tanager; Jamaican
     Spindalis; Jamaican Stripe-headed
     Tanager; Mark-head (WI);
     Cashewbird (WI); Goldfinch (WI);
     Orangebird (WI); Silver-head (WI);
     Spanish Quail (WI); Yam-cutter (WI)

8743   *Ramphocelus nigrogularis*
        Passeriformes - Thraupidae
e   Masked Crimson-Tanager; Masked
     Common Tanager; Masked Tanager;
     Crimson Tanager
f   Tangara masqué
d   Maskentangare
i   Tangara cremisi mascherata

8744   *Ramphocelus passerinii*
        Passeriformes - Thraupidae
e   Scarlet-rumped Tanager; Song
     Tanager; Passerini's Tanager
f   Tangara à croupion rouge
d   Passerin-Tangare; Rotbürzeltangare
i   Tangara grupponescarlatto

8745   *Ramphocelus sanguinolentus*
        Passeriformes - Thraupidae
e   Crimson-collared Tanager
f   Tangara ceinturé
d   Flammentangare; Halsbandtangare
i   Tangara dal collare cremisi

8746   *Ramphocinclus brachyurus*
        Passeriformes - Sturnidae
e   White-breasted Thrasher; White-
     breasted Trembler
f   Moqueur gorge-blanche; Gorge-

blanche (Ants); Gorge-blanc (Ants)
d   Weißbrustspottdrossel
i   Mimo codacorta

8747   *Ramphocoris clotbey*
        Passeriformes - Alaudidae
e   Thick-billed Lark; Clotbey Lark;
     Clot Bey's Lark
f   Alouette de Clotbey
d   Knackerlerche
i   Allodola beccaforte

8748   *Ramphodon dohrnii*
        Apodiformes - Trochilidae
e   Hook-billed Hermit
f   Colibri de Dohrn; Ermite de Dohrn
d   Bronzeschwanzeremit
i   Eremita becco a uncino

8749   *Ramphodon naevius*
        Apodiformes - Trochilidae
e   Saw-billed Hermit
f   Colibri tacheté; Ermite tacheté
d   Sägeschnabeleremit
i   Eremita becco a sega

8750   *Ramphomicron dorsale*
        Apodiformes - Trochilidae
e   Black-backed Thornbill; Black-
     backed-Thornbilled Hummingbird
f   Colibri à dos noir
d   Schwarzrückenkolibri
i   Becco a spina dorsonero

8751   *Ramphomicron microrhynchum*
        Apodiformes - Trochilidae
e   Purple-backed Thornbill; Purple-
     backed Thornbilled-Hummingbird
f   Colibri à petit bec
d   Kleinschnabelkolibri
i   Becco a spina dorsopurpureo

8752   *Ramphotrigon fuscicauda*
        Passeriformes - Tyrannidae
e   Dusky-tailed Flatbill
f   Platyrhynque à queue sombre;
     Tyranneau à queue sombre
d   Chapman-Breitschnabel
i   Beccopiatto codabruna

8753 **Ramphotrigon megacephala**
Passeriformes - Tyrannidae
e Large-headed Flatbill
f Platyrhynque à grosse tête;
Tyranneau à grosse tête
d Brauenbreitschnabel
i Beccopiatto testagrossa

8754 **Ramphotrigon ruficauda**
Passeriformes - Tyrannidae
e Rufous-tailed Flatbill
f Platyrhynque à queue rousse;
Tyranneau à queue rousse
d Rotschwanzbreitschnabel
i Beccopiatto codirosso

8755 **Ramsayornis fasciatus**
Passeriformes - Meliphagidae
e Bar-breasted Honeyeater; White-
breasted Honeyeater; Fasciated
Honeyeater
f Ramsayornis fasciatus; Méliphage
fascié
d Wellenbrusthonigfresseer
i Mangiamiele pettobarrato

8756 **Ramsayornis modestus**
Passeriformes - Meliphagidae
e Brown-backed Honeyeater; Modest
Honeyeater
f Ramsayornis modestus; Méliphage
modeste
d Sumpfhonigfresser
i Mangiamiele dorsobruno

8757 **Randia pseudozosterops**
Passeriformes - Sylviidae
e Rand's Warbler; Maroansetra
Warbler
f Randie malgache; Fauvette deRand
d Randia
i Randia

8758 **Raphus cucullatus**
Columbiformes - Raphidae
e Dodo
f Dronte de Maurice
d Dronte
i Dodo

8759 **Raphus solitarius**
Columbiformes - Raphidae
e Réunion Solitaire; White Dodo
f Dronte de la Réunion
d Réunion-Dronte
i Solitario di Riunione; Solitario
dell'Isola Reunion

8760 **Recurvirostra americana**
Charadriiformes - Recurvirostridae
e American Avocet
f Avocette d'Amérique
d Braunhalssäbelschnäbler;
Amerikanischer Säbelschnäbler
i Avocetta americana

8761 **Recurvirostra andina**
Charadriiformes - Charadriidae
e Andean Avocet
f Avocette des Andes
d Anden-Säbelschäbler
i Avocetta delle Ande

8762 **Recurvirostra avosetta**
Charadriiformes - Recurvirostridae
e Pied Avocet; Avocet; Eurasian
Avocet; Black-capped Avocet
f Avocette élégante; Avocette à tête
noire
d Säbelschnäbler; Eigentlicher
Säbelschnäbler
i Avocetta; Avocetta eurasiatica

8763 **Recurvirostra novaehollandiae**
Charadriiformes - Recurvirostridae
e Red-necked Avocet; Australian Red-
necked Avocet; Australian Avocet
f Avocette d'Australie
d Rotkopfsäbelschnäbler
i Avocetta australiana

8764 **Recurvirostridae**
Charadriiformes -
e Stilts; Avocets
f Échasses; Recurvirostridés
d Stelzenläufer; Säbelschnäbler
i Recurvirostridi

8765 **Regulidae**
Passeriformes
e Kinglets

f   Roitelets; Régulidés
d   Goldhähnchen

8766   *Regulus calendula*
      Passeriformes - Regulidae
e   Ruby-crowned Kinglet
f   Roitelet à couronne rubis; Roitelet
      rubis
d   Rubingoldhähnchen; Rotkrönchen
i   Regolo americano

8767   *Regulus cuvierei*
      Passeriformes - Regulidae
e   Cuvier's Kinglet

8768   *Regulus goodfellowi*
      Passeriformes - Regulidae
e   Flamecrest; Taiwan Kinglet;
      Formosan Kinglet; Taiwan Firecrest;
      Formosan Firecrest
f   Roitelet de Taïwan
d   Formosa-Goldhähnchen
i   Firrancino di Taiwan

8769   *Regulus ignicapillus*
      Passeriformes - Regulidae
e   Firecrest; Eurasian Firecrest
f   Roitelet à triple-bandeau; Roitelet
      triple-bandeau; Roitelet à moustaches
d   Sommergoldhähnchen;
      Augenstreifgoldhähnchen
i   Fiorrancino; Fiorrancino eurasiatico

8770   *Regulus regulus*
      Passeriformes - Regulidae
e   Goldcrest; Common Goldcrest (ISC);
      European Goldcrest; Gold-crested
      Wren
f   Roitelet huppé
d   Wintergoldhähnchen
i   Regolo; Fiorrancino; Fiorrancio;
      Arancino; Regolo eurasiatico

8771   *Regulus satrapa*
      Passeriformes - Regulidae
e   Golden-crowned Kinglet
f   Roitelet à couronne dorée
d   Teneriffa-Goldhähnchen; Satrap
      Goldhähnchen; Satrap
i   Fiorrancino americano

8772   *Regulus teneriffae*
      Passeriformes - Regulidae
e   Canary Islands Kinglet; Canary
      Island Goldcrest; Tenerife Kinglet;
      Tenerife Goldcrest; Orangecrest;
      Tenerifean Goldcrest
f   Roitelet de Ténérife
d   Kanarien-Goldhähnchen
i   Regolo di Tenerife

8773   *Reinwardtipicus validus*
      Piciformes - Picidae
e   Orange-backed Woodpecker
f   Pic vigoureux
d   Reinwardt-Specht;
      Orangerückenspecht
i   Picchio di Reinwardt

8774   *Reinwardtoena browni*
      Columbiformes - Columbidae
e   Pied Cuckoo-Dove; Brown's Long-
      tailed Pigeon; Black-and-white Long-
      tailed Pigeon; Black-and-grey Long-
      tailed Pigeon
f   Phasianelle de Brown
d   Schwarztaube; Schwarze Reinwardt-
      Taube
i   Piccione codalunga di Brown

8775   *Reinwardtoena crassirostris*
      Columbiformes - Columbidae
e   Crested Cuckoo-Dove; Crested
      Long-tailed Pigeon; Crested Pigeon
f   Phasianelle huppée
d   Helmtaube; Salomonen-Schopftaube
i   Piccione codalunga crestata

8776   *Reinwardtoena reinwardtii*
      Columbiformes - Columbidae
e   Great Cuckoo-Dove; Reinwardt's
      Long-tailed Pigeon; Long-tailed
      Cuckoo-Dove; Chestnut-and-grey
      Pigeon; Maroon-and-grey Pigeon
f   Phasianelle de Reinwardt
d   Reinwardt-Taube; Rotbraune
      Reinwardt-Taube
i   Piccione codalunga di Reinwardt

8777   *Remiz consobrinus*
      Passeriformes - Paridae
e   Chinese Penduline-Tit; Eastern

Penduline-Tit
f Rémiz de Chine
i Pendolino della Cina

8778 *Remiz coronatus*
Passeriformes - Paridae
e White-crowned Penduline-Tit;
Black-headed Penduline-Tit
f Rémiz couronnée
i Pendolino coronato

8779 *Remiz macronyx*
Passeriformes - Paridae
e Black-headed Penduline-Tit

8780 *Remiz pendulinus*
Passeriformes - Paridae
e Eurasian Penduline-Tit (NA);
Penduline Tit; Common Penduline-
Tit; Masked Penduline-Tit; White-
throated Penduline-Tit
f Mésange rémiz; Rémiz penduline
d Beutelmeise
i Pendolino eurasiatico; Pendolino

8781 *Rhabdornis grandis*
Passeriformes - Sylviidae
e Long-billed Rhabdornis
f Rhabdornis à long bec
i Rampichino beccolungo

8782 *Rhabdornis inornatus*
Passeriformes - Sylviidae
e Stripe-breasted Rhabdornis; Plain-
headed Creeper; Plain-headed
Rhabdornis
f Rhabdornis à tête brune
d Braunkopfrhabdornis;
Braunkopfbaumläufer
i Rampichino disadorno

8783 *Rhabdornis mystacalis*
Passeriformes - Sylviidae
e Stripe-sided Rhabdornis; Stripe-
headed Creeper; Striped-headed
Creeper; Striped-headed Rhabdornis
f Rhabdornis à tête striée
d Streifenkopfrhabdornis;
Streifenkopfbaumläufer;
Bartbaumläufer
i Rampichino fianchistriati

8784 *Rhagologus leucostigma*
Passeriformes - Corvidae
e Mottled Whistler; Red-vented
Whistler
f Rhagologue maillé
d Wellendickkopf
i Zufolatore macchiettato

8785 *Rhamphomantis megarhynchus*
Cuculiformes - Cuculidae
e Long-billed Cuckoo; Little Long-
billed Cuckoo; Little Koel
f Coucou à long bec; Coucou
d Langschnabelkuckuck
i Cuculo beccolungo

8786 *Rhaphidura leucopygialis*
Apodiformes - Apodidae
e Silver-rumped Spinetail; Silver-
rumped Swift; Silver-rumped
Spinetailed-Swift; White-rumped
Spinetailed-Swift; White-rumped
Spinetail; Silver-rumped Needletail
f Martinet leucopyge
d Weißschwanzsegler
i Rondone codaspinosa dal groppone
bianco

8787 *Rhaphidura sabini*
Apodiformes - Apodidae
e Sabine's Spinetail; Sabine's
Spinetailed-Swift; Sabine's Swift
f Martinet de Sabine
d Sumpfsegler
i Rondone codaspinosa di Sabine

8788 *Rhea americana*
Rheiformes - Rheidae
e Greater Rhea; Greater Common-
Rhea
f Nandou d'Amérique; Nandou
commun; Nandou américain
d Nandu
i Nandù; Nandù comune

8789 *Rhea pennata*
Rheiformes - Rheidae
e Lesser Rhea; Darwin's Rhea; Puna
Rhea
f Nandou de Darwin

d   Darwin-Strauss
i   Nandù di Darwin

8790   *Rhegmatorhina berlepschi*
    Passeriformes - Thamnophilidae
e   Harlequin Antbird; Harlequin
    Antcatcher
f   Fourmilier arlequin
d   Darwin-Strauss
i   Mangiaformiche arlecchino

8791   *Rhegmatorhina cristata*
    Passeriformes - Thamnophilidae
e   Chestnut-crested Antbird; Chestnut-
    crested Antcatcher; Crested
    Antcatcher; Crested Antbird
f   Fourmilier à huppe marron
d   Rotschopfameisenvogel
i   Mangiaformiche crestacastana

8792   *Rhegmatorhina gymnops*
    Passeriformes - Thamnophilidae
e   Bare-eyed Antbird; Santarem
    Antbird; Bare-eyed Antcatcher;
    Sooty Antbird
f   Fourmilier fuligineux
d   Nacktaugenameisenvogel;
    Braunschopfameisenvogel
i   Mangiaformiche occhinudi

8793   *Rhegmatorhina hoffmannsi*
    Passeriformes - Thamnophilidae
e   White-breasted Antbird; White-
    breasted Antcatcher; Hoffmann's
    Antbird; Hoffmann's Antcatcher
f   Fourmilier à poitrine blanche
d   Hoffmanns Ameisenvogel
i   Mangiaformiche pettobianco

8794   *Rhegmatorhina melanosticta*
    Passeriformes - Thamnophilidae
e   Hairy-crested Antbird; Hairy-crested
    Antcatcher
f   Fourmilier chevelu
d   Grauschopfameisenvogel; Hoffmann-
    Ameisenvogel
i   Mangiaformiche crestapelosa

8795   **Rheidae**
    Rheiformes
e   Rheas

f   Rhéidés
d   Nandus; Fleckenrhegmatorhina
i   Reidi

8796   **Rheiformes**
e   Rheas
f   Rhéiformes
d   Nandus
i   Rheiformi

8797   *Rheinardia nigrescens*
    Galliformes - Phasianidae
e   Malaysian Argus
d   Rheinhart-Fasan

8798   *Rheinardia ocellata*
    Galliformes - Phasianidae
e   Crested Argus; Crested Argus-
    Pheasant; Rheinhardt's Argus;
    Ocellated Argus
f   Argus ocellé; Rheinarte ocellé
d   Perlenfasan
i   Argo ocellato

8799   *Rhinocrypta lanceolata*
    Passeriformes - Rhinocryptidae
e   Crested Gallito; Grey Gallito; Gallito
f   Tourco huppé
d   Schopfgallito; Strichelstelzer; Graues
    Hähnchen
i   Gallito crestato

8800   **Rhinocryptidae**
    Passeriformes
e   Tapaculos
f   Rhinocryptidés
d   Tapaculos; Buschschlüpfer;
    Bürzelstelzer
i   Rinocrittidi

8801   *Rhinomyias addita*
    Passeriformes - Muscicapidae
e   Streaky-breasted Jungle-Flycatcher;
    Buru Jungle-Flycatcher; Streak-
    breasted Jungle-Flycatcher; Streaky-
    breasted Flycatcher; Streak-breasted
    Rhinomyias
f   Gobemouche à gorge rayée; Gobe-
    mouches à gorge rayée
d   Buru-Dschungelschnäpper
i   Pigliamosche di Buru

8802 **_Rhinomyias albigularis_**
Passeriformes - Muscicapidae
e White-throated Jungle-Flycatcher;
Negros Jungle-Flycatcher; Negros
White-throated Jungle-Flycatcher
f Gobemouche de Negros; Gobe-
mouches de Negros
d Negros-Dschungelschnäpper
i Pigliamosche golabianca

8803 **_Rhinomyias brunneata_**
Passeriformes - Muscicapidae
e Brown-chested Jungle-Flycatcher;
White-gorgetted Jungle-Flycatcher;
Chinese Olive-Flycatcher; Chinese
White-gorgetted Flycatcher; Brown-
chested Flycatcher; Brown-chested
Rhinomyias; White-gorgetted
Flycatcher; Chinese Jungle-
Flycatcher; Olive Flycatcher;
Migratory Jungle-Flycatcher
f Gobemouche à poitrine brune; Gobe-
mouches à poitrine brune
d Weißkehldschungelschnäpper
i Pigliamosche bruno

8804 **_Rhinomyias colonus_**
Passeriformes - Muscicapidae
e Henna-tailed Jungle-Flycatcher;
Celebes Jungle-Flycatcher; Sula
Jungle-Flycatcher; Gleaning Jungle-
Flycatcher; Henna-tailed Rhinomyias
f Gobemouche à queue henné; Gobe-
mouches à queue henné
d Sula-Dschungelschnäpper
i Pigliamosche di Sula

8805 **_Rhinomyias goodfellowi_**
Passeriformes - Muscicapidae
e Slaty-backed Jungle-Flycatcher;
Goodfellow's Jungle-Flycatcher;
Mindanao Jungle-Flycatcher
f Gobemouche de Goodfellow; Gobe-
mouches de Goodfellow
d Graurückendschungelschnäpper
i Pigliamosche di Mindanao

8806 **_Rhinomyias gularis_**
Passeriformes - Muscicapidae
e Eyebrowed Jungle-Flycatcher;
White-browed Jungle-Flycatcher;

White-throated Jungle-Flycatcher;
Kinabalu Jungle-Flycatcher; White-
browed Rhinomyias
f Gobemouche bridé; Gobe-mouches
bridé
d Kinabalu-Dschungelschnäpper
i Pigliamosche delle Filippine

8807 **_Rhinomyias insignis_**
Passeriformes - Muscicapidae
e White-browed Jungle-Flycatcher;
Insignia Jungle-Flycatcher; Luzon
Jungle-Flycatcher; Rusty-flanked
Jungle-Flycatcher; Lepanto Jungle-
Flycatcher
f Gobemouche de Luçon; Gobe-
mouches de Luçon
d Rotflankendschungelschnäpper
i Pigliamosche di Lepanto

8808 **_Rhinomyias olivacea_**
Passeriformes - Muscicapidae
e Fulvous-chested Jungle-Flycatcher;
Olive-backed Jungle-Flycatcher;
Fulvous-chested Flycatcher; Fulvous-
chested Rhinomyias
f Gobemouche à dos olive; Gobe-
mouches à dos olive; Gobemouche
olive; Gobe-mouches olive
d Olivrückendschungelschnäpper;
Olivrückenmalaienschnäpper
i Pigliamosche dorso oliva

8809 **_Rhinomyias oscillans_**
Passeriformes - Muscicapidae
e Russet-backed Jungle-Flycatcher;
Active Flycatcher; Flores Jungle-
Flycatcher; Russet-backed
Flycatcher; Russet-backed
Rhinomyias
f Gobemouche de Florès; Gobe-
mouches de Florès
d Flores-Dschungelschnäpper
i Pigliamosche di flores

8810 **_Rhinomyias ruficauda_**
Passeriformes - Muscicapidae
e Rufous-tailed Jungle-Flycatcher;
Rufous-tailed Rhinomyias; Chestnut-
tailed Jungle-Flycatcher;
Greybreasted Jungle-Flycatcher

f Gobemouche à queue marron; Gobe-
    mouches à queue marron
d Rotschwanzdschungelschnäpper
i Pigliamosche codirosso

8811 *Rhinomyias umbratilis*
      Passeriformes - Muscicapidae
e Grey-chested Jungle-Flycatcher;
    White-throated Jungle-Flycatcher;
    Grey-chested Flycatcher; Grey
    chested Rhinomyias
f Gobemouche ombré; Gobe-mouches
    ombré
d Graubrustdschungelschnäpper
i Pigliamosche della giungla

8812 *Rhinopomastus aterrimus*
      Coraciiformes - Phoeniculidae
e Black Scimitarbill; Black Wood-
    Hoopoe; Lesser Wood-Hoopoe;
    Scimitar-billed Wood-Hoopoe
f Irrisor noir
i Upupa arboricola nera

8813 *Rhinopomastus cyanomelas*
      Coraciiformes - Phoeniculidae
e Common Scimitarbill; Scimitar-
    billed Wood-Hoopoe (CSA);
    Scimitar-billed Hoopoe; Scimitarbill
f Irrisor namaquois
d Sichelhopf; Baumhopf
i Upupa becco a scimitarra

8814 *Rhinopomastus minor*
      Coraciiformes - Phoeniculidae
e Abyssinian Scimitarbill; Abyssinian
    Scimitar-billed Hoopoe
f Irrisor à cimeterre
i Upupa becco a scimitarra
    dell'Abissinia

8815 *Rhinoptilus africanus*
      Ciconiiformes - Glareolidae
e Double-banded Courser; Two-
    banded Courser (CSA)
f Courvite à double collier
d Doppelbandrennvogel
i Corrione due fasce

8816 *Rhinoptilus bitorquatus*
      Ciconiiformes - Glareolidae

e Jerdon's Courser; Double-banded
    Courser (ISC)
f Courvite de Jerdon
d Goldavari-Rennvogel
i Corrione di Jerdon

8817 *Rhinoptilus chalcopterus*
      Ciconiiformes - Glareolidae
e Bronze-winged Courser; Violet-
    tipped Courser
f Courvite à ailes bronzées
d Bronzeflügelrennvogel;
    Amethystrennvogel
i Corrione alibronzate

8818 *Rhinoptilus cinctus*
      Ciconiiformes - Glareolidae
e Three-banded Courser; Heuglin's
    Courser (CSA); Treble-banded
    Courser; Seebohm's Courser
f Courvite à triple collier
d Bindenrennvogel
i Corrione di Heuglin

8819 *Rhipidura albicollis*
      Passeriformes - Corvidae
e White-throated Fantail; White-
    spotted Fantail-Flycatcher (ISC);
    White-throated Fantail-Flycatcher
f Rhipidure à gorge blanche
d Weißkehlfächerschwanz
i Coda a ventaglio golabianca

8820 *Rhipidura albogularis*
      Passeriformes - Corvidae
e Spot-breasted Fantail; White-spotted
    Fantail-Flycatcher; White-spotted
    Fantail
f Rhipidure à poitrine tachetée

8821 *Rhipidura albolimbata*
      Passeriformes - Corvidae
e Friendly Fantail; White-eared Fantail
f Rhipidure familier
d Graubauchfächerschwanz
i Coda a ventaglio orlata

8822 *Rhipidura atra*
      Passeriformes - Corvidae
e Black Fantail
f Rhipidure noir

d	Mohrenfächerschwanz		*d*	Ceram-Fächerschwanz
i	Coda a ventaglio nera e cannella		*i*	Coda a ventaglio dell'Isola Ceram

8823 *Rhipidura aureola*
Passeriformes - Corvidae
e White-browed Fantail; White-browed Fantail-Flycatcher; White-breasted Fantail-Flycatcher
f Rhipidure à grands sourcils
d Weißstirnfächerschwanz
i Coda a ventaglio coronata

8824 *Rhipidura brachyrhyncha*
Passeriformes - Corvidae
e Dimorphic Fantail; Dimorphic Rufous-Fantail; Rufous Fantail
f Rhipidure dimorphe
d Zweiphasenfächerschwanz
i Coda a ventaglio dorsocastano

8825 *Rhipidura cockerelli*
Passeriformes - Corvidae
e White-winged Fantail; Cockerell's Fantail; Lesser Pied-Fantail
f Rhipidure de Cockerell
d Cockerell-Fächerschwanz
i Coda a ventaglio di Cockerell

8826 *Rhipidura cyaniceps*
Passeriformes - Corvidae
e Blue-headed Fantail
f Rhipidure à tête bleue
d Blaukopffächerschwanz
i Coda a ventaglio testablu

8827 *Rhipidura dahli*
Passeriformes - Corvidae
e Bismarck Fantail; Island Fantail; Island Rufous-Fantail; New Britain Fantail; Dusky-throated Fantail
f Rhipidure des Bismarck
d Graukehlfächerschwanz
i Coda a ventaglio di Dahl

8828 *Rhipidura dedemi*
Passeriformes - Corvidae
e Streaky-breasted Fantail; Van Oort's Fantail; Ceram Rufous Fantail; Ceram Fantail; Streak-breasted Fantail
f Rhipidure de Céram

8829 *Rhipidura diluta*
Passeriformes - Corvidae
e Brown-capped Fantail
f Rhipidure à calotte brune
i Coda a ventaglio capobruno

8830 *Rhipidura drownei*
Passeriformes - Corvidae
e Brown Fantail; Mountain Fantail; Drowne's Fantail
f Rhipidure brun
d Bergfächerschwanz
i Coda a ventaglio di montagna

8831 *Rhipidura euryura*
Passeriformes - Corvidae
e White-bellied Fantail
f Rhipidure à ventre blanc
d Kurzfußfächerschwanz
i Coda a ventaglio ventrebianco

8832 *Rhipidura fuliginosa*
Passeriformes - Corvidae
e Grey Fantail; Collared Grey-Fantail; Grey Collared-Fantail; New Zealand Fantail
f Rhipidure à collier
d Graufächerschwanz
i Coda a ventaglio fuligginosa

8833 *Rhipidura fuliginosa cervinia*
Passeriformes - Corvidae
e Lord Howe Island Grey-Fantail (ANZ)

8834 *Rhipidura fuliginosa pelzelni*
Passeriformes - Corvidae
e Norfolk Island Grey-Fantail (ANZ)

8835 *Rhipidura fuscorufa*
Passeriformes - Corvidae
e Cinnamon-tailed Fantail
f Rhipidure brun-roux
i Coda a ventaglio cannella

8836 *Rhipidura hyperythra*
Passeriformes - Corvidae
e Chestnut-bellied Fantail

f   Rhipidure à ventre roux
d   Braunbauchfächerschwanz
i   Coda a ventaglio ventrecastano

8837   *Rhipidura hypoxantha*
Passeriformes - Corvidae
e   Yellow-bellied Fantail; Yellow-
bellied Fantail-Flycatcher
f   Rhipidure à ventre jaune
d   Goldbauchfächerschwanz
i   Coda a ventaglio ventregiallo

8838   *Rhipidura javanica*
Passeriformes - Corvidae
e   Pied Fantail; Pied Fantail-Flycatcher;
Malaysian Fantail
f   Rhipidure pie
d   Malaien-Fächerschwanz
i   Coda a ventaglio dal collare nero

8839   *Rhipidura kubaryi*
Passeriformes - Corvidae
e   Pohnpei Fantail; Ponapé Fantail
f   Rhipidure de Ponapé
i   Coda a ventaglio di Pohnpei

8840   *Rhipidura lepida*
Passeriformes - Corvidae
e   Palau Fantail; Palau Island Fantail;
Moluccan Fantail
f   Rhipidure des Palau
d   Palau-Fächerschwanz
i   Coda a ventaglio di Palau

8841   *Rhipidura leucophrys*
Passeriformes - Corvidae
e   Willie-Wagtail; Black-and-white
Fantail; White-browed Fantail
f   Rhipidure hochequeue
d   Gartenfächerschwanz;
Schwarzweißfächerschwanz
i   Coda a ventaglio ballerina

8842   *Rhipidura leucophrys melaleuca*
Passeriformes - Corvidae
e   Torres Strait Willie-Wagtail (ANZ)

8843   *Rhipidura leucothorax*
Passeriformes - Corvidae
e   White-bellied Thicket-Fantail;
White-breasted Fantail; White-

breasted Thicket-Fantail; White-
bellied Fantail; Black-throated
Thicket-Fantail
f   Rhipidure à poitrine blanche
d   Dickichtfächerschwanz
i   Coda a ventaglio pettobianco

8844   *Rhipidura maculipectus*
Passeriformes - Corvidae
e   Black Thicket-Fantail; Thicket-
Fantail
f   Rhipidure maculé
d   Sumpffächerschwanz
i   Coda a ventaglio pettomaculato

8845   *Rhipidura malaitae*
Passeriformes - Corvidae
e   Malaita Fantail; Malaita Rufous-
Fantail
f   Rhipidure de Malaita
d   Malaita-Fächerschwanz
i   Coda a ventaglio dell'Isola Malaita

8846   *Rhipidura matthiae*
Passeriformes - Corvidae
e   Matthias Fantail; St. Matthias
Fantail; St. Matthias Rufous-Fantail;
Heinroth's Fantail
f   Rhipidure des Saint-Matthias
d   Schwarzbrustfächerschwanz
i   Coda a ventaglio dell'Isola San
Matthias

8847   *Rhipidura nebulosa*
Passeriformes - Corvidae
e   Samoan Fantail
f   Rhipidure des Samoa
d   Samoa-Fächerschwanz
i   Coda a ventaglio di Samoa

8848   *Rhipidura nigrocinnamomea*
Passeriformes - Corvidae
e   Black-and-cinnamon Fantail
f   Rhipidure noir-et-roux
d   Mindanao-Fächerschwanz
i   Coda a ventaglio nera e cannella

8849   *Rhipidura opistherythra*
Passeriformes - Corvidae
e   Long-tailed Fantail; Tanimbar
Fantail; Tanimbar Rufous-Fantail;

Red-backed Fantail-Flycatcher
f Rhipidure des Tanimbar
d Tanimbar-Fächerschwanz
i Coda a ventaglio codalunga

8850 *Rhipidura perlata*
Passeriformes - Corvidae
e Spotted Fantail; Pearlated Fantail;
Spotted-Fantail-Flycatcher; Pearl-
spotted Fantail
f Rhipidure perlé
d Perlbrustfächerschwanz
i Coda a ventaglio perlata

8851 *Rhipidura personata*
Passeriformes - Corvidae
e Kadavu Fantail
f Rhipidure de Kandavu
d Kandavu-Fächerschwanz
i Coda a ventaglio dell'Isola Kadavu

8852 *Rhipidura phasiana*
Passeriformes - Corvidae
e Mangrove Fantail; Grey Mangrove-
Fantail (ANZ); Mangrove Grey-
Fantail
f Rhipidure des mangroves
i Coda a ventaglio delle mangrovie

8853 *Rhipidura phoenicura*
Passeriformes - Corvidae
e Rufous-tailed Fantail; Red-tailed
Fantail
f Rhipidure rougequeue
d Rotbürzelfächerschwanz
i Coda a ventaglio codirossa

8854 *Rhipidura rennelliana*
Passeriformes - Corvidae
e Rennell Fantail
f Rhipidure de Rennell
d Rennel-Fächerschwanz
i Coda a ventaglio dell'Isola Rennell

8855 *Rhipidura rufidorsa*
Passeriformes - Corvidae
e Rufous-backed Fantail; Grey-
breasted Fantail; Grey-breasted
Rufous-Fantail; Red-backed Fantail
f Rhipidure à dos roux

d Graubrustfächerschwanz
i Coda a ventaglio pettogrigio

8856 *Rhipidura rufifrons*
Passeriformes - Corvidae
e Rufous Fantail; Rufous-fronted
Fantail; Black-breasted Rufous-
Fantail
f Rhipidure roux
d Rotstirnfächerschwanz;
Rotstirnfächerschnäpper
i Coda a ventaglio castana

8857 *Rhipidura rufiventris*
Passeriformes - Corvidae
e Northern Fantail; Red-vented Fantail;
White-throated Fantail
f Rhipidure à ventre chamois
d Witwenfächerschwanz
i Coda a ventaglio dal collare grigio

8858 *Rhipidura rufiventris gularis*
Passeriformes - Corvidae
e Torres Strait Northern-Fantail (ANZ)

8859 *Rhipidura semirubra*
Passeriformes - Corvidae
e Manus Fantail; Mangrove Fantail;
Mangrove Grey-Fantail
f Rhipidure de l'Amirauté
i Coda a ventaglio di Manus

8860 *Rhipidura spilodera*
Passeriformes - Corvidae
e Streaked Fantail; Spotted Fantail
f Rhipidure tacheté
d Fleckenfächerschwanz
i Coda a ventaglio maculata

8861 *Rhipidura superciliaris*
Passeriformes - Corvidae
e Blue Fantail
f Rhipidure bleu
d Blaurückenfächerschwanz
i Coda a ventaglio azzurra

8862 *Rhipidura superflua*
Passeriformes - Corvidae
e Tawny-backed Fantail; Moluccan
Fantail; Moluccan Rufous-Fantail;
Buru Fantail; Buru Rufous-Fantail;

Cinnamon-backed Fantail; Chestnut-backed Fantail; Grey-breasted Fantail
f Rhipidure de Buru
d Buru-Fächerschwanz
i Coda a ventaglio delle Molucche

8863 *Rhipidura tenebrosa*
Passeriformes - Corvidae
e Dusky Fantail; Sombre Fantail
f Rhipidure ombré
d Rußfächerschwanz
i Coda a ventaglio bruna

8864 *Rhipidura teysmanni*
Passeriformes - Corvidae
e Rusty-bellied Fantail; Teysmann's Fantail; Celebes Rufous-Fantail; Celebes Fantail-Flycatcher; Sulawesi Rufous-Fantail; Rusty-flanked Fantail
f Rhipidure des Celebes
d Celebes-Fächerschwanz
i Coda a ventaglio di Sulawesi

8865 *Rhipidura threnothorax*
Passeriformes - Corvidae
e Sooty Thicket-Fantail
f Rhipidure fuligineux
d Rosenberg-Fächerschwanz
i Coda a ventaglio dorsobruno

8866 *Rhizothera longirostris*
Galliformes - Phasianidae
e Long-billed Partridge; Long-billed Wood-Partridge
f Perdrix à long bec
d Langschnabelwachtel
i Pernice beccolungo

8867 *Rhodacanthis flaviceps*
Passeriformes - Fringillidae
e Lesser Koa-Finch; Yellow-headed Koa-Finch
f Petit Psittirostre
d Goldkopfkoagimpel; Kleiner Koafink
i Fringuello minore del koa

8868 *Rhodacanthis palmeri*
Passeriformes - Fringillidae
e Greater Koa-Finch; Orange Koa-Finch; Hopue

f Psittirostre de Palmer
d Orangebrustkoagimpel; Großer Koafink
i Fringuello maggiore del koa

8869 *Rhodinocichla rosea*
Passeriformes - Thraupidae
e Rosy Thrush-Tanager; Rose-breasted Thrush-Tanager; Queo
f Tangara quéo
d Queo; Drosseltangare
i Tangara tordo pettorosa

8870 *Rhodonessa caryophyllacea*
Anseriformes - Anatidae
e Pink-headed Duck
f Nette à cou rose
d Nelkenente
i Anatra testarossa

8871 *Rhodopechys githaginea*
Passeriformes - Fringillidae
e Trumpeter Finch; Trumpeter Bullfinch
f Roselin githagine; Bouvreuil githagine
d Wüstengimpel; Sahara-Wüstengimpel; Wüstentrompeter
i Trombettiere; Trombettiere comune

8872 *Rhodopechys mongolica*
Passeriformes - Fringillidae
e Mongolian Finch; Mongolian Trumpeter Finch; Mongolian Trumpeter Bullfinch; Mongolian Trumpeter
f Roselin de Mongolie; Bouvreuil de Mongolie
d Mongolen-Gimpel
i Trombettiere mongolo

8873 *Rhodopechys obsoleta*
Passeriformes - Fringillidae
e Desert Finch; Lichtenstein's Desert Finch; Lichtenstein's Finch; Black-billed Desert Finch; Black-billed Finch
f Roselin de Lichtenstein; Bouvreuil de Lichtenstein
d Weißflügelgimpel; Schwarzzügelgimpel

i Trombettiere di Lichtenstein;
Trombetiere del deserto

8874 *Rhodopechys sanguinea*
Passeriformes - Fringillidae
e Crimson-winged Finch; Crimson-
winged Desert Finch
f Roselin à ailes roses; Bouvreuil à
ailes roses
d Rotflügelgimpel; Wüstengimpel
i Trombettiere alirosse

8875 *Rhodophoneus cruentus*
Passeriformes - Malaconotidae
e Rosy-patched Bush-Shrike; Red-
patched Shrike; Rosy-patched Shrike
f Tchagra à croupion rose
d Blutfleckbuschwürger
i Averla di macchia rosata

8876 *Rhodopis vesper*
Apodiformes - Trochilidae
e Oasis Hummingbird; Evening
Hummingbird
f Colibri vesper
d Atacama-Kolibri
i Rodope

8877 *Rhodospingus cruentus*
Passeriformes - Fringillidae
e Crimson-breasted Finch; Crimson
Finch; Red-crowned Finch; Crimson
Finch-Tanager
f Rhodospingue ponceau
d Purpurkronfink
i Fringuello cremisi

8878 *Rhodostethia rosea*
Charadriiformes - Laridae
e Ross's Gull; Rosy Gull
f Mouette de Ross; Mouette rosée
d Rosenmöwe
i Gabbiano di Ross

8879 *Rhodothraupis celaeno*
Passeriformes - Fringillidae
e Crimson-collared Grosbeak
f Cardinal à collier
d Halsbandkardinal
i Beccogrosso dal collare rosso

8880 *Rhopocichla atriceps*
Passeriformes - Sylviidae
e Dark-fronted Babbler; Black-headed
Babbler; Black-fronted Babbler
f Timalie à tête noire
d Schwarzkappenbaumtimalie;
Kapuzentimalie
i Garrulo testanera

8881 *Rhopophilus pekinensis*
Passeriformes - Cisticolidae
e White-browed Chinese-Warbler;
Peking Warbler; White-browed
Bush-dweller; Chinese Hill-Warbler;
Chinese Babbler
f Rhopophile de Pékin; Fauvette de
Pékin
d Peking-Sänger
i Ropofilo della Cina

8882 *Rhopornis ardesiaca*
Passeriformes - Thamnophilidae
e Slender Antbird
f Alapi de Bahia
d Bahia-Ameisenschnäpper
i Mangiaformiche snello

8883 *Rhyacornis bicolor*
Passeriformes - Muscicapidae
e Luzon Water-Redstart; Philippine
Water-Redstart; Water Redstart
f Nymphée bicolore
d Luzon-Rötel
i Codirosso delle Filippine

8884 *Rhyacornis fuliginosus*
Passeriformes - Muscicapidae
e Plumbeous Water-Redstart;
Plumbeous Redstart
f Nymphée fuligineuse; Rougequeue
fuligineux
d Wasserrötel
i Codirosso fuligginoso

8885 *Rhynchocyclus brevirostris*
Passeriformes - Tyrannidae
e Short-billed Flatbill
f Tyranneau à bec court
d Brillenkreisschnabel
i Beccopiatto occhibianchi

8886   *Rhynchocyclus fulvipectus*
        Passeriformes - Tyrannidae
e   Fulvous-breasted Flatbill
f   Platyrhynque à poitrine fauve;
        Tyranneau à poitrine fauve
d   Ockerbrustkreisschnabel
i   Beccopiatto pettofulvo

8887   *Rhynchocyclus olivaceus*
        Passeriformes - Tyrannidae
e   Olivaceous Flatbill
f   Platyrhynque olivâtre; Tyranneau
        olivâtre
d   Olivrückenkreisschnabel
i   Beccopiatto olivaceo

8888   *Rhynchocyclus pacificus*
        Passeriformes - Tyrannidae
e   Pacific Flatbill; Eye-ringed Flatbill
f   Platyrhynque de Colombie;
        Platyrhynque à bec court

8889   *Rhynchopsitta pachyrhyncha*
        Psittaciformes - Psittacidae
e   Thick-billed Parrot; Thick-billed
        Parakeet; Maroon-fronted Parrot
f   Conure à gros bec; Perruche à gros
        bec; Conure à front brun
d   Ara-Sittich; Arara-Sittich;
        Maronenstirnstittich
i   Pappagallo beccoforte; Pappagallo
        frontecastano

8890   *Rhynchortyx cinctus*
        Galliformes - Odontophoriidae
e   Tawny-faced Quail; Banded Wood-
        Quail
f   Colin ceinturé
d   Langbeinwachtel
i   Colino facciafulva

8891   *Rhynchostruthus socotranus*
        Passeriformes - Fringillidae
e   Golden-winged Grosbeak
f   Grand-verdier à ailes d'or; Grosbec à
        ailes d'or
d   Maskengimpel; Goldflügelgimpel
i   Beccogrosso alidorate

8892   *Rhynchotus rufescens*
        Tinamiformes - Tinamidae

e   Red-winged Tinamou; Rufous
        Tinamou
f   Tinamou isabelle
d   Pampahuhn; Inambu
i   Pollo delle Pampas

8893   **Rhynochetidae**
        Gruiformes
e   Kagus
f   Rhynochétidés
d   Kagus
i   Rinochetidi

8894   *Rhynochetos jubatus*
        Gruiformes - Rhynochetidae
e   Kagu; Cagu; Cagon
f   Kagou huppé
d   Kagu
i   Kagu

8895   *Rhytipterna holerythra*
        Passeriformes - Tyrannidae
e   Rufous Mourner
f   Aulia roux
d   Rostrhytipterna
i   Piagnone rugginoso

8896   *Rhytipterna immunda*
        Passeriformes - Tyrannidae
e   Pale-bellied Mourner; Cayenne
        Mourner
f   Aulia à ventre pâle
d   Fahlbauchrhytipterna
i   Piagnone beccochiaro

8897   *Rhytipterna simplex*
        Passeriformes - Tyrannidae
e   Greyish Mourner
f   Aulia grisâtre
d   Graurhytipterna
i   Piagnone grigio

8898   *Rimator malacoptilus*
        Passeriformes - Sylviidae
e   Long-billed Wren-Babbler; Long-
        billed Scimitar-Babbler
f   Turdinule à long bec
d   Zwergsäbler; Rimator-Timalie
i   Garrulo scricciolo beccolungo

8899 *Riparia cincta*
Passeriformes - Hirundinidae
e Banded Martin; Banded Sand-Martin
f Hirondelle à collier
d Gebänderte Uferschwalbe;
Weißbrauenschwalbe
i Topino dai sopraccigli bianchi

8900 *Riparia congica*
Passeriformes - Hirundinidae
e Congo Martin; Congo Sand-Martin
f Hirondelle du Congo; Hirondelle de
rivage du Congo
d Afrikanische Uferschwalbe
i Topino del Congo

8901 *Riparia paludicola*
Passeriformes - Hirundinidae
e Plain Martin; African Sand-Martin
(CSA); Brown-throated Martin
(CSA); Plain Sand-Martin; Brown-
throated Sand-Martin; Sand-Martin;
Indian Sand-Martin; Grey-breasted
Sandmartin
f Hirondelle paludicole; Hirondelle des
sables; Hirondelle de Mauretanie
d Uferschwalbe;
Braunkehluferschwalbe
i Topino africano

8902 *Riparia riparia*
Passeriformes - Hirundinidae
e Sand-Martin; Bank Swallow (NA);
European Sand-Martin (CSA);
Common Sand-Martin; Gorgeted
Sand-Martin
f Hirondelle de rivage
d Europäische Uferschwalbe
i Topino; Rondine riparia

8903 *Rissa brevirostris*
Charadriiformes - Laridae
e Red-legged Kittiwake
f Mouette des brumes; Mouette à
pattes rouges
d Klippenmöwe
i Gabbiano tridattilo zamperosse

8904 *Rissa tridactyla*
Charadriiformes - Laridae
e Black-legged Kittiwake; Kittiwake

f Mouette tridactyle
d Dreizehenmöwe
i Gabbiano tridattilo; Gabbiano
terragnolo

8905 *Rollandia microptera*
Podicipediformes - Podicipedidae
e Short-winged Grebe; Flightless
Grebe; Titicaca Flightless-Grebe
f Grèbe microptère
d Titicaca-Taucher
i Rollandia alibreve

8906 *Rollandia rolland*
Podicipediformes - Podicipedidae
e White-tufted Grebe
f Grèbe de Rolland
d Rollandtaucher
i Rolandia ciuffibianchi

8907 *Rollulus rouloul*
Galliformes - Phasianidae
e Crested Partridge; Rouloul; Crested
Wood-Partridge; Rouloul Partridge
f Rouloul couronné
d Straußwachtel
i Rouloul; Rulul

8908 *Roraimia adusta*
Passeriformes - Furnariidae
e Roraiman Barbtail; Roraima Barbtail
f Anabasitte du Roraraima
d Roraima-Stachelschwanz
i Corridore arboricolo di Roraima

8909 *Rostratula benghalensis*
Charadriiformes - Rostratulidae
e Greater Painted-Snipe; Painted Snipe
f Rhynchée peinte
d Goldschnepfe
i Beccaccia dorata; Beccaccia dorata
maggiore

8910 *Rostratula benghalensis australis*
Charadriiformes - Rostratulidae
e Australian Painted-Snipe (ANZ)

8911 *Rostratula semicollaris*
Charadriiformes - Rostratulidae
e American Painted-Snipe; South
American Painted-Snipe

f Rhynchée de Saint-Hilaire
d Weißfleckengoldschnepfe
i Beccaccia dorata del Sudamerica

8912 Rostratulidae
Charadriiformes
e Painted-Snipe
f Rostratulidés
d Goldschnepfen
i Rostratulidi

8913 Rostrhamus hamatus
Falconiformes - Accipitridae
e Slender-billed Kite; Helicolestes;
Slender-billed Helicolestes
f Milan à long bec
d Hakenweihe
i Rostramo del becco sottile; Nibbio
beccosottile; Rostramo

8914 Rostrhamus sociabilis
Falconiformes - Accipitridae
e Snail Kite; Everglade Kite
f Milan des marais; Milan des
Everglades
d Schneckenweihe; Schneckenmilan
i Nibbio delle Everglades; Rostramo
socievole

8915 Rougetius rougetii
Gruiformes - Rallidae
e Rouget's Rail; Abyssinian Rail
f Râle de Rouget
d Braunralle
i Rallo di Rouget

8916 Rowettia goughensis
Passeriformes - Fringillidae
e Gough Finch; Gough Island Finch;
Gough Bunting
f Rowettie de Gough
d Gough-Ammer; Rowettia
i Fringuello dell'Isola Gough

8917 Rukia longirostra
Passeriformes - Zosteropidae
e Long-billed White-eye; Ponapé
White-eye; Large Ponapé White-eye
f Zostérops de Ponapé
d Langschnabelbrillenvogel

8918 Rukia ruki
Passeriformes - Zosteropidae
e Faichuk White-eye; Truck White-
eye; Large Truk White-eye; Great
Truk White-eye; Greater Truk White-
eye
f Zostérops de Truk
d Truk-Brillenvogel; Zimtbrillenvogel
i Occhialino di Truk

8919 Rupicola peruviana
Passeriformes - Tyrannidae
e Andean Cock-of-the-rock; Andean
Red-Cock-of-the-rock; Peruvian
Cock-of-the-rock; Red Cock-of-the-
rock
f Coq-de-roche péruvien; Coq-de-
roche du Pérou
d Anden-Klippenvogel; Anden-
Felsenhahn; Roter Felsenhahn
i Galletto di roccia peruviano; Galletto
di roccia delle Ande; Galletto di
monte del Perù

8920 Rupicola rupicola
Passeriformes - Tyrannidae
e Guianan Cock-of-the-rock; Orange
Cock-of-the-rock
f Coq-de roche orangé; Coq-de-roche
de Guyane
d Cayenne-Klippenvogel;
Klippenvogel; Kammfelsenhahn
i Galletto di roccia; Galletto di roccia
comune; Rupicola; Galletto di roccia
della Guiana; Galletto di monte

8921 Rynchops albicollis
Charadriiformes - Laridae
e Indian Skimmer; Scissorbill
f Bec-en-ciseaux à collier
d Halsbandscherenschnabel
i Becco a cesoie indiano

8922 Rynchops flavirostris
Charadriiformes - Laridae
e African Skimmer; Scissorbill
f Bec-en-ciseaux d'Afrique
d Scherenschnabel; Afrika-
Scherenschnabel;
Braunmantelscherenschnabel
i Becco a cesoie africano

8923 *Rynchops niger*
Charadriiformes - Laridae
- *e* Black Skimmer; American Skimmer
- *f* Bec-en-ciseaux noir
- *d* Schwarzer Scherenschnabel;
  Schwarzmantelscherenschnabel
- *i* Becco a forbice; Becco a cesoie;
  Becco a cesoie americano

S

8924 Sagittariidae
Falconiformes
e Secretarybirds
f Sagittaridés
d Sekretäre
i Sagittaridi

8925 *Sagittarius serpentarius*
Falconiformes - Sagittariidae
e Secretarybird
f Messager sagittaire; Serpentaire
d Sekretär
i Serpentario; Segretario

8926 *Sakesphorus bernardi*
Passeriformes - Thamnophilidae
e Collared Antshrike; White-naped
Antshrike
f Batara de Bernard
d Weißnacken ameisenwürger
i Averla formichiera dal collare

8927 *Sakesphorus canadensis*
Passeriformes - Thamnophilidae
e Black-crested Antshrike
f Batara huppé
d Schwarzhaubenameisenwürger
i Averla formichiera crestanera

8928 *Sakesphorus cristatus*
Passeriformes - Thamnophilidae
e Silvery-cheeked Antshrike
f Batara à joues argent
d Camposameisenwürger
i Averla formichiera guanceargentate

8929 *Sakesphorus luctuosus*
Passeriformes - Thamnophilidae
e Glossy Antshrike
f Batara luisant
d Trauerameisenwürger
i Averla formichiera splendente

8930 *Sakesphorus melanonotus*
Passeriformes - Thamnophilidae
e Black-backed Antshrike
f Batara à dos noir
d Schwarzrückenameisenwürger
i Averla formichiera dorsonero

8931 *Sakesphorus melanothorax*
Passeriformes - Thamnophilidae
e Band-tailed Antshrike; Black-
throated Antshrike
f Batara de Cayenne
d Weißschulterameisenwürger
i Averla formichiera golanera

8932 *Salpinctes obsoletus*
Passeriformes - Certhiidae
e Rock Wren; Wren; American Rock-
Wren
f Troglodyte des rochers
d Felsenzaunkönig
i Scricciolo di roccia

8933 *Salpornis spilonotus*
Passeriformes - Certhiidae
e Spotted Creeper; African Spotted-
Creeper; Spotted-Grey-Creeper
(ISC); Spotted Treecreeper
f Grimpereau tacheté
d Fleckenbaumläufer; Stammsteiger;
Gefleckter Baumsteiger
i Rampichino maculato

8934 *Saltator albicollis*
Passeriformes - Thraupidae
e Lesser Antillean Saltator; Streaked
Saltator
f Grive gros-bec (Ants); Grosbec
(Ants); Saltator gros-bec; Saltator
rayé
d Strichelsaltator
i Beccoforte

8935 *Saltator atriceps*
Passeriformes - Thraupidae
e Black-headed Saltator
f Saltator à tête noire
d Schwarzkappensaltator;
Schwarzkopfhabie
i Beccoforte testanera

8936 **Saltator atricollis**
  Passeriformes - Thraupidae
  e Black-throated Saltator
  f Saltator à gorge noire
  d Schwarzhalssaltator
  i Beccoforte golanera

8937 **Saltator atripennis**
  Passeriformes - Thraupidae
  e Black-winged Saltator
  f Saltator à ailes noires
  d Schwarzschwingensaltator
  i Beccoforte alinere

8938 **Saltator aurantiirostris**
  Passeriformes - Thraupidae
  e Golden-billed Saltator
  f Saltator à bec orangé
  d Goldschnabelsaltator
  i Beccoforte dorato

8939 **Saltator cinctus**
  Passeriformes - Thraupidae
  e Masked Saltator
  f Saltator masqué
  d Maskensaltator
  i Beccoforte mascherto

8940 **Saltator coerulescens**
  Passeriformes - Thraupidae
  e Greyish Saltator; Grey Saltator
  f Saltator gris
  d Grausaltator
  i Beccoforte griastro

8941 **Saltator fuliginosus**
  Passeriformes - Thraupidae
  e Black-throated Grosbeak
  f Cardinal fuligineux; Salttor
    fuligineux
  d Papageischnabelsaltator
  i Beccogrosso golanera; Beccoforte
    golanera

8942 **Saltator grandis**
  Passeriformes - Thraupidae
  e Middle American Saltator

8943 **Saltator grossus**
  Passeriformes - Thraupidae
  e Slate-coloured Grosbeak; White-
    throated Grosbeak
  f Cardinal ardoisé; Saltator ardoisé
  d Rotschnabelsaltator
  i Beccogrosso ardesia; Beccoforte
    ardesia

8944 **Saltator maxillosus**
  Passeriformes - Thraupidae
  e Thick-billed Saltator
  f Saltator à bec épais
  d Dickschnabelsaltator
  i Beccoforte robusto

8945 **Saltator maximus**
  Passeriformes - Thraupidae
  e Buff-throated Saltator
  f Saltator des grands-bois
  d Buntkehlsaltator
  i Beccoforte golachiara

8946 **Saltator nigriceps**
  Passeriformes - Thraupidae
  e Black-cowled Saltator
  f Saltator à capuchon
  i Beccoforte capinero

8947 **Saltator orenocensis**
  Passeriformes - Thraupidae
  e Orinocan Saltator; Orinoco Saltator
  f Saltator de l'Orénoque
  d Zimtflankensaltator
  i Beccoforte dell'Orinoco

8948 **Saltator rufiventris**
  Passeriformes - Thraupidae
  e Rufous-bellied Saltator
  f Saltator à ventre roux
  d Rotbauchsaltator
  i Beccoforte ventreruggionoso

8949 **Saltator similis**
  Passeriformes - Thraupidae
  e Green-winged Saltator
  f Saltator olive; Grand Saltator
  d Grünschwingensaltator
  i Beccoforte aliverdi

8950 **Saltator striatipectus**
  Passeriformis - Thraupidae
  e Streaked Saltator

8951 Saltatricula multicolor
Passeriformes - Fringillidae
e Many-coloured Chaco-Sparrow;
Many-coloured Chaco-Finch
f Saltatricule du Chaco
d Vielfarbenammer
i Fringuello multicolore

8952 Salvadorina waigiuensis
Anseriformes - Anatidae
e Salvadori's Duck; Salvadori's Teal
f Canard de Salvadori
d Salvadori-Ente
i Anatra di Salvadori

8953 Sapayoa aenigma
Passeriformes - Tyrannidae
e Broad-billed Sapayoa; Sapayoa;
Broad-billed Manakin; Enigma
Manakin
f Sapayoa à bec large; Manakin à bec
large
d Breitschnabelpipra
i Manachino beccolargo

8954 Sapheopipo noguchii
Piciformes - Picidae
e Okinawa Woodpecker; Pryer's
Woodpecker; Okinawan Woodpecker
f Pic d'Okinawa
d Okinawa-Specht
i Picchio di Okinawa

8955 Sappho sparganura
Apodiformes - Trochilidae
e Red-tailed Comet; Sappho Comet;
Fire-tailed Comet
f Colibri sapho
d Schleppensylphe; Sapphokolibri
i Colibrì Saffo

8956 Sarcogyps calvus
Falconiformes - Accipitridae
e Red-headed Vulture; Black Vulture
(ISC); King Vulture (ISC); Indian
Black-Vulture; Asiatic King-Vulture;
Asian King-Vulture; Pondicherry
Vulture
f Vautour royal
d Kahlkopfgeier
i Avvoltaio calvo

8957 Sarcops calvus
Passeriformes - Sturnidae
e Coleto; Coleto Mynah; Bald Starling
f Goulin gris; Sarcops chauve
d Kahlkopfatzel
i Storno calvo

8958 Sarcoramphus papa
Falconiformes - Cathartidae
e King Vulture
f Sarcoramphe roi
d Königsgeier
i Avvoltoio papa; Re degli avvoltoi

8959 Sarkidiornis melanotos
Anseriformes - Anatidae
e Comb Duck; Knob-billed Duck
(CSA); African Comb Duck; Nakta
(ISC); Knob-billed Goose; Comb
Nakta; Black-backed Duck
f Canard à bosse; Canard à bosse
bronzée
d Höckerente; Glanzgans; Glanzente
i Anatra dal bernoccolo; Sarkidiornis

8960 Sarkidiornis sylvicola
Anseriformes - Anatidae
e American Comb Duck

8961 Saroglossa aurata
Passeriformes - Sturnidae
e Madagascar Starling
f Étourneau malgache
d Madagaskar-Star
i Storno del Madagascar

8962 Saroglossa spiloptera
Passeriformes - Sturnidae
e Spot-winged Starling; Spot-winged
Stare (ISC); Spotted-wing Starling
f Étourneau à ailes tachetées
d Spleißflügelstar; Braunkehlstar;
Marmorstar
i Storno alimacchiate

8963 Sarothrura affinis
Gruiformes - Rallidae
e Striped Flufftail; Chestnut-tailed
Flufftail; Chestnut-tailed Crake; Red-
tailed Flufftail
f Râle affin; Râle à queue rousse

<table>
<tr><td>d</td><td>Streifenzwergralle; Streifenralle</td></tr>
<tr><td>i</td><td>Schiribilla codacastana</td></tr>
</table>

8964　*Sarothrura ayresi*
Gruiformes - Rallidae

e White-winged Flufftail; White-winged Crake
f Râle à miroir; Räle miroir
d Weißflügelzwergralle; Weißflügelralle
i Schiribilla alibianche

8965　*Sarothrura boehmi*
Gruiformes - Rallidae

e Streaky-breasted Flufftail; Böhm's Flufftail; Böhm's Crake; Streaky-breasted Crake; Streaky-breasted Pygmy-Crake
f Râle de Böhm
d Böhms Zwergralle; Böhm-Ralle
i Schiribilla pettostriato

8966　*Sarothrura elegans*
Gruiformes - Rallidae

e Buff-spotted Flufftail; Buff-spotted Crake; Buff-spotted Pygmy-Crake
f Râle ponctué
d Schmuckzwergralle; Tropfenralle
i Schiribilla macchiefulve

8967　*Sarothrura insularis*
Gruiformes - Rallidae

e Madagascar Flufftail; Madagascar Crake
f Râle insulaire
d Hova-Ralle; Hava-Ralle
i Schiribilla del Madagascar

8968　*Sarothrura lugens*
Gruiformes - Rallidae

e Chestnut-headed Flufftail; African Chestnut-headed Crake; Chestnut-headed Crake; Long-tailed Flufftail
f Râle à tête rousse
d Ugalla-Ralle
i Schiribilla lamentosa

8969　*Sarothrura pulchra*
Gruiformes - Rallidae

e White-spotted Flufftail; White-spotted Crake; White-spotted Pygmy-Crake; White-spotted Pygmy-Rail
f Râle perlé
d Perlenralle
i Schiribilla macchiebianche

8970　*Sarothrura rufa*
Gruiformes - Rallidae

e Red-chested Flufftail; Red-chested Pygmy-Crake; Red-chested Crake
f Râle à camail; Râle camail
d Rotbrustzwergralle; Rotbrustralle
i Schiribilla pettorosso

8971　*Sarothrura watersi*
Gruiformes - Rallidae

e Slender-billed Flufftail; Waters's Flufftail; Waters's Crake
f Râle de Waters
d Lemurenralle
i Schiribilla di Waters

8972　*Sasia abnormis*
Piciformes - Picidae

e Rufous Piculet
f Picumne roux
d Malaien-Mausspecht; Dreizehenmausspecht
i Picchio nano rossiccio

8973　*Sasia africana*
Piciformes - Picidae

e African Piculet; Pygmy Woodpecker; Pygmy Wood Piculet
f Picumne de Verreaux
d Graubauchmausspecht; Afrikanischer Mausspecht
i Picchio nano africano

8974　*Sasia ochracea*
Piciformes - Picidae

e White-browed Piculet; White-browed Rufous-Piculet; Rufous Piculet
f Picumne à sourcils blancs
d Rötelmausspecht; Olivmausspecht
i Picchio nano dai sopraccigli bianchi

8975　*Satrapa icterophrys*
Passeriformes - Tyrannidae

e Yellow-browed Tyrant; Yellow-browed Satrap
f Moucherolle à sourcils jaunes

d   Goldtyrann
i   Wrobell

8976   *Saurothera longirostris*
      Coliformes - Coccyzidae
e   Hispaniolan Lizard-Cuckoo
f   Tacco d'Hispaniola
d   Haiti-Kukuck
i   Mangialucertole di Hispaniola

8977   *Saurothera merlini*
      Coliformes - Coccyzidae
e   Great Lizard-Cuckoo; Rain Crow
      (WI); Big Rain Crow (WI); Kataw
      (WI); Cuban Lizard-Cuckoo
f   Tacco de Cuba; Tacco de Merlin
d   Eidechsenkuckuck
i   Mangialucertole maggiore

8978   *Saurothera vetula*
      Coliformes - Coccyzidae
e   Jamaican Lizard-Cuckoo; Rainbird
      (WI); Old Woman Bird (WI);
      Maybird (WI); Ringtail (WI);
      Sawdering (WI)
f   Tacco de la Jamaïque
d   Jamaika-Kukuck
i   Mangialucertole giamaicano

8979   *Saurothera vieilloti*
      Coliformes - Coccyzidae
e   Puerto Rican Lizard-Cuckoo
f   Tacco de Porto Rico
d   Puertorico-Kukuck
i   Mangialucertole portoricano

8980   *Saxicola aethiops*
      Passeriformes - Muscicapidae
e   Black Bush-Chat

8981   *Saxicola axillaris*
      Passeriformes - Muscicapidae
e   African Stonechat
f   Tarier bifascié

8982   *Saxicola caprata*
      Passeriformes - Muscicapidae
e   Pied Bush-Chat; Pied Stonechat; Pied
      Chat; Ceylon Pied Bush-Chat
f   Tarier pie
d   Mohrenschwarzkehlchen

i   Saltimpalo nero e bianco;
      Saltimpoalo nero

8983   *Saxicola dacotiae*
      Passeriformes - Muscicapidae
e   Canary Islands Chat; Fuerteventura
      Chat; Canary Islands Stonechat;
      Canarian Chat; Meade-Waldo's Chat
f   Tarier des Canaries
d   Kanarien-Schmätzer; Kanarien-
      Schwarzkehlchen
i   Saltimpalo delle Canarie; Stiaccino
      delle Canarie

8984   *Saxicola ferrea*
      Passeriformes - Muscicapidae
e   Grey Bush-Chat; Dark-grey Bush-
      Chat
f   Tarier gris
d   Grauschmätzer
i   Stiaccino grigio

8985   *Saxicola gutturalis*
      Passeriformes - Muscicapidae
e   White-bellied Bush-Chat; Timor
      Bush-Chat; Timor Stonechat; White-
      bellied Chat; Sunda Bush-Chat;
      Sunda Stonechat
f   Tarier à gorge blanche
d   Timor-Schmätzer
i   Sassicola ventrebianco

8986   *Saxicola insignis*
      Passeriformes - Muscicapidae
e   White-throated Bush-Chat;
      Hodgson's Bush-Chat; Hodgson's
      Stonechat; Hodgson's Chat
f   Tarier de Hodgson
d   Hodgson-Schmätzer
i   Sassicola di Hodgson

8987   *Saxicola jerdoni*
      Passeriformes - Muscicapidae
e   Jerdon's Bush-Chat; Jerdon's Chat;
      Rufous-breasted Bush-Chat
f   Tarier de Jerdon
d   Jerdon-Schmätzer
i   Sassicola di Jerdon

8988   *Saxicola leucura*
      Passeriformes - Muscicapidae

   e   White-tailed Stonechat; White-tailed
       Bush-Chat (ISC); Black Stonechat
   f   Tarier à queue blanche
   d   Weißschwanzschwarzkehlchen
   i   Saltimpalo codabianca

8989   *Saxicola macrorhyncha*
       Passeriformes - Muscicapidae
   e   White-browed Bush-Chat; Stoliczka's
       Whinchat; Stoliczka's Chat;
       Stoliczka's Bush-Chat
   f   Tarier de Stoliczka
   d   Wüstenbraunkehlchen
   i   Sassicola di Stoliczka

8990   *Saxicola maura*
       Passeriformes - Muscicapidae
   e   Siberian Stonechat
   f   Tarier de Sibérie; Tarier sibérien;
       Tarrier pâtre sibérien
   d   Sibirisches Schwarzkehlchen
   i   Saltimpalo siberiano

8991   *Saxicola rubetra*
       Passeriformes - Muscicapidae
   e   Whinchat; European Whinchat
   f   Traquet tarier; Tarier des prés; Tarier
       ordinaire; Tarier
   d   Braunkehlchen
   i   Stiaccino; Scrocchino; Saltinvaghile;
       Saltinseccia; Saltancini; Piagnaccia

8992   *Saxicola tectes*
       Passeriformes - Muscicapidae
   e   Réunion Stonechat; Bourbon
       Stonechat
   f   Tarier de la Réunion; Traquet de la
       Réunion
   d   Réunion-Schmätzer
   i   Saltimpalo di Reunion

8993   *Saxicola torquata*
       Passeriformes - Muscicapidae
   e   Common Stonechat; Stonechat;
       Collared Bush-Chat (ISC); Collared
       Stonechat; European Stonechat
   f   Tarier pâtre; Traquet pâtre
   d   Schwarzkehlchen
   i   Saltimpalo; Saltinpunta; Fornaiolo

8994   *Saxicola torquata hibernans*
       Passeriformes - Muscicapidae
   e   Western European Stonechat

8995   *Saxicola torquata rubicola*
       Passeriformes - Muscicapidae
   e   Eastern European Stonechat

8996   *Saxicoloides fulicata*
       Passeriformes - Muscicapidae
   e   Indian Robin; Indian Chat; Black-
       backed Chat; Black Robin; Brown-
       backed Indian Robin; Ceylon Black
       Robin
   f   Pseudotraquet indien
   d   Strauchschmätzer
   i   Pettirosso indiano

8997   *Saxixola macrorhyncha*
       Passeriformes - Muscicapidae
   e   White-browed Bush-Chat; Stoliczka's
       Bush-Chat; Stoliczka's Whinchat
   d   Wüstenbraunkehlchen
   i   Saltimpalo; Stiaccino

8998   *Sayornis nigricans*
       Passeriformes - Tyrannidae
   e   Black Phoebe
   f   Moucherolle noire
   d   Schwarzkopfphoebe
   i   Febe nero

8999   *Sayornis phoebe*
       Passeriformes - Tyrannidae
   e   Eastern Phoebe
   f   Moucherolle phébi
   d   Phoebe; Haustyrann
   i   Febe orientale

9000   *Sayornis saya*
       Passeriformes - Tyrannidae
   e   Say's Phoebe
   f   Moucherolle à ventre roux
   d   Say-Phoebe
   i   Febe di Say

9001   *Scaphidura oryzivora*
       Passeriformes - Icteridae
   e   Giant Cowbird; Rice Grackle; Rice
       Cowbird
   f   Vacher géant

d Riesenkuhstärling
i Molotro gigante

9002 ***Sceloglaux albifacies***
Strigiformes - Strigidae
e Laughing Owl; White-faced Owl
f Ninoxe rieuse
d Lachkauz; Weißwangenkauz
i Gufo facciabianca; Gufo ridente

9003 ***Scelorchilus albicollis***
Passeriformes - Rhinocryptidae
e White-throated Tapaculo
f Tourco à gorge blanche
d Weißkehltapaculo
i Tapacula golabianca

9004 ***Scelorchilus rubecula***
Passeriformes - Rhinocryptidae
e Chucao Tapaculo; Chucao
f Tourco rougegorge
d Rotkehltapaculo
i Tapacula di Chucao

9005 ***Schetba rufa***
Passeriformes - Corvidae
e Rufous Vanga
f Schetbé roux; Artamie rousse
d Rotvanga
i Vanga rossiccia

9006 ***Schiffornis major***
Passeriformes - Pipridae
e Greater Schiffornis; Greater Manakin
f Antriade roussâtre
d Zimtschiffornis
i Manachino maggiore

9007 ***Schiffornis turdinus***
Passeriformes - Pipridae
e Thrush-like Schiffornis; Thrush-like
Manakin; Thrush Shiffornis; Thrush
Manakin; Thrush-like Mourner;
Brown Mourner; Brown Schiffornis
f Antriade turdoïde
d Brozeschiffornis
I Manachino tordo

9008 ***Schiffornis veraepacis***
Passeriformes - Pipridae

e Brown Schiffornis
f Antriade brun

9009 ***Schiffornis virescens***
Passeriformes - Pipridae
e Greenish Schiffornis; Greenish
Manakin
f Antriade verdâtre
d Einfarbschiffornis
i Manachino verdastro

9010 ***Schistochlamys melanopis***
Passeriformes - Fringillidae
e Black-faced Tanager
f Tangara à camail
d Schleiertangare
i Tangara faccianera

9011 ***Schistochlamys ruficapillus***
Passeriformes - Fringillidae
e Cinnamon Tanager
f Tangara cannellé
d Gimpeltangare
i Tangara pettocannella

9012 ***Schizoeaca coryi***
Passeriformes - Furnariidae
e Ochre-browed Thistletail
f Synallaxe de Cory
i Codaspinosa di Cory

9013 ***Schizoeaca fuliginosa***
Passeriformes - Furnariidae
e White-chinned Thistletail; White-
chinned Spinetail; Thistletail;
Andean Thistletail
f Synallaxe à menton blanc; Synallaxe
chardon
d Distelschwanzschlüpfer
i Codaspinosa fuligginoso

9014 ***Schizoeaca griseomurina***
Passeriformes - Furnariidae
e Mouse-coloured Thistletail (NA)
f Synallaxe souris
i Codaspinosa grigiotopo

9015 ***Schizoeaca harterti***
Passeriformes - Furnariidae
e Black-throated Thistletail

f Synallaxe à gorge noire
i Codaspinosa di Hartert

9016 **_Schizoeaca helleri_**
Passeriformes - Furnariidae
e Puna Thistletail
f Synallaxe de Heller
i Codaspinosa della Puna

9017 **_Schizoeaca moreirae_**
Passeriformes - Furnariidae
e Itatiaia Thistletail; Brazilian
Thistletail
f Synallaxe de l'Itatiaia
d Strohschwanzschlüpfer
i Codaspinosa di Itatiaia

9018 **_Schizoeaca palpebralis_**
Passeriformes - Furnariidae
e Eye-ringed Thistletail
f Synallaxe à lunettes
i Codaspinosa del Perù

9019 **_Schizoeaca perijana_**
Passeriformes - Furnariidae
e Perija Thistletail
f Synallaxe des Periya
i Codaspinosa di Perija

9020 **_Schizoeaca vilcabambae_**
Passeriformes - Furnariidae
e Vilcabamba Thistletail
f Synallaxe de Vilcabamba
i Codaspinosa

9021 **_Schoenicola brevirostris_**
Passeriformes - Sylviidae
e Fan-tailed Grassbird; Fan-tailed
Swamp-Warbler; Fan-tailed Grass-
Warbler; Fan-tailed Warbler; Broad-
tailed Warbler
f Graminicole à bec court; Fauvette à
bec court; Fauvette à longue queue
d Breitschwanzsänger
i Forapaglie coda a ventaglio

9022 **_Schoenicola brevirostris alexinae_**
Passeriformes - Sylviidae
e Broad-tailed Warbler (CSA)

9023 **_Schoenicola platyura_**
Passeriformes - Sylviidae
e Broad-tailed Grassbird; Broad-tailed
Grass-Warbler; Broad-tailed Warbler
f Graminicole à queue large; Fauvette
à large queue
d Breitschwanzsänger;
Rundschwanzsänger
i forapaglie codanera

9024 **_Schoeniophylax phryganophila_**
Passeriformes - Furnariidae
e Chotoy Spinetail
f Synallaxe damier
d Weißwangenspitzschwanz;
Weißwangenschlüpfer
i Codaspinosa di Chotoy

9025 **_Schoutedenapus myoptilus_**
Apodiformes - Apodidae
e Scarce Swift
f Martinet de Shoa
d Maussegler
i Rondone dello Shoa

9026 **_Schoutedenapus schoutedeni_**
Apodiformes - Apodidae
e Schouten's Swift; Congo Swift
f Martinet de Schouteden
d Schouteden-Segler
i Rondone di Schoutenden

9027 **_Scissirostrum dubium_**
Passeriformes - Sturnidae
e Finch-billed Mynah; Grosbeak
Starling; Scissorbill Starling; Celebes
Starling; Celebes Mynah; Grosbeak
Mynah; Celebes Mynah
f Scissirostre des Célèbes
d Schmalschnabelstar
i Storno beccogrosso

9028 **_Sclateria naevia_**
Passeriformes - Thamnophilidae
e Silvered Antbird; Silvered
Antcatcher
f Alapi paludicole
d Mangroveameisenvogel

9029 **_Sclerurus albigularis_**
Passeriformes - Furnariidae

e   Grey-throated Leaftosser; Grey-
     throated Leafscraper
f   Sclérure à gorge grise
d   Graukehllaubwender
i   Grattafoglie golachiara

9030   *Sclerurus caudacutus*
          Passeriformes - Furnariidae
e   Black-tailed Leaftosser; Black-tailed
     Leafscraper
f   Sclérure des ombres
d   Schwarzschwanzlaubwender;
     Weißkehlblattwender
i   Grattafoglie codanera

9031   *Sclerurus guatemalensis*
          Passeriformes - Furnariidae
e   Scaly-throated Leaftosser; Scaly-
     throated Leafscraper; Guatemala
     Leafscraper; Guatemalan Leaftosser
f   Sclérure écaillé
d   Schuppenkehllaubwender;
     Buschlaubkratzer
i   Grattafoglie del Guatemala

9032   *Sclerurus mexicanus*
          Passeriformes - Furnariidae
e   Tawny-throated Leaftosser; Tawny-
     throated Leafscraper
f   Sclérure à gorge rousse
d   Rostkehllaubwender
i   Grattafoglie messicano

9033   *Sclerurus rufigularis*
          Passeriformes - Furnariidae
e   Short-billed Leaftosser; Short-billed
     Leafscraper
f   Sclérure à bec court
d   Kurzschnabellaubwender
i   Grattafoglie beccocorto

9034   *Sclerurus scansor*
          Passeriformes - Furnariidae
e   Rufous-breasted Leaftosser; Rufous-
     breasted Leafscraper
f   Sclérure à poitrine rousse
d   Rostbrustlaubwender
i   Grattafoglie pettirosso

9035   Scolopacidae
          Charadriiformes

e   Waders
f   Scolopacidés
d   Schnepfen; Schnepfenvögel
i   Scolopacidi

9036   *Scolopax celebensis*
          Charadriiformes - Scolopacidae
e   Sulawesi Woodcock; Celebes
     Woodcock; Indonesian Woodcock
f   Bécasse des Célèbes
d   Celebes-Schnepfe
i   Beccaccia di Sulawesi

9037   *Scolopax minor*
          Charadriiformes - Scolopacidae
e   American Woodcock; Woodcock
     (NA)
f   Bécasse d'Amérique
d   Amerikanische Waldschnepfe;
     Kanada-Schnepfe
i   Beccaccia americana

9038   *Scolopax mira*
          Charadriiformes - Scolopacidae
e   Amami Woodcock
f   Bécasse d'Amami
d   Amami-Schnepfe
i   Beccaccia di Amami-Oshima

9039   *Scolopax rochussenii*
          Charadriiformes - Scolopacidae
e   Moluccan Woodcock; Obi
     Woodcock; Moluku Woodcock;
     Indonesian Woodcock
f   Bécasse des Moluques
d   Molukken-Schnepfe
i   Beccaccia dell'Isola di Obi

9040   *Scolopax rusticola*
          Charadriiformes - Scolopacidae
e   Eurasian Woodcock; Woodcock;
     European Woodcock; Common
     Woodcock
f   Bécasse des bois
d   Waldschnepfe
i   Beccaccia eurasiatica; Beccaccia
     dorata maggiore

9041   *Scolopax saturata*
          Charadriiformes - Scolopacidae
e   Rufous Woodcock; Dusky

Woodcock; Horsfield's Woodcock;
East Indian Woodcock; Javanese
Woodcock; Indonesian Woodcock;
Rosenberg's Woodcock
f Bécasse de Java
d Malaien-Schnepfe
i Beccaccia delle inde orientale

9042 Scopidae
Ciconiiformes
e Hamerkops; Hammerheads;
Hammer-headed Storks;
Hammerhead Storks
f Scopidés
d Schattenvögel; Hammerköpfe
i Scopidi

9043 *Scopus umbretta*
Ciconiiformes - Scopidae
e Hamerkop; Hammerhead;
Hammerhead Stork
f Ombrette africaine; Ombrette du
Sénégal
d Hammerkopf; Schattenvogel
i Uccello martello; Umbretta

9044 *Scotocerca inquieta*
Passeriformes - Cisticolidae
e Streaked Scrub-Warbler; Scrub-
Warbler
f Dromoique du désert; Dromoïque
vif-argent; Dromoïque du Sahara
d Wüstenprinie; Streifenbuschsänger
i Codascura inquieta; Beccamoschino
dei cespugli

9045 *Scotopelia bouvieri*
Strigiformes - Strigidae
e Vermiculated Fishing-Owl
f Chouette-pêcheuse de Bouvier
d Marmorfischeule
i Civetta pescatrice di Bouvier; Gufo
pescatore di Bouvier

9046 *Scotopelia peli*
Strigiformes - Strigidae
e Pel's Fishing-Owl; African Fishing-
Owl; Fish-Owl
f Chouette-pêcheuse de Pel
d Fischeule; Bindenfischeule

i Civetta pescatrice di Pel; Gufo
pescatore di Pel

9047 *Scotopelia ussheri*
Strigiformes - Strigidae
e Rufous Fishing-Owl; Ussher's
Fishing-Owl
f Chouette-pêcheuse rousse
d Rotrückenfischeule
i Civetta pescatrice rossiccia

9048 *Scytalopus acutirostris*
Passeriformes - Rhinocryptidae
e Sharp-billed Tapaculo
f Mérulaxe argenté

9049 *Scytalopus affinis*
Passeriformes - Rhinocryptidae
e Ancash Tapaculo

9050 *Scytalopus altirostris*
Passeriformes - Rhinocryptidae
e Elfin Forest Tapaculo

9051 *Scytalopus argentifrons*
Passeriformes - Rhinocryptidae
e Silvery-fronted Tapaculo; Silver-
fronted Tapaculo; Chiriquiri
Tapaculo
f Mérulaxe argenté
d Siberstirn-Tapaculo
i Tapaculo fronteargentata

9052 *Scytalopus atratus*
Passeriformes - Rhinocryptidae
e Northern White-crowned Tapaculo

9053 *Scytalopus bolivianus*
Passeriformes - Rhinocryptidae
e Southern White-crowned Tapaculo;
White-crowned Tapaculo; Southern
White-winged Tapaculo
f Mérulaxe de Bolivie

9054 *Scytalopus canus*
Passeriformes - Rhinocryptidae
e Paramo Tapaculo

9055 *Scytalopus caracae*
Passeriformes - Rhinocryptidae
e Caracas Tapaculo

9056   ***Scytalopus chiriquensis***
    Passeriformes - Rhinocryptidae
e   Chiriqui Tapaculo

9057   ***Scytalopus chocoensis***
    Passeriformes - Rhinocryptidae
e   Choco Tapaculo

9058   ***Scytalopus femoralis***
    Passeriformes - Rhinocryptidae
e   Peruvian Rufous-vented Tapaculo;
    Rufous-vented Tapaculo; White-
    crowned Tapaculo
f   Mérulaxe à ventre roux
d Rotbauchtapaculo
i   Tapaculo di Santa Marta

9059   ***Scytalopus fuscus***
    Passeriformes - Rhinocryptidac
e   Dusky Tapaculo

9060   ***Scytalopus griseicollis***
    Passeriformes - Rhinocryptidae
e   Rufous-rumped Tapaculo

9061   ***Scytalopus indigoticus***
    Passeriformes - Rhinocryptidae
e   White-breasted Tapaculo; Indigo
    Tapaculo
f   Mérulaxe à poitrine blanche
d Weißbrusttapaculo
i   Tapaculo pettobianco

9062   ***Scytalopus latebricola***
    Passeriformes - Rhinocryptidae
e   Brown-rumped Tapaculo
f   Mérulaxe à croupion brun; Mérulaxe
    à dos roux
d Braunbürzeltapaculo
i   Tapaculo bruno

9063   ***Scytalopus macropus***
    Passeriformes - Rhinocryptidae
e   Large-footed Tapaculo
f   Mérulaxe à grands pieds
d Großfußtapaculo
i   Tapaculo piedigrandi

9064   ***Scytalopus magellanicus***
    Passeriformes - Rhinocryptidae
e   Magellanic Tapaculo; Andean
    Tapaculo; Churrin Tapaculo
f   Mérulaxe des Andes
d Anden-Tapaculo
i   Tapaculo di Magellano

9065   ***Scytalopus meridanus***
    Passeriformes - Rhinocryptidae
e   Merida Tapaculo

9066   ***Scytalopus micropterus***
    Passeriformes - Rhinocryptidae
e   Equatorial Rufous-vented Tapaculo

9067   ***Scytalopus novacapitalis***
    Passeriformes - Rhinocryptidae
e   Brasilia Tapaculo
f   Mérulaxe de Brasilia
d Brasilia-Tapaculo
i   Tapaculo di Brasilia

9068   ***Scytalopus panamensis***
    Passeriformes - Rhinocryptidae
e   Pale-throated Tapaculo; Tacarcuna;
    Panama Tapaculo; Tarcacuna
    Tapaculo
f   Mérulaxe du Panama
d Schwarzstirntapaculo
i   Tapaculo panamense

9069   ***Scytalopus parkeri***
    Passeriformes - Rhinocryptidae
e   Chusquea Tapaculo

9070   ***Scytalopus parvirostris***
    Passeriformes - Rhinocryptidae
e   Grey Tapaculo

9071   ***Scytalopus psychopompus***
    Passeriformes - Rhinocryptidae
e   Chestnut-sided Tapaculo; Bahia
    Tapaculo
f   Mérulaxe de Bahia

9072   ***Scytalopus robbinsi***
    Passeriformes - Rhinocryptidae
e   Ecuadorian Tapaculo

9073   ***Scytalopus sanctaemartae***
    Passeriformes - Rhinocryptidae
e   Santa Marta Tapaculo

f Mérulaxe des Santa Marta
i Tapaculo di Santa Marta

9074 *Scytalopus schulenbergi*
Passeriformes - Rhinocryptidae
e Diademed Tapaculo

9075 *Scytalopus simonsi*
Passeriformes - Rhinocryptidae
e Andean Tapaculo

9076 *Scytalopus speluncae*
Passeriformes - Rhinocryptidae
e Mouse-coloured Tapaculo
f Mérulaxe souris
d Maustapaculo

9077 *Scytalopus spillmani*
Passeriformes - Rhinocryptidae
e Spillman's Tapaculo

9078 *Scytalopus superciliaris*
Passeriformes - Rhinocryptidae
e White-browed Tapaculo
f Mérulaxe bridé; Mérulaxe à sourcils
d Weißbrauentapaculo
i Tapaculo sopracciglibianchi

9079 *Scytalopus unicolor*
Passeriformes - Rhinocryptidae
e Unicoloured Tapaculo
f Mérulaxe unicolore
d Einfarbtapaculo
i Tapaculo unicolore

9080 *Scytalopus urubambae*
Passeriformes - Rhinocryptidae
e Cuzco Tapaculo

9081 *Scytalopus vicinior*
Passeriformes - Rhinocryptidae
e Narino Tapaculo
f Mérulaxe du Narino
d Fleckenkopftapaculo
i Tapaculo di Narino

9082 *Scytalopus zimmeri*
Passeriformes - Rhinocryptidae
e Zimmer's Tapaculo

9083 *Scythrops novaehollandiae*
Cuculiformes - Cuculidae
e Channel-billed Cuckoo
f Coucou présageur
d Fratzenkuckuck
i Cuculo beccoscanalato

9084 *Seicercus affinis*
Passeriformes - Sylviidae
e White-spectacled Warbler; Allied
Flycatcher-Warbler; Allied Warbler
f Pouillot affin
d Silberbrillenlaubsänger
i Luì pigliamosche dagli occhiali

9085 *Seicercus burkii*
Passeriformes - Sylviidae
e Golden-spectacled Warbler; Yellow-
eyed Flycatcher-Warbler; Golden-
eyed Flycatcher-Warbler; Black-
browed Flycatcher-Warbler; Golden-
spectacled Flycatcher-Warbler;
Yellow-eyed Warbler
f Pouillot de Burke; Gobemouche de
Burke; Fauvette-gobemouches de
Burke; Gobe-mouches de Burke
d Goldbrillenlaubsänger
i Luì pigliamosche occhigialli

9086 *Seicercus castaniceps*
Passeriformes - Sylviidae
e Chestnut-crowned Warbler;
Chestnut-headed Flycatcher-Warbler;
Chestnut-crowned Flycatcher-
Warbler; Chestnut Warbler;
Chestnut-headed Warbler
f Pouillot à couronne marron;
Gobemouche à couronne marron;
Fauvette-gobemouches à couronne
marron; Gobe-mouches à couronne
marron
d Rotkopflaubsänger
i Luì pigliamosche testacastana

9087 *Seicercus grammiceps*
Passeriformes - Sylviidae
e Sunda Warbler; Sunda Flycatcher-
Warbler; White-rumped Warbler
f Pouillot de la Sonde
d Sunda-Laubsänger
i Luì pigliamosche della Sonda

9088   Seicercus montis
Passeriformes - Sylviidae
- *e*   Yellow-breasted Warbler; Yellow-breasted Flycatcher-Warbler
- *f*   Pouillot à poitrine jaune
- *d*   Gelbbauchlaubsänger
- *i*   Luì pigliamosche pettogiallo

9089   Seicercus poliogenys
Passeriformes - Sylviidae
- *e*   Grey-cheeked Warbler; Grey-cheeked Flycatcher-Warbler
- *f*   Pouillot à joues grises; Fauvette à joues grises; Gobemouche à joues grises; Fauvette-gobemouches à joues grises; Gobe-mouches à joues grises
- *d*   Grauwangenlaubsänger
- *i*   Luì pigliamosche guancegrige

9090   Seicercus xanthoschistos
Passeriformes - Sylviidae
- *e*   Grey-hooded Warbler; Grey-headed Flycatcher-Warbler; Grey-headed Warbler
- *f*   Pouillot à tête grise
- *d*   Grauscheitellaubsänger
- *i*   Luì pigliamosche testagrigia

9091   Seiurus aurocapillus
Passeriformes - Fringillidae
- *e*   Ovenbird; Betsy Kick-up (WI)
- *f*   Paruline couronnée; Petite Chitte dorée (Ants); Fauvette couronnée
- *d*   Pieperwaldsänger; Ofenvogel; Goldkopfwaldsänger
- *i*   Seiuro dalla testa d'oro; Tordo corona dorato; Seiuro corona dorata

9092   Seiurus motacilla
Passeriformes - Fringillidae
- *e*   Louisiana Waterthrush
- *f*   Paruline hochequeue; Fauvette hochequeue
- *d*   Stelzenwaldsänger
- *i*   Seiuro della Luisiana; Seiuro della Louisiana

9093   Seiurus noveboracensis
Passeriformes - Fringillidae
- *e*   Antostomus n

- *f*   Paruline des ruisseaux; Fauvette des ruisseaux
- *d*   Drosselwaldsänger; Uferwaldsänger
- *i*   Tordo acquaiolo del Nord; Seiuro del Nord

9094   Selasphorus ardens
Apodiformes - Trochilidae
- *e*   Glow-throated Hummingbird
- *f*   Colibri ardent
- *d*   Glutkehlkolibri
- *i*   Colibrì ardente

9095   Selasphorus flammula
Apodiformes - Trochilidae
- *e*   Volcano Hummingbird; Rose-throated Hummingird; Cerise-throated Hummingbird; Heliotrope-throated Hummingbird
- *f*   Colibri flammule
- *d*   Weinkehlkolibri
- *i*   Colibrì codalarga minore

9096   Selasphorus floresii
Apodiformes - Trochilidae
- *e*   Floresi's Hummingbird

9097   Selasphorus platycercus
Apodiformes - Trochilidae
- *e*   Broad-tailed Hummingbird
- *f*   Colibri à queue large; Colibri tricolore
- *d*   Dreifarbenkolibri
- *i*   Colibrì codalarga

9098   Selasphorus rufus
Apodiformes - Trochilidae
- *e*   Rufous Hummingbird
- *f*   Colibri roux
- *d*   Zimtkolibri
- *i*   Colibrì codalarga rossiccio; Colibrì rossiccio

9099   Selasphorus sasin
Apodiformes - Trochilidae
- *e*   Allen's Hummingbird; Red-backed Woodstar
- *f*   Colibri sasin
- *d*   Allen-Kolibri
- *i*   Colibrì di Allen

9100 *Selasphorus scintilla*
  Apodiformes - Trochilidae
  e Scintillant Hummingbird
  f Colibri scintillant
  d Flämmchenkolibri
  i Colibrì scintillante

9101 *Selasphorus simoni*
  Apodiformes - Trochilidae
  e Cerise-throated Hummingbird;
  Simon's Hummingbird
  d Kirschkehlkolibri

9102 *Selasphorus torridus*
  Apodiformes - Trochilidae
  e Heliotrope-throated Hummingbird;
  Torrid Hummingbird

9103 *Selenidera culik*
  Piciformes - Ramphastidae
  e Guianan Toucanet; Guiana Toucanet
  f Toucanet koulik
  d Pfefferfresser
  i Tucanetto della Guiana

9104 *Selenidera gouldii*
  Piciformes - Ramphastidae
  e Gould's Toucanet
  f Toucanet de Gould
  d Gould-Arassari
  i Tucanetto di Gould

9105 *Selenidera maculirostris*
  Piciformes - Ramphastidae
  e Spot-billed Toucanet
  f Toucanet à bec tacheté
  d Bindenschnabeltukan;
  Goldohrarassari; Fleckenarassari
  i Tucanetto beccomaculato

9106 *Selenidera nattereri*
  Piciformes - Ramphastidae
  e Tawny-tufted Toucanet; Natterer's
  Toucanet
  f Toucanet de Natterer
  d Natterer-Arassari
  i Tucanetto di Natterer

9107 *Selenidera reinwardtii*
  Piciformes - Ramphastidae
  e Golden-collared Toucanet;

  Reinwardt's Toucanet
  f Toucanet de Reinwardt
  d Reinwardt-Arassari
  i Tucanetto di Reinwardt

9108 *Selenidera spectabilis*
  Piciformes - Ramphastidae
  e Yellow-eared Toucanet
  f Toucanet à oreilles d'or; Toucanet à
  oreille d'or
  d Gelbohrarassari
  i Tucanetto orecchiegialle

9109 *Seleucidis melanoleuca*
  Passeriformes - Corvidae
  e Twelve-wired Bird of Paradise
  f Paradisier multifil
  d Fadenhopf
  i Paradisea dai dodici filli

9110 *Semioptera wallacei*
  Passeriformes - Corvidae
  e Standardwing Bird of Paradise;
  Wallace's Standardwing; Standard-
  winged Bird of Paradise;
  Standardwing
  f Paradisier de Wallace
  d Wallace-Paradiesvogel
  i Paradisea di Wallace

9111 *Semnornis frantzii*
  Piciformes - Ramphastidae
  e Prong-billed Barbet
  f Cabézon de Frantzius
  d Aztekenbartvogel
  i Barbuto di Frantz

9112 *Semnornis ramphastinus*
  Piciformes - Ramphastidae
  e Toucan Barbet
  f Cabézon toucan
  d Tukanbartvogel
  i Barbuto tucanetto; Barbuto
  ramfastino

9113 *Sephanoides fernandensis*
  Apodiformes - Trochilidae
  e Juan Fernandez Firecrown;
  Fernandez Firecrown
  f Colibri de Robinson

d   Juan Fernandez-Kolibri
i   Corona di fiamma delle Fernandez

9114   ***Sephanoides sephaniodes***
      Apodiformes - Trochilidae
e   Green-backed Firecrown
f   Colibri du Chili
d   Chile-Kolibri
i   Corona di fiamma dorsoverde

9115   ***Sericornis arfakianus***
      Passeriformes - Pardalotidae
e   Grey-green Scrubwren; Grey-green
      Sericornis
f   Séricorne vert-de-gris
d   Arfak-Sericornis
i   Sericeo grigioverde

9116   ***Sericornis beccarii***
      Passeriformes - Pardalotidae
e   Tropical Scrubwren; Beccari's
      Scrubwren; Little Scrubwren;
      Beccari's Sericornis; Little Sericornis
f   Séricorne de Beccari
d   Beccari-Sericornis
i   Sericeo di Beccari

9117   ***Sericornis citreogularis***
      Passeriformes - Pardalotidae
e   Yellow-throated Scrubwren; Yellow-
      throated Sericornis; Lemon-throated
      Sericornis; Lemon-throated
      Scrubwren
f   Séricorne à gorge jaune
d   Gelbkehlsericornis
i   Sericeo golagialla

9118   ***Sericornis frontalis***
      Passeriformes - Pardalotidae
e   White-browed Scrubwren; White-
      browed Sericornis
f   Séricorne à sourcils blancs
d   Weißbrauensericornis
i   Sericeo dai mustacchi

9119   ***Sericornis frontalis maculatus***
      Passeriformes - Pardalotidae
e   Spotted Sericornis; Spotted
      Scrubwren
f   Séricorne maculé

9120   ***Sericornis humilis***
      Passeriformes - Pardalotidae
e   Tasmanian Scrubwren; Brown
      Scrubwren; Tasmanian Sericornis
f   Séricorne brun
i   Sericeo bruno

9121   ***Sericornis keri***
      Passeriformes - Pardalotidae
e   Atherton Scrubwren; Atherton
      Sericornis
f   Séricorne de l'Atherton
d   Rotstirnsericornis
i   Sericeo di Atherton

9122   ***Sericornis magnirostris***
      Passeriformes - Pardalotidae
e   Large-billed Scrubwren; Large-billed
      Sericornis
f   Séricorne à grand bec
d   Fahlstirnsericornis
i   Sericeo beccogrosso

9123   ***Sericornis nouhuysi***
      Passeriformes - Pardalotidae
e   Large Scrubwren; Large Mountain-
      Scrubwren; Large Mountain-
      Sericornis; Noisy Scrubwren
f   Séricorne montagnard
d   Bergsericornis
i   Sericeo montano

9124   ***Sericornis papuensis***
      Passeriformes - Pardalotidae
e   Papuan Scrubwren; Papuan
      Sericornis; Olive Scrubwren
f   Séricorne papou
d   Papua-Sericornis
i   Sericeo papua

9125   ***Sericornis perspicillatus***
      Passeriformes - Pardalotidae
e   Buff-faced Scrubwren; Buff-faced
      Sericornis; Black-and-green
      Sericornis; Morobe Sericornis
f   Séricorne fardé
d   Brillensericornis
i   Sericeo dagli occhiali

9126   ***Sericornis rufescens***
      Passeriformes - Pardalotidae

e Vogelkop Scrubwren; Arfak Buff-
faced Sericornis; Arfak Buff-faced
Scrubwren; Arfak Scrubwren;
Vogelkop Sericornis; Rufous
Scrubwren
f Séricorne chamois
d Braunohrsericornis
i Sericeo di Vogelkop

9127 *Sericornis spilodera*
Passeriformes - Pardalotidae
e Pale-billed Scrubwren; Pale-billed
Sericornis
f Séricorne à bec blanc
d Fahlschnabelsericornis
i Sericeo beccochiaro

9128 *Sericornis virgatus*
Passeriformes - Pardalotidae
e Perplexing Scrubwren; Perplexing
Sericornis
f Séricorne mystérieux
d Sepik-Sericornis
i Sericeo lineato

9129 *Sericossypha albocristata*
Passeriformes - Fringillidae
e White-capped Tanager
f Tangara à coiffe blanche
d Weißkappentangare;
Schmucktangare
i Tangara capobianco

9130 *Sericulus aureus*
Passeriformes - Ptilonorhynchidae
e Flame Bowerbird; Black-faced
Golden-Bowerbird; New Guinea
Golden-Bowerbird; Flamed
Bowerbird; Flamed Regentbird;
Black-faced Bowerbird; Golden
Regentbird
f Jardinier du prince d'Orange
d Papua Goldvogel; Goldlaubenvogel
i Sericulo di fiamma; Giardiniere di
fiamma

9131 *Sericulus bakeri*
Passeriformes - Ptilonorhynchidae
e Fire-maned Bowerbird; Beck's
Bowerbird; Adelbert's Bowerbird;
Adelbert's Regentbird; Baker's

Bowerbird; Macloud Bowerbird
f Jardinier de Baker
d Rotscheitellaubenvogel
i Sericulo criniera di fiamma;
Giardiniere criniera di fiamma

9132 *Sericulus chrysocephalus*
Passeriformes - Ptilonorhynchidae
e Regent Bowerbird; Australian
Regentbird; Golden Regentbird
f Jardinier prince-régent; Oiseau-
régent
d Samtgoldvogel;
Gelbnackenlaubenvogel
i Sericulo dalla testa gialla;
Giardiniere testadorata; Sericulo a
testa gialla; Sericulo testagialla

9133 *Serilophus lunatus*
Passeriformes - Eurylaimidae
e Silver-breasted Broadbill; Collared
Broadbill; Gould's Broadbill;
Hodgson's Broadbill
f Eurylaime de Gould
d Würgerbreitrachen; Würgerbreitmaul
i Beccolargo pettoargentato

9134 *Serinus alario*
Passeriformes - Fringillidae
e Black-headed Canary; Blackhead
Canary; Mountain Canary; Damara
Canary; Namibia Canary; Blackhead
f Serin alario
d Alariogirlitz; Alario
i Alario; Canarino di Damara

9135 *Serinus albogularis*
Passeriformes - Fringillidae
e White-throated Canary; White-
throated Seedeater
f Serin à gorge blanche
d Weißkehlgirlitz
i Verzellino golabianca

9136 *Serinus ankoberensis*
Passeriformes - Fringillidae
e Ankober Serin; Ankober Seedeater
f Serin d'Ankober
i Canarino di Ankober

9137   *Serinus atrogularis*
        Passeriformes - Fringillidae
   e   Black-throated Canary; Black-
        throated Seedeater (CSA); Peach's
        Canary (CSA); Yellow-rumped
        Seedeater; Southern Yellow-rumped
        Seedeater; Yellow-rumped Serin;
        Black-throated Serin
   f   Serin à gorge noire
   d   Angola-Girlitz
   i   Verzellino golanera

9138   *Serinus buchanani*
        Passeriformes - Fringillidae
   e   Kenya Grosbeak-Canary; Southern
        Grosbeak-Canary (CSA); Grosbeak
        Canary
   f   Serin de Buchanan
   i   Canarino di Buchanan

9139   *Serinus burtoni*
        Passeriformes - Fringillidae
   e   Thick-billed Seedeater; Grosbeak-
        Seedeater; Thick-billed Serin
   f   Serin de Burton
   d   Dickschnabelgirlitz
   i   Canarino beccogrosso

9140   *Serinus canaria*
        Passeriformes - Fringillidae
   e   Island Canary; Canary; Common
        Canary; Canarybird
   f   Serin des Canaries; Canari; Serin
        canari
   d   Kanarien-Girlitz; Kanarienvogel;
        Wilder Kanarienvogel
   i   Canarino; Canarino delle Canarie

9141   *Serinus canicapillus*
        Passeriformes - Fringillidae
   e   West African Seedeater; West
        African Streaky-headed Seedeater;
        West African Serin
   f   Serin ouest-africain
   d   Braunwangengirlitz
   i   Verzellino dell'Africa occidentale

9142   *Serinus canicollis*
        Passeriformes - Fringillidae
   e   Cape Canary; Yellow-crowned
        Canary (CSA); Kenya Canary

   f   Serin du Cap
   d   Gelbscheitelgirlitz
   i   Canarino del Capo

9143   *Serinus canicollis flavivertex*
        Passeriformes - Fringillidae
   f   Serin à calotte jaune

9144   *Serinus capistratus*
        Passeriformes - Fringillidae
   e   Black-faced Canary
   f   Serin à masque noir; Serin à face
        noire
   d   Zügelgirlitz
   i   Canarino faccianera

9145   *Serinus citrinella*
        Passeriformes - Fringillidae
   e   Citril Finch; European Citril-Finch;
        Citril
   f   Venturon montagnard; Venturon
        alpin; Tarin venturon; Serin de
        montagne
   d   Zitronengirlitz; Zitronenzeisig
   i   Venturone; Verzellino europeo

9146   *Serinus citrinelloides*
        Passeriformes - Fringillidae
   e   African Citril; African Citril-Finch;
        Abyssinian Citril-Finch; Abyssinian
        Citril
   f   Serin d'Abyssinie; Venturon africain
   d   Dünnschnabelgirlitz
   i   Venturone africano

9147   *Serinus citrinipectus*
        Passeriformes - Fringillidae
   e   Lemon-breasted Seedeater; Lemon-
        breasted Canary (CSA)
   f   Serin à poitrine citron
   d   Gelbbrustgirlitz
   i   Verzellino pettogiallo

9148   *Serinus corsicanus*
        Passeriformes - Fringillidae
   e   Corsican Citril-Finch
   f   Venturon corse
   d   Korsika-Girlitz; Korsen-Girlitz
   i   Verzellino

9149 **Serinus donaldsoni**
Passeriformes - Fringillidae
e Abyssinian Grosbeak-Canary;
Northern Grosbeak-Canary (CSA);
Grosbeak Canary; Grosbeak Serin
f Serin à gros bec
d Kernbeißergirlitz
i Canarino di Donaldson

9150 **Serinus dorsostriatus**
Passeriformes - Fringillidae
e White-bellied Canary; Somali Canary
f Serin à ventre blanc
d Weißbauchgirlitz; Großer
Mozambique-Girlitz
i Canarino ventrebianco

9151 **Serinus estherae**
Passeriformes - Fringillidae
e Mountain Serin; Malay Goldfinch;
Mindanao Goldfinch; Malay Serin;
Malaysian Finch; Sunda Serin;
Indonesian Serin; Javan Greenfinch
f Serin malais
d Mindanao-Girlitz; Malaien-Girlitz
i Verzellino della Malesia

9152 **Serinus flavigula**
Passeriformes - Fringillidae
e Yellow-throated Seedeater; Yellow-
throated Serin
f Serin à gorge jaune

9153 **Serinus flaviventris**
Passeriformes - Fringillidae
e Yellow Canary; Shell Canary;
Yellow Sysie
f Serin de Sainte-Hélène; Serin jaune
d Gelbbauchgirlitz
i Canarino ventregiallo

9154 **Serinus frontalis**
Passeriformes - Fringillidae
e Western Citril; Yellow-fronted
Canary; Yellow-fronted Citril;
Western Citril-Finch
f Serin à diadème
d Diademgirlitz
i Venturone di Kivu

9155 **Serinus gularis**
Passeriformes - Fringillidae
e Streaky-headed Seedeater; Streaky-
headed Canary (CSA); Streaky-
headed Serin; Streak-headed
Seedeater
f Serin gris; Serin à tête blanche
d Brauengirlitz
i Verzellino capostriato

9156 **Serinus hypostictus**
Passeriformes - Fringillidae
e East African Citril; East African
Citril-Finch; Tanzanian Citril;
Eastern Citril
f Serin est-africain
d Streifengirlitz
i Venturone dell'Africa orientale

9157 **Serinus koliensis**
Passeriformes - Fringillidae
e Papyrus Canary; Koli Canary; Van
Someren's Canary; Papyrus Serin
f Serin du Koli; Serin des papyrus
d Papyrusgirlitz
i Canarino di Van Someren

9158 **Serinus leucopterus**
Passeriformes - Fringillidae
e White-winged Seedeater; Protea;
Protea Canary (CSA); Layard's
Seedeater; Protea Seedeater
f Serin bifascié; Serin des protéas
d Proteagirlitz; Weißbindengirlitz
i Canarino alibianche

9159 **Serinus leucopygius**
Passeriformes - Fringillidae
e White-rumped Seedeater; White-
rumped Canary; Grey Canary;
White-rumped Siskin; White-rumped
Serin
f Serin à croupion blanc; Chanteur
d'Afrique
d Weißbürzelgirlitz; Grauedelsänger
i Verzellino gropponebianco

9160 **Serinus melanochrous**
Passeriformes - Fringillidae
e Kipengere Seedeater; Tanzania
Seedeater

f   Serin des Kipengere
i   Canarino del Kipengere

9161   *Serinus menachensis*
      Passeriformes - Fringillidae
e   Yemen Serin; Yemen Seedeater;
      Yemen Canary; Menacha Seedeater;
      Menacha Serin
f   Serin du Yémen
d   Menacha-Girlitz
i   Verzellino dello Yemen

9162   *Serinus mennelli*
      Passeriformes - Fringillidae
e   Black-eared Seedeater; Black-eared
      Canary (CSA); Black-eared Serin
f   Serin oreillard
d   Schwarzwangengirlitz
i   Verzellino orecchienere

9163   *Serinus mozambicus*
      Passeriformes - Fringillidae
e   Yellow-fronted Canary; Green
      Singing-Finch; Yellow-eyed Canary
      (CSA); Icterine Canary; Green
      Canary
f   Serin du Mozambique; Serin à front
      jaune
d   Mozambique-Girlitz
i   Canarino del Mozambico

9164   *Serinus nigriceps*
      Passeriformes - Fringillidae
e   Abyssinian Siskin; Black-headed
      Siskin; Abyssinian Black-headed
      Siskin; Rüppell's Siskin; Black-
      headed Serin; Abyssinian Canary;
      African Black-headed Siskin;
      Ethiopian Black-headed Siskin
f   Serin à tête noire
d   Schwarzkopfgirlitz
i   Canarino testanera

9165   *Serinus pusillus*
      Passeriformes - Fringillidae
e   Fire-fronted Serin; Red-fronted
      Serin; Gold-fronted Serin; Gold-
      fronted Finch
f   Serin à front rouge; Serin à front d'or
d   Rotstirngirlitz

i   Verzellino testarosso; Verzellino
      fronterossa

9166   *Serinus reichardi*
      Passeriformes - Fringillidae
e   Reichard's Seedeater; Stripe-breasted
      Seedeater (CSA); Reichard's
      Seedeater; Stripe-breasted Serin
d   Miombo-Girlitz
i   Verzellino pettostriato

9167   *Serinus reichenowi*
      Passeriformes - Fringillidae
e   Kenya Yellow-rumped Seedeater;
      Yellow-rumped Seedeater (CSA);
      Kenya Yellow-rumped Canary;
      Reichnow's Serin; Kenyan Yellow-
      rumped Serin
d   Reichenow-Girlitz
i   Verzellino di Reichenow

9168   *Serinus rothschildi*
      Passeriformes - Fringillidae
e   Olive-rumped Serin; Arabian Serin;
      Arabian Yellow-rumped Canary;
      Arabian Yellow-rumped Serin;
      Arabian Canary; tothschild's Serin;
      Rothschild's Canary
f   Serin d'Arabie
d   Rothschild-Girlitz
i   Verzellino d'Arabia; Versellino
      gropponeoliva

9169   *Serinus rufobrunneus*
      Passeriformes - Fringillidae
e   Principe Seedeater; West African
      Island-Seedeater
f   Serin de Principe
d   Prinzengirlitz
i   Canarino dell'Isola Principe

9170   *Serinus scotops*
      Passeriformes - Fringillidae
e   Forest Canary; Striped Canary
f   Serin forestier
d   Schwarzkinngirlitz; Waldgirlitz
i   Canarino delle foreste

9171   *Serinus serinus*
      Passeriformes - Fringillidae
e   European Serin; Serin; Eurasian

Serin; Common Serin
f Serin cini; Bouvreuil cini
d Girlitz
i Verzellino; Verdolino; Raperino; Crespolino; Verzollino europeo

9172 *Serinus striolatus*
Passeriformes - Fringillidae
e Streaky Seedeater; Streaky Serin
f Serin strié
d Strichelgirlitz
i Canarino striato

9173 *Serinus sulphuratus*
Passeriformes - Fringillidae
e Bully Canary; Brimstone Canary; Brimstone Serin; Bully Seedeater
f Serin soufré
d Schwefelgirlitz; Schwefelgelber Girlitz
i Canarino solforato

9174 *Serinus symonsi*
Passeriformes - Fringillidae
e Drakensberg Siskin; Symons's Cape-Siskin; Symons's Syskin
f Serin de Symons; Serin du Drakensberg
d Drakenberg-Girlitz; Basuto-Girlitz
i Verzellino del Drakensberg

9175 *Serinus syriacus*
Passeriformes - Fringillidae
e Syrian Serin; Tristram's Serin
f Serin syriaque
d Zederngirlitz; Syrien-Girlitz
i Verzellino siriano; Verzellino di Siria

9176 *Serinus thibetanus*
Passeriformes - Fringillidae
e Tibetan Serin; Tibetan Siskin
f Serin du Tibet; Tarin du Tibet
d Himalaja-Zeisig
i Verzellino tibetano

9177 *Serinus totta*
Passeriformes - Fringillidae
e Cape Siskin; South African Siskin
f Serin totta
d Hottentotten-Girlitz
i Verzellino del Capo

9178 *Serinus tristriatus*
Passeriformes - Fringillidae
e Brown-rumped Seedeater; Brown-rumped Serin
f Serin à trois raies; Serin à croupion brun
d Rüppells Girlitz
i Verzellino dalle tre strie

9179 *Serinus whytii*
Passeriformes - Fringillidae
e Yellow-browed Seedeater; Southern Streaky-Seedeater
f Serin bridé
d Gelbbrauengirlitz
i Canarino di Whyte

9180 *Serinus xantholaemus*
Passeriformes - Fringillidae
e Salvadori's Seedeater; Salvadori's Serin
f Serin de Salvadori

9181 *Serinus xanthopygius*
Passeriformes - Fringillidae
e Abyssinian Yellow-rumped Seedeater; Abyssinian Yellow-rumped Canary; Yellow-rumped Canary; Yellow-rumped Serin; Yellow-rumped Seedeater
f Serin à croupion jaune
d Gelbbürzelgirlitz
i Verzellino gropponegiallo; Verzellino dal groppone giallo

9182 *Serophaga cinerea*
Passeriformes - Tyrannidae
e Torrent Tyrannulet
f Tyranneau des torrents
d Sturzbachtachuri
i Tiranno piccolo di torrent

9183 *Serophaga griseiceps*
Passeriformes - Tyrannidae
e Grey-crowned Tyrannulet
f Tyranneau de Berlioz
d Ockerbindentachuri

9184 *Serophaga hypoleuca*
Passeriformes - Tyrannidae
e River Tyrannulet

f   Tyranneau des rivières
d   Ufertachuri
i   Tiranno piccolo di fiume

9185   *Serpophaga munda*
      Passeriformes - Tyrannidae
e   White-bellied Tyrannulet;
      Berlepsch's Tyrannulet
f   Tyranneau à ventre blanc
d   Berlepsch-Tachuri
i   Tiranno piccolo pettobianco

9186   *Serpophaga nigricans*
      Passeriformes - Tyrannidae
e   Sooty Tyrannulet
f   Tyranneau noirâtre
d   Maustachuri
i   Tiranno piccolo fuligginoso

9187   *Serpophaga subcristata*
      Passeriformes - Tyrannidae
e   White-crested Tyrannulet
f   Tyranneau à toupet
d   Strichelschopftachuri
i   Tiranno piccolo crestabianca

9188   *Setophaga ruticilla*
      Passeriformes - Fringillidae
e   American Redstart; Christmasbird
      (WI); Butterflybird (WI)
f   Paruline flamboyante; Petite Chitte
      de feu (Ants); Carougette (Ants);
      Gabriel du feu (Ants); Petit du feu
      (Ants); Fauvette flamboyante; Carte
      (Ants)
d   Schnäpperwaldsänger;
      Rotschwanzwaldsänger
i   Codirosso americano

9189   *Setornis criniger*
      Passeriformes - Pycnonotidae
e   Hook-billed Bulbul; Long-billed
      Bulbul
f   Bulbul à long bec; Bulbul à bec long
d   Langschnabelbülbül
i   Bulbul beccolungo

9190   *Sheppardia aequatorialis*
      Passeriformes - Muscicapidae
e   Equatorial Akalat; Jackson's Akalat;
      Equatorial Redstart; Acholi Akalat

f   Rougegorge équatorial; Merle
      rougegorge équatorial
d   Uganda-Akalat
i   Akelat di Jackson

9191   *Sheppardia bocagei*
      Passeriformes - Muscicapidae
e   Bocage's Akalat; Bocage's Chat;
      Rufous-cheeked Robin-Chat;
      Rufous-cheeked Chat, Bocage's
      Robin; Rufous-cheeked Akalat
f   Rougegorge de Bocage; Cossyphe à
      joues rouges
d   Bocage-Rötel
i   Akelat di Bocage

9192   *Sheppardia cyornithopsis*
      Passeriformes - Muscicapidae
e   Lowland Akalat; Akalat; Whiskered
      Akalat; Common Akalat; Whiskered
      Redstart
f   Rougegorge merle; Merle rougegorge
d   Schnäpperrötel; Schnäpperakalat
i   Akelat comune; Akelat

9193   *Sheppardia gabela*
      Passeriformes - Muscicapidae
e   Gabela Akalat; Gabela Robin-Chat;
      Gabela Robin; Gabela Redstart
f   Rougegorge de Gabela
d   Gabela-Rötel; Gabela-Akalat
i   Akelat di Gabela

9194   *Sheppardia gunningi*
      Passeriformes - Muscicapidae
e   East Coast Akalat; Gunning's Robin
      (CSA); Gunning's Robin-Chat;
      Eastern Akalat; Gunning's Akalat
f   Rougegorge de Gunning
d   Blauflügelalethe; Blauflügelakalat;
      Blauflügelrötel
i   Akelat della costa orientale

9195   *Sheppardia lowei*
      Passeriformes - Muscicapidae
e   Iringa Akalat; Iringa Alethe; Iringa
      Robin; Iringa Robin-Chat; Iringa
      Ground-Robin
f   Rougegorge de l'Iringa
d   Njombe-Rötel
i   Akelat dell'Iringa

9196 **Shepppardia montana**
Passeriformes - Muscicapidae
e   Usambara Akalat; Usambara
Ground-Robin (CSA); Usambara
Robin; Usambara Robin-Chat;
Usambara Alethe
f   Rougegorge des Usambara
d   Usambara-Rötel
i   Akelat dell'Iringa

9197 **Shepppardia poensis**
Passeriformes - Muscicapidae
e   Alexander's Akalat; Fernando Po
Robin; Fernando Po Akalat;
Fernando Po Robinchat; Short-tailed
Akalat
f   Rougegorge d'Alexander
d   Kurzschwanzrötel
i   Akelat di Alexander

9198 **Shepppardia sharpei**
Passeriformes - Muscicapidae
e   Sharpe's Akalat; Sharpe's Robin-
Chat; Sharpe's Robin
f   Rougegorge de Sharpe
d   Braunbrustakalat; Braunbruströtel
i   Akelat di Sharpe

9199 **Sialia currucoides**
Passeriformes - Muscicapidae
e   Mountain Bluebird; Rocky Bluebird;
Arctic Bluebird
f   Merlebleu azuré; Merlebleu des
montagnes
d   Berghüttensänger
i   Uccello azzurro montano

9200 **Sialia mexicana**
Passeriformes - Muscicapidae
e   Western Bluebird
f   Merlebleu de l'Ouest; Merlebleu à
dos marron
d   Blaukehlhüttensänger
i   Uccello azzurro occidentale

9201 **Sialia sialis**
Passeriformes - Muscicapidae
e   Eastern Bluebird; Common Bluebird
f   Merlebleu de l'Est; Merlebleu à
poitrine rouge
d   Rotkehlhüttensänger

i   Uccello azzurro comune; Uccello
azzurro; Uccello azzurro orientale

9202 **Sicalis auriventris**
Passeriformes - Emberizidae
e   Greater Yellow-Finch; Yellow-billed
Yellow-Finch
f   Grand Sicale
d   Goldbauchgilbammer
i   Fringuello dorato maggiore

9203 **Sicalis citrina**
Passeriformes - Emberizidae
e   Stripe-tailed Yellow-Finch
f   Sicale citrin
d   Zitronengilbammer
i   Fringuello dorato codastriata

9204 **Sicalis colombiana**
Passeriformes - Emberizidae
e   Orange-fronted Yellow-Finch
f   Sicale à béret
d   Zwergsafranammer; Zwergsafranfink
i   Fringuello dorato frontearancio

9205 **Sicalis flaveola**
Passeriformes - Emberizidae
e   Saffron Finch; Canary (WI); Saffron
Yellow-Finch
f   Sicale bouton-d'or; Serin américain
d   Safranammer; Safranfink
i   Fringuello zafferano; Fringuello
dorato zafferano

9206 **Sicalis lebruni**
Passeriformes - Emberizidae
e   Patagonian Yellow-Finch; Le Brun's
Yellow-Finch
f   Sicale de Patagonie
d   Magellan-Gilbammer
i   Fringuello dorato della Patagonia

9207 **Sicalis lutea**
Passeriformes - Emberizidae
e   Puna Yellow-Finch
f   Sicale jaune
d   Punagilbammer
i   Fringuello dorato della Puna

9208 **Sicalis luteiventris**
Passeriformes - Emberizidae

e   Misto Yellow-Finch
f   Sicale misto
i   Fringuello dorato meridionale

9209   ***Sicalis luteocephala***
Passeriformes - Emberizidae
e   Citron-headed Yellow-Finch
f   Sicale à tête jaune
d   Graunackengilbammer
i   Fringuello dorato testalimone

9210   ***Sicalis luteola***
Passeriformes - Emberizidae
e   Grassland Yellow-Finch; Grass
Sparrow (WI); Grass Canary (WI);
Yellow Grass-Finch; Striped Yellow-
Finch; Yellow-breasted Yellow-
Finch; Montane Yellow-Finch
f   Sicale des savanes; Moisson jaune
(Ants); Petit Serin (Ants)
d   Kurzschnabelgilbammer
i   Fringuello dorato della prateria

9211   ***Sicalis olivascens***
Passeriformes - Emberizidae
e   Greenish Yellow-Finch
f   Sicale olivâtre
d   Olivbrustgilbammer
i   Fringuello dorato verdastro

9212   ***Sicalis pelzelni***
Passeriformes - Emberizidae
e   Pelzeln's Finch

9213   ***Sicalis raimondii***
Passeriformes - Emberizidae
e   Raimondi's Yellow-Finch
f   Sicale de Raimondi
d   Grauflankengilbammer
i   Fringuello dorato di Raimondi

9214   ***Sicalis taczanowskii***
Passeriformes - Emberizidae
e   Sulphur-throated Finch; Sulphur-
breasted Finch; Taczanowski's
Yellow-Finch
f   Sicale de Taczanowski
d   Taczanowski-Gilbammer
i   Fringuello dorato pettosulfureo

9215   ***Sicalis uropygialis***
Passeriformes - Emberizidae
e   Bright-rumped Yellow-Finch
f   Sicale à croupion jaune
d   Goldbürzelgilbammer
i   Fringuello dorato gropponechiaro

9216   ***Sigelus silens***
Passeriformes - Muscicapidae
e   Fiscal Flycatcher
f   Gobemouche fiscal; Gobe-mouches
fiscal
d   Würgerschnäpper
i   Pigliamosche fiscale

9217   ***Silvicultrix diadema***
Passeriformes - Tyrannidae
e   Yellow-bellied Chat-Tyrant;
Hartlaub's Chat-Tyrant
f   Pitajo diadème
d   Diademtyrann
i   Tiranno ventregiallo

9218   ***Silvicultrix frontalis***
Passeriformes - Tyrannidae
e   Crowned Chat-Tyrant
f   Pitajo couronné
d   Braunscheiteltyrann
i   Tiranno coronato

9219   ***Silvicultrix jelskii***
Passeriformes - Tyrannidae
e   Jelski's Chat-Tyrant
f   Pitajo de Jelski
i   Tiranno di Jelski

9220   ***Silvicultrix pulchella***
Passeriformes - Tyrannidae
e   Golden-browed Chat-Tyrant;
Yellow-browed Chat-Tyrant
f   Pitajo à sourcils d'or
d   Goldbrauentyrann
i   Tiranno sopracciglidorati

9221   ***Simoxenops striatus***
Passeriformes - Furnariidae
e   Bolivian Recurvebill; Bolivian
Foliagegleaner
f   Anabate de Bolivie
d   Cochamba-Blattspäher
i   Beccocurvo della Bolivia

9222 **Simoxenops ucayalae**
Passeriformes - Furnariidae
 e Peruvian Recurvebill; Recurvebill;
Peruvian Foliagegleaner
 f Anabate à bec retroussé
 d Ucayali-Blattspäher
 i Beccocurvo del Perù

9223 **Siphonorhis americanus**
Caprimulgiformes - Caprimulgidae
 e Jamaican Poorwill; Jamaican
Pauraque
 f Engoulevent de la Jamaïque
 d Haiti-Nachtschwalbe
 i Succiacapre della Giamaica

9224 **Siphonorhis brewsteri**
Caprimulgiformes - Caprimulgidae
 e Least Poorwill; Least Pauraque
 f Grouiller-corps (Ants); Engoulevent
grouillécor
 i Succiacapre di Brewster

9225 **Sipodotus wallacii**
Passeriformes - Maluridae
 e Wallace's Fairywren; Wallace's
Wren; Blue-capped Fairywren;
Wallace's Wren-Warbler
 f Mérion de Wallace
 d Rostnackenstaffelschwanz
 i Scricciolo splendente di Wallace

9226 **Siptornis striaticollis**
Passeriformes - Furnariidae
 e Spectacled Prickletail
 f Pseudosittine à collier
 d Streifenhalsschlüpfer
 i Codaspinosa dagli occhiali

9227 **Siptornopsis hypochondriacus**
Passeriformes - Furnariidae
 e Great Spinetail
 f Synallaxe à poitrine rayée
 d Salvins Schlüpfer
 i Codaspinosa maggiore

9228 **Sirystes albogriseus**
Passeriformes - Tyrannidae
 e White-rumped Sirystes

9229 **Sirystes sibilator**
Passeriformes - Tyrannidae
 e Sibilant Sirystes; Sirystes (NA)
 f Tyran siffleur
 i Tiranno flautista

9230 **Sitta azurea**
Passeriformes - Sittidae
 e Blue Nuthatch; Azure Nuthatch
 f Sittelle bleue
 d Azurkleiber
 i Picchio muratore azzurra

9231 **Sitta canadensis**
Passeriformes - Sittidae
 e Red-breasted Nuthatch
 f Sittelle à poitrine rousse; Sitelle du
Canada
 d Kanada-Kleiber; Klappenreiher
 i Picchio muratore pettofulvo

9232 **Sitta carolinensis**
Passeriformes - Sittidae
 e White-breasted Nuthatch
 f Sittelle à poitrine blanche
 d Carolina-Kleiber
 i Picchio muratore pettobianco

9233 **Sitta cashmirensis**
Passeriformes - Sittidae
 e Kashmir Nuthatch; Brooks's
Nuthatch
 f Sittelle du Cachemire
 d Kaschmir-Kleiber
 i Picchio muratore del Kashmir

9234 **Sitta castanea**
Passeriformes - Sittidae
 e Chestnut-bellied Nuthatch; Chestnut-
breasted Nuthatch
 f Sittelle à ventre marron
 d Zimtkleiber; Kastanienkleiber
 i Picchio muratore ventrecastano

9235 **Sitta europaea**
Passeriformes - Sittidae
 e Eurasian Nuthatch; Wood Nuthatch;
Nuthatch; European Nuthatch;
Common Nuthatch
 f Sittelle torchepot; Sitelle d'Europe;
Grimpereau bleu

d   Kleiber; Spechtmeise
i   Picchio muratore europeo; Picchio
     muratore

9236   *Sitta formosa*
     Passeriformes - Sittidae
e   Beautiful Nuthatch
f   Sittelle superbe
d   Schönkleiber; Schmuckkleiber
i   Picchio muratore splendente

9237   *Sitta frontalis*
     Passeriformes - Sittidae
e   Velvet-fronted Nuthatch; Velvet-
     fronted Blue-Nuthatch
f   Sittelle veloutée; Sitelle à front
     velouté
d   Samtstirnkleiber
i   Picchio muratore frontevellutata

9238   *Sitta himalayensis*
     Passeriformes - Sittidae
e   White-tailed Nuthatch; Himalayan
     Nuthatch
f   Sittelle de l'Himalaya
d   Weißschwanzkleiber
i   Picchio muratore dell'Himalaya

9239   *Sitta krueperi*
     Passeriformes - Sittidae
e   Krüper's Nuthatch
f   Sittelle de Krüper; Sitelle naine
d   Türkenkleiber; Kleinasien-Kleiber
i   Picchio muratore di Krüper

9240   *Sitta ledanti*
     Passeriformes - Sittidae
e   Algerian Nuthatch; Kabylie
     Nuthatch; Kabylian Nuthatch
f   Sittelle kabyle
d   Kabylenkleiber
i   Picchio muratore della Kabilia

9241   *Sitta leucopsis*
     Passeriformes - Sittidae
e   White-cheeked Nuthatch;
     Przewalski's Nuthatch
f   Sittelle à joues blanches
d   Weißwangenkleiber
i   Picchio muratore guancebianche

9242   *Sitta magna*
     Passeriformes - Sittidae
e   Giant Nuthatch
f   Sittelle géante
d   Riesenkleiber
i   Picchio muratore gigante

9243   *Sitta nagaensis*
     Passeriformes - Sittidae
e   Chestnut-vented Nuthatch; Mountain
     Nuthatch; Naga Nuthatch
f   Sittelle des Naga
i   Picchio muratore di Naga

9244   *Sitta neumayer*
     Passeriformes - Sittidae
e   Western Rock-Nuthatch; Rock
     Nuthatch; Lesser Rock-Nuthatch;
     Neumayer's Rock-Nuthatch; Syrian
     Rock-Nuthatch
f   Sittelle de Neumayer; Sitelle des
     rochers de Neumayer
d   Felsenkleiber
i   Picchio muratore di roccia

9245   *Sitta oenochlamys*
     Passeriformes - Sittidae
e   Sulphur-billed Nuthatch
f   Sittelle des Philippines
i   Picchio muratore beccogiallo

9246   *Sitta pusilla*
     Passeriformes - Sittidae
e   Brown-headed Nuthatch
f   Sittelle à tête brune
d   Braunkopfkleiber
i   Picchio muratore capobruno

9247   *Sitta pygmaea*
     Passeriformes - Sittidae
e   Pygmy Nuthatch
f   Petite Sittelle; Sittelle pygmée
d   Zwergkleiber
i   Picchio muratore pigmeo

9248   *Sitta solangiae*
     Passeriformes - Sittidae
e   Yellow-billed Nuthatch; Lilac
     Nuthatch
f   Sittelle à bec jaune

   d Gelbschnabelkleiber
   i Picchio muratore lilla

9249 *Sitta tephronota*
      Passeriformes - Sittidae
   e Eastern Rock-Nuthatch; Great Rock-Nuthatch; Greater Rock-Nuthatch; Large Rock-Nuthatch; Rock Nuthatch; Persian Nuthatch
   f Sittelle des rochers
   d Klippenkleiber; Steinkleiber
   i Picchio muratore di roccia orientale

9250 *Sitta victoriae*
      Passeriformes - Sittidae
   e White-browed Nuthatch; Victoria Nuthatch
   f Sittelle du Victoria
   i Picchio muratore della regina Victoria

9251 *Sitta villosa*
      Passeriformes - Sittidae
   e Snowy-browed Nuthatch; Chinese Nuthatch; Black-headed Nuthatch
   f Sittelle de Chine
   d Chinesen-Kleiber; Korea-Kleiber
   i Picchio muratore cinese

9252 *Sitta whiteheadi*
      Passeriformes - Sittidae
   e Corsican Nuthatch; Whitehead's Nuthatch
   f Sittelle de Corse; Sitelle corse; Sitelle de Whitehead
   d Korsen-Kleiber; Korsika-Kleiber
   i Picchio muratore corso

9253 *Sitta yunnanensis*
      Passeriformes - Sittidae
   e Yunnan Nuthhatch; Black-masked Nuthatch
   f Sittelle du Yunnan
   d Yunnan-Kleiber
   i Picchio muratore dello Yunnan

9254 *Sittasomus aequatorialis*
      Passeriformes - Dendrocolaptidae
   e Pacific Woodcreeper

9255 *Sittasomus griseicapillus*
      Passeriformes - Dendrocolaptidae
   e Olivaceous Woodcreeper; Grayish Woodcreeper (NA)
   f Grimpar fauvette
   d Olivbaumsteiger; Scheinbaumläufer
   i Rampichino testagrigia

9256 *Sittasomus reiseri*
      Passeriformes - Dendrocolaptidae
   e Reiser's Woodcreeper

9257 *Sittasomus sylviellus*
      Passeriformes - Dendrocolaptidae
   e Olivaceous Woodcreeper

9258 *Sittidae*
      Passeriformes
   e Nuthatches
   f Sittelles; Sittidés
   d Kleiber; Spechtmeisen
   i Sittidi; Picchi muratori

9259 *Skutchia borbae*
      Passeriformes - Thamnophilidae
   e Pale-faced Bare-eye; Pale-faced Antbird
   f Fourmilier à face pâle; Fourmilier à face blanche
   d Samtbrauenameisenvogel
   i Mangiaformiche facciachiara

9260 *Smicrornis brevirostris*
      Passeriformes - Pardalotidae
   e Weebill; Brown Weebill
   f Gérygone à bec court
   d Stutzschnabel
   i Beccominuto

9261 *Smithornis capensis*
      Passeriformes - Eurylaimidae
   e African Broadbill; Black-capped Broadbill; Delacour's Broadbill
   f Eurylaime du Cap
   d Kap-Breitrachen; Spatelbreitrachen; Spatelbreitmaul
   i Beccolargo africano

9262 *Smithornis rufolateralis*
      Passeriformes - Eurylaimidae
   e Rufous-sided Broadbill; Red-sided

Broadbill; Red Broadbill
f Eurylaime à flancs roux
d Kappenbreitrachen;
Rotbrustbreitrachen;
Rotflankenbreitrachen
i Beccolargo fianchirossi

9263 *Smithornis sharpei*
Passeriformes - Eurylaimidae
e Grey-headed Broadbill; Sharpe's
Broadbill
f Eurylaime à tête grise
d Graukopfbreitrachen
i Beccolargo testagrigia

9264 *Somateria fischeri*
Anseriformes - Anatidae
e Spectacled Eider; Fischer's Eider
f Eider à lunettes
d Plüschkopfente
i Edredone degli occhiali

9265 *Somateria mollissima*
Anseriformes - Anatidae
e Common Eider; Eider; Eider Duck
f Eider à duvet
d Eiderente
i Edredone; Edredone comune; Anatra
dal piumino

9266 *Somateria spectabilis*
Anseriformes - Anatidae
e King Eider
f Eider à tête grise
d Prachteiderente
i Re degli edredoni

9267 *Somateria v-nigrum*
Anseriformes - Anatidae
e Pacific Eider

9268 *Spartonoica maluroides*
Passeriformes - Furnariidae
e Bay-capped Wren-Spinetail; Wren-
like Spinetail
f Synallaxe des marais
d Staffelschwanzschlüpfer
i Codaspinosa scricciolo

9269 *Speculipastor bicolor*
Passeriformes - Sturnidae

e Magpie Starling
f Spréo pie
d Spiegelstar
i Storno bicolore

9270 *Speirops brunneus*
Passeriformes - Zosteropidae
e Fernando Po Speirops; Fernando Po
White-eye
f Zostérops de Fernando Po; Speirops
de Fernando Po
d Braunbrillenvogel
i Occhialino del Fernando Po

9271 *Speirops leucophoeus*
Passeriformes - Zosteropidae
e Principe Speirops; Prince's Island
Speirops; Principe Island White-eye;
Prince's Island White-eye
f Zostérops de Principe; Speirops de
Principe
d Silberbrillenvogel
i Occhialino di Principe

9272 *Speirops lugubris*
Passeriformes - Zosteropidae
e Black-capped Speirops; Sao Tomé
Speirops; Sao Tomé White-eye
f Zostérops de Sao Tomé; Speirops à
calotte noire
d Trauerbrillenvogel
i Occhialino capinero africano

9273 *Speirops melanocephalus*
Passeriformes - Zosteropidae
e Cameroon Speirops; Cameroon
Black-capped White-eye; Mount
Cameroon Speirops
f Zostérops du Cameroun
d Mönchsbrillenvogel
i Occhialino del Camerun

9274 *Spelaeornis badeigularis*
Passeriformes - Sylviidae
e Rusty-throated Wren-Babbler;
Mishmi Wren; Mishmi Wren-
Babbler
f Turdinule des Mishmi
i Garrulo scricciolo del Mishmi

9275 **Spelaeornis caudatus**
Passeriformes - Sylviidae
e   Rufous-throated Wren-Babbler;
Short-tailed Wren-Babbler; Tailed
Wren-Babbler
f   Turdinule à gorge rousse
d   Rotkehlzaunkönigstimalie
i   Garrulo scricciolo codabreve

9276 **Spelaeornis chocolatinus**
Passeriformes - Sylviidae
e   Long-tailed Wren-Babbler; Streaked
Long-tailed Wren-Babbler; Godwin-
Austen's Wren-Babbler
f   Turdinule chocolat; Spéléornis
chocolat
d   Langschwanzzaunkönigstimalie
i   Garrulo scricciolo di Godwin-Austin

9277 **Spelaeornis formosus**
Passeriformes - Sylviidae
e   Spotted Wren-Babbler
f   Turdinule tachetée; Spéléornis
tacheté
d   Fleckenbrustzaunkönigstimalie
i   Garrulo scricciolo maculato

9278 **Spelaeornis longicaudatus**
Passeriformes - Sylviidae
e   Tawny-breasted Wren-Babbler;
Long-tailed Wren-Babbler
f   Turdinule à longue queue
d   Khasi-Zaunkönigstimalie
i   Garrulo scricciolo codalunga

9279 **Spelaeornis troglodytoides**
Passeriformes - Sylviidae
e   Bar-winged Wren-Babbler; Barred-
wing Wren-Babbler; Bar-winged
Wren-Babbler; Long-tailed Spotted
Wren-Babbler
f   Turdinule troglodyte; Spéléornis
troglodyte
d   Bindenzaunkönigstimalie
i   Garrulo scricciolo alibarrate

9280 **Speotyto cunicularia**
Strigiformes - Strigidae
e   Burrowing Owl; Cuckoobird (WI)
f   Chevêche des terriers; Chouette à
terrier (Ants)

d   Kaninchenkauz; Kanincheneule
i   Civetta delle tane

9281 **Spermophaga haematina**
Passeriformes - Estrildidae
e   Western Bluebill; Blue-billed
Weaver; Crimson-breasted Bluebill;
Red-breasted Bluebill; Bluebill;
Blue-billed Waxbill; Red-breasted
Forest Weaver; Red-headed Forest
Weaver
f   Sénégali sanguin; Astrild à gros bec
bleu; Grosbec sanguin; Grosbec
rouge-noir
d   Rotbrustsamenknacker
i   Beccoblu occidentale

9282 **Spermophaga poliogenys**
Passeriformes - Estrildidae
e   Grant's Bluebill; Grant's Forest-
Weaver
f   Sénégali à bec bleu; Loxie de Grant;
Grosbec à front rouge
d   Grants Samenknacker
i   Beccoblu di Grant

9283 **Spermophaga ruficapilla**
Passeriformes - Estrildidae
e   Red-headed Bluebill; Red-headed
Bluebill-Weaver; Red-headed Blue-
billed Weaver; Red-billed Bluebill
f   Sénégali à tête rouge; Loxie à tête
rouge; Grosbec à tête rouge
d   Rotkopfsamenknacker
i   Beccoblu testarossa

9284 **Sphecotheres hypoleucus**
Passeriformes - Corvidae
e   Wetar Figbird
f   Sphécothère de Wetar
i   Uccello dei ficchi di Wetar

9285 **Sphecotheres vieilloti**
Passeriformes - Corvidae
e   Green Figbird; Southern Figbird;
Timor Figbird; Yellow Figbird
f   Sphécothère de Vieillot
d   Weißkehlfeigenpirol

9286 **Sphecotheres viridis**
Passeriformes - Corvidae

e   Timor Figbird; Figbird; Green
     Figbird
f   Sphécothère figuier
d   Timor-Feigenpirol
i   Uccello dei fichi verde

9287   *Sphecotheres viridis flaviventris*
        Passeriformes - Corvidae
e   Yellow Figbird
f   Sphécothère à ventre jaune
d   Gelbkehlfeigenpirol

9288   Spheniscidae
        Sphenisciformes
e   Penguins
f   Spheniscidés
d   Pinguine
i   Sfeniscididi

9289   Sphenisciformes
e   Penguins
f   Sphéniciformes; Manchots
d   Flossentaucher; Pinguine
i   Sfenisciformi

9290   *Spheniscus demersus*
        Sphenisciformes - Spheniscidae
e   Jackass Penguin; Black-footed
     Penguin
f   Manchot du Cap
d   Brillenpinguin
i   Sfenisco demerso; Pinguino del
     Capo; Pinguino jackass

9291   *Spheniscus humboldti*
        Sphenisciformes - Spheniscidae
e   Humboldt Penguin; Peruvian
     Penguin; Humboldt's Penguin
f   Manchot de Humboldt
d   Humboldt-Pinguin
i   Sfenisco di Humboldt; Pinguino di
     Humboldt; Pinguino di spatula di
     Humboldt

9292   *Spheniscus magellanicus*
        Sphenisciformes - Spheniscidae
e   Magellanic Penguin
f   Manchot de Magellan
d   Magellan-Pinguin
i   Sfenisco di Magellano; Pinguino di
     Magellano

9293   *Spheniscus mendiculus*
        Sphenisciformes - Spheniscidae
e   Galapagos Penguin
f   Manchot des Galapagos
d   Galapagos-Pinguin
i   Pinguino delle Galapagos; Sfenisco
     delle Galapagos

9294   *Sphenocichla humei*
        Passeriformes - Sylviidae
e   Wedge-billed Wren-Babbler; Hume's
     Wren-Babbler; Wedge-billed Wren;
     Wedge-billed Tree-Babbler
f   Turdinule à gros bec
d   Keilschnabelzaunkönigtimalie;
     Keilschnabeltimalie
i   Garrulo scricciolo beccoconico

9295   *Sphenoeacus afer*
        Passeriformes - Muscicapidae
e   Cape Grass-Warbler; Grass-Warbler;
     Cape Grassbird; Grassbird
f   Fauvette des graminées
d   Kap-Grassänger
i   Codacuta africano delle erbe

9296   *Sphyrapicus nuchalis*
        Piciformes - Picidae
e   Red-naped Sapsucker
f   Pic à nuque rouge
i   Picchio nucarossa

9297   *Sphyrapicus ruber*
        Piciformes - Picidae
e   Red-breasted Sapsucker
f   Pic à poitrine rouge
i   Picchio pettirosso americano

9298   *Sphyrapicus thyroideus*
        Piciformes - Picidae
e   Williamson's Sapsucker
f   Pic de Williamson
d   Kiefernsaftlecker
i   Picchio di Williamson

9299   *Sphyrapicus varius*
        Piciformes - Picidae
e   Yellow-bellied Sapsucker; Common
     Sapsucker
f   Pic maculé
d   Gelbbauchsaftlecker;

Feuerkopfsaftlecker
i Succialinfa comune; Succialinfa;
Picchio ventregiallo

9300 Spiloptila clamans
Passeriformes - Cisticolidae
e Cricket Longtail; Cricket Warbler;
Cricket Scaly-Prinia; Scaly Prinia;
Scaly-fronted Warbler
f Prinia à front écailleux; Fauvette à
front écailleux
d Schuppenkopfsänger
i Beccamoschino vocifero

9301 Spilornis abbotti
Falconiformes - Accipitridae
e Simeulue Serpent-Eagle

9302 Spilornis asturinus
Falconiformes - Accipitridae
e Nias Serpent-Eagle

9303 Spilornis baweanus
Falconiformes - Accipitridae
e Bawean Serpent-Eagle

9304 Spilornis cheela
Falconiformes - Accipitridae
e Crested Serpent-Eagle; Serpent-
Eagle; Ceylon Serpent-Eagle
f Serpentaire bacha
d Schlangenweihe
i Aquila serpentaria crestata; Aquila
serpentaria

9305 Spilornis elgini
Falconiformes - Accipitridae
e Andaman Serpent-Eagle; Andaman
Dark Serpent-Eagle
f Serpentaire des Andaman
d Andamanen-Weihe
i Aquila serpentaria delle Andamane

9306 Spilornis holospilus
Falconiformes - Accipitridae
e Philippine Serpent-Eagle
f Serpentaire des Philippines
d Philippinen-Schlangenweihe
i Aquila serpentaria delle Filippine

9307 Spilornis kinabaluensis
Falconiformes - Accipitridae
e Mountain Serpent-Eagle; Kinabalu
Serpent-Eagle
f Serpentaire des Kinabalu
d Bergschlangenweihe
i Aquila serpentaria di Kinabalu

9308 Spilornis minimus
Falconiformes - Accipitridae
e Nicobar Serpent-Eagle; Small-
crested Serpent-Eagle; Great Nicobar
Serpent-Eagle
f Serpentaire des Nicobar
d Zwergschlangenweihe
i Aquila serpentaria delle Nicobare;
Aquila serpentaria di Nicobar

9309 Spilornis minimus klossi
Falconiformes - Accipitridae
e Nicobar Serpent-Eagle
f Serpentaire menu
d Zwergschlangenweihe

9310 Spilornis natunensis
Falconiformes - Accipitridae
e Natuna Serpent-Eagle

9311 Spilornis perplexus
Falconiformes - Accipitridae
e Ryukyu Serpent-Eagle
f Serpentaire des Ryukyu

9312 Spilornis rufipectus
Falconiformes - Accipitridae
e Sulawesi Serpent-Eagle; Celebes
Serpent-Eagle
f Serpentaire des Célèbes
d Celebes-Schlangenweihe
i Aquila serpentaria di Sulawesi;
Aquila serpentaria di Celebes

9313 Spilornis sipora
Falconiformes - Accipitridae
e Mentawai Serpent-Eagle

9314 Spindalis portoricensis
Passeriformes - Thraupidae
e Puerto Rican Spindalis; Puerto Rican
Stripe-headed Tanager

9315 *Spindalis zena*
  Passeriformes - Thraupidae
 e Stripe-headed Tanager; Western
    Spindalis; Western Stripe-headed
    Tanager; Bastard Cock (WI)
 f Tangara à tête rayée
 d Bahama-Tangare; Dachtangare
 i Tangara testastriata

9316 *Spiza americana*
  Passeriformes - Fringillidae
 e Dickcissel
 f Dickcissel d'Amérique; Dickcissel
 d Dickzissel; Schildammer; Dickcissel
 i Spiza americana

9317 *Spizaetus africanus*
  Falconiformes - Accipitridae
 e Cassin's Hawk-Eagle
 f Aigle de Cassin; Spizaète de Cassin
 d Schwarzachseladler
 i Spizaeto di Cassin; Spizeto di Cassin;
    Spizeto africano

9318 *Spizaetus alboniger*
  Falconiformes - Accipitridae
 e Blyth's Hawk-Eagle; Black-and-
    white Hawk-Eagle; Mountain Hawk-
    Eagle
 f Aigle de Blyth; Spizaète de Blyth
 d Traueradler
 i Spizaeto di Blyth; Spizeto di Blyth

9319 *Spizaetus bartelsi*
  Falconiformes - Accipitridae
 e Javan Hawk-Eagle; Java Hawk-Eagle
 f Aigle de Java; Spizaète de Java
 d Bartel-Adler
 i Spizaeto di Giava; Spizeto di Giava;
    Falco aquila di Giava

9320 *Spizaetus cirrhatus*
  Falconiformes - Accipitridae
 e Changeable Hawk-Eagle; Crested
    Hawk-Eagle (ISC); Ceylon Hawk-
    Eagle; Flores Hawk-Eagle; Sumbarra
    Hawk-Eagle
 f Aigle huppé; Spizaète huppé
 d Haubenadler
 i Spizaeto variabile; Spizeto variabile;
    Falco aquila di Celebes

9321 *Spizaetus lanceolatus*
  Falconiformes - Accipitridae
 e Sulawesi Hawk-Eagle; Celebes
    Hawk-Eagle
 f Aigle des Célèbes; Spizaète des
    Célèbes
 d Celebes-Adler
 i Spizaeto di Sulawesi; Spizeto di
    Sulawesi

9322 *Spizaetus nanus*
  Falconiformes - Accipitridae
 e Wallace's Hawk-Eagle; Small Hawk-
    Eagle
 f Aigle de Wallace; Spizaète de
    Wallace
 d Dschungeladler
 i Spizaeto di Wallace; Spizeto di
    Wallace

9323 *Spizaetus nipalensis*
  Falconiformes - Accipitridae
 e Mountain Hawk-Eagle; Hodgson's
    Hawk-Eagle (ISC); Feather-toes
    Hawk-Eagle (ISC); Feather-toed
    Hawk-Eagle
 f Aigle montagnard; Spizaète
    montagnard
 d Bergadler
 i Spizaeto di Hodgson; Spizeto di
    Hodgson; Falco aquila delle
    montagne

9324 *Spizaetus ornatus*
  Falconiformes - Accipitridae
 e Ornate Hawk-Eagle
 f Aigle orné; Spizaète orné
 d Prachthaubenadler; Prachtadler
 i Spizaeto ornato; Spizeto ornato;
    Falco aquila ornato

9325 *Spizaetus philippensis*
  Falconiformes - Accipitridae
 e Philippine Hawk-Eagle
 f Aigle des Philippines; Spizaète des
    Philippines
 d Philippinen-Adler
 i Spizaeto delle Filippine; Spizeto
    delle Filippine; Falco aquila delle
    Filippine

9326 **Spizaetus tyrannus**
Falconiformes - Accipitridae
e Black Hawk-Eagle
f Aigle tyran; Spizaète tyran
d Tyrannenadler
i Spizaeto nero; Spizeto nero; Tiranno

9327 **Spizastur melanoleucus**
Falconiformes - Accipitridae
e Black-and-white Hawk-Eagle
f Aigle noir-et-blanc
d Elsteradler
I Falco aquila bianco e nero

9328 **Spizella arborea**
Passeriformes - Emberizidae
e American Tree-Sparrow; Tree
Sparrow (NA)
f Bruant hudsonien; Pinson hudsonien
d Bauumammer; Amerikanischer
Baumfink
i Passero arboricola

9329 **Spizella atrogularis**
Passeriformes - Emberizidae
e Black-chinned Sparrow
f Bruant à menton noir
d Schwarzkinnammer
i Passero golanera

9330 **Spizella breweri**
Passeriformes - Emberizidae
e Brewer's Sparrow
f Bruant de Brewer; Pinson de Brewer
d Nevada-Ammer
i Passero di Brewer

9331 **Spizella pallida**
Passeriformes - Emberizidae
e Clay-coloured Sparrow (NA)
f Bruant des plaines; Pinson des
plaines
d Fahlammer
i Passero pallido

9332 **Spizella passerina**
Passeriformes - Emberizidae
e Chipping Sparrow
f Bruant familier; Pinson familier
d Gesellschaftsammer; Kiefernammer;

Schwirrammer
i Passero cinguettante

9333 **Spizella pusilla**
Passeriformes - Emberizidae
e Field Sparrow
f Bruant des champs; Pinson des
champs
d Klapperammer; Feldammer
i Passero dei campi

9334 **Spizella taverneri**
Passeriformes - Emberizidae
e Timberline Sparrow
f Bruant de Taverner
i Passero di Taverner

9335 **Spizella wortheni**
Passeriformes - Emberizidae
e Worthen's Sparrow
f Bruant de Worthen
d Zacatec-Ammer
i Passero di Worthen

9336 **Spiziapteryx circumcinctus**
Falconiformes - Falconidae
e Spot-winged Falconet
f Carnifex à ailes tachetées
d Tropfenfalke
i Falchetto alimacchiate; Falco dalle
ali punteggiate

9337 **Spizixos canifrons**
Passeriformes - Pycnonotidae
e Crested Finchbill; Finch-billed
Bulbul; Crested Finch-billed Bulbul
f Bulbul à gros bec
d Finkenbülbül
i Bulbul crestato

9338 **Spizixos semitorques**
Passeriformes - Pycnonotidae
e Collared Finchbill; Collared Finch-
billed Bulbul; Black-headed
Finchbill; Chinese Finchbill;
Swinhoe's Finch-billed Bulbul
f Bulbul à semi-collier; Bulbul à
collier
d Halsbandbülbül
i Bulbul dal semicollare

9339 *Spizocorys conirostris*
Passeriformes - Alaudidae
e   Pink-billed Lark
f   Alouette à bec rose
d   Rotschnabellerche
i   Calandrella beccorosa

9340 *Spizocorys fringillaris*
Passeriformes - Alaudidae
e   Botha's Lark
f   Alouette de Botha
d   Finkenlerche
i   Calandrella di Botha

9341 *Spizocorys obbiensis*
Passeriformes - Alaudidae
e   Obbia Lark
f   Alouette d'Obbia
d   Obbia-Lerche
i   Calandrella di Obbia

9342 *Spizocorys personata*
Passeriformes - Alaudidae
e   Masked Lark
f   Alouette masquée
d   Maskenlerche
i   Calandrella mascherata

9343 *Spizocorys sclateri*
Passeriformes - Alaudidae
e   Sclater's Lark; Sclater's Short-toed
      Lark
f   Alouette de Sclater
d   Sclaters Kurzhaubenlerche;
      Ammernlerche
i   Calandrella di Sclater

9344 *Sporophila albogularis*
Passeriformes - Emberizidae
e   White-throated Seedeater
f   Sporophile à gorge blanche
d   Weißkehlpfäffchen
i   Beccasemi golabianca

9345 *Sporophila americana*
Passeriformes - Emberizidae
e   Wing-barred Seedeater; Variable
      Seedeater
f   Sporophile à ailes blanches
d   Wechselpfäffchen
i   Beccasemi variabile

9346 *Sporophila americana aurita*
Passeriformes - Emberizidae
e   Variable Seedeater
f   Sporophile variable
i   Beccasemi

9347 *Sporophila ardesiaca*
Passeriformes - Emberizidae
e   Duboi's Seedeater
f   Sporophile de Dubois
d   Dubois-Pfäffchen
i   Beccasemi di Dubois

9348 *Sporophila bouvreuil*
Passeriformes - Emberizidae
e   Capped Seedeater
f   Sporophile bouvreuil
d   Orangepfäffchen
i   Beccasemi capinero

9349 *Sporophila bouvronides*
Passeriformes - Emberizidae
e   Lesson's Seedeater
f   Sporophile faux-bouvron
i   Beccasemi di Lesson

9350 *Sporophila caerulescens*
Passeriformes - Emberizidae
e   Double-collared Seedeater
f   Sporophile à col double
d   Schmuckpfäffchen
i   Beccasemi dal doppio collare

9351 *Sporophila castaneiventris*
Passeriformes - Emberizidae
e   Chestnut-bellied Seedeater
f   Sporophile à ventre châtain
d   Rotbauchpfäffchen
i   Beccasemi ventrecastano

9352 *Sporophila cinnamomea*
Passeriformes - Emberizidae
e   Chestnut Seedeater; Grey-capped
      Chestnut Seedeater
f   Sporophile cannellé
d   Zimtpfäffchen
i   Beccasemi castano

9353 *Sporophila collaris*
Passeriformes - Emberizidae
e   Rusty-collared Seedeater

f Sporophile à col fauve
d Erzpfäffchen; Halsbandpfäffchen
i Beccasemi dal collare rosso

9354 *Sporophila corvina*
Passeriformes - Emberizidae
e Variable Seedeater; Black Seedeater

9355 *Sporophila falcirostris*
Passeriformes - Emberizidae
e Temminck's Seedeater
f Sporophile de Temminck
d Falzschnabelpfäffchen
i Beccasemi di Temminck

9356 *Sporophila frontalis*
Passeriformes - Emberizidae
e Buffy-fronted Seedeater; Buff-fronted Seedeater; Buffy-throated Seedeater
f Sporophile à front blanc
d Riesenpfäffchen
i Beccasemi frontechiara

9357 *Sporophila hypochroma*
Passeriformes - Emberizidae
e Grey-and-chestnut Seedeater; Rufous-rumped Seedeater; Rufous-naped Seedeater
f Sporophile à croupion roux
d Rotbürzelpfäffchen
i Beccasemi grigio e castano

9358 *Sporophila hypoxantha*
Passeriformes - Emberizidae
e Tawny-bellied Seedeater
f Sporophile à ventre fauve
d Ockerbrustpfäffchen
i Beccasemi ventrefulvo

9359 *Sporophila insulata*
Passeriformes - Emberizidae
e Tumaco Seedeater
f Sporophile de Tumaco
d Tumaco-Pfäffchen
i Beccasemi di Tumaco

9360 *Sporophila intermedia*
Passeriformes - Emberizidae
e Grey Seedeater
f Sporophile intermediaire

d Einfarbpfäffchen; Blaupfäffchen
i Beccasemi grigio

9361 *Sporophila leucoptera*
Passeriformes - Emberizidae
e White-bellied Seedeater; Grey-backed Seedeater; Bicoloured Seedeater; Black-backed Seedeater
f Sporophile à ventre blanc
d Zweifarbenpfäffchen
i Beccasemi ventrebianco

9362 *Sporophila lineola*
Passeriformes - Emberizidae
e Lined Seedeater
f Sporophile bouveron
d Diamantpfäffchen
i Beccasemi lineato

9363 *Sporophila luctuosa*
Passeriformes - Emberizidae
e Black-and-white Seedeater
f Sporophile noir-et-blanc
d Trauerpfäffchen
i Beccasemi nero e bianco

9364 *Sporophila melanogaster*
Passeriformes - Emberizidae
e Black-bellied Seedeater
f Sporophile à ventre noir
d Schwarzbauchpfäffchen
i Beccasemi ventrenero

9365 *Sporophila melanops*
Passeriformes - Emberizidae
e Hooded Seedeater; Dark-eyed Seedeater
f Sporophile de I'Araguaia
d Kapuzenpfäffchen
i Beccasemi dal cappuccio

9366 *Sporophila minuta*
Passeriformes - Emberizidae
e Ruddy-breasted Seedeater; Minute Seedeater; Pygmy Seedeater
f Sporophile petit-louis
d Zwergpfäffchen
i Beccasemi pettotugginoso

9367 *Sporophila moroletti*
Passeriformes - Emberizidae

e   White-collared Seedeater; Morellet's
      Seedeater

9368   *Sporophila murallae*
         Passeriformes - Emberizidae
e   Caqueta Seedeater

9369   *Sporophila nigricollis*
         Passeriformes - Emberizidae
e   Yellow-bellied Seedeater; White-
      beak See-see (WI); Black-necked
      Seedeater
f   Sporophile à ventre jaune
d   Gelbbauchpfäffchen;
      Schwarzkappenpfäffchen
i   Beccasemi collonero

9370   *Sporophila nigrorufa*
         Passeriformes - Emberizidae
e   Black-and-tawny Seedeater
f   Sporophile noir-et-roux
d   Schwarzmantelpfäffchen
i   Beccasemi nero e fulvo

9371   *Sporophila palustris*
         Passeriformes - Emberizidae
e   Marsh Seedeater; White-billed
      Seedeater
f   Sporophile des marais
d   Sumpfpfäffchen
i   Beccasemi di palude

9372   *Sporophila peruviana*
         Passeriformes - Emberizidae
e   Parrot-billed Seedeater
f   Sporophile perroquet
d   Papageischnabelpfäffchen
i   Beccasemi peruviano

9373   *Sporophila plumbea*
         Passeriformes - Emberizidae
e   Plumbeous Seedeater
f   Sporophile gris-de-plomb
d   Graupfäffchen
i   Beccasemi plumbeo

9374   *Sporophila ruficollis*
         Passeriformes - Emberizidae
e   Dark-throated Seedeater
f   Sporophile à gorge sombre
i   Beccasemi golascura

9375   *Sporophila schistacea*
         Passeriformes - Emberizidae
e   Slate-coloured Seedeater
f   Sporophile ardoisé
d   Schieferpfäffchen
i   Beccasemi color lavagna

9376   *Sporophila simplex*
         Passeriformes - Emberizidae
e   Drab Seedeater
f   Sporophile simple
d   Dickschnabelpfäffchen
i   Beccasemi modesto

9377   *Sporophila telasco*
         Passeriformes - Emberizidae
e   Chestnut-throated Seedeater
f   Sporophile télasco
d   Braunkehlpfäffchen
i   Beccasemi golacastana

9378   *Sporophila torqueola*
         Passeriformes - Emberizidae
e   White-collared Seedeater;
      Cinnamon-rumped Seedeater;
      Common Collared-Seedeater;
      Collared Seedeater
f   Sporophile à col blanc
d   Braunbürzelpfäffchen
i   Beccasemi gropponecannella

9379   *Sporophila zelichi*
         Passeriformes - Emberizidae
e   Narosky's Seedeater; White-collared
      Seedeater; Entre Rios Seedeater
f   Sporophile de Narosky
i   Beccasemi di Narosky

9380   *Sporopipes frontalis*
         Passeriformes - Ploceidae
e   Speckle-fronted Weaver; Scaly-
      fronted Weaver
f   Sporopipe quadrille; Moineau
      quadrillé
i   Tessitore frontesquamosa

9381   *Sporopipes squamifrons*
         Passeriformes - Ploceidae
e   Scaly Weaver; Scaly-feathered Finch
      (CSA); Scaly-feathered Weaver
f   Sporopipe squameux; Moineau à

front écaillé
d Schurrbärtchen
i Tessitore squamoso

9382 *Spreo albicapillus*
Passeriformes - Sturnidae
e White-crowned Starling
f Spréo à calotte blanche; Étourneau à calotte blanche
d Weißscheitelstar
i Storno capobianco

9383 *Spreo bicolor*
Passeriformes - Sturnidae
e African Pied-Starling; Pied Starling (CSA); Magpie Starling
f Spréo bicolore; Étourneau bicolore
d Zweifarbenstar; Zweifarbstar
i Storno ventrebianco

9384 *Spreo fischeri*
Passeriformes - Sturnidae
e Fischer's Starling
f Spréo de Fischer; Étourneau de Fischer
d Fischer-Glanzstar
i Storno di Fischer

9385 *Stachyris ambigua*
Passeriformes - Sylviidae
e Buff-chested Babbler; Buff-chested Tree-Babbler; Harrington's Tree-Babbler; Equivocal Tree-Babbler; Harrington's Babbler; Yellow-throated Tree-Babbler; Equivocal Jery
f Timalie ambigue
d Harrington-Timalie
i Garrulo pettonocciolo

9386 *Stachyris capitalis*
Passeriformes - Sylviidae
e Rusty-crowned Babbler; Rufous-crowned Tree-Babbler; Philippine Tree-Babbler; Rufous-crowned Babbler; Philippine Babbler
f Timalie mitrée
d Gelbkehlbuschtimalie
i Garrulo arboricolo corona rossiccia

9387 *Stachyris chrysaea*
Passeriformes - Sylviidae
e Golden Babbler; Golden-headed Babbler; Golden-headed Tree-Babbler; Golden Tree-Babbler
f Timalie dorée
d Goldkopftimalie
i Garrulo testadorata

9388 *Stachyris dennistouni*
Passeriformes - Sylviidae
e Golden-crowned Babbler; Golden-crowned Tree-Babbler
f Timalie à calotte dorée
i Garrulo corona dorata

9389 *Stachyris erythroptera*
Passeriformes - Sylviidae
e Chestnut-winged Babbler; Chestnut-winged Tree-Babbler; Red-winged Babbler; Red-winged Tree-Babbler
f Timalie à ailes rousses
d Rotflügeltimalie
i Garrulo arboricolo alicastane

9390 *Stachyris grammiceps*
Passeriformes - Sylviidae
e White-breasted Babbler; White-breasted Tree-Babbler; Javan Babbler
f Timalie à poitrine blanche
d Grauwangenbuschtimalie
i Garrulo di Giava

9391 *Stachyris herberti*
Passeriformes - Sylviidae
e Sooty Babbler; Sooty Tree-Babbler; Laos Dusky Tree-Babbler; Herbert's Babbler
f Timalie de Herbert
d Laos-Buschtimalie
i Garrulo fuligginoso

9392 *Stachyris hypogrammica*
Passeriformes - Sylviidae
e Palawan Striped-Babbler; Palawan Tree-Babbler; Palawan Striped Tree-Babbler; Palawan Babbler; Buff-capped Babbler
f Timalie de Palawan
d Palawan-Timalie
i Garrulo capofulvo

9393 *Stachyris latistriata*
Passeriformes - Sylviidae
e   Panay Striped-Babbler; Panay
Striped Tree-Babbler
f   Timalie à raies larges

9394 *Stachyris leucotis*
Passeriformes - Sylviidae
e   White-necked Babbler; White-
necked Tree-Babbler; White-eared
Tree-Babbler; White-collared Tree-
Babbler; White-eared Babbler
f   Timalie oreillarde
d   Perlhalsbuschtimalie
i   Garrulo arboricolo pettogrigio

9395 *Stachyris maculata*
Passeriformes - Sylviidae
e   Chestnut-rumped Babbler; Chestnut-
rumped Tree-Babbler; Red-rumped
Tree-Babbler; Red-rumped Babbler
f   Timalie maculée
d   Rostbürzeltimalie
i   Garrulo aboricolo pettomaculato

9396 *Stachyris melanothorax*
Passeriformes - Sylviidae
e   Crescent-chested Babbler; Pearl-
cheeked Babbler; Pearl-cheeked
Tree-Babbler; Pearly-cheeked
Babbler
f   Timalie perlée
d   Perlwangentimalie
i   Garrulo arboricolo pettonero

9397 *Stachyris nigriceps*
Passeriformes - Sylviidae
e   Grey-throated Babbler; Black-
throated Tree-Babbler; Black-
throated Babbler; Grey-throated
Tree-Babbler; Black-headed Babbler;
Grey-breasted Tree-Babbler; Grey
Babbler
f   Timalie à tête rayée; Timalie à tête
noire
d   Graukehlbuschtimalie
i   Garrulo golagricia

9398 *Stachyris nigricollis*
Passeriformes - Sylviidae
e   Black-throated Babbler; Black-

throated Tree-Babbler; Black-necked
Tree-Babbler; Black-necked Babbler
f   Timalie à gorge noire
d   Schwarzkehlbuschtimalie
i   Garrulo arboricolo golanera

9399 *Stachyris nigrocapitata*
Passeriformes - Sylviidae
e   Black-crowned Babbler; Black-
crowned Tree-Babbler
f   Timalie à calotte noire
i   Garrulo corona nera

9400 *Stachyris nigrorum*
Passeriformes - Sylviidae
e   Negros Striped-Babbler; Negros
Striped Tree-Babbler; Negros Tree-
Babbler; Black-crowned Tree-
Babbler; Negros Babbler
f   Timalie de Negros
d   Negros-Timalie
i   Garrulo dell'Isola Negros

9401 *Stachyris oglei*
Passeriformes - Sylviidae
e   Snowy-throated Babbler; Austen's
Spotted-Babbler; Austen's Spotted-
Tree-Babbler; Ogle's Snowy-throated
Tree-Babbler; Ogle's Snowy-throated
Babbler; Ogle's Spotted-Babbler
f   Timalie d'Austen
d   Weißkehlbuschtimalie
i   Garrulo di Austen

9402 *Stachyris plateni*
Passeriformes - Sylviidae
e   Pygmy Babbler; Pygmy Tree-
Babbler; Platen's Babbler
f   Timalie pygmée
d   Zwergbuschtimalie
i   Garrulo pigmeo·

9403 *Stachyris poliocephala*
Passeriformes - Sylviidae
e   Grey-headed Babbler; Grey-headed
Tree-Babbler
f   Timalie à tête grise
d   Graukopfbuschtimalie
i   Garrulo testagrigia

9404 *Stachyris pyrrhops*
Passeriformes - Sylviidae
e Black-chinned Babbler; Red-billed
Babbler; Red-billed Tree-Babbler
f Timalie à bec rouge
d Schwarzkinntimalie
i Garrulo beccorosso

9405 *Stachyris rodolphei*
Passeriformes - Sylviidae
e Deignan's Babbler; Rufous-fronted
Babbler
f Timalie de Deignan
d Thai-Timalie
i Garrulo di Deignan

9406 *Stachyris ruficeps*
Passeriformes - Sylviidae
e Rufous-capped Babbler; Rufous-
capped Tree-Babbler; Red-headed
Tree-Babbler; Rufous-crowned
Babbler; Red-headed Babbler;
Rufous-fronted Babbler
f Timalie de Blyth
d Rotkopftimalie
i Garrulo testarossa

9407 *Stachyris rufifrons*
Passeriformes - Sylviidae
e Rufous-fronted Babbler; Red-fronted
Babbler; Red-fronted Tree-Babbler
f Timalie à front roux
d Roststirntimalie; Rotstirnbaumtimalie
i Garrulo fronterosso

9408 *Stachyris speciosa*
Passeriformes - Sylviidae
e Flame-templed Babbler; Rough-
templed Tree-Babbler; Tweeddale's
Babbler; Hume's Babbler; Hume's
Tree-Babbler; Rough-templed
Babbler
f Timalie precieuse
d Goldstirnbuschtimalie
i Garrulo splendido

9409 *Stachyris striata*
Passeriformes - Sylviidae
e Luzon Striped-Babbler; Striped Tree-
Babbler
f Timalie striée

d Streifentimalie
i Garrulo arboricolo striato

9410 *Stachyris striolata*
Passeriformes - Sylviidae
e Spot-necked Babbler; Spot-necked
Tree-Babbler; Spotted-Tree-Babbler;
Spotted-necked Babbler; Spotted-
neck Tree-Babbler
f Timalie à cou tacheté
d Fleckenhalsbuschtimalie
i Garrulo collomacchiato

9411 *Stachyris thoracica*
Passeriformes - Sylviidae
e White-bibbed Babbler; White-
collared Babbler; White-collared
Tree-Babbler; White-bibbed Tree-
Babbler; White-collared Java Tree-
Babbler
f Timalie à col blanc
d Weißbrustbuschtimalie
i Garrulo arboricolo dal collare

9412 *Stachyris whiteheadi*
Passeriformes - Sylviidae
e Chestnut-faced Babbler; Whitehead's
Tree-Babbler; Whitehead's Babbler
f Timalie de Whitehead
d Brillentimalie
i Garrulo di Whitehead

9413 *Stactolaema anchietae*
Piciformes - Lybiidae
e Anchieta's Barbet; Yellow-headed
Barbet
f Barbican à tête jaune
d Strohkopfbartvogel
i Barbuto di Anchieta

9414 *Stactolaema leucotis*
Piciformes - Lybiidae
e White-eared Barbet
f Barbican oreillard
d Weißohrbartvogel
i Barbuto orecchiebianche

9415 *Stactolaema olivacea*
Piciformes - Lybiidae
e Green Barbet; African Green-Barbet;
Olive Barbet

f Barbican olivâtre
d Olivbartvogel
i Barbuto oliva

9416 *Stactolaema whytii*
Piciformes - Lybiidae
e Whyte's Barbet
f Barbican de Whyte
d Spiegelbartvogel
i Barbuto di Whyte

9417 *Stactolaema woodwardi*
Piciformes - Lybiidae
e Woodward's Barbet
f Barbican de Woodward

9418 *Stagonopleura bella*
Passeriformes - Passeridae
e Beautiful Firetail; Tasmanian
Waxbill; Beautiful Firetail-Finch
f Diamant à queue-de-feu
d Feuerschwanzamadine
i Diamante codadifuoco

9419 *Stagonopleura guttata*
Passeriformes - Passeridae
e Diamond Firetail; Firetail-Finch;
Diamond Sparrow; Diamond Firetail-
Finch; Spot-sided Finch; Spotted-
sided Finch
f Diamant à gouttelettes
d Diamantenamadine
i Diamante moscato

9420 *Stagonopleura oculata*
Passeriformes - Passeridae
e Red-eared Firetail; Red-eared
Firetail-Finch; Red-eared Finch
f Diamant oculé; Astrild à oreillons
rouges
d Rotohramadine
i Diamante orecchierosse

9421 *Starnoenas cyanocephala*
Columbiformes - Columbidae
e Blue-headed Quail-Dove; Blue-
headed Ground-Pigeon; Black-
bearded Pigeon
f Colombe à tête bleue
d Kuba-Erdtaube; Kuba-Taube;

Rebhuhntaube
i Colomba pernice testa azzurra

9422 *Steatornis caripensis*
Caprimulgiformes - Steatornithidae
e Oilbird; Guacharo
f Guacharo de Caripe
d Fettschwalm
i Guaciaro; Uccello dell'olio

9423 Steatornithidae
Caprimulgiformes
e Oilbirds; Guacharos
f Steatornithidés
d Fettschwälme
i Steatornitidi

9424 *Stelgidopteryx fucata*
Passeriformes - Hirundinidae
e Tawny-headed Swallow
f Hirondelle fardée
d Fuchsschwalbe
i Rondine testafulva

9425 *Stelgidopteryx ridgwayi*
Passeriformes - Hirundinidae
e Ridgway's Rough-winged Swallow;
Northern Rough-winged Swallow;
Yucatan Rough-winged Swallow
f Hirondelle de Ridgway
i Rondine di Ridgway

9426 *Stelgidopteryx ruficollis*
Passeriformes - Hirundinidae
e Southern Rough-winged Swallow;
American Rough-winged Swallow;
Rough-winged Swallow; Galley
Martin
f Hirondelle à gorge rousse
d Rauhflügelschwalbe
i Rondine aliruvide meridionale

9427 *Stelgidopteryx serripennis*
Passeriformes - Hirundinidae
e Northern Rough-winged Swallow
f Hirondelle à ailes herissées
i Rondine aliruvide settentrionale

9428 *Stellula calliope*
Apodiformes - Trochilidae
e Calliope Hummingbird

f Colibri calliope
d Sternelfe
i Colibrì Calliope

9429 ***Stenostira scita***
Passeriformes - Sylviidae
e Fairy Flycatcher; Fairy Warbler;
Fairy Flycatcher-Warbler
f Mignard enchanteur; Gobemouche
des fées
d Elfenschnäpper; Elfensänger
i Stenostira

9430 ***Stephanoaetus coronatus***
Falconiformes - Accipitridae
e Crowned Hawk-Eagle; Booted Eagle;
Crowned Eagle; African Crowned-
Eagle
f Aigle couronné
d Kronenadler
i Aquila coronata; Spizaeto coronato;
Spizeto coronato; Falco aquila
coronato

9431 ***Stephanophorus diadematus***
Passeriformes - Fringillidae
e Diademed Tanager
f Tangara à diadème
d Diademtangare
i Tangara dal diadema

9432 ***Stephanoxis lalandi***
Apodiformes - Trochilidae
e Plovercrest; Black-breasted
Plovercrest; Delalande's Plovercrest;
Violet-crested Plovercrest
f Colibri de Delalande
d Zopfelfe
i Colibrì crestato di Delalande

9433 **Stercorariidae**
Charadriiformes
e Skuas; Jaegers
f Labbes; Stercorariidés
d Raubmöwen
i Stercoraridi

9434 ***Stercorarius longicaudus***
Charadriiformes - Stercorariidae
e Long-tailed Jaeger; Long-tailed
Skua; Buffon's Skua

f Labbe à longue queue
d Falkenraubmöwe; Kleine Raubmöwe
i Labbo codalunga

9435 ***Stercorarius parasiticus***
Charadriiformes - Stercorariidae
e Parasitic Jaeger; Arctic Jaeger; Arctic
Skua; Parasitic Skua; Richardson's
Skua
f Labbe parasite
d Schmarotzerraubmöwe
i Labbo

9436 ***Stercorarius pomarinus***
Charadriiformes - Stercorariidae
e Pomarine Jaeger; Pomarine Skua;
Pomatorhine Skua; Twist-tailed Skua
f Labbe pomarin
d Spatelraubmöwe; Mittlere
Raubmöwe
i Stercorario mezzano; Gabbiano nero

9437 ***Sterna acuticauda***
Charadriiformes - Laridae
e Black-bellied Tern
d Schwarzbauchseeschwalbe
i Sterna ventrenero

9438 ***Sterna albifrons***
Charadriiformes - Laridae
e Little Tern; Sea-Swallow (CSA);
Least Tern; Black-lored Tern; White-
shafted Ternlet
f Sterne naine
d Zwergseeschwalbe
i Fraticello; Fraticello comune; Sterna
minore

9439 ***Sterna aleutica***
Charadriiformes - Laridae
e Aleutian Tern; Kamchatka Tern
f Sterne des Aléoutiennes; Sterne
aléoute
d Aleuten-Seeschwalbe
i Sterna aleutina; Sterna delle Aleutine

9440 ***Sterna anaethetus***
Charadriiformes - Laridae
e Bridled Tern; Sea-Swallow (CSA);
Eggbird (WI); Hurricanebird; Brown-
winged Tern; Panayan Tern; Smaller

Sooty-Tern
f Sterne bridée; Oiseau fou (Ants);
Thoire (Ants); Sterne à collier;
Dongue (Ants)
d Zügelseeschwalbe
i Sterna dalle redini

9441 *Sterna antillarum*
Charadriiformes - Laridae
e Least Tern; Spratt Gull (WI),
Eggbird (WI)
f Petite Sterne; Pigeon de mer (Ants);
Sterne des Antilles (Ants); Petite
Mauve (Ants)
d Kleinseeschwalbe
i Fraticello minore

9442 *Sterna aurantia*
Charadriiformes - Laridae
e River Tern; Indian River-Tern
f Sterne de rivière
d Hindu-Seeschwalbe
i Sterna di fiume

9443 *Sterna balaenarum*
Charadriiformes - Laridae
e Damara Tern; Sea-Swallow (CSA)
f Sterne des baleiniers
d Damara-Seeschwalbe
i Sterna di Damara

9444 *Sterna bengalensis*
Charadriiformes - Laridae
e Lesser Crested-Tern; Sea-Swallow
(CSA); Crested Tern; Indian Lesser
Crested-Tern; Small Crested-Tern
f Sterne voyageuse
d Rüppelsche Seeschwalbe; Rüppell-
Seeschwalbe
i Rondine di mare di Rüppell; Sterna
di Rüppell

9445 *Sterna bergii*
Charadriiformes - Laridae
e Great Crested-Tern; Crested Tern;
Swift Tern (CSA); Greater Crested-
Tern; Large Crested-Tern (ISC);
Large Tern; Torres Strait Tern; Bass
Straits Tern; Yellow-billed Tern
f Sterne huppée

d Eilseeschwalbe
i Sterna di Berg; Beccapesci maggiore

9446 *Sterna bernsteini*
Charadriiformes - Laridae
e Chinese Crested-Tern; Chinese Tern;
Chinese Lesser Crested-Tern
f Sterne d'Orient
d Schantung-Seeschwalbe; Chinesische
Seeschwalbe
i Sterna di Bernstein

9447 *Sterna caspia*
Charadriiformes - Laridae
e Caspian Tern; Sea-Swallow (CSA)
f Sterne caspienne; Sterne Tchégrava
d Raubseeschwalbe;
Riesenseeschwalbe
i Sterna maggiore

9448 *Sterna dougallii*
Charadriiformes - Laridae
e Roseate Tern; Sea-Swallow (CSA);
Gullie (WI); Eastern Roseate Tern;
Dougall's Tern; Graceful Tern
f Sterne de Dougall; Sterne (Ants);
Petite Fouquette (Ants); Petite
Mauve (Ants); Mauve blanche
(Ants); Mauve (Ants)
d Rosenseeschwalbe
i Sterna di Dougall; Rondine di mare
di Dougall; Sterna del Dougall;
Sterna maggiore; Rondine di mare
maggiore

9449 *Sterna elegans*
Charadriiformes - Laridae
e Elegant Tern
f Sterne élégante
d Schmuckseeschwalbe; Kleine
Königsseeschwalbe
i Sterna elegante

9450 *Sterna eurygnatha*
Charadriiformes - Laridae
e Cayenne Tern
f Sterne de Cayenne
d Cayenne-Seeschwalbe

9451 *Sterna forsteri*
Charadriiformes - Laridae

e Forster's Tern
f Sterne de Forster
d Sumpfseeschwalbe; Forster-
Seeschwalbe
i Sterna di Forster

9452 *Sterna fuscata*
Charadriiformes - Laridae
e Sooty Tern; Sea-Swallow (CSA);
Booby (WI); Eggbird; Wideawake;
Wideawake Tern
f Sterne fuligineuse; Sterne noire
(Ants); Oiseau fou (Ants)
d Rußseeschwalbe
i Sterna scura; Rondine di mare scura;
Sterna fuligginosa

9453 *Sterna hirundinacea*
Charadriiformes - Laridae
e South American Tern; Cassin's Tern
f Sterne hirundinacée
d Falkland-Seeschwalbe
i Sterna del Sudamerica

9454 *Sterna hirundo*
Charadriiformes - Laridae
e Common Tern; Tern; Comic Tern
(CSA); Sea-Swallow (CSA); Gullie
(WI); Black-billed Common Tern;
Asiatic Common Tern; Tibetan
Common Tern
f Sterne pierregarin; Sterne pierre
(Ants); Petite Mauve (Ants)
d Flußseeschwalbe
i Sterna comune; Rondine di mare;
Mignattone; Anima di sbirro grosso;
Sterna

9455 *Sterna lorata*
Charadriiformes - Laridae
e Peruvian Tern; Chilean Tern
f Sterne du Pérou; Sterne pérouvienne
d Peru-Seeschwalbe
i Sterna del Cile

9456 *Sterna lunata*
Charadriiformes - Laridae
e Grey-backed Tern; Spectacled Tern;
Gray-backed Tern (NA)
f Sterne à dos gris

d Brillenseeschwalbe
i Sterna dagli occhiali

9457 *Sterna maxima*
Charadriiformes - Laridae
e Royal Tern; Sea-Swallow (CSA);
Eggbird (WI); Gaby (WI); Seagull
(WI); Gullie (WI); Spratbird (WI)
f Sterne royale; Sterne (Ants); Pigeon
de mer (Ants); Foquette (Ants);
Mauve (Ants)
d Königsseeschwalbe
i Sterna reale

9458 *Sterna nereis*
Charadriiformes - Laridae
e Fairy Tern; Australian Fairy Tern;
Little Tern; Sea-Swallow; Little Sea-
Swallow; White-faced Ternlet;
Nereis Tern
f Sterne néréis
d Australseeschwalbe
i Sterna australiana

9459 *Sterna nilotica*
Charadriiformes - Laridae
e Gull-billed Tern; Sea-Swallow
(CSA); Gullie (WI); Long-legged
Tern
f Sterne hansel; Fou (Ants)
d Lachseeschwalbe
i Sterna zampenere

9460 *Sterna paradisaea*
Charadriiformes - Laridae
e Arctic Tern; Comic Tern (CSA); Sea-
Swallow (CSA)
f Sterne arctique
d Küstenseeschwalbe
i Sterna artica; Rondine di mare
codalunga

9461 *Sterna repressa*
Charadriiformes - Laridae
e White-cheeked Tern; Sea-Swallow
(CSA)
f Sterne à joues blanches
d Weißwangenseeschwalbe
i Sterna guancebianche

9462 **Sterna sandvicensis**
Charadriiformes - Laridae
e   Sandwich Tern; Cabot's Tern; Sea-Swallow (CSA); Yellow-nibbed Tern; Cayenne Tern
f   Sterne caugek
d   Brand-Seeschwalbe
i   Sterna reale

9463 **Sterna saundersi**
Charadriiformes - Laridae
e   Saunders's Tern; Saunders's Little-Tern; Black-shafted Tern; Black-shafted Ternlet; Mexican Coast Little-Tern
f   Sterne de Saunders
d   Orientseeschwalbe
i   Fraticello di Saunders

9464 **Sterna striata**
Charadriiformes - Laridae
e   White-fronted Tern; Black-billed Tern; Southern Tern
f   Sterne tara
d   Tara-Seeschwalbe
i   Sterna frontebianca

9465 **Sterna sumatrana**
Charadriiformes - Laridae
e   Black-naped Tern; Sea-Swallow (CSA)
f   Sterne diamant
d   Schwarznackenseeschwalbe
i   Sterna sumatrana

9466 **Sterna superciliaris**
Charadriiformes - Laridae
e   Yellow-billed Tern; Amazon Tern
f   Sterne argentée
d   Amazonas-Seeschwalbe
i   Sterna dell'Amazzonia

9467 **Sterna trudeaui**
Charadriiformes - Laridae
e   Snowy-crowned Tern; Trudeau's Tern
f   Sterne de Trudeau
d   Weißkopfseeschwalbe
i   Sterna di Trudeau

9468 **Sterna virgata**
Charadriiformes - Laridae
e   Kerguelen Tern; Sea-Swallow (CSA)
f   Sterne de Kerguélen
d   Kerguelen-Seeschwalbe
i   Sterna delle Kerguelen

9469 **Sterna vittata**
Charadriiformes - Laridae
e   Antarctic Tern; Sea-Swallow (CSA); Subantarctic Tern; Wreathed Tern; Tristan Tern
f   Sterne couronnée
d   Gabelschwanzseeschwalbe; Antipoden-Seeschwalbe
i   Sterna antartica

9470 **Sterna vittata bethunei**
Charadriiformes - Laridae
e   New Zealand Antarctic-Tern (ANZ)

9471 **Sternoclyta cyanopectus**
Apodiformes - Trochilidae
e   Violet-chested Hummingbird
f   Colibri à poitrine violette
d   Veilchenbrustkolibri
i   Colibrì pettoviola

9472 **Stictonetta naevosa**
Anseriformes - Anatidae
e   Freckled Duck; Speckled Duck
f   Stictonette tachetée
d   Affenente; Affengans
i   Anatra lentigginosa

9473 **Stigmatura budytoides**
Passeriformes - Tyrannidae
e   Greater Wagtail-Tyrant; Wagtail Tyrant
f   Calandrite bergeronnette
d   Stelzentachuri
i   Tiranno cutrettola maggiore

9474 **Stigmatura napensis**
Passeriformes - Tyrannidae
e   Lesser Wagtail-Tyrant
f   Calandrite du Napo
d   Braunrückentachuri
i   Tiranno cutrettola minore

9475 **Stiltia isabella**
Ciconiiformes' - Glareolidae
e Australian Courser; Australian
Pratincole; Long-legged Pratincole;
Isabelline Pratincole; Nankeen
Plover; Swallow-Plover
f Glaréole isabelle
d Stelzenbrachschwalbe
i Pernice del mare australiana

9476 **Stiphrornis erythrothorax**
Passeriformes - Muscicapidae
e Forest Robin; Forest Robin-Chat;
Forest Akalat
f Rougegorge de forêt
d Waldrötel; Rotkehlakalat;
Rotbrustakalat
i Pettirosso di foresta

9477 **Stipiturus malachurus**
Passeriformes - Maluridae
e Southern Emuwren; Emuwren
f Queue-de-gaze du Sud
d Rotstirnborstenschwanz
i Scricciolo emù meridionale

9478 **Stipiturus malachurus hartogi**
Passeriformes - Maluridae
e Dirk Hartog Issland Southern-
Emuwren (ANZ)

9479 **Stipiturus malachurus intermedius**
Passeriformes - Maluridae
e Fleurieu PeninsulaSouthern-
Emuwren (ANZ)

9480 **Stipiturus malachurus parimeda**
Passeriformes - Maluridae
e Eyre Peninsula Southern-Emuwren
(ANZ)

9481 **Stipiturus mallee**
Passeriformes - Maluridae
e Mallee Emuwren
f Queue-de-gaze du mallée
i Scricciolo emù del mallee

9482 **Stipiturus ruficeps**
Passeriformes - Maluridae
e Rufous-crowned Emuwren

d Rotscheitelborstenschwanz
i Scricciolo emù corona rossa

9483 **Strepera fuliginosa**
Passeriformes – Corvidae
e Black Currawong
f Réveilleur noir
i Gazza chiassosa nera

9484 **Strepera fuliginosa colei**
Passeriformes - Corvidae
e King Island Black Currawong (ANZ)

9485 **Strepera graculina**
Passeriformes - Corvidae
e Pied Currawong; Pied Bell-Magpie;
Scrub Currawong; Black Currawong
f Grand Réveilleur
d Dickschnabelwürgerkrähe
i Gazza chiassosa bianca e nera

9486 **Strepera graculina ashbyi**
Passeriformes - Corvidae
e Western Victoria Pied-Currawong
(ANZ)

9487 **Strepera graculina crissalis**
Passeriformes - Corvidae
e Lord Howe Island Pied Currawong
(ANZ)

9488 **Strepera versicolor**
Passeriformes - Corvidae
e Grey Currawong; Clinking
Currawong; Common Currawong
f Réveilleur cendré
d Schlankschnabelwürgerkrähe
i Gazza chiassosa grigia

9489 **Streptocitta albertinae**
Passeriformes - Sturnidae
e Bare-eyed Myna; Albertina's
Starling; Sula Starling; Sula Mynah;
Sula Magpie; Schlegel's Mynah
f Streptocitte des Sula
d Sula-Atzel
i Maina occhinudi

9490 **Streptocitta albicollis**
Passeriformes - Sturnidae
e White-necked Mynah; Buton

Starling; New Caledonian Starling;
Buton Mynah; Sulawesi Magpie;
Celebes Magpie; Vieillot's Mynah;
Vieillot's Starling
f   Streptocitte à cou blanc
d   Weißhalsatzel
i   Gracula dal collo bianco; Maina
collobianco

9491   Streptopelia bitorquata
Columbiformes - Columbidae
e   Island Collared-Dove; Javanese
Turtle-Dove; Javanese Collared
Dove; Island Turtle-Dove; Javan
Turtle-Dove; Javan Collared-Dove;
Philippine Collared-Dove; Philippine
Turtle-Dove; Java Ring Dove; Red-
necked Ring Dove
f   Tourterelle à double collier
d   Kichertaube
i   Tortora dal doppio collare

9492   Streptopelia capicola
Columbiformes - Columbidae
e   Ring-necked Dove; Cape Turtle-
Dove (CSA); Turtle-Dove; Cape
Ring Dove; Damara Dove; Dark-
eyed Ring Dove
f   Tourterelle du Cap
d   Gurrtaube; Kap-Turteltaube; Kap-
Lachtaube
i   Tortora del Capo

9493   Streptopelia chinensis
Columbiformes - Columbidae
e   Spotted Dove; Spotted Turtle-Dove
(ANZ); Spotted-necked Ddove;
Tigrine Dove; Pearl-necked Dove;
Indian Turtle-Dove; Chinese Turtle-
Dove; Indian Spotted-Dove; Ceylon
Spotted-Dove; Malay Spotted-Dove;
Burmese Spotted-Dove; Necklace
Spotted-Dove
f   Tourterelle tigrine
d   Perlhalstaube
i   Tortora della Cina

9494   Streptopelia decaocto
Columbiformes - Columbidae
e   Eurasian Collared-Dove; Collared
Dove; Collared Turtle-Dove; Ring

Dove (ISC); Indian Ring Dove
f   Tourterelle turque; Tourterelle rieuse;
Tourterelle isabelle
d   Türkentaube
i   Tortora dal collare; Tortora dal
collare orientale

9495   Streptopelia decipiens
Columbiformes - Columbidae
e   Mourning Collared-Dove; African
Mourning Dove (CSA); Mourning
Dove; Angola Dove; Deceptive
Turtle-Dove
f   Tourterelle pleureuse
d   Angola-Brillentaube; Angola-
Turteltaube; Brillentaube; Sudan-
Trauertaube
i   Tortora lamentosa africana

9496   Streptopelia hypopyrrha
Columbiformes - Columbidae
e   Andamawa Turtle-Dove; Pink-
bellied Turtle-Dove
f   Tourterelle de l'Adamaoua
d   Adamaua-Turteltaube
i   Tortora ventrerosato

9497   Streptopelia lugens
Columbiformes - Columbidae
e   Dusky Turtle-Dove; Pink-breasted
Dove; Black Dove
f   Tourterelle à poitrine rose
d   Trauerturteltaube
i   Tortora piangente

9498   Streptopelia orientalis
Columbiformes - Columbidae
e   Oriental Turtle-Dove; Rufous Turtle-
Dove; Eastern Turtle-Dove; Rufous
Dove; Eastern Rufous Turtle-Dove;
Mountain Turtle-Dove; Japanese
Turtle-Dove
f   Tourterelle orientale
d   Meena-Taube; Orientturteltaube
i   Tortora orientale

9499   Streptopelia picturata
Columbiformes - Columbidae
e   Madagascar Turtle-Dove; Painted
Dove; Malagasy Turtle-Dove
f   Tourterelle peinte; Pigeon de

Madagascar
d Madagaskar-Turteltaube
i Colomba del Madagascar

9500 ***Streptopelia reichenowi***
Columbiformes - Columbidae
e White-winged Collared-Dove;
African White-winged Dove (CSA);
Reichenow's Dove; Reichenow's
Ring Dove; White-winged Ring
Dove; White-winged Turtle-Dove
f Tourterelle de Reichenow;
Tourterelle à ailes blanches
d Weißflügelturteltaube
i Tortora di Reichenow

9501 ***Streptopelia risoria***
Columbiformes - Columbidae
e Barbary Dove; Ringed Turtle-Dove
f Tourterelle rieuse
d Lachtaube
i Tortora domestica

9502 ***Streptopelia roseogrisea***
Columbiformes - Columbidae
e African Collared-Dove; Pink-headed
Dove; Barbary Dove; Pink-headed
Turtle-Dove; Rose-grey Turtle-Dove;
Pink-headed Collared Dove; Collared
Dove
f Tourterelle rose-et-grise; Tourterelle
rieuse
d Hauslachtaube; Afrikanische
Lachtaube; Halbmondtaube
i Tortora dal collare africana

9503 ***Streptopelia semitorquata***
Columbiformes - Columbidae
e Red-eyed Dove; Ring Dove (CSA);
Collared-Dove; Half-collared Dove;
Black Dove; Black Pigeon
f Tourterelle à collier
d Palmtaube; Rotaugentaube
i Tortora dal semicollare

9504 ***Streptopelia senegalensis***
Columbiformes - Columbidae
e Laughing Dove; Palm Dove;
Laughing Turtle-Dove (ANZ); Little
Brown-Dove (ISC); Senegal Turtle-
Dove; Senegal Dove; Egyptian Dove;

Egyptian Turtle-Dove; Village Dove;
Town Dove; Garden Dove
f Tourterelle maillée; Tourterelle des
palmiers
d Weinrote Halsringtaube;
Palmentaube; Senegal-Täubchen;
Senegal-Taube
i Tortora delle palme; Tortora
senegalese; Tortora del Senegal

9505 ***Streptopelia tranquebarica***
Columbiformes - Columbidae
e Red Collared-Dove; Red Turtle-
Dove; Indian Red Turtle-Dove; Red
Ring Dove; Dwarf Turtle-Dove
f Tourterelle à tête grise
d Turteltaube; Zwerglachtaube
i Tortora dal collare rossiccio

9506 ***Streptopelia turtur***
Columbiformes - Columbidae
e European Turtle-Dove (NA); Turtle-
Dove; Dove; Eurasian Turtle-Dove;
Western Turtle-Dove; Isabelline
Turtle-Dove
f Tourterelle des bois
d Röteltaube; Orientalische Lachtaube
i Tortora; Tortora selvatica; Torora
comune

9507 ***Streptopelia vinacea***
Columbiformes - Columbidae
e Vinaceous Dove; Vinaceous Turtle-
Dove; Vinaceous Ring Dove
f Tourterelle vineuse
d Weinrote Lachtaube; Weinrote
Turteltaube
i Tortora vinacea

9508 ***Streptoprocne biscutata***
Apodiformes - Apodidae
e Biscutate Swift
f Martinet à collier interrompu
d Schildsegler
i Rondone biscutata

9509 ***Streptoprocne phelpsi***
Apodiformes - Apodidae
e Tepui Swift; Phelps's Swift
f Martinet tepui

d   Phelps-Segler
i   Rondone del Tepui

9510   *Streptoprocne rutila*
     Apodiformes - Apodidae
e   Chestnut-collared Swift
f   Martinet à collier; Martinet à collier
     roux
d   Rothalssegler
i   Rondone dal collare castano

9511   *Streptoprocne semicollaris*
     Apodiformes - Apodidae
e   White-naped Swift
f   Martinet à nuque blanche
d   Weißnackensegler
i   Rondone nucabianca

9512   *Streptoprocne zonaris*
     Apodiformes - Apodidae
e   White-collared Swift; Rainbird (WI);
     Ringed Gowrie (WI); Antillean
     Cloud Swift; Collared Swift; Ringed
     Swift
f   Oiseau de la pluie (Ants); Martinet à
     collier blanc
d   Halsbandsegler
i   Rondone dal collare bianco

9513   *Stresemannia bougainvillei*
     Passeriformes - Meliphagidae
e   Bougainville Honeyeater
f   Stresemannia bougainvillei;
     Méliphage de Bougainville
d   Bougainville-Honigfresser
i   Mangiamiele di Bougainville

9514   Strigidae
     Strigiformes
e   Typical Owls
f   Strigidés; Strigiens; Hiboux
d   Eulen; Eigentliche Eulen; Ohreulen
i   Strigidi

9515   Strigiformes
e   Owls
f   Strigiformes
d   Eulenvögel; Eulen
i   Strigiformi

9516   *Strigops habroptilus*
     Psittaciformes - Psittacidae
e   Kakapo; Owl Parrot
f   Strigops kakapo; Perruche hibou
d   Eulenpapagei; Kakapo
i   Kakapo; Strigope

9517   *Strix albitarsus*
     Strigiformes - Strigidae
e   Rufous-banded Owl
f   Chouette fasciée
d   Rötelkauz
i   Allocco fasciato rossiccio; Gufo dalle
     striscie rosse

9518   *Strix aluco*
     Strigiformes - Strigidae
e   Tawny Owl; Brown Owl; Wood
     Owl; Eurasian Tawny Owl; Tawny
     Wood-Owl
f   Chouette hulotte; Hulotte; Chat-huant
d   Waldkauz; Eigentlicher Waldkauz
i   Allocco; Gufo selvatico; Allocco
     eurasiatico

9519   *Strix butleri*
     Strigiformes - Strigidae
e   Hume's Owl; Hume's Tawny-Owl;
     Hume's Wood-Owl; Desert Owl
f   Chouette de Butler
d   Fahlkauz
i   Alloco di Butler; Allocco di Hume;
     Allocco pallido

9520   *Strix chacoensis*
     Strigiformes - Strigidae
e   Rufous-legged Owl

9521   *Strix davidi*
     Strigiformes - Strigidae
e   Sichuan Wood-Owl; Szechwan
     Wood-Owl; Père David's Owl;
     Szechuan Wood-Owl
f   Chouette du Sitchouan
i   Allocco del Szechwan

9522   *Strix fulvescens*
     Strigiformes - Strigidae
e   Fulvous Owl
f   Chouette fauve

    d   Zebrakauz
    i   Allocco fulvo

9523   ***Strix huhula***
        Strigiformes - Strigidae
    e   Black-banded Owl
    f   Chouette huhul
    i   Allocco fasciato nero; Gufo a righe
        nere

9524   ***Strix hylophila***
        Strigiformes - Strigidae
    e   Rusty-barred Owl; Brazilian Owl
    f   Chouette dryade
    d   Brasil-Kauz
    i   Allocco rossiccio; Gufo brasiliano

9525   ***Strix leptogrammica***
        Strigiformes - Strigidae
    e   Brown Wood-Owl
    f   Chouette leptogramme
    d   Malaien-Kauz
    i   Allocco bruno; Allocco dal petto
        fulvo

9526   ***Strix lucida***
        Strigiformes - Strigidae
    e   Mexican Spotted-Owl

9527   ***Strix nebulosa***
        Strigiformes - Strigidae
    e   Great Grey-Owl; Lapland Owl;
        Lapland Striped-Owl; Lapp Striped-
        Owl; Dark Wood-Owl
    f   Chouette lapone
    d   Bartkauz
    i   Allocco di Lapponia; Gufo della
        Lapponia

9528   ***Strix nigrolineata***
        Strigiformes - Strigidae
    e   Black-and-white Owl
    f   Chouette à lignes noires
    d   Bindenhalskauz
    i   Allocco fasciato bianco e nero; Gufo
        bianco e nero

9529   ***Strix occidentalis***
        Strigiformes - Strigidae
    e   Spotted Owl; California Spotted-
        Owl; Mountain Spotted-Owl;

        Mexican Spotted-Owl
    f   Chouette tachetée
    d   Fleckenkauz
    i   Allocco maculato americano; Gufo
        chazzato

9530   ***Strix ocellata***
        Strigiformes - Strigidae
    e   Mottled Wood-Owl
    f   Chouette ocellée
    d   Mangokauz
    i   Allocco ocellato

9531   ***Strix rufipes***
        Strigiformes - Strigidae
    e   Rufous-legged Owl
    f   Chouette masquée
    d   Rostfußkauz
    i   Allocco zamperosse; Gufo dalle
        gambe rosse

9532   ***Strix seloputo***
        Strigiformes - Strigidae
    e   Spotted Wood-Owl; Seloputo Owl
    f   Chouette des pagodes
    d   Pagodenkauz
    i   Allocco maculato asiatico

9533   ***Strix uralensis***
        Strigiformes - Strigidae
    e   Ural Owl; Ural Wood-Owl
    f   Chouette de l'Oural
    d   Habichtskauz
    i   Allocco degli Urali

9534   ***Strix varia***
        Strigiformes - Strigidae
    e   Barred Owl
    f   Chouette rayée; Chouette barrée
    d   Streifenkauz
    i   Allocco barrato; Civetta rigata;
        Allocco americano

9535   ***Strix virgata***
        Strigiformes - Strigidae
    e   Mottled Owl; American Wood-Owl
    f   Chouette mouchetée
    d   Sprenkelkauz
    i   Allocco marezzato; Gufo screziato

9536   *Strix woodfordii*
Strigiformes - Strigidae
e   African Wood-Owl; Wood Owl
(CSA); Woodford's Wood-Owl; West
African Wood-Owl
f   Chouette africaine; Hulotte africaine
d   Woodford-Kauz
i   Allocco; Gufo dei bosci africano

9537   *Struthidea cinerea*
Passeriformes - Corcoracidae
e   Apostlebird; Grey Jumper
f   Apôtre gris; Graucope gris
d   Grauling; Gimpelhäher
i   Uccello dei dodici apostoli

9538   *Struthio camelus*
Struthiformes - Struthionidae
e   Ostrich; Two-toed Ostrich; Southern
Ostrich
f   Autruche d'Afrique; Autruche
d   Strauß; Afrikanischer Strauß
i   Struzzo

9539   *Struthio molybdophanes*
Struthiformes - Struthionidae
e   Somali Ostrich

9540   **Struthionidae**
Struthiformes
e   Ostriches
f   Struthionidés
d   Strauße; Afrika-Laufvögel
i   Struzionidi

9541   **Struthioniformes**
e   Ostriches
f   Struthioniformes
d   Strauße
i   Struzioniformi

9542   *Sturnella bellicosa*
Passeriformes - Icteridae
e   Peruvian Meadowlark; Peruvian Red-
breasted Meadowlark; White-thighed
Meadowlark; Greater Red-breasted
Meadowlark
f   Sturnelle du Pérou
d   Weißschenkelsoldatenstärling
i   Sturnella pettirosa del Perù

9543   *Sturnella lilianae*
Passeriformes - Icteridae
e   Lilian's Meadowlark
f   Sturnelle de Lilian
i   Sturnella di Lillian

9544   *Sturnella loyca*
Passeriformes - Icteridae
e   Long-tailed Meadowlark; Greater
Red-breasted Meadowlark
f   Sturnelle australe
d   Langschwanzsoldatenstärling
i   Sturnella codalunga

9545   *Sturnella magna*
Passeriformes - Icteridae
e   Eastern Meadowlark
f   Sturnelle des prés
d   Lerchenstärling
i   Sturnella allodola orientale

9546   *Sturnella militaris*
Passeriformes - Icteridae
e   Pampas Meadowlark; Red-breasted
Blackbird; Lesser Red-breasted
Meadowlark; Lesser Meadowlark
f   Sturnelle des pampas
d   Schwarzschenkelsoldatenstärling
i   Sturnella di Defilippi

9547   *Sturnella neglecta*
Passeriformes - Icteridae
e   Western Meadowlark
f   Sturnelle de l'Ouest
d   Wiesenstärling; Westlicher
Lerchenstärling
i   Sturnella negletta; Sturnella allodola
occidentale

9548   *Sturnella superciliaris*
Passeriformes - Icteridae
e   White-browed Blackbird
d   Weißbrauenstärling
i   Sturnella

9549   **Sturnidae**
Passeriformes
e   Starlings
f   Étourneaux; Sturnidés
d   Stare
i   Sturnidi; Storni

9550 **_Sturnus burmannicus_**
Passeriformes - Sturnidae
e Vinous-breasted Starling; Jerdon's Starling
f Étourneau vineux
d Kambodscha-Star
i Storno di Jerdon

9551 **_Sturnus cineraceus_**
Passeriformes - Sturnidae
e White-cheeked Starling; Grey Starling; Ashy Starling; Pale-bellied Starling
f Étourneau gris; Martin gris
d Graustar; Weißwangenstar
i Storno grigio

9552 **_Sturnus contra_**
Passeriformes - Sturnidae
e Asian Pied-Starling; Pied Mynah (ISC); Pied Starling; Asiatic Pied-Starling
f Étourneau pie; Martin pie
d Elsterstar
i Storno bianco e nero

9553 **_Sturnus erythropygius_**
Passeriformes - Sturnidae
e White-headed Starling; Andaman White-headed Mynah
f Étourneau à tête blanche
d Andamanen-Star
i Storno testabianca

9554 **_Sturnus malabaricus_**
Passeriformes - Sturnidae
e Chestnut-tailed Starling; Grey-headed Myhna; Chestnut-faced Starling; Ashy-headed Starling; Grey-headed Starling
f Étourneau à tête grise; Martin à tête grise
d Graukopfstar
i Storno testagrigia

9555 **_Sturnus melanopterus_**
Passeriformes - Sturnidae
e Black-winged Starling
f Étourneau à ailes noires
d Schwarzflügelstar
i Storno alinere

9556 **_Sturnus nigricollis_**
Passeriformes - Sturnidae
e Black-collared Starling; Black-necked Starling
f Étourneau à cou noir; Martin à cou noir
d Schwarzhalsstar
i Storno collonero

9557 **_Sturnus pagodarum_**
Passeriformes - Sturnidae
e Brahminy Starling; Black-headed Mynah (ISC); Pagoda Starling; Brahminy Mynah; Black-headed Starling
f Étourneau des pagodes; Martin des pagodes
d Pagodenstar
i Storno delle pagode

9558 **_Sturnus philippensis_**
Passeriformes - Sturnidae
e Chestnut-cheeked Starling; Red-cheeked Starling; Violet-backed Starling; Violet-backed Starlet; Red-cheeked Mynah
f Étourneau à joues marron; Étourneau philippin
d Violettrückenstar; Rostbackenstar
i Storno dorsovioletto

9559 **_Sturnus roseus_**
Passeriformes - Sturnidae
e Rosy Starling; Rose-coloured Starling; Rosy Pastor (ISC)
f Étourneau roselin; Martin roselin
d Rosenstar
i Storno roseo

9560 **_Sturnus senex_**
Passeriformes - Sturnidae
e White-faced Starling; Ceylon White-faced Starling; Ceylon White-headed Starling; Sri Lanka White-headed Starling; Ceylon White-headed Mynah
f Étourneau de Ceylan
d Greisenstar
i Storno testacanuta

9561   ***Sturnus sericeus***
        Passeriformes - Sturnidae
   e   Red-billed Starling; Silky Starling
   f   Étourneau soyeux; Martin soyeux
   d   Seidenstar; Hellgrauer Star
   i   Storno sericeo

9562   ***Sturnus sinensis***
        Passeriformes - Sturnidae
   e   White-shouldered Starling; Grey-
         backed Starling; Chinese Starling;
         Mandarin Mynah; Grey-backed
         Mynah; Grey-backed Starlet; Chinese
         Mynah
   f   Étourneau mandarin
   d   Mandarinstar
   i   Storno della Cina

9563   ***Sturnus sturninus***
        Passeriformes - Sturnidae
   e   Purple-backed Starling; Daurian
         Starling; Daurian Starlet; Daurian
         Mynah
   f   Étourneau de Daourie
   d   Mongolen-Star; Daurischer Star
   i   Storno daurico

9564   ***Sturnus unicolor***
        Passeriformes - Sturnidae
   e   Spotless Starling; Black Starling;
         Sardinian Starling; Mediterranean
         Starling
   f   Étourneau unicolore
   d   Einfarbstar
   i   Storno nero

9565   ***Sturnus vulgaris***
        Passeriformes - Sturnidae
   e   Common Starling; Starling; Purple-
         winged Starling; Eurasian Starling;
         European Starling; Northern Starling
   f   Étourneau sansonnet; Sansonnet
   d   Star; Gemeiner Star
   i   Storno; Stornello; Storno europeo;
         Storno comune

9566   ***Stymphalornis acutirostris***
        Passeriformes - Thamnophilidae
   e   Black-bellied Starling; Parana Ant-
         Wren

9567   ***Sublegatus arenarum***
        Passeriformes - Tyrannidae
   e   Northern Scrub-Flycatcher
   f   Tyranneau des palétuviers
   i   Tiranno di macchia del Nord

9568   ***Sublegatus glaber***
        Passeriformes - Tyrannidae
   e   Smooth Scrub-Flycatcher

9569   ***Sublegatus modestus***
        Passeriformes - Tyrannidae
   e   Southern Scrub-Flycatcher; Scrub-
         Flycatcher; Short-billed Flycatcher
   f   Tyranneau modeste
   d   Buschfliegenstecher
   i   Tiranno di macchia

9570   ***Sublegatus obscurior***
        Passeriformes - Tyrannidae
   e   Amazonian Scrub-Flycatcher; Dusky
         Flycatcher; Todd's Scrub-Flycatcher;
         Scrub-Flycatcher
   f   Tyranneau ombré

9571   ***Suiriri affinis***
        Passeriformes - Tyrannidae
   e   Campo Suiriri; Campo Suiriri-
         Flycatcher; Campos Scrub-
         Flycatcher; Northern Scrub-
         Flycatcher; Scrub-Flycatcher; Chaco
         Suiriri; Chaco Suiriri-Flycatcher
   f   Tyranneau des campos
   i   Suiriri del Nord

9572   ***Suiriri suiriri***
        Passeriformes - Tyrannidae
   e   Chaco Suiriri; Suiriri Flycatcher;
         Chaco Suiriri-Flycatcher; Southern
         Suiriri
   f   Tyranneau suiriri
   d   Suiriri
   i   Suiriri del Sud

9573   ***Sula dactylatra***
        Pelecaniformes - Sulidae
   e   Masked Booby; Blue-faced Booby;
         White Booby; Booby (WI); Blue-
         faced Gannet (WI); Masked Gannet;
         Whistling Booby
   f   Fou masqué

d   Maskentölpel
i   Sula mascherata

9574   *Sula dactylatra bedouti*
Pelecaniformes - Sulidae
e   Eastern Indian Ocean Masked-Booby
(ANZ)

9575   *Sula dactylatra fullagari*
Pelecaniformes - Sulidae
e   Tasman Sea Masked-Booby (ANZ)

9576   *Sula granti*
Pelecaniformes - Sulidae
e   Nazca Booby

9577   *Sula leucogaster*
Pelecaniformes - Sulidae
e   Brown Booby; White-bellied Booby;
Booby (WI)
f   Fou de Cayenne; Fou brun; Fou
blanc (Ants); Fou à pieds rouges
(Ants)
d   Brauntölpel; Weißbauchtölpel
i   Sula fosca

9578   *Sula nebouxii*
Pelecaniformes - Sulidae
e   Blue-footed Booby
f   Fou à pieds bleus
d   Blaufußtölpel
i   Sula dai piedi azzurri; Sula
piediazzurri

9579   *Sula sula*
Pelecaniformes - Sulidae
e   Red-footed Booby; Booby (WI);
Tree Booby (WI)
f   Fou à pieds rouges; Fou blanc (Ants)
d   Rotfußtölpel
i   Sula piedirossi; Sula dai piedi rossi

9580   *Sula tasmani*
Pelecaniformes - Sulidae
e   Tasman Booby (ANZ)

9581   *Sula variegata*
Pelecaniformes - Sulidae
e   Peruvian Booby; Variegated Booby
f   Fou varié

d   Guanotölpel
i   Sula peruviana

9582   Sulidae
Pelecaniformes
e   Boobies; Sulids; Gannets
f   Fous; Sulidés
d   Tölpel
i   Sulidi; Sule

9583   *Surnia ulula*
Strigiformes - Strigidae
e   Northern Hawk-Owl; Hawk-Owl
f   Chouette épervière; Surnie épervière
d   Sperbereule
i   Ulula

9584   *Surniculus lugubris*
Cuculiformes - Cuculidae
e   Drongo-Cuckoo; Indian Drongo-
Cuckoo (ISC)
f   Coucou surnicou
d   Drongokuckuck
i   Cuculo drongo; Cuculo indiano

9585   *Swynnertonia swynnertoni*
Passeriformes - Muscicapidae
e   Swynnerton's Robin; Swynnerton's
Bush-Robin; Swynnerton's Forest
Robin
f   Rougegorge de Swynnerton
d   Swynnerton-Rötel
i   Pettirosso di Swynnerton

9586   *Sylvia althaea*
Passeriforme - Sylviidae
e   Hume's Whitethroat; Hume's Lesser-
Whitethroat
f   Fauvette de Hume; Fauvette-
babillarde de Hume
d   Eibischgrasmücke
i   Bigia; Bigiarella di Hume

9587   *Sylvia atricapilla*
Passeriformes - Sylviidae
e   Blackcap; European Blackcap
(CSA); Black-capped Warbler
f   Fauvette à tête noire
d   Mönchsgrasmücke;
Schwarzplättchen
i   Capinera

9588 *Sylvia boehmi*
Passeriformes - Sylviidae
e Banded Warbler; Banded Tit-
Warbler; Banded Tit-Babbler;
Banded Tit-Flycatcher; Banded
Parisoma
f Parisome sanglée; Parisome barrée
d Bandmeisensänger
i Bigia fasciata

9589 *Sylvia borin*
Passeriformes - Sylviidae
e Garden Warbler
f Fauvette des jardins
d Gartengrasmücke
i Beccafico; Bigione; Bigia

9590 *Sylvia buryi*
Passeriformes - Sylviidae
e Yemen Warbler; Yemen Tit-Warbler;
Arabian Tit-Warbler; Arabian Tit-
Babbler; Yemen Tit-Flycatcher;
Yemen Tit-Babbler; Yemen
Parisoma; Arabian Warbler
f Parisome du Yémen
d Asir-Meisensänger
i Parisoma dello Yemen

9591 *Sylvia cantillans*
Passeriformes - Sylviidae
e Subalpine Warbler
f Fauvette passerinette; Fauvette
subalpine
d Bartgrasmücke; Weißbartgrasmücke
i Sterpazzolina; Sterpazzuolina

9592 *Sylvia carbonata*
Passeriformes - Sylviidae
e Carbonated Warbler
i Bigia

9593 *Sylvia communis*
Passeriformes - Sylviidae
e Greater Whitethroat; Common
Whitethroat (CSA); Whitethroat;
Common White-throated Warbler;
White-throated Warbler; Eurasian
Whitethroat
f Fauvette grisette
d Dorngrasmücke

i Sterpazzola; Sterpazzuola;
Sterpazzolina

9594 *Sylvia conspicillata*
Passeriformes - Sylviidae
e Spectacled Warbler
f Fauvette à lunettes
d Brillengrasmücke
i Sterpazzola di Sardegna

9595 *Sylvia curruca*
Passeriformes - Sylviidae
e Lesser Whitethroat; Grey
Whitethroat; Indian Lesser
Whitethroat
f Fauvette babillarde
d Klappergrasmücke; Zaungrasmücke
i Bigiarella

9596 *Sylvia deserticola*
Passeriformes - Sylviidae
e Tristram's Warbler
f Fauvette de l'Atlas; Fauvette du
désert
d Atlas-Grasmücke
i Bigia del deserto; Silvia di Tristram

9597 *Sylvia hortensis*
Passeriformes - Sylviidae
e Orphean Warbler
f Fauvette orphée
d Orpheus-Grasmücke
i Bigia grossa

9598 *Sylvia layardi*
Passeriformes - Sylviidae
e Layard's Warbler; Layard's Tit-
Warbler (CSA); Layard's Tit-
Flycatcher; Layard's Tit-Babbler
f Parisome de Layard
d Layards Sänger
i Bigia di Layard

9599 *Sylvia leucomelaena*
Passeriformes - Sylviidae
e Arabian Warbler; Red Sea Warbler;
Blanford's Warbler
f Fauvette d'Arabie; Fauvette arabe
d Akaziengrasmücke; Blandford-
Grasmücke
i Bigia del Mar Rosso

9600 Sylvia lugens
Passeriformes - Sylviidae
e Brown Warbler; Brown Tit-Warbler;
Brown Tit-Babbler; Brown Tit-
Flycatcher; Brown Warbler
f Parisome brune
d Braunmeisensänger
i Bigia bruna

9601 Sylvia melanocephala
Passeriformes - Sylviidae
e Sardinian Warbler
f Fauvette mélanocéphale
d Samtkopfgrasmücke;
Schwarzkopfgrasmücke
i Occhiocotto; Capinera nera;
Occhiorosso

9602 Sylvia melanothorax
Passeriformes - Sylviidae
e Cyprus Warbler
f Fauvette de Chypre
d Schuppengrasmücke
i Bigia di Cipro; Occhiocotto di Cipro

9603 Sylvia minula
Passeriformes - Sylviidae
e Small Whitethroat; Desert Lesser-
Whitethroat; Lesser Whitethroat
f Fauvette minule; Fauvette babillarde
des déserts
i Bigiarella del deserto

9604 Sylvia montana
Passeriformes - Sylviidae
e Blue Mountain-Warbler

9605 Sylvia mystacea
Passeriformes - Sylviidae
e Ménétries's Warbler
f Fauvette de Ménétries
d Tamariskengrasmücke; Östliche
Samtkopfgrasmücke
i Ochiocotta di Ménétries; Bigia di
Ménétries

9606 Sylvia nana
Passeriformes - Sylviidae
e Desert Warbler; Whitethroat; Desert
Whitethroat
f Fauvette naine

d Wüstengrasmücke
i Sterpazzola nana

9607 Sylvia nisoria
Passeriformes - Sylviidae
e Barred Warbler; European Barred
Warbler
f Fauvette éperviére; Fauvette rayée
d Sperbergrasmücke
i Bigia padovana

9608 Sylvia rueppelli
Passeriformes - Sylviidae
e Rüppell's Warbler
f Fauvette de Rüppell; Fauvette
masquée
d Maskengrasmücke
i Silvia di Rüppell; Bigia di Rüppell

9609 Sylvia sarda
Passeriformes - Sylviidae
e Marmora's Warbler
f Fauvette sarde
d Sarden-Grasmücke
i Magnanina sarda

9610 Sylvia subcaeruleum
Passeriformes - Sylviidae
e Rufous-vented Warbler; Tit-Babbler
(CSA); Cape Tit-Warbler; Southern
Tit-Warbler; Southern Warbler; Cape
Warbler; Chestnut-vented Warbler;
Chestnut-vented Tit-Babbler
f Parisome grignette; Parisome à sous-
ventre rouge
d Meisensänger
i Bigia del Capo

9611 Sylvia undata
Passeriformes - Sylviidae
e Dartford Warbler
f Fauvette pitchou
d Provence-Grasmücke
i Magnanina

9612 Sylvietta brachyura
Passeriformes - Sylviidae
e Northern Crombec; Common
Crombec; Senegal Crombec;
Crombec; Northern Sylvietta;
Nuthatch Warbler

f Crombec sittelle; Fauvette crombec; Crombec
d Braunbauchsylvietta
i Silvietta settentrionale

9613 *Sylvietta chapini*
Passeriformes - Sylviidae
e Chapin's Crombec; Chapin's Sylvietta
f Crombec de Chapin
i Silvietta di Chapin

9614 *Sylvietta denti*
Passeriformes - Sylviidae
e Lemon-bellied Crombec; Lemon-bellied Sylvietta
f Crombec à gorge tachetée; Fauvette crombec à gorge tachetée
d Gelbsteißsylvietta
i Silvietta ventregiallo

9615 *Sylvietta isabellina*
Passeriformes - Sylviidae
e Somali Crombec; Somali Long-billed Crombec; Somali Sylvietta; Somali Long-billed Sylvietta
f Crombec isabelle
d Isabellsylvietta
i Silvietta somala beccolungo

9616 *Sylvietta leucophrys*
Passeriformes - Sylviidae
e White-browed Crombec; White-browed Sylvietta
f Crombec à sourcils blancs
d Weißbrauensylvietta
i Silvietta dai sopraccigli bianchi

9617 *Sylvietta philippae*
Passeriformes - Sylviidae
e Short-billed Crombec; Somali Short-billed Crombec; Somali Short-tailed Crombec; Short-tailed Crombec; Somali Short-billed Sylvietta; Somali Crombec
f Crombec de Somalie
d Weißbauchsylvietta
i Silvietta somala beccocorto

9618 *Sylvietta rufescens*
Passeriformes - Sylviidae

e Cape Crombec; Long-billed Crombec (CSA); Long-billed Sylvietta
f Crombec à long bec
d Langschnabelsylvietta; Kurzschwanz-Sylvietta
i Silvietta beccolungo

9619 *Sylvietta ruficapilla*
Passeriformes - Sylviidae
e Red-capped Crombec; Red-capped Sylvietta
f Crombec à calotte rousse; Crombec à joues rousses
d Rotohrsylvietta
i Silvietta capirossa

9620 *Sylvietta virens*
Passeriformes - Sylviidae
e Green Crombec
f Crombec vert; Fauvette crombec verte
d Grünmantelsylvietta
i Silviette verde

9621 *Sylvietta virens flaviventris*
Passeriformes - Sylviidae
f Crombec à ventre jaune

9622 *Sylvietta whytii*
Passeriformes - Sylviidae
e Red-faced Crombec; Red-faced Sylvietta
f Crombec à face rousse; Crobec de Whyte
d Whytes Sylvietta; Whyte-Sylvietta
i Silvietta facciarossa

9623 **Sylviidae**
Passeriformes
e Old World Warblers; Warblers
f Sylviidés
d Zweigsänger; Grasmücken
i Silvie; Silvidi

9624 *Sylviorthorhynchus desmursii*
Passeriformes - Furnariidae
e Des Murs's Wiretail; Desmurs's Spinetail
f Synallaxe de des Murs
d Sechsfedernschlüpfer
i Cada a fili di Des Murs

9625 **Sylviparus modestus**
Passeriformes - Paridae
e Yellow-browed Tit
f Mésange modeste
d Laubsängermeise
i Cincia dai sopraciggli bialli

9626 **Syma megarhyncha**
Coraciiformes - Halcyonidae
e Mountain Kingfisher; Mountain
Greater Yellow-billed Kingfisher;
Mountain Yellow-billed Kingfisher
f Martin-chasseur montagnard
d Bergtorotoro
i Martin pescatore di monte

9627 **Syma torotoro**
Coraciiformes - Halcyonidae
e Yellow-billed Kingfisher; Lesser
Yellow-billed Kingfisher; Lowland
Kingfisher
f Martin-chasseur torotoro
d Torotoro
i Martin pescatore beccogiallo

9628 **Synallaxis albescens**
Passeriformes - Furnariidae
e Pale-breasted Spinetail; Pale-breasted
Castle-builder
f Synallaxe albane
d Temminck-Schlüpfer;
Weißbrustbuschschlüpfer
i Codaspinosa pettopallido

9629 **Synallaxis albigularis**
Passeriformes - Furnariidae
e Dark-breasted Spinetail
f Synallaxe à gorge blanche
d Weißkehlschlüpfer
i Codaspinosa pettoscuro

9630 **Synallaxis albilora**
Passeriformes - Furnariidae
e White-lored Spinetail; Ochre-
breasted Spinetail
f Synallaxe ocré
d Ockerbrustschlüpfer
i Codaspinosa petto ocra

9631 **Synallaxis azarae**
Passeriformes - Furnariidae

e Azara's Spinetail
f Synallaxe d'Azara
d Azara-Schlüpfer
i Codaspinosa di Azara

9632 **Synallaxis brachyura**
Passeriformes - Furnariidae
e Slaty Spinetail; Sooty Spinetail; Slaty
Castle-builder
f Synallaxe ardoisé
d Graukehlschlüpfer
i Codaspinosa ardesia

9633 **Synallaxis cabanisi**
Passeriformes - Furnariidae
e Cabanis's Spinetail
f Synallaxe de Cabanis
d Cabanis-Schlüpfer
i Codaspinosa di Cabanis

9634 **Synallaxis candei**
Passeriformes - Furnariidae
e White-whiskered Spinetail
f Synallaxe à moustaches
d Schwarzohrschlüpfer
i Codaspinosa dai mustacchi bianchi

9635 **Synallaxis castanea**
Passeriformes - Furnariidae
e Black-throated Spinetail; Chestnut
Spinetail
f Synallaxe de Vaurie
d Kastanienschlüpfer
i Codaspinosa castano

9636 **Synallaxis cherriei**
Passeriformes - Furnariidae
e Chestnut-throated Spinetail
f Synallaxe à gorge marron; Synallaxe
à gorge violet
d Braunkehlschlüpfer
i Codaspinosa golacastana

9637 **Synallaxis chinchipensis**
Passeriformes - Furnariidae
e Chinchipe Spinetail; Chinchipe
f Synallaxe du Chinchipe

9638 **Synallaxis cinerascens**
Passeriformes - Furnariidae
e Grey-bellied Spinetail

f   Synallaxe grisin
d   Graubauchschlüpfer
i   Codaspinosa pettogrigio

9639   *Synallaxis cinnamomea*
Passeriformes - Furnariidae
e   Stripe-breasted Spinetail; Striped
Spinetail
f   Synallaxe guiouti
d   Zimtschlüpfer
i   Codaspinosa pettostriato

9640   *Synallaxis courseni*
Passeriformes - Furnariidae
e   Apurimac Spinetail; Coursen's
Spinetail; Blake's Spinetail
f   Synallaxe de Coursen; Synallaxe de
Blake
d   Apurimac-Schlüpfer
i   Codaspinosa di Apurimac

9641   *Synallaxis elegantior*
Passeriformes - Furnariidae
e   Elegant Spinetail
f   Synallaxe élégant
d   Schmuckschlüpfer

9642   *Synallaxis erythrothorax*
Passeriformes - Furnariidae
e   Rufous-breasted Spinetail; Rufous-
breasted Castle-builder
f   Synallaxe à poitrine rousse
d   Rotbrustschlüpfer
i   Codaspinosa pettirosso

9643   *Synallaxis frontalis*
Passeriformes - Furnariidae
e   Sooty-fronted Spinetail; Grey-
browed Spinetail
f   Synallaxe à front sombre
d   Petzeln-Schlüpfer
i   Codaspinosa frontescura

9644   *Synallaxis fuscorufa*
Passeriformes - Furnariidae
e   Rusty-headed Spinetail; Santa Marta
Spinetail
f   Synallaxe des Santa Marta
d   Rotkopfschlüpfer
i   Codaspinosa di Santa Marta

9645   *Synallaxis gujanensis*
Passeriformes - Furnariidae
e   Plain-crowned Spinetail
f   Synallaxe de Cayenne
d   Cayenne-Schlüpfer
i   Codaspinosa della Guana

9646   *Synallaxis hellmayri*
Passeriformes - Furnariidae
e   Red-shouldered Spinetail; Reiser's
Spinetail
f   Synallaxe de Hellmayr
d   Rotshulterschlüpfer
i   Codaspinosa di Reiser

9647   *Synallaxis hypospodia*
Passeriformes - Furnariidae
e   Cinereous-breasted Spinetail
f   Synallaxe cendré
d   Graubrustschlüpfer
i   Codaspinosa pettogrigio

9648   *Synallaxis infuscata*
Passeriformes - Furnariidae
e   Plain Spinetail; Pinto's Spinetail;
Alagoas Spinetail; Pernambuco
Spinetail
f   Synallaxe de Pinto
d   Pinto-Schlüpfer
i   Codaspinosa bruno

9649   *Synallaxis kollari*
Passeriformes - Furnariidae
e   Hoary-throated Spinetail
f   Synallaxe du Roraima
d   Rotohrschlüpfer
i   Codaspinosa di Kollar

9650   *Synallaxis macconnelli*
Passeriformes - Furnariidae
e   MacConnell's Spinetail
f   Synallaxe de McConnell
i   Codaspinosa di McConnell

9651   *Synallaxis maranonica*
Passeriformes - Furnariidae
e   Marañon Spinetail
f   Synallaxe du Marañon
d   Marañon-Schlüpfer
i   Codaspinosa di Maranon

9652 **Synallaxis moesta**
Passeriformes - Furnariidae
e Dusky Spinetail
f Synallaxe obscur
d Trauerschlüpfer
i Codaspinosa scuro

9653 **Synallaxis propinqua**
Passeriformes - Furnariidae
e White-bellied Spinetail
f Synallaxe à ventre blanc
d Weißbauchschlüpfer
i Codaspinosa pettobianco

9654 **Synallaxis ruficapilla**
Passeriformes - Furnariidae
e Rufous-capped Spinetail
f Synallaxe à calotte rousse; Synallaxe
à tête rouge
d Rotkappenschlüpfer
i Codaspinosa testarossiccia

9655 **Synallaxis rutilans**
Passeriformes - Furnariidae
e Ruddy Spinetail
f Synallaxe ardent
d Rötelschlüpfer
i Codaspinosa rutilante

9656 **Synallaxis scutata**
Passeriformes - Furnariidae
e Ochre-cheeked Spinetail; Ochre-
throated Spinetail
f Synallaxe à bavette; Synallaxe
orangé
d Goldohrschlüpfer
i Codaspinosa guanceocra

9657 **Synallaxis spixi**
Passeriformes - Furnariidae
e Spix's Spinetail; Chicli Spinetail
f Synallaxe de Spix
d Spix-Schlüpfer
i Codaspinosa di Spix

9658 **Synallaxis stictothorax**
Passeriformes - Furnariidae
e Necklaced Spinetail
f Synallaxe à collier
d Fleckenbrustschlüpfer
i Codaspinosa dalla collana

9659 **Synallaxis subpudica**
Passeriformes - Furnariidae
e Silvery-throated Spinetail
f Synallaxe à gorge argentée;
Synallaxe à gorge d'argent
d Silberkehlschlüpfer
i Codaspinosa gola argentata

9660 **Synallaxis superciliosa**
Passeriformes - Furnariidae
e Buff-browed Spinetail
f Synallaxe à sourcils fauves
d Brauenschlüpfer

9661 **Synallaxis tithys**
Passeriformes - Furnariidae
e Blackish-headed Spinetail; Black-
faced Spinetail
f Synallaxe tithys
d Taczanowski-Schlüpfer
i Codaspinosa testanera

9662 **Synallaxis unirufa**
Passeriformes - Furnariidae
e Rufous Spinetail
f Synallaxe roux
d Rostschlüpfer
i Codaspinosa rossiccio

9663 **Synallaxis whitneyi**
Passeriformes - Furnariidae
e Bahia Spinetail

9664 **Synallaxis zimmeri**
Passeriformes - Furnariidae
e Russet-bellied Spinetail
f Synallaxe de Zimmer
d Rostbauchschlüpfer
i Codaspinosa di Zimmer

9665 **Syndactyla guttulata**
Passeriformes - Furnariidae
e Guttulated Foliagegleaner;
Guttulated Leafgleaner
f Anabate à gouttelettes
d Zimtbrauenblattspäher
i Ticotico macchiettato

9666 **Syndactyla ruficollis**
Passeriformes - Furnariidae
e Rufous-necked Foliagegleaner; Red-

necked Foliagegleaner; Rufous-
necked Leafgleaner; Red-necked
Leafgleaner
f   Anabate à cou roux
d   Rothalsblattspäher
i   Spigolofoglie collorossiccio; Ticotico
collorossicio

9667   *Syndactyla rufosuperciliata*
Passeriformes - Furnariidae
e   Buff-browed Foliagegleaner; Buff-
browed Leafgleaner
f   Anabate à sourcils fauves
d   Ockerbrauenblattspäher
i   Ticotico dai sopraccigli ruggine

9668   *Syndactyla subalaris*
Passeriformes - Furnariidae
e   Lineated Foliagegleaner; Stripe-
bellied Foliagegleaner; Lineated
Leafgleaner
f   Anabate vergeté
d   Streifenblattspäher
i   Ticotico lineato

9669   *Synthliboramphus antiquus*
Charadriiformes - Alcidae
e   Ancient Murrelet; Ancient Auk;
Ancient Auklet; Short-billed
Guillemot; Black-throated Murrelet;
Grey-headed Murrelet
f   Guillemot à cou blanc; Guillemot
antique; Alque à cou blanc
d   Silberalk
i   Urietta antica

9670   *Synthliboramphus craveri*
Charadriiformes - Alcidae
e   Craver's Murrelet
f   Guillemot de Craveri
d   Craveri-Alk
i   Urietta di Craveri

9671   *Synthliboramphus wumizusume*
Charadriiformes - Alcidae
e   Japanese Murrelet; Temmink's
Murrelet; Japanese Auklet; Crested
Murrelet
f   Guillemot du Japon
d   Japan-Alk
i   Urietta giapponese

9672   *Syntlhliboramphus scrippsi*
Charadriiformes - Alcidae
e   Scripps's Murrelet
f   Guillemot de Scripps

9673   *Syrigma sibilatrix*
Ciconiiformes - Ardeidae
e   Whistling Heron
f   Héron flûte-du-soleil
d   Pfeifreiher
i   Airone fischiatore

9674   *Syrmaticus ellioti*
Galliformes - Phasianidae
e   Elliot's Pheasant; White-necked
Long-tailed Pheasant; Bar-backed
Pheasant; Chinese Bar-backed
Pheasant
f   Faisan d'Elliot
d   Elliot-Fasan
i   Fagiano di Elliot

9675   *Syrmaticus humiae*
Galliformes - Phasianidae
e   Mrs.. Hume's Pheasant; Hume's
Pheasant; Bar-tailed Pheasant; Bar-
backed Pheasant; Black-necked
Pheasant
f   Faisan de Hume
d   Burma-Fasan
i   Fagiano della signora Hume

9676   *Syrmaticus mikado*
Galliformes - Phasianidae
e   Mikado Pheasant; Taiwan Long-
tailed Pheasant
f   Faisan mikado
d   Mikado-Fasan
i   Fagiano mikado

9677   *Syrmaticus reevesii*
Galliformes - Phasianidae
e   Reeves's Pheasant; Bar-tailed
Pheasant; White-crowned Long-
tailed Pheasant
f   Faisan vénéré
d   Königsfasan
i   Fagiano di Reeves; Fagiano venerato

9678   *Syrmaticus soemmerringii*
Galliformes - Phasianidae

e Copper Pheasant; Soemmerring's
   Pheasant
f Faisan scintillant
d Kupferfasan
i Fagiano di Soemmerring

9679 *Syrrhaptes paradoxus*
   Pteroclidiformes - Pteroclididae
e Pallas's Sandgrouse
f Syrrhapte paradoxal; Ganga
   paradoxal
d Steppenhuhn; Steppenflughuhn
i Sirratte; Sirratte del Pallas

9680 *Syrrhaptes tibetanus*
   Pteroclidiformes - Pteroclididae
e Tibetan Sandgrouse
f Syrrhapte du Tibet
d Tibet-Flughuhn
i Sirratte del Tibet

T

9681   *Tachornis furcata*
Apodiformes - Apodidae
e   Pygmy Swift
f   Martinet pygmée
d   Däumlingssegler
i   Rondone delle palme pigmeo

9682   *Tachornis phoenicobia*
Apodiformes - Apodidae
e   Antillean Palm-Swift; Swallow (WI)
f   Petite Rolle (Ants); Hirondelle
(Ants); Jolle jolle (Ants); Martinet
petit-rolle
d   Kuba-Segler
i   Rondone delle palme antilleano

9683   *Tachornis squamata*
Apodiformes - Apodidae
e   Fork-tailed Palm-Swift; Neotropical
Palm-Swift; Fork-tailed Swift
f   Martinet claudia
d   Claudia-Segler
i   Rondone delle palme codaforcuta

9684   *Tachuris rubrigastra*
Passeriformes - Tyrannidae
e   Many-coloured Rush-Tyrant
f   Tyranneau omnicolore
d   Vielfarbentachuri; Königstachuri;
Prachttachuri
i   Tiranno ventrerosso

9685   *Tachybaptus dominicus*
Podicipediformes - Podicipedidae
e   Least Grebe; Duck-and-Teal (WI);
Diving Dapper (WI); Diver (WI);
Least Dabchick; American Dabchick
f   Grèbe minime; Grèbe de Saint-
Domingue; Grèbe (Ant); Ti Plonjon
(Ant)
d   Schwarzkopftaucher
i   Tuffetto minore

9686   *Tachybaptus novaehollandiae*
Podicipediformes - Podicipedidae
e   Australasian Grebe; Australian Little-
Grebe; Australian Dabchick; Black-
throated Little-Grebe
f   Grèbe australasien
d   Neuholland-Taucher
i   Tuffetto australiano

9687   *Tachybaptus pelzelnii*
Podicipediformes - Podicipedidae
e   Madagascar Grebe; Pelzeln's Grebe;
Pelzeln's Dabchick; Madagascar
Little-Grebe
f   Grèbe malgache
d   Pelzeln-Taucher
i   Tuffetto di Madagascar

9688   *Tachybaptus ruficollis*
Podicipediformes - Podicipedidae
e   Little Grebe; Dabchick
f   Grèbe castagneux; Grèbe fluviatile;
Petit Grèbe
d   Zwergtaucher
i   Tuffetto; Tuffolo; Tuffolino; Tuffetto
piccolo; Brinzo; Tuffetto comune

9689   *Tachybaptus rufolavatus*
Podicipediformes - Podicipedidae
e   Alaotra Grebe; Delacour's Grebe;
Delacour's Little-Grebe; Rusty
Grebe; Madagascar Red-necked
Grebe
f   Grèbe roussâtre; Grèbe de Delacour

9690   *Tachycineta albilinea*
Passeriformes - Hirundinidae
e   Mangrove Swallow
f   Hirondelle des mangroves;
Hirondelle des palétuviers
d   Mangroveschwalbe
i   Rondine delle mangrovie

9691   *Tachycineta albiventer*
Passeriformes - Hirundinidae
e   White-winged Swallow
f   Hirondelle à ailes blanches
d   Cayenne-Schwalbe
i   Rondine alibianche

9692 **Tachycineta bicolor**
Passeriformes - Hirundinidae
e Tree Swallow
f Hirondelle bicolore
d Sumpfschwalbe
i Rondine arboricola

9693 **Tachycineta cyaneoviridis**
Passeriformes - Hirundinidae
e Bahama Swallow
f Hirondelle des Bahamas
d Bahama-Schwalbe
i Rondine delle Bahama

9694 **Tachycineta euchrysea**
Passeriformes - Hirundinidae
e Golden Swallow; Swallow (WI);
Rainbird (WI)
f Oiseau de la pluie (Ants); Hirondelle
verte (Ants); Hirondelle dorée
d Antillen-Schwalbe
i Rondine dorata

9695 **Tachycineta leucorrhoa**
Passeriformes - Hirundinidae
e White-rumped Swallow
f Hirondelle à diadème
d Weißbürzelschwalbe
i Rondine dal groppone bianco;
Rondine gropponebianco

9696 **Tachycineta meyeni**
Passeriformes - Hirundinidae
e Chilean Swallow
f Hirondelle du Chili
d Feuerland-Schwalbe
i Rondine del Cile

9697 **Tachycineta stolzmanni**
Passeriformes - Hirundinidae
e Tumbes Swallow; West Peruvian
Swallow
f Hirondelle de Stolzmann
d Veilchenschwalbe

9698 **Tachycineta thalassina**
Passeriformes - Hirundinidae
e Violet-green Swallow
f Hirondelle à face blanche; Hirondelle
émeraudine
i Rondine verde-viola

9699 **Tachyeres brachypterus**
Anseriformes - Anatidae
e Falkland Steamerduck; Falkland
Islands Flightless-Steamerduck
f Brassemer des Malouines; Canard-
vapeur des Malouines
d Falkland-Dampfschiffente
i Anatra vapore delle Falkland

9700 **Tachyeres leucocephalus**
Anseriformes - Anatidae
e Chubut Steamerduck; White-headed
Steamerduck; White-headed
Flightless-Steamerduck
f Brassemer à tête blanche
i Anatra vapore testabianca

9701 **Tachyeres patachonicus**
Anseriformes - Anatidae
e Flying Steamerduck
f Brassemer de Patagonie
d Langflügeldampfschiffente
i Anatra vapore della Patagonia

9702 **Tachyeres pteneres**
Anseriformes - Anatidae
e Flightless Steamerduck; Magellanic
Flightless-Steamerduck
f Brassemer cendré; Canard-vapeur
cendré
d Riesendampfschiffente
i Anatra vapore delle Falkland

9703 **Tachymarptis aequatorialis**
Apodiformes - Apodidae
e Mottled Swift
f Martinet marbré
d Schuppensegler
i Rondone marezzato

9704 **Tachymarptis melba**
Apodiformes - Apodidae
e Alpine Swift; Ceylon White-bellied
Swift
f Martinet alpin; Martinet à ventre
blanc
d Alpensegler
i Rondone; Rondone maggiore;
Rondone alpino; Rondone maggiore
eurasiatico

9705 **Tachyphonus coronatus**
 Passeriformes - Fringillidae
 e Ruby-crowned Tanager
 f Tangara couronné
 d Krontangare
 i Tangara corona di rubino

9706 **Tachyphonus cristatus**
 Passeriformes - Fringillidae
 e Flame-crested Tanager; Scarlet-
 crested Tanager
 f Tangara à huppe ignée
 d Rothaubentangare; Haubentangare
 i Tangara crestadifiamma

9707 **Tachyphonus delatrii**
 Passeriformes - Fringillidae
 e Tawny-crested Tanager
 f Tangara de Delattre
 d Schwarzachseltangare
 i Tangara di Delattre

9708 **Tachyphonus luctuosus**
 Passeriformes - Fringillidae
 e White-shouldered Tanager
 f Tangara à épaulettes blanches
 d Trauertangare
 i Tangara spallebianche

9709 **Tachyphonus nattereri**
 Passeriformes - Fringillidae
 e Natterer's Tanager
 f Tangara de Natterer

9710 **Tachyphonus phoenicius**
 Passeriformes - Fringillidae
 e Red-shouldered Tanager
 f Tangara à galons rouges
 d Rotschultertangare
 i Tangara spallerosse

9711 **Tachyphonus rufiventer**
 Passeriformes - Fringillidae
 e Yellow-crested Tanager
 f Tangara à crête jaune
 d Gelbschopftangare
 i Tangara crestagialla

9712 **Tachyphonus rufus**
 Passeriformes - Fringillidae
 e White-lined Tanager

 f Tangara à galons blancs
 d Schwarztangare
 i Tangara nera

9713 **Tachyphonus surinamus**
 Passeriformes - Fringillidae
 e Fulvous-crested Tanager
 f Tangara à crête fauve
 d Goldschopftangare
 i Tangara crestafulva

9714 **Tadorna cana**
 Anseriformes - Anatidae
 e South African Shelduck; African
 Shelduck; Cape Shelduck
 f Tadorne à tête grise
 d Graukopfrostgans; Graukopfkasarka
 i Casarca sudafricana; Casarca
 africana

9715 **Tadorna cristata**
 Anseriformes - Anatidae
 e Crested Shelduck; Korean Crested-
 Shelduck
 f Tadorne de Corée
 d Schopfkasarka
 i Volpoca crestata

9716 **Tadorna ferruginea**
 Anseriformes - Anatidae
 e Ruddy Shelduck; Brahminy Duck
 (ISC); Ruddy Sheldrake m
 f Tadorne casarca
 d Rostgans
 i Casarca; Germano forestiero;
 Casarca ferruginea

9717 **Tadorna radjah**
 Anseriformes - Anatidae
 e Radjah Shelduck; Rajah Shelduck;
 White-headed Shelduck; Burdekin
 Duck; Black-backed Shelduck
 f Tadorne radjah
 d Radjahgans
 i Casarca testabianca; Casarca di
 Radjah

9718 **Tadorna tadorna**
 Anseriformes - Anatidae
 e Common Shelduck; Shelduck;
 Northern Shelduck; Red-billed

Shelduck
f Tadorne de Belon; Tadorne belon
d Brandgans; Brandente
i Volpoca; Volpoca comune; Tadorna

9719 Tadorna tadornoides
Anseriformes - Anatidae
e Australian Shelduck; Mountain
Duck; Mountain Shelduck; Chestnut-
breasted Shelduck; Chestnut-
coloured Shelduck
f Tadorne d'Australie
d Halsbandkasarka
i Casarca australiano

9720 Tadorna variegata
Anseriformes - Anatidae
e Paradise Shelduck; Paradise Duck;
New Zealand Shelduck; Rangitata
Goose; Painted Duck
f Tadorne de paradis
d Paradieskasarka
i Casarca del Paradiso; Casarca della
Nuova Zelanda

9721 Taeniopygia bichenovii
Passeriformes - Passeridae
e Double-barred Finch; Banded Finch;
Bichenov's Finch; Ringed Finch;
Double-banded Finch
f Diamant de Bichenov
d Ringelastrild;
Weißbürzelringelastrild
i Diamante di Bichenow

9722 Taeniopygia castanotis
Passeriformes - Passeridae
e Chestnut-eared Finch

9723 Taeniopygia guttata
Passeriformes - Passeridae
e Zebra Finch; Spotted-sided Finch;
Chestnut-eared Finch
f Diamant mandarin
d Zebrafink
i Diamante mandarino

9724 Taeniotriccus andrei
Passeriformes - Tyrannidae
e Black-chested Tyrant; Andre's Tyrant
f Microtyran d'André; Tyranneau

d'André
d Schwarzschopftyrann
i Tiranno todo pettonero

9725 Talegalla cuvieri
Galliformes - Megapodiidae
e Red-billed Brush-Turkey; Red-billed
Scrub-Turkey
f Talégalle de Cuvier
d Rotschnabeltalegalla
i Talegalla beccorosso

9726 Talegalla fuscirostris
Galliformes - Megapodiidae
e Black-billed Brush-Turkey; Black-
billed Scrub-Turkey; Yellow-legged
Brush-Turkey
f Talégalle à bec foncé
d Schwarzschnabeltalegalla
i Talegalla becconero

9727 Talegalla jobiensis
Galliformes - Megapodiidae
e Brown-collared Brush-Turkey;
Collared Brush-Turkey; Brown-
billed Brush-Turkey; Red-legged
Brush-Turkey
f Talégalle de Jobi
d Halsbandtalegalla
i Talegalla dal collare bruno

9728 Tangara argyrofenges
Passeriformes - Thraupidae
e Straw-backed Tanager; Green-
throated Tanager
f Calliste à gorge verte
d Grünkehltangare
i Tangara golaverde

9729 Tangara arthus
Passeriformes - Thraupidae
e Golden Tanager
f Calliste doré
d Goldtangare
i Tangara dorata

9730 Tangara brasiliensis
Passeriformes - Thraupidae
e White-bellied Tanager

9731   *Tangara cabanisi*
Passeriformes - Thraupidae
e   Azure-rumped Tanager; Cabanis's
Tanager
f   Calliste azuré
d   Cabanis-Tangare
i   Tangara gropponeblu

9732   *Tangara callophrys*
Passeriformes - Thraupidae
e   Opal-crowned Tanager
f   Calliste à sourcils clairs
d   Opalscheiteltangare
i   Tangara corona opale

9733   *Tangara cayana*
Passeriformes - Thraupidae
e   Burnished-buff Tanager; Rufous-
crowned Tanager
f   Calliste passevert
d   Isabelltangare
i   Tangara della Cayenna

9734   *Tangara chilensis*
Passeriformes - Thraupidae
e   Paradise Tanager
f   Calliste septicolore
d   Siebenfarbentangare
i   Tangara del Paradiso

9735   *Tangara chrysotis*
Passeriformes - Thraupidae
e   Golden-eared Tanager; Blue-browed
Tanager
f   Calliste à oreilles d'or
d   Goldohrtangare
i   Tangara orecchiedorate

9736   *Tangara cucullata*
Passeriformes - Thraupidae
e   Lesser Antillean Tanager; Hooded
Tanager; Golden Tanager (WI);
Princebird (WI); Paw-paw Bird (WI);
Soursopbird (WI); Antillean Tanager
f   Calliste dos-bleu
d   Rotkappentangare
i   Tangara delle Piccole Antille

9737   *Tangara cyanicollis*
Passeriformes - Thraupidae
e   Blue-necked Tanager

f   Calliste à cou bleu
d   Azurkopftangare; Blaukopftangare
i   Tangara colloazzurro

9738   *Tangara cyanocephala*
Passeriformes - Thraupidae
e   Red-necked Tanager; Blue-headed
Tanager
f   Calliste à tête bleue
d   Blaukappentangare
i   Tangara capoazzurro

9739   *Tangara cyanomelaena*
Passeriformes - Thraupidae
e   Silver-breasted Tanager

9740   *Tangara cyanoptera*
Passeriformes - Thraupidae
e   Black-headed Tanager
f   Calliste à tête noire
d   Blauflügeltangare
i   Tangara di monte aliazzurre

9741   *Tangara cyanotis*
Passeriformes - Thraupidae
e   Blue-browed Tanager
f   Calliste à sourcils bleus
d   Silberbrauentangare;
Blauwangentangare
i   Tangara dai sopraccigli azzurri

9742   *Tangara cyanoventris*
Passeriformes - Thraupidae
e   Gilt-edged Tanager
f   Calliste à ventre bleu
d   Blaubrusttangare
i   Tangara pettazzurra

9743   *Tangara desmaresti*
Passeriformes - Thraupidae
e   Brassy-breasted Tanager; Brass-
breasted Tanager
f   Calliste de Desmarest
d   Orangebrusttangare
i   Tangara di Desmarest

9744   *Tangara dowii*
Passeriformes - Thraupidae
e   Spangle-cheeked Tanager; Spangled
Tanager; Dow's Tanager
f   Calliste pailleté

d Glanzfleckentangare
i Tangara di Dow

9745 *Tangara fastuosa*
Passeriformes - Thraupidae
e Seven-coloured Tanager; Superb Tanager
f Calliste superbe
d Vielfarbentangare
i Tangara settecolori

9746 *Tangara florida*
Passeriformes - Thraupidae
e Emerald Tanager
f Calliste émeraude
d Smaragdtangare
i Tangara smeraldo

9747 *Tangara fucosa*
Passeriformes - Thraupidae
e Green-naped Tanager; Pirre Tanager
f Calliste à nuque verte
i Tangara nucaverde

9748 *Tangara guttata*
Passeriformes - Thraupidae
e Speckled Tanager
f Calliste tiqueté
d Tropfentangare
i Tangara maculata

9749 *Tangara gyrola*
Passeriformes - Thraupidae
e Bay-headed Tanager; Gyrola Tanager; Green Tanager
f Calliste rouverdin
d Grüntangare
i Tangara testabaia

9750 *Tangara heinei*
Passeriformes - Thraupidae
e Black-capped Tanager
f Calliste à calotte noire
d Heine-Tangare
i Tangara di Heine

9751 *Tangara icterocephala*
Passeriformes - Thraupidae
e Silver-throated Tanager
f Calliste safran

d Silberkehltangare
i Tangara gialla golabianca

9752 *Tangara inornata*
Passeriformes - Thraupidae
e Plain-coloured Tanager; Plain Tanager
f Calliste gris
d Schlichttangare; Grautangare
i Tangara disadorna

9753 *Tangara johannae*
Passeriformes - Thraupidae
e Blue-whiskered Tanager; Whiskered Tanager
f Calliste moustachu
d Blaubarttangare
i Tangara dai mustacchi azzurri

9754 *Tangara labradorides*
Passeriformes - Thraupidae
e Metallic-green Tanager
f Calliste vert
d Schwarznackentangare
i Tangara verde-matallico

9755 *Tangara larvata*
Passeriformes - Thraupidae
e Golden-hooded Tanager; Golden-masked Tanager; Masked Tanager; Golden-headed Tanager
f Calliste à coiffe d'or
d Purpurmaskentangare
i Tangara maschera dorata

9756 *Tangara lavinia*
Passeriformes - Thraupidae
e Rufous-winged Tanager; Lavinia's Tanager
f Calliste à ailes rousses
d Goldflügeltangare
i Tangara alirosse

9757 *Tangara mexicana*
Passeriformes - Thraupidae
e Turquoise Tanager
f Calliste diable-enrhumé
d Türkis-Tangare
i Tangara turchese

9758 *Tangara meyerdeschauenseei*
Passeriformes - Thraupidae
e Green-capped Tanager
f Calliste de Schauensee
i Tangara di Meyer de Schauensee

9759 *Tangara nigrocincta*
Passeriformes - Thraupidae
e Masked Tanager; Black-tinted
Tanager; Black-banded Tanager;
Blue-headed Tanager
f Calliste masqué
d Schwarzbrusttangare; Maskentangare
i Tangara dal collare nero

9760 *Tangara nigroviridis*
Passeriformes - Thraupidae
e Beryl-spangled Tanager; Green-and-
black Tanager
f Calliste beryl
d Silberfleckentangare
i Tangara verde e nera

9761 *Tangara palmeri*
Passeriformes - Thraupidae
e Grey-and-gold Tanager
f Calliste or-gris
d Palmer-Tangare
i Tangara di Palmer

9762 *Tangara parzudakii*
Passeriformes - Thraupidae
e Flame-faced Tanager; Parzudaki's
Tanager
f Calliste à face rouge
d Rotstirntangare;
Flammengesichttangare
i Tangara facciadifiamma

9763 *Tangara peruviana*
Passeriformes - Thraupidae
e Black-backed Tanager
f Calliste à dos noir
d Schwarzmanteltangare
i Tangara dorsonero

9764 *Tangara phillipsi*
Passeriformes - Thraupidae
e Sira Tanager; Siara Tanager
f Calliste de Phillips
i Tangara di Phillips

9765 *Tangara preciosa*
Passeriformes - Thraupidae
e Chestnut-backed Tanager
f Calliste à dos marron
d Prachttangare
i Tangara dorsocastano

9766 *Tangara pulcherrima*
Passeriformes - Thraupidae
e Yellow-collared Tanager; Golden-
collared Tanager; Yellow-collared
Honeycreeper; Golden-collared
Honeycreeper; Golden-bellied
Honeycreeper
f Calliste sucrier
d Halsbandtangare
i Tangara dal collare giallo

9767 *Tangara punctata*
Passeriformes - Thraupidae
e Spotted Tanager
f Calliste syacou
d Drosseltangare
i Tangara punteggiata

9768 *Tangara ruficervix*
Passeriformes - Thraupidae
e Golden-naped Tanager; Yungas
Tanager; Orange-rumped Tanager
f Calliste à nuque d'or
d Goldnackentangare
i Tangara nucadorata

9769 *Tangara rufigenis*
Passeriformes - Thraupidae
e Rufous-cheeked Tanager
f Calliste à joues rousses
d Rotwangentangare
i Tangara guanceruggine

9770 *Tangara rufigula*
Passeriformes - Thraupidae
e Rufous-throated Tanager
f Calliste à gorge rousse
d Rostkehltangare
i Tangara golarossa

9771 *Tangara schrankii*
Passeriformes - Thraupidae
e Green-and-gold Tanager
f Calliste de Schrank

d Schrank-Tangare; Goldbrusttangare
i Tangara verde e dorato

9772 ***Tangara seledon***
    Passeriformes - Thraupidae
e Green-headed Tanager; Celadon Tanager
f Calliste à tête verte
d Dreifarbentangare; Prachttangare
i Tangara testaverde

9773 ***Tangara varia***
    Passeriformes - Thraupidae
e Dotted Tanager
f Calliste tacheté
d Surinam-Tangare
i Tangara varia

9774 ***Tangara vassorii***
    Passeriformes - Thraupidae
e Blue-and-black Tanager; Buff-naped Tanager
f Calliste bleu-et-noir
d Vassori-Tangare
i Tangara azzurra e nera

9775 ***Tangara velia***
    Passeriformes - Thraupidae
e Opal-rumped Tanager; Opal Tanager
f Calliste varié
d Rotbauchtangare
i Tangara gropponeopale

9776 ***Tangara viridicollis***
    Passeriformes - Thraupidae
e Silver-backed Tanager; Silver Tanager; Silvery Tanager
f Calliste argenté
d Silbertangare
i Tangara colloverde

9777 ***Tangara vitriolina***
    Passeriformes - Thraupidae
e Scrub Tanager
f Calliste vitriolin
d Rotscheiteltangare
i Tangara di macchia

9778 ***Tangara xanthocephala***
    Passeriformes - Thraupidae
e Saffron-crowned Tanager

f Calliste à tête dorée
d Gelbkopftangare
i Tangara corona zafferano

9779 ***Tangara xanthogastra***
    Passeriformes - Thraupidae
e Yellow-bellied Tanager
f Calliste à ventre jaune
d Gelbbauchtangare
i Tangara ventregiallo

9780 ***Tanygnathus gramineus***
    Psittaciformes - Psittacidae
e Black-lored Parrot; Buru Parrot
f Perruche de Buru
d Buru-Papagei; Schwarzstirnedelpapagei
i Pappagallo dai mustacchi neri

9781 ***Tanygnathus lucionensis***
    Psittaciformes - Psittacidae
e Blue-naped Parrot; Blue-crowned Parrot
f Perruche de Luçon
d Blaunackenpapagei; Blauscheiteledelpapagei
i Pappagallo nuca azzurra

9782 ***Tanygnathus megalorynchos***
    Psittaciformes - Psittacidae
e Great-billed Parrot; Large-billed Parrot; Moluccan Parrot; Island Parrot
f Perruche à bec de sang
d Schwarzschulterpapagei
i Pappagallo beccogrosso

9783 ***Tanygnathus sumatranus***
    Psittaciformes - Psittacidae
e Blue-backed Parrot; Müller's Blue-backed Parrot; Everett's Parrot; Burbidge's Parrot; Müller's Parrot; Blue-necked Parrot; Azure-rumped Parrot
f Perruche de Müller
d Everett-Papagei; Müllers Edelpapagei
i Pappagallo di Müller

9784 ***Tanysiptera carolinae***
    Coraciiformes - Halcyonidae

e   Numfor Paradise-Kingfisher
f   Martin-chasseur de Caroline
d   Numfor-Liest
i   Martin pescatore del Paradiso di
    Numfor

9785   *Tanysiptera danae*
       Coraciiformes - Halcyonidae
e   Brown-headed Paradise-Kingfisher;
    Brown-backed Paradise-Kingfisher
f   Martin-chasseur rose
d   Braunmantelliest
i   Martin pescatore del Paradiso
    dorsobruno

9786   *Tanysiptera ellioti*
       Coraciiformes - Halcyonidae
e   Kofiau Paradise-Kingfisher; Elliot's
    Paradise-Kingfisher
f   Martin-chasseur de Kofiau; Martin-
    chasseur d'Elliot
d   Elliot-Liest
i   Martin pescatore del Paradiso di
    Elliot

9787   *Tanysiptera galatea*
       Coraciiformes - Halcyonidae
e   Common Paradise-Kingfisher;
    Paradise-Kingfisher; Beautiful
    Kingfisher
f   Martin-chasseur à longs brins
d   Spatelliest
i   Martin pescatore del Paradiso

9788   *Tanysiptera hydrocharis*
       Coraciiformes - Halcyonidae
e   Little Paradise-Kingfisher; Aru
    Paradise-Kingfisher; Blue-tailed
    Paradise-Kingfisher
f   Martin-chasseur menu; Martin-
    chasseur des forêts
d   Feenlist
i   Martin pescatore minore del Paradiso

9789   *Tanysiptera nympha*
       Coraciiformes - Halcyonidae
e   Red-breasted Paradise-Kingfisher;
    Pink-breasted Paradise-Kingfisher
f   Martin-chasseur nymphe
d   Nymphenliest

i   Martin pescatore del Paradiso
    pettorosa

9790   *Tanysiptera riedelii*
       Coraciiformes - Halcyonidae
e   Biak Paradise-Kingfisher
f   Martin-chasseur de Biak
d   Biak-Liet
i   Martin pescatore del Paradiso di Biak

9791   *Tanysiptera sylvia*
       Coraciiformes - Halcyonidae
e   Buff-breasted Paradise-Kingfisher;
    White-tailed Paradise-Kingfisher;
    White-tailed Kingfisher; Australian
    Paradise-Kingfisher; Black-headed
    Paradise-Kingfisher
f   Martin-chasseur sylvain
d   Paradiesliest
i   Martin pescatore del Paradiso
    codabianca

9792   *Taoniscus nanus*
       Tinamiformes - Tinamidae
e   Dwarf Tinamou; Least Tinamou
f   Tinamou carapé
d   Pfauensteißhuhn
i   Tinamo nano

9793   *Tapera naevia*
       Passeriformes - Neomorphidae
e   Striped Cuckoo
f   Géocoucou tacheté
d   Vierflügelkuckuck
i   Tapera

9794   *Taphrolesbia griseiventris*
       Apodiformes - Trochilidae
e   Grey-bellied Comet
f   Colibri comète
d   Graubauchsylphe
i   Colibrì cometa ventregrigio

9795   *Taphrospilus hypostictus*
       Apodiformes - Trochilidae
e   Many-spotted Hummingbird
f   Colibri grivelé
d   Tropfenkolibri
i   Colibrì ventremacchiato

9796 **Taraba major**
Passeriformes - Formicariidae
e Great Antshrike
f Grand Batara
d Weißbrustameisenwürger; Großer
Ameisenfresser
i Averla formichiera maggiore

9797 **Tarsiger chrysaeus**
Passeriformes - Muscicapidae
e Golden Bush-Robin; Golden Robin
f Rossignol doré
d Blauschwanz
i Usignolo dorato

9798 **Tarsiger cyanurus**
Passeriformes - Muscicapidae
e Red-flanked Bluetail; Orange-
flanked Bush-Robin (ISC); Bluestart;
Red-flanked Bush-Robin; Orange-
flanked Bluetail; Blue-tailed Robin
f Rossignol à flancs roux; Robin à
flancs roux
d Rotbrustblauschwanz
i Codazzurro

9799 **Tarsiger hyperythrus**
Passeriformes - Muscicapidae
e Rufous-breasted Bush-Robin;
Rufous-bellied Bush-Robin; Rufous-
breasted Robin; Rufous-breasted
Bluetail; Rufous-bellied Bluetail
f Rossignol à ventre roux
d Weißbrauenblauschwanz
i Usignolo pettorossiccio

9800 **Tarsiger indicus**
Passeriformes - Muscicapidae
e White-browed Bush-Robin; White-
browed Bluetail; White-browed
Robin
f Rossignol à sourcils blancs;
Rossignol indien
d Taiwan-Rötel
i Usignolo dai sopraccigli bianchi

9801 **Tarsiger johnstoniae**
Passeriformes - Muscicapidae
e Collared Bush-Robin; Collared
Robin; Formosan Bush-Robin;
Taïwan Bush-Robin; Johnston's

Bush-Robin
f Rossignol de Johnston; Rossignol de
Taïwan
d Formosa-Blauschwanz
i Usignolo dal collare

9802 **Tauraco bannermani**
Musophagiformes - Musophagidae
e Bannerman's Turaco; Bannerman's
Touraco
f Touraco doré
d Bannerman-Turaco; Bannerman-
Turako
i Turaco di Bannerman

9803 **Tauraco corythaix**
Musophagiformes - Musophagidae
e Knysna Turaco; Knysna Touraco
f Touraco louri
d Helmturako; Helmturaco
i Turaco di Knysna

9804 **Tauraco erythrolophus**
Musophagiformes - Musophagidae
e Red-crested Turaco; Red-crested
Touraco
f Touraco paruline
d Rothaubenturako; Rothaubenturaco;
Rotschopfturaco; Rotschopfturako
i Turaco crestarossa

9805 **Tauraco fischeri**
Musophagiformes - Musophagidae
e Fischer's Turaco; Fischer's Touraco
f Touraco de Fischer
d Fischer-Turako; Fischer-Turaco
i Turaco di Fischer

9806 **Tauraco hartlaubi**
Musophagiformes - Musophagidae
e Hartlaub's Turaco; Hartlaub's
Touraco; Blue-crested Turaco; Blue-
crested Touraco
f Touraco de Hartlaub
d Seidenhollenturako; Seidenturaco;
Seidenhollenturaco; Seidenturako
i Turaco di Hartlaub; Turaco della
cresta rossa

9807 **Tauraco leucolophus**
Musophagiformes - Musophagidae

e   White-crested Turaco; White-crested
    Touraco
f   Touraco à huppe blanche
d   Weißhaubenturako;
    Weißhaubenturaco
i   Turaco crestabianca

9808   *Tauraco leucotis*
    Musophagiformes - Musophagidae
e   White-cheeked Turaco; White-
    cheeked Touraco
f   Touraco à joues blanches
d   Eißohr; Weißohrturako;
    Weißohrturaco
i   Turaco guancebianche; Turaco dalle
    gote bianche

9809   *Tauraco livingstonii*
    Musophagiformes - Musophagidae
e   Livingstone's Turaco; Livingstone's
    Touraco
f   Touraco de Livingstone
d   Langschopfturako; Langschopfturaco
i   Turaco di Livingstone

9810   *Tauraco macrorhynchus*
    Musophagiformes - Musophagidae
e   Yellow-billed Turaco; Yellow-billed
    Touraco; Black-tipped Turaco;
    Black-tipped Touraco; Red-tipped
    Turaco; Red-tipped Touraco; Crested
    Turaco; Crested Touraco
f   Touraco à gros bec
d   Blaurückenturako; Blaurückenturaco
i   Turaco beccogiallo

9811   *Tauraco persa*
    Musophagiformes - Musophagidae
e   Guinea Turaco; Guinea Touraco
f   Touraco vert
d   Guinea-Turako; Guinea-Turaco
i   Turaco verde

9812   *Tauraco ruspolii*
    Musophagiformes - Musophagidae
e   Ruspoli's Turaco; Ruspoli's Touraco;
    Prince Ruspoli's Turaco; Prince
    Ruspoli's Touraco
f   Touraco de Ruspoli; Touraco de
    prince Ruspoli

d   Ruspoli-Turako; Ruspoli-Turaco
i   Turaco del principe Ruspoli

9813   *Tauraco schalowi*
    Musophagiformes - Musophagidae
e   Schalow's Turaco; Schalow's
    Touraco
f   Touraco de Schalow
d   Federhelmturako; Federhelmturaco
i   Turaco di Schallow

9814   *Tauraco schuettii*
    Musophagiformes - Musophagidae
e   Black-billed Turaco; Black-billed
    Touraco
f   Touraco à bec noir
d   Schwarzschnabelturako
i   Turaco becconero; Turaco dal becco
    nero

9815   *Tchagra anchietae*
    Passeriformes - Malaconotidae
e   Southern Blackcap; Anchieta's
    Tchagra
f   Tchagra d'Anchieta
i   Ciagra di Anchieta

9816   *Tchagra australis*
    Passeriformes - Malaconotidae
e   Brown-crowned Tchagra; Three-
    streaked Tchagra (CSA); Brown-
    headed Tchagra; Black-crowned
    Bush-Shrike; Brown-headed Bush-
    Shrike; Souza's Tchagra
f   Tchagra à tête brune
d   Damara-Tschagra; Dorntschagra
i   Ciagra testabruna

9817   *Tchagra jamesi*
    Passeriformes - Malaconotidae
e   Three-streaked Tchagra; James's
    Tchagra; Three-streaked Bush-Shrike
f   Tchagra de James
d   Somali-Tchagra
i   Ciagra di James

9818   *Tchagra minuta*
    Passeriformes - Malaconotidae
e   Marsh Tchagra; Blackcap Tchagra;
    Black-capped Bush-Shrike; Three-
    streaked Bush-Shrike; Blackcap

Bush-Shrike; Lesser Tchagra; Little
Tchagra
f Tchagra des marais; Petit téléphone
d Sumpftschagra
i Ciagra capinera

9819 *Tchagra senegala*
Passeriformes - Malaconotidae
e Black-crowned Tchagra; Black-
crowned Bush-Shrike; Black-headed
Bush-Shrike; Bush Shrike; Hooded
Tchagra; Black-capped Bush-Shrike
f Tchagra à tète noire; Téléphone
tchagra
d Senegal-Tschagra;
Schwarzkopftschagra
i Ciagra corone nera

9820 *Tchagra tchagra*
Passeriformes - Malaconotidae
e Southern Tchagra; Tchagra Bush-
Strike; Levaillant's Bush-Shrike;
Levaillant's Tchagra; Tchagra Shrike
f Tchagra du Cap
d Kap-Tschagra
i Ciagra meridionale

9821 *Telacanthura melanopygia*
Apodiformes - Apodidae
e Black Spinetail; Chapin's Swift;
Chapin's Spinetailed-Swift; Ituri
Mottle-throated Spinetailed-Swift;
Ituri Mottle-throated Spinetail
f Martinet de Chapin
d Ituri-Segler
i Rondone codaspinosa dell'Ituri

9822 *Telacanthura ussheri*
Apodiformes - Apodidae
e Mottled Spinetail; Mottle-throated
Spinetailed-Swift
f Martinet d'Ussher
d Baobabsegler
i Rondone codaspinosa di Ussher

9823 *Teledromas fuscus*
Passeriformes - Rhinocryptidae
e Sandy Gallito; Barrancolino Gallito;
Barrancolino
f Tourco sable

d Braungallito; Pampabürzelstelzer
i Gallito color sabbia

9824 *Telespiza cantans*
Passeriformes - Fringillidae
e Laysan Finch; Laysan Honeycreeper
f Psittirostre de Laysan
d Laysan-Gimpel; Ou
i Fringuello di Laysan

9825 *Telespiza ultima*
Passeriformes - Fringillidae
e Nihoa Finch; Nihoa Honeycreeper
f Psittirostre de Nihoa
i Fringuello di Nihoa

9826 *Telophorus bocagei*
Passeriformes - Corvidae
e Grey-green Bush-Shrike; Bocage's
Bush-Shrike; Grey Bush-Shrike
f Gladiateur à front blanc
d Bocage-Würger; Grauer
Buschwürger
i Averla di macchia grigioverde

9827 *Telophorus kupeensis*
Passeriformes - Corvidae
e Serle's Bush-Shrike; Mount Kupé
Bush-Shrike; Kupé Mountain Bush-
Shrike; Kupe Bush-Shrike
f Gladiateur du Kupé; Gladiateur du
Mont Kupé
d Halsbandwürger
i Averla di macchia del Monte Kupè

9828 *Telophorus multicolor*
Passeriformes - Corvidae
e Many-coloured Bush-Shrike
f Gladiateur multicolore
d Vielfarbenwürger
i Averla di macchia multicolore

9829 *Telophorus nigrifrons*
Passeriformes - Corvidae
e Black-fronted Bush-Shrike; Many-
coloured Bush-Shrike
f Gladiateur à front noir
d Schwarzstirnbuschwürger;
Graustirnwürger
i Averla di macchia frontenera

9830   *Telophorus olivaceus*
Passeriformes - Corvidae
e   Olive Bush-Shrike; Rufous-breated Bush-Shrike
f   Gladiateur olive
d   Olivwürger; Olivbuschwürger
i   Averla di macchia oliva

9831   *Telophorus quadricolor*
Passeriformes - Corvidae
e   Four-coloured Bush-Shrike; Gorgeous Bush-Shrike; Konkoit (CSA)
f   Gladiateur quadricolore
d   Vierfarbenwürger
i   Averla di macchia quadricolore

9832   *Telophorus sulfureopectus*
Passeriformes - Corvidae
e   Sulphur-breasted Bush-Shrike; Orange-breasted Bush Shrike (CSA)
f   Gladiateur soufré; Pie-grièche soufrée
d   Orangebrustwürger; Orangewürger; Orangebrustbuschwürger
i   Averla di macchia pettosulfureo

9833   *Telophorus viridis*
Passeriformes - Corvidae
e   Perrin's Bush-Shrike; Perrin's Shrike; Gorgeous Bush-Shrike; Four-coloured Bush-Shrike
f   Gladiateur vert
d   Vierfarbenwürger
i   Averla di macchia di Perrin

9834   *Telophorus zeylonus*
Passeriformes - Corvidae
e   Bokmakierie; Bokmakierie Shrike; Bacbakiri; Bokmakiere Bush-Shrike
f   Gladiateur bacbakiri
d   Bokmakiri
i   Averla di macchia Bok-Makirì

9835   *Temnurus temnurus*
Passeriformes - Corvidae
e   Ratchet-tailed Treepie; Notch-tailed Treepie; Black Treepie
f   Temia temnure; Pie temnure
d   Buchtschwanzelster
i   Gazza codadentata

9836   *Tephrodornis gularis*
Passeriformes - Corvidae
e   Large Wood-Shrike; Brown-tailed Wood-Shrike; Hook-billed Wood-Shrike; Hook-billed Greybird
f   Téphrodorne bridé; Échenilleur bridé brun
d   Braunschwanztephradornis
i   Averla di bosca codabruna

9837   *Tephrodornis pondicerianus*
Passeriformes - Corvidae
e   Common Wood-Shrike; Wood-Shrike; Lesser Wood-Shrike; Ceylon Wood-Shrike
f   Téphrodorne de Pondichéry; Échenilleur de Pondichéry
d   Weißbrauentephradornis
i   Averla di bosca comune; Averla di bosca

9838   *Tephrozosterops stalkeri*
Passeriformes - Zosteropidae
e   Bicoloured White-eye; Two-coloured White-eye; Kakopi White-eye; Ceram White-eye; Stalker's White-eye; Bicoloured Darkeye; Rufescent White-eye; Rufescent Darkeye
f   Zostérops de Stalker
d   Kakopi-Brillenvogel
i   Occhialino bicolore

9839   *Terathopius ecaudatus*
Falconiformes - Accipitridae
e   Bateleur
f   Bateleur des savanes; Aigle bateleur
d   Gaukler
i   Falco giocoliere

9840   *Terenura callinota*
Passeriformes - Formicariidae
e   Rufous-rumped Ant-Wren
f   Grisin à croupion roux
d   Grauwangenameisenfänger
i   Mangiaformiche gropponerossiccio

9841   *Terenura humeralis*
Passeriformes - Formicariidae
e   Chestnut-shouldered Ant-Wren
f   Grisin à épaules rousses

d Rotschulterameisenfänger
i Mangiaformiche spallecastane

9842 ***Terenura maculata***
Passeriformes - Formicariidae
e Streak-capped Ant-Wren
f Grisin à tête rayée
d Streifenkopfameisenfänger
i Mangiaformiche testacastana

9843 ***Terenura sharpei***
Passeriformes - Formicariidae
e Yellow-rumped Ant-Wren
f Grisin à croupion jaune
d Gelbbürzelameisenfänger
i Mangiaformiche di Sharp

9844 ***Terenura sicki***
Passeriformes - Formicariidae
e Orange-bellied Ant-Wren; Alagoas
Ant-Wren
f Grisin de Sick
i Mangiaformiche di Sick

9845 ***Terenura spodioptila***
Passeriformes - Formicariidae
e Ash-winged Ant-Wren
f Grisin spodioptile
d Grauschwingenameisenfänger
i Mangiaformiche alicenerine

9846 ***Terenura venezuelana***
Passeriformes - Formicariidae
e Perija Ant-Wren
f Grisin des Periya

9847 ***Teretistris fernandinae***
Passeriformes - Fringillidae
e Yellow-headed Warbler
f Paruline de Fernandina
d Gelbkopfwaldsänger
i Parula testagialla

9848 ***Teretistris fornsi***
Passeriformes - Fringillidae
e Orient Warbler
f Paruline d'Oriente
d Forns-Waldsänger
i Parula di Cuba

9849 ***Terpsiphone atrocaudata***
Passeriformes - Monarchidae
e Japanese Paradise-Flycatcher; Black
Paradise-Flycatcher; Black-tailed
Paradise-Flycatcher
f Tchitrec du Japon; Gobemouche du
paradis du Japon; Gobe-mouches du
paradis du Japon
d Prinzen-Paradiesschnäpper;
Japanischer Paradiesschnäpper
i Pigliamosche del Paradiso nero

9850 ***Terpsiphone atrochalybeia***
Passeriformes - Monarchidae
e Sao Tomé Paradise-Flycatcher
f Tchitrec de Sao Tomé; Gobemouche
paradis; Gobemouche de paradis de
San Tomé; Gobe-mouches paradis;
Gobe-mouches paradis de San
Thomé; Moucherolle de Sao Tomé
d Stahlparadiesschnäpper
i Pigliamosche del Paradiso di Sao
Tomè

9851 ***Terpsiphone batesi***
Passeriformes - Monarchidae
e Bates's Paradise-Flycatcher
f Tchitrec de Bates

9852 ***Terpsiphone bedfordi***
Passeriformes - Monarchidae
e Bedford's Paradise-Flycatcher; Red-
bellied Paradise-Flycatcher
f Tchitrec de Bedford; Moucherolle de
Bedford
i Pigliamosche del Paradiso di Bedford

9853 ***Terpsiphone bourbonnensis***
Passeriformes - Monarchidae
e Mascarene Paradise-Flycatcher
f Tchitrec des Mascareignes;
Gobemouche de paradis des îles
Bourbon; Moucherolle de l'île
Maurice; Gobe-mouches de paradis
des îles Bourbon; Gobe-mouches de
paradis de la Vierge; Gobemouche de
paradis de la Vierge
d Maskarenen-Paradiesschnäpper
i Pigliamosche del Paradiso delle
Mascarene

9854   Terpsiphone cinnamomea
Passeriformes - Monarchidae
e   Rufous Paradise-Flycatcher; Alut
Paradise-Flycatcher; Philippine
Paradise-Flycatcher
f   Tchitrec roux
d   Zimtparadiesschnäpper
i   Pigliamosche del Paradiso rossiccio

9855   Terpsiphone corvina
Passeriformes - Monarchidae
e   Seychelles Paradise-Flycatcher;
Seychelles Black Paradise-Flycatcher
f   Tchitrec des Seychelles; Moucherolle
noire
d   Seychellen-Paradiesschnäpper
i   Pigliamosche del Paradiso delle
Seychelle

9856   Terpsiphone cyanescens
Passeriformes - Monarchidae
e   Blue Paradise-Flycatcher
f   Tchitrec de Palawan
d   Kobaltparadiesschnäpper
i   Pigliamosche del Paradiso azzurro

9857   Terpsiphone cyanomelas
Passeriformes - Monarchidae
f   Tchitrec du Cap

9858   Terpsiphone mutata
Passeriformes - Monarchidae
e   Madagascar Paradise-Flycatcher
f   Tchitrec malgache; Gobemouche de
paradis malgache; Gobemouche de
paradis de Madagascar; Gobe-
mouches de paradis malgache; Gobe-
mouches de paradis de Madagascar;
Moucherolle malgache
d   Rotbrustparadiesschnäpper
i   Pigliamosche del Paradiso del
Madagascar

9859   Terpsiphone nitens
Passeriformes - Monarchidae
f   Tchitrec noir
i   Pigliamosche del Paradiso

9860   Terpsiphone paradisi
Passeriformes - Monarchidae
e   Asian Paradise-Flycatcher; Paradise

Flycatcher (ISC); Asiatic Paradise-
Flycatcher; Common Paradise-
Flycatcher (ISC); Ceylon Paradise-
Flycatcher; Indian Paradise-
Flycatcher
f   Tchitrec de paradis; Gobemouche du
paradis noir; Gobe-mouches du
paradis noir; Gobemouche de
paradis; Gobemouche de paradis
indien; Gobe-mouches du paradis
asiatique; Gobe-mouches de paradis;
Gobe-mouches de paradis indien;
Gobe-mouches du paradis
d   Fahlbauchparadiesschnäpper;
Indischer Paradiesschnäpper;
Hainparadiesschnäpper
i   Pigliamosche del Paradiso asiatico

9861   Terpsiphone rufiventer
Passeriformes - Monarchidae
e   Black-headed Paradise-Flycatcher;
Red-bellied Paradise-Flycatcher;
Red-bellied Paradise-Monarch;
Annobon Paradise-Flycatcher
f   Tchitrec à ventre roux; Moucherolle
à ventre roux; Gobemouche du
paradis à ventre roux; Gobe-mouches
paradis à ventre roux
d   Senegal-Paradiesschnäpper
i   Pigliamosche del Paradiso
ventrerosso

9862   Terpsiphone rufiventer emini
Passeriformes - Monarchidae
e   Red-bellied Paradise-Flycatcher
(CSA)
d   Emin-Paradiesschnäpper

9863   Terpsiphone rufocinerea
Passeriformes - Monarchidae
e   Rufous-vented Paradise-Flycatcher;
Rufous-vented Paradise-Monarch
f   Tchitrec du Congo; Gobemouche
paradis du Congo; Gobe-mouches
paradis du Congo; Moucherolle du
Congo
i   Pigliamosche del Paradiso dal
sottocoda rosso

9864   Terpsiphone smithii
Passeriformes - Monarchidae

e   Annobon Paradise-Flycatcher
i   Pigliamosche del Paradiso

9865 *Terpsiphone viridis*
Passeriformes - Monarchidae
e   African Paradise-Flycatcher;
Paradise Flycatcher (CSA); African
Paradise-Monarch
ƒ   Tchitrec d'Afrique; Gobemouche
paradis; Gobemouche de paradis à
long bec; Moucherolle paradis;
Moucherolle de paradis; Gobe-
mouches de paradis à long bec
d   Graubrustparadiesschnäpper;
Arikanischer Paradiesschnäpper
i   Pigliamosche del Paradiso africano

9866 *Tersina viridis*
Passeriformes - Fringillidae
e   Swallow-Tanager
ƒ   Tersine hirondelle
d   Schwalbentangare
i   Tangara rondine

9867 *Tesia castaneocoronata*
Passeriformes - Sylviidae
e   Chestnut-headed Tesia; Chestnut-
headed Ground-Warbler
ƒ   Tésie à tête marron; Tesia à tête
marron
d   Rotkopftesia; Gelbbauchtesia
i   Tesia corona castana

9868 *Tesia cyaniventer*
Passeriformes - Sylviidae
e   Grey-bellied Tesia; Grey-bellied
Ground-Warbler; Dull Ground-
Warbler; Yellow-browed Tesia;
Slaty-bellied Ground-Warbler; Slaty-
bellied Tesia; Dull Tesia
ƒ   Tésie à sourcils jaunes; Tesia à
sourcils jaunes
d   Olivscheiteltesia; Graubauchtesia
i   Tesia ventregrigio

9869 *Tesia everetti*
Passeriformes - Sylviidae
e   Russet-capped Tesia; Russet-capped
Stubtail; Russet-capped Bush-
Warbler
ƒ   Tésie d'Everett

d   Rotscheiteltesia
i   Tesia di Everett

9870 *Tesia olivea*
Passeriformes - Sylviidae
e   Slaty-bellied Tesia; Slaty-bellied
Ground-Warbler; Bright Slaty-bellied
Tesia; Bright Slaty-bellied Ground-
Warbler; Bright Tesia; Olive
Ground-Warbler
ƒ   Tésie à ventre ardoisé; Tesia à ventre
ardoisé
d   Goldscheiteltesia
i   Tesia ventreardesia

9871 *Tesia superciliaris*
Passeriformes - Sylviidae
e   Javan Tesia; Java Ground-Warbler;
Javan Ground-Warbler; Malaysian
Tesia; Malaysian Eyebrowed Tesia
ƒ   Tésie de Java
d   Brauentesia
i   Tesia di Giava

9872 *Tetrao mlokosiewiczi*
Galliformes - Phasianidae
e   Caucasian Grouse; Caucasian Black-
Grouse; Caucasian Blackcock
ƒ   Tétras du Caucase
d   Kaukasus-Birkhuhn
i   Fagiano di monte del Caucaso

9873 *Tetrao parvirostris*
Galliformes - Phasianidae
e   Black-billed Capercaillie; Rock
Capercaillie; Small-billed
Capercaillie; Short-billed
Capercaillie; Spotted Capercaillie
ƒ   Tétras à bec noir; Grand Tétrao
d   Felsenauerhuhn
i   Gallo cedrone becconero

9874 *Tetrao tetrix*
Galliformes - Phasianidae
e   Black Grouse; Heath Fowl; Black
Game; Grey Hen f; Black Cock m;
Brood (col); Eurasian Black-Grouse;
Northern Black-Grouse
ƒ   Tétras lyre
d   Birkhuhn
i   Fagiano di monte; Gallo forcello

9875   ***Tetrao urogallus***
        Galliformes - Phasianidae
e   Western Capercaillie; Wood Grouse;
        Great Grouse; Capercaillie; Eurasian
        Capercaillie
f   Grand Tétras
d   Auerhuhn
i   Gallo cedrone; Urogallo

9876   ***Tetraogallus altaicus***
        Galliformes - Phasianidae
e   Altai Snowcock
f   Tétraogalle de I'Altai
d   Altai-Königshuhn
i   Tetraogallo del Altai

9877   ***Tetraogallus caspius***
        Galliformes - Phasianidae
e   Caspian Snowcock
f   Tétrogalle de Perse
d   Kaspisches Königshuhn; Kaspik-
        Königshuhn
i   Tetraogallo del Caspio

9878   ***Tetraogallus caucasicus***
        Galliformes - Phasianidae
e   Caucasian Snowcock
f   Tétrogalle du Caucase
d   Kaukasus-Königshuhn
i   Tetraogallo del Caucaso

9879   ***Tetraogallus himalayensis***
        Galliformes - Phasianidae
e   Himalayan Snowcock
f   Tétraogalle de l'Himalaya
d   Himalaja-Königshuhn
i   Tetraogallo dell'Himalaya

9880   ***Tetraogallus tibetanus***
        Galliformes - Phasianidae
e   Tibetan Snowcock
f   Tétraogalle du Tibet
d   Tibetaner Königshuhn
i   Tetraogallo del Tibet

9881   ***Tetraophasis obscurus***
        Galliformes - Phasianidae
e   Verreaux's Monal Partridge;
        Verreaux's Pheasant-Grouse;
        Verreaux's Partridge; Chestnut-
        throated Partridge; Chestnut-throated

        Monal Partridge; Pheasant-Partridge
f   Tétraophase de Verreaux; Tétraogalle
d   Braunkehlkeilschwanzhuhn
i   Monal di Verreaux

9882   ***Tetraophasis szechenyii***
        Galliformes - Phasianidae
e   Buff-throated Partridge; Szecheny's
        Monal Partridge; Szecheny's
        Pheasant-Grouse; Szecheny's
        Pheasant-Partridge
f   Tétraophase de Szecheny
d   Rostkehlkeilschwanz
i   Monal di Szecheny

9883   ***Tetrax tetrax***
        Gruiformes - Otididae
e   Little Bustard
f   Outarde canepetière
d   Zwergtrappe
i   Gallina prataiola

9884   ***Thalassarche carteri***
        Procellariiformes - Diomedeidae
e   Indian Yellow-Albatross (ANZ)

9885   ***Thalassarche eremita***
        Procellariiformes - Diomedeidae
e   Chatham Albatross (ANZ)

9886   ***Thalassarche impavida***
        Procellariiformes - Diomedeidae
e   Campbell Albatross (ANZ)

9887   ***Thalassarche steadi***
        Procellariiformes - Diomedeidae
e   White-capped Albatross (ANZ)

9888   ***Thalassoica antarctica***
        Procellariiformes - Procellariidae
e   Antarctic Petrel; Antarctic Fulmar
f   Pétrel antarctique
d   Weißflügelsturmvogel; Antarktik-
        Sturmvogel; Grauweißer Sturmvogel
i   Procellaria dell'Antartide; Procellaria
        antartica

9889   ***Thalassornis leuconotus***
        Anseriformes - Dendrocygnidae
e   White-backed Duck; White-backed
        Whistling-Duck

f Canard à dos blanc; Dendrocygne à dos blanc
d Weißrückenpfeifgans; Weißrückenente

9890 ***Thalurania colombica***
Apodiformes - Trochilidae
e Blue-crowned Woodnymph; Colombian Woodnymph; Violet-crowned Woodnymph (NA); Crowned Woodnymph
f Dryade couronnée
i Ninfa coronata

9891 ***Thalurania fannyi***
Apodiformes - Trochilidae
e Green-crowned Woodnymph
f Dryade de Fanny

9892 ***Thalurania furcata***
Apodiformes - Trochilidae
e Fork-tailed Woodnymph; Princess Woodnymph; Woodnymph; Common Woodnymph; Cayenne Woodnymph
f Dryade à queue fourchue
d Schwalbennymphe
i Ninfa codafurcuta

9893 ***Thalurania glaucopis***
Apodiformes - Trochilidae
e Violet-capped Woodnymph; Blue-crowned Woodnymph; Blue-headed Woodnymph; Brazilian Woodnymph
f Dryade glaucope
d Veilchenkopfnymphe
i Ninfa corona violetta

9894 ***Thalurania hypochlora***
Apodiformes - Trochilidae
e Emerald-bellied Woodnymph

9895 ***Thalurania ridgwayi***
Apodiformes - Trochilidae
e Mexican Woodnymph
f Dryade du Mexique

9896 ***Thalurania townsendi***
Apodiformes - Trochilidae
e Violet-crowned Woodnymph

9897 ***Thalurania watertonii***
Apodiformes - Trochilidae
e Long-tailed Woodnymph; Waterton's Woodnymph
f Dryade de Waterton
d Langschwanznymphe
i Ninfa codalunga

9898 ***Thamnistes anabatinus***
Passeriformes - Thamnophilidae
e Russet Antshrike; Tawny Antshrike
f Batara rousset
d Rostwürgerling
i Averla formichiera rossiccio

9899 ***Thamnistes rufescens***
Passeriformes - Thamnophilidae
e Peruvian Antshrike

9900 ***Thamnolaea cinnamomeiventris***
Passeriformes - Muscicapidae
e Mocking Cliff-Chat; White-shouldered Cliffchat (CSA); Mocking Chat (CSA); Cliff-Chat; Common Clif-Chat
f Traquet à ventre roux; Traquet de roche à ventre roux
d Rotbauchschmätzer
i Sassicola ventrecannella

9901 ***Thamnolaea coronata***
Passeriformes - Muscicapidae
e White-crowned Cliff-Chat; Crowned Cliff-Chat
f Traquet couronné
i Sassicola rupestre corona bianca

9902 ***Thamnolaea semirufa***
Passeriformes - Muscicapidae
e White-winged Cliff-Chat
f Traquet demi-roux; Traquet de roche à ventre blanc
d Spiegelschmätzer
i Sassicola rupestre alibianchi

9903 ***Thamnomanes ardesiacus***
Passeriformes - Thamnophilidae
e Dusky-throated Antshrike; Saturnine Antshrike; Grey-throated Antvireo; Saturnine Antvireo
f Batara ardoisé

d   Grauwürgerling
i   Averela formichiera ardesia

9904   ***Thamnomanes caesius***
Passeriformes - Thamnophilidae
e   Cinereous Antshrike
f   Batara cendré
d   Buschwürgerling
i   Averela formichiera cinerea

9905   ***Thamnomanes saturninus***
Passeriformes - Thamnophilidae
e   Saturnine Antshrike
f   Batara saturnin
d   Schwarzkehlwürgerling

9906   ***Thamnomanes schistogynus***
Passeriformes - Thamnophilidae
e   Bluish-slate Antshrike
f   Batara bleu-gris
d   Blauwürgerling
i   Averela formichiera blu-lavagna

9907   **Thamnophilidae**
Passseriformes
e   Antbirds
f   Thamnophilidés
d   Wollrücken
i   Tamnofilidi

9908   ***Thamnophilus aethiops***
Passeriformes - Thamnophilidae
e   White-shouldered Antshrike
f   Batara à épaulettes blanches; Batara à
ailes blanches
d   Trauerwollrücken
i   Averla formichiera spallebianche

9909   ***Thamnophilus amazonicus***
Passeriformes - Thamnophilidae
e   Amazonian Antshrike
f   Batara d'Amazonie; Batara Amazone
d   Zwillingswollrücken
i   Averla formichiera dell'Amazzonia

9910   ***Thamnophilus aroyae***
Passeriformes - Thamnophilidae
e   Upland Antshrike
f   Batara montagnard
d   Marcapala-Wollrücken
i   Averla formichiera dei monti

9911   ***Thamnophilus asperiventer***
Passeriformes - Thamnophilidae
e   Andean Antshrike
d   Anden-Wollrücken
i   Averla formichiera

9912   ***Thamnophilus atrinucha***
Passeriformes - Thamnophilidae
e   Western Slaty-Antshrike; Slaty
Antshrike
f   Batara à nuque noire

9913   ***Thamnophilus bridgesi***
Passeriformes - Thamnophilidae
e   Black-hooded Antshrike; Bridges's
Antshrike
f   Batara capucin
d   Kapuzenwollrücken
i   Averla formichiera di Bridges

9914   ***Thamnophilus caerulescens***
Passeriformes - Thamnophilidae
e   Variable Antshrike
f   Batara bleuâtre
d   Grauscheitel;
Grauscheitelwollrücken; Grauer
Ameisenvogel; Dunkler
Ameisenwürger
i   Averla formichiera variabile

9915   ***Thamnophilus cryptoleucus***
Passeriformes - Thamnophilidae
e   Castelnau's Antshrike
f   Batara de Castelnau
d   Castelnau-Wollrücken
i   Averla formichiera di Castelnau

9916   ***Thamnophilus doliatus***
Passeriformes - Thamnophilidae
e   Barred Antshrike
f   Batara rayé
d   Bindenwollrücken;
Ringelameisenwürger
i   Averla formichiera barrata

9917   ***Thamnophilus insignis***
Passeriformes - Thamnophilidae
e   Streak-backed Antshrike; Tepui
Antshrike
f   Batara à dos rayé; Batara du
Venezuela

   d   Roraima-Wollrücken
   i   Averla formichiera dorsostriato

9918   ***Thamnophilus multistriatus***
         Passeriformes - Thamnophilidae
   e   Bar-crested Antshrike
   f   Batara de Lafresnaye
   d   Streifenwollrücken

9919   ***Thamnophilus murinus***
         Passeriformes - Thamnophilidae
   e   Mouse-coloured Antshrike
   f   Batara souris
   d   Silberwollrücken
   i   Averla formichiera grigiotopo

9920   ***Thamnophilus nigriceps***
         Passeriformes - Thamnophilidae
   e   Black Antshrike
   f   Batara noir; Batara du Nechi
   d   Mohrenwollrücken
   i   Averla formichiera nera

9921   ***Thamnophilus nigrocinereus***
         Passeriformes - Thamnophilidae
   e   Blackish-grey Antshrike; Grey
         Antshrike
   f   Batara demi-deuil
   d   Graubauchwollrücken
   i   Averla formichiera grigionera

9922   ***Thamnophilus palliatus***
         Passeriformes - Thamnophilidae
   e   Chestnut-backed Antshrike; Lined
         Antshrike
   f   Batara mantelé
   d   Mantelwollrücken
   i   Averla formichiera lineata

9923   ***Thamnophilus praecox***
         Passeriformes - Thamnophilidae
   e   Cocha Antshrike
   f   Batara du Cocha
   d   Cocha-Wollrücken
   i   Averla formichiera di Cocha

9924   ***Thamnophilus punctatus***
         Passeriformes - Thamnophilidae
   e   Eastern Slaty-Antshrike; Slaty-
         Antshrike
   f   Batara tacheté

   d   Tüpfelwollrücken
   i   Averla formichiera ardesia

9925   ***Thamnophilus ruficapillus***
         Passeriformes - Thamnophilidae
   e   Rufous-capped Antshrike
   f   Batara à tête rousse
   d   Rotscheitelwollrücken; Gesperberter
         Ameisenvogel
   i   Averla formichiera capirossa

9926   ***Thamnophilus ruficapillus***
         marcapatae
         Passeriformes - Thamnophilidae
   f   Batara subandin

9927   ***Thamnophilus schistaceus***
         Passeriformes - Thamnophilidae
   e   Plain-winged Antshrike; Black-
         capped Antshrike
   f   Batara à ailes unies; Batara
         fuligineux
   d   Kappenwollrücken
   i   Averla formichiera scistacea

9928   ***Thamnophilus tenuepunctatus***
         Passeriformes - Thamnophilidae
   e   Lined Antshrike
   f   Batara vermiculé
   d   Rotschwingenwollrücken

9929   ***Thamnophilus torquatus***
         Passeriformes - Thamnophilidae
   e   Rufous-winged Antshrike; Ringed
         Antshrike
   f   Batara à ailes rousses
   d   Rotschwingenwollrücken
   i   Averla formichiera alirosse

9930   ***Thamnophilus unicolor***
         Passeriformes - Thamnophilidae
   e   Uniform Antshrike
   f   Batara unicolore
   d   Einfarbwollrücken
   i   Averla formichiera unicolore

9931   ***Thamnophilus zarumae***
         Passeriformes - Thamnophilidae
   e   Chapman's Antshrike
   f   Batara de Chapman

9932 **_Thamnornis chloropetoides_**
Passeriformes - Sylviidae
e Thamnornis Warbler; Kirikita;
Kirikita Warbler
f Nésille kiritika; Kirikita; Fauvette
chloropétoïde
d Kirikita
i Kiritika

9933 **_Thaumastura cora_**
Apodiformes - Trochilidae
e Peruvian Sheartail; Cora Sheartail
f Colibri cora
d Corakolibri
i Colibri di Cora

9934 **_Theristicus branickii_**
Ciconiiformes - Threskiornithidae
e Andean Ibis; Puna Buff-necked Ibis;
Branicki's Ibis
f Ibis des Andes
i Ibis delle Ande

9935 **_Theristicus caerulescens_**
Ciconiiformes - Threskiornithidae
e Plumbeous Ibis; Blue Ibis
f Ibis plombé
d Stirnbandibis
i Ibis plumbeo

9936 **_Theristicus caudatus_**
Ciconiiformes - Threskiornithidae
e Buff-necked Ibis; White-throated
Ibis; Black-faced Ibis
f Ibis mandore
d Weißhalsibis
i Ibis collochiaro

9937 **_Theristicus melanopis_**
Ciconiiformes - Threskiornithidae
e Black-faced Ibis
f Ibis à face noire
d Schwarzflügelibis
i Ibis faccianera

9938 **_Thescelocichla leucopleura_**
Passeriformes - Pycnonotidae
e Swamp Greenbul; Swamp Palm-
Bulbul; Swamp Bulbul; Swamp
Palm-Greenbul; White-tailed
Greenbul

f Bulbul des raphias; Bulbul à queue
tachetée
d Raphiabülbül
i Bulbul delle paludi

9939 **Thinocoridae**
Charadriiformes
e Seedsnipe; American Seed-Snipe
f Thinocoridés
d Höhenläufer
i Tinocoridi

9940 **_Thinocorus orbignyianus_**
Charadriiformes - Thinocoridae
e Grey-breasted Seedsnipe; D'orbigny's
Seedsnipe
f Thinocore d'Orbigny
d Graukehlhöhenläufer
i Tinocoride pettogrigio

9941 **_Thinocorus rumicivorus_**
Charadriiformes - Thinocoridae
e Least Seedsnipe; Patagonian
Seedsnipe; Pygmy Seedsnipe
f Thinocore de Patagonie
d Zwerghöhenläufer
i Tinocoride minor; Tinocoride della
Patagonia

9942 **_Thlypopsis fulviceps_**
Passeriformes - Fringillidae
e Fulvous-headed Tanager
f Tangara à tête fauve
d Rostkopftangare
i Tangara testafulva

9943 **_Thlypopsis inornata_**
Passeriformes - Fringillidae
e Buff-bellied Tanager
f Tangara à ventre roux
d Braunbauchtangare
i Tangara pettochiaro

9944 **_Thlypopsis ornata_**
Passeriformes - Fringillidae
e Rufous-chested Tanager; Fulvous-
chested Tanager
f Tangara à flancs roux
d Zimtbrusttangare
i Tangara pettirossa

9945 *Thlypopsis pectoralis*
Passeriformes - Fringillidae
e   Brown-flanked Tanager
f   Tangara à flancs bruns
d   Braunflankentangare
i   Tangara fianchibruni

9946 *Thlypopsis ruficeps*
Passeriformes - Fringillidae
e   Rust-and-yellow Tanager
f   Tangara à ventre jaune
d   Goldkappentangare
i   Tangara ruggine e gialla

9947 *Thlypopsis sordida*
Passeriformes - Fringillidae
e   Orange-headed Tanager
f   Tangara à tête orangé
d   Orangekopftangare
i   Tangara testa aranciata

9948 **Thraupidae**
Passeriformes
e   Tanagers; Thraupids
f   Tanagridés; Thraupidés
d   Tangaren
i   Traupidi

9949 *Thraupis abbas*
Passeriformes - Thraupidae
e   Yellow-winged Tanager; Abbot's
Tanager
f   Tangara à miroir jaune
d   Abt-Tangare
i   Tangara aligialli

9950 *Thraupis bonariensis*
Passeriformes - Thraupidae
e   Blue-and-yellow Tanager
f   Tangara fourchu
d   Furchentangare
i   Tangara di Buenos Aires

9951 *Thraupis cyanocephala*
Passeriformes - Thraupidae
e   Blue-capped Tanager
f   Tangara à tête bleue
d   Gelbschenkeltangare
i   Tangara testablu

9952 *Thraupis cyanoptera*
Passeriformes - Thraupidae
e   Azure-shouldered Tanager
f   Tangara à épaulettes bleues
d   Violettschultertangare
i   Tangara aliazzurre

9953 *Thraupis episcopus*
Passeriformes - Thraupidae
e   Blue-grey Tanager; Blue Tanager
f   Tangara evêque
d   Bischofstangare; Blautangare;
Blaugraue Tangare
i   Tangara grigio-azzurra

9954 *Thraupis glaucocolpa*
Passeriformes - Thraupidae
e   Glaucous Tanager
f   Tangara glauque
i   Tangara glauca

9955 *Thraupis ornata*
Passeriformes - Thraupidae
e   Golden-chevroned Tanager
f   Tangara orné
d   Schmucktangare
i   Tangara spalledorate

9956 *Thraupis palmarum*
Passeriformes - Thraupidae
e   Palm Tanager; Common Palm-
Tanager
f   Tangara des palmiers
d   Palmentangare; Palmtangare
i   Tangara delle palme

9957 *Thraupis sayaca*
Passeriformes - Thraupidae
e   Sayaca Tanager; Blue Tanager
f   Tangara sayaca
d   Prälattangare; Meerblaue Tangare;
Blauflügeltangare
i   Tangara sayaca

9958 *Threnetes leucurus*
Apodiformes - Trochilidae
e   Pale-tailed Barbthroat; White-tailed
Barbthroat; Bronze-tailed Barbthroat;
Christina's Barbthroat
f   Ermite à queue blanche

d   Hellschwanzeremit
i   Colibrì barbuto codabianca

9959   *Threnetes leucurus loehkeni*
    Apodiformes - Trochilidae
e   Bronze-tailed Barbthroat
f   Ermite de Loehken
d   Loehken-Eremit

9960   *Threnetes niger*
    Apodiformes - Trochilidae
e   Sooty Barbthroat
f   Ermite d'Antonie
d   Mohreneremit
i   Colibrì barbuto fuligginoso

9961   *Threnetes ruckeri*
    Apodiformes - Trochilidae
e   Band-tailed Barbthroat; Rucker's
    Bandthroat; Rucker's Fantailed-
    Bandthroat; Blue-Fantailed-
    Bandthroat
f   Ermite de Rucker
d   Bindenschwanzeremit
i   Colibrì barbuto di Rucker

9962   *Threskiornis aethiopicus*
    Ciconiiformes - Threskiornithidae
e   Sacred Ibis; White Ibis (ISC); Black-
    headed Ibis (ISC); Madagascar Ibis
f   Ibis sacré
d   Heiliger Ibis
i   Ibis sacro

9963   *Threskiornis bernieri*
    Ciconiiformes - Threskiornithidae
e   Madagascar Ibis
f   Ibis malgache
i   Ibis sacro del Madagascar

9964   *Threskiornis melanocephalus*
    Ciconiiformes - Threskiornithidae
e   Black-headed Ibis; Oriental Ibis;
    Indian Ibis; Black-necked Ibis; White
    Ibis; Asiatic White Ibis; Indian
    Black-necked Ibis; Oriental White
    Ibis
f   Ibis à tête noire
d   Schwarzhals Ibis
i   Ibis orientale; Ibis sacro orientale

9965   *Threskiornis molucca*
    Ciconiiformes - Threskiornithidae
e   Australian Ibis; Australian Black-
    necked Ibis; Australian White Ibis
    (ANZ); Black-headed Ibis; Black-
    necked Ibis
f   Ibis à cou noir; Ibis leucon
d   Molukken-Ibis
i   Ibis sacro australiano

9966   *Threskiornis spinicollis*
    Ciconiiformes - Threskiornithidae
e   Straw-necked Ibis
f   Ibis d'Australie
d   Stachelibis
i   Ibis collospinoso

9967   **Threskiornithidae**
    Ciconiiformes
e   Ibises, Spoonbills
f   Threskiornithides; Plataleides
d   Ibise; Ibisvögel
i   Treschiornitidi

9968   *Thripadectes flammulatus*
    Passeriformes - Furnariidae
e   Flammulated Treehunter
f   Anabate flammé
d   Strichelbaumspäher
i   Grattafoglie flammulato

9969   *Thripadectes holostictus*
    Passeriformes - Furnariidae
e   Striped Treehunter
f   Anabate strié
d   Streifenbaumspäher
i   Grattafoglie striato

9970   *Thripadectes ignobilis*
    Passeriformes - Furnariidae
e   Uniform Treehunter
f   Anabate uniforme
d   Einfarbbaumspäher
i   Grattafoglie uniforme

9971   *Thripadectes melanorhynchus*
    Passeriformes - Furnariidae
e   Black-billed Treehunter
f   Anabate à bec noir
d   Tschudi-Baumspäher
i   Grattafoglie becconero

9972 **Thripadectes rufobrunneus**
Passeriformes - Furnariidae
e Streak-breasted Treehunter;
Streaked-breasted Treehunter
f Anabate des ravins
d Streifenbrustbaumspäher
i Grattafoglie bruno-rossiccio

9973 **Thripadectes scrutator**
Passeriformes - Furnariidae
e Peruvian Treehunter; Buff-throated
Treehunter
f Anabate inca
d Schuppenkehlbaumspäher
i Grattafoglie scrutatore

9974 **Thripadectes virgaticeps**
Passeriformes - Furnariidae
e Streak-capped Treehunter; Streaked-
capped Treehunter
f Anabate à tête striée
d Streifenkopfbaumspäher
i Grattafoglie capostriato

9975 **Thripophaga cherriei**
Passeriformes - Furnariidae
e Orinoco Softtail
f Synallaxe de l'Orénoque
d Orinoco-Schlüpfer
i Codamorbida dell'Orinoco

9976 **Thripophaga fusciceps**
Passeriformes - Furnariidae
e Plain Softtail
f Synallaxe terne
d Braunkopfbündelnister
i Codamorbida testabruna

9977 **Thripophaga macroura**
Passeriformes - Furnariidae
e Striated Softtail
f Synallaxe rayé
d Breitschwanzschlüpfer
i Codamorbida striata

9978 **Thryomanes bewickii**
Passeriformes - Certhiidae
e Bewick's Wren
f Troglodyte de Bewick
d Buschzaunkönig
i Scricciolo di Bewick

9979 **Thryomanes sissonii**
Passeriformes - Certhiidae
e Socorro Wren
f Troglodyte de Socorro
i Scricciolo di Revillagigedo

9980 **Thryorchilus browni**
Passeriformes - Certhiidae
e Timberline Wren; Irazú Wren
f Troglodyte des volcans
d Bergzaunkönig
i Scricciolo di Brown

9981 **Thryothorus albinucha**
Passeriformes - Certhiidae
e White-browed Wren; Cabot's Wren

9982 **Thryothorus atrogularis**
Passeriformes - Certhiidae
e Black-throated Wren; Northern
Black-throated Wren
f Troglodyte à gorge noire
d Schwarzkehlzaunkönig
i Scricciolo golanera

9983 **Thryothorus coraya**
Passeriformes - Certhiidae
e Coraya Wren
f Troglodyte coraya; Troglodyte de
Coraya
d Coraya-Zaunkönig
i Scricciolo coraya

9984 **Thryothorus eisenmanni**
Passeriformes - Certhiidae
e Inca Wren
f Troglodyte inca
i Scricciolo inca

9985 **Thryothorus euophrys**
Passeriformes - Certhiidae
e Plain-tailed Wren; Spot-chested
Wren
f Troglodyte maculé
d Fraser-Zaunkönig
i Scricciolo codasemplice

9986 **Thryothorus fasciatoventris**
Passeriformes - Certhiidae
e Black-bellied Wren
f Troglodyte à ventre noir

d   Bindenbauchzaunkönig
i   Scricciolo ventrenero

9987   *Thryothorus felix*
      Passeriformes - Certhiidae
e   Happy Wren
f   Troglodyte joyeux
d   Buntwangenzaunkönig
i   Scricciolo guancestriate

9988   *Thryothorus genibarbis*
      Passeriformes - Certhiidae
e   Moustached Wren
f   Troglodyte à moustaches
d   Wangenstreifzaunkönig
i   Scricciolo dai mustacchi

9989   *Thryothorus griseus*
      Passeriformes - Certhiidae
e   Grey Wren; Amazon Wren
f   Troglodyte gris
d   Grauzaunkönig
i   Scricciolo grigio

9990   *Thryothorus guarayanus*
      Passeriformes - Certhiidae
e   Fawn-breasted Wren
f   Troglodyte des Guarayos
d   Gurayos-Zaunkönig
i   Scricciolo di Guaraya

9991   *Thryothorus leucopogon*
      Passeriformes - Certhiidae
e   Stripe-throated Wren
f   Troglodyte balafre
i   Scricciolo golastriata

9992   *Thryothorus leucotis*
      Passeriformes - Certhiidae
e   Buff-breasted Wren; Buff-bellied
      Wren
f   Troglodyte à face pâle
d   Weißohrzaunkönig
i   Scricciolo guancebianche

9993   *Thryothorus longirostris*
      Passeriformes - Certhiidae
e   Long-billed Wren; Gray Wren (NA);
      Buff-breasted Wren
f   Troglodyte à long bec

d   Langschnabelzaunkönig
i   Scricciolo beccolungo

9994   *Thryothorus ludovicianus*
      Passeriformes - Certhiidae
e   Carolina Wren; White-browed Wren
f   Troglodyte de Caroline
d   Flötenvogel; Carolina-Zaunkönig
i   Scricciolo della Carolina

9995   *Thryothorus maculipectus*
      Passeriformes - Certhiidae
e   Spot-breasted Wren
f   Troglodyte à poitrine tachetée
d   Fleckenbrustzaunkönig
i   Scricciolo pettomaculato

9996   *Thryothorus modestus*
      Passeriformes - Certhiidae
e   Plain Wren; Modest Wren
f   Troglodyte modeste
d   Cabanis-Zaunkönig
i   Scricciolo modesto

9997   *Thryothorus mystacalis*
      Passeriformes - Certhiidae
e   Whiskered Wren
f   Troglodyte à favoris
i   Scricciolo delle redini

9998   *Thryothorus nicefori*
      Passeriformes - Certhiidae
e   Niceforo's Wren
f   Troglodyte de Niceforo
d   Niceforo-Zaunkönig
i   Scricciolo di Niceforo

9999   *Thryothorus nigricapillus*
      Passeriformes - Certhiidae
e   Bay Wren; Black-capped Wren
f   Troglodyte à calotte noire;
      Troglodyte à tête noire
d   Kastanienzaunkönig; Uferzaunkönig
i   Scricciolo capinero

10000   *Thryothorus pleurostictus*
      Passeriformes - Certhiidae
e   Banded Wren
f   Troglodyte barré
d   Akazienzaunkönig
i   Scricciolo pettobianco

10001   *Thryothorus rufalbus*
Passeriformes - Certhiidae
e   Rufous-and-white Wren
f   Troglodyte rufalbin; Troglodyte de Branicki
d   Rotrückenzaunkönig
i   Scricciolo bianco e ruggine

10002   *Thryothorus rutilus*
Passeriformes - Certhiidae
e   Rufous-breasted Wren
f   Troglodyte des halliers
d   Rotbrustzaunkönig
i   Scricciolo pettirosso

10003   *Thryothorus sclateri*
Passeriformes - Certhiidae
e   Speckle-breasted Wren; Speckled Wren
f   Troglodyte de Sclater
i   Scricciolo di Sclater

10004   *Thryothorus semibadius*
Passeriformes - Certhiidae
e   Riverside Wren; Salvin's Wren
f   Troglodyte des ruisseaux
i   Scricciolo di Salvin

10005   *Thryothorus sinaloa*
Passeriformes - Certhiidae
e   Sinaloa Wren; Bar-vented Wren
f   Troglodyte du Sinaloa
d   Sinaloa -Zaunkönig
i   Scricciolo di Sinaloa

10006   *Thryothorus spadix*
Passeriformes - Certhiidae
e   Sooty-headed Wren; Smoky-headed Wren; Southern Black-throated Wren
f   Troglodyte moine
d   Rußkopfzaunkönig
i   Scricciolo testafuligginosa

10007   *Thryothorus superciliaris*
Passeriformes - Certhiidae
e   Superciliated Wren
f   Troglodyte bridé; Troglodyte à sourcils blancs
d   Küstenzaunkönig
i   Scricciolo dai sopraccigli bianchi

10008   *Thryothorus thoracicus*
Passeriformes - Certhiidae
e   Stripe-breasted Wren; Stripe-throated Wren
f   Troglodyte flammé; Troglodyte à poitrine striée
d   Streifenkehlzaunkönig
i   Scricciolo pettostriato

10009   *Thryothorus zeledoni*
Passeriformes - Certhiidae
e   Canebreak Wren

10010   *Tiaris bicolor*
Passeriformes - Emberizidae
e   Black-faced Grassquit; Black Sparrow (WI); Tobaccobird (WI); Grassbird (WI); Cheecheebird (WI); Tobacco-seed (WI); Groundquit (WI); Grass Sparrow (WI); Sinbird (WI)
f   Sporophile cici; Sporophile à face noire; Mangeur d'herbe (Ants)
d   Schwarzgesichtchen; Jamaika-Fink
i   Fringuello cantore faccianera

10011   *Tiaris canora*
Passeriformes - Emberizidae
e   Cuban Grassquit; Melodious Grassquit
f   Sporophile petit-chanteur
d   Kuba-Fink; Kleiner Kuba-Fink
i   Fringuello cantore di Cuba

10012   *Tiaris fuliginosa*
Passeriformes - Emberizidae
e   Sooty Grassquit
f   Sporophile fuligineux
d   Schwarzbrüstchen
i   Fringuello cantore fuligginoso

10013   *Tiaris obscura*
Passeriformes - Emberizidae
e   Dull-coloured Grassquit
f   Sporophile obscur
i   Fringuello cantore bruno

10014   *Tiaris olivacea*
Passeriformes - Emberizidae
e   Yellow-faced Grassquit; Squit (WI); Grassbird (WI)

f Sporophile grand-chanteur
d Großer Kuba-Fink
i Fringuello cantore facciagialla

10015 *Tichodroma muraria*
Passeriformes - Sittidac
e Wallcreeper; Red-winged
Wallcreeper
f Tichodrome échelette; Sporophile
échelette
d Mauerläufer
i Picchio muraiolo

10016 *Tickellia hodgsoni*
Passeriformes - Sylviidae
e Broad-billed Warbler; Broad-billed
Flycatcher-Warbler; Tickell's
Flycatcher-Warbler; Tickell's
Warbler
f Pouillot de Hodgson; Fauvette-
gobemouches de Hodgson
d Breitschnabellaubsänger
i Luì pigliamosche di Hodgson

10017 *Tigriornis leucolophus*
Ciconiiformes - Ardeidae
e White-crested Tiger-Heron; African
Tiger-Heron
f Onoré à huppe blanche; Butor à crête
d Weißschopfreiher
i Airone tigrato crestabianco

10018 *Tigrisoma fasciatum*
Ciconiiformes - Ardeidae
e Fasciated Tiger-Heron; Salmon's
Tiger-Heron
f Onoré fascié
d Salmon-Reiher
i Airone tigrato fasciato

10019 *Tigrisoma lineatum*
Ciconiiformes - Ardeidae
e Rufescent Tiger-Heron
f Onoré rayé
d Marmorreiher
i Airone tigrato rossicio

10020 *Tigrisoma mexicanum*
Ciconiiformes - Ardeidae
e Bare-throated Tiger-Heron; Cabanis's
Tiger-Heron; Mexican Tiger-Heron;

Nexican Tiger Bittern
f Onoré du Mexique
d Nacktkehlreiher
i Airone tigrato golanuda

10021 *Tijuca atra*
Passeriformes - Tyrannidae
e Black-and-gold Cotinga; Black-and-
gold Tijuca
f Cotinga noir
d Tijuca
i Cotinga nera e dorata

10022 *Tijuca condita*
Passeriformes - Tyrannidae
e Grey-winged Cotinga; Orgaos
Cotinga; Snow's Cotinga
f Cotinga à ailes grises
d Braunpfäffchen
i Cotinga aligrige

10023 *Tilmatura dupontii*
Apodiformes - Trochilidae
e Sparkling-tailed Hummingbird;
Dupont's Humingbird; Sparkling-
tailed Woodstar
f Colibri zémès
d Dupont-Kolibri
i Colibrì di Dupont

10024 *Timalia pileata*
Passeriformes - Sylviidae
e Chestnut-capped Babbler; Red-
capped Babbler (ISC)
f Timalie coiffée; Timalie à tête rousse
d Rotkäppchentimalie
i Garrulo capocastano; Garrulo dal
ciuffo rosso

10025 *Timeliopsis fulvigula*
Passeriformes - Meliphagidae
e Olive Straightbill; Mountain Straight-
billed Honeyeater; Mountain
Straightbill
f Méliphage olivâtre
d Buschhonigfresser
i Mangiamiele beccodiritto di
montagna

10026 *Timeliopsis griseigula*
Passeriformes - Meliphagidae

e   Tawny Straightbill; Lowland
Straight-billed Honeyeater; Grey-
throated Straightbill
f   Méliphage chamois
d   Geradschnabelhonigfresser
i   Mangiamiele beccodiritto di pianura

10027   Tinamidae
Tinamiformes
e   Tinamous
f   Tinamous; Tinamidés
d   Steißhühner; Tinamus
i   Tinamidi

10028   Tinamiformes
e   Tinamous
f   Tinamiformes
d   Tinamus; Steißhühner
i   Tinamiformi

10029   *Tinamotis ingoufi*
Tinamiformes - Tinamidae
e   Patagonian Tinamou; Ingouf's
Tinamou
f   Tinamou de Patagonie
d   Patagonien-Steißhuhn
i   Tinamo della Patagonia

10030   *Tinamotis pentlandii*
Tinamiformes - Tinamidae
e   Puna Tinamou; Pentland's Tinamou
f   Tinamou quioula
d   Punasteißhuhn
i   Tinamo della Puna

10031   *Tinamus guttatus*
Tinamiformes - Tinamidae
e   White-throated Tinamou
f   Tinamou à gorge blanche
d   Weißkehltinamu
i   Tinamo golabianca

10032   *Tinamus major*
Tinamiformes - Tinamidae
e   Great Tinamou
f   Grand Tinamou; Tinamou magoua
d   Großtinamu
i   Tinamo gigante

10033   *Tinamus osgoodi*
Tinamiformes - Tinamidae

e   Black Tinamou
f   Tinamou noir
d   Schwarztinamu
i   Tinamo nero

10034   *Tinamus solitarius*
Tinamiformes - Tinamidae
e   Solitary Tinamou
f   Tinamou solitaire
d   Grausteißtinamu
i   Tinamo solitario

10035   *Tinamus tao*
Tinamiformes - Tinamidae
e   Grey Tinamou; Tao Tinamou
f   Tinamou tao
d   Tao
i   Tinamo grigio

10036   *Tityra cayana*
Passeriformes - Tyrannidae
e   Black-tailed Tityra; Cayenne Tityra
f   Tityre gris; Tityra gris
d   Schwarzschwanztityra
i   Titira della Cayenna

10037   *Tityra inquisitor*
Passeriformes - Tyrannidae
e   Black-crowned Tityra; Inquisitor
Tityra
f   Tityre à tête noire; Tityra à tête noire
d   Kappentityra
i   Titira testanera

10038   *Tityra semifasciata*
Passeriformes - Tyrannidae
e   Masked Tityra
f   Tityre masqué; Tityra masquée
d   Maskentityra; Maskenbekarde
i   Titira mascherata

10039   *Tmetothylacus tenellus*
Passeriformes - Passeridae
e   Golden Pipit
f   Pipit doré
d   Goldpieper
i   Pispola dorata

10040   *Tockus albocristatus*
Coraciiformes - Bucerotidae
e   White-crested Hornbill; African

White-crested Hornbill; White-
crowned Hornbill; Crowned
Hornbill; Long-tailed Hornbill
f   Calao à huppe blanche
d   Weißschopftoko;
Wcißschopfhornvogel;
Perückenhornvogel
i   Bucero crestabianca

10041   *Tockus alboterminatus*
Coraciiformes - Bucerotidae
e   Crowned Hornbill; African Crowned-
Hornbill
f   Calao couronné
d   Kronentoko
i   Bucero codabianca

10042   *Tockus bradfieldi*
Coraciiformes - Bucerotidae
e   Bradfield's Hornbill
f   Calao de Bradfield
d   Bradfield-Toko; Felsentoko
i   Bucero di Bradfield

10043   *Tockus camurus*
Coraciiformes - Bucerotidae
e   Red-billed Dwarf Hornbill; Dwarf
Hornbill
f   Calao pygmée
d   Zwergtoko
i   Bucero nano

10044   *Tockus deckeni*
Coraciiformes - Bucerotidae
e   Von der Decken's Hornbill; Decken's
Hornbill
f   Calao de Decken
d   Von der Decken-Toko; Jackson-Toko
i   Bucero di von der Decken; Lofocera
di Van Decken; Rincacero di Decken

10045   *Tockus erythrorhynchus*
Coraciiformes - Bucerotidae
e   Red-billed Hornbill; African Red-
billed Hornbill; Red-beaked Hornbill
f   Calao à bec rouge
d   Rotschnabeltoko
i   Bucero beccorosso; Lofocera dal
becco rosso; Rincacero comune;
Rincacero; Tok

10046   *Tockus fasciatus*
Coraciiformes - Bucerotidae
e   African Pied-Hornbill; Pied Hornbill;
Zande Hornbill; Allied Hornbill;
Black-and-white-tailed Hornbill
f   Calao longibande
d   Elstertoko
i   Bucero bianco e nero

10047   *Tockus flavirostris*
Coraciiformes - Bucerotidae
e   Eastern Yellow-billed Hornbill;
Yellow-billed Hornbill
f   Calao à bec jaune
d   Gelbschnabeltoko
i   Bucero beccogiallo orientale;
Lofocera dal becco giallo

10048   *Tockus hartlaubi*
Coraciiformes - Bucerotidae
e   Black Dwarf Hornbill
f   Calao de Hartlaub; Calao pygmée à
bec noir
d   Schwarzer Zwergtoko
i   Bucero di Hartlaub

10049   *Tockus hemprichii*
Coraciiformes - Bucerotidae
e   Hemprich's Hornbill
f   Calao de Hemprich
d   Hemprich-Toko
i   Bucero di Hemprich

10050   *Tockus jacksoni*
Coraciiformes - Bucerotidae
e   Jackson's Hornbill
f   Calao de Jackson
i   Bucero di Jackson

10051   *Tockus leucomelas*
Coraciiformes - Bucerotidae
e   Southern Yellow-billed Hornbill
f   Calao leucomèle
d   Gelbschnabeltoko
i   Bucero beccogiallo meridionale

10052   *Tockus monteiri*
Coraciiformes - Bucerotidae
e   Monteiro's Hornbill
f   Calao de Monteiro

d   Monteiro-Toko
i   Bucero di Monteiro

10053   ***Tockus nasutus***
       Coraciiformes - Bucerotidae
e   African Grey-Hornbill; Grey Hornbill
f   Calao à bec noir; Calao nasique
d   Grautoko; Weißschafttoko
i   Bucero grigio africano; Bucero nasuto

10054   ***Tockus pallidirostris***
       Coraciiformes - Bucerotidae
e   Pale-billed Hornbill
f   Calao à bec pâle
d   Blaßschnabeltoko
i   Bucero beccopallido

10055   **Todidae**
       Coraciiformes
e   Todies
f   Todidés
d   Todis
i   Todidi

10056   ***Todiramphus albonotatus***
       Coraciiformes - Halcyonidae
e   New Britain Kingfisher; White-backed Kingfisher; White-mantled Kingfisher
f   Martin-chasseur à dos blanc
d   Bismarck-Liest
i   Martin pescatore dorsobianco

10057   ***Todiramphus australasius***
       Coraciiformes - Halcyonidae
e   Cinnamon-banded Kingfisher; Australasian Kingfisher; Lesser Sunda Kingfisher; Timor Kingfisher; Cinnamon-backed Kingfisher; Cinnamon-collared Kingfisher; Sunda Kingfisher
f   Martin-chasseur couronné
d   Timor-Liest
i   Martin pescatore di Timor

10058   ***Todiramphus chloris***
       Coraciiformes - Halcyonidae
e   Collared Kingfisher; White-collared Kingfisher; Mangrove Kingfisher

f   Martin-chasseur à collier blanc
d   Halsbandliest; Grünkopfliest
i   Martin pescatore dal collare bianco

10059   ***Todiramphus cinnamominus***
       Coraciiformes - Halcyonidae
e   Micronesian Kingfisher
f   Martin-chasseur cannellé
d   Zimtkopfliest
i   Martin pescatore di Ryukyu

10060   ***Todiramphus cinnamominus miyakoensis***
       Coraciiformes - Halcyonidae
e   Ryukyu Kingfisher
f   Martin-chasseur de Miyaco
i   Martin pescatore

10061   ***Todiramphus diops***
       Coraciiformes - Halcyonidae
e   Blue-and-white Kingfisher; Moluccan Kingfisher
f   Martin-chasseur des Moluques; Martin-chasseur de Temminck
d   Lasurliest
i   Martin pescatore delle Molucche

10062   ***Todiramphus enigma***
       Coraciiformes - Halcyonidae
e   Talaud Kingfisher; Obscure Kingfisher
f   Martin-chasseur des Talaud
i   Martin pescatore enigmatico

10063   ***Todiramphus farquhari***
       Coraciiformes - Halcyonidae
e   Chestnut-bellied Kingfisher
f   Martin-chasseur à ventre roux
d   Braunbauchliest
i   Martin pescatore ventrecastano

10064   ***Todiramphus funebris***
       Coraciiformes - Halcyonidae
e   Sombre Kingfisher; Halmahera Kingfisher; Funereal Kingfisher; Olive-backed Kingfisher
f   Martin-chasseur funèbre
d   Molukken-Liest
i   Martin pescatore bruno

10065   *Todiramphus gambieri*
         Coraciiformes - Halcyonidae
   e   Tuamotu Kingfisher; Mangareva
         Kingfisher; Niau Kingfisher
   f   Martin-chasseur des Gambier
   d   Tuamotu-Liest
   i   Martin pescatore di Tuamotu

10066   *Todiramphus godeffroyi*
         Coraciiformes - Halcyonidae
   e   Marquesan Kingfisher; Marquesas
         Kingfisher; Marquesa Kingfisher
   f   Martin-chasseur des Marquises
   d   Marquesas-Liest
   i   Martin pescatore delle Marquesas

10067   *Todiramphus lazuli*
         Coraciiformes - Halcyonidae
   e   Lazuli Kingfisher; South Moluccan
         Kingfisher
   f   Martin-chasseur lazuli
   i   Martin pescatore lapislazzuli

10068   *Todiramphus leucopygius*
         Coraciiformes - Halcyonidae
   e   Ultramarine Kingfisher
   f   Martin-chasseur outremer
   d   Ultramarinliest
   i   Martin pescatore ultramarino

10069   *Todiramphus macleayii*
         Coraciiformes - Halcyonidae
   e   Forest Kingfisher; Australian Forest-
         Kingfisher
   f   Martin-chasseur forestier; Martin-
         chasseur de Jacquinot
   d   Spiegelliest
   i   Martin pescatore di foresta

10070   *Todiramphus nigrocyaneus*
         Coraciiformes - Halcyonidae
   e   Blue-black Kingfisher; Blue-backed
         Kingfisher; Blue-sided Kingfisher
   f   Martin-chasseur bleu-noir
   d   Dunkelliest
   i   Martin pescatore blu-nero

10071   *Todiramphus pyrrhopygia*
         Coraciiformes - Halcyonidae
   e   Red-backed Kingfisher
   f   Martin-chasseur à dos de feu

   d   Rotbürzelliest
   i   Martin pescatore dal groppone rosso

10072   *Todiramphus recurvirostris*
         Coraciiformes - Halcyonidae
   e   Flat-billed Kingfisher
   f   Martin-chasseur des Samoa
   d   Flachschnabelliest
   i   Martin pescatore di Samoa

10073   *Todiramphus ruficollaris*
         Coraciiformes - Halcyonidae
   e   Mangaia Kingfisher; Tanga'eo
   f   Martin-chasseur de Mangaia
   i   Martin pescatore di Mangaia

10074   *Todiramphus sanctus*
         Coraciiformes - Halcyonidae
   e   Sacred Kingfisher
   f   Martin-chasseur sacré
   d   Götzenliest
   i   Martin pescatore sacro

10075   *Todiramphus sanctus norfolkiensis*
         Coraciiformes - Halcyonidae
   e   Norfolk Island Sacred-Kingfisher
         (ANZ)

10076   *Todiramphus sanctus vagans*
         Coraciiformes - Halcyonidae
   e   Tasman Sea Sacred-Kingfisher
         (ANZ)

10077   *Todiramphus saurophaga*
         Coraciiformes - Halcyonidae
   e   Beach Kingfisher; White-headed
         Kingfisher
   f   Martin-chasseur à tête blanche
   d   Eidechsenliest
   i   Martin pescatore testabianca

10078   *Todiramphus tuta*
         Coraciiformes - Halcyonidae
   e   Chattering Kingfisher; Borabora
         Kingfisher; Respected Kingfisher;
         Polynesian Kingfisher; Pacific
         Kingfisher
   f   Martin-chasseur respecté
   d   Borabora-Liest
   i   Martin pescatore di Borabora

10079 **Todiramphus veneratus**
Coraciiformes - Halcyonidae
e Tahiti Kingfisher; Venerated
Kingfisher; Tahitian Kingfisher
f Martin-chasseur vénéré
d Tahiti-Liest
i Martin pescatore di Tahiti

10080 **Todiramphus winchelli**
Coraciiformes - Halcyonidae
e Rufous-lored Kingfisher; Winchell's
Kingfisher
f Martin-chasseur de Winchell;
Martin-chasseur à collier roux
d Rotnackenliest
i Martin pescatore di Winchell

10081 **Todirostrum calopterum**
Passeriformes - Tyrannidae
e Golden-winged Tody-Flycatcher;
Jardine's Tody-Flycatcher; Jardine's
Golden-winged Tody-Tyrant; Black-
backed Tody-Flycatcher
f Todirostre à ailes d'or
d Spateltyrann
i Becco di todo alidorate

10082 **Todirostrum chrysocrotaphum**
Passeriformes - Tyrannidae
e Yellow-browed Tody-Flycatcher;
Painted Tody-Flycatcher; Painted
Tody-Tyrant; Black-headed Tody-
Flycatcher
f Todirostre bridé; Todirostre à
sourcils
d Goldflügelspateltyrann
i Becco di todo dai sopraccigli gialli

10083 **Todirostrum cinereum**
Passeriformes - Tyrannidae
e Common Tody-Flycatcher; Tody-
Flycatcher; Common Tody-Tyrant;
Tody-Tyrant; Black-fronted Tody-
Flycatcher
f Todirostre familier; Todirrostre
commun; Todirostre
d Gelbbauchspateltyrann
i Becco di todo cinereo

10084 **Todirostrum fumifrons**
Passeriformes - Tyrannidae

e Smoky-fronted Tody-Flycatcher;
Smoky-fronted Tody-Tyrant
f Todirostre à front gris
d Graustirnspateltyrann
i Becco di todo frontescura

10085 **Todirostrum latirostre**
Passeriformes - Tyrannidae
e Rusty-fronted Tody-Flycatcher;
Rusty-faced Tody-Tyrant
f Todirostre à front roux
d Rotflügelspateltyrann
i Becco di todo beccolargo

10086 **Todirostrum maculatum**
Passeriformes - Tyrannidae
e Spotted Tody-Flycatcher; Spotted
Tody-Tyrant
f Todirostre tacheté
d Tropfenbrustspateltyrann
i Becco di todo maculato

10087 **Todirostrum nigriceps**
Passeriformes - Tyrannidae
e Black-headed Tody-Flycatcher;
Blackheaded Tody-Tyrant
f Todirostre à tête noire
d Schwarzkopfspateltyrann
i Becco di todo testanera

10088 **Todirostrum pictum**
Passeriformes - Tyrannidae
e Painted Tody-Flycatcher
f Todirostre peint; Todirostre à collier
d Weißkehlspateltyrann
i Becco di todo variopinto

10089 **Todirostrum plumbeiceps**
Passeriformes - Tyrannidae
e Ochre-faced Tody-Flycatcher; Lead-
crowned Tody-Tyrant
f Todirostre gorgeret
d Zimtkehlspateltyrann
i Becco di todo testaplumbea

10090 **Todirostrum poliocephalum**
Passeriformes - Tyrannidae
e Yellow-lored Tody-Flycatcher; Grey-
headed Tody-Tyrant; Grey-headed
Tody-Flycatcher
f Todirostre à tête grise

d Graukopfspateltyrann
i Becco di todo testagrigia

10091 *Todirostrum pulchellum*
Passeriformes - Tyrannidae
e Black-backed Tody-Flycatcher
f Todirostre à dos noir
i Becco di todo dorsonero

10092 *Todirostrum russatum*
Passeriformes - Tyrannidae
e Ruddy Tody-Flycatcher; Ruddy
Tody-Tyrant
f Todirostre roussâtre
d Rostkopfspateltyrann
i Becco di todo rossiccio

10093 *Todirostrum senex*
Passeriformes - Tyrannidae
e Buff-cheeked Tody-Flycatcher;
Plumbeous-crowned Tody-Tyrant;
Ancient Tody-Tyrant; Barba Tody-
Flycatcher; Todybill
f Todirostre à joues rousses
d Braunwangenspateltyrann
i Becco di todo corona plumbea

10094 *Todirostrum sylvia*
Passeriformes - Tyrannidae
e Slate-headed Tody-Flycatcher; Slate-
headed Tody-Tyrant; Desmarest's
Tody-Tyrant; Slaty Tody-Flycatcher
f Todirostre de Desmarest
d Graukehlspateltyrann
i Becco di todo capolavagna

10095 *Todirostrum viridanum*
Passeriformes - Tyrannidae
e Maracaibo Tody-Flycatcher; Short-
tailed Tody-Flycatcher
f Todirostre du Maracalbo
i Becco di todo codacorta

10096 *Todus angustirostris*
Coraciiformes - Todidae
e Narrow-billed Tody
f Colibri (Ants); Chicorette (Ants);
Todier à bec étroit
d Schmalschnabeltodi
i Todo beccosottile

10097 *Todus mexicanus*
Coraciiformes - Todidae
e Puerto Rican Tody; Hypochondriac
Tody
f Todier de Porto Rico
d Gelbflankentodi
i Todo di Portorico

10098 *Todus multicolor*
Coraciiformes - Todidae
e Cuban Tody
f Todier de Cuba
d Vielfarbentodi
i Todo di Cuba

10099 *Todus subulatus*
Coraciiformes - Todidae
e Broad-billed Tody; Hispaniolan Tody
f Perroquet de terre (Ants); Colibri
(Ants); Todier à bec large
d Breitschnabeltodi
i Todo beccolargo

10100 *Todus todus*
Coraciiformes - Todidae
e Jamaican Tody; Rastabird (WI);
Robin (WI); Robin-Redbreast (WI)
f Todier de la Jamaïque
d Grüntodi
i Todo della Giamaica; Todo

10101 *Tolmomyias assimilis*
Passeriformes - Tyrannidae
e Yellow-margined Flycatcher; Similar
Flycatcher; Yellow-margined Flatbill
f Platyrhynque à miroir; Tyranneau à
miroir
d Spiegelbreitschnabel
i Tiranno marginato

10102 *Tolmomyias flaviventris*
Passeriformes - Tyrannidae
e Yellow-breasted Flycatcher; Yellow-
breasted Flatbill
f Platyrhynque à poitrine jaune;
Tyranneau à poitrine jaune
d Gelbbauchbreitschnabel
i Tiranno pettogiallo

10103 *Tolmomyias poliocephalus*
Passeriformes - Tyrannidae

e   Grey-crowned Flycatcher; Grey-
     crowned Flatbill
f   Platyrhynque poliocéphale;
     Tyranneau poloicéphale
d   Schieferkopfbreitschnabel
i   Tiranno grigio

10104   *Tolmomyias sulphurescens*
            Passeriformes - Tyrannidae
e   Yellow-olive Flycatcher; White-eyed
     Flycatcher; Yellow-olive Flatbill
f   Platyrhynque jaune-olive; Tyranneau
     jaune-olive
d   Flachschnabeltyrann;
     Olivscheitelbreitschnabel
i   Tiranno gialla-oliva

10105   *Topaza pella*
            Apodiformes - Trochilidae
e   Crimson Topaz; Beautiful Topaz;
     Topaz-throated Hummingbird; King
     Hummingbird
f   Colibri topaze
d   Topazkolibri
i   Colibrì topazio cremisi

10106   *Topaza pyra*
            Apodiformes - Trochilidae
e   Fiery Topaz; Inca Topaz
f   Colibri flamboyant
d   Flammenkolibri
i   Colibrì topazio di fiamma

10107   *Torgos tracheliotus*
            Falconiformes - Accipitridae
e   Lappet-faced Vulture; Nubian
     Vulture (CSA); African Black-
     Vulture; King Vulture
f   Vautour oricou
d   Ohrengeier
i   Avvoltoio orecchiuto; Avvoltaio
     della Nubia

10108   *Torgos tracheliotus negevensis*
            Falconiformes - Accipitridae
e   Arabian Lappet-faced Vulture; La
f   Vautour d'Arabie
d   Ohrengeier

10109   *Torreornis inexpectata*
            Passeriformes - Fringillidae

e   Zapata Sparrow; Cuban Sparrow
f   Bruant de Zapata
e   Zapata-Ammer
i   Passero di Zapata

10110   *Touit batavica*
            Psittaciformes - Psittacidae
e   Lilac-tailed Parrotlet; Seven-coloured
     Parrotlet
f   Toui à sept couleurs; Perruche à dos
     noir
d   Siebenfarbenpapagei; Trinidad-
     Papagei
i   Tui settecolori

10111   *Touit costaricensis*
            Psittaciformes - Psittacidae
e   Red-fronted Parrotlet
f   Toui du Costa Rica
i   Tui della Costarica

10112   *Touit dilectissima*
            Psittaciformes - Psittacidae
e   Blue-fronted Parrotlet; Red-winged
     Parrotlet
f   Toui à front bleu
d   Kronenpapagei
i   Tui ali rosse-gialle

10113   *Touit huetii*
            Psittaciformes - Psittacidae
e   Scarlet-shouldered Parrotlet; Huet's
     Parrotlet
f   Toui de Huet
d   Schwarzstirnpapagei
i   Tui spallescarlatte

10114   *Touit melanonotus*
            Psittaciformes - Psittacidae
e   Brown-backed Parrotlet; Black-eared
     Parrotlet
f   Toui à dos noir
d   Braunrückenpapagei;
     Schwarzrückenpapagei
i   Tui orecchienere

10115   *Touit purpurata*
            Psittaciformes - Psittacidae
e   Sapphire-rumped Parrotlet; Green-
     banded Parrot
f   Perroquet à croupion bleu

d Purpurschwanzpapagei
i Tui gropponezaffiro

10116 ***Touit stictoptera***
Psittaciformes - Psittacidae
e Spot-winged Parrotlet; Spotted-
winged Parrotlet
f Toui tacheté; Toui d'Emma
d Tüpfelpapagei; Braunschulterpapagei
i Tui alimacchiate

10117 ***Touit surda***
Psittaciformes - Psittacidae
e Golden-tailed Parrotlet
f Toui à queue d'or
d Goldschwanzpapagei
i Tui codadorata

10118 ***Toxorhamphus novaeguineae***
Passeriformes - Melanocharitidae
e Green-crowned Longbill; Yellow-
bellied Longbill; New Guinea
Longbill; Canary Longbill; Canary
False-Sunbird
f Toxoramphe à ventre jaune
d Gelbbauchpfriemschnabel
i Beccolungo corona verde

10119 ***Toxorhamphus poliopterus***
Passeriformes - Melanocharitidae
e Grey-winged Longbill; Slaty-chinned
Longbill; Grey-winged False-Sunbird
f Toxoramphe à tête grise
d Graukinnpfriemschnabel
i Beccolungo aligrige

10120 ***Toxostoma arenicola***
Passeriformes - Sturnidae
e Vizcaino Thrasher; Rosalia Thrasher

10121 ***Toxostoma bendirei***
Passeriformes - Sturnidae
e Bendire's Thrasher; Bendire Thrasher
f Moqueur de Bendire; Moqueur à bec
droit
d Kaktusspottdrossel
i Mimo di Bendire

10122 ***Toxostoma cinereum***
Passeriformes - Sturnidae
e Grey Thrasher; San Lucas Thrasher

f Moqueur gris
d Mausspottdrossel
i Mimo cenerino

10123 ***Toxostoma crissale***
Passeriformes - Sturnidae
e Crissal Thrasher
f Moqueur cul-roux
d Rotsteißspottdrossel
i Mimo dal sottocoda rosso

10124 ***Toxostoma curvirostre***
Passeriformes - Sturnidae
e Curve-billed Thrasher
f Moqueur à bec courbé
d Krummschnabelspottdrossel
i Mimo beccocurvo

10125 ***Toxostoma guttatum***
Passeriformes - Sturnidae
e Cozumel Thrasher
f Moqueur de Cozumel
d Cozumel-Spottdrossel
i Mimo del Cozumel

10126 ***Toxostoma lecontei***
Passeriformes - Sturnidae
e Le Conte's Thrasher
f Moqueur de Le Conte
d Wüstenspottdrossel
i Mimo de LeConte

10127 ***Toxostoma longirostre***
Passeriformes - Sturnidae
e Long-billed Thrasher
f Moqueur à long bec
d Langschnabelspottdrossel
i Mimo beccolungo

10128 ***Toxostoma ocellatum***
Passeriformes - Sturnidae
e Ocellated Thrasher
f Moqueur ocellé
d Tropfenspottdrossel
i Mimo ocellato

10129 ***Toxostoma redivivum***
Passeriformes - Sturnidae
e California Thrasher
f Moqueur de Californie
d Sichelspottdrossel; Kalifornischer

Sichelspötter
i Mimo della California

10130 Toxostoma rufum
Passeriformes - Sturnidae
e Brown Thrasher
f Moqueur roux
d Rote Spottdrossel;
Rotrückenspotdrossel;
Rotrückensichler; Braune
Spottdrossel
i Mimo rossiccio; Mimo rosso

10131 Trachyphonus darnaudii
Piciformes - Lybiidae
e D'Arnaud's Barbet
f Barbican d'Arnaud
d D'Arnaud-Bartvogel;
Ohrfleckbartvogel
i Barbuto di D'Arnaud; Trachyfono
d'Arnaud

10132 Trachyphonus erythrocephalus
Piciformes - Lybiidae
e Red-and-yellow Barbet
f Barbican à tête rouge
d Flammenkopfbartvogel
i Trachyfono dalla testa rossa; Barbuto
testarossa

10133 Trachyphonus margaritatus
Piciformes - Lybiidae
e Yellow-breasted Barbet
f Barbican perlé
d Perlenbartvogel
i Barbuto perlato; Trachyfono perlato

10134 Trachyphonus purpuratus
Piciformes - Lybiidae
e Yellow-billed Barbet
f Barbican pourpré
d Gelbschnabelbartvogel;
Rotbandbartvogel
i Barbuto beccogiallo; Trachyfono dal
becco giallo

10135 Trachyphonus usambiro
Piciformes - Lybiidae
e Usambiro Barbet; Masai Barbet
f Barbican masai

d Haubenbartvogel
i Barbuto dell'Usambiro

10136 Trachyphonus vaillantii
Piciformes - Lybiidae
e Crested Barbet
f Barbican promépic; Barbican huppé
d Schwarzrückenbartvogel
i Barbuto di Levaillant; Trachyfono di
Levaillant

10137 Tragopan blythii
Galliformes - Phasianidae
e Blyth's Tragopan; Grey-bellied
Tragopan
f Tragopan de Blyth
d Graubauchsatyrhuhn;
Graubauchtragopan
i Tragopan di Blyth; Trogopano di
Blyth

10138 Tragopan caboti
Galliformes - Phasianidae
e Cabot's Tragopan; Yellow-bellied
Tragopan; Chinese Tragopan
f Tragopan de Cabot
d Braunbauchsatyrhuhn;
Braunbauchtragopan
i Tragopan di Temminck; Tragopano
di Temminck

10139 Tragopan melanocephalus
Galliformes - Phasianidae
e Western Tragopan; Western Horned-
Tragopan; Black-headed Tragopan
f Tragopan de Hastings
d Schwarzkopfsatyrhuhn;
Schwarzkopftragopan
i Tragopan occidentale; Trogopano
occidentale

10140 Tragopan satyra
Galliformes - Phasianidae
e Satyr Tragopan; Crimson Tragopan;
Indian Tragopan; Crimson Horned-
Pheasant
f Tragopan satyre
d Rotes Satyrhuhn; Satyrtragopan
i Tragopan satiro; Tragopano satiro

10141   *Tragopan temminckii*
Galliformes - Phasianidae
e   Temminck's Tragopan; Crimson-
bellied Tragopan
f   Tragopan de Temminck
d   Temminck-Satyrhuhn; Temminck-
Tragopan
i   Tragopan di Temminck

10142   *Tregellasia capito*
Passeriformes - Petroicidae
e   Pale-yellow Robin-Flycatcher; Pale-
yellow Robin
f   Miro jaunâtre
d   Fahlgesichtschnäpper
i   Pigliamosche giallo pallido

10143   *Tregellasia leucops*
Passeriformes - Petroicidae
e   White-faced Robin; White-faced
Robin Flycatcher; White-faced
Yellow Robin-Flycatcher; White-
faced Yellow-Robin
f   Miro à face blanche
d   Weißgesichtschnäpper
i   Pigliamosche facciabianca

10144   *Treron apicauda*
Columbiformes - Columbidae
e   Pin-tailed Green-Pigeon; Pin-tailed
Pigeon; Pintail Green-Pigeon; Long-
tailed Green-Pigeon
f   Colombar à longue queue
d   Himalaja-Spitzschwanztaube;
Spitzschwanzgrüntaube
i   Piccione verde codacuta

10145   *Treron australis*
Columbiformes - Columbidae
e   Madagascar Green-Pigeon; Green
Pigeon; African Green-Pigeon
f   Colombar maitsou; Pigeon vert
d   Graunasengrüntaube; Grüntaube;
Madagaskar-Grüntaube;
Rotbuggrüntaube
i   Piccione verde del Madagascar

10146   *Treron bicincta*
Columbiformes - Columbidae
e   Orange-breasted Green-Pigeon;
Orange-breated Pigeon; Ceylon

Orange-breasted Green-Pigeon
f   Colombar à double collier
d   Bindengrüntaube
i   Piccione verde pettoarancio

10147   *Treron calva*
Columbiformes - Columbidae
e   African Green-Pigeon; Green Pigeon
(CSA); Fruit Pigeon (CSA); African
Fruit-Pigeon; Green Fruit-Pigeon
f   Colombar à front nu
d   Grüne Fruchttaube
i   Piccione verde africano

10148   *Treron capellei*
Columbiformes - Columbidae
e   Large Green-Pigeon; Great Green-
Pigeon; Large Thick-billed Green-
Pigeon
f   Colombar de Capelle
d   Dickschnabeltaube, Große
Grüntaube; Großschnabelgrüntaube
i   Piccione verde maggiore

10149   *Treron curvirostra*
Columbiformes - Columbidae
e   Thick-billed Green-Pigeon
f   Colombar à gros bec
d   Dickschnabeltaube;
Papageischnabeltaube
i   Piccione verde beccogrosso

10150   *Treron floris*
Columbiformes - Columbidae
e   Flores Green-Pigeon
f   Colombar de Florès
d   Flores-Grüntaube
i   Piccione verde di Flores

10151   *Treron formosae*
Columbiformes - Columbidae
e   Whistling Green-Pigeon; Formosan
Green-Pigeon; Red-capped Green-
Pigeon; Ryukyu Green-Pigeon
f   Colombar de Formose
d   Formosa-Grüntaube; Taiwan-
Fruchttaube
i   Piccione verde di Formosa

10152   *Treron fulvicollis*
Columbiformes - Columbidae

e Cinnamon-headed Green-Pigeon;
  Cinnamon-headed Pigeon; Cinnamon
  Green-Pigeon; Chestnut-headed
  Green-Pigeon
f Colombar à cou roux
d Zimtkopfgrüntaube
i Piccione verde collofulvo

10153 *Treron griseicauda*
Columbiformes - Columbidae
e Grey-cheeked Green-Pigeon; Grey-
  faced Green-Pigeon; Grey-faced
  Thick-billed Green-Pigeon; Green
  Pigeon; Grey-fronted Green-Pigeon
f Colombar à face grise
d Graumaskentaube
i Piccione verde facciagrigia

10154 *Treron olax*
Columbiformes - Columbidae
e Little Green-Pigeon
f Colombar odorifère
d Graukopfgrüntaube; Kleine
  Grüntaube
i Piccione verde minore

10155 *Treron oxyura*
Columbiformes - Columbidae
e Sumatran Green-Pigeon; Yellow-
  bellied Pin-tailed Pigeon; Yellow-
  bellied Pin-tailed Green-Pigeon;
  Green Spectacled Pigeon; Yellow-
  bellied Green-Pigeon; Sunda Pin-
  tailed Pigeon; Pin-tailed Green-
  Pigeon
f Colombar à queue pointue
d Gelbbauchgrüntaube; Sumatra-
  Spitzschwanztaube
i Piccione verde ventregiallo

10156 *Treron pembaensis*
Columbiformes - Columbidae
e Pemba Green-Pigeon; Pemba Island
  Green-Pigeon
f Colombar de Pemba

10157 *Treron phoenicoptera*
Columbiformes - Columbidae
e Yellow-footed Green-Pigeon;
  Common Green-Pigeon (ISC); Green
  Pigeon (ISC); Yellow-legged Green-

Pigeon; Yellow-footed Pigeon;
Ceylon Green-Pigeon; Ceylon
Yellow-legged Green-Pigeon; Bengal
Green-Pigeon
f Colombar commandeur
d Gelbfußtaube; Rotschultertaube
i Piccione verde zampegialle

10158 *Treron pompadora*
Columbiformes - Columbidae
e Pompadour Green-Pigeon; Grey-
  fronted Green-Pigeon (ISC);
  Pompadour Pigeon; Ashy-headed
  Green-Pigeon; Andaman Green-
  Pigeon
f Colombar pompadour
d Andamanen-Grüntaube;
  Graustirngrüntaube; Pompadourtaube
i Piccione verde della Pompadour

10159 *Treron psittacea*
Columbiformes - Columbidae
e Timor Green-Pigeon
f Colombar unicolore
d Timor-Grüntaube
i Piccione verde di Timor

10160 *Treron sanctithomae*
Columbiformes - Columbidae
e Sao Tomé Green-Pigeon
f Colombar de Sao Tomé
d St. Thomas-Grüntaube
i Piccione verde di Sao Tomè

10161 *Treron seimundi*
Columbiformes - Columbidae
e Yellow-vented Green-Pigeon; White-
  bellied Pin-tailed Pigeon; White-
  bellied Pin-tailed Green-Pigeon;
  Yellow-vented Pigeon; Seimund's
  Pintail Green-Pigeon; Seimund's
  Green-Pigeon; White-bellied Pin-
  tailed Green-Pigeon
f Colombar de Seimund
d Malacca-Spitzschwanztaube;
  Weißbauchgrüntaube
i Piccione verde ventrebianco

10162 *Treron sieboldii*
Columbiformes - Columbidae
e White-bellied Green-Pigeon; Green

Pigeon; White-bellied Wedge-tailed
Pigeon; Siebold's Pigeon; White-
bellied Pigeon; White-bellied
Wedge-tailed Green-Pigeon;
Japanese Green-Pigeon; Siebold's
Green-Pigeon
f Colombar de Siebold
d Siebold-Taube
i Piccione verde giapponese

10163 *Treron sphenura*
Columbiformes - Columbidae
e Wedge-tailed Green-Pigeon; Kokla
(ISC); Singing Green-Pigeon;
Korthasi's Pigeon
f Colombar chanteur
d Keilschwanzgrüntaube
i Piccione verde codacuneata

10164 *Treron teysmannii*
Columbiformes - Columbidae
e Sumba Green-Pigeon; Sumba Island
Green-Pigeon
f Colombar de Sumba
d Teysmann-Taube
i Piccione verde di Sumba

10165 *Treron vernans*
Columbiformes - Columbidae
e Pink-necked Green-Pigeon; Pink-
necked Pigeon
f Colombar giouanne
d Frühlingstaube;
Halsfleckengrüntaube
i Piccione verde collorosa

10166 *Treron waalia*
Columbiformes - Columbidae
e Bruce's Green-Pigeon; Yellow-
bellied Pigeon; Yellow-bellied
Green-Pigeon; Yellow-bellied Fruit-
Pigeon
f Colombar waalia; Pigeon à épaulettes
violettes
d Waalia-Taube
i Piccione verde pettogiallo

10167 *Trichastoma bicolor*
Passeriformes - Sylviidae
e Ferruginous Babbler; Ferruginous
Jungle-Babbler

f Akalat ferrugineux
d Weißwangenmaustimalie
i Garrulo ferrugineo

10168 *Trichastoma celebense*
Passeriformes - Sylviidae
e Sulawesi Babbler; Celebes Jungle-
Babbler; Sulawesi Jungle-Babbler;
Celebes Babbler; Strickland's
Babbler
f Akalat des Célèbes
d Weißkehlmaustimalie
i Garrulo di Sulawesi

10169 *Trichastoma rostratum*
Passeriformes - Sylviidae
e White-chested Babbler; Blyth's
Babbler; Blyth's Jungle-Babbler;
White-chested Jungle-Babbler;
Mangrove Brown-Babbler
f Akalat à front noir
d Mangrovemaustimalie;
Schwarzstirndschungeltimalie
i Garrulo pettobianco

10170 *Trichastoma woodi*
Passeriformes - Sylviidae
e Bagobo Babbler; Bagobo Jungle-
Babbler
f Akalat de Wood
d Leonardina
i Garrulo di Wood

10171 *Trichixos pyrropyga*
Passeriformes - Muscicapidae
e Rufous-tailed Shama; Orange-tailed
Shama
f Shama à queue rousse
d Feuerschwanzschama
i Shama codirosso

10172 *Trichocichla rufa*
Passeriformes - Sylviidae
e Long-legged Thicketbird; Long-
legged Warbler; Greater Forest
Warbler
f Mégalure des Fidji
d Langbeinbuschsänger
i Cantore di macchia zampelunghe

10173 *Trichodere cockerelli*
Passeriformes - Meliphagidae
e White-streaked Honeyeater; Brush-
throated Honeyeater; Cockerell's
Honeyeater
f Trichodere cockerelli; Méliphage de
Cockerell
d Uferhonigfresser
i Mangiamiele di Cockerell

10174 *Trichoglossus chlorolepidotus*
Psittaciformes - Loriidae
e Scaly-breasted Lorikeet; Green
Lorikeet; Green-and-yellow Lorikeet;
Green Keet; Green Leek; Green
Parrot; Scaly-breasted Lory; Green-
and-gold Lorikeet
f Loriquet vert; Loriquet écaillé;
Perruche écaillée
d Schuppenlori
i Lorichetto pettosquamato

10175 *Trichoglossus euteles*
Psittaciformes - Loriidae
e Olive-headed Lorikeet; Perfect
Lorikeet; Yellow-headed Lorikeet
f Loriquet eutèle; Loriquet à tête jaune
d Gelbkopflori
i Lorichetto testagialla

10176 *Trichoglossus flavoviridis*
Psittaciformes - Loriidae
e Yellow-and-green Lorikeet; Yellow-
and-green Lory
f Loriquet jaune-et-vert
d Gelbgrüner Lori; Celebes Lori
i Lorichetto giallo e verde

10177 *Trichoglossus haematodus*
Psittaciformes - Loriidae
e Rainbow Lorikeet; Coconut Lory;
Blue-bellied Parrot; Swainson's Blue-
bellied Lorikeet; Blue Mountain
Lorikeet; Green-naped Lorikeet; Blue
Mountain Parrot; Rainbow Lory;
Red-breasted Lorikeet; Blue-bellied
Lorikeet; Swainson's Lorikeet
f Loriquet à tête bleue
d Allfarblori; Gebirgslori
i Tricoglosso arcobaleno; Lorichetto
arcobaleno

10178 *Trichoglossus johnstoniae*
Psittaciformes - Loriidae
e Mindanao Lorikeet; Johnstone's
Lorikeet; Mrs.. Johnstone's Lorikeet;
Mount Apo Lory; Apo Lorikeet
f Loriquet de Johnstone
d Mindanao-Lori
i Lorichetto della signora Johnstone

10179 *Trichoglossus ornatus*
Psittaciformes - Loriidae
e Ornate Lorikeet; Ornamented
Lorikeet; Ornate Lory
f Loriquet orné
d Schmucklori
i Lorichetto ornato

10180 *Trichoglossus rubiginosus*
Psittaciformes - Loriidae
e Pohnpei Lorikeet; Ponapé Lory;
Cherry Lorikeet; Cherry-red Lorikeet
f Loriquet de Ponapé
d Kirschlori
i Lorichetto di Ponapè

10181 *Trichoglossus rubritorquis*
Psittaciformes - Loriidae
e Red-collared Lorikeet
f Loriquet à col rouge
i Tricoglosso del collare rosso;
Lorichetto dal collare rosso

10182 *Tricholaema diademata*
Piciformes - Lybiidae
e Red-fronted Barbet
f Barbican à diadème
d Diadembartvogel
i Barbuto diademato

10183 *Tricholaema frontata*
Piciformes - Lybiidae
e Miombo Barbet; Miombo Pied-
Barbet
f Barbican du Miombo
i Barbuto del Miombo

10184 *Tricholaema hirsuta*
Piciformes - Lybiidae
e Hairy-breasted Barbet; Black-
throated Barbet
f Barbican herissé

d   Fleckenbartvogel
i   Barbuto irsuto

10185   Tricholaema hirsuta flavipunctata
Piciformes - Lybiidae
f   Barbican de Verreaux

10186   Tricholaema lacrymosa
Piciformes - Lybiidae
e   Spot-flanked Barbet; Spotted-flanked
Barbet
f   Barbican tacheté; Barbican funèbre
d   Tränenbartvogel
i   Barbuto fianchimacchiati

10187   Tricholaema leucomelas
Piciformes - Lybiidae
e   Pied Barbet; Acacia Pied-Barbet;
Acacia Barbet
f   Barbican pie
d   Rotstirnbartvogel
i   Barbuto bianco e nero

10188   Tricholaema melanocephala
Piciformes - Lybiidae
e   Black-throated Barbet; African
Black-throated Barbet
f   Barbican à tête noire
d   Schwarzkopfbartvogel
i   Barbuto africano golanera

10189   Trichothraupis melanops
Passeriformes - Fringillidae
e   Black-goggled Tanager
f   Tangara à front noir
d   Haarschopftangare
i   Tangara dagli occhiali nere

10190   Triclaria malachitacea
Psittaciformes - Psittacidae
e   Blue-bellied Parrot; Purple-bellied
Parrot; Violet-bellied Parrot
f   Crick à ventre bleu; Caïque à ventre
bleu
d   Blaubauch
i   Pappagallo ventre malachite

10191   Trigonoceps occipitalis
Falconiformes - Accipitridae
e   White-headed Vulture
f   Vautour à tête blanche

d   Wollkopfgeier
i   Avvoltoio testabianca; Avvoltaio
dalla testa bianca

10192   Tringa brevipes
Charadriiformes - Scolopacidae
e   Grey-tailed Tattler; Grey-rumped
Sandpiper; Grey-tailed Sandpiper;
Grey-rumped Tattler; Grey
Sandpiper; Siberian Tattler;
Polynesian Tattler; Ashen Tringine-
Sandpiper
f   Chevalier de Sibérie
d   Graubürzelwasserläufer;
Wanderwasserläufer
i   Piro-piro asiatico

10193   Tringa cinerea
Charadriiformes - Scolopacidae
e   Terek Sandpiper; Avocet Sandpiper
(ISC)
f   Chevalier bargette; Bargette de
Térek; Chevalier de Térek
d   Térek-Wasserläufer; Isländischer
Strandläufer
i   Piro-piro becco torto; Terecchio;
Piro-piro térek

10194   Tringa cooperi
Charadriiformes - Scolopacidae
e   Cooper's Sandpiper
d   Dunkelwasserläufer

10195   Tringa erythropus
Charadriiformes - Scolopacidae
e   Spotted Redshank; Dusky Redshank
(ISC)
f   Chevalier arlequin; Petit Chevalier
d   Dunkler Wasserläufer; Großer
Rotschenkel
i   Totano moro

10196   Tringa flavipes
Charadriiformes - Scolopacidae
e   Lesser Yellowlegs; Yellowshank;
Snipe (WI); Pondbird (WI)
f   Chevalier à pattes jaunes; Chevalier
pattejaune; Bécassine à pattes jaunes
(Ants); Patte-jaune (Ants); Petit
Chevalier; Petit chevalier à pattes
jaunes

d Gelbschenkel; Kleiner Gelbschenkel
i Totano zampegialle minore

10197   ***Tringa glareola***
Charadriiformes - Scolopacidae
e Wood Sandpiper; Spotted Sandpiper (ISC)
f Chevalier sylvain
d Bruchwasserläufer; Wasserläufer
i Piro-piro boschereccio

10198   ***Tringa guttifer***
Charadriiformes - Scolopacidae
e Nordmann's Greenshank; Armstrong's Sandpiper; Spotted-Greenshank; Okhotsk Tringine-Sandpiper
f Chevalier tacheté
d Kurzfußwasserläufer
i Pantana macchiata

10199   ***Tringa hypoleucos***
Charadriiformes - Scolopacidae
e Common Sandpiper; Sandpiper; Carrier Sandpiper; Eurasian Sandpiper
f Chevalier guignette
d Flußuferläufer; Wasserläufer
i Piro-piro piccolo

10200   ***Tringa incana***
Charadriiformes - Scolopacidae
e Wandering Tattler; Polynesian Tattler; American Ashen Tringine-Sandpiper; American Grey-rumped Sandpiper
f Chevalier errant
d Wanderwasserläufer
i Piro-piro vagabondo

10201   ***Tringa macularia***
Charadriiformes - Scolopacidae
e Spotted Sandpiper; Weatherbird (WI); Tip-up (WI); Dipper (WI)
f Chevalier grivelé; Chevalier branlequeue (Ants)
d Amerikanischer Drosseluferläufer; Drosseluferläufer; Amerikanischer Uferläufer
i Piro-piro americano; Piro-piro macchiato; Piro-piro machiettato

10202   ***Tringa melanoleuca***
Charadriiformes - Scolopacidae
e Greater Yellowlegs; Snipe (WI); Pondbird (WI); Greater Yellowshank
f Chevalier criard; Bécasse à pattes jaunes (Ants); Grand Chevalier; Clin (Ants); Clinclin (Ants); Grand chevalier à pattes jaunes
d Großer Gelbschenkel
i Totano zampegialle maggiore

10203   ***Tringa nebularia***
Charadriiformes - Scolopacidae
e Common Greenshank; Greenshank; Greater Greenshank; Large Tringine-Sandpiper
f Chevalier aboyeur
d Grünschenkel; Heller Wasserläufer
i Pantana; Pantana comune

10204   ***Tringa ochropus***
Charadriiformes - Scolopacidae
e Green Sandpiper
f Chevalier cul-blanc
d Waldwasserläufer
i Piro-piro culbianco

10205   ***Tringa solitaria***
Charadriiformes - Scolopacidae
e Solitary Sandpiper; Pondbird (WI)
f Chevalier solitaire; Grande-aile (Ants); Chevalier à aile noire (Ants); Aile-noire
d Einsiedelwasserläufer; Einsamer Wasserläufer
i Piro-piro solitario

10206   ***Tringa stagnatilis***
Charadriiformes - Scolopacidae
e Marsh Sandpiper
f Chevalier stagnatile
d Teichwasserläufer
i Piro-piro gambe lunghe; Albastrello

10207   ***Tringa totanus***
Charadriiformes - Scolopacidae
e Common Redshank; Redshank
f Chevalier gambette
d Rotschenkel
i Pettegola; Gambetta

10208 Trochilidae
Apodiformes
e Hummingbirds
f Trochilidés; Colibris
d Kolibris
i Trochilidi

10209 *Trochilus polytmus*
Apodiformes - Trochilidae
e Streamertail; Western Streamertail;
Godbird (WI); Doctor-Bird (WI);
Jamaican Doctor-Bird (WI);
Scissors-tail (WI); Long-tailed
Hummingbird (WI); Red-billed
Streamertail
f Colibri à tête noire
d Wimpelschwanz
i Colibrì coda a festone; Colibrì dalla
testa nera

10210 *Trochilus scitulus*
Apodiformes - Trochilidae
e Eastern Streamertail; Black-billed
Streamertail; Godbird (WI); Doctor-
Bird (WI); Jamaican Doctor-Bird
(WI); Scissors-tail (WI); Longtail
(WI); Long-tailed Doctor-Bird (WI);
Streamertail

10211 *Trochilus violajugulum*
Apodiformes - Trochilidae
e Violet-throated Hummingbird

10212 *Trochocercus cyanomelas*
Passeriformes - Monarchidae
e African Crested-Flycatcher; Crested-
Flycatcher; Cape Crested-Flycatcher;
Blue-mantled Flycatcher; Blue-
mantled Crested-Flycatcher; Blue-
mantled Crested-Monarch
d Blaumantelhaubenschnäpper
i Pigliamosche crestato del Capo

10213 *Trochocercus cyanomelas bivittatus*
Passeriformes - Monarchidae
e Blue-mantled Crested-Flycatcher
(CSA); Blue-mantled Flycatcher
(CSA)
d Blaumantelschopfschnäpper

10214 *Trochocercus nitens*
Passeriformes - Monarchidae
e Blue-headed Crested-Flycatcher;
Blue-headed Crested-Monarch
f Gobemouche noir huppé
d Glanzhaubenschnäpper
i Pigliamosche crestato testablu

10215 *Troglodytes aedon*
Passeriformes - Certhiidae
e House Wren; Wren; Northern House-
Wren; Godbird (WI); Rockbird (WI);
Wallbird (WI); Brown-throated
Wren; Brown-throated House-Wren;
Southern House-Wren; Cozumel
House-Wren; Cozumel Wren
f Troglodyte familier; Rossignol
(Ants); Oiseau Bon-Dieu (Ants);
Troglodyte commun; Troglodyte
d Hauszaunkönig
i Scricciolo delle case

10216 *Troglodytes beani*
Passeriformes - Certhiidae
e Cozumel Wren
f Troglodyte de Cozumel

10217 *Troglodytes brunneicollis*
Passeriformes - Certhiidae
e Brown-throated Wren
f Troglodyte à gorge brune

10218 *Troglodytes cobbi*
Passeriformes - Certhiidae
e Cobb's Wren

10219 *Troglodytes martinicensis*
Passeriformes - Certhiidae
e Antillean Wren; Antillean House-
Wren
f Troglodyte des Antilles

10220 *Troglodytes monticola*
Passeriformes - Certhiidae
e Santa Marta Wren; Paramo Wren
f Troglodyte des Santa Marta
i Scricciolo di Santa Marta

10221 *Troglodytes musculus*
Passeriformes - Certhiidae

e   Southern House-Wren
f   Troglodyte austral

10222   *Troglodytes ochraceus*
Passeriformes - Certhiidae
e   Ochraceous Wren
f   Troglodyte ocré
i   Scricciolo ocraceo

10223   *Troglodytes rufociliatus*
Passeriformes - Certhiidae
e   Rufous-browed Wren; House Wren
f   Troglodyte à sourcils roux
i   Scricciolo dai sopraccigli rossicci

10224   *Troglodytes rufulus*
Passeriformes - Certhiidae
e   Tepui Wren
f   Troglodyte des tépuis
d   Roraima-Zaunkönig
i   Scricciolo del Tepui

10225   *Troglodytes sissonii*
Passeriformes - Certhiidae
e   Socorro Wren; Revillagigedo Wren

10226   *Troglodytes solstitialis*
Passeriformes - Certhiidae
e   Mountain Wren; Sun Wren; Rufous-
browed Wren
f   Troglodyte montagnard
d   Rostbrauenzaunkönig
i   Scricciolo di montagna

10227   *Troglodytes tanneri*
Passeriformes - Certhiidae
e   Clarion Wren; Clarion Island Wren
f   Troglodyte de Clarion
i   Scricciolo di Tanner

10228   *Troglodytes troglodytes*
Passeriformes - Certhiidae
e   Winter Wren; Wren; Northern Wren;
Common Wren; European Wren;
Holarctic Wren
f   Troglodyte mignon; Troglodyte des
forêts; Troglodyte; Oiseau béni
d   Zaunkönig
i   Scricciolo; Re di macchia; Reatino;
Regillo; Reillo; Foramacchie;
Forasieppe; Scricciolo comune

10229   *Trogon ambiguus*
Trogoniformes - Trogonidae
e   Coppery-tailed Trogon

10230   *Trogon aurantiiventris*
Trogoniformes - Trogonidae
e   Orange-bellied Trogon
f   Trogon à ventre orangé
d   Goldbauchtrogon
i   Trogone ventrearancino

10231   *Trogon australis*
Trogoniformes - Trogonidae
e   Chapman's Trogon

10232   *Trogon bairdii*
Trogoniformes - Trogonidae
e   Baird's Trogon
f   Trogon de Baird
d   Baird-Tangare
i   Trogone di Baird

10233   *Trogon caligatus*
Trogoniformes - Trogonidae
e   Gartered Trogon

10234   *Trogon citreolus*
Trogoniformes - Trogonidae
e   Citreoline Trogon
f   Trogon citrin
d   Graukopftrogon
i   Trogone citrino

10235   *Trogon clathratus*
Trogoniformes - Trogonidae
e   Lattice-tailed Trogon
f   Trogon échelette
d   Sperberschwanz-Trogon
i   Trogone codareticolata

10236   *Trogon collaris*
Trogoniformes - Trogonidae
e   Collared Trogon; Bar-tailed Trogon
f   Trogon rosalba
d   Jungferntrogon
i   Trogone dal collare

10237   *Trogon comptus*
Trogoniformes - Trogonidae
e   White-eyed Trogon; Blue-tailed
Trogon

f	Trogon aux yeux blancs; Trogon à
	queue bleue
d	Blauschwanztrogon
i	Trogone codablu

10238   Trogon curucui
Trogoniformes - Trogonidae
e   Blue-crowned Trogon
f   Trogon couroucou; Trogon de Linné
d   Blauscheiteltrogon
i   Trogone corona blu

10239   Trogon elegans
Trogoniformes - Trogonidae
e   Elegant Trogon; Coppery-tailed
Trogon
f   Trogon élégant
d   Kupfertrogon
i   Trogone elegante

10240   Trogon macroura
Trogoniformes - Trogonidae
e   Large-tailed Trogon

10241   Trogon massena
Trogoniformes - Trogonidae
e   Slaty-tailed Trogon; Massena
Trogon; Chapman's Trogon
f   Trogon de Masséna
d   Massena-Trogon
i   Trogone di Massena

10242   Trogon melanocephalus
Trogoniformes - Trogonidae
e   Black-headed Trogon; Citreoline
Trogon
f   Trogon à tête noire
d   Schwarzkopftrogon
i   Trogone testanera

10243   Trogon melanurus
Trogoniformes - Trogonidae
e   Black-tailed Trogon
f   Trogon à queue noire
d   Schwarzschwanztrogon
i   Trogone codanera

10244   Trogon mexicanus
Trogoniformes - Trogonidae
e   Mountain Trogon; Mexican Trogon
f   Trogon montagnard

d   Bronzetrogon
i   Trogone messicano

10245   Trogon personatus
Trogoniformes - Trogonidae
e   Masked Trogon
f   Trogon masqué
d   Maskentrogon
i   Trogone mascherato

10246   Trogon puella
Trogoniformes - Trogonidae
e   Bar-tailed Trogon

10247   Trogon rufus
Trogoniformes - Trogonidae
e   Black-throated Trogon; Graceful
Trogon
f   Trogon aurore
d   Schwarzkehltrogon
i   Trogone golanera

10248   Trogon surrucura
Trogoniformes - Trogonidae
e   Surucua Trogon; Brazilian Trogon
f   Trogon surucua
d   Surucua-Trogon
i   Trogone surucua

10249   Trogon violaceus
Trogoniformes - Trogonidae
e   Violaceous Trogon; Amazonian
Trogon
f   Trogon violacé
d   Veilchentrogon
i   Trogone violaceo

10250   Trogon viridis
Trogoniformes - Trogonidae
e   White-tailed Trogon
f   Trogon à queue blanche; Trogon à
ventre jaune
d   Grüntrogon; Weißschwanztrogon
i   Trogone codabianca

10251   Trogonidae
Trogoniformes
e   Trogons
f   Trogonidés; Couroucous
d   Trogone; Trogonten
i   Trogonidi

10252 Trogoniformes
e Trogons
f Trogoniformes
d Trogone; Trogonten; Verkehrtfüßler
i Trogoniformi

10253 *Trugon terrestris*
Columbiformes - Columbidae
e Thick-billed Ground-Pigeon; Slaty Ground-Pigeon; Grey Ground-Pigeon
f Trugon terrestre
d Neuguinea-Erdtaube
i Piccione terricolo beccogrosso

10254 *Tryngites subruficollis*
Charadriiformes - Scolopacidae
e Buff-breasted Sandpiper
f Bécasseau rousset; Bécasseau rousse; Bécassine rousse; Bécasseau roussâtre
d Grasläufer
i Piro-piro fulvo

10255 *Turacoena manadensis*
Columbiformes - Columbidae
e White-faced Cuckoo-Dove; White-faced Pigeon; White-faced Black Cuckoo-Dove; White-faced Black-; Sulawesi Black-
f Phasianelle de Manado
d Manado-Taube
i Piccione facciabianca

10256 *Turacoena modesta*
Columbiformes - Columbidae
e Black Cuckoo-Dove; Timor Black-Pigeon; Slaty Cuckoo-Dove; Slate-coloured Cuckoo-Dove
f Phasianelle modeste
d Timor-Taube
i Piccione modesta di Timor

10257 Turdidae
Passeriformes
e Thrushes
f Turdidés
d Drosselvögel; Drosseln
i Turdidi; Tordi

10258 *Turdoides affinis*
Passeriformes - Sylviidae

e Yellow-billed Babbler; White-headed Babbler (ISC); Indian White-headed Babbler; White-billed Babbler; Southern Common Babbler
f Cratérope affin
d Gelbschnabeldrossling
i Garrulo testabianca

10259 *Turdoides altirostris*
Passeriformes - Sylviidae
e Iraq Babbler
f Cratérope d'Irak
d Rieddrossling
i Garrulo dell'Iraq

10260 *Turdoides aylmeri*
Passeriformes - Sylviidae
e Scaly Chatterer; Scaly Babbler; Aylner's Babbler
f Cratérope ardoisé
d Schuppenbrustdrossling; Braunkopfdrossling
i Garrulo di Aylmer

10261 *Turdoides bicolor*
Passeriformes - Sylviidae
e Southern Pied-Babbler; Pied Babbler (CSA); Bicoloured Babbler
f Cratérope bicolore
d Elsterdrossling
i Garrulo bicolore

10262 *Turdoides caudatus*
Passeriformes - Sylviidae
e Common Babbler
f Cratérope de l'Inde; Cratérope indien
d Langschwanzdrossling
i Garrulo; Garrulo comune; Timalide; Timalide comune

10263 *Turdoides earlei*
Passeriformes - Sylviidae
e Striated Babbler; Earle's Babbler
f Cratérope strié
d Streifendrossling
i Garrulo di Earle

10264 *Turdoides fulvus*
Passeriformes - Sylviidae
e Fulvous Chatterer; Fulvous Babbler
f Cratérope fauve

d Akaziendrossling; Sahara-
Langschwanzdrossling
i Garrulo fulvo; Timalide nordafricano

10265 *Turdoides gularis*
Passeriformes - Sylviidae
e White-throated Babbler; Burmese
White-throated Babbler; Burmese
Babbler
f Cratérope à gorge blanche
d Weißkehldrossling; Burma-Drossling
i Garrulo golabianca

10266 *Turdoides gymnogenys*
Passeriformes - Sylviidae
e Bare-cheeked Babbler
f Cratérope à joues nues
d Nacktohrdrossling;
Nacktwangendrossling
i Garrulo guancenude

10267 *Turdoides hartlaubii*
Passeriformes - Sylviidae
e Angola Babbler; Hartlaub's Babbler
(CSA); White-rumped Babbler
(CSA); Angola Wite-rumped Babbler
f Cratérope de Hartlaub
d Weißbürzeldrossling; Hartlaub-
Drossling
i Garrulo dell'Angola

10268 *Turdoides hindei*
Passeriformes - Sylviidae
e Hinde's Pied-Babbler; Hinde's
Babbler
f Cratérope de Hinde; Cratérope pie de
Hinde
i Garrulo di Hinde

10269 *Turdoides hypoleucus*
Passeriformes - Sylviidae
e Northern Pied-Babbler; Pied Babbler
(CSA)
f Cratérope bigarré; Cratérope pie
d Bronzedrossling
i Garrulo bicolore settentrionale

10270 *Turdoides jardineii*
Passeriformes - Sylviidae
e Arrow-marked Babbler; Jardine's
Babbler

f Cratérope flèché; Cratérope de
Jardine
d Braundrossling
i Garrulo di Jardine

10271 *Turdoides leucocephalus*
Passeriformes - Sylviidae
e Cretschmar's Babbler; White-headed
Babbler
f Cratérope à tête blanche
d Weißkopfdrossling
i Garrulo di Cretzschmar

10272 *Turdoides leucopygius*
Passeriformes - Sylviidae
e White-rumped Babbler; Abyssinian
Babbler
f Cratérope à croupion blanc
d Weißbürzeldrossling
i Garrulo gropponebianco; Tordo
garrulo leucopigio; Cratropodo

10273 *Turdoides longirostris*
Passeriformes - Sylviidae
e Slender-billed Babbler
f Cratérope à bec fin
d Schlankschnabeldrossling
i Garrulo beccolungo

10274 *Turdoides malcolmi*
Passeriformes - Sylviidae
e Large Grey-Babbler; Great Grey-
Babbler; Malcolm's Babbler
f Cratérope gris
d Malcolm-Drossling
i Garrulo di Malcolm

10275 *Turdoides melanops*
Passeriformes - Sylviidae
e Black-lored Babbler; Black-faced
Babbler (CSA)
f Cratérope masqué
d Dunkler Drossling;
Schwarzflügeldrossling;
Schwarzzügeldrossling
i Garrulo occhigrigi

10276 *Turdoides nipalensis*
Passeriformes - Sylviidae
e Spiny Babbler
f Cratérope du Nepal

d Idel-Drossling
i Garrulo del Nepal

10277 ***Turdoides plebejus***
Passeriformes - Sylviidae
e Brown Babbler; African Brown-Babbler; Sudan Babbler; Sudan Brown Babbler
f Cratérope brun
d Schuppendrossling; Sudan-Drossling
i Garrulo bruno

10278 ***Turdoides reinwardtii***
Passeriformes - Sylviidae
e Blackcap Babbler; Western Dusky Babbler; Black-eye Babbler; Black-eyed Babbler
f Cratérope à tête noire
d Weißaugendrossling
i Garrulo di Reinwardt

10279 ***Turdoides rubiginosus***
Passeriformes - Sylviidae
e Rufous Chatterer; Rufous Babbler
f Cratérope rubigineux
d Rüppels Drossling
i Garrulo rossiccio africano

10280 ***Turdoides rufescens***
Passeriformes - Sylviidae
e Orange-billed Babbler; Ceylon Babbler; Ceylon Jungle-Babbler; Ceylon Rufous Babbler; Sri Lanka Orange-billed Babbler
f Cratérope de Ceylan
d Ceylon-Drossling
i Garrulo rossiccio di Ceylon; Gracula rossiccio di Sri-Lanka

10281 ***Turdoides sharpei***
Passeriformes - Sylviidae
e Sharpe's Pied-Babbler; Black-lored Babbler (CSA); Tabora Black-faced Babbler
f Cratérope de Sharpe
d Tabora-Drossling
i Garrulo di Sharpe

10282 ***Turdoides squamiceps***
Passeriformes - Sylviidae
e Arabian Babbler; Arabian Brown-Babbler; Brown Babbler
f Cratérope écaillé
d Graudrossling; Palästina-Langschwanzdrossling
i Garrulo arabo; Timalide d'Arabia

10283 ***Turdoides squamulatus***
Passeriformes - Sylviidae
e Scaly Babbler; Squamulated Babbler
f Cratérope maillé
d Schwarzkopfdrossling
i Garrulo squamoso

10284 ***Turdoides striatus***
Passeriformes - Sylviidae
e Jungle Babbler; Striated Jungle-Babbler; Striated Babbler; Deccan Babbler; Whit-headed Jungle-Babbler; White-headed Babbler
f Cratérope de brousse
d Dschungeldrossling
i Garrulo striato

10285 ***Turdoides subrufus***
Passeriformes - Sylviidae
e Rufous Babbler; Rufous-backed Babbler
f Cratérope roussâtre
d Graustirndrossling
i Garrulo rossiccio indiano

10286 ***Turdoides tenebrosus***
Passeriformes - Sylviidae
e Dusky Babbler
f Cratérope ombré; Cratérope fuligineux
d Uferdrossling
i Garrulo fosco

10287 ***Turdus abyssinicus***
Passeriformes - Turdidae
e Abyssinian Thrush; Mountain Thrush; Northern Olive-Thrush; African Mountain-Thrush
d Gmelin-Drossel

10288 ***Turdus albicollis***
Passeriformes - Turdidae
e White-necked Thrush
f Merle à col blanc

d Trauerdrossel
i Tordo collobianco

10289 *Turdus albocinctus*
Passeriformes - Turdidae
e White-collared Blackbird; White-collared Thrush; White-throated Thrush
f Merle à collier blanc
d Weißhalsamsel
i Merlo dal collare orientale

10290 *Turdus amaurochalinus*
Passeriformes - Turdidae
e Creamy-bellied Thrush; Cream-bellied Thrush; Dusky Thrush
f Merle à ventre clair
d Rahmbauchdrossel
i Tordo ventrechiaro

10291 *Turdus ardosiaceus*
Passeriformes - Turdidae
e Eastern Red-legged Thrush

10292 *Turdus assimilis*
Passeriformes - Turdidae
e White-throated Thrush; White-throated Robin
f Merle à gorge blanche
i Tordo golabianca

10293 *Turdus aurantius*
Passeriformes - Turdidae
e White-chinned Thrush; Hopping-Dick; Jumping-Dick (WI); Twopenny Chick (WI); Chick-me-chick (WI); Chap-man-chick (WI)
f Merle à miroir
d Weißkinndrossel
i Tordo golabianca

10294 *Turdus bewsheri*
Passeriformes - Turdidae
e Comoro Thrush; Comoros Thrush; Johamma Thrush
f Merle des Comores; Grive des Comores
d Komoro-Drossel
i Tordo delle Comore

10295 *Turdus boulboul*
Passeriformes - Turdidae
e Grey-winged Blackbird
f Merle à ailes grises
d Bülbüldrossel; Bülbülamsel
i Merlo aligrige

10296 *Turdus cardis*
Passeriformes - Turdidae
e Japanese Thrush; Grey Thrush; Japanese Grey Thrush
f Merle du Japon; Grive du Japon
d Scheckendrossel
i Tordo grigio giapponese

10297 *Turdus celaenops*
Passeriformes - Turdidae
e Izu Thrush; Seven Islands Thrush; Izu Islands Thrush
f Merle des Izu
d Kurodas-Drossel
i Tordo delle Sette Isole

10298 *Turdus chiguanco*
Passeriformes - Turdidae
e Chiguanco Thrush
f Merle chiguanco
d Chiguanco-Drossel
i Tordo chiguanco

10299 *Turdus chrysolaus*
Passeriformes - Turdidae
e Brown-headed Thrush; Red-bellied Thrush; Red-billed Thrush; Brown Thrush; Japanese Thrush
f Merle à flancs roux; Grive à flancs roux
d Rotkopfdrossel
i Tordo beccorosso

10300 *Turdus confinis*.
Passeriformes - Turdidae
e San Lucas Robin

10301 *Turdus daguae*
Passeriformes - Turdidae
e Dagua Thrush

10302 *Turdus dissimilis*
Passeriformes - Turdidae
e Black-breasted Thrush

Migratory Thrush
- *f*   Merle migrateur; Merle d'Amérique; Merle américain; Grive de Canada
- *d*   Wanderdrossel
- *i*   Tordo migratore americano; Tordo migratore

10330   ***Turdus mupinensis***
Passeriformes - Turdidae
- *e*   Chinese Thrush; Chinese Song-Thrush; Verreaux's Song-Thrush; Mongolian Song-Thrush; Eastern Song-Thrush
- *f*   Grive de Verreaux
- *d*   Mupin-Drossel
- *i*   Tordo della Mongolia

10331   ***Turdus naumanni***
Passeriformes - Turdidae
- *e*   Naumann's Thrush; Dusky Thrush; Rufous-tailed Thrush
- *f*   Grive de Naumann; Merle de Naumann; Grive à ailes rousses
- *d*   Rostschwanzdrossel; Naumann-Drossel; Rotschwanzdrossel; Rostflügeldrossel
- *i*   Tordo di Naumann; Cesena di Nauman; Cesena fosca

10332   ***Turdus nigrescens***
Passeriformes - Turdidae
- *e*   Sooty Thrush; Sooty Robin; Sooty Ouzel
- *f*   Merle fuligineux
- *d*   Rußdrossel
- *i*   Merlo occhigialli

10333   ***Turdus nigriceps***
Passeriformes - Turdidae
- *e*   Andean Slaty-Thrush; Slaty Thrush; Black-headed Thrush; Black-capped Thrush
- *f*   Merle ardoisé
- *d*   Weißachseldrossel
- *i*   Tordo testanera

10334   ***Turdus nudigenis***
Passeriformes - Turdidae
- *e*   Yellow-eyed Thrush; Bare-eyed Thrush (NA); Bare-eyed Robin (NA); Naked-eye Thrush; Yellow-eyed Grieve (WI); American Bare-eyed Thrush
- *f*   Grive à lunettes (Ants); Grive à paupières (Ants); Grive chat (Ants); Grive à lunettes (Ants); Merle à lunettes
- *d*   Nacktaugendrossel
- *i*   Tordo occhinudi americano

10335   ***Turdus obscurus***
Passeriformes - Turdidae
- *e*   Eyebrowed Thrush; Dark Thrush; White-browed Thrush; Dusky Thrush; Grey-headed Thrush
- *f*   Grive obscure; Merle obscure; Merle pâle
- *d*   Weißbrauendrossel
- *i*   Tordo oscuro

10336   ***Turdus obsoletus***
Passeriformes - Turdidae
- *e*   Pale-vented Thrush
- *f*   Merle cul-blanc
- *d*   Blaßbauchdrossel
- *i*   Tordo scolorito

10337   ***Turdus olivaceofuscus***
Passeriformes - Turdidae
- *e*   Olivaceous Thrush; Gulf of Guinea Thrush; Sao Tomé Thrush
- *f*   Merle de Sao Tomé; Grive de Sao Tomé
- *d*   St. Thomas-Drossel
- *i*   Tordo di Sao Tomè

10338   ***Turdus olivaceus***
Passeriformes - Turdidae
- *e*   Olive Thrush; Cape Thrush (CSA); Northern Olive-Thrush
- *f*   Merle olivâtre
- *d*   Olivdrossel; Kap-Drossel; Kap-Amsel
- *i*   Tordo olivaceo

10339   ***Turdus olivaceus smithi***
Passeriformes - Turdidae
- *e*   Namaqua Thrush (CSA)

10340   ***Turdus olivater***
Passeriformes - Turdidae
- *e*   Black-hooded Thrush; Black-hooded

Ouzel; Olive-backed Thrush
f   Merle à froc noir
d   Kapuzendrossel
i   Tordo dal cappuccio nero

10341   *Turdus pallidus*
Passeriformes - Turdidae
e   Pale Thrush
f   Merle pâle; Grive pâle
d   Fahldrossel
i   Tordo pallido

10342   *Turdus pelios*
Passeriformes - Turdidae
e   African Thrush; West African Thrush
f   Merle africain; Grive grisâtre
d   Pelius-Drossel; Pelio-Drossel; Pelio-Amsel
i   Tordo africano

10343   *Turdus personus*
Passeriformes - Turdidae
e   Lesser Antillean Thrush
i   Tordo

10344   *Turdus philomelos*
Passeriformes - Turdidae
e   Song-Thrush; Common Song-Thrush; European Song-Thrush
f   Grive musicienne; Grive chanteuse; Grive commune; Grive
d   Singdrossel
i   Tordo bottaccio

10345   *Turdus pilaris*
Passeriformes - Turdidae
e   Fieldfare
f   Grive litorne
d   Wacholderdrossel; Krammetsvogel
i   Cesena

10346   *Turdus plebejus*
Passeriformes - Turdidae
e   American Mountain-Thrush; Mountain Thrush; Mountain Robin (NA); Cabanis's Thrush; American Mountain-Robin
f   Merle de montagne
d   Cabanis-Drossel
i   Tordo plebeo

10347   *Turdus plumbeus*
Passeriformes - Turdidae
e   Red-legged Thrush; Western Red-legged Thrush; Bahama Thrush (WI)
f   Merle vantard
d   Rotfußdrossel
i   Tordo plumbeo

10348   *Turdus poliocephalus*
Passeriformes - Turdidae
e   Island Thrush; Mountain Blackbird
f   Merle des îles
d   Südseedrossel
i   Tordo isolano

10349   *Turdus poliocephalus erythropleurus*
Passeriformes - Turdidae
e   Christmas Island Island-Thrush (ANZ)

10350   *Turdus poliocephalus poliocephalus*
Passeriformes - Turdidae
e   Grey-headed Blackbird (ANZ)

10351   *Turdus poliocephalus vinitinctus*
Passeriformes - Turdidae
e   Vinous-tinted Thrush (ANZ)

10352   *Turdus ravidus*
Passeriformes - Turdidae
e   Grand Cayman Thrush; Thrush (WI)
f   Merle de Grande Caiman
d   Rotaugendrossel
i   Tordo di Grand Cayman

10353   *Turdus reevei*
Passeriformes - Turdidae
e   Plumbeous-backed Thrush; Reeve's Thrush
f   Merle de Reeve
d   Mausdrossel
i   Tordo di Reeve

10354   *Turdus rubrocanus*
Passeriformes - Turdidae
e   Chestnut Thrush; Grey-headed Thrush
f   Merle à tête grise; Grive à tête grise
d   Kastaniendrossel
i   Tordo golarosso; Tordo testagrigia

10355   ***Turdus ruficollis***
Passeriformes - Turdidae
e   Red-throated Thrush; Dark-throated
Thrush (ISC)
f   Grive à gorge rousse; Merle à gorge
rousse
d   Rotkehldrossel; Bechsteindrossel
i   Tordo golanera

10356   ***Turdus ruficollis atrogularis***
Passeriformes - Turdidae
e   Dark-throated Thrush; Black-
throated Thrush
f   Grive à gorge noire; Merle à gorge
noire
d   Schwarzkehldrossel

10357   ***Turdus rufitorques***
Passeriformes - Turdidae
e   Rufous-collared Robin; Rufous-
collared Thrush
f   Merle à col roux
d   Rotnackendrossel
i   Tordo dal collare rossiccio

10358   ***Turdus rufiventris***
Passeriformes - Turdidae
e   Rufous-bellied Thrush
f   Merle à ventre roux
d   Rotbauchdrossel
i   Tordo rufiventre; Tordo dal ventre
rossiccio

10359   ***Turdus rufopalliatus***
Passeriformes - Turdidae
e   Rufous-backed Thrush; Rufous-
backed Robin (NA)
f   Merle à dos roux
d   Rotmanteldrossel
i   Tordo dorsocastano

10360   ***Turdus serranus***
Passeriformes - Turdidae
e   Black Thrush; Black Robin; Glossy-
black Thrush
d   Samtdrossel; Guatemala-Drossel
i   Merlo splendente

10361   ***Turdus subalaris***
Passeriformes - Turdidae
e   Eastern Slaty Thrush; Slaty-capped

Thrush; Behn's Thrush
f   Merle à calotte grise
i   Tordo di Behn

10362   ***Turdus swalesi***
Passeriformes - Turdidae
e   La Selle Thrush; Swale's Thrush
f   Merle (Ants); Merle de La Selle
d   Haiti-Drossel
i   Tordo di La Selle

10363   ***Turdus tephronotus***
Passeriformes - Turdidae
e   Bare-eyed Thrush; African Bare-eyed
Thrush
f   Merle cendré
d   Brillendrossel
i   Merlo col petto bianco; Tordo
occhinudi africano

10364   ***Turdus torquatus***
Passeriformes - Turdidae
e   Ring Ouzel
f   Merle à plastron; Merle des
montagnes; Merle à collier
d   Ringdrossel; Ringamsel
i   Merlo dal collare

10365   ***Turdus unicolor***
Passeriformes - Turdidae
e   Tickell's Thrush; Indian Grey-Thrush
(ISC)
f   Merle unicolore; Grive unicolore
d   Einfarbdrossel
i   Tordo di Tickell

10366   ***Turdus viscivorus***
Passeriformes - Turdidae
e   Mistlethrush
f   Grive draine
d   Misteldrossel
i   Tordela

10367   ***Turnagra capensis***
Passeriformes - Corvidae
e   Piopio; New Zealand Thrush; New
Zealand Piopio
f   Piopio de Nouvelle-Zélande
d   Piopio
i   Turnagra

10368   **Turnicidae**
            Gruiformes
   e   Button-Quails
   d   Laufhühnchen

10369   *Turnix castanota*
            Gruiformes - Turnicidae
   e   Chestnut-backed Button-Quail;
            Chestnut-backed Quail; Thick-billed
            Quail
   f   Turnix castanote
   d   Rotrückenlaufhühnchen

10370   *Turnix everetti*
            Gruiformes - Turnicidae
   e   Sumba Button-Quail, Everett's
            Button-Quail
   f   Turnix de Sumba
   d   Sumba-Laufhühnchen

10371   *Turnix hottentotta*
            Gruiformes - Turnicidae
   e   Hottentot Button-Quail; Black-
            rumped Button-Quail; Dwarf Button-
            Quail; South African Button-Quail
   f   Turnix hottentot
   d   Hottentotten-Laufhühnchen

10372   *Turnix maculosa*
            Gruiformes - Turnicidae
   e   Red-backed Button-Quail; Australian
            Hemipode; Black-backed Button-
            Quail; Black-spotted Button-Quail;
            Orange-breasted Button-Quail; Red-
            collared Button-Quail; Spotted
            Button-Quail
   f   Turnix moucheté
   d   Fleckenlaufhühnchen

10373   *Turnix melanogaster*
            Gruiformes - Turnicidae
   e   Black-breasted Button-Quail; Black-
            fronted Quail
   f   Turnix à poitrine noire
   d   Schwarzbrustlaufhühnchen

10374   *Turnix nana*
            Gruiformes - Turnicidae
   e   Black-rumped Button-Quail; Dwarf
            Button-Quail; African Button-Quail;

            Natal Button-Quail
   f   Turnix nain

10375   *Turnix nigricollis*
            Gruiformes - Turnicidae
   e   Madagascar Button-Quail; Black-
            necked Button-Quail
   f   Turnix de Madagascar
   d   Schwarzkehllaufhühnchen

10376   *Turnix ocellata*
            Gruiformes - Turnicidae
   e   Spotted Button-Quail; Ocellated
            Button-Quail; Philippine Button-
            Quail; Chestnut-breasted Button-
            Quail
   f   Turnix de Luçon
   d   Riesenlaufhühnchen

10377   *Turnix olivii*
            Gruiformes - Turnicidae
   e   Buff-breasted Button-Quail; Buff-
            backed Button-Quail; Olive's Quail;
            Robinson's Button-Quail
   f   Turnix de Robinson
   d   Robinson-Laufhühnchen

10378   *Turnix pyrrhothorax*
            Gruiformes - Turnicidae
   e   Red-chested Button-Quail; Red-
            breasted Button-Quail; Rufous-
            breasted Button-Quail; Chestnut-
            breasted Button-Quail; Yellow
            Button-Quail; Yellow Quail
   f   Turnix à poitrine rousse
   d   Rotbrustlaufhühnchen

10379   *Turnix suscitator*
            Gruiformes - Turnicidae
   e   Barred Button-Quail; Common
            Bustard-Quail (ISC); Blue-legged
            Bustard-Quail (ISC); Bustard-Quail;
            Common Bustard-Quail; Indian
            Bustard-Quail; Ceylon Bustard-
            Quail; Dusky Button-Quail; Powell's
            Button-Quail
   f   Turnix combattant
   d   Bindenlaufhühnchen

10380   *Turnix sylvatica*
            Gruiformes - Turnicidae

e Small Button-Quail; Kurrichane
Button-Quail; Andalusian Hemipode;
Little Bustard-Quail (ISC); Button-
Quail; Little Button-Quail;
Kurrichane Bustard-Quail; Common
Button-Quail; Bustard-Quail;
Bustard-Quail
f Turnix mugissant; Turnix
d'Andalousie; Turnix d'Afrique;
Turnix sylvatique
d Laufhühnchen;
Spitzschwanzlaufhühnchen;
Rotkehllaufhühnchen
i Quaglia tridattila; Quaglia tridattila
eurasiatica

10381 *Turnix tanki*
Gruiformes - Turnicidae
e Yellow-legged Button-Quail; Indian
Button-Quail (ISC)
f Turnix indien
d Rotnackenlaufhühnchen

10382 *Turnix varia*
Gruiformes - Turnicidae
e Painted Button-Quail; Varied Button-
Quail; Speckled Button-Quail; Scrub
Button-Quail
f Turnix bariolé
d Buntlaufhühnchen

10383 *Turnix velox*
Gruiformes - Turnicidae
e Little Button-Quail; Little Quail;
Australian Little Button-Quail; Swift-
flying Quail; White-bellied Quail
f Petit Turnix
d Zwerglaufhühnchen

10384 *Turnix worcesteri*
Gruiformes - Turnicidae
e Worcester's Button-Quail; Luzon
Button-Quail
d Luzonlaufhühnchen
i Quaglia tridattila di Worcester

10385 *Turtur abyssinicus*
Columbiformes - Columbidae
e Black-billed Wood-Dove; Black-
billed Blue-spotted Wood-Dove;
Abyssinian Wood-Dove

f Tourtelette d'Abyssinie; Émerauldine
à bec noir
d Abessinisches Waldtäubchen;
Erzflecktaube
i Tortora abissina

10386 *Turtur afer*
Columbiformes - Columbidae
e Blue-spotted Wood-Dove; Blue-
spotted Dove (CSA); Red-billed
Wood-Dove; Sapphire-spotted Dove;
Red-billed Blue-spotted Wood-Dove
f Tourtelette améthystine; Émeraudine
à bec rouge
d Stahlflecktaube
i Tortora macchiata di blu

10387 *Turtur brehmeri*
Columbiformes - Columbidae
e Blue-headed Wood-Dove; Blue-
headed Dove; Maiden Dove
f Tourtelette demoiselle; Tourtelette à
tête bleue
d Maidtaube
i Tortora testablu

10388 *Turtur chalcospilos*
Columbiformes - Columbidae
e Emerald-spotted Wood-Dove;
Emerald-spotted Dove (CSA);
Green-spotted Dove; Green-spotted
Wood-Dove
f Tourtelette émeraudine
d Bronzeflecktaube
I Tortora macchiata smeraldo

10389 *Tylas eduardi*
Passeriformes - Corvidae
e Tylas Vanga; Kinkimavo; Tylas
f Tylas à tête noire; Vanga à tête noire;
Tylas; Vanga tylas
d Bülbülvanga
i Kinkimavo

10390 *Tympanistria tympanistria*
Columbiformes - Columbidae
e Tambourine Dove; White-breasted
Wood-Dove; Forest Dove; White-
breasted Pigeon
f Tourterelle tambourette

d   Tamburintäubchen
i   Tortora tamburina

10391   *Tympanuchus cupido*
      Galliformes - Phasianidae
e   Greater Prairie-Chicken; Prairie-
      Chicken; Heath Hen; Pinnated
      Grouse
f   Tétras des prairies; Grande Poule-
      des-prairies (Qué); Cupidon des
      prairies; Poule des prairies
d   Präriehuhn
i   Tetraone della prateria; Tetraone
      delle praterie; Cupidonia

10392   *Tympanuchus pallidicinctus*
      Galliformes - Phasianidae
e   Lesser Prairie-Chicken; Prairie-
      Chicken; Pinnated Grouse; Lesser
      Prairie-Grouse; Lesser Pinnated
      Grouse
f   Tétras pâle
d   Schweifhuhn
i   Tetraone minore delle praterie

10393   *Tyranneutes stolzmanni*
      Passeriformes - Tyrannidae
e   Dwarf Tyrant-Manakin; Dwarf
      Manakin
f   Manakin nain
d   Zwergpipra; Zwergschnurvogel

10394   *Tyranneutes virescens*
      Passeriformes - Tyrannidae
e   Tiny Tyrant-Manakin; Tiny Manakin
f   Manakin minuscule
d   Tyrann

10395   Tyrannidae
      Passeriformes
e   Tyrant-Flycatchers
f   Tyrannidés
d   Tyranne

10396   *Tyranniscus australis*
      Passeriformes - Tyrannidae
e   Olrog's Tyrannulet
f   Tyranneau d'Olrog
d   Gelbscheitelfliegenstecher

10397   *Tyrannopsis sulphurea*
      Passeriformes - Tyrannidae
e   Sulphury Flycatcher
f   Tyran des palmiers
d   Fliegenstecher; Haschvogel;
      Schwefeltyrann
i   Tiranno sulfureo

10398   *Tyrannulus elatus*
      Passeriformes - Tyrannidae
e   Yellow-crowned Tyrannulet
f   Tyranneau roitelet
d   Tyrann
i   Tiranno piccolo corona gialla

10399   *Tyrannus albogularis*
      Passeriformes - Tyrannidae
e   White-throated Kingbird
f   Tyran à gorge blanche
d   Mönchstyrann
i   Tiranno golabianca

10400   *Tyrannus caudifasciatus*
      Passeriformes - Tyrannidae
e   Loggerhead Kingbird; Loggerhead
      Flycatcher; Petchary
f   Tyran tête-police
d   Texas-Tyrann
i   Tiranno codafasciata

10401   *Tyrannus couchii*
      Passeriformes - Tyrannidae
e   Couch's Kingbird
f   Tyran de Couch
d   Dickschnabeltyrann
i   Tiranno di Couch

10402   *Tyrannus crassirostris*
      Passeriformes - Tyrannidae
e   Thick-billed Kingbird
f   Tyran à bec épais; Tyran à gros bec
      (Qué); Tyranneau gros-bec
d   Kuba-Tyrann
i   Tiranno beccoforte

10403   *Tyrannus cubensis*
      Passeriformes - Tyrannidae
e   Giant Kingbird
f   Tyran géant
d   Grautyrann
i   Tiranno gigante di Cuba

10404 **Tyrannus dominicensis**
Passeriformes - Tyrannidae
e Grey Kingbird
f Tyran gris
d Scherentyrann
i Tiranno di Dominica

10405 **Tyrannus forficatus**
Passeriformes - Tyrannidae
e Scissor-tailed Flycatcher; Fork-tailed Flycatcher
f Tyran à longue queue; Moucherolle à longue queue; Moucherolle à queue-en-ciseaux
d Rotscheiteltyrann
i Tiranno codaforcuta meridionale

10406 **Tyrannus melancholicus**
Passeriformes - Tyrannidae
e Tropical Kingbird; Yellow Pippiree (WI); Azara's Kingbird; West Mexican Kingbird
f Tyran mélancolique
d Schneekehltyrann; Trauertyrann
i Tiranno tropicale

10407 **Tyrannus niveigularis**
Passeriformes - Tyrannidae
e Snowy-throated Kingbird
f Tyran chimu
d Gabeltyrann
i Tiranno goladineve

10408 **Tyrannus savana**
Passeriformes - Tyrannidae
e Fork-tailed Flycatcher; Scissor-tail (WI); Swallow-tailed Flycatcher
f Tyran des savanes; Tyran à queue fourchue
d Königstyrann
i Tiranno codaforcuta meridionale

10409 **Tyrannus tyrannus**
Passeriformes - Tyrannidae
e Eastern Kingbird; Kingbird
f Tyran tritri; Tyran oriental
d Arkansas-Tyrann; Königsvogel; Königssatrap
i Tiranno orientale; Tiranno

10410 **Tyrannus verticalis**
Passeriformes - Tyrannidae
e Western Kingbird; Arkansas Kingbird
f Tyran de l'Ouest
d Cassin-Tyrann
i Tiranno occidentale

10411 **Tyrannus vociferans**
Passeriformes - Tyrannidae
e Cassin's Kingbird
f Tyran de Cassin
d Eule
i Tiranno di Cassin

10412 **Tyto alba**
Strigiformes - Tytonidae
e Barn Owl; Common Barn-Owl; Jumbiebird; Patoo (WI); Scritch Owl (WI); Kritch Owl (WI); White Owl (WI); Screech-Owl (WI) (ISC); Jumpiebird (WI)
f Effraie des clochers; Effraie; Chouette des clochers; Chat-houant (Ants); Frize (Ants); Chouette effraie
d Schleiereule
i Barbagianni; Barbagianni selvatico; Gufo reale; Alloco bianco; Barbaianni; Barbagianni comune

10413 **Tyto aurantia**
Strigiformes - Tytonidae
e Bismarck Masked-Owl; New Britain Barn-Owl; Golden Barn-Owl
f Effraie dorée
d Goldeule
i Barbagianni della Nuova Britannia

10414 **Tyto capensis**
Strigiformes - Tytonidae
e African Grass-Owl; Grass-Owl; Cape Grass-Owl; Cape Owl; Abyssinian Grass-Owl
f Effraie du Cap; Chouette effraie du Cap
d Graseule; Kap-Eule
i Barbagianni delle erbe africano; Gufo delle praterie

10415 **Tyto castanops**
Strigiformes - Tytonidae

e   Tasmanian Masked-Owl; Tasmanian
      Owl
f   Effraie de Tasmanie
i   Barbagianni della Tasmania

10416   *Tyto detorta*
      Strigiformes - Tytonidae
e   Cape Verde Barn-Owl

10417   *Tyto glaucops*
      Strigiformes - Tytonidae
e   Ashy-faced Owl; Hispaniolan Barn-
      Owl; Ashy-faced Barn-Owl
f   Frize (Ants); Effraie d'Hispaniola
i   Barbagianni facciagrigia

10418   *Tyto inexpectata*
      Strigiformes - Tytonidae
e   Mynahassa Masked-Owl;
      Unexpected Barn-Owl; Mynahassa
      Barn-Owl; Unexpected Owl
f   Effraie de Mynahassa
d   Mynahassa-Eule
i   Barbagianni di Minahassa

10419   *Tyto longimembris*
      Strigiformes - Tytonidae
e   Eastern Grass-Owl; Grass-Owl;
      Australasian Grass-Owl
f   Effraie de prairie
d   Graseule
i   Barbagianni delle erbe orientale

10420   *Tyto manusi*
      Strigiformes - Tytonidae
e   Manus Masked-Owl; Manus Owl
f   Effraie de Manus
i   Barbagianni di Manus

10421   *Tyto multipunctata*
      Strigiformes - Tytonidae
e   Lesser Sooty-Owl; Sooty Owl
f   Effraie piquetée
i   Barbagianni fuligginoso minore

10422   *Tyto nigrobrunnea*
      Strigiformes - Tytonidae
e   Taliabu Masked-Owl; Taliabu Owl;
      Sula Owl; Sula Masked-Owl; Black-
      brown Owl; Taliabu Barn-Owl
f   Effraie de Taliabu

d   Taliabu-Eule
i   Barbagianni di Sula

10423   *Tyto novaehollandiae*
      Strigiformes - Tytonidae
e   Australian Masked-Owl; Masked
      Owl
f   Effraie masquée
d   Neuholland-Eule
i   Barbagianni australiano; Gufo
      mascherato

10424   *Tyto novaehollandiae castanops*
      Strigiformes - Tytonidae
e   Tasmanian Masked-Owl (ANZ)

10425   *Tyto novaehollandiae kimberli*
      Strigiformes - Tytonidae
e   Northern Masked-Owl (ANZ)

10426   *Tyto novaehollandiae melvillensis*
      Strigiformes - Tytonidae
e   Twi Islands Masked-Owl (ANZ)

10427   *Tyto rosenbergii*
      Strigiformes - Tytonidae
e   Sulawesi Owl; Rosenberg's Owl;
      Celebes Barn-Owl; Sulawesi
      Masked-Owl; Celebes Masked-Owl
f   Effraie des Célèbes
i   Barbagianni di Rosenberg;
      Barbagianni di Celebes

10428   *Tyto sororcula*
      Strigiformes - Tytonidae
e   Lesser Masked-Owl; Tanimbar Owl
f   Effraie des Tanimbar
i   Barbagianni minore

10429   *Tyto soumagnei*
      Strigiformes - Tytonidae
e   Madagascar Red-Owl; Madagascar
      Grass-Owl; Madagascar Masked-
      Owl
f   Effraie de Soumagne; Effraie jaune;
      Chouette-effraie jaune
d   Malegassen-Eule
i   Barbagianni del Madagascar

10430   *Tyto tenebricosa*
      Strigiformes - Tytonidae

e   Greater Sooty-Owl; Sooty Owl
f   Effraie ombrée; Effraie géante
d   Rußeule
i   Barbagianni fuligginoso minore;
    Barbagianni fuligginoso

10431   **Tytonidae**
      Strigiformes
e   Barn-Owls
f   Effraies; Tytonidés
d   Schleiereulen
i   Titonidi

U

10432   *Upucerthia albigula*
Passeriformes - Furnariidae
e   White-throated Earthcreeper
f   Upucerthie à gorge blanche
d   Weißkehlerdhacker
i   Rampichino di terra golabianca

10433   *Upucerthia andaecola*
Passeriformes - Furnariidae
e   Rock Earthcreeper
f   Upucerthie des rochers
d   Felsenerdhacker
i   Rampichino di terra delle Ande

10434   *Upucerthia certhioides*
Passeriformes - Furnariidae
e   Chaco Earthcreeper
f   Upucerthie du Chaco
d   Chaco-Erdhacker
i   Rampichino di terra di Chaco

10435   *Upucerthia dumetaria*
Passeriformes - Furnariidae
e   Scale-throated Earthcreeper
f   Upucerthie des buissons
d   Schuppenkehlerdhacker
i   Rampichino di terra golascagliosa

10436   *Upucerthia harterti*
Passeriformes - Furnariidae
e   Bolivian Earthcreeper
f   Upucerthie de Bolivie; Upucerthie de Hartert
d   Braunkappenerdhacker
i   Rampichino di terra della Bolivia

10437   *Upucerthia jelskii*
Passeriformes - Furnariidae
e   Plain-breasted Earthcreeper
f   Upucerthie de Jelski
d   Buscherdhacker
i   Rampichino di terra di Jelski

10438   *Upucerthia ruficauda*
Passeriformes - Furnariidae
e   Straight-billed Earthcreeper
f   Upucerthie à bec droit
d   Geradschnabelerdhacker
i   Rampichino di terra codaruggine

10439   *Upucerthia serrana*
Passeriformes - Furnariidae
e   Striated Earthcreeper
f   Upucerthie striée
d   Streifenerdhacker; Wüstenhacker
i   Rampichino di terra striato

10440   *Upucerthia validirostris*
Passeriformes - Furnariidae
e   Buff-breasted Earthcreeper
f   Upucerthie fauve
d   Braunbaucherdhacker
i   Rampichino di terra beccoforte

10441   *Upupa africana*
Coraciiformes - Upupidae
e   African Hoopoe; Hoopoe; Madagascar Hoopoe
f   Huppe d'Afrique; Huppe malgache
i   Upupa africana

10442   *Upupa epops*
Coraciiformes - Upupidae
e   Eurasian Hoopoe; Hoopoe; Common Hoopoe; Ceylon Hoopoe
f   Huppe fasciée
d   Wiedehopf
i   Upupa; Upupa eurasiatica

10443   *Upupa marginata*
Coraciiformes - Upupidae
e   Madagascar Hoopoe
f   Huppe malgache

10444   *Upupa senegalensis*
Coraciiformes - Upupidae
e   Central African Hoopoe

10445   **Upupidae**
Coraciiformes
e   Hoopoes
f   Huppes; Upupidés
d   Wiedehöpfe
i   Upupidi; Upupi

10446 **_Uraeginthus angolensis_**
Passeriformes - Estrildidae
e Blue-breasted Cordonbleu; Angola
Cordonbleu; Angolan Cordonbleu;
Blue-breasted Waxbill; Blue
Waxbill; Blue-cheeked Cordonbleu;
Cordonbleu; Southern Cordonbleu
f Cordonbleu de I'Angola; Astrild
bleu; Cordonbleu d'Angola
d Angola Schmetterlingsfink;
Blauastrild
i Cordon blu dell'Angola

10447 **_Uraeginthus angolensis niassensis_**
Passeriformes - Estrildidae
e Southern Cordonbleu; Blue Waxbill
(CSA)
d Schmetterlingsfink

10448 **_Uraeginthus bengalus_**
Passeriformes - Estrildidae
e Red-cheeked Cordonbleu; Red-
cheeked Waxbill; Red-cheeked Blue-
Waxbill; Cordonbleu
f Cordonbleu à joues rouges; Bengali
cordonbleu
d Schmetterlingsfink;
Schmetterlingsastrild
i Cordon blu guancerosso

10449 **_Uraeginthus cyanocephalus_**
Passeriformes - Estrildidae
e Blue-capped Cordonbleu; Blue-
headed Cordonbleu; Blue-capped
Waxbill; Blue-headed Waxbill
f Cordonbleu cyanocéphale; Astrild à
tête bleue; Cordonbleu a calotte bleue
d Blaukopfschmetterlingsfink;
Blauköpfiger Schmetterlingsfink
i Cordon blu dal cappuccio

10450 **_Uraeginthus granatina_**
Passeriformes - Estrildidae
e Common Grenadier; Grenadier;
Violet-eared Cordonbleu; Violet-
eared Waxbill (CSA); Grenadier
Waxbill
f Cordonbleu grenadin; Grenadin;
Astrild grenadin; Cordonbleu à joues
violettes

d Granatastrild; Blaubäckchen
i Granatino comune; Granatino

10451 **_Uraeginthus ianthinogaster_**
Passeriformes - Estrildidae
e Purple Grenadier; Purple Waxbill;
Purple-bellied Waxbill
f Cordonbleu violacé; Grenadin à
poitrine bleue
d Veilchenastrild; Purpurgranatastrild;
Blaubäuchiger Granatastrild
i Nidiaceo dal petto azzurro; Granatino
purpureo

10452 **_Uragus sibiricus_**
Passeriformes - Fringillidae
e Long-tailed Rosefinch; Siberian
Rosefinch
f Roselin à queue longue; Roselin à
longue queue
d Meisengimpel
i Ciuffolotto siberiano

10453 **_Uratelornis chimaera_**
Coraciiformes - Brachypteraciidae
e Long-tailed Ground-Roller
f Brachyptérolle à longue queue
d Langschwanzerdracke
i Coracia terricola codalunga

10454 **_Uria aalge_**
Charadriiformes - Alcidae
e Common Murre (NA); Guillemot;
Thin-billed Murre; Murre; Common
Guillemot; Atlantic Murre
f Guillemot de Troil; Guillemot
marmette; Marmette de Troil
d Trottellumme
i Uria; Uria comune

10455 **_Uria lomvia_**
Charadriiformes - Alcidae
e Thick-billed Murre (NA); Brünnich's
Guillemot; Arctic Guillemot
f Guillemot de Brünnich; Marmette de
Brünnich
d Dickschnabellumme
i Uria di Brünnich

10456 **_Urochroa bougueri_**
Apodiformes - Trochilidae

e   White-tailed Hillstar
f   Colibri de Bouguer
d   Glanzfleckenkolibri
i   Stella dei monti codabianca

10457   ***Urocissa caerulea***
Passeriformes - Corvidae
e   Formosan Magpie; Formosan Blue-Pie; Formosan Blue-Magpie; Taiwan Magpie; Taiwan Blue-Magpie
f   Pirolle de Taïwan; Pie bleue de Taïwan; Pie bleue de Formose
d   Dickschnabelkitta
i   Gazza azzurra di Taiwan

10458   ***Urocissa erythrorhyncha***
Passeriformes - Corvidae
e   Blue Magpie; Red-billed Magpie (ISC); Red-billed Blue-Magpie; Occipital Blue-Pie; Chinese Blue-Pie
f   Pirolle à bec rouge; Pirolle de la Chine; Pie bleue à bec rouge
d   Rotschnabelkitta; Rotschnabelschweifkitta
i   Gazza azzurra beccorosso

10459   ***Urocissa flavirostris***
Passeriformes - Corvidae
e   Gold-billed Magpie; Gold-billed Blue-Magpie; Yellow-billed Blue-Magpie; Golden-billed Magpie; Golden-billed Blue-Magpie; Black-headed Magpie
f   Pirolle à bec jaune; Pie bleue à bec jaune
d   Gelbschnabelkitta; Gelbschnabelschweifkitta
i   Gazza azzurra beccogiallo

10460   ***Urocissa ornata***
Passeriformes - Corvidae
e   Ceylon Magpie; Sri Lanka Magpie; Ceylon Blue-Magpie; Ceylon Jay; Ceylon Hunting-Crow; Sri Lanka Blue-Magpie
f   Pirolle de Ceylan; Pie bleue ornée
d   Schmuckkitta
i   Gazza azzurra di Sri-Lanka

10461   ***Urocissa whiteheadi***
Passeriformes - Corvidae

e   White-winged Magpie; Whitehead's Magpie; Grey Magpie; Whitehead's Hunting-Crow; Whitehead's Blue-Magpie; White-winged Blue-Magpie
f   Pirolle de Whitehead; Pie bleue de Whitehead
d   Graubauchkitta
i   Gazza azzurra alibianche

10462   ***Urocolius indicus***
Coliiformes - Coliidae
e   Red-faced Mousebird
f   Coliou quiriva
d   Rotzügelmausvogel; Brillenmausvogel
i   Uccello topo facciarossa

10463   ***Urocolius macrourus***
Coliiformes - Coliidae
e   Blue-naped Mousebird
f   Coliou huppé
d   Blaunackenmausvogel
i   Uccello topo nucablu

10464   ***Urocynchramus pylzowi***
Passeriformes - Fringillidae
e   Przewalski's Finch; Przewalski's Rosefinch; Pink-tailed Bunting; Rose Bunting
f   Bruselin de Przewalski
d   Rosenschwanz
i   Zigolo di Przewalski

10465   ***Uroglaux dimorpha***
Strigiformes - Strigidae
e   Papuan Boobook; Papuan Hawk-Owl
f   Ninoxe papoue
d   Rundflügelkauz
i   Ulula della Nuova Guinea; Civetta sparviero papuana

10466   ***Urolais epichlora***
Passeriformes - Cisticolidae
e   Green Longtail
f   Prinia verte
d   Langschwanzprinie
i   Codalunga verde

10467   ***Uromyias agilis***
Passeriformes - Tyrannidae
e   Agile Tit-Tyrant

e   White-tailed Lapwing; White-tailed
    Plover
f   Vanneau à queue blanche
d   Weißschwanzkiebitz;
    Weißschwanzsteppenkiebitz
i   Pavoncella codabianca

10491   *Vanellus lugubris*
    Charadriiformes - Charadriidae
e   Senegal Lapwing; Senegal Plover
    (CSA); Lesser Black-winged Plover
    (CSA); Lesser Black-winged
    Lapwing
f   Vanneau terne; Vanneau demideuil
d   Trauerkiebitz
i   Pavoncella lugubre

10492   *Vanellus macropterus*
    Charadriiformes - Charadriidae
e   Javanese Lapwing; Javanese
    Wattled-Plover; Javan Lapwing;
    Black-thighed Wattled-Lapwing;
    Sunda Lapwing
f   Vanneau hirondelle
d   Schwarzbauchkiebitz
i   Pavoncella di Giava

10493   *Vanellus malarbaricus*
    Charadriiformes - Charadriidae
e   Yellow-wattled Lapwing; Yellow-
    wattled Plover
f   Vanneau de Malabar
d   Malabar-Kiebitz
i   Pavoncella del Malabar

10494   *Vanellus melanocephalus*
    Charadriiformes - Charadriidae
e   Spot-breasted Lapwing; Spot-
    breasted Plover
f   Vanneau d'Abyssinie
d   Strichelkiebitz
i   Pavoncella abissina

10495   *Vanellus melanopterus*
    Charadriiformes - Charadriidae
e   Black-winged Lapwing; Black-
    winged Plover (CSA)
f   Vanneau à ailes noires
d   Schwarzflügelkiebitz
i   Pavoncella alinere

10496   *Vanellus miles*
    Charadriiformes - Charadriidae
e   Masked Lapwing; Masked Plover;
    Northern Masked-Plover; Spur-
    winged Plover; Australian Spur-
    winged Plover; Wattled Plover
f   Vanneau soldat
d   Soldatenkiebitz;
    Schwarznackenkiebitz
i   Pavoncella mascherata

10497   *Vanellus resplendens*
    Charadriiformes - Charadriidae
e   Andean Lapwing
f   Vanneau des Andes
d   Anden-Kiebitz
i   Pavoncella delle Ande

10498   *Vanellus senegallus*
    Charadriiformes - Charadriidae
e   Wattled Lapwing; Wattled Plover
    (CSA); Senegal Wattled-Plover;
    Senegal Wattled-Lapwing; African
    Wattled-Plover
f   Vanneau du Sénégal
d   Senegal-Kiebitz
i   Pavoncella del Senegal

10499   *Vanellus senegallus lateralis*
    Charadriiformes - Charadriidae
e   African Wattled-Plover (CSA)

10500   *Vanellus spinosus*
    Charadriiformes - Charadriidae
e   Spur-winged Lapwing; Spur-winged
    Plover; Spurwing Plover
f   Vanneau éperonné; Vanneau à
    éperons
d   Spornkiebitz; Sporenkiebitz
i   Pavoncella armata; Pavoncella
    spinosa

10501   *Vanellus superciliosus*
    Charadriiformes - Charadriidae
e   Brown-chested Lapwing; Brown-
    chested Plover (CSA); Brown-
    chested Wattled-Plover
f   Vanneau à poitrine châtaine;
    Vanneau caronculé
d   Rotbrustkiebitz
i   Pavoncella pettobruno

10502 *Vanellus tectus*
Charadriiformes - Charadriidae
- *e* Black-headed Lapwing; Black-headed Plover; Crested Wattled-Plover
- *f* Vanneau à tête noire; Vanneau coiffé
- *d* Schwarzkopfkiebitz; Schwarzschopfkiebitz
- *i* Pavoncella testanera

10503 *Vanellus tricolor*
Charadriiformes - Charadriidae
- *e* Banded Lapwing; Banded Plover; Plain Plover; Plain Banded-Plover; Black-breasted Plover
- *f* Vanneau tricolore
- *d* Schildkiebitz
- *i* Pavoncella tricolore

10504 *Vanellus vanellus*
Charadriiformes - Charadriidae
- *e* Northern Lapwing; Plover; Green Plover; Peewit; Lapwing; Common Northern Lapwing; Common Lapwing; Eurasian Lapwing
- *f* Vanneau huppé
- *d* Kiebitz
- *i* Pavoncella; Fifa; Miciola; Mivola; Pavoncella paleartica

10505 *Vanga curvirostris*
Passeriformes - Corvidae
- *e* Hook-billed Vanga
- *f* Vanga écorcheur
- *d* Hakenvanga
- *i* Vanga beccouncinato

10506 **Vangidae**
Passeriformes
- *e* Vangas; Vanga-Shrikes
- *f* Vangidés
- *d* Vangas; Vangawürger; Blauwürger
- *i* Vangidi

10507 *Veniliornis affinis*
Piciformes - Picidae
- *e* Red-stained Woodpecker
- *f* Pic affin; Pic de Selys
- *d* Blutflügelspecht
- *i* Picchio di Selys

10508 *Veniliornis callonotus*
Piciformes - Picidae
- *e* Scarlet-backed Woodpecker
- *f* Pic rubin
- *d* Scharlachrückenspecht
- *i* Picchio dorsoscarlatto

10509 *Veniliornis cassini*
Piciformes - Picidae
- *e* Golden-collared Woodpecker
- *f* Pic de Cassin
- *d* Goldnackenspecht
- *i* Picchio di Cassin

10510 *Veniliornis chocoensis*
Piciformes - Picidae
- *e* Choco Woodpecker
- *f* Pic du Choco
- *i* Picchio del Choco

10511 *Veniliornis dignus*
Piciformes - Picidae
- *e* Yellow-vented Woodpecker; Yellow-fronted Woodpecker
- *f* Pic à ventre jaune
- *d* Gelbbauchspecht
- *i* Picchio dal sottocoda giallo

10512 *Veniliornis frontalis*
Piciformes - Picidae
- *e* Dot-fronted Woodpecker
- *f* Pic étoilé
- *d* Perlstirnspecht
- *i* Picchio stellato

10513 *Veniliornis fumigatus*
Piciformes - Picidae
- *e* Smoky-brown Woodpecker; Brown Woodpecker
- *f* Pic enfumé
- *d* Rußspecht
- *i* Picchio fumigato

10514 *Veniliornis kirkii*
Piciformes - Picidae
- *e* Red-rumped Woodpecker
- *f* Pic à croupion rouge; Pic de Kirk
- *d* Blutbürzelspecht
- *i* Picchio di Kirk

10515   *Veniliornis maculifrons*
        Piciformes - Picidae
  e   Yellow-eared Woodpecker
  f   Pic à oreilles d'or; Pic à oreille d'or
  d   Goldohrspecht
  i   Picchio frontemaculata

10516   *Veniliornis nigriceps*
        Piciformes - Picidae
  e   Bar-bellied Woodpecker
  f   Pic à ventre barré; Pic d'Orbigny
  d   Bindenbauchspecht
  i   Picchio di D'Orbigny

10517   *Veniliornis passerinus*
        Piciformes - Picidae
  e   Little Woodpecker
  f   Pic passerin
  d   Sperlingsspecht
  i   Picchio passerino

10518   *Veniliornis sanguineus*
        Piciformes - Picidae
  e   Blood-coloured Woodpecker (NA)
  f   Pic rougeâtre
  d   Blutrückenspecht
  i   Picchio sanguigno

10519   *Veniliornis spilogaster*
        Piciformes - Picidae
  e   White-spotted Woodpecker
  f   Pic aspergé
  d   Perlbauchspecht
  i   Picchio ventremacchiato

10520   *Vermivora bachmanii*
        Passeriformes - Fringillidae
  e   Bachman's Warbler
  f   Paruline de Bachman
  d   Gelbstirnwaldsänger
  i   Parula di Bachman

10521   *Vermivora celata*
        Passeriformes - Fringillidae
  e   Orange-crowned Warbler
  f   Paruline verdâtre; Fauvette verdâtre
  d   Orangefleckwaldsänger
  i   Parula corona arancio

10522   *Vermivora chrysoptera*
        Passeriformes - Fringillidae

  e   Golden-winged Warbler
  f   Paruline à ailes dorées; Paruline aile-
        d'or; Fauvette à ailes dorées
  d   Goldflügelraupenfresser;
        Goldflügelwaldsänger
  i   Parula alidorate

10523   *Vermivora crissalis*
        Passeriformes - Fringillidae
  e   Colima Warbler
  f   Paruline de Colima
  d   Colima-Waldsänger
  i   Parula di Colima

10524   *Vermivora lawrencii*
        Passeriformes - Fringillidae
  e   Lawrence's Warbler

10525   *Vermivora leucobronchialis*
        Passeriformes - Fringillidae
  e   Brewster's Warbler

10526   *Vermivora luciae*
        Passeriformes - Fringillidae
  e   Lucy's Warbler
  f   Paruline de Lucy
  d   Rotbürzelwaldsänger
  i   Parula di Lucy

10527   *Vermivora peregrina*
        Passeriformes - Fringillidae
  e   Tennessee Warbler
  f   Paruline obscure; Fauvette obscure
  d   Brauenwaldsänger
  i   Parula di Tenesee; Beccaverme del
        Tennessee

10528   *Vermivora pinus*
        Passeriformes - Fringillidae
  e   Blue-winged Warbler
  f   Paruline à ailes bleues; Petite Chitte
        aile-bleue (Ants); Fauvette à ailes
        bleues
  d   Blauflügelwaldsänger
  i   Parula aliazzurre

10529   *Vermivora ruficapilla*
        Passeriformes - Fringillidae
  e   Nashville Warbler; Grey-headed
        Warbler
  f   Paruline à joues grises; Fauvette à

joues grises
d   Rubinfleckwaldsänger
i   Parula di Nashville

10530   *Vermivora virginiae*
Passeriformes - Fringillidae
e   Virginia's Warbler
f   Paruline de Virginia; Fauvette de
Virginia
d   Virgininia-Waldsänger
i   Parula di Virginia

10531   *Vestiaria coccinea*
Passeriformes - Fringillidae
e   Iiwi
f   Iiwi rouge
d   Iiwi; Scharlachkleidervogel
i   Iiwi

10532   *Vidua camerunensis*
Passeriformes - Estrildidae
e   Cameroon Indigobird

10533   *Vidua chalybeata*
Passeriformes - Estrildidae
e   Village Indigobird; Common
Indigobird (CSA); Indigobird (CSA);
Steel-blue Widow-Finch (CSA);
Senegal Indigo Finch; Village
Widowfinch; Green Indigobird; Red-
billed Firefinch-Indigobird; Purple
Indigobird; South African Indigobird
f   Combassou du Sénégal
d   Rotschnabelige Atlaswitwe;
Rotfußatlaswitwe
i   Uccello indaco dei villaggi

10534   *Vidua codringtoni*
Passeriformes - Estrildidae
e   Green Indigobird; Green Widow-
Finch (CSA); Indigobird (CSA);
Twinspot Indigobird; Codrington's
Indigobird
f   Combassou vert
d   Grüne Atlaswitwe

10535   *Vidua fischeri*
Passeriformes - Estrildidae
e   Straw-tailed Whydah; Fischer's
Whydah; Straw-tailed Widow
f   Veuve de Fischer; Veuve à queue de

paille
d   Strohwitwe; Fischer-Witwe
i   Vedova di Fischer

10536   *Vidua funerea*
Passeriformes - Estrildidae
e   Variable Indigobird; Black Widow-
Finch (CSA); Indigobird (CSA);
Black Indigobird; African Firefinch-
Indigobird; Dusky Indigobird
f   Combassou noir; Combassou
variable
d   Purpuratlaswitwe;
Weißfußatlaswitwe
i   Uccello indaco variabile

10537   *Vidua hypocherina*
Passeriformes - Estrildidae
e   Steel-blue Whydah; Steel-blue
Widow
f   Veuve métallique
d   Glanzwitwe
i   Vedova azzurro acciaio

10538   *Vidua interjecta*
Passeriformes - Estrildidae
e   Long-tailed Paradise-Whydah;
Nigeria Paradise-Widow; Nigeria
Paradise-Whydah; Uelle Paradise-
Whydah; Uelle Paradise-Widow
f   Veuve nigérienne; Veuve d'Uelle
d   Langschwanzparadieswitwe
i   Vedova codalunga

10539   *Vidua larvaticola*
Passeriformes - Estrildidae
e   Black-faced Firefinch-Indigobird
f   Combassou de Baka; Combassou
baka
i   Uccello indaco di Baka

10540   *Vidua macroura*
Passeriformes - Estrildidae
e   Pin-tailed Whydah; King of Six
(CSA); Pin-tailed Widow
f   Veuve dominicaine
d   Dominikanerwitwe
i   Vedova coda a spilli

10541   *Vidua maryae*
Passeriformes - Estrildidae

e   Thick-billed Vireo; Chick-of-the-village (WI); Shear-bark (WI)
f   Oiseau canne (Ants); Viréo à bec fort
d   Dickschnabelvireo
i   Vireo beccoforte

10569   *Vireo flavifrons*
Passeriformes - Vireonidae
e   Yellow-throated Vireo
f   Viréo à gorge jaune
d   Gelbkehlvireo
i   Vireo frontegialla

10570   *Vireo flavoviridis*
Passeriformes - Vireonidae
e   Yellow-green Vireo
f   Viréo jaune-verdâtre
i   Vireo giallo-verde

10571   *Vireo gilvus*
Passeriformes - Vireonidae
e   Eastern Warbling-Vireo (NA); Warbling Vireo
f   Viréo mélodieux
d   Sängervireo
i   Vireo canoro

10572   *Vireo gracilirostris*
Passeriformes - Vireonidae
e   Noronha Vireo
f   Viréo de Noronha
i   Vireo di Noronha

10573   *Vireo griseus*
Passeriformes - Vireonidae
e   White-eyed Vireo
f   Viréo aux yeux blancs
d   Weißaugenvireo
i   Vireo occhibianchi

10574   *Vireo gundlachii*
Passeriformes - Vireonidae
e   Cuban Vireo
f   Viréo de Cuba
d   Grundlach-Vireo
i   Vireo di Cuba

10575   *Vireo huttoni*
Passeriformes - Vireonidae
e   Hutton's Vireo
f   Viréo de Hutton

d   Hutton-Vireo
i   Vireo di Hutton

10576   *Vireo hypochryseus*
Passeriformes - Vireonidae
e   Golden Vireo
f   Viréo doré
d   Goldbauchvireo
i   Vireo dorato

10577   *Vireo latimeri*
Passeriformes - Vireonidae
e   Puerto Rican Vireo; Latimer's Vireo (WI)
f   Viréo de Porto Rico
d   Braunscheitelvireo
i   Vireo di Portorico

10578   *Vireo leucophrys*
Passeriformes - Vireonidae
e   Brown-capped Vireo; Warbling Vireo
f   Viréo à calotte brune
d   Braunkappenvireo
i   Vireo capobruno

10579   *Vireo magister*
Passeriformes - Vireonidae
e   Yucatan Vireo; Sweet Bridget (WI)
f   Viréo du Yucatan
d   Yucatan-Vireo
i   Vireo dello Yucatan

10580   *Vireo masteri*
Passeriformes - Vireonidae
e   Choco Vireo

10581   *Vireo modestus*
Passeriformes - Vireonidae
e   Jamaican Vireo; Jamaican White-eyed Vireo; White-eyed Vireo (WI); Sewi-sewi (WI)
f   Viréo de la Jamaïque
d   Jamaika-Vireo
i   Vireo della Giamaica

10582   *Vireo nanus*
Passeriformes - Vireonidae
e   Flat-billed Vireo
f   Viréo d'Hispaniola

d Schnäppervireo
i Vireo nano

10583 *Vireo nelsoni*
Passeriformes - Vireonidae
e Dwarf Vireo; Nelson's Vireo
f Viréo nain
d Zwergvireo
i Vireo di Nelson

10584 *Vireo olivaceus*
Passeriformes - Vireonidae
e Red-eyed Vireo; Chivi Vireo
f Viréo à oeil rouge; Viréo aux yeux
      rouges
d Rotaugenvireo
i Vireo occhirossi

10585 *Vireo osburni*
Passeriformes - Vireonidae
e Blue Mountains Vireo; Blue
      Mountain Vireo (NA)
f Viréo d'Osburn
d Osburn-Vireo
i Vireo blu

10586 *Vireo pallens*
Passeriformes - Vireonidae
e Mangrove Vireo; Pale Vireo
f Viréo des mangroves
d Mangrovevireo
i Vireo delle mangrovie

10587 *Vireo perquisitor*
Passeriformes - Vireonidae
e Veracruz Vireo

10588 *Vireo philadelphicus*
Passeriformes - Vireonidae
e Philadelphia Vireo
f Viréo de Philadelphie
d Philadelphia-Vireo; Schlichtvireo
i Vireo Filadelfia

10589 *Vireo plumbeus*
Passeriformes - Vireonidae
e Plumbeous Vireo; Solitary Vireo
f Viréo plombé
i Vireo plumbeo

10590 *Vireo semiflavus*
Passeriformes - Vireonidae
e Maya Vireo

10591 *Vireo solitarius*
Passeriformes - Vireonidae
e Blue-headed Vireo; Solitary Vireo
f Viréo à tête bleue
d Graukopfvireo
i Vireo solitario

10592 *Vireo swainsonii*
Passeriformes - Vireonidae
e Western Warbling-Vireo; Warbling
      Vireo
f Viréo de Swainson
i Vireo di Swainson

10593 *Vireo vicinior*
Passeriformes - Vireonidae
e Grey Vireo
f Viréo gris
d Grauvireo
i Vireo grigio

10594 *Vireo victoriae*
Passeriformes - Vireonidae
e Cape Warbling-Vireo

10595 Vireolaniidae
Passeriformes
e Shrike-Vireos; Vireos
f Vireolaniidés
d Laubwürger; Vireos

10596 *Vireolanius eximius*
Passeriformes - Vireolaniidae
e Yellow-browed Shrike-Vireo;
      Venezuelan Shrike-Vireo
f Smaragdan à sourcils jaunes
i Vireo averla dai sopraccigli gialli

10597 *Vireolanius leucotis*
Passeriformes - Vireolaniidae
e Slaty-capped Shrike-Vireo; Grey-
      capped Shrike-Vireo
f Smaragdan oreillard
d Schieferkopfvireo
i Vireo averla capoardesia

10598 *Vireolanius melitophrys*
Passeriformes - Vireolaniidae
e Chestnut-sided Shrike-Vireo;
Highland Shrike-Vireo
f Smaragdan ceinturé
d Brustbandvireo; Bindenlaubwürger
i Vireo averla fianchicastani

10599 *Vireolanius pulchellus*
Passeriformes - Vireolaniidae
e Green Shrike-Vireo
f Smaragdan émeraude
d Smaragdvireo; Smaragdlaubwürger
i Vireo averla verde

10600 **Vireonidae**
Passeriformes
e Vireos; Shrike-Vireos; Peppershrikes
f Viréos; Viréonidés
d Laubwürger; Vireos
i Vireonidi

10601 *Vireosylva propinqua*
Passeriformes - Vireonidae
e Vera Paz Vireo

10602 *Viridonia stejnegeri*
Passeriformes - Drepanididae
e Kauai Amakihi
f Amakihi de Stejneger
i Amakihi di Kauai

10603 *Volatinia jacarina*
Passeriformes - Emberizidae
e Blue-black Grassquit; Blue-black
See-see (WI); Johnny-jump-up (WI);
Blue-black Seedeater
f Jacarini noir
d Jacarini; Jacarinifink
i Fringuello negrillo

10604 *Vultur gryphus*
Falconiformes - Cathartidae
e Andean Condor; Condor
f Condor des Andes
d Kondor
i Condor delle Ande; Condor

W

e   Bare-eyed White-eye; Woodford's
    White-eye; Rennel White-eye;
    Eyebrowed White-eye
f   Zostérops de Woodford
d   Woodfords Brillenvogel

10605   *Wetmorethraupis sterrhopteron*
          Passeriformes - Fringillidae
e   Orange-throated Tanager, Wetmore's
    Tanager
f   Tangara à gorge orangée
d   Veilchenschultertangare
i   Tangara aliazzure di Wetmore

10606   *Wilsonia canadensis*
          Passeriformes - Parulidae
e   Canada Warbler
f   Paruline du Canada; Fauvette du
    Canada
d   Kanada-Waldsänger
i   Parula canadese

10607   *Wilsonia citrina*
          Passeriformes - Parulidae
e   Hooded Warbler
f   Paruline à capuchon; Fauvette à
    capuchon
d   Kapuzenwaldsänger;
    Kappenwaldsänger
i   Parula dal cappuccio

10608   *Wilsonia pusilla*
          Passeriformes - Parulidae
e   Wilson's Warbler; Black-capped
    Warbler; Pileolated Warbler
f   Paruline à calotte noire; Fauvette à
    calotte noire
d   Mönchswaldsänger
i   Parula di Wilson

10609   *Woodfordia lacertosa*
          Passeriformes - Zosteropidae
e   Sanford's White-eye
f   Zostérops de Sanford
d   Sanford-Brillenvogel
i   Occhialino di Sanford

10610   *Woodfordia superciliosa*
          Passeriformes - Zosteropidae

X

10611 Xanthocephalus xanthocephalus
Passeriformes - Fringillidae
e Yellow-headed Blackbird
f Carouge à tête jaune; Étourneau à
tête jaune
d Gelbkopfstarling; Brillenstärling
i Ittero testagialla

10612 Xanthomyza phrygia
Passeriformes - Meliphagidae
e Regent Honeyeater; Embroidered
Honeyeater; Warty-faced Honeyeater
f Xanthomyza phrygia; Méliphage
régent
d Warzenhonigfresser;
Warzenhonigesser
i Mangiamiele frigio

10613 Xanthotis flaviventer
Passeriformes - Meliphagidae
e Tawny-breasted Honeyeater; Tawny-
breasted Xanthotis; Streaked
Honeyeater; Brown Honeyeater;
Brown Xanthotis
f Xanthotis flaviventer; Méliphage à
ventre fauve
d Ockerbrusthonigfresser
i Mangiamiele pettogiallo

10614 Xanthotis flaviventer saturatior
Passeriformes - Meliphagidae
e Torres Strait Tawny-breasted
Honeyeater (ANZ)

10615 Xanthotis macleayana
Passeriformes - Meliphagidae
e Macleay's Honeyeater; Yellow-
streaked Honeyeater; Macleay's
Xanthotis; Buff-striped Honeyeater
f Xanthotis macleayana; Méliphage de
Macleay
d Kappenhonigfresser
i Mangiamiele di MacLeay

10616 Xanthotis polygramma
Passeriformes - Meliphagidae
e Spotted Honeyeater; Spotted
Xanthotis; Many-spotted Honeyeater;
Gray's Honeyeater
f Xanthotis polygramma; Méliphage
moucheté
d Drosselhonigfresser
i Mangiamiele maculato

10617 Xanthotis provocator
Passeriformes - Meliphagidae
e Kadavu Honeyeater; Yellow-faced
Honeyeater; Kadavu Xanthotis;
Yellow-faced Xanthotis; Kadavu
Honeyeater; Kadavu Xanthotis
f Xanthotis provocator; Foulehaio
provocator; Méliphage de Kandavu
d Goldaugenhonigfresser
i Mangiamiele di Kadavu

10618 Xenerpestes minlosi
Passeriformes - Furnariidae
e Double-banded Greytail; Double-
banded Softtail
f Queue-grise des feuilles
d Bindengrauschwanz
i Codagrigia bifasciato

10619 Xenerpestes singularis
Passeriformes - Furnariidae
e Equatorial Greytail; Singular
Greytail; Equatorial Softtail; Singular
Softtail
f Queue-grise d'Équateur
d Streifengrauschwanz
i Codagrigia equatoriale

10620 Xenicidae
Passeriformes
e New Zealand Wrens
f Xenicidés
d Maorischlüpfer; Neuseeland-Pittas;
Neuseeland-Schlüpfer
i Xenicidi

10621 Xenicus gilviventris
Passeriformes - Xenicidae
e South Island Wren; Rock Wren
f Xénique des rochers

d   Felsschlüpfer
i   Xenico di roccia

10622   *Xenicus longipes*
Passeriformes - Xenicidae
e   Bush Wren
f   Xénique des buissons
d   Neuseeland-Schlüpfer;
Baumscheinzaunkönig
i   Xenico dei cespugli

10623   *Xenicus lyalli*
Passeriformes - Xenicidae
e   Stephen Island Wren; Stephen's
Island Rock-Wren; Travers's Wren
f   Xénique de Stephen
d   Stephen-Schlüpfer; Scheinzaunkönig
i   Xenico dell'Isola Stephen

10624   *Xenocopsychus ansorgei*
Passeriformes - Muscicapidae
e   Angola Cave-Chat; Ansorge's Robin;
Cave-Chat; Angola Robin-Chat;
Ansorge's Cave-Chat; Angolan Cave-
Chat
f   Cossyphe des grottes
d   Höhlenrötel
i   Pettirosso di Ansorge

10625   *Xenodacnis parina*
Passeriformes - Fringillidae
e   Tit-like Dacnis; Tit Dacnis
f   Xénodacnis mésange
d   Meisenpitpit
i   Dacne cincia

10626   *Xenoglaux loweryi*
Strigiformes - Strigidae
e   Long-whiskered Owlet
f   Chevêchette nimbée
d   Peruaner Kauz
i   Civetta di Lowery

10627   *Xenoligea montana*
Passeriformes - Fringillidae
e   White-winged Warbler; White-
winged Ground-Warbler; Chapman's
Warbler
f   Petite Chitte (Ants); Petit Quatre-
yeux (Ants); Paruline quatre-yeux

d   Spiegelwaldsänger
i   Parula alibianchi

10628   *Xenoperdix udzungwensis*
Galliformes - Phasianidae
e   Udzungwa Forest Partridge; Udzunga
Partridge

10629   *Xenopipo atronitens*
Passeriformes - Tyrannidae
e   Black Manakin
f   Manakin noir
d   Mohrenpipra
i   Manachino nero

10630   *Xenopirostris damii*
Passeriformes - Corvidae
e   Van Dam's Vanga
f   Vanga de Van Dam
d   Schlegel-Vanga
i   Vanga di Van Dam

10631   *Xenopirostris polleni*
Passeriformes - Corvidae
e   Pollen's Vanga
f   Vanga de Pollen
d   Pollen-Vanga
i   Vanga di Pollen

10632   *Xenopirostris xenopirostris*
Passeriformes - Corvidae
e   Lafresnaye's Vanga
f   Vanga de Lafresnaye
d   Schmalschnabelvanga
i   Vanga di Lafresnaye

10633   *Xenops milleri*
Passeriformes - Furnariidae
e   Rufous-tailed Xenops
f   Sittine à queue rousse
d   Rotschwanzbaumspäher
i   Xenope di Miller

10634   *Xenops minutus*
Passeriformes - Furnariidae
e   Plain Xenops
f   Sittine brune
d   Sparrman-Steigschnabel
i   Xenope minuto

10635 **Xenops rutilans**
Passeriformes - Furnariidae
e Streaked Xenops
f Sittine striée
d Rötelsteigschnabel
i Xenope striato

10636 **Xenops tenuirostris**
Passeriformes - Furnariidae
e Slender-billed Xenops
f Sittine des rameaux
d Streifenschwanzsteigschnabel
i Xenope beccosottile

10637 **Xenopsaris albinucha**
Passeriformes - Tyrannidae
e White-naped Xenopsaris; Xenopsaris
f Bécarde à nuque blanche; Xenopsaris
à nuque blanche
d Kappentachuri
i Tiranno nucabianca

10638 **Xenornis setifrons**
Passeriformes - Thamnophilidae
e Speckled Antshrike; Speckle-
breasted Antshrike; Spiny-faced
Antshrike; Grey-faced Antshrike
f Batara masqué
d Streifenbrustwürgerling
i Averla formichiera pettomacciato

10639 **Xenospingus concolor**
Passeriformes - Fringillidae
e Slender-billed Finch
f Xénospingue uniforme
d Feinschnabelämmerling
i Fringuello beccofino

10640 **Xenospiza baileyi**
Passeriformes - Fringillidae
e Sierra Madre Sparrow
f Bruant des sierras
d Sierraammer
i Passero della Sierra Madre

10641 **Xenotriccus callizonus**
Passeriformes - Tyrannidae
e Belted Flycatcher
f Moucherolle ceinturée
d Gürteltyrann
i Tiranno ricurvo

10642 **Xenotriccus mexicanus**
Passeriformes - Tyrannidae
e Pileated Flycatcher; Crested Wood
Pewee
f Moucherolle aztèque
d Schopftyrann
i Tiranno pileato

10643 **Xiphidiopicus percussus**
Piciformes - Picidae
e Cuban Green-Woodpecker; Cuban
Woodpecker
f Pic poignardé
d Blutfleckspecht
i Picchio verde di Cuba

10644 **Xiphirhynchus superciliaris**
Passeriformes - Sylviidae
e Slender-billed Scimitar-Babbler
f Pomatorhin à bec fin
d Dünnschnabelsäbler; Sicheltimalie
i

10645 **Xiphocolaptes albicollis**
Passeriformes - Dendrocolaptidae
e White-throated Woodcreeper; White-
throated Creeper; Snethlage's
Woodcreeper
f Grimpar à gorge blanche
d Weißkehlbaumsteiger
i Garrulo scimitarra beccofine

10646 **Xiphocolaptes falcirostris**
Passeriformes - Dendrocolaptidae
e Moustached Woodcreeper;
Moustached Creeper
f Grimpar à moustaches
d Bartbaumsteiger
i Rampichino dai mustacchi

10647 **Xiphocolaptes franciscanus**
Passeriformes - Dendrocolaptidae
e Snethlage's Woodcreeper;
Snethlage's Creeper
f Grimpar de Snethlage; Grimpar du
Sao Francisco
d Grauschnabelbaumsteiger
i Rampichino di Snethlage

10648 **Xiphocolaptes major**
Passeriformes - Dendrocolaptidae

e   Great Rufous Woodcreeper; Great
      Rufous Creeper
f   Grand Grimpar
d   Riesenbaumsteiger
i   Rampichino rossiccio maggiore

10649   *Xiphocolaptes orenocensis*
      Passeriformes - Dendrocolaptidae
e   Great-billed Woodcreeper; Rusty-
      breasted Woodcreeper
d   Rotbrustbaumsteiger

10650   *Xiphocolaptes promeropirhynchus*
      Passeriformes - Dendrocolaptidae
e   Strong-billed Woodcreeper; Giant
      Creeper; Giant Woodcreeper; Strong-
      billed Creeper
f   Grimpar géant
d   Strichelkopfbaumsteiger
i   Rampichino beccoforte

10651   *Xiphocolaptes villanovae*
      Passeriformes - Dendrocolaptidae
e   Villa Nova Woodcreeper
f   Grimpar de Villa Nova
d   Bahia-Baumsteiger
i   Rampichino

10652   *Xipholena atropurpurea*
      Passeriformes - Tyrannidae
e   White-winged Cotinga; Wied's
      Cotinga
f   Cotinga porphyrion
d   Weißflügelkotinga
i   cotinga alibianche

10653   *Xipholena lamellipennis*
      Passeriformes - Tyrannidae
e   White-tailed Cotinga; Lafresnaye's
      Cotinga
f   Cotinga à queue blanche
d   Weißschwanzkotinga
i   Cotinga codabianca

10654   *Xipholena punicea*
      Passeriformes - Tyrannidae
e   Pompadour Cotinga
f   Cotinga pompadour
d   Pompadourkotinga;
      Pompadourschmuckvogel
i   Cotinga vinacea

10655   *Xiphorhynchus aequatorialis*
      Passeriformes - Tyrannidae
e   Spot-throated Woodcreeper

10656   *Xiphorhynchus elegans*
      Passeriformes - Dendrocolaptidae
e   Elegant Woodcreeper; Elegant
      Creeper
f   Grimpar élégant
d   Schmuckbaumsteiger
t   Rampichino elegante

10657   *Xiphorhynchus erythropygius*
      Passeriformes - Dendrocolaptidae
e   Spotted Woodcreeper; Spotted
      Creeper
f   Grimpar tacheté
d   Sternfleckenbaumsteiger

10658   *Xiphorhynchus eytoni*
      Passeriformes - Dendrocolaptidae
e   Dusky-billed Creeper; Dusky-billed
      Woodcreeper; Eyton's Woodcreeper;
      Eyton's Creeper
f   Grimpar d'Eyton
d   Streifenmantelbaumsteiger
i   Rampichino di Eyton

10659   *Xiphorhynchus flavigaster*
      Passeriformes - Dendrocolaptidae
e   Ivory-billed Woodcreeper; Ivory-
      billed Creeper; Laughing
      Woodcreeper; Laughing Creeper;
      Stripe-throated Woodcreeper
f   Grimpar à bec ivoire; Grimpar à bec
      d'ivoire
d   Lachbaumsteiger
i   Rampichino ventregiallo

10660   *Xiphorhynchus guttatus*
      Passeriformes - Dendrocolaptidae
e   Buff-throated Woodcreeper; Buff-
      throated Creeper
f   Grimpar des cabosses
d   Tropfenstirnbaumsteiger
i   Rampichino gocciolato

10661   *Xiphorhynchus lachrymosus*
      Passeriformes - Dendrocolaptidae
e   Black-striped Woodcreeper; Black-
      striped Creeper

f Grimpar maillé
d Tränenbaumsteiger
i Rampichino lacrimoso

10662 *Xiphorhynchus necopinus*
  Passeriformes - Dendrocolaptidae
e Zimmer's Woodcreeper; Zimmer's
  Creeper
f Grimpar de Zimmer
d Amazonas-Baumsteiger
i Rampichino di Zimmer

10663 *Xiphorhynchus obsoletus*
  Passeriformes - Dendrocolaptidae
e Striped Woodcreeper; Striped
  Creeper
f Grimpar strié
d Streifenbaumsteiger
i Rampichino striato

10664 *Xiphorhynchus ocellatus*
  Passeriformes - Dendrocolaptidae
e Ocellated Woodcreeper; Ocellated
  Creeper
f Grimpar ocellé
d Augenbaumsteiger
i Rampichino ocellato

10665 *Xiphorhynchus pardalotus*
  Passeriformes - Dendrocolaptidae
e Chestnut-rumped Woodcreeper;
  Chestnut-rumped Creeper
f Grimpar flambé
d Rostkehlbaumsteiger
i Rampichino dal groppone castano

10666 *Xiphorhynchus picus*
  Passeriformes - Dendrocolaptidae
e Straight-billed Woodcreeper;
  Straight-billed Creeper
f Grimpar talapiot
d Geradschnabelbaumsteiger;
  Spechtbaumhacker
i Rampichino picchio

10667 *Xiphorhynchus spixii*
  Passeriformes - Dendrocolaptidae
e Spix's Woodcreeper; Spix's Creeper
f Grimpar de Spix
d Spix-Baumsteiger
i Rampichino di Spix

10668 *Xiphorhynchus striatigularis*
  Passeriformes - Dendrocolaptidae
e Stripe-throated Woodcreeper; Stripe-
  throated Creeper; Richmond's
  Creeper
f Grimpar d'Altamira
d Tamaulipas-Baumsteiger

10669 *Xiphorhynchus susurrans*
  Passeriformes - Dendrocolaptidae
e Cocoa Woodcreeper

10670 *Xiphorhynchus triangularis*
  Passeriformes - Dendrocolaptidae
e Olive-backed Woodcreeper; Olive-
  backed Creeper; Spotted-
  Woodcreeper
f Grimpar à dos olive; Grimpar de
  Lafresnaye
d Schuppenbrustbaumsteiger
i Rampichino dorso oliva

10671 *Xolmis cinerea*
  Passeriformes - Tyrannidae
e Grey Monjita; Grey Pepoaza
f Pépoaza cendré
d Bartmonjita
i Monjita cenerina

10672 *Xolmis coronata*
  Passeriformes - Tyrannidae
e Black-crowned Monjita
f Pépoaza couronné
d Kronenmonjita
i Monjita corona nera

10673 *Xolmis dominicana*
  Passeriformes - Tyrannidae
e Black-and-white Monjita; Dominican
  Monjita
f Pépoaza dominicain
d Dominikanermonjita
i Monjita bianca e nera

10674 *Xolmis irupero*
  Passeriformes - Tyrannidae
e White Monjita; Widow Monjita
f Pépoaza irupero
d Pepoaza; Witwenmonjita
i Monjita bianca

10675   *Xolmis pyrope*
Passeriformes - Tyrannidae
e   Fire-eyed Diucon
f   Pépoaza oeil-de-feu; Pépoaza à oeil
de feu
d   Feueraugentyrann
i   Tiranno occhiodifuoco

10676   *Xolmis rubetra*
Passeriformes - Tyrannidae
e   Rusty-backed Monjita; Chat like
Monjita
f   Pépoaza traquet
d   Rotrückenmonjita
i   Monjita dorsoruggine

10677   *Xolmis salinarum*
Passeriformes - Tyrannidae
e   Salinas Monjita
f   Pépoaza de Salinas
i   Monjita delle saline

10678   *Xolmis velata*
Passeriformes - Tyrannidae
e   White-rumped Monjita; Veiled
Monjita
f   Pépoaza voilé
d   Weißbürzelmonjita
i   Monjita gropponebianco

Y

10679 Yuhina bakeri
Passeriformes - Sylviidae
e White-naped Yuhina; Baker's
Chestnut-headed Yuhina; Chestnut-
headed Yuhina; Blyth's Yuhina;
White-naped Ixulus
f Yuhina à nuque blanche
d Rotkopfyuhina
i Yuhina testacastana

10680 Yuhina brunneiceps
Passeriformes - Sylviidae
e Formosan Yuhina; Taiwan Yuhina
f Yuhina de Taïwan; Yuhina de
Formose
d Braunkopfyuhina; Haubenbräunling
i Yuhina di Taiwan

10681 Yuhina castaniceps
Passeriformes - Sylviidae
e Striated Yuhina; Chestnut-headed
Yuhina; White-browed Yuhina;
Chestnut-headed Staphida; Collared
Siva
f Yuhina à tête marron
d Rotohryuhina
i Yuhina dorsostriato

10682 Yuhina diademata
Passeriformes - Sylviidae
e White-collared Yuhina; Diademed
Yuhina
f Yuhina à diadème
d Diademyuhina
i Yuhina coronata

10683 Yuhina everetti
Passeriformes - Sylviidae
e Chestnut-crested Yuhina; Chestnut-
headed Siva; Chestnut-headed
Yuhina; Chestnut-crested Babbler
f Yuhina de Bornéo

d Rotschopfyuhinia
i Yuhina di Everett

10684 Yuhina flavicollis
Passeriformes - Sylviidae
e Whiskered Yuhina; Yellow-naped
Yuhina; Whiskered Ixulus
f Yuhina à cou roux
d Gelbnackenyuhina
i Yuhina dai mustacchi

10685 Yuhina gularis
Passeriformes - Sylviidae
e Stripe-throated Yuhina; Striped-
throated Yuhina; Striped Yuhina;
Stripe-throated Ixulus; Striped-
throated Ixulus
f Yuhina à gorge striée
d Kehlstreifenyuhina
i Yuhina golastriata

10686 Yuhina humilis
Passeriformes - Sylviidae
e Burmese Yuhina; Burmese Ixulus
f Yuhina de Birmanie
i Yuhina birmana

10687 Yuhina nigrimenta
Passeriformes - Sylviidae
e Black-chinned Yuhina; Pale Yuhina
f Yuhina à menton noir
d Meisenyuhina; Zwergtimalie
i Yuhina mentenero

10688 Yuhina occipitalis
Passeriformes - Sylviidae
e Rufous-vented Yuhina; Slaty-headed
Yuhina
f Yuhina à ventre roux
d Roststeißyuhina
i Yuhina nucarossiccia

10689 Yuhina zantholeuca
Passeriformes - Sylviidae
e White-bellied Yuhina; Erpornis;
White-bellied Tree-Babbler; White-
bellied Crested-Babbler; White-
breasted Erpornis
f Yuhina à ventre blanc
d Grünrückenyuhina
i Yuhina pettobianco

Z

10690 *Zaratornis stresemanni*
Passeriformes - Tyrannidae
e White-cheeked Cotinga
f Cotinga à joues blanches
d Mönchszuser; Mönchskotinga
i Cotinga di Stresemann; Corvide di
Zavatteri

10691 *Zavattariornis stresemanni*
Passeriformes - Corvidae
e Stresemann's Bush-Crow;
Zavattariornis; Abyssinian Pie; Bush-
Crow; Abyssinian Bush-crow;
Ethiopian Bush-crow
f Corbin de Stresemann; Corbeau de
Stresemann
d Dornhäher; Zavattarivogel
i Corvide di Zavattari

10692 *Zebrilus undulatus*
Ciconiiformes - Ardeidae
e Zigzag Heron; Vermiculated Heron
f Onoré zigzag
d Zickzackreiher
i Tarabusino zebrato

10693 *Zeledonia coronata*
Passeriformes - Fringillidae
e Wrenthrush; Zeledonia
f Paruline de Zeledon
d Zaunkönigsdrossel; Zeledonie
i Zeledonia coronata

10694 *Zenaida asiatica*
Columbiformes - Columbidae
e White-winged Dove; Lapwing (WI);
Whitewing (WI); Singing Dove;
Mesquitre Dove
f Tourterelle à ailes blanches; Barbarin
(Ants)
d Weißflügeltaube
i Tortora dalle ali bianche

10695 *Zenaida auriculata*
Columbiformes - Columbidae
e Eared Dove; Trinidad Ground Dove
(WI); Violet-eared Dove; Blue-eared
Dove; Torpedo Dove; Bronze-necked
Dove; Gold-necked Dove; Wood-
Dove; Pea Dove; Seaside Dove
f Tourterelle oreillarde; Tourterelle
Ortolan (Ants)
d Chile-Trauertaube; Ohrflecktaube
i Tortora auricolato

10696 *Zenaida aurita*
Columbiformes - Columbidae
e Zenaida Dove; Seaside Dove (WI);
Pea Dove (WI); Wood Dove (WI);
Mountain Dove (WI)
f Tourterelle à queue carrée; Grosse
Tourterelle (Ants); Tourterelle
(Ants); Toutrelle (Ants)
d Liebestaube
i Tortora zenaida

10697 *Zenaida galapagoensis*
Columbiformes - Columbidae
e Galapagos Dove
f Tourterelle des Galapagos
d Galapagos-Taube
i Tortora delle Galapagos

10698 *Zenaida graysoni*
Columbiformes - Columbidae
e Socorro Dove
f Tourterelle de Socorro
i Tortora di Socorro

10699 *Zenaida macroura*
Columbiformes - Columbidae
e Mourning Dove; Paloma (WI); Longtail (WI); Pea Dove (WI); Carolina
Dove
f Tourterelle triste; Tourterelle (Ants);
Queue-fine (Ants); Tourterelle (Ants)
d Trauertaube; Carolina-Taube
i Totora lamentosa americana;
Colomba della Carolina

10700 *Zenaida meloda*
Columbiformes - Columbidae
e Pacific Dove

10701 **Zimmerius bolivianus**
Passeriformes - Tyrannidae
e Bolivian Tyrannulet
f Tyranneau de Bolivie; Tyranneau
bolivien
d Olivfliegenstecher
i Tiranno piccolo della Bolivia

10702 **Zimmerius chrysops**
Passeriformes - Tyrannidae
e Golden-faced Tyrannulet
f Tyranneau à face d'or
i Tiranno piccolo facciadorata

10703 **Zimmerius cinereicapillus**
Passeriformes - Tyrannidae
e Red-billed Tyrannulet
f Tyranneau à bec rouge
d Rotschnabelfliegenstecher
i Tiranno piccolo beccogrosso

10704 **Zimmerius gracilipes**
Passeriformes - Tyrannidae
e Slender-footed Tyrannulet
f Tyranneau à petits pieds
d Schlankfußfliegenstecher
i Tiranno piccolo piedigracili

10705 **Zimmerius improbus**
Passeriformes - Tyrannidae
e Paltry Tyrannulet; Venezuelan
Tyrannulet
f Tyranneau trompeur
i Tiranno piccolo del Venezuela;
Tiranno piccolo di Venezuela

10706 **Zimmerius vilissimus**
Passeriformes - Tyrannidae
e Paltry Tyrannulet; Mistletoe
Tyrannulet
f Tyranneau gobemoucheron
d Augenringfliegenstecher
i Tiranno piccolo modesto

10707 **Zimmerius viridiflavus**
Passeriformes - Tyrannidae
e Golden-faced Tyrannulet; Peruvian
Tyrannulet; Tschudi's Tyrannulet
f Tyranneau à face jaune
d Goldstirnfliegenstecher
i Tiranno piccolo del Perù

10708 **Zonerodius heliosylus**
Ciconiiformes - Ardeidae
e Forest Heron; Forest Bittern; Zebra
Bittern; New Guinea Tiger-Heron;
New Guinea Tiger Bittern
f Onoré phaeton
d Bindenreiher
i Tarabusco delle foreste

10709 **Zonotrichia albicollis**
Passeriformes - Emberizidae
e White-throated Sparrow
f Bruant à gorge blanche; Pinson à
gorge blanche
d Weißkehlammer;
Weißkehlammerfink
i Zonotrichia collobianco; Passero
golabianca

10710 **Zonotrichia atricapilla**
Passeriformes - Emberizidae
e Golden-crowned Sparrow
f Bruant à couronne dorée; Pinson à
couronne dorée
d Kronenammer
i Zonotrichia corona dorata; Passero
corona dorato

10711 **Zonotrichia capensis**
Passeriformes - Emberizidae
e Rufous-collared Sparrow; Andean
Sparrow
f Bruant chingolo; Bruant brésilien
d Morgenammer; Braunnackenammer
i Zonotrichia dal collare rossiccio;
Passero dal collare rossiccio

10712 **Zonotrichia leucophrys**
Passeriformes - Emberizidae
e White-crowned Sparrow
f Bruant à couronne blanche; Pinson à
couronne blanche
d Dachsammer
i Passero corona bianco; Zonotrichia
corona bianca

10713 **Zonotrichia querula**
Passeriformes - Emberizidae
e Harris's Sparrow; White-rumped
Spinetail (ISC)
f Bruant de Harris; Bruant à face noire;

Pinson à face noire
d   Harris-Ammer
i   Passero di Harris

10714   *Zoonavena grandidieri*
Apodiformes - Apodidae
e   Madagascar Spinetailed-Swift;
Grandidier's Spinetailed-Swift;
Madagascar Spinetail; Malagasy
Spinetail; Malagasy Needletail
f   Martinet de Grandidier
d   Malagassen-Segler
i   Rondone codaspinosa del
Madagascar

10715   *Zoonavena sylvatica*
Apodiformes - Apodidae
e   Indian White-rumped Spinetailed-
Swift; Indian White-rumped
Spinetail; White-rumped Spinetail;
White-rumped Needletail
f   Martinet indien; Martinet hindou
d   Hindu-Segler
i   Rondone codaspinosa indiano

10716   *Zoonavena thomensis*
Apodiformes - Apodidae
e   Sao Tomé Spinetailed-Swift; Sao
Tomé Spinetail
f   Martinet de Sao Tomé
s   St. Thomas-Segler
i   Rondone codaspinosa di Sao Tomé

10717   *Zoothera andromeda*
Passeriformes - Muscicapidae
e   Sunda Ground-Thrush; Sunda Long-
billed Thrush; Andromeda Thrush;
Sunda Thrush
f   Grive andromède
d   Andromedadrossel
i   Tordo della Sonda

10718   *Zoothera cameronensis*
Passeriformes - Muscicapidae
e   Black-eared Ground-Thrush;
Cameroon Ground-Thrush
f   Grive du Cameroun; Grive terrestre
du Cameroun
d   Kamerun-Drossel
i   Tordo del Camerun

10719   *Zoothera cinerea*
Passeriformes - Muscicapidae
e   Ashy Thrush; Ashy Ground-Thrush
f   Grive cendrée
d   Mindoro-Drossel
i   Tordo cenerino

10720   *Zoothera citrina*
Passeriformes - Muscicapidae
e   Orange-headed Thrush (ISC);
Orange-headed Ground-Thrush
(ISC); Ground-Thrush; Northern
Orange-headed Ground-Thrush;
White-throated Thrush
f   Grive à tête orangé
d   Dama-Drossel
i   Tordo testa aranciata

10721   *Zoothera citrina cyanotis*
Passeriformes - Muscicapidae
e   White-throated Ground-Thrush (ISC)

10722   *Zoothera crossleyi*
Passeriformes - Muscicapidae
e   Crossley's Ground-Thrush
f   Grive de Crossley; Grive terrestre de
Crossley
i   Tordo di Crossley

10723   *Zoothera dauma*
Passeriformes - Muscicapidae
e   White's Thrush; Scaly Thrush (ISC);
Golden Mountain-Thrush; White's
Ground-Thrush; Scaly Ground-
Thrush; White's Scaly-Thrush;
Speckled Mountain-Thrush; Ceylon
Scaly-Thrush
f   Grive dorée; Merle doré
d   Erddrossel; Buntdrossel
i   Tordo dorato

10724   *Zoothera dixoni*
Passeriformes - Muscicapidae
e   Long-tailed Thrush; Long-tailed
Mountain-Thrush
f   Grive de Dixon
d   Dixon-Drossel
i   Tordo di Dixon

10725 **Zoothera dumasi**
Passeriformes - Muscicapidae
e Moluccan Ground-Thrush; Buru
Ground-Thrush; Buru Thrush; Ceram
Thrush; Ceram Thrush; Ceram
Ground-Thrush
f Grive de Dumas
d Molukken-Drossel
i Tordo delle Molucche

10726 **Zoothera erythronota**
Passeriformes - Muscicapidae
e Celebes Ground-Thrush; Red-backed
Ground-Thrush; Lombok Thrush;
Chestnut-headed Thrush; Chestnut-
backed Ground-Thrush
f Grive à dos roux
d Rotrückendrossel
i Tordo dorsorosso

10727 **Zoothera everetti**
Passeriformes - Muscicapidae
e Everett's Thrush; Everett's Ground-
Thrush
f Grive d'Everett
d Everett-Drossel
i Tordo di Everrett

10728 **Zoothera gurneyi**
Passeriformes - Muscicapidae
e Orange Thrush (CSA); Gurney's
Thrush; Orange Ground-Thrush;
Guerney's Ground-Thrush
f Grive de Gurney; Grive terrestre
orangée
d Gurneys Drossel
i Tordo di Gurney

10729 **Zoothera guttata**
Passeriformes - Muscicapidae
e Spotted Thrush; Natal Thrush (CSA)
f Grive tachetée
d Fleckengrunddrossel
i Tordo maculato

10730 **Zoothera guttata fischeri**
Passeriformes - Muscicapidae
e Spotted Ground-Thrush; Natal
Ground-Thrush; Spotted Thrush;
Fischer's Thrush
f Grive de Fischer; Grive terrestre

tachetée
d Natal-Drossel

10731 **Zoothera heinei**
Passeriformes - Muscicapidae
e Russet-tailed Thrush; Heine's
Ground-Thrush; Heine's Thrush
f Grive de Heine
i Tordo del Queensland

10732 **Zoothera horsfieldi**
Passeriformes - Muscicapidae
e Horsfield's Thrush; Spot-winged
Thrush; Spotted-wing Thrush
f Grive de Horsfield
i Tordo di Horsfield

10733 **Zoothera interpres**
Passeriformes - Muscicapidae
e Chestnut-capped Thrush; Chestnut-
capped Ground-Thrush; Chestnut-
headed Thrush; Kühl's Ground-
Thrush
f Grive de Kuhl
d Rotkappendrossel
i Tordo capocastano

10734 **Zoothera kibalensis**
Passeriformes - Muscicapidae
e Kibale Ground-Thrush; Prigogine's
Ground-Thrush
f Grive de Kibale
d Kibale-Drossel
i Tordo di Prigogine

10735 **Zoothera lunulata**
Passeriformes - Muscicapidae
e Olive-tailed Thrush; Bassian Thrush;
Australian Ground-Thrush
f Grive à lunules
i Tordo dorato di Bass

10736 **Zoothera lunulata halmaturina**
Passeriformes - Muscicapidae
e South Australian Bassian-Thrush
(ANZ)

10737 **Zoothera machiki**
Passeriformes - Muscicapidae
e Fawn-breasted Thrush

f   Grive à poitrine fauve
i   Tordo dorato di Tanimbar

10738   *Zoothera major*
      Passeriformes - Muscicapidae
e   Amami Thrush
f   Grive d'Amami
i   Tordo di Amami

10739   *Zoothera margaretae*
      Passeriformes - Muscicapidae
e   San Cristobal Thrush; San Cristobal
      Ground-Thrush
f   Grive des Salomon
i   Tordo di San Cristobal

10740   *Zoothera marginata*
      Passeriformes - Muscicapidae
e   Dark-sided Thrush; Lesser Long-
      billed Thrush; Lesser Brown-Thrush;
      Long billed Thrush
f   Grive à grand bec
d   Langschnabeldrossel
i   Tordo beccolungo minore

10741   *Zoothera mollissima*
      Passeriformes - Muscicapidae
e   Plain-backed Thrush; Plain-backed
      Mountain-Thrush
f   Grive de Hodgson
d   Himalaja-Drossel
i   Tordo di montagna

10742   *Zoothera monticola*
      Passeriformes - Muscicapidae
e   Long-billed Thrush; Greater Long-
      billed Thrush; Large Brown-Thrush;
      Brown Thrush; Large Long-billed
      Thrush
f   Grive montagnarde
d   Bergdrossel
i   Tordo beccolungo maggiore

10743   *Zoothera naevia*
      Passeriformes - Muscicapidae
e   Varied Thrush
f   Grive à collier; Merle à collier
d   Halsbanddrossel
i   Tordo variopinto; Tordo vario

10744   *Zoothera oberlaenderi*
      Passeriformes - Muscicapidae
e   Oberländer's Ground-Thrush; Congo
      Thrush; Forest Ground-Thrush;
      Oberländer's Forest Ground-Thrush
f   Grive d'Oberländer; Grive terrestre
      d'Oberländer
d   Oberländer-Drossel
i   Tordo di Oberländer

10745   *Zoothera peronii*
      Passeriformes - Muscicapidae
e   Orange-sided Thrush; Orange-
      banded Thrush; Péron's Ground-
      Thrush; Péron's Timor Ground-
      Thrush; Timor Thrush; Orange-
      banded Ground-Thrush; Orange-
      banded Thrush
f   Grive de Péron
d   Timor-Drossel
i   Tordo di Peron

10746   *Zoothera piaggiae*
      Passeriformes - Muscicapidae
e   Abyssinian Ground-Thrush; Orange
      Thrush; Orange Ground-Thrush
f   Grive de Piaggia
d   Orangedrossel
i   Tordo abissino

10747   *Zoothera pinicola*
      Passeriformes - Muscicapidae
e   Aztec Thrush
f   Grive aztèque
d   Aztekendrossel
i   Tordo azteco

10748   *Zoothera princei*
      Passeriformes - Muscicapidae
e   Grey Ground-Thrush
f   Grive olivâtre; Grive terrestre grise
d   Ghana-Drossel; Grauerddrossel
i   Tordo di Prince

10749   *Zoothera schistacea*
      Passeriformes - Muscicapidae
e   Slaty-backed Thrush; Tanimbar
      Thrush; Tanimbar Ground-Thrush;
      Slaty-backed Ground-Thrush; White-
      eared Thrush; White-eared Ground-
      Thrush

f Grive schistacée
d Weißohrdrossel
i Tordo dorsoardesia

10750 *Zoothera sibirica*
Passeriformes - Muscicapidae
e Siberian Thrush; Siberian Ground-Thrush
f Grive de Sibérie; Merle sibérien; Merle de Sibérie; Grive sibérienne
d Schieferdrossel; Sibirische Drossel
i Tordo siberiano

10751 *Zoothera spiloptera*
Passeriformes - Muscicapidae
e Spot-winged Thrush; Spotted-wing Thrush; Spotted-wing Ground-Thrush; Spot-winged Ground-Thrush; Sri-Lanka Spot-winged Ground-Thrush
f Grive à ailes tachetées
d Ceylon-Drossel
i Tordo alimacchiate

10752 *Zoothera talaseae*
Passeriformes - Muscicapidae
e New Britain Thrush; New Britain Ground-Thrush; Talaseae Thrush; Northern Melanesian Ground-Thrush; Melanesian Ground-Thrush; Black-backed Ground-Thrush
f Grive de Nouvelle-Bretagne
i Tordo delle Bismarck

10753 *Zoothera tanganjicae*
Passeriformes - Muscicapidae
e Kivu Ground-Thrush; Orange Ground-Thrush; Uganda Ground-Thrush; Western Ground-Thrush
f Grive du Kivu
d Tanganika-Drossel
i Tordo del Tanganica

10754 *Zoothera terrestris*
Passeriformes - Muscicapidae
e Bonin Thrush; Kittlitz's Thrush; Bonin Islands Thrush
f Grive des Bonin
d Bonin-Drossel
i Tordo dell'Isola Bonin

10755 *Zoothera wardii*
Passeriformes - Muscicapidae
e Pied Thrush; Pied Ground-Thrush
f Grive de Ward
d Elsterdrossel
i Tordo di Ward

10756 Zosteropidae
Passeriformes
e White-eyes
f Zosteropidés; Oiseaux-lunettes
d Brillenvögel
i Zosteropidi

10757 *Zosterops abyssinicus*
Passeriformes - Zosteropidae
e White-breasted White-eye; Abyssinian White-eye (CSA); African White-breasted White-eye; Yellow White-eye
f Zostérops à front noir; Zosterops d'Abyssinie
d Somali-Brillenvogel; Abessinischer Brillenvogel
i Occhialino pettobianco

10758 *Zosterops albogularis*
Passeriformes - Zosteropidae
e White-chested White-eye; Norfolk White-eye; Norfolk Island White-eye
f Zostérops à flancs jaunes
d Norfolk-Brillenvogel
i Occhialino golabianca

10759 *Zosterops anomalus*
Passeriformes - Zosteropidae
e Lemon-throated White-eye; Black-ringed White-eye; Makassar White-eye; Celebes White-eye
f Zostérops à poitrine blanche
i Occhialino di Sulawesi

10760 *Zosterops atricapillus*
Passeriformes - Zosteropidae
e Black-capped White-eye
f Zostérops à gorge citron
d Schwarzstirnbrillenvogel
i Occhialino di Beccari

10761 *Zosterops atriceps*
Passeriformes - Zosteropidae

e   Creamy-throated White-eye; Batjan White-eye; Moluccan White-eye; Cream-throated White-eye; Halmahera Black-fronted White-eye; Black-fronted White-eye
ƒ   Zostérops à calotte noire
d   Braunscheitelbrillenvogel
i   Occhialino capinero

10762   *Zosterops atrifrons*
Passeriformes - Zosteropidae
e   Black-crowned White-eye; Black-fronted White-eye; Moluccan Black-fronted White-eye; Moluccan White-eye
ƒ   Zostérops à gorge crème
d   Nehrkorn-Brillenvogel
i   Occhialino delle Molucche

10763   *Zosterops borbonicus*
Passeriformes - Zosteropidae
e   Mascarene Grey White-eye; Grey White-eye; Bourbon White-eye; Mascarene White-eye; Réunion White-eye
ƒ   Zostérops des Mascareignes; Zosterops de Bourbon
d   Mascarenen-Brillenvogel
i   Occhialino delle Mascarene

10764   *Zosterops buruensis*
Passeriformes - Zosteropidae
e   Buru Yellow White-eye; Buru Island White-eye
ƒ   Zostérops de Buru
d   Buru-Brillenvogel
i   Occhialino di Buru

10765   *Zosterops ceylonensis*
Passeriformes - Zosteropidae
e   Sri Lanka White-eye; Large Sri Lanka White-eye; Ceylon White-eye; Ceylon Hill White-eye
ƒ   Zostérops de Ceylan
d   Ceylon-Brillenvogel
i   Occhialino di Sri-Lanka

10766   *Zosterops chloris*
Passeriformes - Zosteropidae
e   Lemon-bellied White-eye; Moluccan White-eye; Mangrove White-eye;

Yellow-bellied White-eye; Yellow White-eye; Large Brindled White-eye
ƒ   Zostérops à ventre citron
d   Molukken-Brillenvogel
i   Occhialino ventregiallo

10767   *Zosterops chloronothos*
Passeriformes - Zosteropidae
e   Mauritius Olive White-eye
ƒ   Zostérops de Maurice; Zosterops de l'île Maurice
i   Occhialino di Mauritius

10768   *Zosterops cinereus*
Passeriformes - Zosteropidae
e   Grey-brown White-eye; Caroline White-eye; Grey White-eye
ƒ   Zostérops cendré
d   Kittlitzs Brillenvogel
i   Occhialino grigio bruno

10769   *Zosterops citrinellus*
Passeriformes - Zosteropidae
e   Pale White-eye; Ash-bellied White-eye; Australian Pale White-eye; Australian Silvereye; Ashy-bellied Silvereye
ƒ   Zostérops pâle
d   Zitronenbrillenvogel
i   Occhialino pallido

10770   *Zosterops consobrinorum*
Passeriformes - Zosteropidae
e   Pale-bellied White-eye; Peninsular White-eye; Celebes White-eye; Sulawesi White-eye; Laloumera White-eye
ƒ   Zostérops à ventre pâle
d   Celebes-Brillenvogel
i   Occhialino ventrepallido

10771   *Zosterops conspicillatus*
Passeriformes - Zosteropidae
e   Bridled White-eye; Rota White-eye; Saipau White-eye
ƒ   Zostérops bridé
d   Semper-Brillenvogel
i   Occhialino dalle redini

10772   ***Zosterops erythropleurus***
Passeriformes - Zosteropidae
e   Chestnut-flanked White-eye
f   Zostérops à flancs marron
d   Rotflankenbrillenvogel;
     Goldkinnbrillenvogel
i   Occhialino fianchicastani

10773   ***Zosterops everetti***
Passeriformes - Zosteropidae
e   Everett's White-eye
f   Zostérops d'Everett
d   Everett-Brillenvogel
i   Occhialino di Everett

10774   ***Zosterops explorator***
Passeriformes - Zosteropidae
e   Layard's White-eye; Fiji White-eye
f   Zostérops des Fidji
d   Layard-Brillenvogel
i   Occhialino di Layard

10775   ***Zosterops ficedulinus***
Passeriformes - Zosteropidae
e   Principe White-eye; Principe Island
     White-eye; Sao Tomé White-eye
f   Zostérops becfigue
d   Fahlbrillenvogel
i   Occhialino di Hartlaub

10776   ***Zosterops finschii***
Passeriformes - Zosteropidae
e   Dusky White-eye; Finsch's White-
     eye
f   Zostérops de Finsch
d   Finsch-Brillenvogel
i   Occhialino bruno

10777   ***Zosterops flavifrons***
Passeriformes - Zosteropidae
e   Yellow-fronted White-eye; Yellow-
     White-eye
f   Zostérops à front jaune
d   Gelbstirnbrillenvogel
i   Occhialino frontegiallo

10778   ***Zosterops flavus***
Passeriformes - Zosteropidae
e   Javan White-eye; Yellow White-eye;
     Philippine White-eye
f   Zostérops flavescent

d   Horsfield-Brillenvogel
i   Occhialino di Giava

10779   ***Zosterops fuscicapillus***
Passeriformes - Zosteropidae
e   Capped White-eye; Yellow-bellied
     Mountain-White-eye; Yellow-bellied
     White-eye; Western Mountain-
     White-eye
f   Zostérops mitré
d   Arfak-Brillenvogel
i   Occhialino capobruno

10780   ***Zosterops grayi***
Passeriformes - Zosteropidae
e   Pearl-bellied White-eye; Great Kai
     Island White-eye; Kai Island White-
     eye; Kai White-eye
f   Zostérops de la Grande Kai
d   Grays Brillenvogel
i   Occhialino ventreperlato

10781   ***Zosterops griseotinctus***
Passeriformes - Zosteropidae
e   Louisiade White-eye; Louisiades
     White-eye; Islet White-eye; Island
     White-eye; Dull-coloured White-eye
f   Zostérops de la Louisiade
d   Louisiaden-Brillenvogel
i   Occhialino delle Luisiade

10782   ***Zosterops griseovirescens***
Passeriformes - Zosteropidae
e   Annobon White-eye
f   Zostérops d'Annobon
d   Annobon-Brillenvogel
i   Occhialino di Annobon

10783   ***Zosterops hypolais***
Passeriformes - Zosteropidae
e   Plain White-eye
f   Zostérops hypolais
i   Occhialino disadorno

10784   ***Zosterops hypoxanthus***
Passeriformes - Zosteropidae
e   Black-headed White-eye; Bismarck
     Yellow-bellied White-eye; Bismarck
     Blackfronted White-eye; Bismarck
     White-eye
f   Zostérops des Bismarck

d Bismarck-Brillenvogel
i Occhialino testanera

10785 *Zosterops inornatus*
Passeriformes - Zosteropidae
e Large Lifou White-eye; Large Lifu
White-eye
f Zostérops de Lifu
d Lifu-Brillenvogel
i Occhialino disadorno

10786 *Zosterops japonicus*
Passeriformes - Zosteropidae
e Japanese White-eye; Chinese White-
eye
f Zostérops du Japon
d Japanischer Brillenvogel; Japan-
Brillenvogel
i Occhialino giapponese

10787 *Zosterops kirki*
Passeriformes - Zosteropidae
e Kirk's White-eye; Grand Comoro
White-eye
f Zostérops de Kirk
d Kirk-Brillenvogel
i Occhialino di Kirk

10788 *Zosterops kuehni*
Passeriformes - Zosteropidae
e Ambon Yellow White-eye; Amboina
White-eye; Kühn's White-eye;
Ambon White-eye
f Zostérops d'Ambon
d Amboina-Brillenvogel
i Occhialino di Ambon

10789 *Zosterops kulalensis*
Passeriformes - Zosteropidae
e Kulal White-eye
d Brillenvogel

10790 *Zosterops kulambangrae*
Passeriformes - Zosteropidae
e Solomon Islands White-eye;
Solomons White-eye; Rendova
White-eye
f Zostérops des Salomon
d Salomonen-Brillenvogel
i Occhialino delle Isole Salomone

10791 *Zosterops lateralis*
Passeriformes - Zosteropidae
e Silvereye; Grey-breasted White-eye;
Grey-backed White-eye; Grey-
breasted Silver-eye
f Zostérops à dos gris
d Mantelbrillenbogel; Silberauge;
Graurückenbrillenvogel;
Silberbrillenvogel; Australischer
Brillenvogel
i Occhialino dorsogrigio

10792 *Zosterops luteirostris*
Passeriformes - Zosteropidae
e Ghizo White-eye; Splendid White-
eye; Hartert's White-eye
f Zostérops de Gizo
d Ganonga-Brillenvogel
i Occhialino di Gizo

10793 *Zosterops luteus*
Passeriformes - Zosteropidae
e Australian Yellow White-eye;
Yellow Silver-eye; Mangrove White-
eye; Gulliver's White-eye
f Zostérops à ventre jaune
d Mangrovebrillenvogel
i Occhialino giallo africano

10794 *Zosterops maderaspatanus*
Passeriformes - Zosteropidae
e Malagasy White-eye; Madagascar
White-eye; Mascarene White-eye;
Hove Grey-backed White-eye
f Zostérops malgache
d Madagaskar-Brillenvogel
i Occhialino del Madagascar

10795 *Zosterops mayottensis*
Passeriformes - Zosteropidae
e Chestnut-sided White-eye; Chestnut-
flanked White-eye; Mayotte Chesnut-
flanked White-eye; Mayotte White-
eye
f Zostérops de Mayotte
d Dotterbrustbrillenvogel
i Occhialino fianchicastane delle
Seychelle

10796 *Zosterops meeki*
Passeriformes - Zosteropidae

e White-throated White-eye; Meek's
White-eye; Louisiade Black-fronted
White-eye; Tagula White-eye;
Tagula Black-fronted White-eye;
Louisiades Black-fronted White-eye
f Zostérops à gorge blanche
d Tagula-Brillenvogel
i Occhialino golabianca

10797 *Zosterops metcalfii*
Passeriformes - Zosteropidae
e Yellow-throated White-eye;
Metcalfe's White-eye; Bukida White-
eye
f Zostérops à gorge jaune
d Goldkehlbrillenvogel
i Occhialino golagialla

10798 *Zosterops meyeni*
Passeriformes - Zosteropidae
e Lowland White-eye; Luzon White-
eye; Philippine White-eye
f Zostérops des Philippines
d Luzon-Brillenvogel
i Occhialino delle Filippine

10799 *Zosterops minor*
Passeriformes - Zosteropidae
e Black-fronted White-eye; New
Guinea Black-fronted White-eye;
Papuan Black-fronted White-eye;
Papuan White-eye; Papuan
Mountain-White-eye
f Zostérops mineur
i Occhialino minore

10800 *Zosterops minutus*
Passeriformes - Zosteropidae
e Small Lifou White-eye; Small Lifu
White-eye
f Zostérops minute
d Ameisenbrillenvogel
i Occhialino piccolo

10801 *Zosterops modestus*
Passeriformes - Zosteropidae
e Seychelles Grey White-eye;
Seychelles Brown White-eye
f Zostérops des Seychelles
d Mahé-Brillenvogel
i Occhialino delle Seychelle

10802 *Zosterops montanus*
Passeriformes - Zosteropidae
e Mountain White-eye
f Zostérops montagnard
d Gebirgsbrillenvogel;
Bergbrillenvogel
i Occhialino di montagna

10803 *Zosterops mouroniensis*
Passeriformes - Zosteropidae
e Mount Karthala White-eye; Comoro
White-eye; Grand Comoro White-
eye; Mount Karthala Green White-
eye; Karthala White-eye
f Zostérops du Karthala; Zosterops du
Mont Karthala
d Karthala-Brillenvogel
i Occhialino di Grand Comoro

10804 *Zosterops murphyi*
Passeriformes - Zosteropidae
e Hermit White-eye; Kulambangra
White-eye; Murphy's White-eye;
Kulambangra Mountain-White-eye
f Zostérops de Murphy
d Murphy-Brillenvogel
i Occhialino dei Monti Kulambangra

10805 *Zosterops mysorensis*
Passeriformes - Zosteropidae
e Biak White-eye; Soepiori White-eye
f Zostérops de Biak
d Biak-Brillenvogel
i Occhialino di Biak

10806 *Zosterops natalis*
Passeriformes - Zosteropidae
e Christmas Island White-eye;
Christmas White-eye
f Zostérops de Christmas
d Weißstirnbrillenvogel
i Occhialino di Natale

10807 *Zosterops nehrkorni*
Passeriformes - Zosteropidae
e Sangihi White-eye

10808 *Zosterops nigrorum*
Passeriformes - Zosteropidae
e Golden-green White-eye; Philippine
Yellow White-eye; Philippine White-

eye; Golden-yellow White-eye;
Yellow White-eye; Yellowish White-
eye
f Zostérops jaunâtre
d Philippinen-Brillenvogel
i Occhialino delle Filippine

10809 Zosterops novaeguineae
Passeriformes - Zosteropidae
e New Guinea White-eye; New Guinea
Mountain-White-eye
f Zostérops de Nouvelle-Guinée
d Papua-Brillenvogel; Neuguinea-
Brillenvogel
i Occhialino della Nuova Guinea

10810 Zosterops oleagineus
Passeriformes - Zosteropidae
e Yap Olive White-eye; Yap White-
eye; Large Yap White-eye; Yap
Greater White-eye
f Zostérops de Yap
d Yap-Brillenvogel
i Occhialino di Yap

10811 Zosterops olivaceus
Passeriformes - Zosteropidae
e Réunion Olive White-eye; Olive
White-eye; Olivaceous White-eye;
Mascarene Olive White-eye
f Zostérops de la Réunion
d Olivbrillenvogel
i Occhialino di Reunion

10812 Zosterops pallidus
Passeriformes - Zosteropidae
e Pale White-eye (CSA); Cape White-
eye; African White-eye; Pallid
White-eye
d Kap-Brillenvogel
i Occhialino del Capo

10813 Zosterops pallidus capensis
Passeriformes - Zosteropidae
e Cape White-eye (CSA)
f Zostérops du Cap
d Oranjebrillenvogel; Kap-Brillenvogel

10814 Zosterops palpebrosus
Passeriformes - Zosteropidae
e Oriental White-eye; White-eye (ISC);

Indian White-eye; Small White-eye;
Ceylon White-eye; Yellow-bellied
White-eye
f Zostérops oriental
d Ganges-Brillenvogel; Indien-
Brillenvogel
i Uccello dagli occhiali orientale;
Occhialino orientale

10815 Zosterops poliogaster
Passeriformes - Zosteropidae
e Broad-ringed White-eye; Montane
White-eye; Yellow-bellied White-
eye; Heuglin's White-eye; African
Mountain-White-eye; Highland
White-eye; South Pare White-eye
f Zostérops alticole; Zosterops des
montagnes
d Bergbrillenvogel
i Occhialino di monte africano

10816 Zosterops poliogaster eurycricotus
Passeriformes - Zosteropidae
f Zostérops du Kilimandjaro

10817 Zosterops poliogaster kikuyensis
Passeriformes - Zosteropidae
e Kikuyu White-eye (CSA)
f Zostérops du Kikuyu

10818 Zosterops poliogaster winifreda
Passeriformes - Zosteropidae
e South Pare White-eye (CSA)

10819 Zosterops rendovae
Passeriformes - Zosteropidae
e Grey-throated White-eye; Central
Solomons White-eye; Solomon
Islands White-eye; Variable White-
eye
f Zostérops à gorge grise
d Graukehlbrillenvogel
i Occhialino golagrigia

10820 Zosterops rennellianus
Passeriformes - Zosteropidae
e Rennell White-eye; Rennel Island
White-eye; Grey-throated White-eye
f Zostérops de Rennell
d Rennell-Brillenvogel
i Occhialino di Rennell

English Index

African Marsh-Warbler
(CSA) 112
African Mosque-Swallow
4684
African Mountain-Buzzard
1405
African Mountain-Thrush
10287
African Mountain-White-
eye 10815
African Mourning Dove
(CSA) 9495
African Moustached-
Warbler (CSA) 5775
African Nicator 6538
African Olivaceous
Alseonax 6071
African Olive-Pigeon 2478
African Openbill 576
African Open-billed Stork
(CSA) 576
African Orange-bellied
Parrot (CSA) 8064
African Oriole 6779
African Owl 1008
African Oystercatcher 4425
African Palm-Swift 3020
African Paradise-
Flycatcher 9865
African Paradise-Monarch
9865
African Paradise-Whydah
10545
African Parsonfinch 5138
African Peafowl 7181
African Penduline-Tit
651,653
African Piculet 8973
African Pied-Crow 2679
African Pied-Hornbill
10046
African Pied-Starling 9383
African Pied-Wagtail 6035
African Pigeon 2478
African Pipit 702
African Pitta 7842
African Pygmy-Cormorant
7407
African Pygmy-Falcon
8068

African Pygmy-Goose
6530
African Pygmy-Kingfisher
4886
African Quailfinch 6842
African Rail (CSA) 8711
African Raven 2678
African Red-billed Hornbill
10045
African Red-eyed Bulbul
8612
African Red-faced Apalis
765
African Red-tailed Buzzard
1391
African Red-tailed Hawk
1391
African Red-winged
Starling 6734
African Reed-Warbler 112
African Reef-Heron 3404
African River-Martin 8326
African Rock-Bunting
3485
African Rock-Martin 4667
African Rock-Pipit 704
African Rook 2683
African Sand-Martin
(CSA) 8901
African Scops-Owl 6913
African Scrub-Warbler
1292
African Sedge-Warbler
1290
African Serpent-Eagle
3341
African Shelduck 9714
African Short-toed Lark
1483
African Shrike-Flycatcher
1230
African Silverbill 5292
African Singing Bush-Lark
5925
African Skimmer 8922
African Snipe 4028
African Sooty-Flycatcher
6067
African Spoonbill 7878
African Spotted-Creeper
8933

African Stonechat 8981
African Striated-Swallow
4663
African Striped-Cuckoo
2367
African Swallow-tailed
Kite (CSA) 2072
African Swamphen 8160
African Swamp-Warbler
121
African Swift 834
African Tailorbird 6834
African Tawny-Eagle 857
African Thrush 10342
African Thrush-Babbler
8557
African Tiger-Heron 10017
African Violet-backed
Sunbird 677
African Wagtail 6035
African Warbler 2334
African Warblers 2355
African Water-Rail 8711
African Wattled-Plover
(CSA) 10499
African Wattled-Plover
10498
African White-backed
Griffon-Vulture 4401
African White-breasted
White-eye 10757
African White-crested
Hornbill 10040
African White-eared
Bulbul 8597
African White-eye 10812
African White-naped
Raven 2678
African White-necked
Raven 2678
African White-rumped
Swift 838
African White-tailed
Nightjar (CSA) 1685
African White-throated
Bulbul 7547
African White-winged
Dove (CSA) 9500
African Wood-Owl 9536
African Wood-Pigeon 2525
African Wryneck 4926

Asiatic Golden-Weaver 7956
Asiatic Grasshopper-Warbler 5286
Asiatic House-Martin 3054
Asiatic King-Vulture 8956
Asiatic Knot 1513
Asiatic Lark 5704
Asiatic Long-billed Lark 5704
Asiatic Martin 3054
Asiatic Migratory Quail 2772
Asiatic Openbill 577
Asiatic Open-billed Stork 577
Asiatic Paradise-Flycatcher 9860
Asiatic Pied-Starling 9552
Asiatic Scoter 5692
Asiatic Short-toed Lark 1482
Asiatic Snipe 4032
Asiatic Sparrowhawk 49
Asiatic Whimbrel 6617
Asiatic White Crane 4353
Asiatic White Ibis 9964
Asiatic White-crested Hornbill 83
Asiatic White-eyed Pochard 1130
Asities 7468
Assam Brown-backed Hornbill 624
Assam Bush-Quail 7246
Assam Quail 7246
Astley's Leiothrix 5134
Astrild 3614
Atherton Scrubwren 9121
Atherton Sericornis 9121
Athi Short-toed Lark 1479
Atitlan Grebe 8023
Atiu Island Swiftlet 213
Atiu Swiftlet 213
Atjeh Pheasant 5363
Atlantic Fairy Tern 4368
Atlantic Fulmar 3965
Atlantic Gannet 6032
Atlantic Murre 10454
Atlantic Petrel 8438
Atlantic Puffin 3934

Atlantic Royal-Flycatcher 6743
Atlantic Yellow-nosed Albatross 3268
Atoll Fruit-Dove 8489
Atoll Starling 798
Atoll Warbler 110
Auckland Cormorant 7385
Auckland Island Merganser 5790
Auckland Island Teal 525
Auckland Islands Cormorant 7385
Auckland Islands Merganser 5790
Auckland Islands Rail 5217
Auckland Islands Shag 7385
Auckland Islands Snipe 2422
Auckland Merganser 5790
Auckland Shag 7385
Auckland Snipe 2422
Auckland Teal 525
Audebert's Hummingbird 2106
Audouin's Gull 5073
Audubon's Black-headed Oriole 4813
Audubon's Oriole 4813
Audubon's Shearwater 8573
Audubon's Warbler 3124, 3129
Augur Buzzard 1390,1412
August Amazon-Parrot 449
Aukland Rail 5217
Auks 2017,325
Auntie Katie (WI) 4820
Aurora Finch 8684,8685, 8687
Aurora Waxbill 8687
Austen's Barwing 156
Austen's Brown-Hornbill 622
Austen's Laughingthrush 4075,4089
Austen's Spotted-Babbler 9401
Austen's Spotted-Tree-Babbler 9401

Austral Blackbird 2902
Austral Canastero 1016
Austral Conure 3517
Austral Flowerpecker 3180,3205
Austral Hyliota 4713
Austral Negrito 5191
Austral Parakeet 3517
Austral Pygmy-Owl 4261
Austral Rail 8709
Austral Screech-Owl 6920
Austral Thrush 10304
Australasian Bittern 1263
Australasian Bush-Lark 5934
Australasian Emu 3309
Australasian Gannet 6034
Australasian Goshawk 45
Australasian Grass-Owl 10419
Australasian Grebe 9686
Australasian Grey-Teal 550
Australasian Harrier 2297
Australasian Kingfisher 10057
Australasian Koel 3644
Australasian Lark 5934
Australasian Little Bittern (ANZ) 4894
Australasian Magpie 4392
Australasian Marsh-Harrier 2297
Australasian Pipit 721
Australasian Pochard 1129
Australasian Robins 7285
Australasian Shoveler 564
Australasian Wild Duck 572
Australasian Wrens 5528
Australian Avocet 8763
Australian Banded Fruit-Dove (ANZ) 8488
Australian Bittern 1263
Australian Black-fronted Plover 3445
Australian Black-necked Ibis 9965
Australian Black-shouldered Kite 3432
Australian Blue-billed Duck 6928

Avocet Sandpiper (ISC)
10193
Avocetbill 6747
Avocets 8764
Aylner's Babbler 10260
Aymara Parakeet 1245
Ayres's Cisticola (CSA)
2310
Ayres's Cloud-Cisticola
2310
Ayres's Eagle 4623
Ayres's Hawk-Eagle 4623
Azara Aracari 8460
Azara Cuckoo 2402
Azara's Bittern 4892
Azara's Elaenia 6197
Azara's Grass-Finch 3492
Azara's Kingbird 10406
Azara's Sandplover 2024
Azara's Spinetail 9631
Azores Bullfinch 8661
Aztec Parakeet 902
Aztec Thrush 10747
Aztek Conure 902
Azure Gallinule 8157
Azure Jay 2923
Azure Kingfisher 308
Azure Nuthatch 9230
Azure Roller 3766
Azure Tit 7091
Azure-breasted Pitta 7869
Azure-crown 428
Azure-crowned
Hummingbird 408
Azure-hooded Jay 2947
Azure-naped Jay 2930
Azure-rumped Parrot 9783
Azure-rumped Tanager
9731
Azure-shouldered Tanager
9952
Azure-winged Magpie
2955
Babax 1140
Babbling Starling 6454
Bacbakiri 9834
Bachman's Sparrow 280
Bachman's Warbler 10520
Baer's Pochard 1130
Baglafecht Weaver 7936
Bagobo Babbler 10170

Bagobo Jungle-Babbler
10170
Bahama Bananaquit 2424
Bahama Duck 527
Bahama Flycatcher 6160
Bahama Mockingbird 5897
Bahama Parrot (WI) 451
Bahama Pintail 527
Bahama Swallow 9693
Bahama Thrush (WI)
10347
Bahama Woodpecker 5685
Bahama Woodstar 1524
Bahama Yellowthroat 4187
Bahaman Mockingbird
5897
Bahaman Oriole (WI) 4809
Bahaman Yellowthroat
4187
Bahia Ant-Wren 4594
Bahia Spinetail 9663
Bahia Tapaculo 9071
Bahia Tyrannulet 7586
Baikal Bulfinch 8657
Baikal Bullfinch 8663
Baikal Teal 544
Baillon's Crake 8175
Baillon's Toucan 1144
Baird's Cormorant 7403
Baird's Creeper 6757
Baird's Flycatcher 6186
Baird's Junco 4916
Baird's Sandpiper 1497
Baird's Sparrow 475
Baird's Trogon 10232
Baker's Bowerbird 9131
Baker's Chestnut-headed
Yuhina 10679
Baker's Imperial-Pigeon
3346
Baker's Pigeon 3346
Balabac Blue-Flycatcher
2996
Bald Coot 3957
Bald Eagle 4445
Bald Friarbird 7454
Bald Ibis 4213,4214
Bald Starling 8957
Bald-faced Rail 4377
Bald-headed Wood-Shrike
7877

Baldpate (WI) 2497
Baldpate 524
Balearic Shearwater 8574
Bali Mynah 5198
Bali Starling 5198
Bali Tailorbird 6839
Balicassiao 3221
Balicassiao Drongo 3221
Ballivan's Quail 6684
Ballivan's Wood-Quail
6684
Ballmann's Malimbe 5517
Balsas Screech-Owl 6911
Balsas Woodpecker 5676
Baltimore Oriole 4811
Bamboo Antshrike 2984
Bamboo-Partridge 1149,
1150
Bamboo-Wren 8354
Bamenda Apalis 745,768
Bamenda Wattle-eye 7914
Banana Katie (WI) 4820
Bananabird (WI)
2425,3142,4809,4820
Bananal Antbird 1896
Bananal Tyrannulet 6192
Bananaquit 2425
Bananaquits 2426
Banda Honeyeater 6311
Banda Myzomela 6311,
6325
Banda Sea Honeyeater
5250
Band-backed Cactus-Wren
1633
Band-backed Wren 1633
Band-bellied Crake 8173
Band-bellied Owl 8584
Banded Antbird 3213
Banded Antcatcher 3213
Banded Ant-Wren 3213
Banded Barbet 5433
Banded Bay-Cuckoo 1464
Banded Broadbill 3758
Banded Cactus-Wren 1633
Banded Cotinga 2759
Banded Crake 8702
Banded Crane-Hawk 4211
Banded Curassow 2822
Banded Dotterel 2023
Banded Eagle-Owl 1342

Barred Warbler 9607
Barred Waxbill 3614
Barred Woodcreeper 3080
Barred Woodpecker 5655
Barred Wren 1633
Barred Wren-Warbler 1473
Barred-wing Wren-Babbler
9279
Barred-winged Rail 6511
Barrot's Fairy 4513
Barrow Island White-
winged Fairywren
(ANZ) 5542
Barrow's Bustard 3732,
3743
Barrow's Goldeneye 1356
Bar-rumped Godwit 5260
Bar-shouldered Dove 4150
Bar-shouldered Ground-
Dove 4150
Bar-shouldered Mangrove
Dove 4150
Bar-tailed Cuckoo-Dove
(ISC) 5477
Bar-tailed Desert Lark 485
Bar-tailed Godwit 5260
Bar-tailed Lark 485,488
Bar-tailed Pheasant 9675,
9677
Bar-tailed Sandlark 485
Bar-tailed Snowfinch 6029
Bar-tailed Thornbill 18
Bar-tailed Treecreeper
1931
Bar-tailed Trogon 10236,
10246,773
Bar-throated Apalis 769
Bar-throated Minla 5907
Bar-throated Siva 5907
Bar-throated Wreathed-
Hornbill 91
Bartlett's Bleeding-heart
3997
Bartlett's Blood-breasted
Pigeon 3997
Bartlett's Pigeon 3997
Bartlett's Punalada 3997
Bartlett's Tinamou 2858
Bartram's Sandpiper 1157
Bar-vented Wren 10005

Bar-winged Cinclodes
2246
Bar-winged Flycatcher-
Shrike 4536
Bar-winged Hemispingus
4536
Bar-winged Oriole 4821
Bar-winged Prinia 8190
Bar-winged Pygmy-Triller
4536
Bar-winged Rail 6511
Bar-winged Weaver 7932
Bar-winged Wood-Wren
4581
Bar-winged Wren-Babbler
9279,9279
Bar-winged Wren-Warbler
8190
Basilica Pigeon 3347
Basra Reed-Warbler 122
Bass Straits Tern 9445
Bassian Thrush 10735
Bastard Cock (WI) 9315
Bastard Grieve (WI) 10558
Bastard Hawk (WI) 3815
Bat Falcon 3811,5444
Bat Hawk 5444
Bat Kite 5444
Bat Lorikeet 5371
Bat-eating Buzzard 5444
Bat-eating Hawk 5444
Bateleur 9839
Bate's Olive-Sunbird 6368
Bates's Black-Swift 835
Bates's Forest-Nightjar
1655
Bates's Nightjar 1655
Bates's Paradise-Flycatcher
9851
Bates's Sunbird 6368
Bates's Swift 835
Bates's Weaver 7938
Batjan White-eye 10761
Bat-like Spinetail 6356
Bat-like Spinetailed-Swift
6356
Baudin's Black-Cockatoo
(ANZ) 1549
Baudo Guan 7222
Baudo Oropendola 4395
Bauer's Parakeet 1156

Bauer's Parrot 1156
Baumann's Bulbul 7550
Baumann's Greenbul 7550
Baumann's Olive-Bulbul
7550
Baumann's Olive-Greenbul
7550
Bawean Serpent-Eagle
9303
Bay Ant-Pitta 4301
Bay Coucal 1849
Bay Hornero 3973
Bay Owl 7496
Bay Woodpecker 1238
Bay Wren 9999
Baya 7974
Baya Weaver 7974
Baya Weaverbird (ISC)
7974
Bay-backed Ant-Pitta 4314
Bay-backed Shrike 5065
Bay-breasted Cuckoo 4701
Bay-breasted Warbler 3126
Bay-breasted Warbling-
Finch 8147
Bay-capped Wren-Spinetail
9268
Bay-chested Warbling-
Finch 8147
Bay-crowned Brush-Finch
1077
Bay-crowned Finch 1077
Bay-headed Bee-eater 5806
Bay-headed Tanager 9749
Bay-Owl (ISC) 7495
Bay-ringed Tyrannulet
7604
Bay-ringed Tyrant 7604
Bay-vented Cotinga
3291,3292
Bay-winged Cowbird 5965
Bay-winged Hawk 7012
Beach Dikkop 1375
Beach Goose 632
Beach Kingfisher 10077
Beach Stone-Curlew 1375
Beach Thick-knee 1373,
1375
Bean Goose 636
Bearded Barbet 5424
Bearded Bee-eater 5796

Bearded Bellbird 8251
Bearded Bulbul 2834
Bearded Flycatcher 6170
Bearded Greenbul 2834
Bearded Guan 7214
Bearded Helmetcrest 6926
Bearded Manakin 5550
Bearded Melidectes 5731
Bearded Mountaineer 6759
Bearded Parrotbill 7008
Bearded Partridge 3170,
    7247
Bearded Penguin 8638
Bearded Reedling 7008
Bearded Robin (CSA) 1922
Bearded Screech-Owl 6862
Bearded Scrub-Robin 1913,
    1922
Bearded Tachuri 8101
Bearded Tit 7008
Bearded Tit-Babbler 7008
Bearded Tree-Partridge
    3170
Bearded Tree-Quail 3170
Bearded Vulture 4398
Bearded Wood-Partridge
    3170
Bearded Woodpecker 3163
Beaudoin's Snake-Eagle
    2282
Beaudouin's Harrier-Eagle
    2282
Beautiful Conure 8665
Beautiful Firetail 9418
Beautiful Firetail-Finch
    9418
Beautiful Flowerpecker
    3189
Beautiful Fruit-Dove
    8520,8525
Beautiful Grass-Parakeet
    6485
Beautiful Hummingbird
    1537
Beautiful Jay 2950
Beautiful Kingfisher 9787
Beautiful Long-tailed
    Sunbird 6416
Beautiful Lorikeet 2062
Beautiful Myzomela 6333
Beautiful Niltava 6550

Beautiful Nuthatch 9236
Beautiful Parakeet 8316
Beautiful Parrot 6485,7761
Beautiful Rosefinch 1764
Beautiful Sibia 4620
Beautiful Sunbird 6416
Beautiful Topaz 10105
Beautiful Treerunner 5568
Beautiful Woodpecker
    5680
Beavan's Bullfinch 8658
Beavan's Wren-Warbler
    8205
Beccari's Scops-Owl 6863
Beccari's Scrubwren 9116
Beccari's Sericornis 9116
Beck's Bowerbird 9131
Beck's Petrel 8426
Bedford's Paradise-
    Flycatcher 9852
Bee Hummingbird (WI)
    5774
Bee Hummingbird 5773
Beechey Jay 2922
Beechey's Jay 2922
Bee-eater 5798
Bee-eaters 5795
Beenybird (WI) 2425
Behn's Thrush 10361
Beijing Flycatcher 3830
Beijing Laughingthrush
    4084
Belcher's Gull 5074
Belding's Jay 781
Belding's Yellowthroat
    4181
Belford's Honeyeater 5727
Belford's Melidectes 5727
Bell Miner 5556
Bellbird 650
Bell-Magpie 4392
Bell's Sage-Sparrow 496
Bell's Sparrow 496
Bell's Vireo 10562
Bell's Warbler 1162
Bellshrike 5019,5028
Belted Flycatcher 10641
Belted Kingfisher 5586
Bendire Thrasher 10121
Bendire's Thrasher 10121
Bengal Bush-Lark 5924

Bengal Florican 3733
Bengal Green-Pigeon
    10157
Bengal Lark 5924
Bengal Pitta 7845
Bengal Vulture (ISC) 4402
Bengal Weaver 7939
Bengalese Finch 5321
Bennett's Cassowary 1793
Bennett's Crow 2680
Bennett's Woodpecker
    1588,1595
Bensch's Mesite 6011
Bensch's Monia 6011
Bensch's Rail 6011
Benson's Robin-Chat 8335
Benson's Rock-Thrush
    8335
Bentbill 6728,6729
Bequia Sweet (WI) 8692
Berard's Diving-Petrel
    7193
Berlepsch's Antbird 6235
Berlepsch's Ant-Pitta 4731
Berlepsch's Canastero 1019
Berlepsch's Conure 8674
Berlepsch's Dacnis 3029
Berlepsch's Emrald 2169
Berlepsch's Gnatcatcher
    8076
Berlepsch's Parakeet 8674
Berlepsch's Pigeon
    2479,2521
Berlepsch's Softtail 1019
Berlepsch's Tinamou 2859
Berlepsch's Tyrannulet
    9185
Berlepsch's Woodstar 94
Berlioz's Black-Flycatcher
    5657
Berlioz's Flycatcher 5657
Berlioz's Sunbird 684
Berlioz's Swift 836
Bermuda Petrel 8428
Bernier's Teal 528
Bernier's Vanga 6776
Bernstein's Coucal 1847
Berrypeckers 5699
Berthelot's Pipit 693
Bertoni's Antbird 3323
Bertram's Weaver 7940

Black-and-white Tanager
2571
Black-and-white Tody-
Flycatcher 8035
Black-and-white Tody-
Tyrant 8035
Black-and-white Triller
4974
Black-and-white Vanga-
Flycatcher 1231
Black-and-white Warbler
5953
Black-and-white Wren
5529
Black-and-white Wren-
Warbler 5529
Black-and-white-casqued
Hornbill 1888
Black-and-white-tailed
Hornbill 10046
Black-and-yellow Bishop
3716
Black-and-yellow Broadbill
3759
Black-and-yellow Crested-
Flycatcher 4344
Black-and-yellow
Flycatcher 3844
Black-and-yellow
Grosbeak 6118
Black-and-yellow
Hawfinch 6118
Black-and-yellow Monarch
5983
Black-and-yellow
Monarch-Flycatcher
5983
Black-and-yellow Silky-
Flycatcher 7366
Black-and-yellow Tanager
2213
Black-backed Antshrike
8930
Black-backed Apalis 766
Black-backed Barbet 5428
Black-backed Bittern
1263,4895
Black-backed Blue-Wren
5545
Black-backed Brushfinch
10478

Black-backed Bush-
Tanager 10478
Black-backed Butcherbird
2790
Black-backed Button-Quail
10372
Black-backed Chat 8996
Black-backed Cisticola
(CSA) 2328,2330
Black-backed Cloud-
Cisticola 2328
Black-backed Duck 8959
Black-backed Finch 10478
Black-backed Forktail 3519
Black-backed Fruit-Dove
8487
Black-backed Grosbeak
7440
Black-backed Ground-
Thrush 10752
Black-backed Honeyeater
8546
Black-backed Imperial-
Pigeon 3359
Black-backed Kingfisher
(ISC) 1966
Black-backed Magpie 4392
Black-backed Monarch
5991,5991
Black-backed Mouse-
Babbler 2816
Black-backed Mouse-
Warbler 2816
Black-backed Nightingale-
Thrush 1820
Black-backed Oriole
4800,4804
Black-backed Pitta 7870
Black-backed Puffback
3335
Black-backed Puffback-
Shrike 3335
Black-backed Seedeater
9361
Black-backed Shelduck
9717
Black-backed Sibia 4618
Black-backed Streaked-
Honeyeater 8546
Black-backed Tanager
9763

Black-backed Thornbill
8750
Black-backed Three-toed
Woodpecker 7683
Black-backed Tody
Flycatcher 0081,10091
Black-backed Triller 4969
Black-backed Wagtail 6048
Black-backed Water-Tyrant
3860
Black-backed Woodpecker
(ISC) 2205
Black-backed Woodpecker
7683
Black-backed Wren 5545
Black-backed Yellow
Woodpecker 2205
Black-backed-Thornbilled
Hummingbird 8750
Black-banded Barbet 5604
Black-banded Crake 740
Black-banded Creeper
3085
Black-banded
Flowerpecker 3194
Black-banded Flycatcher
3855
Black-banded Fruit-Dove
8482
Black-banded Owl 9523
Black-banded Pigeon 8482
Black-banded Plover 2048
Black-banded Rail 740
Black-banded Sandplover
2048
Black-banded Tanager
9759
Black-banded Woodcreeper
3085
Black-barred Cuckoo-
Shrike 2622,2622
Black-beaked Bronze-
Mannikin 5305
Black-bearded Pigeon 9421
Black-bellied Ant-Wren
3874
Black-bellied Bustard 3739
Black-bellied Cuckoo 7671
Black-bellied Cuckoo-
Shrike 2640

Black-faced Monarch 5996
Black-faced Munia 5311
Black-faced Parrotfinch
  3605
Black-faced Pitta 7841
Black-faced Prinia 8201
Black-faced Quailfinch
  6843
Black-faced Quelea 8690
Black-faced Rufous-
  Warbler 1193,1194
Black-faced Sandgrouse
  8409
Black-faced Shag 7388
Black-faced Sheathbill
  2081
Black-faced Shrikebill
  2383
Black-faced Solitaire 6106
Black-faced Spinetail 9661
Black-faced Spoonbill
  7881
Black-faced Tanager 9010
Black-faced Treepie 3076
Black-faced Warbler 3
Black-faced Waxbill (CSA)
  3618
Black-faced Waxbill 3622
Black-faced Weaver 7990
Black-faced Woodswallow
  994
Black-fced Oriole 6780
Black-footed Albatross
  3277
Black-footed Penguin 9290
Black-fronted Babbler
  8880
Black-fronted Bulbul 8612
Black-fronted Bush-Shrike
  9829
Black-fronted Dotterel
  3445
Black-fronted
  Flowerpecker 3194
Black-fronted Ground-
  Tyrant 6092
Black-fronted Guan 7773
Black-fronted
  Hummingbird 4724
Black-fronted
  Laughingthrush 4119

Black-fronted Nunbird
  6010
Black-fronted Nunlet 6010
Black-fronted Parakeet
  2966
Black-fronted Parrotbill
  7033
Black-fronted Piping-Guan
  7773
Black-fronted Plover 3445
Black-fronted Quail 10373,
  6683
Black-fronted Tara 2097
Black-fronted Tern 2097,
  2098
Black-fronted Tody-
  Flycatcher 10083
Black-fronted Tyrannulet
  7596
Black-fronted Weaver 7990
Black-fronted White-eye
  10761,10762,10799
Black-fronted Wood-Quail
  6683
Black-girdled Barbet 1639
Black-goggled Tanager
  10189
Black-gorgeted
  Laughingthrush 4108
Black-head (WI) 1127
Blackhead 9134
Blackhead Canary 9134
Black-headed Antbird 7241
Black-headed Ant-Thrush
  3867
Black-headed Apalis 759
Black-headed Babbler
  8880,9397
Black-headed Batis 1203
Black-headed Bee-eater
  5800
Black-headed Berryeater
  1784
Blackheaded Brush-Finch
  1058
Black-headed Bulbul
  8588,8626
Black-headed Bunting
  3472
Black-headed Bush-Shrike
  5018,9819

Black-headed Butcherbird
  2788
Black-headed Caique 7756
Black-headed Canary 9134
Black-headed Cloud-
  Cisticola 2328
Black-headed Conure 6343
Black-headed Cotinga 1784
Black-headed Cuckoo-
  Shrike 2636
Black-headed Diamondbird
  7048
Black-headed Duck 4612
Black-headed Finch 1058
Black-headed Finchbill
  9338
Black-headed Flyeater
  4232
Black-headed Forest Oriole
  6793
Black-headed Fruit-Dove
  8508
Black-headed Fruiteater
  7814
Black-headed Gerygone
  4232
Black-headed Gonolek
  5018
Black-headed Greenfinch
  1716,1743
Black-headed Grey-Jay
  7266
Black-headed Grosbeak
  7443,7444
Black-headed Gull 5110
Black-headed Hemispingus
  4547
Black-headed Heron 958
Black-headed Honeyeater
  5764,6328
Black-headed Ibis (ISC)
  9962
Black-headed Ibis 9964,
  9965
Black-headed Jay
  2930,4129,7266
Black-headed Lapwing
  10502
Black-headed Logrunner
  6824

Blossom-headed Parakeet
(ISC) 8366
Blossom-headed Parakeet
8376
Blue Baize (WI) 3672
Blue Bird of Paradise 7021
Blue Bunting 2920,5131
Blue Bustard 3734
Blue Chaffinch 3951
Blue Chat (ISC) 5410
Blue Chlorophonia 2127
Blue Cotinga 2758,2761
Blue Coua 2777
Blue Crane 3405,4356
Blue Crow 4393
Blue Crowned-Goura 4289
Blue Crowned-Pigeon 4289
Blue Cuckoo-Shrike 2612
Blue Dacnis 3030
Blue Dove (WI) 4208
Blue Dove 2371
Blue Duck 2074,4763
Blue Eared-Pheasant 2842
Blue Fairy Flycatcher 3443
Blue Fairywren 5535
Blue Fantail 8861
Blue Finch 4364,8162
Blue Flycatcher 3831
Blue Frog-Hawk 72
Blue Gauldin (WI) 3400
Blue Gaulin (WI) 3400,954
Blue Gay (WI) 3672
Blue Goose 631
Blue Grosbeak 4364
Blue Ground-Dove 2371
Blue Grouse 3062
Blue Hawk 3797
Blue Heron 3405
Blue Hill-Pigeon 2517
Blue Honeycreeper 2911
Blue Ibis 9935
Blue Jay (CSA) 2601
Blue Jay (ISC) 2600
Blue Jay 2916
Blue Jewel-Babbler 8552
Blue Korhaan (CSA) 3734
Blue Lorikeet 10555
Blue Lory 10555
Blue Madagascar Coua
2777

Blue Madagascar Coucal
2777
Blue Magpie 10458,7674
Blue Manakin 2084
Blue Mockingbird 5720
Blue Mountain Duck 8429
Blue Mountain Lorikeet
10177
Blue Mountain Parrot
10177
Blue Mountain Vireo (NA)
10585
Blue Mountain-Duck 4763
Blue Mountains Vireo
10585
Blue Mountain-Warbler
9604
Blue Noddy 8249
Blue Nuthatch 9230
Blue Paradise-Flycatcher
9856
Blue Partridge (WI) 4208
Blue Peafowl 7180
Blue Penguin 3656
Blue Petrel 4454,6982
Blue Pitta 7847
Blue Quail 1522,2767,2768
Blue Quit (WI) 3697
Blue Rail 8178
Blue Reef-Egret 3407
Blue Reef-Heron 3407
Blue Robin 2238
Blue Rock-Pigeon (ISC)
2500
Blue Rock-Thrush 6022
Blue Roller 2603
Blue Rosella 7884
Blue Roughwing 8297
Blue Rough-winged
Swallow 8297
Blue Saw-wing 8297
Blue Saw-winged Swallow
8297
Blue Shag 7406
Blue Shortwing 1276
Blue Swallow 4658
Blue Tanager 9953,9957
Blue Ternlet 8249
Blue Thrasher (WI) 3383
Blue Thrush (WI) 3383
Blue Tit 7085

Blue Titmouse 7085
Blue Vanga 2941
Blue Waxbill (CSA) 10447
Blue Waxbill 10446
Blue Whistling-Thrush
6226
Blue Wren-Warbler 5535,
5536
Blue-and-black Jay
2934,2940
Blue-and-black Tanager
9774
Blue-and-gold Macaw 862
Blue-and-gold Tanager
1151,1152
Blue-and-grey
Sparrowhawk 56
Blue-and-orange Niltava
6550
Blue-and-white Flycatcher
2957,5541
Blue-and-white Kingfisher
10061,1752
Blue-and-white
Mockingbird 5721
Blue-and-white Swallow
8635
Blue-and-white Wren-
Warbler 5541
Blue-and-yellow Macaw
862
Blue-and-yellow Tanager
9950
Blue-backed Conebill 2550
Blue-backed Fairy-bluebird
4877
Blue-backed Jay 2937
Blue-backed Kingfisher
10070
Blue-backed Manakin
2083,2087
Blue-backed Niltava 2995
Blue-backed Parrot 9783
Blue-backed Pitta 7868
Blue-backed Tanager 2914
Blue-banded Grass-
Parakeet 6482
Blue-banded Grass-Parrot
6482
Blue-banded Kingfisher
313

Brazilian Woodnymph
9893
Brazza's Martin 7437
Brazza's Swallow 7437
Brehm's Parrot 8356
Brehm's Tiger-Parrot 8356
Brent 1312
Brent Goose 1312
Brewer's Blackbird 3677
Brewer's Sparrow 9330
Brewster's Warbler 10525
Briar Warbler 8204
Bridges's Antshrike 9913
Bridges's Woodcreeper
3325
Bridled Chickadee (NA)
7136
Bridled Guillemot 1877
Bridled Honeyeater 5225
Bridled Quail-Dove 4204
Bridled Screech-Owl 6862
Bridled Sparrow 285
Bridled Tern 9440
Bridled Tit 7136
Bridled Titmouse 7136
Bridled White-eye 10771
Brier Warbler (CSA) 8204
Bright Slaty-bellied
Ground-Warbler 9870
Bright Slaty-bellied Tesia
9870
Bright Tesia 9870
Bright-capped Cisticola
2327
Bright-green Warbler 7640
Bright-headed Cisticola
2327
Bright-rumped Attila 1098
Bright-rumped Yellow-
Finch 9215
Brigida's Woodcreeper
4708
Brimstone Canary 9173
Brimstone Serin 9173
Bristlebill 1237
Bristle-crowned Starling
6736
Bristled Grassbird 1977
Bristled Grass-Warbler
(ISC) 1977
Bristled Shrike 7877

Bristled Shrike-Starling
7877
Bristlehead 8388
Bristle-necked Brownbul
7569
Bristle-nosed Barbet 4373
Bristle-thighed Curlew
6618
Bristle-Tyrant 7593
British Pied-Wagtail (NA)
6037
British Pied-Wagtail 6036
British Rock-Pipit 724
British Storm-Petrel 4694
British Yellow-Wagtail
6044
Broadbill 1135
Broadbilled Dove-Petrel
6982
Broad-billed Fairywren
5538
Broad-billed Flycatcher
6142
Broad-billed Flycatcher-
Warbler 10016
Broad-billed Hummingbird
2986,2987
Broad-billed Manakin 8953
Broad-billed Monarch 6142
Broad-billed Motmot 3437
Broad-billed Myiagra 6142
Broad-billed Myiagra-
Flycatcher 6142
Broad-billed Prion 6982
Broad-billed Roller (ISC)
3769
Broad-billed Roller 3767
Broad-billed Sandpiper
5251
Broad-billed Sapayoa 8953
Broad-billed Tody 10099
Broad-billed Warbler
10016
Broad-billed Wren 5538
Broad-billed Wren-Warbler
5538
Broadbills 3757
Broad-ringed White-eye
10815
Broad-tailed Camaroptera
1562

Broad-tailed Grassbird
9023
Broad-tailed Grass-Warbler
9023
Broad-tailed Hummingbird
9097
Broad-tailed Paradise-
Whydah 10543,10544
Broad-tailed Paradise-
Widow 10543
Broad-tailed Petrel 3945
Broad-tailed Thornbill 9
Broad-tailed Warbler
(CSA) 9022
Broad-tailed Warbler
9021,9023
Broad-tipped Hermit 7339
Broad-winged Hawk 1406
Broad-zoned Kingfisher
313
Brolga 4357
Bronze Cuckoo 2189
Bronze Drongo (CSI) 3216
Bronze Euphonia 3701
Bronze Ground-Dove 3995
Bronze Hermit 4275
Bronze Mannikin 5294
Bronze Munia 5294
Bronze Pygmy-Tyrant
8350
Bronze Shag 7384
Bronze Sunbird 6395
Bronze-backed
Flowerpecker 3195
Bronze-backed Imperial-
Pigeon 3345
Bronze-brown Cowbird
5963
Bronze-collared Pigeon
6853
Bronzed Cowbird 5962
Bronzed Drongo 3216
Bronzed Eastern Pygmy-
Sunbird 679
Bronzed Grackle
8698,8699
Bronzed Pygmy-Sunbird
679
Bronzed Racket-tailed
Treepie 2850

Buff-throated Partridge 9882

Buff-throated Purpletuft 4869

Buff-throated Saltator 8945

Buff-throated Sunbird 6360

Buff-throated Thickhead 4725

Buff-throated Tody-Tyrant 4570

Buff-throated Treehunter 9973

Buff-throated Warbler 7657

Buff-throated Willow-Warbler 7657

Buff-throated Woodcreeper 10660

Buff-vented Bulbul 4871

Buff-winged Starfrontlet 2415

Buffy Fish-Owl 4933

Buffy Hummingbird 5194

Buffy Pipit 736

Buffy Rockjumper 1973

Buffy Tuftedcheek 8330

Buffy-crested Hornbill 623

Buffy-crowned Tree-Partridge 3171

Buffy-crowned Tree-Quail 3171

Buffy-crowned Wood-Partridge 3171

Buffy-fronted Seedeater 9356

Buffy-fronted Wood-Partridge 3171

Buffy-throated Seedeater 9356

Bukida White-eye 10797

Bulbul (CSA) 8590

Bulbul 2836

Bulbuls 8587

Buller's Albatross 3266

Buller's Gull 5076

Buller's Mollymawk 3266

Buller's Shearwater 8565

Bullfinch 8663

Bull-headed Shrike 5038

Bulloak Parrot 6583

Bullock's Oriole 4804

Bull-of-the-bog 1264

Bulloo Grey-Grasswren (ANZ) 501

Bully Canary 9173

Bully Seedeater 9173

Bullybird 5293

Bulo Burti Boubou 5023

Bulwer's Petrel 1366

Bulwer's Pheasant 5358

Bumblebee Hummingbird 1090

Buntings 3490

Burbidge's Parrot 9783

Burchell's Coucal 1848

Burchell's Courser 2906

Burchell's Glossy-Starling 4992

Burchell's Gonolek 5013

Burchell's Sandgrouse 8407

Burchell's Starling 4992

Burdekin Duck 9717

Bürger's Goshawk 3599

Bürger's Sparrowhawk 3599

Burmeister's Ground-Tyrant 6088

Burmeister's Seriema 2217

Burmeister's Woodstar 5886

Burmese Babbler 10265

Burmese Francolin 3920

Burmese Hornbill 90

Burmese Ixulus 10686

Burmese Jay 4128

Burmese Marsh-Tit 7119

Burmese Peacock-Pheasant 8092

Burmese Scaly-bellied Woodpecker 7742

Burmese Shrike 5042

Burmese Spottbill (ISC) 560

Burmese Spotted-Dove 9493

Burmese White-throated Babbler 10265

Burmese Yuhina 10686

Burnished-buff Tanager 9733

Burnt-neck Eremomela 3571

Burnt-necked Eremomela 3571

Burrowing Owl 9280

Burrowing Parakeet 2943

Burrowing Parrot 2943

Burton's Finch 1516

Buru Cuckoo-Shrike 2624

Buru Fantail 8862

Buru Flowerpecker 3187

Buru Flycatcher 3832

Buru Ground-Thrush 10725

Buru Honeyeater 5242

Buru Island Cuckoo-Shrike 2624

Buru Island White-eye 10764

Buru Islands Cuckoo-Shrike 2624

Buru Jungle-Flycatcher 8801

Buru Lorikeet 2066

Buru Monarch 5994

Buru Mountain-White-eye 5483

Buru Oriole 6780

Buru Parrot 9780

Buru Racket-tail 8214

Buru Racket-tailed Parakeet 8214

Buru Racket-tailed Parrot 8214

Buru Racquet-tail 8214

Buru Racquet-tailed Parakeet 8214

Buru Racquet-tailed Parrot 8214

Buru Red-Lory 3539

Buru Rufous-Fantail 8862

Buru Thrush 10725

Buru Yellow White-eye 10764

Bush Blackcap (CSA) 5266

Bush Bustard 3741

Bush Fowl 3897

Bush Hawk 3807

Bush Lark (ISC) 5924

Bush Petronia 7286

Bush Pipit 696

Bush Shrike 9819

Bush Snipe 2423

Bush Sparrow 7286
Bush Stone-Curlew (ANZ)
    1374
Bush Tanager 2150
Bush Thick-knee 1374
Bush Tit 8300
Bush Wren 10622
Bush-Crow 10691
Bush-hen 391
Bush-Lark 5934
Bush-Quail 7245
Bush-Robin 8056
Bush-Shrikes 5490
Bushveld Pipit (CSA) 697
Bushveld Pipit 696
Bushveld Tree Pipit 696
Bush-Warbler 1474,1955
Bushy-crested Hornbill 623
Bushy-crested Jay 2931
Bustard-Quail 10379,10380
Bustards 6852
Butcherbird 5040
Butembo Greenbul 2120
Buton Hornbill 82
Buton Mynah 9490
Buton Starling 9490
Butterbird (WI) 3290
Butterflybird (WI) 9188
Büttikofer's Babbler
    5487,7209
Button-grass Parrot 7297
Button-Quail 10380
Button-Quails 10368
Buzzard 1395,1396
Buzzing Flowerpecker
    3193
Caatinga Ant-Wren 4597
Caatinga Black-Tyrant
    4939
Caatinga Cachalote 8346
Caatinga Catchalote 8346
Caatinga Nighthawk 2188
Caatinga Parakeet 906
Cabanis's Bunting 3452
Cabanis's Emerald 2169
Cabanis's Greenbul 7551
Cabanis's Ground-Sparrow
    5782
Cabanis's Spinetail 9633
Cabanis's Tanager 9731
Cabanis's Thrush 10346

Cabanis's Tiger-Heron
    10020
Cabanis's Tyrant 4937
Cabanis's Warbler 1164,
    1170
Cabanis's Yellow Bunting
    3452
Cabot's Tern 9462
Cabot's Tragopan 10138
Cabot's Wren 9981
Cacique 1445,1447,
    1448,1451,1456
Cackling Goose 1318
Cackling-Falcon 3784
Cactus Canastero 1020
Cactus Conure 906
Cactus Ground-Finch 4178
Cactus Parakeet 906
Cactus Wren 1620
Cactus-Finch 4178
Cadet Hummingbird 584
Caffer Swift 838
Cagon 8894
Cagu 8894
Cahow 8428
Cahow Petrel 8428
Caica Parrot 7758
Caique 7756
Calandra Lark 5702
Caledonian Myiagra 6128
Calfbird 7268
California Condor 4379
California Gnatcatcher
    8075
California Gull 5078
California Jay 781
California Quail 1519
California Spotted-Owl
    9529
California Thrasher 10129
California Towhee 7778
Californian Condor 4379
Californian Scrub Jay 781
Califoronian Pygmy-Owl
    4248
Calliope Hummingbird
    9428
Cambodian Hill-Partridge
    928
Cameroon Apalis 753

Cameroon Bare-headed
    Rockfowl 7678
Cameroon Black-capped
    White-eye 9273
Cameroon Blue-headed
    Sunbird 6410
Cameroon Bulbul 597,7564
Cameroon Cliff-Swallow
    4666
Cameroon Francolin 8403
Cameroon Greenbul 587,
    597,7564
Cameroon Ground-Thrush
    10718
Cameroon Indigobird
    10532
Cameroon Montane
    Greenbul 597
Cameroon Mountain Saw-
    wing 8289
Cameroon Mountain-
    Bulbul 597
Cameroon Mountain-Bush-
    shrike 5496
Cameroon Mountain-Chat
    2750
Cameroon Mountain-
    Francolin 8403
Cameroon Mountain-
    Greenbul 597
Cameroon Mountain-
    Warbler 1300
Cameroon Olive-Greenbul
    7564
Cameroon Olive-Pigeon
    2518
Cameroon Pipit 697
Cameroon Rameraon
    Pigeon 2518
Cameroon Robin-Chat
    2750
Cameroon Rockfowl 7678
Cameroon Roughwing
    8289
Cameroon Rough-winged
    Swallow 8289
Cameroon Saw-wing 8289
Cameroon Scrub-Warbler
    1300
Cameroon Sombre
    Greenbul (CSA) 587

Chinese Pipit 734
Chinese Pitta 7862
Chinese Pond-Heron 964
Chinese Reed-Bunting 3489
Chinese Reed-Warbler 136
Chinese Rusty-cheeked Scimitar-Babbler 8115
Chinese Scrub-Warbler 1308
Chinese Sedge-Warbler 136
Chinese Shortwing 4691
Chinese Shrike 5061
Chinese Snipe 4025
Chinese Song-Thrush 10330
Chinese Sparrowhawk 72
Chinese Spot-breasted Scimitar-Babbler 8115
Chinese Starling 9562
Chinese Striated-Swallow 4687
Chinese Swiftlet 194
Chinese Tern 9446
Chinese Thrush 10330, 4080
Chinese Tragopan 10138
Chinese Turtle-Dove 9493
Chinese Water-Pheasant 4697
Chinese White-eye 10786
Chinese White-gorgetted Flycatcher 8803
Chinese Willow Tit 7127
Chinese Willow-Warbler 7657
Chinese Yellow-bellied Leaf-Warbler 7657
Chinese Yellow-Tit 7128
Ching-ching (WI) 8696
Chinquis Peacock-Pheasant 8092
Chinspot Batis 1206,1214
Chin-spot Flycatcher 1206
Chin-spot Puffback 1206
Chin-spot Puffback-Flycatcher 1206
Chinstrap Penguin 8638
Chip-chip (WI) 3144,3145, 3153

Chipping Sparrow 9332
Chipwillow 1701
Chir Pheasant 1828
Chiribiquete Emerald 2168
Chirinda Apalis 749
Chiriqui Pigeon 4195
Chiriqui Quail-Dove 4195
Chiriqui Tapaculo 9056
Chiriqui Yellowthroat 4183
Chiriquiri Dove 4195
Chiriquiri Quail-Dove 4201
Chiriquiri Tapaculo 9051
Chiriri Parrotlet 1321
Chirping Cisticola 2344
Chirruping Wedgebill 8394
Chivi Vireo 10567,10584
Chloridops 2107
Chlorotic Euphonia 3686
Chocho Quit (WI) 3697
Chockallot 1435
Choco Poorwill 6639
Choco Tapaculo 9057
Choco Tinamou 2869
Choco Toucan 8722
Choco Vireo 10580
Choco Warbler 1165
Choco Woodpecker 10510
Chocolate Flycatcher 3282, 6080
Chocolate Tyrant 6499
Chocolate-backed Kingfisher 4432
Chocolate-vented Tyrant 6499
Choiseul Pigeon 5867
Choliba Scops-Owl 6867
Cholo Alethe 360
Cholo Mountain Alethe 360
Chopi Blackbird 4282
Chopi Grackle 4282
Chorister Chat 2746
Chorister Robin (CSA) 2746
Chorister Robin-Chat 2746
Chotoy Spinetail 9024
Choucador Glossy-Starling 5003
Chough 8652
Chowchilla 6824
Chowchillas (ANZ) 6823

Christina's Barbthroat 9958
Christmas Boobook 6558
Christmas Frigatebird (ISC) 3939
Christmas Imperial-Pigeon 3379
Christmas Island Emerald-Dove (ANZ) 1992
Christmas Island Frigatebird 3939
Christmas Island Glossy Swiftlet (ANZ) 2457
Christmas Island Hawk-Owl (ANZ) 6558
Christmas Island Imperial-Pigeon 3379
Christmas Island Island-Thrush (ANZ) 10349
Christmas Island Reed-Warbler 107
Christmas Island Shearwater 8575
Christmas Island White-eye 10806
Christmas Island White-tailed Tropicbird (ANZ) 7330
Christmas Pigeon 3379
Christmas Reed-Warbler 107
Christmas Shearwater 8575
Christmas Warbler 107
Christmas White-eye 10806
Christmasbird (WI) 3121, 3703,4682,9188
Chubb's Cisticola 2319
Chubut Steamerduck 9700
Chucao 9004
Chucao Tapaculo 9004
Chuck-will's-widow 1658
Chuck-will's-widow Nightjar 1658
Chukar 346
Chukar Partridge 346
Churrin Tapaculo 9064
Churring Cisticola 2343
Chusquea Tapaculo 9069
Cicada Cuckoo-Shrike 2626

Cipo Canastero 1028
Cirl Bunting 3460
Citreoline Trogon 10234,
    10242
Citril 9145
Citril Finch 9145
Citrine Canary-Flycatcher
    2900
Citrine Flycatcher 2900
Citrine Wagtail 6040
Citrine Warbler 1180
Citron-bellied Attila 1095
Citron-crested Cockatoo
    1442
Citron-headed Yellow-
    Finch 9209
Citron-throated Toucan
    8723
Clamorous Reed-Warbler
    137
Clapper Bush-Lark 5920
Clapper Rail 8714
Clapper-Lark 5920
Clapperton's Francolin
    3899
Claret-breasted Fruit-Dove
    8532
Clarion Island Wren 10227
Clarion Wren 10227
Clarke's Weaver 7953
Clark's Crow 6610
Clark's Grebe 160
Clark's Nutcracker 6610
Clay-coloured Robin (NA)
    10309
Clay-coloured Sparrow
    (NA) 9331
Clay-coloured Thrush
    10309
Clicking Peltops 7210
Clicking Peltops-Flycatcher
    7210
Cliff Flycatcher 4650
Cliff Swallow 7272
Cliff-Chat 9900
Cling-cling (WI) 8696
Clink Parakeet 5128
Clinking Currawong 9488
Close-barred Sandgrouse
    8412,8413
Clot Bey's Lark 8747

Clotbey Lark 8747
Cloud Cisticola (CSA)
    2351
Cloud Forest Swallow 6605
Cloud-forest Screech-Owl
    6878,6895
Cloudscraper (CSA)
    2310,2334
Cloudscraper Cisticola
    2321
Cloud-scraping Cisticola
    2321
Cloven-feathered Dove
    3303
Club-winged Manakin
    5441
Clucking Hen (WI) 900
Coal Tit 7080
Coal Titmouse 7080
Coal-black Flowerpiercer
    3243
Coal-crested Finch 2053
Coast Bush-Tit 8300
Coast Swallow 4688
Coastal Flycatcher 3503
Coastal Miner 4168
Coastal Screech-Owl 6905
Cobalt-winged Parakeet
    1323
Coban Swallow 6605
Cobb's Wren 10218
Cocha Antshrike 9923
Cochabamba Mountain-
    Finch 8139
Cockatiel 6644
Cockatoo Parrot 6644
Cockatoos 1444
Cockerell's Fantail 8825
Cockerell's Honeyeater
    10173
Cock-tailed Tyrant 357
Cocoa Thrush 10307
Cocoa Woodcreeper 10669
Cocoi Heron 952
Coconut Lory 10177
Coconutbird (WI) 4809
Cocos Cuckoo 2400
Cocos Fairy Tern 4368
Cocos Finch 7750
Cocos Flycatcher 6523
Cocos Islands Cuckoo 2400

Cocos Islands Finch 7750
Cocos Islands Flycatcher
    6523
Cocos-Keeling Islands
    Buff-banded Rail
    (ANZ) 4055
Codrington's Indigobird
    10534
Coe's Honeyguide 5743
Coffinbird (WI) 2403
Coiba Spinetail 2801
Colasisi 5377
Colazizi 5377
Coleto 8957
Coleto Mynah 8957
Colies 2446,2447
Colima Pygmy-Owl 4260,
    4263
Colima Warbler 10523
Collared Accentor 8273
Collared Antshrike 8926
Collared Apalis 763,767
Collared Aracari 8470
Collared Broadbill 9133
Collared Brush-Turkey
    9727
Collared Bush-Chat (ISC)
    8993
Collared Bush-Lark 5927
Collared Bush-Robin 9801
Collared Crescentchest
    5716
Collared Crow 2720
Collared Dove 9494,9502
Collared Falconet 5868
Collared Finchbill 9338
Collared Finch-billed
    Bulbul 9338
Collared Fishing Buzzard
    1381
Collared Fishing Hawk
    1381
Collared Flycatcher 3828
Collared Forest-Falcon
    5848
Collared Gnatwren 5851
Collared Grey-Fantail 8832
Collared Grosbeak 6116
Collared Hemipode 7183
Collared Hill-Partridge 934

Common Oystercatcher
4426
Common Palm-Tanager
9956
Common Paradise-
Flycatcher (ISC) 9860
Common Paradise-
Kingfisher 9787
Common Pariah-Kite (ISC)
5892
Common Parrotlet 3881
Common Partridge 7249
Common Pauraque 6636
Common Peafowl 7180
Common Penduline-Tit
8780
Common Pheasant 7433
Common Pied-
Oystercatcher 4426
Common Pied-Wagtail
6036
Common Pied-Wheatear
6724
Common Pigeon 2501
Common Pintail 523
Common Piping-Guan
7771
Common Pipit 721
Common Pochard 1132
Common Poorwill 7413
Common Potoo 6630
Common Pratincole 4242
Common Prion 6982
Common Puffback 3337
Common Puffin 3934
Common Quail 2770
Common Quailfinch 6842
Common Quaker-Babbler
334
Common Quelea 8690
Common Raven 2685
Common Redpoll 1727
Common Redshank 10207
Common Redstart 7524
Common Reed-Bunting
(NA) 3479
Common Reed-Warbler
135
Common Ringed-Plover
2028
Common Robin-Chat 2744

Common Rock-Thrush
(CSA) 6021
Common Roller 2603
Common Rosefinch 1760
Common Rubythroat 5411
Common Sand-Martin
8902
Common Sandpiper 10199
Common Sapsucker 9299
Common Scaly-
Woodpecker 7740
Common Scimitarbill 8813
Common Scops-Owl 6910
Common Scoter 5690
Common Screech-Owl
6858
Common Scrubfowl 5639
Common Scrubhen 5639
Common Scythebill 1618
Common Seaside-Sparrow
480
Common Serin 9171
Common Shag 7370
Common Shama 2594
Common Shelduck 9718
Common Short-toed Lark
1486
Common Shoveler 532
Common Shrike-Flycatcher
1230
Common Sicklebill 3775
Common Sicklewing Chat
1910
Common Sickle-winged
Chat 1910
Common Silverbeak-
Tanager 8736
Common Silverbill 5308
Common Siskin 1744
Common Skylark 299
Common Slender-billed
Weaver 7973
Common Snipe 4019
Common Snowfinch 6026
Common Social-Weaver
7469
Common Song-Thrush
10344
Common Spoonbill 7880
Common Squacco Heron
967

Common Starling 9565
Common Stilt 4629
Common Stonechat 8993
Common Streaked-Bulbul
4901
Common Swallow 4682
Common Swift 832
Common Tailorbird 6841
Common Teal 534
Common Tern 9454
Common Tetraka 7562
Common Thick-knee 1376
Common Thornbird 7304
Common Tit-Babbler 334
Common Tody-Flycatcher
10083
Common Tody-Tyrant
10083
Common Treecreeper 1930
Common Treepie 3079
Common Troupial 4817
Common Trumpeter 8389
Common Tsikirity 6506
Common Tufted Flycatcher
5946
Common Turkey 5724
Common Wattlebird 645
Common Wattle-eye
(CSA) 7912
Common Waxbill 3614
Common Waxwing 1250
Common Weaver 7974
Common Wheatear 6720
Common White Noddy
4368
Common White Tern 4368
Common White-eye 1137
Common Whiteface 778
Common Whitethroat
(CSA) 9593
Common White-throated
Warbler 9593
Common Woodcock 9040
Common Woodnymph
9892
Common Wood-Pigeon
2511
Common Wood-Shrike
9837
Common Wren 10228

Common Yellowthroat
(NA) 4191
Common Yellowthroat
4190
Comores Bulbul 4787
Comoro Blue-Pigeon 355
Comoro Brush-Warbler
6503
Comoro Bulbul 4791
Comoro Drongo 3227
Comoro Fody 3885
Comoro Fruit-Dove 355
Comoro Olive-Pigeon 2514
Comoro Scops-Owl 6902
Comoro Thrush 10294
Comoro Tsikirity 6505
Comoro Warbler 6503
Comoro White-eye 10803
Comoro Wood-Pigeon
2514
Comoros Black-Bulbul
4791
Comoros Brush-Warbler
6503
Comoros Bulbul 4791
Comoros Thrush 10294
Compact Weaver 7986
Comte de Paris's
Starfrontlet 2415
Concolor Creeper 3081
Concolor Woodcreeper
3081
Concolored Woodcreeper
3081
Condamine's Sicklebill
3776
Condor 10604,4379
Cone-billed Tanager 2570
Congo Bay-Owl 7496
Congo Bearded-Bulbul
2836
Congo Black-bellied
Sunbird 6379
Congo Bulbul 2836
Congo Greenbul 2120
Congo Martin 7437,8900
Congo Moorchat 6260
Congo Owl 7496
Congo Peacock 239,7181
Congo Peafowl 239,7181
Congo Sand-Martin 8900

Congo Saw-wing 8291
Congo Serpent-Eagle 3341
Congo Snake-Eagle 3341
Congo Sunbird 6379
Congo Swift 9026
Congo Thrush 10744
Congo Wood-Pigeon 2525
Connecticut Warbler 6748
Conover's Dove 5178
Constant's Starthroat 4505
Cook Islands Flycatcher
8110
Cook Islands Fruit-Dove
8522
Cook Islands Monarch-
Flycatcher 8110
Cook Islands Reed-Warbler
123
Cook Islands Starling 794
Cook Islands Swift 213
Cook Islands Warbler 123
Cook Pigeon 2498
Cook Straights Cormorant
7383
Cookacheea 1696
Cook's Petrel 8431
Cooper's Hawk 41
Cooper's Sandpiper 10194
Cooper's Screech-Owl
6869
Coot 3957
Copper Pheasant 9678
Copper Sunbird 6381
Copperhead 1129
Copper-rumped
Hummingbird 425
Coppersmith (ISC) 5601
Coppersmith Barbet 5601
Copper-tailed Coucal 1852
Copper-tailed Glossy-
Starling 4998
Copper-throated Sunbird
6373
Copper-vented Puffleg
3584
Copper-winged
Hummingbird 266
Coppery Emerald 2171
Coppery Metaltail 5834
Coppery Sunbird (CSA)
6381

Coppery Thorntail 8152
Coppery-bellied Puffleg
3584
Coppery-chested Jacamar
3985
Coppery-headed Emerald
3447
Coppery-necked Dove
4150
Coppery-tailed Coucal
1852
Coppery-tailed Glossy-
Starling 4998
Coppery-tailed Trogon
10229,10239
Coquerel's Coua 2778
Coqui Frankolin 3900
Cora Sheartail 9933
Coral Fruit-Dove 8489
Coralbill 4779
Coral-billed Ground-
Cuckoo 1754
Coral-billed Nuthatch 4779
Coral-billed Scimitar-
Babbler 8117
Coraya Wren 9983
Corby Crow 2686
Cordillera Canastero 1030
Cordilleran Canastero 1030
Cordilleran Flycatcher
3511
Cordilleran Snipe 4022,
4033
Cordilleran Woodcreeper
3083
Cordoba Canastero 1036
Cordoba Cinclodes 2244
Cordonbleu 10446,10448
Corella 1441
Cormorant 7381
Cormorants 7367
Corn Bunting 3454
Cornbird (WI) 5966
Corncrake 2831
Cornish Chough 8652
Coroneted Fruit-Dove 8490
Coroneted Redstart 7522
Coroneted Sandgrouse
8408
Corporalbird (WI) 6519
Correndera Pipit 703

Corsican Citril-Finch 9148
Corsican Nuthatch 9252
Cory's Shearwater 1531,
1532
Coscoroba 2738
Coscoroba Swan 2738
Costa Rica Hummingbird
3658
Costa Rica Quail-Dove
4197
Costa Rica Woodstar 7470
Costa Rican Hummingbird
3658
Costa Rican Quail 3171
Costa Rican Quail-Dove
4197
Costa Rican Woodstar
7470
Costa's Hummingbird 1539
Cotingas 2763
Cotta's Elaenia 6193
Cotton Pygmy-Goose 6531
Cotton Teal (ISC) 6531
Cottonbird (WI) 8074
Cotton-tree Plover (WI)
1157
Cotton-tree Sparrow (WI)
5397
Coucal (ISC) 1865
Couch's Jay 783,786
Couch's Kingbird 10401
Coue's Flycatcher 2585
Coue's Gadwall 533
Coulibri (WI) 6826
Coulon's Macaw 866
Count Raggi's Bird of
Paradise 7019
Courol 5174
Courol Roller 5174
Courols 5173
Coursen's Spinetail 9640
Coursers 4243
Courtois's Laughingthrush
4089
Cowbird (WI) 1347,2403
Cowbird 5966
Coxen's Fif-Parrot (ANZ)
2973
Cox's Sandpiper 1507
Cozumel Emerald 2163

Cozumel House-Wren
10215
Cozumel Thrasher 10125
Cozumel Vireo 10561
Cozumel Wren 10215,
10216
Crab Hawk (WI) 1418
Crab Plover 3311
Crabcatcher (WI) 6624
Crabcracker (WI)
6624,6635
Crab-Plovers 3304
Crackpot Soldier (WI)
4634
Crag Chestnut-winged
Starling 6734
Crag Chilia 2078
Crag Martin 4681
Crakes 8700
Crane 4351
Crane Hawk 4210
Cranes 4345,4346
Craver's Murrelet 9670
Cream-backed Woodpecker
1579
Cream-bellied Fruit-Dove
8510
Cream-bellied Gnatcatcher
8078
Cream-bellied Thrush
10290
Cream-breasted Canastero
1021
Cream-breasted Fruit-Dove
8510
Cream-coloured Courser
2905
Cream-coloured
Gnatcatcher 8078
Cream-coloured Pratincole
4236
Cream-coloured
Woodpecker (NA)
1833
Cream-rumped Miner 4166
Cream-striped Bulbul 8605
Cream-throated White-eye
10761
Cream-vented Bulbul 8616
Creamy Gnatcatcher 8078

Creamy-bellied Ant-Wren
4591
Creamy-bellied
Gnatcatcher 8078
Creamy-bellied Thrush
10290
Creamy-breasted Canastero
1021
Creamy-crested Spinetail
2795
Creamy-rumped Miner
4166
Creamy-throated White-eye
10761
Creepers 1936
Crescent Honeyeater 7544
Crescent-chested Babbler
9396
Crescent-chested Puffbird
5513
Crescent-chested Warbler
7075
Crescented Ant-Pitta 4329
Crescent-eyed Pewee (WI)
2575
Crescent-faced Ant-Pitta
4329
Crested Antbird 8791
Crested Antcatcher 8791
Crested Ant-Tanager 4409
Crested Argus 8798
Crested Argus-Pheasant
8798
Crested Auklet 222
Crested Barbet 10136
Crested Baza 1125
Crested Becard 6996
Crested Bellbird 6754
Crested Berrypecker 7041
Crested Bird of Paradise
2389
Crested Black Tit 7112
Crested Black-Tyrant 4941
Crested Bobwhite 2448
Crested Bowerbird 470
Crested Bronzewing 4155
Crested Brown-Bulbul
8599
Crested Bulbul 8603
Crested Bunting 5776

Crowned Pigeon 4289
Crowned Plover (CSA)
 10485
Crowned Plover 10480
Crowned Sandgrouse 8408
Crowned Slaty-Flycatcher
 4344
Crowned Solitary-Eagle
 4479
Crowned Warbler 7621,
 7649
Crowned Willow-Warbler
 7621
Crowned Woodnymph
 9890
Crow-Pheasant (ISC) 1865
Crows 2675
Crozet Shag 7396
Cryptic Ant-Thrush 2011
Cryptic Flycatcher 3833
Cryptic Warbler 2856
Cuban Amazon 451
Cuban Amazon-Parrot 451
Cuban Blackbird 3286
Cuban Black-Hawk 1419
Cuban Bullfinch 5778
Cuban Conure 910
Cuban Crested-Flycatcher
 6160
Cuban Crow 2708
Cuban Emerald 2170
Cuban Flicker 2432,2434
Cuban Gnatcatcher 8079
Cuban Grassquit 10011
Cuban Green-Woodpecker
 10643
Cuban Ivory-billed
 Woodpecker 1574
Cuban Kite 2182
Cuban Lizard-Cuckoo 8977
Cuban Macaw 867,880
Cuban Martin 8258
Cuban Nightjar 1665
Cuban Parakeet 910
Cuban Parrot 451
Cuban Pewee 2575
Cuban Pygmy-Owl 4271
Cuban Red-headed
 Woodpecker 5685
Cuban Red-winged
 Blackbird 250

Cuban Screech-Owl 6887
Cuban Solitaire 6103
Cuban Sparrow 10109
Cuban Swallow 6607
Cuban Tody 10098
Cuban Trogon 8238
Cuban Vireo 10574
Cuban Whistling-Duck
 3112
Cuban Woodpecker 10643,
 2434
Cuckoo 2883
Cuckoo Hawk 1121
Cuckoo Owl 4254
Cuckoo Owlet 4252,4254
Cuckoobird (WI)
 6901,9280
Cuckoo-Finch (CSA) 621
Cuckoo-Finches (CSA)
 7925
Cuckoo-Hawk 1120
Cuckoo-Roller 5174
Cuckoo-Rollers 5173
Cuckoos 2882
Cuckoo-Shrikes 1573
Cuckoo-Weaver 621
Cuming's Scrubfowl 5636
Cundinamarca Ant-Pitta
 4315
Curassows 2787
Curl-breasted Manucode
 5561
Curl-crested Aracari 8461
Curl-crested Jay 2926
Curl-crested Manucode
 5561
Curl-crested Toucan 8461
Curlew (WI)
 3640,6617,7923
Curlew 6612
Curlew Sandpiper 1499
Curlew Stint 1499
Curly-crested Manucode
 5561
Curve-billed Cacique 1456
Curve-billed Reedhaunter
 5255
Curve-billed Scythebill
 1615
Curve-billed Thrasher
 10124

Curve-billed Tinamou 6594
Curve-winged Sabrewing
 1603
Cutia 2909
Cut-throat 378
Cut-throat Finch (CSA)
 378
Cut-throat Weaver 378
Cuvier's Hummingbird
 7323
Cuvier's Kinglet 8767
Cuvier's Nightjar 1692
Cuvier's Rail 3333
Cuvier's Sabrewing 7323
Cuvier's Scaly-breasted
 Hummingbird 7323
Cuvier's Toucan 8725
Cuzco Mountain-Finch
 8136
Cuzco Tapaculo 9080
Cuzco Warbler 1166
Cyprus Pied-Wheatear
 6706
Cyprus Warbler 9602
Cyprus Wheatear 6706
Dabbene's Guan 7215
Dabchick 9688
Dagua Thrush 10301
Dainty Honeyeater 6333
Dalmatian Pelican 7196
Damar Blue-Flycatcher
 3838
Damar Flycatcher 3838
Damara Canary 9134
Damara Dove 9492
Damara Rockjumper 98
Damara Tern 9443
Damaraland Rockjumper
 98
Danish Crow 2688
Danjou's Babbler 4907
Dapper (WI) 8024
Dappled Bulbul 944
Dappled Greenbul 944
Dappled Illadopsis 944
Dappled Mountain-Bulbul
 944
Dappled Mountain-Robin
 944
Dappled Robin 944
Dapple-throat 944

Dusky-capped Greenlet 4744
Dusky-capped Vireo 4744
Dusky-chested Flycatcher 6222
Dusky-faced Tanager 5947
Dusky-green Fulvetta 330
Dusky-green Oropendola 8303
Dusky-green Tit-Babbler 330
Dusky-grey Heron 962
Dusky-headed Brush-Finch 1064
Dusky-headed Conure 924
Dusky-headed Finch 1064
Dusky-headed Parakeet 924
Dusky-legged Guan 7220
Dusky-tailed Antbird 3321
Dusky-tailed Ant-Tanager 4410
Dusky-tailed Canastero 1026
Dusky-tailed Flatbill 8752
Dusky-throated Antshrike 9903
Dusky-throated Fantail 8827
Dusky-throated Hermit 7357
Dusky-vented Storm-Petrel 3947
Dusky-winged Foliagegleaner 7481
Duyvenbode's Sunbird 227
Dwarf Bearded-Bulbul 371
Dwarf Bittern 4897
Dwarf Button-Quail 10371, 10374
Dwarf Cassowary 1793
Dwarf Cuckoo 2397,2404
Dwarf Emu 3307,3308
Dwarf Forest-Kingfisher 1969
Dwarf Fruit-Dove 8512
Dwarf Honeyeater 6700
Dwarf Honeyguide 4859
Dwarf Hornbill 10043
Dwarf Jay 2949

Dwarf Kingfisher 1969, 4884
Dwarf Koel 5859
Dwarf Lory 4279
Dwarf Manakin 10393
Dwarf Mannikin 5138
Dwarf Olive-Ibis 1256
Dwarf Pugmy Goose 6530
Dwarf Quail 2768
Dwarf Rail 8175
Dwarf Raven (CSA) 2714
Dwarf River-Kingfisher 312
Dwarf Sparrowhawk 62
Dwarf Tinamou 9792
Dwarf Turtle-Dove 9505
Dwarf Tyrant-Manakin 10393
Dwarf Vireo 10583
Dwarf Whistler 6935
Dyal 2596
Dyal-Thrush 2596
Dybowski's Dusky Twinspot 3773
Dybowski's Twinspot 3773
Eagle-Hawk 4453,848
Eagle-Owl 1334
Eagles 81
Eared Dove 10695
Eared Grebe (NA) 8018
Eared Honeyeater 5240
Eared Pitta 7864
Eared Poorwill 6637
Eared Pygmy-Tyrant 6213
Eared Quetzal 3745
Eared Trogon 3745
Eared-Nightjars 3748
Earl of Derby's Parakeet 7892
Earle's Babbler 10263
East African Batis 1205
East African Citril 9156
East African Citril-Finch 9156
East African Puffback 1205
East African Swee (CSA) 3628
East African Swee-Waxbill 3628
East African Wheatear 6716

East Asian Swallow 4682
East Coast Akalat 9194
East Coast Batis (CSA) 1214
East Indian Woodcock 9041
East Indies Bush-Warbler 1294
East Siberian Sandpiper 1498
Eastern Akalat 9194
Eastern Alpine-Mannikin 5313
Eastern Alpine-Munia 5313
Eastern Barred Bush-Warbler 1475
Eastern Barred Owlet 4270
Eastern Bearded Scrub-Robin 1922
Eastern Bearded Tit 7008
Eastern Bearded-Bulbul 2836
Eastern Black-chinned Honeyeater (ANZ) 5768
Eastern Black-headed Oriole 6790
Eastern Black-Wheatear 6716
Eastern Blossom-headed Parakeet 8376
Eastern Bluebird 9201
Eastern Blue-Swallow 4658
Eastern Bonelli's Warbler 7643
Eastern Bowerbird 471
Eastern Bristlebird 3045
Eastern Broad-billed Roller 3769
Eastern Bronze-naped Pigeon 2484
Eastern Bush-hen 391
Eastern Bush-Lark 5934
Eastern Calandra 5701
Eastern Calandra-Lark 5701
Eastern Cattle-Egret 1346
Eastern Chachalaca 6820
Eastern Chanting-Goshawk 5741,5742

Firewood-gatherer 738
Fiscal (CSA) 5040
Fiscal Flycatcher 9216
Fiscal-Shrike (CSA) 5040
Fischer's Bulbul 7555
Fischer's Eider 9264
Fischer's Finch-Lark 3576
Fischer's Fruit-Dove 8494
Fischer's Greenbul 7555
Fischer's Lovebird 242
Fischer's Sparrow-Lark 3576
Fischer's Starling 9384
Fischer's Thrush 10730
Fischer's Touraco 9805
Fischer's Turaco 9805
Fischer's Whydah 10535
Fish Crow 2711
Fish Eagle (WI) 7006
Fish Hawk 7006
Fish-eye (WI) 10318
Fishing Buzzard 1381
Fish-Owl 9046
Five-colored Munia (NA) 5319
Five-coloured Barbet 1643
Five-coloured Mannikin 5319
Five-striped Sparrow 499
Fjordland Crested-Penuin 3651
Flame Bowerbird 9130
Flame Dove 8531
Flame Minivet 7256
Flame Robin 7281
Flame Robin-Flycatcher 7281
Flame-breasted Flowerpecker 3187
Flame-breasted Fruit-Dove 8507
Flame-breasted Sunbird 6427
Flame-capped Parakeet 904
Flame-chested Flowerpecker 3187
Flame-coloured Minivet 7255
Flame-coloured Tanager 7820
Flamecrest 8768

Flame-crested Manakin 4605
Flame-crested Tanager 9706
Flame-crowned Flowerpecker 3178
Flame-crowned Manakin 4605
Flamed Bowerbird 9130
Flamed Regentbird 9130
Flame-faced Tanager 9762
Flame-fronted Barbet 5592
Flame-headed Oriole 4828
Flame-rumped Cacique 1452
Flame-rumped Sapphire 4722
Flame-rumped Tanager 8739
Flame-templed Babbler 9408
Flame-throated Warbler 7073
Flame-winged Conure 8665
Flame-winged Parakeet 8665
Flaming Parakeet 915
Flaming Sunbird 229
Flamingo (ISC) 7506
Flamingo 7507
Flamingos 7502
Flammulated Bamboo-Tyrant 4556
Flammulated Flycatcher 3058
Flammulated Owl 6872
Flammulated Pygmy-Tyrant 4556
Flammulated Screech-Owl 6872
Flammulated Stock-Tyrant 4556
Flammulated Treehunter 9968
Flappet Bush-Lark 5940
Flappet Lark 5940
Flat-billed Kingfisher 10072
Flat-billed Vireo 10582

Flat-chested Curassow 6592
Flat-chested Urumutum 6592
Flat-clawed Petrel 6650
Flat-nosed Phalarope 7414
Flavescent Bulbul 8601
Flavescent Flycatcher 6202
Flavescent Warbler 1172
Flesh-footed Petrel 8566
Flesh-footed Shearwater 8566
Fleshy-fronted Petrel 8566
Fleshy-fronted Shearwater 8566
Fleurieu PeninsulaSouthern-Emuwren (ANZ) 9479
Flicker 2428,2435
Flightless Cormorant 7392
Flightless Grebe 8905
Flightless Steamerduck 9702
Flightless Teal 525,531
Flinders Ranges Chestnut-rumped Heathwren (ANZ) 4707
Flint-billed Woodpecker 1576
Flock Bronzewing 7426
Flock Fruit-Pigeon 5329
Flock Pigeon 5329,7426
Flop (CSA) 3727
Floreana Mockingbird 6518
Flores Crow 2694
Flores Flowerpecker 3177
Flores Green-Pigeon 10150
Flores Hanging-Parakeet 5375
Flores Hawk-Eagle 9320
Flores Jungle-Flycatcher 8809
Flores Minivet 7258
Flores Monarch 6002
Flores Mountain-Monarch 6002
Flores Scops-Owl 6856, 6892,6914
Flores White-eye 4483

Fulvous-rumped
  Woodpecker 5654
Fulvous-streaked Prinia
  8194
Fulvous-vented Euphonia
  3692
Funereal Cockatoo 1550
Funereal Kingfisher 10064
Furtive Flycatcher 3834
Fuscous Flycatcher 2391
Fuscous Honeyeater 5226
Gabar Goshawk 5740
Gabela Akalat 9193
Gabela Bush-Shrike 5012
Gabela Helmet-Shrike
  8230
Gabela Redstart 9193
Gabela Robin 9193
Gabela Robin-Chat 9193
Gabon Batis 1202
Gabon Boubou 5016
Gabon Chestnut-breasted
  Helkmetshrike 8235
Gabon Coucal 1844
Gabon Helmet-Shrike 8235
Gabon Nightjar (CSA)
  1672
Gabon Nightjar 1672
Gabon Quailfinch 6843
Gabon Woodpecker 3159
Gaboon Coucal 1844
Gaboon Nightjar 1672
Gaby (WI) 9457
Gadfly-Petrel (CSA) 1366,
  1367,8428,8444
Gadfly-Petrel 8430
Gadwall 571
Gaimard's Cormorant 7390
Gaimard's Elaenia 6195
Galah 1440
Galapagos Albatross 3275
Galapagos Cormorant 7392
Galapagos Crake 5125
Galapagos Dove 10697
Galapagos Flightless-
  Cormorant 7392
Galapagos Flycatcher 6154,
  8647
Galapagos Hawk 1397
Galapagos Heron 1428

Galapagos Mockingbird
  6517
Galapagos Penguin 9293
Galapagos Petrel 6663,
  8453
Galapagos Pintail 546
Galapagos Rail 5125
Galapagos Storm-Petrel
  6663
Galeated Curassow 7178
Galeated Jay 2938
Galebird (WI) 8259
Galley Martin 9426
Gallinaceous Birds 4016
Gallinule 4036,4036
Gallito 8799
Gambaga Flycatcher 6065
Gambaga Spotted-
  Flycatcher 6065
Gambage Dusky-
  Flycatcher 6065
Gambel's Chickadee 7100
Gambel's Quail 1521
Gambel's Tit 7100
Gambia Barn-Swallow
  4671
Gambia Swallow 4671
Gambian Barn-Swallow
  4671
Gambian Puffback-Shrike
  3337
Game Birds 4016
Gang (col) 5724
Gang-gang 1525
Gang-Gang Cockatoo 1525
Gannet 6032,6033,6034
Gannets 9582
Ganongga White-eye
  10828
Gansu Leaf-Warbler 7632
Ganuche (WI) 7198
Garden Bulbul 8590
Garden Dove 9504
Garden Emerald 2158
Garden Thrush 10309
Garden Warbler 9589
Gardner Bowerbird 469
Garganey 563
Garnet Flycatcher 3659
Garnet Flyeater 3659
Garnet Hummingbird 3664

Garnet Pitta 7853
Garnet Robin 3659
Garnet-throated
  Hummingbird 3664,
  4987
Gartered Trogon 10233
Gaudichaud's Kingfisher
  3024
Gaudy Barbet 5609
Gaulin (WI) 1347,1429,
  3400,3406,3408,3409,
  4890
Gay's Sierra-Finch 7533
Geelvink Flowerpecker
  3190
Geelvink Pygmy-Parrot
  5879
Geelvink Scrubfowl 5640
Geese 578
Gentian Kingfisher 1969
Gentoo Penguin 8639
Geoffrey's Plover 2030
Geoffroy's Dotterel 2030
Geoffroy's Parrot 4144
Geoffroy's Sandplover
  2030
Geomalia 4148
Georgia Shag 7391
Georgian Diving-Petrel
  7191
Georgian Pintail 547
Georgian Teal 547
Gerfalcon 3813
Germain's Peacock-
  Pheasant 8095
German's Swiftlet 198,2459
Geronimo Swift 7010
Geronimo Swiftlet 7010
Gestroi's Fruit-Dove 8514
Ghana Cuckoo-Shrike 1568
Ghizo White-eye 10792
Giant Ant-Pitta 4310
Giant Antshrike 1192
Giant Babax 1141
Giant Brown-Needletail
  4648
Giant Bush-Warbler 1298
Giant Cactus-Wren 1622
Giant Conebill 6756
Giant Coot 3961
Giant Coua 2782

Greater Pied-Woodpecker
3105
Greater Pitohui 7839
Greater Prairie-Chicken
10391
Greater Racket-tailed
Drongo 3235
Greater Racquet-tailed
Drongo 3235
Greater Red-breasted
Meadowlark 9542,9544
Greater Red-headed
Babbler 5505
Greater Red-headed
Parrotbill 7036
Greater Rhea 8788
Greater Ringed-Plover
2028
Greater Roadrunner 4141
Greater Rock-Nuthatch
9249
Greater Rufous-headed
Parrotbill 7036
Greater Sage-Grouse 1842
Greater Sandplover 2030
Greater Scaly Wren-
Babbler 8002
Greater Scaup 1135
Greater Scaup Duck 1135
Greater Schiffornis 9006
Greater Scimitar-Babbler
8119
Greater Scythebill 1616
Greater Seed-Finch 6850
Greater Shearwater (NA)
8569
Greater Sheathbill 2080
Greater Short-toed Lark
1481,1483
Greater Shrike-Babbler
8478
Greater Sicklebill 3549
Greater Snow Goose 631
Greater Snow-Petrel 7003
Greater Sooty-Owl 10430
Greater Spotted-Eagle 851
Greater Spotted-
Woodpecker 3105
Greater Streaked-
Honeyeater 6307

Greater Streaked-Lory
1998
Greater Striated-Swallow
4687
Greater Striped- Swallow
4660
Greater Sulawesi
Honeyeater 6307
Greater Sulphur-crested
Cackatoo 1432
Greater Swamp-Warbler
133
Greater Thornbird 7303
Greater Truk White-eye
8918
Greater Vasa-Parrot 2659
Greater Wagtail-Tyrant
9473
Greater Waxwing 1250
Greater White-crested
Cockatoo 1430
Greater White-eared Fruit-
Pigeon 7421
Greater White-fronted
Goose 628
Greater White-tailed Leaf-
Warbler 7649
Greater Whitethroat 9593
Greater Wood-Shrike 7839
Greater Woodswallow
1000
Greater Yellow-crested
Cockatoo 1432
Greater Yellow-eared
Spiderhunter 886
Greater Yellow-Finch 9202
Greater Yellow-headed
Vulture 1809
Greater Yellowlegs 10202
Greater Yellownape 7735
Greater Yellow-naped
Woodpecker 7735
Greater Yellowshank
10202
Great-tailed Grackle 8694
Great-winged Petrel 8445
Great-winged Wren 1628
Grebes 8021,8022
Gree-gree (WI) 1406
Greek Partridge 347
Green Aracari 8471

Green Avadavat 382
Green Barbet 5606,
5618,9415
Green Bee-eater 5811
Green Berrypecker 5697
Green Broadbill 1542,8325
Green Bulbul (ISC) 2135
Green Canary 9163
Green Carib 3663
Green Catbird 277
Green Cochoa 2410
Green Conure 913
Green Cormorant 7370
Green Coucal 1872,1965
Green Crombec 9620
Green Doctorbird (WI)
3663
Green Dove 8503
Green Dwarf Goose 6533
Green Figbird 9285,9286
Green Flycatcher-Warbler
7661
Green Fruit-Pigeon 10147
Green Glossy-Starling
4995
Green Grosbeak 1786
Green Hanging-Parakeet
5374
Green Hanging-Parrot 5374
Green Hermit 7342
Green Heron 1427,1429
Green Honeycreeper 2125
Green Honeyeater 4383
Green Hummingbird (WI)
3663
Green Hunting-Crow 2299
Green Hylia 4712
Green Hylie 4712
Green Ibis 1259,5823
Green Imperial-Pigeon
3343
Green Indigobird
10533,10534
Green Jay
2299,2936,2938,2939
Green Jery 6469
Green Junglefowl 4069
Green Keet 10174,
4277,4279
Green Kingfisher 2110
Green Leaf-Warbler 7640

Horseshoe Honeyeater
 7544
Horsfield's Babbler 5488
Horsfield's Bronze-Cuckoo
 2189
Horsfield's Bush-Lark 5934
Horsfield's Cuckoo 2189
Horsfield's Goshawk 72
Horsfield's Jungle-Babbler
 5488
Horsfield's Pheasant 5367
Horsfield's Scimitar-
 Babbler 8118
Horsfield's Sparrowhawk
 72
Horsfield's Thrush 10732
Horsfield's Woodcock 9041
Horus Swift 839
Hose's Broadbill 1541
Hose's Magnificent
 Broadbill 1541
Hottentot Button-Quail
 10371
Hottentot Teal 551
Houbara 2096
Houbara Bustard 2096
Houhuysi 5731
House Bunting 3483
House Crow 2717
House Finch 1762
House Martin 3056
House Mynah 105
House Pigeon 2501
House Sparrow 7141
House Swallow 4664,4688
House Swift 830,841
House Wren 10215,10223
Hova Lark 5932
Hove Grey-backed White-
 eye 10794
Hoy's Screech-Owl 6877
Huahuna Flycatcher 8111
Huallaga Tanager 8741
Hudson Bay Bird 7264
Hudsonian Chickadee 7104
Hudsonian Curlew 6614,
 6617
Hudsonian Godwit 5259
Hudsonian Tit 7104
Hudson's Black-Tyrant
 4940

Hudson's Canastero 1025
Huet-Huet 8474
Huet-huet 8476
Huet's Parrotlet 10113
Huia 4602
Humble Canastero 1027
Humblot's Flycatcher 4693
Humblot's Heron 955
Humblot's Sunbird 6390
Humboldt Penguin 9291
Humboldt's Penguin 9291
Humboldt's Sapphire 4720
Hume's Babbler 9408
Hume's Blue-throated
 Barbet 5603
Hume's Chat 6703
Humes Ground-Chough
 8344
Hume's Ground-Jay 8344
Hume's Groundpecker
 8344
Hume's Large-billed Reed-
 Warbler 138
Hume's Lark 1478
Hume's Leaf-Warbler 7629
Hume's Lesser-Whitethroat
 9586
Hume's Owl 9519
Hume's Parakeet 8371
Hume's Pheasant 9675
Hume's Short-toed Lark
 1478
Hume's Swiftlet 197
Hume's Tawny-Owl 9519
Hume's Tree-Babbler 9408
Hume's Warbler 7629
Hume's Wheatear 6703
Hume's Whitethroat 9586
Hume's Wood-Owl 9519
Hume's Wren-Babbler
 9294
Hume's Yellow-browed
 Warbler 7629
Hummingbird (WI)
 1524,2170
Hummingbirds 10208
Hungarian Partridge (NA)
 7249
Hunstein's Mannikin 5302
Hunstein's Munia 5302
Hunter (WI) 4700

Hunter's Cisticola 2332
Hunter's Sunbird 6391
Hunting Cissa 2299
Hunting Crow 2299
Huon Astrapia 1043
Huon Bird of Paradise
 1043
Huon Honeyeater 5728,
 5761
Huon Melidectes 5728
Huon Melipotes 5761
Huon Parotia 7070
Huon Wattled-Honeyeater
 5728
Huon Wattled-Melidectes
 5728
Hurricanebird (WI) 3942
Hurricanebird 9440
Hutchins's Goose 1315
Hutton's Fruit-Dove 8497
Hutton's Shearwater 8572
Hutton's Vireo 10575
Hwamei 4080
Hwamei Laughingthrush
 4080
Hyacinth Macaw 617
Hyacinth Visorbearer 1102
Hyacinthine Flycatcher
 2995
Hyacinthine Niltava 2995
Hyacynthine Macaw 617
Hylocitrea 4725
Hypochondriac Tody
 10097
Hypocolius 4774
Hyrcanian Tit 7106
Iago Sparrow 7148,7152
Ibadan Malimbe 5522
Iberian Chiffchaff 7614
Ibis (WI) 1347
Ibis 3640,3641,6576
Ibisbill 4794
Ibisbills 4795
Ibises 9967
Iceland Gull 5088
Iceland Sandpiper 1498
Icterids 4799
Icterine Bulbul 7559
Icterine Canary 9163
Icterine Greenbul 7559
Icterine Warbler 4639

Java Finch 7001
Java Frogmouth 1221
Java Ground-Warbler 9871
Java Hawk-Eagle 9319
Java Kingfisher 4435
Java Mannikin 5295
Java Munia 5295,5305
Java Ring Dove 9491
Java Sandplover 2029
Java Sparrow 7001
Java White-bellied Munia
   5305
Javan Babbler 9390
Javan Barbet 5596,5604
Javan Brown Hill-Warbler
   8203
Javan Brown-Prinia 8203
Javan Brown-throated
   Barbet 5596
Javan Bush-Warbler 1304
Javan Cochoa 2407
Javan Collared-Dove 9491
Javan Cormorant 7399
Javan Coucal 1861
Javan Cuckoo-Shrike 2628
Javan Fire-breasted
   Flowerpecker 3207
Javan Flowerpecker 3207
Javan Frogmouth 1221
Javan Fulvetta 338
Javan Greenfinch 9151
Javan Grey-throated White-
   eye 5354
Javan Ground-Warbler
   9871
Javan Hanging-Parrot 5378
Javan Hawk-Eagle 9319
Javan Hill-Mynah 4294
Javan Hill-Partridge 936
Javan Junglefowl 4069
Javan Kingfisher 4435
Javan Lapwing 10492
Javan Lark 5934
Javan Mannikin 5305
Javan Munia 5305
Javan Mynah 104
Javan Nun-Babbler 338
Javan Owlet 4252
Javan Parakeet 8362
Javan Plover 2029
Javan Pond-Heron 969

Javan Quaker-Babbler 338
Javan Scops-Owl 6857
Javan Streaked-Bulbul
   4793
Javan Sunbird 232
Javan Tailorbird 6839
Javan Tesia 9871
Javan Thrush 2407
Javan Tit-Babbler 5455
Javan Turtle-Dove 9491
Javan White-eye 10778,
   5354
Javanese Collared Dove
   9491
Javanese Cormorant 7399
Javanese Fulvetta 338
Javanese Hanging Parakeet
   5378
Javanese Imperial-Pigeon
   3359
Javanese Lapwing 10492
Javanese Mannikin 5305
Javanese Mountain-Pigeon
   3359
Javanese Munia 5305
Javanese Mynah 104
Javanese Pond-Heron 969
Javanese Turtle-Dove 9491
Javanese Wattled-Plover
   10492
Javanese White-bellied
   Munia 5305
Javanese Woodcock 9041
Jay 4124
Jelski's Bush-Tyrant 4946
Jelski's Chat-Tyrant 9219
Jelski's Tanager 4880
Jerdon's Leafbird 2136
Jerdon's Babbler 2210
Jerdon's Baza 1122
Jerdon's Bush-Chat 8987
Jerdon's Chat 8987
Jerdon's Chloropsis 2136
Jerdon's Courser 8816
Jerdon's Imperial-Pigeon
   3345
Jerdon's Laughingthrush
   4092
Jerdon's Leaf-Warbler 7626
Jerdon's Mannikin 5303
Jerdon's Minivet 7254

Jerdon's Moupinia 2210
Jerdon's Nightjar 1653
Jerdon's Reed-Warbler 108
Jerdon's Starling 9550
Jerdon's Willow-Warbler
   7626
Jerfalcon 3813
Jerryang 4279
Jery 6468
Jet Antbird 1900
Jet Manakin 2133
Jewelfront 4496
Jobi Island Dove 4001
Jobi Manucode 5562
Johamma Thrush 10294
Johanna's Sunbird 6392
Johannes's Tody-Tyrant
   4560
John Crow (WI) 1807
John Philip (WI) 10558
John-chew-it (WI) 10558
Johnny-jump-up (WI)
   10603
Johnson's Warbler 2552
Johnstone's Lorikeet 10178
Johnstone's Touraco 6097
Johnston's Bush-Robin
   9801
Johnston's Red-tufted
   Sunbird 6393
John-to-whit (WI) 10558
Joly's Jay 2953
Jos Plateau Indigobird
   10541
Josephine's Lorikeet 2056
Josephine's Tody-Tyrant
   4561
Jouanin's Gadfly Petrel
   1367
Jouanin's Petrel 1367
Jourdain's Woodstar 1972
Jouyi's Wood-Pigeon 2495
Joyful Bulbul 2119
Joyful Greenbul 2119
Ju River Scrub-Warbler
   1298
Ju River Warbler 1298
Juan Fernandez Firecrown
   9113
Juan Fernandez Petrel 8433

Juan Fernandez Tit-Tyrant
516
Juba Weaver 7950
Jubaland Weaver 7950
Judy f (WI) 6991
Julie's Hummingbird 3038
Jumbiebird 10412
Jumpiebird (WI) 10412
Jumping-Dick (WI) 10293
Jungle Ant-Tanager 4411,
4412
Jungle Babbler 10284
Jungle Boobook 6574
Jungle Bush-Quail 7244
Jungle Coucal 1857
Jungle Crow 2702,2703
Jungle Long-tailed Warbler
8210
Jungle Mynah 101,103
Jungle Nightjar 1675
Jungle Owlet 4268
Jungle Pigeon 2491
Jungle Prinia 8210
Jungle Quail 7244
Jungle Sparrow 7156
Jungle Tanager 4410
Jungle Wren-Warbler (ISC)
8210
Junglefowl 4066
Junin Canastero 1039
Junin Grebe 8020
Junin Rail 5120,5126
Juniper Finch 7754
Juniper Titmouse 7120
Kabobo Apalis 756
Kabylian Nuthatch 9240
Kabylie Nuthatch 9240
Kadavu Dove 8503
Kadavu Fantail 8851
Kadavu Honeyeater
10617,10617
Kadavu Shining Parrot
8268
Kadavu Xanthotis 10617
Kaempfer's Tody-Tyrant
4562
Kaffervink (CSA) 3712,
3713,3715,3716,
3718,3720
Kaffir Rail 8711
Kaffra Saw-wing 8294

Kagu 8894
Kagus 8893
Kai Cicadabird 2621
Kai Coucal 1866
Kai Cuckoo-Shrike 2621,
2650
Kai Greybird 2621
Kai Island Cuckoo-Shrike
2650
Kai Island Pygmy-Parrot
5880
Kai Island White-eye
10780
Kai Kuckooshrike 2648
Kai Monarch 5993
Kai Monarch-Flycatcher
5993
Kai White-eye 10780
Kaka (ANZ) 6524
Kakamega Greenbul 593
Kakapo 9516
Kakariki 2958
Kakatoe 1435
Kakawahie 7063
Kakawahie Creeper 7063
Kakelaar (CSA) 7513
Kakopi White-eye 9838
Kalabat Nightjar 3751
Kalahari Robin (CSA)
1920
Kalahari Sandy Scrub-
Robin 1920
Kalahari Scrub-Robin 1920
Kalanga 3394
Kalij 5367
Kalij Pheasant 5367
Kalinowski's Tinamou
6595
Kallej Pheasant 5367
Kamao 6107
Kamchatka Gull 5079,5113
Kamchatka Pied-Wagtail
6048
Kamchatka Tern 9439
Kamchatkan Pied-Wagtail
6048
Kamchatkan Sea-Eagle
4448
Kandt's Waxbill 3619
Kangaroo Island Emu
(ANZ) 3308

Kangaroo Island Glossy
Black-Cockatoo (ANZ)
1552
Kangaroo Island Western-
Whipbird (ANZ) 8396
Kapoc Penduline-Tit 656
Kapul Eagle 4481
Karamajoa Warbler 756
Karamoja Apalis 757
Karoo Bustard 3744
Karoo Chat 1907
Karoo Eremomela (CSA)
3565
Karoo Green Warbler 3565
Karoo Korhaan (CSA)
3744
Karoo Lark 1937
Karoo Prinia 8200
Karoo Robin (CSA) 1914
Karoo Scrub-Robin 1914
Karthala White-eye 10803
Kashmir Flycatcher 3853
Kashmir House-Martin
3054
Kashmir Nuthatch 9233
Kashmir Red-breasted
Flycatcher 3853
Kashmir Tit 168
Katanga Masked-Weaver
7960
Kataw (WI) 8977
Kathala Scops-Owl 6902
Kauai Akepa 5400
Kauai Akiola 4524
Kauai Amakihi 10602
Kauai Creeper 6757
Kauai Oo 5957
Kawall's Parrot 450
Kea 6525
Keay's Bleeding-heart 4002
Kedong Cisticola 2323
Keel-billed Motmot 3436
Keel-billed Toucan 8726,
8727
Keeog 8652
Keg 8652
Kelley's Tit-Babbler 5457
Kelp Goose 2101
Kelp Gull 5083
Kemp's Bushcreeper 5480

Lemon-bellied Flyrobin
5861
Lemon-bellied Sylvietta
9614
Lemon-bellied White-eye
10766
Lemon-breased Flyrobin
5861
Lemon-breasted
Berrypecker 5695
Lemon-breasted Canary
(CSA) 9147
Lemon-breasted Flycatcher
5861
Lemon-breasted Microeca
5861
Lemon-breasted Seedeater
9147
Lemon-browed Flycatcher
2554
Lemon-browed Tanager
2178
Lemon-cheeked
Honeyeater 5232
Lemon-chested Greenlet
4743,4755
Lemon-chested Vireo 4755
Lemon-crested Cockatoo
1442
Lemon-rumped Tanager
8739
Lemon-rumped Tinkerbird
8046
Lemon-rumped Warbler
7619,7647
Lemon-spectacled Tanager
2178
Lemon-throated Barbet
3635
Lemon-throated Leaf-
Warbler 7618
Lemon-throated Scrubwren
9117
Lemon-throated Sericornis
9117
Lemon-throated Warbler
7618
Lemon-throated White-eye
10759
Lendu Alseonax 6069
Lendu Flycatcher 6069

Lepanto Jungle-Flycatcher
8807
Lepe Cisticola 2337
Leschenault's Forktail 3520
Lesser Adjutant 5171
Lesser Adjutant-Stork 5171
Lesser African Jacana 5876
Lesser Amakihi 4523
Lesser Andean Flamingo
7505
Lesser Antillean Bullfinch
5395
Lesser Antillean Flycatcher
6157
Lesser Antillean Grackle
8692
Lesser Antillean Pewee
2580
Lesser Antillean Saltator
8934
Lesser Antillean Swift
1984
Lesser Antillean Tanager
9736
Lesser Antillean Thrush
10343
Lesser Bar-tailed Cuckoo-
Dove 5472
Lesser Bay-Woodpecker
1239
Lesser Becard 6990
Lesser Bird of Paradise
7018
Lesser Black-backed
Cisticola 2352
Lesser Black-backed Gull
5085
Lesser Black-Coucal 1847
Lesser Black-winged
Lapwing 10491
Lesser Black-winged
Plover (CSA) 10491
Lesser Blue-eared Glossy-
Starling 4996
Lesser Blue-eared Starling
(CSA) 4996
Lesser Blue-Pitta 7847
Lesser Blue-winged Pitta
7860,7862
Lesser Bristlebill 1236

Lesser Broad-billed Petrel
6979
Lesser Brown Fruit-Dove
7423
Lesser Brown Fruit-Pigeon
7423
Lesser Brown Wren-
Warbler 8205
Lesser Brown-Bulbul 8598
Lesser Brown-Prinia 8205
Lesser Brown-Thrush
10740
Lesser Canastero 1035
Lesser Cape Verde Kestrel
3805
Lesser Coucal 1846
Lesser Creeper 5142
Lesser Crested-Tern 9444
Lesser Cuckoo 2892
Lesser Cuckoo-Shrike 2623
Lesser Double-collared
Sunbird (CSA) 6375
Lesser Eagle-Owl 6875
Lesser Eared-Nightjar 3755
Lesser Egret 3403,957
Lesser Elaenia 3414
Lesser Fish-Eagle 4796
Lesser Fishing-Eagle 4796
Lesser Flameback 3259
Lesser Flamingo 7501
Lesser Florican 3738
Lesser Forktail 3525
Lesser Frigatebird 3941
Lesser Fulvous Whistling-
Duck 3113
Lesser Gallinule (CSA)
8155
Lesser Glossy-Starling 807
Lesser Golden-backed
Woodpecker (ISC)
3259
Lesser Golden-Plover
7997,7998
Lesser Goldfinch 1739
Lesser Goshawk 45
Lesser Grass-Finch 3493
Lesser Green-billed
Malkoha 7312
Lesser Green-Broadbill
1542

Lita Woodpecker 7702
Little Tern 4370
Little Amazon Parrot 4456
Little Auk 368
Little Banded-Goshawk
  (CSA) 31
Little Barbet 5594
Little Bee-eater 5815
Little Bee-Hummingbird
  (WI) 5774
Little Bishop 3725
Little Bittern 4889,4893
Little Black-and-white
  Cormorant 7397
Little Black-Bustard 3730
Little Black-Cormorant
  7409
Little Black-Petrel 5404
Little Black-Shag 7409
Little Blue Macaw 2956
Little Bluebill 1127
Little Blue-Heron 3400
Little Blue-Kingfisher 307
Little Blue-Penguin 3656
Little Blue-Petrel 8249
Little Blue-Swallow 4675
Little Blue-winged Pitta
  7860
Little Broadbill 1127
Little Bronze-Cuckoo 2199
Little Bronze-Drongo 3216
Little Bronze-Pigeon 7425
Little Brown-Bustard 3737
Little Brown-Crane 4349
Little Brown-Dove (ISC)
  9504
Little Brown-Korhaan 3737
Little Bunting 3475
Little Bustard 9883
Little Bustard-Quail (ISC)
  10380
Little Button-Quail
  10380,10383
Little Capped Fruit-Dove
  8490
Little Chachalaca 6816
Little Cockatoo 1441
Little Corella 1441
Little Cormorant 7397,
  7399,7407

Little Coroneted Fruit-
  Dove 8490
Little Crabier (WI) 1429
Little Crake 8172,8175
Little Crow 2680,2693
Little Cuckoo 2892,7672
Little Cuckoo-Dove 5474
Little Cuckoo-Shrike 2645
Little Curlew 6616,6617
Little Devil (WI) 8573
Little Doctor-Bird (WI)
  5774,6826
Little Dove 4149
Little Duck Hawk 3800
Little Eagle 4626
Little Egret 3402,3403
Little Fairy Tern 4370
Little Falcon 3800
Little Fieldwren 2216
Little Forktail 3524
Little Friarbird 7452
Little Fruit-Dove 8512
Little Gallinule 8157
Little Grassbird 5627
Little Grass-Warbler 5627
Little Grebe 9688
Little Green Barbet 5617
Little Green Bee-eater 5811
Little Green Heron (ISC)
  1427
Little Greenbul 602
Little Green-Bulbul 602
Little Green-Pigeon 10154,
  1991
Little Greenshank 1494
Little Green-Sunbird 6424
Little Green-Woodpecker
  1591
Little Grey-Alseonax 6063
Little Grey-Bulbul 590
Little Grey-Flycatcher
  1284,1286,6063
Little Grey-Greenbul 590
Little Grey-Kiwi 828
Little Grey-Tern 8249
Little Grey-Woodpecker
  3156
Little Ground-Chough
  8344
Little Ground-Dove 2535
Little Ground-Jay 8344

Little Ground-Tyrant 6091
Little Guan 7233
Little Gull 5101
Little Hanging-Parrot 5378
Little Hawk 39
Little Hermit 7347
Little Heron 1427
Little Inca-Finch 4851
Little Inca-Sparrow 4851
Little Indian Nightjar 1652
Little Kai White-eye 10833
Little Keet 4279
Little King Bird of Paradise
  2225
Little Kingfisher 307,318
Little Koel 5859,8785
Little Long-billed Cuckoo
  8785
Little Lorikeet 4279
Little Lory 4279
Little Malay Parrot 8386
Little Mangrove Heron
  1427
Little Marshbird 5627
Little Masked-Weaver
  7961
Little Minivet 7252,7258
Little Moorhen 4035
Little Nightjar 1691
Little Oliveback 6509
Little Olive-Sunbird 6424
Little Olive-Waxback 6509
Little Olive-Waxbill 6509
Little Olive-Weaver 6509
Little Owl 1053
Little Papuan Frogmouth
  8007
Little Paradise-Kingfisher
  9788
Little Penelopina 7233
Little Penguin 3656
Little Petrel 8578
Little Pied Blue-Flycatcher
  3857
Little Pied-Cormorant 7397
Little Pied-Flycatcher 3857
Little Pied-Goshawk 55
Little Pied-Sparrowhawk
  55
Little Plover 2025

Lord Howe Island White-throated Pigeon (ANZ) 2528
Lord Howe Island Woodrail 4060
Lord Howe Swamphen 8154
Lord Howe White-eye 10829,10832
Lord Howe Woodhen (ANZ) 4060
Lord Howe Woodrail 4060
Lord Howe's Rail 4060
Lord Rothschild's Bird of Paradise 1043
Lorentz's Whistler 6952
Lorenz's Bulbul 7561
Loria's Bird of Paradise 2388
Lories 5382
Lorikect (ISC) 5381
Lorikeets 5382
Loten's Sunbird 6396
Lotusbird 4875
Louisiade Black-fronted White-eye 10796
Louisiade Butcherbird 2789
Louisiade Flowerpecker 3201
Louisiade Honeyeater 5759
Louisiade Lory 5389
Louisiade Meliphaga 5759
Louisiade White-eye 10781
Louisiades Black-fronted White-eye 10796
Louisiades Butcherbird 2789
Louisiades Flowerpecker 3201
Louisiades Honeyeater 5759
Louisiades Lory 5389
Louisiades Meliphaga 5759
Louisiades White-eye 10781
Louisiana Heron 3409
Louisiana Waterthrush 9092
Loveliest Sunbird 235
Lovely Cotinga 2756

Lovely Fairywren 5530
Lovely Hummingbird 397
Lovely Sunbird 237
Lovely Wren 5530
Lovely Wren-Warbler 5530
Loveridge's Sunbird 6397
Lower Amazonian Antwren 5885
Lowland Akalat 9192
Lowland Chat-Tanager 1544
Lowland Eupates 8552
Lowland Hepatic Tanager 7822
Lowland Jewel-Babbler 8552
Lowland Kingfisher 9627
Lowland Lorikeet 2062
Lowland Mouse-Babbler 2815
Lowland Mouse-Warbler 2815
Lowland Peltops 7210
Lowland Peltops-Flycatcher 7210
Lowland Quail-Thrush 8552
Lowland Rail-Babbler 8552
Lowland Sicklebill 3548
Lowland Straight-billed Honeyeater 10026
Lowland Swallow 4663
Lowland Swiftlet 217
Lowland White-eye 10798
Lowland Wood-Wren 4582
Low's Swiftlet 203
Loyalty Honeyeater 5244
Loyalty Islands Honeyeater 5244
Lucifer Hummingbird 1536
Lucy's Emerald 418
Lucy's Warbler 10526
Ludlow's Fulvetta 333
Ludwig's Bustard 6497
Lufira Masked-Weaver 7979
Lühder's Bush-Shrike (CSA) 5024
Lühder's Bush-Shrike 5024
Lunulated Antbird 4389

Lunulated Antcatcher 4389
Lunulated Honeycreeper 5770
Luzon Bleeding-heart 4004
Luzon Bleeding-heart Pigeon 4004
Luzon Bush-Warbler 1963
Luzon Button-Quail 10384
Luzon Duck 553
Luzon Flycatcher 3833
Luzon Guaiabero 1243
Luzon Hanging-Parrot 5377
Luzon Hornbill 7229
Luzon Jungle-Flycatcher 8807
Luzon Racket-tail 8215
Luzon Racket-tailed Parakeet 8213
Luzon Racquet-tail 8215
Luzon Racquet-tailed Parrot 8213
Luzon Rail 5216
Luzon Scops-Owl 6890
Luzon Striped-Babbler 9409
Luzon Tailorbid 6831
Luzon Tarictic-Hornbill 7229
Luzon Water-Redstart 8883
Luzon White-eye 10798
Luzon Wren-Babbler 6353
Lyne's Bustard 3742
Lynes's Cisticola 2323
Lyrebirds 5787
Lyre-tailed Honeyguide 5726
Lyre-tailed Nightjar 10470
Macaroni Penguin 3650
MacClounie's Barbet 5429
Maccoa Duck 6933
MacConnell's Flycatcher 5910
MacConnell's Spinetail 9650
MacCormick's Skua 1804
MacGillivray's Warbler 6751
MacGregor's Bird of Paradise 5438

Madagascar Spinetailed-
Swift 10714
Madagascar Squacco 966
Madagascar Squacco
Heron (CSA) 966
Madagascar Starling 8961
Madagascar Sunbird 6408
Madagascar Swamp-
Warbler 127
Madagascar Swift 833
Madagascar Teal 528
Madagascar Tetraka 7562
Madagascar Tsikirity 6506
Madagascar Turtle-Dove
9499
Madagascar Wagtail 6046
Madagascar Weaver 3887
Madagascar White-eye
10794,1134
Madagascar White-throated
Rail 3333
Madagascar Woodrail 1634
Madagascar Yellowbrow
2841
Madame Verreaux's
Sunbird 6392
Madanga White-eye 5483
Madarasz's Parrot 8357
Madarasz's Tiger-Parrot
8357
Madeira Petrel 8446
Madeira Storm-Petrel 6652
Madeiran Fork-tailed Petrel
6652
Madeiran Petrel 6652
Madeiran Storm-Petrel
6652
Mademoiselle Thura's
Rosefinch 1777
Maes's Laughingthrush
4097
Mafor Pygmy-Parrot 5879
Magdalena Parakeet 8674
Magdalena Tinamou 2874
Magellan Blue-eyed Shag
7371
Magellan Conure 3517
Magellan Cormorant 7395
Magellan Diving-Petrel
7192
Magellan Goose 2103

Magellan Gull 5115
Magellan Shag 7395
Magellan Snipe 4030
Magellanic Diving-Petrel
7192
Magellanic Flightless-
Steamerduck 9702
Magellanic Oystercatcher
4422
Magellanic Penguin 9292
Magellanic Plover 8000
Magellanic Tapaculo 9064
Magellanic Woodpecker
1580
Magellan's Diving-Petrel
7192
Magenta Petrel 8447
Magenta-throated
Woodstar 7470
Magnificent Bird of
Paradise 2224
Magnificent Broadbill 1541
Magnificent Dove 8506
Magnificent Frigatebird
3942
Magnificent Fruit-Dove
8506
Magnificent Fruit-Pigeon
8506
Magnificent Green-
Broadbill 1541
Magnificent Hummingbird
3657
Magnificent Night-Heron
4287
Magnificent Quetzal 7430
Magnificent Riflebird 8549
Magnolia Warbler 3137
Magpie (ANZ) 4392
Magpie 7675
Magpie Geese 642
Magpie Goose 641
Magpie Lark 4335
Magpie Mannikin (CSA)
5298
Magpie Munia 5298
Magpie Shrike 2677
Magpie Starling 9269,9383
Magpie Tanager 2303
Magpie-Jay 1528,1529
Magpie-Robin 2596

Maguari Stork 2231
Mahali Sparrow-Weaver
7927
Mahatea Fruit-Dove 8486
Mahratta Woodpecker
(ISC) 3104
Maiden Dove 10387
Major Hall's Babbler 8125
Major Mitchell's Cockatoo
1435
Makassar White-eye 10759
Makatea Fruit-Dove 8486
Malabar Blue-winged
Parakeet 8365
Malabar Crested-Lark 3992
Malabar Grey-Hornbill
6681
Malabar Lark 3992
Malabar Parakeet 8365
Malabar Pied-Hornbill 659
Malabar Trogon 4469
Malabar Whistling-Thrush
6229
Malacca Parakeet 8375
Malachite Kingfisher 311,
323
Malachite Sunbird 6385
Malagasy Brush-Warbler
6506
Malagasy Heron 955
Malagasy Jacana 158
Malagasy Kingfisher 323
Malagasy Marsh-Harrier
2292
Malagasy Needletail 10714
Malagasy Pond-Heron
(CSA) 966
Malagasy Scops Owl 6896
Malagasy Scops-Owl 6907
Malagasy Snipe 4023
Malagasy Spinetail 10714
Malagasy Turtle-Dove
9499
Malagasy White-eye 10794
Malaita Fantail 8845
Malaita Honeyeater 6327
Malaita Myzomela 6327
Malaita Rufous-Fantail
8845
Malaita White-eye 10830
Malawi Batis 1197

Malawi Greenbul 7548
Malawi Puffback-
 Flycatcher 1197
Malay Banded-Crake 8703
Malay Banded-Rail 8703
Malay Black-Hornbill 660
Malay Bronze-Cuckoo
 2199
Malay Crested-Jay 7897
Malay Cuckoo 2199
Malay Eagle-Owl 1343
Malay Falconet 5870
Malay Fish-Owl 4933
Malay Forest-Kingfisher
 1971
Malay Fowl 5644
Malay Goldfinch 9151
Malay Green-Cuckoo 2199
Malay Ground-Cuckoo
 1753
Malay Hanging Lorkeet
 5376
Malay Honeyguide 4852
Malay House-Swift 841
Malay Lorikeet 5376
Malay Night-Heron 4288
Malay Peacock-Pheasant
 8098
Malay Pitta 7859
Malay Plover 2042
Malay Pond-Heron 969
Malay Rail-Babbler 3674
Malay Sandplover 2042
Malay Scrub-Robin 3674
Malay Serin 9151
Malay Spotted-Dove 9493
Malayan Barbet 5610
Malayan Crested Peacock-
 Pheasant 8098
Malayan Ground-Cuckoo
 1753
Malayan Night-Heron 4288
Malayan Peacock-Pheasant
 8096,8098
Malayan Scops-Owl 6908
Malayan Whistling-Thrush
 6232
Malaysian Argus 8797
Malaysian Banded-Crake
 8703

Malaysian Black-Hornbill
 660
Malaysian Blue-Flycatcher
 3006
Malaysian Blue-wattled
 Bulbul 8611
Malaysian Bronze-Cuckoo
 2197,2199
Malaysian Cochoa 2407
Malaysian Cockooshrike
 2628
Malaysian Crested
 Peacock-Pheasant 8098
Malaysian Eagle-Owl 1343
Malaysian Eared-Nightjar
 3755
Malaysian Eyebrowed
 Tesia 9871
Malaysian Fantail 8838
Malaysian Finch 9151
Malaysian Fish-Owl 4933
Malaysian Fulvetta 327
Malaysian Grey-breasted
 Woodpecker 4518
Malaysian Ground-Cuckoo
 1753
Malaysian Hanging
 Lorikeet 5376
Malaysian Honeyguide
 4852
Malaysian Imperial-Pigeon
 3369
Malaysian Night-Heron
 4288
Malaysian Nightjar 3755
Malaysian Niltava
 3006,6549
Malaysian Nun-Babbler
 327
Malaysian Oriole 6803
Malaysian Peacock-
 Pheasant 8096,8098
Malaysian Pied-Hornbill
 658
Malaysian Plover 2042
Malaysian Pygmy-
 Woodpecker 3108
Malaysian Rail-Babbler
 3674
Malaysian Sandplover
 2042

Malaysian Tesia 9871
Malaysian Treepie 3074,
 3078
Malaysian Wattled-Bulbul
 8611
Malaysian Woodpecker
 3108
Malborough Sound Shag
 7383
Malcolm's Babbler 10274
Malee Hen 5135
Maleo 5450
Maleo Fowl 5450
Malgas 6034
Mali Firefinch 4963
Malia 5516
Malindi Pipit 717
Mallard 558
Mallard Duck 558
Mallee Emuwren 9481
Mallee Honeyeater 5233
Mallee Hylacola 4702
Mallee Parrot 7885
Mallee Ringneck 7885
Mallee Ring-necked Parrot
 7885
Mallee Wren 4702
Malleefowl 5135
Mamo 3302
Mamos 3300
Manakins 7816
Manchurian Bush-Warbler
 1952
Manchurian Crane 4352
Manchurian Eared-
 Pheasant 2845
Manchurian Paddyfield
 Warbler 141
Manchurian Red-footed
 Falcon 3819
Manchurian Reed-Warbler
 141
Mandarin 294
Mandarin Duck 294
Mandarin Mynah 9562
Mandell's Hill-Partridge
 937
Maned Duck 2074
Maned Goose 2074
Maned Owl 4913
Maned Wood Duck 2074

Marico Flycatcher (CSA) 1283
Marico Sunbird 6400
Marion Island Petrel 6979
Marion Island Shag 7396
Marion Shag 7396
Mariqua Flycatcher 1283
Mariqua Sunbird 6400
Markham's Petrel 6658
Markham's Storm-Petrel 6658
Mark-head (WI) 8742
Marlock Parakeet 8104
Marmora's Warbler 9609
Marmy Dove (WI) 4204
Maroansetra Warbler 8757
Maroon Oriole 6802
Maroon Woodpecker 1239
Maroon-and-grey Pigeon 8776
Maroon-backed Accentor 8277
Maroon-backed Imperial-Pigeon (ISC) 3345
Maroon-backed Whistler 2660
Maroon-bellied Conure 8669
Maroon-bellied Parakeet 8669
Maroon-breasted Crowned-Pigeon 4290
Maroon-breasted Flycatcher 7465
Maroon-breasted Monarch 7465
Maroon-breasted Philentoma 7465
Maroon-breasted Sunbird (ISC) 6396
Maroon-breasted Whistler 2660
Maroon-chested Chat-Tyrant 6674
Maroon-chested Ground-Dove 2370
Maroon-chinned Fruit-Dove 8528
Maroon-faced Conure 8673
Maroon-faced Parakeet 8673

Maroon-fronted Parrot 8889
Maroon-tailed Conure 8674
Maroon-tailed Parakeet 8674
Marquesa Flycatcher 8112
Marquesa Fruit-Dove 8492, 8509
Marquesa Ground-Dove 4008
Marquesa Imperial-Pigeon 3357
Marquesa Kingfisher 10066
Marquesa Lory 10557
Marquesa Monarch 8112
Marquesa Pigeon 3357
Marquesa Reed-Warbler 126
Marquesa Swiftlet 202,206
Marquesan Flycatcher 8112
Marquesan Fruit-Dove 8492,8509
Marquesan Ground-Dove 4008,4008
Marquesan Imperial-Pigeon 3357
Marquesan Kingfisher 10066
Marquesan Lory 10557
Marquesan Monarch 8112
Marquesan Pigeon 3357
Marquesan Reed-Warbler 126
Marquesan Swiftlet 202,206
Marquesas Flycatcher 8112
Marquesas Fruit-Dove 8492,8509
Marquesas Ground-Dove 4008,4008
Marquesas Imperial-Pigeon 3357
Marquesas Kingfisher 10066
Marquesas Lory 10557
Marquesas Monarch 8112
Marquesas Pigeon 3357
Marquesas Reed-Warbler 126

Marquesas Swiftlet 202, 206
Marsabit Lark 5944
Marsh Babbler 7206
Marsh Coucal 1852
Marsh Crake 8175
Marsh Finch 6844
Marsh Harrier 2283,2297
Marsh Hawk 2288,2289
Marsh Hen (WI) 8714
Marsh Owl 1009
Marsh Sandpiper 10206
Marsh Seedeater 9371
Marsh Snipe 4025
Marsh Spotted-Babbler 7206
Marsh Tchagra 9818
Marsh Tern 2098
Marsh Tit 7119
Marsh Titmouse 7119
Marsh Warbler 130
Marsh Waxbill 3625
Marsh Whydah 3720
Marsh Widowbird 3720
Marsh Wren 2356,2359
Marshall's Fig-Parrot 2972
Marshall's Iora 171
Martial Eagle 8066
Martial Hawk-Eagle 8066
Martinique Macaw 619
Martinique Oriole 4803
Martinique Quail-Dove 4202
Marvellous Hummingbird 5288
Marvelous Spatuletail 5288
Mary Perk (WI) 1429
Maryland Yellowthroat 4191
Mas a Tierra Petrel 8432
Mas a Tierra Tit-Tyrant 516
Mas Afuera Rayadito 789
Mas Atierra Petrel 8432
Masafuera Island Rayadito 789
Masafuera Rayadito 789
Masai Barbet 10135
Mascarene Black-Petrel 8422

McGregor's House-Finch
  1761
McKay's Bunting 7919
McLelland's Mountain-
  Bulbul 4788
Meade-Waldo's Chat 8983
Meade-Waldo's
  Oystercatcher 4424
Meadow Bunting 3459
Meadow Pipit 725
Meadow Warbler 4186
Mealy Amazon 445
Mealy Parrot 445
Mealy Redpoll 1727
Mealy Rosella 7884
Mearn's Quail 3022
Meatbird 7264
Mechow's Long-tailed
  Cuckoo 1890
Median Egret (ISC) 957
Mediterranean Black-
  headed Gull 5100
Mediterranean Gull 5100
Mediterranean Shearwater
  1531,8574,8582
Mediterranean Starling
  9564
Medium Ground-Finch
  4175
Medium Hermit 7357
Medium Tree-Finch 1560
Medium-billed Prion 6979
Meek's Hawk-Owl 6557
Meek's Honeyeater 8545
Meek's Lorikeet 2058
Meek's Pigeon 5867
Meek's Pygmy-Parrot 5881
Meek's Streaked-
  Honeyeater 8545
Meek's White-eye 10796
Mees's Monarch 6002
Mees's Monarch-Flycatcher
  6002
Megapode 5636,5639
Megapodes 5633
Meinertzhagen's Snowfinch
  6029
Melancholy Woodpecker
  3162
Melanesian Broadbill 6128

Melanesian Cuckoo-Shrike
  2617
Melanesian Flycatcher
  6128
Melanesian Greybird 2617
Melanesian Ground-Thrush
  10752
Melanesian Myiagra 6128
Melanesian Myiagra-
  Flycatcher 6138
Melanesian Scrubfowl
  5637
Melba Finch (CSA) 8686
Melba Waxbill 8686
Melipotes 5762
Meller's Duck 554
Mell's Maroon Oriole 6792
Mell's Oriole 6792
Melodious Babbler 5506
Melodious Blackbird 3287
Melodious Grassquit 10011
Melodious Lark 5926
Melodious Laughingthrush
  4080
Melodious Warbler 4643
Melsetter Apalis 749
Melville Island Firebird
  7462
Menacha Seedeater 9161
Menacha Serin 9161
Menando Starling 3526
Menbek's Coucal 1857
Mencke's Monarch 5997
Mencke's Monarch-
  Flycatcher 5997
Mendoza Flycatcher 8112
Ménétries's Ant-Wren 6292
Ménétries's Warbler 9605
Mengkoka Honeyeater
  6307
Mentaur Scops-Owl 6919
Mentaur Screech-Owl 6919
Mentawai Scops-Owl 6898
Mentawai Serpent-Eagle
  9313
Mentawi Scops-Owl 6898
Menzbier's Pipit 718
Mercenary Amazon-Parrot
  452
Merganser 5791
Merida Blue-Jay 2946

Merida Flowerpecker 3245
Merida Flowerpiercer 3245
Merida Hemispingus 4539
Merida Sunangel 4491
Merida Tapaculo 9065
Merida Teal 542
Merida Wren 2357
Merle (WI) 8692
Merlin 3789
Merrill's Fruit-Dove 8510
Mesapotamian Crow 2686
Mesites 5826
Mesquitre Dove 10694
Metallic Pigeon 2527
Metallic Starling 806
Metallic Sunbird 679
Metallic Wood-Pigeon
  2527
Metallic-green Tanager
  9754
Metallic-winged Sunbird
  235
Metaltail 2002
Metcalfe's White-eye
  10797
Meve's Starling 5001
Meves's Glossy-Starling
  5001
Meves's Long-tailed
  Glossy-Starling 5001
Mew Gull 5079
Mexican Ant-Thrush 3866
Mexican Becard 6988
Mexican Cacique 1451
Mexican Chachalaca 6820
Mexican Chickadee 7125
Mexican Coast Little-Tern
  9463
Mexican Crow 2698,2716
Mexican Curassow 2824
Mexican Dipper 2267
Mexican Duck 537,558
Mexican Green-Conure
  913
Mexican Green-Parakeet
  913
Mexican Hermit 7351,7361
Mexican Jay 786
Mexican Junco 4924
Mexican Long-tailed
  Partridge 3172

Mexican Parrotlet 3880
Mexican Pygmy-Owl 4255
Mexican Rail 8714
Mexican Sheartail 3296
Mexican Spadebill 7898
Mexican Spotted-Owl
    9526,9529
Mexican Spotted-Wren
    1625
Mexican Thick-knee 1371
Mexican Tiger-Heron
    10020
Mexican Trogon 10244
Mexican Violet-ear 2445
Mexican Woodnymph
    9895
Mexican Wren 1625
Mexican Yellow Grosbeak
    7442
Meyer's Bronze-Cuckoo
    2198
Meyer's Cuckoo 2198
Meyer's Friarbird 7460
Meyer's Goshawk 60
Meyer's Myza 6306
Meyer's Parrot 8061
Meyer's Sicklebill 3550
Meyer's Whistler 6954
Micronesian Broadbill
    6138
Micronesian Cardinal-
    Honeyeater 6335
Micronesian Flycatcher
    6131,6138
Micronesian Honeyeater
    6335
Micronesian Imperial-
    Pigeon 3366
Micronesian Kingfisher
    10059
Micronesian Megapode
    5641
Micronesian Myiagra 6138
Micronesian Myzomela
    6335
Micronesian Pigeon 3366
Micronesian Scrubfowl
    5641
Micronesian Starling 810
Micronesian Swiftlet
    193,201

Middendorf's Grasshopper-
    Warbler 5286
Middendorf's Stint 1511
Middendorf's Warbler 5286
Middle American Jacana
    4911
Middle American Saltator
    8942
Middle American Screech-
    Owl 6874
Middle Spotted-
    Woodpecker 3106
Midget Flowerpecker 3174
Mid-mountain Berrypecker
    5695
Mid-mountain Eupates
    8553
Mid-mountain Honeyeater
    5732
Mid-mountain Jewel-
    Babbler 8553
Mid-mountain Melidectes
    5732
Mid-mountain Mouse-
    Babbler 2816
Mid-mountain Mouse-
    Warbler 2816
Mid-mountain Rail-Babbler
    8553
Mid-mountain Scrub-Robin
    8553
Migrant Shrike 5050
Migratory Jungle-
    Flycatcher 8803
Migratory Pigeon 3397
Migratory Thrush 10329
Mikado Pheasant 9676
Military Macaw 876
Milky Eagle-Owl 1337
Milky Pratincole 4237
Milky Stork 6121
Millerbird 119
Miller's Ant-Pitta 4316
Millet's Laughingtrush
    4100
Milne-Edward's Leaf-
    Warbler 7610
Milne-Edward's Warbler
    7610
Milne-Edward's Willow-
    Warbler 7610

Mimetic Honeyeater 5748
Mimic Honeyeater 5748
Mimic Meliphaga 5748
Mimic Starling 792
Minas Gerais Tyrannulet
    7602
Mindanao Bleeding-heart
    3997
Mindanao Eagle-Owl 6875
Mindanao Goldfinch 9151
Mindanao Hornbill 7227
Mindanao Jungle-
    Flycatcher 8805
Mindanao Lorikeet 10178
Mindanao Parrotfinch 3601
Mindanao Racket-tail 8219
Mindanao Racket-tailed
    Parrot 8219
Mindanao Racquet-tail
    8219
Mindanao Racquet-tailed
    Parrot 8219
Mindanao Scops-Owl 6900
Mindanao Tarictic-Hornbill
    7227
Mindanao Wattled-
    Broadbill 3761
Mindanao White-eye 5353
Mindanao Wrinkled-
    Hornbill 86
Mindoro Bleeding-heart
    4007
Mindoro Flowerpecker
    3206
Mindoro Hornbill 7230
Mindoro Imperial-Pigeon
    3363
Mindoro Pigeon 3363,4007
Mindoro Punalada 4007
Mindoro Scops-Owl 6899
Mindoro Tarictic-Hornbill
    7230
Mindoro Zone-tailed
    Pigeon 3363
Miner 4165
Miniature Greenbul 7554
Miniature Kingfisher 4886
Miniature Tit-Babbler 5874
Minute Bittern 4893
Minute Cuckoo 7672
Minute Hermit 7344

Minute Seedeater 9366
Miombo Barbet 10183
Miombo Bearded Scrub-
Robin 1913
Miombo Bush-Warbler
1476
Miombo Double-collared
Sunbird 6399
Miombo Grey-Tit 7101
Miombo Pied-Barbet
10183
Miombo Rock-Thrush
6012
Miombo Scrub-Robin 1913
Miombo Sunbird 6399
Miombo Tit 7101
Miranda's Tody-Tyrant
4566
Mirror Peacock Pheasant
8096
Mishmi Wren 9274
Mishmi Wren-Babbler
9274
Mississippi Kite 4833
Mistlethrush 10366
Mistletoe Flowerpecker
3192
Mistletoe Tyrannulet 10706
Mistletoebird (WI) 3703
Mistletoebird 3192
Misto Yellow-Finch 9208
Mitchell's Cockatoo (ANZ)
1435
Mitchell's Woodstar 7471
Mitred Comure 916
Mitred Grebe 8015
Mitred Parakeet 916
Mocking Chat (CSA) 9900
Mocking Cliff-Chat 9900
Mockingbird 5901
Modest Grassfinch 6446
Modest Honeyeater 8756
Modest Parrot 8358
Modest Tiger-Parrot 8358
Modest Wren 9996
Moheli Brush-Warbler
6505
Moheli Tsikirity 6505
Moho 4767,8158
Mohua 5961
Molina's Conure 8675

Molokai Creeper 7063
Molokai Mano 3301
Molokai Oo 5956
Molucca Fruit-Dove 8511
Moluccan Barred
Sparrowhawk 52
Moluccan Black-fronted
White-eye 10762
Moluccan Boobook 6571
Moluccan Bulbul 369
Moluccan Cicadabird 2618
Moluccan Cockatoo 1436
Moluccan Collared
Sparrowhawk 42
Moluccan Coucal 1866
Moluccan Crow 2725
Moluccan Cuckoo 1460
Moluccan Cuckoo-Shrike
2611,2641
Moluccan Fantail
8840,8862
Moluccan Flowerpecker
3212
Moluccan Flycatcher 6134
Moluccan Friarbird 7457,
7461
Moluccan Fruit-Dove 8511
Moluccan Fruit-Pigeon
3347
Moluccan Goshawk 52
Moluccan Ground-Thrush
10725
Moluccan Hanging-Parrot
5370
Moluccan Hawk-Owl 6571
Moluccan Honeyeater 5248
Moluccan Kestrel 3803
Moluccan Kingfisher
10061
Moluccan King-Parrot 365
Moluccan Mannikin 5311
Moluccan Megapode 5647
Moluccan Monarch 4780,
5999
Moluccan Monarch-
Flycatcher 4780
Moluccan Munia 5311
Moluccan Myiagra-
Flycatcher 6134
Moluccan Oriole 6797

Moluccan Owlet-Nightjar
181
Moluccan Parrot 9782
Moluccan Pitta 7858,7860
Moluccan Red-Lory 3544
Moluccan Rufous-bellied
Fruit-Pigeon 3347
Moluccan Rufous-bellied
Pigeon 3347
Moluccan Rufous-Fantail
8862
Moluccan Rufous-tailed
Moorhen 390
Moluccan Scops-Owl 6892
Moluccan Scrubfowl 5647
Moluccan Scrubhen 5647
Moluccan Shrike-Robin
3832
Moluccan Sparrowhawk
42,52
Moluccan Starling 1189,
808
Moluccan Swiftlet 200
Moluccan Triller 4969
Moluccan White-eye
10761,10762,10766
Moluccan Woodcock 9039
Moluccas Friarbird 7461
Moluccas Scrubfowl 5647
Moluccas Scrubhen 5647
Molucccan Greybird 2641
Moluku Woodcock 9039
Mombasa Woodpecker
1592
Monarch-Flycatcher 5997
Mondetour's Dove 2370
Mongalla Cisticola 2339
Mongolian Accentor 8278
Mongolian Bunting 3473
Mongolian Buzzard 1399
Mongolian Desert Jay 8026
Mongolian Dotterel 2034
Mongolian Eared-Pheasant
2842
Mongolian Finch 8872
Mongolian Ground-Jay
8026
Mongolian Gull 5109
Mongolian Lark 5705
Mongolian Pheasant 7434
Mongolian Plover 2034

Moustached Green-
Tinkerbird (CSA) 8050
Moustached Hawk-Cuckoo
2897
Moustached Jay 2933
Moustached Kingfisher 145
Moustached
Laughingthrush 4082
Moustached Parakeet 8362
Moustached Puffbird 5509
Moustached Quail-Dove
4193
Moustached Scrub-Robin
1919
Moustached Swift 4533
Moustached Tinkerbird
8050
Moustached Tree-Babbler
5504
Moustached Treeswift
4533
Moustached Turca 8475
Moustached Turco 8475
Moustached Warbler
(CSA) 5775
Moustached Warbler 125
Moustached Woodcreeper
10646
Moustached Wren 9988
Moutain Greybird 2609
Mozambique Batis (CSA)
1214
Mozambique Nightjar
(CSA) 1672
Mrs. Boulton's Warbler
7634
Mrs. Boulton's Woodland-
Warbler 7634
Mrs. Moreau's Rufous-
Warbler 1195
Mrs.. Benson's Brush-
Warbler 6505
Mrs.. Benson's Warbler
6505
Mrs.. Forbes-Watson's
Black-Flycatcher 5656
Mrs.. Gould's Sunbird 230
Mrs.. Hall's Greenbul 591
Mrs.. Hume's Pheasant
9675

Mrs.. Johnstone's Lorikeet
10178
Mrs.. Morden's Owlet 6883
Mrs.. Moreau's Warbler
1195
Mt. Loft Ranges Chestnut-
rumped Heathwren
(ANZ) 4706
Mudlark (ANZ) 4334,4335
Mud-nesters 2665
Mueller's Barbet 5610
Mufumbiri Gonolek 5025
Mufumbri Shrike 5025
Mugimaki Flycatcher 3843
Muir's Corella (ANZ) 1439
Mulga Parrot 8317
Müller's Imperial-Pigeon
3364
Müller's Barbet 5610
Müller's Blue-backed
Parrot 9783
Müller's Bush-Warbler
1964
Müller's Cuckoo-Shrike
2641
Müller's Fruit-Pigeon 3364
Müller's Greybird 2641
Müller's Parrot 9783
Müller's White-eye 4484
Müller's Wren-Babbler
6352
Mulsant's Woodstar 97
Multicoloured Tanager
2114
Multicrested Bird of
Paradise 2389
Muna Cuckoo-Shrike 2630
Muna Greybird 2613
Mundane Monarch 5998
Munia Cuckoo-Shrike 2613
Murky Sandpiper 1501
Murphy's Petrel 8458
Murphy's White-eye 10804
Murray Smoker 7891
Murre 10454
Murumbidgee Lory 7891
Murumbidgee Rosella 7891
Murumbidgee Swamp-Lory
7891
Musao Island Monarch
5997

Muschenbroek's Lorikeet
6490
Muscovy 1466
Muscovy Duck 1466
Musician Wren 3008
Musk Duck 1233,1466
Musk Lorikeet 4277
Musk Lory 4277
Musky Lorikeet 4277
Mute Swan 2982
Muttonbird (CSA) 8559,
8566,8569,8570,8573,
8578,8580
Muttonbird 8581
Mynah 105
Mynahassa Barn-Owl
10418
Mynahassa Flowerpecker
3179
Mynahassa Masked-Owl
10418
Mynas Gerais Tyrannulet
7602
Myrtle Warbler (NA) 3129
Mysterious Scops-Owl
6892
Mysterious Starling 805
Nabipur Laughingthrush
4122
Nacunda Nighthawk 8005
Naga Nuthatch 9243
Nahan's Francolin 3915
Naked-eye Thrush 10334
Naked-eyed Partridge-
Bronzewing 4160
Naked-faced Barbet 4372
Naked-faced Spiderhunter
883
Naked-legged Swiftlet 205
Nakta (ISC) 8959
Namaqua Dove 6702
Namaqua Prinia 8209
Namaqua Sandgrouse 8414
Namaqua Thrush (CSA)
10339
Namaqua Warbler 8209
Namibia Canary 9134
Namibia Sparrow 7154
Namuli Apalis 758
Nanday Conure 6343
Nanday Parakeet 6343

Nankeen Heron 6634
Nankeen Kestrel 3786
Nankeen Night-Heron 6634
Nankeen Plover 9475
Napo Sabrewing 1613
Napoleon Bishop 3711
Napoleon Weaver 3711
Napoleon's Peacock-
    Pheasant 8094
Narcissina Flycatcher 3844
Narcissus Flycatcher 3844
Narcodoman Hornbill 87
Narcondam Hornbill 87
Narcondom Wreathed
    Hornbill 87
Narina Trogon 772
Narino Tapaculo 9081
Narosky's Seedeater 9379
Narrow-billed Ant-Wren
    3872
Narrow-billed Bronze-
    Cuckoo 2189
Narrow-billed Creeper
    5141
Narrow-billed Prion 6975
Narrow-billed Tody 10096
Narrow-billed
    Woodcreeper 5141
Narrow-tailed Emerald
    2173
Narrow-tailed Starling
    8039
Narrow-tailed Tyrant 8101
Nashville Warbler 10529
Natal Button-Quail 10374
Natal Chat 2751
Natal Francolin 3916
Natal Ground-Thrush
    10730
Natal Honeybird 8264
Natal Nightjar (CSA) 1685
Natal Robin (CSA) 2751
Natal Robin-Chat 2751
Natal Thrush (CSA) 10729
Natal Yellow-Warbler
    2123
Native Pigeon 5329
Natterer's Cotinga 2761
Natterer's Hermit 7352
Natterer's Manakin 7797
Natterer's Piculet 7713

Natterer's Tanager 9709
Natterer's Toucanet 9106
Natterer's Tyrant 7590
Natuna Serpent-Eagle 9310
Naumann's Thrush 10303,
    10331
Nauru Reed-Warbler 131
Nava's Wren 4761
Nazca Booby 9576
Nduk Eagle-Owl 1345
Neblina Foliagegleaner
    1111
Neblina Leafgleaner 1111
Neblina Metaltail 5832
Nechisar Nightjar 1704
Necklace Spotted-Dove
    9493
Necklaced Hill-Partridge
    943
Necklaced Laughingthrush
    (ISC) 4103
Necklaced Laughingthrush
    4108
Necklaced Spinetail 9658
Neddicky (CSA) 2329
Neddicky Cisticola 2329
Needle-billed Hermit 7353
Neergaard's Sunbird 6406
Neergard's Double-collared
    Sunbird 6406
Negrito Flycatcher 5191
Negros Babbler 9400
Negros Bleeding-heart
    4002
Negros Blood-breasted
    Pigeon 4002
Negros Fruit-Dove 8483
Negros Jungle-Flycatcher
    8802
Negros Pigeon 4002
Negros Punalada 4002
Negros Striped Tree-
    Babbler 9400
Negros Striped-Babbler
    9400
Negros Tree-Babbler 9400
Negros White-throated
    Jungle-Flycatcher 8802
Nehrkorn's Flowerpecker
    3199
Nelicourvi Weaver 7966

Nellie (CSA) 5453
Nelson's Flycatcher 775
Nelson's Gull 5103
Nelson's Sharp-tailed
    Sparrow 482
Nelson's Vireo 10583
Néné 1320
Néné Goose 1320
Neospiza 6492
Neotropic Black-headed
    Siskin 1736
Neotropic Cormorant
    7377,7401
Neotropic River-Warbler
    1183
Neotropical Fan-tailed
    Warbler 3774
Neotropical Palm-Swift
    9683
Neotropical River-Warbler
    1183
Nepal Babbler 335
Nepal Barwing 152
Nepal Cutia 2909
Nepal Fulvetta 335
Nepal House-Martin 3055
Nepal Martin 3055
Nepal Parrotbill 7033
Nepal Quaker-Babbler 335
Nepal Rosefinch 1763
Nepal Sunbird 233
Nepal Treecreeper 1932
Nepal Wood-Pigeon 2515
Nepal Wren-Babbler 8003
Nepal Yellow-backed
    Sunbird 233
Nepalese Treecreeper 1932
Nereis Tern 9458
Neumann's Bush-Warbler
    4549
Neumann's Coucal 1860
Neumann's Short-tailed
    Warbler 4549
Neumann's Warbler 4549
Neumann's Waxbill 3631
Neumayer's Rock-Nuthatch
    9244
Nevada Cowbird 5964
New Britain Babbler-
    Warbler 5623

New Guinea Robin 8030
New Guinea Scrubfowl
  5634
New Guinea Spinetail 5578
New Guinea Spinetailed-
  Swift 5578
New Guinea Swiftlet 205
New Guinea Thornbill 18
New Guinea Tiger Bittern
  10708
New Guinea Tiger-Heron
  10708
New Guinea Treecreeper
  2674
New Guinea White-eye
  10809
New Guinea Woodswallow
  1000
New Guinea Wreathed-
  Hornbill 89
New Guinea Wren 5535
New Guinean Brown-
  Honeyeater 8631
New Hanover Mannikin
  5315
New Hanover Munia 5315
New Hebrides Broadbill
  6128
New Hebrides Flycatcher
  6464
New Hebrides Fruit-Dove
  8530
New Hebrides Honeyeater
  7542
New Hebrides Monarch
  6464
New Hebrides Scrubfowl
  5642
New Hebrides Starling 816
New Holland Honeyeater
  7543
New Holland Snipe 4020
New Ireland Drongo 3232
New Ireland Finch 5297
New Ireland Flowerpecker
  3189
New Ireland Friarbird 7456
New Ireland Hawk-Owl
  6575
New Ireland Honeyeater
  6333

New Ireland Mannikin
  5297
New Ireland Munia 5297
New World Warblers 7076
New World Blackbirds
  4799
New World Cuckoos 2395
New World Quails 6682
New World Vulture 2661
New World Vultures 1810
New Zealand Antarctic-
  Tern (ANZ) 9470
New Zealand Bellbird 650
New Zealand Bittern 1263
New Zealand Brown-
  Creeper 5960
New Zealand Bush-Falcon
  3807
New Zealand Creeper 5960
New Zealand Crested-
  Penuin 3651
New Zealand Dabchick
  8070
New Zealand Dotterel 2038
New Zealand Falcon 3807
New Zealand Fantail 8832
New Zealand Fernbird
  5630
New Zealand Flycater 4223
New Zealand Fruit-Pigeon
  4528
New Zealand Gerygone
  4223
New Zealand Grebe 8070
New Zealand Grey
  Gerygone 4223
New Zealand Grey-
  Flyeater 4223
New Zealand Hobby 3807
New Zealand Kaka 6524
New Zealand King-
  Cormorant 7383
New Zealand King-Shag
  7383
New Zealand Little-Bittern
  4895
New Zealand Parakeet
  2958,2961
New Zealand Pigeon 4528
New Zealand Piopio 10367
New Zealand Pipit 721

New Zealand Plover 2038
New Zealand Quail 2773
New Zealand Quail-Hawk
  3807
New Zealand Red-breasted
  Dotterel 2038
New Zealand Robin 7275
New Zealand Robin-
  Flycatcher 7275
New Zealand Scaup 1136
New Zealand Semi-
  woodcock 2422
New Zealand Shag 7383
New Zealand Shearwater
  8565
New Zealand Shelduck
  9720
New Zealand Shore-Plover
  2037
New Zealand Shoveler 564
New Zealand Snipe
  2422,2423
New Zealand Sooty-
  Oystercatcher 4428
New Zealand Stilt 4635
New Zealand Teal 525
New Zealand Thrush 10367
New Zealand Tit 7278
New Zealand Tomtit 7279
New Zealand Wattlebirds
  1518
New Zealand Woodrail
  4045
New Zealand Wrens 10620
Newell's Shearwater 8576
Newtonia 6536
Newton's Bowerbird 8226
Newton's Cuckoo-Shrike
  2642
Newton's Fiscal 5055
Newton's Fiscal-Shrike
  5055
Newton's Gardner-
  Bowerbird 8226
Newton's Golden-
  Bowerbird 8226
Newton's Kestrel 3806
Newton's Parakeet 8370
Newton's Scrub-Warbler
  1288
Newton's Sunbird 6407

North Atlantic Gannet 6032
North Atlantic Shearwater
  1531
North Island Oystercatcher
  4428
North Island Pied-
  Oystercatcher 4428
North Island Woodhen
  4045
Northern Anteater Chat
  6254
Northern Ant-eating Chat
  6254
Northern Bald Ibis 4214
Northern Barred-
  Woodcreeper 3087
Northern Bearded Scrub-
  Robin 1919
Northern Beardless-
  Flycastcher 1601
Northern Beardless-
  Tyrannulet 1601
Northern Bentbill 6728
Northern Black Tit 7108
Northern Black-Flycatcher
  5658
Northern Black-Grouse
  9874
Northern Black-headed
  Gull 5110
Northern Black-Korhaan
  3731
Northern Black-throated
  Wren 9982
Northern Black-Tit (CSA)
  7102
Northern Boat-billed Heron
  2406
Northern Bobwhite 2451
Northern Brownbul 7567
Northern Brown-Bulbul
  7567
Northern Brown-throated
  Weaver 7946
Northern Brown-Tit 7094
Northern Bullfinch 8663
Northern Cactus-Wren
  1620
Northern Cardinal 1712
Northern Carmine Bee-
  eater 5809

Northern Cassowary 1796
Northern Catbird 3383
Northern Chowchilla 6824
Northern Crag-Martin 4681
Northern Creepers 1936
Northern Crested Shrike-
  Tit (ANZ) 3826
Northern Crombec 9612
Northern Crowned-Crane
  1147
Northern Diver 4133
Northern Double-collared
  Sunbird 6414,6420
Northern Eagle-Owl 1334
Northern Eastern-
  Bristlebird (ANZ) 3046
Northern Fairy-Flycatcher
  3443
Northern Fantail 8857
Northern Fig-Parrot 2972
Northern Flicker 2428,2435
Northern Fulmar (NA)
  3965
Northern Fulmar-Petrel
  3965
Northern Gannet (NA)
  6032
Northern Gerygone 4230
Northern Giant-Petrel
  (ANZ) 5454
Northern Goshawk 47
Northern Grey-Tit (CSA)
  7101
Northern Grey-Tit 7131
Northern Grosbeak-Canary
  (CSA) 9149
Northern Ground Hornbill
  1363
Northern Harrier 2288
Northern Hawk-Cuckoo
  2888
Northern Hawk-Owl 9583
Northern Hazelhen 1253
Northern Helmeted-
  Curassow 7178
Northern Hen-Harrier 2288
Northern Hepatic Tanager
  7823
Northern Hobby (NA) 3816
Northern House-Martin
  3056

Northern House-Wren
  10215
Northern Ivory-billed
  Woodpecker 1583
Northern Jacana 4911
Northern Jay 7265
Northern Jery 6468
Northern Lapwing 10504
Northern Logrunner 6824
Northern Long-tailed
  Glossy-Starling 4993
Northern Marsh-Harrier
  2283
Northern Marsh-
  Meadowlark 5136
Northern Masked-Owl
  (ANZ) 10425
Northern Masked-Plover
  10496
Northern Masked-Weaver
  7987
Northern Melanesian
  Ground-Thrush 10752
Northern Mockingbird
  5901
Northern Mountain-
  Cacique 1450
Northern Mountain-
  Greenbul 588
Northern Musical Wren
  3009
Northern Musician-Wren
  3009
Northern Needletail 4645
Northern Needle-tailed
  Swift 4645
Northern Nightingale-Wren
  5855
Northern Nightjar 1673
Northern Olive-Thrush
  10287,10338
Northern Orange-headed
  Ground-Thrush 10720
Northern Orange-tufted
  Sunbird 6411
Northern Oriole 4804,4811
Northern Paradise-Whydah
  10544
Northern Parula 7071
Northern Phalarope (NA)
  7415

Pale-billed Hornero 3973
Pale-billed Scrubwren 9127
Pale-billed Sericornis 9127
Pale-billed Sicklebill 3548
Pale-billed Woodcreeper
   3084
Pale-billed Woodpecker
   1576
Pale-blue Niltava 3007
Pale-breasted Castle-
   builder 9628
Pale-breasted Illadopsis
   4846
Pale-breasted Robin 10321
Pale-breasted Spinetail
   9628
Pale-breasted Thrush
   10321
Pale-breasted Thrush-
   Babbler 4846
Pale-browed Tinamou 2878
Pale-browed Treehunter
   2222
Pale-brown Shrike 5048
Pale-capped Pigeon 2516
Pale-chinned Flycatcher
   2999
Pale-coloured Prinia 8207
Pale-crested Woodpecker
   1837
Pale-crowned Cisticola
   (CSA) 2313
Pale-crowned Cloud-
   Cisticola 2313
Pale-edged Flycatcher 6149
Pale-eyed Blackbird 261
Pale-eyed Pygmy-Tyrant
   1048
Pale-eyed Thrush 7896
Pale-faced Antbird 9259
Pale-faced Bare-eye 9259
Pale-faced Bulbul 8601
Pale-faced Sheathbill 2080
Pale-footed Bush-Warbler
   1959
Pale-footed Cettia 1959
Pale-footed Shearwater
   8566
Pale-footed Stubtail 1959
Pale-footed Swallow 6605
Pale-fronted Dove 5185

Pale-fronted Negrofinch
   6544
Pale-grey Cicadabird 2618
Pale-grey Cuckoo-Shrike
   2618
Pale-headed Brush-Finch
   1070
Pale-headed Finch 1070
Pale-headed Frogmouth
   1224
Pale-headed Jacamar 1267
Pale-headed Mannikin
   5307,5317
Pale-headed Munia 5317
Pale-headed Nun 5317
Pale-headed Rosella 7884
Pale-headed Woodpecker
   4138
Pale-legged Hornero 3970
Pale-legged Leaf Warbler
   7659
Pale-legged Ovenbird 3970
Pale-legged Warbler 1185
Pale-legged Willow-
   Warbler 7659
Pale-mandibled Aracari
   8464
Pale-mandibled Toucan
   8464
Pale-naped Brush-Finch
   1071
Pale-naped Finch 1071
Pale-olive Bulbul 7557
Pale-olive Greenbul 7557
Paler Chinspot Puffback-
   Flycatcher 1214
Pale-rumped Swift 1982
Pale-rumped Warbler 7619
Palestine Sunbird 6411
Pale-tailed Barbthroat 9958
Pale-tailed Canastero 1024
Pale-throated Flycatcher
   6156
Pale-throated Pampa-Finch
   3494
Pale-throated Sierra-Finch
   3494
Pale-throated Tapaculo
   9068
Pale-tipped Inezia 4865

Pale-tipped Tyrannulet
   4865
Pale-vented Euphonia 3693
Pale-vented Pigeon 2482
Pale-vented Thrush 10336
Pale-winged Ant-Wren
   6274
Pale-winged Indigobird
   10550
Pale-winged Starling 6735
Pale-winged Trumpeter
   8390
Pale-yellow Robin 10142
Pale-yellow Robin-
   Flycatcher 10142
Palila 5398
Pallas's Bunting 3473
Pallas's Cormorant 7405
Pallas's Crake 8175
Pallas's Dipper 2268
Pallas's Eagle 4447
Pallas's Eared-Pheasant
   2842
Pallas's Fish-Eagle 4447
Pallas's Fishing-Eagle
   (ISC) 4447
Pallas's Grasshopper-
   Warbler 5280
Pallas's Gull 5094
Pallas's Leaf-Warbler 7647
Pallas's Reed-Bunting 3473
Pallas's Rosefinch 1772
Pallas's Rosy-Finch 1772
Pallas's Sandgrouse 9679
Pallas's Sea-Eagle (ISC)
   4447
Pallas's Warbler 5280,7647
Pallas's Willow-Warbler
   7647
Pallatanga Elaenia 3424
Pallid Cuckoo 2890
Pallid Dove 5182
Pallid Falcon 3799
Pallid Finch 5317
Pallid Flycatcher (CSA)
   1285
Pallid Harrier 2291
Pallid Honeyguide 4857
Pallid Mannikin 5317
Pallid Munia 5317
Pallid Scops-Owl 6865

Rheas 8795,8796
Rheinhardt's Argus 8798
Rhinoceros Auk 1926
Rhinoceros Auklet 1926
Rhinoceros Hornbill 1359
Ribbon Finch 378
Ribbon-tailed Astrapia 1041
Ribbon-tailed Drongo 3232
Rice Cowbird 9001
Rice Grackle 9001
Rice Munia 7001
Ricebird (WI) 3290
Ricebird 7001
Richard's Fruit-Dove 8524
Richard's Monarch 6000
Richard's Pipit (CSA) 702
Richard's Pipit 721
Richardson's Goose 1315
Richardson's Skua 9435
Richmond's Creeper 10668
Rickery-Dick (WI) 2184
Rickett's Hill-Partridge 934
Rickett's Parrotbill 7025
Rickett's Partridge 934
Rickett's Willow-Warbler 7650
Ridgeway's Whip-poor-will 1696
Ridgway's Cotinga 2762
Ridgway's Hawk 1410
Ridgway's Rough-winged Swallow 9425
Ridgway's Titmouse 7120
Rieffer's Hummingbird 426
Rifleman 8
Rimatara Reed-Warbler 132
Ring Dove (CSA) 9503
Ring Dove (ISC) 9494
Ring Dove 2511
Ring Ouzel 10364
Ringbill 1131
Ring-billed Duck 1131
Ring-billed Gull 5082
Ringed Antpipit 2737
Ringed Antshrike 9929
Ringed Dotterel 2028
Ringed Finch 9721
Ringed Gowrie (WI) 9512
Ringed Kingfisher 5589

Ringed Penguin 8638
Ringed Plover 2028
Ringed Storm-Petrel 6655
Ringed Swift 9512
Ringed Teal 1526
Ringed Turtle-Dove 9501
Ringed Warbling-Finch 8148
Ringed Woodpecker 1839
Ring-neck (WI) 2047
Ringneck Parrot 1156
Ring-necked Crow 2720
Ring-necked Dove 9492
Ring-necked Duck 1131
Ring-necked Francolin 3929
Ring-necked Parakeet 8374
Ring-necked Parrot 1156, 7885
Ring-necked Pheasant (NA) 7433
Ring-necked Teal 1526
Ringtail (WI) 2481,8978
Ring-tailed Fishing-Eagle (ISC) 4447
Ring-tailed Pigeon 2481
Rio Branco Antbird 1894
Rio de Janeiro Antbird 1893,6278
Rio de Janeiro Ant-Wren 6278
Rio de Janeiro Greenlet 4755
Rio Grande Jay 2939
Rio Suno Ant-Wren 6300
Ripley's Fruit-Dove 8483
Ripley'sDove 8483
Riroriro 4223
River Chat 1990
River Chink (WI) 4911
River Eagle 4450
River Flycatcher 5975
River Honeyeater 5237
River Kingfisher 307,310
River Lapwing 10487
River Prinia 8193
River Redstart 1990
River Robin 5975
River Tern 9442
River Tyrannulet 9184

Riverina Shy Heathwren (ANZ) 4703
River-Martin 8326
Riverside Tyrant 4943
Riverside Wren 10004
River-Warbler 1183,5282
Rivoli's Hummingbird 3657
Roadrunner 4141
Roadside Hawk 1404
Roaroa Kiwi 827
Roatelo 5824
Robert's Chat 2754
Robert's Forest-Prinia 8204
Robert's Hummingbird 7324
Roberts's Prinia (CSA) 8204
Robin (NA) 10329
Robin (WI) 10100,5395
Robin 3594
Robin Accentor 8282
Robin Dyal 2596
Robin Hedge-Sparrow 8282
Robin Redbreast 3594
Robin-Chat 2744
Robin-Flycatcher 3843, 7275
Robin-Redbreast (WI) 10100
Robinson's Button-Quail 10377
Robinson's Whistling-Thrush 6232
Roborate Screech-Owl 6905
Roborowski's Rosefinch 1769
Robust Thornbill 23
Robust White-eye 10829
Robust Woodpecker 1584
Rock Accentor 8273
Rock Bunting (CSA) 3485
Rock Bunting 3457
Rock Bush-Quail 7243
Rock Capercaillie 9873
Rock Cisticola (CSA) 2305
Rock Conure 8681
Rock Cormorant 7395
Rock Dove 2500

Sind Pied-Woodpecker 3090
Sind Sparrow 7156
Sind Woodpecker 3090
Singing Blackbird 3287
Singing Bush-Lark (CSA) 5926
Singing Bush-Lark 5925,5934
Singing Bush-Warbler 1955
Singing Cisticola 2315
Singing Dove 10694
Singing Green-Pigeon 10163
Singing Honeyeater 5239
Singing Lark 5925
Singing Parrot 4146
Singing Peltops 7211
Singing Quail 3037
Singing Starling 793
Single-wattled Cassowary 1796
Singular Greytail 10619
Singular Softtail 10619
Sinhalese Frogmouth 1223
Sinhalese Hanging-Parrot 5372
Sinkiang Ground-Jay 8025
Sipora Scops-Owl 6898
Siquijor Bulbul 4905
Sira Tanager 9764
Sirkeer 7314
Sirkeer Cuckoo (ISC) 7314
Sirkeer Malkoha 7314
Sirystes (NA) 9229
Siskin 1738,1744
Six-plumed Bird of Paradise 7068
Sjostedt's Greenbul 1142
Sjostedt's Barred Owlet 4272
Sjostedt's Honeyguide Greenbul 1142
Sjostedt's Owlet 4272
Sjostedt's Pigeon 2518
Sjostedt's White-tailed Bulbul 1142
Sjostedt's White-tailed Greenbul 1142
Skuas 9433

Skylark 299
Sladen's Barbet 4374
Slate-and-gold Warbler 1173
Slate-and-rufous Warbling-Finch 8139
Slate-backed Thornbill 23
Slate-blue Seedeater 396
Slate-breasted Rail 5218
Slate-colored Junco (NA) 4920
Slate-coloured Antbird 7242
Slate-coloured Boubou 5021
Slate-coloured Boubou-Shrike 5021
Slate-coloured Bush-Shrike 5021
Slate-coloured Coot 3955
Slate-coloured Cuckoo-Dove 10256
Slate-coloured Fox-Sparrow 7167
Slate-coloured Grosbeak 8943
Slate-coloured Hawk 5207
Slate-coloured Seedeater 9375
Slate-coloured Solitaire 6115
Slate-crowned Ant-Pitta 4331
Slate-headed Tody-Flycatcher 10094
Slate-headed Tody-Tyrant 10094
Slater's Leaf-Warbler 7650
Slate-throated Gnatcatcher 8085
Slate-throated Redstart 6181
Slate-throated Whitestart 6181
Slaty Antshrike 9912
Slaty Ant-Wren 6296
Slaty Becard 6994
Slaty Blue-Flycatcher 3856
Slaty Blue-Robin 7236
Slaty Bristlefront 5821
Slaty Brush-Finch 1076

Slaty Bunting 5131
Slaty Castle-builder 9632
Slaty Cuckoo-Dove 10256
Slaty Cuckoo-Shrike 2651, 2652
Slaty Egret 3410
Slaty Elaenia 3430
Slaty Finch 1076,4463
Slaty Flowerpecker 3241, 3251,3252
Slaty Flowerpiercer 3241, 3251,3252
Slaty Flycatcher 3282, 5575,5576,6134
Slaty Gnateater 2557
Slaty Ground-Pigeon 10253
Slaty Highland-Honeycreeper 3241
Slaty Monarch 5575,6134
Slaty Robin-Flycatcher 7236
Slaty Solitaire 6115
Slaty Spinetail 9632
Slaty Tanager 2829
Slaty Thicket-Flycatcher 7236
Slaty Thrush 10333
Slaty Tody-Flycatcher 10094
Slaty Vireo 10563
Slaty-Antshrike 9924
Slaty-backed Blue-Flycatcher 3839
Slaty-backed Chat-Tyrant 6665
Slaty-backed Flycatcher 3839
Slaty-backed Forest-Falcon 5845
Slaty-backed Forktail 3523
Slaty-backed Goshawk 56
Slaty-backed Ground-Thrush 10749
Slaty-backed Gull 5113
Slaty-backed Hemispingus 4540
Slaty-backed Jungle-Flycatcher 8805
Slaty-backed Nightingale-Thrush 1815

Solomons Broadbill 6132
Solomons Broadbill-
  Flycatcher 6132
Solomons Cockatoo 1431
Solomons Corella 1431
Solomons Crow 2704
Solomons Flowerpecker
  3174
Solomons Flycatcher 6132
Solomons Mountain-
  Whistler 6947
Solomons Pied-Monarch
  5977
Solomons Rail 6512
Solomons Satin Flycatcher
  6132
Solomons Thicketbird 5624
Solomons Whistler 6947
Solomons White-eye 10790
Somali Bee-eater 5816
Somali Blackbird 10324
Somali Black-throated
  Bustard 3737
Somali Bulbul 8618
Somali Bunting 3474
Somali Bush-Lark 5942
Somali Canary 9150
Somali Chanting-Goshawk
  5742
Somali Chestnut-winged
  Starling 6731
Somali Courser 2907
Somali Crombec
  9615,9617
Somali Fiscal 5059
Somali Fiscal-Shrike 5059
Somali Golden-breasted
  Bunting 3474
Somali Grey-Tit 7131
Somali Lark
  4606,5942,5943
Somali Long-billed Bush-
  Lark 5943
Somali Long-billed
  Crombec 9615
Somali Long-billed Lark
  5943
Somali Long-billed
  Sylvietta 9615
Somali Long-clawed Lark
  4606

Somali Ostrich 9539
Somali Pigeon 2508
Somali Raven 2714
Somali Rock-Pigeon 2508
Somali Short-billed
  Crombec 9617
Somali Short-billed
  Sylvietta 9617
Somali Short-tailed
  Crombec 9617
Somali Short-toed Lark
  1487
Somali Shrike 5059
Somali Sparrow 7139
Somali Starling 6731
Somali Stock-Dove 2508
Somali Stock-Pigeon 2508
Somali Sylvietta 9615
Somali Thrush 10324
Somali Tit 7131
Somali Wheatear 6721
Sombre Bulbul (CSA) 592
Sombre Chat 1903
Sombre Fantail 8863
Sombre Gallinule 4043
Sombre Greenbul 587,
  592,597
Sombre Hummingbird 777
Sombre Kingfisher 10064
Sombre Leaf-Warbler 7609
Sombre Monarch 5990
Sombre Moorhen 4043
Sombre Nightjar 1673
Sombre Pigeon 2851
Sombre Rock-Chat 1903
Sombre Tit 7110
Song Parrot 4146
Song Sparrow 5781
Song Tanager 8744
Song Wren 3008,3009
Songar Tit 7127
Song-Thrush 10344
Sonnerat's Junglefowl 4068
Sook (WI) 7694
Sooty Albatross 7497
Sooty Antbird 6239,8792
Sooty Anteater Chat 6259
Sooty Ant-Tanager 4411
Sooty Babbler 7751,9391
Sooty Barbthroat 9960
Sooty Boubou 5020,5022

Sooty Boubou-Shrike 5022
Sooty Bush-Shrike 5022
Sooty Chat 1903,
  6259,7751
Sooty Crake 8178
Sooty Eagle-Owl 1338
Sooty Falcon 3790
Sooty Finch 1076
Sooty Flycatcher 6067,
  6076
Sooty Fox-Sparrow 7168
Sooty Grassquit 10012
Sooty Grouse 3061,3062
Sooty Guillemot 1877
Sooty Gull 5091
Sooty Honeyeater 5729
Sooty Jay 7266
Sooty Long-tailed Tit 166
Sooty Melidectes 5729
Sooty Myzomela 6338
Sooty Nightjar 1702
Sooty Ouzel 10332
Sooty Owl 10421,10430
Sooty Oystercatcher 4421
Sooty Rail 8178
Sooty Robin 10332
Sooty Rock-Chat
  1903,7751
Sooty Shearwater 8570
Sooty Shrike-Thrush 2469
Sooty Spinetail 9632
Sooty Storm-Petrel 6658,
  6659,6664
Sooty Swift 3013
Sooty Tanager 4411
Sooty Tern 9452
Sooty Thicket-Fantail 8865
Sooty Thrush 10332
Sooty Tit 166
Sooty Tree-Babbler 9391
Sooty Tyrannulet 9186
Sooty Whistler 2469
Sooty Woodpecker 6054
Sooty-backed Hawk-Owl
  6574
Sooty-backed Owl 6574
Sooty-capped Babbler 5501
Sooty-capped Bush-
  Tanager 2152
Sooty-capped Hermit 7335
Sooty-capped Petrel 8441

Southern Bristle-Tyrant
7590
Southern Brown Fruit-
Dove 7422
Southern Brown-throated
Weaver 7995
Southern Cape Petrel
(ANZ) 3042
Southern Caracara 8090
Southern Carmine Bee-
eater 5817
Southern Cassowary 1794
Southern Common Babbler
10258
Southern Common Indian
Nightjar 1652
Southern Cordonbleu
10446,10447
Southern Crested Bellbird
(ANZ) 6755
Southern Crested
Madagascar Coucal
2786
Southern Crested-Grebe
8014
Southern Crowned-Crane
1148
Southern Crowned-Pigeon
4290
Southern Double-collared
Sunbird 6375
Southern Emuwren 9477
Southern Fairy-Flycatcher
3440,3443
Southern Fairy-Prion
(ANZ) 6981
Southern Figbird 9285
Southern Fulmar 3966
Southern Fulmar-Prion
(ANZ) 6977
Southern Giant-Fulmar
5453
Southern Giant-Petrel 5453
Southern Godwit 5260
Southern Great Reed-
Warbler 137
Southern Great-Skua 1802
Southern Green-Jery 6469
Southern Grey Wren-
Warbler 1476

Southern Grey-headed
Sparrow 7140
Southern Grey-Shrike 5053
Southern Grey-Tit (CSA)
7077
Southern Grosbeak-Canary
(CSA) 9138
Southern Ground Hornbill
1364
Southern Helmeted-
Curassow 7179
Southern Helmet-Shrike
8231
Southern Hill-Mynah 4292
Southern Honeyeater 5747
Southern Hooded Sierra-
Finch 7533
Southern House-Wren
10215,10221
Southern Hyliota 4713
Southern Island Snipe 2422
Southern Kapoc Penduline-
Tit 653
Southern Lapwing 10483
Southern Lesser Blue-eared
Glossy-Starling 4999
Southern Logrunner 6825
Southern Long-tailed
Glossy-Starling 5001
Southern Magpie-Robin
2596
Southern Martin 8260
Southern Masked-Weaver
7990
Southern Mockingbird
5896
Southern Mountain-
Greenbul 586
Southern Musician-Wren
3008
Southern Nightingale
Whistler-Wren 5854
Southern Nightingale-Wren
5854
Southern Orange-tufted
Sunbird 6371
Southern Ostrich 9538
Southern Painted Stork
6121
Southern Pale Chanting-
Goshawk 5741

Southern Palm Cockatoo
(ANZ) 8240
Southern Penduline-Tit 653
Southern Penguin 3656
Southern Pied-Babbler
10261
Southern Pintail 539
Southern Pochard 6527
Southern Puffback 3335
Southern Pygmy-Sunbird
683
Southern Red-Bishop
(CSA) 3726
Southern Rock-Bunting
3455
Southern Rough-winged
Swallow 9426
Southern Royal-Albatross
(ANZ) 3271
Southern Rufous Scrub-
bird (ANZ) 1085
Southern Rufous-rumped
Shrike 5057
Southern Rufous-Sparrow
7154
Southern Screamer 2070
Southern Screech-Owl
6920
Southern Scrub-Flycatcher
9569
Southern Scrub-Robin
3315
Southern Serkeer 7314
Southern Shoveler 564
Southern Shrikebill 2384
Southern Singing Bush-
Lark 5926
Southern Singing Lark
5926
Southern Skua 1802
Southern Snake-Eagle 2280
Southern Snipe 4033
Southern Spinetailed
Chowchila 6825
Southern Squatter-Pigeon
(ANZ) 4159
Southern Star Finch (ANZ)
6451
Southern Stone-Curlew
1374

Southern Streaky-Seedeater
9179
Southern Suiriri 9572
Southern Tchagra 9820
Southern Tern 9464
Southern Thick-knee 1374
Southern Tit-Warbler 9610
Southern Treepie 3077
Southern Trogon (ISC)
4469
Southern Violet Wood-
Hoopoe 7511
Southern Warbler 9610
Southern White-bellied
Sunbird 6433
Southern White-crowned
Shrike 3746
Southern White-crowned
Tapaculo 9053
Southern White-eared
Honeyeater 5750
Southern White-eared
Mountain-Honeyeater
5750
Southern Whiteface 778
Southern White-winged
Tapaculo 9053
Southern Wigeon 566
Southern Yellow Grosbeak
7441
Southern Yellow-bellied
Flycatcher 4713
Southern Yellow-bellied
Hyliota 4713
Southern Yellow-bellied
Warbler 4713
Southern Yellow-billed
Hornbill 10051
Southern Yellow-Robin
3533
Southern Yellow-rumped
Seedeater 9137
Southern Yellowthroat
4191
Southern Yellow-throated
Sparrow 7289
South-western Cape Barren
Goose (ANZ) 1925
South-Western Red-tailed
Black-Cockatoo
(ANZ) 1548

Souza's Tchagra 9816
Spadiced Owlet 4252
Spain-Spain (WI) 8074
Spalding's Logrunner 6824
Spangle-cheeked Tanager
9744
Spangled Coquette 5343
Spangled Cotinga 2757
Spangled Drongo 3222,
3228
Spangled Honeyeater 5761
Spangled Kookaburra 3027
Spangled Tanager 9744
Spanish Imperial-Eagle 847
Spanish Nightingale (WI)
5897
Spanish Quail (WI) 8742
Spanish Sparrow 7147
Spanish Wheatear 6710
Spanish Woodpecker (WI)
7694
Sparkling Violet-ear 2441
Sparkling-tailed
Hummingbird 10023
Sparkling-tailed Woodstar
10023
Sparrow (WI) 5395
Sparrow 7141
Sparrowhawk 3786,3807,
3815,63,65
Sparrows 7169
Sparrow-Weaver 7929
Speckle-breasted Ant-Pitta
4735,4736
Speckle-breasted Antshrike
10638
Speckle-breasted
Woodpecker 3165
Speckle-breasted Wren
10003
Speckle-chested Piculet
7727
Speckled Ant-Pitta 4736
Speckled Antshrike 10638
Speckled Boobook 6565
Speckled Button-Quail
10382
Speckled Chachalaca 6814
Speckled Crake 2765
Speckled Duck 9472
Speckled Firefinch 4959

Speckled Ground-Dove
2536
Speckled Hawk-Owl 6565
Speckled Honeyeater 5248
Speckled Hummingbird
159
Speckled Mountain-Thrush
10723
Speckled Mourner 5036
Speckled Mousebird 2455
Speckled Piculet 7715
Speckled Pigeon 2490
Speckled Rail 2765
Speckled Reed-Warbler
136
Speckled Spinetail 2803
Speckled Tanager 9748
Speckled Teal 542
Speckled Tinker-Barbet
8053
Speckled Tinkerbird 8053
Speckled Warbler 2216
Speckled Wood-Pigeon
2478,2491
Speckled Wren 10003
Speckled-marked Darwin's
Rail 2765
Speckle-faced Parrot 7770
Speckle-fronted Weaver
9380
Speckle-necked Pigeon
2519
Speckle-throated
Woodpecker 1597
Spectacled Amazon 436
Spectacled Amazon-Parrot
436
Spectacled Ant-Pitta 4737
Spectacled Barwing
151,154
Spectacled Bristle-Tyrant
7598
Spectacled Bulbul 8598
Spectacled Cormorant 7405
Spectacled Dove 5840
Spectacled Duck 569
Spectacled Eider 9264
Spectacled Finch 1516
Spectacled Foliagegleaner
513
Spectacled Fulvetta 339

Spectacled Greenbul 7571
Spectacled Guillemot 1877
Spectacled Imerial-Pigeon 3368
Spectacled Laughingthrush 4109
Spectacled Monarch 6004
Spectacled Monarch-Flycatcher 6004
Spectacled Owl 8585
Spectacled Parrotbill 7026
Spectacled Parrotlet 3879
Spectacled Pelican 7195
Spectacled Petrel 8241, 8243
Spectacled Pigeon 4386
Spectacled Prickletail 9226
Spectacled Redstart 6180
Spectacled Reed-Warbler 136
Spectacled Spiderhunter 886
Spectacled Tanager 2178
Spectacled Tern 9456
Spectacled Tinkerbird 8053
Spectacled Tyrant 4764
Spectacled Warbler 6180, 9594
Spectacled Weaver 7969, 7971
Spectacled Whitestart 6180
Speke's Weaver 7982
Spence's Sunangel 4491
Spice Finch (WI) 5318
Spice Imperial-Pigeon 3365
Spice Mannikin 5318
Spice Pigeon 3348
Spicebird 5318
Spike-heeled Lark 2076
Spillman's Tapaculo 9077
Spine-tailed Chowchilla 6825
Spine-tailed Logrunner 6825
Spinetailed-Swift 4645
Spinifex Parrot 7296
Spinifex Pigeon 4156
Spinifexbird 3559
Spiny Babbler 10276

Spiny-cheeked Honeyeater 6
Spiny-cheeked Wattlebird 6
Spiny-faced Antshrike 10638
Spix's Creeper 10667
Spix's Guan 7216
Spix's Macaw 2956
Spix's Spinetail 9657
Spix's Woodcreeper 10667
Spix's Woodrail 895
Splendid Astrapia 1044
Splendid Bird of Paradise 1044
Splendid Blue Wren-Warbler 5547
Splendid Blue-Wren 5547
Splendid Coquette 5341
Splendid Fairywren 5547
Splendid Fruit-Dove 8527
Splendid Glossy-Starling 5009
Splendid Grass-Parakeet 6486
Splendid Grass-Parrot 6486
Splendid Pigeon 2519
Splendid Sabrewing 1613
Splendid Starling 5009
Splendid Sunbird 6377
Splendid White-eye 10792,10828
Splendid Woodpecker 1586
Splendid Wren 5547
Splendid Wren-Warbler 5547
Spoonbill (WI) 532
Spoonbill 7880
Spoonbill Sandpiper 3762
Spoon-billed Sandpiper 3762
Spoonbills 9967
Spoot-throated Accentor 8281
Spot-backed Antbird 4757
Spot-backed Antshrike 4776
Spot-backed Ant-Wren 4588
Spot-backed Puffbird 6646

Spot-bellied Bobwhite 2449
Spot-bellied Crested-Bobwhite 2449
Spot-bellied Eagle-Owl 1339
Spotbill (ISC) 559
Spotbill Duck 559
Spot-billed Duck 559
Spot-billed Grey-Duck 559
Spot-billed Ground-Tyrant 6095
Spot-billed Pelican 7200
Spot-billed Toucanet 9105
Spot-breasted Antbird 6253
Spot-breasted Antvireo 3390
Spot-breasted Cuckoo-Dove 5470
Spot-breasted Darkeye 4484
Spot-breasted Fantail 8820
Spot-breasted Flycatcher 6066
Spot-breasted Foliagegleaner 512
Spot-breasted Honeyeater 5754
Spot-breasted Ibis 1260
Spot-breasted Lapwing 10494
Spot-breasted Laughingthrush 4099
Spot-breasted Oriole 4826, 4828
Spot-breasted Parrotbill 7031
Spot-breasted Plover 10494
Spot-breasted Scimitar-Babbler 8115
Spot-breasted Thornbird 7302
Spot-breasted Warbler 7075
Spot-breasted White-eye 4484
Spot-breasted Woodpecker 2439
Spot-breasted Wren 9995
Spot-chested Aracari 8470
Spot-chested Wren 9985

Sri Lanka White-headed
   Starling 9560
Sri Lanka Wood-Pigeon
   2523
Sri Lanka Yellow-eared
   Bulbul 8613
Sri Lanka Yellow-fronted
   Barbet 5599
Sri Lankan Dull Blue
   Flycatcher 3670
Sri Lankan Dusky blue
   Flycatcher 3670
Sri Lankan Grey-Hornbill
   6680
Sri-Lanka Spot-winged
   Ground-Thrush 10751
Sri-Lankan Blue-Flycatcher
   3670
Sspectacled Crowtit 7026
St. Ana Myzomela 6338
St. Andrew Mockingbird
   5899
St. Andrew Vireo 10564
St. Helena Plover 2046
St. Helena Sandplover
   2046
St. Helena Waxbill 3614
St. Lucia Amazon-Parrot
   459
St. Lucia Amzon 459
St. Lucia Black-Finch 5719
St. Lucia Nightjar 1690
St. Lucia Oriole 4819
St. Lucia Parrot 459
St. Lucia Pewee (WI) 6157
St. Lucia Warbler 3130
St. Lucian Nightjar 1690
St. Matthias Fantail 8846
St. Matthias Monarch 5997
St. Matthias Rufous-Fantail
   8846
St. Thomas Canary 6492
St. Thomas Conure 919
St. Vincent Amazon 448
St. Vincent Parrot 448
St. Vincent Solitaire 6113
St. Vincent's Gulf Slender-
   billed Thornbird (ANZ)
   15
Stagemaker 278
Stager's Piculet 7728

Stalker's White-eye 9838
Standardwing 9110
Standardwing Bird of
   Paradise 9110
Standard-winged Bird of
   Paradise 9110
Standard-winged Nightjar
   5451
Stanley Crane 4356
Stanley's Bustard 6495
Stanley's Crane 4356
Stanley's Parakeet 7892
Stanley's Rosella 7892
Star Finch 6449
Starchy Thrush 6510
Stark's Lark 3558
Stark's Short-toed Lark
   3558
Starling 9565
Starlings 9549
Starred Antshrike 8634
Starred Bush-Robin 8056
Starred Quail 6696
Starred Robin (CSA) 8056
Starred Wood-Quail 6696
Star-spotted Nightjar 1705
Star-throated Ant-Wren
   6281
Steamer Duck 1233
Steel-barred Ground-Dove
   2537
Steel-blue Flycatcher 6132
Steel-blue Myiagra 6132
Steel-blue Whydah 10537
Steel-blue Widow 10537
Steel-blue Widow-Finch
   (CSA) 10533
Steely-vented
   Hummingbird 424
Steere's Babbler 5265
Steere's Broadbill 3761
Steere's Coucal 1867
Steere's Liocichla 5265
Steere's Pitta 7869
Steere's Tinamou 2878
Steinbach's Canastero 1037
Stejneger's Petrel 8443
Stejneger's Storm-Petrel
   6664
Stella's Lorikeet 2061
Steller's Albatross 3263

Steller's Cormorant 7405
Steller's Eider 8100
Steller's Jay 2917
Steller's Sea-Eagle 4448
Stepanyan's Grasshopper-
   Warbler 5279
Stepanyan's Warbler 5279
Stephanie's Astrapia 1045
Stephan's Dove 1993
Stephan's Emerald Dove
   1993
Stephan's Ground-Dove
   1993
Stephan's Pigeon 1993
Stephen Island Wren 10623
Stephen's Island Rock-
   Wren 10623
Stephen's Lorikeet 10556
Stephen's Lory 10556
Steppe Buzzard (CSA)
   1396
Steppe Buzzzard 1395
Steppe Eagle 855
Steppe Grey-Shrike 5053
Stewart Cormorant 7384
Stewart Island Shag 7384
Stewart Shag 7384
Stewart's Bunting 3482
Stierling's Barred Wren-
   Warbler 1475
Stierling's Barred-Warbler
   1475
Stierling's Woodpecker
   3168
Stierling's Wren-Warbler
   1475
Stifftail 6932
Stiff-tailed Duck 6928
Stilt 4629
Stilt Sandpiper 1501
Stilts 8764
Stinker Petrel 5453
Stippled Ant-Wren 6284
Stipple-throated Ant-Wren
   6284
Stitchbird 6608
Stock Antshrike 2983
Stock Dove 2506
Stock Pigeon 2506
Stock-Flycatcher-Warbler 4

Sydney Firetail 6453
Sydney Waxbill 6453
Sykes Crested-Lark 3990
Sykes's Lark 3990
Sykes's Nightjar 1682
Sykes's Warbler 4644
Symons's Cape-Siskin 9174
Symons's Syskin 9174
Syrian Rock-Nuthatch
  9244
Syrian Serin 9175
Syrian Woodpecker 3110
Szecheny's Pheasant-
  Partridge 9882
Szecheny's Monal Partridge
  9882
Szecheny's Pheasant-
  Grouse 9882
Szechuan Grey-Jay 7266
Szechuan Liocichla 5263
Szechuan Wood-Owl 9521
Szechwan Grey-Jay 7266
Szechwan Liocichla 5263
Szechwan Wood-Owl 9521
Tabity Newtonia 6535
Tabon 5636
Tabon Scrubfowl 5636
Tabora Black-faced
  Babbler 10281
Tabora Cisticola 2307
Tabora Grey-Tit 7101
Tabuan Crake 8178
Tabuan Parakeet 8269
Tacarcuna 9068
Tacarcuna Bush-Tanager
  2155
Tacarcuna Quail 6687
Tacarcuna Wood-Quail
  6687
Tacazze Sunbird 6432
Tachira Ant-Pitta 4303
Tachira Emerald 411
Taczanowski's Cinclodes
  2253
Taczanowski's Finch 6028
Taczanowski's Flycatcher
  6149
Taczanowski's Grebe 8020
Taczanowski's Nightjar
  5407

Taczanowski's Snowfinch
  6028
Taczanowski's Tinamou
  6599
Taczanowski's Yellow-
  Finch 9214
Tadjourna Francolin 3918
Tagula Black-fronted
  White-eye 10796
Tagula Butcherbird 2789
Tagula Honeyeater 5759
Tagula White-eye 10796
Taha Bishop 3729
Taha Weaver 3711
Tahiti Blue-Lory 10555
Tahiti Flycatcher 8113
Tahiti Fruit-Dove 8521
Tahiti Kingfisher 10079
Tahiti Monarch 8113
Tahiti Monarch-Flycatcher
  8113
Tahiti Parakeet 2966
Tahiti Petrel 8455
Tahiti Rail 4053
Tahiti Reed-Warbler 116
Tahiti Swiftlet 202
Tahiti Warbler 116
Tahitian Fruit-Dove 8521
Tahitian Kingfisher 10079
Tahitian Lory 10555
Tahitian Pigeon 3344
Tahitian Sandpiper 8266
Tahitian Swiftlet 202
Tai Malimbe 5517
Taiga Bean-Goose 636
Taiko 8447
Tailed Wren-Babbler 9275
Tailorbird 6841
Taita Falcon 3795
Taita Fiscal 5044
Taita Fiscal-Shrike 5044
Taita Olive-Thrush 10313
Taita Thrush 10313
Taita Tit 7088
Taita White-eye 10827
Taiwan Bamboo-Partridge
  1150
Taiwan Barwing 152
Taiwan Blue-Magpie
  10457

Taiwan Blue-Pheasant
  5369
Taiwan Bulbul 8621
Taiwan Bush-Robin 9801
Taiwan Bush-Warbler 1289
Taiwan Firecrest 8768
Taiwan Hill-Partridge 931
Taiwan Kinglet 8768
Taiwan Laughingthrush
  4104
Taiwan Long-tailed
Pheasant 9676
Taiwan Magpie 10457
Taiwan Partridge 931
Taiwan Sibia 4614
Taiwan Whistling-Thrush
  6230
Taiwan Yuhina 10680
Takahe 8158
Talamanca Jay 2921
Talaseae Thrush 10752
Talaud Kingfisher 10062
Taliabu Barn-Owl 10422
Taliabu Masked-Owl
  10422
Taliabu Owl 10422
Talking Mynah 4294
Tallman's Fruiteater 7814
Talpacoti Dove 2539
Tamarugo Conebill 2552
Tamaulipas Crow 2698
Tamaulipas Pygmy-Owl
  4269
Tambourine Dove 10390
Tana River Cisticola 2345
Tanager Finch 6769
Tanagers 9948
Tanga'eo 10073
Tanganyika Masked-
  Weaver 7977
Tanimbar Bush-Warbler
  1953
Tanimbar Cockatoo 1433
Tanimbar Corella 1433
Tanimbar Crow 2709
Tanimbar Fantail 8849
Tanimbar Flycatcher 5865
Tanimbar Gerygone 4219
Tanimbar Ground-Thrush
  10749
Tanimbar Honeyeater 5250

Tanimbar Megapode 5646
Tanimbar Microeca 5865
Tanimbar Microeca-
Flycatcher 5865
Tanimbar Monarch 5998
Tanimbar Monarch-
Flycatcher 5998
Tanimbar Oriole 6780
Tanimbar Owl 10428
Tanimbar Parrotfinch 3612
Tanimbar Rufous-Fantail
8849
Tanimbar Scrubfowl 5646
Tanimbar Starling 796
Tanimbar Thrush 10749
Tanimbar Triller 4975
Tanna Fruit-Dove 8530
Tanna Ground-Dove 3999
Tanniabird (WI) 4824
Tanzania Masked-Weaver
7977
Tanzania Seedeater 9160
Tanzanian Bay-Owl 7496
Tanzanian Citril 9156
Tanzanian Mountain-
Weaver 7967
Tao Tinamou 10035
Tapaculos 8800
Tara 2097
Tarcacuna Tapaculo 9068
Tarictic Hornbill 7227,
7229,7231
Tasman Booby (ANZ)
9580
Tasman Island Starling 799
Tasman Parrot 7886
Tasman Sea Little
Shearwater (ANZ)
8560
Tasman Sea Masked-
Booby (ANZ) 9575
Tasman Sea Sacred-
Kingfisher (ANZ)
10076
Tasman Sea White-bellied
Storm-Petrel (ANZ)
3946
Tasman Starling 799
Tasmanian Australian
Owlet-Nightjar 183

Tasmanian Azure
Kingfisher (ANZ) 309
Tasmanian Brown-Quail
2775
Tasmanian Eastern Rosella
(ANZ) 7890
Tasmanian Emu (ANZ)
3310
Tasmanian Honeyeater
7544
Tasmanian Masked-Owl
(ANZ) 10424
Tasmanian Masked-Owl
10415
Tasmanian Muttonbird
8581
Tasmanian Native-hen
4039
Tasmanian Owl 10415
Tasmanian Raven 2718
Tasmanian Rosella 7886
Tasmanian Scrubwren
9120
Tasmanian Sericornis 9120
Tasmanian Thornbill 11
Tasmanian Waterhen 4039
Tasmanian Waxbill 9418
Tasmanian Wedge-tailed
Eagle (ANZ) 849
Tataki Thrush 7043
Tataupa Tinamou 2877
Tate's Barbtail 8180
Taumolu Ground-Dove
3998
Taveta Golden-Weaver
7945
Taveta Weaver 7945
Tawitawi Bleeding-heart
4005
Tawitawi Pigeon 4005
Tawny Ant-Pitta 4319
Tawny Antshrike 9898
Tawny Crested-Lark 3990
Tawny Eagle (ISC) 859
Tawny Eagle 857
Tawny Fish-Owl 4932
Tawny Frogmouth 8010
Tawny Grassbird 5632
Tawny Grass-Warbler 5632
Tawny Lark 3990
Tawny Marshbird 5632

Tawny Owl 9518
Tawny Piculet 7712
Tawny Pipit 698
Tawny Straightbill 10026
Tawny Tit-Spinetail 5159
Tawny Wood-Owl 9518
Tawny-backed Fantail
8862
Tawny-bellied Babbler
3384
Tawny-bellied Euphonia
3696
Tawny-bellied Hermit 7362
Tawny-bellied Motmot
5974
Tawny-bellied Screech-
Owl 6923
Tawny-bellied Seedeater
9358
Tawny-breasted Flycatcher
6173
Tawny-breasted
Honeyeater 10613
Tawny-breasted Parrotfinch
3604
Tawny-breasted Tinamou
6590
Tawny-breasted Wren-
Babbler 9278
Tawny-breasted Xanthotis
10613
Tawny-browed Owl 8583
Tawny-capped Euphonia
3683
Tawny-chested Flycatcher
776
Tawny-collared Nightjar
1701
Tawny-crested Tanager
9707
Tawny-crowned Greenlet
4747
Tawny-crowned
Honeyeater 7540
Tawny-crowned Pygmy-
Tyrant 3770
Tawny-crowned Vireo
4747
Tawny-faced Gnatwren
5850
Tawny-faced Quail 8890

Thick-billed Murre (NA)
10455
Thick-billed Nutcracker
6609
Thick-billed Parakeet 8889
Thick-billed Parrot 8889
Thick-billed Penguin 3651
Thick-billed Plover 2052
Thick-billed Prion 6976
Thick-billed Quail 10369
Thick-billed Raven 2690
Thick-billed Reed-Warbler
106
Thick-billed Saltator 8944
Thick-billed Seedeater
9139
Thick-billed Seed-Finch
6849
Thick-billed Serin 9139
Thick-billed Shrike 5063
Thick-billed Siskin 1723
Thick-billed Spiderhunter
884
Thick-billed Vireo 10568
Thick-billed Warbler 106
Thick-billed Weaver (CSA)
472
Thick-billed White-eye
4483
Thick-billed Willow-
Warbler 7653
Thickbird (CSA) 1347
Thicket Ant-Pitta 4732
Thicket Flycatcher 3840
Thicket Tinamou 2864
Thicket Warbler 5624
Thicket-Fantail 8844
Thick-knee 1376
Thick-knees (CSA) 1372
Thick-knees 1370
Thin-billed Flycatcher
2122
Thin-billed Flycatcher-
Warbler 2122
Thin-billed Murre 10454
Thin-billed Prion 6975
Thin-billed Starling 6737
Thin-billed Yellow-
Warbler 2122
Thistletail 9013
Thollon's Moorchat 6260

Thornbird 7304
Thornbush Barred Wren-
Warbler 1473
Thorn-tailed Rayadito 790
Thraupids 9948
Three-banded Courser
8818
Three-banded Mockingbird
6518
Three-banded Plover 2024,
2049
Three-banded Rosefinch
1778
Three-banded Warbler
1186
Three-coloured Mannikin
5309
Three-coloured Parrotfinch
3612
Three-streaked Bush-
Shrike 9817,9818
Three-streaked Tchagra
(CSA) 9816
Three-streaked Tchagra
9817
Three-striped Flycatcher
2556
Three-striped Hemispingus
4546
Three-striped Tanager 4546
Three-striped Warbler 1187
Three-toed Crowtit 7034
Three-toed Forest-
Kingfisher 1966
Three-toed Golden-backed
Woodpecker 3262
Three-toed Jacamar 4908
Three-toed Kingfisher
(ISC) 1966
Three-toed Lapwing 10482
Three-toed Parrotbill 7034
Three-toed Plover 10482
Three-toed Swiftlet 209
Three-toed Woodpecker
7693
Three-wattled Bellbird
8253
Thrush (WI) 10352
Thrush Ant-Pitta 6269
Thrush Manakin 9007
Thrush Shiffornis 9007

Thrush-Babbler 8557
Thrushes 10257
Thrush-like Ant-Pitta 6269
Thrush-like Cactus-Wren
1631
Thrush-like Manakin 9007
Thrush-like Mourner 9007
Thrush-like Schiffornis
9007
Thrush-like Woodcreeper
3071
Thrush-like Wren 1631
Thrush-Nightingale 5413
Thrush-Wren 1631
Thunberg's Swiftlet 197
Thura's Rosefinch 1777
Thyolo Alethe 360
Tibet Rosefinch 1769
Tibet Snipe 4031
Tibet Snowfinch 6023
Tibetan Babax 1139
Tibetan Bunting 3470
Tibetan Common Tern
9454
Tibetan Eared-Pheasant
2843,2844
Tibetan Greenfinch 1716
Tibetan Ground-Chough
8344
Tibetan Ground-Jay 8344
Tibetan Lark 5704
Tibetan Owl 1052
Tibetan Partridge 7248
Tibetan Plover 2034
Tibetan Rosefinch 1769
Tibetan Sandgrouse 9680
Tibetan Scrub-Robin 7515
Tibetan Serin 9176
Tibetan Shrike 5062
Tibetan Siskin 9176
Tibetan Snipe 4031
Tibetan Snowcock 9880
Tibetan Snowfinch 6023
Tibetan Twite 1730
Tickbird (WI) 2846
Tickbird 1368,1369
Tickell's Babbler 7209
Tickell's Blue-Flycatcher
3005
Tickell's Flowerpecker
(ISC) 3186

Topknot 5329
Top-knot Judas (WI) 3421
Topknot Pigeon 4155,5329
Toppie (CSA) 8590,8595
Toreana Tree-Finch 1560
Toro Greenbul 7558
Toro Olive-Bulbul 7558
Toro Olive-Greenbul 7558
Torotoroka Scops-Owl
    6891
Torpedo Dove 10695
Torrent Duck 5788
Torrent Lark 4334
Torrent Robin 5975
Torrent Tyrannulet 9182
Torres Straits Torresian
    crow (ANZ) 2710
Torres strait Black
    Butcherbird (ANZ)
    2793
Torres Strait Dusky-
    Honeyeater (ANZ)
    6331
Torres Strait Frilled
    Monarch (ANZ) 991
Torres Strait Imperial-
    Pigeon 3377
Torres Strait Large-billed
    Gerygone (ANZ) 4228
Torres Strait Leaden
    Flycatcher (ANZ) 6141
Torres Strait Little
    Kingfisher (ANZ) 319
Torres Strait Northern-
    Fantail (ANZ) 8858
Torres Strait Peaceful Dove
    (ANZ) 4153
Torres Strait Pensinsula
    Eclectus Parrot (ANZ)
    3396
Torres Strait Pigeon 3377
Torres Strait Red-headed
    Honeyeater (ANZ)
    6322
Torres Strait Spangled
    Drongo (ANZ) 3223
Torres Strait Tawny-
    breasted Honeyeater
    (ANZ) 10614
Torres Strait Tern 9445

Torres Strait Trumpet
    Manucode (ANZ) 5564
Torres Strait Willie-
    Wagtail (ANZ) 8842
Torresian Crow 2709
Torresian Imperial-Pigeon
    3377
Torrid Hummingbird 9102
Tothschild's Serin 9168
Toucan Barbet 9112
Toucans 8720
Toulson's Swift 845
Tourati Bush-Shrike 5028
Tourmaline Sunangel 4487
Toutouwai 7275
Tovi Parakeet 1324
Town Crow 2717
Town Dove 9504
Town Pigeon (CSA) 2501
Townsend's Bunting 3486
Townsend's Cormorant
    7404
Townsend's Shearwater
    8562
Townsend's Solitaire 6114
Townsend's Warbler 3151
Tractrac Chat 1912
Traill's Flycatcher 3512
Train-bearing Hermit 7362
Transvaal Rock-Thrush
    6017
Travancore Scimitar-
    Babbler 8118
Travers's Wren 10623
Treble-banded Courser
    8818
Treble-banded Sandplover
    2049
Tree Booby (WI) 9579
Tree Martin 4674
Tree Pipit 735
Tree Sparrow (NA) 9328
Tree Sparrow 7153
Tree Swallow 9692
Tree Wagtail 3154
Treecreeper 1927,1930
Treecreepers 1936
Tree-Ducks 3120
Treefern Flyeater 4233
Treefern Gerygone 4233

Treefern Gerygone-
    Warbler 4233
Treepie (ISC) 3079
Treeswift 4531
Treeswifts 4534
Trembler 2240
Tres Marias Oriole 4814
Tri-collared Plover 2049
Tri-collared Sandplover
    2049
Tricolour Hornero 3974
Tricoloured Blackbird 259
Tricoloured Brush-Finch
    1081
Tricoloured Bulbul 8622
Tricoloured Finch 1081
Tricoloured Flycatcher
    3858
Tricoloured Heron 3409
Tricoloured Lory 5390
Tricoloured Mannikin 5309
Tricoloured Munia 5309
Tricoloured Nun 5309
Tricoloured Parrotfinch
    3612
Trilling Cisticola 2354
Trilling Nighthawk 2183
Trinidad Euphonia 3708
Trinidad Ground Dove
    (WI) 10695
Trinidad Piping-Guan 7774
Trinidade Island Petrel
    8421
Trinidade Petrel 8421
Tristan Albatross (ANZ)
    3270
Tristan Bunting 6521
Tristan Diving-Petrel 7193
Tristan Finch 6521,6522
Tristan Gallinule 4040
Tristan Grosbeak 6522
Tristan Mooorhen 4040
Tristan Rail 4040
Tristan Skua 1802
Tristan Tern 9469
Tristan Thrush 6510
Tristram's Bunting 3487
Tristram's Flowerpecker
    3209
Tristram's Grackle 6738
Tristram's Honeyeater 6338

Wedge-tailed Shearwater 8578
Wedge-tailed Shrike 5061
Wee Juggler Cocklerina 1435
Weebill 9260
Weero 6644
Weid's Tinamou 2870
Weigall's Roller 2605
Weigeu Bird of Paradise 2226
Weka 4045
Weka Rail 4045
Welcome Swallow 4673
Wells's Dove 5186
West African Batis 1204, 1207
West African Black-Flycatcher 5656
West African Cuckoo-Falcon 1120
West African Freckled Nightjar 1706
West African Guineafowl 6620
West African Island-Seedeater 9169
West African Little-Sparrowhawk 44
West African Nicator 6538
West African Penduline-Tit 655
West African Prinia 8208
West African Red-Bishop 3718
West African Reef-Heron 3404
West African River-Eagle 4450
West African Seedeater 9141
West African Serin 9141
West African Streaky-headed Seedeater 9141
West African Swallow 4663,4676
West African Thrush 10342
West African Wood-Owl 9536
West Andean Emerald 2166

West Indian Red-bellied Woodpecker 5685
West Indian Robin (WI) 5395
West Indian Tree-Duck 3112
West Indian Whistling-Duck 3112
West Indian Woodpecker 5685
West Mexican Chachalaca 6817
West Mexican Kingbird 10406
West Peruvian Screech-Owl 6905
West Peruvian Swallow 9697
West Polynesian Ground-Dove 4012
Westermann's Flycatcher 3857
Western Alpine-Mannikin 5312
Western Antshrike 3387
Western Antvireo 3387
Western Australian Lewins-Rail (ANZ) 8716
Western Banded Snake-Eagle (CSA) 2278
Western Bearded Scrub-Robin 1913,1919
Western Bearded-Greenbul 2834
Western Black Flycatcher 5658
Western Black-capped Lory 5390
Western Black-headed Oriole 6781
Western Bluebill 9281
Western Bluebird 9200
Western Bonelli's Warbler 7611
Western Bowerbird 2089
Western Bristlebird 3050
Western Bronze-naped Pigeon 2493
Western Capercaillie 9875

Western Chat-Tanager 1545
Western Citril 9154
Western Citril-Finch 9154
Western Corella 1438
Western Crested Shrike-Tit (ANZ) 3825
Western Crow (NA) 2682
Western Crowned Leaf-Warbler 7641
Western Crowned-Pigeon 4289
Western Crowned-Warbler 7641
Western Cuckoo-Shrike 1568
Western Curlew 6612
Western Dusky Babbler 10278
Western European Stonechat 8994
Western Flycatcher 3503
Western Gerygone 4221
Western Gerygone-Warbler 4221
Western Golden-backed Weaver 7975
Western Golden-backed Woodpecker 1591
Western Golden-Weaver 7985
Western Goshawk 45
Western Grasshopper-Warbler 5285
Western Grasswren 2360
Western Grebe 161
Western Green Tinker-Barbet 8049
Western Green-backed Honeybird 8254
Western Greenfinch 1722
Western Green-Tinkerbird 8049
Western Grey-Plantain-eater 2832
Western Ground Parrot 7296
Western Ground-Parrot (ANZ) 7298
Western Ground-Thrush 10753

Whip-poor-will Nightjar
1707
Whiskered Akalat 9192
Whiskered Arfak-Lorikeet
6763
Whiskered Auklet 224
Whiskered Fig-Parrot 8381
Whiskered Flowerpecker
3203
Whiskered Ixulus 10684
Whiskered Lorikeet 6763
Whiskered Owl 6918
Whiskered Pitta 7857
Whiskered Redstart 9192
Whiskered Screech-Owl
6918
Whiskered Tanager 9753
Whiskered Tern 2098
Whiskered Tinker-Barbet
8050
Whiskered Treeswift
4530,4533
Whiskered Wren 9997
Whiskered Yuhina 10684
Whisky Jack 7264
Whispering Ibis 7488
Whistler (WI) 3112,3421
Whistling Bird (WI) 1806
Whistling Booby 9573
Whistling Cisticola 2336
Whistling Dove (WI) 5186
Whistling Dove 8503
Whistling Duck 3113
Whistling Eagle 4453
Whistling Green-Pigeon
10151
Whistling Hawk 4453
Whistling Heron 9673
Whistling Hornbill 1887
Whistling Kite 4453
Whistling Nightingale 5419
Whistling Swan 2977
Whistling Teal 3118
Whistling Tree-Duck 3113,
3116
Whistling Warbler 1806
Whistling Wren 5853
Whistling-Ducks 3120
Whistling-Thrush 6226
Whistling-winged Pigeon
4155

White Booby 9573
White Buzzard 5199
White Cockatoo 1430,
1431,1432,1441
White Cotinga 1781
White Dodo 8759
White Eared-Pheasant 2843
White Fruit-Pigeon 3361,
3377
White Gallinule (ANZ)
8154
White Goshawk 64
White Hawk 5199
White Helmet-Shrike 8231
White Ibis (ISC) 9962
White Ibis 3640,9964
White Imperial-Pigeon
3361
White Laughingthrush
4106
White Monjita 10674
White Mynah 5198
White Noddy 4368
White Owl (WI) 10412
White Pelican 7199
White Reef-Egret 3407
White Reef-Heron 3407
White Sparrow 7159
White Spoonbill 7880
White Stork 2229
White Wagtail 6036
White Woodpecker 5666
White-and-grey Warbling-
Finch 8143
White-and-red Eagle-Kite
4452
White-backed Ant-Wren
6272
White-backed Black-Tit
7109
White-backed Duck 9889
White-backed Fire-eye
8645
White-backed Griffon-
Vulture 4401
White-backed Kingfisher
10056
White-backed Mousebird
2453
White-backed Munia (ISC)
5321

White-backed Night-Heron
4286
White-backed Rock-Thrush
6021
White-backed Stilt 4633
White-backed Swallow
1091,2075
White-backed Thrush
10319
White-backed Tit 7109
White-backed Vulture
(ISC) 4402
White-backed Vulture 4401
White-backed Wheatear
6708
White-backed Whistling-
Duck 9889
White-backed Woodpecker
3101
White-backed
Woodswallow 1003
White-banded Hornero
3969
White-banded Mockingbird
5904
White-banded Swallow
1091
White-banded Tanager
6493
White-banded Tyrannulet
5585
White-barred Piculet 7710
White-beak See-see (WI)
9369
White-bearded Antshrike
1232
White-bearded Bulbul 2837
White-bearded Flycatcher
7439
White-bearded Greenbul
2835,2837
White-bearded Hermit
7343
White-bearded Honeyeater
7543
White-bearded Manakin
5550
White-Bellbird 8250
White-bellied Akalat 2754
White-bellied Amethyst-
Starling 2275

Yamdena Bush-Warbler 1953
Yangtse Crowtit 7032
Yangtse Parrotbill 7032
Yap Greater White-eye 10810
Yap Island Monarch 5988
Yap Monarch 5988
Yap Olive White-eye 10810
Yap Quail-Dove 4014
Yap White-eye 10810
Yapacana Antbird 6236
Yarrell's Curassow 2823
Yarrell's Siskin 1748
Yarrell's White-Wagtail 6037
Yarrell's Woodstar 3665
Yaruquian Hermit 7363
Yelkouan Shearwater 8582
Yellow Bishop 3716
Yellow Bittern 4896
Yellow Bulbul 8608
Yellow Bunting 3452, 3461,3484
Yellow Button-Quail 10378
Yellow Canary 9153
Yellow Cardinal 4361
Yellow Chat 3553
Yellow Crake 2766
Yellow Dove 8505
Yellow Elaenia 6193
Yellow Figbird 9285,9287
Yellow Finch (CSA) 7949
Yellow Flycatcher 2123, 3595
Yellow Flycatcher-Warbler 2123
Yellow Fody 3886
Yellow Grass-Finch 9210
Yellow Grosbeak 7441, 7442
Yellow Hawaii Finch 8387
Yellow Honeyeater 5224
Yellow Junglefowl 4067
Yellow Laysan-Finch 8387
Yellow Longbill 5479
Yellow Oriole 4823,6786
Yellow Parrot 7891
Yellow Penduline-Tit 655

Yellow Pippiree (WI) 10406
Yellow Quail 10378
Yellow Rail 2766
Yellow Robin 3533
Yellow Rosella 7891
Yellow Samp Flycatcher-Warbler 2122
Yellow Silver-eye 10793
Yellow Streaked- Lorikeet 2059
Yellow Swamp-Warbler 2122
Yellow Sysie 9153
Yellow Thicket-Flycatcher 8032
Yellow Thornbill 19
Yellow Tit 7103
Yellow Tyrannulet 1709
Yellow Wagtail 6043
Yellow Warbler 2123, 3122,3142
Yellow Wattlebird 648
Yellow Weaver (ISC) 7963
Yellow Weaver 7984
Yellow White-eye (CSA) 10826
Yellow White-eye 10757, 10766,10778,10808
Yellow-and-green Lorikeet 10176
Yellow-and-green Lory 10176
Yellow-back (WI) 5399
Yellow-backed Eremomela 3566
Yellow-backed Finch 5399
Yellow-backed Grassbird (WI) 5399
Yellow-backed Grassquit 5399
Yellow-backed Lory 5388
Yellow-backed Oriole 4806
Yellow-backed Peppershrike 2970
Yellow-backed Pyrtilia 8683
Yellow-backed Sunbird (ISC) 238
Yellow-backed Tanager 4550

Yellow-backed Weaver 7950,7964
Yellow-backed Whistler 6938
Yellow-backed Widow (CSA) 3723
Yellow-bearded Bulbul 2838
Yellow-bearded Greenbul 2838
Yellow-bellied Ant-Wren 4591
Yellow-bellied Asity 6463
Yellow-bellied Bristle-Tyrant 7591
Yellow-bellied Bulbul 2118,372,375,7556, 7565
Yellow-bellied Bunting 3463
Yellow-bellied Bush-Warbler 1961
Yellow-bellied Chat-Tyrant 9217
Yellow-bellied Chough 8651
Yellow-bellied Dacnis 3032
Yellow-bellied Elaenia 3418
Yellow-bellied Eremomela 3566
Yellow-bellied Fantail 8837
Yellow-bellied Fantail-Flycatcher 8837
Yellow-bellied Flowerpecker 3197
Yellow-bellied Flycatcher 3505,4714
Yellow-bellied Flycatcher-Warbler 4,4714
Yellow-bellied Flyeater 4217
Yellow-bellied Fruit-Dove 8509
Yellow-bellied Fruit-Pigeon 10166
Yellow-bellied Gerygone 4217

Yellowbird (WI) 3142
Yellowbreast (WI) 2425
Yellow-breasted Ant-Pitta
4309
Yellow-breasted Ant-Wren
4587
Yellow-breasted Apalis
751
Yellow-breasted Azure Tit
7097
Yellow-breasted Babbler
(ISC) 5456
Yellow-breasted Barbet
10133
Yellow-breasted Bird of
Paradise 5277
Yellow-breasted Boatbill
5439
Yellow-breasted Boat-
billed Flycatcher 5439
Yellow-breasted Boubou
5014
Yellow-breasted Bowerbird
2091
Yellow-breasted Bulbul
2118
Yellow-breasted Bunting
3449,3463
Yellow-breasted Chat 4798
Yellow-breasted Crake
8167
Yellow-breasted Flatbill
10102,5439
Yellow-breasted Flatbill-
Flycatcher 5439
Yellow-breasted
Flowerpecker 8220
Yellow-breasted Flycatcher
10102,5439
Yellow-breasted
Flycatcher-Warbler
9088
Yellow-breasted Flyeater
4234
Yellow-breasted Fruit-
Dove 8513,8527
Yellow-breasted Fruit-
Pigeon 8513
Yellow-breasted Gerygone-
Warbler 4234

Yellow-breasted Green
Magpie 2301
Yellow-breasted
Greenfinch 1743
Yellow-breasted Hylotia
(CSA) 4714
Yellow-breasted Magpie
2301
Yellow-breasted Marshbird
8340
Yellow-breasted Musk-
Parakeet 8267
Yellow-breasted Pipit 701
Yellow-breasted Prinia
8187
Yellow-breasted Racket-
tail 8212
Yellow-breasted Racquet-
tail 8212
Yellow-breasted Shining
Parrot 8267
Yellow-breasted Shrike
5014
Yellow-breasted Sunbird
230,6394,6407,6436
Yellow-breasted Tailorbird
6838
Yellow-breasted Tit 7097
Yellow-breasted Toucan
8721,8728
Yellow-breasted Warbler
9088
Yellow-breasted Willow-
Warbler 7650
Yellow-breasted Wren-
Warbler 4234
Yellow-breasted Yellow-
Finch 9210
Yellow-breated Gerygone
4234
Yellow-bridled Finch 5708
Yellow-browed Antbird
4771
Yellow-browed Bulbul
4870
Yellow-browed Bunting
3456
Yellow-browed
Camaroptera 1566
Yellow-browed Chat-
Tyrant 9220

Yellow-browed Darkeye
5357
Yellow-browed Foditany
2841
Yellow-browed Honeyeater
5734
Yellow-browed
Laughingthrush 4090
Yellow-browed Leaf-
Warbler 7631
Yellow-browed Melidectes
5734
Yellow-browed Oxylabes
2841,6305
Yellow-browed Satrap
8975
Yellow-browed Seedeater
9179
Yellow-browed Shrike-
Vireo 10596
Yellow-browed Sparrow
474
Yellow-browed Tanager
2177,2178
Yellow-browed Tesia 9868
Yellow-browed Tetraka
2841
Yellow-browed Tit 9625
Yellow-browed Tody-
Flycatcher 10082
Yellow-browed Toucanet
1107
Yellow-browed Tyrant
8975
Yellow-browed Warbler
7631
Yellow-browed White-eye
5357
Yellow-browed
Woodpecker 7697
Yellow-brown Quail-Dove
4009
Yellow-capped Pygmy-
Parrot 5880
Yellow-capped Weaver
7951
Yellow-carpalled Sparrow
287
Yellow-casqued Hornbill
1886

Index Français

Caique à face rouge 4456
Caique à joues roses 7761
Caique à queue courte 4343
Caique à tête noire 7758
Caique à ventre blanc 7755
Caïque à ventre bleu 10190
Caïque à ventre noir 7756
Caique de Barraband 7757
Caique de Bonaparte 7762
Caique de Fuertes 4457
Caïque de Salvin 4459
Caique malpourri 7756
Caique mitré 7760
Caique vautourin 4400
Calandre de Mongolie 5705
Calandre leucoptère 5703
Calandre orientale 5701
Calandrelle des sables 1485
Calandres 303
Calandrite bergeronnette 9473
Calandrite du Napo 9474
Calao à bec jaune 10047
Calao à bec noir 10053
Calao à bec pâle 10054
Calao à bec rouge 10045
Calao à cannelures 7228
Calao à casque jaune 1886
Calao à casque noir 1882
Calao à casque plat 1358
Calao à casque rond 1360
Calao à casque rouge 84
Calao à cimier 82
Calao à cou roux 88
Calao à cuisses blanches 1881
Calao à gorge claire 90
Calao à huppe blanche 10040
Calao à joues argent 1883
Calao à joues brunes 1885
Calao à joues grises 1888
Calao bicorne 1357
Calao brun 624
Calao charbonnier 660
Calao coiffé 83
Calao couronné 10041
Calao d'Abyssinie 1363
Calao de Bradfield 10042
Calao de Decken 10044
Calao de Gingi 6680

Calao de Hartlaub 10048
Calao de Hemprich 10049
Calao de Jackson 10050
Calao de Malabar 659
Calao de Manille 7229
Calao de Mindanao 7227
Calao de Mindoro 7230
Calao de Monteiro 10052
Calao de Narcondam 87
Calao de Palawan 661
Calao de Samar 7232
Calao de Sumba 85
Calao de Vieillot 86
Calao de Walden 92
Calao des Célèbes 7228
Calao des Sulu 662
Calao festonné 91
Calao gris 6681
Calao largup 623
Calao leucomèle 10051
Calao longibande 10046
Calao malais 658
Calao nasique 10053
Calao papou 89
Calao pie 659
Calao pygmée 10043
Calao pygmée à bec noir 10048
Calao rhinocéros 1359
Calao siffleur 1887
Calao taritic 7227
Calao terrestre 1364
Calao trompette 1884
Caleçon rouge (Ants) 8237
Calfat 7001
Calicalic malgache 1492
Calléatidés 1518
Calliope de Sibérie 5411
Calliope sibérienne 5411
Calliste à ailes rousses 9756
Calliste à calotte noire 9750
Calliste à coiffe d'or 9755
Calliste à cou bleu 9737
Calliste à dos marron 9765
Calliste à dos noir 9763
Calliste à face rouge 9762
Calliste à gorge rousse 9770
Calliste à gorge verte 9728

Calliste à joues rousses 9769
Calliste à nuque d'or 9768
Calliste à nuque verte 9747
Calliste à oreilles d'or 9735
Calliste à sourcils bleus 9741
Calliste à sourcils clairs 9732
Calliste à tête bleue 9738
Calliste à tête dorée 9778
Calliste à tête noire 9740
Calliste à tête verte 9772
Calliste à ventre bleu 9742
Calliste à ventre jaune 9779
Calliste argenté 9776
Calliste azuré 9731
Calliste beryl 9760
Calliste bleu-et-noir 9774
Calliste de Desmarest 9743
Calliste de Phillips 9764
Calliste de Schauensee 9758
Calliste de Schrank 9771
Calliste diable-enrhumé 9757
Calliste doré 9729
Calliste dos-bleu 9736
Calliste émeraude 9746
Calliste étincelant 2115
Calliste gris 9752
Calliste masqué 9759
Calliste moustachu 9753
Calliste multicolore 2114
Calliste oreillard 2113
Calliste or-gris 9761
Calliste pailleté 9744
Calliste passevert 9733
Calliste rouverdin 9749
Calliste safran 9751
Calliste septicolore 9734
Calliste sucrier 9766
Calliste superbe 9745
Calliste syacou 9767
Calliste tacheté 9773
Calliste tiqueté 9748
Calliste varié 9775
Calliste vert 9754
Calliste vitriolin 9777
Calobate de l'Annam 1754
Calobate radieux 1753

Capucin à ventre blanc 5304
Capucin à ventre roux 5303
Capucin bec-d'argent 5292
Capucin bec-de-plomb 5308
Capucin bicolore 5290
Capucin coloré 5319
Capucin damier 5318
Capucin damier muscade 5318
Capucin de Java 5305
Capucin de Hunstein 5302
Capucin de l'Inde 5309
Capucin de Madagascar 5138
Capucin de Nevermann 5314
Capucin de Nouvelle-Bretagne 5310
Capucin de Nouvelle-Hanovre 5315
Capucin de Nouvelle-Irlande 5297
Capucin des Arfak 5325
Capucin des Moluques 5311
Capucin des montagnes 5303,5313
Capucin des Snow 5312
Capucin domino 5321
Capucin donacole 5293
Capucin gris 5291
Capucin jacobin 5311
Capucin javanais 5305
Capucin marron 5295
Capucin noir 5322
Capucin nonnette 5294
Capucin pâle 5317
Capucin pie 5298
Capucin ponctué (Ants) 5318
Capucin sombre 5299
Capucin tacheté 5306
Capucin tricolore 5309
Capucin triste 5324
Caracara à gorge blanche 7417
Caracara à gorge rouge 3043
Caracara à tête jaune 5887

Caracara à ventre blanc 3043
Caracara austral 7418
Caracara caronculé 7419
Caracara chimango 5888
Caracara de Guadalupe 8089
Caracara huppé 8090
Caracara montagnard 7420
Caracara noir 3044
Caramoptère barrée 1476
Cardinal à collier 8879
Cardinal à cuisses noires 7445
Cardinal à dos noir 7440
Cardinal à épaulettes 1787
Cardinal à poitrine rose 7443
Cardinal à poitrine rouge 7443
Cardinal à tête jaune 7441
Cardinal à tête noire 7444
Cardinal à ventre blanc 1788
Cardinal ardoisé 8943
Cardinal de virginie 1712
Cardinal érythromèle 7263
Cardinal flavert 1786
Cardinal fuligineux 8941
Cardinal gris 7059
Cardinal huppé 1712
Cardinal jaune 7442
Cardinal pyrrhuloxia 1715
Cardinal rouge 1712
Cardinal vermillon 1714
Cardinal vert 4361
Cardinalidés 1711
Cardinaux 1711
Cariama de Burmeister 2217
Cariama huppé 1750
Cariamidés 1751
Carillonneur huppé 6754
Carnifex à ailes tachetées 9336
Carnifex à collier 5848
Carnifex à gorge cendrée 5844
Carnifex ardoisé 5845
Carnifex barré 5847
Carnifex de Buckley 5843

Carnifex plombé 5846
Carouge (Ants) 3703,4803, 4809,4819
Carouge à calotte rousse 257
Carouge à capuchon 255
Carouge à épaulettes 256
Carouge à oeil clair 261
Carouge à tête jaune 10611
Carouge de Californie 259
Carouge de la Jamaïque 6519
Carouge de Porto Rico 260
Carouge galonné 258
Carouge loriot 4380
Carouge safran 252
Carouge unicolore 251
Carougette (Ants) 9188
Carpophage à cire rouge 3375
Carpophage à double huppe 5329
Carpophage à lunettes 3368
Carpophage à manteau brun 3345
Carpophage à queue barrée 3373
Carpophage à queue bleue 3353
Carpophage à tête rose 3374
Carpophage à ventre rose 3372
Carpophage à ventre roux 3376
Carpophage argenté 3377
Carpophage blanc 3348
Carpophage brillant 3351
Carpophage cendrillon 3352
Carpophage charlotte 3350
Carpophage cuivré 3365
Carpophage d'Albertis 4384
Carpophage d'Aurora 3344
Carpophage de Baker 3346
Carpophage de Brenchley 3349
Carpophage de Finsch 3355
Carpophage de Forsten 3356

Carpophage de la Societé 3344

Carpophage de Micronésie 3366

Carpophage de Mindoro 3363

Carpophage de Müller 3364

Carpophage de Nouvelle-Zélande 4528

Carpophage de Peale 3360

Carpophage de Pickering 3369

Carpophage de Wharton 3379

Carpophage de Zoé 3380

Carpophage des Célèbes 2851

Carpophage des Marquises 3357

Carpophage des Moluques 3347

Carpophage des Salomon 4386

Carpophage géant 3358

Carpophage luctuose 3361

Carpophage mada 4385

Carpophage mantelé 3359

Carpophage meunier 3371

Carpophage müllerien 3364

Carpophage noir 3362

Carpophage océanique 3366

Carpophage pacifique 3367

Carpophage pauline 3343

Carpophage pinon 3370

Carte (Ants) 9188

Casiorne à dos brun 1789

Casiorne rouge 1790

Casiorne roux 1790

Casoar à casque 1794

Casoar de Bennett 1793

Casoar unicaronculé 1796

Casoars 1791

Cassenoix d'Amérique 6610

Cassenoix moucheté 6609

Cassican à collier 2794

Cassican à dos noir 2790

Cassican à gorge noire 2791

Cassican à tête noire 2788

Cassican de Tagula 2789

Cassican des mangroves 2792

Cassican flûteur 4392

Cassique à ailes jaunes 1451

Cassique à bec blanc 1450

Cassique à bec mince 1452

Cassique à dos rouge 1456

Cassique à épaulettes 1447

Cassique à queue frangée 6678

Cassique à tête brune 8310

Cassique bicolore 4397

Cassique casqué 8308

Cassique cul-jaune 1445

Cassique cul-rouge 1448

Cassique de Cassin 4395

Cassique de Koepcke 1449

Cassique de Montézuma 4396

Cassique d'Équateur 1454

Cassique du Para 4394

Cassique huppé 8306

Cassique montagnard 1446

Cassique noir 8307

Cassique olivâtre 8303

Cassique roussâtre 8302

Cassique solitaire 1455

Cassique vert 8309

Cassseur de Burgau (Ants) 4427

Casuariformes 1792

Casuariidés 1791

Cataménie du paramo 1799

Cataménie maculée 1798

Cataménie terne 1800

Cathartidés 1810

Caurale soleil 3764

Cent-coups-de couteau (Ants) 1699

Céréopse cendré 1924

Certhiidés 1936

Cerylidés 1948

Chanteur d'Afrique 9159

Charadridés 2016

Charadriiformes 2017

Chardonneret 1721

Chardonneret à bec épais 1723

Chardonneret à croupion jaune 1746

Chardonneret à menton noir 1719

Chardonneret à tête noire 1736

Chardonneret à ventre jaune 1747

Chardonneret commun 1721

Chardonneret de Magellan 1734

Chardonneret de Yarrell 1748

Chardonneret des Andes 1742

Chardonneret des Antilles 1725

Chardonneret des pins 1738

Chardonneret élégant 1721

Chardonneret gris 1733

Chardonneret jaune 1745

Chardonneret mineur 1739

Chardonneret noir 1717

Chardonneret olivâtre 1737

Chardonneret rouge 1724

Chardonneret safran 1740

Charitospize charbonnier 2053

Charpentier (Ants) 5684, 6513

Charpentier camelle (Ants) 6513

Charpentier-bois (Ants) 6513

Chat-houant (Ants) 10412

Chat-huant 9518

Chat-huant (Ants) 1011, 6631

Chétopse bridé 1974

Chétopse doré 1973

Chétusie social 10488

Chevalier à aile noire (Ants) 10205

Chevalier à ailes blanches 8266

Chevalier à pattes jaunes 10196

Chevalier aboyeur 10203

Chevalier arlequin 10195

Chevalier bargette 10193

Chevalier branlequeue
(Ants) 10201
Chevalier combattant 7473
Chevalier criard 10202
Chevalier cul-blanc 10204
Chevalier de Sibérie 10192
Chevalier de Térek 10193
Chevalier de terre (Ants)
2051
Chevalier des Touamotou
8265
Chevalier errant 10200
Chevalier gambette 10207
Chevalier grivelé 10201
Chevalier guignette 10199
Chevalier pattejaune 10196
Chevalier pied-vert (Ants)
1501
Chevalier semipalmé 1827
Chevalier solitaire 10205
Chevalier stagnatile 10206
Chevalier sylvain 10197
Chevalier tacheté 10198
Chevêche 1053
Chevêche brame 1052
Chevêche commune 1053
Chevêche d'Athéna 1053
Chevêche des terriers 9280
Chevêche forestière 1051
Chevêchette à collier 4247
Chevêchette à dos marron
4251
Chevêchette à pieds jaunes
4273
Chevêchette à queue barrée
4272
Chevêchette australe 4261
Chevêchette brune 4246
Chevêchette cabouré 4260
Chevêchette châtaine 4250
Chevêchette cuculolde
4254
Chevêchette d'Europe 4265
Chevêchette d'Amazonie
4257
Chevêchette de Cuba 4271
Chevêchette de Graben
4245
Chevêchette de Ngami
4262

Chevêchette de Scheffler
4270
Chevêchette des Andes
4259
Chevêchette des rocheuses
4248
Chevêchette des saguaros
5849
Chevêchette du Cap 4249
Chevêchette du jungle 4268
Chevêchette du Pérou 4267
Chevêchette elfe 5849
Chevêchette naine 4255
Chevêchette nimbée 10626
Chevêchette perlée 4266
Chevêchette spadicée 4252
Chicorette (Ants) 10096
Chilia des rochers 2078
Chionidés 2079
Chionis blanc 2080
Chipiu à capuchon 8143
Chipiu à col noir 8140
Chipiu à croupion roux
8142
Chipiu à flancs roux 8141
Chipiu à moustaches 4847
Chipiu à poitrine baie 8147
Chipiu à sourcils roux 8138
Chipiu à tête cendrée 8137
Chipiu alticole 8133
Chipiu cannellé 8145
Chipiu césar 8136
Chipiu costumé 4849
Chipiu de Bolivie 8135
Chipiu de Cochabamba
8139
Chipiu de Tucuman 8134
Chipiu de Watkins 4851
Chipiu d'Ortiz 4848
Chipiu noiron 8149
Chipiu noiroux 8144
Chipiu remarquable 4850
Chipiu rougegorge 8146
Chipiu sanglé 8148
Chlamydère tacheté 2092
Chlamydochère à poitrine
noire 2094
Chloropète aquatique 2122
Chloropète de montagne
2124
Chloropète jaune 2123

Chloropsis de Hardwicke
2139
Chocard à bec jaune 8651
Choucador à épaulettes
rouges 5002
Choucador à longue queue
4993
Choucador à oreillons bleus
4995
Choucador à queue bronzée
4998
Choucador à queue fine
4991
Choucador à queue violette
4994
Choucador à tête pourprée
5005
Choucador à ventre noir
4997
Choucador à ventre roux
5004
Choucador de Burchell
4992
Choucador de Hildebrandt
5000
Choucador de Meves 5001
Choucador de Principe
5003
Choucador de Rüppell
5007
Choucador de Shelley 5008
Choucador de Swainson
4996
Choucador elisabeth 4999
Choucador iris 2394
Choucador pourpre 5006
Choucador splendide 5009
Choucador superbe 5010
Choucas à collier 2692
Choucas de Daourie 2692
Choucas des tours 2706
Chouette (Ants) 1014,1658,
1665
Chouette à collier 8584
Chouette à lignes noires
9528
Chouette à lunettes 8585
Chouette à sourcils jaunes
8583
Chouette à terrier (Ants)
9280

Cormoran de Pallas 7405
Cormoran de Socotra 7400
Cormoran de Tasmanie
  7388
Cormoran de Temminck
  7380
Cormoran de Vieillot 7399
Cormoran des Auckland
  7385
Cormoran des bancs 7398
Cormoran des Chatham
  7402
Cormoran du Cap 7379
Cormoran géorgien 7391
Cormoran huppé 7370
Cormoran impérial 7371
Cormoran largup 7370
Cormoran magellanique
  7395
Cormoran moucheté 7406
Cormoran noir 7381,7409
Cormoran ordinaire 7381
Cormoran pélagique 7403
Cormoran pie 7397
Cormoran pygmée 7407
Cormoran varié 7411
Cormoran vigua 7377
Cormorans 7367
Corneille à bec blanc 2726
Corneille à bec fin 2693
Corneille à collier 2720
Corneille à gros bec 2703
Corneille à tête brune 2696
Corneille d'Alaska 2684
Corneille d'Amérique 2682
Corneille de Cuba 2708
Corneille de Florès 2694
Corneille de Guam 2700
Corneille de la Jamaïque
  2699
Corneille de Meek 2704
Corneille de rivage 2711
Corneille des Banggai 2724
Corneille des Célèbes 2723
Corneille des Moluques
  2725
Corneille d'Hawai 2697
Corneille d'Hispaniola
  2701
Corneille du Cap 2683
Corneille du Mexique 2698

Corneille du Sinaloa 2716
Corneille grise 2721
Corneille mantelée 2688
Corneille noire 2686
Corneille palmiste 2712
Cornichon (Ants) 1544
Corvidés 2675
Corvinelle 2676
Corvinelle à bec jaune
  2676
Corvinelle noir-et-blanc
  2677
Coryllis à front orangé
  5371
Coryllis à gorge jaune 5378
Coryllis à tête bleue 5376
Coryllis de Ceylan 5372
Coryllis de Wallace 5375
Coryllis des Bismarck 5380
Coryllis des Célèbes 5379
Coryllis des Moluques
  5370
Coryllis des Philippines
  5377
Coryllis des Sangi 5373
Coryllis vernal 5381
Coryllis vert 5374
Coryphaspize à joues
  noires 2728
Corythopis à collier 2737
Corythopis de Delalande
  2736
Coscoroba blanc 2738
Cossyphe à ailes bleues
  2745
Cossyphe à calotte blanche
  2741
Cossyphe à calotte
  neigeuse 2752
Cossyphe à calotte rousse
  2751
Cossyphe à flancs olives
  2742
Cossyphe à gorge blanche
  2749
Cossyphe à joues rouges
  9191
Cossyphe à sourcils blancs
  2753
Cossyphe à tête blanche
  2747

Cossyphe à ventre blanc
  2754
Cossyphe choriste 2746
Cossyphe d'Archer 2743
Cossyphe de Heuglin 2748
Cossyphe de Rüppell 2755
Cossyphe des grottes 10624
Cossyphe d'Isabelle 2750
Cossyphe du Cap 2744
Cotinga à ailes grises
  10022
Cotinga à bec jaune 1780,
  7446
Cotinga à col noir 7500
Cotinga à collier 7810
Cotinga à face noire 2542
Cotinga à gorge mauve
  8161
Cotinga à gorge rouge 7807
Cotinga à huppe rouge 493
Cotinga à joues blanches
  10690
Cotinga à poitrine d'or
  7806
Cotinga à queue blanche
  10653
Cotinga à queue fourchue
  7446
Cotinga à queue rayée 7810
Cotinga à tête noire 1784
Cotinga à tête rousse 494
Cotinga barré 7805
Cotinga blanc 1781
Cotinga bleu 2761
Cotinga brun 4867
Cotinga celeste 2756
Cotinga chevalier 7809
Cotinga coqueluchon 1783
Cotinga cordonbleu 2759
Cotinga cordon-rouge 7815
Cotinga d'Antonia 1780
Cotinga de Cayenne 2757
Cotinga de Daubenton
  2758
Cotinga de Lubomirsk
  7812
Cotinga de Ridgway 2762
Cotinga de Sclater 3292
Cotinga des Maynas 2760
Cotinga d'Isabelle 4868
Cotinga écaillé 492

Dicée à sous-caudales jaunes 3183
Dicée à tête écarlate 3210
Dicée à tête rouge 3199
Dicée à ventre blanc 3193
Dicée à ventre jaune 3197
Dicée à ventre orangé 3208
Dicée bicolore 3181
Dicée cendré 3212
Dicée concolore 3184
Dicée couronné 3178
Dicée cul-d'or 3183
Dicée de Bornéo 3198
Dicée de Ceylan 3211
Dicée de Geelvink 3190
Dicée de la Louisiade 3201
Dicée de la Sonde 3177
Dicée de Mauge 3196
Dicée de Mindoro 3206
Dicée de Palawan 8223
Dicée de San Cristobal 3209
Dicée des Bismarck 3189
Dicée des Célèbes 3182
Dicée des Philippines 3180
Dicée des Salomon 3174
Dicée d'Everett 3188
Dicée hirondelle 3192
Dicée olive 8221
Dicée poignarde 8222
Dicée porte-flamme 3194
Dicée pygmée 3204
Dicée quadricolore 3205
Dicée rayé 3175
Dicée sanglant 3207
Dicée tacheté 8220
Dickcissel 9316
Dickcissel d'Amérique 9316
Dicruridés 3214
Diduncule strigirostre 3239
Dindon 5724
Dindon bronzé 5724
Dindon commun 5724
Dindon ocellée 5725
Dindon sauvage 5724
Dindons sauvages 5723
Diomédéidés 3280
Diuca gris 3284
Diuca leucoptère 3285

Domino à longue queue 5321
Donacole 5293
Donacole à grosse tête 5310
Donacole à poitrine blanche 4609
Donacole à poitrine jaune 5296
Donacole à poitrine noire 5323
Donacole à tête grise 5296
Donacole commun 5293
Donacole des hauteurs 5312
Donacole des montagnes 5313
Donacole pectorale 4609
Donacospize des marais 3295
Dongue (Ants) 9440
Doradite à ailes pointues 8331
Doradite babillarde 8333
Doradite de Dinelli 8332
Doradite de Sclater 8334
Dormilon à bec jaune 6095
Dormilon à bec maculé 6095
Dormilon à calotte rousse 6096
Dormilon à front blanc 6085
Dormilon à front noir 6092
Dormilon à grands sourcils 6087
Dormilon à nuque jaune 6090
Dormilon à queue courte 6083
Dormilon à sourcils blancs 6086
Dormilon à tête rousse 6096
Dormilon à ventre roux 6088
Dormilon andin 6087
Dormilon bistré 6094
Dormilon cendré 6089
Dormilon de Junin 6093

Dormilon des Maloines 6094
Dormilon fluviatile 6091
Dos-olive de Shelley 6509
Dos-rouge (Ants) 1504
Dos-vert à collier 6507
Dos-vert à joues blanches 6508
Dos-vert à tête noire 6509
Double Collier (Ants) 2051
Drépanide mamo 3302
Drépanide noir 3301
Drépanidés 3300
Dromadidés 3304
Dromaiidés 3306
Drome 3311
Drome ardéole 3311
Droméocerque brun 3305
Droméocerque tacheté 495
Dromoïque du désert 9044
Dromoïque du Sahara 9044
Dromoïque vif-argent 9044
Drongo à crinière 3228
Drongo à dos brillant 3215
Drongo à gros bec 3219
Drongo à rames 3236
Drongo à raquettes 3235
Drongo à ventre blanc 3224
Drongo balicassio 3221
Drongo brillant 3215
Drongo bronzé 3216
Drongo cendré 3229
Drongo d'Aldabra 3217
Drongo de forêt 3220
Drongo de la Grande Comore 3227
Drongo de la Sonde 3225
Drongo de Ludwig 3230
Drongo de Mayotte 3238
Drongo de Nouvelle-Irlande 3232
Drongo de Sumatra 3237
Drongo des Aldabra 3217
Drongo des Andaman 3218
Drongo des Célèbes 3234
Drongo malgache 3226
Drongo modeste 3233
Drongo pailleté 3222
Drongo papou 1976
Drongo royal 3231
Dronte de la Réunion 8759

Gobemouche de Hodgson 3839

Gobemouche de l'Angola 3281

Gobemouche de Lendu 6069

Gobemouche de Livingstone 3596

Gobemouche de l'Itombwe 6068

Gobemouche de Lompobattang 3831

Gobemouche de Luçon 8807

Gobemouche de Marico 1283

Gobemouche de McGrigor 6548

Gobemouche de Negros 8802

Gobemouche de Palawan 3847

Gobemouche de paradis 9860

Gobemouche de paradis à long bec 9865

Gobemouche de paradis de la Vierge 9853

Gobemouche de paradis de Madagascar 9858

Gobemouche de paradis de San Tomé 9850

Gobemouche de paradis des îles Bourbon 9853

Gobemouche de paradis indien 9860

Gobemouche de paradis malgache 9858

Gobemouche de Rueck 3001

Gobemouche de Sanford 3003

Gobemouche de Seth-Smith 6075

Gobemouche de Sharpe 1286

Gobemouche de Sibérie 6076

Gobemouche de Sumatra 6549

Gobemouche de Sumba 6074

Gobemouche de Tessmann 6078

Gobemouche de Tickell 3005

Gobemouche de Timor 3855

Gobemouche de Vaurie 3833

Gobemouche de ward 8324

Gobemouche de Williamson 6080

Gobemouche des Célèbes 2997

Gobemouche des collines 2989

Gobemouche des Comores 4693

Gobemouche des fées 9429

Gobemouche des îles 3669

Gobemouche des mangroves 3002

Gobemouche des marais 6057

Gobemouche des Nilgiri 3667

Gobemouche drongo 5658

Gobemouche du paradis du Japon 9849

Gobemouche du Cachemire 3853

Gobemouche du Libéria 5656

Gobemouche du Marico 1283

Gobemouche du paradis à ventre roux 9861

Gobemouche du paradis noir 9860

Gobemouche d'Ussher 6079

Gobemouche écorcheur 1230

Gobemouche enfumé 6067

Gobemouche ferrugineux 6064

Gobemouche fiscal 9216

Gobemouche forestier 3933

Gobemouche fuligineux 6076

Gobemouche givré 3840

Gobemouche gris 6077

Gobemouche grisâtre 1284

Gobemouche huppé à tête blanche 3441

Gobe-mouche huppée (Ants) 6157

Gobemouche hyacinthe 2995

Gobemouche indigo 3668

Gobemouche malais 3006

Gobemouche mésange 6199

Gobemouche mugimaki 3843

Gobemouche muttui 6070

Gobemouche nain 3846

Gobemouche narcisse 3844

Gobemouche noir 3841

Gobemouche noir huppé 10214,3444

Gobemouche olivâtre 6071

Gobemouche olive 8808

Gobemouche ombré 8811

Gobemouche orangé-et-noir 3845

Gobemouche pâle 1285

Gobemouche paradis 9850, 9865

Gobemouche paradis du Congo 9863

Gobemouche pie 3857

Gobemouche pygmée 6081

Gobemouche roux 6456

Gobemouche saphir 3849

Gobemouche sombre 6056

Gobemouche soyeux à tête grise 1202

Gobemouche soyeux à joues noires 1203

Gobemouche soyeux du Sénégal 1213

Gobemouche sud-africain 5660

Gobemouche sundara 6550

Gobemouche traquet 1282

Gobemouche ultramarin 3854

Gobemouche vert-de-gris 3671

Gobe-mouches de Daourie
6062
Gobe-mouches de David
6546
Gobe-mouches de Dohrn
4692
Gobe-mouches de Fischer
5659
Gobe-mouches de Florès
8809
Gobe-mouches de
Gambaga 6065
Gobe-mouches de
Goodfellow 8805
Gobe-mouches de Hainan
2992
Gobe-mouches de Hartert
3837
Gobe-mouches de Hodgson
3839
Gobe-mouches de Lendu
6069
Gobe-mouches de
l'Itombwe 6068
Gobe-mouches de
Lompobattang 3831
Gobe-mouches de Luçon
8807
Gobe-mouches de Marico
1283
Gobe-mouches de
McGrigor 6548
Gobe-mouches de Negros
8802
Gobe-mouches de Palawan
3847
Gobe-mouches de paradis
9860
Gobe-mouches de paradis à
long bec 9865
Gobe-mouches de paradis
de la Vierge 9853
Gobe-mouches de paradis
de Madagascar 9858
Gobe-mouches de paradis
indien 9860
Gobe-mouches de paradis
malgache 9858
Gobe-mouches de Rueck
3001

Gobe-mouches de Sanford
3003
Gobe-mouches de Seth-
Smith 6075
Gobe-mouches de Sharpe
1286
Gobe-mouches de Sumatra
6549
Gobe-mouches de Sumba
6074
Gobe-mouches de
Tessmann 6078
Gobe-mouches de Tickell
3005
Gobe-mouches de Timor
3855
Gobe-mouches de Vaurie
3833
Gobe-mouches de
Williamson 6080
Gobe-mouches del'Angola
3281
Gobe-mouches des Célèbes
2997
Gobe-mouches des collines
2989
Gobe-mouches des
Comores 4693
Gobe-mouches des îles
3669
Gobe-mouches des
mangroves 3002
Gobe-mouches des marais
6057
Gobe-mouches des Nilgiri
3667
Gobe-mouches drongo
5658
Gobe-mouches du
Cachemire 3853
Gobe-mouches du Liberia
5656
Gobe-mouches du Marico
1283
Gobe-mouches du paradis
9860
Gobe-mouches du paradis
asiatique 9860
Gobe-mouches du paradis
du Japon 9849

Gobe-mouches du paradis
noir 9860
Gobe-mouches d'Ussher
6079
Gobe-mouches enfumé
6067
Gobe-mouches ferrugineux
6064
Gobe-mouches fiscal 9216
Gobe-mouches forestier
3933
Gobe-mouches fuligineux
6076
Gobe-mouches givré 3840
Gobe-mouches gris 6077
Gobe-mouches grisâtre
1284
Gobe-mouches gris-bleu
8074
Gobe-mouches huppé à tête
blanche 3441
Gobe-mouches hyacinthe
2995
Gobe-mouches indigo 3668
Gobe-mouches malais 3006
Gobe-mouches mésange
6199
Gobe-mouches monarque
azuré 4780
Gobe-mouches mugimaki
3843
Gobe-mouches muttui 6070
Gobe-mouches nain 3846
Gobe-mouches narcisse
3844
Gobe-mouches noir 3841
Gobe-mouches noir huppé
3444
Gobe-mouches olivâtre
6071
Gobe-mouches olive 8808
Gobe-mouches ombré 8811
Gobe-mouches orangé-et-
noir 3845
Gobe-mouches pâle 1285
Gobe-mouches paradis
9850
Gobe-mouches paradis à
ventre roux 9861
Gobe-mouches paradis de
San Thomé 9850

Grallaire de Carriker 4302
Grallaire de Cuzco 4306
Grallaire de Kaestner 4315
Grallaire de Natterer 4735
Grallaire de Przewalski
   4318
Grallaire de Quito 4319
Grallaire de Watkins 4325
Grallaire demi-lune 4329
Grallaire des Andes 4298
Grallaire des Santa Marta
   4299
Grallaire du Pérou 4333
Grallaire du Tachira 4303
Grallaire ecaillée 4312
Grallaire flammée 4304
Grallaire géante 4310
Grallaire grand-beffroi
   6269
Grallaire maillée 4330
Grallaire masquée 4307
Grallaire naine 4331
Grallaire ocrée 4328
Grallaire ondée 4323
Grallaire roi 4324
Grallaire rousse 4322
Grallaire secrète 4305
Grallaire sobre 6270
Grallaire tachetée 4734
Grallaire teguy 4736
Gralline papoue 4334
Gralline pie 4335
Graminicole à bec court
   9021
Graminicole à queue large
   9023
Graminicole rayée 1977
Grand Arachnothère 889
Grand Barbu 5616
Grand Batara 9796
Grand Bec-scie 5791
Grand Bulbul 2117
Grand Bulbul huppé 2834
Grand Butor 1264
Grand Cacatoés à crète
   jaune 1432
Grand Calao d'Abyssinie
   1363
Grand Calao terrestre 1364
Grand Capucin 5300
Grand Chevalier 10202

Grand chevalier à pattes
   jaunes 10202
Grand Corbeau 2685
Grand Cormoran 7381
Grand Cormoran
   continental 7382
Grand Cossyphe à tête
   blanche 2741
Grand Coucal 1865
Grand Échenilleur 2643
Grand Eclectus 3394
Grand Égothèle 184
Grand Gaucho 272
Grand Géocoucou 4141
Grand Gobemouche 6547
Grand Gobe-mouches 6547
Grand Gosier (Ants) 7198
Grand Gravelot 2028
Grand Grèbe 8017
Grand Grimpar 10648
Grand Harle 5791
Grand Héron 954
Grand Hocco 2824
Grand Ibijau 6629
Grand Indicateur 4855
Grand Jacamar 4909
Grand Labbe 1805
Grand Langrayen 1000
Grand Mainate 103
Grand Manchot 823
Grand Martin 103
Grand Minivet 7256
Grand Moineau 7154
Grand Morillon 1135
Grand Niltava 6547
Grand Oedicnème 1377
Grand Paradisier 7015
Grand Percefleur 3249
Grand Phaéton 7327
Grand Pic 3331
Grand Pingouin 7752
Grand Pinson terrestre
   4177
Grand Plongeon 4133,8024
Grand Puffin 8569
Grand Quiscale 8694
Grand Réveilleur 9485
Grand Roselin strié 1775
Grand Roselin tacheté 1774
Grand Saltator 8949
Grand Sicale 9202

Grand Tétrao 9873
Grand Tétras 9875
Grand Tinamou 10032
Grand Urubu 1809
Grandala bleu 4340
Grandala bleu-ciel 4340
Grand-duc à aigrettes 1341
Grand-duc africain 1331
Grand-duc ascalaphe 1332
Grand-duc bruyant 1343
Grand-duc d'Europe 1334
Grand-duc d'Amérique
   1344
Grand-duc de Coromandel
   1336
Grand-duc de Shelley 1342
Grand-duc de Verreaux
   1337
Grand-duc de Virginie
   1344
Grand-duc des Philippines
   1340
Grand-duc des Usambara
   1345
Grand-duc du Cap 1335
Grand-duc du désert 1332
Grand-duc du Népal 1339
Grand-duc indien 1333
Grand-duc tacheté 1338
Grande Aigrette 3398
Grande Bouscarle 1958
Grande Calandre 5704
Grande Éréonesse 6467
Grande Fauvette des marais
   5628
Grande Grallaire 4308
Grande Graminicole 4336
Grande Mélampitte 5661
Grande Mésange
   charbonnière 7111
Grande Nonne 5298
Grande Outarde 6854
Grande Panure 2569
Grande Phasianelle 5471
Grande Poule-des-prairies
   (Qué) 10391
Grande Pseudobrève 380
Grande Turdinule 6351
Grande-aile (Ants) 10205
Grand-verdier à ailes d'or
   8891

Graucope gris 9537
Grauérie striée 4342
Gravelot à collier
    interrompu 2018
Gravelot à front blanc 2031
Gravelot à triple collier
    2049
Gravelot de Leschenault
    2030
Gravelot de Madagascar
    2045
Gravelot du désert 2030
Gravelot kildir 2051
Gravelot mongol 2034
Gravelot pâtre 2041
Gravelot patte-noire 2018
Gravelot semipalmé 2047
Grèbe (Ant) 9685
Grèbe à bec bigarré 8024
Grèbe à bec cerclé 8024
Grèbe à cou brun 8013
Grèbe à cou noir 8018
Grèbe à crète 8013
Grèbe à face blanche 160
Grèbe à joues grises 8016
Grèbe argenté 8069
Grèbe australasien 9686
Grèbe aux belles joues
    8019
Grèbe castagneux 9688
Grèbe cornu 8013
Grèbe de Delacour 9689
Grèbe de l'Atitlan 8023
Grèbe de l'Ouest 161
Grèbe de Nouvelle-Zélande
    8070
Grèbe de Rolland 8906
Grèbe de Saint-Domingue
    9685
Grèbe de Taczanowski
    8020
Grèbe des Andes 8012
Grèbe du Lac Atitlan 8023
Grèbe élégant 161
Grèbe esclavon 8013
Grèbe fluviatile 9688
Grèbe huppé 8014
Grèbe jougris 8016
Grèbe malgache 9687
Grèbe microptère 8905
Grèbe minime 9685

Grèbe mitré 8015
Grèbe oreillard 8013
Grèbe roussâtre 9689
Grèbes 8021,8022
Grébifoulque d'Afrique
    8011
Grébifoulque d'Amérique
    4510
Grébifoulque d'Asie 4509
Grébifoulque du Sénégal
    8011
Grèbifoulques 4511
Grenadier (Ants) 3718
Grenadin 10450
Grenadin à poitrine bleue
    10451
Grigri (Ants) 3815
Grigri morne 3789
Grigri poulet (Ants) 3815
Grimpar à ailes rousses
    3064
Grimpar à bec brun 1617
Grimpar à bec courbe 1615
Grimpar à bec d'ivoire
    10659
Grimpar à bec étroit 5141
Grimpar à bec ivoire 10659
Grimpar à bec rouge 1618
Grimpar à bec-en-faux
    1614
Grimpar à collier 3063
Grimpar à dos olive 10670
Grimpar à gorge blanche
    10645
Grimpar à gorge rousse
    3063
Grimpar à gorge tachetée
    3052
Grimpar à longue queue
    3051
Grimpar à menton blanc
    3070
Grimpar à moustaches
    10646
Grimpar barré 3080
Grimpar bec-en-coin 4281
Grimpar brun 5142
Grimpar concolore 3081
Grimpar d'Altamira 10668
Grimpar de Hoffmanns
    3082

Grimpar de Lafresnaye
    10670
Grimpar de Perrot 4709
Grimpar de Pucheran 1616
Grimpar de Snethlage
    10647
Grimpar de Souleyet 5145
Grimpar de Spix 10667
Grimpar de Stresemann
    4710
Grimpar de Villa Nova
    10651
Grimpar de Zimmer 10662
Grimpar des cabosses
    10660
Grimpar des plateaux 3086
Grimpar d'Eyton 10658
Grimpar du Sao Francisco
    10647
Grimpar écaillé 5146
Grimpar élégant 10656
Grimpar enfumé 3066
Grimpar fauvette 9255
Grimpar flambé 10665
Grimpar géant 10650
Grimpar givré 5144
Grimpar grive 3071
Grimpar lancéolé 5140
Grimpar maillé 10661
Grimpar montagnard 5139
Grimpar moucheté 5139
Grimpar nasican 6355
Grimpar ocellé 10664
Grimpar roux 3067
Grimpar strié 10663
Grimpar tacheté 10657
Grimpar talapiot 10666
Grimpar tyran 3072
Grimpar varié 3085
Grimperaux 1936
Grimpereau à doigts courts
    1928
Grimpereau bleu 9235
Grimpereau brachydactyle
    1928
Grimpereau brun 1927
Grimpereau de I'Himalaya
    1931
Grimpereau des arbres
    1928
Grimpereau des bois 1930

Hirondelle 4682
Hirondelle (Ants)
1984,3015,8259,9682
Hirondelle à ailes blanches 9691
Hirondelle à ailes herissées 9427
Hirondelle à ailes tachetées 4670
Hirondelle à bande rousse 7273
Hirondelle à bavette 4675
Hirondelle à ceinture blanche 1091
Hirondelle à collier 8899
Hirondelle à croupion gris 4669
Hirondelle à cuisses blanches 6444
Hirondelle à diadème 9695
Hirondelle à dos blanc 2075
Hirondelle à face blanche 9698
Hirondelle à front blanc 7272
Hirondelle à front brun 4668
Hirondelle à gorge blanche 4654
Hirondelle à gorge fauve 4680
Hirondelle à gorge perlée 4662
Hirondelle à gorge rousse 9426
Hirondelle à gorge striée 4652
Hirondelle à longs brins 4685
Hirondelle à queue blanche 4672
Hirondelle à queue courte 8292
Hirondelle à queue fourchue 8288
Hirondelle à tête blanche 8285
Hirondelle à tête noire 6607

Hirondelle à tête rousse 4660
Hirondelle à ventre blanc 8259
Hirondelle à ventre brun 6606
Hirondelle à ventre roux 4683
Hirondelle ariel 4657
Hirondelle bicolore 9692
Hirondelle blanc (Ants) 7272
Hirondelle bleue 4658
Hirondelle bleue-et-blanche 8635
Hirondelle brune 4667, 8289
Hirondelle chalybée 8257
Hirondelle concolore 4659
Hirondelle d'Angola 4656
Hirondelle de Blanford 8287
Hirondelle de Bocage 4680
Hirondelle de Bonaparte 3054
Hirondelle de Brazza 7437
Hirondelle de Chapman 6605
Hirondelle de cheminée 4682
Hirondelle de Cuba 8258
Hirondelle de Fanti 8293
Hirondelle de fenêtre 3056
Hirondelle de forêt 4666
Hirondelle de Guinée 4671
Hirondelle de Hodgson 3055
Hirondelle de l'Angola 4656
Hirondelle de Kafa 8294
Hirondelle de la Mer Rouge 4678
Hirondelle de Mauretanie 8901
Hirondelle de montagne (Ants) 3015
Hirondelle de Petit 8296
Hirondelle de Preuss 4679
Hirondelle de Reichenow 8295

Hirondelle de Ridgway 9425
Hirondelle de rivage 8902
Hirondelle de rivage du Congo 8900
Hirondelle de rivière 8326
Hirondelle de Salvadori 8286
Hirondelle de Stolzmann 9697
Hirondelle de Tahiti 4688
Hirondelle des Andes 4655
Hirondelle des arbres 4674
Hirondelle des Bahamas 9693
Hirondelle des Galapagos 8260
Hirondelle des granges 4682
Hirondelle des Mangbetu 8291
Hirondelle des mangroves 9690
Hirondelle des Mascareignes 7436
Hirondelle des mosquées 4684
Hirondelle des Nilgiri 4664
Hirondelle des palétuviers 9690
Hirondelle des rochers 4681
Hirondelle des roches 4681
Hirondelle des sables 8901
Hirondelle des torrents 1092
Hirondelle d'Éthiopie 4653
Hirondelle domestique 3056
Hirondelle dorée 9694
Hirondelle du Cachemire 3054
Hirondelle du Chili 9696
Hirondelle du Congo 8900
Hirondelle du désert 4677
Hirondelle du Nepal 3055
Hirondelle du Ruwenzori 8290
Hirondelle du Sinaloa 8261
Hirondelle émeraudine 9698

Hirondelle fanti 8293
Hirondelle fanti hérissée
   8293
Hirondelle fardée 9424
Hirondelle fauve (Ants)
   4668
Hirondelle fluviatile 4665
Hirondelle gracieuse 8260
Hirondelle hérissée 8297
Hirondelle hérissée à queue
   courte 8292
Hirondelle isabelline 4667
Hirondelle messagère 4673
Hirondelle morne (Ants)
   3015
Hirondelle noire 4675,8262
Hirondelle ouest-africaine
   4663
Hirondelle paludicole 8901
Hirondelle pourprée 8262
Hirondelle rouge-et-noire
   4676
Hirondelle rousseline 4661
Hirondelle roux (Ants)
   4682
Hirondelle roux-et-noir
   4676
Hirondelle rustique 4682
Hirondelle sombre 8260
Hirondelle striée 4652,4687
Hirondelle striolée 4687
Hirondelle sud-africaine
   4686
Hirondelle tapère 7326
Hirondelle verte (Ants)
   9694
Hirondelles 4651
Hirondelles de tempête
   4695
Hirundinidés 4651
Histurgopse à queue rouge
   4690
Hoazin huppé 6745
Hoazins 6744
Hobereau africain 3791
Hocco à face nue 2822
Hocco à pierre 7178
Hocco alector 2819
Hocco d'Albert 2818
Hocco de Blumenbach
   2820

Hocco de Daubenton 2821
Hocco de Salvin 5950
Hocco de Spix 5951
Hocco globuleux 2823
Hocco mitou 5949
Hocco nocturne 5952,6592
Hocco unicorne 7179
Hokki blanc 2843
Hokki bleu 2842
Hokki brun 2845
Hokki du Tibet 2844
Houbara de Macqueen
   2095
Houbara ondulée 2096
Huart à bec jaune 4131
Huart à collier 4133
Huart à gorge rousse 4135
Huart arctique 4132
Huart du Pacifique 4134
Huia dimorphe 4602
Huîtrier (Ants) 4427
Huîtrier à long bec 4423
Huîtrier américain 4427
Huîtrier d'Amérique 4427
Huîtrier de Bachman 4418
Huîtrier de Finsch 4420
Huîtrier de Garnot 4422
Huîtrier de Moquin 4425
Huîtrier des Chatham 4419
Huîtrier fuligineux 4421
Huîtrier noir 4417
Huîtrier pie 4426
Huîtrier variable 4428
Huîtriers 4416
Hulotte 9518
Hulotte africaine 9536
Hupolaïs botté 4638
Hupolaïs des oliviers 4641
Hupolaïs d'Upcher 4640
Hupolaïs ictérine 4639
Hupolaïs pâle 4642
Hupolaïs polyglotte 4643
Hupolaïs rama 4644
Huppe d'Afrique 10441
Huppe fasciée 10442
Huppe malgache
   10441,10443
Huppes 10445
Hydrobatidés 4695
Hylia verte 4712
Hyliote à dos violet 4715

Hyliote à ventre jaune 4714
Hyliote australe 4713
Hypocolius gris 4774
Hypolaïs russe 4638
Hypositte malgache 4779
Ibidorhynchidés 4795
Ibijau à ailes blanches 6632
Ibijau à longue queue 6627
Ibijau des Andes 6633
Ibijau gris 6630
Ibijau jamaïcain 6631
Ibijau roux 6628
Ibis à cou noir 9965
Ibis à face blanche 7922
Ibis à face noire 9937
Ibis à face nue 7488
Ibis à queue pointue 1889
Ibis à tête noire 9964
Ibis blanc 3640
Ibis caronculé 1257
Ibis chauve 4214
Ibis d'Australie 9966
Ibis de Davison 8321
Ibis de Ridgeway 7924
Ibis des Andes 9934
Ibis du Cap 4213
Ibis falcinelle 7923
Ibis géant 8322
Ibis guarana 7922
Ibis hagedash 1258
Ibis huppé 5347
Ibis leucon 9963
Ibis malgache 9963
Ibis mandore 9936
Ibis nippon 6576
Ibis noir 8323
Ibis noir (Ants) 7923
Ibis olivâtre 1259
Ibis olive 1259
Ibis peché (Ants) 7923
Ibis plombé 9935
Ibis rouge 3641
Ibis sacré 9962
Ibis vermiculé 1260
Ibis vert 5823
Ictéridés 4799
Idiopsar à queue courte
   4835
Ifrita de Kowald 4836
Ignicolore 3718,3725
Inca à collier 2419

Manakin à dos bleu 7795
Manakin à fraise 2662
Manakin à front blanc 7800
Manakin à gorge blanche
2663
Manakin à longue queue
2084
Manakin à moustaches
4605
Manakin à panache doré
6477
Manakin à queue barrée
7792
Manakin à queue ronde
7787
Manakin à tête blanche
7797,7798
Manakin à tête bleue 7789
Manakin à tête d'opale
7794
Manakin à tête d'or 7790
Manakin à tête grise 7818
Manakin à tête jaune 2131
Manakin à tête rouge 7799
Manakin à ventre blanc
6478,6479
Manakin à ventre jaune
6480
Manakin auréole 7785
Manakin aux ailes d'or
5574
Manakin cannellé 6488
Manakin casqué 737
Manakin casse-noisette
5550
Manakin céruléen 7786
Manakin de Sclater 2664
Manakin des yungas 2083
Manakin deuil 2133
Manakin doré 7803
Manakin fastueux 2086
Manakin filifère 7793
Manakin lancéolé 2085
Manakin militaire 4837
Manakin minuscule 10394
Manakin nain 10393
Manakin neigeux 7797
Manakin noir 10629
Manakin olive 2134
Manakin orné 2664
Manakin rubis 5443

Manakin tête-de-feu 5442
Manakin tijé 2087
Manakin tyran 6476
Manakin unicolore 2133
Manakin vert 2132
Manchot à jugulaire 8638
Manchot antipode 5591
Manchot d'Adélie 8637
Manchot de Humboldt
9291
Manchot de Magellan 9292
Manchot des Galapagos
9293
Manchot du Cap 9290
Manchot empéreur 822
Manchot jugulaire 8638
Manchot papou 8639
Manchot pygmée 3656
Manchot royal 823
Manchots 9289
Manger-poulet (Ants) 1406
Mangeur d'herbe (Ants)
10010
Mango à cravate noire 666
Mango à cravate verte 669
Mango de la Jamaïque 665
Mango de Prévost 667
Mango doré 663
Mango vert 670
Manucaude royal 2225
Marabou d'Afrique 5169
Marabout argala 5170
Marabout chevelu 5171
Marangouin (Ants) 1509
Marmaronette marbrée
5572
Marmette de Brünnich
10455
Marmette de Troil 10454
Marouette à bec jaune 388
Marouette à poitrine
blanche 393
Marouette à sourcils blancs
8167
Marouette akool 386
Marouette australienne
8168
Marouette bicolore 387
Marouette brune 8169
Marouette caroline 8165
Marouette d' Elwes 387

Marouette d'Australie 8168
Marouette de Baillon 8175
Marouette de Caroline
8165
Marouette de Henderson
8164
Marouette de Kusaie 8170
Marouette de la Caroline
8165
Marouette de Laysan 8171
Marouette des Hawai 8176
Marouette des Philippines
391
Marouette d'Olivier 392
Marouette fuligineuse 8178
Marouette grise 8166
Marouette isabelle 389
Marouette maillée 8177
Marouette mandarin 8173
Marouette plombée 8163
Marouette ponctuée 8174
Marouette poussin 8172
Marouette rayée 189
Martin à collier 99
Martin à cou noir 9556
Martin à tête grise 9554
Martin à ventre blanc 104
Martin caronculé 2827
Martin couronné 491
Martin de Rothschild 5198
Martin des berges 102
Martin des pagodes 9557
Martin des rivages 102
Martin forestier 101
Martin gris 9551
Martin huppé 100
Martin pie 9552
Martin roselin 9559
Martin soyeux 9561
Martin triste 105
Martin-chasseur à ailes
bleues 3025
Martin-chasseur à ailes
brunes 7185
Martin-chasseur à bec noir
7187
Martin-chasseur à coiffe
noire 4438
Martin-chasseur à collier
blanc 10058

Martin-chasseur à collier
 roux 10080
Martin-chasseur à dos
 blanc 10056
Martin-chasseur à dos de
 feu 10071
Martin-chasseur à longs
 brins 9787
Martin-chasseur à
 moustaches 145
Martin-chasseur à poitrine
 bleue 4437
Martin-chasseur à tête
 blanche 10077
Martin-chasseur à tête
 brune 4431
Martin-chasseur à tête grise
 4436
Martin-chasseur à tête
 rouge 4884
Martin-chasseur à tête
 rousse 4884
Martin-chasseur à ventre
 blanc 315
Martin-chasseur à ventre
 roux 10063
Martin-chasseur bec-en-
 cuillère 2377
Martin-chasseur bleu-et-
 blanc 4440
Martin-chasseur bleu-noir
 10070
Martin-chasseur cannellé
 10059
Martin-chasseur couronné
 10057
Martin-chasseur de Biak
 9790
Martin-chasseur de
 Caroline 9784
Martin-chasseur de
 Gaudichaud 3024
Martin-chasseur de
 Hombron 147
Martin-chasseur de
 Jacquinot 10069
Martin-chasseur de Java
 4435
Martin-chasseur de Kofiau
 9786

Martin-chasseur de
 Mangaia 10073
Martin-chasseur de Miyaco
 10060
Martin-chasseur de Smyrne
 4442
Martin-chasseur de
 Temminck 10061
Martin-chasseur de
 Winchell 10080
Martin-chasseur d'Elliot
 9786
Martin-chasseur des forêts
 9788
Martin-chasseur des
 Gambier 10065
Martin-chasseur des
 mangroves 4439,4441
Martin-chasseur des
 Marquises 10066
Martin-chasseur des
 Moluques 10061
Martin-chasseur des Samoa
 10072
Martin-chasseur des Talaud
 10062
Martin-chasseur
 d'Euphrosine 5738
Martin-chasseur du Sénégal
 4439
Martin-chasseur étincelant
 1752
Martin-chasseur forestier
 10069
Martin-chasseur funèbre
 10064
Martin-chasseur géant 3026
Martin-chasseur gurial
 7186
Martin-chasseur lazuli
 10067
Martin-chasseur marron
 4432
Martin-chasseur menu
 9788
Martin-chasseur mignon
 4951
Martin-chasseur moine 149
Martin-chasseur
 montagnard 9626

Martin-chasseur nymphe
 9789
Martin-chasseur oreillard
 2363
Martin-chasseur oreillon-
 bleu 2363
Martin-chasseur outremer
 10068
Martin-chasseur pailleté
 3027
Martin-chasseur pygmé
 4886
Martin-chasseur respecté
 10078
Martin-chasseur rose 9785
Martin-chasseur roux 4885
Martin-chasseur royal 150
Martin-chasseur sacré
 10074
Martin-chasseur strié 4433
Martin-chasseur sylvain
 9791
Martin-chasseur tacheté
 148
Martin-chasseur torotoro
 9627
Martin-chasseur trapu 146
Martin-chasseur tyro 3027
Martin-chasseur vénéré
 10079
Martin-chasseur violet
 4434
Martinet (Ants) 4890
Martinet à collier 9510
Martinet à collier blanc
 9512
Martinet à collier
 interrompu 9508
Martinet à collier roux
 9510
Martinet à croupion blanc
 830
Martinet à croupion gris
 1981
Martinet à gorge blanche
 221
Martinet à menton blanc
 3012
Martinet à nuque blanche
 9511

Méliphage de Lombok 5247
Méliphage de Macleay 10615
Méliphage de Mayr 8544
Méliphage de Meek 8545
Méliphage de Mimika 5754
Méliphage de Nouvelle-Hollande 7543
Méliphage de Reichenow 5734
Méliphage de San Cristobal 5735
Méliphage de Tagula 5759
Méliphage de Timor 5243
Méliphage de Wetar 5249
Méliphage de White 2568
Méliphage de Whiteman 5737
Méliphage des Bonin 774
Méliphage des Célèbes 6306
Méliphage des mangroves 5221
Méliphage des Nouvelles-Hébrides 7542
Méliphage des Snow 6764
Méliphage enfumé 5762
Méliphage fardé 7541
Méliphage fascié 8755
Méliphage festonné 25
Méliphage flavescent 5222
Méliphage forestier 5755
Méliphage foulehaio 3891
Méliphage frangé 5246
Méliphage fuligineux 5729
Méliphage gracile 5752
Méliphage grimé 5220
Méliphage grisâtre 5226
Méliphage gris-de-plomb 8547
Méliphage grivelé 5240
Méliphage hihi 6608
Méliphage jaune 5224
Méliphage kioéa 1975
Méliphage lancéolé 7918
Méliphage leucotique 5229
Méliphage mao 4382
Méliphage maquille 5736
Méliphage marbré 8630
Méliphage marqué 5756

Méliphage mineur 647
Méliphage modeste 8756
Méliphage montagnard 5757
Méliphage moucheté 10616
Méliphage obscur 5232
Méliphage ocré 8631
Méliphage olivâtre 10025
Méliphage orné 5233
Méliphage pailleté 5761
Méliphage peint 4341
Méliphage régent 10612
Méliphage reticulé 5758
Méliphage serti 5234
Méliphage sosie 5748
Méliphage souriant 5751
Méliphage strié 8546
Méliphage toulou 4381
Méliphage trompeur 4280
Méliphage unicolore 5237
Méliphage versicolore 5238
Méliphage vert 4383
Meliphagidés 5760
Mélocichle à moustaches 5775
Ménure d'Albert 5785
Ménure du prince Albert 5785
Ménure superbe 5786
Menuridés 5787
Mère des cailles 2831
Merganette des torrents 5788
Mergule nain 368
Mérion à bec large 5538
Mérion à dos rouge 5544
Mérion à épaulettes 5529
Mérion à gorge bleue 5546
Mérion à tête rousse 2381
Mérion couronné 5533
Mérion de Campbell 5532
Mérion de Lambert 5539
Mérion de Wallace 9225
Mérion élégant 5537
Mérion empéreur 5536
Mérion leucoptère 5541
Mérion ravissant 5530
Mérion splendide 5547
Mérion superbe 5535
Mérion turquoise 5531

Merle des rochers 7751
Merle 10328
Merle (Ants) 10362,254, 2846,8692
Merle à ailes grises 10295
Merle à bec noir 10315
Merle à calotte grise 10361
Merle à col blanc 10288
Merle à col roux 10357
Merle à collier 10364, 10743
Merle à collier blanc 10289
Merle à dos gris 10314
Merle à dos roux 10359
Merle à flancs roux 10299
Merle à froc noir 10340
Merle à gorge blanche 10292
Merle à gorge noire 10356
Merle à gorge rousse 10355
Merle à lunettes 10334
Merle à miroir 10293
Merle à oeil clair 7896
Merle à pattes jaunes 7895
Merle à plastron 10364
Merle à poitrine noire 10302
Merle à tête grise 10354
Merle à ventre clair 10290
Merle à ventre fauve 10306
Merle à ventre roux 10358
Merle africain 10342
Merle américain 10329
Merle améthyste 2275
Merle ardoisé 10333
Merle austral 10304
Merle aux yeux blancs 10318
Merle bleu 6022
Merle bleu de Taïwan 6230
Merle bleu siffleur 6226
Merle bleu siffleur de Taïwan 6230
Merle bronzé à longue queue 4993
Merle bronzé pourpré 5006
Merle brun 10303
Merle cacao 10307
Merle cendré 10363
Merle chiguanco 10298
Merle Corbeau (Ants) 2846

Mesia à oreillons argentés 5133
Mésite monias 6011
Mésite unicolore 5824
Mésite variée 5825
Mesitornithidés 5826
Messager sagittaire 8925
Métallure à gorge feu 5830
Métallure à queue bronzée 2000
Métallure à queue d'airain 5828
Métallure à tête rousse 2002
Métallure arc-en-ciel 1999
Métallure de Baron 5829
Métallure de Stanley 2003
Métallure de Thérèse 5834
Métallure dorée 5831
Métallure du Chinguela 5832
Métallure émeraude 5835
Métallure olivâtre 2001
Métallure phébé 5833
Métallure verte 5836
Mézette à moustaches 7008
Microbate à collier 5851
Microbate à long bec 8733
Microbate cendré 5850
Microgoura de Choiseul 5867
Micropsitte à tête fauve 5882
Micropsitte de Bruijn 5877
Micropsitte de Finsch 5878
Micropsitte de Geelvink 5879
Micropsitte de Meek 5881
Micropsitte pygmée 5880
Microtyran à calotte noire 6212
Microtyran à face blanche 8034
Microtyran à queue courte 6214
Microtyran à ventre blanc 6211
Microtyran bariolé 8037
Microtyran bifascié 5351
Microtyran casqué 5349
Microtyran chevelu 5350

Microtyran coiffé 1048
Microtyran d'André 9724
Microtyran eulophe 5348
Microtyran noir-et-blanc 8035
Microtyran oreillard 6213
Microtyran tricolore 8036
Mi-deuil (Ants) 5953
Mignard enchanteur 9429
Milan à long bec 8913
Milan à plastron 4455
Milan à queue carrée 5328
Milan à queue fourchue 3431
Milan bec-en-croc 2181
Milan bidenté 4476
Milan bleuâtre 4834
Milan brun 5891
Milan de Cayenne 5160
Milan de Cuba 2182
Milan de Forbes 5161
Milan de Riocour 2072
Milan d'Égypte 5889
Milan des chauves-souris 5444
Milan des Everglades 8914
Milan des marais 8914
Milan diodon 4477
Milan du Mississippi 4833
Milan noir 5892
Milan royal 5893
Milan sacré 4452
Milan siffleur 4453
Minivet à bec court 7250
Minivet à ventre blanc 7254
Minivet cendré 7253
Minivet de Sumbawa 7258
Minivet de Swinhoe 7251
Minivet des Ryukyu 7262
Minivet flamboyant 7257
Minivet mandarin 7261
Minivet montagnard 7261
Minivet oranor 7252
Minivet rose 7260
Minivet rouge 7255
Minivet vermillon 7259
Minla à aile bleue 5905
Minla à ailes bleues 5905
Minla à gorge striée 5907
Minla à queue rousse 5906

Mino anais 5908
Mino de Dumont 5909
Miro à ailes blanches 7237
Miro à capuchon 5709
Miro à croupion blanc 7234
Miro à dos vert 6972
Miro à face blanche 10143
Miro à flancs noirs 8031
Miro à front rouge 7277
Miro à gorge noire 8029
Miro à menton noir 8030
Miro à pattes jaunes 5864
Miro à poitrine blanche 3535
Miro à poitrine grise 3536
Miro à poitrine jaune 3533
Miro à queue brune 5862
Miro à tête grise 4611
Miro à ventre citron 5861
Miro à ventre jaune 3534
Miro aux yeux blancs 6973
Miro bridé 8033
Miro ceinturé 8032
Miro cendré 4610
Miro de Tasmanie 5712
Miro des Chatham 7284
Miro des mangroves 3537
Miro des rochers 7274
Miro des Tanimbar 5865
Miro des torrents 5975
Miro écarlate 7279
Miro embrasé 7281
Miro enchanteur 5860
Miro grenat 3659
Miro gris-bleu 7236
Miro incarnat 7282
Miro jaunâtre 10142
Miro mésange 7278
Miro montagnard 7276
Miro olive 5863
Miro ombré 7235
Miro papou 5866
Miro rose 7283
Miro rubisole 7275
Modeste 6446
Modulatrice à lunettes 5954
Modulatrice grivelée 944
Moho à tête noire 4767
Moho de Bishop 5956
Moho de Kauai 5957
Moho d'Hawai 5958

Moho d'Oahu 5955
Mohoua à tête blanche 5959
Mohoua à tête jaune 5961
Mohoua pipipi 5960
Moien (Ants) 626
Moineau 7141
Moineau à dos roux 7148
Moineau à front écaillé 9381
Moineau à gorge jaune 7289
Moineau à point jaune 7288
Moineau à sourcils 7289
Moineau à tête grise 7146
Moineau blanc 7159
Moineau bridé 7289
Moineau d'Arabie 7143
Moineau de la mer Morte 7152
Moineau de Socotra 7149
Moineau de Somalie 7139
Moineau de Swainson 7161
Moineau d'Emin 7142
Moineau d'Emin Bey 7142
Moineau d'Emin Pacha 7142
Moineau des herbes (Ants) 484
Moineau des îles du Cap-Vert 7148
Moineau des saxaouls 7138
Moineau domestique 7141
Moineau doré 7150
Moineau doré d'Arabie 7143
Moineau doré d'Emin 7142
Moineau du Cap 7151
Moineau du Cap-Vert 7148
Moineau du Paradis 377
Moineau du Sind 7156
Moineau espagnol 7147
Moineau flavéole 7144
Moineau friquet 7153
Moineau gris 7146
Moineau gris austral 7140
Moineau mélanure 7151
Moineau moabite 7152
Moineau pâle 1785
Moineau perroquet 7145

Moineau quadrillé 9380
Moineau roux 7157
Moineau rutilant 7158
Moineau sociale à tête grise 8342
Moineau sociale à tête noire 8343
Moineau soulcie 7287
Moineau soulcie à point jaune 7288
Moineau soulcie austral 7289
Moineau soulcie pâle 1785, 7286
Moineau sud-africain 7140
Moineau swahili 7160
Moineau tisserin 7929
Moineau-tisserin à calotte maronne 7929
Moineau-tisserin à dos roux 7928
Moineau-tisserin à sourcils blancs 7927
Moineau-tisserin de Donaldson 7926
Moineaux 7169
Moinelette à dos gris 3580
Moinelette à front blanc 3578
Moinelette à oreillons blancs 3577
Moinelette à oreillons noirs 3574
Moinelette croisée 3575
Moinelette de Fischer 3576
Moinelette d'Oustalet 3579
Moisson jaune (Ants) 9210
Moisson pied-blanc 5719
Molothre noir 5964
Momot à bec large 3437
Momot caréné 3436
Momotidés 5968
Monarque à ailes noires 5987
Monarque à ailes tachetées 5989
Monarque à bec large 6142
Monarque à capuchon 5995
Monarque à col blanc 6006
Monarque à collerette 990

Monarque à crête bleue 6127
Monarque à face noire 5996
Monarque à froc roux 988
Monarque à gorge noire 2383
Monarque à lunettes 6004
Monarque à menton noir 5998
Monarque à nuque blanche 5999
Monarque à poitrine jaune 5439
Monarque à poitrine noire 5440
Monarque à ventre marron 5981
Monarque acier 6132
Monarque brun 2384
Monarque de Biak 6126
Monarque de Boano 5978
Monarque de Bougainville 5985
Monarque de Brehm 5979
Monarque de Brown 5980
Monarque de Buru 5994
Monarque de Fatuhiva 8114
Monarque de Florès 6002
Monarque de Guam 6133
Monarque de Kofiau 5991
Monarque de Lesson 5575
Monarque de Ponapé 6139
Monarque de Rarotonga 8110
Monarque de Rennell 2382
Monarque de Richards 6000
Monarque de San Cristobal 6129
Monarque de Tahiti 8113
Monarque de Tinian 6003
Monarque de Truk 5827
Monarque de Vanikoro 6143
Monarque de Yap 5988
Monarque des Banks 6464
Monarque des Bismarck 6005
Monarque des Fidji 2385

Océanite à ventre blanc 3945
Océanite à ventre noir 3947
Océanite cendré 6654
Océanite cul-blanc 6656
Océanite de Castro 6652
Océanite de Guadeloupe 6657
Océanite de Hornby 6655
Océanite de Markham 6658
Océanite de Matsudaira 6659
Océanite de Swinhoe 6662
Océanite de tempête 4694
Océanite de Tristram 6664
Océanite de Wilson 6650
Océanite d'Elliot 6649
Océanite frégate 7184
Océanite minute 6661
Océanite nain 6661
Océanite néréide 4072
Océanite noir 6660
Océanite tempête 4694
Océanite tethys 6663
Océanites 4695
Odontophoridés 6682
Oedicnème bistrié 1371
Oedicnème bridé 1374
Oedicnème criard 1376
Oedicnème des récifs 1373
Oedicnème du Pérou 1379
Oedicnème du Sénégal 1378
Oedicnème tachard 1372
Oedicnème vermiculé 1380
Oedicnème vocifère 1371
Oedicnèmes 1370
Oie à bec court 630
Oie à tête barrée 639
Oie cendrée 629
Oie cygnoïde 633
Oie de Groenland 637
Oie de Ross 640
Oie d'Égypte 376
Oie des moissons 636
Oie des neiges 631
Oie empereur 632
Oie naine 635
Oie rieuse 628
Oie-armée de Gambie 7921

Oiseau à berceau satiné 8540
Oiseau à berceau tacheté 2092
Oiseau à berceau vert 277
Oiseau béni 10228
Oiseau Bon-Dieu (Ants) 10215
Oiseau canne (Ants) 10558, 10568
Oiseau de la pluie (Ants) 1981,3015,9512,9694
Oiseau détiqueur 1347
Oiseau fou (Ants) 9440,9452
Oiseau grandpère (Ants) 3703
Oiseau musicien (Ants) 6104
Oiseau palmiste (Ants) 3382
Oiseau-bleu des fàes 4877
Oiseau-chat 3383
Oiseau-lunettes jaune 10826
Oiseau-lyre du prince Albert 5785
Oiseau-mouche (Ants) 663
Oiseau-régent 9132
Oiseaux de tempête 4695
Oiseaux des tempêtes 8246
Oiseaux-lunettes 10756
Oiseaux-lyres 5787
Ombrette africaine 9043
Ombrette du Sénégal 9043
Onoré à huppe blanche 10017
Onoré du Mexique 10020
Onoré fascié 10018
Onoré phaeton 10708
Onoré rayé 10019
Onoré zigzag 10692
Ophistocomidés 6744
Ophrysie de I.Himalaya 6746
Oréochare des Arfak 6753
Oréophase cornu 6760
Oréotangara élégant 6769
Organiste à bec épais 3698
Organiste à calotte bleue 2129

Organiste à calotte d'or 3707
Organiste à calotte jaune 3699
Organiste à capuchon 3690
Organiste à col jaune 2128
Organiste à couronne rousse 3683
Organiste à gorge jaune 3695
Organiste à nuque bleue 2127
Organiste à sourcils jaunes 2126
Organiste à ventre brun 2130
Organiste à ventre marron 3704
Organiste à ventre orangé 3710
Organiste à ventre roux 3706
Organiste chalybée 3685
Organiste chlorotique 3686
Organiste cul-blanc 3702
Organiste cul-roux 3692
Organiste de brousse 3682
Organiste de Finsch 3691
Organiste de la Jamaïque 3697
Organiste de la Magdalena 3688
Organiste de Trinidad 3708
Organiste doré 3689
Organiste fardé 3687
Organiste louis-d'or 3703
Organiste mordoré 3701
Organiste moucheté 3696
Organiste nègre 3684
Organiste olive 3694
Organiste plombé 3705
Organiste téité 3709
Origma des rochers 6775
Oriole à ailes blanches 4804
Oriole à capuchon 4809
Oriole à dos rayé 4828
Oriole à épaulettes 4805
Oriole à gros bec 4815
Oriole à tête d'or 4802
Oriole cul-noir 4830

Oriole d'Abeillé 4800
Oriole d'Audubon 4813
Oriole de Baltimore 4811
Oriole de la Martinique 4803
Oriole de Sainte-Lucie 4819
Oriole des campos 4818
Oriole des vergers 4829
Oriole du Nord 4800,4811
Oriole jaune-verdâtre 4825
Oriole leucoptère 4812
Oriole masqué 4808
Oriole moriche 4807
Oriole noir-et-or 4806
Oriole orangé 4801,4811
Oriole troupiale 4817
Oriolidés 6777
Oriolie de Bernier 6776
Ortalide à sourcils 6819
Ortalide à tête grise 6811
Ortalide à tête rousse 6812
Ortalide à ventre blanc 6815
Ortalide à ventre marron 6821
Ortalide à ventre roux 6818
Ortalide babillarde 6813
Ortalide chacamel 6820
Ortalide de Wagler 6817
Ortalide du Chaco 6810
Ortalide maillée 6814
Ortalide motmot 6816
Orthonychidés 6823
Orthonyx de Spalding 6824
Orthonyx de Temminck 6825
Ortolan (Ants) 2536
Otididés 6852
Otidiphaps noble 6853
Ouanga négresse (Ants) 5774
Ouette à ailes bleues 2915
Ouette à tête grise 2104
Ouette à tête rousse 2105
Ouette de l'Orénoque 6445
Ouette de Magellan 2103
Ouette d'Égypte 376
Ouette des Andes 2102
Ouette marine 2101
Outarde à collerette 2096

Outarde à miroir blanc 3731
Outarde à tête noire 973
Outarde à ventre noir 3739
Outarde arabe 970
Outarde barbue 6854
Outarde canepetière 9883
Outarde d'Australie 971
Outarde de Denham 6494
Outarde de Hartlaub 3736
Outarde de Heuglin 6496
Outarde de Ludwig 6497
Outarde de Rüppell 3740
Outarde de Stanley 6495
Outarde de Vigors 3744
Outarde d'Oustalet 3735
Outarde du Bengale 3733
Outarde du Sénégal 3743
Outarde houbara 2096
Outarde houppette 3741
Outarde korhaan 3730
Outarde kori 973
Outarde nubienne 6498
Outarde passarage 3738
Outarde plombée 3734
Outarde somalienne 3737
Outardes 6852
Oxylabe à gorge blanche 6924
Oxylabe à sourcils jaunes 2841
Oxyrhynque en feu 6927
Oxyrhynque huppé 6927
Pachycare nain 6935
Padda 7001
Padda brun 7000
Padda de Java 7001
Padda de riz 7001
Pagophile blanche 7004
Paille en queue (Ants) 7327
Paille en queue à bec rouge (Ants) 7327
Palette à couronne bleue 8211
Palette à manteau d'or 8217
Palette de Buru 8214
Palette de Cassin 8212
Palette de Mindanao 8219
Palette de Palawan 8216
Palette des Sulu 8218
Palette momot 8215

Palette verte 8213
Palicour de Cayenne 6268
Palmérie huppée 7005
Palmiste africain 4399
Panure à moustaches 7008
Paon bleu 7180
Paon du Congo 239
Paon spicifère 7181
Pape de Manille 3613
Pape de Nouméa 3609
Pape des prairies 3608
Papegai maillé 3173
Paradiséidé 7022
Paradisier à bec blanc 3548
Paradisier à gorge noire 1042
Paradisier à queue courte 7013
Paradisier à rubans 1041
Paradisier bleu 7021
Paradisier caronculé 7014
Paradisier corvin 5435
Paradisier d'Albertis 3547
Paradisier de Carola 7066
Paradisier de Goldie 7016
Paradisier de Guillaume 7017
Paradisier de Jobi 5562
Paradisier de Keraudren 5563
Paradisier de Lawes 7068
Paradisier de Loria 2388
Paradisier de Macgregor 5438
Paradisier de Meyer 3550
Paradisier de Raggi 7019
Paradisier de Rodolphe 7021
Paradisier de Rothschild 1043
Paradisier de Rudolph 7021
Paradisier de Stéphanie 1045
Paradisier de Victoria 8551
Paradisier de Wahnes 7070
Paradisier de Wallace 9110
Paradisier d'Entrecasteaux 5561
Paradisier d'Helena 7067
Paradisier du prince Albert 8402

Platyrhynque jaune-olive 10104
Platyrhynque olivâtre 8887
Platyrhynque poliocéphale 10103
Platystéiridés 7917
Plectrophane blanc 7919
Plectrophane des neiges 7920
Plocéidés 7925
Plongeon (Ants) 8024
Plongeon à bec blanc 4131
Plongeon à gorge noire 4132
Plongeon arctique 4132
Plongeon catmarin 4135
Plongeon damier 4132
Plongeon du Pacifique 4134
Plongeon glacial 4133
Plongeon huard 4133
Plongeon imbrin 4133
Plongeon lumme 4132
Plongeons 4136
Pluvian d'Égypte 8001
Pluvian fluviatile 8001
Pluvianelle magéllanique 8000
Pluvier (Ants) 7999
Pluvier à bandeau noir 2048
Pluvier à camail 2044
Pluvier à collier interrompu 2018
Pluvier à double collier 2023
Pluvier à face noire 3445
Pluvier à front blanc 2031
Pluvier à long bec 2043
Pluvier à tête rousse 2045
Pluvier à triple collier 2049
Pluvier anarhynque 522
Pluvier argenté 7999
Pluvier asiatique 2021
Pluvier australien 2022
Pluvier bronzé 7996
Pluvier bronzé (Ants) 7997
Pluvier ceinturé 3598
Pluvier d'Azara 2024
Pluvier de Forbes 2027
Pluvier de Java 2029

Pluvier de Leschenault 2030
Pluvier de Mongolie 2034
Pluvier de Nouvelle-Zélande 2037
Pluvier de Péron 2042
Pluvier de Sainte-Hélène 2046
Pluvier de Wilson 2052
Pluvier des Andes 7438
Pluvier des Falkland 2026
Pluvier des salines (Ants) 974
Pluvier dominicain 7997
Pluvier doré 7996
Pluvier doré (Ants) 7997
Pluvier doré américain 7997
Pluvier doré d'Amérique 7997
Pluvier doré d'Eurasie (Qué) 7996
Pluvier du désert 2030
Pluvier du Puna 2020
Pluvier d'Urville 2033
Pluvier élégant 2040
Pluvier facou (Ants) 974
Pluvier fauve 7998
Pluvier grand-gravelot 2028
Pluvier gris (Ants) 7999
Pluvier grosse tête (Ants) 7999
Pluvier guignard 2036
Pluvier kildir 2051
Pluvier montagnard 2035
Pluvier neigeux 2019
Pluvier oréophile 6761
Pluvier oriental 2050
Pluvier pâtre 2041
Pluvier petit-gravelot 2025
Pluvier roux 2038
Pluvier semipalmé 2047
Pluvier sibérien 7998
Pluvier siffleur 2032
Pluvier sociable 10488
Pluviers 2016
Pnoepyga à poitrine maillée 8004
Pnoepyga à ventre blanc 8002

Podarge à tête grise 1224
Podarge cornu 1218
Podarge de Blyth 1216
Podarge de Bornéo 1222
Podarge de Ceylan 1223
Podarge de Hartert 1219
Podarge de Hodgson 1220
Podarge de Java 1221
Podarge des Philippines 1225
Podarge étoilé 1226
Podarge gris 8010
Podarge ocellé 8007
Podarge oreillard 1217
Podarge oriental 1220
Podarge papou 8009
Podargidé 8006
Podicipédidés 8021
Podicipediformes 8022
Podoce de Biddulph 8025
Podoce de Henderson 8026
Podoce de Hume 8344
Podoce de Pander 8027
Podoce de Pleske 8028
Poliolais à queue blanche 8071
Polochion à menton jaune 7452
Polochion à nuque blanche 7448
Polochion casqué 7451
Polochion couronné 7449
Polochion criard 7454
Polochion de Brass 7450
Polochion de Céram 7463
Polochion de Kisar 7459
Polochion de Meyer 7460
Polochion de Nouvelle-Bretagne 7453
Polochion de Nouvelle-Guinée 7462
Polochion de Nouvelle-Irlande 7456
Polochion des Moluques 7461
Polochion moine 7455
Polochion sobre 7458
Polochion sombre 7457
Polochion strié 5772
Pomatorhin à bec corail 8117

Prinia à face noire 8201
Prinia à front écailleux
 9300
Prinia à front roux 8184
Prinia à gorge blanche
 8199
Prinia à gorge noire 8182
Prinia à joues rousses 5515
Prinia à oreilles rousses
 5515
Prinia à plastron 8191
Prinia à poitrine canelle
 8196
Prinia à ventre jaune 8192
Prinia aquatique 8193
Prinia bifasciée 8190
Prinia cendrée 8206
Prinia crinigère 8188
Prinia de Burnes 8185
Prinia de Hodgson 8195
Prinia de Roberts 8204
Prinia de Sao Tomé 8202
Prinia des marais 8186
Prinia des montagnes 8203
Prinia du Drakensberg
 8196
Prinia du Karroo 8200
Prinia du Namaqua 8209
Prinia du Sierra Leone
 8198
Prinia forestière 8210
Prinia gracile 8194
Prinia grise 3314
Prinia modeste 8208
Prinia pâle 8207
Prinia rayée 8183
Prinia roussâtre 8205
Prinia simple 8197
Prinia verte 10466
Prion à bec epais 6976
Prion bleu 4454
Prion colombe 6980
Prion de Belcher 6975
Prion de Forster 6982
Prion de la Désolation 6978
Prion de Salvin 6979
Prionopidés 8227
Ririt à collier 7912
Pririt à front blanc 7907
Pririt à gorge noire 7915
Pririt à joues noires 1203

Pririt à queue courte 1205
Pririt à taches blanches
 7916
Pririt à tête grise 1208
Pririt à ventre dorée 7911
Pririt chalybée 7910
Pririt châtain 7909
Pririt de Blisset 7908
Pririt de Boulton 1201
Pririt de Fernando Po 1210
Pririt de l'Angola 1204
Pririt de Jameson 7913
Pririt de Lawson 1207
Pririt de l'Ituri 1200
Pririt de Reichenow 1212
Pririt de Verreaux 1202
Pririt de Vieillot 1211
Pririt de Woodward 1199
Pririt du Bamenda 7914
Pririt du Cap 1196
Pririt du Malawi 1197
Pririt du Ruwenzori 1198
Pririt du Sénégal 1213
Pririt molitor 1206
Pririt pâle 1214
Pririt pygmée 1209
Procéllaridés 8246
Procellariformes 8247
Promérops de Gurney 8264
Promérops du Cap 8263
Pseudolangrayen à lunettes
 8327
Pseudolangrayen d'Afrique
 8326
Pseudolangrayen d'Asie
 8327
Pseudolangrayen de rivière
 8326
Pseudosittine à collier 9226
Pseudotraquet indien 8996
Psitacule de Malacca 8386
Psittacidés 8360
Psittaciformes 8361
Psittacule à poitrine orangé
 2974
Psittacule de Desmarest
 8379
Psittacule de Salvadori
 8381
Psittacule d'Edwards 8380
Psittacule double-oeil 2972

Psittirostre à gros bec 2107
Psittirostre de Laysan 9824
Psittirostre de Maui 8341
Psittirostre de Nihoa 9825
Psittirostre de Palmer 8868
Psittirostre palila 5398
Psittirostre psittacin 8387
Psittrichas de Pesquet 8388
Psophidés 8392
Psophode à menton noir
 8395
Psophode à tête noire 7538
Psophode babillard 8394
Psophode carillonneur
 8400
Pstittaculidés 8378
Ptéroclidés 8419
Ptéroclididés 8419
Ptilogon à longue queue
 8537
Ptilogon cendré 8538
Ptilonorhynchidés 8539
Ptilonorhynque satiné 8540
Ptilope à ceinture 8487
Ptilope à couronne lilas
 8490
Ptilope à diadème 8523
Ptilope à front d'or 8484
Ptilope à poitrine écarlate
 8485
Ptilope à ventre orangé
 8501
Ptilope batilde 8513
Ptilope caronculé 8495
Ptilope casqué 8499
Ptilope de Clementine 8518
Ptilope de Dupetit-Thouars
 8492
Ptilope de Fischer 8494
Ptilope de Grey 8496
Ptilope de Henderson 8500
Ptilope de Hutton 8497
Ptilope de l'Alligator 8482
Ptilope de La Pérouse 8517
Ptilope de la Société 8521
Ptilope de Layard 8503
Ptilope de Leclancher 8504
Ptilope de Makatea 8486
Ptilope de Marché 8507
Ptilope de Mercier 8509
Ptilope de Merrill 8510

Serpentaire des Philippines 9306
Serpentaire des Ryukyu 9311
Serpentaire du Congo 3341
Serpentaire menu 9309
Shama à croupion blanc 2594
Shama à queue rousse 10171
Shama bridé 2593
Shama dayal 2596
Shama de Cebu 2592
Shama de Madagascar 2591
Shama de Strickland 2598
Shama des Seychelles 2597
Shama noir 2595
Sibia à dos marron 4613
Sibia à longue queue 4619
Sibia à tête noire 4618
Sibia casquée 4615
Sibia de Formose 4614
Sibia de Taïwan 4614
Sibia du Langbian 2840
Sibia gracile 4617
Sibia grise 4617
Sibia superbe 4620
Sibia tachetée 2839
Sicale à béret 9204
Sicale à croupion jaune 9215
Sicale à tête jaune 9209
Sicale bouton-d'or 9205
Sicale citrin 9203
Sicale de Patagonie 9206
Sicale de Raimondi 9213
Sicale de Taczanowski 9214
Sicale des savanes 9210
Sicale jaune 9207
Sicale misto 9208
Sicale olivâtre 9211
Siffleur (Ants) 3115,3421, 6157
Siffleur à bavette blanche 6950
Siffleur à cape jaune 6938
Siffleur à dos marron 2660
Siffleur à dos vert 6936
Siffleur à face rousse 6967

Siffleur à flancs jaunes 4725
Siffleur à gorge nue 6957
Siffleur à nuque rousse 344
Siffleur à queue noire 6953
Siffleur à tête grise 6941
Siffleur à ventre blanc 6951
Siffleur à ventre jaune 6971
Siffleur américain (Ants) 524
Siffleur blanc (Ants) 3421
Siffleur calédonien 6939
Siffleur cendré 6943
Siffleur d'Amérique 524
Siffleur de Blasius 6944
Siffleur de Bornéo 6946
Siffleur de Gilbert 6948
Siffleur de Lorentz 6952
Siffleur de montagne (Ants) 6104
Siffleur de Schlegel 6968
Siffleur de Sclater 6970
Siffleur de Wallace 6937
Siffleur des Moluques 6964
Siffleur des Philippines 6965
Siffleur des Salomon 6947
Siffleur des Samoa 6940
Siffleur des Tonga 6949
Siffleur doré 6961
Siffleur du Vogelkop 6954
Siffleur huppé (Ants) 6157
Siffleur itchong 6966
Siffleur modeste 6955
Siffleur moine 6956
Siffleur morne (Ants) 6104
Siffleur olivâtre 6958
Siffleur orphée 6960
Siffleur rouilleux 6945
Siffleur sobre 6969
Siffleur terne 6942
Sifilet de Lawes 7068
Sirli de Dupont 2077
Sirli de Witherby 298
Sirli des déserts 297
Sirli du désert 297
Sirli ricoti 2077,297
Sitelle à front velouté 9237
Sitelle corse 9252
Sitelle de Whitehead 9252

Sitelle des rochers de Neumayer 9244
Sitelle d'Europe 9235
Sitelle du Canada 9231
Sitelle naine 9239
Sittelle à bec jaune 9248
Sittelle à joues blanches 9241
Sittelle à poitrine blanche 9232
Sittelle à poitrine rousse 9231
Sittelle à tête brune 9246
Sittelle à ventre marron 9234
Sittelle bleue 9230
Sittelle de Chine 9251
Sittelle de Corse 9252
Sittelle de l'Himalaya 9238
Sittelle de Krüper 9239
Sittelle de Neumayer 9244
Sittelle des Naga 9243
Sittelle des Philippines 9245
Sittelle des rochers 9249
Sittelle du Cachemire 9233
Sittelle du Victoria 9250
Sittelle du Yunnan 9253
Sittelle géante 9242
Sittelle kabyle 9240
Sittelle pygmée 9247
Sittelle superbe 9236
Sittelle torchepot 9235
Sittelle veloutée 9237
Sittelles 9258
Sittidés 9258
Sittine à bec fin 4495
Sittine à queue rousse 10633
Sittine brune 10634
Sittine des rameaux 10636
Sittine striée 10635
Sizerin à tête rouge 1727
Sizerin blanchâtre 1731
Sizerin boréal 1727
Sizerin flammé 1727
Sizerin groenlandais 1731
Smaragdan à sourcils jaunes 10596
Smaragdan ceinturé 10598

Stizorhin de Fraser 6456
Stourne à longue queue 804
Stourne à queue courte 807
Stourne aux yeux blancs 792
Stourne bronzé 811
Stourne calédonien 814
Stourne chanteur 793
Stourne de Grant 809
Stourne de Kusaie 795
Stourne de Micronésie 810
Stourne de Norfolk 799
Stourne de Polynésie 815
Stourne de Ponapé 812
Stourne de Rarotonga 794
Stourne de Rennell 803
Stourne de Samoa 791
Stourne de San Cristobal 797
Stourne des Fead 798
Stourne des Moluques 808
Stourne des Salomon 802
Stourne des Tanimbar 796
Stourne d'Espiritu Santo 813
Stourne luisant 806
Stourne mélanésien 816
Stourne métallique 806
Stourne mystérieux 805
Streptocitte à cou blanc 9490
Streptocitte des Sula 9489
Stresemannia bougainvillei 9513
Strigidés 9514
Strigiens 9514
Strigiformes 9515
Strigops kakapo 9516
Struthionidés 9540
Struthioniformes 9541
Sturnelle à sourcils blancs 5137
Sturnelle australe 9544
Sturnelle de Lilian 9543
Sturnelle de l'Ouest 9547
Sturnelle des pampas 9546
Sturnelle des prés 9545
Sturnelle du Pérou 9542
Sturnelle militaire 5136
Sturnidés 9549
Suce-fleur (Ants) 5774

Sucrier (Ants) 2425
Sucrier à poitrine jaune (Ants) 2425
Sucrier à ventre jaune 2425
Sucrier mang (Ants) 3142
Sucrier mangle (Ants) 3142
Sulidés 9582
Surnie épervière 9583
Sylphe à longue queue 270
Sylphe à queue violette 269
Sylviidés 9623
Synallaxe à bandeaux 2796
Synallaxe à bavette 9656
Synallaxe à bec courbé 5255
Synallaxe à bec court 1018
Synallaxe à bec droit 5256
Synallaxe à calotte blanche 2795
Synallaxe à calotte rayée 2809
Synallaxe à calotte rousse 9654
Synallaxe à collier 9658
Synallaxe à dos marron 7300
Synallaxe à face rouge 2801
Synallaxe à filets 5155
Synallaxe à front rouille 1031
Synallaxe à front roux 7304
Synallaxe à front sombre 9643
Synallaxe à gorge argentée 9659
Synallaxe à gorge blanche 9629
Synallaxe à gorge d'argent 9659
Synallaxe à gorge jaune 1933
Synallaxe à gorge marron 9636
Synallaxe à gorge noire 9015
Synallaxe à gorge rayée 5158
Synallaxe à gorge violet 9636
Synallaxe à lunettes 9018

Synallaxe à menton blanc 9013
Synallaxe à moustaches 9634
Synallaxe à poitrine rayée 9227
Synallaxe à poitrine rousse 9642
Synallaxe à queue marbrée 1029
Synallaxe à queue noire 1026
Synallaxe à queue pâle 1024
Synallaxe à sourcils blancs 4514
Synallaxe à sourcils fauves 9660
Synallaxe à sourcils gris 2799
Synallaxe à tête brune 5152
Synallaxe à tête grise 2810
Synallaxe à tête rouge 9654
Synallaxe à ventre blanc 9653
Synallaxe albane 9628
Synallaxe ardent 9655
Synallaxe ardoisé 9632
Synallaxe austral 1016
Synallaxe aux yeux rouges 7301
Synallaxe belette 1934
Synallaxe casqué 2795
Synallaxe cendré 9647
Synallaxe chardon 9013
Synallaxe couronné 5153
Synallaxe damier 9024
Synallaxe d'Azara 9631
Synallaxe de Baron 2798
Synallaxe de Berlepsch 1019
Synallaxe de Blake 9640
Synallaxe de Cabanis 9633
Synallaxe de Cayenne 9645
Synallaxe de Coiba 2800
Synallaxe de Cordoba 1036
Synallaxe de Cory 9012
Synallaxe de Coursen 9640
Synallaxe de des Murs 9624
Synallaxe de Garlepp 1031

Synallaxe de Heller 9016
Synallaxe de Hellmayr 9646
Synallaxe de Hudson 1025
Synallaxe de Junin 1039
Synallaxe de la Plata 5154
Synallaxe de Lafresnaye 7306
Synallaxe de l'Itatiaia 9017
Synallaxe de l'Orénoque 9975
Synallaxe de Marcapata 2805
Synallaxe de Masafuera 789
Synallaxe de McConnell 9650
Synallaxe de Patagonie 1032
Synallaxe de Pinto 9648
Synallaxe de Spix 9657
Synallaxe de Vaurie 9635
Synallaxe de Vilcabamba 9020
Synallaxe de Wyatt 1040
Synallaxe de Zimmer 9664
Synallaxe des Andes 5151
Synallaxe des bambous 2373
Synallaxe des broméliades 2803
Synallaxe des cactus 1020
Synallaxe des cañons 1033
Synallaxe des canyons 1033
Synallaxe des épines 7307
Synallaxe des joncs 7492
Synallaxe des marais 9268
Synallaxe des Periya 9019
Synallaxe des rocailles 1030
Synallaxe des roseaux 5255
Synallaxe des Santa Marta 9644
Synallaxe d'Iquico 1023
Synallaxe d'Orbigny 1021
Synallaxe du Chinchipe 9637
Synallaxe du Cipo 1028
Synallaxe du Marañon 9651

Synallaxe du puna 1034
Synallaxe du Roraima 9649
Synallaxe écaillé 2806
Synallaxe élégant 9641
Synallaxe fauve 5159
Synallaxe flammé 1022
Synallaxe grimpeur 2797
Synallaxe grisin 9638
Synallaxe guiouti 9639
Synallaxe huppé 2811
Synallaxe inca 1038
Synallaxe maculé 7302
Synallaxe mantelé 1019
Synallaxe marron 1037
Synallaxe mésange 5150
Synallaxe montagnard 5159
Synallaxe obscur 9652
Synallaxe ocré 9630
Synallaxe olive 2807
Synallaxe orangé 9656
Synallaxe pâle 2808
Synallaxe ponctué 2802
Synallaxe Rayadito 790
Synallaxe rayé 9977
Synallaxe renard 2814
Synallaxe rouge 7303
Synallaxe rousselé 7307
Synallaxe rousserole 5256
Synallaxe roux 9662
Synallaxe siffleur 7305
Synallaxe soufré 2812
Synallaxe souris 9014
Synallaxe strié 5156
Synallaxe striolé 5157
Synallaxe terne 9976
Synallaxe terrestre 1027
Synallaxe tithys 9661
Synallaxe vannier 1035
Syrrhapte du Tibet 9680
Syrrhapte paradoxal 9679
Tacco de Cuba 8977
Tacco de la Jamaïque 8978
Tacco de Merlin 8977
Tacco de Porto Rico 8979
Tacco d'Hispaniola 8976
Tadorne à tête grise 9714
Tadorne belon 9718
Tadorne casarca 9716
Tadorne d'Australie 9719
Tadorne de Belon 9718

Tadorne de Corée 9715
Tadorne de paradis 9720
Tadorne radjah 9717
Talégalle à bec foncé 9726
Talégalle de Bruijn 191
Talégalle de Cuvier 9725
Talégalle de Jobi 9727
Talégalle de Latham 358
Talégalle des Arfak 190
Talève d'Allen 8155
Talève de Lord Howe 8154
Talève favorite 8157
Talève poule-sultane 8160
Talève pourprée 8159
Talève sultane 8160
Talève takahé 8158
Talève violacée 8159
Tamatia à col roux 5511
Tamatia à collier 1349
Tamatia à gorge fauve 5507
Tamatia à gorge rousse 4769
Tamatia à gros bec 6584
Tamatia à moustaches 5509
Tamatia à plastron 6586
Tamatia à semi-collier 5512
Tamatia barré 6647
Tamatia brun 5508
Tamatia chacuru 6645
Tamatia de Cassin 6585
Tamatia de Colombie 1351
Tamatia de Lafresnaye 5510
Tamatia macrodactyle 1350
Tamatia pie 6588
Tamatia rayé 5513
Tamatia soussâtre 5511
Tamatia striolé 6648
Tamatia tacheté 1352
Tamatia tamajac 6646
Tanagra écarlate 7827
Tanagridés 9948
Tangara à bavette jaune 4879
Tangara à bec court 2151
Tangara à bec d'argent 8736
Tangara à boucles d'or 1152

Tétras pâle 10392
Tétras sombre 3062
Tétrogalle de Perse 9877
Tétrogalle du Caucase 9878
Thamnophilidés 9907
Thinocore de Patagonie
   9941
Thinocore d'Orbigny 9940
Thinocoridés 9939
Thoire (Ants) 9440
Thraupidés 9948
Threskiornithidés 9967
Ti Plonjon (Ant) 9685
Tichodrome échelette
   10015
Tic-tic (Ants) 3146
Timalie à ailes rousses
   9389
Timalie à bec rouge 9404
Timalie à calotte dorée
   9388
Timalie à calotte noire
   9399
Timalie à col blanc 9411
Timalie à collier roux 4950
Timalie à cou tacheté 9410
Timalie à couronne rousse
   2211
Timalie à face grise 5455
Timalie à front roux 9407
Timalie à gorge blanche
   4949
Timalie à gorge jaune 5456
Timalie à gorge noire 9398
Timalie à gorge striée 5456
Timalie à poitrine blanche
   9390
Timalie à poitrine tachetée
   7208
Timalie à raies larges 9393
Timalie à tête blanche 4071
Timalie à tête grise 9403
Timalie à tête noire
   4767,8880,9397
Timalie à tête rayée 9397
Timalie à tête rousse 10024
Timalie à ventre roux 3384
Timalie ambigue 9385
Timalie aux yeux d'or 2212
Timalie brune 5459
Timalie chamasa 5458

Timalie coiffée 10024
Timalie d'Abyssinie 7062
Timalie d'Austen 9401
Timalie de Blyth 9406
Timalie de Chapin 4948
Timalie de Deignan 9405
Timalie de Herbert 9391
Timalie de Jerdon 2210
Timalie de Kelley 5457
Timalie de Negros 9400
Timalie de Palawan 9392
Timalie de Tickell 7209
Timalie de Whitehead 9412
Timalie dorée 9387
Timalie maculée 9395
Timalie miniature 5874
Timalie mitrée 9386
Timalie oreillarde 9394
Timalie perlée 9396
Timalie precieuse 9408
Timalie pygmée 9402
Timalie striée 9409
Tinamidés 10027
Tinamiformes 10028
Tinamou tepui 2873
Tinamou à calotte noire
   2857
Tinamou à capuchon 6591
Tinamou à gorge blanche
   10031
Tinamou à grands sourcils
   2878
Tinamou à petit bec 2872
Tinamou à pieds rouge
   2857
Tinamou à pieds rouges
   2866
Tinamou à tête rousse 6590
Tinamou barré 2862
Tinamou boraquira 6600
Tinamou brun 2871
Tinamou cannellé 2864
Tinamou carapé 9792
Tinamou cendré 2863
Tinamou curvirostre 6594
Tinamou de Bartlett 2858
Tinamou de Berlepsch
   2859
Tinamou de Bonaparte
   6589
Tinamou de Boucard 2860

Tinamou de Colombie
   2867
Tinamou de Darwin 6602
Tinamou de Kalinowski
   6595
Tinamou de Kerr 2869
Tinamou de la Magdalena
   2874
Tinamou de Patagonie
   10029
Tinamou de Taczanowski
   6599
Tinamou de Zimmer 2865
Tinamou des Andes 6597
Tinamou des Santa Marta
   2868
Tinamou des tépuis 2873
Tinamou du Chaco 6601
Tinamou élégant 3642
Tinamou forestier 2874
Tinamou isabelle 8892
Tinamou magoua 10032
Tinamou noctivague 2870
Tinamou noir 10033
Tinamou oariana 2876
Tinamou orné 6596
Tinamou perdrix 6598
Tinamou quioula 10030
Tinamou rubigineux 2861
Tinamou sauvageon 6593
Tinamou solitaire 10034
Tinamou soui 2875
Tinamou superbe 3643
Tinamou tacheté 6603
Tinamou tao 10035
Tinamou tataupa 2877
Tinamou varié 2880
Tinamou vermiculé 2879
Tinamous 10027
Tiqueur (Ants) 1347
Tisserin à ailes rouges 521
Tisserin à bec grêle 7985
Tisserin à cape brune 7957
Tisserin à cape jaune 7951
Tisserin à cou noir 7969
Tisserin à dos d'or 7959
Tisserin à front noir 7990
Tisserin à gorge brune
   7995
Tisserin à gorge noire 7946

Toucanet à oreilles d'or 9108
Toucanet à sourcils jaunes 1107
Toucanet de Derby 1105
Toucanet de Gould 9104
Toucanet de Natterer 9106
Toucanet de Reinwardt 9107
Toucanet émeraude 1108
Toucanet koulik 9103
Toui à ailes jaunes 1321
Toui à ailes variées 1328
Toui à bandeau jaune 1244
Toui à dos noir 10114
Toui à front bleu 10112
Toui à front d'or 1326
Toui à front roux 1246
Toui à lunettes 3879
Toui à menton d'or 1324
Toui à queue d'or 10117
Toui à sept couleurs 10110
Toui à tête jaune 3883
Toui aymara 1245
Toui catherine 1247
Toui céleste 3878
Toui de Catherine 1247
Toui de D'Achille 6344
Toui de Deville 1323
Toui de Huet 10113
Toui de Sclater 3882
Toui de Spix 3884
Toui d'Emma 10116
Toui des tépuis 6345
Toui d'Orbigny 1248
Toui du Costa Rica 10111
Toui du Mexique 3880
Toui été 3881
Toui flamboyant 1325
Toui para 1322
Toui tacheté 10116
Toui tirica 1327
Touraco à bec noir 9814
Touraco à gros bec 9810
Touraco à huppe blanche 9807
Touraco à huppe splendide 6098
Touraco à joues blanches 9808

Touraco à queue barrée 2833
Touraco a ventre blanc 2734
Touraco concolore 2733
Touraco de Fischer 9805
Touraco de Hartlaub 9806
Touraco de Lady Ross 6099
Touraco de Livingstone 9809
Touraco de prince Ruspoli 9812
Touraco de Ruspoli 9812
Touraco de Schalow 9813
Touraco doré 9802
Touraco du Ruwenzori 6097
Touraco géant 2732
Touraco gris 2832
Touraco gris d'Abyssinie 2833
Touraco louri 9803
Touraco masqué 2735
Touraco paruline 9804
Touraco vert 9811
Touraco violet 6100
Tourco à gorge blanche 9003
Tourco à gorge marron 8474
Tourco à moustaches 8475
Tourco ceinturé 5267
Tourco huet-huet 8476
Tourco huppé 8799
Tourco rougegorge 9004
Tourco sable 9823
Tournepierre (Ants) 974
Tournepierre à collier 974
Tournepierre noir 975
Tourte (Qué) 3397
Tourte voyageuse 3397
Tourtelette à tête bleue 10387
Tourtelette améthystine 10386
Tourtelette d'Abyssinie 10385
Tourtelette demoiselle 10387

Tourtelette émeraudine 10388
Tourterelle (Ants) 10696,10699
Tourterelle à ailes blanches 10694,9500
Tourterelle à collier 9503
Tourterelle à double collier 9491
Tourterelle à masque blanc 817
Tourterelle à masque de fer 6702
Tourterelle à poitrine rose 9497
Tourterelle à queue carrée 10696
Tourterelle à tête grise 9505
Tourterelle de l'Adamaoua 9496
Tourterelle de Reichenow 9500
Tourterelle de Socorro 10698
Tourterelle des bois 9506
Tourterelle des Galapagos 10697
Tourterelle des palmiers 9504
Tourterelle du Cap 9492
Tourterelle isabelle 9494
Tourterelle maillée 9504
Tourterelle masquée 6702
Tourterelle oreillarde 10695
Tourterelle orientale 9498
Tourterelle Ortolan (Ants) 10695
Tourterelle peinte 9499
Tourterelle pleureuse 9495
Tourterelle rieuse 9494,9501,9502
Tourterelle rose-et-grise 9502
Tourterelle tambourette 10390
Tourterelle tigrine 9493
Tourterelle triste 10699
Tourterelle turque 9494
Tourterelle vineuse 9507

Toutrelle (Ants) 10696
Toxoramphe à tête grise 10119
Toxoramphe à ventre gris 6700
Toxoramphe à ventre jaune 10118
Toxoramphe pygmée 6701
Tragopan de Blyth 10137
Tragopan de Cabot 10138
Tragopan de Hastings 10139
Tragopan de Temminck 10141
Tragopan satyre 10140
Traîne-buisson 8279
Traquet à capuchon 6718
Traquet à front blanc 6255
Traquet à poitrine rousse 6705
Traquet à queue brune 1908
Traquet à queue noire 1906
Traquet à queue rousse 6726
Traquet à tête blanche 6712
Traquet à tête grise 6717
Traquet à ventre roux 9900
Traquet afroalpin 1911
Traquet aile-en-faux 1910
Traquet bistré 1905
Traquet brun 6254
Traquet commandeur 6259
Traquet couronné 9901
Traquet d'Arabie 6715
Traquet d'Arnott 6256
Traquet de Chypre 6706
Traquet de Finsch 6708
Traquet de Heuglin 6709
Traquet de Hume 6703
Traquet de la Réunion 8992
Traquet de roche à queue noire 1906
Traquet de roche à queue rousse 1904
Traquet de roche à ventre blanc 9902
Traquet de roche à ventre roux 9900
Traquet de Rüppell 6258
Traquet de Somalie 6721

Traquet demi-roux 9902
Traquet des Héréros 6342
Traquet deuil 6714
Traquet du Cap 6723
Traquet du Congo 6260
Traquet du désert 6707
Traquet du Karoo 1907
Traquet familier 1904
Traquet fourmilier 6257
Traquet isabelle 6711
Traquet leucomèle 6724
Traquet montagnard 6719
Traquet motteux 6720
Traquet noir 6713
Traquet noir à front blanc 6255
Traquet noir d'Abyssinie 6258
Traquet noir d'Arnott 6256
Traquet oreillard 6710
Traquet pâtre 8993
Traquet pie 6724
Traquet pie d'Orient 6722
Traquet rieur 6713
Traquet sombre 1903
Traquet stapazin 6710
Traquet tarier 8991
Traquet tractrac 1912
Traquet traîne-charrue 6720
Traquet variable 6722
Traquet-fourmilier brun 6254
Traquet-fourmilier brun du nord 6254
Traquet-fourmilier brun du sud 6257
Traquet-fourmilier du Congo 6260
Traquet-fourmilier noir 6259
Travailleur à bec rouge 8690
Travailleur à queue courte 1265
Travailleur à tête rouge 8689
Travailleur cardinal 8688
Trembleur brun 2240
Trembleur gris 2239

Trichodere cockerelli 10173
Trochilidés 10208
Troglodyte 10215,10228
Troglodyte à ailes blanches 4581
Troglodyte à bec court 2360
Troglodyte à bec fin 4762
Troglodyte à calotte noire 9999
Troglodyte à face pâle 9992
Troglodyte à favoris 9997
Troglodyte à gorge brune 10217
Troglodyte à gorge noire 9982
Troglodyte à long bec 9993
Troglodyte à miroir 3294
Troglodyte à moustaches 9988
Troglodyte à nuque rousse 1630
Troglodyte à poitrine blanche 4582
Troglodyte à poitrine grise 4580
Troglodyte à poitrine striée 10008
Troglodyte à poitrine tachetée 9995
Troglodyte à sourcils blancs 10007
Troglodyte à sourcils roux 10223
Troglodyte à tête blanche 1619
Troglodyte à tête noire 9999
Troglodyte à ventre blanc 10472
Troglodyte à ventre noir 9986
Troglodyte arada 3008
Troglodyte austral 10221
Troglodyte balafre 9991
Troglodyte bambla 5852
Troglodyte barré 10000
Troglodyte bicolore 1624
Troglodyte bridé 10007

Tyranneau à queue courte
6214
Tyranneau à queue rousse
8754
Tyranneau à queue sombre
8752
Tyranneau à queue-en-
aiguille 2901
Tyranneau à sourcils blancs
5585
Tyranneau à sourcils roux
7603
Tyranneau à tête brune
6806
Tyranneau à tête cendrée
7574
Tyranneau à tête grise 7577
Tyranneau à tête noire
7578
Tyranneau à tête rousse
8351
Tyranneau à toupet 9187
Tyranneau à ventre blanc
6211,9185
Tyranneau à ventre jaune
6808
Tyranneau barbu 8101
Tyranneau bariolé 8037
Tyranneau bolivien 10701
Tyranneau bridé 8102
Tyranneau bronzé 8350
Tyranneau brun 2392
Tyranneau casqué 5349
Tyranneau chevelu 5350
Tyranneau coiffé 5351
Tyranneau d'André 9724
Tyranneau de Panama
7592
Tyranneau de Berlioz 9183
Tyranneau de Bolivie
10701
Tyranneau de Burmeister
7573
Tyranneau de Cecilia 7587
Tyranneau de Chapman
7588
Tyranneau de Hellmayr
5581
Tyranneau de Krone 7594
Tyranneau de Lanyon 7595

Tyranneau de l'Araguaya
6192
Tyranneau de Lawrence
6729
Tyranneau de Mynas
Gerais 7602
Tyranneau de Reiser 7580
Tyranneau de Salvadori
4864
Tyranneau de Sao Paulo
7600
Tyranneau de Sclater 7581
Tyranneau de Zeledon
7585
Tyranneau d'Équateur 7593
Tyranneau des campos
8102,9571
Tyranneau des Cocos 6523
Tyranneau des palétuviers
9567
Tyranneau des rivières
9184
Tyranneau des torrents
9182
Tyranneau distingué 7590
Tyranneau d'Ihering 7589
Tyranneau d'Olrog 10396
Tyranneau d'Oustalet 7599
Tyranneau du Paraguay
8101
Tyranneau du Venezuela
7605
Tyranneau eulophe 5348
Tyranneau fascié 7575
Tyranneau flavéole 1709
Tyranneau givré 4865
Tyranneau gobemoucheron
10706
Tyranneau gris-et-jaune
7575
Tyranneau gros-bec 10402
Tyranneau imberbe 1601
Tyranneau inézia 4865
Tyranneau jaune-olive
10104
Tyranneau jaune-vert 7592
Tyranneau marbré 7597
Tyranneau masqué 7591
Tyranneau minute 6807
Tyranneau modeste 9569

Tyranneau montagnard
7579
Tyranneau nain 7576
Tyranneau noirâtre 9186
Tyranneau noir-et-blanc
8035
Tyranneau olivâtre 8887
Tyranneau ombré 9570
Tyranneau omnicolore
9684
Tyranneau passegris 1602
Tyranneau pattu 7573
Tyranneau plombé 7579
Tyranneau poloicéphale
10103
Tyranneau roitelet 10398
Tyranneau soufré 5583
Tyranneau souris 7325
Tyranneau suiriri 9572
Tyranneau sylvain 7604
Tyranneau terne 4864
Tyranneau trompeur 10705
Tyranneau varié 7601
Tyranneau ventru 7606
Tyranneau verdâtre 7607
Tyranneau verdin 7584
Tyrannidés 10395
Tytonidés 10431
Upucerthie à bec droit
10438
Upucerthie à gorge blanche
10432
Upucerthie de Bolivie
10436
Upucerthie de Hartert
10436
Upucerthie de Jelski 10437
Upucerthie des buissons
10435
Upucerthie des rochers
10433
Upucerthie du Chaco
10434
Upucerthie fauve 10440
Upucerthie striée 10439
Upupidés 10445
Urotangara de Stolzmann
10478
Urubu à tête jaune 1808
Urubu à tête rouge 1807
Urubu noir 2661

Urubus 1810
Vacher 5964
Vacher à ailes baies 5965
Vacher à tête brune 5964
Vacher bronzé 5962
Vacher brun 5963
Vacher criard 5967
Vacher géant 9001
Vacher luisant 5966
Vachers 4799
Valet de Caïman (Ants) 1429
Vanga à queue rousse 1492
Vanga à tête noire 10389
Vanga de Lafresnaye 10632
Vanga de Pollen 10631
Vanga de Van Dam 10630
Vanga écorcheur 10505
Vanga tylas 10389
Vanga-sitelle malgache 4779
Vangidés 10506
Vanneau à ailes blanches 10486
Vanneau à ailes noires 10495
Vanneau à éperons 10500
Vanneau à poitrine châtaine 10501
Vanneau à queue blanche 10490
Vanneau à tête blanche 10480
Vanneau à tête grise 10484
Vanneau à tête noire 10502
Vanneau armé 10481
Vanneau caronculé 10501
Vanneau coiffé 10502
Vanneau couronné 10485
Vanneau d'Abyssinie 10494
Vanneau de Cayenne 10482
Vanneau de Malabar 10493
Vanneau demideuil 10491
Vanneau des Andes 10497
Vanneau du Sénégal 10498
Vanneau éperonné 10500
Vanneau hirondelle 10492
Vanneau huppé 10504

Vanneau indien 10489
Vanneau pie 10487
Vanneau sociable 10488
Vanneau soldat 10496
Vanneau terne 10491
Vanneau téro 10483
Vanneau tricolore 10503
Vanneau-pluvier social 10488
Vautour (Ants) 1807
Vautour à tête blanche 10191
Vautour africain 4401
Vautour charognard 6359
Vautour chassefiente 4403
Vautour chaugoun 4402
Vautour d'Arabie 10108
Vautour de l'Himalaya 4405
Vautour de Rüppell 4407
Vautour fauve 2281,4404
Vautour indien 4406
Vautour moine 188
Vautour oricou 10107
Vautour palmiste 4399
Vautour percnoptère 6487
Vautour royal 8956
Venturon africain 9146
Venturon alpin 9145
Venturon corse 9148
Venturon montagnard 9145
Verdier 1722
Verdier commun 1722
Verdier de Chine 1741
Verdier de l'Himalaya 1743
Verdier d'Europe 1722
Verdier d'Oustalet 1716
Verdier du Vietnam 1735
Verdin à ailes jaunes 2138
Verdin à front bleu 2142
Verdin à front d'or 2135
Verdin à tête jaune 2136
Verdin barbe-bleue 2137
Verdin de Hardwicke 2139
Verdin de Palawan 2140
Verdin de Sonnerat 2141
Veuve à collier d'or 10544
Veuve à dos d'or 3723
Veuve à épaulettes orangées 3715

Veuve à queue de paille 10535
Veuve de Chapin 10543
Veuve de Fischer 10535
Veuve de paradis 10545
Veuve dominicaine 10540
Veuve du Togo 10549
Veuve d'Uelle 10538
Veuve métallique 10537
Veuve nigérienne 10538
Veuve noire 3713
Veuve noire en feu 3713
Veuve royale 10548
Viduidés 10552
Viréo à ailes jaunes 10565
Viréo à bec fort 10568
Viréo à calotte brune 10578
Viréo à gorge jaune 10569
Viréo à moustaches 10558
Viréo à oeil rouge 10584
Viréo à tête bleue 10591
Viréo à tête noire 10560
Viréo ardoisé 10563
Viréo aux yeux blancs 10573
Viréo aux yeux rouges 10584
Viréo de Bell 10562
Viréo de Cassin 10566
Viréo de Cozumel 10561
Viréo de Cuba 10574
Viréo de Hutton 10575
Viréo de la Jamaïque 10581
Viréo de Noronha 10572
Viréo de Philadelphie 10588
Viréo de Porto Rico 10577
Viréo de San Andrés 10564
Viréo de Swainson 10592
Viréo des mangroves 10586
Viréo d'Hispaniola 10582
Viréo doré 10576
Viréo d'Osburn 10585
Viréo du Yucatan 10579
Viréo gris 10593
Viréo jaune-verdâtre 10570
Viréo mélodieux 10571
Viréo nain 10583
Viréo plombé 10589

Deutsches Register

Aaskrähe 2686
Abbotts Raupenfänger
　2609
Abbott-Star 2274
Abbott-Tölpel 7011
Abdim-Storch 2227
Abeillé-Kernbeißer 4600
Abeillé-Kolibri 1
Abendammer 8132
Abendfalke 3819
Abendkernbeißer 4601
Aberdare-Cistensänger
　2304
Abessinien-Nektarvogel
　6388
Abessinische Felsentaube
　2475
Abessinischer Brillenvogel
　10757
Abessinisches Wald-
　täubchen 10385
Abt-Tangare 9949
Aburri 5
Acapulco-Blaurabe 2934
Adamaua-Turteltaube 9496
Adams-Schneefink 6023
Adams-Schneesperling
　6023
Adanson-Wachtel 2767
Adelaide-Rosella 7883
Adelaide-Sittich 7883
Adela-Kolibri 6770
Adelie-Pinguin 8637
Adlerbussard 1411
Adlerfregattvogel 3940
Adlerschnabel 3775
Affenadler 7832
Affenente 9472
Affengans 9472
Afghanen-Schneefink 6029
Afghanen-Sperling 6029
Afrika-Habicht 75
Afrika-Laufvögel 9540
Afrikanische Bartvögel
　5421
Afrikanische Bekassine
　4028
Afrikanische Binsenralle
　8011
Afrikanische Grasmücken
　2355

Afrikanische Graumeise
　7101
Afrikanische Lachtaube
　9502
Afrikanische Ohrenlerche
　3573
Afrikanische Pitta 7842
Afrikanische Uferschwalbe
　8900
Afrikanische Zwerggans
　6530
Afrikanische Zwergohreule
　6910
Afrikanische Zwergwachtel
　2767
Afrikanischer Bandwürger
　5046
Afrikanischer Baumfalke
　3791
Afrikanischer Blaß-
　schnäpper 1285
Afrikanischer
　Halsbandsittich 8374
Afrikanischer Hecken-
　sänger 1915
Afrikanischer Kuckuck
　2887
Afrikanischer Löffler 7878
Afrikanischer Mausspecht
　8973
Afrikanischer Nimmersatt
　6122
Afrikanischer Rohrweihe
　2296
Afrikanischer Schlangen-
　halsvogel 607
Afrikanischer Sperber 75
Afrikanischer Strauß 9538
Afrikanisches Sultanshuhn
　8155
Afrika-Ruderente 6933
Afrika-Scherenschnabel
　8922
Afrika-Sultanshuhn 8155
Agapane 4637
Aguja 4209
Ägyptischer Ziegenmelker
　1648
Ahanta-Frankolin 3895
Ährenpfau 7181
Ährenschopftyrann 5351

Ajaxflöter 2256
Akaziendrossel 10323
Akaziendrossling 10264
Akaziengrasmücke 9599
Akaziensänger 7572
Akazienzaunkönig 10000
Akekee 5400
Akepa 5401
Akiola 4522
Akiopalaau 4527
Alario 9134
Alariogirlitz 9134
Alauwahio 7064
Albatros 3272
Albatrosse 3280
Albertis-Taube 4384
Albert-Paradiesvogel 8402
Aldabra-Buschsänger 6502
Aldabra-Drongo 3217
Alekto-Weber 1329
Aleuten-Alk 8556
Aleuten-Seeschwalbe 9439
Alexander-Bülbül 587
Alexander-Schnäpper 1210
Alexander-Segler 831
Alexander-Sittich 8369
Alexandra-Sittich 8103
Alice-Kolibri 2157
Alken 325
Alkenvögel 2017,325
Allen-Kolibri 9099
Allfarblori 10177
Almenschmätzer 1911
Almora-Specht 7740
Alpenbraunelle 8273
Alpendohle 8651
Alpengimpel 1775
Alpenkrähe 8652
Alpenmeise 7113
Alpenrotschwanz 7520
Alpenschneehuhn 4967
Alpensegler 9704
Alpensittich 2960
Alpenstrandläufer 1496
Altai-Königshuhn 9876
Altweltgeier 188
Amakihi 4526
Amami-Schnepfe 9038
Amani-Nektarvogel 682
Amarant 4960

Angola-Pitta 7842
Angola-Rötel 6012
Angola-Schnäpper 1204
Angola-Schwalbe 4656
Angola-Schwarzbäckchen
 3620
Angola-Turteltaube 9495
Angola-Weber 7988
Anhinga 606
Anianiau 4523
Anjouan-Nektarvogel 6378
Anna-Kleidervogel 2298
Annas Kolibri 1538
Annobon-Brillenvogel
 10782
Ansell-Kuckuck 1844
Ansorge-Bülbül 585
Antarktik-Skua 1804
Antarktik-Sturmvogel 9888
Antarktische Raubmöwe
 1804
Antillen-Drossel 2221
Antillen-Elaenie 3417
Antillen-Grackel 8696
Antillen-Krähe 2701
Antillen-Nachtschwalbe
 2184
Antillen-Schwalbe 9694
Antillen-Segler 1984
Antillen-Taube 2520
Antillen-Trupial 4809
Antillen-Tyrann 6144
Antillen-Waldsänger 3121
Antinori-Würger 5059
Antipoden-Seeschwalbe
 9469
Anumbi 738
Apalopteron 774
Aplomado-Falke 3796
Apolinar-Zaunkönig 2356
Appert-Bülbül 7549
Apurimac-Schlüpfer 9640
Araberspecht 3096
Arabertrappe 970
Arabien-Steinsperling 1785
Arabischer Seidenschwanz
 4774
Arabisches Wüstenhuhn
 490
Ara-Kakadu 8239
Arakanga 873

Araponga 8251
Arara-Kakadu 8239
Arara-Sittich 8889
Ara-Sittich 8889
Arassari 8459
Araucaner-Taube 2476
Araukaner Kauz 4261
Araukarien-Schlüpfer 5155
Arauna 862
Archbold-Laubenvogel 945
Archbold-Nachtschwalbe
 3749
Archbold-Sperber 62
Arfak-Beerenpicker 5693
Arfak-Brillenvogel 10779
Arfak-Gelbwangenvogel
 6753
Arfak-Honigfresser 6308
Arfak-Lori 6763
Arfak-Nonne 5325
Arfak-Paradiesvogel 7069
Arfak-Sericornis 9115
Arfak-Strahlenparadies-
 vogel 7069
Argala 5170
Argentinische Ruderente
 6934
Argusfasan 976
Argusnachtschwalbe 3750
Arielfregattvogel 3941
Arielschwalbe 4657
Arikanischer Paradies-
 schnäpper 9865
Arizona-Waldsänger 3135
Arkansas-Tyrann 10409
Arkansas-Zeisig 1739
Armenien-Möwe 5070
Arnott-Schmätzer 6256
Aru-Honigfresser 5749
Aru-Liest 3027
Aru-Schnäpper 5863
Aschbrustämmerling 7535
Aschbrusthonigfresser
 5762
Aschenmeise 7089
Aschgrauer Wellenläufer
 6654
Aschkehlkrummschnabel
 6728
Aschkopfameisenfänger
 1895

Aschkopfammer 3481
Aschkopfhabicht 66
Aschkopfvireo 4749
Asiatische Bartvögel 5619
Asiatischer Trauerdrongo
 3231
Asir-Meisensänger 9590
Atacama-Kolibri 8876
Athi-Lerche 1479
Atita-aucher 8023
Atiu-Salangane 213
Atjeh-Maustimalie 5489
Atlantis-Ralle 1054
Atlas-Grasmücke 9596
Atlas-Grünspecht 7741
Atlas-Kuckuck 1850
Atlas-Liest 1752
Atlas-Pipra 2133
Atollstar 798
Auckland-Ente 525
Auckland-Ralle 5217
Auckland-Säger 5790
Auckland-Schnepfe 1827,
 2422
Audubon-Sturmtaucher
 8573
Auerhuhn 9875
Augenbaumsteiger 10664
Augenbrauenameisen-
 schnäpper 6262
Augenbrauenbusch-
 drossling 4841
Augenbrauenente 572
Augenbrauenhäherling
 4080
Augenbrauenhemispingus
 4545
Augenbrauenmahali 7927
Augenbrauensperling 7289
Augenbrauenweber 7986
Augennachtschwalbe 6638
Augenralle 1635
Augenringfliegenstecher
 10706
Augenringspateltyrann
 4569
Augenringwaldsänger 6748
Augenstreifgoldhähnchen
 8769
Augenstreifscheindrossel
 3316

Bergpapagei 249
Bergpapapageiamadine
  3601
Bergpeltops 7211
Bergpieper 732
Bergpirol 6796
Bergprinie 8188
Bergpurpurfink 1757
Bergraupenfänger 2640
Bergregenpfeifer 2035
Bergrötel 6014
Bergrubinkehlchen 5417
Bergschilffink 5313
Bergschlangenweihe 9307
Bergschmätzer 6719
Bergschneidervogel 6830
Bergsegler 220
Bergsericornis 9123
Bergshwalm 178
Bergsittich 8104
Bergspinnenjäger 887
Bergspottdrossel 6766
Bergstachelschwanz 5569
Bergstar 6735
Bergstelze 6039
Bergstelzenkrähe 7678
Bergstrandläufer 1503
Bergtaube 4203
Bergtinamu 6589
Bergtorotoro 9626
Bergtrogon 773
Bergtyrann 274
Berguhu 1331
Bergwachtel 6765,7474
Bergwaldbülbül 601
Bergwaldflöter 8554
Bergwaldhonigfresser 5755
Bergwaldschnäpper 7276
Bergzaunkönig 9980
Bergzipzalp 7655
Bering-Strandläufer 1508
Berlepsch-Amazilie 429
Berlepsch-Ameisenpitta
  4731
Berlepsch-Elfe 94
Berlepsch-Schlüpfer 1019
Berlepsch-Tachuri 9185
Bermuda-Sturmvogel 8428
Bernier-Ente 528
Bernstein-Kuckuck 1847
Bertrand-Weber 7940

Beryll-Amazilie 400
Besra-Sperber 80
Betschuanen-Lerche 1941
Beuteljahoo 7795
Beutelmeise 8780
Beutelsäbler 8126
Biak-Brillenvogel 10805
Biak-Liet 9790
Bienenelfe 5773
Bienenfresser 5795,5798
Bigua-Scharbe 7377,7401
Bindenameisendrossel
  2012
Bindenameisenwürger
  1192
Bindenbartvogel 8046
Bindenbauchspecht 10516
Bindenbauchzaunkönig
  9986
Bindenbaumsteiger 3080
Bindenbuschsänger 1293
Bindenerdracke 1270
Bindenfalke 3820
Bindenfischeule 9046
Bindenflughuhn 8412
Bindenfregattvogel 3943
Bindengimpel 1778
Bindengrauschwanz 10618
Bindengrundkuckuck 6473
Bindengrünling 1741
Bindengrüntaube 10146
Bindenhalskauz 9528
Bindenkauz 8584
Bindenkreuzschnabel 5392
Bindenlärmvogel 2833
Bindenlaubwürger 10598
Bindenlaufhühnchen 10379
Bindennektarvogel 6400
Bindenralle 4054
Bindenraupenfänger 2653
Bindenreiher 10708
Bindenrennvogel 8818
Bindenruderente 6934
Bindensänger 1473
Bindenschmuckvogel 7805
Bindenschnabeltukan 9105
Bindenschwanzeremit 9961
Bindenschwanzguan 7213
Bindenschwanznacht-
  schwalbe 6641
Bindenschwanztaube 5477

Bindenseeadler 4447
Bindenstrandläufer 1501
Bindentaube 2487
Bindentaucher 8024
Bindentinamu 2862
Bindentöpfer 3969
Bindentrupial 4821
Bindenuferwipper 2246
Bindenuhu 1342
Bindenwachtel 6765,7474
Bindenwaldsänger 1163
Bindenwollrücken 9916
Bindenzaunkönig 1623
Bindenzaunkönigstimalie
  9279
Binsenastrild 6449
Binsenhühner 4511
Binsenralle 8011
Binsenrallen 4511
Binsenrohrsänger 129
Birkenmeise 7094
Birkentyrann 3505
Birkenzeisig 1727,1728,
  1729
Birkhuhn 9874
Bischof 7443
Bischofstangare 9953
Bischofstaube 4208
Bischofsweber 3719
Bismarck-Brillenvogel
  10784
Bismarck-Fischer 324
Bismarck-Kauz 6563
Bismarck-Liest 10056
Bismarck-Mistelfresser
  3189
Bismarck-Papagei 4146
Bismarck-Schwalbenstar
  998
Bismarck-Weihe 4576
Blandford-Grasmücke 9599
Blanford-Bülbül 8592
Blanford-Gimpel 1773
Blanford-Lerche 1480
Blanford-Schneefink 6024
Blanford-Sperling 6024
Blaßbauchbülbül 374
Blaßbauchcistensänger
  2337
Blaßbauchdrossel 10336
Blaßbürzelsegler 1982

Blutkotinga 4415,7499
Blutohrpapagei 7759
Blutohrrotschwanzsittich
  8670
Blutohrsittich 8670
Blutpirol 6802
Blutrückenspecht 10518
Blutrumpfsittich 8315
Blutschnabelmöwe 5115
Blutschnabelweber 8690
Blutschwingenfruchttaube
  8507
Blutseidenschwanz 1251
Blutspecht 3110
Blutstirnkardinal 7057
Bluttangare 7820
Bobolink 3290
Bocage-Nektarvogel 6370
Bocage-Rötel 9191
Bocage-Weber 7988
Bocage-Würger 9826
Bogota-Ralle 8718
Böhm-Ralle 8965
Böhms Zwergralle 8965
Böhm-Schnäpper 6058
Böhm-Spint 5799
Bojers Weber 7942
Bokmakiri 9834
Bolivianischer Helmhokko
  7179
Bolivia-Rotschwanzsittich
  8667
Bolivien-Rotschwanzsittich
  8667
Bolles Lorbeertaube 2480
Bonaparte-Möwe 5107
Bonaparte-Taube 5183
Bonaparte-Trupial 4825
Bonaparte-Waldsänger
  1180
Bonin-Drossel 10754
Bonin-Fink 2071
Bonin-Gimpel 2071
Bonin-Honigfresser 774
Bonin-Sturmvogel 8437
Bonin-Taube 2526
Bootschwanzgrackel 8693
Bootschwanznacht-
  schwalbe 1656
Borabora-Liest 10078
Boran-Cistensänger 2311

Borbabaumsteiger 3081
Borneo-Bartvogel 5608
Borneo-Bronzemännchen
  5299
Borneo-Buschsänger 1287
Borneo-Buschwachtel 935
Borneo-Dickkopf 6946
Borneo-Mistelfresser 3198
Borneo-Stutzschwanz
  10475
Borneo-Wolltimalie 8535
Borstenbartvogel 4373
Borstenbrachvogel 6618
Borstenhäherling 4094
Borstenkopf 8388
Borstenmantelbülbül 4786
Borstenrabe 2713
Borstentyrann 6173
Bothain-Schnäpper 3831
Botteri-Ammer 281
Bougainville-Honigfresser
  9513
Bougainville-Krähe 2704
Boulton-Buschwachtel 941
Boultons Laubsänger 7634
Bourke-Sittich 6489
Bouvier-Nektarvogel 6371
Brachpieper 698
Brachschwalbe 4242
Brachschwalbenartige
  Vögel 4243
Bradfield-Toko 10042
Brahmakauz 1052
Brahminenmilan 4452
Brahminenweihe 4452
Brandente 9718
Brandgans 9718
Brand-Seeschwalbe 9462
Brandtaube 3997
Brandweber 3724
Brasilia-Tapaculo 9067
Brasil-Kauz 9524
Brauenbreitschnabel 8753
Brauenbrilliant 4504
Brauenbülbül 8607
Brauenfliegenstecher 7325
Brauengirlitz 9155
Brauengrasschlüpfer 500
Brauenmeise 7129
Brauenrohrsänger 114
Brauensäbler 8128

Brauenschama 2593
Brauenschlüpfer 9660
Brauenschnäpper 3854
Brauenschopftangare 4622
Brauenspecht 7742
Brauenstelze 6049
Brauentesia 9871
Brauenwaldsänger 10527
Brauenwaldschnäpper 3932
Brauenzaunkönig 1624
Braunachselstärling 5449
Braunaugenvireo 10562
Braunbartbuschammer
  1063
Braunbartvogel 1535
Braunbauchamazilie 403
Braunbauchameisen-
  schlüpfer 6283
Braunbauchbaumrutscher
  2375
Braunbauchbergtangare
  3057
Braunbauchbrilliant 4502
Braunbauchbülül 4871
Braunbauchbuschdrossling
  4842
Braunbauchdickichtvogel
  1083
Braunbaucherdhacker
  10440
Braunbaucheremit 7362
Braunbauchfächerschwanz
  8836
Braunbauchfaulvogel 4460
Braunbauchflughuhn 8410
Braunbauchfruchttaube
  3349
Braunbauchkielralle 386
Braunbauchkleinralle 386
Braunbauchlaubenvogel
  2088
Braunbauchliest 10063
Braunbauchorganist 2130,
  3704
Braunbauchsatyrhuhn
  10138
Braunbauchsylvietta 9612
Braunbauchtangare 9943
Braunbauchtragopan 10138
Braunbauchtyrann 2575

Braunbauchwaldhuscher 2817

Braunbindenfaulvogel 6585

Braunbrillenvogel 9270

Braunbrustakalat 9198

Braunbrustalethe 363

Braunbrustbartvogel 5427

Braunbrustblauschnäpper 3005

Braunbrustbülbül 8627

Braunbrustgudilang 2470

Braunbruströtel 9198

Braunbrustschilffink 5293

Braunbrustschlangenadler 2278

Braunbrustschmätzer 6705

Braunbrustschwalbe 7326

Braunbrustsittich 8665

Braunbrustspateltyrann 4570

Braunbrusttalegalla 191

Braunbrusttyrann 3506

Braunbrustwaldsänger 3126

Braunbrustweißstirnchen 780

Braunbrustwürgertangare 5032

Braunbrustzaunkönig 3010

Braunbürzelamarant 4955

Braunbürzelammer 3448

Braunbürzelbaumspäher 1115

Braunbürzeldornschnabel 24

Braunbürzelpfäffchen 9378

Braunbürzelspecht 5654

Braunbürzeltapaculo 9062

Braunbürzelweber 7986

Brauncachalote 8348

Braundrossling 10270

Braundrosssel 10320

Braune Maustimalie 4842

Braune Spottdrossel 10130

Braune Wasseramsel 2268

Braunelaenie 3426

Brauner Bartheckensänger 1922

Brauner Cistensänger 2329

Brauner Faulvogel 6578

Brauner Fliegenschnäpper 6062

Brauner Ibis 7923

Brauner Ohrfasan 2845

Brauner Pfaufasan 8095

Brauner Reisfink 7000

Brauner Schlangenadler 2279

Brauner Sichler 7923

Brauner Tropfenastrild 2386

Braunes Hähnchen 8474

Braunflankenrötel 2742

Braunflankentangare 9945

Braunfleckenwaldwächter 4757

Braunflügelbrachschwalbe 4242

Braunflügelguan 6820

Braunflügelgurial 7185

Braunflügelige Brach-schwalbe 4242

Braunflügelkuhstärling 5965

Braunflügelmausvogel 2455

Braunflügelspechtdrossel 3064

Braungallito 9823

Braunhäher 8355

Braunhäherling 4086

Braunhalsnachtschwalbe 1696

Braunhalspecht 3262

Braunhalssäbelschnäbler 8760

Braunhaubenstärling 8302

Braunhonigfresser 5245

Braunkappenerdhacker 10436

Braunkappenerdtimalie 7205

Braunkappenfaulvogel 1350

Braunkappenhäherling 4075

Braunkappenmeise 7104

Braunkappenralle 892

Braunkappenschnäpper 3597

Braunkappenvireo 10578

Braunkappenwaldsänger 6176

Braunkappenweber 7957

Braunkehl Huëthuët 8474

Braunkehlbartvogel 5596

Braunkehlbaumläufer 1929

Braunkehlbaumspäher 1116

Braunkehlchen 8991

Braunkehleremit 7344

Braunkehlfaulvogel 5507

Braunkehlglanzvogel 1268

Braunkehlgoldweber 7995

Braunkehlhonigfresser 6

Braunkehlkardinal 7061

Braunkehlkeilschwanzhuhn 9881

Braunkehllappenschnäpper 7912

Braunkehlnektarvogel 678

Braunkehlpfäffchen 9377

Braunkehlreiher 3410

Braunkehlschlüpfer 9636

Braunkehlspecht 3098

Braunkehlstar 8962

Braunkehluferschwalbe 8901

Braunkehlweber 7995

Braunkehlwendehals 4926

Braunkinnfruchttaube 8528

Braunkinnsittich 1322

Braunkopfalcippe 331

Braunkopfammer 3450

Braunkopfbartvogel 5618

Braunkopfbaumläufer 8782

Braunkopfbreitrachen 3758

Braunkopfbündelnister 9976

Braunkopfbuschammer 1059

Braunkopfdrossel 10305

Braunkopfdrossling 10260

Braunkopffeinsänger 743

Braunkopffliegenstecher 5164

Braunkopfhonigschmecker 5766

Braunköpfiger Raben-kakadu 1551

Braunkopfinezia 4865

Braunkopfkakadu 1551

Braunkopfkleiber 9246
Braunkopfkolibri 2988
Braunkopfkrähe 2696
Braunkopfkuhstärling 5964
Braunkopflackvogel 3045
Braunkopfliest 4431
Braunkopfmöwe 5075
Braunkopfmusendrossel
1824
Braunkopfnicator 6539
Braunkopfpapagei 8058
Braunkopfpapageimeise
7039
Braunkopfpapageischnabel
7039
Braunkopfpitta 7863
Braunkopfprinie 8197
Braunkopfrhabdornis 8782
Braunkopfsittich 2964,924
Braunkopfspint 5806
Braunkopfstärling 257
Braunkopftangare 2150
Braunkopftrupial 4803
Braunkopfvireo 4740
Braunkopfyuhina 10680
Braunkopfzuser 494
Braunkopfzwergfisher
4884
Braunkrausschwanz 5957
Braunkuhstärling 5965
Braunlätzchen 3672
Braunlaubsänger 7625
Braunliest 4439,4442
Bräuling 384
Braunlori 1996
Braunmaina 101
Braunmantelämmerling
7530
Braunmantelausternfischer
4427
Braunmantelbeerenfresser
1783
Braunmantelliest 9785
Braunmantelscheren-
schnabel 8922
Braunmantelstirnvogel
4395,8305
Braunmeisensänger 9600
Braunmotmot 3666
Braunnackenammer 10711
Braunnackenfrankolin 8404

Braunnackenrabe 2714
Braunohrammer 3464
Braunohrarassari 8463
Braunohrbülbül 4575
Braunohrbunttangare 2113
Braunohrdickkopf 6954
Braunohrsericornis 9126
Braunohrsittich 8669
Braunohrspecht 1590
Braunohrzwergspecht 7724
Braunpelikan 7198
Braunpfäffchen 10022
Braunralle 8915
Braunrückenameisenpitta
4313
Braunrückenameisenvogel
6237
Braunrückendickkopf 6955
Braunrückenelsterchen
5290
Braunrückengerygone 4235
Braunrückengoldsperling
7143,7150
Braunrückengrundammer
7780
Braunrückenklarino 6110
Braunrückenkurzflügel
1277
Braunrückenlalage 4977
Braunrückenleierschwanz
5785
Braunrückenmeise 7122
Braunrückenmistelfresser
3188
Braunrückennektarvogel
6382
Braunrückenpapagei 10114
Braunrückenpieper 713
Braunrückenrötel 2755
Braunrückenspälteltyrann
4562
Braunrückenspecht 3164
Braunrückenspottdrossel
5895
Braunrückentachuri 9474
Braunrückentukan 8728
Braunrückenwaldhuscher
2815
Braunrückenzwergspecht
7714,7719

Braunscheitelameisenvogel
6251
Braunscheitelbrillenvogel
10761
Braunscheitelfinkenlerche
3576
Braunscheitelhonigesser
7540
Braunscheitellerche 3576
Braunscheitelmotmot 5972
Braunscheitelorganist 3683
Braunscheitelpapagei 8359
Braunscheitelschwalbe
4665
Braunscheitelspecht 3108
Braunscheiteltyrann 9218
Braunscheitelvireo 10577
Braunscheitelwürger 5024
Braunschenkelweihe 4477
Braunschmätzer 1905
Braunschnäpper 6062
Braunschopfameisenvogel
8792
Braunschopftyrann 6153
Braunschulterpapagei
10116
Braunschulterstärling 254
Braunschwanz 1908
Braunschwanzalcippe 327
Braunschwanzamazilie 426
Braunschwanzamazone 449
Braunschwanzfruchttaube
3360
Braunschwanzschnäpper
5862
Braunschwanzseeadler
4796
Braunschwanzsichelhopf
3548
Braunschwanzsittich 8674
Braunschwanztachuri 5581
Braunschwanztephradornis
9836
Braunschwingendrongo
3227
Braunschwingenmusketier
2415
Braunschwingenspecht
7705
Braunsegler 840
Braunsensenschnabel 1617

Brustbandtapaculo 5267
Brustbandtyran 2737
Brustbandvireo 10598
Brustfleckenpapagei-
  schnabel 7031
Brustfleckentangare 6520
Brustfleckentimalie 6350
Brustfleckflachschnabel
  5440
Brustschildsteinschmätzer
  6723
Bubu 7309
Buchendickkopf 6958
Buchentyrann 3513
Buchfink 3949
Buchschwanztyrann 4557
Buchstabentaube 4158
Buchtschwanzelster 9835
Budongo-Laubsänger 7615
Büffelkopfente 1354
Büffelweber 1330
Büffelwürger 5038
Buffon-Kolibri 2004
Bülbülamsel 10295
Bülbüldrossel 10295
Bülbülhonigfresser 8631
Bülbüls 8587
Bülbültimalie 944
Bülbülvanga 10389
Bülbülwürger 6538
Bullabulla-Sittich 7885
Buller-Albatros 3266
Bulwer-Fasan 5358
Bulwer-Sturmschwalbe
  1366
Bulwer-Sturmvogel 1366
Buntastrild 8686
Buntbartvogel 3637
Buntbauchnektarvogel 671
Buntbaumsteiger 5144
Buntbrustzwergpapagei
  8379
Buntbürzelameisenfänger
  3213
Buntdrossel 10723
Bunter Waldsänger 6184
Buntfalke 3815
Buntfasan 7435
Buntflöter 8553
Buntflügelameisen-
  schlüpfer 6292

Buntflügelhäherling 4120
Buntfüssige Sturmschwalbe
  6650
Buntfußsturmschwalbe
  6650
Bunthöschen 4461
Buntkehlsaltator 8945
Buntkopffalcippe 342
Buntkopfbartvogel 5609
Buntkopffeldhüpfer 7678
Buntkopfhonigfresser 25
Buntkopfpapageiamadine
  3601
Buntkopfspecht 5668
Buntlaufhühnchen 10382
Buntlori 8385
Buntmeise 7134
Buntpipra 5442
Buntscharbe 7390
Buntschnabelhonigfresser
  5225
Buntschnabelkrähe 2726
Buntschnabelkuckuck 7308
Buntschwanzdegenflügel
  1612
Buntschwanzschmuckvogel
  7810
Buntsittich 7889
Buntspecht 3105
Buntstorch 6123
Bunttukan 8726
Buntwachtel 7245
Buntwangenzaunkönig
  9987
Buntwarzenhonigfresser
  5729
Burgundergimpel 1779
Buriti-Segler 1978
Burma-Bülbül 4873
Burma-Drossling 10265
Burma-Fasan 9675
Burma-Würger 5042
Burmeister-Kolibri 5886
Bürstentapaculo 5821
Burton-Gimpel 1516
Buru-Brillenvogel 10764
Buru-Dschungelschnäpper
  8801
Buru-Fächerschwanz 8862
Buru-Honigfresser 5242
Buru-Lederkopf 7461

Buru-Lori 2066
Buru-Mistelfresser 3187
Buru-Monarch 5994
Buru-Papagei 9780
Buru-Pirol 6780
Buru-Raupenfänger 2624
Bürzelstelzer 8800
Buschdrossling 8557
Buschelaenie 3411
Büscheleule 6889
Buschelster 2302
Buscherdhacker 10437
Buscheule 6905
Buschfliegenstecher 9569
Buschflöter 8394
Buschflughuhn 8417
Buschhäher 782
Buschhonigfresser 10025
Buschhuhn 358
Buschkletterer 2446,2447
Buschkuckuck 1465
Buschlaubkratzer 9031
Buschläufer 2729
Buschlerche 5925,5934
Buschmeise 8300
Buschorganist 3682
Buschpieper 696
Buschrohrsänger 118
Buschschlüpfer 8800
Buschschwarzkäppchen
  5266
Buschsperling 7286
Buschspötter 4638
Buschstärling 3288
Buschtaube 7425
Buschtinamu 2864
Buschtyrann 3510
Buschvireo 4742
Buschwaldsittich 7888
Buschwürger 5490
Buschwürgerling 9904
Buschzaunkönig 9978
Bussard 1395
Bussardmilan 4455
Büttelwürger 5055
Caatinga-Spateltyrann 4566
Cabanis-Ammer 3452
Cabanis-Bülbül 7551
Cabanis-Drossel 10346
Cabanis-Schlüpfer 9633
Cabanis-Tangare 9731

Flußseeschwalbe 9454
Flußsumpfhuhn 8168
Flußuferläufer 10199
Flußwaldsänger 1183
Flußwasseramsel 2268
Forbes-Nonne 5297
Forbes-Papageiamadine
3612
Forbes-Stärling 2903
Forbes-Weihe 5161
Formosa-Bartvogel 5612
Formosa-Blauschwanz
9801
Formosa-Bülbül 8621
Formosa-Buschwachtel 931
Formosa-Goldhähnchen
8768
Formosa-Grüntaube 10151
Formosa-Häherling 5265
Formosa-Meise 7103
Formosa-Pfeifdrossel 6230
Formosa-Sibia 152
Forns-Waldsänger 9848
Forsten-Pirol 6787
Forster-Seeschwalbe 9451
Foxs Weber 7983
Franklin-Möwe 5108
Franklins Möwe 5108
Frankolinwachtel 7244
Fraser-Zaunkönig 9985
Fratzeneule 7495
Fratzenkuckuck 9083
Frauenlori 5390
Frauflankenuferwipper
2249
Fregattensturmschwalbe
7184
Fregattvogel 3944
Friedenstäubchen 4152
Friedmann-Lerche 5938
Froschmäuler 1215
Froschschnabel 2377
Fruchtpicker 2094
Frühlingsfruchttaube 8532
Frühlingspapageichen 5381
Frühlingstaube 10165
Fuchsammer 7166
Fuchscistensänger 2353
Fuchseule 6915
Fuchsfalke 3780
Fuchshabicht 3600

Fuchslerche 5918
Fuchslöffelente 557
Fuchsmonarch 6001
Fuchsrote Kuckuckstaube
5470
Fuchssänger 1193
Fuchsscheitelvireo 4747
Fuchsschlüpfer 2814
Fuchsschnäpper 6080
Fuchsschwalbe 9424
Fuchstangare 6752
Fuchsweber 7945
Fuelleborns Würger 5020
Fülleborn-Nektarvogel
6401
Fülleborn-Pieper 5465
Fünffarbenbartvogel 1643
Fünffarbennonne 5319
Fünfstreifenammer 499
Funkenkehlchen 4505
Furchenjahrvogel 91
Furchentangare 9950
Furchenvogel 2389
Gabarhabicht 5740
Gabela-Akalat 9193
Gabela-Brillenwürger 8230
Gabela-Rötel 9193
Gabela-Würger 8230
Gabeldrongo 3226
Gabelkuckuck 7319
Gabelnachtschwalbe 10471
Gabelracke 2601
Gabelschwalbe 8260
Gabelschwanzfasan 5361
Gabelschwanzhuhn 4069
Gabelschwanzkolibri 2162
Gabelschwanzkotinga 7446
Gabelschwanzmöwe 2826
Gabelschwanzraupenfänger
2634
Gabelschwanzseeschwalbe
9469
Gabelschwanzspint 5804
Gabelschwanzwellenläufer
6653
Gabelschwanzzuser 7446
Gabeltyrann 10407
Gabelweihe 5893
Gabun-Buschsänger 1298
Gabun-Nachtschwalbe
1672

Gabun-Nektarvogel 676
Gabun-Pirol 6794
Gabun-Schnäpper 1202
Gabun-Specht 3159
Gackelkuckuck 2892
Gackeltrappe 3730,3731
Galapagos-Albatros 3275
Galapagos-Bussard 1397
Galapagos-Kuckuck 2402
Galapagos-Pinguin 9293
Galapagos-Ralle 5125
Galapagos-Sharbe 7392
Galapagos-Spottdrossel
6517
Galapagos-Taube 10697
Galapagos-Tyrann 6154
Galapagos-Wellenläufer
6663
Gambaga-Schnäpper 6065
Gambel-Meise 7100
Gambel-Wachtel 1521
Gambia-Schneeballwürger
3337
Ganga-Kakadu 1525
Ganges-Brillenvogel 10814
Ganonga-Brillenvogel
10792
Gans 634
Gänsegeier 4404
Gänsesäger 5791
Garlepp-Ammerfink 8139
Garlepp-Schlüpfer 1031
Garnot-Sturmvogel 7190
Gartenammer 3466
Gartenbaumläufer 1928
Gartenbülbül 8617
Gartenfächerschwanz 8841
Gartengrasmücke 9589
Gartenlaubvogel 4639
Gartenmausvogel 2453
Gartenraupenfänger 2649
Gartenrohrsänger 112
Gartenrotschwanz 7524
Gartenspötter 4639
Gartentrupial 4829
Gartentyrann 3509
Gaukler 9839
Gebänderte Fruchttaube
3380
Gebänderte Uferschwalbe
8899

Goldflügelspateltyrann
10082
Goldflügelspecht 7695
Goldflügeltangare 9756
Goldflügelwaldsänger
10522
Goldglanzschwänzchen
5831
Goldhähnchen 8765
Goldhähnchenblau-
schnäpper 6081
Goldhähnchendornschnabel
22
Goldhähnchenlaubsänger
7647
Goldhähnchenwaldsänger
1170
Goldhalspieper 5464
Goldhaubengärtner 470
Goldhelmhornvogel 1886
Goldhonigfresser 2372
Goldkappensittich 904
Goldkappentangare 9946
Goldkehlbartvogel 5600
Goldkehlbrillenvogel
10797
Goldkehlbülbül 8625
Goldkehleremomela 3561
Goldkehlnektarvogel 237
Goldkehlpitta 7855
Goldkehltukan 8721
Goldkehltyrann 6205
Goldkehlwaldsänger 3132
Goldkernbeißer 6118
Goldkinnbrillenvogel
10772
Goldkinnsittich 1324
Goldkopfbreitschnabel
7899
Goldkopfbuschammer 1062
Goldköpfchen 1110
Goldkopfcistensänger 2327
Goldkopfkoagimpel 8867
Goldkopfpapagei 7762
Goldkopfpipra 7790
Goldkopfsittich 1326
Goldkopftimalie 9387
Goldkopftrogon 7428
Goldkopfwaldsänger 9091
Goldkronenmistelfresser
3178

Goldkrontyrann 6187
Goldkuckuck 2190
Goldlaubenvogel 9130
Goldman-Kolibri 4284
Goldmann-Taube 4199
Goldmann-Wachteltaube
4199
Goldmans Kolibri 4284
Goldmantelspecht 1591
Goldmantelweber 7987
Goldmaskenspecht 5672
Goldmeise 1110
Goldmonarch 5983
Goldnackenara 864
Goldnackenarara 864
Goldnackenspecht 10509,
5668
Goldnackentangare 9768
Goldnackenweber 7934
Goldohrarassari 9105
Goldohrhonigfresser 5753
Goldohrschlüpfer 9656
Goldohrspecht 10515
Goldohrtangare 9735
Goldpieper 10039
Goldraupenesser 1599
Goldraupenfresser 1599
Goldregenpfeifer 7996
Goldringtangare 1152
Goldrückenbergtangare
1423
Goldrückenfledermaus-
papagei 5377
Goldrückengimpel 8656
Goldrückenhonig-
schmecker 5769
Goldrückenspecht 3161,
3259
Goldrückenspreizschwanz
3595
Goldrückenweber 3714
Goldsaphir 4716
Goldscheitelelaenie 6194
Goldscheitelgelbkehlchen
4184
Goldscheitelhonigfresser
7540
Goldscheitellaubsänger
7650
Goldscheiteltangare 4883
Goldscheiteltaucher 8015

Goldscheiteltesia 9870
Goldscheiteltyrann 6206
Goldscheitelwaldsänger
1169
Goldscheitelwürger 5015
Goldschlüpfer 5838
Goldschnabelammer 4847
Goldschnabelmusendrossel
1811
Goldschnabelralle 6460
Goldschnabelruderammer
979
Goldschnabelsaltator 8938
Goldschnabeltäubchen
2532
Goldschnäpper 3858
Goldschnepfe 8909
Goldschnepfen 8912
Goldschopfpinguin 3650
Goldschopfspecht 5670
Goldschopftangare 9713
Goldschulterfruchttaube
8533
Goldschulterkassike 1447
Goldschultersittich 8313
Goldschulterspecht 2205
Goldschulterstärling 258
Goldschwanzbülbül 369
Goldschwanzpapagei
10117
Goldschwanzsaphir 4718
Goldschwanzspecht 1587
Goldschwingenhäherling
4118
Goldschwingenpipra 5571
Goldsittich 912
Goldspecht 2428
Goldstirnbeutelmeise 652
Goldstirnblattspäher 7487
Goldstirnblattvogel 2135
Goldstirnbuschtimalie 9408
Goldstirndickkopf 6935
Goldstirnfaulvogel 5511
Goldstirnfliegenstecher
10707
Goldstirnfruchttaube 8484
Goldstirnpapageichen 5371
Goldstirnpipra 7785
Goldstirnsittich 903
Goldstirnspecht 5664

Goldstirntrugschmätzer
3552
Goldstirnvireo 4739
Goldstirnzwergspecht 7707
Goldstreifenhonigfresser
5236
Goldstreifenwaldsänger
1162
Goldstrichellori 2063
Goldtangare 9729
Goldtaube 8505
Goldtrupial 4801
Goldtukan 1144
Goldtyrann 8975
Goldwaldsänger 3142
Goldwangenbartvogel 5595
Goldwangenhonigfresser
6764
Goldwangenpapagei 7757
Goldwangenspecht 5669
Goldwangenwaldsänger
3128
Goldweber 7984
Goldzeisig 1745
Goldzügelamazone 463
Goldzügelbülbül 8591
Goldzügeltimalie 4930
Goliathkuckuck 1853
Goliathreiher 953
Goodson-Taube 2489
Gosling-Feinsänger 754
Göttervogel 7019
Götzenliest 10074
Goudot-Kolibri 5148
Gough-Ammer 8916
Gough-Teichhuhn 4037
Gould-Amadine 3603
Gould-Arassari 9104
Gould-Elfe 5338
Gould-Nektarvogel 230
Gould-Scharbe 7389
Gould-Sturmschwalbe
8441
Graf von Paris 2415
Granada-Amazone 444
Granatastrild 10450
Granatkehlnymphe 4987
Granatkolibri 3664
Granatpitta 7853
Grandala 4340
Grant-Honigfresser 4341

Grant-Pirol 6800
Grants Samenknacker 9282
Grasammer 7165
Graseule 10414,10419
Grasklapperlerche 5920
Grasläufer 10254
Grasmücken 9623
Grauammer 3454
Grauämmerling 7746
Grauastrild 3632
Grauaugenbülbül 4872
Graubartfalke 3786
Graubauchameisenpitta
6268
Graubauchameisen-
schlüpfer 6285
Graubauchbülbül 8596
Graubauchdegenschnäbler
5850
Graubauchdickicht-
schnäpper 7236
Graubauchfächerschwanz
8821
Graubauchfruchttaube 8482
Graubauchfuchssänger
1194
Graubauchhabicht 67
Graubauchkitta 10461
Graubauchlaubpicker 8256
Graubauchmausspecht
8973
Graubauchmeise 7101
Graubauchnymphe 4492
Graubauchpfriemschnabel
6700
Graubauchsatyrhuhn 10137
Graubauchschlüpfer 9638
Graubauchsegler 1989
Graubauchsylphe 9794
Graubauchtesia 9868
Graubauchtimalie 6351
Graubauchtragopan 10137
Graubauchwollrücken 9921
Graubekarde 6993
Graubeutelmeise 654
Graubindenzaunkönig 1628
Graublauers Täubchen
2371
Graubrauenbambushuhn
1150

Graubrustameisenjäger
6270
Graubrustameisenpitta
4296
Graubrustameisenschlüpfer
6284
Graubrustammer 1489
Graubrustbaumelster 3075
Graubrustbuschammer
1076
Graubrustbuschsänger 1288
Graubrustbuschtangare
2154
Graubrustdegenflügel 1610
Graubrustdrongo 3224
Graubrustdrosseltimalie
4928
Graubrustdschungel-
schnäpper 8811
Graubrustelaenie 3429
Graubrusterdtaube 3995
Graubrusteremit 7342
Graubrustfächerschwanz
8855
Graubrustfrankolin 3924
Graubrustguan 6817
Graubrustgudilang 2465
Graubrusthäher 786
Graubrusthäherling 4092
Graubrustmeise 7112
Graubrustmistelfresser
3203
Graubrustmückenfresser
2557
Graubrustnektarvogel 6435
Graubrustparadies-
schnäpper 9865
Graubrustpflanzenmäher
7666
Graubrustprinie 8195
Graubrustralle 4059
Graubruströtel 2220
Graubrustschlangenadler
2280
Graubrustschlüpfer 9647
Graubrustschwalbe 8257
Graubrustspecht 3160
Graubrustspinnenjäger 881
Graubruststrandläufer 1504
Graubrusttachuri 515
Graubrusttaube 3995,5177

Kleiner Gelbstreifen-
　laubbülbül 7554
Kleiner Goldregenpfeifer
　7997
Kleiner Grundfink 4176
Kleiner Halsband-
　nektarvogel 6375
Kleiner Honiganzeiger
　4858
Kleiner Hyazintharara 618
Kleiner Kehlflecksperling
　7150
Kleiner Koafink 8867
Kleiner Kuba-Fink 10011
Kleiner Organist 3702
Kleiner Paradiesvogel 7018
Kleiner Purpurastrild 8641
Kleiner Rindenspalter 1239
Kleiner Schlammläufer
　5252
Kleiner Singhabicht 5741
Kleiner Soldatenara 876
Kleiner Soldatenarara 876
Kleiner Sturmtaucher 8559
Kleiner Textor 7964
Kleiner Vasa 2658
Kleiner Vasapapagei 2658
Kleines Sumpfhuhn 8172
Kleingrundfink 4176
Kleinschmidts Papagei-
　amadine 3605
Kleinschnabeldarwinfink
　1560
Kleinschnabelkolibri 1100,
　8751
Kleinschnabeltinamu 2872
Kleinschnäpper 7917
Kleinseeschwalbe 9441
Kleinspecht 3107
Kleinster Zwergspecht
　7722
Kleinsumpfhuhn 8172
Kleintöpfer 3971
Kletterhonigfresser 5771
Kletterkleidervogel 7064
Klettertöpfer 7476
Kletterwaldsänger 5953
Kletterweber 5526
Kletterwürgertimalie 4613
Klingelpipra 5442
Klingelschwatzvogel 5556

Klippenausternfischer 4421
Klippenhuhn 345
Klippenkleiber 9249
Klippenmöwe 8903
Klippenpieper 704
Klippenregenpfeifer 6761
Klippenrötel 6020
Klippensänger 98
Klippenschwalbe 4686
Klippensittich 6484
Klippenstrandläufer 1502
Klippentaube 2517
Klippenvogel 8920
Klufttyrann 6669
Klunkerhonigesser 646
Klunkeribis 1257
Klunkerkranich 4350
Knackente 563
Knackerlerche 8747
Knackerpapageischnäbler
　2107
Knäkente 563
Knarrtrappe 3744
Knopffruchttaube 8499
Knutt 1498
Koakleidervogel 8341
Kobaltämmerling 8162
Kobalteisvogel 321
Kobaltflügelsittich 1323
Kobaltirene 4876
Kobaltniltava 6547
Kobaltparadiesschnäpper
　9856
Koël 3646
Koepke-Eremit 7345
Kohala-Kleidervogel 2298
Köhler-Drossel 7895
Kohletyrann 4944
Kohlmeise 7111
Kokardenspecht 7685
Kokil 7318
Koklas-Fasan 8558
Kokos-Fink 7750
Kokos-Honigschmecker
　6337
Kolbenente 6529
Kolibri 2986
Kolibrifink 381
Kolibris 10208
Kolkrabe 2685
Kolumbia-Sittich 923

Komoren-Fruchttaube 355
Komoren-Taube 2514
Komoren-Weber 3885
Komoro-Drossel 10294
Kona-Gimpel 2107
Kona-Papageischnäbler
　2107
Kondor 10604
Kongo-Kuckuckswürger
　1570
Kongo-Nektarvogel 6379
Kongo-Papagei 8060
Kongo-Pfau 239
Kongo-Raupenfresser 1570
Kongo-Schmätzer 6260
Kongo-Schnäpper 1203
Kongo-Taube 2525
Kongo-Waldtaube 2525
Königsalbatros 3271
Königsamazone 448
Königsameisenpittta 4324
Königsameisenstelzer 4324
Königsatzel 1188
Königsbussard 1409
Königsdrongo 3231
Königsfasan 9677
Königsfruchttaube 8523
Königsgeier 8958
Königsglanzstar 2739
Königskleidervogel 3302
Königsliest 150
Königslöffler 7882
Königsmeise 7128
Königsmusketier 2421
Königsnektarvogel 6418
Königsnymphe 4490
Königspapageiamadine
　3610
Königsparadiesvogel 2225
Königspinguin 823
Königsralle 8712
Königssatrap 10409
Königsscharbe 7369,7371
Königsseeschwalbe 9457
Königssittich 367
Königsspecht 1576
Königstachuri 9684
Königstyrann 10408,6740
Königsvogel 10409
Königsweber 7933
Königswitwe 10548

Mangokauz 9530
Mangroveamazilie 401
Mangroveameisenvogel
9028
Mangroveblauschnäpper
3002
Mangrovebrillenvogel
10793
Mangrovedarwinfink 1557
Mangrovedickkopf 6953
Mangroveeule 6869,6884
Mangrovefink 1557
Mangrovegerygone 4226
Mangrovehonigfresser
5221
Mangroveliest 4441
Mangrovemaustimalie
10169
Mangrovemonarch 5984
Mangrovenhonigfresser
5221
Mangrovenkuckuck 2403
Mangrovenwürgatzel 2792
Mangrovepirol 6786
Mangrovepitta 7859
Mangroveralle 3661
Mangrovereiher 1427
Mangroveschnäpper 3537
Mangroveschwalbe 9690
Mangrovevireo 10586
Manila-Papageiamadine
3613
Manipur-Wachtel 7246
Mantelbrillenbogel 10791
Mantelbuschschlüpfer 1021
Mantelbussard 5205
Mantelhabicht 58
Mantelkardinal 7058
Mantelmöwe 5099
Mantelschwärzling 6543
Mantelwollrücken 9922
Manus-Kauz 6557
Manus-Lederkopf 7448
Manyar-Weber 7962
Mao 4382
Maoriente 1136
Maorifalke 3807
Maorifruchttaube 4528
Maorigerygone 4223
Maorigrasmücke 5961
Maorimöwe 5076

Maoriregenpfeifer 2038
Maorischlüpfer 10620
Maorischnäpper 7278
Maoritaucher 8070
Marabu 5169
Marakana 875
Maranhao-Eremit 7350
Marañon-Bandvogel 5714
Maranon-Drossel 10326
Marañon-Schlüpfer 9651
Marcapala-Wollrücken
9910
Marcapata-Schlüpfer 2805
Margareten-Eremit 7349
Margareten-Lori 2057
Margareten-Schnäpper
1201
Marianen-Erdtaube 4014
Marianen-Fruchttaube 8526
Marianen-Myiagra 6133
Marico-Blaßschnäpper
1283
Marico-Schnäpper 1283
Marmelalk 1280
Marmelente 5572
Marmorameisenwürger
3937
Marmorfischeule 9045
Marmorhonigfresser 8630
Marmorkehlameisen-
schlüpfer 6279
Marmornachtschwalbe
1676
Marmorreiher 10019
Marmorschnepfe 5258
Marmorschwalm 8007
Marmorspätzling 8342
Marmorstar 8962
Marmortimalie 6352
Marmorweber 8342
Maronenbauchhaken-
schnabel 3246
Maronenbrustammerfink
8147
Maronenspecht 1239
Maronensperling 7142
Maronenstirnstittich 8889
Maronentaube 5473
Marquesas-Erdtaube 4008
Marquesas-Fliegen-
schnäpper 8112

Marquesas-Fruchttaube
3357
Marquesas-Liest 10066
Marquesas-Monarch 8112
Marquesas-Taube 4008
Marula-Schnäpper 5656
Maryland-Waldsänger
4190
Masafuera-Schlüpfer 789
Masatierra-Sturmvogel
8432
Mascarenen-Brillenvogel
10763
Maschona-Hyliota 4713
Maskarenen-Fluchtvogel
4783
Maskarenen-Papagei 5573
Maskarenen-Paradies-
schnäpper 9853
Maskarenen-Purpurhuhn
8156
Maskarenen-Schwalbe
7436
Maskarenen-Sturmvogel
8422,8426
Maskarenschwalbe 7436
Maskenamadine 8044
Maskenammer 3481
Maskenbaumelster 3076
Maskenbekarde 10038
Maskenbülbül 8593,8612
Maskenbuschsänger 5620
Maskenente 6929
Maskeneule 7495
Maskenfeinsänger 746
Maskenflughuhn 8416
Maskengelbkehlchen 4179
Maskengimpel 8658,8891
Maskengrasmücke 9608
Maskenhäherling 4109
Maskenhakenschnabel
3255
Maskenkernbeißer 3532
Maskenköpfchen 245
Maskenkuckuck 1856
Maskenkuhstärling 5964
Maskenlerche 9342
Maskenlori 4279
Maskenmonarch 5996
Maskenmückenfänger 8076
Maskenpfuhlhuhn 4038

Mittelmeer-Steinschmätzer 6710
Mittelmeer-Sturmtaucher 8574,8582
Mittelreiher 957
Mittelsäger 5793
Mittelspecht 3106
Mittlere Raubmöwe 9436
Mitu 5949
Moab-Sperling 7152
Moheli-Buschsänger 6505
Mohrenameisenfänger 1901
Mohrenbussard 1388
Mohrendajal 2595
Mohreneremit 9960
Mohrenfächerschwanz 8822
Mohrengimpel 8655
Mohrenguan 2009
Mohrenhabicht 59
Mohrenhonigfresser 6329
Mohrenibis 7488
Mohrenklaffschnabel 576
Mohrenklarino 3528
Mohrenkopf 8065
Mohrenkopfameisenvogel 7241
Mohrenkopfpapagei 8065
Mohrenkuckuck 1857
Mohrenlerche 5706
Mohrenmeise 7115
Mohrenmusketier 2418
Mohrennonne 5315
Mohrenpipra 10629
Mohrenpirol 6788
Mohrenpitohui 7840
Mohrenpitta 7870
Mohrenralle 388
Mohrenraupenfresser 1571
Mohrenreisknacker 6848
Mohrenscharbe 7399
Mohrenschwalbe 4675
Mohrenschwarzkehlchen 8982
Mohrensegler 835
Mohrenstirnvogel 8307
Mohrentrappist 6007
Mohrenweber 7968
Mohrenweihe 2293
Mohrenwollrücken 9920

Mohrenwürger 5026
Mohrenwürgerling 6461
Molina-Sittich 8675
Mollymauk 3276
Molokai-Krausschwanz 5956
Molukken-Adler 852
Molukken-Atzel 1189
Molukken-Brillenvogel 10766
Molukken-Bronzefrucht-taube 3353
Molukken-Buschsänger 1294
Molukken-Drossel 10725
Molukken-Falke 3803
Molukken-Grundschnäpper 3832
Molukken-Habicht 52
Molukken-Huhn 5647
Molukken-Ibis 9965
Molukken-Kakadu 1436
Molukken-Kauz 6571
Molukken-Kielralle 390
Molukken-Krähe 2725
Molukken-Kuckuck 1460
Molukken-Liest 10064
Molukken-Myiagra 6134
Molukken-Raupenfänger 2611
Molukken-Schnepfe 9039
Molukken-Schwalm 181
Molukken-Star 808
Mönchsalcippe 4838
Mönchsbrillenvogel 9273
Mönchsfliegenstecher 7578
Mönchsgeier 188
Mönchsgrasmücke 9587
Mönchsittich 6210
Mönchskotinga 10690
Mönchskranich 4354
Mönchskuckuck 1859
Mönchsmeise 7113
Mönchspirol 6793
Mönchsraupenfänger 2656
Mönchsregenpfeifer 2044
Mönchssittich 6210
Mönchstyrann 10399
Mönchswaldsänger 10608
Mönchsweber 7973
Mönchszuser 10690

Mondetour-Täubchen 2370
Mondstreifhonigschmecker 5770
Mongolei-Regenpfeifer 2034
Mongolen-Bussard 1399
Mongolen-Gimpel 8872
Mongolen-Häher 8026
Mongolen-Lerche 5705
Mongolen-Regenpfeifer 2034
Mongolen-Star 9563
Moniasralle 6011
Monteiro-Astrild 2386
Monteiro-Toko 10052
Monteiro-Würger 5498
Montezuma-Stirnvogel 4396
Montserrat-Trupial 4824
Moorente 1137
Moorhuhn 4965
Moorschneehuhn 4964
Mooshacker 5516
Moosnestsalangane 217
Moostimalie 8004
Mooswaldtimalie 5516
Moreau-Nektarvogel 6404
Moreno-Täubchen 5842
Morgenammer 10711
Morgenrötel 2218
Morinell-Regenpfeifer 2036
Morio-Raupenfänger 2641
Morotai-Lederkopf 7457
Morrison-Häherling 4104
Moschusente 1466
Moschuslori 4277
Moskitokolibri 2207
Motmotpapagei 8215
Möwen 5066
Möwenbussard 5208
Möwenvögel 2017,5066
Mozambique-Girlitz 9163
Mückenesser 2565
Mückenfänger 2565
Mückenfresser 2565
Mugimaki-Schnäpper 3843
Müller-Amazone 445
Müllers Edelpapagei 9783
Mupin-Drossel 10330

Murphy-Brillenvogel
  10804
Murphy-Sturmvogel 8458
Muskatamadine 5318
Muskatfink 5318
Mynahassa-Eule 10418
Mynahassa-Mistelfresser
  3179
Myrtensänger 3129
Myrtenwaldsänger 3129
Mystacornis 6305
Nachtfalke 2185
Nachtflughuhn 8406
Nachtigall 5414
Nachtigallkolibri 1603
Nachtigallrohrsänger 5284
Nachtigallschwirl 5284
Nachtigalzaunkönig 5854
Nachtreiher 6635
Nachtschattenfresser 4774
Nachtschwalbe 1669
Nachtschwalben 1646,3748
Nachtschwalben-artige
  Vögel 1647
Nachtsittich 7296
Nacktaugenameisenvogel
  8792
Nacktaugenbrillenvogel
  4483
Nacktaugendrossel 10334
Nacktaugenfruchttaube
  3370
Nacktaugenhonigesser
  5736
Nacktaugenkakadu 1441
Nacktaugenralle 4377
Nacktaugentaube 2483
Nacktaugentrupial 4380
Nacktbeineule 6868
Nacktfußeule 6901
Nacktgesichthokko 2822
Nacktgesichttäubchen 5840
Nackthalsschmuckvogel
  4378
Nacktkehldickkopf 6957
Nacktkehlfrankolin 3895
Nacktkehlglockenvogel
  8252
Nacktkehlglöckner 8252
Nacktkehllärmvogel 2735
Nacktkehlreiher 10020

Nacktkehlschirmvogel
  1873
Nacktkopfameisenvogel
  4375
Nacktkopfameisenwürger
  4375
Nacktohrdrossling 10266
Nacktschnabelhäher 4393
Nacktwangenblaurabe 2933
Nacktwangendrossling
  10266
Nacktwangenkakadu 1441
Nacktwangenspinnenjäger
  883
Nackzügelsänger 1977
Naka 6011
Nama-Flughuhn 8414
Nama-Prinie 8209
Nama-Specht 3163
Nama-Trappe 3744
Namib-Lerche 487
Namib-Schmätzer 1910
Namib-Schnäpper 6342
Nanday-Sittich 6343
Nandu 8788
Nandus 8795,8796
Napo-Degenflügel 1613
Napoleon-Fasan 8094
Napoleon-Weber 3711
Narcondam-Jahrvogel 87
Narina-Trogon 772
Narzissenschnäpper 3844
Nasenkakadu 1443
Nasenstreifhoniganzeiger
  4858
Nashornalk 1926
Nashornpelikan 7197
Nashornvögel 1361
Natal-Drossel 10730
Natal-Felsenspringer 1973
Natal-Frankolin 3916
Natal-Glanzköpchen 6425
Natal-Heckensänger 1923
Natal-Honigfresser 8264
Natal-Nachtschwalbe 1685
Natal-Rötel 2751
Natal-Specht 1594
Natal-Zwergeisvogel 4886
Natal-Zwergfischer 4886
Natterer-Arassari 9106
Naumann-Drossel 10331

Nebelkrähe 2688,2712
Neblina-Blattspäher 1111
Neblina-Glanz-
  schwänzchen 5832
Neddicky 2329
Neergard-Nektarvogel
  6406
Neergards Nektarvogel
  6406
Neevermanns Nonne 5314
Negerralle 388
Negros-Dschungel-
  schnäpper 8802
Negros-Fruchttaube 8483
Negros-Taube 4002
Negros-Timalie 9400
Nehrkorn-Brillenvogel
  10762
Nehrkorn-Mistelfresser
  3199
Nektarpitta 6462
Nektarvögel 6441
Nelkenente 8870
Nelkenfruchttaube 3368
Nelson-Gelbkehlchen 4185
Nelson-Tyrann 775
Nemosia 6442
Nepal-Alcippe 335
Nepal-Baumläufer 1932
Nepal-Bekassine 4027
Nepal-Hornvogel 88
Nepal-Pitta 7862
Nepal-Schwalbe 3055
Nepal-Sibia 153
Nepal-Uhu 1339
Nepar-Taube 2515
Netzbauchspecht 7744
Neubrittanien-Bronze-
  flügeltaube 4579
Neuguinea-Brillenvogel
  10809
Neuguinea-Erdtaube 10253
Neuguinea-Habicht 5650
Neuguinea-Krähe 2721
Neuholland-Dommel 4895
Neuholland-Eule 10423
Neuholland-Habicht 64
Neuholland-Krähe 2689
Neuholland-Schwalbe 4673
Neuholland-Taucher 9686

Philippinen-Segler 5579
Philippinen-Specht 6054
Philippinen-Trogon 4465
Philippinen-Wolltimalie
 8536
Philippinen-Würger 5064
Phoebe 8999
Phoenixsturmvogel 8420
Piapia 8555
Piccazurtaube 2512
Pickering-Fruchttaube 3369
Picui-Täubchen 2537
Piepertyrann 6091
Pieperwaldsänger 9091
Pinaia-Brillenvogel 5355
Pinguine 9288,9289
Pinkpink 2351
Pinselscharbe 7404
Pinselsittich 5172
Pinseltangare 3639
Pinto-Schlüpfer 9648
Pioho 8691
Piopio 10367
Pipra 7784
Pipras 7816
Pirol 6795
Pirole 6777
Pirolgimpel 5262
Pirolhonigfresser 5238
Pirolsänger 4767
Piroltrupial 4828
Pirolweber 7952
Pissaca 6596
Pitcairn-Rohrsänger 142
Pittadrossel 379
Pittas 7876
Pittatimalie 4836
Piura-Stelzling 4333
Piura-Tyrann 6670
Piwih 2576,2588
Planalto-Eremit 7354
Platen-Taube 4007
Plättchentukan 8461
Plattschnabelmotmot 3437
Pleske-Häher 8028
Plüschkopfente 9264
Plüschkopftangare 1797
Polarbirkenzeisig 1731
Polarmöwe 5088
Polartaucher 4132
Pollen-Vanga 10631

Pompadourkotinga 10654
Pompadourschmuckvogel
 10654
Pompadoursittich 8268,
 8269
Pompadourtaube 10158
Ponapé-Myiagra 6139
Ponderosa-Eule 6872
Poortman-Kolibri 2169
Poorwill 7413
Porphyrkopflori 4278
Port Lincoln-Sittich 7886
Portorico-Taube 2520
Pracht Astrapia 1044
Prachtadler 9324
Prachtamazone 455
Prachtammer 3456
Prachtatzel 1191
Prachtbartvogel 5613
Prachtblauschnäpper 3004
Prachteiderente 9266
Prachtelfe 5340
Prachtente 544
Prachtfink 381
Prachtfinken 3633
Prachtflügelsittich 8665
Prachtfregattvogel 3942
Prachtfruchttaube 8529
Prachtglanzstar 5009
Prachthabicht 3599
Prachthäher 4130
Prachthäherling 4088,4101
Prachthaubenadler 9324
Prachtkauz 4272
Prachtkleiber 3040
Prachtkuckuck 2196
Prachtleierschwanz 5786
Prachtliest 4432
Prachtlori 5388
Prachtmoho 5958
Prachtnachtschwalbe 1670
Prachtnektarvogel 6431
Prachtnonne 5320
Prachtparadieselster 1044
Prachtparadiesvogel 8549
Prachtpfeifente 566
Prachtpipra 2087
Prachtreiher 969
Prachtsittich 8316
Prachtstaffelschwanz 5535
Prachttachuri 9684

Prachttangare 9765,9772
Prachttaube 2519
Prachttaucher 4132
Prairiepieper 733
Prälatfasan 5359
Prälattangare 9957
Prarieammer 1477
Präriebussard 1414
Präriefalke 3802
Präriehuhn 10391
Prärieläufer 1157
Präriemöwe 5108
Prärieregenpfeifer 2035
Preuss-Nektarvogel 6414
Preuss-Weber 7975
Priestervogel 8270
Prigogine-Bülbül 2120
Prigogine-Eule 7496
Princess-of-Wales Sittich
 8103
Prinzenbussard 5206
Prinzendrossling 4692
Prinzengirlitz 9169
Prinzen-Glanzstar 5003
Prinzenhabicht 68
Prinzenhäherling 4091
Prinzen-Paradiesschnäpper
 9849
Prinzen-Weber 7976
Prinzen-Zwergfischer 317
Prinzessin Stephanie-
 Paradiesvogel 1045
Priritschnäpper 1211
Pritchard-Huhn 5644
Proteagirlitz 9158
Protonotarsänger 8271
Protonotarwaldsänger 8271
Provence-Grasmücke 9611
Przewalski-Papagei-
 schnabel 7035
Pseudokoa 8341
Pseudonektarvogel 6462
Puderspecht 6055
Puertorico-Amazone 462
Puertorico-Kukuck 8979
Puertorico-Nachtschwalbe
 1688
Punabekassine 4017
Punabussard 1407
Punaente 562
Punaerdhacker 4169

Punagilbammer 9207
Punaregenpfeifer 2020
Punaschlüpfer 1034
Punasteißhuhn 10030
Punataucher 8020
Punatyrann 6093
Pünktchenente 551
Pünktchenkauz 6565
Pünktchenspecht 1596
Purpurastrild 8642
Purpuratlaswitwe 10536
Purpurbauchnektarvogel
    6417
Purpurbindentäubchen
    2369
Purpurblaurabe 2927
Purpurbrustamazilie 422
Purpurbrustfruchttaube
    8506
Purpurbrustkotinga 2758
Purpurbrusttäubchen 2370
Purpurbürzelglanzköpchen
    6391
Purpurdegenflügel 1608
Purpurfink 1766
Purpurgimpel 1766
Purpurglanzstar 5006
Purpurgrackel 8698
Purpurgranatastrild 10451
Purpurhähnchen 5163
Purpurhuhn 8160
Purpurkappenfruchttaube
    8521
Purpurkardinal 1714
Purpurkehlglanz-
    schwänzchen 5829
Purpurkehlkotinga 8161
Purpurkehlnektarvogel
    6429
Purpurkehlorganist 3686
Purpurkopfelfe 4513
Purpurkronfink 8877
Purpurkuckuck 1871
Purpurkuckuckswürger
    1572
Purpurmanteltangare 4881
Purpurmaskenbartvogel
    5425
Purpurmaskentangare 9755
Purpurmistelfresser 3207
Purpurnaschvogel 2910

Purpurnektarvogel 6364
Purpurpfeifdrossel 6226
Purpurraupenfresser 1572
Purpurreiher 961
Purpurrückenkolibri 264
Purpurrückentaube 4200
Purpurrückenwachteltaube
    4200
Purpurscheitelfruchttaube
    8518
Purpurschnäpperdrossel
    2409
Purpurschultertaube 4012
Purpurschwalbe 8262
Purpurschwanzpapagei
    10115
Purpurspint 5803
Purpurstaffelschwanz 5533
Purpurstärling 3677
Purpurstirnpapagei 7770
Purpurtangare 8736
Purpurtaube 2521
Purpurtyrann 8648
Purpurwaldsänger 3581
Purpurwaldtaube 2516
Purpurzügelhonigfresser
    5220
Puvels Buschdrossling
    4843
Pycroft-Sturmvogel 8454
Queo 8869
Quesal 7430
Quetzal 7430
Rabaul-Habicht 56
Rabengeier 2661
Rabenkakadu 1546
Rabenkrähe 2686,2713
Rabenpapagei 2658
Rabenstar 795
Rabenvögel 2675
Rabors Timalie 6353
Rachel-Weber 5525
Rachenvögel 3757
Racken 2607
Rackenvögel 2607,2608
Rackettschwanzelster 2850
Radjaheule 6864
Radjahgans 9717
Rahmbauchdrossel 10290
Rahmbauchmückenfänger
    8078

Rahmbrustprinie 8208
Rahmkopffruchttaube 8491
Rainammer 2180
Raketenschwanzpapagei
    8213
Rallen 8700
Rallenklarino 6112
Rallenkranich 900
Rallenkraniche 891
Rallenläuter 3674
Rallenreiher 967
Rallenvögel 4346
Rand-Honigfresser 5750
Randia 8757
Rapa-Fruchttaube 8497
Raphiabülbül 9938
Rara 7667
Rarita 7667
Rarotonga-Fruchttaube
    8522
Rarotonga-Monarch 8110
Rarotonga-Star 794
Raubadler 857
Raubmöwe 1804
Raubmöwen 9433
Raubseeschwalbe 9447
Raubvögel 3822
Raubwürger 5045
Rauchschwalbe 4682
Rauchsegler 3013
Rauchspechtdrossel 3066
Rauchtyrann 6667
Rauhflügelschwalbe 9426
Rauhfußbussard 1401
Rauhfußhuhn 1255
Rauhfußkauz 175
Raupenfänger 2637,2648
Raupenfresser 1573
Raytallerche 1485
Razolerche 302
Rebhuhn 7249
Rebhuhnastrild 6842
Rebhuhnfrankolin 3914
Rebhuhntaube 9421
Regenbogenpitta 7856
Regenbogenspint 5812
Regenbogen-Tukan 8727
Regenbrachvogel 6617
Regenkuckuck 4700
Regenkuckucke 2395
Regenpfeifer 2016

Regenstorch 2227
Regensturmvogel 8439
Regenwachtel 2769
Reichard-Weber 7977
Reichenow-Bergastrild
  2853
Reichenow-Buschdrossling
  4845
Reichenow-Girlitz 9167
Reichenows Bergastild
  2853
Reichenow-Weber 7936
Reichsvogel 5013
Reiher 963
Reiherente 1133
Reiherläufer 3304,3311
Reinwardt-Arassari 9107
Reinwardt-Specht 8773
Reinwardt-Taube 8776
Reinwardt-Trogon 4473
Reiser-Grünrücken-
  fliegenstecher 7580
Reisfink 7001
Reisstärling 3290
Reisvogel 7001
Renauld-Kuckuck 1754
Rennel-Fächerschwanz
  8854
Rennell-Brillenvogel 10820
Rennell-Würgermonarch
  2382
Rennel-Star 803
Rennkuckuck 4142
Renntaucher 160,161
Rennvogel 2905
Réunion-Dronte 8759
Réunion-Schmätzer 8992
Rheinhart-Fasan 8797
Rhinozerusvogel 1359
Rhododendronbuschsänger
  1958
Rhododendrondrossel
  10319
Rhododendrongimpel 7754
Ribeiro-Tyrann 4568
Ricord-Kolibri 2170
Ridgway-Kauz 177
Ridgway-Kotinga 2762
Riedammerfink 3295
Rieddrossling 10259
Riedsänger 5629

Riedscharbe 7368
Riedweber 7946
Riesenadler 4448
Riesenalk 7752
Riesenameisenpitta 4310
Riesenani 2847
Riesenbabax 1141
Riesenbaumsteiger 10648
Riesenbekassine 4034
Riesenbläßhuhn 3961
Riesenbreitrachen 2727
Riesencoua 2782
Riesendampfschiffente
  9702
Riesendrossel 10308
Riesenelsterchen 5298
Riesenfischer 5588
Riesenfischuhu 4931
Riesenfroschmaul 1217
Riesenfroschschwalm 1217
Riesenfruchttaube 3358
Riesenglanzstar 4992
Riesenhäherling 4098
Riesenibis 8322
Riesenkauz 6572
Riesenkleiber 9242
Riesenkolibri 7177
Riesenkrähe 2696
Riesenkuhstärling 9001
Riesenlaufhühnchen 10376
Riesenlerche 5933
Riesennachtschwalbe 3752
Riesennektarvogel 6434
Riesenpfäffchen 9356
Riesenpieper 691
Riesenpinguin 822
Riesenpitta 7846,7859
Riesenralle 900
Riesenrallen 891
Riesenrotschwanz 7518
Riesensäbler 8119
Riesensalangane 4696
Riesenschnabelani 2848
Riesenschwalbenstar 1000
Riesenschwalk 6629
Riesenschwirl 5281
Riesenseeschwalbe 9447
Riesensperling 7154
Riesenspitzschnabel 6756
Riesensteigschnabel 5651
Riesenstorch 3545

Riesensturmvogel 5453,
  5454
Riesensumpflerche 5704
Riesentafelente 1138
Riesentrappe 972
Riesentukan 8729
Riesenturaco 2732
Riesenweber 7954
Riesenzaunkönig 1622
Riffenreiher 3407
Riff-Triel 1373
Rimator-Timalie 8898
Rindenfink 1561
Rindenpicker 4281
Ringamsel 10364
Ringdrossel 10364
Ringelameisenwürger 9916
Ringelastrild 9721
Ringelgans 1312
Ringeltaube 2511
Ringfasan 7433
Ringrennvogel 2022
Ringschnabelente 1131
Ringschnabelmöwe 5082
Ringschnäpper 990
Ringsittich 1156
Ringstar 4295
Ripley-Fruchttaube 8483
Rivoli-Kolibri 3657
Roberts-Prinie 8204
Robinson-Laufhühnchen
  10377
Roborowski-Gimpel 1769
Rodriguez-Rohrsänger
  1227
Rodriguez-Sittich 8370
Rodriguez-Weber 3886
Rohrammer 3479
Rohrdommel 1264
Rohrdoradito 8333
Röhrennasen 8246,8247
Rohrschlüpfer 7492
Rohrschwirl 5284
Rohrsittich 8666
Rohrspottdrossel 3294
Rohrweihe 2283
Rollandtaucher 8906
Roraima-Ameisenfänger
  4595
Roraima-Nachtschwalbe
  1708

Rostkehleremomela 3571
Rostkehlfaulvogel 4769
Rostkehlhonigfresser 2567
Rostkehlkeilschwanz 9882
Rostkehllaubwender 9032
Rostkehlmeise 7098
Rostkehlnachtigall 5409
Rostkehlrotschwanz 7515
Rostkehltangare 9770
Rostkehlwasseramsel 2269
Rostkinnhäherling 4113
Rostkopffuchssänger 1195
Rostkopfschilfsteiger 5632
Rostkopfspateltyrann
    10092
Rostkopfstärling 257
Rostkopftangare 9942
Rostlerche 5939
Rostmantelblattspäher 7477
Rostmantelbündelnister
    1019
Rostmantelwürger 5060
Rostnachtschwalbe 1699
Rostnackenammer 5783
Rostnackenstaffelschwanz
    9225
Rostnackenvireo 4753
Rostnackenzwergspecht
    7713
Rostohrhonigfresser 5732
Rostpiha 5273
Rostprinie 8205
Rostrenvogel 2906
Rostrhytipterna 8895
Rostrote Pfuhlschnepfe
    5260
Rostroter Faulvogel 5511
Rostrückenameisenvogel
    6238
Rostrückenammer 288
Rostrückenklarino 2223
Rostscheitelammer 290
Rostscheitellaubsänger
    7651
Rostscheitelwaldsänger
    3131
Rostschlüpfer 9662
Rostschwanzammer 289
Rostschwanzdrossel 10331
Rostschwanzhonigfresser
    5735

Rostschwanzschmätzer
    1904
Rostschwanzschnäpper
    6073
Rostschwanztyrann 4945
Rostschwingenbusch-
    drossling 4845
Rostschwingenralle 4376
Rostschwingenstachel-
    schwanz 8181
Rostsperling 7148,7154
Roststärling 3676
Roststeißyuhina 10688
Roststirndornschnabel 20
Roststirntimalie 9407
Rosttäubchen 2539
Rosttinamu 2861
Rosttöpfer 3972
Rosttyrann 6201
Rostwangenlaubsänger 2
Rostwangenmaustimalie
    7207
Rostwangennachtschwalbe
    1698
Rostwangenschneidervogel
    6839
Rostwangenschwanzmeise
    167
Rostwürgerling 9898
Rotachselkuhstärling 5967
Rotachselpapagei 8386
Rotachselraupenfänger
    2614
Rotachseltaube 5184
Rotaugenameisenvogel
    7491
Rotaugenbülbül 8593
Rotaugenbündelnister 7301
Rotaugendrossel 10352
Rotaugenente 6527
Rotaugenfruchttaube 3370
Rotaugenkuhstärling 5962
Rotaugentaube 9503
Rotaugenvireo 10584
Rotbackensänger 5515
Rotbandbartvogel 10134
Rotbandregenpfeiffer 2040
Rotbartfruchttaube 8509
Rotbartspint 6642
Rotbauchadler 4625

Rotbauchameisenschlüpfer
    6282
Rotbauchara 874
Rotbaucharara 874
Rotbauchbergfruchttaube
    3351
Rotbauchbuschwachtel
    927,936
Rotbauchbussard 1417
Rotbauchdrossel 10358
Rotbaucheremit 7355
Rotbauchglanzstar 5004
Rotbauchkurzflügel 1273
Rotbauchlalage 4969
Rotbauchliest 3024
Rotbauchmeise 7123
Rotbauchmistelfresser 3180
Rotbauchmohrenkopf 8064
Rotbauchmyiagra 6143
Rotbauchnektarvogel 6377
Rotbauchniltava 6550
Rotbauchnymphe 4982
Rotbauchorganist 3706
Rotbauchpapagei 8064
Rotbauchpfäffchen 9351
Rotbauchpipra 7785
Rotbauchpitta 7852
Rotbauchreiher 968
Rotbauchsaltator 8948
Rotbauchschmätzer 9900
Rotbauchschwalbe 4683
Rotbauchsittich 8680
Rotbauchspecht 3166
Rotbauchsperber 71
Rotbauchstärling 4778
Rotbauchtangare 611,9775
Rotbauchtapaculo 9058
Rotbauchtimalie 3384
Rotbauchweber 5520
Rotbauchwürger 5013
Rotbauchwürgertimalie
    8480
Rotbauchzwergspecht 7723
Rotbrauenpanthervogel
    7046
Rotbrauenspitzschnabel
    2549
Rotbrauenstar 3526
Rotbraune Reinwardt-
    Taube 8776
Rotbrauner Sperling 7153

Rotbrauner Weber 7978
Rotbrillentaube 4160
Rotbrustakalat 9476
Rotbrustameisendrossel
  3869
Rotbrustammerfink 8144
Rotbrustbaumsteiger 10649
Rotbrustblauschwanz 9798
Rotbrustbrachschwalbe
  4238
Rotbrustbraunelle 8282
Rotbrustbreitrachen 9262
Rotbrustbülbül 4903
Rotbrustbuschsänger 5623
Rotbrustcoua 2785
Rotbrustfalke 3792
Rotbrustfischer 5589
Rotbrustfliegenstecher
  5165
Rotbrustglanzköpchen
  6425
Rotbrustglanzschwänzchen
  1999
Rotbrustgrundschnäpper
  3840
Rotbrustguan 7221
Rotbrüstiger Schönsittich
  6486
Rotbrustkauz 4273
Rotbrustkiebitz 10501
Rotbrustkrontaube 4290
Rotbrustlaufhühnchen
  10378
Rotbrustmaustimalie 7209
Rotbrustmayrornis 5577
Rotbrustmeise 7121
Rotbrustnektarvogel 6425
Rotbrustparadiesschnäpper
  9858
Rotbrustpirol 6785
Rotbrustpitpit 3029
Rotbrustpitta 7853
Rotbrustralle 8970
Rotbrustregenpfeifer 2033
Rotbrustsamenknacker
  9281
Rotbrustschlüpfer 9642
Rotbrustspecht 3094
Rotbrustspechtpapagei
  5877
Rotbruststärling 5136

Rotbruststelzling 4327
Rotbrusttangare 2540
Rotbrusttinamu 2880
Rotbrustwachtel 6695
Rotbrustwaldsänger 6184
Rotbrustzaunkönig 10002
Rotbrustzwerggans 6530
Rotbrustzwergralle 8970
Rotbugamazone 434
Rotbugara 879
Rotbugarara 879
Rotbuggrüntaube 10145
Rotbürstelwürgerling 3392
Rotburzelameisenfänger
  3322
Rotbürzelammerfink 8142
Rotbürzelbartvogel 8045
Rotbürzelblattspäher 7478
Rotbürzelfächerschwanz
  8833
Rotbürzelhonigfresser 6319
Rotbürzelkassike 1448
Rotbürzellerche 7748
Rotbürzelliest 10071
Rotbürzellori 2065
Rotbürzelpfäffchen 9357
Rotbürzelpiha 5274
Rotbürzelspecht 7734
Rotbürzelsteinschmätzer
  6726
Rotbürzeltangare 8744
Rotbürzeltrogon 4467
Rotbürzelwaldsänger
  10526
Rotbürzelwürger 5047
Rotbüschelaulia 5035
Rotbüschelbartvogel 8353
Rotdrossel 10316
Rote Erdtaube 4203
Rote Fidschi-Flaumfuß-
  taube 8531
Rote Spottdrossel 10130
Rötelammer 3478
Rötelbraunelle 8283
Röteldickichtvogel 1084
Röteleule 6906
Rötelfalke 3804
Rötelgrundammer 7779
Rötelkauz 9517
Rötelkuckuck 7672
Rötelmausspecht 8974

Rötelmerle 6018
Rötelpelikan 7201
Rötelschlüpfer 9655
Rötelschwalbe 4661
Rötelsperling 7158
Rötelsteigschnabel 10635
Röteltaube 9506
Röteltyrann 6671
Roter Felsenhahn 8919
Roter Flamingo 7507
Roter Hokko 2824
Roter Kardinal 1712
Roter Kronfink 2730
Roter Milan 5893
Roter Paradiesvogel 7020
Roter Sichelschnabel 3549
Rotes Satyrhuhn 10140
Rotes Spornhuhn 4065
Rotflankenammerfink 8141
Rotflankenbreitrachen 9262
Rotflankenbrillenvogel
  10772
Rotflankendschungel-
  schnäpper 8807
Rotflankenfruchttaube
  3373
Rotflankenhabicht 36
Rotflankenhonigfresser
  8542
Rotflügelbrachschwalbe
  4242
Rotflügelbreitschnabel
  7901
Rotflügelbussard 1383
Rotflügelfrankolin 3913
Rotflügelgimpel 8874
Rotflügelguan 6813
Rotflügellerche 5930
Rotflügelralle 894
Rotflügelsittich 820
Rotflügelspateltyrann
  10085
Rotflügelspecht 7738
Rotflügelstärling 256
Rotflügeltaube 4579
Rotflügeltimalie 9389
Rotflügelwachtelastrild
  6844
Rotfußatlaswitwe 10533
Rotfußdrossel 10347
Rotfußente 565

Indice Italiano

Allocco bruno 9525
Allocco dal petto fulvo 9525
Allocco degli Urali 9533
Allocco del Szechwan 9521
Allocco di Hume 9519
Allocco di Lapponia 9527
Allocco eurasiatico 9518
Allocco fasciato bianco e nero 9528
Allocco fasciato nero 9523
Allocco fasciato rossiccio 9517
Allocco fulvo 9522
Allocco maculato americano 9529
Allocco maculato asiatico 9532
Allocco marezzato 9535
Allocco ocellato 9530
Allocco pallido 9519
Allocco rossiccio 9524
Allocco zamperosse 9531
Alloco bianco 10412
Alloco di Butler 9519
Allode 303
Allodola 299
Allodola abissina 5918
Allodola alirosse 5930
Allodola australe 3574
Allodola beccaforte 8747
Allodola beccocurvo 297
Allodola beccolunga 1942
Allodola calandra 5702
Allodola canora 5925
Allodola capocenerino 3575
Allodola cappeluluta 3989
Allodola castana 5916
Allodola codafaciata 485
Allodola comune 299
Allodola dagli speroni 2076
Allodola dal collare 5927
Allodola del Chuana 1941
Allodola del deserto 486
Allodola del deserto codafasciata 485
Allodola del deserto codarossa 488
Allodola del Dupont 2077
Allodola del Karoo 1937

Allodola del Kordofan 5928
Allodola del Marsabit 5944
Allodola del Sidamo 4608
Allodola dell'Angola 5919
Allodola dell'Assam 5924
Allodola delle nuvole 5920
Allodola dell'Isola Razo 302
Allodola di Archer 4606
Allodola di Ash 5923
Allodola di Bradfield 5935
Allodola di Dunn 3557
Allodola di Dupont 2077
Allodola di Erard 5929
Allodola di Fischer 3576
Allodola di Fremantle 8318
Allodola di Friedmann 5938
Allodola di Giava 5925,5934
Allodola di Gillett 5931
Allodola di Gray 487
Allodola di Hamerton 298
Allodola di macchia alirosse 5933
Allodola di macchia bruna 7749
Allodola di macchia codarossa 7748
Allodola di macchia meridionale 5926
Allodola di Rudd 4607
Allodola di Salvadori 5940
Allodola di Sharpe 5942
Allodola di Stark 3558
Allodola di Temminck 3573
Allodola dimacchia settentrionale 5917
Allodola dorsogrigio 3580
Allodola Erard 5937
Allodola ferruginea 1939
Allodola fringuello 3578
Allodola giapponese 301
Allodola golagialla 3572
Allodola hova 5932
Allodola mattolina 5405
Allodola nucarossicia 5915
Allodola orecchiebianche 3577

Allodola orientale 300
Allodola panterana 299
Allodola passerina 5936
Allodola passerina capinera 3578
Allodola rossa 1943
Allodola rugginosa 5939
Allodola sabota 5941
Allodola somala 5943
Allodola testacastana 3579
Allodole 303
Allodolina 299
Alzavola 534
Alzavola aliverdi 535
Alzavola anellata 1526
Alzavola argentina 542
Alzavola asiatica 544
Alzavola brasiliana 1526, 465
Alzavola comune 534
Alzavola del Capo 529
Alzavola del Madagascar 528
Alzavola della Sonda 548
Alzavola delle Filippine 553
Alzavola grigia 550
Alzavola macchiettata 542
Alzavole aliverdi 535
Amadina testarossa 377
Amakihi 4526
Amakihi comune 4526
Amakihi di Kauai 10602
Amakihi maggiore 4525
Amaranto africano 4958
Amaranto beccochiaro 4953
Amaranto beccorosso 4960
Amaranto bruno 4955
Amaranto di Dybowski 3773
Amaranto di Jameson 4957
Amaranto di Kuli Koro 4963
Amaranto di Monteiro 2386
Amaranto di Reichenow 4961
Amaranto fiammante 4766
Amaranto mascherato 4954

Colibrì magnifico 3657
Colibrì minimo 5774
Colibrì mirabile 5288
Colibrì modesto 777
Colibrì nero 5722
Colibrì orecchiebianche 4721
Colibrì orecchieviola scintillante 2441
Colibrì orechieviola bruno 2443
Colibrì orechieviola talassino 2445
Colibrì pettoviola 9471
Colibrì rossiccio 9098
Colibrì rubino del Brasil 2380
Colibrì rubino-topazio 2207
Colibrì Saffo 8955
Colibrì scintillante 9100
Colibrì smeraldo 3447
Colibrì smeraldo codablu 2167
Colibrì smeraldo codaforcuta 2162
Colibrì smeraldo codastretta 2173
Colibrì smeraldo codaverde 2157
Colibrì smeraldo dei giardini 2158
Colibrì smeraldo di Brace 2161
Colibrì smeraldo di Cuba 2170
Colibrì smeraldo di Gibson 2164
Colibrì smeraldo di Hispaniola 2174
Colibrì smeraldo di Poortman 2169
Colibrì smeraldo di Portorico 2165
Colibrì smeraldo ramato 2171
Colibrì smeraldo ventredorato 2159
Colibrì testablu 2954
Colibrì testavioletta 4935

Colibrì topazio cremisi 10105
Colibrì topazio di fiamma 10106
Colibrì ventremacchiato 9795
Colibrì ventrenero 3680
Colibrì ventreviola 3038
Colibrì ventrezaffiro 5149
Colibrì verde splendente 5148
Colibrì violetto 2443
Colibrì zaffiro golabianca 4717
Colibrì zaffiro golarossiccia 4723
Colibrì zafiro codadorato 2215
Colibrì zampepiumose canuto 4462
Colibrì zampepiumose coda di zaffiro 3588
Colibrì zampepiumose corona blu 3586
Colibrì zampepiumose di Derby 3585
Colibrì zampepiumose gola turchese 3587
Colibrì zampepiumose multicolore 3589
Colibrì zampepiumose pettodorato 3590
Colibrì zampepiumose pettonero 3591
Colibrì zampepiumose pettosmeraldo 3583
Colibrì zampepiumose smagliante 3592
Colibrì zampepiumose ventreramato 3584
Colibrì zampepiumose verdastro 4461
Colidi 2446
Coliformi 2447
Colino 2451
Colino alimacchiate 6685
Colino barbuto 3170
Colino barrato 7474
Colino canoro 3037
Colino castano 6691
Colino codalunga 3172

Colino crestato 2448
Colino dal collare 6697
Colino del Venezuela 6686
Colino della California 1519
Colino della Virginia 2451
Colino di Douglas 1520
Colino di Gambel 1521
Colino di Montezuma 3022
Colino di Tacarcuana 6687
Colino dorsonero 6693
Colino facciafulva 8890
Colino fasciastrata 6684
Colino frontenera 6683
Colino fronterosso 6688
Colino golabianca 6692
Colino golanera 2450
Colino macchiato 6690
Colino marmorizzato 6689
Colino ocellato 3023
Colino orecchienere 6694
Colino orechiebianche 3171
Colino pettorosso 6695
Colino plumifero 6765
Colino squamato 1522
Colino stellato 6696
Collotagliato 378
Colloverde 558
Colomba alimacchiate 2502
Colomba argentata 2477
Colomba azzurra del Madagascar 352
Colomba azzurra delle Comore 355
Colomba azzurra delle Seychelle 354
Colomba azzurra di Mauritius 353
Colomba beccogiallo 2488
Colomba becconero 2505
Colomba capochiaro 2516
Colomba collorosso 2520, 2538
Colomba corona bianca 2497
Colomba coronata 4289
Colomba coronata di Vittoria 4291
Colomba cruentata 4004

Fringuello cantore di Cuba
10011
Fringuello cantore
facciagialla 10014
Fringuello cantore
faccianera 10010
Fringuello cantore
fianchirossi 8141
Fringuello cantore
fuligginoso 10012
Fringuello cantore grigio
dal collare 8148
Fringuello cantore grigio e
bianco 8137
Fringuello cantore
gropponerosso 8142
Fringuello cantore nero e
castano 8149
Fringuello cantore nero e
ruggine 8144
Fringuello cantore ornato
8145
Fringuello cantore
pettirosso 8146
Fringuello cantore
pettocannella 8147
Fringuello cantore
pettocastano 8136
Fringuello cantore
tucumano 8134
Fringuello cenerino del
Perù 7746
Fringuello codabreve delle
Ande 4835
Fringuello codalungo 3295
Fringuello comune 3949
Fringuello cremisi 8877
Fringuello crestagrigia
5344
Fringuello crestanera 5345
Fringuello crestarossa 2730
Fringuello dai mustacchi
gialli 5708
Fringuello dal diadema
1797
Fringuello della Patagonia
7534
Fringuello della Sierra
codafasciata 7527
Fringuello della Sierra del
Perù 7536

Fringuello della Sierra della
Patagonia 7534
Fringuello della Sierra
dorsocastano 7530
Fringuello della Sierra
fuligginoso 7529
Fringuello della Sierra
golabianca 7531
Fringuello della Sierra
lamentoso 7532
Fringuello della Sierra
pettocenerino 7535
Fringuello della Sierra
plumbeo 7537
Fringuello della Sierra
testagrigia 7533
Fringuello della Sierra
testanera 7528
Fringuello delle Canarie
3951
Fringuello delle Isole
Cocos 7750
Fringuello dell'Isola Gough
8916
Fringuello dell'Isola Tristan
da Cunha 6521
Fringuello di fuoco 3725
Fringuello di Laysan 9824
Fringuello di macchia
alibianchi 1067
Fringuello di macchia
ardesia 1076
Fringuello di macchia
capinero 1058
Fringuello di macchia
capirosso 1073
Fringuello di macchia
capocastano 1059
Fringuello di macchia dagli
occhiali 1066
Fringuello di macchia dai
mustacchi 1057
Fringuello di macchia dalle
strie gialle 1061
Fringuello di macchia dalle
strie verdi 1082
Fringuello di macchia del
Tepui 1072
Fringuello di macchia di
Santa Marta 1068

Fringuello di macchia
golagialla 1065
Fringuello di macchia
nucabianca 1056
Fringuello di macchia
nucapallida 1071
Fringuello di macchia
nucarossa 1075
Fringuello di macchia
orecchierosse 1074
Fringuello di macchia petto
ocra 1078
Fringuello di macchia si
Seebohm 1077
Fringuello di macchia testa
oliva 1062
Fringuello di macchia
testabianca 1055
Fringuello di macchia
testafulva 1063
Fringuello di macchia
testapallida 1070
Fringuello di macchia
testascura 1064
Fringuello di macchia
testastriata 1079
Fringuello di macchia
tricolore 1081
Fringuello di macchia
ventrerosso 1069
Fringuello di Nihoa 9825
Fringuello di Wilkins 6522
Fringuello dorato
codastriata 9203
Fringuello dorato della
Patagonia 9206
Fringuello dorato della
prateria 9210
Fringuello dorato della
Puna 9207
Fringuello dorato di
Raimondi 9213
Fringuello dorato
frontearancio 9204
Fringuello dorato
gropponechiaro 9215
Fringuello dorato maggiore
9202
Fringuello dorato
meridionale 9208

Mangiamiele barbalunga
5733
Mangiamiele barrato 7545
Mangiamiele becco a spina
occidentale 25
Mangiamiele becco a spina
orientale 26
Mangiamiele beccodiritto
di montagna 10025
Mangiamiele beccodiritto
di pianura 10026
Mangiamiele beccoforte
5771
Mangiamiele beccolungo
5745
Mangiamiele boccagialla
5751
Mangiamiele
boccapurpurea 5220
Mangiamiele bruno scuro
5244
Mangiamiele cantore 5239
Mangiamiele capobruno
5766
Mangiamiele capomaculato
8632
Mangiamiele
cappucciorosso 6325
Mangiamiele cardinale
6313
Mangiamiele caruncolato
3891
Mangiamiele collogiallo
5223
Mangiamiele
collomacchiato 5234
Mangiamiele corona fulva
7540
Mangiamiele dai mustacchi
bianchi 5746
Mangiamiele dai mustacchi
gialli 5230
Mangiamiele dai mustacchi
neri 5225
Mangiamiele dai
sopraccigli bianchi
5765
Mangiamiele dai
sopraccigli rossi 5770
Mangiamiele dal collare
1945

Mangiamiele dal groppone
rosso 6339
Mangiamiele dalle
caruncole rosse 645
Mangiamiele del Mallee
5233
Mangiamiele della
Micronesia 6335
Mangiamiele della Nuova
Caledonia 6312
Mangiamiele della Nuova
Irlanda 6333
Mangiamiele della Nuova
Olanda 7543
Mangiamiele delle colline
5732
Mangiamiele delle foreste
5755
Mangiamiele delle Isole
Bismarck 6332
Mangiamiele delle Luisiade
5759
Mangiamiele delle
mangrovie 5221
Mangiamiele dell'Isola
Bonin 774
Mangiamiele dell'Isola
Sumba 6318
Mangiamiele di Adolfina
6308
Mangiamiele di Amboina
6310
Mangiamiele di Aru 5749
Mangiamiele di Banda
6311
Mangiamiele di Belford
5727
Mangiamiele di
Bougainville 9513
Mangiamiele di Buru 5242
Mangiamiele di Cockerell
10173
Mangiamiele di Eichhorn
6319
Mangiamiele di Foerster
5728
Mangiamiele di Gilliard
5737
Mangiamiele di
Guadalcanal 4360

Mangiamiele di Hindwood
5227
Mangiamiele di Kadavu
10617
Mangiamiele di Lombok
5247
Mangiamiele di macchia
5747
Mangiamiele di MacLeay
10615
Mangiamiele di Malaita
6327
Mangiamiele di Mayr 8544
Mangiamiele di Reichenow
5734
Mangiamiele di Rosenberg
6334
Mangiamiele di Rotuma
6314
Mangiamiele di Salvadori
8631
Mangiamiele di San
Cristobal 5735
Mangiamiele di Sclater
6337
Mangiamiele di Seram
5248
Mangiamiele di Sulawesi
6315
Mangiamiele di Timor
5243
Mangiamiele di Tristram
6338
Mangiamiele di Vogelkop
5730
Mangiamiele di Wakolo
6340
Mangiamiele dorsobruno
8756
Mangiamiele dorsocastano
8543
Mangiamiele dorsodorato
5769
Mangiamiele dorsonero
1944
Mangiamiele dorsoscuro
8546
Mangiamiele facciadorata
5219
Mangiamiele facciarossa
4381

Merlo di Brewer 3677
Merlo di Cebu 2592
Merlo di Strickland 2598
Merlo europeo 10327
Merlo indiano 4294
Merlo m 10327
Merlo nero 10327,2595
Merlo occhibianchi 7896
Merlo occhigialli 10332
Merlo podobè 1921
Merlo shama 2594
Merlo splendente 10360
Merlo zampegialle 7895
Merope 5798
Meropidi 5795
Mesena bruna 5824
Mesena pettobianco 5825
Mesitornitidi 5826
Mestolino 532
Mestolone 531,532
Mestolone argentino 557
Mestolone australiano 564
Mestolone comune 532
Mestolone del Capo 567
Miagra blu-acciaio 6132
Miagra capoazzurro 6127
Miagra dal sottocoda
 bianco 6124
Miagra dall'elmo 6134
Miagra della Micronesia
 6138
Miagra della Nuova
 Caledonia 6128
Miagra di Guam 6133
Miagra di Palau 6131
Miagra di Ponapè 6139
Miagra di San Cristobal
 6129
Miagra di seta 6130
Miagra maggiore 6136
Miagra modesta 6135
Miagra nera 6126
Miagra pettirossa 6142
Miagra splendente 6125
Miagra ventrerosso 6143
Miciola 10504
Microgura di Meek 5867
Miglarino 3479
Migliarino del Pallas 3473
Migliarino di palude 3479

Migliarino di palude del
 Giappone 3489
Migliarino polare 3473
Mignattaio 7923
Mignattino 2100
Mignattino alibianche 2099
Mignattino bigio 2098
Mignattino comune 2100
Mignattino piombato 2098
Mignattone 9454
Mimo alifasciate 5904
Mimo azzurro 5720
Mimo beccocurvo 10124
Mimo beccolungo 10127
Mimo bianco e azzurro
 5721
Mimo blu del Messico
 5720
Mimo cenerino 10122
Mimo codacorta 8746
Mimo codalunga 5898
Mimo dal sottocoda rosso
 10123
Mimo de LeConte 10126
Mimo dei campi 5902
Mimo del Cile 5903
Mimo del Cozumel 10125
Mimo della California
 10129
Mimo della Carolina 3383
Mimo della Patagonia 5900
Mimo della salvia 6766
Mimo delle Bahama 5897
Mimo delle Galapagos
 6517
Mimo di Bendire 10121
Mimo di McDonald 6515
Mimo di San Cristobal
 6516
Mimo di Socorro 5894
Mimo do Darwin 6518
Mimo dorsobruno 5895
Mimo grigio delle Antille
 2239
Mimo occhidiperla 5566
Mimo ocellato 10128
Mimo pettoscaglioso 5567
Mimo poliglotto 5901
Mimo rossiccio 10130
Mimo rosso 10130
Mimo ruficauda 2240

Mimo tropiccale 5896
Minatore 4165
Minatore alibrune 4171
Minatore beccocorto 4163
Minatore beccogrosso 4164
Minatore beccosottile 4172
Minatore comune 4165
Minatore del Campo 4140
Minatore del Perù 4168
Minatore della Puna 4169
Minatore grigiastro 4167
Minatore isabellino 4166
Minatore rufipenne 4170
Ministro 7173
Minla aliazzurre 5905
Minla codacastana 5907
Minla codarossa 5906
Mitteria africana 6122
Mitteria americana 6120
Mitteria cinerea 6121
Mitteria del Senegal 3546
Mitteria indiana 6123
Mitu 5949
Mivola 10504
Mniotilta bianca e nera
 5953
Moho delle Hawaii 5958
Moho di Bishop 5956
Moho di Kauai 5957
Moho di Oahu 5955
Molotra splendente 5966
Molotro alicastane 5965
Molotro ascelle castane
 5967
Molotro badio 5965
Molotro bonariense 5966
Molotro bronzato 5962
Molotro gigante 9001
Molotro nero 5964
Molotro testabruna 5963,
 5964
Momotidi 5968
Monaca beccogiallo 6008
Monaca frontebrianca 6009
Monaca frontenera 6010
Monaca nera 6007
Monacchia 2706
Monachella 6710
Monachella araba 6715
Monachella bella
 testabianca 6712

Otarda di Smith 3731
Otarda di Stanley 6495
Otarda di Vigors 3744
Otarda korhaan 3730
Otarda kori 972
Otarda maggiore 6854
Otarda maggiore indiana
    973
Otarda minore indiana
    3738
Otarda nubiana 6498
Otarda umile 3737
Otarda ventrenero 3739
Otarde 6852
Otididi 6852
Ou 8387
Oubara 2096
Padda 7001
Padda bruno di Timor 7000
Paglianculo 164
Pagliarolo 129
Palamedea cornuta 604
Palila 5398
Palombella 2506
Pantana 10203
Pantana comune 10203
Pantana macchiata 10198
Papa della Louisiana 7172
Papa della Luisiana 7172
Papa di Leclancher 7174
Papa lazuli 7171
Papa pettorosa 7175
Papa variabile 7176
Papagallo del Niam-Niam
    8057
Papagallo del Senegal 8065
Papagallo di Jardine 8060
Papagallo di Meyer 8061
Papagallo facciagialla 8059
Papagallo robusto 8062
Papagallo rufiventre 8064
Papagallo testabruna 8058
Pappagallino aliblu 3884
Pappagallino dagli occhiali
    3879
Pappagallino dal groppone
    turchese 3880
Pappagallino dal groppone
    verde 3881
Pappagallino dalla testa
    nera 6343

Pappagallino del Pacifico
    3878
Pappagallino del Tepui
    6345
Pappagallino della signora
    D'Achille 6344
Pappagallino di Sclater
    3882
Pappagallino facciagialla
    3883
Pappagallino ondulato
    5777
Pappagallo accipitrino
    3173
Pappagallo aliazzurre 4457
Pappagallo alibronzate
    7763
Pappagallo alinere 4458
Pappagallo becco di corallo
    7769
Pappagallo beccoforte 8889
Pappagallo beccogrosso
    9782
Pappagallo capobruno 7759
Pappagallo caporosso 7760
Pappagallo cenerino 8382
Pappagallo coda a racchetta
    corona blu 8211
Pappagallo coda a racchetta
    corona rossa 8212
Pappagallo coda a racchetta
    di Buru 8214
Pappagallo coda a racchetta
    di Mindanao 8219
Pappagallo coda a racchetta
    di montagna 8215
Pappagallo coda a racchetta
    di Palawan 8216
Pappagallo coda a racchetta
    di Sulu 8218
Pappagallo coda a racchetta
    dorsodorato 8217
Pappagallo coda a racchetta
    verde 8213
Pappagallo codacorta 4343
Pappagallo corona bianca
    7767
Pappagallo corona di
    prugna 7770
Pappagallo dai mustacchi
    neri 9780

Pappagallo dal collare blu
    4147
Pappagallo dal petto giallo
    6210
Pappagallo dei Caica 7758
Pappagallo dei ficchi dagli
    occhiali 2972
Pappagallo dei ficchi
    pettoarancia 2974
Pappagallo dei fichi di
    Desmarest 8379
Pappagallo dei fichi di
    Edwards 8380
Pappagallo dei fichi di
    Salvadori 8381
Pappagallo del Congo 8062
Pappagallo delle
    Mascarene 5573
Pappagallo dell'Ecuador
    4459
Pappagallo di Barraband
    7757
Pappagallo di Brehm 8356
Pappagallo di Madarasz
    8357
Pappagallo di Massimiliano
    7765
Pappagallo di Müller 9783
Pappagallo di Pennant 7888
Pappagallo di Pesquet 8388
Pappagallo di Rüppell 8063
Pappagallo facciarossa
    7761
Pappagallo facciarugginosa
    4456
Pappagallo fosco 7764
Pappagallo frontecastano
    8889
Pappagallo gropponeblu
    8386
Pappagallo guancerosse
    4144
Pappagallo modesto 8358
Pappagallo nuca azzurra
    9781
Pappagallo pigmeo delle
    Kai 5880
Pappagallo pigmeo di
    Brujin 5877
Pappagallo pigmeo
    facciafulva 5882

Rampichino 10651,1928
Rampichino alifulve 3064
Rampichino alpestre 1930
Rampichino americano
  1927
Rampichino bandenere
  3085
Rampichino barrato 3080
Rampichino becco a cuneo
  4281
Rampichino becco a
  scimitarra 3325
Rampichino beccoforte
  10650
Rampichino beccolungo
  6355,8781
Rampichino beccorosso
  4709
Rampichino beccosottile
  5141
Rampichino bruno
  australiano 2671
Rampichino bruno minore
  5142
Rampichino codalunga
  3051
Rampichino comune 1928
Rampichino corona
  macchiata 5139
Rampichino dai mustacchi
  10646
Rampichino dai sopraccigli
  bianchi 2667
Rampichino dai sopraccigli
  rossi 2669
Rampichino dal groppone
  castano 10665
Rampichino del Planalto
  3086
Rampichino della Nuova
  Zelanda 5960
Rampichino delle Hawaii
  6758
Rampichino delle palme
  1228
Rampichino dell'Himalaya
  1931
Rampichino di Eyton
  10658
Rampichino di Hoffmann
  3082

Rampichino di Kauai 6757
Rampichino di Maui 7065
Rampichino di Molokai
  7063
Rampichino di Oahu 7064
Rampichino di Snethlage
  10647
Rampichino di Spix 10667
Rampichino di Stoliczka
  1932
Rampichino di Stresemann
  4710
Rampichino di terra
  beccoforte 10440
Rampichino di terra
  codaruggine 10438
Rampichino di terra della
  Bolivia 10436
Rampichino di terra delle
  Ande 10433
Rampichino di terra di
  Chaco 10434
Rampichino di terra di
  Jelski 10437
Rampichino di terra
  golabianca 10432
Rampichino di terra
  golascagliosa 10435
Rampichino di terra striato
  10439
Rampichino di Zimmer
  10662
Rampichino disadorno
  8782
Rampichino dorso oliva
  10670
Rampichino elegante
  10656
Rampichino fianchistriati
  8783
Rampichino formichiere
  codamarginata 4772
Rampichino formichiere
  codanera 6264
Rampichino formichiere
  dai sopraccigli bianchi
  6262
Rampichino formichiere
  faccianera 6265
Rampichino formichiere
  mentonero 4773

Rampichino formichiere
  pettocenerino 6263
Rampichino fuligginoso
  3066
Rampichino gocciolato
  10660
Rampichino golabianca
  2671
Rampichino golabruna
  1929
Rampichino golamacchiata
  3052
Rampichino golarossiccia
  3063
Rampichino lacrimoso
  10661
Rampichino lineato 5140
Rampichino maculato 8933
Rampichino mentobianco
  3070
Rampichino ocellato 10664
Rampichino papua 2674
Rampichino picchio 10666
Rampichino rossiccio 3067
Rampichino rossiccio
  maggiore 10648
Rampichino rugginoso
  2376
Rampichino squamato
  5146
Rampichino striato 10663
Rampichino terriccola 3560
Rampichino testagrigia
  9255
Rampichino testastriata
  5145
Rampichino tiranno 3072
Rampichino tordo 3071
Rampichino ventrebianco
  5144
Rampichino ventregiallo
  10659
Randia 8757
Raperino 9171
Rayadito codaspinosa 790
Rayadito di Masafuera 789
Re degli avvoltoi 8958
Re degli edredoni 9266
Re di macchia 10228
Re di quaglie 2831
Reatino 10228

Scricciolo formichiere alirosse 4596
Scricciolo formichiere beccocorto 6294
Scricciolo formichiere beccolungo 4590
Scricciolo formichiere capinero 4586
Scricciolo formichiere codafasciata 6303
Scricciolo formichiere codalunga 6290
Scricciolo formichiere codamacchiata 4599
Scricciolo formichiere codarossiccia 6277
Scricciolo formichiere delle colline 6299
Scricciolo formichiere di Bahia 4594
Scricciolo formichiere di Cherrie 6276
Scricciolo formichiere di Hauxwell 6285
Scricciolo formichiere di Ihering 6287
Scricciolo formichiere di Klages 6288
Scricciolo formichiere di Menetries 6292
Scricciolo formichiere di Parker 4592
Scricciolo formichiere di Quixos 5885
Scricciolo formichiere di Rio de Janeiro 6278
Scricciolo formichiere di Rio Suna 6300
Scricciolo formichiere di Roraima 4595
Scricciolo formichiere di Salvadori 6293
Scricciolo formichiere di Sclater 6297
Scricciolo formichiere di Todd 4598
Scricciolo formichiere dorsomacchiato 4588
Scricciolo formichiere fianchibianchi 6273

Scricciolo formichiere golagialla 6271
Scricciolo formichiere golamacchiata 6284
Scricciolo formichiere golastellata 6281
Scricciolo formichiere grigio 6280
Scricciolo formichiere lavagna 6296
Scricciolo formichiere macchiettato 6282
Scricciolo formichiere occhibianchi 6289
Scricciolo formichiere ornato 6295
Scricciolo formichiere pettogiallo 4587
Scricciolo formichiere pettorale 4593
Scricciolo formichiere pigmeo 6275
Scricciolo formichiere striato 6301
Scricciolo formichiere unicolore 6302
Scricciolo formichiere ventrebruno 6283
Scricciolo formichiere ventrecastano 4591
Scricciolo formichiere ventrefulvo 6279
Scricciolo gigante 1622
Scricciolo golanera 9982
Scricciolo golastriata 9991
Scricciolo grigio 9989
Scricciolo grigio barrato 1628
Scricciolo guancebianche 9992
Scricciolo guancestriate 9987
Scricciolo inca 9984
Scricciolo macchiato 1625
Scricciolo mimo 3294
Scricciolo modesto 9996
Scricciolo nucarossiccia 1630
Scricciolo ocraceo 10222
Scricciolo pettirosso 10002

Scricciolo pettobianco 10000
Scricciolo pettocastano 3010
Scricciolo pettomaculato 9995
Scricciolo pettoscaglioso 5854
Scricciolo pettostriato 10008
Scricciolo rossiccio 2273
Scricciolo splendente 5542
Scricciolo splendente alibianche 5541
Scricciolo splendente amabile 5530
Scricciolo splendente azzurro 5547
Scricciolo splendente corona aranciata 2381
Scricciolo splendente coronato 5533
Scricciolo splendente di Campbell 5532
Scricciolo splendente di Gray 5538
Scricciolo splendente di palude 5537
Scricciolo splendente di Wallace 9225
Scricciolo splendente dorsorosso 5544
Scricciolo splendente imperatore 5536
Scricciolo splendente pettoblu 5546
Scricciolo splendente spallebianche 5529
Scricciolo splendente variegato 5539
Scricciolo testabianca 1619
Scricciolo testafuligginosa 10006
Scricciolo tordo 1631
Scricciolo usignolo 5855
Scricciolo ventrebianco 10472
Scricciolo ventrenero 9986
Scrocchino 8991
Segretario 8925
Seiuro corona dorata 9091

Stella dei boschi del Cile
3665
Stella dei boschi della
Colombia 93
Stella dei boschi di
Esmeralda 94
Stella dei boschi di Fanny
6304
Stella dei boschi di Jourdan
1972
Stella dei boschi di
Mitchell 7471
Stella dei boschi
golamagenta 7470
Stella dei boschi
pettobianco 97
Stella dei boschi piccola 95
Stella dei monti andina
6772
Stella dei monti codabianca
10456
Stella dei monti
codacuneata 6770
Stella dei monti
dell'Ecuador 6771
Stella dei monti
fianchibianchi 6773
Stella dei monti pettonero
6774
Stenostira 9429
Stercoraridi 9433
Stercorario antartico 1802
Stercorario del Cile 1803
Stercorario di McCormick
1804
Stercorario maggiore 1805
Stercorario mezzano 9436
Sterna 9454
Sterna aleutina 9439
Sterna antartica 9469
Sterna artica 9460
Sterna australiana 9458
Sterna beccogrosso 7364
Sterna bianca 4368
Sterna bianca minore 4370
Sterna comune 9454
Sterna dagli occhiali 9456
Sterna dalle redini 9440
Sterna del Cile 9455
Sterna del Dougall 9448

Sterna del Sudamerica
9453
Sterna dell'Amazzonia
9466
Sterna delle Aleutine 9439
Sterna delle Kerguelen
9468
Sterna di Berg 9445
Sterna di Bernstein 9446
Sterna di Damara 9443
Sterna di Dougall 9448
Sterna di fiume 9442
Sterna di Forster 9451
Sterna di Rüppell 9444
Sterna di Trudeau 9467
Sterna elegante 9449
Sterna frontebianca 9464
Sterna frontenera 2097
Sterna fuligginosa 9452
Sterna guancebianche 9461
Sterna inca 5067
Sterna maggiore 9447,9448
Sterna minore 9438
Sterna reale 9457,9462
Sterna scura 9452
Sterna stolida 626
Sterna stolida bruna 626
Sterna stolida cerulea 8249
Sterna stolida grigia 8248
Sterna stolida minore 627
Sterna stolida nera 625
Sterna sumatrana 9465
Sterna ventrenero 9437
Sterna zampenere 9459
Sterpazzola 9593
Sterpazzola di Sardegna
9594
Sterpazzola nana 9606
Sterpazzolina 9591,9593
Sterpazzuola 9593
Sterpazzuolina 9591
Stiaccino 8991,8997
Stiaccino delle Canarie
8983
Stiaccino grigio 8984
Stornello 9565
Storni 9549
Storno 9565
Storno alibianchi 6454
Storno alicastane 6734
Storno alichiare 6735

Storno alimacchiate 8962
Storno alinere 9555
Storno ametista 2275
Storno beccobianco 6730
Storno beccofino 6737
Storno beccogrosso 9027
Storno bianco e nero 9552
Storno bicolore 9269
Storno calvo 8957
Storno canoro 793
Storno capobianco 9382
Storno caruncolato 2827
Storno cenerino 2740
Storno codabreve 807
Storno codagraduata 8039
Storno codalunga 804
Storno collonero 9556
Storno color d'accaio 4995
Storno comune 9565
Storno dal collare bianco
4295
Storno daurico 9563
Storno del Madagascar
8961
Storno della Cina 9562
Storno della Micronesia
810
Storno delle Filippine 811
Storno delle Isole Fead 798
Storno delle Isole Samoa
791
Storno delle Molucche 808
Storno delle Nuove Ebridi
816
Storno delle pagode 9557
Storno dell'Isola Norfolk
799
Storno dell'Isola Rennell
803
Storno di Abbott 2274
Storno di Bali 5198
Storno di Blyth 6731
Storno di Burchell 4992
Storno di Fischer 9384
Storno di Grant 809
Storno di Hildebrandt 5000
Storno di Jerdon 9550
Storno di Kenrick 8038
Storno di Kosrae 795
Storno di Meves 5001
Storno di Ponapè 812

Tangara vermiglia 1527
Tangra olivacea 2150
Tantalo 6120,6122
Tantalo africano 6122
Tantalo americano 6120
Tantalo cinereo 6121
Tantalo indiano 6123
Tapacula di Chucao 9004
Tapacula golabianca 9003
Tapaculo bruno 9062
Tapaculo cenerino 6233
Tapaculo cintura rossiccia 5267
Tapaculo di Brasilia 9067
Tapaculo di Magellano 9064
Tapaculo di Narino 9081
Tapaculo di Santa Marta 9058,9073
Tapaculo fianchi ocra 3660
Tapaculo fronteargentata 9051
Tapaculo ocellato 143
Tapaculo panamense 9068
Tapaculo pettobianco 9061
Tapaculo piedigrandi 9063
Tapaculo sopracciglibianchi 9078
Tapaculo unicolore 9079
Tapera 9793
Tarabugino 109
Tarabugio 1264
Tarabusco delle foreste 10708
Tarabusi 963
Tarabusino 4893
Tarabusino americano 4890
Tarabusino cannella 4888
Tarabusino cinese 4896
Tarabusino comune 4893
Tarabusino dorsonero 4895
Tarabusino dorsostriato 4892
Tarabusino nano africano 4897
Tarabusino nero 4891
Tarabusino orientale 4889
Tarabusino zebrato 10692
Tarabuso 1264
Tarabuso amazzonico 1262
Tarabuso americano 1261

Tarabuso australiano 1263, 1264
Tarrabuso 1264
Tassolo 6910
Teccola 8174
Terecchio 10193
Tesia corona castana 9867
Tesia di Everett 9869
Tesia di Giava 9871
Tesia ventreardesia 9870
Tesia ventregrigio 9868
Tessitore 7947
Tessitore alibarrate 7932
Tessitore arancio 7933
Tessitore baglafecht 7936
Tessitore baya 7974
Tessitore beccogrosso 472
Tessitore capobruno 7957
Tessitore capogiallo 7951
Tessitore castano 7978
Tessitore codabreve 1265
Tessitore codarossa 4690
Tessitore color cannella 7935
Tessitore color di fuoco 3725
Tessitore compatto 7986
Tessitore dagli occhiali 7971
Tessitore dei bufali beccobianco 1329
Tessitore dei bufali beccorosso 1330
Tessitore dei bufali testabianca 3258
Tessitore del Bengala 7939
Tessitore del Capo 7944
Tessitore del Lago Victoria 7992
Tessitore delle palme 7942
Tessitore delle Sychelle 3890
Tessitore dell'Usambara 7967
Tessitore di Bannerman 7937
Tessitore di Bates 7938
Tessitore di Bertrand 7940
Tessitore di Bocage 7988
Tessitore di Clarke 7953
Tessitore di Finn 7963

Tessitore di foresta 3888, 7941
Tessitore di Fox 7983
Tessitore di Jackson 7959
Tessitore di Loanga 7985
Tessitore di Mauritius 3889
Tessitore di Pelzeln 7973
Tessitore di Rodriguez 3886
Tessitore di Rüppell 7952
Tessitore di Salvadori 7950
Tessitore di Sao Tomè 7981
Tessitore di Speke 7982
Tessitore di Weyns 7993
Tessitore dorato 7984
Tessitore dorato asiatico 7956
Tessitore dorato dell'Isola Principe 7976
Tessitore dorato di Holub 7994
Tessitore dorato di Taveta 7945
Tessitore dorato golabruna 7995
Tessitore dorato testaoliva 7972
Tessitore dorsodorato occidentale 7975
Tessitore fiammante 3887
Tessitore frontesquamosa 9380
Tessitore giallo 7952
Tessitore giallo e nero 7969
Tessitore gigante 7954
Tessitore golabruna settentrionale 7946
Tessitore mascherato 7961
Tessitore mascherato africano 7990
Tessitore mascherato del Katanga 7960
Tessitore mascherato della Tanzania 7977
Tessitore mascherato di Lufira 7979
Tessitore mascherato minore 7958
Tessitore mascherato settentrionale 7987

Trombetiere del deserto 8873

Trombettiere 8871

Trombettiere alirosse 8874

Trombettiere comune 8871

Trombettiere di Lichtenstein 8873

Trombettiere mongolo 8872

Trupiale 4817

Trupiale dalle ali bianchi 4817

Tucanetto beccomaculato 9105

Tucanetto beccoscanalato 1109

Tucanetto dai sopracigglie gialli 1107

Tucanetto dal groppone rosso 1106

Tucanetto dalla banda cerulea 1104

Tucanetto della Guiana 9103

Tucanetto di Derby 1105

Tucanetto di Gould 9104

Tucanetto di Natterer 9106

Tucanetto di Reinwardt 9107

Tucanetto orecchiegialle 9108

Tucanetto smeraldo 1108

Tucanetto zafferano 1144

Tucano 8730

Tucano ambiguo 8721

Tucano beccoscanalato 8731

Tucano bicolore 8726

Tucano cioco 8722

Tucano crestagialla 8724

Tucano di Cuvier 8725

Tucano di monte becconero 583

Tucano di monte beccopiatto 582

Tucano di monte dl cappuccio 580

Tucano di monte pettogrigio 581

Tucano di Swainson 8728

Tucano golagialla 8723

Tucano solforato 8727

Tucano sulfureo 8727

Tucano toco 8729

Tuffetto 9688

Tuffetto australiano 9686

Tuffetto comune 9688

Tuffetto di Madagascar 9687

Tuffetto minore 9685

Tuffetto piccolo 9688

Tuffolino 9688

Tuffolo 9688

Tui 8270

Tui ali rosse-gialle 10112

Tui alimacchiate 10116

Tui codadorata 10117

Tui della Costarica 10111

Tui gropponezaffiro 10115

Tui orecchienere 10114

Tui settecolori 10110

Tui spallescarlatte 10113

Turaco azzurro gigante 2732

Turaco beccogiallo 9810

Turaco becconero 9814

Turaco crestabianca 9807

Turaco crestarossa 9804

Turaco crestavioletta 6098

Turaco dal becco nero 9814

Turaco dalle gote bianche 9808

Turaco del principe Ruspoli 9812

Turaco del Ruwenzori 6097

Turaco della cresta rossa 9806

Turaco di Bannerman 9802

Turaco di Fischer 9805

Turaco di Hartlaub 9806

Turaco di Knysna 9803

turaco di Lady Ross 6099

Turaco di Livingstone 9809

Turaco di Schallow 9813

Turaco gigante 2732

Turaco grigio occidentale 2832

Turaco grigio orentale 2833

Turaco guancebianche 9808

Turaco mascherato 2735

Turaco unicolore 2733

Turaco ventrebianco 2734

Turaco verde 9811

Turaco violetto 6100

Turdidi 10257

Turka dai mustacchi 8475

Turnagra 10367

Ubara 2096

Uccelli da preda 3822

Uccelli delle tempeste 4695

Uccellino degli spinifex 3559

Uccello azzurro 9201

Uccello azzurro comune 9201

Uccello azzurro montano 9199

Uccello azzurro occidentale 9200

Uccello azzurro orientale 9201

Uccello campanello australiano 5556

Uccello campanello della Nuova Zelanda 650

Uccello cappuccino 7268

Uccello dagli occhiali orientale 10814

Uccello d'argento 3516

Uccello dei cespugi rossiccio 1084

Uccello dei cespugli occidentale 1083

Uccello dei dodici apostoli 9537

Uccello dei fichi di Wetar 9284

Uccello dei fichi verde 9286

Uccello del sole 5134

Uccello del sole comune 5134

Uccello del vischio 3192

Uccello delle palme 3382

Uccello delle risaie 7001

Uccello delle tempeste 4694

Uccello delle tempeste cinereo 6654

Uccello delle tempeste codaforcuta 6656